This volume presents a comprehensive and up-to-the-minute review of all aspects of the physics of the aurorae australis and borealis. The broad topics covered relate to the different sections of a conference which was held in Cambridge in July 1988 to celebrate the centenary of the birth of Professor Sydney Chapman, F.R.S, who can be considered the founder of the subject in its modern form. Contributions from an international group of experts in the field discuss current thinking on the physical causes and effects of the aurora, the nightly show of dancing lights in the atmosphere, at heights above 100 km.
The book is aimed primarily at students and researchers in auroral physics, but will also be of interest to magnetospheric, ionospheric and atmospheric physicists.

AURORAL PHYSICS

AURORAL PHYSICS

Editors

Ching -I. Meng
The John Hopkins University
Applied Physics Laboratory

Michael J. Rycroft
British Antarctic Survey

Louis A. Frank
University of Iowa

The right of the
University of Cambridge
to print and sell
all manner of books
was granted by
Henry VIII in 1534.
The University has printed
and published continuously
since 1584.

CAMBRIDGE UNIVERSITY PRESS

Cambridge

New York   Port Chester

Melbourne   Sydney

CAMBRIDGE UNIVERSITY PRESS
Cambridge, New York, Melbourne, Madrid, Cape Town, Singapore,
São Paulo, Delhi, Dubai, Tokyo, Mexico City

Cambridge University Press
The Edinburgh Building, Cambridge CB2 8RU, UK

Published in the United States of America by Cambridge University Press, New York

www.cambridge.org
Information on this title: www.cambridge.org/9780521157414

First published 1991
First paperback edition 2010

A catalogue record for this publication is available from the British Library

ISBN 978-0-521-38049-2 Hardback
ISBN 978-0-521-15741-4 Paperback

Additional resources for this publication at www.cambridge.org/9780521157414

AURORAL PHYSICS

Contents

Preface xv
Participants xvii

I. INTRODUCTORY OBSERVATIONS 1

I-1. Auroral Phenomena 3

S.-I. Akasofu

1. Introduction 3
2. Power Generation for Auroral Phenomena 4
3. Magnetosphere/Ionosphere Interaction 4
4. Auroral Potential Structure 5
5. Auroral Phenomena at the Ionospheric Level 5
6. Auroral Substorms and Magnetospheric Substorms 7
7. Polar Cap During Periods of Northward IMF 10
8. Need for a Global Imaging of the Magnetosphere 10
9. Concluding Remarks 10

II. AURORAL SPECTROSCOPY AND THERMOSPHERE 13

II-1. Overview of Auroral Spectroscopy 15

A. Vallance Jones

1. Introduction 15
2. Spectroscopic Techniques 16
3. Atomic and Molecular Species and Their Energy Levels 16
4. Lifetimes of Expected States 16
5. The Auroral Spectrum 16
6. Far Infrared Spectrum 26
7. Aurora on Other Planets 26
8. Analysis of Spectra and Applications 26
9. Conclusions 26

II-2. Auroral Excitation Processes 29

M. H. Rees and D. Lummerzheim

1. Introduction 29
2. Electron Impact 29
3. Excitation by Electron Impact 31
4. Results of Model Computations 32
5. Proton Impact 34
6. Chemical-Ionic Reactions 34
7. Cascading 35
8. Concluding Remarks 35

II-3. Auroral Emission Processes and Remote Sensing..37

R. R. Meier and D. J. Strickland

    1. Introduction..37
    2. Energy Partitioning and Emission Features..37
    3. Emission Parameters..40
    4. Emission Diagnostics..42
    5. Conclusions..48

II-4. Thermospheric Response and Feedback to Auroral Inputs.................................51

D. Rees and T. J. Fuller-Rowell

    1. Introduction..51
    2. The Coupled Global Thermosphere/Polar Ionosphere Model..................54
    3. Coupled Model Simulations...55
    4. Upper Thermospheric Neutral Wind and Temperature, Composition, and Ion Density,
       as Functions of Geomagnetic Activity...55
    5. Lower Thermospheric Neutral Wind and Temperature, Composition, Plasma Density,
       and Auroral Ionization Rate, as a Function of Geomagnetic Activity.....58
    6. Electrodynamics..60
    7. Summary..62

II-5. Thermospheric Dynamics, Energetics, and Composition at Auroral Latitudes.........67

T. L. Killeen, F. G. McCormac, A.G. Burns, and R. G. Roble

    1. Introduction..67
    2. Dynamics Explorer Instrumentation and Observations........................68
    3. The NCAR Thermosphere General Circulation Model..........................72
    4. Auroral Forcing Mechanisms..73
    5. Concluding Remarks..80

III. AURORAL PARTICLES AND ACCELERATION MECHANISMS........83

III-1. Overview of Electron and Ion Precipitation in the Auroral Oval.......................85

P. T. Newell, C.-I. Meng, and D. A. Hardy

    1. Introduction..85
    2. Global Maps of Auroral Precipitation Binned by MLAT and MLT.............86
    3. Morphology and Dynamics of the Auroral Boundaries........................89
    4. Statistical Studies Involving Recognition of Special Features................91
    5. Summary: A Tour of the Auroral Oval..92

III-2. Diagnosis of Auroral Acceleration Mechanisms by Particle Measurements...........97

J. L. Burch

    1. Introduction..97
    2. Acceleration by Quasi-Static Field-Aligned Potential Differences............98
    3. Electron Burst-Type Distributions..102
    4. Summary and Conclusions...105

III-3. Characteristics of Magnetic-Field Aligned Electric Fields in the Auroral Acceleration Region 109

L. P. Block and C.-G. Fälthammar

1. Introduction...109
2. A Concise Definition of DC Electric Fields...110
3. Examples of Observed DC Fields...110
4. Double Layers..111
5. Spectra of the Electric Field Fluctuations...113
6. Particle Observations...113
7. Simultaneous Upward Acceleration of Ions and Electrons.........................116
8. Waves with Higher Frequencies..117
9. The Cause of Maximum Power Spectral Density Below 1 Hz.....................117
10. Summary...117

III-4. Auroral Electron Acceleration: A Case for the Stochastic Alternative.................119

D. A. Bryant, D. S. Hall, and R. Bingham

1. Introduction...119
2. Acceleration Model...121
3. Predictions...123
4. Energetics..126
5. Conclusions..127

III-5. Auroral Ion Acceleration and Its Relationship to Ion Composition.....................129

E. G. Shelley and H. L. Collin

1. Introduction...129
2. Accelerated Ionospheric Ions...129
3. Acceleration at Low Altitude...130
4. Perpendicular Acceleration at Higher Altitudes..131
5. Parallel Acceleration and Ion Beams..135
6. Summary and Status..140

III-6. Ion Precipitation and the Transport of Ions Accelerated by Auroral Processes....143

J. M. Bosqued

1. Introduction...143
2. Transport of Ionospheric Ions: Experimental Aspects................................144
3. Modeling Ion Transport from the Ionosphere..150
4. Concluding Remarks...153

IV. AURORAS AND MAGNETOSPHERIC CONFIGURATION.............157

IV-1. What Determines the Size of the Auroral Oval?...159

G. L. Siscoe

1. An Introduction to Auroras and Magnetospheric Configuration...................159
2. The Auroral Oval and Predictive Geospace Science..................................161

3. Observed Properties of the Oval of Visual Auroras............162
4. Observed Properties of the Oval of Diffuse Auroras............166
5. Relation of the Observations to the Models............167
6. Superposition Models............167
7. Convection Models............171
8. Conclusion............174

IV-2. The Quiet-Time Aurora and the Magnetospheric Configuration............177
R. Lundin, L. Eliasson, and J. S. Murphree

1. Introduction............177
2. Polar Region Aurora............179
3. Boundary Layer Linkage to the Oval............188
4. Conclusions............191

IV-3. Discrete Auroras and Magnetotail Processes............195
L. R. Lyons

1. Introduction............195
2. Magnetotail Region of Arc Generation............195
3. Magnetospheric Region of Substorm Initiation............199
4. Effects of Variation in the Polar Cap Area............200
5. Conclusions............203

IV-4. Auroral Luminosity and Its Relationship to Magnetospheric Plasma Domains.....207
Yu. I. Galperin and Ya. I. Feldstein

1. Introduction............207
2. Basic Data and Definitions of Auroral Oval and Plasma Sheet............209
3. Evidence that Supports Mapping the Region of Diffuse Luminosity Equatorward of
   the Auroral Oval to the Region Between the Inner Boundary of the Plasma Sheet
   and the Plasmapause............213
4. Evidence that Supports Mapping the Auroral Oval to the Low-Latitude
   (Central, Main) Plasma Sheet............214
5. Identification of the Soft Diffuse Electron Precipitation and Luminosity Region
   Poleward of the Auroral Oval as a Separate Magnetospheric Plasma Domain
   Mapped from the High-Latitude Magnetotail Boundary Plasma Sheet............216
6. Some Implications of Mapping for the Substorm Onset Location and Development.........217
7. Conclusions............219

IV-5. The Aurora and Middle Magnetospheric Processes............223
B. H. Mauk and C.-I. Meng

1. Introduction............223
2. Field-Aligned Electromagnetic Phenomena............226
3. Discharge Mechanisms............231
4. Discussion............236

IV-6. Auroral Plasma Waves...241

D. A. Gurnett

    1. Introduction...241
    2. Plasma Wave Modes..242
    3. Electromagnetic Waves.......................................243
    4. Electrostatic Waves...250
    5. Conclusion..252

V.   AURORAL SUBSTORMS AND DYNAMICS.................255

V-1.   Overview of Observations and Models of Auroral Substorms..........257

G. Rostoker

    1. Introduction...257
    2. Evolution of the Definition of a Substorm.................260
    3. Substorm Features that Any Model Must Explain.........263
    4. An Evaluation of the Two Models.........................266
    5. Conclusions..269

V-2.   Diagnosis of Auroral Dynamics Using Global Aurora Imaging with Emphasis
      on Large-Scale Evolutions...............................273

J. D. Craven and L. A. Frank

    1. Introduction...273
    2. Large-Scale Spatial Distributions..........................274
    3. Motion of the Theta Aurora.................................278
    4. Response to Shocks in the Interplanetary Medium.......280
    5. Auroral Substorms..282

V-3.   Diagnosis of Auroral Dynamics Using Global Auroral Imaging with Emphasis
      on Localized and Transient Features.....................289

G. G. Shepherd and J. S. Murphree

    1. Introduction...289
    2. Introduction to Viking UVI.................................289
    3. Auroral Behavior Prior to Substorm Onset...............292
    4. The Location of Substorm Onset..........................294
    5. Multiple "Simultaneous" Intensifications in a Poleward Arc.........295
    6. Summary and Conclusions..................................296

V-4.   Poleward Motions of Auroral Structures............................299

E. W. Hones, Jr., A. B. Galvin, and P. R. Higbie

    1. Introduction...299
    2. Observations of a Substorm on May 4, 1986............300
    3. Discussion of the May 4, 1986 Event....................303
    4. Observations of a Substorm on March 28, 1983........306

    5. Discussion of the March 28, 1983 Event 308

    6. Conclusions 308

V-5. A Magnetosphere-Ionosphere Coupling Theory of Substorms Including
      Magnetotail Dynamics 311

    J. R. Kan, L. Zhu, A. T. Y. Lui, and S.-I. Akasofu

    1. Introduction 311

    2. Enhancement of Magnetospheric Convection 312

    3. M-I Coupling Model of Substorms 313

    4. Closure of Field-Aligned Currents in the Magnetosphere 316

    5. Magnetotail Dynamics During Substorms 317

    6. Summary 318

VI. AURORAL STRUCTURES 323

VI-1. Overview of Auroral Spatial Scales 325

    D. J. Gorney

    1. Introduction 325

    2. The "Characteristic" Scale Size 326

    3. The Scale Spectrum 329

    4. A Case Study 331

    5. Summary 333

VI-2. Mesoscale Structures in Auroral Phenomena 335

    O. A. Troshichev

    1. Introduction 335

    2. Mesoscale Structures in the Auroral Oval 335

    3. Mesoscale Auroral Structures and Electric Current Systems 338

    4. Mesoscale Structures in the Polar Caps 342

    5. Conclusions 347

VI-3. The Pulsating Aurora and Its Relationship to Fields and Charged-Particle
      Precipitation 351

    P. J. Tanskanen

    1. Aurora 351

    2. Auroral Structures 351

    3. Pulsing and Pulsating Aurora 351

    4. Structure and Motion of Pulsating Aurora 352

    5. Altitude of Pulsating Aurora 353

    6. Where and When Do Auroral Pulsations Occur? 354

    7. Black Aurora 356

    8. Electron Precipitation, Pitch-Angle Distributions, and Energy Spectra 356

    9. Relaxation Oscillator Mechanism 357

  10. Summary 358

VI-4. Electrodynamics of Active Auroral Forms: Westward Traveling Surges and Omega Bands 361

W. Baumjohann

1. Introduction 361
2. Westward Traveling Surges 362
3. Omega Bands 364
4. Conclusions 366

VI-5. Large-Scale Distribution of Discrete Auroras and Field-Aligned Currents 369

E. Friis-Christensen and K. Lassen

1. Introduction 369
2. Observations 370
3. Statistical Distribution of Field-Aligned Currents and Aurora 373
4. Region 1a and Region 1b Currents 378
5. Discussion 378
6. Summary 380

VII. AURORA AND IONOSPHERE 383

VII-1. The Auroral Electrojets: Relative Importance of Ionospheric Conductivities and Electric Fields 385

Y. Kamide

1. Introduction 385
2. The Auroral Electrojet 386
3. Conductance and Electric Field in the Auroral Electrojet 388
4. Discussion 394

VII-2. Large-Scale Currents Connecting the Polar Ionosphere with the Magnetosphere ...401

T. Iijima

1. Introduction 401
2. Basic Patterns of Polar Geomagnetic Disturbances 401
3. Basic Patterns of Large-Scale Birkeland Current Systems 404
4. Concluding Remarks 407

VII-3. Ionosphere-Magnetosphere Mapping of Dynamic Auroral Structures During Substorms 409

E. Nielsen

1. Introduction 409
2. Observations and Results 410
3. Discussion 415

VII-4. Incoherent Scatter Observations of the Auroral Ionosphere with the
EISCAT Radar Facility...419

J. Röttger

1. Introduction..419
2. Incoherent Scatter Radar Research of the Polar Upper Atmosphere...........419
3. A Brief Description of EISCAT...420
4. The EISCAT Common and Unusual Programmes and the Special Campaign
   Operations...421
5. Magnetosphere-Ionosphere Coupling..423
6. Anisotropic and Non-Maxwellian Ion Velocity Distributions, Large
   Electric Fields, and Frictional Heating...424
7. Field-Aligned Plasma Velocities, Magnetospheric Convection, and Flux
   Transfer Events...425
8. Ionospheric Composition...425
9. Investigations of the Aurora with ISR and Other Ground-Based Instruments
   and Satellites...426
10. Auroral Substorms..426
11. Auroral Particle Precipitation..427
12. Electric Fields, Currents, and Intensifications of Auroral Arcs..............427
13. Incoherent Scatter and Magnetometer Observations of Pulsations.............430
14. Damping of Pulsations by Joule Heating..431
15. Ionospheric Conductivities...432
16. E-Region Irregularities and Plasma Waves..432
17. Extension of E-Region Observations into the D-Region and Observations of the
    Lower Thermosphere and Mesosphere...433
18. Coupling with the Neutral Atmosphere: Mean Winds, Tides, and Atmospheric
    Gravity Waves..434
19. Conclusion...435

VII-5. Ground-Based Measurements of Joule Heating Rates...........................439

O. de la Beaujardière, R. Johnson, and V. B. Wickwar

1. Introduction..439
2. Seasonal Dependence of High-Latitude Ionospheric Convection.................444
3. Conductivities from Solar-Produced Ionization..................................444
4. Neutral Wind...446
5. Summary of Observations...447
6. Discussion..447

VII-6. Global Observations: A Future Research Thrust in Auroral and
Magnetospheric Research..449

D. J. Williams

1. Introduction..449
2. Global Observations..450
3. Low-Altitude Regions...450
4. High-Altitude Regions..453
5. Summary...456

INDEX **457**

PLATES 1-50 *
* These plates are available in colour for download from www.cambridge.org/9780521157414

PREFACE

This book is the final outcome of an international conference on Auroral Physics that was held at St. John's College, Cambridge, from 11 to 15 July, 1988. The conference marked the centenary of the birth of Professor Sydney Chapman, FRS, on 29 January, 1988.

Sydney Chapman, a most eminent mathematician and physicist, carried out much fundamental research on the kinetic theory of gases, upper atmospheric physics, geomagnetism and solar-terrestrial physics. Auroras hold a special place in the field of solar-terrestrial physics. Auroras play a role in solar-terrestrial phenomena analogous to the role Sydney Chapman played in the development of the field.

More than 150 scientists from 16 countries travelled to attend the conference. (See listing and photograph on following pages.) The conference's objectives were to bring together auroral research scientists, to have a focussed interdisciplinary forum on auroral physics, to review the state-of-the-art understanding of auroral phenomena, to stimulate future research in auroral research, and to interest young scientists in further auroral investigations. Almost all the invited review papers that were presented in Cambridge are published here. The papers are appropriately grouped within lengthy chapters on auroral spectroscopy, acceleration processes, relationships to the magnetosphere, substorms, structures, ionosphere, thermosphere, and future prospects. We are most grateful to our colleagues who reviewed these chapters in considerable detail.

Ching –I. Meng (Johns Hopkins University)
Michael J. Rycroft (British Antarctic Survey)
Louis A. Frank (Univesity of Iowa)

September 1989

The publication of this volume is supported by The Johns Hopkins University Applied Physics Laboratory (JHU/APL). Special thanks are given to my colleagues in space physics who reviewed the material and provided many helpful comments. Thanks are also given to John W. Kaufman and John B. Moffett, at JHU/APL, who managed the publication.

Auroral Physics Conference, St. Johns College, Cambridge, England
July 11-15, 1988

PARTICIPANTS

for

CHAPMAN CONFERENCE ON AURORAL PHYSICS

St. John's College, Cambridge, 11–14 July 1988

AKASOFU, S.-I.	University of Alaska, USA.
ALPERT, J.	c/o St. John's College, Cambridge, UK.
ATKINSON, G.	NRC, Herzberg Inst. of Astrophysics, Canada.
BAKER, K.	Johns Hopkins Applied Physics Laboratory, USA.
BARKER, M.	University of York, UK.
BARROW, C.	Max-Planck Institut für Aeronomie, FRG.
BAUMJOHANN, W.	Max-Planck Institut für Extraterrestriche Physik, FRG.
BIRMINGHAM, T.	NASA Goddard Space Flight Center, USA.
BLOCK, L.	The Royal Institute of Technology, Sweden.
BLOMBERG, L.	The Royal Institute of Technology, Sweden.
BOEHM, M.	Max-Planck Institut für Extraterrestriche Physik, FRG.
BOSQUED, J.-M.	CESR, Toulouse, France.
BRUNING, K.	The Royal Institute of Technology, Sweden.
BRYANT, D.	Rutherford Appleton Laboratory, UK.
BURCH, J. L.	Southwest Research Institute, USA.
CANDIDI, M.	IFFI-CNR, Frascati, Italy.
CARLSON, C.	University of California, USA.
CAROVILLANO, R.	Boston College, USA.
CHAMBERLAIN, J.	Rice University, USA.
CHATURVEDI, P.	c/o Dr. S. L. Ossakow, NRL, Washington, DC, USA.
CHIAN, A.	Institute for Space Research, Sao Paulo, Brazil.
COLE, K.	La Trobe University, Australia.
CRAVEN, J.	University of Iowa, USA.
CROOKER, N.	University of California, USA.
DANBOURAS, J.	CESR, Toulouse, France.
DANIELSEN, C.	Danish Meteorological Institute, Denmark.
DUDENEY, J. R.	British Antarctic Survey, UK.
DE LA BEAU JARDIERE, O.	SRI International, Menlo Park, CA, USA.
EATHER, R.	Boston College, USA.
EGELAND, A.	University of Oslo, Norway.
ERGUN, A.	University of Califorina, USA.
ERLANDSON, R.	Johns Hopkins Applied Physics Laboratory, USA.
ESPY, P.	Utah State University, USA.
EVANS, D.	Space Environment Laboratory, Boulder, CO, USA.
FALTHAMMAR, C.-G.	The Royal Institute of Technology, Sweden.
FARMER, A.	Rutherford Appleton Laboratory, UK.
FRANK, L.	The University of Iowa, USA.
FRASER, B.	University of Newcastle, Australia.

FREEMAN, K.   Rutherford Appleton Laboratory, UK.
FRIIS-CHRISTENSEN, E.   Danish Meteorological Institute, Denmark.
GALPERIN, Y.   IKI, Moscow, USSR.
GARBE, G.   University of New Hampshire, USA.
GORNEY, D.   The Aerospace Corporation, USA.
GREENWALD, R.   Johns Hopkins Applied Physics Laboratory, USA.
GURNETT, D.   University of Iowa, USA.
GUSTAFSSON, G.   Swedish Institute of Space Physics, Sweden.
HALL, D. S.   Rutherford Appleton Laboratory, UK.
HARDY, D.   AFGL/PHP, MA, USA.
HONES, E.   Los Alamos, NM, USA.
HORNE, R. B.   British Antarctic Survey, UK.
HRUSKA, A.   NRC Herzberg Institute of Astrophysics, Canada.
HUGHES, A.   University of Natal, S. Africa.
HULTQVIST, B.   The Swedish Institute of Space Physics, Sweden.
IIJIMA, T.   The University of Tokyo, Japan.
ISHIMOTO, M.   Johns Hopkins Applied Physics Laboratory, USA.
JARVIS, M.   British Antarctic Survey, UK.
JOHNSON, R.   Geoscience and Engineering Center, USA.
JONES, D.   British Antarctic Survey, UK.
KAMIDE, Y.   Kyoto Sangyo University, Japan.
KAN, J.   Geophysical Institute, USA.
KENDALL, D.   NRC, Ottawa, Canada.
KIDD, S.   University of Alberta, Canada.
KILLEEN, T.   University of Michigan, USA.
KOSCH, M.   University of Natal, S. Africa.
KRIGE, D.   University of Natal, S. Africa.
LAUDER, M.   Oxford Computer Services, UK.
LLEWELLYN, E.   University of Saskatchewan, Canada.
LOCKWOOD, M.   Rutherford Appleton Laboratory, UK.
LUI, A.   Johns Hopkins Applied Physics Laboratory, USA.
LUMMERZHEIM, D.   University of Alaska, USA.
LUNDIN, R.   The Swedish Institute of Space Physics, Sweden.
LYONS, L.   The Aerospace Corporation, Los Angeles, USA.
MALINGRE, M.   Centre de Recherches en Physique de l'Environnement, France.
MANUEL, J.   University of Alberta, USA.
MARKLUND, G.   Royal Institute of Technology, USA.
MATTIN, N.   British Antarctic Survey, UK.
MAUK, B.   Johns Hopkins Applied Physics Laboratory, USA.
McCOMAS, D.   Los Alamos National Laboratory, USA.
McEWAN, D.   University of Saskatchewan, Canada.
McFADDEN, J.   University of California, USA.
McILWAIN, C.   University of California, USA.
McKENZIE, J. F.   University of Natal, S. Africa.
McPHERRON, R.   University of California, USA.
MEIER, R.   Naval Research Laboratory, Washington, USA.
MENG, C. I.   Johns Hopkins Applied Physics Laboratory, USA.

MORRISON, K.	British Antarctic Survey, UK.
NAKAMURA, M.	Max-Planck Institut für Physik und Astrophysik, FRG.
NAKAMURA, R.	University of Tokyo, Japan.
NEWELL, P.	Johns Hopkins Applied Physics Laboratory, USA.
NIELSEN, E.	Max-Planck Institut für Aeronomie, FRG.
NISHIDA, A.	ISAS, Kanagawa, Japan.
NISHIKAWA, K.-I.	University of Iowa, USA.
NISHITANI, N.	University of Tokyo, Japan.
OHTA, A.	Oslo University, Norway.
OMURA, Y.	University of Kyoto, Japan.
ORR, D.	University of York, UK.
PARROT, M.	LPCE/CNRS, Orleans, France.
PERRAUT, S.	CRPE/CNET, Issy-les-Moulineaux, France.
PETERSON, W.	Lockheed Palo Alto Laboratories, USA.
PINNOCK, M.	British Antarctic Survey, UK.
POTEMRA, T.	Johns Hopkins Applied Physics Laboratory, USA.
POTELETTE, R.	CRPE, Saint-Maur des Fosses, France.
REES, D.	University College London, UK.
REES, M.	University of Alaska, USA.
RETTERER, J.	Boston College, USA.
RIDLEY, C.	High Altitude Observatory, National Center for Atmospheric Research, Boulder, USA.
RODGER, A.	British Antarctic Survey, UK.
ROETTGER, J.	EISCAT Scientific Association, Sweden.
ROMICK, G.	University of Alaska, USA.
ROSTOKER, G.	University of Alberta, Canada.
ROTHWELL, P. S.	AFGL/PHG, MA, USA.
ROTHWELL, Pamela	University of Sussex, UK.
RYCROFT, M. J.	British Antarctic Survey, UK.
SAGDEEV, R.	IKI, Moscow, USSR.
SAMSON, J.	University of Alberta, Canada.
SAUNDERS, M.	Imperial College London, UK.
SAZHIN, S.	University of Sheffield, UK.
SCHRODER, W.	Geophysical Station, D-2820 Bremen-Roennebeck, Hechelstrasse 8, FRG.
SCHULZ, M.	Aerospace Corporation, Los Angeles, USA.
SCOURFIELD, M.	University of Natal, S. Africa.
SHELLEY, E.	Lockheed Palo Alto Research Laboratories, USA.
SHEPHERD, G.	Cress York University, Canada.
SHIBAJI, T.	University of Tokyo, Japan.
SHUKLA, A.	Royal Aerospace Establishment, UK.
SILEVITCH, M.	North-eastern University, Boston, USA.
SINGER, H.	AFGL/PHG, MA, USA.
SISCOE, G.	University of California, USA.
SMITH, A.	British Antarctic Survey, UK.
SOLHEIM, B.	York University, Ontario, Canada.
SOUTHWOOD, D.	Imperial College London, UK.

STEEN, A.	The Swedish Institute of Space Physics, Sweden.
STERN, D.	NASA Goddard Space Flight Center, USA.
TANSKANEN, P.	University of Oulu, Finland.
TEMERIN, M.	University of California, USA.
TINSLEY, B.	National Science Foundation, Washington, DC, USA.
TREILHOU, J.-P.	CESR, Toulouse, France.
TROSHICHEV, O.	Arctic and Antarctic Institute, 33 Bering St., 199226 Leningrad, USSR.
VALLANCE-JONES, A.	NRC, Herzberg Institute of Astrophysics, Ottawa, Canada.
VASYLUNAS, V.	Max-Planck Institut für Aeronomie, FRG.
WALKER, A. D.	University of Natal, S. Africa.
WALKER, J.	GSC/EMR, Ottawa, Canada.
WARD, W.	Department of Applied and Theoretical Physics, University of Cambridge, UK.
WERDEN, W.	University of Washington, USA.
WHITAKER, W.	P. O. Box 3036, McLean, VA, USA.
WHITEHEAD, J.	Queensland University, Australia.
WILLIAMS, D.	Johns Hopkins Applied Physics Laboratory, USA.
WILLIS, D.	Rutherford Appleton Laboratory, UK.
WINCKLER, J. R.	University of Minnesota, USA.
WINGLEE, R.	University of Colorado, USA.
WINSER, K.	Rutherford Appleton Laboratory, UK.
WRIGHT, W.	British Antarctic Survey, UK.
YEOMAN, T.	University of York, UK.

I. INTRODUCTORY OBSERVATIONS

I-1. AURORAL PHENOMENA

S.-I. Akasofu*

The great progress in auroral physics during the last few decades has created the awareness that most of what we call auroral phenomena are various manifestations of dissipation processes associated with the discharge of electrical power generated by the solar-wind/magnetosphere interaction. Here we review briefly the progress that has been made in understanding some of the basic processes in the solar-wind/magnetosphere/thermosphere/ionosphere interaction.

1. INTRODUCTION

As a star, the Sun is continuously emitting enormous amounts of energy into space. This energy emission takes several forms, the first of which is the familiar black-body radiation. The second mode of energy emission is the *solar wind*, which consists of protons with energies of about 1 kV and an equal number of electrons with energies of a few hundred electron volts. They stream out from the Sun at supersonic speeds. The solar wind tends to confine the Earth and its magnetic field into a comet-shaped cavity called the *magnetosphere*. As the solar wind interacts with the magnetosphere, as much as 10^6 MW of electrical power is generated, discharged, and subsequently dissipated, partly through that portion of the upper atmosphere called the polar *ionosphere*. Both the solar X-ray and ultraviolet radiations, the third mode of energy emission, are responsible for producing the ionosphere. Most of what we call *auroral phenomena* are various manifestations of dissipation processes associated with this discharge.

The discharge process produces, among many fascinating phenomena, visible emissions that we recognize as the *aurora*. In fact, of all the manifestations, the aurora is the only visible phenomenon. As described in the following chapters, a great variety of other manifestations occur and can be detected by specific instruments, such as magnetometers, ionosondes, and many satellite-borne instruments.

Three systems—the solar wind, the magnetosphere, and the ionosphere—interact, transmitting and transforming the solar-wind energy into energies of auroral phenomena, and eventually depositing most of it as heat energy in the ionosphere (Fig. 1). So far, the electrical connection between the magnetosphere and the upper atmosphere has been considered. Thus, its ionized component, the ionosphere, has been emphasized. However, the neutral component of the upper atmosphere also responds to the discharge process. Thus, the term *thermosphere* includes both the ionized and neutral component of the upper atmosphere above the mesosphere (to about 80 km). The importance of the thermosphere in auroral phenomena is emphasized in Chapter 2.

Sydney Chapman played the most important role in establishing the foundations of the scientific discipline to which all the participants of the International Conference on Auroral Physics belong. However, it is impossible to describe Chapman's contributions to auroral science in a short paper. Thus, we confine ourselves to making a few remarks, and we limit the subject area to auroral science:

1. Chapman established the present concept of geomagnetic storms in terms of the initial phase and the main phase (1918).
2. He obtained the storm-time current system in terms of the equivalent (two-dimensional) currents (1918–35).
3. He published a theory of the night airglow and the formation of the ozone layer (1930).
4. With V. C. A. Ferraro, he published a theory of magnetosphere formation by proposing that solar-wind particles constitute a plasma and are not a cloud of individual particles (1931).
5. He published a theory of the formation of the ionosphere (1931).
6. With T. G. Cowling, he obtained the standard formulas for the ionospheric conductivities (1939) and published the classic treatise "The Mathematical Theory of Non-Uniform Gases" (1953).
7. With J. Bartels, he published the *magnum opus* "Geomagnetism" in 1940. It served as the basic reference and treatise until about 1970.

Each of these seven contributions can be regarded as fundamental. Even one such contribution may be considered to be sufficient by any single researcher in his entire scientific career. Those who are interested in Chapman's contributions, not only in auroral physics but also in other fields, should refer to *Sydney Chapman, Eighty, From his Friends*, by Akasofu et al. [1968].

---

*Geophysical Institute, University of Alaska, Fairbanks, Alaska 99775-0800.

Auroral Physics, edited by C.-I. Meng, M. J. Rycroft and L. A. Frank. © Cambridge UP 1991

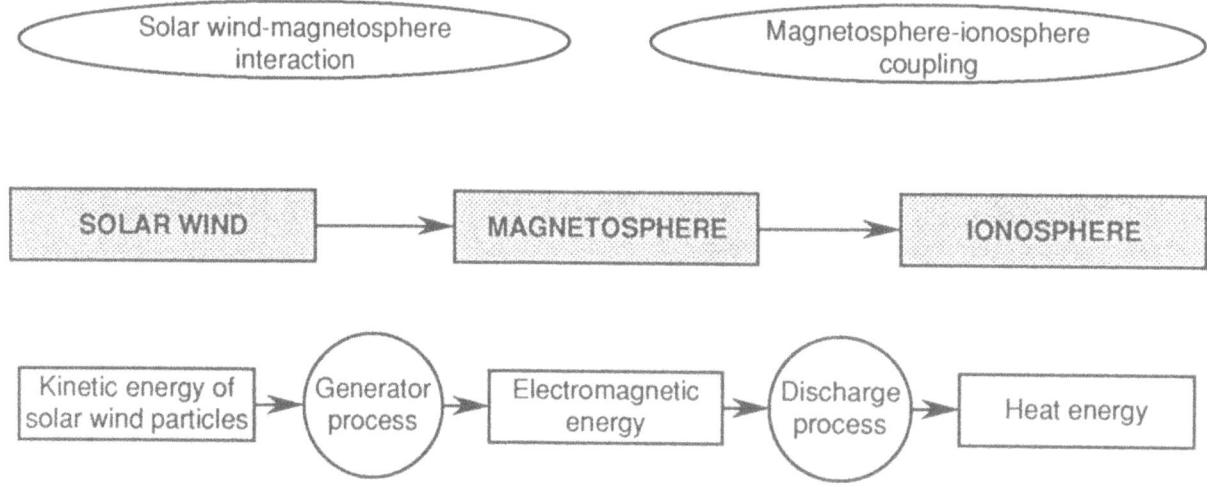

Figure 1—Flow chart for solar-wind/magnetosphere/ionosphere coupling. It shows the energy flow, energy conversion, and associated processes.

2. POWER GENERATION FOR AURORAL PHENOMENA

If the solar wind were a nonmagnetized plasma, we would not expect much more than the formation of the magnetospheric cavity. However, since the solar wind is a magnetized plasma (carrying the solar magnetic field, generally called the interplanetary magnetic field (IMF)), the interaction between the solar wind and the magnetosphere becomes very complex. Thus, progress has been slow in understanding this particular interaction process, which is called dayside magnetic *reconnection*. The simplest situation occurs when the IMF is directed southward, so that it is antiparallel to the Earth's magnetic field near the nose of the magnetosphere. It is a recent finding by *Russell and Elphic* [1979] that this interaction is not a steady process. Indeed, a more recent computer-simulation study confirms that it is basically a nonsteady process (*Lee and Fu* [1985]).

It is through dayside magnetic reconnection that some magnetic field lines from the magnetosphere are connected to solar-wind magnetic field lines across the magnetopause. We are indebted to *Dungey* [1961] for the present concept of the so called *open magnetosphere*. It is understood that solar-wind particles flow along the magnetopause, crossing the newly connected magnetic field lines, although the details have not yet been fully understood. This process is basically the same as that of a magnetohydrodynamic (MHD) generator. Thus, the entire magnetopause constitutes a gigantic natural generator that we call the solar-wind/magnetosphere generator. Individual solar-wind ions

lose only a very small fraction of their kinetic energy by this interaction, but it is through this process that more than 10^6 MW of power is generated. This amount is estimated on the basis of the total energy-deposition rate in the polar ionosphere. The total potential drop generated, about 100 kV, is estimated from the potential difference between the dawnside and duskside of the auroral oval.

3. MAGNETOSPHERE/IONOSPHERE INTERACTION

The dynamo process described above would have little significance if the ionosphere did not exist. Without the ionosphere, the dynamo has to power an open circuit. Many auroral phenomena occur because the power is transmitted to the ionosphere from the magnetosphere. Furthermore, the ionosphere is not simply a passive load, and the magnetosphere and the ionosphere constitute a complex feedback system.

As the magnetosphere is filled with a very rarefied plasma, which is permeated by the Earth's magnetic field, electric currents tend to flow along the magnetic field lines. Called Birkeland currents, these *field-aligned currents* (FAC) transmit the power generated by the dynamo to the ionosphere. Actually, the current system that connects the magnetosphere and the ionosphere is very complex, consisting of the primary (region 1) and secondary (region 2) currents [*Iijima and Potemra*, 1976]. The aurora is the result of this discharge process.

When the aurora can be seen from well above the northern polar region, it appears as a ring of luminosity around the geomagnetic pole, the *auroral oval*. The

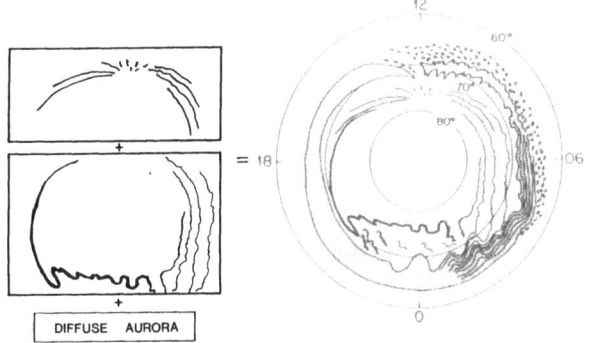

Figure 2—Schematic representation of the local time distribution of the aurora.

auroral oval delineates approximately the area called the *polar cap*; the geomagnetic field lines that anchor to the polar cap are connected to the IMF lines across the magnetopause, and such field lines are termed *open field lines*. On the other hand, other geomagnetic field lines connect two points (one in each hemisphere) across the equatorial plane; such field lines are called *closed field lines*. The auroral oval consists of two parts: the first is the oval of discrete (curtain-like) aurora, and the second is the oval of the diffuse aurora (Fig. 2). In spite of the great progress made in auroral physics in recent years, one of the long-standing unsolved problems is how the region of discrete aurora and the diffuse aurora are connected by geomagnetic field lines from different plasma regimes in the tail of the magnetosphere. This is a subject of Chapter 4.

4. AURORAL POTENTIAL STRUCTURE

The magnetosphere and the ionosphere together constitute a complex interactive system. Alfvén waves and the FAC carry information between the magnetosphere and the ionosphere. However, in conducting the upward electric current from the ionosphere, the magnetic field lines have only a limited capacity because, like all charged particles in the Van Allen radiation belt, the current-carrying electrons have a helical motion along geomagnetic field lines. As the electrons approach the Earth (move into a region of stronger field), the *pitch* of the helical motion increases. As a result, the electron motion becomes completely circular at a certain height. At this mirror point, the electrons are *reflected back* and start to move upward, so that they cannot reach the ionosphere. However, it has been suggested that when the generator power and the FAC density become high enough, an interesting potential distribution develops in this rarefied plasma environment at an altitude of 10,000–20,000 km above the ground. It appears that the structure is a sort of electrical double layer, but its exact nature in the Earth's environment is presently a controversial issue among

auroral scientists [*Akasofu and Kan*, 1981]. The *auroral potential structure* appears to have a U-shaped geometry in its north–south cross section in a gross time-average sense; it is a source of intense kilometric radio emissions [*Gurnett and Inan*, 1988]. An electron moving downward along the center of the structure is accelerated toward the Earth (ionosphere) by the upward-directed electric field, increasing its velocity component along the magnetic field line. Thus, its pitch decreases, allowing the electron to reach the ionosphere. The potential drop in the structure is estimated to be a few kilovolts, so that the electron has acquired a few kilo-electron volts of energy by the time it emerges from the bottom of the structure. The energy spectral characteristics of auroral electrons have been studied extensively by rocket- and satellite-borne instruments in the past [*Burch*, 1988].

Mechanisms for accelerating energetic charged particles in natural conditions have greatly concerned astrophysicists, solar physicists, and auroral (magnetospheric) physicists, because such particles are common in cosmic, solar, and magnetospheric environments. It has been widely believed that it was impossible to maintain a significant electric field along magnetic field lines in a very rarefied plasma, making it impossible to accelerate charged particles by an electric field along the field line. Thus, as alternatives, a variety of MHD processes has been conceived. It seems, however, that, at least in the magnetosphere, an electric field parallel to the geomagnetic field lines can be produced and maintained in a limited region when an intense electric current flows along the field lines, as first suggested by Alfvén [1950]. There is no doubt that there are other processes involved in the acceleration of auroral particles. This important subject is discussed in Chapter 3.

The auroral potential structure and other acceleration processes are crucial in producing the aurora and associated auroral phenomena. A large number of current-carrying electrons reach the top of the ionosphere after being accelerated to a few kiloelectron volts. As a result, the electrons are capable of ionizing and dissociating atoms and molecules in the polar upper atmosphere. The potential structure also accelerates positive ions upward, producing upward-streaming ions. Complex plasma-wave/particle interaction processes also occur in the auroral potential structure, generating intense radio emissions in the kilometric range (0.6–1.5 MHz). These subjects are discussed particularly in Chapter 4, and also in all other chapters.

5. AURORAL PHENOMENA AT THE IONOSPHERIC LEVEL

As the electrons penetrate downward, they collide with atmospheric atoms and molecules, losing about

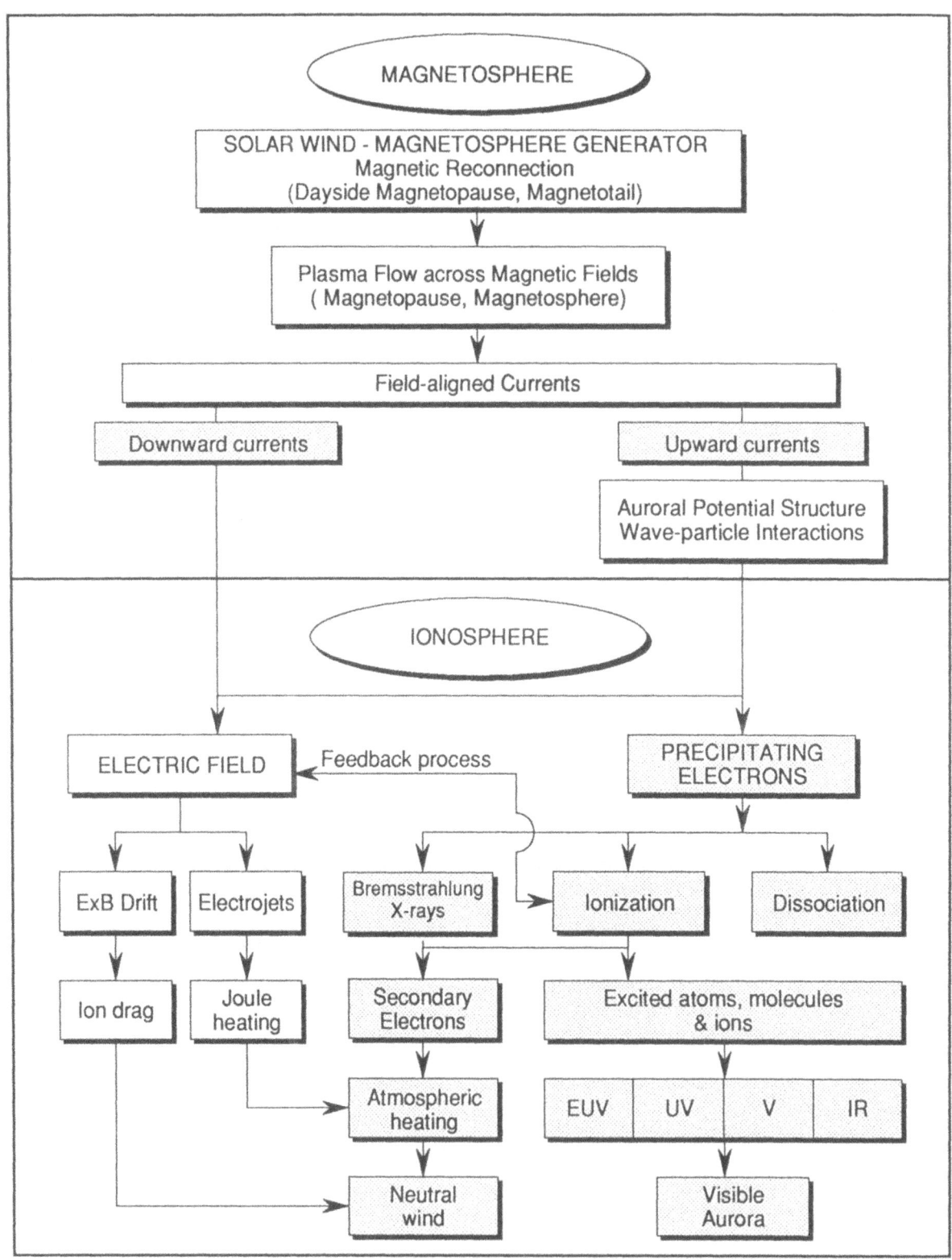

Figure 3—Flow chart for magnetosphere/ionosphere coupling and production of the aurora.

30 eV at each collision. The kinetic energy of a few kiloelectron volts is needed for auroral electrons to penetrate to an altitude of 100 km, where the atmospheric density is high enough that the optical emissions from the excited atoms and molecules can be detected by the naked eye (see Fig. 3). By colliding with atmospheric atoms and molecules, the precipitating electrons ionize and excite them, and dissociate molecules. Subsequently, a complex chain of chemical reactions takes place, including excitation by secondary electrons. The excited ions, atoms, and molecules are responsible for electromagnetic radiation over a wide spectral range extending from the extreme ultraviolet (EUV) to the infrared (IR). The most common emission from the aurora is a whitish-green light (5577 Å) that is emitted by oxygen atoms, excited in part by colliding secondary electrons. A red emission near the bottom of the auroral curtain comes partly from excited nitrogen molecules. The EUV (from excited oxygen and hydrogen atoms, ionized oxygen atoms, and others), UV (from ionized nitrogen molecules and others), and IR (from oxygen, nitrogen molecules and ionized nitrogen molecules, and others) radiations cannot be observed on the ground because they are absorbed by the intervening atmosphere. These topics are dealt with in Chapters 2 and 7.

The voltage produced by the dynamo process varies between 25 and 150 kV. This potential drop is transmitted to the ionosphere by FACs. The resulting electric field in the ionosphere and the energetic electrons produced by the auroral potential structure are responsible for most of the phenomena associated with the aurora.

The aurora has a thin, curtain-like form; its thickness (north–south) is about 1 km or less, while its lateral extent (east–west) is thousands of kilometers. The reason for this particular auroral form is simply that the accelerated electrons are confined to a thin sheet. However, it is not known at present why the FACs tend to develop such a thin sheet. Multiple curtains (two or more) can form, but the cause is not well understood. The bottom of the curtain is at about 100 km altitude, because most of the precipitating electrons lose their penetrating power at that height as the atmospheric density increases rapidly downward. The thin-electron-sheet beam exhibits a variety of instabilities, including a series of vortices and curls, as well as large-scale structures (such as westward traveling surges). This subject is discussed in Chapter 6.

In the polar upper atmosphere surrounded by the auroral oval, the electric field lies in the dawn-to-dusk direction, causing the ionospheric plasma to have an $E \times B$ drift motion from the dayside to the nightside. Just outside the oval, the drift motion reverses direction (namely, from the nightside to the dayside in both the dawn and dusk hemispheres), resulting in two large-scale vortex motions that are, as a whole, called *convection*. There are several methods of observing the $E \times B$ drift motion, including direct measurement by satellite-borne instruments. Barium-ion clouds released from a rocket at an altitude of a few hundred kilometers participate in the same $E \times B$ drift motion and provide a method that has been used extensively to map the electric field in the auroral upper atmosphere [*Heppner and Maynard*, 1987; *Heelis*, 1988; *Fälthammer*, 1989]. An incoherent scatter radar is a powerful ground-based observing device for detecting ion motions. The drifting ions impart their momentum to the neutral particles in the upper ionosphere, causing them to be dragged in the same direction, i.e., from the dayside to the nightside. The resulting motion is another type of atmospheric wind [*Killeen and Roble*, 1988].

In the lower ionosphere, only electrons can participate in the convection motion. As a result, two large-scale vortex currents occur in the polar ionosphere. They are particularly concentrated along the oval and are called the westward and eastward electrojets. Intense Joule heat is produced along the auroral oval. This heating is another cause of large-scale winds in the upper atmosphere. A number of researchers have made a detailed study of the atmospheric motions resulting from such heating [e.g., *Rees and Fuller-Rowell*, 1987]. The hot secondary-electron gas resulting from the ionization heats the atmospheric atoms and molecules, and the interaction becomes an important cause of large-scale winds in the polar upper atmosphere. As the energetic electrons are decelerated by the collisions, X-rays are generated that can be detected by a balloon-borne X-ray detector at an altitude of 30 km and by satellite-borne detectors from above. For details of this subject, see new textbooks by *Akasofu and Kamide* [1987], *Kamide* [1988], and *Rees* [1989].

6. AURORAL SUBSTORMS AND MAGNETOSPHERIC SUBSTORMS

It was the all-sky camera operation and the subsequent analyses during the International Geophysical Year (1957/58) that revealed systematic auroral activity over the entire polar region, called the *auroral substorm*. A series of typical auroral features is as follows. The first indication of an auroral substorm is a sudden brightening of the auroral curtain in the midnight or late evening sector. This brightening spreads rapidly along the curtain, so that in a matter of several minutes the entire section of the curtain in the dark hemisphere becomes bright. The bright curtain begins to move poleward in the midnight sector with a speed of a few hundred meters per second. At the same time, a large-scale wavy motion is generated near the western end of the poleward motion. This wavy motion, called the westward traveling surge, propagates westward (toward the dusk-sunset line) with a speed of about 1 km s^{-1}. In the morning sector, auroral curtains appear to disintegrate into many patches. The poleward motion in the midnight sector lasts typically for about 30 min to 1 hour. After this poleward advancing curtain reaches its highest latitude, auroral activity begins to subside; however, the westward traveling surge often continues to propagate along the dayside part of the oval. During the last decade, excellent auroral images have been taken from satellites, with the result that many global features of the auroral substorm have been clarified. These findings are reported in Chapter 5.

The auroral substorm is the only visible manifestation of what we call the *magnetospheric substorm*. There are many other different manifestations of the magnetospheric substorm [*Akasofu*, 1977]. The electrojets are greatly intensified during the auroral substorm, causing intense geomagnetic disturbances. This phenomenon is called the polar magnetic substorm. The flow patterns of the electrojets have been studied extensively by *Kamide* [1988], *Baumjohann et al.* [1981], and others.

The causes of the magnetospheric substorm have been one of the major topics among magnetospheric physicists during the last two decades. Many theorists have speculated that magnetic reconnection in the *magnetotail* is responsible for the energy supply. In the magnetotail, the magnetic field is directed toward Earth in the northern half and away from Earth in the southern half; the magnetotail can be considered to consist of two solenoids, producing antiparallel magnetic fields. They have speculated that such an antiparallel magnetic-field system is intrinsically unstable and that the fields can spontaneously and explosively annihilate themselves. Thus, they have hypothesized that the magnetic energy accumulated and stored in the magnetotail would be suddenly converted into energy for the magnetospheric substorm by a process that is intrinsic to the magnetosphere. In fact, it has long been said that the magnetotail contains enough energy for many intense substorms and that a search should be made for internal processes that could trigger magnetic reconnection explosively.

However, it has become increasingly clear that the occurrence of magnetospheric substorms is at least partially controlled by the solar wind and the IMF. In other words, the growth and decay of magnetospheric substorms are controlled by the rise and fall of the power generated by the solar-wind/magnetosphere generator, which is a function of at least the speed (**V**) of the solar wind, the magnitude (**B**) of the magnetic field,

and the polar angle (θ) of the magnetic field vector. Time variations of the power equation are similar to those of the rate of total energy dissipation in the inner magnetosphere (which includes the ring-current injection rate, the Joule heat production rate, the kinetic energy injection rate of auroral electrons, etc.). The exact dependence of the power on these quantities is, however, a matter of great controversy. One empirical formula suggested by *Perreault and Akasofu* [1978] is given by

$$\text{Power (MW)} = 20 \, \mathbf{V} \, (\text{km s}^{-1}) \times \mathbf{B}^2 \, (\text{nT})$$

$$\times \sin^4 (\theta/2)$$

where θ is approximately the polar angle ($\theta = 0°$ for a northward-directed field, and $\theta = 180°$ for a southward-directed field). The equation has been theoretically confirmed by *Pudovkin et al.* [1986]. *Reiff et al.* [1981] have also shown that the total potential drop across the polar cap is closely related to the power given in the equation above.

Among the solar-wind quantities that control the power of the solar-wind/magnetosphere generator, the angle θ is, on the average, the most variable. Consider a simple situation in which circularly polarized Alfvén waves propagate along the IMF (which lie approximately in the equatorial plane); in this situation the angle θ is most effective in modulating the power (since V and B do not vary in this situation). Obviously, the power will be highest when $\theta = 180°$, namely, when the IMF is directed southward. Therefore, the occurrence of magnetospheric substorms is most often associated with the southward turning of the IMF vector. When $\theta \sim 180°$, the intensity of substorms depends on the magnitude of the magnetic field; the greater the magnitude, the more intense is the magnetospheric substorm.

At present, we are still far from a firm understanding of processes that lead to the onset of auroral substorms. If magnetic reconnection is involved in substorm processes, it is not certain whether or not it is the primary cause [*Hones*, 1984] or an effect [*Kan et al.*, 1988; *Akasofu*, 1989]. It is important to note that the onset is signaled by the sudden brightening of an auroral curtain in the midnight sector. Thus, one of the most interesting problems in this regard is how an intensified FAC along a narrow strip in the midnight sector can arise after the magnetospheric convection becomes enhanced. Several ideas on onset processes are presented in Chapter 5.

The size of the auroral oval is a function of the power (or the north–south component of the IMF). In the midnight sector, the latitude of the oval is about 67° or above when the power is less than 10^5 MW. As the

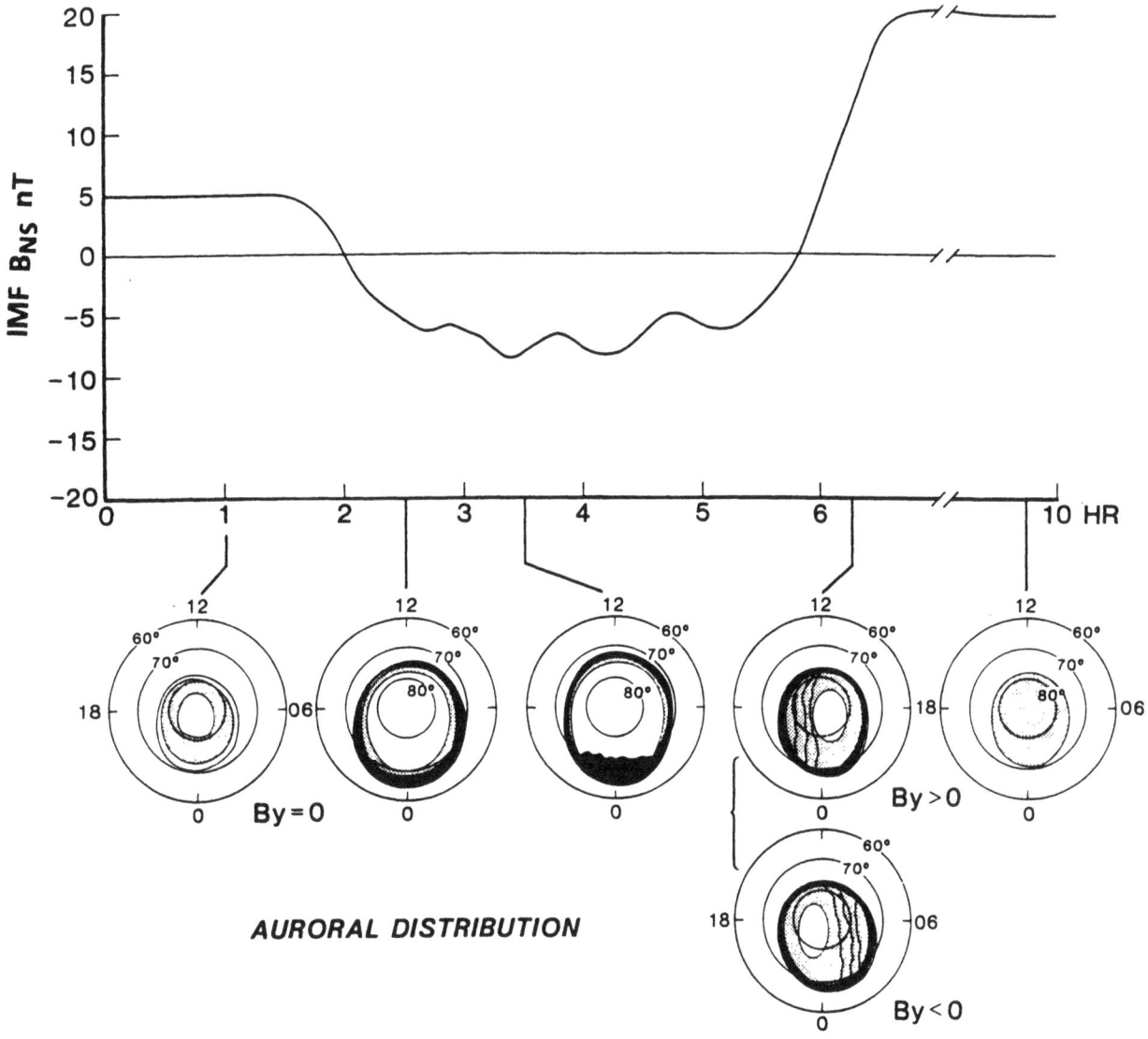

Figure 4—Schematic representation of changes of the northern auroral distribution as changes occur in the north–south component of the interplanetary magnetic field.

power increases to 10^6 MW, the oval expands to $65°$ or lower. Figure 4 shows schematically how the auroral distribution varies as the north–south component $\mathbf{B}_{NS}$ of the IMF varies from $+5$ nT to a larger negative value (-7 nT) and then to a very large positive value ($\sim +20$ nT). At the beginning, the oval is small and much of the polar cap is covered by a weak glow (mostly subvisual) except in the vicinity of the geomagnetic pole. As $\mathbf{B}_{NS}$ becomes negative (namely, as the power of the auroral generator increases), the oval expands rapidly, and the subvisual glow disappears except for a narrow belt just to the poleward side of the oval.

Then, a substorm begins. Bright and active auroral curtains advance toward higher latitudes. As a result, the area enclosed by the auroral oval contracts. *Frank and Craven* [1988] suggest that this is an indication that magnetic energy accumulated in the magnetotail (during the initial expansion of the oval) is released. When $\mathbf{B}_{NS}$ becomes positive again, the substorm begins to subside and the oval begins to contract poleward. At the same time, a subvisual glow starts to fill the polar cap.

Many other interesting phenomena also occur in the magnetosphere during magnetospheric substorms. Particles with energies of a few hundred kilovolts are produced in the magnetotail and stream along the magnetic field lines [*Williams et al.*, 1985]. The distributions of plasmas in the magnetosphere undergo drastic changes. Together with magnetotail reconnection, these changes are expected to be directly or indirectly related to auroral dynamics during auroral substorms.

7. POLAR CAP DURING PERIODS OF NORTHWARD IMF

One of the important questions among magnetospheric physicists today is: what happens to the aurora when θ becomes almost 0° for an extended period? As the power of the solar-wind/magnetosphere generator decreases, the aurora becomes dim and the auroral electrojets become weak. However, an unexpectedly interesting auroral phenomenon takes place in this situation. There appear a number of auroral curtains and subvisual patches across the auroral oval that are parallel to the noon–midnight meridian. Such auroras are called polar-cap auroras and have recently been studied extensively by *Frank and Craven* [1988], *Meng and Lundin* [1986], and others. This phenomenon and many others associated with it (field-aligned currents, convection, etc.) cannot be simply understood in terms of the decreasing power of the generator. *Lassen and Danielson* [1978] showed that the azimuthal angle (or the east–west component) of the IMF plays an important role in determining the distribution of the aurora when the angle θ becomes small. The convection pattern also becomes significantly asymmetric with respect to the noon–midnight meridian.

8. NEED FOR A GLOBAL IMAGING OF THE MAGNETOSPHERE

High-time resolution global imaging of the aurora has made a major contribution in advancing magnetospheric physics during the last decade. It provides a "visible frame of reference" in studying individual substorms and in dealing with very complex magnetospheric phenomena, as well as with a number of vital quantitative parameters. In the past, despite the fact that the concept of the auroral oval has been useful as a natural frame of reference (rather than geomagnetic latitude) in sorting out a great variety of ground-based and satellite observations, auroral imaging had not necessarily been the top-priority project, compared with observations of "invisible" physical quantities, such as electric fields, magnetic fields, particle fluxes, etc., and even "visible" components, including auroral spectra. The present success in imaging the aurora by spacecraft has made it clear that imaging has been established as a truly vital tool in auroral and magnetospheric physics.

It may not be an exaggeration to say that imaging of the entire magnetosphere is needed for the future advancement of magnetospheric substorm studies. The complexity of magnetospheric phenomena and their three-dimensional aspects cannot easily be studied completely by observations made at single points by a few satellites. Although past satellite observations have suggested many interesting phenomena, proving their validity is not an easy task. There have been many two- and three- dimensional simulation studies to explain these phenomena. However, we must be cautious in inferring magnetospheric phenomena on the basis of two- or three-dimensional computer simulation studies alone. The simulation studies must be tested by theoretical analyses and by global observations.

In this context, like auroral imaging, it is expected that the global observation of the magnetosphere could make a major advance in magnetospheric substorm studies. It must be stressed that such global imaging does not reduce the importance of in situ observations of the magnetosphere by satellites and from ground-based stations. On the contrary, these observations will be truly complementary, just as satellite and ground-based observations are. Such a complementary effort was dramatically demonstrated by both in situ spacecraft (ICE) and ground-based imaging observation of P/Giacobini-Zinner comet. Figure 5 shows an anticipated view of the magnetosphere at a distance of the orbit of the Moon, i.e., at 60 Earth radii (R_E). The figure was made by computing the line-of-sight densities, along different view directions, of sunlight resonantly scattered by oxygen ions.

9. CONCLUDING REMARKS

Perhaps, the research for truth follows the geometry of a polyhedron. Often, a researcher stands on one surface of it, while another researcher stands on another surface. For the first, everything on his surface appears to be consistent with his own model (a paradigm), while the same holds true for the second on his surface. Unfortunately, the first does not understand why the second attempts to understand a phenomena differently from his, or vice versa. Such a difference becomes a cause of controversy. Eventually, however, both will come to the realization that each has been studying only one aspect of a multisurface phenomenon and that the phenomenon in which they are interested has at least two surfaces. Often, this task has been left to a new generation. This is what the present generation has done; it is why we have made such good progress in this field. Often, it is in the polyhedron that understanding of a natural phenomenon deepens, by revealing one new surface after another, which collectively constitute the truth. Alfvén, Birkeland, Chapman, Dungey, and many others have established

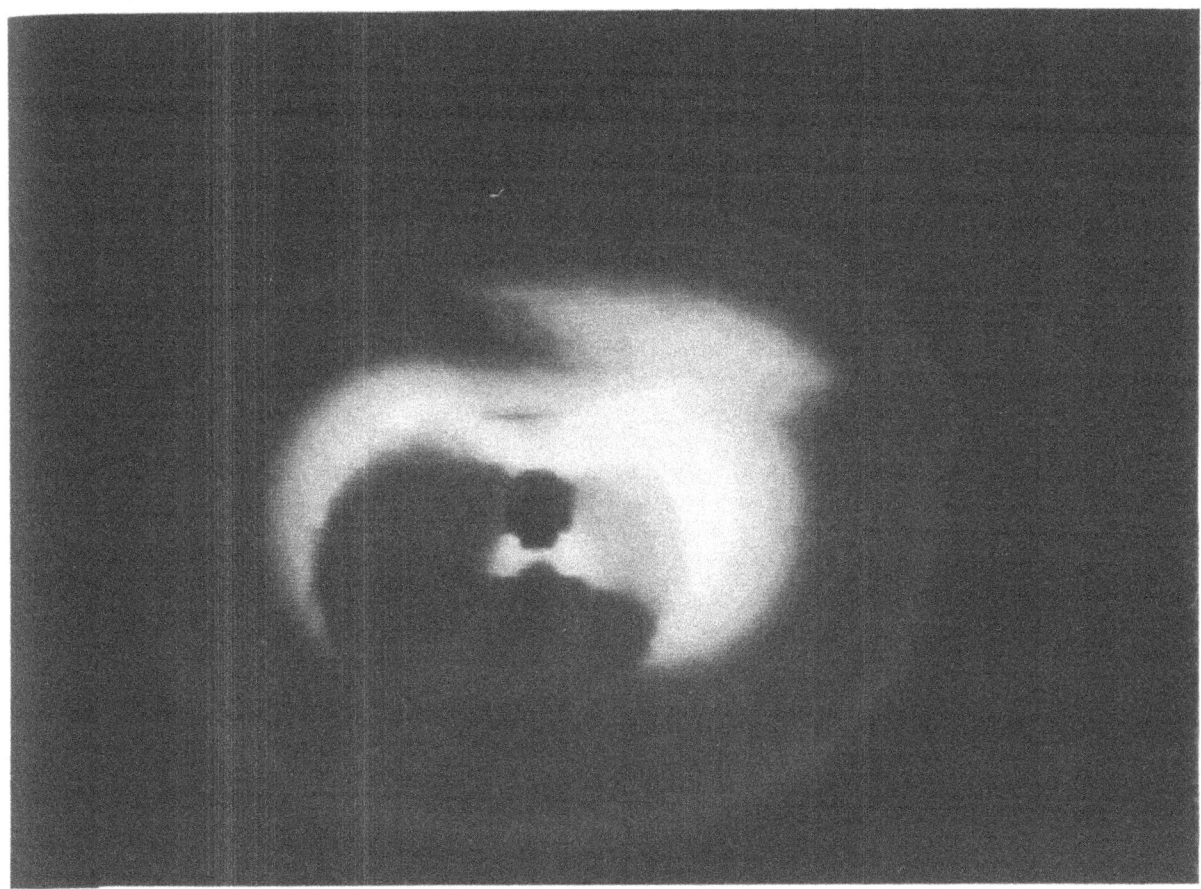

Figure 5—Results of model calculations, showing an image of the magnetosphere in resonantly scattered sunlight from O⁺. (This figure also appears in color: Plate 1.)

different surfaces of a polyhedron. During the last three decades, we have also discovered several surfaces of the conceptual polyhedron of the magnetosphere and linked them to the earlier surfaces. There is no doubt that new generations of auroral physicists will discover many other surfaces.

ACKNOWLEDGMENT—The author would like to thank C.-I. Meng, M. Rycroft, and L. Frank for their valuable suggestions in improving the manuscript.

The research reported here is supported in part by National Science Foundation grant ATM 88-03133.

REFERENCES

Akasofu, S.-I., B. Fogle, and B. Haurwitz, *Sydney Chapman, Eighty, From his Friends*, University of Colorado Press (1968).
Akasofu, S.-I., *Physics of Magnetospheric Substorms*, D. Reidel Pub. Co., Dordrecht, Holland (1977).
Akasofu, S.-I., and J. R. Kan, *Physics of Auroral Arc Formation*, American Geophysical Union, Washington, DC (1981).
Akasofu, S.-I., and Y. Kamide (ed.), *The Solar Wind and the Earth*, Terra Science Pub. Co. and D. Reidel Pub. Co., Dordrecht/Boston/Lancaster/Tokyo (1987).
Akasofu, S.-I., "Substorms: Future of Magnetospheric Substorm-Storm Research," *EOS Trans. AGU*, **70**, 529 (1989).
Alfvén, H., *Cosmical Electrodynamics*, Oxford University Press, Oxford (1950).
Baumjohann, W., R. J. Pellinen, H. J. Opgenoorth, and E. Nielsen, "Joint Two-Dimensional Observations of Ground Magnetic and Ionospheric Electric Fields Associated with Auroral Zone Currents: Current Systems Associated with Local Auroral Breakups," *Planet. Space Sci.*, **29**, 431 (1981).

Burch, J. L., "Energetic Particles and Currents: Results from Dynamics Explorer," *Rev. Geophys.*, **26**, 215 (1988).
Chapman, S., "An Outline of a Theory of Magnetic Storms," *Proc. R. Soc.*, **97**, 61 (1918).
Chapman, S., "On Ozone and Atomic Oxygen in the Upper Atmosphere," *Phil. Mag.*, **10**, 369 (1930).
Chapman, S., "The Absorption and Dissociative or Ionizing Effect of Monochromatic Radiation in an Atmosphere on a Rotating Earth," *Proc. Phys. Soc. London*, **43**, 26 (1931).
Chapman, S., and V. C. A. Ferraro, "A New Theory of Magnetic Storms, Part I, The Initial Phase," *Terr. Magn.*, **36**, 77 (1931).
Chapman, S., "The Electric Current Systems of Magnetic Storms," *Terr. Magn.*, **40**, 349 (1935).
Chapman, S., and T. G. Cowling, *The Mathematical Theory of Non-Uniform Gases*, Cambridge University Press, Cambridge (1939).
Dungey, J. W., "Inteplanetary Magnetic Field and the Auroral Zone," *Phys. Res. Lett.*, **6**, 47 (1961).

Fälthammer, C.-G., "Electric Fields in the Magnetosphere—The Evidence from ISEE and Viking," *Plasma Sci.*, **17**, 174 (1989).

Frank, L. A., and J. D. Craven, "Imaging Results from Dynamics Explorer I," *Rev. Geophys.*, **26**, 249 (1988).

Gurnett, D. A., and U. S. Inan, "Plasma Wave Observations with the Dynamics Explorer I Spacecraft," *Rev. Geophys.*, **26**, 285 (1988).

Heelis, R. A., "Studies of Ionospheric Plasma and Electrodynamics and Their Application to Ionosphere-Magnetosphere Coupling," *Rev. Geophys.*, **20**, 317 (1988).

Heppner, J. P., and N. C. Maynard, "Empirical High-Latitude Electric Field Models," *J. Geophys. Res.*, **92**, 4467 (1987).

Hones, E. W., Jr., "Plasma Sheet Behavior During Substorms," in *Magnetic Reconnection in Space and Laboratory Plasmas*, Geophysical Monograph 30, E. W. Hones, Jr., ed., American Geophysical Union, Washington, DC (1984).

Iijima, T., and T. A. Potemra, "Large-Scale Characteristics of Field-Aligned Currents Associated with Substorms," *J. Geophys. Res.*, **81**, 3999 (1976).

Kamide, Y., *Electrodynamic Processes in the Earth's Ionosphere and Magnetosphere*, Kyoto Sangyo University Press, Kyoto (1988).

Kan, J. R., L. Zhu, and S.-I. Akasofu, "A Theory of Substorms: Onset and Subsidence," *J. Geophys. Res.*, **93**, 5024 (1988).

Killeen, T. L., and R. G. Roble, "Thermosphere Dynamics: Contributions from the First 5 Years of the Dynamics Explorer Program," *Rev. Geophys.*, **26**, 329 (1988).

Lassen, K., and C. Danielson, "Quiet Time Pattern of Auroral Arcs for Different Directions of Interplanetary Magnetic Fields in the Y-Z Plane," *J. Geophys. Res.*, **83**, 5277 (1978).

Lee, L. C., and Z. F. Fu, "A Theory of Magnetic Flux Transfer at the Earth's Magnetopause," *Geophys. Res. Lett.*, **12**, 105 (1985).

Meng, C.-I., and R. Lundin, "Auroral Morphology of the Midday Oval," *J. Geophys. Res.*, **91**, 1572 (1986).

Perreault, P., and S.-I. Akasofu, "A Study of Geomagnetic Storms," *Geophys. J. R. Astron. Soc.*, **54**, 547 (1978).

Pudovkin, M. I., V. S. Semenov, M. F. Iteyu, and K. H. Biernat, "Implications of the Stagnation Line Model for Energy Input Through the Dayside Magnetopause," *Geophys. Res. Lett.*, **13**, 213 (1986).

Rees, D., and T. J. Fuller-Rowell, "Global Thermospheric Modeling," *Phys. Scr.*, **T18**, 212 (1987).

Rees, M., *Physics and Chemistry of the Upper Atmosphere*, Cambridge University Press, Cambridge (1989).

Reiff, P. H., R. W. Spiro, and T. W. Hill, "Dependence of Polar Cap Potential on Interplanetary Parameters," *J. Geophys. Res.*, **86**, 7639 (1981).

Russell, C. T., and R. C. Elphic, "ISEE Observations of Flux Transfer Events at the Dayside Magnetopause," *Geophys. Res. Lett.*, **6**, 33 (1979).

Williams, D. J., D. G. Mitchell, T. E. Eastman, and L. A. Frank, "Energetic Particle Observations in the Low-Latitude Boundary Layer," *J. Geophys. Res.*, **90**, 5097 (1985).

II. AURORAL SPECTROSCOPY AND THERMOSPHERE

II-1. OVERVIEW OF AURORAL SPECTROSCOPY*

A. Vallance Jones[†]

Historically, spectroscopy provided the first information about conditions and composition of the upper atmosphere. A set of spectra from 400–16,000 Å is presented and the main lines and band systems of the spectrum discussed. Progress in obtaining and interpreting auroral spectra is reviewed. References are given to detailed studies that have been made of classical nightside aurora, as well as of polar-cap, cleft, and equatorial aurora.

1. INTRODUCTION

Until the first rocket measurements were made of the fluxes of electrons above aurora in 1958 by *McIlwain* [1960] and by *Meredith, Davis, Heppner, and Berg* [1958], the domain of present interest was as inaccessible to direct study as the other planets of the solar system or the stars and galaxies beyond. Optical observations, magnetometry, and reflections of radio waves provided the main sources of information about the upper atmosphere and the regions beyond. The optical phenomena of interest were meteor trails, the feeble "light of the night sky," and, in polar regions, the mysterious and exciting northern lights or aurora. The object of the study of the latter was then, as it is now, partly to understand the nature of the phenomenon and partly to obtain knowledge about the inaccessible upper atmosphere of our planet. In 1951 when a celebrated earlier Conference on Auroral Physics was held at the University of Western Ontario, one was still in the era when the interests of the scientists involved encompassed the whole field from spectra to the theories of the origin of the exciting particles. Sydney Chapman participated at that conference and entered into a wide range of discussions including the influence of upper atmospheric composition on spectral emissions. Chapman's name is, of course, closely associated with the spectroscopy of the nightglow.

The development of means of direct exploration of the upper atmosphere and the magnetosphere has led to definitive advances in both objectives so that spectroscopic studies are no longer the only method of ob-taining information about the higher atmosphere and beyond. Nevertheless, optical observations have an important role to play because they provide the means to make remote observations over a wide spatial region. Spectroscopic observations from spacecraft, as well as from the ground, play a key role in present and future studies. Both for this reason and for historic reasons it is appropriate to begin with papers on spectroscopic and optical studies of aurora.

Spectroscopic studies of aurora began slowly in the last century with the help of visual spectroscopes. Such studies were difficult because of the very low intensity of emissions even for bright aurora. The advent of the photographic plate combined with high-aperture prism spectrographs provided increasingly better results between 1920 and 1940 (e.g., *Vegard*, [1933]). A further great improvement occurred in the early 1950s when good, large diffraction gratings were combined with excellent flat-field Schmidt cameras by *Meinel* [1950]. Instruments of this type were used to map the spectrum from 3150–10,500 Å at resolutions down to 1 Å (see *Vallance Jones* [1971]). Image converters and photoconductive detectors made it possible to obtain spectra further into the infrared. Scanning spectrometers with photomultipliers as detectors greatly improved quantitative intensity measurements and made it possible to carry out exact comparisons with theoretically calculated spectra. However, it may be noted that, in comparison to laboratory spectroscopy, the low intensity of aurora still does not permit the high resolutions that are necessary to resolve fully the many overlapping bands and lines in the spectrum.

We here present a brief introduction rather than an exhaustive review of this topic. The references given are intended as useful entry points into the more recent literature. More detailed accounts of auroral spectroscopy may be found in the books by *Chamberlain* [1961], *Omholt* [1971], and *Vallance Jones* [1974].

*National Research Council of Canada, paper No. 30608
[†]Herzberg Institute of Astrophysics, National Research Council of Canada, Ottawa, Canada K1A 0R6.

2. SPECTROSCOPIC TECHNIQUES

The three basic devices used in auroral spectroscopy are the diffraction grating, the Michelson interferometer, and the Fabry-Perot etalon. The last device includes the interference filter, which is a fixed low-resolution etalon. Table 1 summarizes the techniques most commonly used in terms of spectral resolution.

TABLE 1. Summary of Spectroscopic Instruments

Device	Resolution	Remarks
Low Resolution		
Transmission filters	100–1000 Å	Imagers
Interference filters	2–100 Å	Narrow-band photometers
Medium resolution		
Grating spectrographs	1–10 Å	With area or linear detectors
Spectrometers	1–10 Å	With exit slit and photomultipliers
High resolution		
Fabry-Perot etalon Michelson interferometer	0.01–2 Å	Line profiles IR or special visible

The information in Table 1 provides but a brief summary. In general, the interference-filter instruments are most useful in sampling the spectrum with high time or spatial resolution at a variety of wavelengths. The grating instruments are most useful in covering major portions of the spectrum at medium resolution while the Fabry-Perot systems [*Hernandez*, 1978; *Meriwether et al.*, 1983; *Rees et al.*, 1984] facilitate high resolution studies of very short spectral ranges, particularly for the study of Doppler profiles and shifts of strong spectral lines. Michelson interferometry [*Shepherd et al.*, 1984] is another powerful way of measuring the widths and Doppler shifts of spectral lines.

In comparison with a grating spectrometer, for a given spectral bandwidth, the Fabry-Perot (and interference-filter devices) provide a considerably larger photon flux at the detector (for a dispersing element of the same size). However, the growing availability of efficient linear and area detectors such as resistive anodes, anode arrays, and charge-coupled devices (CCDs) makes it possible to recoup much of this loss by observing all the elements of a wide spectral range simultaneously. This possibility is particularly useful in the visible and ultraviolet regions [e.g. *Broadfoot et al.*,

1977; *Bowyer et al.*, 1981; *Torr and Torr*, 1984; *Siskind and Barth*, 1987]. For the infrared, Fourier-transform spectroscopy with Michelson interferometers [e.g. *Espy et al.*, 1987] is the most generally useful technique.

3. ATOMIC AND MOLECULAR SPECIES AND THEIR ENERGY LEVELS

The principal atomic and molecular species appearing in the electronic auroral spectrum are O, O^+, N, N^+, N_2, N_2^+, O_2, O_2^+ and H, i.e., almost all the possible simple species that can be produced by energetic electron excitation or by dissociative excitation of the major atmospheric constituents. The spectrum of aurora is quite complicated because many transitions are possible between the ground state and excited states of these species. A few selected simplified energy-level diagrams are reproduced here. A more complete set is to be found in *Vallance Jones* [1974].

4. LIFETIMES OF EXCITED STATES

An important factor in understanding and making use of the spectra of aurora is the absence of container walls and the decreasing frequency with height of the deactivation (or quenching) in collisions of excited atoms or molecules. This makes it possible for radiation to be observed from metastable species with radiative lifetimes much longer than the typical 10^{-8} seconds of the allowed transitions normally observed in the laboratory. (These transitions are referred to as "forbidden" because they violate the spectroscopic selection rules for electric dipole radiation.)

The long lifetimes have two important consequences. First, atoms or molecules excited to these metastable states may transfer their energy to other species before they can radiate and so excite other atoms or molecules indirectly. Secondly, below the height where the quenching rate equals the radiative emission rate for a particular metastable species, most of the excited molecules lose their energy in collisions rather than by radiating.

5. THE AURORAL SPECTRUM

Studies during the last 30 years have led to an increasingly accurate knowledge of the spectrum. Figures 1 to 10 are a reproduction of a set of spectra summarizing our present knowledge of the spectrum from 400 Å in the extreme ultraviolet to 16,000 Å in the infrared. References to the sources from which the spectral data have been taken are given in the figure captions. Some of the more important features in the spectrum are briefly discussed in the following notes. It is important to note that below 3100 Å the spectra can be observed only from rockets or satellites and that below 1700 Å there may be absorption of radiation in the atmosphere above the emitting region. In the far ultraviolet (FUV) and extreme ultraviolet (EUV) regions, there is serious absorption of the emitted radiation by O_2 and the observed spectrum is very sensitive to the conditions of observation [e.g., *Feldman and Gentieu*, 1982].

O I Forbidden Lines

The low-lying levels of O derived from the ground-state configuration are shown schematically in Figure 11. The approximate radiative lifetimes of the atom in the excited state are indicated on this and other energy-level diagrams. The 2972 Å line is a strong feature in the near ultraviolet spectrum of Figure 3. The 5577 Å line or "green line" (Figure 6) is the strongest line in the visible auroral spectrum. It has a lifetime of about 0.75 s [*Froese Fischer and Saha*, 1983; *Baluja and Zeippen*, 1988]. The 6300–6364 Å multiplet or "red line(s)" are notable because of the very long lifetime of their upper state, the metastable $O(^1D)$ level [*Kernahan and Pang*, 1975; *Baluja and Zeippen*, 1988]. Quenching, mostly by N_2, deactivates most of the $O(^1D)$ produced by harder electrons. The softer the primary excitation flux, the greater the fraction of excited atoms that can radiate, so the ratio of λ6300 to N_2^+ emission provides the basis for optical estimates of primary particle energy.

Spectacular high-altitude red aurora (type-A red aurora), which is occasionally seen during great magnetic storms, owes its red color to a strong enhancement of the 6300–6364 Å multiplet due in part to the height effect just discussed and likely also to thermal excitation of the $O(^1D)$ level by high electron temperatures. Similar red aurora is seen in the cusp region [*Wickwar and Kofman*, 1984]. Red lower bordered aurora (type B red aurora) appears to be associated with excitation by higher energy auroral electrons. It may be partly due to weakening of the green line because of quenching of $O(^1S)$ below 100 km and to decreasing concentrations of O in the same region. Shifts in the vibrational populations of the upper state of the N_2 first-positive bands could produce a further reddening effect below 85 km. This is further discussed in the paragraph on the N_2 first-positive bands.

N_2^+ Band Systems

The two excited electronic levels of the N_2^+ ion give rise to the first-negative bands from the $B^2\Sigma_u^+$ state and the Meinel bands from the lower energy $A^2\Pi_u$ state.

N_2^+ First-Negative (1N) Bands. Bands of this system are prominent in the near ultraviolet and blue regions of the spectrum shown in Figures 4, 5, and 6. The strongest are those at 3914, 4278, and 4709 Å. The transitions involved are fully allowed so that quenching is negligible. The intensity of the bands pro-

vides a good measure of the energy input to the atmosphere arising from auroral electrons. The rotational structure of these bands is very simple and is very sensitive to the atmospheric temperature in the emission region. Even if the rotational lines are not resolved, the band shape (as in Figures 4 and 5) provides a means of estimating atmospheric neutral temperature profiles. The rotational temperature is much higher in the spectrum of type-A aurora [*Vallance Jones*, 1967], which is excited much higher in the atmosphere by soft electrons. Similar effects are observed in cleft aurora [*Sivjee*, 1983; *Henriksen*, 1984], although these aurora are often partly sunlit so that more complex resonance processes [*Degen*, 1987] may be important.

Theoretically, the rotational temperatures from these and other bands should be close to the ambient neutral atmospheric temperature when the bands are excited by electrons. This is usually the case in normal bright aurora. In the case of dominant excitation by heavier particles, the apparent rotational temperature may be enhanced. Such enhancements have been observed in low-latitude aurora of which one type is probably excited by O^+ ions as discussed by *Tinsley et al.* [1986] and *Ishimoto et al.* [1986].

To the extent that the atmosphere is well modeled, it is possible to invert the procedure and, by measuring zenith rotational temperatures, get estimates of primary electron energies for auroral zone electron aurora [*Vallance Jones et al.*, 1987]. However, the density and composition of the neutral atmosphere can vary substantially with time and location, as noted by *Meier and Strickland* [this volume].

N_2^+ Meinel Bands. The distinctive triple-peaked structure of the strong bands of this system is clearly seen in Figures 8, 9, and 10 in the infrared. Derivation of atmospheric temperature is also possible from these bands, but is more difficult than for the first-negative band because the band structure is more complex.

N_2 Band Systems

A simplified energy level diagram for N_2 is shown in Figure 12. The band systems arising from transitions between these levels are prominent in the spectrum from the far UV to IR. The three main triplet levels, A, B, and C are excited both by electron impact from the X ground state and also by cascade as indicated in the figure. Inspection of a complete energy-level diagram with potential curves and vibrational levels shows that more complicated transitions are possible and must be taken into account in understanding the vibrational populations of these excited states [*Cartwright*, 1978].

N_2 First-Positive System. This is a strong, allowed, band system with bands of very complex rotational fine structure not well suited for temperature determination.

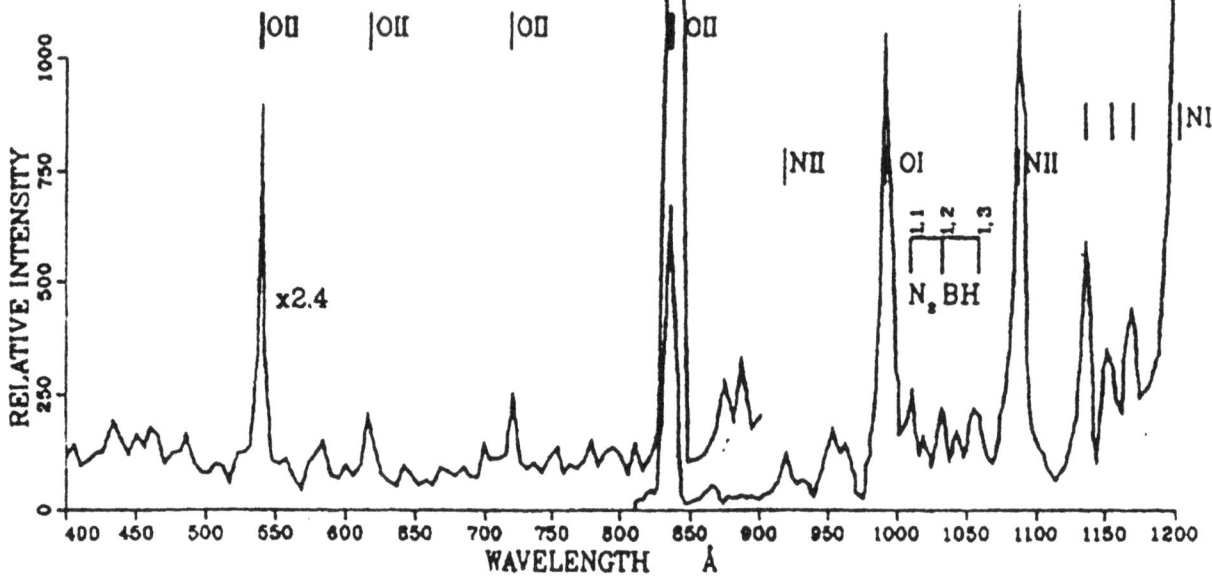

Figure 1—Spectrum of aurora 400–1200 Å. The left-hand curve is from data of *Paresce et al.* [1983a], obtained with a satellite spectrograph with a resolution of about 8 Å from a height of about 600 km. The right-hand curve was plotted from data of *Feldman and Gentieu* [1982], obtained with a rocket spectrometer at 160–180 km, viewing horizontally with a resolution of 6.5 Å.

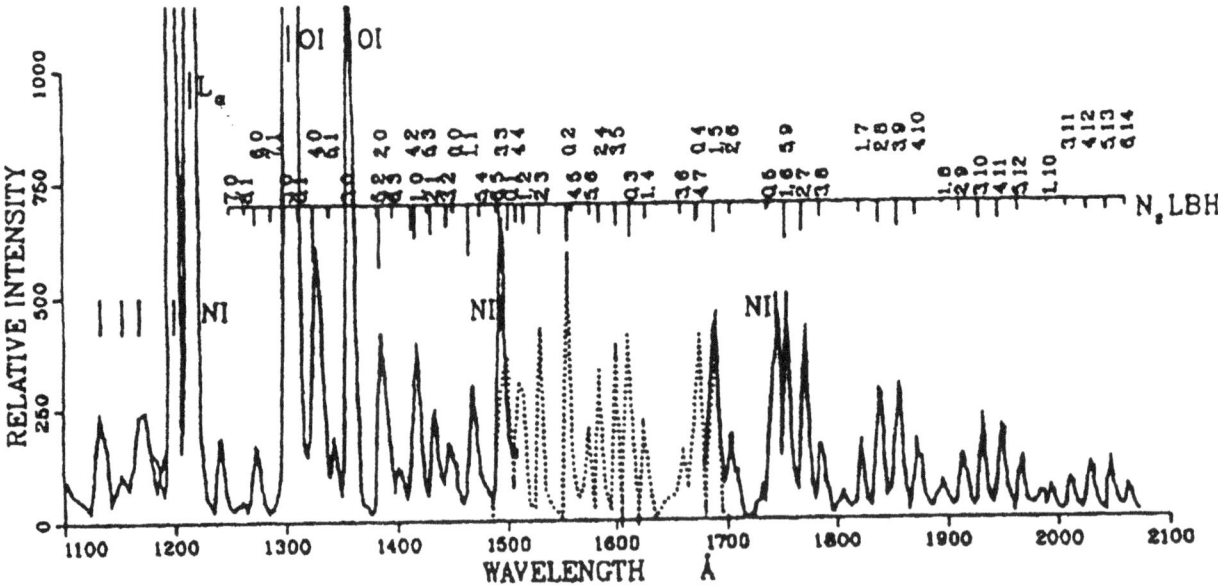

Figure 2—Spectrum of aurora 1100–2100 Å. For 1100–1190 Å, the plot is from data of *Paresce et al.* [1983b], obtained with the same satellite instrument as was used by *Paresce et al.* [1983a]. The plot from 1200–1500 Å is from the data of *Feldman and Gentieu* [1982] as described for Figure 1. From 1500–1700 Å the dashed curve shows synthetic N_2 LBH bands calculated by *R. L. Gattinger* [private communication] at a resolution of 5 Å (no higher resolution spectrum was available over this range). From 1675–2075 Å the curve is from data of *Eastes and Sharp* [1987] obtained at 110 km with a rocket spectrometer viewing horizontally with a resolution of 4 Å.

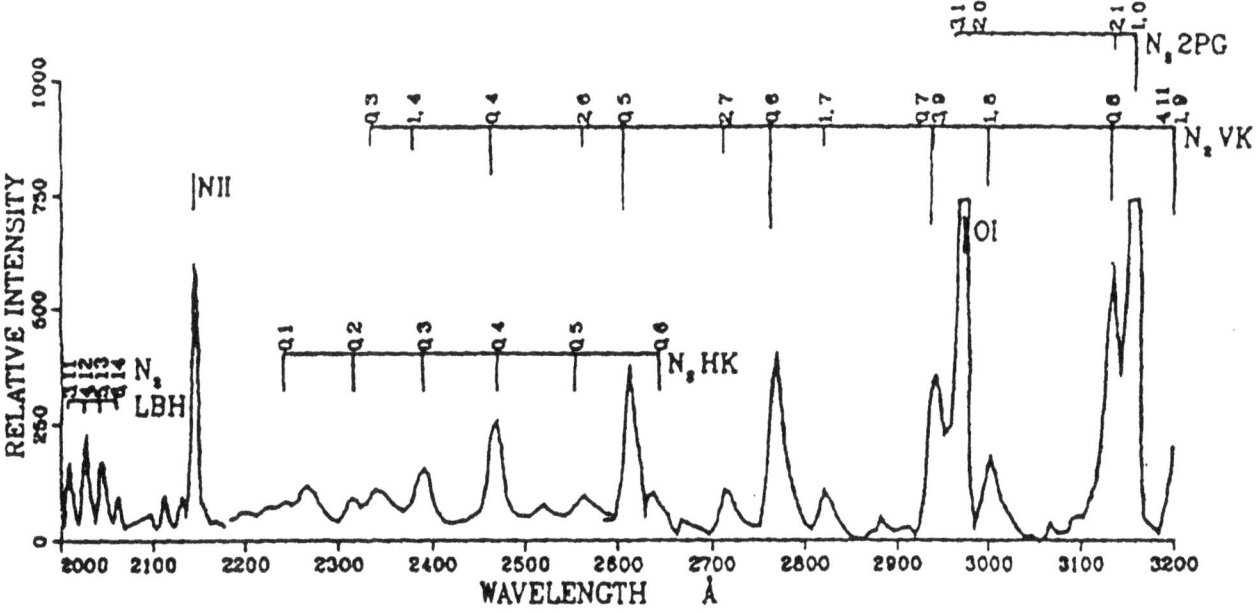

Figure 3—Spectrum of aurora 2000–3200 Å; 2000–2180 Å from data of *Siskind and Barth* [1987], obtained at a resolution of 6.7 Å with a rocket spectrograph; 2180–2600 Å from data of *Beiting and Feldman* [1979] with a rocket spectrometer with a resolution of 15 Å; 2600–3200 Å from data of *Sharp* [1971] with a rocket-borne spectrometer with a resolution of 7.9 Å, looking upward.

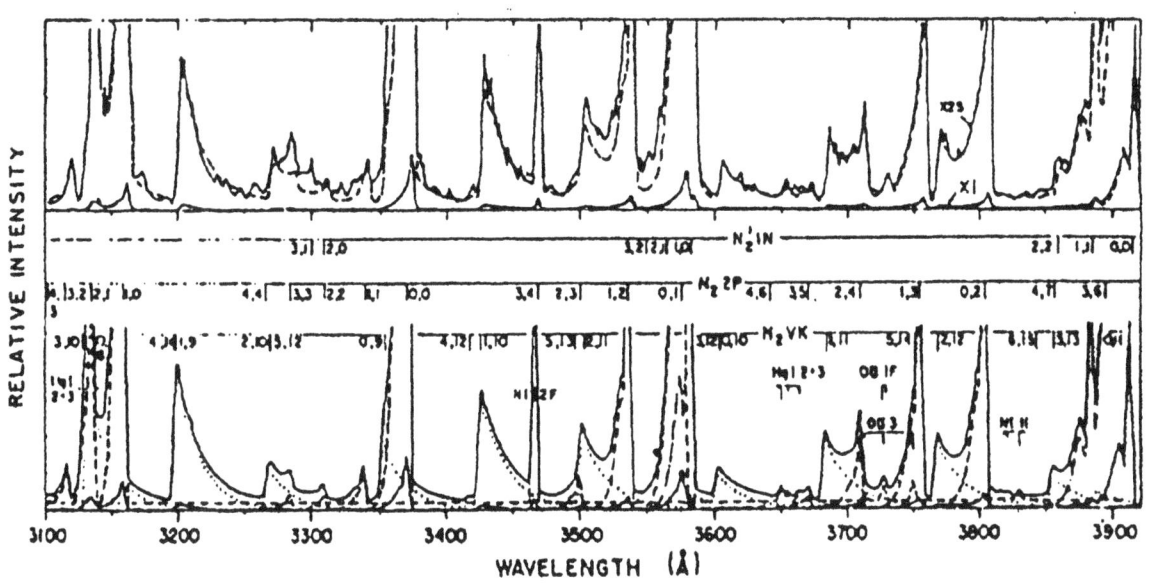

Figure 4—Spectrum of the aurora 3100–3920 Å obtained with a ground-based spectrometer with a resolution of 3.5 Å, from *Vallance Jones and Gattinger* [1975]. Upper panel: observed and synthetic spectra (dashed curve); lower panel: components of synthetic spectrum. (Courtesy Canadian Journal of Physics.)

19

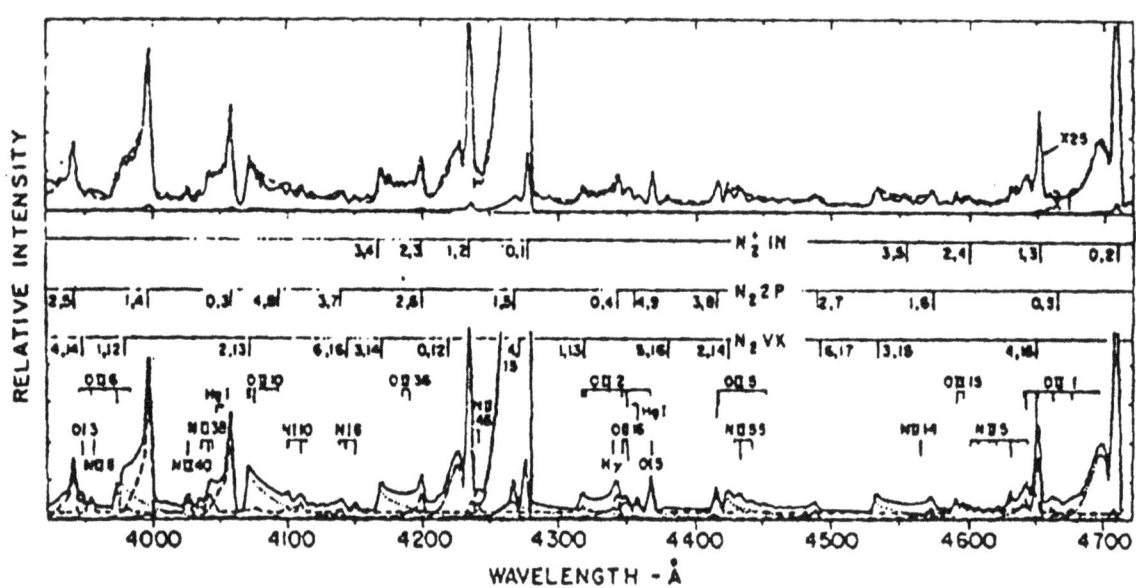

Figure 5—Spectrum of aurora 3920–4720 Å; otherwise the same as Figure 4. (Courtesy Canadian Journal of Physics.)

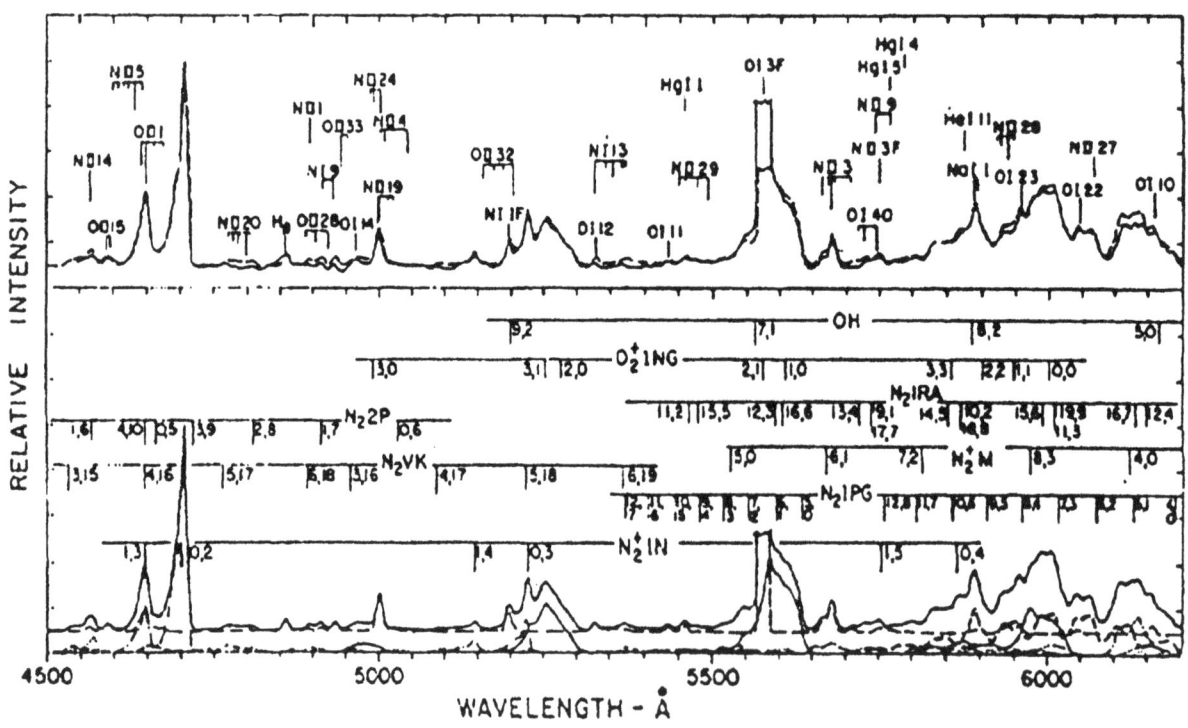

Figure 6—Spectrum of aurora 4500–6200 Å from *Gattinger and Vallance Jones* [1974] from a ground-based spectrometer with a resolution of about 10 Å; otherwise similar to Figure 4. (Courtesy Canadian Journal of Physics.)

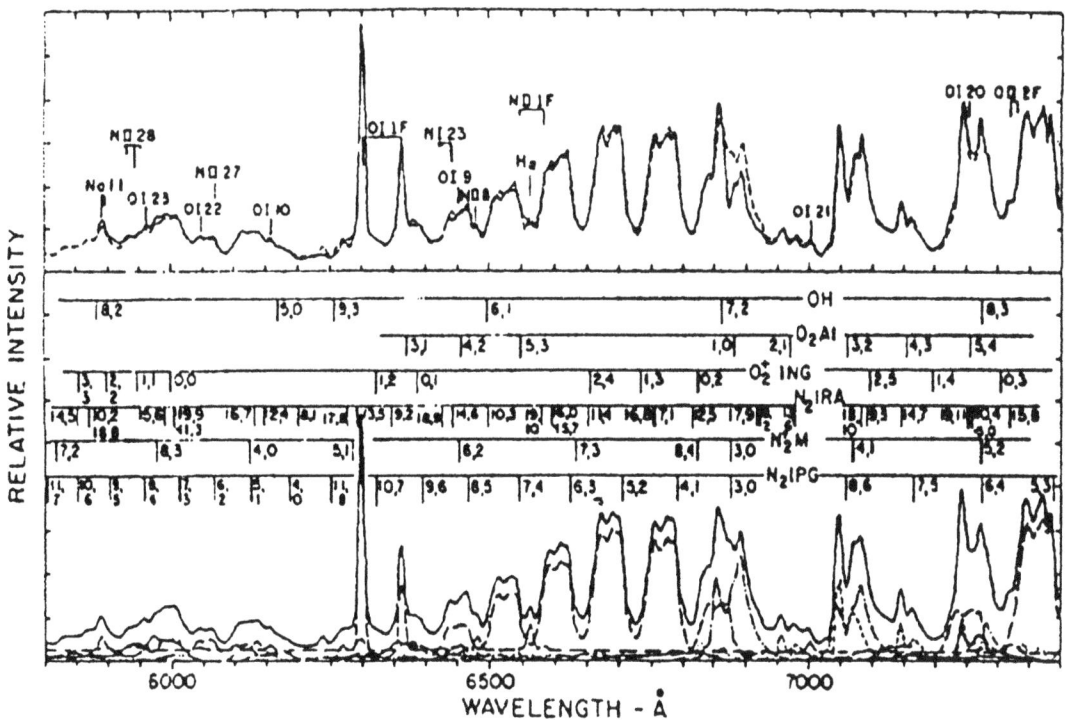

Figure 7—Spectrum of aurora 5800-7400 Å; continuation of Figure 6. (Courtesy Canadian Journal of Physics.)

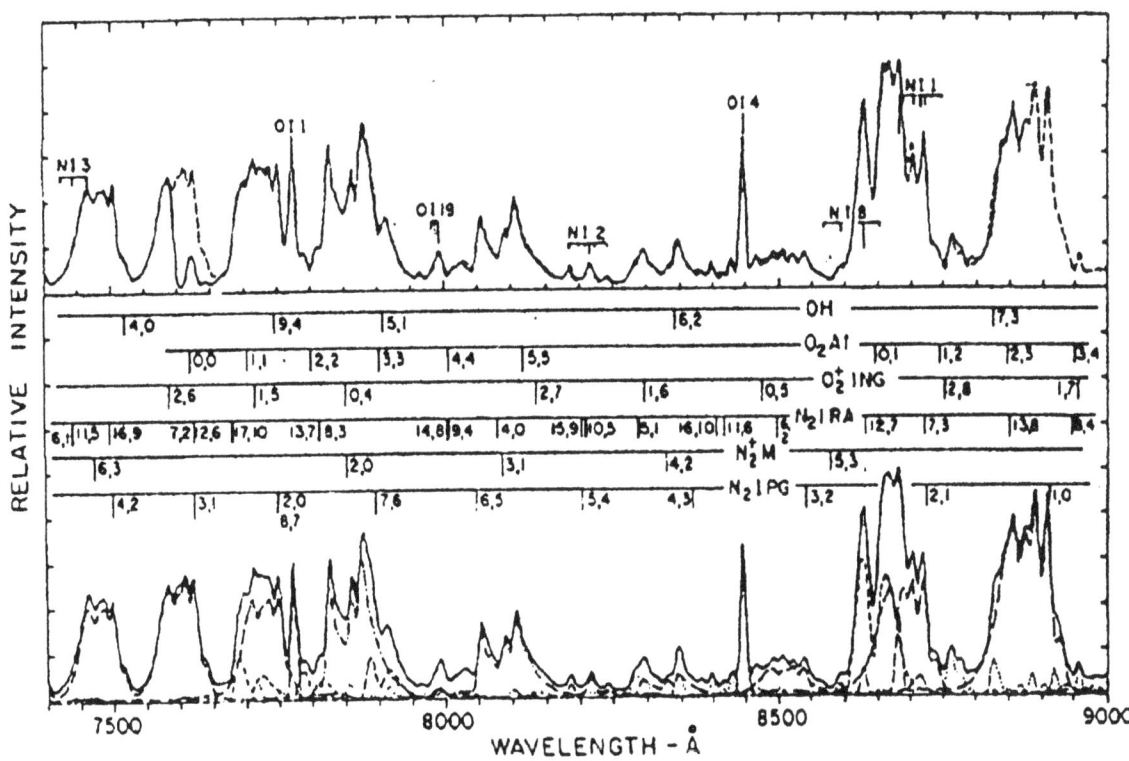

Figure 8—Spectrum of aurora 7400-9000 Å; further continuation of Figure 6. (Courtesy Canadian Journal of Physics.)

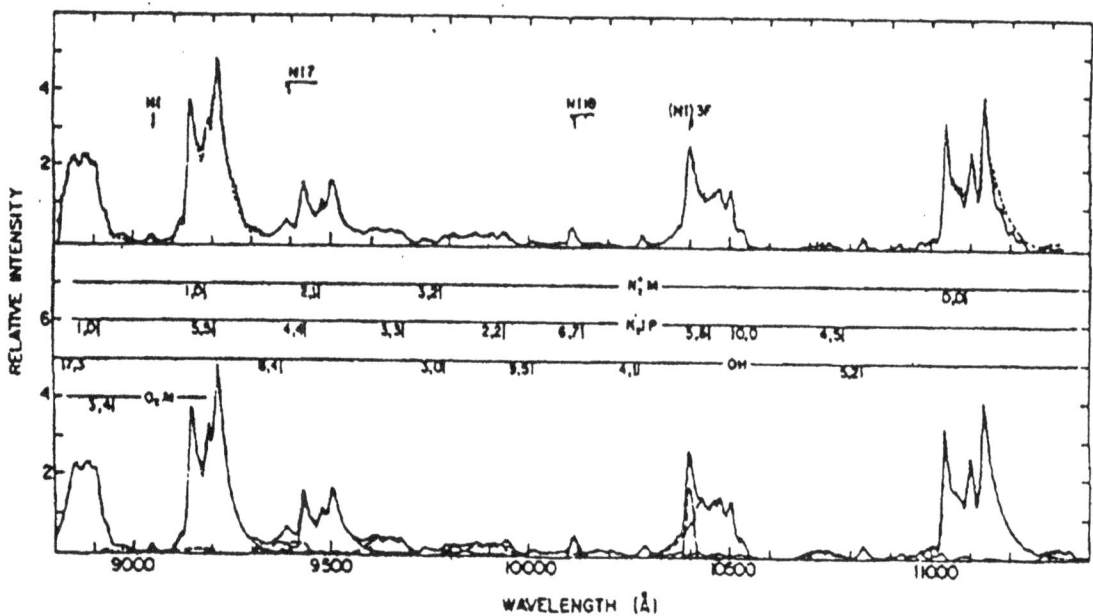

Figure 9—Spectrum of aurora 8800–11400 Å from *Vallance Jones and Gattinger* [1976] from a ground-based spectrometer with a resolution of 15 Å; the format is similar to that of Figure 4. (Courtesy Canadian Journal of Physics.)

The unresolved bands are very prominent in the 6400–6800 Å region in Figure 7 and are important features right out to the very strong 0,0 band near 10,500 Å in Figures 9 and 10. *Benesch* [1981] has suggested a mechanism whereby collisions at heights below 85 km could produce perturbations in the vibrational population in the B state and so cause the visual reddening effect typical of type-B red aurora. (See also *Gattinger et al.* [1985] and *Kaila* [1986].)

N$_2$ Second-Positive Bands. This strong allowed system in the blue and near UV is prominent in Figures 3 and 4, especially the 0,0 band at 3370 Å and the $\Delta v = -1$ sequence in the 3700–3800 Å region. The rotational structure is complex and not normally resolved by auroral spectrometers.

N$_2$ Vegard-Kaplan Bands. These bands arise from the forbidden $A \rightarrow X$ transition (Figure 12). The radiative lifetime of the upper state is about 1–2 s so that quenching is important, and the relative intensity of the system depends on primary electron energy. The rotational structure is relatively simple and sensitive to atmospheric temperature in the emitting region. The bands are prominent in the ultraviolet and blue regions of the spectrum (Figures 3, 4, and 5). A method of deriving primary electron energy fluxes and mean energies has been described recently by *Ishimoto et al.* [1988].

N$_2$ Lyman-Birge-Hopfield Bands. This system is dominant in the far UV (Figure 2). The numerous bands between 1300 and 2100 Å each have complex rotational structure that is not resolved in auroral spectra. The bands in the 1350–1600 Å region are significantly absorbed by overlying molecular oxygen in normal night time aurora.

Other N$_2$ Band Systems. Several other band systems are weakly present. The $b \rightarrow X$ Birge-Hopfield system (Figure 12) appears in the spectrum of *Feldman and Gentieu* [1982], reproduced in Figure 1. Only the $v' = 1$ progression is found; most other vibrational levels of the upper state predissociate to atomic N instead of radiating [*Zipf and Gorman,* 1980]. Other transitions from higher level singlet states, of which some are found in the laboratory, probably predissociate under optically thick conditions in aurora [*Zipf and McLaughlin,* 1978; see also *Roncin et al.* 1987]. Several Herman-Kaplan bands (from a higher triplet state) were identified in the *Beiting and Feldman* [1979] spectrum forming part of Figure 3.

O$_2^+$ First-Negative Band System

This is a strong allowed system of complex bands due to transitions from the $b\ ^4\Sigma_g^-$ excited quartet states of the O$_2^+$ ion down to the lower $a\ ^4\Pi_u$ quartet state. The a state is metastable because the transition to the $X\ ^2\Pi_g$ ground state of the ion is forbidden. The radiative lifetime of the a state is different for different components of the state with the longest being of the order of 100 ms [*Bustamente et al.* 1987]. The strong bands of the system are seen in Figure 6 near 5250, 5600, and 6000 Å. The relative intensity

22

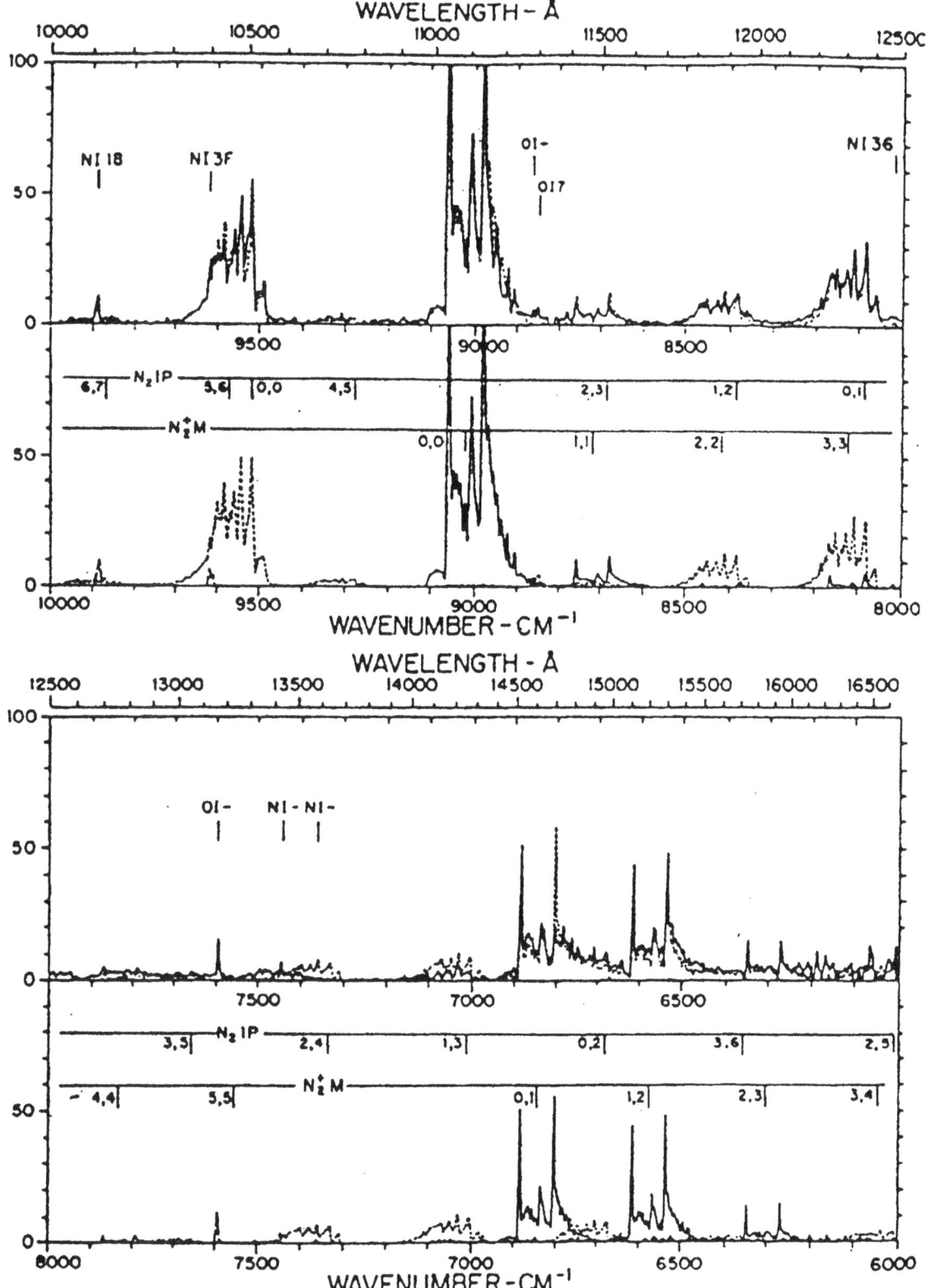

Figure 10—Spectrum of aurora 10,000–16,000 Å obtained with a Michelson interferometer with a resolution of 12 Å by *Gattinger and Vallance Jones* [1981]. The format is similar to that of Figure 4. (Courtesy Canadian Journal of Physics.)

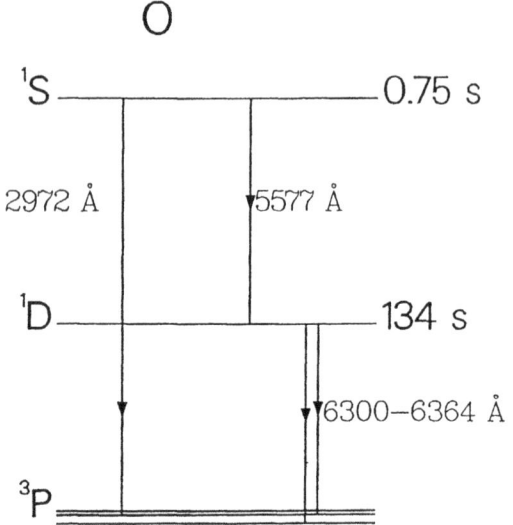

Figure 11—Schematic diagram of the energy levels derived from the ground-state configuration for atomic oxygen.

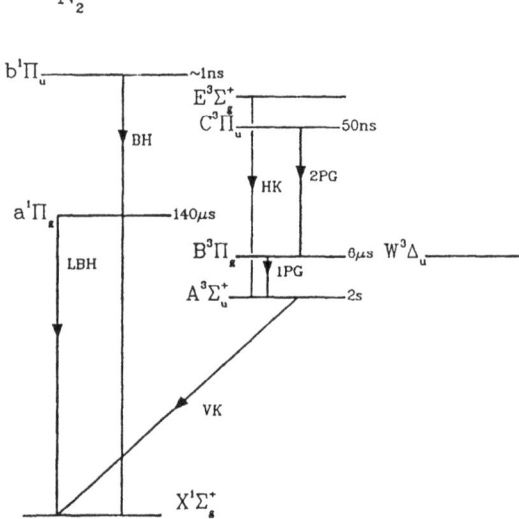

Figure 12—Simplified schematic energy-level diagram for the N_2 molecule.

of this system is sensitive to primary particle energy and atmospheric composition [*Niciejewski et al.*, 1989].

O_2 Band Systems

The most prominent band system in aurora is the atmospheric system corresponding to the $b \rightarrow X$ transition as defined in the energy-level diagram of Figure 13 (*Slanger*, [private communication, 1988]). As with the ground-state configuration transitions of atomic oxygen, transitions between all the levels shown in Figure 13 are forbidden. The strongest band of the atmospheric system, as observed from the ground is the 0,1 band, centered at 8645 Å in Figure 8, although the 0,0 band at 7619 Å is even stronger above the absorbing O_2 of the lower atmosphere. The radiative lifetime of the *b* state is about 12 s.

The infrared atmospheric system $(a \rightarrow X)$ is remarkable in that its upper state lifetime is about 60 min [*Badger et al.* 1965]. This state is also extremely resistant to quenching. The 0,0 band can appear strongly in aurora but is subject to severe time-lag effects because of the long radiative lifetime [*Gattinger and Vallance Jones,* 1973]. It is also strongly absorbed by ground-state molecular oxygen and is consequently best observed from aircraft or balloon altitudes.

None of the other O_2 band systems seen in the nightglow have been reported in aurora, even though their upper states have quite large electron-impact cross sections. This is likely because electron-impact excitation leads to dissociation of the upper state and would be expected according to the *Frank-Condon* principle if the energy-level diagram of Figure 13 is correct.

Atomic Hydrogen Emissions

Lines of the Balmer and Lyman series appear in the auroral spectrum when the auroral emission is partly or wholly excited by incoming energetic protons. The Lα line seen in Figure 2 may be largely due to the geocoronal glow, although it does appear in spectra of proton aurora. The Balmer Hβ line appears as a weak feature at 4861 Å in Figure 6. When the excitation by

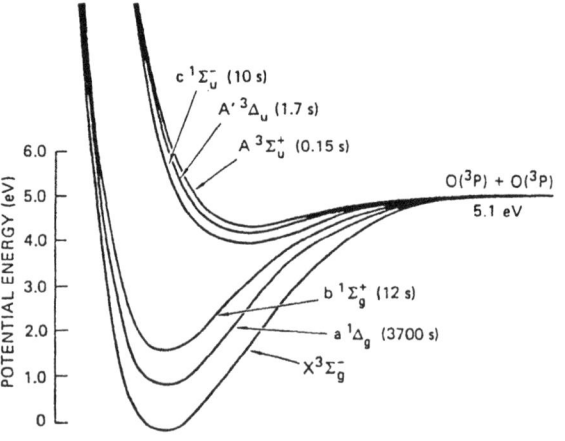

Figure 13—Potential energy curves for the states of O_2 derived from ground-state oxygen atoms [*Slanger*, 1988]. Lifetimes 50–60% shorter have recently been calculated for the *c,A'* and *A* states by *Bates* [1988]. (Courtesy T.G. Slanger.)

protons is dominant, the intensity of the Hβ line becomes comparable to that of neighboring 4709 Å N_2^+ band. Frequently the Hα line at 6563 Å is also prominent, although it is blended into the N_2 1PG bands in Figure 7.

Historically the Balmer lines have been of great interest because they afforded the first direct proof that aurora was excited by high-velocity incoming particles. *Meinel* [1951] demonstrated conclusively that the line profile of Hα observed in the magnetic zenith shows a Doppler shift, indicating that the emitting protons are traveling with high downward velocity, parallel to the terrestrial magnetic field. The analysis of the Doppler profile of the Hα and Hβ lines provided the possibility of inferring the angular and velocity distribution of the incoming protons. The history of these endeavors has been reviewed by *Eather* [1967].

Multiplets Due to Higher Transitions of Atomic Oxygen. The very strong multiplet of O at 1304 Å, which appears in Figure 2, arises from the transition shown on Figure 14. This is the so-called "resonance" line of atomic oxygen, i.e., the first permitted transition connecting to the ground state. Because the emitted

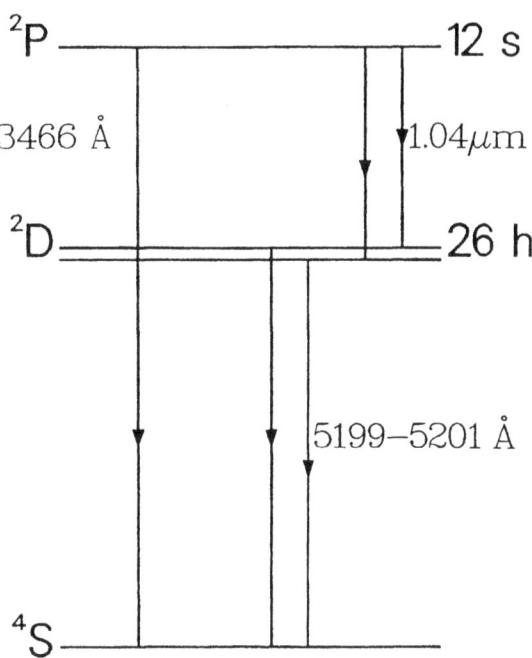

Figure 15—Schematic diagram of the energy levels derived from the ground-state configuration for atomic nitrogen.

photons are strongly reabsorbed and re-emitted thousands of times by atmospheric O at auroral heights, the emission is spatially diffused away from the original region of excitation and may also be strongly absorbed by other, normally weakly, absorbing atmospheric components such as O_2. The weakly forbidden 1356 Å multiplet (Figure 2) involves the transition to the ground state from the lowest quintet level. The interpretation of the spectrum has been discussed by *Meier et al.* [1982, 1985]. The strong infrared multiplets at 8446 and 7774 Å (Figure 8) result from cascade emissions from higher levels shown in Figure 14 to the upper levels of the 1304 and 1356 Å multiplets. The relation between these and other cascading transitions in the infrared and far ultraviolet (as shown in Figure 14) is a subject of considerable interest [*Christensen et al.*, 1983].

Forbidden Atomic Nitrogen Lines

The low-lying levels of the atomic nitrogen are shown schematically in Figure 15. As it is in the case of atomic oxygen, transitions between these levels are highly forbidden, with the ^{2}P and ^{2}D levels having lifetimes of about 12 seconds and 26 hours, respectively. The

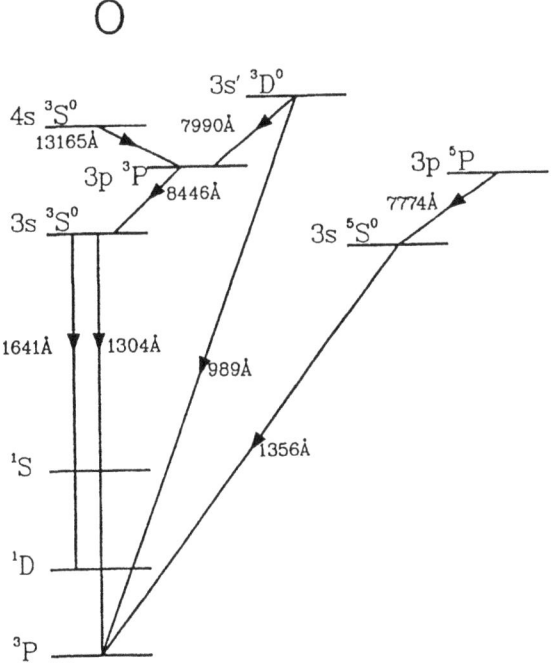

Figure 14—Partial atomic oxygen energy-level diagram showing transitions that give rise to important auroral OI lines. The prime on the 3s′ ^{3}D^0 state indicates that the O$^+$ core is excited to the 2p^3 ^{2}D state. The unprimed levels have the ion core in the ground state.

10,400 Å doublet is clearly seen in Figure 9 while the 5199–5200 Å appears in Figure 6. The 3466 Å line is very prominent in Figure 4. The long lifetimes and consequent susceptibility to quenching of these metastable upper states makes them of considerable interest as potential indicators of particle energy and composition [*Rees and Romick, 1985*].

Multiplets Due to Higher Transitions of N and N^+. There are many, quite strong allowed N and N^+ multiplets to be seen throughout the spectrum. No doubt these multiplets arise from dissociative excitation of N_2. The strong line in the far UV at 1200 Å (Figure 2) and the visible region multiplet at 5000 Å are prominent examples.

The N^+ feature at 2143 Å (Figure 3) is historically and spectroscopically interesting. Discovered first in lower-resolution rocket spectra, it was originally thought to be the 1,0 band of the γ-system of NO. It was subsequently identified as a transition from the somewhat metastable $2s2p^3\ ^5S$ state of N^+ to the ground state. Details have been well summarized by *Siskind and Barth* [1987] from whose data the spectral curve from 2000–2180 Å in Figure 3 was derived.

6. FAR INFRARED SPECTRUM

Beyond 2 microns, one gradually enters the region of pure vibration-rotation spectra of NO, NO^+, O_3, etc. that can be excited in aurora by energy transfer (see review by *Gordiets* [1986]).

7. AURORA ON OTHER PLANETS

Aurora on other planets is now an important topic, recently reviewed by *Fox* [1986]. The absence of magnetic fields in Venus and Mars means that localized charged particle precipitation like that in the Earth's auroral ovals is not to be expected. Excess 1304 Å and 1356 Å oxygen emission is ascribed to the effects of low-energy electron precipitation on Venus. In the case of Jupiter, there is localized aurora due to electron or ion precipitation from the Io torus [*Waite et al., 1988*]. The emission detected is in the far UV and comprises H and H_2 features. The visible region spectrum remains to be explored. The situation for Saturn is similar.

8. ANALYSIS OF SPECTRA AND APPLICATIONS

A further phase in the basic science is to understand how the different excited states of the emitting species are populated as observed. This involves the theoretical calculation of excitation rates (as a function of height) and the loss rates by emission and other loss mechanisms. This topic is covered by *Rees and Lummerzheim*

[this volume]. Once this understanding has been attained it may be applied to remote sensing measurements of a number of important quantities. These measurements are of particular value in that they can give information over a wide spatial region, particularly when the observations are made from satellites.

The main applications of spectroscopic observations may be summarized as follows:

1. Identification of precipitating charged particles as is discussed above under atomic hydrogen emissions and the rotational structure of the N_2^+ first-negative bands.
2. Incident energy flux in aurora, also discussed in the section on N_2^+ bands.
3. Energy spectrum of primary electrons from the ratios of forbidden and allowed spectral emissions as is discussed in the section on the 6300 Å line and the N_2^+ 1N bands.
4. Atmospheric temperatures from band rotational temperatures.
5. Atmospheric composition from spectral ratios.
6. Measurements of winds from the Doppler shifts of spectral lines.
7. Inferences about the ion and neutral chemistry of long-lived excited states of atmospheric constituents.

Some examples of these applications have been given in the sections noted above. Elsewhere in this volume, *Rees and Lummerzheim* and *Meier and Strickland*, give more detailed discussions of particular applications. The measurement of winds and temperatures from the Doppler width and shifts of spectral lines is a powerful tool, both in ground-based and satellite applications. Most measurements have been made with the O I 6300 Å line [e.g., *Killeen and Roble*, 1988; *Meriwether et al.*, 1988; *Rees et al.*, 1984], some with the O I 5577 Å line [*Wiens et al.*, 1988] and the O^+ 7320 Å line [*Smith et al.*, 1985]. The last reference permits direct observations of ion drifts. These different emission features give a mean wind, weighted according to the emission height profile of the observed line.

9. CONCLUSIONS

The study of the spectroscopy of aurora is a fascinating topic in its own right and there remain many interesting problems in understanding the spectrum, particularly in the less well-explored ultraviolet and infrared regions. At the same time optical observations of aurora provide an extremely valuable remote sensing tool in atmospheric and magnetospheric studies.

REFERENCES

Abreu, V. J., J. H. Yee, S. C. Solomon, and A. Dalgarno, "The Quenching Rate of $O(^1D)$ by $O(^3P)$," *Planet. Space Sci., 34,* 1143–1145 (1986).

Badger, R., A. C. Wright, and F. R. Whitlock, "The Absolute Intensities of the Discrete and Continuous Absorption Bands of Oxygen Gas at 1.26 μ and 1.075 μ and the Radiative Lifetime of the $^1\Delta_g$ State of Oxygen," *J. Chem. Phys., 43,* 4345–4350 (1965).

Bates, D. R., "Transition Probabilities of the Bands of the Oxygen Systems of the Nightglow," *Planet. Space Sci., 36,* 869, 1988.

Beiting, E. J. and P. D. Feldman, "Ultraviolet Spectrum of the Aurora (2000–2800 Å)," *J. Geophys. Res., 84,* 1287–1296 (1979).

Beluja, K. L., and C. J. Zeippen, "M1 and E2 Transition Probabilities for States within the $2p^4$ Configuration of the OI Isoelectronic Sequence," *J. Phys. B., 21,* 1455 (1988).

Benesch, W. "Mechanism for the Auroral Red Lower Border," *J. Geophys. Res., 86,* 9065–9072 (1981).

Broadfoot, A. L., B. R. Sandel, D. E. Shemansky, S. K. Atreya, T. M. Donahue, H. W. Moos, J. L. Bertaux, J. E. Blamont, R. Goody, M. B. McElroy, and Y. L. Yung, "Ultraviolet Spectrometer Experiment on the Voyager Mission," *Space Sci. Rev., 21,* 183–205 (1977).

Bowyer, S., R. Kimble, F. Paresce, M. Lampton, and G. Penegor, "Continuous-Readout Extreme-Ultraviolet Airglow Spectrometer," *App. Opt., 20,* 477–486 (1981).

Bustamante, S. W., H. S. Kwok, L. R. Carlson, and Y. T. Lee, "Spin-Forbidden Radiative Decay of the a $^4\Pi_u$ State of O_2^+," *J. Chem. Phys., 86,* 508 (1987).

Cartwright, D. C., "Vibrational Populations of the Excited States of N_2 Under Auroral Conditions," *J. Geophys. Res., 83,* 517–531 (1978).

Chamberlain, J. W., *Physics of the Aurora and Airglow,* 704 pp, Academic Press, New York (1961).

Christensen, A. B., G. G. Sivjee, and J. M. Hecht, "OI(7990 Å) Emission and Radiative Entrapment of Auroral EUV," *J. Geophys. Res., 88,* 4911–4917 (1983).

Degen, V., "Modelling of the N_2^+ First Negative Bands in Sunlit Aurora," *Planet. Space Sci., 35,* 1061–1066 (1987).

Eastes, R. W. and W. E. Sharp, "Rocket-Borne Spectroscopic Measurements in the Ultraviolet Aurora: The Lyman-Birge-Hopfield Bands," *J. Geophys. Res., 92,* 10,095–10,100 (1987).

Eather, R. H., "Auroral Proton Precipitation and Hydrogen Emissions," *Rev. Geophys., 5,* 207–285 (1967).

Espy, P. J., W. R. Pendleton, Jr., G. G. Sivjee, and M. P. Fetrow, "Vibrational Development of the N_2^+ Meinel Band System in Aurora," *J. Geophys. Res., 92,* 11257–11261 (1987).

Feldman, P. D. and E. P. Gentieu, "The Ultraviolet Spectrum of an Aurora 530–1520 Å", *J. Geophys. Res., 87,* 2453–2458 (1982).

Fox, J. L., "Models for Aurora and Airglow Emissions from Other Planetary Atmosphere," *Can. J. Phys., 64,* 1631–1656 (1986).

Froese Fischer, C., and H. P. Saha, "Multiconfiguration Hartree-Fock Results with Breit-Pauli Corrections for Forbidden Transitions in the $2p^4$ Configuration," *Phys. Rev. A, 28,* 3169–3178 (1983).

Gattinger, R. L., F. R. Harris, and A. Vallance Jones, "The Height, Spectrum and Mechanism of Type-B Red Aurora and Its Bearing on the Excitation of $O(^1S)$ in Aurora," *Planet. Space Sci., 33,* 207–221 (1985).

Gattinger, R. L. and A. Vallance Jones, "Observation and Interpretation of O_2 1.27μ Emission Enhancements in Aurora," *J. Geophys. Res., 78,* 8305–8313 (1973).

Gattinger, R. L. and A. Vallance Jones, "Quantitative Spectroscopy of the Aurora. II. The Spectrum of Medium Intensity Aurora Between 4500 and 8900 Å," *Can. J. Phys., 52,* 2343–2356 (1974).

Gattinger, R. L., and A. Vallance Jones, "Quantitative Spectroscopy of the Aurora. V. The Spectrum of Strong Aurora Between 10,000 and 16,000 Å," *Can. J. Phys., 59,* 480–487 (1981).

Gordiets, B. F., "Excitation of Vibrational Emission Bands in Aurora and Airglow," *Can. J. Phys., 64,* 1673–1678 (1986).

Henriksen, K., "N_2^+ Emissions in Sunlit Cusp and Night-Side Aurora," *Ann. Geophys., 2,* 457–462 (1984).

Hernandez, G., "Analytical Description of Fabry-Perot Spectrometer 4. Signal Noise Limitations in Data Retrieval; Winds, Temperature, and Emission Rate," *Applied Optics, 17,* 2967–2972 (1978).

Ishimoto, M., C.-I. Meng, G. J. Romick, and R. E. Huffman, "Auroral Electron Energy and Flux from Molecular Nitrogen Ultraviolet Emissions Observed by the S3-4 Satellite," *J. Geophys. Res., 93,* 9854–9866 (1988).

Ishimoto, M., M. R. Torr, P. G. Richards, and D. G. Torr, "The Role of Energetic O^+ Precipitation in a Mid-Latitude Aurora," *J. Geophys. Res., 91,* 5793–5802 (1986).

Kaila, K., "Altitude of the Red Lower Border in Auroral Forms," in *Proc. 13th Annual Meeting on Upper Atmosphere Studies by Optical Methods,* Report 86-28, K. Máseide, ed., Department of Physics, Univ. of Oslo, pp. 347–352 (1986).

Kernahan, J. A. and P. H.-L. Pang, "Experimental Determinations of Absolute A Coefficients for 'Forbidden Atomic Oxygen Lines'," *Can. J. Phys., 53,* 455 (1975).

Killeen, T. L., and R. G. Roble, "Thermospheric Dynamics: Contributions from the First 5 Years of Dynamics Explorer," *Rev. Geophys., 26,* 329–367 (1988).

McIlwain, C. E., "Direct Measurement of Particles Producing Visible Auroras," *J. Geophys. Res., 65,* 2727–2747 (1960).

Meier, R. R., R. R. Conway, D. E. Anderson, P. D. Feldman, R. W. Eastes, E. P. Gentieu, and A. B. Christensen, "The Ultraviolet Dayglow at Solar Maximum, 3. Photoelectron Excited Emissions of N_2 and O," *J. Geophys. Res., 90,* 6608–6616 (1985).

Meier, R. R., R. R. Conway, P. D. Feldman, D. J. Strickland, and E. P. Gentieu, "Analysis of Nitrogen and Oxygen Far Ultraviolet Auroral Emissions," *J. Geophys. Res., 87,* 2444–2452 (1982).

Meinel, A. B. "OH Emission Bands in the Spectrum of the Night Sky, I," *Astrophys. J., 111,* 555–564 (1950).

Meinel, A. B., "Doppler Shifted Auroral Emission," *Astrophys. J., 113,* 50–54 (1951).

Meridith L. H., L. R. Davis, J. P. Heppner, and O. E. Berg, "Rocket Auroral Investigations." in *Experimental Results of the U.S. Rocket Program for the IGY to* 1 *July* 1958, J. Hanessian, Jr. and I. Guttmacker, eds., IGY Rocket Report Series, No. 1, National Academy of Sciences, pp. 169–178 (1958).

Meriwether, Jr., J. W., T. L. Killeen, F. G. McCormac, A. G. Burns, and R. G. Roble, "Thermospheric Winds in the Geomagnetic Polar Cap for Solar Minimum Conditions," *J. Geophys. Res., 93,* 7478–7492 (1988).

Meriwether, Jr., J. W., C. A. Tepley, S. A. Price, and P. B. Hays, "Remote Ground-Based Observations of Terrestrial Airglow Emissions and Thermospheric Dynamics at Calgary, Alberta, Canada," *Opt. Eng., 22,* 128–131 (1983).

Niciejewskski, R., J. W. Meriwether, Jr., A. Vallance Jones, R. L. Gattinger, C. E. Valladares, V. B. Wickwar, and J. Kelly, "Ground-Based Observations of the O_2^+ 1N Band Enhancements Relative to N_2^+ 1N Band Emission," in *Planet. Space Sci., 37,* 131–143 (1989).

Omholt, A., *The Optical Aurora,* 198 pp., Springer, New York, Heidelberg, Berlin (1971).

Paresce, F., S. Chakrabarti, S. Bowyer, and R. Kimble, "The Extreme Ultraviolet Spectrum of Day and Nightside Aurorae: 800–1400 Å," *J. Geophys. Res., 88,* 4905–4910 (1983b).

Paresce, F., S. Chakrabarti, R. Kimble, and S. Bowyer, "The 300–900 Å Spectrum of Nightside Aurora," *J. Geophys. Res., 88,* 10247–10252 (1983a).

Rees, D., A. H. Greenaway, I. McWhirter, P. J. Charleton, and Åke Steen, "The Doppler Imaging System: Initial Observations of the Auroral Thermosphere," *Planet. Space Sci., 32,* 273–285 (1984).

Rees, M. H. and G. J. Romick, "Atomic Nitrogen in Aurora: Production, Chemistry and Optical Emissions," *J. Geophys. Res., 90,* 9871–9879 (1985).

Roncin, J.-Y., F. Launay, and K. Yoshino, "New Emission Bands in the High Resolution Vacuum Ultraviolet Spectrum of Molecular Nitrogen," *Planet. Space Sci., 35,* 267–269 (1987).

Sharp, W. E., "Rocket-Borne Spectroscopic Measurements in the Ultraviolet Aurora: Nitrogen Vegard-Kaplan Bands," *J. Geophys. Res., 76,* 987–1005 (1971).

Shepherd, G. G., W. A. Gault, R. A. Koehler, J. C. McConnell, K. V. Paulson, E. J. Llewellyn, C. D. Anger, L. L. Cogger, J. W. Haslett, D. R. Moorcroft, and R. L. Gattinger, "Doppler Imaging of the Aurora Borealis," *Geophys. Res. Let., 11,* 1003–1005 (1984).

Siskind, D. E., and C. A. Barth, "Rocket Observations of the NII 2143 Å Emission in an Aurora," *Geophys. Res. Lett., 14,* 479–482 (1987).

Sivjee, G. G., "Differences in Near UV (~3400–4300 Å) Optical Emissions from Midday Cusp and Nighttime Auroras," *J. Geophys. Res., 88,* 435–441 (1983).

Smith, R. W., K. J. Winser, A. P. Van Eyken, S. Quegan, and B. T. Allen, "Observation and Theoretical Modelling of a Region of Downward Field Aligned Flow of O^+ in the Winter Dayside Polar Cap," *J. Atmos. Terr. Phys., 47,* 489–495 (1985).

Tinsley, B. A., R. Rohrbaugh, H. Rassoul, Y. Sahai, N. R. Teixeira, and D. Slater, "Low-Latitude Aurorae and Storm Time Current Systems," *J. Geophys. Res.,* **91**, 11257-11269 (1986).

Torr, M. R., and D. G. Torr, "Atmospheric Spectral Imaging," *Science,* **225**, 169-171 (1984).

Vallance Jones, A., "Auroral Observations from West-Central Canada", in *Aurora and Airglow,* B.M. McCormac, ed., Reinhold, New York (1967).

Vallance Jones, A., "Auroral Spectroscopy," *Space Sci. Rev.,* **11**, 776-826 (1971).

Vallance Jones, A., *Aurora,* 301 pp., D. Reidel Pub. Co., Dordrecht-Holland (1974).

Vallance Jones, A., and R. L. Gattinger, "Quantitative Spectroscopy of the Aurora. III. The Spectrum of Medium Intensity Aurora Between 3100 and 4700 Å, *Can. J. Phys.,* **53**, 1806-1813 (1975).

Vallance Jones, A., and R. L. Gattinger, "Quantitative Spectroscopy of the Aurora. IV. The Spectrum of Medium Intensity Aurora Between 8800 and 11400 Å," *Can. J. Phys.,* **54**, 2128-2133 (1976).

Vallance Jones, A., R. L. Gattinger, P. Shih, J. W. Meriwether, V. B. Wickwar, and J. Kelly, "Optical and Radar Characterization of a Short-Lived Auroral Event at High Latitude," *J. Geophys. Res.,* **92**, 4575-4589 (1987).

Vegard, L., "Investigations of the Auroral Spectrum Based on Observations from the Auroral Observatory, Tromso," *Geofys. Pub.,* **10**, no.4 (1933).

Waite, J. H., J. T. Clarke, T. E. Cravens, and C. M. Hammond, The Jovian Aurora: Electron or Ion Precipitation," *J. Geophys. Res.,* **93**, 7244-7250 (1988).

Wickwar, V. B., and W. Kofman, "Dayside Red Auroras at Very High Latitudes: The Importance of Thermal Excitation," *Geophys. Res. Lett.,* **1**, 923-926 (1984).

Wiens, R. H., G. G. Shepherd, W. A. Gault, and P. R. Kosteniuk, "Optical Measurements of Winds in the Lower Thermosphere," *J. Geophys. Res.,* **93**, 5973-5980 (1988).

Zipf, E. C., and M. R. Gorman, "Electron-Impact Excitation of the Singlet States of N_2. I. The Birge-Hopfield system ($b^1\Pi_u - X^1\Sigma_g^+$)," *J. Chem. Phys.,* **73**, 813-926 (1980).

Zipf, E. C. and R. W. McLaughlin, "On the Dissociation of Nitrogen by Electron Impact and by EUV Photo-Absorption," *Planet. Space Sci.,* **26**, 449-462 (1978).

II-2. AURORAL EXCITATION PROCESSES

M. H. Rees* and D. Lummerzheim*

1. INTRODUCTION

Vallance Jones [this volume] presents the emission spectrum of the aurora in the wavelength range of 400–16,500 Å. Atomic lines and molecular bands are identified and the application of auroral spectroscopy to atmospheric and magnetospheric physics is indicated. The continuing improvements in spectroscopic instruments with respect to sensitivity and resolution have led to the discovery of previously unobserved features and to correcting previously incorrectly identified ones. Weak spectral features currently labeled as an unresolved background emission will, no doubt, be identified as the quest for better instruments continues.

Here we describe the mechanisms, or processes, that give rise to the excited states of atoms, molecules, and ions, from which the spectral emissions originate. Excited states that do not lead to observable spectral features are also discussed because of their importance in auroral chemistry and energetics.

For convenience in the presentation, auroral excitation processes may be classified into direct and indirect ones. Energetic electron and heavy charged particle impact are considered to be direct excitation sources, while a variety of chemical-ionic reactions, including cascading, are labeled indirect auroral excitation mechanisms. Some chemical processes operate in the atmosphere to produce the airglow radiations. Auroral spectra, therefore, always include nightglow and dayglow emissions. This unavoidable contamination must be taken into account in quantitative analyses of auroral spectra, especially at high resolution and in weak aurora. We identify indirect auroral processes as those in which at least one of the reactants is produced by a direct particle impact reaction.

2. ELECTRON IMPACT

Precipitation of energetic electrons into the atmosphere is the principal energy source of the aurora. Scattering and slowing down of the electron beam occurs by collisional interactions with the molecules and atoms in the atmosphere, collisions that have a high probability of leaving the atom, molecule, or ion in an excited state.

Atomic oxygen may be excited into one of several electronic states,

$$e + O(^3P) \rightarrow e + O(^1D, ^1S, ^3S, ^5S, \dots)$$

Partial energy-level diagrams are given by *Vallance Jones* [this volume] and a more complete listing of this and all other species may be found in the monograph by *Rees* [1989]. The excited $3s^3S^0$ state is the upper state of the allowed transition at 1304 Å, a bright feature in the ultraviolet aurora.

Ionization of atomic oxygen may leave the ion in an excited state,

$$e + O(^3P) \rightarrow 2e + O^+(^2D, ^2P, ^4P, \dots)$$

Excited states in the ground configuration of the ion, $2p^3\,^2D$ and $2p^3\,^2P$, are metastable. Many allowed states are excited and contribute to lines observed in the ultraviolet auroral spectrum.

Excited states of atomic oxygen are also produced by electron impact dissociation and dissociative ionization,

$$e + O_2(X^3\Sigma_g^-) \rightarrow e + O(^3P) + O(^1D, ^1S, ^3S, ^5S, \dots)$$

Both atomic and molecular sources need to be included when the excitation rate of oxygen atoms and ions is computed.

Electron impact on O_2 leads to excitation of the molecule,

$$e + O_2(X^3\Sigma_g^-) \rightarrow e + O_2(a^1\Delta_g, b^1\Sigma_g^+, \dots)$$

and to ionization excitation,

$$e + O_2(X^3\Sigma_g^-) \rightarrow 2e + O_2^+(a^4\Pi_u, A^2\Pi_u, b^4\Sigma_g^-, \dots)$$

The $O_2^+(a^4\Pi)$ state is highly metastable, with a radiative lifetime of about 10^4 seconds. Radiation from this state has not been detected, but it has an important role in auroral ion chemistry.

Molecular nitrogen is the most abundant neutral species in the Earth's atmosphere below about 200 km. Electron impact produces excitation of the molecule,

$$e + N_2(X^1\Sigma_g) \rightarrow e + N_2(A^3\Sigma,$$
$$B^3\Pi, C^3\Pi, a^1\Pi, b^1\Pi \dots)$$

---

*Geophysical Institute, University of Alaska Fairbanks, Fairbanks, Alaska 99701.

Auroral Physics, edited by C.-I. Meng, M. J. Rycroft and L. A. Frank. © Cambridge UP 1991

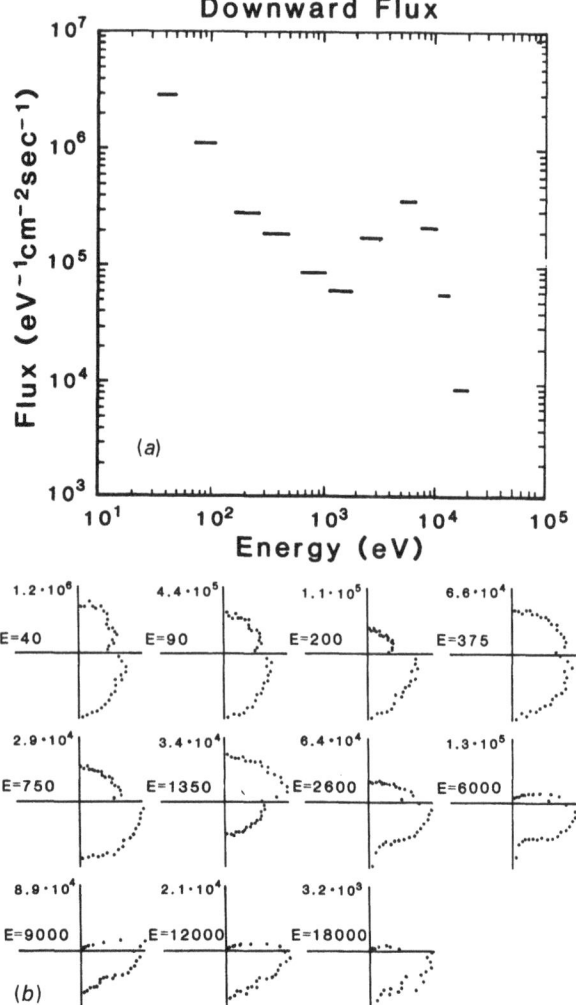

Figure 1—(a) The downward electron flux as a function of energy and (b) the angular distribution of the intensity measured by electron spectrometers on board a rocket traversing over an auroral arc at 340 km altitude at Poker Flat, Alaska. The length of the bars in the energy plot shows the range of the individual sensors. The angular distribution plots are normalized individually for each energy bin with the normalization factor shown on the ordinate with units eV^{-1} cm^{-2} s^{-1}. Each plot is identified by the center energy (in eV) of the corresponding energy bin. The abscissae are oriented normal to the magnetic field. Each data point represents a 4-second average in a 5° pitch-angle bin. The radii at which the points are located give a linear measure of the intensity [*Lummerzheim et al.*, 1989].

Some states lead to allowed transitions (e.g., $C^3\Pi \rightarrow B^3\Pi$) while others are metastable (e.g., $A^3\Sigma$). Several high-lying states of N$_2$ (e.g., $b^1\Pi$) lead to dissociation of the molecule and are named predissociation states,

$$e + N_2(X^1\Sigma_g) \rightarrow e + N_2^*(b^1\Pi) \rightarrow e + N(^4S) + N(^2D,^2P)$$

The product nitrogen atoms may be in excited states with the excess energy going into kinetic energy of the products. The energy-level diagram in Figure 12 of *Vallance Jones* [this volume] shows various electronic states of N$_2$.

Ionization of N$_2$ leads to several important excited states,

$$e + N_2(X^1\Sigma_g) \rightarrow 2e + N_2^+ \ (A^2\Pi,B^2\Sigma)$$

as well as to dissociative ionization,

$$e + N_2(X^1\Sigma_g) \rightarrow 2e + N(^4S) + N^+(^1D)$$
$$e + N_2(X^1\Sigma_g) \rightarrow 2e + N(^2D) + N^+(^3P)$$

Evidently, either of the products may be in an excited state, the ion, N$^+$(^1D), or the atom, N(^2D).

Atomic nitrogen is a minor constituent in the neutral atmosphere and most of the excited atomic states are produced through dissociative channels. However, the atomic nitrogen that is present in the Earth's atmosphere is subject to electron impact excitation and ionization, providing an additional source of excited species. This situation is opposite that encountered earlier with oxygen where electron impact on the atom is the dominant source of excited atoms and ions, while molecular dissociation is the minor source.

There are other minor species in the Earth's atmosphere (e.g., NO, CO$_2$) that are subject to electron bombardment that could cause excitation, ionization, and dissociation, but there is no observational evidence of excited electronic states being produced in the minor molecular species by electron impact. On the other hand, vibrational excitation of minor and major species by electron impact is a very important process, accounting for much of the energy loss of auroral electrons below the excitation threshold of electronic states,

$$e + M(v' = 0) \rightarrow e + M(v' > 0)$$

where M stands for N$_2$, O$_2$, NO, and CO$_2$. The vibrational distribution may not correspond to a Boltzmann distribution at the kinetic temperature of the ambient gas. The rotational distribution is in quasi-equilibrium and follows the kinetic temperature.

A contribution to the thermalization of auroral electrons is the excitation of the fine-structure levels in atomic oxygen,

$$e + O(^3P_2) \rightarrow e + O(^3P_{1,0})$$

Enhancement over airglow of the lines in the far infrared, corresponding to the transitions between fine structure levels, has not yet been detected in aurora.

3. EXCITATION BY ELECTRON IMPACT

The volume excitation rate of species j into state l at altitude z is

$$\eta_j^l(z) = n_j(z) \int_{E_{thr}}^{E_{max}} \sigma_j^l(E) I(E,z) dE \quad (cm^{-3}s^{-1})$$

where $\sigma_j^l(E)$ is the energy-dependent inelastic collision cross section (in cm^2), $n_j(z)$ is the concentration of the target atom or molecule at altitude z (in cm^{-3}), and $I(E,z)$ is the sum of the upward- and downward-directed hemispheric electron intensity at altitude z (in $cm^{-2} s^{-1} eV^{-1}$).

To evaluate the excitation rate, therefore, requires a neutral atmosphere obtained from measurements or models, collision cross sections obtained from laboratory experiments or theory, and the electron flux obtained from measurements and computations. The electron flux is usually measured above the aurora by satellite or rocket-borne instruments. The intensity as a function of altitude is obtained by solving the transport equation,

$$\mu \frac{dI(z,E,\mu)}{dz} = -n(z)\sigma(E,\mu)I(z,E,\mu)$$

$$+ n(z) \int_E^\infty \int_{-1}^{+1} \sigma(E' \rightarrow E, \mu' \rightarrow \mu)I(z,E',\mu')dE' \, d\mu'$$

$$+ n_e(z) \frac{\partial[L(E)I(z,E,\mu)]}{\partial E}$$

where μ is the cosine of the scattering angle. The boundary conditions are specified at an altitude below the deepest penetration of auroral electrons, $I(z_0,E,\mu) = 0$ and above the aurora $I(z_{max},E,\mu)$, by measurements or synthetic precipitation spectra. All excited and ionized states of all species must be included in the inelastic scattering processes, although the subscript j and superscript l have been dropped in the equation. The first term on the right side represents absorption, analogous to the optical case, while the second term includes the production of secondary electrons by ionization, the production of energy-degraded primary electrons, and elastic scattering by neutrals. The last term describes the energy loss to the ambient electrons of density $n_e(z)$ by means of a loss function $L(E)$. The

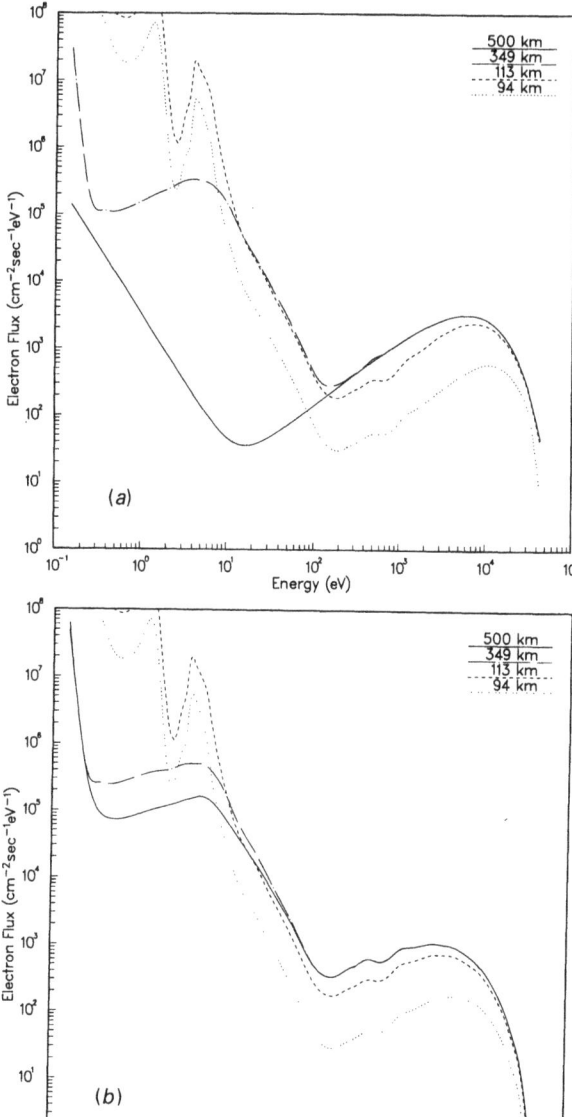

Figure 2—The downward flux (*a*) and the upward directed flux (*b*) computed at four altitudes with the transport model. An isotropic Maxwellian input spectrum with $E_{char} = 6$ keV was assumed at the upper boundary (500 km), supplemented by a power law spectrum, E^{-2}, below about 10 eV.

hemispheric electron intensity is derived from the intensity,

$$I(z,E) = 2\pi \int_{-1}^{+1} I(z,E,\mu)d\mu \quad (cm^{-2}s^{-1}eV^{-1})$$

Figure 1 presents the downward electron flux as a

function of energy measured at an altitude of 340 km by a rocket-borne spectrometer and the electron flux as a function of pitch angle at eleven energies. This event may be typical of electron precipitation associated with discrete aurora, but no two events are ever identical. The sensitivity of excitation rates of various states in atmospheric species to the characteristics of the auroral electron spectrum is a topic of current research activity.

The solution of the transport equation for an assumed auroral spectrum at 500 km altitude is presented in Figure 2. The flux is shown as a function of energy at four altitudes, including the input spectrum. Degraded primaries and secondary electrons make up the large flux of electrons below 100 eV. It is noted that a substantial fraction of the electron flux is scattered back out of the atmosphere. It is difficult to verify observationally the electron flux height profiles computed from models because this would require that the electron spectrometer be flown along a magnetic field line. Yet, knowledge of $I(z,E)$ is crucial for correctly predicting the rate of excitation of a multitude of states in atmospheric gases.

4. RESULTS OF MODEL COMPUTATIONS

The excitation rates of a few selected states are presented in Figures 3 through 6. Inelastic and elastic cross sections for the production of all the states in-

cluded in the transport calculations are given by *Lummerzheim* [1987]. Electron spectra can be specified quite generally by a number flux and an energy flux.

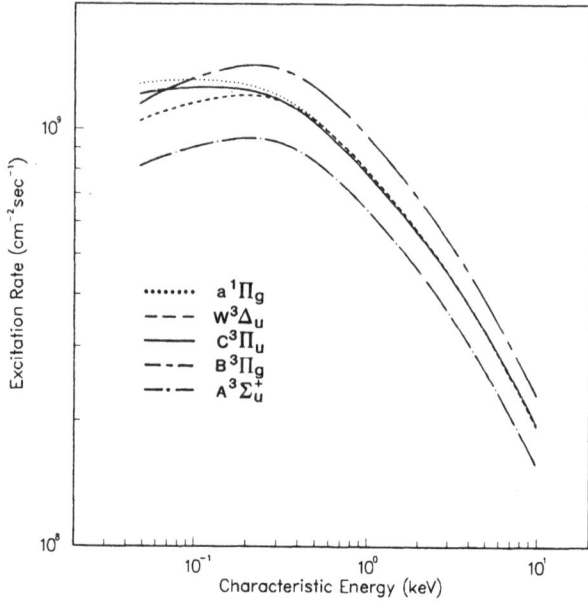

Figure 4—Column excitation rates of five states of N_2 by electron impact on N_2 as a function of the characteristic energy of an electron energy flux of 1 erg cm^{-2}s^{-1}.

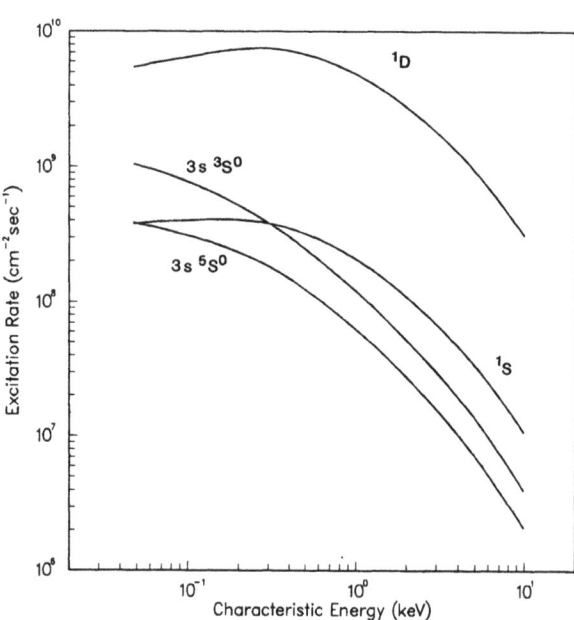

Figure 3—Column excitation rates of O(^1D), O(^1S), O($3s\,^3S^0$), and O($3s\,^5S^0$) by electron impact on O as a function of the characteristic energy of an electron energy flux of 1 erg cm^{-2} s^{-1}.

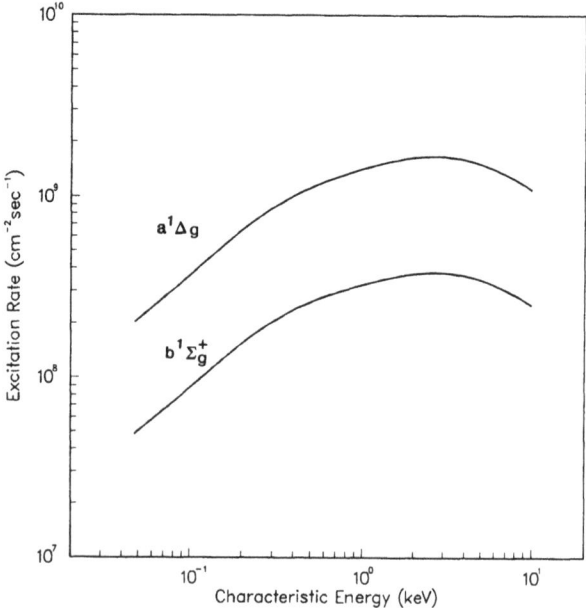

Figure 5—Column excitation rates of O$_2(a^1\Delta_g)$ and of O$_2(b^1\Sigma_g^+)$ by electron impact on O$_2$ as a function of the characteristic energy of an electron flux of 1 erg cm^{-2}s^{-1}.

The downward component of the number flux is

$$F_N^-(z) = 2\pi \int_0^\infty dE \int_{-1}^0 I(z,E,\mu)\mu d\mu \ (\text{cm}^{-2}\text{s}^{-1})$$

and the downward energy flux is

$$F_E^-(z) = 2\pi \int_0^\infty EdE \int_{-1}^0 I(z,E,\mu)\mu d\mu \ (\text{eV cm}^{-2}\text{s}^{-1})$$

The characteristic energy is defined at ∞ (or a z_{max} in practice),

$$E_{char} = \frac{1}{2}\frac{F_E^-(\infty)}{F_N^-(\infty)} \ (\text{eV})$$

For example, a Maxwellian spectrum that is proportional to $E \exp(-E/E_0)$ has a characteristic energy $E_{char} = E_0$. Real auroral spectra are seldom exactly Maxwellian, but may quite generally be described by a characteristic energy, as defined above. The quantities that are measured are not volume emission rates but column emission rates. Model results are therefore presented as height-integrated production rates of selected excited states,

$$H_j^l = \int_{z_{min}}^{z_{max}} \eta_j^l(z)dz \ (\text{cm}^{-2}\text{s}^{-1})$$

Emission rates are presented by *Meier and Strickland* [this volume].

The column excitation rates of four states in atomic oxygen from which prominent auroral radiations originate are presented in Figure 3. Excitation rates of five states of N_2 are given in Figure 4. Radiation from $W^3\Delta_u$ is not observed, but the state is believed to contribute to the excitation of the B and/or the A state by energy transfer. The $a^1\Pi_g$ is a weak predissociation state with most of the energy radiated in the ultraviolet. The column excitation rates of the two lowest excited states of O_2 are shown in Figure 5. Both states are metastable. Although the radiative lifetime of $a^1\Delta_g$ is about one hour, the state is almost immune to collisional deactivation and the infrared atmospheric band system is a prominent feature of the auroral spectrum, as noted by *Vallance Jones* [this volume].

An important energy degradation process for energetic auroral electrons is the collisional transfer of energy to the ambient electron gas in the auroral ionosphere. The resulting electron temperature is always higher than the neutral gas temperature and may reach 4000–6000 K in the F region. The Boltzmann distribution of an electron gas includes a substantial popu-

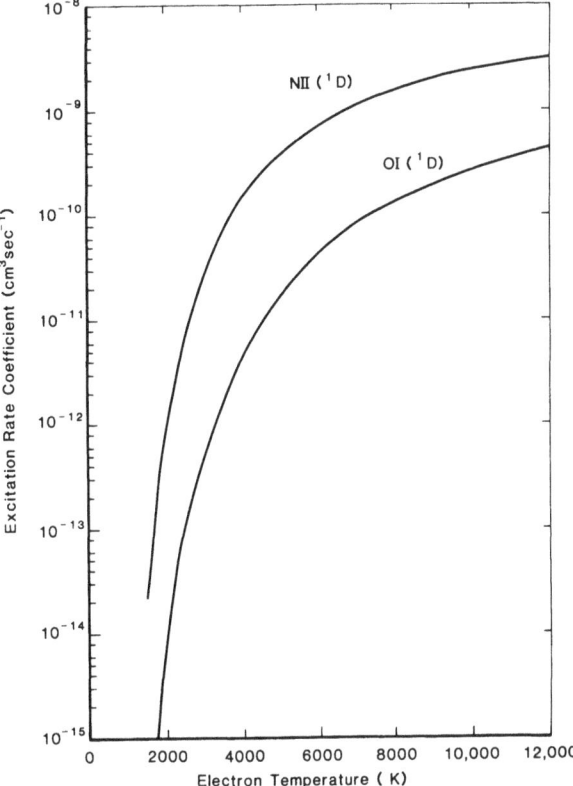

Figure 6—Excitation rate coefficients of O(^1D) and N$^+$(^1D) by electron impact as a function of the electron temperature of the ambient thermal population.

lation in the high-energy tail, and these electrons can excite species with sufficiently low excitation thresholds. This excitation mechanism with a threshold energy of 1.97 eV is known to contribute to the production of O(^1D) by electron impact. The mechanism may also contribute to the excitation of atomic nitrogen ions at high altitudes,

$$e + N^+(^3P) \rightarrow e + N^+(^1D)$$

for which the excitation threshold is 1.89 eV. The excitation rate coefficients for the two species are given in Figure 6 as a function of electron temperature.

Thresholds for excitation of electronic states in neutral species are in the energy range from a few to a few tens of electron volts. The largest number flux of auroral electrons occurs at these lower energies (see Figure 2). Since photoelectrons produced by energetic solar UV photons have a similar energy spectrum, excitation of the dayglow and the aurora occurs through similar processes. The enhancement of excited states of ions is a signature of auroral excitation.

5. PROTON IMPACT

The latitude distribution of proton and electron precipitation has been examined spectroscopically from ground-based observations and by energetic-particle detectors carried on polar orbiting satellites. Although electron precipitation is usually the dominant source of auroral excitation, regions have been identified, not infrequently, where proton fluxes are large in an absolute sense (> 1 erg cm^{-2} s^{-1}) and also large relative to electron precipitation. Proton bombardment produces auroral spectral emissions and the excitation mechanisms need to be investigated.

Excitation by proton impact with atmospheric species may occur in charge-exchange collisions,

$$H^+ + N_2 \rightarrow H(n=2,3,4\ldots) + N_2^+(A^2\Pi, B^2\Sigma, \ldots)$$

or by ionizing collisions,

$$H^+ + N_2 \rightarrow H^+ + N_2^+(A^2\Pi, B^2\Sigma, \ldots) + e$$

Hydrogen atoms and molecular nitrogen ions may be produced in excited states and the cross sections associated with these processes are large [*Van Zyl et al.*, 1984]. The hydrogen atoms produced in charge-exchange collisions retain a large fraction of the proton's velocity (producing the Doppler shifted hydrogen lines) and may produce excitation in the atmosphere,

$$H + N_2 \rightarrow H(n=2,3,4, \ldots) + N_2(C^3\Pi, B^3\Pi, \ldots)$$

The analogous excitation processes occur in collisions of the proton/hydrogen beam with O_2 and with O.

Volume excitation rates are computed as they are for electron impact, requiring the density-height profiles of atmospheric neutral species, the excitation cross sections, and the energetic-particle flux. Both the proton flux and the hydrogen atom flux produce excitation but with different cross sections. The transport equations for protons and hydrogen atoms are coupled by the charge exchange and ionization reactions [*Basu et al.*, 1987],

$$H^+ + M \rightarrow H + M^+$$

$$H + M \rightarrow H^+ + M + e$$

$$H + M \rightarrow H + M^+ + e$$

The electrons ejected by ionizing collisions have sufficient energy to produce additional excitation of atmospheric atoms and molecules by processes described earlier.

Fluxes of energetic O^+ ions precipitating into the atmosphere have been detected with satellite-borne instruments. Modeling the excitation of atmospheric species associated with an energetic O^+/O beam has, most recently, been discussed by *Ishimoto et al.* [1986].

6. CHEMICAL-IONIC REACTIONS

Chemical-ionic reactions occur in the atmosphere without aurora. The reactants are the atmospheric gases and the ions and atoms that are produced by the absorption of solar ultraviolet radiation in the thermosphere. This is a subject to which Sydney Chapman made fundamental contributions. While the basic excitation reactions are the same in the dayglow as in aurora, the latter produces larger excitation and ionization rates that are limited in spatial extent and in duration. The auroral ionosphere is therefore characterized by density gradients and non-steady-state conditions.

Several types of chemical reactions lead to the production of excited states and one example of each is given. Dissociative recombination of molecular nitrogen ions leads to the production of two excited nitrogen atoms,

$$N_2^+ + e \rightarrow N(^2D) + N(^2D).$$

This reaction and the corresponding ones for O_2^+ and NO^+ account for the decay of auroral ionization. The atom-atom interchange reaction,

$$N(^2D) + O_2 \rightarrow NO + O(^1D)$$

is an important source of NO molecules as well as oxygen atoms excited to the ^1D state. The loss process for NO is slow, and substantial concentrations of this species can build up in the auroral atmosphere. The ion-atom interchange reaction

$$N^+ + O_2 \rightarrow NO^+ + O(^1D)$$

is also a source of excited oxygen, $O(^1D)$, as well as of NO^+ ions. Chemical-ionic reactions are sources of excited states that have a low threshold energy, e.g., $O(^1D)$ at 1.97 eV, $N(^2D)$ at 2.38 eV, and $O(^1S)$ at 4.13 eV. The volume production rate of species j in state l is

$$\eta_j^l(z) = K(l' \rightarrow l)n_a(z)n_b(z) \ (\text{cm}^{-3}\text{s}^{-1})$$

where the $n_x(z)$'s are the concentrations of the reactants at altitude z and $K(l' \rightarrow l)$ is the reaction rate coefficient (cm^{3}s^{-1}) which may be a function of temperature. The continuity equations that specify the concentrations of atmospheric neutral and ionized species, including excited states, are coupled and time dependent. There are local source and loss terms, and transport terms [*Rees*, 1989].

7. CASCADING

Excited states may be produced by radiative transitions from higher-lying states. An example in N_2 is

$$N_2(C^3\Pi) \rightarrow N_2(B^3\Pi) + h\nu$$

$$N_2(B^3\Pi) \rightarrow N_2(A^3\Sigma) + h\nu$$

where all three states, *C*, *B*, and *A*, are excited states of N_2.

An example in atomic oxygen is

$$O(3p\ ^3S) \rightarrow O(3s\ ^3S) + h\nu$$

$$O(3s\ ^3S) \rightarrow O(2p\ ^3P) + h\nu$$

Energy-level diagrams of atmospheric species are good sources for identifying states that could be excited by cascading.

8. CONCLUDING REMARKS

Only a small fraction of the energy carried by auroral particles that precipitate into the atmosphere goes into excitation of electronic states of atoms, molecules, and ions. These excitation processes are the source of the spectroscopic aurora discussed by *Vallance Jones* [this volume]. Estimates of the energy budget associated with auroral particle precipitation show that about one half of the energy goes into heating the neutral gas and the

plasma. The remainder is shared between enhancement of the ionization, increased dissociation, chemical energy stored in long-lived species, and energy stored in vibrational enhancement. Depending on the energy and pitch-angle distribution of the downward particle flux, 15–30% of the energy may be scattered back out of the atmosphere.

The processes responsible for excitation of the spectroscopic aurora have mostly been identified. Many, but not all impact excitation cross sections have been measured in the laboratory. There are uncertainties in some reaction-rate coefficients, particularly regarding the applicability to the atmospheric environment of results obtained under laboratory conditions. Finally, solutions of the electron and proton transport equation have yet to be fully tested and verified.

ACKNOWLEDGMENT—The National Science Foundation provided partial support for this work under grant ATM 8701192.

REFERENCES

Basu, B., J. R. Jasperse, R. M. Robinson, R. R. Vondrak, and D. S. Evans, "Linear Transport Theory of Auroral Proton Precipitation: A Comparison with Observations," *J. Geophys. Res.*, **92**, 5920–5932 (1987).

Ishimoto, M., M. R. Torr, P. G. Richards, and D. G. Torr, "The Role of Energetic O⁺ Precipitation in a Mid-Latitude Aurora," *J. Geophys. Res.*, **91**, 5793 (1986).

Lummerzheim, D., "Electron Transport and Optical Emissions in the Aurora," Ph.D. Thesis, 98 pp., University of Alaska, Fairbanks (1987).

Lummerzheim, D., M. H. Rees, and H. R. Anderson, "Angular Dependent Transport of Auroral Electrons in the Upper Atmosphere," *Planet. Space Sci.*, **37**, 109 (1989).

Rees, M. H., *Physics and Chemistry of the Upper Atmosphere*, 289 pp., Cambridge University Press (1989).

Van Zyl, B., M. W. Gealy, and H. Neumann, "Prediction of Photon Yields for Proton Aurorae in an N_2 Atmosphere," *J. Geophys., Res.*, **89**, 1701–1710 (1984).

II-3. AURORAL EMISSION PROCESSES AND REMOTE SENSING

R. R. Meier* and D. J. Strickland[†]

Optical observations of the aurora have evolved over the years from somewhat imprecise qualitative research into quantitative remote sensing of the characteristics of the energetic particles and the state of the atmosphere and ionosphere. In reviewing the latest spectroscopic approaches to the study of the aurora, the intent here is to provide continuity following the paper by Rees and Lummerzheim on excitation processes and the papers on atmospheric composition. The discussion is mainly from a theoretical perspective, describing the auroral emission spectrum resulting from energetic electron precipitation into a nominal atmosphere.

1. INTRODUCTION

An important fraction of the energy carried into the auroral zone by precipitating electrons appears as optical emissions. For electrons with characteristic energies of 8 keV (in a Maxwellian energy distribution), the percentage is 16% while, at 100 eV, radiation accounts for 25–30% of the incident energy. Thus, emission processes play an important role in the energetics of auroras. They also provide important diagnostics for determining the characteristics of the energetic electron spectrum as well as of atmospheric composition, temperature, dynamics, and chemistry. Here we focus on applications of auroral spectroscopy to the remote sensing of precipitating electrons and atmospheric composition. The approach is largely theoretical, mainly discussing physical concepts and recent progress. Previous reviews of the auroral spectroscopic emissions and relevant physical processes have been given by *Vallance Jones* [1974] and *Bates* [1982]. Work since then has been covered by *Rees* [1983], *Vallance Jones et al.* [1985], and *Meier* [1987].

Historically, ground-based observing systems, and more recently imaging instruments on board orbiting satellites, have provided spectroscopic data mainly in the form of column emission rates, without direct knowledge of the altitude profiles of the emitters. This is the problem addressed here: to obtain quantitative information about the precipitating particles and the atmosphere from selected line-of-sight spectroscopic observations. Of course, if altitude information is available from a rocket or spinning satellite, spectroscopic techniques become even more powerful. First, auroral spectroscopy is discussed briefly from a theoretical perspective. Next, principles of obtaining quantitative information from spectra are reviewed. Finally, various observational approaches are examined.

2. ENERGY PARTITIONING AND EMISSION FEATURES

Rees and Lummerzheim [this volume] have discussed the fate of energetic electrons precipitating into the auroral zone. For a Maxwellian distribution of electron energies with 1-keV characteristic energy, nearly half of the energy goes into ionization, the rest resulting in molecular dissociation and excitation of atomic and molecular states. (See Figures 11 through 15 of *Vallance Jones* [this volume] for energy-level diagrams showing the locations of some of these states.) According to the electron transport model of *Strickland et al.* [1976, 1983, 1989], using a *Jacchia* [1977] atmospheric model at 1000 K and adjusting the O concentration downward by 0.5 for the auroral zone, some 31% of the total incident energy causes excitation of N_2, about 12% going into triplet states (A, B, C, W, and Rydbergs) and 19% into singlet states (a, a', a'', b, b', c_4', and w). Of that 31% which goes into N_2 excitation, 14% results in radiation from the triplet and $a^1\Pi$ states, with much of the remaining 17% resulting in predissociation. Even for the $a^1\Pi$ state (and other far ultraviolet) emissions, half of the radiation is emitted downward and primarily causes additional dissociation of O_2. Excitation of states of atomic oxygen accounts for nearly 6% of the total incident energy at $E_0 = 1$ keV (E_0 being the characteristic energy of primary particles with a Maxwellian distribution). Most of that ends up in radiation. For the optically thick ultraviolet transitions of O, most of the radiation escapes from the top of the atmosphere after undergoing multiple scattering (unless the electron characteristic energy is greater than 5 keV, in which case the auroral excitation occurs so low in the atmosphere that much of the original emission is absorbed by O_2). As noted in the introduction, a larger portion of the energy in softer auroras is emitted from the atmosphere, because of high-altitude excitation of atomic oxygen. For harder auroras, relatively more of the energy is dissipated through dissociation and ionization processes.

*E. O. Hulburt Center For Space Research, Naval Research Laboratory, Washington, DC 20375.
[†]Computational Physics Incorporated, P.O. Box 360, Annandale, VA 22003.

Auroral Physics, edited by C.-I. Meng, M. J. Rycroft and L. A. Frank. © Cambridge UP 1991

TABLE 1. Some important auroral emission features within wavelengths bands: model with $E_0 = 1$ keV and $Q = 1$ erg cm^{-2} s^{-1}

Band	Wavelength (Å)	Emission line source	Excitation mechanism	Column emission rate (R)	Notes
EUV[†]	834	O II	e + O, O_2	286,17*	Excitation only
	989	O I	e + O, O_2	111,5*	Excitation only
	1027	O I	e + O	57,6*	Excitation only
	1085	N II	e + N_2	57	
	1134	N I	e + N_2	32	
FUV[†]	1200	N I	e + N_2	167*	Excitation only
	1216	H I	H$^{+\prime}$ + H, H$^\prime$ + X	–	Proton events
	1304	O I	e + O, O_2	403,11*	Excitation only
	1356	O I	e + O, O_2	84,19	Excitation only
	1493	N I	e + N_2	61	
	1743	N I	e + N_2	31	
	1000–2600	N_2 LBH	e + N_2	53	1383 Å band
MUV[†]	2143	N II	e + N_2	–	Not in model
	2972	O I	e + O, chem.	43	
	1250–5325	N_2 VK	e + N_2	46	3199 Å band
NUV	2680–5460	N_2 2*PG*	e + N_2	188	3371 Å band
	3727	O II	e + O, chem.	0.1	Preliminary
	3466	N I	e + N_2, chem.	78	Preliminary
VIS-NIR	2860–5870	N_2^+ 1*NG*	e + N_2	782	3914 Å band
	4780–25310	N_2 1*PG*	e + N_2	304	7754 Å band
	5200	N I	e + N_2, chem.	29	Preliminary
	5500–17700	N_2^+ Meinel	e + N_2, chem.	–	Not in model
	4990–8530	O_2^+ 1*NG*	e + O_2	–	Not in model
	5577	O I	e + O, chem.	1000	
	6300	O I	e + O, chem.	383	
	5380–9970	O_2 ATM	e + O_2, chem.	–	
	7320	O II	e + O, chem.	85	Preliminary
	7774	O I	e + O, O_2	60,17	
	7990	O I	e + O	–	Not in model
	8446	O I	e + O, O_2	277,8	

*Optically thick; only column excitation rates are given.

[†]Nadir viewing from 400 km for EUV, FUV, and MUV bands; pure absorption included.

In his overview of auroral spectroscopy, *Vallance Jones* provides a comprehensive set of spectra covering the range from 400–16,500 Å along with line and band identifications. He compiled a number of independent observations to show the entire optical spectrum. Because of differing observing conditions of the datasets in his compilation, the relative line and band strengths are different from spectrum to spectrum. In order to place the auroral spectrum on a common intensity scale, a theoretical model can be employed using a specific set of conditions. Several models have been developed that predict auroral emission rates (e.g., *Strickland et al.* [1976, 1989], *Solomon et al.* [1988], and *Rees and Lummerzheim* [1989]). For the results presented here we use the model of Strickland and his colleagues. Space limitations preclude describing the model in detail; many of the details of electron transport processes, cross sections, and chemistry can be found in the above papers and in *Daniell and Strickland* [1986].

Table 1 lists important emission features in the auroral spectrum for the extreme (EUV), far (FUV), middle (MUV), near (NUV) ultraviolet, visible (VIS),

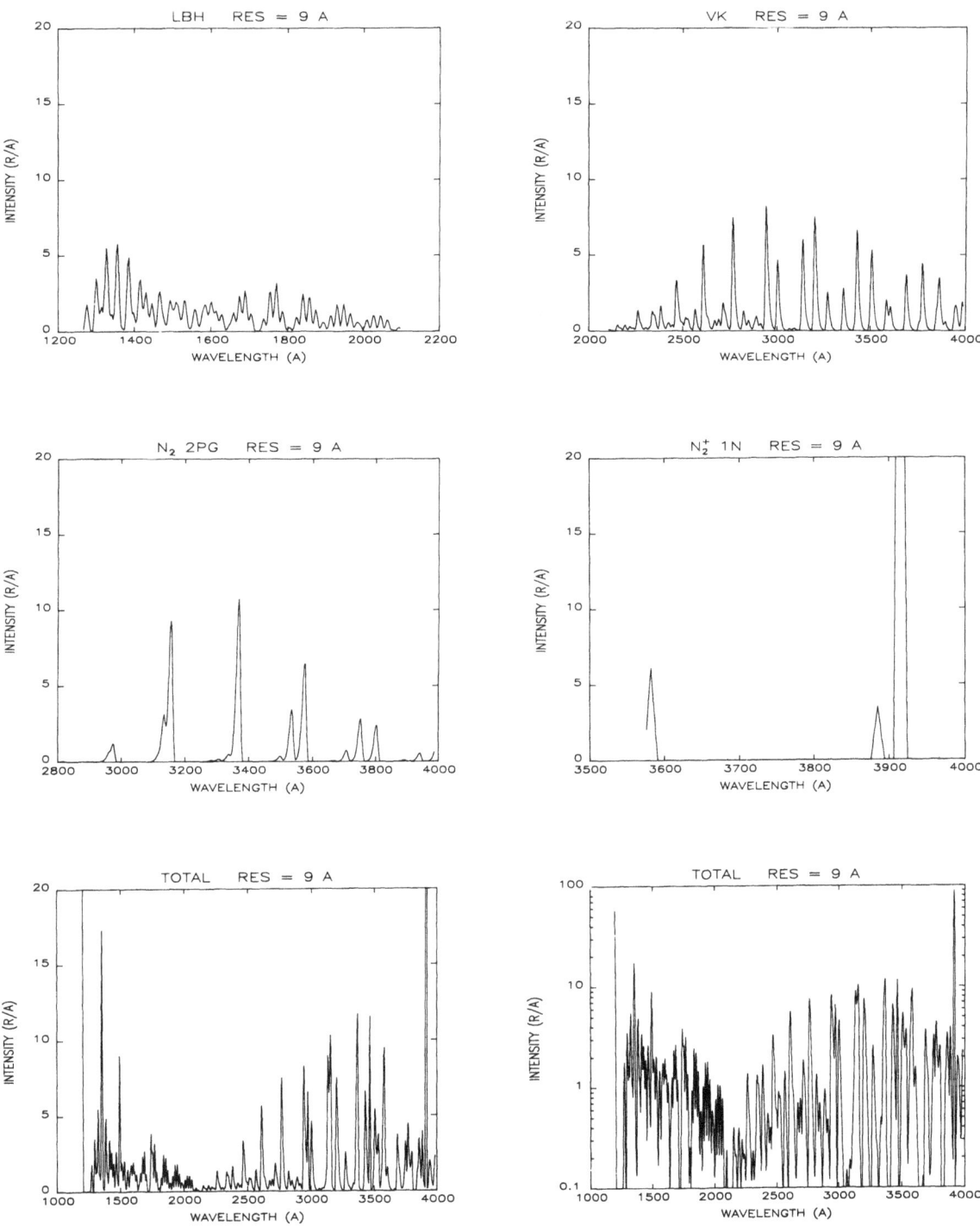

Figure 1—Auroral UV spectrum at 9 Å spectral resolution for E_0 = 1 keV and Q = 1 erg cm^{-2} s^{-1}. *Upper left*, the N$_2$ Lyman-Birge-Hopfield bands; *upper right*, N$_2$ Vegard-Kaplan bands; *middle left*, N$_2$ second positive bands; *middle right*, N$_2^+$ first negative bands (band heads only); *lower left*, total spectrum including atomic lines at 1200, 1356, 1493, 1743, 2972, and 3466 Å; *lower right*, same as lower left, except on logarithmic scale. A rotational temperature of 400K is assumed for the LBH, VK, and 2PG bands.

and near infrared (NIR) bands of the optical spectrum. The viewing direction is assumed to be along the magnetic zenith, except for the UV emissions that are along the magnetic nadir from above the atmosphere, since atmospheric extinction prevents viewing from the ground. Some of the features are optically thick so that multiple-scattering calculations must be performed in order to determine the emission rate. This was not done for Table 1, although a later section addresses the use of optically thick features and relevant branching transitions to determine auroral atmospheric characteristics. Other spectral features have recently been included in the model, but have not been compared extensively with observation; their emission rates must be considered with caution. They are indicated in the table as preliminary. The papers cited by Vallance Jones discuss production and loss processes of many of the spectral features.

The ultraviolet portion of the spectrum is shown graphically in Figure 1. A resolution of 9 Å is assumed for the spectrum, since that is typical of many spectrographs. Only atomic lines at 1200, 1356, 1493, 1743, 2972, and 3466 Å are included. Some important lines not included in the synthetic spectrum are H I 1216, O I 1304, and N II 2143 Å. The spectrum is dominated by N_2 Lyman-Birge-Hopfield (LBH) bands in the FUV, Vegard-Kaplan and 2PG bands in the MUV and NUV, and N_2^+ 1NG bands in the NUV. Detailed identifications are given in Figures 2 through 4 of *Vallance Jones* [this volume].

Table 1 and Figure 1 are meant to illustrate the relative emission rates of spectral features for a typical nighttime electron-excited aurora. There will be important changes in the spectrum for the much softer daytime aurora, where excitation of atomic oxygen dominates and pure absorption (extinction due to molecular absorption) becomes negligible. Similarly for harder auroras, the molecular features increase relative to the atomic emissions. For very hard auroras, absorption or quenching can cause the brightness of some features to decrease because of the low altitude of peak excitation.

3. EMISSION PARAMETERS

There is a variety of parameters that affect auroral emissions: energy distribution and type of precipitating particles, atmospheric and ionospheric composition and opacity, dynamics, chemistry, and atomic and molecular constants (cross sections, transition probabilities, reaction rates). Here, we intend to concentrate on optical emissions that depend on the parameters of the precipitating electron flux and atmospheric composition. The chosen parameters of a Maxwellian energy distribution for the electron flux are E_0 (the characteristic energy in keV), and Q (the incident energy

flux in ergs cm^{-2} s^{-1}). A low energy "tail" (LET) is included to account for escaping secondary electrons that are reflected by potential barriers or by wave-particle interactions at high altitudes and reenter the atmosphere as part of the primary particle distribution. The LET does not add significantly to Q, but does influence emissions produced at high altitude, such as O I 6300 Å. Analytic representations of LETs and references to observations of primary spectra are given by *Meier et al.* [1989]. For atmospheric composition, a standard model is assumed (*Jacchia* [1977]), but linear multipliers are included as a first-order attempt to delineate departures from standard conditions. We recognize that it is not physically realistic to adjust atmospheric composition in an altitude-independent manner in response to energy or momentum inputs. *D. Rees and Fuller-Rowell* [this volume] and *Killeen et al.* [this volume] discuss more realistic thermospheric global circulation models. Apart from those changes caused by variable photodissociation or auroral dissociation, there are strong responses to intense magnetospheric energy deposition. Systematic large-scale upwelling, outflow from heated regions, combined with horizontal advection, and distant downwelling can modify gas composition significantly. Heated regions display enhancement of heavier molecular species and a depletion of lighter atomic species, with the converse in regions of horizontal convergence and downwelling. This spectrum of compositional changes is not reflected particularly well in hydrostatic diffusive equilibrium models, such as that used here.

However, parameterization is not available for these global models in a way that can be used in developing algorithms for extraction of composition information from spectral data. On the other hand, as discussed by *Strickland et al.* [1989], the details of the model atmosphere used in developing algorithms for extracting compositional information are not particularly critical since the analysis of optical data will yield results that can be used to infer departures from the baseline model. For example, *Hecht et al.* [1989] found that in deducing O multipliers for the *Jacchia* [1977] model from analysis of spectroscopic data (see next section for a discussion of the algorithm), the O multiplier was found to increase with characteristic energy of the incident primary electrons, showing that the O scale height in the baseline model was too large and that a modification was required for internal consistency. The algorithms described in the next section are also very useful in identifying other trends. *Hecht et al.* [1989] actually observe the depletion in atomic oxygen relative to molecular nitrogen as auroral heating increases. This finding was true for both the Jacchia and the MSIS models. In the future, more refined techniques involving iteration to the actual altitude distribution of

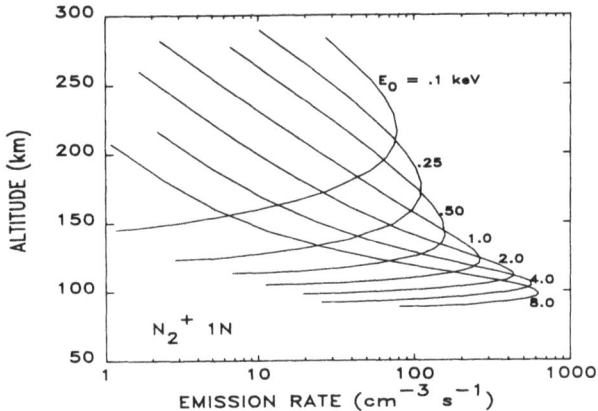

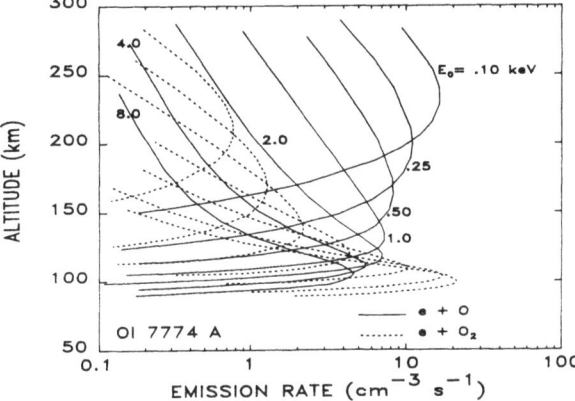

Figure 2—Volume emission rates from all levels of N_2^+ 1N state (upper panel) and for O I 7774 Å emission from O and O_2 sources (lower panel). Characteristic energy of the primary electrons is indicated, except for the e + O_2 curves. (From *Meier et al.* [1989].)

the atmospheric species are expected to be developed. In the meantime, a standard model that is easy to parameterize with exospheric temperature and constant multipliers has been used to illustrate the salient features of the spectral emissions and to develop analysis algorithms.

Spectroscopic techniques take advantage of the fact that, as the characteristic energy of the primary electrons increases, the altitude of maximum energy deposition decreases. Combinations of spectral features whose brightnesses depend on the height of emission in different ways can provide information about E_0. The ratio of atomic to molecular emissions also decreases with increasing energy, so compositional parameters can be deduced. These points are illustrated in Figure 2 (from *Strickland et al.* [1989]), which shows volume excitation rates of N_2^+ ($B\ ^2\Sigma$) (the parent state of 3914 and 4278 Å 1NG bands) and the O I

$(3p\ ^5P)$ state which emits at 7774 Å. Electron impact excitation of ground state N_2, O_2 and O is the source of these emissions. They are allowed transitions and consequently are prompt (lifetimes $\sim 10^{-8}$ s), so that emission from excited states is more rapid than collisional processes.

Column integrals of the functions in Figure 2 yield observables. But other emissions, which are absorbed or quenched, offer alternative methods of parameter delineation. Table 2 illustrates various ways in which the column emission rates of selected features depend on altitude and characteristic energy. The generic quantities g and B represent, respectively, the excitation rate per second per atom (or molecule) and the branching ratio into the desired rotational line and vibrational band of a particular transition. The "g-factor" is formally defined as the integral of the electron impact excitation cross section times the energetic electron flux over energy. Square brackets indicate concentration. The product of g times the concentration is the volume excitation rate, as shown by the curves in Figure 2. Simple integrals of the volume emission rates yield the O I 7774 Å column emission rate or the N_2 4278 Å (0,1) band. *Strickland et al.* [1989] have shown that for E_0, ranging from 0.1–8 keV, 4278 Å increases by about a factor of 2, 7774 Å (e + O) decreases by 15, and 7774 Å (e + O_2) increases by about 6. The precise variations, however, were found to depend on the O concentration.

Excited states that are quenched, such as O(^1D), (parent state of O I 6300 Å) and $N_2(A)$ (parent state of the Vegard-Kaplan band)), have dependencies on atmospheric composition that are included in the β terms in Table 2. β is the ratio of the quenching coefficient times the concentration of the quenching species to the transition rate. O(^1D) is quenched mainly by N_2 and possibly O (*Abreu et al.* [1986]), thereby restricting the emission to higher altitudes. The upper-case G for 6300 Å in Table 2 is meant to represent the fact that there are chemical sources of O(^1D), although electron impact excitation dominates in the emission region according to *Solomon et al.* [1988] or *Meier et al.* [1989]. While *Sharp et al.* [1979] and *Rees and Roble* [1986] have argued that below 200 km, an additional excitation source is required, *Meier et al.* [1989] have reanalyzed the *Sharp et al.* data and found no compelling need for the additional source.

Figure 3 shows the volume emission rate at 6300 Å for various E_0s. The column emission rate of 6300 Å decreases by about two orders of magnitude for E_0 increasing from 0.1–8 keV (*Strickland et al.* [1989]). The red line is also strongly dependent on the O concentration.

Another way to determine E_0 is to take advantage of the effect of pure absorption on the emission rate.

41

TABLE 2. Examples of emissions.

N$_2$

LBH $\qquad\qquad I \;=\; \int B\, g_{LBH}\, [N_2]\, e^{-\tau(O_2)}\, dz$

VK $\qquad\qquad I \;=\; \int \dfrac{B\, g_{VK}\, [N_2]\, dz}{1 + \beta_{VK}\,(O)}$

2 *PG* (1*PG*, 1*NG*, Meinel) $\quad I \;=\; \int B\, g\, [N_2]\, dz$

O$_2$

1*NG*, ATM $\qquad\qquad I \;=\; \int B\, g\, [O_2]\, dz$

O I 7774 $\qquad\qquad I_{O_2} \;=\; \int g_{O_a}\, [O_2]\, dz$

O

6300 $\qquad\qquad I_R \;=\; \int \dfrac{G_R\, [O]\, dz}{1 + \beta(N_2, O)}$

7774 $\qquad\qquad I_0 \;=\; \int g_0\, [O]\, dz$

For example, the N$_2$ LBH bands are absorbed strongly by the O$_2$ Schumann-Runge continuum near 1400 Å. Absorption is greater for larger E_0 since the excitation occurs at altitudes with higher O$_2$ concentration. The functional dependence on O$_2$ optical depth τ (O$_2$) is shown in Table 2.

There are other spectroscopic techniques for obtaining the abundance of atomic oxygen using measurements of temperature to obtain the altitude of the emitting species (and consequently E_0 through the altitude of peak energy deposition), or using ratios of optically thick to thin features. Some of these are discussed next.

4. EMISSION DIAGNOSTICS

Red-to-Blue and Other Spectroscopic Ratios

Perhaps the most widely used spectroscopic emissions for determining characteristic energy and energy flux in auroras are O I 6300 Å (red) and N$_2^+$ 4278 Å (blue). They are among the brightest auroral emissions observable from the ground. *Rees and Luckey* [1974] presented a series of curves that allowed the deduction of E_0 and Q from the ratio and the magnitude of the emission rates. The sensitivity of the red-to-blue ratio on E_0 is because, as E_0 increases, the altitude of peak energy deposition decreases. For a fixed energy flux, a relatively smaller amount of red line is produced as E_0 increases because quenching dominates over emission as the main loss of O(^1D) atoms. Thus the large decrease in the volume emission rate with increasing E_0 in Figure 3 can be understood. Since the blue emission is much less dependent on E_0, and since electron impact excitation dominates both emissions,

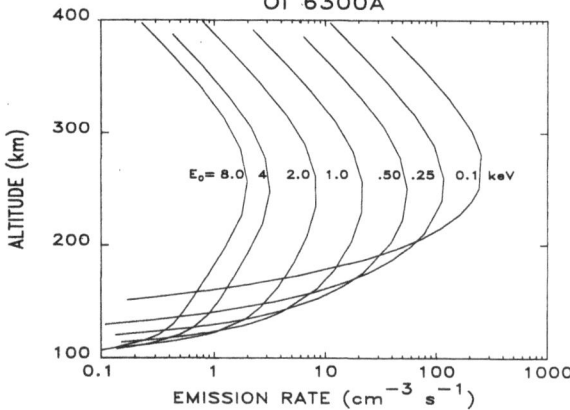

Figure 3—Volume emission rate of O (^1D) at 6300 Å for various characteristic energies.

42

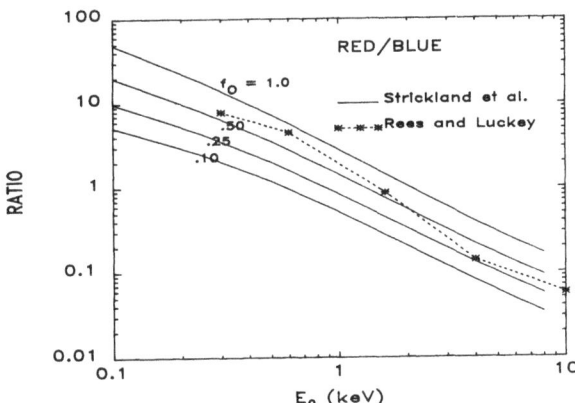

Figure 4—Red-to-blue ratio vs. characteristic energy for various multipliers of the atomic oxygen concentration. Values from *Rees and Luckey* [1974] are shown for a 3914 Å emission rate of 1 kR. The theoretical curves are taken from *Strickland et al.* [1989].

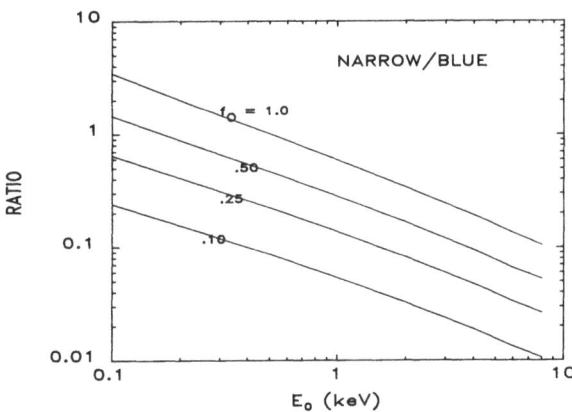

Figure 5—Ratio of O I 7774 Å (narrow)-to-blue emission vs. characteristic energy for various multipliers of the atomic oxygen concentration. (From *Strickland et al.* [1989].)

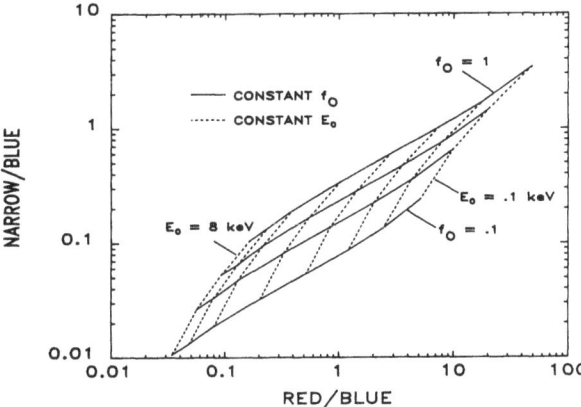

Figure 6—Line ratios plotted for various characteristic energies E_0 (with constant O multipliers) and for various O concentration multipliers f_0 (with constant characteristic energies). (From *Strickland et al.* [1989].)

the red-to-blue ratio is sensitive to E_0 and to first-order, not sensitive to Q. However, as pointed out by *Rees and Luckey* [1974], the ratio can depend on the total energy input in the aurora, since as Q increases, the ionospheric electron density increases, and loss of energy of the secondary electrons to the plasma increases, thereby reducing the $O(^1D)$ production.

A more serious difficulty facing the use of the red-to-blue ratio is its dependence on the O concentration. Figure 4 shows the dependence of the ratio on E_0 for various constant multipliers (f_0) of the O concentration in the Jacchia 1000 K model. This figure is taken from *Strickland et al.* [1989] and assumes a nominal F region electron density profile to account for energy loss to the plasma; LETs are included in the Maxwellian velocity distributions. The Rees and Luckey ratio is shown for a blue emission rate of 1 kR. The different slopes of the curves at different energies could be due to several factors. The Strickland et al. results use the Strickland electron transport model while a simple range-energy analysis was used by Rees and Luckey. Another difference is that the Strickland representation incorporates a low energy component with the Maxwellian distribution, while Rees and Luckey use pure Maxwellian functions. The two computations, moreover, use very different model atmospheres.

The results in Figure 4 show that to separate compositional from other effects, it is necessary to measure additional emission features that also depend on E_0 and f_0. One candidate for this task is the O I 7774 Å emission line [*Hecht et al.*, 1985]. This feature has two spectral components, a "narrow" line due to electron impact on atomic oxygen and a "broad" component due to impact on O_2. The spectral width of the narrow component (~ 0.03 Å) is due to the thermal motion of O atoms; the broad component (~ 0.2 Å) is due to the excess kinetic energy of the excited fragment atoms produced by dissociation of O_2. The two lines can be resolved with a Fabry-Perot interferometer [*Hecht et al.*, 1985]. The narrow-to-blue ratio, shown in Figure 5, varies with both E_0 and f_0, but in a different way than red-to-blue, since quenching is not involved in the production of the O I 7774 Å line. An alternative way of displaying the two sets of ratios is to plot one against the other for specific values of E_0 and f_0. This is shown in Figure 6, taken from *Strickland et al.* [1989]. An observation of the narrow-to-blue and red-to-blue ratios yields a data point that can be plotted directly on Figure 6 to find E_0 and f_0. This has been done by *Hecht et al.* [1989] for a variety of observations. It is encouraging that most of the data points obtained by them fall within the model boundaries in Figure 6, demonstrating that the model is doing reasonably well. *Hecht et al.* [1989] also plotted the derived f_0 as a function of geomag-

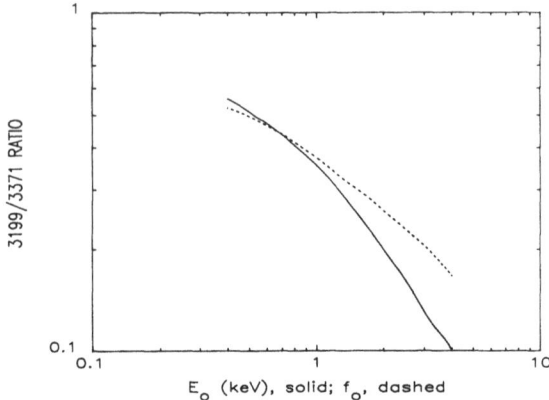

3199/3371 RATIO

E_o (keV), solid; f_o, dashed

Figure 7—Ratio of VK-to-2*PG* bands vs. characteristic energy E_0 for an O multiplier of unity (solid curve) and vs. O concentration multiplier f_0 for $E_0 = 1$ keV. The curves were scaled from the figures of *Daniell and Strickland* [1986].

netic activity. They found a systematic decrease of atomic oxygen as magnetic activity increased thereby supporting the view that increased heating results in depletion of O in the auroral zone.

The plots in Figures 4 through 6 are relatively insensitive to the O_2 abundance. Similar plots using the broad-to-blue ratio can yield information about the concentration of O_2 [*Strickland et al.*, 1989; *Hecht et al.*, 1989]. Application of such plots, however, is of limited use for soft auroras since the broad component becomes rather weak.

N_2 Vegard-Kaplan (VK) and Second Positive Bands

Quenching of the VK bands plays a role similar to $O(^1D)$ quenching, so the ratio of a forbidden VK emission to an allowed N_2 emission yields information on the characteristic energy of the aurora (e.g., *Sharp* [1971]). A recent analysis of the VK problem has been published by *Daniell and Strickland* [1986]. The solid line in Figure 7 shows a plot of the ratio of the (1,9) VK band at 3199 Å to the N_2 second positive (2*PG*) (0,0) band at 3371 Å as a function of characteristic energy. The individual intensities were scaled from the curves of Daniell and Strickland for a Maxwellian primary spectrum and $f_0 = 1$ in the *Jacchia* [1977] model. However, since atomic oxygen is the primary quencher, there is a strong dependence on f_0. This is shown as the dashed curve in Figure 7, which is the 3199/3371 ratio as a function of f_0 for $E_0 = 1$ keV.

Thus VK bands could fulfill the same role as O I 6300 Å in deducing characteristic energy (see *Ishimoto et al.* [1988]). In principle, a family of curves similar to Figure 6 could be constructed using an appropriate allowed O I emission along with the VK and 3371 Å emissions. The advantage of using VK instead of O I 6300 Å is that there are no chemical mechanisms to complicate the excitation rate. The disadvantage is that there are cascade transitions from the B, C, and W states that populate the A state. *Conway and Christensen* [1985] and *Daniell and Strickland* [1986] have taken cascade into account by combining the appropriate cross sections of *Cartwright et al.* [1977].

A great deal of controversy has accompanied the VK problem. The quenching coefficient of the $N_2(A)$ state by O has been measured several times in the laboratory and has been found to be of order 3.4×10^{-11} cm^3 s^{-1} for $v' = 1$ [*Piper et al.*, 1981; *Thomas and Kaufman*, 1985; *De Souza et al.*, 1985]. On the other hand, some rocket observations have supported a value of 2×10^{-10} cm^3 s^{-1} (*Sharp et al.* [1979] as corrected by *Torr and Sharp* [1979]; *Beiting and Feldman* [1979]). *McDade and Llewellyn* [1984], *Shepherd* [1984], and *Meier et al.* [1989] have argued that, if the concentration of atomic oxygen were adjusted appropriately, the laboratory measurements could be brought into agreement with the rocket observations. Until this issue is completely resolved, VK quenching must be used with caution.

Pure Absorption of N_2 Lyman-Birge-Hopfield (LBH) Bands

There is substantial absorption of N_2 Lyman-Birge-Hopfield bands near 1400 Å by the O_2 Schumann-Runge absorption continuum. As the primary electron characteristic energy increases, more absorption occurs due to the lower altitude of energy deposition. *Strickland et al.* [1983] suggested that LBH band absorption could be used to deduce E_0. To illustrate the concept, LBH synthetic spectra are plotted in Figure 8 for characteristic energies of 1 and 5 keV (see also Figure 1 of *Strickland et al.* [1983]). The individual band features are indentified in Figure 2 of *Vallance Jones* [this volume] paper. The lack of absorption outside the O_2 continuum is evident at shorter and longer wavelengths. The sensitivity to characteristic energy is clearly seen from 1350–1600 Å.

The energy dependencies of the various band emissions were calculated by *Strickland et al.* [1983] for a standard *Jacchia* [1977] model atmosphere. Figure 9 shows the ratios of several bands as functions of E_0; individual intensities were scaled from the curves of *Strickland et al.* [1983]. The viewing was assumed to be along magnetic nadir from the top of the atmosphere. While the 1273 Å band was used as the reference emission in Figure 9 due to weak O_2 absorption at that wavelength, a brighter N_2 feature

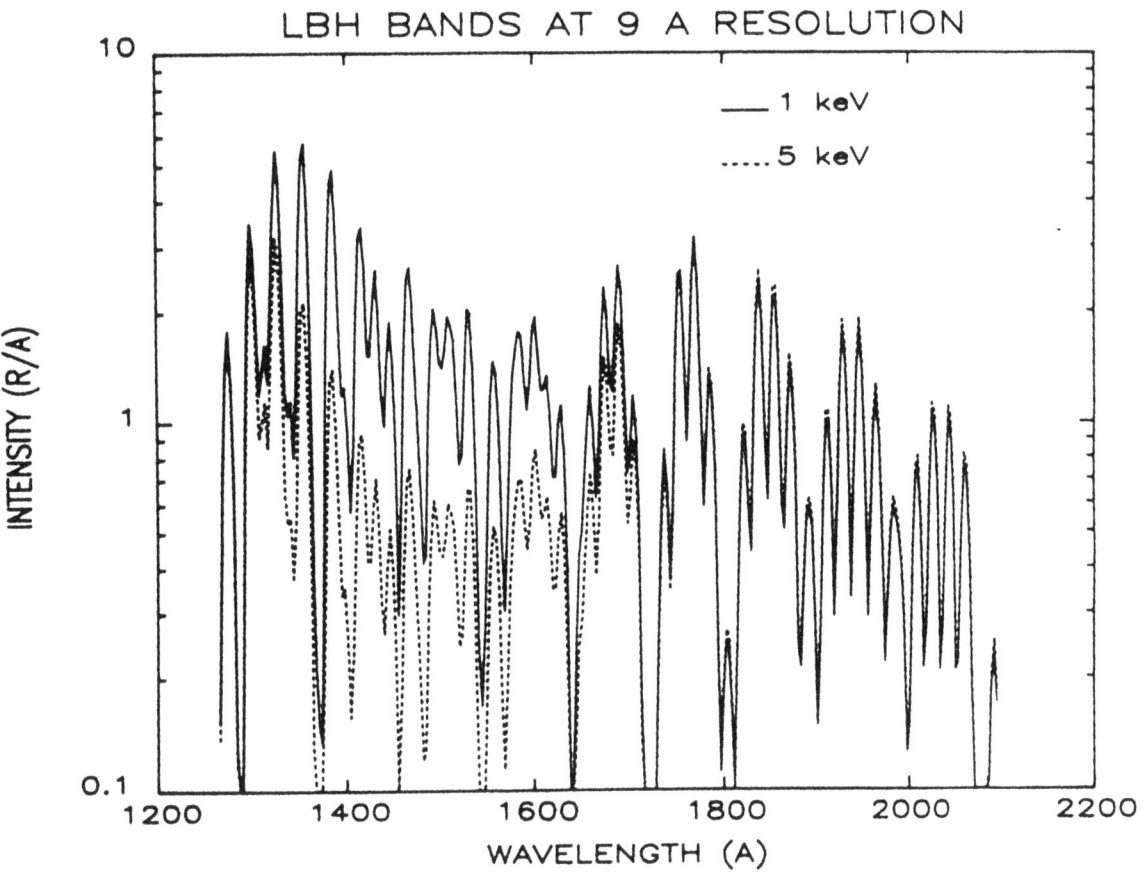

Figure 8—N_2 LBH band spectrum at 9 Å resolution for characteristic energies of 1 and 5 keV.

such as the 2*PG* 3371 Å band could also be used. As seen in the figure, the absorption method is most useful above 1 keV, where altitudes of greater O_2 absorption are reached by the energetic electrons. This technique was employed recently by *Ishimoto et al.* [1988] in the interpretation of satellite observations of the FUV aurora.

There is a lesser dependence of the LBH band ratios on the O_2 abundance: for example at 3 keV, multiplying and dividing the O_2 concentration by 1.5 leads to a ±10–15% uncertainty in E_0 derived from the 1383 Å band. Since the band ratios vary differently with O_2 than with E_0, it should be possible to estimate the O_2 concentration if the measurements are of sufficient precision and accuracy. The O multiplier can then be obtained using, say, the O I 1356 Å emission produced by electron impact excitation. The ratio of intensities at 1356 Å to 1275 Å is also shown in Figure 9 (after downward adjustment of the e + O cross section used by *Strickland et al.* [1983] by a factor of nearly 3 [see *Zipf and Erdman*, 1985 and *Morrison and Meier*, 1988]). The dependence of the O I 1356 Å emission on the O concentration is nearly

linear; a weak effect due to multiple scattering should be taken into account [*Strickland et al.*, 1983].

E_0 and Thermospheric Temperature

If the temperature of an emission source is measured, the altitude of the emission and consequently the characteristic energy can be estimated if the height-dependent thermospheric temperature is known. Two observational techniques that employ this principle use measurements of molecular rotational lines and atomic line widths. Recent applications of the former method are reported by *Vallance Jones et al.* [1987] and *Espy et al.* [1987]; *Hecht et al.* [1989] have attempted the latter. *Strickland et al.* [1989] discuss the various issues from a theoretical standpoint.

The simplest approach is to derive a temperature from a measurement, to assume that all of the emission originates from a single altitude, to look up the altitude from a model of thermospheric temperature vs. altitude, and then use a plot of volume emission rates such as Figure 2 to obtain E_0. However, since the emission is spread out in altitude, large errors are likely to result from this simple approach.

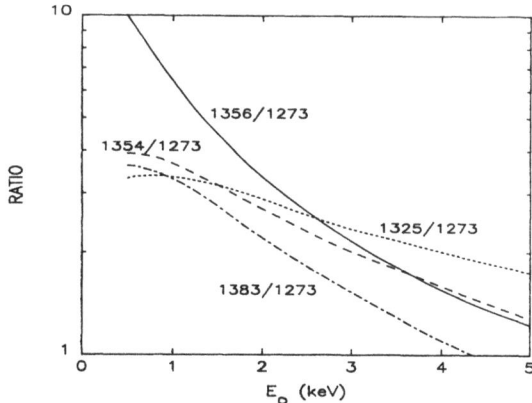

Figure 9—N$_2$ LBH band ratios and O I 1356 Å to N$_2$ 1273 Å ratio vs. characteristic energy. Emission rates were taken from *Strickland et al.* [1983].

In order to account for the finite width of emission layers, an effective temperature can be defined:

$$\langle T \rangle = \int \frac{T(z)\, j(z)\, dz}{j(z)\, dz}$$

where j is the volume emission rate (cm^{-3} s^{-1}) of the particular feature. A value of $\langle T \rangle$ can be calculated for each E_0. This was used by *Vallance Jones et al.* [1987] to derive a characteristic energy from the N$_2^+$ 1NG band at 4278 Å, whose rotational temperature they measured. This procedure will be accurate only when the rotational line distribution varies linearly with temperature. The exact analysis procedure would be to fit the observations with a synthetic spectrum that allows for temperature variations throughout the emission layer. Different synthetic spectra would be required for each E_0.

If an emission line profile were measured, and were assumed to have a Doppler (Gaussian) profile, the de-

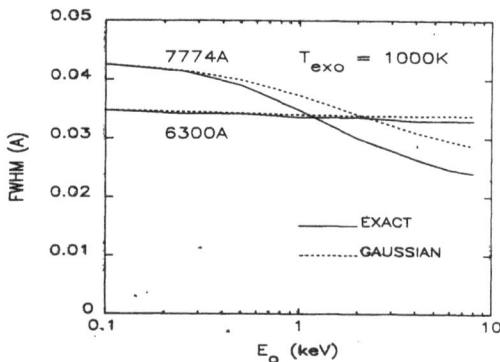

Figure 10—Full widths at half maximum intensity for O I 7774 and 6300 Å emission lines vs. characteristic energy. (From *Strickland et al.* [1989].)

rived temperature could be related to E_0 using $\langle T \rangle$. However, use of $\langle T \rangle$ introduces serious errors since the altitude weighting is not linear [*Strickland et al.*, 1989]. The correct procedure would be to calculate the line profile, $4\pi I$ (photons cm^{-2} s^{-1} Å^{-1}) using the Doppler profile ψ (Å^{-1}), appropriate to each altitude:

$$4\pi I(\lambda) = \int j(z)\psi(\lambda, z)\, dz$$

$$\psi = \frac{1}{\sqrt{\pi}\, \Delta\lambda_D(z)} \exp -\left[\frac{\lambda - \lambda_0}{\Delta\lambda_D(z)} \right]^2$$

$$\Delta\lambda_D = \frac{\lambda}{c} \sqrt{\frac{2kT(z)}{m}}$$

The various quantities have their usual meanings. The altitude integral must be performed at each wavelength within the line. The resulting profile is not Gaussian, as shown by *Strickland et al.* [1989]. Figure 10, taken from the paper by those authors, shows the full width at half maximum (FWHM) of the e + O component of O I 7774 Å for the exact profile, and for the assumption of a Gaussian line calculated with $\langle T \rangle$. As an example of the error involved in the Gaussian assumption, if a line width of 0.03 Å were actually observed, the approximate scheme would result in an E_0 too large by more than a factor of 2.

Figure 10 also shows line widths for O I 6300 Å, a commonly used feature for obtaining temperatures and winds. The red line is not useful as an E_0 diagnostic because quenching restricts the main emission profile to 200–300 km (Figure 3). There is not much difference between the exact and approximate calculations of the FWHM since the variation of the thermospheric temperature is not large at those altitudes.

Optically Thick Lines and Branching in Atomic Oxygen

Calculations of emission lines with small atmospheric opacity for resonant scattering can be made with relative ease by integrating the volume emission rate along the line of sight. The volume emission rate is a linear function of concentration for electron impact excitation (Table 2). As the optical depth (defined as the integral along the line of sight of the resonant scattering cross section at line center times the concentration) increases toward unity, the column emission rate increases more rapidly due to enhanced multiple scattering. But the rate of increase is moderated because the atmospheric transmission decreases with increasing optical depth. For large optical depths, the emission rate again approaches a linear dependence on

TABLE 3. O I optically thick emissions ($Q = 1$ erg cm^{-2} s^{-1}).

		e + O			e + O$_2$
	E_0	I_0	I	Ratio	
1304 Å					
	1 keV	403R	788R	1.9	11R
	5 keV	131	119	0.9	15
1356 Å					
	1 keV	84	76	0.9	19
	5 keV	27	14	0.52	14
1641 Å					
	1 keV	9.7×10^{-4}	4.9	5100	2.6×10^{-5}
	5 keV	2.9×10^{-4}	1.6	5500	3.6×10^{-5}

the concentration (with a proportionality determined by the actual atmospheric conditions), primarily through the initial volume excitation rate. This is caused by the approximate balance between multiple scattering and atmospheric transmission. These comments assume viewing from outside with the excitation source inside the atmosphere [e.g., *Meier and Lee*, 1981]. Consequently, the emission-rate ratio of two emission lines with low and high opacity, such as O I (^3P-^5S) 1356 Å and (^3P-^3S) 1304 Å, will not be sensitive to the concentration of atomic oxygen. But the ratio is somewhat sensitive to E_0 due to greater O$_2$ pure absorption effects for 1304 Å than for 1356 Å (see Table 3).

On the other hand, weak intercombination transitions from the upper states of optically thick lines can be enhanced dramatically through the multiple scattering process. An example of this is the atomic oxygen transition from the upper state of 1304 (^3S) to (^1D), resulting in the emission of a photon at 1641 Å. (See the energy level diagram from $3s\ ^3S^0$ in Figure 14 of *Vallance Jones* [this volume].) The branching ratio to ^1D is about 2.4×10^{-6} [*Conway et al.*, 1988]. This would be the value of the ratio of the 1641 Å to 1304 Å emission rates under optically thin conditions. However, at 1304 Å, the vertical optical depth above the O$_2$ absorption altitude near 100 km approaches 10^5, causing thousands of scatterings before a photon escapes from the top of the auroral atmosphere. (For E_0 greater than about 5 keV, the low altitude of excitation results in the absorption of a substantial fraction of the photons.) Entrapment of radiation causes the 1641 Å/1304 Å ratio to increase to near 10^{-2} [*Meier and Conway*, 1985; *Conway et al.*, 1988]. Those authors have shown that in the dayglow the ratio has a power-law dependence on the O concentration, the power depending on the observation altitude and the altitude of the excitation source. The low branching ratio prevents a significant decrease in the 1304 Å radiation field, so the O sensitivity comes mainly via the

1641 Å line. This result needs to be investigated for auroras, where O$_2$ absorption becomes increasingly important for hard electron spectra. It is likely that in principle, if the characteristic energy is known, the 1641 Å to 1304 Å ratio will give the O concentration, but detailed computations are needed for verification.

Table 3 shows the column emission rates of O I 1304 Å, 1641 Å, and 1356 Å, for characteristic energies of 1 and 5 keV and the Jacchia 1000 K atmosphere with $f_0 = 0.5$. The multiple scattering computations were performed using the Monte Carlo model of *Meier and Lee* [1982]. The contributions from e + O$_2$ dissociative excitation by electron impact are also shown for general interest. The quantity I_0 is the initial column excitation rate for e + O and I includes multiple scattering. At 1 keV, multiple scattering nearly doubles the 1304 Å emission rate, while at 5 keV enhanced pure absorption at lower altitudes actually causes a reduction of the emission rate to 90% of the initial excitation rate. Note the strong enhancement of the still weak 1641 Å emission rates; for 1641 Å, the ratio of I to I_0 equals the mean number of scatterings of a 1304 Å photon. An intense aurora should show this feature. Radiation at 1641 Å is much stronger in the dayglow [*Conway et al.*, 1988]. For 1356 Å, I is lower than I_0, again primarily because of O$_2$ absorption.

Transitions with larger branching ratios offer even greater sensitivity to the O concentration because the radiation field and consequently the strength of the parent emission can be reduced by the branching losses. One example that has received some attention is O I 989 Å and its infrared branch at 7990 Å [*Christensen et al.*, 1977; *Meier*, 1982; *Gladstone et al.*, 1987; *Link et al.*, 1988; *Morrison and Meier*, 1988]. (See Figure 14 of *Vallance Jones* [this volume].) The 7990 Å feature is attractive in that it can be measured from the ground. *Meier* [1982] showed that the ratios of lines in the multiplet depend on the O concentration. Also the ratio of 7990 Å to an allowed emission such as 7774 Å should be sensitive to the amount of O. However, the upper state of 989 Å ($3s'$ $^3D_0$) also branches to ^1D, emitting 1173 Å radiation [*Morrison,* 1985]. The total branching from the $^3D_0$ state appears to be a few times 10^{-4}, although *Erdman and Zipf* [1983] place that of 7990 Å at $< 2.5 \times 10^{-5}$. Various unsettled issues about the $^3D_0$ branching problem are addressed in the above papers. The confident use of 7990 Å emission awaits their resolution.

Yet another transition that has diagnostic potential is O I 1027 Å and the branch from its upper state ($3d$ $^3D_0$) to the intermediate state ($3p$ ^3P), accompanied by 11290 Å photons. The branching ratio is 0.29. Since the atmosphere is optically thick to 1027 Å photons, multiple scattering will quickly convert every UV photon into one at 11290 Å. That is not the case in the dayglow, where an accidental resonance with the solar Lyman β line causes excitation of high-altitude oxygen atoms [*Meier et al.,* 1987]. The 11290 Å line should play the same diagnostic role as O I 7990 Å. It ap-

pears as a weak feature in the auroral IR spectrum (see Figure 10 of *Vallance Jones* [this volume]).

5. CONCLUSIONS

In the past several years, there has been substantial progress in the modeling of auroral emission features. Important aspects of classical emission processes have been delineated. New and innovative techniques have been proposed for quantitative analysis of auroral excitation processes and atmospheric conditions. Perhaps the most important need is for experimental validation. Comprehensive observing campaigns are required to obtain data under many different observing conditions. The highly variable aurora requires large databases. Especially useful would be campaigns that include rockets and radars, as well as space-based observations for complete validation. We can look forward in the not too distant future to the implementation of on-line algorithms that routinely provide quantitative assessment of atmospheric conditions from field sites.

REFERENCES

Abreu, V. J., J. H. Yee, S. C. Solomon, "The Quenching Rate of O(^1D) by O(^3P)," *Planet. Space Sci.,* **54**, 1143 (1986).

Bates, D. R., "Airglow and Auroras," in *Applied Atomic Collision Physics I,* H. S. W. Massey, B. Baderson, and E. W. McDaniel, eds., Academic Press, New York, p. 149 (1982).

Beiting, E. J. III, and P. D. Feldman, "Ultraviolet Spectrum of the Aurora (2000-2800 Å)," *J. Geophys. Res.,* **84**, 1287 (1979).

Cartwright, D. C., S. Trajman, A. Chutjian, and W. Williams, "Electron Impact Excitation of the Electronic States of N$_2$, II, Integral Cross Sections at Incident Energies from 10 to 50 eV," *Phys. Res.,* **A16**, 1014-1051 (1977).

Christensen, A. B., G. J. Romick, and G. G. Sivjee, "Auroral O I (989 Å) and O I (1027 Å) Emissions," *J. Geophys. Res.,* **82**, 4997 (1977).

Conway, R. R., and A. B. Christensen, "The Ultraviolet Dayglow at Solar Maximum 2. Photometer Observations of N$_2$ Second Positive (0,0) Band Emission," *J. Geophys. Res.,* **90**, 6601 (1985).

Conway, R. R., R. R. Meier, and R. E. Huffman, "Abundance of Atomic Oxygen in the Lower Thermosphere from Satellite Observations of the O I 1641 Å Dayglow," *Planet. Space Sci.,* **36**, 963 (1988).

Daniell, R. E., Jr., and D. J. Strickland, "Dependence of Auroral Middle UV Emissions on the Incident Electron Spectrum and Neutral Atmosphere," *J. Geophys. Res.,* **91**, 321 (1986).

De Souza, G., Gousset, M. Touzeau, and Tu Khiet, "Note on the Determination of the Efficiency of the Reaction N$_2$(A) + O (^3P) → N$_2$ + O (^1S)," *J. Phys. B: At. Mol. Phys.,* **18**, L661 (1985).

Erdman, P. W., and E. C. Zipf, "Electron Impact Excitation of the O I 7990 Å Multiplet," *J. Geophys. Res.,* **88**, 7245 (1983).

Espy, P. J., W. R. Pendleton, Jr., G. G. Sivjee, and M. P. Fetrow, "Vibrational Development of the N$_2^+$ Meinel Band System in Aurora," *J. Geophys. Res.,* **92**, 11257 (1987).

Gladstone, G. R., R. Link, S. Chakrabarti, and J. C. McConnell, "Modeling of the O I 989-Å to 1173-Å Ratio in the Terrestrial Dayglow," *J. Geophys. Res.,* 12,445 (1987).

Hecht, J. H., A. B. Christensen, and J. B. Pranke, "High-Resolution Auroral Observations of the O I (7774) and O I (8446) Multiplets," *Geophys. Res. Lett.,* **12**, 605-608 (1985).

Hecht, J. H., A. B. Christensen, D. J. Strickland, and R. R. Meier, "Deducing Composition and Incident Electron Spectra from Ground-Based Auroral Optical Measurements: Variations in Oxygen Density," *J. Geophys. Res.,* **94**, 13553 (1989).

Ishimoto, M., C.-I. Meng, G. J. Romick, and R. E. Huffman, "Auroral Electron Energy and Flux from Molecular Nitrogen Ultraviolet Emissions Observed by the S3-4 Satellite," *J. Geophys. Res.,* **93**, 9854 (1988).

Jacchia, L. G., "Thermospheric Temperature, Density and Composition: New Models," *Smithsonian Astrophys. Obs. Spec. Rept.,* **375**, Cambridge, MA (1977).

Link, R., S. Chakrabarti, G. R. Gladstone, and J. C. McConnell, "An Analysis of Satellite Observations of the O I EUV Dayglow," *J. Geophys. Res.,* **92**, 2693 (1988).

Lummerzheim, D., M. H. Rees, and H. R. Anderson, "Angular Dependent Transport of Auroral Electrons in the Upper Atmosphere," *Planet. Space Sci.,* in press (1989).

McDade, I. C., and E. J. Llewellyn, "Atomic Oxygen Concentrations in the Auroral Ionosphere," *Geophys. Res. Lett.,* **11**, 247 (1984).

Meier, R. R., "Spectroscopy of the O I 989- and 7990-Å Multiplets in the Dayglow Airglow," *J. Geophys. Res.,* **87**, 6307 (1982).

Meier, R. R., "Thermospheric Aurora and Airglow," *Rev. Geophys. Space Phys.,* **25**, 471 (1987).

Meier, R. R., and R. R. Conway, "The ^1D-^3S Transition in Atomic Oxygen: A New Method of Measuring the O Abundance in Planetary Thermospheres," *Geophys. Res. Lett.,* **12**, 601 (1985).

Meier, R. R., and J.-S. Lee, "Angle-Dependent Frequency Redistribution: Internal Source Case," *Astrophys. J.,* **250**, 376 (1981).

Meier, R. R., and J.-S. Lee, "An Analysis of the O I 1304 Å Dayglow Using a Monte Carlo Resonant Scattering Model with Partial Frequency Redistribution," *Planet. Space Sci.,* **30**, 439 (1982).

Meier, R. R., D. E. Anderson, Jr., L. J. Paxton, and R. P. McCoy, "The O I 3d $^3D^0$ – 2p^4 ^3P Transition at 1026 Å in the Day Airglow," *J. Geophys. Res.,* **92**, 8767 (1987).

Meier, R. R., D. J. Strickland, J. H. Hecht, and A. B. Christensen, "Deducing Composition and Incident Electron Spectra from Ground-Based Auroral Optical Measurements: A Study of Auroral Red Line Processes," *J. Geophys. Res.,* **94**, 13541 (1989).

Morrison, M. D., "Laboratory Measurement of the O I 1173/989 Å Branching Ratio," *Planet. Space Sci.,* **33**, 135-139 (1985).

Morrison, M. D. and R. R. Meier, "The O I 989 and 1173 Å Multiplets in the Dayglow," *Planet. Space Sci.,* **36**, 987 (1988).

Piper, L. G., G. E. Caledonia, and J. P. Keneally, "Rate Constants for Deactivation of N$_2$ (A$^3\Sigma_u^+$, v' = 0,1) by O," *J. Chem. Phys.,* **10**, 3365 (1981).

Rees, M. H., "Auroral Excitation and Energy Dissipation," *Sol. Terr. Phys.,* R. L. Carovillano and J. M. Forbes, eds., D. Reidel Publishing, Co., Dordrecht, Holland, p. 753 (1983).

Rees, M. H., and D. Luckey, "Auroral Electron Energy Derived from Ratio of Spectroscopic Emissions, 1, Model Computations," *J. Geophys. Res.,* **79**, 5181 (1974).

Rees, M. H., and D. Lummerzheim, "Characteristics of Auroral Electron Precipitation Derived from Optical Spectroscopy," *J. Geophys. Res.*, **94**, 6799 (1989).

Rees, M. H., and R. G. Roble, "Excitation of $O(^1D)$ Atoms in Aurorae and Emission of the [O I]6300-Å Line," *Can. J. Phys.*, **64**, 1608 (1986).

Sharp, W. E., "Rocket-Borne Spectroscopic Measurements in the Ultraviolet Aurora; Nitrogen Vegard-Kaplan Bands," *J. Geophys. Res.*, **76**, 987–1005 (1971).

Sharp, W. E., M. H. Rees, and A. E. Stewart, "Coordinated Rocket and Satellite Measurements of an Auroral Event 2. The Rocket Observations and Analysis," *J. Geophys. Res.*, **84**, 1977 (1979).

Shepherd, G. G., "Atomic Oxygen Concentrations in the Auroral Thermosphere: Application of a Thermospheric Temperature Criterion," *Geophys. Res. Lett.*, **11**, 1117 (1984).

Solomon, S. C., P. B. Hays, and V. J. Abreu, "The Auroral 6300 Å Emission: Observations and Modelling," *J. Geophys. Res.*, **93**, 9867 (1988).

Strickland, D. J., J. R. Jasperse, and J. A. Whalen, "Dependence of Auroral FUV Emissions on the Incident Electron Spectrum and Neutral Atmosphere," *J. Geophys. Res.*, **88**, 8051 (1983).

Strickland, D. J., D. L. Book, T. P. Coffey, and J. A. Fedder, "Transport Equation Techniques for the Deposition of Auroral Electrons," *J. Geophys. Res.*, **812**, 2755 (1976).

Strickland, D. J., R. R. Meier, J. H. Hecht, and A. B. Christensen, "Deducing Composition and Incident Electron Spectra from Ground-Based Auroral Optical Measurements: Theory and Model Results," *J. Geophys. Res.*, **94**, 13527 (1989).

Thomas, J. M., and F. Kaufman, "Rate Constants of the Reaction of Metastable $N_2(A^3\Sigma_u^+)$ in ν = 0,1,2, and 3 with Ground State O_2 and O," *J. Chem. Phys.*, **83**, 2900 (1985).

Torr, D. G., and W. E. Sharp, "The Concentration of Atomic Oxygen in the Auroral Lower Thermosphere," *Geophys. Res. Lett.*, **6**, 860 (1979).

Vallance Jones, A., *Aurora*, D. Reidel Publishing Co., Dordrecht, Holland (1974).

Vallance Jones, A., R. R. Meier, and N. N. Shefov, "Atmospheric Quantal Emissions: A Review of Recent Results," *J. Atmos. Terr. Phys.*, **47**, 623 (1985).

Vallance Jones, A., R. L. Gattinger, P. Shih, J. W. Meriwether, V. B. Wickwar, and J. Kelly, "Optical and Radar Characterization of a Short-Lived Auroral Event at High Latitude," *J. Geophys. Res.*, **92**, 4575 (1987).

Zipf, E. C., and P. W. Erdman, "Electron Impact Excitation of Atomic Oxygen: Revised Cross Sections," *J. Geophys. Res.*, **90**, 11,087 (1985).

II-4. THERMOSPHERIC RESPONSE AND FEEDBACK TO AURORAL INPUTS

D. Rees* and T. J. Fuller-Rowell*

Three predominant mechanisms cause the thermosphere to respond dramatically to intense forcing by auroral processes: direct heat deposition from auroral electrons and ions, ionization of the neutral atmosphere, and dissociation of molecular nitrogen. The direct heating effect of auroral electrons is generally small, except at high altitudes, and in the dayside cusp. However, the additional auroral ionization has a paramount role by increasing the Pedersen conductivity. The combination of the high values of Pedersen conductivity and magnetospheric convection electric field greatly increases Joule heating, generating the dominant energy source at high latitudes during geomagnetically disturbed periods. When the Pedersen conductivity is enhanced, momentum transfer from convecting ions to the neutral gas is similarly enhanced. The momentum transfer and effects of enhanced Joule heating together cause the intense wind systems observed at high latitudes during disturbed periods. The dissociation of molecular nitrogen, through the odd nitrogen chemistry, enhances the number density of nitric oxide, which then plays very important roles in ionospheric chemistry and in the neutral energy balance, due to its infrared radiative properties. Feedback between induced thermospheric changes and the ionosphere and magnetosphere occurs via four routes:

(i) Strong induced neutral winds create a dynamo, generally opposing the magnetospheric convection field, reducing ionospheric currents and Joule heating.

(ii) The strong horizontal winds also cause rapid vertical movements of ionization; these vertical movements can transport ionization upward into regions of lower effective recombination rates, or downward into regions of higher effective recombination rates.

(iii) Ionospheric changes induced by the enhanced auroral production of nitric oxide modulate the ionospheric conductivity, Pedersen currents, and Joule dissipation.

(iv) Large-scale neutral chemical compositional changes, responding to thermospheric convection and advection induced by Joule and particle heating, cause further changes in the plasma density distribution via recombination rate changes. Locally, in regions of very high plasma velocity, this effect is further enhanced by the impact of fast atomic oxygen ions on neutral molecular nitrogen. The divergence of horizontal ionospheric currents reflect the ionospheric closure of the field-aligned currents coupling the magnetosphere and polar ionosphere. Changes of conductivity, or of the neutral wind dynamo, may be expected to modify either the field-aligned current system or the effective magnetospheric electric field. Some illustrations of these effects, using a coupled thermospheric/ionospheric model, are presented and discussed.

1. INTRODUCTION

This review demonstrates the response of the polar thermosphere and ionosphere to magnetospheric inputs, as the level of geomagnetic activity increases from low ($Kp = 2$), to moderately disturbed ($Kp = 4$) conditions. A fully coupled numerical model of the terrestrial thermosphere and ionosphere system will be used to explore the increasing feedback between the ionosphere and thermosphere as the level of geomagnetic activity rises. The behavior of the high-latitude regions will be the major topic, since that is where the influences of the auroral precipitation and magnetospheric electric fields are most strongly imprinted.

At all times and at all geomagnetic activity levels, the magnetosphere imprints unmistakable signatures on the high-latitude thermosphere and ionosphere. These disturbances of the thermosphere and ionosphere are always co-located with signatures of energetic particle precipitation and convective electric fields, or else rep-

resent "fossils" of strong forcing during the previous 1–24 hour period. Under quiet magnetospheric conditions, the regions of such imprints are contracted poleward, away from areas of historical observations. However, recent global observations from satellites and new ground-based polar cap observatories have shown clearly that ionospheric structures, and thermospheric winds, temperature, and composition are persistently disturbed in the vicinity of the auroral oval and within the geomagnetic polar cap, even under the most quiet conditions.

As geomagnetic and magnetospheric activity levels increase, the regions of magnetospheric inputs expand away from the geomagnetic poles. The energy and momentum deposition rates increase greatly and are strongly space and time dependent. This intensification of auroral precipitation is well shown in statistical surveys and analyses of the energetic electron precipitation [*Fuller-Rowell and Evans*, 1987; *Hardy et al.*, 1985]. These statistical surveys complement the impression obtained from individual observations. As shown by analyses of polar plasma convection [*Heppner*,

*Department of Physics and Astronomy, University College London, Gower St., London WC1 E6BT, England.

Auroral Physics, edited by C.-I. Meng, M. J. Rycroft and L. A. Frank. © Cambridge UP 1991

1977; *Heppner and Maynard,* 1987; *Foster et al.,* 1986], the regions of strong magnetospheric convection electric fields imprinted on the polar ionosphere undergo a similar and closely related equatorward expansion and intensification as geomagnetic activity increases. From the point of view of numerical modeling, it is critically important that convection and precipitation boundaries match realistically, particularly if time-dependent simulations for variable geomagnetic activity are to be performed.

Convection electric fields drive ionospheric plasma of the auroral oval and polar cap to velocities of the order of 1 km/s. The ions impart momentum to the neutral gas via "ion drag," at the same time losing a little of their net ($\mathbf{E} \times \mathbf{B}$) velocity, creating the dissipative Pedersen ionospheric current component, which causes Joule heating. If there is to be any effective energy or momentum transfer from the solar wind, via the magnetopause and magnetosphere to the ionosphere and thermosphere, AC and DC components of the field-aligned current (or Birkeland current) are required. Such magnetospheric currents are associated with the dissipative Pedersen current within the auroral ionosphere (component parallel to the ionospheric electric field). Both the field-aligned current and the Pedersen currents within the auroral ionosphere intensify sharply as geomagnetic activity increases. The efficiency of momentum transfer and the Joule heating both increase linearly with ionospheric plasma density. Since Joule heating almost always considerably exceeds direct particle precipitation, knowledge of the ionospheric plasma response to precipitation is particularly important [*Rees et al.,* 1983b].

The polar regions that display the imprints of these important magnetospheric phenomena also show a wide range of other disturbances. These disturbances produce a range of characteristic signatures in the charged and energetic particle populations, the AC and DC electric and magnetic fields, and the optical aurora. However, for the moment, we will concentrate on those phenomena that have the most direct connection to excitation of the thermosphere.

It is very important to recognize that classical ground-based signatures have been used historically to identify intense auroral substorms. These are short, strong, negative magnetic excursions; brilliant aurora, and riometer absorption events. However, they are only a small part of the sequence of phenomena that cause the ionospheric and thermospheric response to geomagnetic disturbances. Indeed the signatures of the most important momentum and energy sources for the thermosphere are difficult to sense. For these reasons, major thermospheric and ionospheric disturbances are often poorly related to classical magnetospheric activity indices such as *Kp* and *AE*, which are dominated by the magnetic effects of Hall currents during auroral substorms. Pedersen currents more accurately reflect the intensity of momentum and energy transport from the magnetosphere. However, the Pedersen current and the field-aligned-current system are poorly reflected in ground-based magnetic perturbations within the auroral oval.

A signature of increasing geomagnetic activity is the intensification of convective electric fields and auroral precipitation, with an equatorward expansion of the auroral oval. The field-aligned current, Joule heating, and ion-drag acceleration of thermospheric winds all increase. The temporal and spatial variability of all of the previously mentioned terms—"geomagnetic forcing"—increases sharply, particularly at very high activity levels ($6 < Kp < 9$). Disturbances of the lower ionospheric regions, in the E-region up to 150 km, respond directly to "auroral" inputs and ion production.

At higher thermospheric and ionospheric altitudes, the situation becomes much more complex. Large-scale advection and convection forced upon the thermosphere by geomagnetic heating causes the polar F-region neutral gas composition to change dramatically [*Rees et al.,* 1985a; *Fuller-Rowell et al.,* 1988]. Very strong enhancement of molecular nitrogen density, and a corresponding depletion of atomic oxygen density occurs, particularly in the disturbed summertime polar cap [*Fuller-Rowell et al.,* 1988]. Enhanced concentrations of molecular nitrogen cause significant depletions of F-region plasma density by greatly increasing the effective recombination rate, while ionization rates, due to the combination of solar photoionization and auroral precipitation, are only slightly changed. The combination of induced ionospheric chemistry changes and dynamical effects on the F-region plasma, resulting from strong induced horizontal winds, cause the ionospheric response to magnetospheric forcing to be very complex and nonlinear.

Global-scale disturbances within the thermosphere follow the initial high-latitude geomagnetic forcing. Propagating waves, and the consequences of gross wind-driven compositional changes, have truly global consequences for the thermosphere, and force some very large, and long-lasting, disturbances of the ionosphere—the ionospheric F-region storm [*Martyn,* 1953]. During major geomagnetic disturbances, the decay of energetic particles from the ring current, probably mainly energetic ions [*Tinsley et al.,* 1988] may directly cause the negative phase of the low-latitude ionospheric storm. This process is, however, still rather difficult to include in a coherent and self-consistent model of the entire coupled solar-terrestrial system.

Numerical models of the thermosphere provide a

global framework incorporating the well-understood basic mechanisms and phenomena, and a means of predicting both the mean structure, and the qualitative and quantitative variations caused by seasonal changes and by solar activity variations. The general form of large-scale thermospheric disturbances resulting from magnetospheric activity can also be simulated quite well. Detailed prediction of localized and short-lived disturbances created at times of high geomagnetic activity will not be as reliable, since we currently have no corresponding temporal and spatial description of the energy and momentum inputs.

Numerical simulation of the thermosphere from first principles requires that the most important physical processes are properly treated [*Fuller-Rowell and Rees*, 1980, 1983; *Roble et al.*, 1982; *Dickenson et al.*, 1984]. It is assumed that most of the energy and momentum sources driving the thermosphere are predetermined, and invariant to the response of the thermosphere. The thermosphere does not determine the nature of the solar UV and EUV inputs that provide important heat and ionization sources. However, the thermosphere does react strongly to forcing. The major responses in wind, temperature, and composition of the polar thermosphere to ion convection and heating within the auroral oval and polar cap are now well documented by ground-based and spaceborne observation [*Hays et al.*, 1984; *Rees et al.*, 1983a, 1986]. Some of these thermospheric responses may change the nature or magnitude of the forcing itself.

While this review will concentrate on external thermospheric forcing from the magnetosphere, it should be noted that significant effects, particularly in the lower thermosphere, occur from internal forcing from the lower atmosphere, as a result of the combination of tidal, gravity, and planetary waves. The first can be handled numerically within a thermospheric model [*Fesen et al.*, 1986; *Parish et al.*, 1989] by introducing a "flexible" lower boundary. Self-consistent wind and temperature amplitude and phase changes corresponding to specific propagating tidal modes can be adjusted, by numerical experiment, until the tides within the lower thermosphere correspond to observed tidal variations as functions of altitude, season, and latitude. The relatively large amplitudes of observed tidal winds in the lower thermosphere can be successfully simulated by introducing such propagating tides. This is not possible, if only the in situ generated tides are considered.

Ion-neutral frictional drag [*Rishbeth and Garriot*, 1969] in regions of rapid convective ion flow causes direct heating of both ions and neutrals, commonly known as Joule heating [*Cole*, 1962, 1971]. Induced winds may increase or decrease (but generally decrease) the ion drag, and the resulting frictional heating. The

induced winds (or more correctly, changed winds, since there is always a complex wind system in existence prior to a given geomagnetic disturbance) may induce a "back-EMF," opposing the initial magnetospheric convective electric field, and decreasing the electromotive force, i.e., $(\mathbf{E} + \mathbf{V}_n \times \mathbf{B})$. This also limits the maximum induced winds, the local electrojet current (at all heights), and the Joule heating. These processes are independent of any plasma density and conductivity modifications, however. The thermosphere will respond to the reduction in both electromotive force and electrojet current and consequent Joule heating.

The wind system induced by ion drag, subtly modified by gas pressure changes from neutral heating, will also force ion motions parallel to the local magnetic field. Such "parallel" ion drifts will also induce a field-aligned electron flow, to maintain quasi-charge neutrality. Thus the entire vertical plasma distribution will respond to wind changes, an effect that becomes increasingly important at greater altitudes. This change of ion density distribution will modify the consequent ion drag on the neutrals, and thus the wind acceleration terms, and finally the winds themselves. Since, in the vicinity of the auroral oval and polar cap, there are always various contra-flowing streams of field-aligned thermal and suprathermal particles, it is difficult to identify the net field-aligned current by direct observation. Yet it is the net flow that powers the magnetosphere-thermosphere forcing process, and variations of this field-aligned flow caused by feedback processes are important, but necessarily second-order changes, and thus difficult to observe directly.

Modifications of the horizontal current (usually decreases) due to the induced winds affect the capacity of the ionosphere to carry field-aligned current connecting to the magnetosphere. Intuitively, the feedback effects of the induced winds on the total electromotive force, and the modified capacity of the auroral ionosphere to transmit or connect the field-aligned current might be expected to cause some significant effects on the magnetosphere at times of large disturbances, when the E-region winds are known to reach 50% of the $\mathbf{E} \times \mathbf{B}$ ion drift velocity, driven by magnetospheric electric fields [*Rees*, 1971, 1973; *Pereira et al.*, 1980].

Although some numerical experiments in these areas have been carried out [*Harel et al.*, 1981], theoretical and experimental exploration of these problems is still at a very preliminary phase. We do not understand whether limits of the availability of charge carriers (ionospheric or magnetospheric) are important. Slight changes of the field-aligned current may be matched by compensating changes in field-aligned potentials. Alternatively, slight shifts may occur in the patterns of overall magnetospheric convection, or their mapping to the ionosphere, to compensate for thermospheric or ionospheric feedback processes [*Fuller-Rowell et al.,* 1987a]. Feedback processes within the thermosphere and ionosphere probably affect a number of magnetospheric processes that have historically been thought of as purely magnetospheric/plasma physics phenomena.

2. THE COUPLED GLOBAL THERMOSPHERE/POLAR IONOSPHERE MODEL

The development of the UCL Three-Dimensional Time-Dependent Thermospheric Model is well documented in previous publications [*Fuller-Rowell and Rees,* 1980, 1983], as is the Sheffield ionospheric model [*Quegan et al.,* 1982]. The development of this coupled model has been described in *Quegan et al.* [1982, 1986]; *Fuller-Rowell et al.,* [1984, 1987b]; and *Allen et al.,* [1986].

The UCL Three-Dimensional Time-Dependent Thermospheric Model (or GCM) simulates the time-dependent structure of the vector wind, temperature, density, and composition of the neutral atmosphere by numerically solving the nonlinear equations of momentum, energy, and continuity, and a time-dependent mean mass equation. The global atmosphere is divided into a series of elements in geographic latitude, longitude, and pressure. In a Eulerian system, each grid point rotates with the Earth to define a noninertial frame of reference in a spherical-polar-coordinate system. The latitude resolution is 2°, the longitude resolution is 18°, and each longitude slice sweeps through all local times, with a 1-min time step. In the vertical direction the atmosphere is divided into 15 levels in log (pressure), each layer is equivalent to one scale-height thickness, from a lower boundary of 1 Pascal at 80-km height.

The time-dependent variables of southward and eastward neutral wind, total energy density, and mean molecular mass are evaluated at each grid point by an explicit time-stepping numerical technique. After each iteration the vertical wind is derived, together with temperature, heights of pressure surfaces, density, and atomic oxygen and molecular nitrogen concentrations. The data can be interpolated to fixed heights for comparison with experimental data or with empirical models. The momentum equation is nonlinear and the solutions fully describe the horizontal and vertical advection, i.e., the transport of momentum.

The initial versions of the global three-dimensional time-dependent (3-D T-D) numerical thermospheric models used theoretical models or the simple empirical *Chiu* [1975] global model of the ionosphere to calculate ion drag and Joule heating. However, the lack of any response at high latitudes to geomagnetic processes (precipitation, convection) within the Chiu ionospheric model caused a gross underestimate of the magnitude of ion drag and of Joule/frictional heating at E-region altitudes in the auroral oval. In the F region, the Chiu model did not so seriously underestimate plasma densities. When the Chiu model was used in the 3-D T-D (or GCM) models, it was possible to simulate F-region winds and temperatures within the upper thermosphere (which were realistic for quiet and slightly disturbed geomagnetic conditions). However, under disturbed conditions, and in the E-region at all times, simulations using the *Chiu* [1975] ionospheric model generated winds, currents, and heating that were all unreasonably low [*Rees et al.,* 1983a, 1986].

In the first interactive model for the polar ionosphere and thermosphere [*Quegan et al.,* 1982], datasets from the UCL global thermosphere and the "Sheffield" polar ionosphere model (UT-independent) were iteratively exchanged until stability was achieved. The effects of the model iterations showed that significant changes in plasma density were caused by the effects of induced winds. The auroral oval plasma densities were greatly enhanced compared with those of the *Chiu* [1975] model. As a result, induced thermospheric winds and heating were generally greatly increased compared with previous simulations using the global *Chiu* [1975] model.

This "simple" coupled model could not, however, be universally applied to study UT variations, let alone to the effects of variable solar, geomagnetic, and seasonal conditions. The next stage, was to develop a fully interactive thermosphere and polar ionosphere model [*Fuller-Rowell et al.,* 1987b, 1988]. The fully coupled model exchanges ionospheric and thermospheric parameters throughout the region poleward of 40° geomagnetic latitude. At lower latitudes, the numerical model is currently still dependent on empirical ionospheric descriptions. When the physics of the major ionospheric-thermospheric interactions are included within the coupled model, many of the additional "geomagnetic" energy sources, required previously to explain observations, are unnecessary [*Fuller-Rowell et al.,* 1987b, 1988].

The approach taken in the development of the coupled model described above differs from that employed by the Utah State University group [*Sojka and Schunk,* 1983]. In the latter modelling, the structure, dynamics, and composition of the neutral atmosphere is assumed to be invariant to the response of the ionosphere to solar and geomagnetic forcing. In this work we show that induced neutral atmosphere winds, temperature, and composition changes do, in fact, cause major feedback changes of the ionosphere.

The neutral atmosphere numerical model uses an Eulerian approach. Earlier versions of the "Sheffield" ionospheric code [*Watkins* 1978; *Allen et al.,* 1986] were evaluated in a Lagrangian system, however. For the fully coupled model, it is necessary to couple the ionospheric and thermospheric grids very closely and accurately for appropriate data exchange. A hybrid system has thus been developed. The ionospheric plasma is sampled in a grid that corresponds to the Eulerian grid of the thermospheric model. To follow the evolution of the ionospheric plasma, a second Lagrangian system has been developed, including the appropriate mathematical terms to cross-couple the two grids. The complex convection patterns imposed by a magnetospheric electric field on plasma movements within the polar regions are referenced to a fixed Sun-Earth frame, assuming pure $\mathbf{E} \times \mathbf{B}$ drifts. The electric field is derived by merging a model of magnetospheric convection [e.g., *Heppner and Maynard,* 1987] with the corotation potential (induced by Earth rotation). Parcels of plasma are traced along their convections paths, which are often complex, particularly if the convection field is time-dependent.

The recently developed NCAR Ionosphere-Thermosphere General Circulation Model (ITGCM) [*Roble et al.,* 1988] is believed to function in a similar way to interchange ionospheric and thermospheric datasets, although the results are not yet available for direct intercomparison of the simulated behavior of the respective regions with the simulations using the UCL/Sheffield model.

In the ionospheric code, atomic (H^+ and O^+) and molecular ion concentrations are evaluated over the height range from 100–1500 km, and used in the thermospheric code poleward of 40° magnetic latitude. The use of the self-consistent ionosphere at high- and mid-latitudes and an empirical description at low-latitudes can result in a discontinuity at the boundary. The ionospheric code is being extended to include the self-consistent calculation at low latitudes, including computation of the equatorial anomaly, and allowance for interhemispheric flow, but these new results will not be discussed here.

3. COUPLED MODEL SIMULATIONS

Four simulations of the coupled thermosphere and ionosphere for solar illumination corresponding to May 12, using the UCL/Sheffield coupled model have been performed. These simulations have been generated for geomagnetic activity levels corresponding to approximately $Kp = 1, 2, 3,$ and 4, respectively, and for low solar activity ($F_{10.7}$ cm = 90). Only those for $Kp = 2$ and $Kp = 4$ will be explicitly discussed here. The simulations are time-dependent, that is they are UT-dependent, and the results are diurnally reproducible. However, the external solar and geomagnetic inputs are

time-independent. These two simulations use a full International Geomagnetic Reference Field (IGRF) representation of the geomagnetic field, to define the mapping of convection and precipitation boundaries in a geographic frame. It should be noted that these simulations include lower-boundary forcing, representative of lower atmosphere propagating tides. Some of the effects of these tides are obvious in the illustrations of the E-region response.

The characteristic UT variations of the summer and winter polar regions are dependent on the offset of the geomagnetic poles from the geographic poles. During the UT day, at all seasons, the geomagnetic polar caps are carried into and out of sunlight. There is, therefore, a large diurnal modulation of the solar photoionization and UV/EUV heating of the geomagnetic polar regions. This also causes large UT variations in plasma density, conductivity, ion drag, and Joule and solar heating of the polar thermosphere. There are consequent large UT modulations of the thermospheric and ionospheric response. Seasonal and latitudinal variations of insolation and solar photoionization cause variations of a factor of 2 in the hemispherically integrated Joule heating rates between the summer and winter hemisphere near solstice. For prescribed electric field and particle precipitation patterns, the higher background conductivity of the summer polar cap and auroral regions allows larger Pedersen and field-aligned currents to flow in those regions, increasing magnetospheric dissipation. Obviously, if the magnetosphere cannot support the currents, or provide the necessary transfer of energy or momentum, the hemispheric dissipation ratio may be smaller. The present models do not shed light on this interesting subject, neither do experimental observations.

4. UPPER THERMOSPHERIC NEUTRAL WIND AND TEMPERATURE, COMPOSITION, AND ION DENSITY, AS FUNCTIONS OF GEOMAGNETIC ACTIVITY

Figure 1 illustrates, in four panels, the distribution of neutral wind and temperature, mean molecular mass, atomic oxygen ion number density (O^+), and molecular ion number density, respectively, within the northern polar region at 1800 UT. The simulation uses NOAA/TIROS precipitation and the Millstone Hill convection field corresponding to low geomagnetic activity, $Kp = 2$, for low solar activity, $F_{10.7} = 90$, and solar illumination conditions corresponding to May 12. The neutral parameters are shown at pressure level 12, which is close to the 300-km altitude selected for the ionospheric parameters. Figure 2 illustrates the same

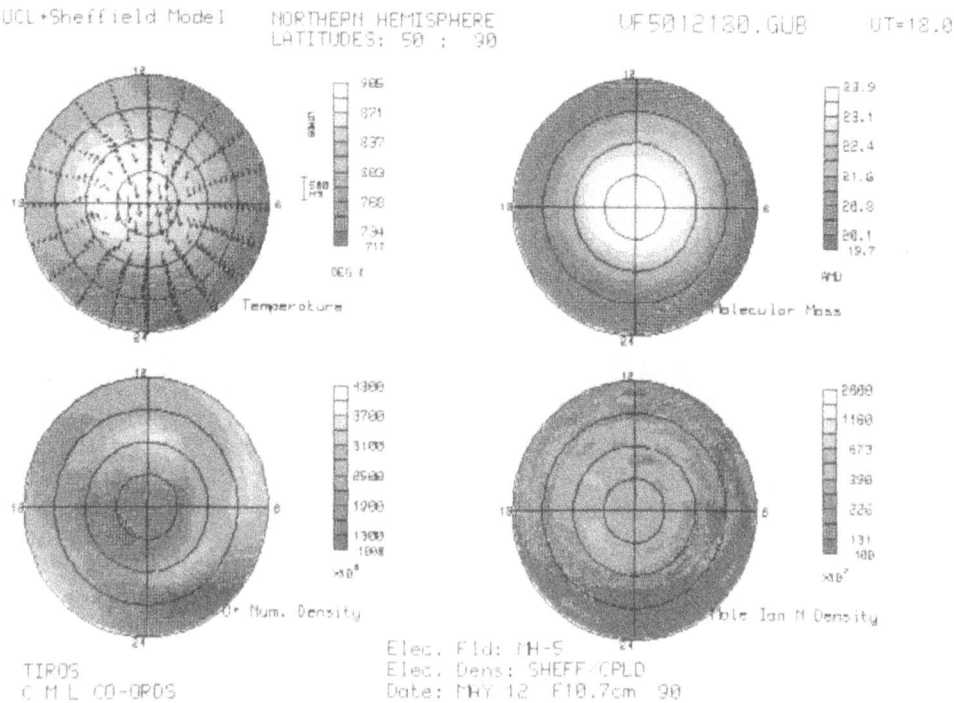

Figure 1—Northern polar distributions of upper thermospheric temperature and wind velocity, mean molecular mass, electron density, and ion temperature at pressure level 12, computed for May 12, by the coupled ionosphere-thermosphere model, using the Millstone Hill convection field, low solar activity, $F_{10.7} = 90$, low geomagnetic activity, $Kp = 2$. The simulations include the effects of lower atmosphere propagating semidiurnal tides. (This figure also appears in color: Plate 2.)

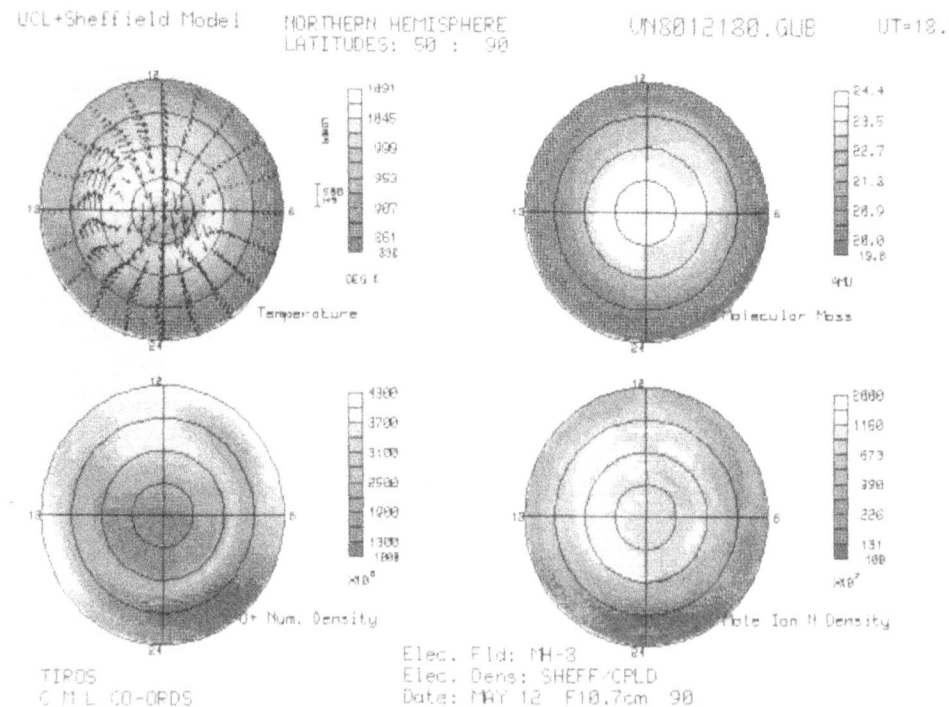

Figure 2—Same as Figure 1, for a higher level of geomagnetic activity, $Kp = 4$. (This figure also appears in color: Plate 3.)

TABLE 1. Characteristics of inputs and response

Kp	1	2	3	4
TIROS/NOAA Auroral Activity Index	3	5	7	8

Hemispherically Integrated Power Inputs (GW)

Particle energy influx	5	13	32	50
Particle heating	2	5	13	20
Northern hemisphere Joule heating	14	28	70	89
Southern hemisphere Joule heating	7	15	41	61
Northern hemisphere solar UV	592	592	592	592
Southern hemisphere solar UV	336	336	336	336
Northern hemisphere solar EUV	102	102	102	102
Southern hemisphere solar EUV	55	55	55	55

Height-Integrated Quantities

Peak Joule heating rate (mWm^{-2})	2.2	4.9	7.8	11.9
Peak upward field-aligned current (μAm^{-2})	0.3	0.9	1.1	1.4
Peak downward field-aligned current (μAm^{-2})	0.3	0.4	0.7	1.0
Peak Pedersen conductivity (Siemens)	5	9	14	19

Neutral Parameters at Pressure Level 12 (~ 300 km)

Peak neutral temperature (K)	863	905	1000	1091
Peak mean molecular mass (amu)	23.1	23.9	24.4	24.4
Peak sunward neutral wind in dusk sector auroral oval (ms^{-1})	108	197	355	436
Peak antisunward neutral wind over polar cap (ms^{-1})	298	330	419	450

Parameters at Pressure Level 7 (~ 125 km)

Peak neutral temperature (K)	370	392	466	554
Peak mean molecular mass (amu)	26.79	26.85	26.88	26.90
Peak sunward neutral wind in dusk sector auroral oval (ms^{-1})	70	104	243	353
Peak ion density ($\times 10^{11}$ m^{-3})	1.2	2.0	3.1	3.7
Peak ion temperature at 300 km (K)	1033	1116	1434	1505
$[NO^+]/[O^+]$ ratio (300 km)	3%	5%	7%	9%

set of parameters as does Figure 1, but for moderately disturbed geomagnetic activity ($Kp = 4$). Some of the characteristics of the inputs and response are summarized in Table 1.

At high latitudes the magnetospheric momentum and energy sources resulting from the convection electric field, auroral particle precipitation, and Joule heating have increased the neutral wind velocities and neutral temperature, with the additional heat source coming primarily from Joule heating.

At $Kp = 1$, the wind circulation at F-region altitudes is only slightly different from the antisolar flow over the polar cap that would be produced without any magnetospheric forcing. At $Kp = 2$, there is a distinct sunward wind, 200 m s^{-1} in the dusk auroral oval, and the antisunward polar cap winds are 300 m s^{-1}. At $Kp = 4$, these wind speeds increase to over 400 and 450 m s^{-1}, respectively.

The sunward winds of the dawn auroral oval are always weaker than the strong sunward winds of the dusk auroral oval. The primary reasons for this dusk/dawn asymmetry have been discussed previously [*Fuller-Rowell and Rees,* 1984]. The asymmetric response is caused by a natural atmospheric "resonance" in which the clockwise wind vortex forced by sunward ion convection in the dusk auroral oval and antisunward ion convection over the polar cap is preferentially excited, given the sense and rate of rotation of the Earth.

Coriolis and curvature accelerations balance within the clockwise vortex, of which the dusk auroral oval is part. In the dawn auroral oval, the conditions for this resonance do not exist. As a result, for the same plasma densities and convection velocities, there is a much smaller effective sunward wind acceleration. This effect was noted in earlier rocket wind measurements, and has been well observed by the Dynamics Explorer 2 (DE 2) satellite [*Fuller-Rowell and Rees*, 1983]. The effect was predicted by much simpler simulations [*Rees et al.*, 1985b] than those we are describing here. However, it is perhaps reassuring to see that the asymmetric dusk/dawn wind response is still observed in the present, rather more sophisticated, simulations, where the possibility of a complex feedback process between the thermosphere and ionosphere, which might change the nature or magnitude of the wind acceleration and response can now be discounted. The coupled model accounts for all the major feedback mechanisms, with the possible exception of feedback processes involving magnetosphere/ionosphere coupling.

The increased geomagnetic heating in the northern polar cap creates, in addition to the high neutral gas temperature, a plateau of raised mean molecular mass, caused by systematic upwelling and outflow to the high midlatitudes. The value of mean molecular mass increases steadily with sustained geomagnetic forcing and heating (enhancement of molecular nitrogen, depletion of atomic oxygen). The consequences are that within the polar cap, as the activity level increases, there is an increase of molecular ion density, increasing the ratio of molecular to atomic ions from about 3–9%, as Kp increases from 1 to 4 (Table 1). The atomic oxygen ion number density at 300 km increases fairly uniformly as activity increases. The number density in the polar region increases, in spite of the neutral chemistry, specifically an increase in the N_2/O ratio, due to the increase in the ionization rate from precipitating particles. At midlatitudes, the increase is due to a combination of two effects. The first is the decrease in the N_2/O ratio, and the second is due to the stronger equatorward wind circulation, driving plasma up the field lines to regions where recombination is slower. The evidence of the underlying changes in polar thermospheric composition, related to the geomagnetic heating and momentum exchange can be found in [*Rees et al.*, 1983a, 1985a, 1986]. Recent review of data from the DE 2 satellite can be found in *Killeen and Roble* [1988] and the accompanying review by *Killeen et al.* [this volume].

5. LOWER THERMOSPHERIC NEUTRAL WIND AND TEMPERATURE, COMPOSITION, PLASMA DENSITY, AND AURORAL IONIZATION RATE, AS A FUNCTION OF GEOMAGNETIC ACTIVITY

Figure 3 illustrates the neutral wind and temperature, mean molecular mass and the plasma density, and auroral ionization rate in the lower thermosphere, at pressure level 7, about 125-km altitude, close to the peak in the Pedersen conductivity. The conditions of the simulation are the same as those used in Figure 1. Figure 4 has a display similar to that of Figure 3, but for a level of geomagnetic activity corresponding to $Kp = 4$.

In Figure 3, the neutral circulation around the polar region (near the summer solstice) is anticyclonic, best seen at high midlatitudes. Maxima in the temperature distribution around 0400 and 1600 LT at midlatitudes are the result of the propagating lower atmosphere tide included in the simulation. Some modest wind changes, around 30 m s^{-1}, are also induced, although these are difficult to distinguish against the winds induced by ion drag forcing.

At higher latitudes, the major wind features result from ion drag, and are thus responsive to convection changes and plasma density enhancements (resulting from combined particle precipitation and solar photoionization). The basic auroral oval wind pattern follows the ion convection, as in the F region. The E-region wind velocities are typically a factor of about 2–3 lower than at F-region altitudes. The E-region winds increase in magnitude as the activity increases rather more strongly than do the F-region winds. At $Kp = 4$, sunward winds within the dusk auroral oval reach 300–400 m s^{-1}.

The self-consistent simulations show E-region winds, at times of moderate geomagnetic disturbances, that are of similar magnitudes and structures to those reported in experimental studies of E-region winds [e.g., *Rees et al.*, 1971; *Johnson et al.*, 1987; *Killeen et al.*, 1989]. Until the present self-consistent thermosphere-ionosphere code was developed, such correspondence of simulated E-region winds with data could only be obtained by extensive and artificial tuning of E-region plasma density distributions [*Rees et al.*, 1985a, 1985b].

The asymmetry of the wind response in the dusk and dawn parts of the auroral oval is again related to the conditions required for resonance. The E-region winds follow the $\mathbf{E} \times \mathbf{B}$ direction more directly than would be expected, considering the rotation of the ion drift vector toward the electric field direction at these alti-

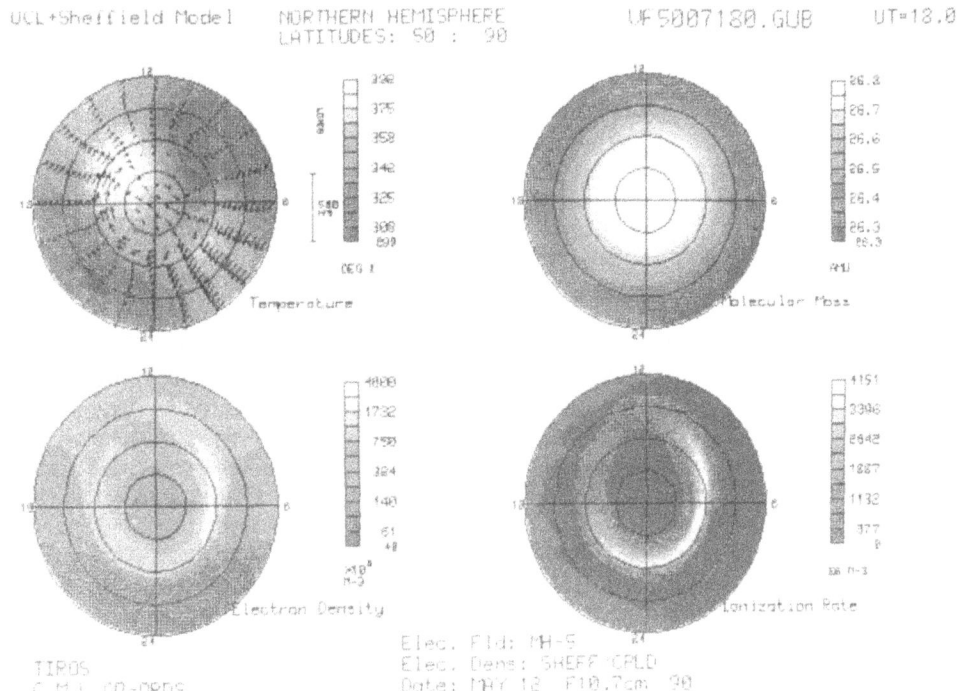

Figure 3—Northern polar distributions of lower thermospheric temperature and wind velocity, mean molecular mass, electron density and auroral ionization rate at pressure level 7 (E-region, approx. 125 km), computed for May 12, by the coupled ionosphere-thermosphere model, using the Millstone Hill convection field, low solar activity, $F_{10.7} = 90$, low geomagnetic activity, $Kp = 2$. The simulations include the effects of lower atmosphere propagating semidiurnal tides. (This figure also appears in color: Plate 4.)

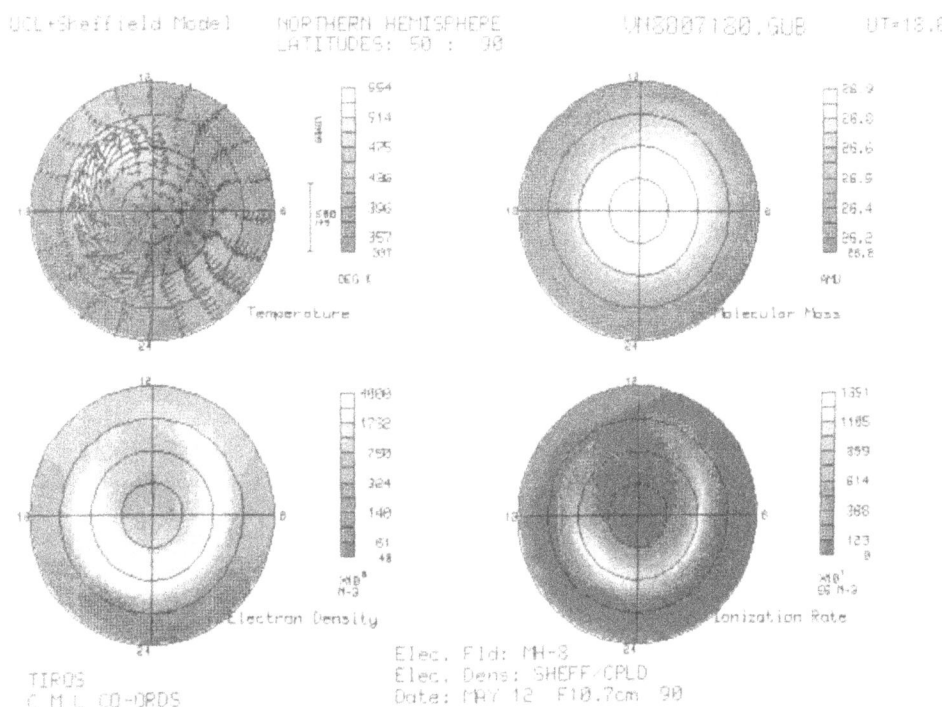

Figure 4—Same as Figure 3, for $Kp = 4$, and the data are displayed for pressure level 7, about 125-km altitude. (This figure also appears in color: Plate 5.)

tudes. This is the result of the stable, steady-state, response of the atmosphere, where pressure gradients restrict the convergence or divergence of the component of winds, induced by ion drag acceleration in the direction of the electric field vector at these relatively low altitudes. A short-period impulse would have rather different consequences, with the initial winds following more closely the mean ion vectors. Only later, as the winds respond to changes in pressure gradients, Coriolis, advection, and viscosity terms, do the patterns shown in Figures 3 and 4 develop.

The E-region plasma densities show a direct response to auroral production, with values increasing with activity. The background E-region values correspond to solar production. Within the auroral oval E-region, the seasonal modulation of plasma density and the effects of plasma transport by ion drifts or winds are generally of little importance. Vertical plasma transport due to the horizontal winds and the electric fields are likely to be the most important single effect, and this may be particularly important in connection with the formation and subsequent behavior of sporadic E layers, resulting from metallic ions. Such long-lived ions are not subject to the rapid destruction of the E-region molecular ions that, alone, are considered in this study. For metallic E-region ions, horizontal as well as vertical transport effects will be much more significant [*Nygren et al.,* 1984].

In combination with the convection electric field distributions, plasma density enhancements caused by auroral precipitation are a critical factor in producing the high observed values of ion drag wind acceleration terms, ionospheric conductivities, and electric currents. Implicitly, the distribution of magnetospheric field-aligned current, which power the entire high-latitude system during geomagnetic disturbances, are therefore modulated by precipitation-induced conductivity enhancements. Typically, in the summer period, peak average E-region ion density values within the auroral oval during moderate geomagnetic disturbances are a factor of 3 greater than those of the surrounding sunlit ionosphere, which is not affected by auroral precipitation. In winter time, the corresponding conductivity enhancement factor due to auroral electron precipitation is between a factor of 10 and 100 times above the non-sunlit E-region polar ionosphere.

6. ELECTRODYNAMICS

In the following section, the key electrodynamic parameters obtained from the numerical simulations will be discussed and compared. These parameters are the height-integrated Joule heating rate and the field-aligned current. Unlike the previous figures, which have been referenced to a fixed pressure level, the electro-

dynamic parameters are height-integrated through all the levels of the thermosphere and ionosphere. Generally both these quantities have their maximum contribution from the lower thermosphere, but there are times and locations when this is not the case, and a significant contribution can come from the upper levels.

The height-integrated Joule heating rate is very sensitive to model inputs. This is due to its dependence on the square of the electric field, and also due to its critical dependence on the co-location of boundaries of conductivity enhancements through electron precipitation and maximum ion drifts. Obviously, one must treat with caution the idea of using the product of two independently produced models, particularly when one expects a strong causal relationship between the two sources via their magnetospheric origin. Ideally, the models of precipitation and convection should be assembled "self-consistently." Any correlation or anti-correlations (as have been reported) between the two datasets can then be maintained. However, no convenient empirical or statistical model of the combined magnetospheric sources is available, so we must rely on using, with some caution, those currently at our disposal. The two models used are convenient in that they have at least been constructed by a binning process using an identical auroral activity index [*Foster et al.,* 1986; *Fuller-Rowell and Evans,* 1987].

Figure 5 shows the northern polar region distributions of the height-integrated Joule heating rate and pressure level 12 (F region) ion flow vectors, the ion temperature (300 km altitude), the height-integrated Pedersen conductivity and the field-aligned currents, calculated from horizontal current convergence or divergence. These values are all computed for May 12, by the coupled ionosphere-thermosphere model, using NOAA/TIROS precipitation and the Millstone Hill convection field corresponding to low geomagnetic activity, $Kp = 2$, and for low solar activity, $F_{10.7} = 90$. The simulations include the effects of lower atmosphere propagating semidiurnal tides.

Figure 6 has the same presentation as that used for Figure 5, but shows more active geomagnetic conditions for $Kp = 4$. Due to its dependence on the square of the neutral wind-ion velocity difference, the peak Joule heating rates are associated with the peaks in the ion drift vectors.

There are two broad regions of peak Joule heating. One is, as expected, throughout the dusk sector auroral oval, corresponding to maximum sunward ion drifts. However, there is a second region of high Joule heating that is associated with enhanced antisunward ion flow over the dusk side of the polar cap. This second region continues into the dawn auroral oval. There is a quite distinct gap between these two regions, follow-

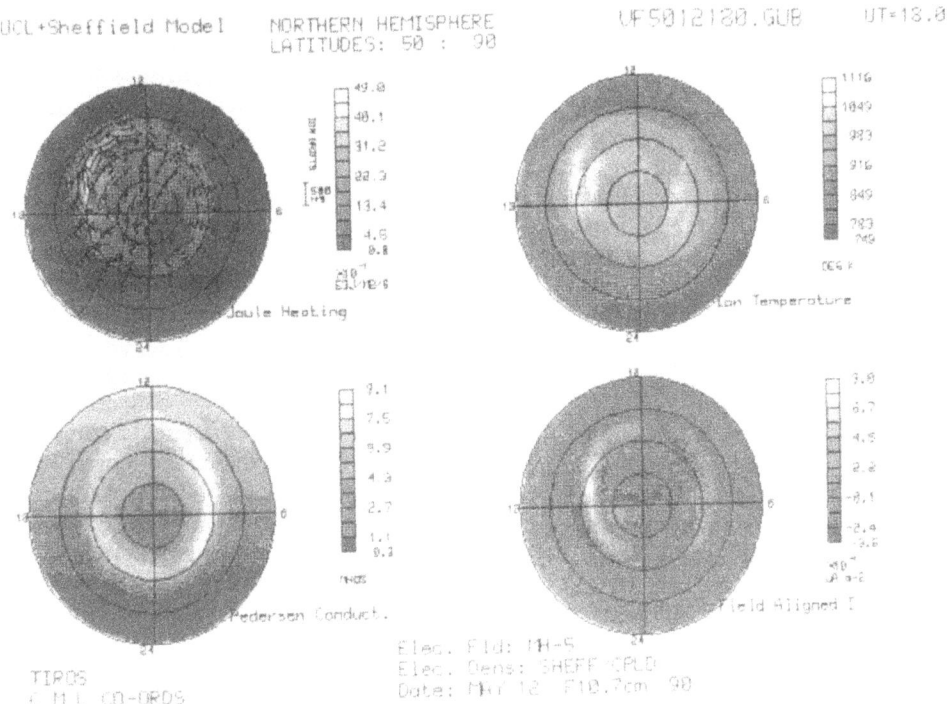

Figure 5—The northern polar region distributions are shown for the height-integrated Joule heating rate and ion flow vectors, the ion temperature (pressure level 12), the height-integrated Pedersen conductivity, and the field-aligned currents, calculated from horizontal current convergence or divergence. The conditions are otherwise as for Figures 1 and 3. (This figure also appears in color: Plate 6.)

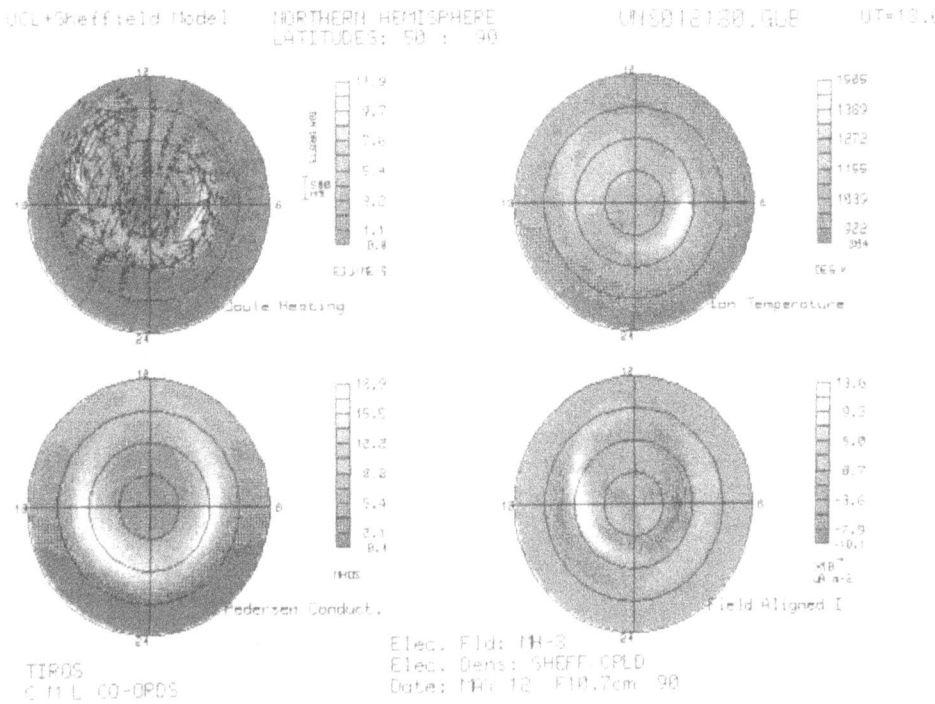

Figure 6—Same as Figure 5, for a higher level of geomagnetic activity, *Kp* = 4, corresponding to the simulations shown in Figures 2 and 4. (This figure also appears in color: Plate 7.)

ing the Harang discontinuity. The patterns are generally similar at all levels of geomagnetic activity, but the magnitudes increase substantially with Kp (Table 1). Joule heating rates peak in the dusk sector of the auroral oval under quiet conditions. As activity increases, the dusk sector values increase by a factor of about 2 to 3, whereas the dawn sector heating rates increase substantially, to values in excess of those in the dusk sector. For asymmetric convection patterns [*Rees et al.*, 1986; *Heppner and Maynard*, 1987], there will be a distinct change of the location of the highest Joule heating rates between the dusk and dawn parts of the polar cap for B_y negative or positive, respectively, in the northern hemisphere.

In Table 1, we have summarized a number of the parameters, including the hemispherically integrated heating rates from solar UV and EUV sources, from particle heating and from Joule heating from each of the four simulations used in this study: $Kp = 1, 2, 3,$ and 4. At $Kp = 4$, the integrated, as well as peak, values are more than 10 times higher than at $Kp = 1$. As the area of the polar cap and auroral oval expands, integrated Joule heating rates increase more rapidly with increasing activity than the local peak rates.

The neutral winds induced by ion drag, heating, and the natural resonance of the thermosphere considerably influence the height-integrated Joule heating rate. For example, Joule heating in the dusk auroral oval is decreased by 30–40% as a result of the induced wind dynamo feedback, as neutral winds follow the ion motion [*Rees and Fuller-Rowell*, 1988]. The preferential excitation of sunward winds in the dusk auroral oval, following ion drifts, is established both from observation [*Rees*, 1973; *Hays et al.*, 1984] and from simulations [*Rees et al.*, 1985b, 1986], as discussed earlier. One consequence of the larger induced winds in the dusk sector due to this inertial resonance effect, is that when the $\mathbf{V}_n \times \mathbf{B}$ component is removed from the full expression for the Joule heating rate the main effect is an increase of Joule heating in the dusk sector.

We should note that the Pedersen conductivity in the winter polar region in the lower thermosphere is very small, except where there is significant auroral precipitation. A significant contribution to the winter polar height-integrated conductivity, and consequently also the Joule heating rate, comes from the upper thermosphere. The winter F-region plasma concentrations are at least as high as summer values.

The ion temperature (F-region) distributions correspond roughly to the Joule heating patterns, when the baseline F-region neutral gas temperature is considered (Figures 1 and 2). As the general level of geomagnetic activity rises, this baseline neutral gas temperature also increases. In these models, using sim-

ple and smooth functions for convection patterns and precipitation, the neutral gas temperature distribution is relatively smooth. However, there are large localized increases of neutral temperature in the polar region that are not represented by semi-empirical thermospheric models such as MSIS 1986 [*Hedin*, 1987]. Thus some care is required to interpret ion temperature data properly in isolation when attempting to infer either Joule heating rates or the neutral gas temperature. The Joule real heating rates are considerably more structured, and with locally much higher values, than these model simulations for mean conditions would indicate.

The field-aligned current distributions derived from the statistical models of convection and precipitation, even using the fully coupled thermosphere-ionosphere model, have to be used cautiously. The height-integrated horizontal ionospheric current system is obtained from a product of the conductivity and total electric field, and the field-aligned current system is calculated from the divergence of the horizontal current. This condition then satisfies current continuity. Combining this calculation with the previously derived quantities, we then have an indication of the total demand made by the ionosphere on the magnetospheric sources of plasma, charge, momentum and energy. For the horizontal current to flow in the ionosphere, current continuity demands that either the magnetosphere is able to supply the necessary field-aligned current or polarization charges build up within the ionosphere, (or via instabilities are triggered by large field-aligned current or "double-layers," etc.). All these tend to reduce the magnitude of the electric field imposed upon the upper atmosphere. The sensitivity of the magnetosphere to the feedback from the ionospheric current system is as yet difficult to quantify experimentally or theoretically.

Figures 5 and 6 show that the field-aligned current (FAC) progressively increases as the level of geomagnetic activity increases. The region 1 and region 2 Birkeland current systems are clearly defined, even at the lowest activity level, and generally expand and intensify as the activity increases. In Table 1, the peak FAC values are summarized for the four simulations corresponding to $Kp = 1, 2, 3,$ and 4. The peak values increase approximately linearly with Kp. We should note, however, that the integrated values are likely to increase more rapidly, due to the expansion of area of the polar cap and auroral oval, with increasing activity.

7. SUMMARY

The model data from two numerical simulations of the coupled thermosphere and ionosphere, which have been presented, are taken from a series of four simulations, for solar illumination conditions corresponding to May 12. Each simulation was at low solar activity and at levels of geomagnetic energy and momentum input corresponding approximately to $Kp = 1, 2, 3,$ and 4. The data have been presented for the northern polar region and at 1800 UT. These simulations illustrate ways in which relatively localized auroral processes affect the behavior of the thermosphere and ionosphere.

Thermospheric Winds

The major effect of auroral precipitation is to increase the Pedersen conductivity and the ion drag wind acceleration, via enhanced E-region ion production. The coupled model simulations show very clearly that it is only the additional effect of auroral electron precipitation, together with enhanced ion convection, that can generate the observed E-region neutral winds of the order of 400 m s^{-1} during disturbed periods. In the F-region, plasma densities do not increase, and may even decrease due to thermospheric compositional changes following intense Joule heating. Therefore F-region winds are not so dependent on auroral electron precipitation and ionization. In all cases, however, it is the patterns of magnetospheric convection and ion flows that are imparted onto the neutral gas motions. The effect of direct auroral heating on winds is minimal, and it is even hard to detect the direct effects on winds of Joule heating.

Thermospheric Composition

The global temperature and pressure fields create a prevailing summer to winter flow at mid and high latitudes. This is particularly pronounced under quiet geomagnetic conditions, and is already well established by May 12, although this is not at the full solstice. For average or high geomagnetic activity, the geomagnetic heating in the summer hemisphere enhances this seasonal interhemispheric flow, whereas in winter it opposes the solar forcing. At upper thermospheric altitudes, a mean wind flows from both the summer pole and from the winter pole to winter midlatitudes. The global pattern of mean molecular mass in the upper thermosphere is largely a result of these large-scale horizontal (and associated vertical) wind fields, with competing flow in the upper mesosphere. High values of mean molecular mass (above 22 amu at F-region altitudes) are found over the summer geomagnetic polar region, even under quiet conditions. As the geomagnetic activity rises, the summer polar values increase further, as high as 24–25 amu, while values of above 20 amu can be found over the winter geomagnetic pole under active geomagnetic conditions. At high winter midlatitudes, the mean molecular mass is close to 16 amu, indicating nearly pure atomic oxygen.

Ion Density

In the upper thermosphere, there is a general increase in the ion density as activity increases. At high latitudes, the enhancement is due to increased auroral precipitation. At midlatitudes, the increase is due to stronger equatorward winds. Holes may appear within the summer polar cap under very disturbed geomagnetic activity conditions, in regions of intense Joule heating, due to strongly enhanced molecular nitrogen concentrations at F-region altitudes. The recombination rate is then increased, decreasing the plasma density despite high values of solar and auroral production. In the auroral E-region, at all levels of geomagnetic activity, the ion density distribution is dominated by the effects of auroral precipitation. For the simulations discussed here, the precipitation has no dependence on the IMF B_y component or season. Additional solar ionization in summer causes a relatively small E-region plasma density increase within the auroral oval, compared with the surrounding polar cap and subauroral regions. In the winter polar E-region, on the other hand, plasma densities can fall to very low values, in the absence of solar or particle ionization sources.

Ionospheric Composition

The seasonal variation of neutral thermospheric composition at midlatitudes, causes enhanced concentrations of [N$_2$] relative to [O] in the summer hemisphere between 250 and 350 km, compared with the equivalent winter regions. This changed summertime neutral composition feeds back into the ionospheric chemistry by increasing the recombination of the dominant ion O$^+$, and by decreasing the N$_m$F$_2$ (peak electron density at the peak of the F$_2$ layer). Conversely, in the winter hemisphere, the enhanced [O/N$_2$] ratio decreases the recombination rate of the F region O$^+$ ions, and causes a general increase in N$_m$F$_2$ in sunlit regions, despite the overall decrease of solar ionization. The ratio of molecular to atomic ions at F-region altitudes has a strong seasonal dependence at a given level of geomagnetic activity, reflecting the strong seasonal difference in the neutral composition. In winter O$^+$ dominates, in summer the concentration of molecular ions increases dramatically. We also see in these simulations that, as Kp increases from 1 to 4, the proportion of molecular to atomic ions increases from about 3% to about 9%, although the atomic oxygen ion concentrations are not greatly changed.

Joule Heating

Joule heating is normally 3–4 times larger than direct particle heating [*Evans et al.,* 1988], although the role of particle ionization is crucial in the production of the high E-region plasma densities and Pedersen conductivity. There are two broad regions of peak Joule heating. One is within the dusk auroral oval, corresponding to the region of maximum sunward ion drifts. A second region of high Joule heating is associated with enhanced antisunward ion flow within the polar cap. This second region continues into the dawn auroral oval. There is a distinct gap between these two regions, aligned with the Harang discontinuity. The patterns are generally similar at all levels of geomagnetic activity. Joule heating rates peak in the dusk sector of the auroral oval for low geomagnetic activity, but are roughly similar for the dusk and dawn sectors for the more active conditions. We should note that the distributions of Joule heating are very dependent on the plasma convection patterns, and are also dependent on precipitation patterns.

Neutral winds induced by ion drag, heating, and the natural resonance of the thermosphere considerably influence the height-integrated Joule heating rate. For example, Joule heating in the dusk auroral oval is decreased by 30–40% as a result of the induced wind dynamo feedback, as neutral winds follow the ion motion, due to the inertial resonance effect. Therefore the effect of ignoring the $\mathbf{V}_n \times \mathbf{B}$ component in the full expression for the Joule heating rate is to overestimate the Joule heating rate in the dusk sector by 30–40%.

F-Region Ion Temperature

The F-region ion temperature distributions correspond roughly to the Joule heating patterns, when the baseline F-region neutral gas temperature is also considered. As the general level of geomagnetic activity rises, this baseline neutral gas temperature also increases, while the neutral temperature distribution becomes more complex. Some considerable care is required to interpret ion temperature data properly when attempting to infer either Joule heating rates or the neutral gas temperature. The Joule heating rates are highly structured, and with locally very high values. Similarly,

neutral gas temperatures, the baseline for the ion temperature, can be highly structured (though less so than the ion temperature). The model simulations shown here for mean conditions only indicate a general guide to Joule heating and to ion temperature distributions.

Pedersen Conductivity

Peak values of Pedersen conductivity outside the auroral oval reflect the seasonal differences in the conductivity of the ionosphere. In the summer sunlit polar region, the Pedersen conductivity is relatively high throughout the polar cap, and also at subauroral latitudes. Hemispherically integrated Joule heating rates are about a factor of 2–3 larger in the summer hemisphere than in the winter hemisphere. As geomagnetic activity level increases, the peak values of Pedersen conductivity increase, while the regions of intensified Pedersen conductivity expand equatorward.

Field-Aligned Currents

The field-aligned currents progressively increase within the summer northern polar region as the level of geomagnetic activity increases. The region 1 and region 2 Birkeland current/field-aligned current systems are clearly defined, even at the lowest activity level, and generally expand equatorward and intensify as the activity increases. While the peak field-aligned current values increase approximately linearly with Kp, the integrated values increase more rapidly, as the area of the polar cap and auroral oval expands with increasing activity. The patterns of field-aligned current are closely related to conductivity distributions and changes.

ACKNOWLEDGMENT—We express our particular thanks to J. Harmer and H. Hughes for their assistance in preparing, running, and processing the computer simulations using the UCL/Sheffield coupled ionospheric/thermospheric model. Computer time was made available by the University of London Computer Center (CRAY 1-S), and the CRAY-XMP-48 at the Rutherford Appleton Laboratory. The research was supported by grants from the United Kingdom Science and Engineering Research Council, and from the European Office of Aerospace Research and Development (AFOSR-86-341).

REFERENCES

Allen, B. T., G. J. Bailey, and R. J. Moffett, "Ion Distributions in the High-Latitude Topside Ionosphere," *Ann. Geophys.,* **4A**, 97–106 (1986).

Cole, K. D., "Joule Heating of the Upper Atmosphere," *Aust. J. Phys.,* **15**, 223–235 (1962).

Cole, K. D., "Electrodynamic Heating and Movement of the Thermosphere," *Planet, Space Sci.,* **19**, 59–75 (1971).

Chiu, Y. T., "An Improved Phenomenological Model of Ionospheric Density," *J. Atmos. Terr. Phys.,* **37**, 1563–1570 (1975).

Dickinson, R. E., E. C. Ridley, and R. G. Roble, "Thermospheric General Circulation with Coupled Dynamics and Composition," *J. Atmos. Terr. Phys.,* **41**, 205–219 (1984).

Evans, D. S., T. J. Fuller-Rowell, S. Maeda, and J. Foster, "Specification of the Heat Input to the Thermosphere from Magnetospheric Processes Using TIROS/NOAA Auroral Particles Observations," *Adv. Astron. Sci.,* **65**, 1649 (1988).

Fesen C., R. G. Roble, and E. C. Ridley, "Simulations of Thermospheric Tides at Equinox with the NCAR Thermospheric General Circulation Model," *J. Geophys. Res.,* **91**, 4471–4489 (1986).

Foster, J. C., J. M. Holt, R. G. Musgrove, and D. S. Evans, "Ionospheric Convection Associated with Discrete Levels of Particle Precipitation," *Geophys. Res. Lett.,* **13**, 656–659 (1986).

Fuller-Rowell, T. J., and D. S. Evans, "Height-Integrated Pedersen and Hall Conductivity Patterns Inferred from the NOAA/TIROS Satellite Data," *J. Geophys. Res.,* **92**, 7606–7618 (1987).

Fuller-Rowell, T. J., and D. Rees, "A Three-Dimensional, Time-Dependent, Global Model of the Thermosphere," *J. Atmos. Sci.,* **37**, 2545–2567 (1980).

Fuller-Rowell, T. J., and D. Rees, "Derivation of a Conservative Equation for Mean Molecular Weight for a Two Constituent Gas within a Three-Dimensional, Time-Dependent Model of the Thermosphere," *Planet. Space Sci.,* **31**, 1209–1222 (1983).

Fuller-Rowell, T. J., and D. Rees "Interpretation of an Anticipated Long-Lived Vortex in the Lower Thermosphere Following Simulation of an Isolated Substorm," *Planet. Space Sci., 32*, 69–85 (1984).

Fuller-Rowell, T. J., D. Rees, S. Quegan, G. J. Bailey, and R. J. Moffett, "The Effect of Realistic Conductivities on the High-Latitude Thermospheric Circulation," *Planet. Space Sci., 32*, 469–480 (1984).

Fuller-Rowell, T. J., D. Rees, S. Quegan, R. J. Moffett, and G. J. Bailey, "The Thermospheric Response and Feedback to Magnetospheric Forcing" (extended abstract), Symposium on Quantitative Modeling of Magnetosphere-Ionosphere Coupling Processes, Convenors: Y. Kamide and R. A. Wolf, March 9–13, 1987, Kyoto Sangyo University, p. 20 (1987*a*).

Fuller-Rowell, T. J., D. Rees, S. Quegan, R. J. Moffett, and G. J. Bailey, "Simulations of the Seasonal and Universal Time Variations of the Thermosphere and Ionosphere Using a Coupled, Three-Dimensional, Global Model," *PAGEOPHSY, 127*, 189–217 (1988).

Fuller-Rowell, T. J., S. Quegan, D. Rees, R. J. Moffett, and G. J. Bailey, "Interactions Between Neutral Thermospheric Composition and the Polar Ionosphere Using a Coupled Ionosphere-Thermosphere Model," *J. Geophys. Res., 92*, 7744–7748 (1987*b*).

Hardy, D., M. S. Gussenhoven, E. Holeman, "A Statistical Model of Auroral Electron Precipitation," *J. Geophys. Res., 90*, 4229–4248 (1985).

Harel, M., R. A. Wolf, P. H. Reiff, R. W. Spiro, W. J. Burke, F. J. Rich, and M. Smiddy, "Quantitative Simulation of a Magnetospheric Substorm, 1, Model Logic and Overview," *J. Geophys. Res., 86*, 2217–2241 (1981).

Hays, P. B., T. L. Killeen, N. W. Spencer, L. E. Wharton, R. G. Roble, B. A. Emery, T. J. Fuller-Rowell, D. Rees, L. A. Frank, and J. D. Craven, "Observations of the Dynamics of the Polar Thermosphere," *J. Geophys. Res., 89*, 5547–5612 (1984).

Hedin, A. E., "MSIS-86 Thermospheric Model," *J. Geophys. Res., 92*, 4649 (1987).

Heppner, J. P., "Empirical Models of High Latitude Electric Field," *J. Geophys. Res., 82*, 1115–1125 (1977).

Heppner, J. P., and N. C. Maynard, "Empirical High-Latitude Electric Field Models," *J. Geophys. Res., 92*, 4467–4490 (1987).

Johnson, R. M., V. B. Wickwar, R. G. Roble, and J. G. Luhmann, "Lower Thermospheric Winds at High Latitudes: Chatanika Radar Observations," *Ann. Geophys., 5*, 383–404 (1987).

Killeen, T. L., and R. G. Roble, "Thermosphere Dynamics: Contributions from the First 5 Years of the Dynamics Explorer Program," *Rev. Geophys., 26*, 329–367 (1988).

Killeen, T. L., F. G. McCormac, A. G. Burns, and R. G. Roble, "Thermospheric Dynamics, Energetics and Composition at Auroral Latitudes," *this volume.*

Killeen, T. L., B. Nardi, F. G. McCormac, J. W. Meriwether, Jr., J. P. Thayer, R. G. Roble, T. J. Fuller-Rowell, and D. Rees, "Lower Thermospheric Structure and Dynamics Inferred from Satellite and Ground-Based Fabry-Perot Observations of the O(1S) Green Line Emission," *Adv. Space Res.*, in press (1989).

Martyn, D. F., "Geo-Morphology of F2-Region Ionospheric Storms," *Nature, 171*, 14–16 (1953).

Nygren, T., L. Jalonen, J. Oskman, and T. Turinen, "The Role of Electric Field and Neutral Wind Direction in the Formation of Sporadic-E Layers," *J. Atmos. Terr. Phys., 46*, 373 (1984).

Parish, H., T. J. Fuller-Rowell, D. Rees, T. S. Virdi, and P. S. J. Williams, "Numerical Simulations of the Seasonal Response of the Thermosphere to Propagating Tides," *Adv. Space Res.*, in press (1989).

Pereira, E., M. C. Kelley, D. Rees, I. S. Mikkelson, T. S. Jorgensen, and T. J. Fuller-Rowell, "Observations of Neutral Wind Profiles Between 115 and 176 km Altitude in the Dayside Auroral Oval," *J. Geophys. Res., 85*, 2935–2940 (1980).

Quegan, S., G. J. Bailey, R. J. Moffett, and L. C. Wilkinson, "Universal Time Effects on the Plasma Convection in the Geomagnetic Frame," *J. Atmos. Terr. Phys., 48*, 25–40 (1986).

Quegan, S., G. J. Bailey, R. J. Moffett, R. A. Heelis, T. J. Fuller-Rowell, D. Rees, and R. W. Spiro, "Theoretical Study of the Distribution of Ionization in the High-Latitude Ionosphere and the Plasmasphere: First Results on the Mid-Latitude Trough and the Light-Ion Trough," *J. Atmos. Terr. Phys., 44*, 619–640 (1982).

Rees, D., "Ionospheric Winds in the Auroral Zone," *J. Brit. Interplanet. Soc., 24*, 233–346 (1971).

Rees, D., "Neutral Wind Structure in the Thermosphere During Quiet and Disturbed Geomagnetic Periods," *Physics and Chemistry of Upper Atmospheres*, B. M. McCormac, ed., Reidel, Dortecht, pp. 11–23 (1973).

Rees, D. and T. J. Fuller-Rowell, "Seasonal and Universal Time Variations of the Geomagnetic Response of the Thermosphere and Ionosphere," *Proc. AGARD/NATO Symp.*, Munich (May 1988).

Rees, D., R. Gordon, T. J. Fuller-Rowell, M. F. Smith, G. R. Carignan, T. L. Killeen, P. B. Hayes, and N. W. Spencer, "The Composition, Structure, Temperature and Dynamics of the Upper Thermosphere in the Polar Regions During October to December 1981," *Planet. Space Sci., 33*, 617–666 (1985*a*).

Rees, D., T. J. Fuller-Rowell, M. F. Smith, R. Gordon, T. L. Killeen, P. B. Hays, N. W. Spencer, L. E. Wharton, and N. C. Maynard, "The Westward Thermospheric Jet Stream of the Evening Auroral Oval," *Planet. Space Sci., 33*, 425–456 (1985*b*).

Rees, D., T. J. Fuller-Rowell, R. Gordon, M. F. Smith, J. P. Heppner, N. C. Maynard, N. W. Spencer, L. E. Wharton, P. B. Hays, and T. L. Killeen, "A Theoretical and Empirical Study of the Response of the High-Latitude Thermosphere to the Sense of the 'Y' Component of the Inerplanetary Magnetic Field," *Planet. Space Sci., 34*, 1–40 (1986).

Rees, D., T. J. Fuller-Rowell, R. Gordon, T. L. Killeen, P. B. Hays, L. E. Wharton, and N. W. Spencer, "A Comparison of the Wind Observations from the Dynamics Explorer Satellite and the Predictions of a Global Time-Dependent Model," *Planet. Space Sci., 31*, 1299–1314 (1983*a*).

Rees, M. H., B. A. Emery, R. G. Roble, and K. Stamnes, "Neutral and Ion Gas Heating by Auroral Electron Precipitation," *J. Geophys. Res., 88*, 6289–6300 (1983*b*).

Rishbeth, H., and O .K. Garriot, *Introduction to Ionospheric Physics*, Academic Press, New York and London (1969).

Roble, R. G., E. C. Ridley, A. D. Richmond, and R. E. Dickinson, "A Coupled Thermosphere/Ionosphere Model," *Geophys. Res. Lett., 15*, 1325–1328 (1988).

Roble, R. G., R. E. Dickinson, and E. C. Ridley, "The Global Circulation and Temperature Structure of the Thermosphere with High-Latitude Plasma Convection," *J. Geophys. Res., 87*, 1599–1614 (1982).

Sojka, J. J., and R. W. Schunk, "A Theoretical Study of the High-Latitude F-Region Response to Magnetospheric Storm Inputs," *J. Geophys. Res., 88*, 2112–2122 (1983).

Tinsley, B. A., Y. Sahai, M. A. Biondi, and J. W. Meriwether, "Equatorial Particle Precipitation During Geomagnetic Storms and Relationship to Equatorial Thermospheric Heating," *J. Geophys. Res.,, 93*, 270–276 (1988).

Watkins, B. J., "A Numerical Computer Investigation of the Polar F-Region," *Planet. Space Sci., 26*, 559–569 (1978).

II-5. THERMOSPHERIC DYNAMICS, ENERGETICS, AND COMPOSITION AT AURORAL LATITUDES

T. L. Killeen,* F. G. McCormac,* A. G. Burns,* and R. G. Roble[†]

Instrumentation on the Dynamics Explorer-2 spacecraft (DE 2) was capable of measuring the three-dimensional vector neutral wind and ion drift in the thermosphere along the orbital track. These direct measurements of dynamics for both the neutral and ionized species, together with simultaneous observations of electric and magnetic fields, precipitating charged particle fluxes, auroral luminosity distributions, and constituent densities and temperatures, have led to an improved description and quantitative understanding of global-scale motions in the high-latitude thermosphere. The comprehensive nature of the DE 2 observations has also enabled stringent experimental constraints to be placed on the numerical models of the region (thermospheric general circulation models, TGCMs), leading to an improved theoretical understanding of the important physical processes that control thermospheric circulation and variability. In particular, DE 2 observations have amply demonstrated the close coupling that exists between the magnetosphere and the thermosphere, with magnetospheric energy and momentum sources playing a key role in establishing thermospheric dynamical and thermal structure via ion-neutral collisions. The mean thermospheric dynamical, thermal, and compositional response to auroral-region forcings has been investigated using averages from many orbital passes of DE 2 across the polar regions. In addition, individual passes have been studied to provide measurements of the detailed local response at high spatial resolution. Results from the spacecraft observations have been extensively compared with calculations from the NCAR thermosphere general circulation model. This paper presents a review of the progress made in this area over the past several years, with emphasis on the identification and quantification of the major physical processes coupling the thermosphere to the ionosphere and magnetosphere in the auroral region.

1. INTRODUCTION

The dynamics, thermodynamics, and compositional structure of the neutral gas in the high-latitude thermosphere are strongly controlled by ion-neutral collision processes that are intimately associated with the aurora. Magnetospheric convection electric fields, for example, map down into the ionosphere along equipotential geomagnetic field lines and drive the charged particles there into cellular motion. Rapidly moving ions in these ionospheric convection cells can readily transfer momentum to the neutral gas via ion-drag forcing and can convert a portion of their kinetic energy of motion into internal energy of the neutral gas via Joule or collisional heating. In addition, the enhanced concentrations of ions associated with the aurora serve to increase the rates of exchange of energy and momentum between the neutral and ionized species by increasing ion-neutral collision frequencies. The effect of these auroral ion-neutral collision processes is to modify the neutral wind, temperature, and compositional structure away from that which would be expected if the only source of energy to the thermosphere were solar UV and EUV insolation.

Many examples of the effects of auroral/magnetospheric forcing of the neutral thermosphere at high lati-

tudes have been reported in the literature over the past ten years (e.g., see the recent review articles of *Meriwether* [1983]; *Roble* [1983]; *Mayr et al.* [1985]; *Killeen* [1987]; *Killeen and Roble* [1988]). Rapid experimental and theoretical progress has been sustained due, in part, to the comprehensive nature of new datasets provided by the NASA Dynamics Explorer-2 spacecraft (DE 2) and the ground-based network of optical interferometer observatories and incoherent scatter radar facilities, as well as to the maturity of the numerical, three-dimensional, time-dependent general circulation models (TGCMs) of the thermosphere. The two best-developed examples of the latter are the National Center for Atmospheric Research model (NCAR-TGCM) of *Dickinson et al.* [1981] and *Roble et al.* [1982, 1988] and the University College London model (UCL-TGCM) of *Fuller-Rowell and Rees* [1980] and *Fuller-Rowell et al.* [1987]. The UCL-TGCM uses a mixed Eulerian-Lagrangian approach to the specification of the interaction between the ionosphere and the thermosphere, whereas the latest version of the NCAR model uses a fully Eulerian scheme for the calculations. While the two models differ in the detailed approach adopted and in the type of input parameterizations used, they are functionally equivalent and have both been extensively validated through comparisons with experimental data of various kinds.

We here summarize some of the significant high-latitude results obtained using the DE 2 thermospheric neutral wind, temperature, and composition data, together with the theoretical predictions of the NCAR-

---

*Space Physics Research Laboratory, Department of Atmospheric, Oceanic and Space Sciences, The University of Michigan, Ann Arbor, Michigan 48109.
[†]High Altitude Observatory, National Center for Atmospheric Research, Boulder, Colorado 80307.

Auroral Physics, edited by C.-I. Meng, M. J. Rycroft and L. A. Frank. © Cambridge UP 1991

TGCM. A full discussion of the results from the UCL-TGCM is presented by *Rees and Fuller-Rowell* [this volume]. The emphasis of the discussion here will be on perturbations to the solar-driven upper thermospheric winds, temperatures, and compositional structure due to forcings associated with the aurora. In particular, we use both experimental and theoretical results to discuss the relative importance of the individual terms in the equations that govern the behavior of the thermosphere (namely, the conservation of momentum, conservation of energy, and single-species continuity equations). In the following sections we (1) describe briefly the nature of the experimental measurements made on the DE 2 spacecraft, (2) describe briefly the NCAR model and some sample results, (3) present the governing equations and discuss experimental and theoretical considerations pertaining to the individual terms, and (4) conclude with some statements regarding the current state of the field.

2. DYNAMICS EXPLORER INSTRUMENTATION AND OBSERVATIONS

The Dynamics Explorer 2 spacecraft payload complement included the Fabry-Perot interferometer (FPI) [*Hays et al.*, 1981], the Wind and Temperature Spectrometer (WATS) [*Spencer et al.*, 1981], and the Neutral Atmosphere Composition Spectrometer (NACS) [*Carignan et al.*, 1981]. These instruments measured the meridional component of the neutral wind, the zonal component of the neutral wind, and neutral constituent abundances, respectively, along the track of the polar-orbiting satellite. Both the FPI and WATS instruments also measured neutral kinetic temperatures. In addition to these instruments, the Langmuir probe (LANG) [*Krehbiel et al.*, 1981], the Ion Drift Meter (IDM) [*Heelis et al.*, 1981], and the Retarding Potential Analyzer (RPA) [*Hanson et al.*, 1981] measured the ion density, the zonal component of the horizontal ion drift, and the meridional component of the ion drift, respectively. The RPA and LANG also enabled measurements of ion and electron temperatures, respectively. The comprehensive nature of the DE 2 dataset has enabled various studies of thermospheric ion-neutral coupling at high latitudes (see review by *Killeen and Roble* [1988]).

An example of the DE 2 coverage for a single orbital segment at southern high latitudes is shown in Figure 1 (taken from *Killeen et al.* [1984]). This figure shows, as a function of various geophysical parameters, the following measured or derived observables (from top to bottom): (1) the ion drift vector from IDM and RPA; (2) the neutral wind vector from FPI and WATS; (3) the electron, ion, and neutral kinetic temperatures from LANG, RPA, and FPI, respectively; (4) the

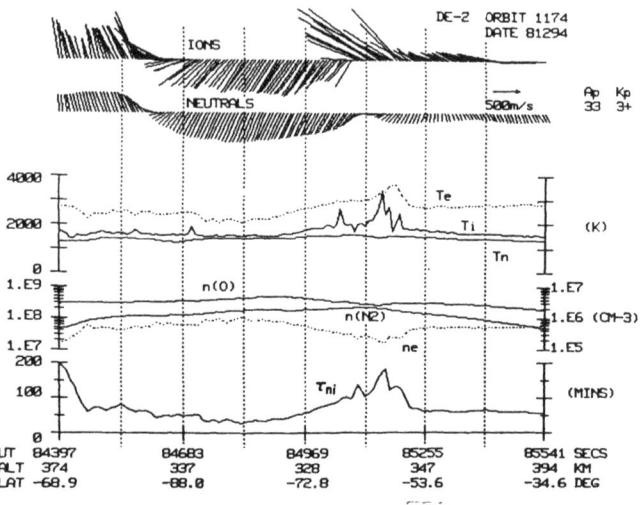

Figure 1—Geophysical observables measured along the track of Dynamics Explorer 2 during orbit 1174. The ion drifts and the neutral winds are shown in the top two traces plotted against time, altitude, and latitude of the spacecraft. The second panel shows the electron, ion, and neutral temperatures measured along the track, and the third panel shows the atomic oxygen and molecular nitrogen densities (left-hand scale) and the electron (ion) density (right-hand scale). The bottom trace shows the ion-neutral momentum transfer time constant evaluated along the track (see text). (From *Killeen et al.* [1984].)

atomic oxygen, molecular nitrogen, and electron number densities from NACS and LANG; and (5) a calculated time constant, τ_{ni}, representing the e-folding time for ion drag to modify the neutral wind in response to an abrupt change in ionospheric convection (see Eq. 3, below). The vectors are plotted such that local noon is to the top of the diagram and local dawn to the right. Thus, the downward-directed ion drift vectors are in the antisunward direction and are associated with the central region of the geomagnetic polar cap, while the two regions of upward-directed ion drift vectors are in the sunward direction and are associated with the dawn and dusk sectors of the auroral zone. An analysis of this and many other passes of DE 2 over the high latitude region near solar maximum (1981–1983) has indicated that the neutral winds generally follow, but lag behind, the pattern of ionospheric convection with some important differences [e.g., *Killeen et al.*, 1982, 1984; *Rees et al.*, 1983, 1985; *Roble et al.*, 1983; *Hays et al.*, 1984]. Thus, in Figure 1, for example, the neutrals and ions have similar velocities (speed and direction) throughout the high-latitude region of the polar dial, except in the vicinity of the dawn convection channel near 62°S latitude (~ 85,040s UT). In the dusk auroral zone (near ~ 78°S latitude, ~ 84,540s UT) and in the central polar cap region (at ~ 84,820s UT) the neutral wind vectors resemble the

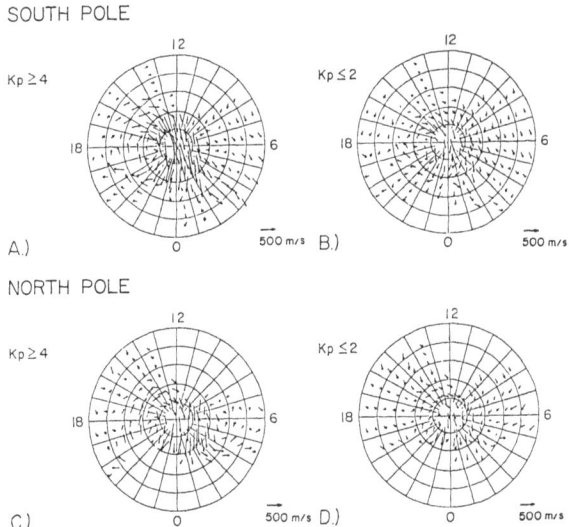

SOUTH POLE

NORTH POLE

Figure 2—Averaged upper thermospheric neutral wind measurements at ~300 km altitude for (a) south pole high *Kp*; (b) south pole low *Kp*; (c) north pole high *Kp*; (d) north pole low *Kp*. Data collected between November 1981 and January 1982 and between November 1982 and January 1983 were averaged according to the given range of *Kp* and plotted in geomagnetic polar coordinates (magnetic latitude and magnetic local time). The outer circle of each polar dial is at 40° geomagnetic latitude. (From *McCormac et al.* [1987].)

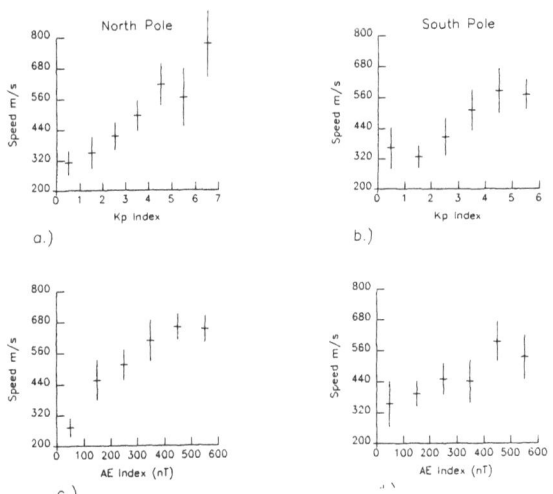

Figure 3—Dependence of the maximum geomagnetic polar cap antisunward neutral wind (~300 km altitude) as a function of the geomagnetic indices *Kp* (top) and *AE* (bottom). North pole observations are given in (a) and (c); South pole observations are given in (b) and (d). (From *McCormac et al.* [1987].)

ion drift vectors closely, demonstrating the effectiveness of the ion-drag force in altering the neutral circulation pattern, away from the simple day-to-night flow that would be expected from considerations of solar insolation only, to one that more closely mimics the twin-cell pattern of ion convection.

The FPI and WATS measurements of the two components of the upper thermospheric horizontal neutral wind vector have been binned and averaged according to various geophysical indices in order to investigate the mean response of the thermospheric neutral wind to forcings associated with the aurora. These DE studies of high-latitude neutral circulation have demonstrated the important role played by the high-latitude sources of momentum (ion drag) and energy (Joule heating) in the establishment of the neutral wind system, at least for the solar maximum conditions pertaining to the DE mission period. A primary conclusion is that above 60° latitude, the neutral wind pattern in the upper thermosphere is best ordered in a geomagnetic coordinate system [*Killeen et al.*, 1983, 1988; *Hays et al.*, 1984].

McCormac et al. [1987] averaged neutral wind vector information from many orbits of DE 2 over a six-month period in geomagnetic coordinates to investigate quantitatively the dependence of the circulation at high latitudes on geomagnetic activity. Figures 2 and 3 show

results from their study. In Figure 2, the average neutral wind pattern for all available southern hemisphere polar passes for (a) active and (b) quiet geomagnetic conditions are compared with the respective northern hemisphere data (c and d). The *Kp* geomagnetic index was used to order these data. In all four cases, the mean neutral circulation shows the imprint of momentum transferred from the twin-cell ionospheric convection pattern, with strong antisunward winds over the geomagnetic polar cap bounded by strong sunward winds in the dusk auroral sector and much weaker sunward or antisunward winds in the dawn auroral sector. The magnitudes of the neutral winds are generally greater for the high *Kp* case than for the low *Kp* case as would be expected from the stronger auroral forcings. Figure 3 shows the measured statistical dependence of the maximum value of the mean antisunward wind in the central polar cap as a function of the indices *Kp* and *AE*. There is a clear positive correlation between the magnitudes of the maximum mean wind speed and these indices, implying that geomagnetic forcing is directly responsible for the enhancement in neutral wind speed. The values shown in Tables 1 and 2 are least-squares fits to the specified regression relationships among the various averaged high-latitude neutral wind features and the geomagnetic indices.

Thayer et al. [1987] investigated the dependency of the high-latitude neutral circulation pattern on the east-west (B_y) component of the interplanetary magnetic field (IMF). They used average neutral wind measurements from the same data base as the *McCormac et*

69

TABLE 1. Relationship between the neutral winds in the polar region and *Kp* for the regression line v(m s^{-1}) = *a* + *b* × *Kp*.

Direction	a	b	r (measured)	Correlation Coefficient (95% confidence interval)
Antisunward				
Northern	261 (±87)	68.7 (±30)	0.97	0.80 ≤ R ≤ 0.99
Southern	269 (±58)	101.6 (±30)	0.94	0.54 ≤ R ≤ 0.99
Dusk sunward				
Northern	158 (±74)	45.3 (±21)	0.98	0.82 ≤ R ≤ 0.99
Southern	113 (±77)	30.4 (±19)	0.94	0.54 ≤ R ≤ 0.99
Dawn sunward				
Northern	68 (±64)	21.0 (±27)	0.91	0.14 ≤ R ≤ 0.99
Antisunward width*	24 (± 4)	3.2 (±1.3)	0.98	0.82 ≤ R ≤ 0.99

*The distance between the sunward/antisunward reversal boundaries of the neutral wind in the polar cap, expressed in degrees latitude.

TABLE 2. Relationship between the neutral winds in the polar region and *AE* for the regression line v(m s^{-1}) = *a* + *b* × *AE*.

Direction	a	b	r (measured)	Correlation Coefficient (95% confidence interval)
Antisunward				
Northern	261 (± 71)	0.82 (±0.20)	0.96	0.67 ≤ R ≤ 0.99
Southern	330 (±112)	0.45 (±0.37)	0.91	0.38 ≤ R ≤ 0.99
Dusk sunward				
Northern	129 (± 58)	0.62 (±0.20)	0.94	0.54 ≤ R ≤ 0.99
Southern	153 (± 78)	0.30 (±0.26)	0.92	0.42 ≤ R ≤ 0.99
Dawn sunward				
Northern	43 (± 47)	0.38 (±0.21)	0.98	0.72 ≤ R ≤ 0.99

al. [1987] study, discussed above, to illustrate changes in the configuration of the neutral wind pattern that were clearly ordered by the sign of B_y. Simultaneous measurements of B_y from the ISEE 3 spacecraft were used to select from and order the DE 2 neutral wind data. For this study, only data for which the north-south (B_z) component of the IMF was ≤ +1 nT were used to avoid complications arising from the multicellular ion and neutral patterns known to exist for strongly northward IMF. Results from this study are shown in Figure 4, where data from many hundreds of DE 2 orbital passes have been separated according to the sign of B_y and averaged into bins of geomagnetic latitude and geomagnetic local time for the two hemispheres.

The criteria used to separate the data sets according to B_y involved the preexistence of a definite positive or negative value for B_y as measured by ISEE 3 for 1 hour prior to the DE orbital pass. Passes occurring either during very high or very low levels of geomagnetic activity were excluded from the study and allowance was made for the propagation time to the Earth of the IMF measured at ISEE 3 altitudes. As can be seen from Figure 4, there are significant differences in the mean thermospheric circulation patterns for B_y positive and negative. To some extent the effects are mirrored between hemispheres; that is, a B_y positive (negative) signature in the northern hemisphere resembles a B_y negative (positive) signature in the southern hemisphere. In all cases, the average circula-

B_y- DEPENDENCE OF THE AVERAGE THERMOSPHERIC NEUTRAL CIRCULATION

(IN GEOMAGNETIC COORDINATES)

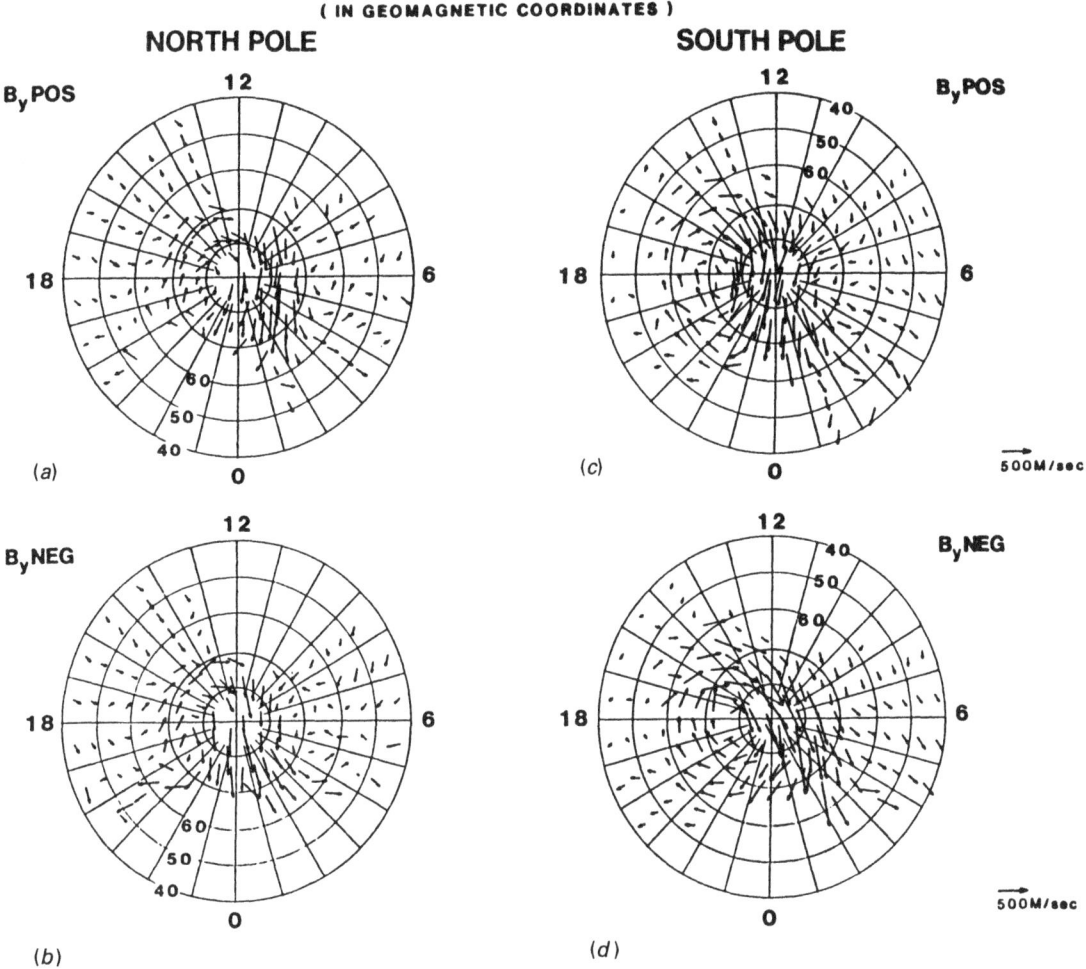

Figure 4—Averaged upper thermospheric wind measurements for (a) north pole B_y positive; (b) north pole B_y negative; (c) south pole B_y positive; (d) south pole B_y negative. Data collected during the same period as for Figure 2 were averaged according to the sign of the B_y component of the IMF and plotted in geomagnetic polar coordinates (magnetic latitude and local time). The outer circle of each polar dial is at 40° geomagnetic latitude. (From *Thayer et al.* [1987].)

tion patterns shown illustrate the dominance of the dusk anticyclonic neutral vortex mentioned above. The spatial magnitude of this vortex, however, is greater for B_y positive conditions in the northern hemisphere and for B_y negative conditions in the southern hemisphere. A distinct rotation in the direction of the polar cap antisunward flow that was dependent on the sign of B_y was also noted.

While the above studies demonstrated quantitative relationships among the various geomagnetic and solar wind parameters (often used to describe auroral conditions) and high-latitude neutral winds, the later study by *Killeen et al.* [1988] unambiguously demonstrated a clear geometric relationship between thermospheric winds and the visible aurora. Figure 5 shows

simultaneous measurements of global-scale auroral luminosity distributions and vector neutral winds over the northern (winter) polar cap, using data from the spin-scan auroral imager (SAI) [*Frank et al.*, 1981] on DE 1 and the FPI and WATS instruments on DE 2, respectively. The purpose of this work was to illustrate the spatial relationship between large-scale morphological features of the F-region neutral wind field in the winter polar region and the location and spatial extent of the aurora. A definite correlation was found to exist between reversals and boundaries in the neutral wind field and the location of the visible auroral oval. The dusk sector reversal to sunward flow in the neutral gas was found to be more pronounced than that occurring in the dawn sector. Examples of such simultaneous

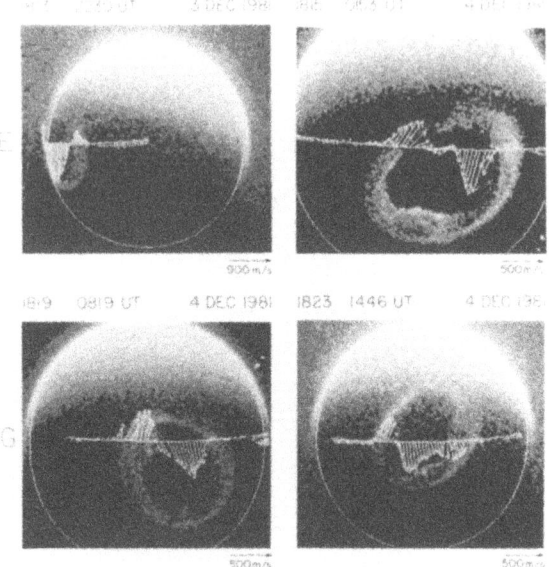

Figure 5—Simultaneously measured neutral wind vectors from DE 2 (orbit 1813) and DE 1 auroral image (courtesy L. A. Frank and J. D. Craven, University of Iowa). The images were obtained using the SAI instrument viewing at ultraviolet wavelengths. The images are oriented such that the direction towards the Sun is to the top of the figure, dusk to the left. The solar terminator is evident, running roughly horizontally across each image, as is the entire auroral oval located just to the nightside of the terminator. The neutral wind vectors are denoted by the arrows whose origins are positioned along the DE 2 orbital track. The wind scale is given at lower right. (From *Killeen et al.* [1988].) (This figure also appears in color: Plate 8.)

datasets indicated that the neutral wind boundaries follow the substorm-dependent expansion and contraction of the auroral oval, consistent with the measured variation of the *Kp*-dependent width of the antisunward neutral wind region given in Table 1. The superposition of vector neutral-wind measurements from DE 2 and auroral imagery from DE 1 demonstrates clearly the power of the DE mission to determine parameters associated with the auroral input of energy, momentum, and charged particles to the upper atmosphere, as well as its dynamical response to such inputs.

3. THE NCAR THERMOSPHERE GENERAL CIRCULATION MODEL

The NCAR-TGCM solves the primitive equations of dynamic meteorology adapted to thermospheric altitudes, including equations for the conservation of energy and momentum and the individual species continuity equations. The physics inherent in the model's parameterizations and input prescriptions are those appropriate to thermospheric altitudes. The basic model and subsequent major developments have been described in detail by *Dickinson et al.* [1981, 1984], *Roble et al.* [1982], *Fesen et al.* [1986] and *Roble and Rid-*

ley [1987]. Most recently, the model was further extended [*Roble et al.*, 1988] to include a self-consistent aeronomic scheme for the thermosphere and ionosphere using a fully Eulerian approach. This new version of the NCAR model is called the Thermosphere-Ionosphere General Circulation Model (TIGCM).

The earlier versions of the TGCM have been used for numerous studies involving ground-based and satellite measurements (see the review of *Killeen and Roble* [1988] for a discussion of work done relative to the DE 2 database). As discussed above, the DE measurements clearly showed a high-latitude thermospheric wind system that was strongly controlled by the momentum source due to magnetospheric convection, with strong antisunward winds in the magnetic polar cap and strong sunward winds equatorward of the polar cap in the auroral zone. *Roble et al.* [1983, 1984] and *Hays et al.* [1984] were able to obtain reasonable agreement between the DE 2 wind measurements and the NCAR-TGCM calculations. In this work, the auroral forcings were calculated within the model by assuming a polar convection electric field distribution as specified by the model of *Heelis et al.* [1982] and an auroral particle distribution as specified by the description of *Roble and Ridley* [1987]. Similar experiment-theory comparisons have also been made with the UCL-TGCM (e.g., *Rees et al.* [1983, 1985, 1986]); results from the UCL model are discussed in greater detail in the accompanying paper by *Rees and Fuller-Rowell* [1989].

Figure 6 shows the results of one of the experimental comparisons made with the NCAR model. In this study, *Killeen et al.* [1986] collated all available F-region neutral wind data from seven ground-based Fabry-Perot interferometers, as well as from several hundred DE 2 orbital passes taken during December 1981, in order to compare the fullest possible experimental description of the mean polar circulation with NCAR-TGCM calculations. Since weather conditions at the ground stations limited the number of occasions when simultaneous observations were possible, the data were averaged by UT to describe the mean diurnal behavior of the thermospheric wind pattern in the northern winter hemisphere. The TGCM was run for a diurnally reproducible case, i.e., with fixed but diurnally modulated auroral forcings. Good agreement was obtained as can be seen by examining the data for three UT bins shown in Figure 6. Both the experimental measurements and the model predictions show the characteristic twin-cell neutral wind pattern driven by ion drag. This pattern is seen to rotate diurnally about the geographic pole in much the same way as the auroral zone rotates about the geographic pole when plotted in a latitude/solar local time polar coordinate system. This type of successful comparison was important since it

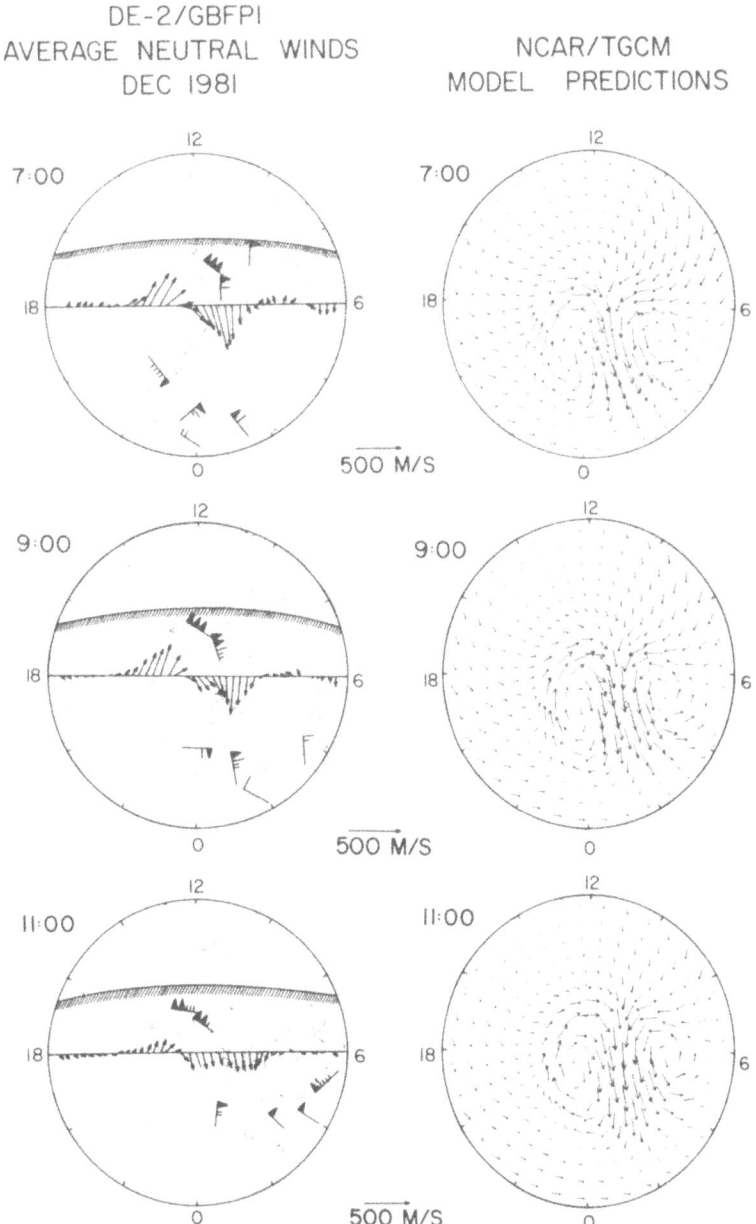

DE-2/GBFPI
AVERAGE NEUTRAL WINDS
DEC 1981

NCAR/TGCM
MODEL PREDICTIONS

Figure 6—Averaged thermospheric wind field for bins 0700, 0900, and 1100 UT (top to bottom). The left-hand side shows the DE 2 and ground-based average measurements plotted in geographic polar coordinates (latitude and local time). The satellite winds are given by the arrows according to the wind scale at lower right. The ground-based Fabry-Perot interferometer wind measurements are plotted as standard meteorological symbols with a barb, 100m s^{-1}; long line, 50 m s^{-1}; short line, 10 m s^{-1}. The curved line is the solar terminator. The left-hand side illustrates the NCAR-TGCM predictions for the midpoint of the particular UT bin. Model winds are plotted according to the same scale as the satellite winds. (From *Killeen et al.* [1986].)

implied that the TGCM could be used with confidence to study quantitatively the forcing processes responsible for the mean circulation and, furthermore, could be used to predict the mean circulation in areas where measurements were not available.

4. AURORAL FORCING MECHANISMS

We now describe briefly the basic governing equations of the neutral thermosphere and discuss the relative importance of the various terms at high latitudes using experimental observations from DE 2 and theoretical calculations from the NCAR TGCM. The equations given here are not the actual equations used within the TGCM, but are simplified forms useful for diagnostic purposes.

Momentum Equation

The equation for the conservation of momentum for an individual thermospheric species, i, can be given in the following form using standard approximations:

$$\rho_i \left[\frac{\partial \mathbf{V}_i}{\partial t} + (\mathbf{V}_i \cdot \nabla) \mathbf{V}_i \right] = \rho_i \mathbf{F}_i - \nabla P_i$$

$$+ \eta \frac{\partial^2 \mathbf{V}_H}{\partial z^2} + 2 \rho_i \mathbf{\Omega} \times \mathbf{V}_i + \rho_i \sum_{j \neq i}$$

$$\left\{ \frac{\mu_{ij} \nu_{ij}}{m_i} (\mathbf{V}_j - \mathbf{V}_i) + m_i (\mathbf{V}_j P_{ij} - \mathbf{V}_i P_{ij})_{\text{chem}} \right\}$$

(1)

where the subscript i refers to the ith species; ρ is the mass density; $\mathbf{V}$ is the velocity; t is time; $\mathbf{\Omega}$ is the angular velocity of the Earth; $\mathbf{F}_i$ refers to all external body forces, including gravity and electrostatic forces; P is the scalar pressure; η is the coefficient of viscosity; $\mathbf{V}_H$ is the horizontal vector velocity; ν_{ij} is the collision frequency of the ith species with the jth species; μ_{ij} is the reduced mass; m_i is the mass of the ith species; P_{ij} is the chemical production rate of the ith species in interaction with the jth species. The left-hand term represents the total time derivative, including the partial time derivative and advective terms. The right-hand terms represent the external body forces, the pressure gradient force, the horizontal force due to vertical viscosity, the Coriolis force, the summation term involving frictional drag and chemistry. When the summation is carried out over all species, the chemical terms cancel and the summation reduces to the familiar ion-drag term. If ion drag is the dominant term and nonlinear advection is neglected, then the momentum equation for the neutral gas reduces to

$$\frac{\partial \mathbf{V}_n}{\partial t} = \frac{n_i}{n_n} \nu_{in} (\mathbf{V}_i - \mathbf{V}_n)$$

(2)

where here the subscripts n and i refer to neutrals and ions, respectively, and n represents number density. This equation may be solved to derive a time constant, τ_{ni}, which represents the e-folding time taken for the velocity of the neutral gas to approach the velocity of the ions following an instantaneous change in the ion drift. τ_{ni} may be written as

$$\tau_{ni} = \frac{(n(\text{N}_2) + n(\text{O}))}{(n_i \, \nu_{ni} (T_n, T_i, n(\text{N}_2), n(\text{O}))}$$

(3)

where $n(\text{N}_2)$ and $n(\text{O})$ are the number densities for the two principal neutral thermospheric constituents at these altitudes, namely, molecular nitrogen and atomic oxygen, and the neutral-ion collision frequency is seen to be a function of ion and neutral temperature and neutral density.

Killeen and Roble [1984] have used the TGCM to investigate the magnitude and direction of the various terms in the above momentum equation at high latitudes. Figure 7 shows some of the results of their study for the southern (summer) hemisphere polar region at F-region altitudes. In this figure (panels *b* through *f*), the various individual terms operating within the TGCM giving rise to the predicted neutral wind shown in panel (*a*) are shown as vectors. The model run was for diurnally reproducible conditions, with moderate and steady geomagnetic conditions. From a careful examination of the directions and magnitudes of these individual forces, it can be seen that the ion-drag force provides the strongest local forcing of the thermosphere, but that the largest ion-drag force vectors are restricted to the dawnside and duskside of the auroral oval/polar-cap boundary, where sunward-drifting ions impart momentum to the neutrals through collisions. *Killeen and Roble* [1982] showed that there was a UT dependence of thermospheric momentum forcing due to the diurnal oscillation of the convection pattern with respect to the solar terminator. At F-region altitudes away from the auroral zone, the basic balance is between ion drag and the pressure gradient force although, at various latitudes and times, significant contributions can be made by the Coriolis, momentum advection and viscous forces. In the auroral zone, however, the neutral flow is strongly controlled by the large ion-drag forces, which can exceed magnitudes of 0.6 m s^{-2} at times. It can be seen in Figure 7 that the pressure gradient force in the polar regions is modified from the basic day-to-night forcing expected from solar heating only. These variations from simple cross-polar-cap pressure-gradient forcing arise due to the strong Joule heating discussed below and the consequent large temperature gradients.

The large ion-drag forces associated with the auroral zone predicted by the TGCM have been confirmed by DE 2 observations. Figure 8 shows the calculated ion-drag force vectors along an orbital pass of the spacecraft [*Killeen et al.*, 1988]. The format shows a composite of neutral winds from FPI and WATS and the simultaneously measured auroral luminosity distribution from SAI. Directly below the composite image are plotted the neutral wind vectors, the zonal ion convec-

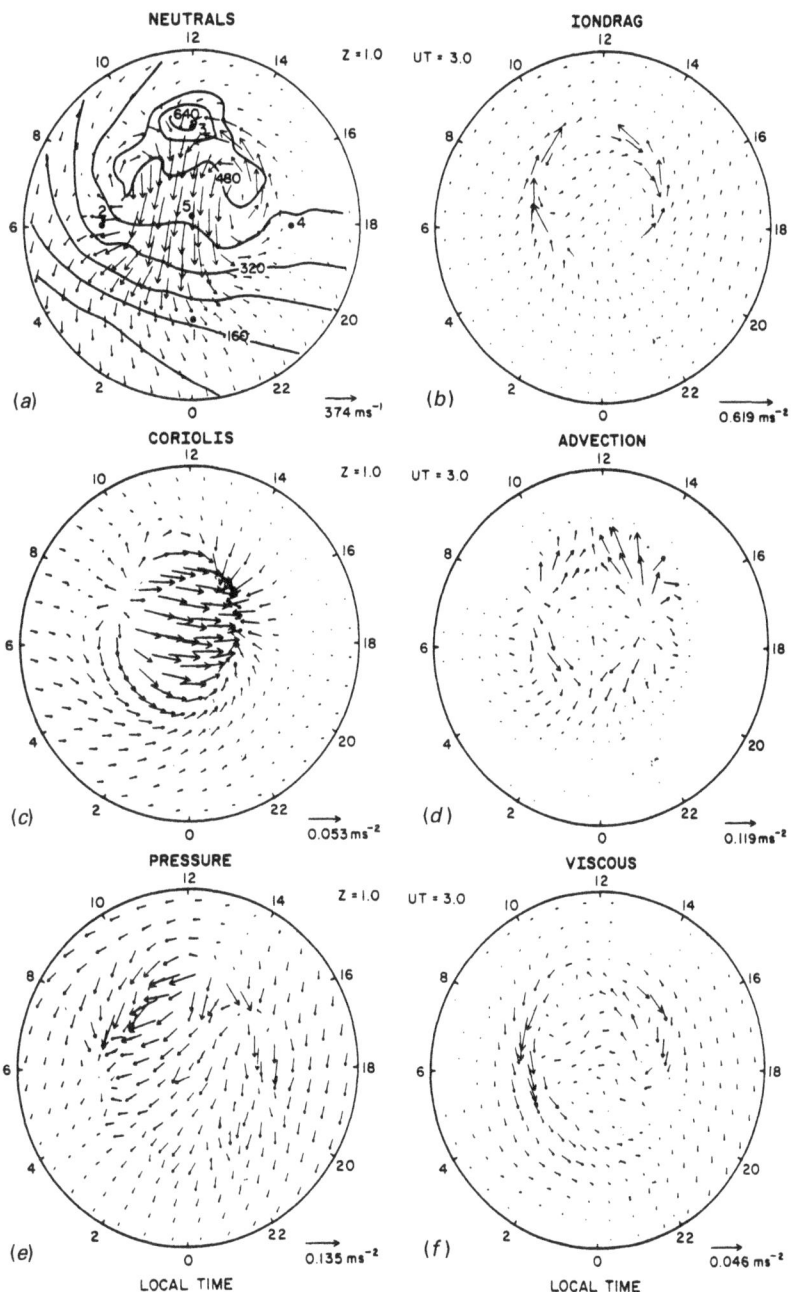

Figure 7—The calculated (a) wind vectors and temperature contours along the $z = +1$ constant pressure surface (~320 km altitude) over the southern hemisphere polar cap at 0300 UT on October 21, 1981 from the NCAR-TGCM. The maximum arrow in (a) represents a wind velocity of 375 m s^{-1}. Vectors of the various forces acting on the neutral wind are given in (b) through (f): (b) the ion drag, (c) Coriolis, (d) advection, (e) pressure, and (f) viscous forces. Note that the length of the maximum arrow represents a different magnitude (m s^{-1}) in each of the figures: (b) 0.62, (c) 0.05, (d) 0.12, (e) 0.14, and (f) 0.05, respectively. (From *Killeen and Roble* [1984].)

tion speeds, the derived Coriolis force vectors of the neutral gas, the ion-drag force vectors, and the ion-neutral momentum-transfer time constant. These parameters are plotted as a function of time and spacecraft coordinates. Note that time increases from right to left in order to match the dawn-to-dusk motion of DE 2 across the polar region of the accompanying auroral image. The polar dial to the left is used to present the ion densities from LANG (solid fill) and, once again, the FPI/WATS neutral wind vectors, for reference purposes.

In this example, the ion-drag force can be seen to be highly variable in the high-latitude region but particularly large in the auroral zone, reaching magnitudes of ~0.4–0.5 ms^{-2}. The magnitudes of the measured force vectors are roughly equivalent to those from the TGCM calculations, although there is much more structure evident in the observations. It can also be seen that the time constant for ion drag to be effective is also highly variable, but is typically on the order of tens of minutes for this high-latitude F-region orbital pass near solar maximum. In general, the theoretical and experimental studies of the thermosphere at auroral latitudes have demonstrated that the neutral-wind configuration is highly dependent on the characteristics of the ion-drag force associated with ionospheric convection.

Energy Equation

The energy equation for the ith neutral thermospheric species is often written as

$$\frac{3}{2} \frac{DP_i}{Dt} + \frac{5}{2} P_i \mathbf{\nabla} \cdot \mathbf{V}_i - \bar{\bar{\tau}} : \nabla \mathbf{V}_i + \mathbf{\nabla} \cdot \mathbf{q}$$

$$+ \rho Q_{\text{solar}} - \rho Q_{\text{rad}} = \sum_j \frac{n_i \mu_{ij} \nu_{ij}}{(m_i + m_j)}$$

$$\left\{ 3k \, (T_j - T_i) + m_j (\mathbf{V}_i - \mathbf{V}_j)^2 \right\} \qquad (4)$$

where the scalar equation is couched in terms of the total derivative of the partial pressure P_i of the ith species; $\mathbf{q}$ is the heat flux vector; Q_{solar} is the heat due to the absorption of solar UV and EUV radiation; Q_{rad} is the heat lost due to radiational cooling; $\bar{\bar{\tau}}$ is the viscous stress tensor and other parameters are as before. The major heating terms include adiabatic heating (included in second term on left-hand side), viscous heating, heat conduction, solar insolation, radiational cooling, and the collision terms on the right-hand side, including heat exchange and frictional heating. For typical F-region conditions, where the neutrals and ions have different velocities, the term within the summation is equivalent to the classical Joule heating

term [*Cole*, 1962; *St. Maurice and Schunk*, 1981], with roughly half the Joule heating coming from the ion-neutral frictional heat term (proportional to the square of the ion-neutral difference velocity) and the other half coming to the neutrals from the frictionally heated ions via the ion-neutral heat exchange term (directly proportional to the ion-neutral temperature difference).

While a full analysis of the individual heating terms within the NCAR-TGCM at high latitudes has not yet been completed, *Roble et al.* [1988] have used the aeronomical scheme first developed by *Roble et al.* [1987] within the new version of the TGCM to calculate successfully the basic global mean thermal structure of the thermosphere for both solar maximum and solar minimum conditions. The major heating and cooling processes that appear to be responsible for the global mean structure appear to have been identified [*Roble et al.*, 1987]. At high latitudes, the predominance of the Joule heating term in establishing the thermal and dynamical structure has been clearly demonstrated and various theoretical studies [*Roble et al.*, 1983; *Rees et al.*, 1985] have provided estimates for the global Joule heating rate in the approximate range ~0.3–1.5 × 10^{11} watts, depending on the level of geomagnetic activity and value for the cross-cap potential used in the calculations. *Emery et al.* [1985] have performed a case study using DE 2 measurements and TGCM calculations and found that the height-integrated Joule heating rate over a typical satellite track in the southern hemisphere magnetic polar cap exceeded both the particle and solar heating rate.

Figure 9 shows results from DE 2 of the calculated Joule heat rate to the neutral gas (per particle) along the track of the spacecraft as it crossed the northern and southern hemisphere polar regions on orbit 7366. The figure is taken from the study by *Killeen et al.* [1989] and shows, from top to bottom, the following measured and derived parameters: (1) the neutral wind vector; (2) the ion drift vector; (3) the electron, ion, and neutral temperatures; (4) the in-situ Joule heating rate per neutral particle; and (5) the measured ratio of molecular nitrogen to atomic oxygen mapped down to a reference altitude of 300 km using a simple diffusive equilibrium approximation. The Joule heating rates are plotted on a per-particle basis to remove, to first order, large changes due to variations in the spacecraft altitude. These data illustrate several characteristic features of thermospheric Joule heating. First, it can be noted that the largest Joule heating rates per particle are generally confined to the high-latitude regions. This is to be expected as the ion-neutral difference velocity maximizes in regions of strong ion convection such as the auroral oval and polar cap. Secondly, the Joule heating rate is seen to be both episodic and highly structured. This fact is due, in part, to the nonlinear (quad-

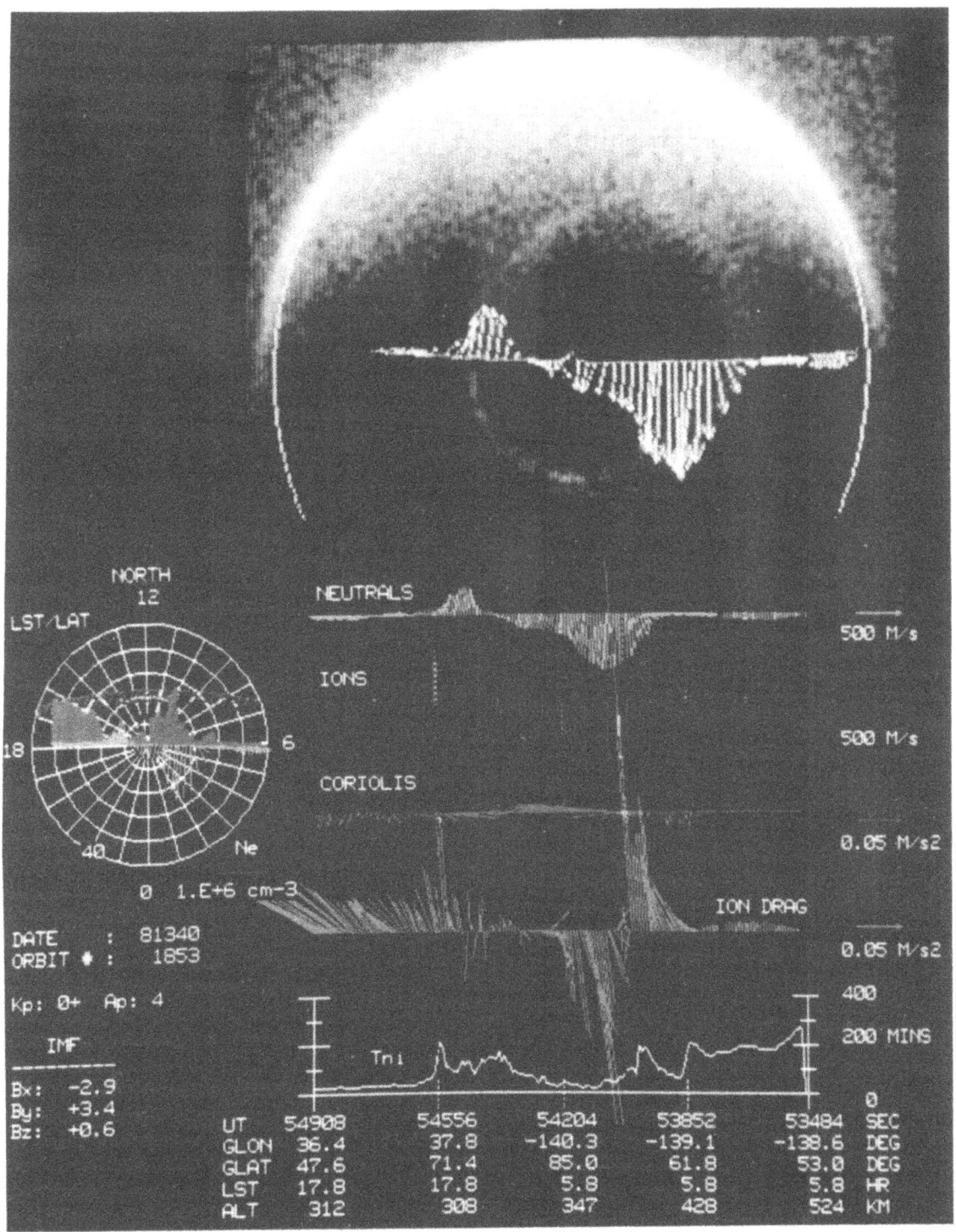

Figure 8—DE 1 auroral image for 1459 UT on December 1981 (top). Measured and derived parameters for orbit 1853 (bottom) are plotted as a function of time along the orbital track of DE 2. These parameters are from top to bottom: the neutral wind vector; the zonal (cross track) component of the ion drift vector; the Coriolis force; the ion-drag force, and the calculated ion-neutral momentum time constant. The scales for the various parameters are to the right of the diagram. The polar dial indicates the track of the spacecraft across the northern polar cap in geographic polar coordinates. The bars on the polar dial represent the neutral wind measurements and the solid fill represents the measured ion densities according to the scale given below the dial. The solar terminator is also shown on the polar dial. (From *Killeen et al.* [1988].) (This figure also appears in color: Plate 9.)

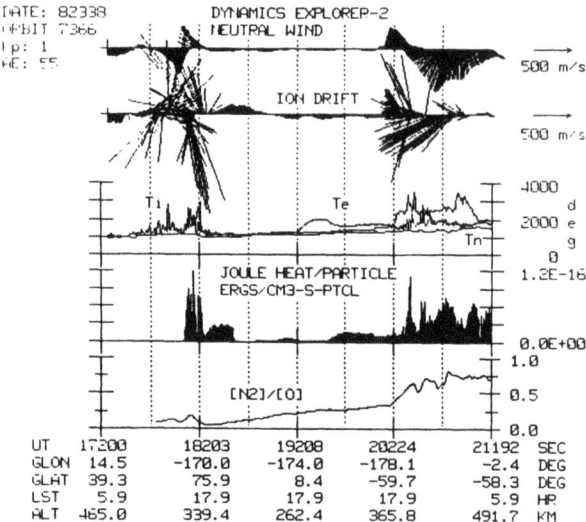

DATE: 82338
ORBIT 7366
Ip: 1
KE: 55

Figure 9—Measured geophysical observables and derived parameters along the track of DE 2 during orbit 7366. The neutral winds and the ion drifts are shown in the top two traces plotted against time, altitude, and latitude of the spacecraft. The second panel shows the electron, ion, and neutral temperatures measured along the track, and the third panel calculated Joule heating rate per neutral particle. The bottom trace shows the molecular nitrogen-to-atomic oxygen ratio referenced to an altitude of 300 km. The ratio was adjusted downward in altitude, using a diffusive equilibrium assumption. (From *Killeen et al.* [1989].)

ratic) nature of dependence of the Joule heat rates on the ion-neutral difference velocity as seen in Eq. 4. The compositional response of the neutral thermosphere at F-region heights to the high-latitude Joule heating is evident in the variations of the molecular nitrogen-to-atomic oxygen ratio plotted in the bottom panel of Figure 9. This ratio increases in response to the upwelling of the atmosphere (upward vertical winds) induced by the Joule heating. Such increases occur at high latitudes and are superimposed upon the general seasonal increase from the winter (northern) hemisphere to the summer (southern) hemisphere (left-to right on the diagram).

Continuity Equation for Individual Species

The single species continuity equation for the major neutral thermospheric species at F-region altitudes may be written

$$\frac{\partial n_i}{\partial t} + \mathbf{U}_n \cdot \nabla n_i = P_i - L_i - \nabla \cdot \mathbf{F}_i$$
$$- n_i \nabla \cdot \mathbf{U}_n \qquad (5)$$

where $\mathbf{U}_n$ is the mass-averaged bulk velocity of the neutral gas, P_i and L_i are chemical production and

loss terms, and $\mathbf{F}_i$ is the total (molecular plus eddy) diffusion velocity of the ith species. The second term on the left-hand side of the equation is the advection term, which can be broken down into vertical and horizontal components. For the constituents N_2 and O, at upper thermospheric altitudes in the auroral region, the eddy diffusion and chemical terms are negligible and the number densities are controlled by the three major terms: molecular diffusion, vertical advection, and horizontal advection.

Recently, *Burns et al.* [1989] performed a term analysis of the single-species continuity equations solved within a diurnally reproducible model run of the NCAR-TGCM for atomic oxygen and molecular nitrogen. They used a version of the continuity equation that describes compositional variations in terms of mass mixing ratios rather than number densities as in Eq. 5. The version of the NCAR model used did not incorporate the self-consistent scheme for ionosphere/thermosphere interactions contained within the TIGCM. Sample results from their study of compositional forcing at high southern latitudes are given in Figure 10, which shows individual compositional forcing terms for molecular nitrogen at ~320 km altitude over the southern hemisphere polar cap. The horizontal advection, vertical advection, and molecular diffusion terms, in units of s^{-1}, are shown in Figures 10a, b and c, respectively, while Figure 10d illustrates the instantaneous total rate of change of the mass mixing ratio for N_2, given essentially by the sum of these three terms. Note that the contour labels have to be multiplied by the factor 10^{-7}, shown to the right of each figure, in order to be converted to the correct units. For comparison with the forcing terms, the calculated instantaneous mass mixing ratio of N_2 is given in Figure 10f and the neutral horizontal wind (arrows, with scale at lower right in m s^{-1}) and vertical wind (contours, with units of 10^{-6} s^{-1}) calculated by the TGCM at the same time are shown in Figure 10e. The vertical "winds" plotted here are not winds in the normal sense, but have been normalized to the local scale height in order to simplify the relationship with vertical advection on a constant-pressure surface. The units are therefore s^{-1} (see *Burns et al.* [1989] for a full discussion).

The pattern of compositional forcing that is seen at high latitudes results from both horizontal and vertical advection, where the horizontal and vertical winds are due primarily to ion drag and Joule heating, respectively. From Figure 10, it can be inferred that the high-latitude thermospheric wind and composition pattern has a strong UT dependence, governed by changes in the relative position of the geomagnetic pole with respect to the geographic pole. For the 0300 UT shown, the geomagnetic pole, and consequently the entire

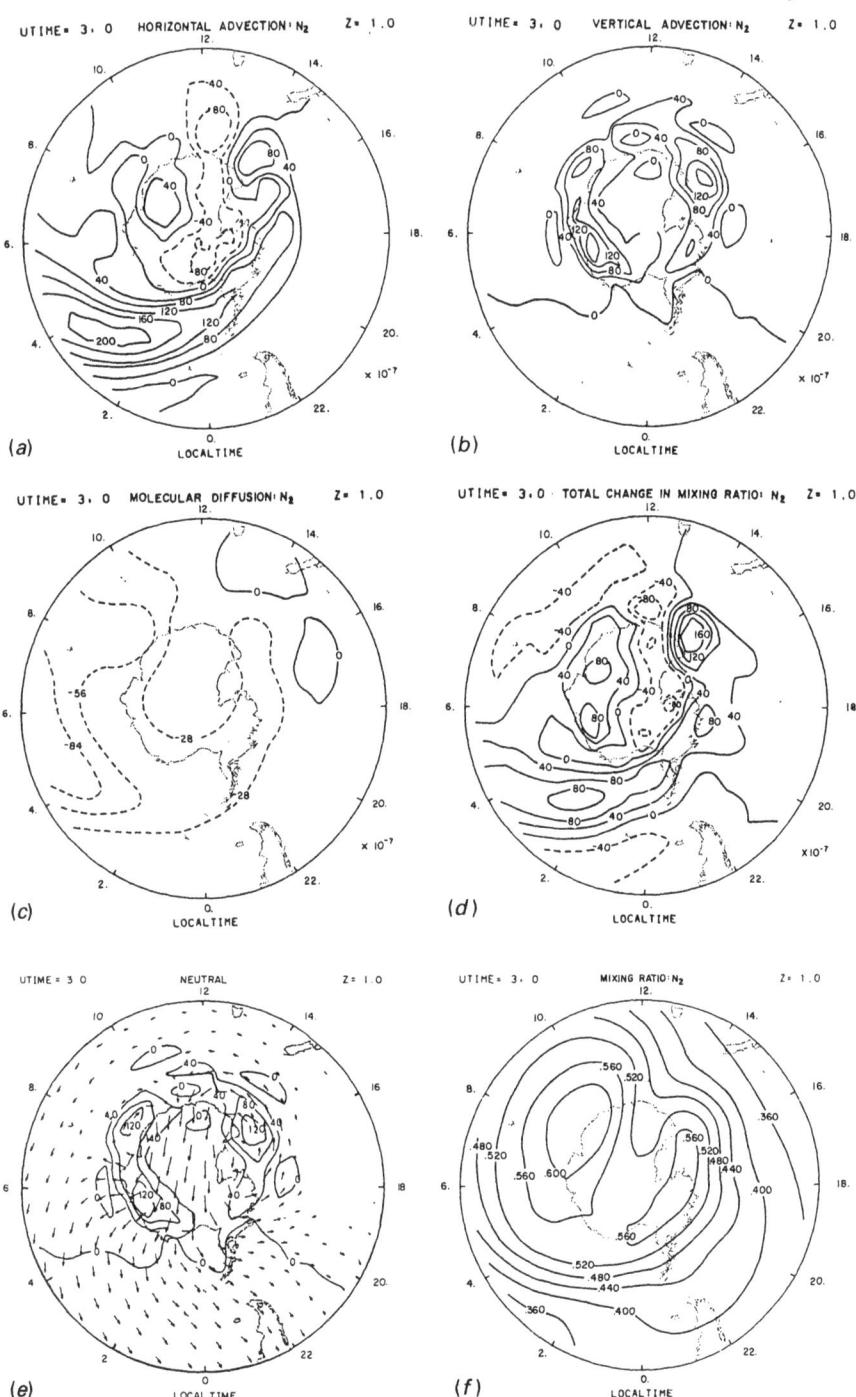

Figure 10—Global polar plots (latitude and local solar time) of the major compositional forcing terms controlling the mixing ratio of N_2 at the z = 1 (~300 km) pressure level at 0300 UT. The outer circle is at −40° latitude and the symbol "S" refers to the position of the geomagnetic pole. The units used for the forcing terms are s^{-1} and the terms are horizontal advection (a), vertical advection (b), and molecular diffusion (c), respectively. The total rate of change in mixing ratio (in units of s^{-1}), which is given by the summation of the three terms shown in (a), (b), and (c), is shown in (d). Note that the contour labels for these first four plots should be multiplied by the factor 10^{-7} as noted to the right of each figure. The TGCM calculated global horizontal wind field ($m\ s^{-1}$) superimposed on vertical "wind" contours (units: $10^{-6}\ s^{-1}$, see text) is shown in (e), and the calculated N_2 mixing ratio is shown in (f). (From *Burns et al.* [1989].)

auroral oval, is offset from the geographic pole toward the noon sector. Joule and particle heating in the auroral zone cause upward vertical winds, especially in the morning and evening sectors (Figure 10e), leading to large increases in the N_2 mass mixing ratio (Ψ_{N2}), that result from this upward advection of nitrogen-rich air (Figure 10b). Vertical advection is much weaker in the auroral oval near midnight and noon and in the central region of the geomagnetic polar cap. Values of Ψ_{N2} are consequently smaller over the magnetic pole than they are in the morning and evening sectors of the auroral oval (Figure 10f). Molecular diffusion (Figure 10c) provides a negative forcing term over much of the summer polar region as it attempts to restore the diffusive equilibrium that has been disturbed by the (generally positive) effects of the vertical and, in particular, the horizontal advection terms (Figures 10b and 10a, respectively).

In general, we may conclude from this study that the thermospheric neutral composition at auroral latitudes is most strongly influenced by dynamical perturbations (vertical and horizontal winds) associated with regions of heating and momentum forcing. While these general results are not new [e.g., *Hays et al.*, 1973; *Mayr et al.*, 1978, 1985; *Roble et al.*, 1984; *Rees et al.*, 1985], the term analysis capability provides a novel quantitative basis for the discussion of individual forcing processes.

5. CONCLUDING REMARKS

Recent experimental results from the Dynamics Explorer-2 spacecraft and theoretical calculations from the NCAR-TGCM have illustrated the manner by which auroral processes control the behavior of the high-latitude thermosphere. The neutral winds at high altitudes typically tend to follow the twin-cell pattern of ionospheric convection, as momentum is transferred from the ions to the neutrals at a rate that is dependent on the ambient concentrations of the two species. As a consequence of this tight ion-neutral momentum coupling, the neutral-wind pattern displays strong dependencies on the level of geomagnetic activity and on the orientation of the IMF, much in the same way as does the ionospheric convection pattern itself. Thermospheric temperatures are directly controlled by the Joule and particle heat sources associated with auroral latitudes. The auroral heating also serves to modulate the thermospheric circulation through the generation of strong pressure-gradient forces and through local upwelling of the atmosphere. Changes in the vertical and horizontal winds at high latitudes lead, in turn, to neutral compositional perturbations. The various dynamical, energetic and compositional variations associated with auroral processes are thus highly coupled with each other.

Significant progress has been made recently in the identification and quantification of the important terms within the governing equations for the thermosphere through combined use of data from Dynamics Explorer 2 and calculations of the numerical models of the thermosphere. Future planned work using the new version of the NCAR-TIGCM, which incorporates a self-consistent thermospheric and ionospheric aeronomic scheme, will lead to additional insights into the coupling among the magnetosphere, ionosphere, and thermosphere in the auroral regions.

ACKNOWLEDGMENT—This work was supported by the NASA grants NAG5-465 and NAG5-482 and by NSF grant ATM-8610085 to the University of Michigan.

REFERENCES

Burns, A. G., T. L. Killeen, and R. G. Roble, "Processes Responsible for the Compositional Structure of the Thermosphere," *J. Geophys. Res.*, 94, 3670–3686 (1989).

Carignan, G. R., B. P. Block, J. C. Maurer, A. E. Hedin, C. A. Reber, and N. W. Spencer, "The Neutral Mass Spectrometer on Dynamics Explorer," *Space Sci. Instrum.*, 5, 493 (1981).

Cole, K. D., "Joule Heating of the Upper Atmosphere," *Aust. J. Phys.*, 15, 223–235 (1962).

Dickinson, R. E., E. C. Ridley, and R. G. Roble, "A Three-Dimensional, Time-Dependent General Circulation Model of the Thermosphere," *J. Geophys. Res.*, 86, 1499–1512 (1981).

Dickinson, R. E., E. C. Ridley, and R. G. Roble, "Thermospheric General Circulation With Coupled Dynamics and Composition," *J. Atmos. Sci.*, 41, 205–219 (1984).

Emery, B. A., R. G. Roble, E. C. Ridley, T. L. Killeen, M. H. Rees, J. D. Winningham, G. R. Carignan, P. B. Hays, R. A. Heelis, W. B. Hanson, N. W. Spencer, L. H. Brace, and M. Sugiura, "Thermospheric and Ionospheric Structure of the Southern Hemisphere Polar Cap on October 21, 1981, as Determined from Dynamics Explorer 2 Satellite Data," *J. Geophys. Res.*, 90, 6553–6566 (1985).

Fesen, C. S., R. E. Dickinson, and R. G. Roble, "Simulation of Thermospheric Tides at Equinox with the NCAR Thermospheric General Circulation model," *J. Geophys. Res.*, 91, 4471–4489 (1986).

Frank, L. A., J. D. Craven, K. L. Ackerson, M. R. English, R. H. Eather, and R. L. Carovillano, "Global Auroral Imaging Instrumentation for the Dynamics Explorer Mission," *Space Sci. Instrum.*, 5, 369–393 (1981).

Fuller-Rowell, T. J., and D. Rees, "A Three-Dimensional, Time-Dependent Global Model of the Thermosphere," *J. Atmos. Sci.*, 37, 2545–2657 (1980).

Fuller-Rowell, T. J., S. Quegan, D. Rees, R. J. Moffett, and G. J. Bailey, "Interactions Between Neutral Thermospheric Composition and the Polar Ionosphere using a Coupled Global Model," *J. Geophys. Res.*, 92, 7744–7748 (1987).

Hanson, W. B., R. A. Heelis, R. A. Power, C. R. Lippincott, D. R. Zuccaro, B. J. Holt, L. H. Harmon, and S. Sanatani, "The Retarding Potential Analyzer for Dynamics Explorer-B," *Space Sci. Instrum.*, 5, 503–510 (1981).

Hays, P. B., R. A. Jones, and M. H. Rees, "Auroral heating and the Composition of the Neutral Atmosphere, *Planet. Space Sci.*, 21, 559–573 (1973).

Hays, P. B., T. L. Killeen, and B. C. Kennedy, "The Fabry-Perot Interferometer on Dynamics Explorer," *Space Sci. Instrum.*, 5, 395–416 (1981).

Hays, P. B., T. L. Killeen, N. W. Spencer, L. E. Wharton, R. G. Roble, B. E. Emery, T. J. Fuller-Rowell, D. Rees, L. A. Frank, and J. D. Craven, "Observations of the Dynamics of the Polar Thermosphere," *J. Geophys. Res.*, 89, 5597–5612 (1984).

Heelis, R. A., W. B. Hanson, C. R. Lippincott, D. R. Zuccaro, L. H. Harmon, B. J. Holt, J. E. Doherty, and R. A. Power, "The Ion Drift Meter for Dynamics Explorer-B," *Space Sci. Instrum.*, 5, 511–521 (1981).

Heelis, R. A., J. K. Lowell, and R. W. Spiro, "A Model of the High-Latitude Ionospheric Convection Pattern," *J. Geophys. Res.*, 87, 6339–6345 (1982).

Killeen, T. L., "Energetics and Dynamics of the Earth's Thermosphere," *Rev. Geophys.*, 25, 433–454 (1987).

Killeen, T. L., and R. G. Roble, "An Analysis of the High Thermospheric Wind Pattern Calculated by a Thermospheric General Circulation Model, 1, Momentum forcing," *J. Geophys. Res.*, **89**, 7509–7522 (1984).

Killeen, T. L., and R. G. Roble, "Thermosphere Dynamics: Contributions From the First 5 Years of the Dynamics Explorer Program," *Rev. Geophys.*, **26**, 329–367 (1988).

Killeen, T. L., P. B. Hays, N. W. Spencer, and L. E. Wharton, "Neutral Winds in the Polar Thermosphere as Measured from Dynamics Explorer," *Geophys. Res. Lett.*, **9**, 957–960 (1982).

Killeen, T. L., P. B. Hays, N. W. Spencer, and L. E. Wharton, "Neutral Winds in the Polar Thermosphere as Measured from Dynamics Explorer," *Adv. Space Res.*, **2**(10) 133–136 (1983).

Killeen, T. L., P. B. Hays, G. R. Carnignan, R. A. Heelis, W. B. Hanson, N. W. Spencer, and L. H. Brace, "Ion-Neutral Coupling in the High Latitude F-region: Evaluation of Ion Heating Terms from Dynamics Explorer 2, *J. Geophys. Res.*, **89**, 7495–7508 (1984).

Killeen, T. L., R. W. Smith, N. W. Spencer, J. W. Meriwether, D. Rees, G. Hernandez, P. B. Hays, L. L. Cogger, D. P. Sipler, M. A. Biondi, and C. A. Tepley, "Mean Neutral Circulation in the Winter Polar F-Region," *J. Geophys. Res.*, **91**, 1633–1649 (1986).

Killeen, T. L., J. D. Craven, L. A. Frank, J. J. Ponthieu, N. W. Spencer, R. A. Heelis, L. H. Brace, R. G. Roble, P. B. Hays, and G. R. Carignan, "On the Relationship Between the Dynamics of the Polar Thermosphere and the Morphology of the Aurora: Global-Scale Observations from Dynamics Explorers 1 and 2," *J. Geophys. Res.*, **93**, 2675–2692 (1988).

Killeen, T. L., R. A. Heelis, L. H. Brace, G. R. Carignan, and N. W. Spencer, "In Situ Joule Heating Rates from Dynamics Explorer 2," *J. Geophys. Res.*, manuscript in preparation (1989).

Krehbiel, J. P., L. H. Brace, R. F. Theis, W. H. Pinkus, and R. B. Kaplan, "The Dynamics Explorer Langmuir Probe Instrument," *Space Sci. Instrum.*, **5**, 493–502 (1981).

Mayr, H. G., I. Harris, and N. W. Spencer, "Some Properties of Atmospheric Dynamics," *Rev. Geophys.*, **16**, 539–565 (1978).

Mayr, H. G., I. Harris, F. Varosi, F. A. Herrero, H. Volland, N. W. Spencer, A. E. Hedin, R. E. Hartle, H. A. Taylor, Jr., L. E. Wharton, and G. R. Carignan, "On the Structure and Dynamics of the Thermosphere," *Adv. Space Res.*, **5**, 283–288 (1985).

McCormac, F. G., T. L. Killeen, J. P. Thayer, C. R. Tschan, G. Hernandez, C. R. Tschan, J. J. Ponthieu, and N. W. Spencer, "Circulation of the Polar Thermosphere During Geomagnetically Quiet and Active Times as Observed from DE 2," *J. Geophys. Res.*, **92**, 10133–10139 (1987).

Meriwether, J. W. Jr., "Observations of Thermospheric Dynamics at High Latitudes from Ground and Space," *Radio Sci.*, **18**, 1035–1052 (1983).

Rees, D., T. J. Fuller-Rowell, R. Gordon, T. L. Killeen, P. B. Hays, L. E. Wharton, and N. W. Spencer, "A Comparison of Wind Observations of the Upper Thermosphere from the Dynamics Explorer Satellite with the Predictions of a Global Time-Dependent Model," *Planet. Space Sci.*, **31**, 1299–1314 (1983).

Rees, D., R. Gordon, T. J. Fuller-Rowell, M. Smith, G. R. Carignan, T. L. Killeen, P. B. Hays, and N. W. Spencer, "The Composition, Structure, Temperature and Dynamics of the Upper Thermosphere in the Polar Regions During October to December, 1981, *Planet. Space Sci.*, **33**, 617–666 (1985).

Rees, D., T. J. Fuller-Rowell, R. Gordon, J. P. Heppner, N. C. Maynard, N. W. Spencer, L. E. Wharton, P. B. Hays, and T. L. Killeen, "A Theoretical and Empirical Study of the Response of the High-Latitude Thermosphere to the Sense of the RYS Component of the Interplanetary Magnetic Field," *Planet. Space Sci.*, **34**, 1–40 (1986).

Roble, R. G., "Dynamics of the Earth's Thermosphere," *Rev. Geophys. Space Phys.*, **21**, 217–233 (1983).

Roble, R. G., and E. C. Ridley, "An Auroral Model for the NCAR Thermospheric General Circulation Model (TGCM)," *Ann. Geophys.*, **5A**, 369–382 (1987).

Roble, R. G., R. E. Dickinson, and E. C. Ridley, "Global Circulation and Temperature Structure of the Thermosphere with High-Latitude Plasma Convection," *J. Geophys. Res.*, **87**, 1599–1614 (1982).

Roble, R. G., R. E. Dickinson, E. C. Ridley, B. A. Emery, P. B. Hays, T. L. Killeen, and N. W. Spencer, "The High Latitude Circulation and Temperature Structure of the Thermosphere Near Solstice," *Planet. Space Sci.*, **31**, 1479–1499 (1983).

Roble, R. G., B. A. Emery, R. E. Dickinson, E. C. Ridley, T. L. Killeen, P. B. Hays, G. R. Carignan, and N. W. Spencer, "Thermospheric Circulation, Temperature and Compositional Structure of the Southern Hemisphere Polar Cap During October-November, 1981," *J. Geophys. Res.*, **89**, 9057–9068 (1984).

Roble, R. G., E. C. Ridley, and R. E. Dickinson, "On the Global Mean Structure of the Thermosphere," *J. Geophys. Res.*, **92**, 8745–8758 (1987).

Roble, R. G., E. C. Ridley, A. D. Richmond, and R. E. Dickinson, "A Coupled Thermosphere/Ionosphere General Circulation Model," *Geophys. Res. Lett.*, **15**, 1325–1328 (1988).

Spencer, N. W., L. E. Wharton, H. B. Niemann, A. E. Hedin, G. R. Carignan, and J. C. Maurer, "The Dynamics Explorer Wind and Temperature Spectrometer," *Space Sci. Instrum.*, **5**, 417–428 (1981).

St. Maurice, J.-P., and R. W. Schunk, "Ion-Neutral Momentum Coupling Near Discrete High-Latitude Ionospheric Features," *J. Geophys. Res.*, **86**, 11299–11321 (1981).

Thayer, J. P., T. L. Killeen, F. G. McCormac, C. R. Tschan, J.-J. Ponthieu, and N. W. Spencer, "Interplanetary Magnetic Field-Dependent Thermospheric Neutral Wind Signatures for Northern and Southern Hemispheres from Dynamics Explorer-2 Data," *Ann. Geophys.*, **5A**, 363–368 (1987).

III. AURORAL PARTICLES AND ACCELERATION MECHANISMS

III-1. OVERVIEW OF ELECTRON AND ION PRECIPITATION IN THE AURORAL OVAL

P. T. Newell,* C.-I. Meng,* and D. A. Hardy[†]

A review is given of statistical results on the regions of electron and ion precipitation within the auroral oval. Generally the precipitating number flux of electrons exceeds that of the ions by a factor of about 50, while the electron energy flux exceeds the ion energy flux by only a factor of 6–9. Progress has been made in understanding the largest-scale auroral morphology, as the tour of the auroral oval in the final section makes clear. The morphology of the equatorward boundary of precipitation is now fairly well determined; for example, it has been established that this boundary as defined by number flux is well fitted by an offset circle in magnetic local time and latitude coordinates. The poleward boundary has been less thoroughly investigated, perhaps partly due to its often irregular configuration. Less is known about dynamics: the equatorward expansion of the auroral oval during substorm activity is well documented; but the quantitative effects of the various factors (such as the interplanetary magnetic field (IMF), substorm activity, and the strength of the ring current) have not been disentangled.

1. INTRODUCTION

Three decades have passed since the first direct measurements by rocket-borne detectors of the precipitating particles responsible for the aurora [*Meridith et al.*, 1958; *McIlwain*, 1960; *Davis et al.*, 1960]. Naturally, over so much time, considerable advances have been made in particle detector instrumentation (particularly for ions), and in the ability to gather, process, and organize large amounts of data. Our purpose here is to give an overview of the global morphology of electron and ion precipitation into the auroral oval in an average sense. Therefore, we concentrate heavily on statistical studies, but original references are also given for many historical "firsts."

More general reviews (including imaging results) have been given recently by *Gorney* [1987] and *Feldstein and Galperin* [1985], while earlier auroral work was reviewed by *Meng* [1978] and *Eather* [1973]. Other papers in this volume review the physical mechanisms involved in particle precipitation. Thus mechanisms of auroral (electron) acceleration are discussed by Burch and by Bryant and Hall; the characteristics of the associated field-aligned electric fields are reviewed by Block and Fälthammar; auroral ion acceleration and composition is considered by Shelley and Collin; while Bosqued considers the effects of transport on ion precipitation.

Recently there has been considerable interest in times of very quiet magnetic activity at the auroral and middle latitudes, which also tend to be times of strong polar cap activity. These phenomena are addressed by *Lundin* [this volume]. Thus, although the very-high-latitude precipitation may represent the expansion of the quiet time oval into the polar cap [*Meng*, 1981], we will not discuss it here. Similarly statistical work on other polar cap phenomena such as polar rain are not discussed.

Here we discuss the global maps in which the average charged particle precipitation is plotted against magnetic latitude (MLAT) and magnetic local time (MLT), review work in the phenomenology and dynamics of the statistical auroral boundaries, consider work on the promising but comparatively undeveloped field of statistical studies that involve pattern recognition, and finally summarize auroral precipitation as we follow the oval around in MLT.

Roughly speaking, most auroral precipitations can be divided into two broad categories. The "hard zone" is characterized by relatively structureless precipitation, with a typical electron energy of a few kiloelectron volts. At slightly higher latitudes there is a "soft zone," with lower average energies, generally below 600 eV, in which there is embedded usually bursty structures and/or inverted-Vs with higher energies. Early reports of a soft zone/hard zone distinction were made by *Johnson et al.* [1966] and *Burch* [1968]; inverted-Vs are defined and discussed in a later section. The hard zone is generally believed to map to the central plasma sheet, and the soft zone to either the plasma sheet boundary layer (on the nightside and extending past the dawn-dusk meridian) or the low-latitude boundary layer (within a few hours of noon). Cusp precipitation, which is fairly closely confined near noon, and has still lower average energies and very high number fluxes representing almost direct magnetosheath particle entry, is not considered here. The relationship between auroral

---

*The Johns Hopkins University Applied Physics Laboratory, Laurel, MD 20707.
[†]Air Force Geophysical Laboratory, Hanscom Air Force Base, MA 01731.

Auroral Physics, edited by C.-I. Meng, M. J. Rycroft and L. A. Frank. © Cambridge UP 1991

precipitation zones and magnetospheric configuration is given careful review by *Galperin and Feldstein* [this volume].

2. GLOBAL MAPS OF AURORAL PRECIPITATION BINNED BY MLAT AND MLT

Low-altitude polar orbiting satellites cut rapidly across polar region field lines, with the observed precipitation thus affording a projection screen view of the distant magnetosphere. A technique for creating synoptic global precipitation maps that has been useful on account of its simplicity takes a grid of MLAT by MLT bins within which the precipitation is averaged. Often there is an additional binning parameter, usually the Kp index of geomagnetic activity, but sometimes the auroral electrojet index AE or interplanetary conditions are used. *Reidler* [1972] used ESRO IA data to produce gridded electron (1.3-keV) and ion (5.8-keV) maps at a fixed energy. *McDiarmid et al.* [1975] used about 1100 Isis 2 passes to create a 1° by 2 hour grid of electron precipitation, for several Kp ranges. The energy range covered was primarily from 150 eV–10 keV. *McDiarmid et al.* [1975] sought to apply the results to the question of the local time extent of the entry of magnetosheathlike particles into the polar regions; they felt a very wide region of entry was indicated.

Spiro et al. [1982] used observations from the AE-C and AE-D satellites to make similar mappings, and used the results to estimate Pedersen and Hall conductances. *Candidi et al.* [1983] produced contours of dayside soft electron fluxes with a similar binning. Such maps have also been produced, using data from the NOAA satellites [e.g., *Foster et al.*, 1986]. The most comprehensive maps, both in terms of energy range covered and sample size (as well as what are apparently the first multi-energy ion maps), have been produced by the U.S. Air Force Geophysics Laboratory (AFGL) group using observations from the Defense Meteorological Satellite Program (DMSP) satellites [*Hardy et al.*, 1985, 1989]. We now concentrate on those results.

Figures 1 (*a-d*) and Figures 2 (*a-d*) depict the synoptic spatial distribution of precipitating electron and ion energy fluxes for $Kp = 2$ and $Kp \geq 6$, respectively. These global maps make it clear that the auroral oval is well named, at least in the statistical sense. The offset of the center of the oval from the geomagnetic pole is also clear. Comparison of the left and right halves of Figures 1 and 2 illustrates the dramatic nature of the nightside precipitation increase during disturbed periods.

There are many useful results that can be derived from such statistical maps. We consider here Figures

3 (*a,b*) and 4 (*a,b*), which give the average electron and ion energy (a) and number fluxes (b) (integrated over the globe, over all volumes of MLAT) as a function of MLT and Kp. A second integration (over MLT) gives the best available values for the total energy input to the upper atmosphere from precipitating particles. As expected, with increasing Kp, the hemispheric energy flux associated with precipitation increases; for electrons it rises from 8.1×10^9 W for $Kp = 1$ to 7.8×10^{10} W for $Kp \geq 6-$. Most of the number flux is carried by low-energy electrons (below 660 eV) and most of the energy flux by high-energy electrons, except for $Kp \leq 1$.

Multi-energy global ion precipitation maps such as shown in Figure 2 have only appeared recently [*Hardy et al.*, 1989]. One result of a precise confirmation of earlier measurements of auroral particles (e.g., *McIlwain* [1960]), namely that the number flux of precipitating electrons, far exceeds that of ions. The ion number flux into the ionosphere is on the average, only 2.4% to 1.5% (for $Kp = 0$ and $Kp \geq 6-$, respectively) that of the electron number flux. Because the ions tend to have a higher average energy than the electrons, the global ion energy flux is about 1/6 to 1/9 that of the electron energy flux (the hemispheric ion energy flux is 1.4×10^9 W and 1.0×10^{10} W for $Kp = 2$ and $Kp \geq 6-$, respectively). At times, on the equatorward side of the evening-sector auroral oval, the ion energy flux can actually exceed the electron energy flux. This is also the sector where the ion average energy is greatest, sometimes apparently exceeding 30 keV.

Figures 3 and 4 show that the nightside energy flux due to precipitating particles of both species is higher than that on the dayside for all levels of magnetic activity. However, for low values of Kp, the number flux on the dayside can be as large as, or larger than, the nightside values. The ion number flux on the dayside shows the peak associated with the cusp at noon more clearly than does the electron number flux. It should be emphasized, however, that there are not separate "electron" and "ion" cusps; in individual passes, the regions of electron and ion cusp precipitation generally coincide.

Although the global statistical precipitation maps based on the MLAT/MLT binning technique are quite valuable, they also have severe limitations. It is not only fine structure such as is associated with discrete arcs that are blurred out by this technique. The auroral boundaries are highly dynamic, and can vary considerably within a 3-hour Kp interval, depending on substorm activity. Other factors such as seasonal effects can be quite strong, as both theoretical [*Choe et al.*, 1973] and empirical [*Mead and Fairfield*, 1975] magnetic field models indicate. Finer binning over a larger

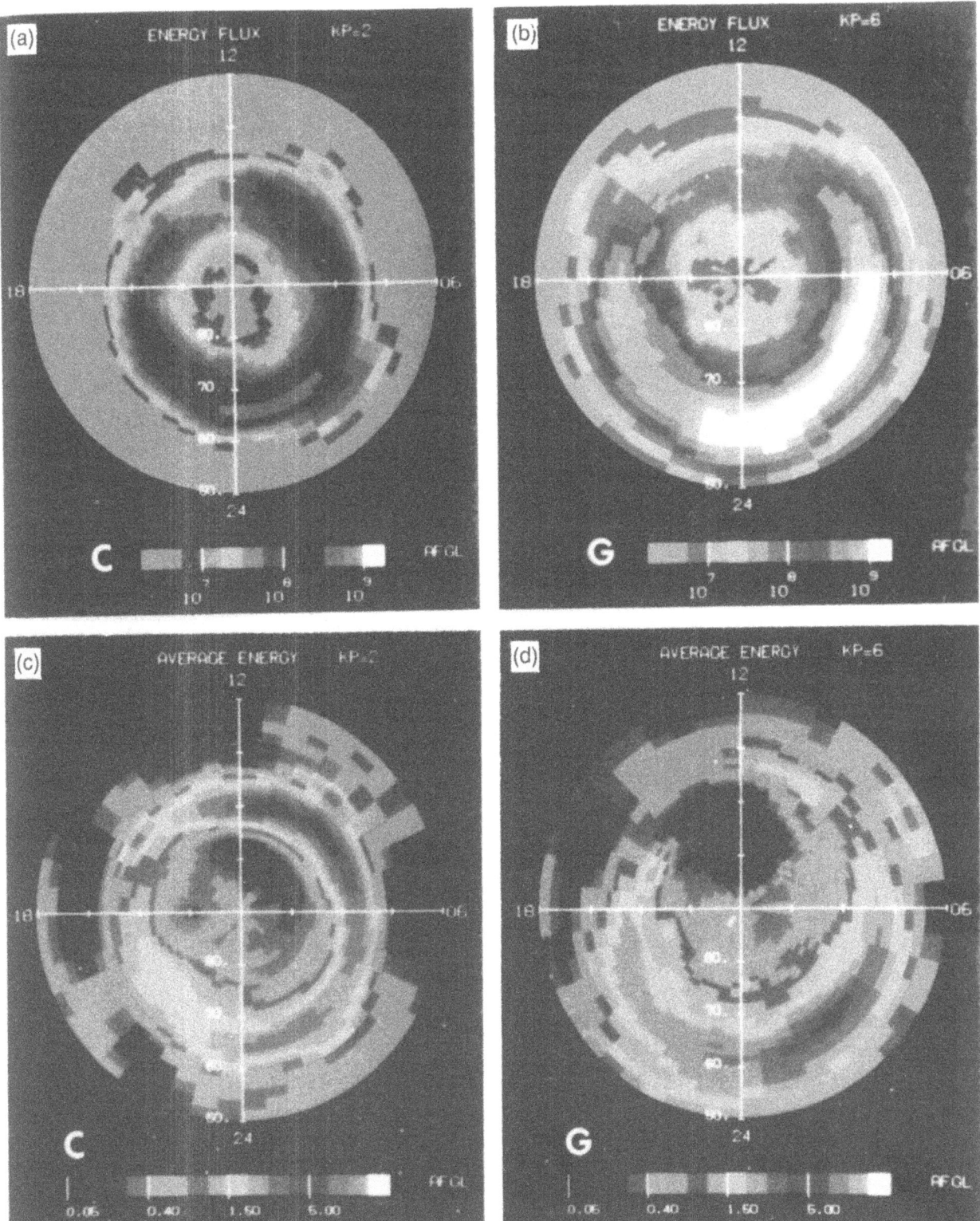

Figure 1—Polar projection map, from 50° magnetic latitude to the magnetic pole, for electron precipitations under *Kp* = 2 (left side) and *Kp* = 6 (right side) conditions of energy flux (top) and average energy (bottom). (From Hardy et al. [1985; 1989]. (This figure also appears in color: Plate 10.)

87

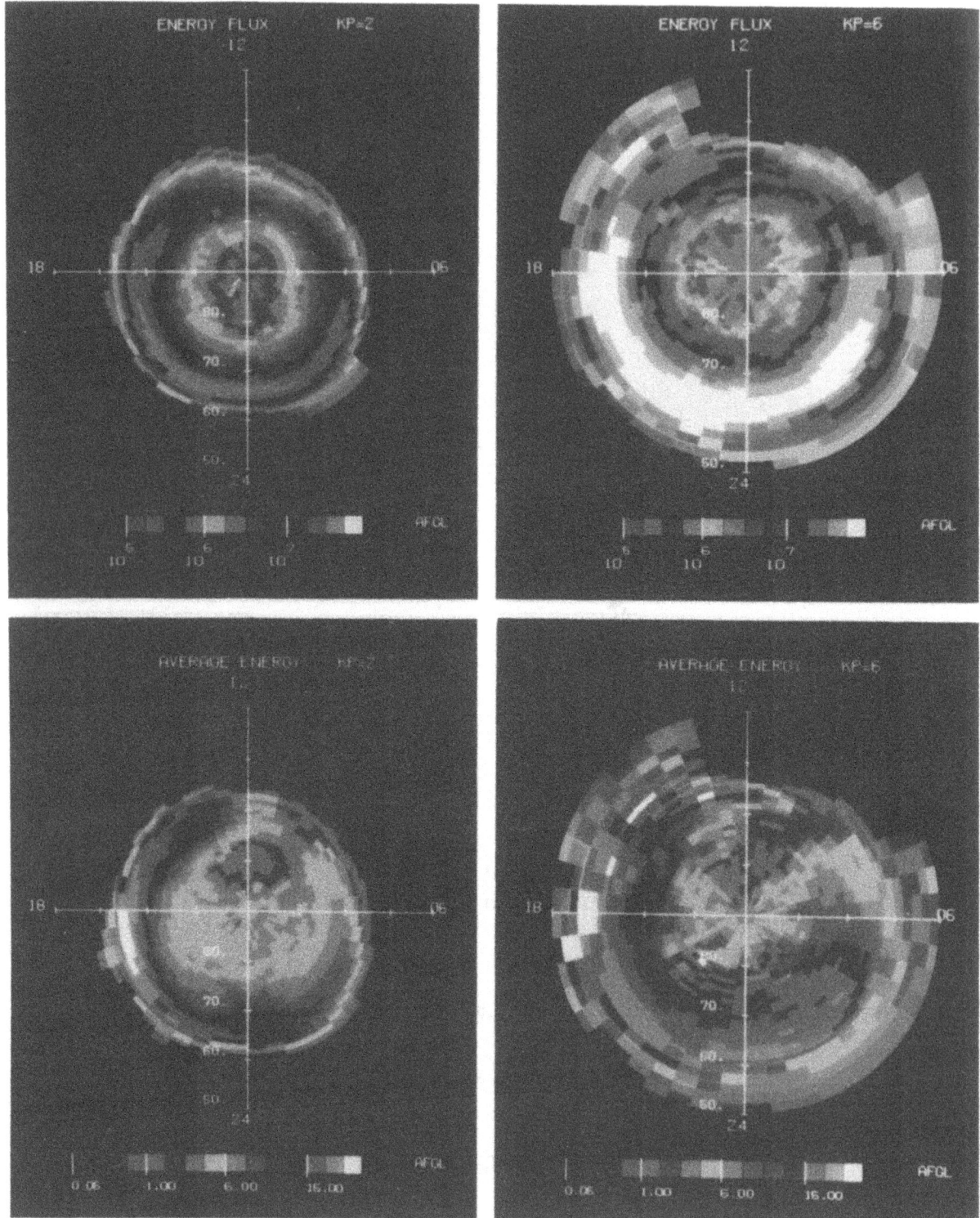

Figure 2—Same as Figure 1, except for ion precipitations. (This figure also appears in color: Plate 11.)

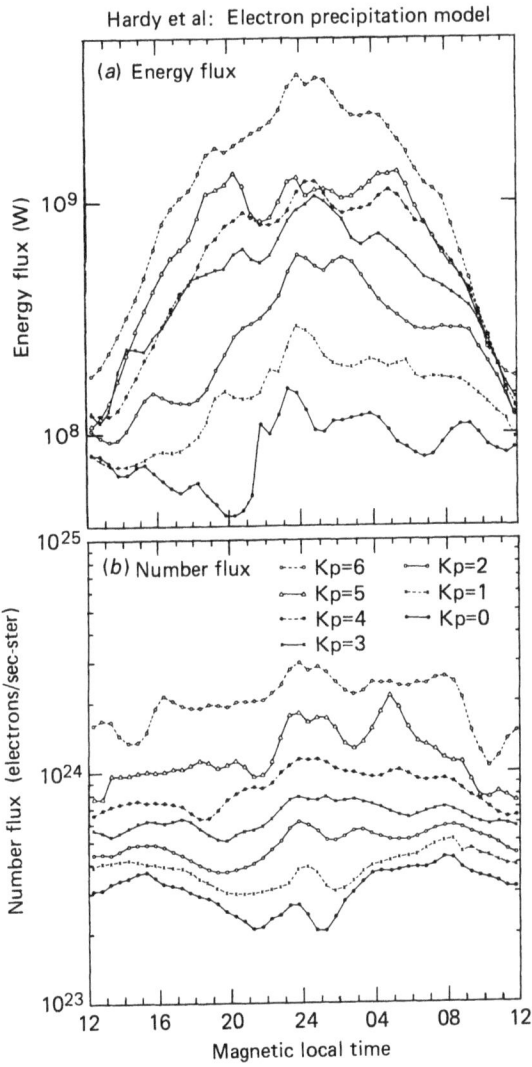

Figure 3—Distribution with magnetic local time (MLT) of the global precipitating electron (*a*) energy flux, and (*b*) number flux, as a function of MLT, for different *Kp* values. (From *Hardy et al.* [1985].)

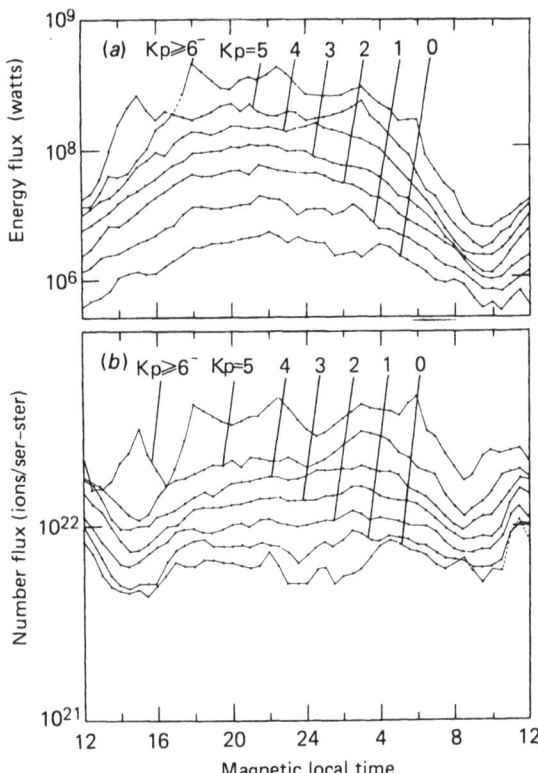

Figure 4—Distribution with magnetic local time (MLT) of the global precipitating ion (*a*) energy flux, and (*b*) number flux as a function of MLT, for different *Kp* values. (From *Hardy et al.* [1989].)

3. MORPHOLOGY AND DYNAMICS OF THE AURORAL BOUNDARIES

We are concerned here with statistical studies of the phenomenology of the auroral precipitation boundaries. Studies based on imaging data lie outside the scope of the present review, although there has been some statistical research on the morphology and dynamics of the auroral oval. The most work in the field of particle precipitation has been done on the equatorward boundaries, and far more has been done on electron than on ion boundaries. *Kamide and Winningham* [1977] used 351 passes of the Isis 1 and 2 satellites to investigate the equatorward boundary of the diffuse auroral electron precipitation as a function of the Interplanetary Magnetic Field (IMF) and substorm activity in the 2000–0400 MLT sector. They reported that the IMF predicted the boundary position better than substorm activity did, although both seemed to have an influence. The correlation coefficient in various local time sectors ranged from 0.42–0.71 (all with fairly small sample sizes). Examining their plots, it appears that the correlation is much better for southward than

number of factors improves resolution, but there are severe limitations on how far even the largest datasets can be subdivided in this way. In the next section, we discuss a promising but hitherto little utilized technique in which the individual regions are first identified, and the average properties are calculated subsequently. One special class of this "pattern recognition" technique has long been employed; namely, the identification and phenomenology of the boundaries of the auroral oval. One can learn, for example, very little about the dynamics of the auroral boundaries from the synoptic maps of the average precipitation.

northward B_z; however, Kamide and Winningham felt that the equatorward boundary did depend upon the magnitude of a northward B_z. They did not report correlation coefficients of the electron precipitation boundary positions with magnetic activity indices.

In a series of articles, *Gussenhoven et al.* [1981, 1983, 1987] have conducted a massive study of the position of the equatorward boundary, as defined by the total number flux of auroral oval precipitation obtained using DMSP data. They studied the precipitating electron number flux boundaries in the dawn and dusk sectors [*Gussenhoven et al.*, 1981] and in the near noon and midnight sectors [*Gussenhoven et al.*, 1983]. With the advent of DMSP ion detectors, the work was extended to include the ion number flux boundaries in the dawn and dusk sectors [*Gussenhoven et al.*, 1987]. Combining the results from the different local time bins, they found that for all activity levels the electron number flux equatorward boundary was well fitted by a circle that was offset (toward midnight) from the geomagnetic poles, a result that is in agreement with the earlier optical observations of *Holzworth and Meng* [1975]. They found very high correlation coefficients between the boundary positions and *Kp*, ranging from -0.82 in the evening sector to -0.35 near noon. The poorer correlation near noon is not unexpected; indeed, there is often no diffuse auroral precipitation in that sector. However, the excellent correlations with *Kp* reported in the nightside sector cast some doubt on the claim of *Kamide and Winningham* [1977] that IMF B_z correlates best with the equatorward boundary of diffuse precipitation. *Gussenhoven et al.* [1987] reported that the ion boundary was equatorward of the electron boundary in the evening sector by an average of 1.4°, while in the morning sector the electron boundary reached further equatorward, by 2.6° on average.

There has been comparatively little work done on variations of the positions of the poleward precipitation boundaries. *Makita et al.* [1983] investigated the poleward, equatorward, and transition boundaries in the DMSP electron data for the dawn and dusk sectors. The transition boundary is the usually sharp boundary between the equatorward, hard, central plasma-sheet-like precipitation, and the poleward softer boundary layer type precipitation (which contains most of the discrete auroral structures). Generally, the boundary positions correlated with the IMF much better for IMF B_z southward than for B_z northward, particularly the equatorward boundary. All the correlation coefficients (typically ~ 0.5) were lower than the results of *Kamide and Winningham* [1977]. *Makita et al.* [1983] found that as B_z turned from northward to increasingly southward, both the equatorward and poleward boundaries shifted equatorward, but that the

poleward boundary shifted further, so that the region of auroral oval precipitation was thinner for southward than for northward B_z. Although this would appear at first to contradict the well-known observation that the optically observable auroral oval is latitudinally much wider in active times, it should be pointed out that the poleward boundary of *Makita et al.* [1983] was the poleward boundary of soft, nearly subvisual, "drizzle," which during very quiet times can fill much of the high-latitude region [*Meng*, 1981; *Lassen et al.*, 1988; *Lundin et al.*, this volume].

Recently attention has turned to the energy-dependent structure of the auroral precipitation cutoffs. Assuming that the equatorward edge of diffuse precipitation maps in the equatorial plane to the inner edge of the plasma sheet, a definite energy and species dependence in the latitude of the precipitation boundaries is predicted by the steady-state convection model by a cross-tail electric field from the distant tail. It is predicted [*Kennel*, 1969] that lower energy electrons will convect further earthward than higher energy electrons in the equatorial plane, which is in fact observed [*Vasyliunas*, 1968; *Schield and Frank*, 1970; *Fairfield and Vinas*, 1984], and therefore lower energy electrons precipitate equatorward of higher energy electrons [e.g., *Kamide and Winningham*, 1977; *Horwitz et al.*, 1986]. *Horwitz et al.* [1986] studied the 100 eV, 1000 eV, and 10,000 eV electron equatorward boundaries observed by DE 2, as well as the equatorial earthward plasma sheet cutoffs at the same energies from DE 1. They found good correspondence between the high-altitude equatorial and the low-altitude precipitation boundaries, with the lower energies reaching further earthward (equatorward) in each region. The dynamics of the precipitation boundaries also matched well with those of the equatorial boundaries during substorm activity.

Newell and Meng [1987a; 1988a] have studied the latitudinal dispersion curves of the equatorward cutoffs as a function of energy for ion as well as for electron precipitation. Figure 5 illustrates an example of such a dispersion curve observed during a polar region pass by DMSP F7 on December 20, 1983. In the near midnight sector of the auroral oval (the right side of the figure), the ions form a C-shaped pattern, in which ions of a few kiloelectron volts cutoff furthest poleward, and lower and higher energy ions extend further equatorward. The zero energy electron and ion precipitation cutoff boundaries coincide, as would be expected from a convection signature. *Newell and Meng* [1988a], using 503 auroral oval crossings by the DMSP F6 and F7 spacecraft, have found that at dusk and near midnight the sense of ion dispersion shown in Figure 5 was by far the most common, accounting for about half

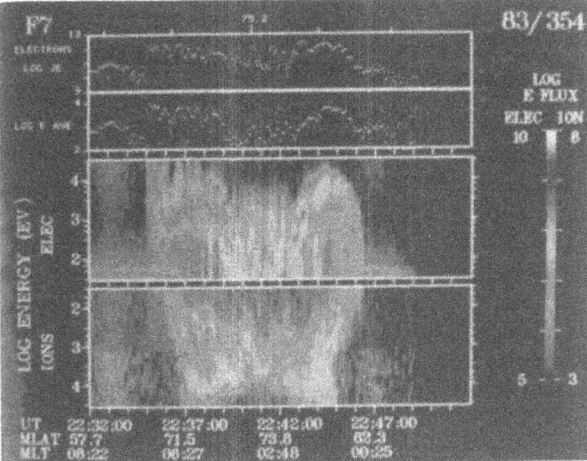

Figure 5—DMSP F7 measurements of precipitating electrons and ions for 20 min on day 354 of 1983. The top line plot is energy flux (eV/cm² s sr), and the bottom line plot is average energy (eV). The lower two spectrograms for electrons and ions, show differential energy flux values in eV/(cm² s sr eV). Notice the curve formed by the equatorward edge of the diffuse aurora on the night (right hand) side: the ions cutoffs form a "C" shape, which is the type of curve most commonly observed. (This figure also appears in color: Plate 12.)

of all ion dispersion curves. (For electrons the curves are almost always monotonic, with lower energy electrons reaching further equatorward [*Horwitz et al.*, 1986; *Newell and Meng*, 1987*a*]). Shortly (within about 10 min) after the onset of substorm activity, the electron and ion curves are frequently dispersionless, i.e., all energies cut off at the same latitude [*Newell and Meng*, 1987*b*], associated with dispersionless plasma injections near geosynchronous orbit [*McIlwain*, 1974].

The finding that in the evening and midnight sectors the lowest energy electron and ion cutoffs coincide modifies earlier findings that, as defined by a number flux threshold, the ion boundaries extend equatorward of the electron boundaries in the evening sector. The general tendency in the ion dispersion curves for low-energy ions to extend equatorward of ions of a few kiloelectron volts energy may be explained in part by ions of ionospheric origin, as discussed by *Bosqued* [this volume].

The time scales for the expansion and contraction of the auroral oval have been investigated by *Nakai et al.* [1986] using the large dataset of electron number flux equatorward boundaries compiled by *Gussenhoven et al.* [1983]. *Nakai et al.* [1986] found that the time scale for the expansion of the auroral oval following a southward turning of the IMF was about 45 min, while the time scale for the contraction after a northward turning was about 8 hours. The 45-min time scale for expansion reported by *Nakai et al.* [1986] is

quite consistent with time scales calculated earlier by other means [e.g., *Meng et al.*, 1973].

4. STATISTICAL STUDIES INVOLVING RECOGNITION OF SPECIAL FEATURES

Many auroral features and boundaries are latitudinally sharp, and most are highly dynamic. Although binning all precipitation (regardless of the plasma source region) by MLAT and MLT is a powerful tool because of its simplicity, much of the discreteness of the auroral precipitation structure is blurred or lost altogether. An alternative approach is first to identify the features of interest, and then carry out the statistical analysis. For very-large-scale statistical studies this involves some sort of automated pattern recognition algorithm. This extra work has, however, the promise of corresponding scientific benefits. The special features that have been most often identified are the boundaries of precipitation associated with the auroral oval, as the previous section made clear. The usual approach, however, has been to carry out smaller studies, and manually identifying the features of interest. One such feature of considerable interest to auroral researchers is the presence of inverted-V precipitation.

In inverted-V structures the precipitating electron spectrum has a "monoenergetic" peak energy that rises and falls as the satellite transverses the feature, thus forming an inverted-V on an energy-time spectrogram. The term "monoenergetic" is something of a misnomer, since any peak in a plot of differential energy flux versus energy is generally called a monoenergetic component. Monoenergetic electron spectra have been seen from the earliest rocket flights [*McIlwain*, 1960] and their importance was soon emphasized [*Albert*, 1967; *Evans*, 1968]. Although it was recognized that the monoenergetic peak energy changed, apparently with latitude, over the course of a rocket flight through a discrete aurora [*Evans*, 1968]; the full structure of inverted-Vs was not apparent until the first color energy-time spectrograms of auroral precipitation were published [*Frank and Ackerson*, 1971].

Lin and Hoffman [1982] studied 430 inverted-V precipitation events using electron data (over the range 200 eV to 25 keV) from the AE-D satellite. They found that the mean latitudinal width of inverted-Vs in their data was about 0.5°, although because of a cutoff in the width distribution at 0.2° associated with the identification technique, the inverted-Vs may actually be a significantly narrower average. Occasionally, inverted-Vs can have latitudinal widths of several degrees.

Lin and Hoffman [1982] found that the maximum probability of observing inverted-V events was in the 2000–0200 MLT sector (although above 80° MLAT the probability was independent of MLT). The lowest latitude at which they observed an inverted-V was 62° MLAT. In accordance with widely held expectations, they found that most inverted-Vs occurred near the poleward edge of the auroral oval. The electron precipitation associated with inverted-Vs is a mixture of a monoenergetic component and secondary and backscattered electrons; *Burch* [this volume] discusses the physical processes involved.

Lyons et al. [1988], using 247 auroral zone transversals of the S3-3 satellite, studied the association between inverted-V events (as defined by the presence of a monoenergetic component) and regions of isotropic energetic ion precipitation, which they interpret as containing field lines that thread the tail current sheet. The coincidence of the inverted-Vs and the isotropic ions led them to suggest that discrete arcs are associated with lines that map to the tail current sheet.

There have been several statistical studies of precipitation associated with the polar cusp, primarily based on electron data [e.g., *Burch*, 1973; *Candidi and Meng*, 1984; *Carbary and Meng*, 1986]. For the purposes of this review, we do not consider cusp precipitation to be part of the auroral oval. A distinction can be made between the dayside extension of the auroral oval (the cleft) and the cusp proper [*Heikkila*, 1985; *Newell and Meng*, 1988b]. The cusp proper is the region of fairly direct magnetosheath entry (the cusp, less strictly defined, was first identified by *Heikkila and Winningham* [1971] and *Frank* [1971]); the cleft encompasses both the low-latitude boundary layer (LLBL) and the plasma sheet boundary layer (PSBL), which are topologically connected [*Vasyliunas*, 1979]. The LLBL was first identified on the flanks of the magnetosphere by *Hones et al.*, [1972]. Discrete auroral arcs appear on the dayside, primarily in the boundary layer regions (the cleft).

Newell and Meng [1988b] have developed a simple algorithm based on the spectral properties of the precipitation to distinguish the two regions (in the cusp the energies are closer to energies in the magnetosheath and the number fluxes are higher). They used the algorithm to determine some statistical properties of the cusp (which is not of direct interest here) and the cleft (which, as the location of dayside aurora, is of interest). The cleft was found to be latitudinally thinnest near noon (about 1.5° MLAT) and to widen rapidly away from noon to 2–3°, with no further widening after about 1 hour away from noon. The ion energy in the boundary layer was found to vary from a spectral peak near magnetosheath values (moderately above 1 keV) near noon to PSBL values (with a spectral peak of several kiloelectron volts) near 0600 MLT. The average flux in the cleft has little dependence on IMF conditions [*Newell and Meng*, 1989]. The electron precipitation in the cleft typically has an average energy of a few hundred electron volts, and tends to be quite variable (spatially "bursty"), apparently associated with discrete auroras.

A special type of precipitation phenomena that appears not to have been studied statistically but is too important to not be mentioned is the precipitation of energetic heavy ions during magnetic storms [*Shelley et al.*, 1972]. The heavy ions, primarily oxygen, tend to be concentrated at the equatorward edge of the oval precipitation; and in disturbed times can exceed proton precipitation at latitudes just below the traditional oval. The discovery by *Shelley et al.* [1972] of large fluxes of precipitating oxygen ions during a major storm was of historical importance in establishing the ionosphere as a major plasma source for the magnetosphere. Related topics such as upflowing ions are discussed by *Shelley and Collin* [this volume].

5. SUMMARY: A TOUR OF THE AURORAL OVAL

After three decades of direct measurements of precipitating particles in the auroral oval, much is known about the large-scale morphology of auroral precipitation. We summarize this knowledge with a tour of the auroral oval. Some readers may be interested in comparing this tour with that given by *Frank and Ackerson* [1972] 17 years ago. Generally, many important features of the electron precipitation patterns as a function of MLT were already known at that date. However, as a comparison of the spectrogram published in this paper with those early spectrograms shows, the quality of the ion detectors has greatly improved; and although knowledge of ion precipitation still lags that of electron precipitation, much progress has been made in this area. As a final cautionary note, there are still basic questions of field-line mapping to the equatorial magnetosphere that have not been fully resolved, as a reading of the article by *Galperin and Feldstein* [this volume] should make clear.

We begin the tour in the evening sector of the auroral oval. Globally, the energy flux associated with electron precipitation is 6–9 times greater than that associated with ion precipitation [*Hardy et al.*, 1989]. Because high-energy ions, convected earthward from the tail, drift westward, the ions are very energetic at the equatorward edge of the evening sector (the average energy can exceed 30 keV), and the ion energy flux can at times exceed the electron energy flux. The ion number flux boundary extends poleward of the electron boundary in the evening sector [*Gussenhoven et al.*, 1987], but zero-energy electron and ion boundaries coincide [*Newell and Meng*, 1988a]. Electrons with energies above a few kiloelectron volts (which drift eastward) are less likely to be present, except during magnetic disturbances. The poleward portion of both the electron and ion precipitation is generally softer than the equatorward portion, a distinction that is generally believed to correspond to the PSBL and the CPS, respectively. Inverted-V events are generally more energetic and have a greater spatial extent in the evening sector than elsewhere [*Frank and Ackerson*, 1972; *Lin and Hoffman*, 1982].

In the midnight sector, the oval precipitation reaches to lower latitudes; on average the precipitation around 0100 extends furthest equatorward [*Gussenhoven et al.*, 1983]. The electrons are a little more energetic and the ions a little less so than in the evening sector. The distinction between the CPS precipitation and the boundary layer precipitation, while still normally resolvable, is less dramatic in the midnight sector than at other local times.

In the morning sector, there is a sharper distinction between the hard precipitation of the diffuse aurora (the continuation of the CPS) and the softer, erratic, poleward boundary layer precipitation associated with most discrete aurorae. In the hard zone, the ion precipitation begins to fade (i.e., the flux levels drop), while the electron precipitation hardens. Electrons below about 1 keV begin to disappear altogether from the hard zone. Inverted-V events in the dawn sector are often of lower energy and have a smaller spatial scale.

Moving from dawn toward noon, the hard zone of precipitation (which is often almost entirely electrons by this point) often begins to fade [*Hardy et al.*, 1985]. Following a southward turning of B_z, the hard zone is usually absent altogether near noon [*Gorney and Evans*, 1987]. In the softer boundary layer region, the ion precipitation begins to soften, and at times precipitation from the low-latitude boundary layer precipitation can be observed instead of (or in addition to) PSBL precipitation. Within about 1.5 hours of noon, the cusp is usually observed [*Newell and Meng*, 1988b], and the boundary layer region, if it is observed at all, is thinnest.

Continuing past noon toward the afternoon sector, there is generally no longer a continuation of the hard zone (CPS) electron precipitation [*Meng and Akasofu*, 1983], although energetic ion precipitation can often be observed. Near 1400 MLT, there are often bursty enhancements of the electron precipitation, usually of comparatively low energy. Around 1700 MLT the precipitation begins to become characteristic of the evening sector, as described previously.

The overall morphology of the equatorward boundary of the auroral oval as defined by particle precipitation is fairly well fitted by an offset circle [*Gussenhoven et al.*, 1983] as had long been supposed [see the review by *Feldstein and Galperin*, 1985]. Work on the poleward portion of the auroral oval, which can be more irregular in shape, is less advanced. At quiet times it appears that boundary-layer-type plasma may fill the oval [*Lundin*, this volume]. The dynamics of the oval are insufficiently investigated. Certainly the oval moves equatorward for southward B_z [*Kamide and Winningham*, 1975]. Apart from the fact that the oval generally expands much more rapidly than it contracts [*Nakai et al.*, 1986], comparatively little is known about the dynamics of the oval. The respective effects of the north-south component of the IMF B_z, substorm activity, and Dst are not yet really disentangled.

The point has been reached where future progress in statistical work lies less in the use of larger datasets and more in the more sophisticated use of datasets. We believe that the most promising means of improving the accuracy of statistical precipitation studies lies in the development of more intelligent pattern recognition algorithms. This will allow more accurate statistical studies to be made of the morphology and dynamics of both individual auroral regions and average auroral precipitation features.

ACKNOWLEDGMENT—The work at APL was supported by the Atmospheric Sciences Division, National Science Foundation grant ATM-8713212, and by the Air Force Office of Scientific Research grant 88-0101.

REFERENCES

Albert, R. D., "Energy and Flux Variations of Nearly Monoenergetic Auroral Electrons," *J. Geophys. Res.*, **72**, 5811-5815 (1967).

Block, L. P., and C.-G. Fälthammar, "Characteristics of Magnetic-Field Aligned Electric Fields in the Auroral Acceleration Region," *this volume.*

Bosqued, J. M., "Ion Precipitation and the Transport of Ions Accelerated by Auroral Processes," *this volume.*

Bryant, D. A., and D. S. Hall, "The Auroral Electron Accelerator: Manifestation of a Universal Ion/Wave/Electron Interaction," *this volume.*

Burch, J. L., "Low-Energy Electrons Fluxes at Latitudes Above the Auroral Zone," *J. Geophys. Res.*, **73**, 3585-3591 (1968).

Burch, J. L., "Rate of Erosion of Dayside Magnetic Flux Based on a Quantitative Study of the Dependence of Polar Cusp Latitude on the Interplanetary Magnetic Field," *Radio Sci.*, **8**, 955-961 (1973).

Burch, J. L., "Overview of the Diagnosis of Auroral Acceleration Mechanisms by Particle Measurements," *this volume.*

Candidi, M., H. W. Kroehl, and C.-I. Meng, "Intensity Distribution of Dayside Polar Soft Electron Precipitation and the IMF," *Planet. Space Sci.*, **31**, 489-498 (1983).

Candidi, M., and C.-I. Meng, "The Relation of the Cusp Precipitating Electron Flux to the Solar Wind and Interplanetary Magnetic Field," *J. Geophys. Res.*, **89**, 9741-9751 (1984).

Carbary, J. F., and C.-I. Meng, "Correlation of Cusp Latitude with B_z and AE(12) Using Nearly One Year's Data," *J. Geophys. Res.*, **91**, 10,047-10,054 (1986).

Choe, J. Y., D. B. Beard, and E. C. Sullivan, "Precise Calculation of the Magnetosphere Surface for a Tilted Dipole," *Planet. Space Sci.*, **21**, 485-498 (1973).

Davis, L. R., O. E. Berg, and L. H. Meredith, "Direct Measurements of Particle Fluxes in and Near Auroras," in *Proc. Cospar Space Sci. Symposium*, North Holland Publishing Company, Amsterdam (1960).

Eather, R. H., "The Auroral Oval—A Reevaluation," *Rev. Geophys. Space. Phys.*, **11**, 155-167 (1973).

Evans, D. S., "The Observations of a Near Monoenergetic Flux of Auroral Electrons," *J. Geophys. Res.*, **73**, 2315-2323 (1968).

Fairfield, D. H., and A. F. Vinas, "The Inner Edge of the Plasma Sheet and the Diffuse Aurora," *J. Geophys. Res.*, **89**, 841-854 (1984).

Feldstein, Ya. I., and Yu. I. Galperin, "The Auroral Luminosity Structure in the High-Latitude Upper Atmosphere: Its Dynamics and Relationship to the Large-Scale Structure of the Earth's Magnetosphere," *Rev. Geophys.*, **23**, 217-275 (1985).

Foster, J. C., J. M. Holt, R. G. Musgrove, and D. S. Evans, "Solar Wind Dependencies of High-Latitude Convection and Precipitation," in *Solar Wind-Magnetosphere Coupling*, Y. Kamide and J. A. Slavin, eds., pp. 477-494 (1986).

Frank, L. A., "Plasma in the Earth's Polar Magnetosphere," *J. Geophys. Res.*, **76**, 5202-5219 (1971).

Frank, L. A., and K. L. Ackerson, "Observations of Charged Particle Precipitation into the Auroral Zone," *J. Geophys. Res.*, **76**, 3612-3643 (1971).

Frank, L. A., and K. L. Ackerson, "Local-Time Survey of Plasma at Low Altitudes Over the Auroral Zones," *J. Geophys. Res.*, **77**, 4116-4127 (1972).

Galperin, Yu. I., and Ya. I. Feldstein, "Auroral Luminosity and Its Relationship to Magnetospheric Plasma Domains," *this volume.*

Gorney, D. J., "U.S. Progress in Auroral Research: 1983-1986," *Rev. Geophys.*, **25**, 555-569 (1987).

Gorney, D. J., and D. S. Evans, "The Low-Latitude Auroral Boundary: Steady-State and Time-Dependent Representations," *J. Geophys. Res.*, **92**, 13,537-13,545 (1987).

Gussenhoven, M. S., D. A. Hardy, and W. J. Burke, "DMSP/F2 Electron Observations of Equatorward Auroral Boundaries and Their Relationship to Magnetospheric Electric Fields," *J. Geophys. Res.*, **86**, 768-778 (1981).

Gussenhoven, M. S., D. A. Hardy, and N. Heinemann, "Systematics of the Equatorward Diffuse Auroral Boundaries," *J. Geophys. Res.*, **88**, 5692-5708 (1983).

Gussenhoven, M. S., D. A. Hardy, and N. Heinemann, "The Equatorward Boundary of Auroral Ion Precipitation," *J. Geophys. Res.*, **92**, 3273-3283 (1987).

Hardy, D. A., M. S. Gussenhoven, and E. Holeman, "A Statistical Model of Auroral Electron Precipitation," *J. Geophys. Res.*, **90**, 4229-4248 (1985).

Hardy, D. A., M. S. Gussenhoven, and D. Brautigam, "A Statistical Model of Auroral Ion Precipitation," *J. Geophys. Res.*, **94**, 370-392 (1989).

Heikkila, W. J., "Definition of the Cusp," in *The Polar Cusp*, J. A. Holtet and A. Egeland, eds., D. Reidel, Hingham, MA, pp. 387-395 (1985).

Heikkila, W. J., and J. D. Winningham, "Penetration of Magnetosheath Plasma to Low Altitudes Through the Dayside Magnetospheric Cusps," *J. Geophys. Res.*, **76**, 883-891 (1971).

Holzworth, R. H., and C.-I. Meng, "Mathematical Representation of the Auroral Oval," *Geophys. Res. Lett.*, **2**, 377-380 (1975).

Hones, Jr., E. W., J. R. Asbridge, S. J. Bame, M. D. Montgomery, S. Singer, and S.-I. Akasofu, "Measurements of Magnetotail Plasma Flow Made with Vela 4B," *J. Geophys. Res.*, **77**, 5503-5522 (1972).

Horwitz, J. L., S. Menteer, J. Turnley, J. L. Burch, J. D. Winningham, C. R. Chappell, J. D. Craven, L. A. Frank, and D. W. Slater, "Plasma Boundaries in the Inner Magnetosphere," *J. Geophys. Res.*, **91**, 8861-8882 (1986).

Johnson, R. G., R. D. Sharp, M. F. Shea, and G. B. Shook, "Satellite Observations of Two Distinct Dayside Zones of Auroral Precipitation," (abstract) *Eos Trans. AGU*, **47**, 64 (1966).

Kamide, Y., and J. D. Winningham, "A Statistical Study of the 'Instantaneous' Nightside Auroral Oval: The Equatorward Boundary of Electron Precipitation as Observed By the Isis 1 and 2 Satellites," *J. Geophys. Res.*, 5573-5588 (1977).

Kennel, C. F., "Consequences of a Magnetospheric Plasma," *Rev. Geophys.*, **7**, 379-419 (1969).

Lassen, K., C. Danielsen, and C.-I. Meng, "Quiet-Time Average Auroral Configuration," *Planet. Space Sci.*, **36**, 791-799 (1988).

Lin, C. S., and R. A. Hoffman, "Observations of Inverted-V Electron Precipitation," *Space Sci. Rev.*, **33**, 415-457 (1982).

Lundin, R., L. Eliasson, and J. S. Murphree, "The Quiet-Time Aurora and the Magnetosphere Configuration," *this volume.*

Lyons, L. R., J. F. Fennell, and A. L. Vampola, "A General Association Between Discrete Auroras and Ion Precipitation from the Tail," *J. Geophys. Res.*, **93**, 12,932-12,940 (1988).

Makita, K., C.-I. Meng, and S.-I. Akasofu, "The Shift of the Auroral Electron Precipitation Boundaries in the Dawn-Dusk Sector in Association with Geomagnetic Activity and Interplanetary Magnetic Field," *J. Geophys. Res.*, **88**, 7967-7981 (1983).

McDiarmid, I. B., J. R. Burrows, and E. E. Budzinski, "Average Characteristics of Magnetospheric Electrons (150 eV to 200 keV) at 1400 km," *J. Geophys. Res.*, **80**, 73-79 (1975).

McIlwain, C. E., "Direct Measurement of Particles Producing Visible Auroras," *J. Geophys. Res.*, **65**, 2727-2747 (1960).

McIlwain, C. E., "Substorm Injection Boundaries," in *Magnetospheric Physics*, B. M. McCormac, ed., D. Reidel, Hingham, MA, pp. 143-154 (1974).

Mead, G. D., and D. H. Fairfield, "A Quantitative Magnetospheric Model Derived from Spacecraft Magnetometer Data," *J. Geophys. Res.*, **80**, 523-534 (1975).

Meng, C.-I., "Electron Precipitation and Polar Auroras," *Space Sci. Rev.*, **22**, 223-300 (1978).

Meng, C.-I., "Polar Caps Arcs and the Plasma Sheet," *Geophys. Res. Lett.*, **8**, 273-276 (1981).

Meng, C.-I., and S.-I. Akasofu, "Electron Precipitation Equatorward of the Midday Oval and the Mantle Aurora," *Planet. Space Sci.*, **31**, 889 (1983).

Meng, C.-I., B. Tsurutani, K. Kawasaki, and S.-I. Akasofu, "Cross-Correlation Analysis of the AE Index and the Interplanetary Magnetic Field B_z Component," *J. Geophys. Res.*, **78**, 617-629 (1973).

Meredith, L. H., L. R. Davis, J. P. Heppner, and O. E. Berg, *IGY Rocket Rep. Ser. 1*, National Academy of Sciences, Washington, DC (1958).

Nakai, H., Y. Kamide, D. A. Hardy, and M. S. Gussenhoven, "Time Scales of Expansion and Contraction of the Auroral Oval," *J. Geophys. Res.*, **91**, 4437-4450 (1986).

Newell, P. T., and C.-I. Meng, "Energy Dependence of the Equatorward Cutoffs in Auroral Electron and Ion Precipitation," *J. Geophys. Res.*, **92**, 7519-7530 (1987a).

Newell, P. T., and C.-I. Meng, "Low Altitude Observations of Dispersionless Substorm Plasma Injections," *J. Geophys. Res.*, **92**, 10,063-10,072 (1987b).

Newell, P. T., and C.-I. Meng, "Categorization of Dispersion Curves in the Equatorward Edge of the Diffuse Aurora," *Planet. Space Sci.*, **36**, 1031-1038 (1988a).

94

Newell, P. T., and C.-I. Meng, "The Cusp and the Cleft/Boundary Layer: Low-Altitude Identification and Statistical Local Time Variation," *J. Geophys. Res.*, **93**, 14,549-14,556 (1988*b*).

Newell, P. T., and C.-I. Meng, "On Quantifying the Distinction Between the Cusp and the Cleft/LLBL," in *NATO Advanced Workshop on Electromagnetic Coupling in the Polar Cleft and Cusps*, Lillehammer, Norway (1989).

Reidler, W., "Auroral Particle Precipitation Patterns," in *Earth's Magnetospheric Processes*, B. M. McCormac, ed., D. Reidel Publishing, Dordrecht-Holland, pp. 133-140 (1972).

Schield, M. A., and L. A. Frank, "Electron Observations Between the Inner Edge of the Plasma Sheet and the Plasmasphere," *J. Geophys. Res.*, **75**, 5401-5414 (1970).

Shelley, E. G., R. G. Johnson, and R. D. Sharp, "Satellite Observations of Energetic Heavy Ions During a Geomagnetic Storm," *J. Geophys. Res.*, **77**, 6104-6110 (1972).

Shelley, E. G., and H. L. Collin, "Auroral Ion Acceleration and Its Relationship to Ion Composition," *this volume*.

Spiro, R. W., P. H. Reiff, and L. J. Maher, "Precipitating Electron Energy Flux and Auroral Zone Conductances—An Empirical Model," *J. Geophys. Res.*, **87**, 8215-8227 (1982).

Vasyliunas, V. M., "A Survey of Low-Energy Electrons in the Evening Sector of the Magnetosphere with OGO 1 and OGO 3," *J. Geophys. Res.*, **73**, 2839-2878 (1968).

Vasyliunas, V. M., "Interaction Between the Magnetospheric Boundary Layers and the Ionosphere," *Proc. Magnetospheric Boundary Layers Conference*, Alpback, 11-15 June, 1979 (ESA SP-148).

III-2. DIAGNOSIS OF AURORAL ACCELERATION MECHANISMS BY PARTICLE MEASUREMENTS

J. L. Burch*

Measurements of suprathermal auroral particles from sounding rockets and satellites at various altitudes have been made with some regularity since about 1960. Although many recurring features of the particle distributions are noted, we have not yet arrived at a universal auroral particle acceleration process or even at a set of independent processes that can explain the data unambiguously. It is popular now to separate auroral acceleration mechanisms into two classes. One involves single-particle adiabatic motion of electrons and ions through a field-aligned quasi-static electric potential structure. This class of model can easily explain a field-aligned primary auroral electron beam and an isotropic secondary and backscattered-primary electron distribution at lower energies; it also offers a natural explanation for the observed highly collimated upward-moving ion beams. The second class of auroral acceleration mechanism involves nonadiabatic, or diffusive, processes that are needed to explain electron distribution functions that are field-aligned over a broad range of (generally low) energies and that are often associated with conical ion distributions. Electron distributions with designations such as field-aligned bursts, counterstreaming electrons, type-1 counterstreaming electrons, suprathermal bursts, and edge precipitation belong to this second category. Contemporary models in the second class are said to be nonadiabatic or diffusive because they involve either wave-particle interactions or diffusive entry of thermal electrons into an extended field-aligned potential structure, or both. After a brief historical review, some of the more recent satellite and sounding-rocket measurements that bear on the question of auroral acceleration mechanisms, particularly those that are crucial to the success of contemporary models, are discussed.

1. INTRODUCTION

Early sounding-rocket flights into visible auroras [McIlwain, 1960] revealed that most of the auroral light could be explained by the observed influx of electrons in the energy range from 5–10 keV. Later sounding-rocket measurements (e.g., Albert [1967]; Evans [1968]) above auroral arcs indicated that the bulk of the incoming electron energy flux resides in a "near-monoenergetic beam," and that the beam is superimposed on a continuum spectrum that extends to very low energies [Westerlund, 1969]. Although a sharply peaked energy spectrum with fluxes peaked along the magnetic field line is readily explained by a field-aligned potential difference, the existence of the low-energy continuum and the "anomalously high" backscatter ratio within it remained as serious "sticking points" in models of auroral acceleration by parallel electric fields. Several years later Evans [1974], with remarkable insight, proposed a simple explanation for this apparent dilemma. Secondary electrons and backscattered primary electrons, if reflected by a parallel potential difference, produce a precipitating low-energy component equal to the upgoing electrons of the same energies because they are one and the same distribution. The work of Evans, along with the theoretical analysis by Knight [1973], set the stage for rapid advances in our understanding of auroral electron and ion energization, using further sounding-rocket measurements and satellite data (e.g., DMSP, S3-2, S3-3, DE 1, DE 2). Chiu and coworkers (see review by Chiu et al. [1983]) focused on the data from the USAF satellites, which became available in the mid to late 1970s (e.g., Fennell et al. [1981]), and developed a detailed theoretical model [Chiu and Schulz, 1978; Chiu and Cornwall, 1980] incorporating the ideas of Alfvén and Fälthammer [1963], Knight [1973], and Lennartsson [1976] on the maintenance of parallel electric fields through the principle of quasi-neutrality as applied to plasma-sheet electron and ion distributions. Although the more localized double layers are not ruled out by the recent observations of S3-3 and DE 1, many such smaller-scale phemonena would have to be distributed along auroral field lines in order to explain the data. At altitudes between 4000 and 8000 km with S3-3, and on out to 15,000 km with DE 1, the most general result contains evidence of a potential difference below the spacecraft and a (generally smaller) potential drop above the spacecraft. In fact, Gurgiolo and Burch [1988] have shown that the population of the trapping region of velocity space, which Chiu et al. [1983] regard as incontrovertible evidence of the existence of a field-aligned potential difference, can occur adiabatically for preexisting plasma-sheet electrons only if the spacecraft is in an electric-field-free region with potential drops located below and above the point of observation.

Another commonly observed type of auroral electron distribution is not readily explained by adiabatic mo-

*Southwest Research Institute, San Antonio, Texas 78228.

tion in a quasi-static field-aligned potential. This type of distribution is referred to variously as field-aligned bursts [*Hoffman and Evans*, 1968], suprathermal bursts [*Johnstone and Winningham*, 1982], counterstreaming electrons [*Sharp et al.*, 1980], type-1 counterstreaming electrons [*Lin et al.*, 1984], and edge precipitation [*Lotko*, 1986]. Common features of these distributions include a field alignment over a broad range of energies, small latitudinal extent of a few kilometers, and temporal variability on a several-second time scale. There is some indication that downward moving distributions of this type are concentrated toward low altitudes of a few thousand kilometers or less, although the upward moving and counterstreaming distributions are seen more often at higher altitudes of an Earth radius or more. The burst-type electron distributions are often observed simultaneously with conical ion distributions, suggesting that electric-field fluctuations on the scale of the ion gyrofrequency may play a role in the acceleration of both ions, which can resonate with the waves in their gyromotion, and electrons, which can resonate with the waves in their field-aligned motion. A number of models of acceleration by wave-particle interactions have been suggested (e.g., *Lotko* [1986]; *Temerin et al.* [1986]; *McFadden et al.* [1987]; *Robinson et al.* [1989]), but they generally can explain only the downward moving electron bursts. *Sharp et al.* [1980] suggested that a set of multiple flickering double layers may be able to explain the counterstreaming electron distributions.

In the following sections, the existing experimental data on auroral plasma distributions will be reviewed and the current models of acceleration by both adiabatic and diffusive processes will be evaluated. The data needed for definitive results will be very difficult to obtain because they will require the measurement of weak, turbulent electric fields and thermal electron distributions with high spatial and temporal resolution.

2. ACCELERATION BY QUASI-STATIC FIELD-ALIGNED POTENTIAL DIFFERENCES

Advances in low-energy charged-particle detection technology during the 1960s led to the realization that the primary influx of particle energy into the auroral atmosphere is carried by electrons with energies between 1 and 10 keV. The integrated energy fluxes in bright auroral forms often reach values between 10 and 100 erg cm^{-2} s^{-1}, with most of the flux in a relatively narrow range of electron energies. *Albert* [1967] and *Evans* [1968] noted the existence of the sharply peaked energy spectra, both referring to them as monoenergetic, with Evans suggesting that they may be consistent with field-aligned potential differences. *Westerlund*

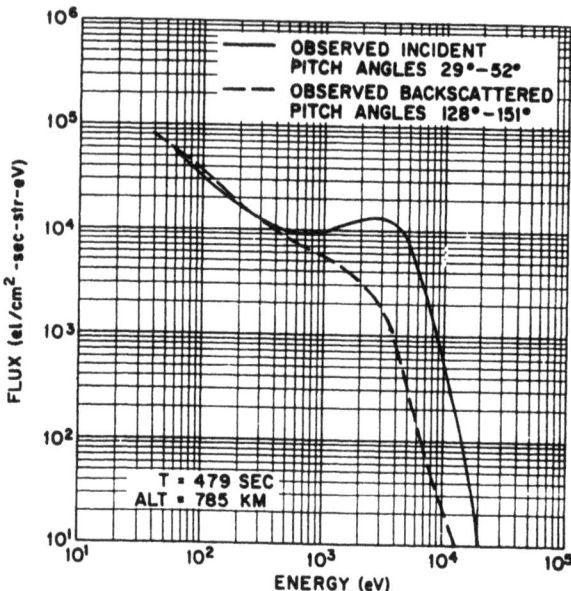

Figure 1—Incident and backscattered auroral electron energy spectra. (From *Reasoner and Chappell* [1973].)

[1969] extended the auroral electron spectral measurements to lower energies and noted that the sharply peaked auroral primary electron distribution is superimposed on a continuum that has another peak at lower energies of about 100 eV. The existence of this continuum and the fact that at energies below the auroral peak the upward and downward fluxes are about equal [*Reasoner and Chappell*, 1973] cast doubt on the applicability of acceleration models involving field-aligned potential differences. Examples of precipitating and backscattered auroral electron energy spectra measured by a sounding-rocket experiment are shown in Figure 1. For some time the existence of a downward moving continuum extending to low energies below the primary spectral peak could not be reconciled with the proposed existence of field-aligned potential differences (e.g., *O'Brien* [1970]). The anomalous backscatter ratio (~1) at low energies also was considered to be puzzling and perhaps required a reverse parallel electric field at lower altitudes [*Reasoner and Chappell*, 1973]. Both of these problems with parallel electric field models were solved by the work of *Evans* [1974].

Evans [1974] modeled the auroral electron distribution with a primary Maxwellian population, which is accelerated through a field-aligned potential difference, and a secondary population, which is trapped at low altitudes by the same potential difference. In Evans' model the composite backscattered primary and secondary electron distribution is isotropic in pitch angle because all of these electrons are reflected by the field-aligned potential difference; that is, the upward and

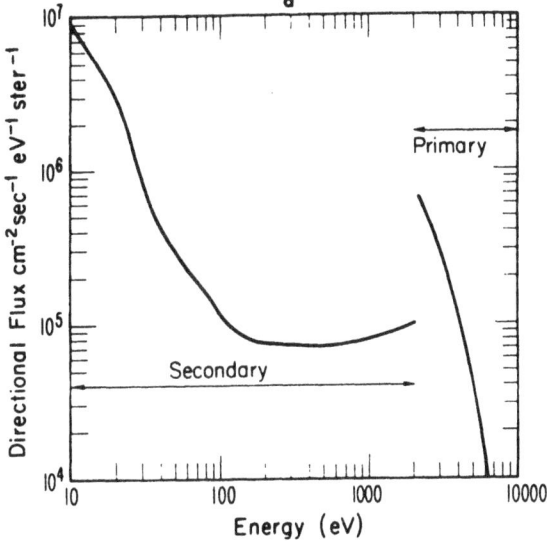

Figure 2—Model energy spectrum for 0° pitch angle precipitating electrons observed just above the atmosphere. The discontinuity separates electrons of atmospheric origin from those of magnetospheric origin. The electrons were assumed to have originated from an 800-eV plasma that had been accelerated through a 2000-V potential drop located at 2000-km altitude. (From *Evans* [1974].)

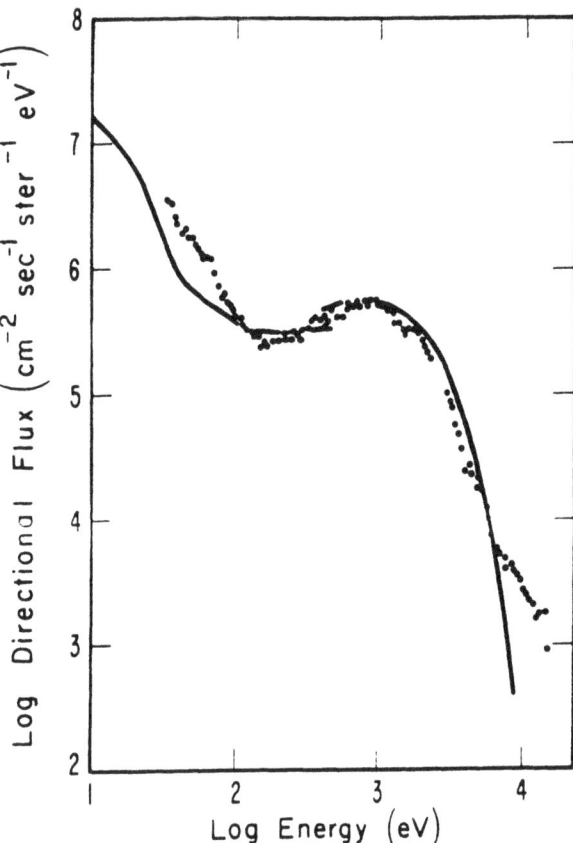

Figure 3—Model electron energy spectrum computed by assuming a 400-V potential difference along a magnetic field line and an unenergized Maxwellian electron distribution of temperature 800 eV and density 5 cm $^{-3}$. The data represent an electron spectrum observed by *Frank and Ackerson* [1971]. (From *Evans* [1974].)

downward distributions are one and the same. Examples of the primary and secondary distributions at 0° pitch angle that are produced by the model of *Evans* [1974] are shown in Figure 2. Figure 3 demonstrates how Evans' model succeeds in reproducing satellite data acquired on Injun 5 [*Frank and Ackerson*, 1971]. One discrepancy between the Evans model and measured spectra is the sharp discontinuity in the model between the primary spectral peak and the secondary electrons. Evans suggested that the discontinuity may just be smoothed out by a real physical detector, although another explanation may be in order. Much more detailed comparisons of the model with sounding-rocket data are presented by *Pulliam et al.* [1981].

Knight [1973] published a theoretical model of the regions of access in velocity space to ionospheric and plasma-sheet plasmas in the presence of a potential difference between the two regions. Knight's model also produced a theoretical relationship between the inferred potential drop and the field-aligned current that later was verified experimentally by *Lyons et al.* [1979] and others. Although *Knight* [1973], unlike *Evans* [1974], did not apply his model to spacecraft data, the two models are very consistent. In fact, Knight's partitioning of velocity space with conic sections derived from conservation of energy and magnetic moment set the stage for rapid interpretation of the data that were to be obtained in 1977 by the S3-3 satellite on electron

and ion distributions at altitudes near and above one Earth radius along auroral field lines. Plots and the equations of the hyperbolas, parabolas, and ellipses that divide velocity space into regions of access of ions and electrons originating in the ionosphere (I) and magnetosphere (M) are shown in Figure 4, which is taken from *Chiu and Schulz* [1978]. Also shown in Figure 4 are the regions of access to secondary electrons (S) and a forbidden or trapping region (T) to which access is denied to electrons from both the magnetosphere and the ionosphere. Chiu and Schulz extended the work of *Knight* [1973] by presenting self-consistent calculations of not only the response of auroral particles to field-aligned potential differences but also of the formation of the potential difference through the principle of quasi-neutrality.

Figure 5, taken from *Croley et al.* [1978], shows the remarkable agreement that was obtained between the S3-3 electron data and the velocity-space partitioning of the type derived by *Knight* [1973] and *Chiu and*

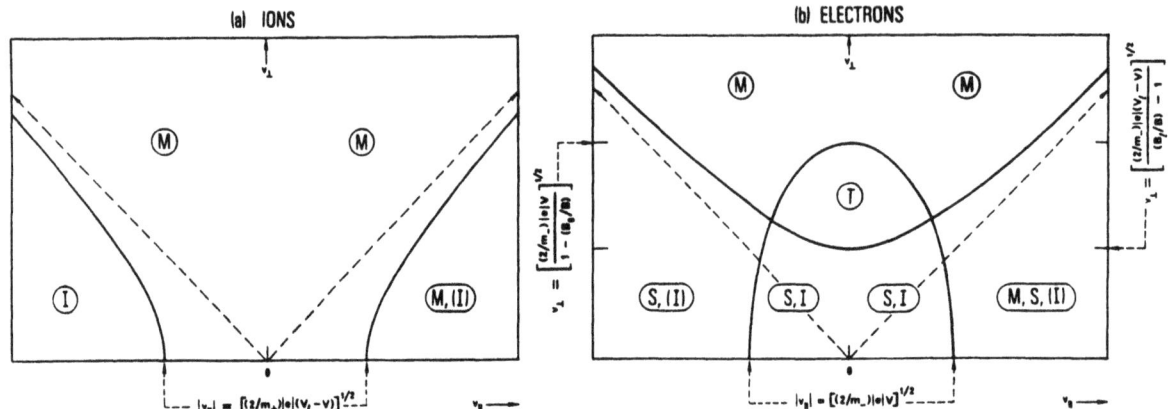

Figure 4—Regions of velocity space accessible to various particle populations. The letters M and I refer to magnetospheric and ionospheric particles; the letter S indicates secondary electrons. The solid curves are the domain boundaries computed from conservation of energy and magnetic moment. The parameters V and B are the electrostatic potential and the magnetic induction at the point of observation, with zero potential assumed at the equatorial intersection of the field line. The subscript I on these parameters refers to the low-altitude limit of this model, which is at 2000 km. (From *Chiu and Schulz* [1978].)

Schulz [1978]. Similar agreement was also found by *Croley et al.* [1978] for the ion data. In Figure 5 the ellipse, beyond which all magnetospheric electrons must reside, is very nearly circular, indicating that the top of the potential-drop region must be far above the point of observation. In fact, for the ellipse shown in Figure 5 the zero potential point was taken to be at the equator.

The solid hyperbolic traces in Figure 5 denote the theoretical widening of the loss cone that results from a potential difference between the spacecraft and the atmosphere. The edge of the measured loss cone is clearly well represented by the hyperbolas along the $-v_\parallel$ axis. In addition, along the $+v_\parallel$ axis, the region of velocity space between the hyperbolic traces and just inside the ellipse contains a depression, which is known as the electron "hole" distribution [*Omidi et al.*, 1984]. With the background of the models of *Evans* [1974] and *Knight* [1973], the hole distribution is easily explainable as the electrostatic reflection of the loss cone. Not so easily understood, and in fact forbidden by the model of *Knight* [1973], is the appearance of electron fluxes in the electron trapping region, which lies outside the ellipse but inside the hyperbola in Figure 5. Another example of the electron hole distribution and the population of the electron-trapping region is shown in Figure 6a. In Figure 6a the hole distribution is observed to lie just inside the ellipse (dotted line) and the contours defining the hole continue smoothly into the region between the ellipse and the hyperbola (dashed line), which is the trapping region.

It was suggested by *Whipple* [1977] that the trapping region may be populated by plasma-sheet electrons that reside at mid-altitudes along auroral field lines

when the field-aligned potential difference is formed. The implication that population of this forbidden region results from the trapping of electrons toward low altitudes perhaps led *Chui et al.* [1983] to consider it as "incontrovertible evidence for the existence of a parallel potential."

Figure 6b shows the simultaneous measurement of an upward ion beam along with the electron distribution discussed above. The upward ion beam is centered near $v_\perp = -0.6 \times 10^3$ km s^{-1} (with the ion velocity determined using the assumption of hydrogen ions). The simultaneous observation of the widening of the electron loss cone, the electron hole distribution, the population of the electron-trapping region, and an upward ion beam constitutes very strong evidence for the existence of field-aligned potential differences. No other competing model of auroral particle acceleration can explain all of these observations.

The electron hole distribution and the population of the electron-trapping region both imply potential drops above the point of observation; widening of the electron loss cone and the appearance of upward ion beams both imply potential drops below it. Taken together these observations would tend to indicate that the S3-3 and the later DE 1 observations were made in regions of parallel electric fields and that the parallel potentials are distributed over a large range of altitudes. Two additional sets of data must be considered, however, before even a preliminary estimate of the altitude dependence of the parallel potential differences can be made. These data sets have been published by *Ghielmetti et al.* [1978] and *Gurgiolo and Burch* [1988].

Ghielmetti et al. [1978] studied the occurrence of upward ion fluxes at energies above 500 eV, with most

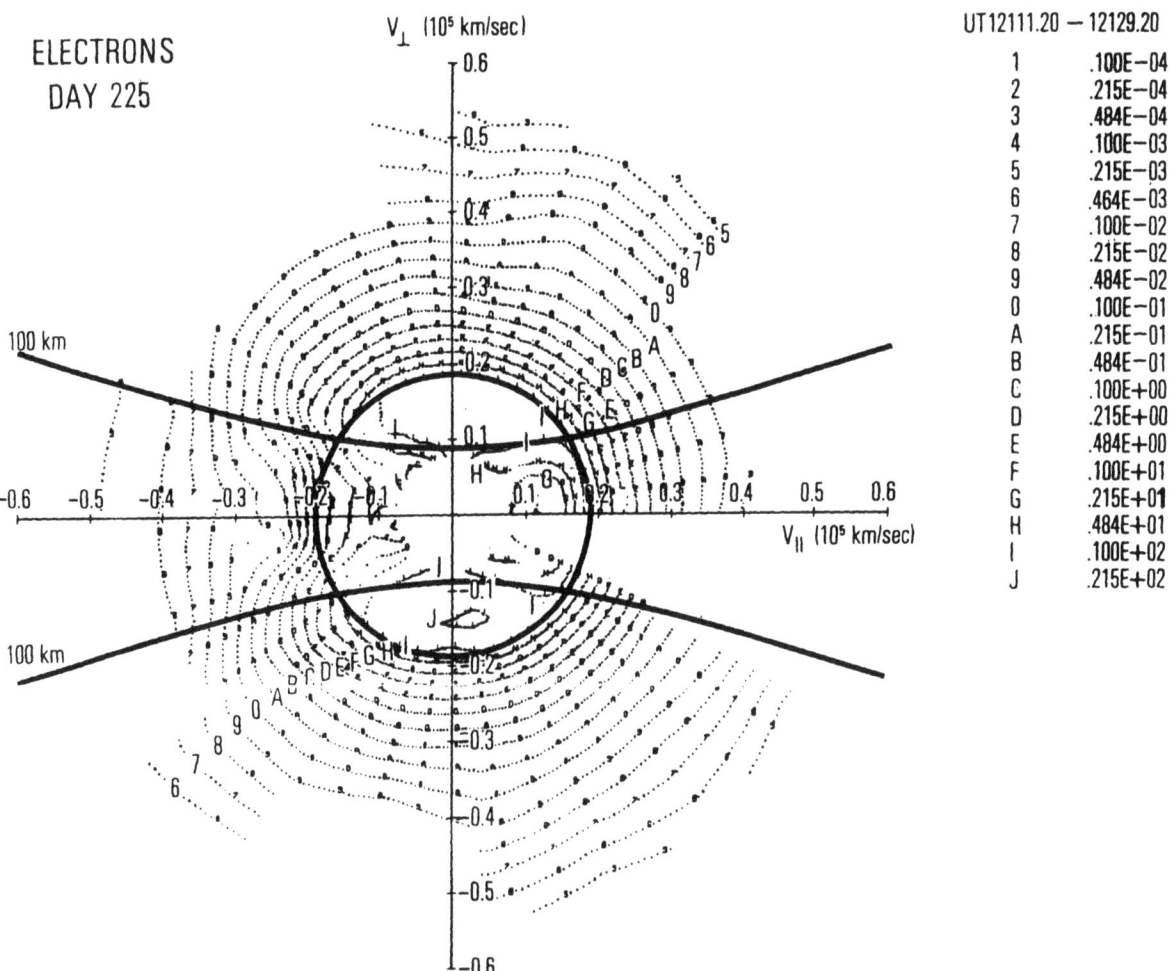

ELECTRONS
DAY 225

$V_\perp$ (10⁵ km/sec)

UT 12111.20 — 12129.20

1	.100E−04
2	.215E−04
3	.484E−04
4	.100E−03
5	.215E−03
6	.464E−03
7	.100E−02
8	.215E−02
9	.484E−02
0	.100E−01
A	.215E−01
B	.484E−01
C	.100E+00
D	.215E+00
E	.484E+00
F	.100E+01
G	.215E+01
H	.484E+01
I	.100E+02
J	.215E+02

100 km

100 km

$V_\parallel$ (10⁵ km/sec)

Figure 5—Values of constant velocity-space distribution function plotted for electron measurements made by the S3-3 space-craft and first reported by *Mizera and Fennel* [1977]. The solid curves are domain boundaries that were derived using a lower boundary potential of 3000 V and a potential at the satellite altitude (7330 km) of 1000 V. The lower boundary in this case was taken as 100 km,. (From *Croley et al.* [1978].)

of the events containing upward accelerated ion beams as would be associated with parallel potential differences below the spacecraft (S3-3). They found a rather rapid onset of normalized probability of occurrence at about 4000-km altitude and a continuing increase with altitude to a peak at 7000 km followed by a slight decrease up to the spacecraft apogee of about 8000 km. This result indicates that at least the lower-altitude reaches of the parallel potentials are often found at altitudes near one Earth radius and at times down to 4000 km. In itself, the result says nothing about the altitude of the top of the parallel potential; however, the fact that the electron hole distribution and population of the electron-trapping region usually appear along with the upward ion beams suggests that the parallel potential extends well above 8000 km.

Gurgiolo and Burch [1988] used computer simulation to attempt to reproduce major features of the electron distribution functions observed by DE 1 in the altitude range from 10,000–15,000 km along auroral field lines. An example of an electron distribution measured in this region is shown in Figure 7. The velocity-space plot in Figure 7 is rotated 90° from those of Figures 5 and 6 with the positive $v_\parallel$ axis (for downward moving electrons) oriented upward instead of to the right. Clearly evident in Figure 7 are the widened loss cone, the hole distribution along the $+v_\parallel$ axis, a nearly circular boundary beyond which the magnetospheric electron population has been accelerated, and a populated trapping region with weak maxima along the $\pm v_\perp$ axes. The trapping region population, which peaks near 90° pitch angles, has been termed the "bump" by, for

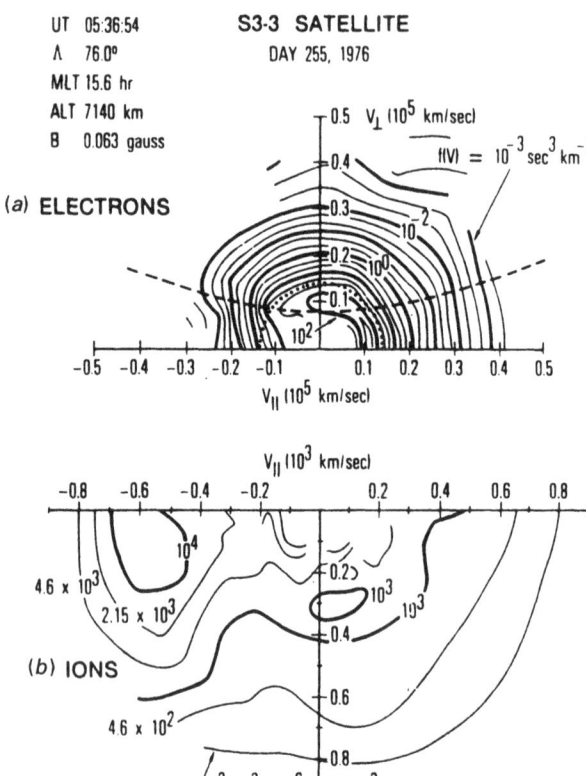

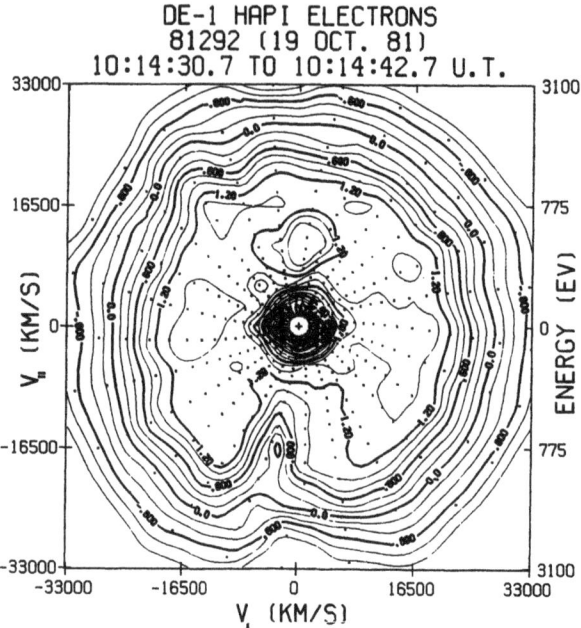

Figure 7—Phase-space contours of constant electron distribution function measured by DE 1 on October 19, 1981. Contours are labeled in units of the base-10 logarithm of the distribution function in units of s^3 km^{-6}. Positive values of $v_\parallel$ indicate downward moving electrons. (From *Lin et al.* [1986].)

Figure 6—Examples of electron (a) and ion (b) distribution function contour plots in ($v_\parallel$, $v_\perp$)-space. Positive $v_\parallel$ marks the downward direction. The electron plot is marked by fitted adiabatic boundaries (dashed and dotted curves). The ion plot shows an upgoing (negative $v_\parallel$) ion beam. (From *Chiu et al.*, [1983].)

example, *Omidi et al.* [1984]. A major objective of the *Gurgiolo and Burch* [1988] study was to investigate the formation of the trapping region and any relationship it may have to the structure of the potential drop. The computer simulation used by Gurgiolo and Burch allowed the full particle velocity-space distribution to be followed from the onset of the field-aligned acceleration. As shown in Figure 8 a simulation using two regions of field-aligned potential differences that are separated in altitude with no parallel electric field at the observing point reproduces well the population of the trapping region, the electron hole distribution, and the evacuation of a widened loss cone. On the other hand, distributed potential drops, either linear or exponential with altitude, which passed through the point of observation, did not produce distributions that agreed nearly as well with DE 1 data. As shown in *Gurgiolo and Burch* [1988], such potential structures tended to produce distribution functions with minima along the $v_\perp$ axes rather than maxima, as observed in the "bump" distributions.

Definitive results on the altitude dependence of the auroral potential structure will require multipoint measurements separated in altitude along with a more reliable method of measuring parallel electric fields, perhaps with a test-particle approach. Neither the results of *Ghielmetti* [1978] nor those of *Gurgiolo and Burch* [1988] rule out either distributed potentials or the more localized double layers.

3. ELECTRON BURST-TYPE DISTRIBUTIONS

A second class of auroral acceleration mechanisms involves nonadiabatic, or diffusive, processes that lead to burst-type electron distributions that are field-aligned over a broad range of energies and are often associated with conical ion distributions. Electron distributions with designations such as field-aligned bursts, counterstreaming electrons, type-1 counterstreaming electrons, suprathermal bursts, and edge precipitation belong to this second category. The burst-type distributions are also characterized by spatial confinement to a few kilometers and temporal variations with periodicities of a few seconds. Although the burst-type distributions are generally ascribed to wave-particle acceleration processes rather than to acceleration by quasi-static parallel electric fields, virtually all contemporary models of the bursts involve regions of such fields.

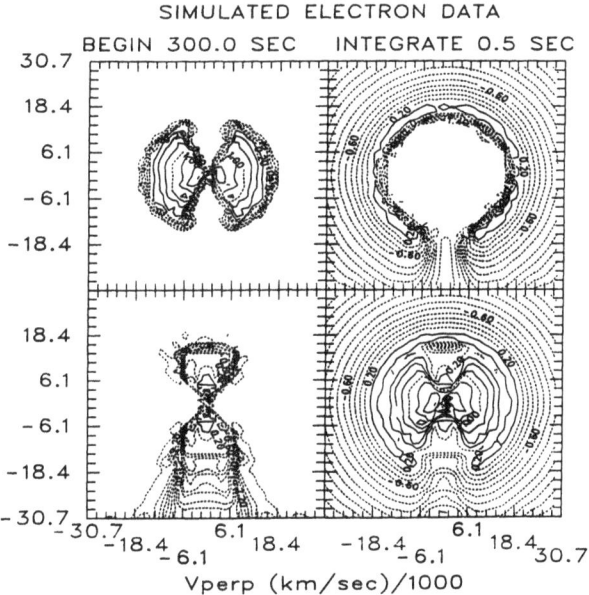

RUN #3
SIMULATED ELECTRON DATA
BEGIN 300.0 SEC   INTEGRATE 0.5 SEC

Vperp (km/sec)/1000

Figure 8—A series of contour plots generated by a computer simulation using separated field-aligned potential-drop regions each of which occupied 1000 km in altitude centered at altitudes of 7500 and 40,500 km, respectively. The total upper potential drop was 700 V and the total lower potential drop was 2500 V. The plots represent equilibrium distributions sampled at the satellite altitude (14,400 km). The upper right panel shows the distribution of electrons initially occupying the field line (the trapped distribution). The upper left contour plot is the injected plasma-sheet electron distribution. The lower left contour plot shows the distribution of secondary and backscattered primary electrons. The lower right panel contains the combination of the trapped, plasma-sheet, and secondary plus backscattered electron distributions,. (From *Gurgiolo and Burch* [1988].)

Many observations of electron bursts have been reported. One of the earliest reports was that of *Hoffman and Evans* [1968] using OGO 4 data. In fact, Hoffman and Evans proposed a parallel potential drop to explain their bursts. However, later measurements with more extensive energy spectral information showed that the field-aligned bursts extended over a wide range of energies and often occurred in the energy regime below the primary peak of inverted-V type electron distributions.

Several years later *Arnoldy* [1974] reported on sounding-rocket measurements above a discrete aurora in which he observed a second, strongly field-aligned energy spectral peak at energies below an approximately 6-keV "monoenergetic peak" that was isotropic in pitch angle. The field-aligned peaks were fairly broad in energy and were described by Arnoldy as at times seeming to fill in the spectrum below the monoenergetic peak. *Arnoldy* [1974] commented further that the

field-aligned electron distributions could be of importance in Birkeland current systems. With Atmosphere Explorer satellite data, *Burch et al.* [1979] and *Lin and Hoffman* [1979] reported similar field-aligned bursts within inverted-V structures. *Burch et al.* [1979] suggested that the bursts originated from thermal electrons trapped within regions of parallel potential drops. *Lin and Hoffman* [1979] employed fixed-energy detectors to obtain high time-resolution data that could be interpreted in terms of time variations and found periodicities at frequencies near 1.75 Hz.

Johnstone and Winningham [1982] investigated field-aligned electron bursts with the Isis 2 data and referred to the distributions as suprathermal bursts. The suprathermal bursts were primarily directed downward, although some upward bursts were also reported. *Johnstone and Winningham* [1982] also noted a strong field alignment over a broad range of energies and a tendency for the bursts to occur at low altitudes in the central plasma sheet (CPS) region or in the boundary plasma sheet (BPS) but in the regions surrounding inverted-V structures rather than within them. This apparent discrepancy with the measurements described above is not yet resolved. Nevertheless, *Johnstone and Winningham* [1982] considered the suprathermal burst process to be closely related to the inverted-V mechanism because the upper-energy limit of the bursts seemed to rise and then fall again with latitude or, when an inverted V was present, to fall off with increasing distance from the edge of the inverted V.

Klumpar and Heikkila [1982] examined the upward suprathermal bursts along with magnetometer data from Isis 2 and determined that the upward directed field-aligned electron distributions are responsible for downward Birkeland currents. With DE 1 at altitudes above 10,000 km, *Burch et al.* [1983] identified upward electron beams as the primary charge carriers of the morning-sector region-1 Birkeland currents and showed that the distribution functions were consistent with downward directed parallel electric fields at altitudes near one Earth radius with total potential drops of a few tens of electron volts. Later work by *Menietti and Burch* [1987] showed that both upward and downward Birkeland currents were carried primarily by field-aligned low-energy electron beams in regions in and around theta auroras in the polar cap.

Sharp et al. [1980], using some of the first data obtained within the auroral acceleration region near altitudes of one Earth radius with S3-3, observed counter-streaming electron distributions, which they suggested could result from fluctuating double layers. The model proposed by *Sharp et al.* [1980] followed a proposal made by *Hall and Bryant* [1974] and by several others who are referenced in *Sharp et al.* [1980] and included *Whalen and Daly* [1979] who suggested turbulent ac-

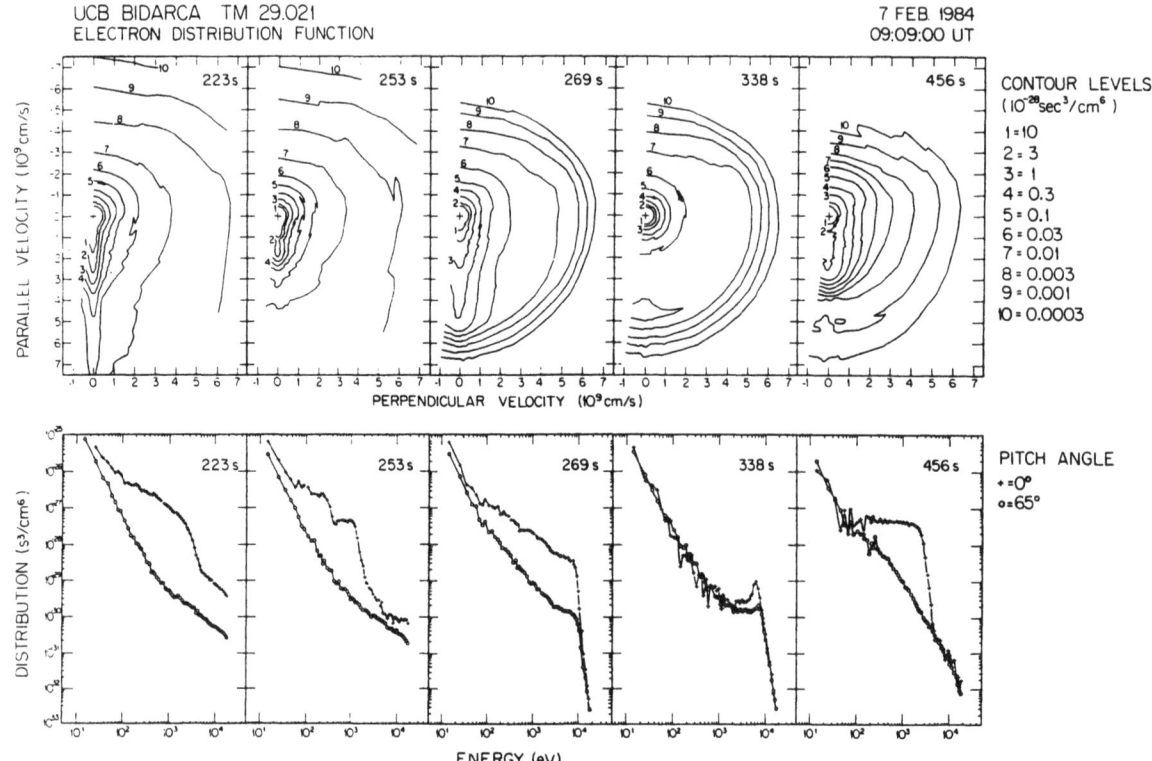

Figure 9—The top panel shows contour plots of the electron distribution function measured in a broad auroral arc. Downward parallel electron velocities are plotted along the upward vertical axis as in Figures 7 and 8. The bottom panel shows slices through the five distribution functions at 0° and 65° pitch angles. (From *McFadden et al.* [1987].)

celeration by broadband electrostatic waves as a possible mechanism for producing field-aligned beams of auroral electrons with broad energy distributions. *Lin et al.* [1982, 1984] investigated similar counterstreaming electron distributions, which they referred to as type-1 counterstreaming electrons and which showed a clear association with conical ion distributions. Type-2 counterstreaming electrons appeared to be a different phenomenon resulting from fine structure in the auroral potential distribution. *Lin et al.* [1984] showed further that most of the Birkeland current density could be accounted for by the counterstreaming distributions, in which either the upward or the downward branch usually dominates the field-aligned flux.

Recent sounding-rocket measurements reported by *McFadden et al.* [1987] suggest that burst-type electron distributions are responsible for the flickering aurora. Figure 9 shows typical electron distributions from *McFadden et al* [1987], in which field-aligned bursts are superimposed on inverted-V type accelerated electron spectra. *McFadden et al.* [1987] note that the burst spectra often extended up to the primary energy peak, which at times reached 18 keV. A major advance in particle instrumentation, an electron spectrograph, allowed McFadden et al. to detect order-of-magnitude

coherent flux oscillations in the 2–20 Hz frequency range, associated with ground-based television recordings of flickering aurora. Analysis of velocity dispersions in the flux oscillations indicated source altitudes between 4000 and 8000 km, corresponding to the expected altitudes of at least the lower reaches of auroral potential structures.

In general, then, the electron bursts seem to be characterized by strong magnetic field alignment and energy spectra that are often fairly broad up to a rather sharp drop-off at a well-defined upper energy limit. They can occur within regions of primary auroral acceleration, at the edges of such regions, or totally outside them. In any case, the mechanism producing the bursts appears to be a part of, or closely related to, the primary auroral acceleration process, which leads to inverted-V precipitation. Outside inverted-V regions, downward bursts seem to predominate at low altitudes (below a few thousand kilometers), while upward and downward bursts occur with nearly equal probability, and sometimes together as counterstreaming distributions, at the higher altitudes (near and above one Earth radius).

Most of the recent models of electron bursts involve interactions with low-frequency waves with frequencies near the ion gyrofrequency. Such waves can lead to

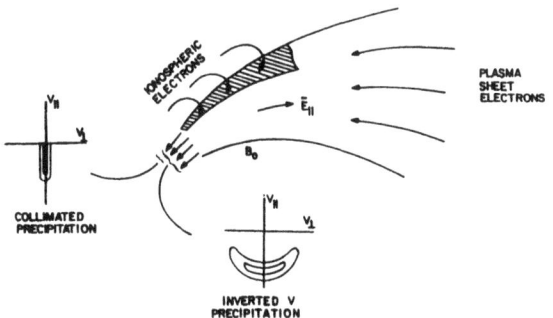

Figure 10—Schematic diagram of primary electron precipitation in a region of upward parallel electric field of finite extent in latitude and altitude. Plasma-sheet electrons enter at the top. Ionospheric electrons enter at the edges through static or turbulent convective motion. The resulting spectra below the acceleration region are identified, respectively, as inverted-V or collimated. (From *Lotko* [1986].)

the transverse acceleration of ions, which can resonate with the waves in their cyclotron motion, and to the parallel acceleration of electrons, which can be trapped and accelerated by the field-aligned motion of the waves. However, virtually all contemporary models of electron bursts also involve a region of quasi-static parallel electric fields at altitudes of several thousand kilometers.

The model of *Temerin* [1986] considers resonant wave acceleration by ion cyclotron waves propagating downward from a parallel electric field region at an altitude of about one Earth radius. Temerin's model was designed primarily to explain the long-standing observations of periodicities in precipitating auroral electrons associated with the flickering aurora [*Evans*, 1967; *Spiger and Anderson*, 1985]. In Temerin's model, the ion cyclotron waves are generated in the auroral potential structure with the oxygen gyrofrequency of about 4 Hz accounting for the observed periodicities in the auroral electron flux. At the higher altitudes the phase velocities of the electromagnetic ion cyclotron waves in the model are fairly low, and thermal electrons can be trapped and accelerated by them. Owing to the increasing magnetic field, the phase velocity increases toward lower altitudes, allowing acceleration to higher energies of up to several kiloelectron volts. However, the model does not predict acceleration up to the primary auroral energy peak as observed by *Arnoldy* [1974] and *McFadden et al.* [1987].

McFadden et al. [1987] presented a conceptually different model, which also involves an auroral parallel potential structure and ion cyclotron waves generated by the turbulence within it. However, in this case the oxygen cyclotron waves, as they propagate downward through the potential structure, perturb and release thermal electrons that have either been trapped above

and within the turbulent acceleration region when it was formed or have diffused or convected into the acceleration region [*Lotko*, 1986; *McFadden et al.*, 1986]. The energy of the accelerated electrons depends on the altitude within the potential drop where they become untrapped, and the untrapping is expected to occur in phase with the waves, leading to the observed periodicities. This model is similar in some respects to the suggestion made by *Burch et al.* [1979], in which atmospheric secondary electrons populate the potential-drop region and are later accelerated through a portion of the total potential difference.

Lotko [1986] has presented a comprehensive model of the trapping of cold electrons in a turbulent acceleration region followed by their diffusion to field-aligned motion through which they gain energy from the portion of the potential difference which they traverse. Lotko's model focuses on edge precipitation by postulating entry of cold electrons through the "sides" of the potential structure either by convection or diffusion. Figure 10 is a schematic of Lotko's model, showing the entry of plasma-sheet electrons from the top of the potential structure, leading to inverted-V distributions at low altitudes, and the entry of cold electrons from the sides of the acceleration region, leading to the production of a field-aligned lower-energy population. Since thermal electrons are trapped in the potential for a time, the effects of turbulence in the parallel potential are important for them, whereas these effects would tend to average out for the inverted-V electrons.

4. SUMMARY AND CONCLUSIONS

Two general classes of auroral electron acceleration have been identified from sounding-rocket and spacecraft measurements performed over the past 28 years. The first class resembles adiabatic acceleration in field-aligned potential structures and produces inverted-V events and the associated upward ion beams. The second type of mechanism is thought to involve diffusive acceleration by wave-particle interactions and/or diffusive transport of cold electrons into parallel potential structures. This second class of acceleration is understood as the likely cause of field-aligned electron bursts, with the same wave modes seen as responsible for the transverse ion acceleration that produces conical ion distributions, which are often observed along with the electron bursts.

Although quasi-static parallel potential structures are the most favored explanation for inverted-V type of auroral particle fluxes, they are not universally accepted. Other proposals have been made, for example by *Bryant and Hall* [this volume]. The mechanism favored by Bryant and Hall (see also *Bingham et al.* [1984]) involves low-frequency waves (e.g., at the lower hybrid frequency) with a parallel electric field component.

Such waves can reproduce many of the observed features of auroral electron distributions; however, they have not been shown to be capable of simultaneously producing the observed upward ion beams.

Recent advances in particle detector technology have led to a greater understanding of the nature of the accelerated electron distributions, particularly the burst-type distributions for which coherent oscillations have been observed, using spectrographic detectors that can sample a wide range of energy steps simultaneously. For significant future progress other new measurement technologies will be needed. A partial list of required measurements includes:

1. Thermal electron distributions in the acceleration region (altitudes of about one Earth radius) with high spatial resolution (<1 km),
2. Electromagnetic and electrostatic ion cyclotron and lower hybrid wave forms and $k_\parallel$ and $k_\perp$ in the acceleration region,
3. $E_\parallel$ of a few millivolts per meter,
4. Birkeland current densities in the acceleration region with spatial resolution of <1 km, and
5. Complete suprathermal plasma distributions in the acceleration region with high spatial resolution (<1 km).

Even with the new measurement technologies our knowledge of the temporal evolution of the auroral potential structure and its distribution with altitude will be severly limited until multipoint measurements of the auroral acceleration regions are made. Appropriate clusters of spacecraft making in situ measurements, along with the imaging of visible ion gas releases in the acceleration region, will pave the way for an eventual full understanding of auroral particle acceleration.

ACKNOWLEDGMENT—This work was supported by NASA Contract NAS5-28711 at Southwest Research Institute.

REFERENCES

Albert, R. D., "Energy and Flux Variations of Nearly Monoenergetic Auroral Electrons," *J. Geophys. Res.*, **72**, 5811 (1967).

Alfvén, H., and C.-G. Fälthammar, *Cosmical Electrodynamics*, Second Edition, Oxford University Press (1963).

Arnoldy, R. L., "Auroral Particle Precipitation and Birkeland Currents," *Rev. Geophys. Space Phys.*, **12**, 217-231 (1974).

Bingham, R., D. A. Bryant, and D. S. Hall, "A Wave Model of the Aurora," *Geophys. Res. Lett.*, **11**, 327-330 (1984).

Burch, J. L., S. A. Fields, and R. A. Heelis, "Polar Cap Electron Acceleration Regions," *J. Geophys. Res.*, **84**, 5863-5874 (1979).

Burch, J. L., P. H. Reiff, and M. Sugiura, "Upward Electron Beams Measured by DE-1: A Primary Source of Dayside Region-1 Birkeland Currents," *Geophys. Res. Lett.*, **10**, 753-756 (1983).

Chiu, Y. T., and J. M. Cornwall, "Electrostatic Model of a Quiet Auroral Arc," *J. Geophys Res.*, **85**, 543 (1980).

Chiu, Y. T., and M. Schulz, "Self Consistent Particle and Parallel Electrostatic Field Distributions in the Magnetospheric-Ionospheric Auroral Region," *J. Geophys. Res.*, **83**, 629 (1978).

Chiu, Y. T., J. M. Cornwall, J. F. Fennell, D. J. Gorney, and P. F. Mizera, "Auroral Plasmas in the Evening Sector: Satellite Observations and Theoretical Interpretations," *Space Sci. Rev.*, **35**, 211-257 (1983).

Collin, H. L., R. D. Sharp, and E. G. Shelley, "The Occurrence and Characteristics of Electron Beams over the Polar Regions," *J. Geophys. Res.*, **87**, 7504 (1982).

Croley, D. R., Jr., P. F. Mizera, and J. F. Fennell, "Signature of a Parallel Electric Field in Ion and Electron Distributions in Velocity Space," *J. Geophys. Res.*, **83**, 2701-2705 (1978).

Evans, D. S., "A 10-cps Periodicity in the Precipitation of Auroral-Zone Electrons," *J. Geophys. Res.*, **72**, 4281 (1967).

Evans, D. S., "The Observations of a Near Monoenergetic Flux of Auroral Electrons," *J. Geophys. Res.*, **73**, 2315-2323 (1968).

Evans, D. S., "Fine Structure in the Energy Spectrum of Low Energy Auroral Electrons," in *Atmospheric Emissions*, B. M. McCormac and A. Omholt, eds., Van Nostrand-Reinhold, New York, p. 107 (1969).

Evans, D. S., "Precipitating Electron Fluxes Formed by a Magnetic Field Aligned Potential Difference," *J. Geophys. Res.*, **79**, 2853-2858 (1974).

Fennell, J. F., D. J. Gorney, and P. F. Mizera, "Auroral Particle Distribution Functions and their Relationship to Inverted Vs and Auroral Arcs," in *Physics of Auroral Arc Formation*, Geophysical Monograph 25, S.-I. Akasofu and J. R. Kan, eds., American Geophysical Union, Washington, DC, pp. 91-102 (1981).

Frank, L. A., and K. L. Ackerson, "Observations of Charged Particle Precipitation into the Auroral Zone," *J. Geophys. Res.*, **76**, 3612-3643 (1971).

Ghielmetti, A. G., R. G. Johnson, R. D. Sharp, and E. G. Shelley, "The Latitudinal, Diurnal, and Altitudinal Distributions of Upward Flowing Energetic Ions of Ionospheric Origin," *Geophys. Res. Lett.*, **5**, 59-62 (1978).

Gurgiolo, C., and J. L. Burch, "Simulation of Electron Distributions within Auroral Acceleration Regions," *J. Geophys. Res.*, **93**, 3989-4003 (1988).

Hall, D. S., and D. A. Bryant, "Collimation of Auroral Particles by Time Varying Acceleration," *Nature*, **251**, 402 (1974).

Hoffman, R. A., and D. S. Evans, "Field-Aligned Electron Bursts at High Latitudes Observed by OGO 4," *J. Geophys. Res.*, **73**, 6201-6214 (1968).

Johnstone, A. D., and J. D. Winningham, "Satellite Observations of Suprathermal Bursts," *J. Geophys. Res.*, **87**, 2321-2329 (1982).

Klumpar, D. M., and W. J. Heikkila, "Electrons in the Ionospheric Source Cone: Evidence for Runaway Electrons as Carriers of Downward Birkeland Currents," *Geophys. Res. Lett.*, **9**, 873-876 (1982).

Knight, S., "Parallel Electric Fields," *Planet. Space Sci.*, **21**, 741 (1973).

Lennartsson, W., "On the Magnetic Mirroring as the Basic Cause of Parallel Electric Fields," *J. Geophys. Res.*, **81**, 5583-5586 (1976).

Lin, C. S., and R. A. Hoffman, "Fluctuations of Inverted V Electron Fluxes," *J. Geophys. Res.*, **84**, 6547-6553 (1979).

Lin, C. S., J. L. Burch, J. D. Winningham, J. D. Menietti, and R. A. Hoffman, "DE-1 Observations of Counterstreaming Electrons at High Altitudes," *Geophys. Res. Lett.*, **9**, 925-928 (1982).

Lin, C. S., M. Sugiura, J. L. Burch, J. N. Barfield, and E. Nielsen, "DE 1 Observations of Type 1 Counterstreaming Electrons and Field-Aligned Currents," *J. Geophys. Res.*, **89**, 8907-8917 (1984).

Lotko, W., "Diffusive Acceleration of Auroral Primaries," *J. Geophys. Res.*, **91**, 191 (1986).

Lyons, L. R., D. S. Evans, and R. Lundin, "An Observed Relation Between Magnetic Field-Aligned Electric Fields and Downward Electron Energy Fluxes in the Vicinity of Auroral Forms," *J. Geophys. Res.*, **84**, 457 (1979).

McFadden, J. P., C. W. Carlson, and M. H. Boehm, "Field-Aligned Electron Precipitation at the Edge of an Arc," *J. Geophys. Res.*, **91**, 1723 (1986).

McFadden, J. P., C. W. Carlson, M. H. Boehm, and T. J. Hallinan, "Field-Aligned Electron Flux Oscillations that Produce Flickering Aurora," *J. Geophys. Res.*, **92**, 11,133-11,148 (1987).

McIlwain, C. E., "Direct Measurement of Particles Producing Visible Aurorae," *J. Geophys. Res.*, **65**, 2727 (1960).

Menietti, J. D., and J. L. Burch, "DE 1 Observations of Theta Aurora Plasma Source Regions and Birkeland Current Charge Carriers," *J. Geophys. Res.*, **92**, 7503-7518 (1987).

Mizera, P. F., and J. F. Fennell, "Signatures of Electric Fields from High and Low Altitude Particle Distributions," *Geophys. Res. Lett.*, **4**, 311-314 (1977).

O'Brien, B.J., "Considerations that the Source of Auroral Energetic Particles Is Not a Parallel Electrostatic Field," *Planet. Space Sci.*, **18**, 1821-1827 (1970).

106

Omidi, N., C. S. Wu, and D. A. Gurnett, "Generation of Auroral Kilometric and Z Mode Radiation by the Cyclotron Maser Mechanism," *J. Geophys. Res.,* **89,** 883-895 (1984).

Pulliam, D. M., H. R. Anderson, K. Stamnes, and M. H. Rees, "Auroral Electron Acceleration and Atmospheric Interactions: (1) Rocket-Borne Observations and (2) Scattering Calculations," *J. Geophys. Res.,* **86,** 2397-2404 (1981).

Reasoner, D. L., and C. R. Chappell, "Twin Payload Observations of Incident and Backscattered Auroral Electrons," *J. Geophys. Res.,* **78,** 2176-2186 (1973).

Robinson, R. M. J. D. Winningham, J. Sharber, J. L. Burch, and R. Heelis, "Plasma and Field Properties of Suprathermal Electron Bursts," *J. Geophys. Res.,* in press (1989).

Sharp, R. D., E. G. Shelley, R. G. Johnson, and A. G. Ghielmetti, "Counterstreaming Electron Beams at Altitudes of 1 R_E Over the Auroral Zone," *J. Geophys Res.,* **85,** 92-100 (1980).

Spiger, R. J., and H. R. Anderson, "Fluctuations of Precipitated Electron Intensity in Flickering Auroral Arcs," *J. Geophys. Res.,* **90,** 6647 (1985).

Temerin, M., J. McFadden, M. Boehm, C. W. Carlson, and W. Lotko, "Production of Flickering Aurora and Field-Aligned Electron Flux by Electromagnetic Ion Cyclotron Waves," *J. Geophys. Res.,* **91,** 5769-5792 (1986).

Westerlund, L. H., "The Auroral Energy Spectrum Extended to 45 eV," *J. Geophys. Res.,* **74,** 351-354 (1969).

Whalen, B. A., and P. W. Daly, "Do Field-Aligned Auroral Particle Distributions Imply Acceleration by Quasi-Static Parallel Electric Fields?", *J. Geophys. Res.,* **84,** 4175-4182 (1979).

Whipple, E. C., Jr., "The Signature of Parallel Electric Fields in a Collisionless Plasma," *J. Geophys. Res.,* **82,** 1525 (1977).

III-3. CHARACTERISTICS OF MAGNETIC-FIELD ALIGNED ELECTRIC FIELDS IN THE AURORAL ACCELERATION REGION

L. P. Block* and C.-G. Fälthammar*

Electric fields in the auroral acceleration region are strongly fluctuating, both transverse and parallel to the magnetic field. Measurements on the Swedish satellite Viking have confirmed and extended earlier observations on S3-3. The results shed additional light on the role of magnetic-field aligned electric fields for auroral acceleration. In terms of their effects on charged particles, these electric fields can be considered as DC up to a finite frequency that depends on the length of the acceleration region and is different for different particles. For an acceleration length of a few thousand kilometers, this frequency is a few hertz for electrons and of the order of 0.1 Hz for ions. According to Viking observations the frequency spectrum of the electric field variations has a broad maximum just below 1 Hz. This holds both for the parallel and the perpendicular components, although the height of the maximum is greater for the latter by typically more than a power of ten. In addition, the electric field spectrum sometimes has a sharp peak near one-quarter, one-half, or three-quarters of the local proton gyrofrequency. This peak is identified as due to Alfvén waves, associated with magnetic-field aligned electric field components. The power density maximum below 1 Hz may be due to standing waves trapped within a plasma density cavity in the altitude range between about one-half and a few Earth radii where the Alfvén velocity is much higher than in the ionosphere and near the equatorial plane.

The small (about a volt) double layers known from S3-3 are also seen on Viking. They appear to be numerous enough that their combined potential drops, in series along an auroral flux tube, can be large enough to account for the energy of auroral electrons. Potential drops below the satellite, inferred from upward ion beams, are about equal to those computed from widened electron loss cones, as expected for DC electric field acceleration.

1. INTRODUCTION

As described by *Burch* [this volume] and in reviews by, for example, *Fälthammar* [1983, 1988], a wide variety of observations, including observations of both upward-flowing ions and downward-accelerated electrons have been interpreted in terms of potential drops along the magnetic field (henceforth called parallel potential drops or $V_\parallel$), although some important features of the particle spectra require additional mechanisms, such as wave-particle interactions [e.g., *Bryant*, 1987].

The magnetic-field aligned potential drop is usually estimated to be in the range a fraction of a kilovolt to several kilovolts. Recent determinations of $V_\parallel$, by three independent methods, were found to give similar results [*Burch*, 1988; *Reiff et al.*, 1988]. *Weimer* [1988] shows that large-scale (> 100 km) electric fields map fairly well between DE 1 and DE 2, but smaller-scale features do not. He also shows that large-scale, nearly sinusoidal electric-field variations are spatial rather than temporal.

Direct measurements of electric fields in the acceleration region are technically very difficult [*Mozer et al.*, 1978; *Block et al.*, 1987] and have until now been performed on only a very few missions. This is particu-

larly true for the component $E_\parallel$. However, direct measurements of $E_\parallel$ have been obtained with the satellites S3-3 and Viking and will be discussed below. The magnetic-field aligned as well as the transverse component of the electric field has been found to have strong time variations in the auroral acceleration region. Thus, the fluctuating fields in the 0.1–100 Hz range have amplitudes that exceed the DC (< 0.1 Hz) field by a power of ten or more. It is therefore important to take these fields into account. We shall show below that electric fields in the lower part of this frequency range affect the electrons, but not the ions, as if they were truly DC (this has important consequences in terms of selective acceleration). Here, we shall use some recently analyzed Viking data to modify the original concept of DC acceleration by taking into account the time variations up to a few hertz. Certain higher frequencies observed on Viking seem to be generated by the accelerated particles, thus contributing not so much to particle acceleration but rather to deceleration and scattering.

This distinction between high and low frequencies is based on a definition of DC fields in terms of the time required for a typical particle to pass through the acceleration region. This time depends on the kind of particle, its velocity along the magnetic field, and the length of the acceleration region.

---

*Department of Plasma Physics, The Royal Institute of Technology, S-100 44 Stockholm, Sweden.

Auroral Physics, edited by C.-I. Meng, M. J. Rycroft and L. A. Frank. © Cambridge UP 1991

The problems we must address are:

1. How can we distinguish between DC and AC electric fields, from the point of view of the particle?

2. Do DC fields exist in this sense?

3. Are DC fields sufficiently extended and coherent within the acceleration region to accelerate particles, as observed?

4. What is the role of AC electric fields? Do they enhance or impede DC acceleration, or do they merely scatter the particles, in pitch angle and/or energy?

5. How are DC and AC fields generated?

2. A CONCISE DEFINITION OF DC ELECTRIC FIELDS

Consider a particle moving through an acceleration region with $E_\parallel$ having frequencies from zero up to some arbitrary value. Select a frequency band from zero to an upper frequency f. The corresponding electric field $E_\parallel(f)$ integrated over this frequency band may be regarded as a DC field for all particles that can pass through the acceleration region between P_1 and P_2 (Figure 1) in a time $\Delta t << 1/f$. The kinetic energy W of the particle will then change by about

$$\Delta W = \pm \int_{p_1}^{p_2} qE_\parallel(f)dz \qquad (1)$$

due to this "DC" field. Wave-particle interaction at higher frequencies may of course add or subtract from this, depending on which way the energy transfer goes.

As a rough measure of the demarcation between AC and DC fields as experienced by the particles, we take $f = v_\parallel/2h$.

Figure 2 shows the dependence of f on the length h of the magnetic field line through the acceleration region ($p_1 \rightarrow p_2$), and on the particle energy

$$W = mv_\parallel^2/2 \qquad (2)$$

where $V_\parallel$ is the average "parallel" velocity within the acceleration region. It is seen in Figure 2 that the critical frequency for typical (keV) electrons is of the order of a few hertz for an acceleration length of a few megameters. For ions of comparable energy, it is a fraction of a hertz.

It should be noted that this definition does not necessarily imply that the DC fields, with finite but low frequency, are potential fields in an Earth-fixed or any other coordinate system. No doubt, potential fields exist since several charge-separation processes are known to exist in the magnetosphere, e.g., double-layer for-

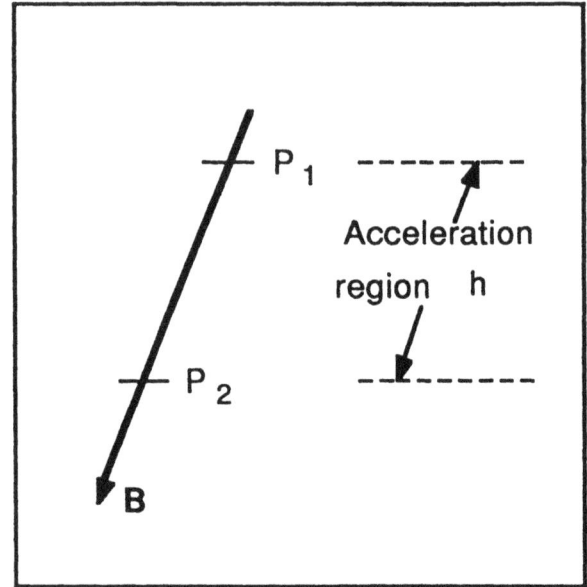

Figure 1—Explanation of symbols used in Eq. 1 for defining DC electric-field components along the magnetic field.

mation and differential gradient and curvature drifts. For examples of charge-separation fields, see e.g., *Alfvén* [1950] §6.2, *Alfvén and Fälthammar* [1963] §§2.6.1 and 5.1.3, *Block* [1966, 1972, 1983], *Vasyliunas* [1970], *Taylor and Perkins* [1971], *Rostoker and Boström* [1976]. Quantitatively, the importance of DC fields does not depend only on how slowly they vary in time, but also on their spatial variations, i.e., how large the integral in Eq. 1 is. Maximum $\Delta W_\parallel$ is obtained if Δz is half a wavelength, assuming a sinusoidal variation.

For the balance of this presentation DC fields will mean electric fields with sufficiently low frequencies for all relevant particles of a certain species. More specifically, $E_\parallel(A, b)$ will be used to denote fields of type A (DC or AC) for particles of kind b (e or p for electron or proton, respectively).

3. EXAMPLES OF OBSERVED DC FIELDS

Since measurements on a single satellite cannot discriminate between spatial and temporal variations, it is in principle impossible to determine whether an observed field is DC or AC in the above sense. At best, with presently available datasets, one can make judicious assumptions about the spatial/temporal character of the observed variations.

With the above caution we shall here present a few examples of data from Viking. Figure 3 shows two components, E_1 and E_2, of the electric field, derived by sine fitting from the 20-second spin-modulated raw data. They therefore represent a least-squares fit to a DC field with 20-second time resolution. E_1 and E_2

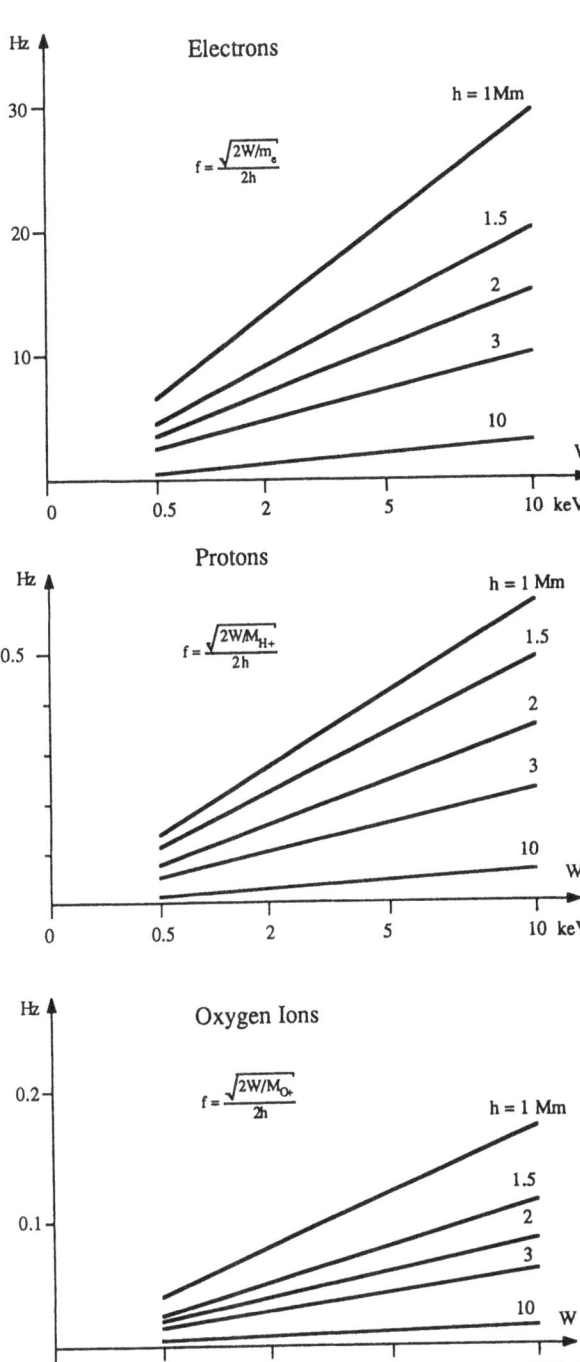

Figure 2—Maximum frequency for sinusoidally varying DC electric fields from the point of view of electrons and protons at different average kinetic energies and lengths of acceleration region.

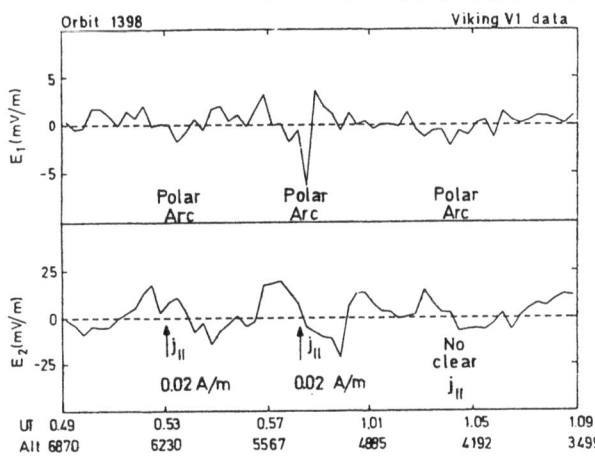

Figure 3—Viking observations of parallel (E_1) and perpendicular (E_2) electric field components. An upward field (negative E_1) is observed at auroral arcs, where E_2 is directed toward the arc on both sides.

are the two components in the spin plane, E_1 being almost parallel to the magnetic field (off by $<7°$) positive downward. E_2 is precisely perpendicular to the magnetic field, positive forward along the satellite trajectory. Since E_1 is not exactly parallel to **B** it is somewhat contaminated by the component perpendicular to both E_2 and the magnetic field. For a more detailed description, see *Block et al.* [1987].

Figure 3 is an example of data from a passage over three polar cap arcs as observed by the Viking UV camera. It shows that above each of the arcs, E_1 is more negative than elsewhere, indicating an upward DC field. Furthermore, E_2 is pointing toward the polar arcs on both sides, implying upward currents above at least two of the three arcs.

4. DOUBLE LAYERS

The existence of double layers on auroral field lines was indirectly inferred from particle velocity distributions observed long ago [*Albert and Lindstrom*, 1970]. The first direct in situ measurements were made on the S3-3 satellite [*Temerin et al.*, 1982; *Mozer and Temerin*, 1983]. They found a large number of small double layers with potential drops of the order of a volt. Their frequency of occurrence indicated that there are thousands of them in series along auroral field lines. They can therefore account for the kilovolt potential drops required for electrostatic acceleration of auroral electrons and upward ion beams. The same kind of double layers have been observed in more detail on Viking [*Boström et al.*, 1987, 1988]. They are associated with density depletions as required by theory (see Figures 4 and 5). They move upward along the magnetic field with velocities of the order of 10–15 km s^{-1}.

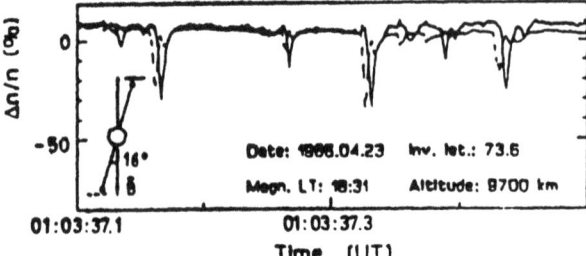

Figure 4—Rarefaction solitary structures recorded by two spatially separated plasma density probes on Viking [*Boström et al.*, 1988].

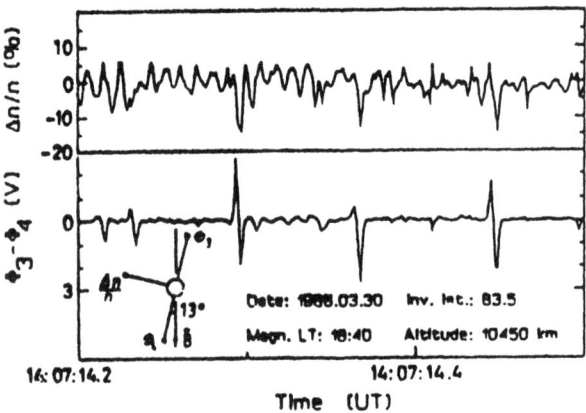

Figure 5—Simultaneous density and electrostatic potential difference variations during the passage of solitary structures. Some of them have a net potential difference (predominantly negative potential difference, i.e., upward field), which means that they are double layers [*Boström et al.*, 1988].

It is possible that the DC fields responsible for charged-particle acceleration are made up of a multitude of these small-scale double layers. Since the internal field of a double layer originates in two layers of opposite charge, the field must be potential on a scale comparable to the size of the double layer. The question then arises: How are the equipotential surfaces closed? If they are closed in the immmediate neighborhood they cannot produce any net acceleration. They must therefore couple to a large-scale field that may be either inductive or potential, or a combination of both. Large-scale inductive fields can only exist during large-scale magnetic field variations such as substorm breakup. During more quiet conditions the coupling should be with the large-scale magnetospheric field, which is potential, at least on the scale of the acceleration region. For a discussion of large-scale equipotential surfaces, see, for example, *Atkinson* [1982] and *Block* [1983].

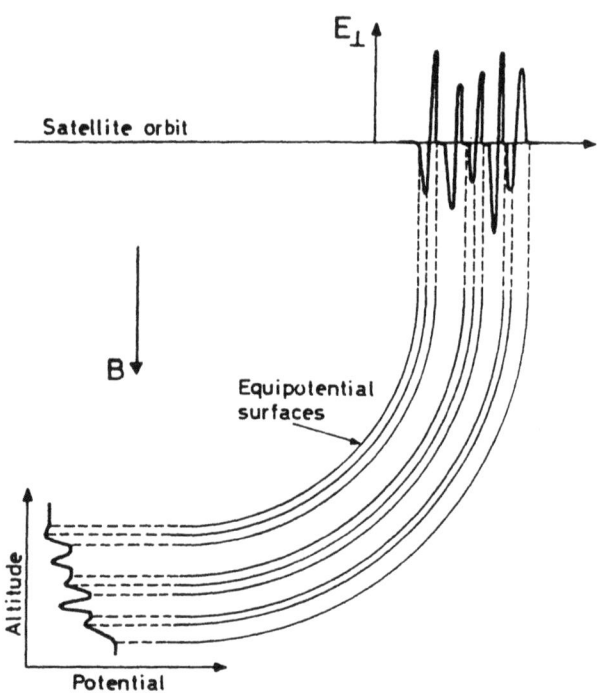

Figure 6—Many small double layers and other solitary structures produce a potential variation along the magnetic field (lower left). If the corresponding parallel fields couple to U-shaped equipotential surfaces on a larger scale, they will also couple to strongly fluctuating perpendicular fields at a higher altitude.

If the double layers are part of large-scale U-shaped structures associated with auroral arcs, or even inverted-V events, the equipotential surfaces in the neighborhood may appear as shown in Figure 6, which also includes small-scale solitary structures with no net potential drop, as described by *Temerin et al.* [1982] and *Boström et al.* [1987, 1988]. Some of these solitary structures are seen in Figure 5. If the equipotential surfaces are as shown in Figure 6 on a scale of the order of the distance between double layers, then the fluctuations should appear similar in both $E_\parallel$ and $E_\perp$ in the appropriate frequency range, i.e., some tens of hertz (see time scale of Figure 5, which shows that the separation between individual double layers and solitary structures is less than 0.1 second). However, at lower frequencies the amplitudes of $E_\parallel$ should be smaller than those of $E_\perp$ by an order of magnitude since the parallel potential drop appears to be extended over thousands of kilometers, whereas the corresponding distance perpendicular to the magnetic field is of the order of 100 km or less [*Mozer*, 1981; *Temerin et al.*, 1982]. This means that the small-scale fluctuations are compressed and merge into larger-scale structures. Evidence in favor of this picture is presented next.

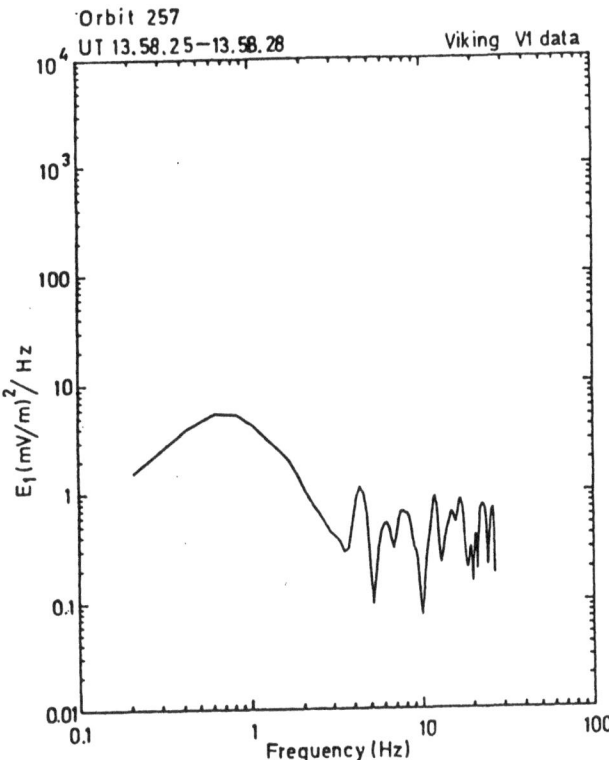

Figure 7—Power density spectrum for parallel electric field variations observed on Viking.

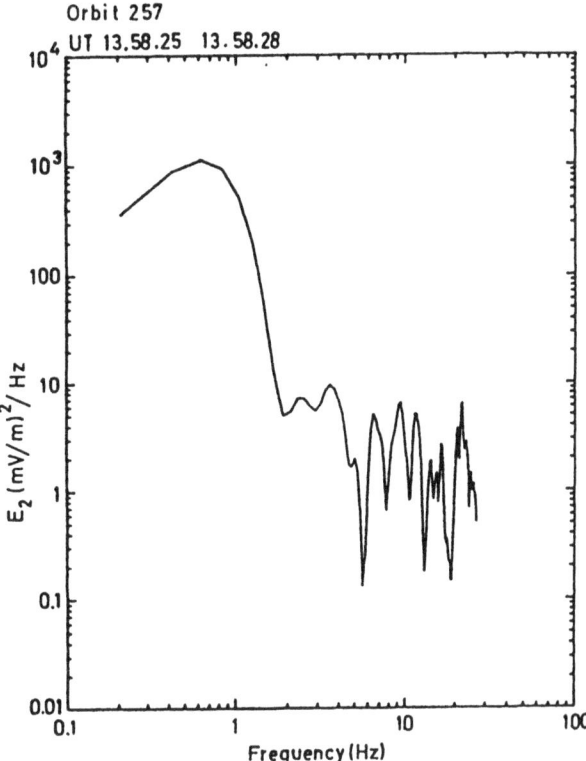

Figure 8—Power density spectrum for perpendicular electric field variations observed simultaneously with those displayed in Figure 7.

5. SPECTRA OF THE ELECTRIC FIELD FLUCTUATIONS

Figures 7 and 8 show power spectral densities (PSD) for E_1 and E_2, respectively, which we assume to be representative of $E_\parallel$ and $E_\perp$ as described above. The maximum PSD is below 1 Hz for both components, but $PSD(E_\perp) \gg PSD(E_\parallel)$ below about 2 Hz, by about two orders of magnitude. Above about 5 Hz, the ratio is only a factor 1–3, which is consistent with the hypothesis presented above about equipotential surfaces like those shown in Figure 6. The high power density at low frequencies suggests that $E_\parallel(DC,e)$ could play an important role for the electrons while having only small effects on the ions. This is further elaborated in the following section.

6. PARTICLE OBSERVATIONS

If DC fields play a dominant role in the acceleration process, the energy of an upward ion beam, observed on a satellite within or above the acceleration region, should correspond to the same $v_\parallel$ as that computed from simultaneously observed widened loss cones for upward flowing electrons. Figure 9 shows Viking

observations of ions and electrons in the eveningside auroral oval. Qualitatively, it is clear that the width of the loss cone increases with increasing ion-beam energy. Table 1 shows several quantitative comparisons of the potential drop below the satellite, derived from the ion-beam energy versus the electron loss cone width. The loss-cone boundary is, of course, not infinitely sharp. The readings are taken in the middle of the boundary. The ion beam is not monoenergetic. Following *Reiff et al.* [1988], we take the average ion energy within the beam as the acceleration potential for the ions.

The results show a good consistency among the potential drops derived at different energies within the same loss cone, indicating that most electrons have been energized by the same amount. The ion data indicate potential drops that agree well with those derived from the electrons for all cases except two where they are lower.

Error bars are difficult to determine because of assumptions behind the calculations. The relatively small observed differences between the potential drops determined from ions and electrons may be entirely due

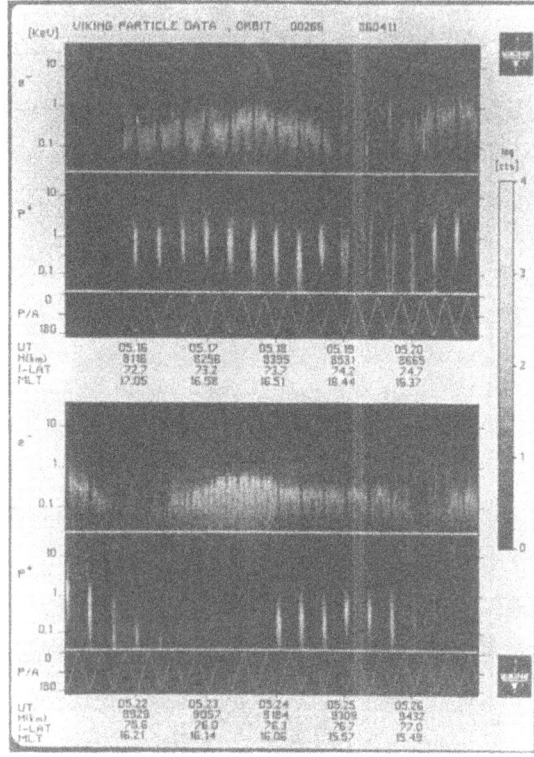

Figure 9—Particle data from Viking, showing that higher upward ion-beam energies correspond to wider loss cones for electrons in the upward direction. Corresponding quantitative results are given in Table 1. (Courtesy, R. Lundin and L. Eliasson.)

to the errors inherent in the method. A much larger statistical sample is required to determine whether a systematic difference exists. This is important since it may provide a clue to the spatial structure throughout the acceleration region of $E_\parallel(\mathrm{DC},e)$ near maximum PSD (0.1–1 Hz).

Suppose there is an upward directed $E_\parallel(\mathrm{DC},p)$ that can account for the upward ion beams. An $E_\parallel(\mathrm{DC},e)$ that is fluctuating too fast for the ions should give the precipitating electrons additional energy if the integral (Eq. 1) is significant, i.e., if $E_\parallel(\mathrm{DC},e)$ is predominantly of one sign. Such is the case, for example, if the wavelength is large compared with the length of the acceleration region, while short wavelengths severely limit the integrated potential drop. $E_\parallel(\mathrm{DC},p)$ inhibits upward acceleration of cold ionospheric electrons when the fluctuating $E_\parallel(\mathrm{DC},e)$ is directed downward, especially if the latter has a node in the more conductive ionosphere. No such effect can inhibit plasma sheet electrons from being accelerated downward when $E_\parallel(\mathrm{DC},e)$ is directed upward. The frequency-integrated PSD for $E_\parallel(\mathrm{DC},e)$ corresponds to an am-

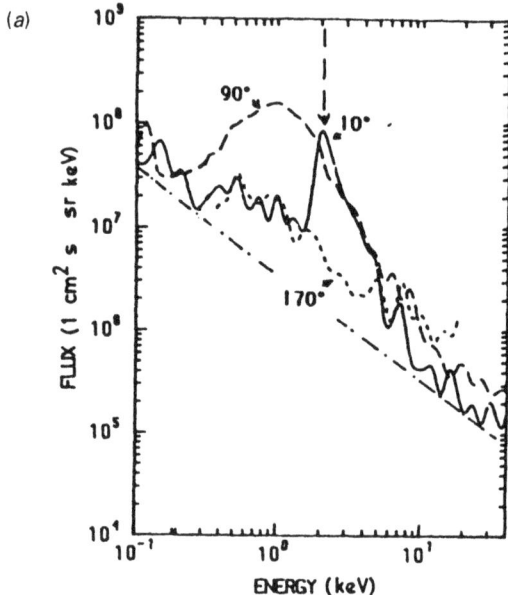

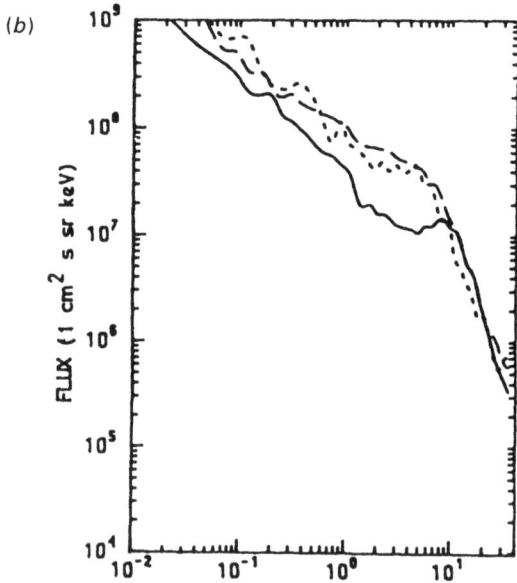

Figure 10—Examples of electron energy spectra for 10°, 90°, and 170° pitch angle taken during two different passes of inverted-V electrons. (*a*) Electron spectra taken within a field-aligned acceleration region with ≈2 kV above and ≈2 kV below the satellite; (*b*) Electron energy spectra taken below a field aligned acceleration region of ≈10 kV. (From *Lundin*, 1988.)

114

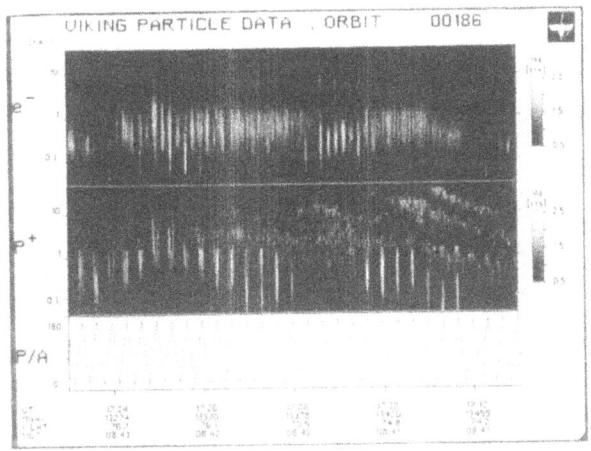

Figure 11—Particle data from Viking orbit 186 on March 17, 1986. During the period from 1724:00–1726:30 UT there are simultaneous upward "elevated" ion conics and upward and downward electron beams. A tentative explanation is proposed in terms of strong fluctuating "parallel" electric fields superimposed on a weaker DC field [*Hultqvist*, 1988]. (Courtesy, R. Lundin.)

plitude of one to a few millivolts per meter (Figure 7), which can easily account for the discrepancies (if real) between the potential drops computed from the electrons and from the ions in Table 1, provided the conditions described are reasonably well fulfilled.

The electron spectra generally indicate acceleration above the satellite when Viking is on field lines threading auroral arcs. Figure 10 shows two examples of spectra over auroral arcs [*Lundin*, 1988]. The sharp peak at 2 keV for the 10° pitch angle in Figure 10a was observed within an acceleration region. The much less pronounced peak at 10 keV in Figure 10b indicates that the electrons have suffered considerable scattering through wave-particle interaction. Viking was then below the acceleration region. The two spectra are, of course, taken at different times so they cannot be directly compared. A statistical analysis of a large number of events is required to draw any general conclusions. However, it seems reasonable to expect that the evidence for DC field acceleration should be cleaner within the acceleration region than below it, where all

TABLE 1. Parallel potential drops calculated from electron loss cones, $V_{\parallel}(e)$, and from upward ion beams, $V_{\parallel}(i)$, observed during different orbits of the Viking satellite.

Orbit	UT	Altitude (km)	E(keV)/LC°	$V_{\parallel}^{(e)}$ (kV)	$V_{\parallel}^{(i)}$ (kV)
266	0516:17	8157	0.15/41	0.58	
			0.25/33	0.59	0.85
			0.28/32	0.60	
			0.18/38	0.59	
343	0537:49	11788	0.755/36	4.9	4.6
			0.37/55	5.0	
343	0538:49	11865	0.73/21	1.34	1.4
			0.37/29	1.55	
1169	0950:09	9192	0.97/18	0.30	0.4
			0.52/22	0.48	
1169	0950:28	9151	0.52/28	1.0	0.65
			0.28/37	1.1	
1169	0955:26	8496	0.91/32	2.1	1.9
			0.57/36	1.8	
1169	0955:46	8451	0.91/32	2.1	1.6
			0.57/37	1.9	
1169	0956:04	8410	0.85/21	0.43	0.55
			0.52/25	0.57	
1169	0956:25	8361	0.85/21	0.42	0.5

Double or multiple values of $V_{\parallel}(e)$ obtained for the same particle velocity distribution correspond to loss cone widths (LC°) measured at different energies E(keV) as indicated at the top.

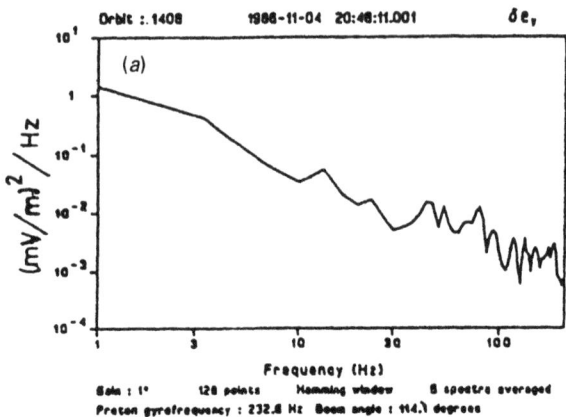
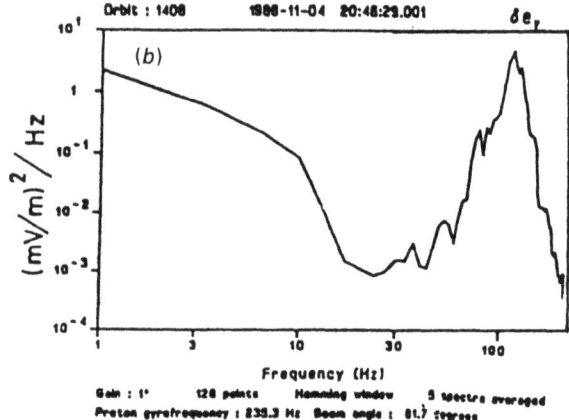

Figure 12—Power density spectra in the frequency range from 1–214 Hz at two different times during a Viking pass over an auroral arc. Sometimes the power spectral density decreases monotonically with increasing frequency (panel *a*). At other times peaks occur at some frequency below the proton gyrofrequency. Panel *b* shows a peak near half the proton gyrofrequency.

sorts of particle-plasma-wave interactions may have smeared out the expected spectral features. Particle heating within the acceleration region has been discussed by *Reiff et al.* [1988].

7. SIMULTANEOUS UPWARD ACCELERATION OF IONS AND ELECTRONS

Figure 11 shows particle data from Viking orbit 186. Of particular interest is the period 1724:00–1726:30 UT when there are upward ion conics and both upward and downward electron beams, all with comparable energies, mainly in the 0.1–1 keV range. The ion conics are "elevated," especially around 1725 UT. This means that there are no ions below a certain energy, which approaches 1 keV in the middle of the event, as if the ions in the conic have been accelerated upward by a DC electric field in addition to the magnetic mirror force responsible for the conic itself. A study of this event is reported by *Hultqvist et al.* [1988a]. *Hultqvist* [1988] has proposed an explanation in terms of an $E_\parallel(DC,p)$ producing the ion conic elevation. The upward and downward electron beams are accelerated by fluctuating and substantially stronger, spatially coherent $E_\parallel(DC,e)$ with frequencies that appear as AC for the protons.

If $E_\parallel(DC,e) > E_\parallel(DC,p)$ in the ionosphere (at the topside F-layer, say), upward acceleration of ionospheric electrons will not be inhibited when the fluctuating field is directed downward. The Viking magnetometer indicated a downward Birkeland current at this time. That current will certainly help in explaining the upward electron beams, but it also means that there is

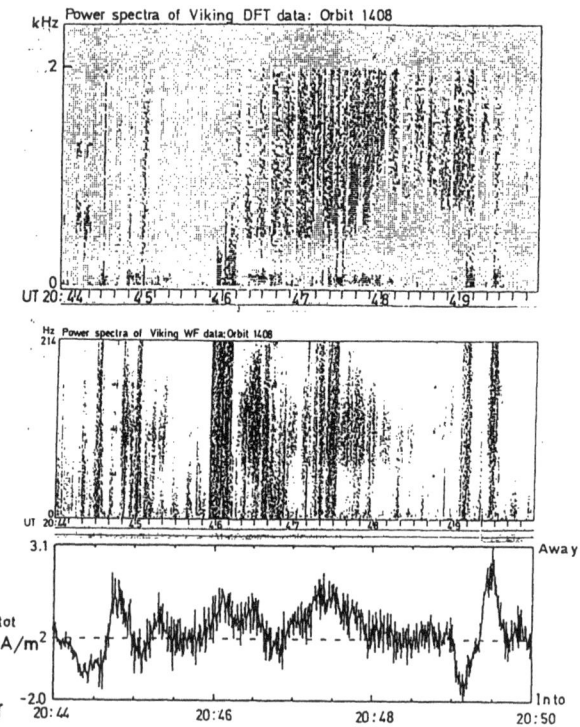

Figure 13—Wave spectrograms for the event shown in Figure 12. The top panel extends to 2000 Hz, the middle panel is an expanded view of the lower part of the frequency band (1–214 Hz). The bottom panel shows the total field-aligned current density calculated from simultaneous Viking magnetometer data. The pronounced spectral peaks around 100–150 Hz (as in Figure 12*b*) do not always occur at maximum current density but more often where the current is relatively low, e.g., near 2047:00, at 2047:30–2047:40, and at 2048:00–2048:10 UT.

116

a generator for $E_\parallel(DC,p)$ that is directed upward. If Hultqvist's explanation is correct, it presents a challenge for theoreticians to explain this generator.

8. WAVES WITH HIGHER FREQUENCIES

Figure 12 shows electric field power spectra (PSD) extending up to 214 Hz taken at two different times during a passage through the auroral acceleration region. One of the spectra shows a pronounced peak near half of the local proton gyrofrequency. (Similar peaks near one-quarter or three-quarters of the local proton gyrofrequency have also been observed.) These have been attributed to shear Alfvén waves that are driven by a precipitating electron beam with field-aligned velocities (in this event ~ 60 Mm s^{-1}) below the local Alfvén velocity (~ 80 Mm s^{-1}) [*Gustafsson et al.*, 1989]. It is interesting that the energy flows from the particles to the waves and not the converse. Thus the current cannot be driven by the waves.

Figure 13 shows a wave spectrogram for the same event that is shown in Figure 12 and includes the two spectra in Figure 12. The bottom panel in Figure 13 shows the Birkeland current density during the same period. The shear Alfvén waves have frequencies around 100–150 Hz. Their correlation with the current is not very high. The high peak at 2044:40 UT has virtually nothing of them, while the peak at 2049:30 UT displays high intensity over a broader frequency range.

If the waves were driving the currents, a more obvious correlation should be expected. However, if the waves are driven by the current-carrying particles only under certain conditions on their particle distributions, a lower correlation may be expected.

9. THE CAUSE OF MAXIMUM POWER SPECTRAL DENSITY BELOW 1 Hz

The PSD maximum below 1 Hz may be caused by a resonance cavity between the topside ionosphere and the equator. The Alfvén velocity is of the order of 1 Mm s^{-1}, both in the ionosphere and at geosynchronous orbit and even lower further away from Earth. However, at about 3000-km altitude the Alfvén velocity can occasionally be as high 100 Mm s^{-1}, i.e., one-third of the velocity of light, due to low particle density of the order of 10–20 protons cm^{-3}. At Viking's apogee the Alfvén velocity is usually of the order of 10–30 Mm s^{-1}. It then falls off toward the equator because of a weakening magnetic field. The reason for the high Alfvén velocity at intermediate altitude is that the plasma density decays much faster with increasing altitude than does the magnetic field.

The resulting Alfvén velocity profile along the geomagnetic field lines provides a resonance cavity for waves in the 0.1–1 Hz frequency range. Normally the

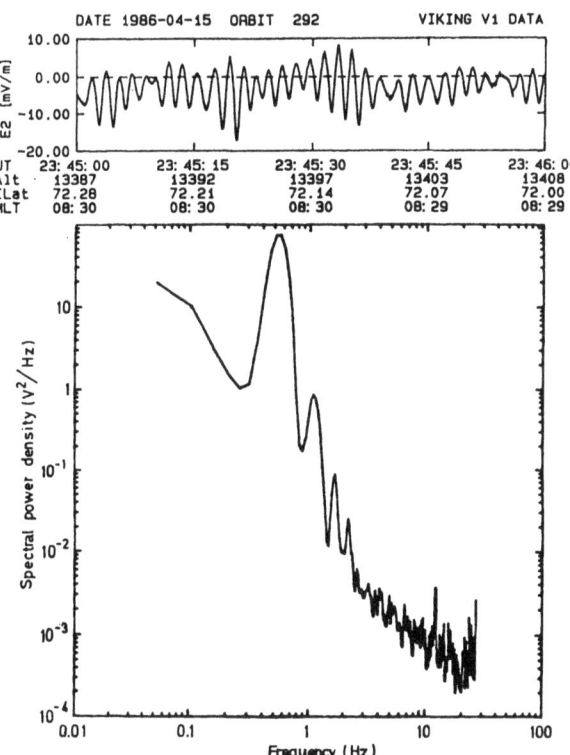

Figure 14—Nearly monochromatic waves with frequency around 0.5–1 Hz are often observed in the late morning sector. The top panel gives an example of the associated electric field. The corresponding spectrum with a strong peak around 0.6 Hz is shown in the lower panel. It is hypothesized that this peak is due to a resonance in a cavity of high Alfvén velocity at an altitude of a few Earth radii.

resonant cavity is not very well defined since the variation of Alfvén velocity does not have clear steps from low to high velocity and vice versa. The PSD maximum is therefore quite broad as in Figures 7 and 8. In the late morning sector, however, very monochromatic waves with frequency around 0.4–0.6 Hz are often seen [see e.g., *Block et al.*, 1987]. Figure 14 shows a spectrum of such a wave with a sharp peak at about 0.55 Hz. We suggest that it is due to a very steep density decay in the topside ionosphere.

10. SUMMARY

The results reported here may be summarized as follows:

1. DC "parallel" electric fields ($E_\parallel$) are defined, from the point of view of the particles, in terms of a maximum frequency that depends on the energy and charge-to-mass ratio, and on the length of the acceleration region. For example: f_{max} is a few hertz for a 1-keV electron, and less than 0.1 Hz for a 1-keV proton.

2. The acceleration region is usually below $3R_E$ geocentric distance.

3. DC parallel electric fields with duration greater than 10 seconds and of the order of 1–10 mV m^{-1} appear to be common in the acceleration region. They are probably distributed over a multitude of small double layers.

4. The power spectral densities (PSD) of $E_\parallel$ and $E_\perp$ have a broad maximum at 0.1–1 Hz. In the spectra so far examined we have found no exception.

5. At frequencies below 1 Hz, $E_\perp \gg E_\parallel$. However, at 20–30 Hz, $E_\perp \geq E_\parallel$ (with increasing frequency, there is a steeper decay for $E_\perp$).

6. The PSD is much lower outside than inside the acceleration region, but both have a maximum in the same frequency range (0.1–1 Hz).

7. For upward Birkeland currents, the field-aligned potential drop computed from the electron loss cone is nearly equal to or, in a few cases, larger than the same derived from the upward ion-beam energy. The differences may be within the error bars but, if they are real, a possible explanation may be that electrons are energized more than ions because of the broader frequency range that appears as DC fields for electrons.

8. Simultaneous upward electron and ion beams, with about the same energy, are sometimes observed when the Birkeland current is directed downward, as reported by *Hultqvist et al.* [1988]. These are explained by *Hultqvist* [1988] in terms of a DC field and strong fluctuating fields that appear as DC fields to the electrons but not to the ions.

9. Sometimes, the PSD has a secondary peak at about one-quarter, one-half, or three-quarters of the proton gyrofrequency. The corresponding waves are identified as shear Alfvén waves, driven by electron beams with a field-aligned velocity below the Alfvén velocity.

10. The maximum PSD at 0.1–1 Hz may be due to a resonance in the altitude region between the topside ionosphere and the low-latitude plasma sheet (2,000–30,000 km or so) where the Alfvén velocity peaks.

ACKNOWLEDGMENT—The authors thank G. Gustafsson for detailed discussions of the wave data from Viking, and to L. Eliasson for supplying the particle data.

REFERENCES

Albert, R. D., and P. J. Lindstrom, "Auroral Particle Precipitation and Trapping Caused by Electrostatic Double Layers in the Ionosphere," *Science*, **170**, 1938 (1970).

Alfvén, H., *Cosmical Electrodynamics*, Clarendon Press, Oxford (1950).

Alfvén, H., and C.-G. Fälthammar, *Cosmical Electrodynamics, Fundamental Principles*, 2nd ed., Oxford University Press, New York (1963).

Atkinson, G., "Inverted V's and/or Discrete Arcs: A Three-dimensional Phenomenon at Boundaries between Magnetic Flux Tubes," *J. Geophys. Res.*, **87**, 1528 (1982).

Block, L. P., "On the Distribution of Electric Fields in the Magnetosphere," *J. Geophys. Res.*, **71**, 855 (1966).

Block, L. P., "Potential Double Layers in the Ionosphere," *Cosmic Electrodynam.*, **3**, 349 (1972).

Block, L. P., "Three-dimensional Potential Structure Associated with Birkeland Currents," in *Magnetospheric Currents*, Geophysical Monograph 28, T. A. Potemra, ed., American Geophysical Union, Washington, DC, p. 315 (1983).

Block, L. P., C.-G. Fälthammar, P. A. Lindqvist, G. T. Marklund, F. S. Mozer, and A. Pedersen, "Electric Field Measurements on Viking: First Results," *Geophys. Res. Lett.*, **14**, 435 (1987).

Boström, R., H. Koskinen, and B. Holback, "Low Frequency Waves and Solitary Structures Observed by Viking," in *Small Scale Plasma Processes, Proc. 21st ESLAB Symposium*, Bolkesjö, Norway, 22–25 Jun 1987, p. 185 (1987).

Boström, R., G. Gustafsson, B. Holback, G. Holmgren, H. Koskinen, and P. Kintner, "Characteristics of Solitary Waves and Weak Double Layers in the Magnetospheric Plasma," *Phys. Rev. Lett.*, **61**, 82 (1988).

Bryant, D. A., "Wave Acceleration of Auroral Electrons," *Proc. 8th ESA Symposium on European Rocket and Balloon Programmes and Related Research*, Sunne, Sweden, 17–23 May 1987, ESA SP-270, p. 273 (1987).

Burch, J. L., "Simultaneous Plasma Observations with DE-1 and DE-2," Paper No. 8.5.2, presented at XXVI COSPAR Meeting, 18–29 July 1988, Helsinki, Finland, *Adv. Space Res.*, **7** (1988).

Fälthammar, C.-G., "Magnetic-Field-Aligned Electric Fields," *ESA J.*, **7**, 385 (1983).

Fälthammar, C.-G., "Magnetosphere-Ionosphere Interactions—Near-Earth Manifestations of the Plasma Universe," *Astrophys. Space Sci.*, **144**, 105–133 (1988).

Gustafsson, G., L. Matson, and M. André, "Waves Below the Local Proton Gyrofrequency in Auroral Acceleration Region," submitted to *J. Geophys. Res.* (1989).

Hultqvist, B., "On the Acceleration of Electrons and Positive Ions in the Same Direction Along Magnetic Field Lines by Parallel Electric Fields," *J. Geophys. Res.*, **93**, 9777–9784 (1988).

Hultqvist, B., R. Lundin, K. Stasiewicz, L. P. Block, P.-A. Lindqvist, G. Gustafsson, H. Koskinen, A. Bahnsen, T. A. Potemra, and L. J. Zanetti, "Simultaneous Observation of Upward Moving Field-aligned Energetic Electrons and Ions on Auroral Zone Field Lines," *J. Geophys. Res.*, **93**, 9766–9776 (1988).

Lundin, R., "Acceleration and Heating of Plasma on Auroral Field Lines: Preliminary Results from the Viking Satellite," *Ann. Geophys.*, **88**, 143 (1988).

Mozer, F. S., "The Low Altitude Electric Field Structure of Discrete Auroral Arcs," in *Physics of Auroral Arc Formation*, Geophysical Monograph 25, S.-I. Akasofu and J. R. Kan, eds., American Geophysical Union, Washington, DC, p. 136 (1981).

Mozer, F. S., and M. Temerin, "Solitary Waves and Double Layers as the Source of Parallel Electric Fields in the Auroral Acceleration Region," in *High-Latitude Space Plasma Physics*, B. Hultqvist and T. Hagfors, eds., Plenum Press, New York and London, pp. 453–468 (1983).

Mozer, F. S., R. B. Torbert, U. V. Fahleson, C.-G. Fälthammar, A. Golfalone, and A. Pedersen, "Measurement of Quasi-Static and Low-Frequency Electric Fields with Spherical Double Probes on the ISEE-1 Spacecraft," *IEEE Trans. Geosci. Electron.*, **GE-16**, 258 (1978).

Reiff, P. H., H. L. Collin, J. D. Craven, J. L. Burch, J. D. Winningham, E. G. Shelley, L. A. Frank, and M. A. Friedman, "Determination of Auroral Electrostatic Potentials Using High- and Low-Altitude Particle Distributions," *J. Geophys. Res.*, **93**, 7441–7465 (1988).

Rostoker, G., and R. Boström, "A Mechanism for Driving the Gross Birkeland Current Configuration in the Auroral Oval," *J. Geophys. Res.*, **81**, 235–244 (1976).

Taylor, H. E., and F. W. Perkins, "Auroral Phenomena Driven by the Magnetospheric Plasma," *J. Geophys. Res.*, **76**, 272 (1971).

Temerin, M., K. Cerny, W. Lotko, and F. S. Mozer, "Observations of Double Layers and Solitary Waves in the Auroral Plasma," *Phys. Rev. Lett.*, **48**, 1175–1179 (1982).

Vasyliunas, V. M., "Mathematical Models of Magnetospheric Convection and Its Coupling to the Ionosphere," in *Particles and Fields*, B. M. McCormac, ed., D. Reidel, Dordrecht-Holland, p. 60 (1970).

Weimer, D. R., "Auroral E-fields from DE-1 and -2 at Magnetically Conjugate Points," Paper No. 8.5.4, presented at XXVI COSPAR Meeting, 18–29 July 1988, Helsinki, Finland, *Adv. Space Res.*, **7** (1988).

III-4. AURORAL ELECTRON ACCELERATION:
A CASE FOR THE STOCHASTIC ALTERNATIVE

D. A. Bryant,* D. S. Hall,* and R. Bingham*

A model of the auroral electron acceleration process in which the electrons are accelerated resonantly by lower-hybrid waves is presented. The essentially stochastic acceleration process is approximated, for the purposes of computation, by a deterministic model involving an empirically derived energy transfer function. The empirical function, which is consistent with all that is known of electron energization by lower-hybrid waves, allows many, possibly all, observed features of the electron distribution to be reproduced.

1. INTRODUCTION

It is well established, following the pioneering experiments of *McIlwain* [1960], *Albert* [1967], and *Evans* [1968], that electrons of the terrestrial plasma sheet or plasma-sheet boundary layer become accelerated in the course of being precipitated into the atmosphere to produce the more structured and generally brighter forms of aurora such as auroral arcs and rays. The energies affected and the energies gained during acceleration are typically 1–10 keV, just above the mean energy of the frequently power-law initial-velocity distribution. An often present magnetic-field alignment shows that the acceleration is preferentially, if not totally, parallel to **B**, and that it takes place at geocentric distances ranging from just above the atmosphere, to $3R_E$, where R_E is the Earth radius of 6370 km. It is clear, therefore, that the accelerator has a component of electric field **E** which, in the northern hemisphere, is antiparallel to **B** such that

$$\mathbf{E} \cdot \mathbf{B} < 0$$

The key question is, what gives rise to this electric field? Is it the (conservative) field of a space charge, static or varying, or is it an induced field? We here explore the possibilities for a conservative field, and find that, if it derives from a particular form of electrostatic wave, it is fully able to account for the vast majority, possibly all, of the features observed in auroral electron velocity distributions. The lack of systematic oscillations, with alternate phases of acceleration and retardation, argues immediately, in any case, against an induced field being primarily responsible.

Electron Motion in a Space-Charge Field

We recall first the basic modes of interaction of an electrically charged particle with a finite configuration of space charge. Without loss of generality, we con-

*Rutherford Appleton Laboratory, Chilton Didcot, Oxfordshire, OX110QX, United Kingdom.

Auroral Physics, edited by C.-I. Meng, M. J. Rycroft and L. A. Frank. © Cambridge UP 1991

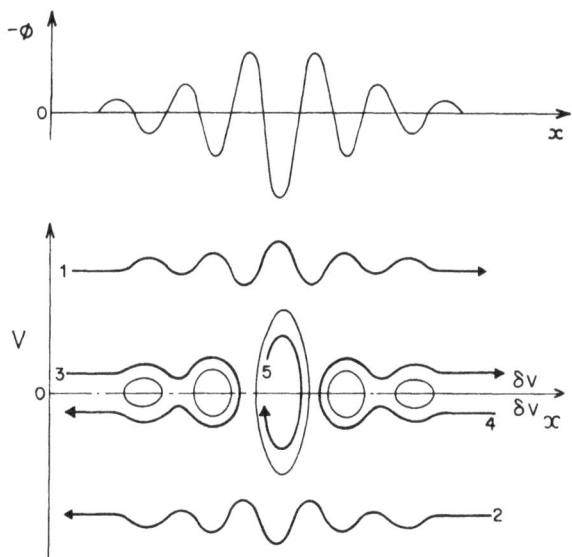

Figure 1—Trajectories (schematic) in two-dimensional phase space (lower panel) experienced by electrons in the potential distribution, $\phi(x)$, of an electrostatic wave packet (upper panel). In the frame of reference of the wave packet, resonant electrons experience simply a reversal in velocity. In other frames they gain or lose energy. Nonresonant and trapped electrons suffer no net change in velocity in any frame.

sider a charge distribution giving a one-dimensional distribution of potential barriers and potential wells, as shown in the upper panel of Figure 1. In the lower panel we illustrate in two-dimensional phase space the types of trajectory experienced by electrons encountering such a potential distribution. In the frame of reference in which the space charge (assumed to have an effective mass very much greater than that of a single electron) is at rest, high-speed electrons (labeled 1 and 2), whether traveling in the $+x$ or $-x$ direction, will negotiate the potential distribution and emerge, after some temporary modulation, with no net change in velocity. Electrons (labeled 3 and 4) approaching with speeds insufficient to surmount the barriers will be reflected without change in speed, and electrons trapped in potential wells (labeled 5) will undergo oscillations, but will again experience no net change in speed. This is, of course, no more than a particular confirmation that there is no exchange of energy in the center-of-mass frame.

In a frame of reference in which the space charge is moving and becomes an electrostatic wave packet at velocity v, an electron approaching at $v \pm \delta v$ will, if δv is less than the critical value required to surmount the wave, be reflected in the space-charge frame and suffer an energy change

$$\Delta E = \pm 2mv\delta v$$

where m is the electron mass. Overtaking electrons will lose energy to the wave packet on reflection, while overtaken electrons will gain energy. Such electrons may be said to be resonant. Nonresonant and trapped electrons will experience no net change of energy in this, or any other frame.

Resonant Acceleration by Lower-Hybrid Waves

One type of electrostatic wave that can resonate with auroral electrons is the lower-hybrid wave. Following earlier indications that a wave process was in operation [*Bryant et al.*, 1978; *Whalen and Daly*, 1979; *Bryant*, 1981, 1983; *Hall*, 1983; *Hall et al.*, 1984], a theory of auroral electron acceleration by such waves was advanced by *Bingham et al.* [1984] and further developed by *Bryant* [1987] and *Bingham et al.* [1988]. The process was invoked also by *Chernikov et al.* [1987]. Lower-hybrid waves propagate almost perpendicular to **B** with a phase velocity approximately equal to the ion sound speed. However, their phase velocity parallel to **B**, by virtue of the small angle of attack (square root of the electron-to-proton mass ratio), is $\geq v_{Te}$, where v_{Te} is the electron thermal speed in a Maxwellian plasma, as illustrated in Figure 2. *McBride et al.* [1972] have shown that the wave spectral density is negligible below v_{Te} and that it peaks at 3–$5v_{Te}$, decreasing continuously toward the speed of light. This characteristic was discussed by *Papadopoulos* [1981] in relation to electron acceleration in magnetosonic shock fronts. We shall assume that for non-Maxwellian electron distributions there is equivalent behavior with respect to the mean energy. It may be seen from *Bingham et al.* [1988] that the parallel group velocity is also high, being approximately equal to the parallel phase velocity divided by $\sqrt{2}$.

Such waves not only abound on auroral-zone flux tubes [*Scarf et al.*, 1973; *Gurnett and Frank*, 1977; *Mozer et al.*, 1980; *Pottelette et al.*, 1988] but are ubiquitous in space-plasma and laboratory-plasma interactions. While lower-hybrid waves will appear to have the necessary properties to account for auroral elec-

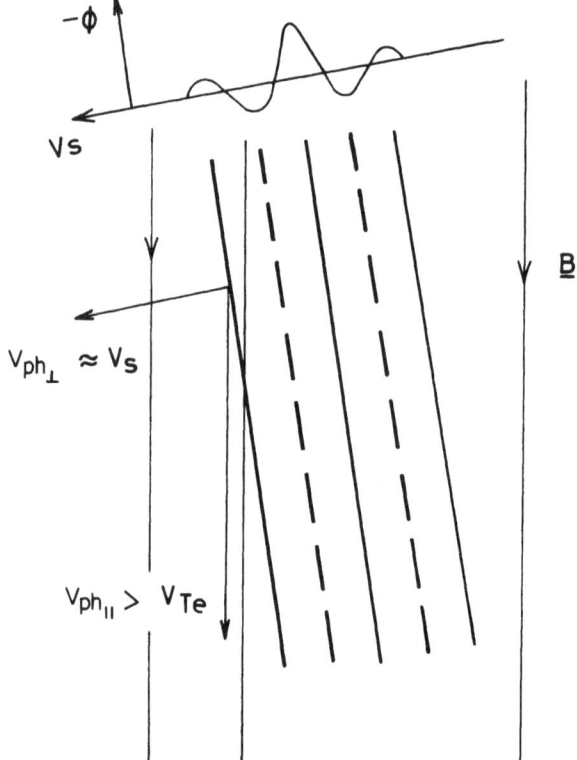

Figure 2—Construction showing how the phase velocity of a lower-hybrid wave packet exceeds the electron thermal speed v_{Te} parallel to the magnetic vector **B**, while the phase velocity perpendicular to $\mathbf{B} \approx v_s$, the ion sound speed.

tron acceleration, we note here in passing that any other electrostatic wave modes with similar properties could also perform the same function.

In view of the small angle of attack, there is very little in the present application to distinguish between upward- and downward-moving wave packets. It is to be expected, therefore, that at least on occasions, the accelerator may be bidirectional though not necessarily symmetric.

Stochastic Acceleration

The factors that influence the flow of energy between electrons and wave packets may readily be gauged from Figure 3. Consider two wave packets WP_1 and WP_2 moving at speeds v_1 and v_2 in the x direction. A resonant electron initially at point 1 in phase space will be accelerated by WP_1 to point 2, where it might subsequently be overtaken by WP_2 and further accelerated to point 3. Alternatively, the same electron after acceleration to point 2 might (from point 2*) overtake another wave packet of the same speed as WP_1 and be retarded to point 1*, thereby returning to its initial velocity at 1. The probability that an electron gains energy (point 2 to point 3) or loses energy (2* to 1*)

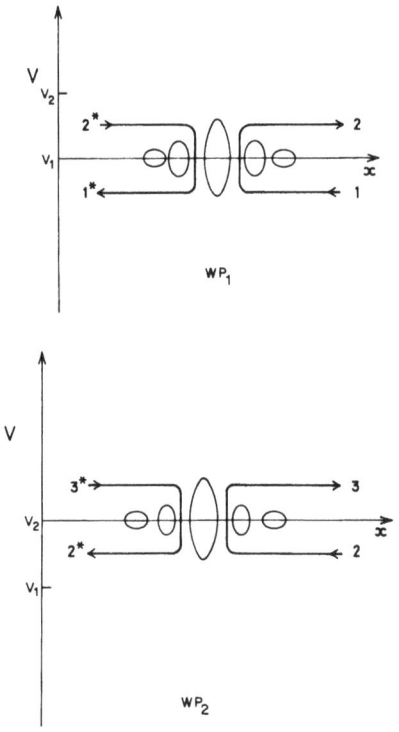

Figure 3—Framework (schematic) for discussion of stochastic energy exchange between lower-hybrid waves and resonant electrons. Successive interactions of an electron with wave packets WP_1 and WP_2 will accelerate an electron from point 1 to point 3. Interactions in the reverse order will cause a retardation from 3* to 1*. Electrons at 2 or 2* will gain or lose energy depending on whether electrons at this velocity encounter more or fewer wave packets of type WP_2 than WP_1 (i.e., on whether the wave-packet velocity distribution has a positive or negative slope). Similar considerations apply to the electron velocity distribution from the standpoint of a given wave packet. For further clarification see text.

will clearly depend on the relative numbers of head-on or overtaking collisions, respectively, that in turn depend on whether it encounters more or fewer wave packets of type WP_2, than WP_1, respectively, i.e., on whether df_w/dv is positive or negative at the velocity in question, f_w being the wave-packet velocity distribution function.

By the same token, the probability of a given wave packet gaining energy from, or losing energy to, the electron distribution depends on whether df_e/dv is positive or negative. Energy flow from waves to electrons is, therefore, promoted by $df_w/dv > 0$ and $df_e/dv < 0$. Flow is from electrons to waves if the inequalities are reversed. A rigorous assessment of the rate of flow of energy and the effect of competing inequalities awaits a full statistical treatment.

2. ACCELERATION MODEL

Pending a rigorous treatment of the problem, we take below an intuitive step leading to a quantitative hypothesis that can be tested against the wealth of observations.

Our hypothesis is that the properties of lower-hybrid waves (outlined above) and the shapes of the velocity distribution functions of wave packets and electrons combine to produce a velocity-dependent degree of acceleration $a(v)$, v being the velocity parallel to **B**, which operates on an initial electron velocity distribution over a series of small increments in time, δt, to modify the distribution through

$$\delta v = a(v)\delta t$$

The function itself will, when used quantitatively, be expressed as a notional energy gain at v and termed the energy transfer function,

$$\Delta E(v) = a(v)dm$$

where d is the length of the acceleration region in the direction of **B**.

The effect of velocity-dependent acceleration on an electron velocity distribution is shown schematically in Figure 4. The velocity distribution becomes distorted over a limited range of velocities. Distortion is greatest at velocities close to, but below, those where the transfer function peaks, at ΔE_{max}, while the low-energy portion of the distribution is unaffected and the high-energy portion is only weakly affected. A similar approach to the problem was adopted by *Karney and Fisch* [1979] who, in considering electron heating by lower-hybrid waves in tokamaks (a phenomenon demonstrated earlier by *Boyd et al.* [1976]), used a rectangular function rather than the continuous curve used here.

A vital consideration in any acceleration process is the continuity of velocity-space density. As a step in the standard proof of the Liouville theorem for the invariance of velocity-space density f in a conservative Hamiltonian system (e.g., *Marion* [1970]), it is shown that

$$(df/dt) + f\mathrm{div}_c v + f\mathrm{div}_v a = 0$$

where div_c is the divergence in configuration space, and div_v is the divergence in velocity space. In a Hamiltonian system, each term of the divergencies exactly cancels its opposite number (in fact in most common systems, both divergencies are independently zero also), leading to the familiar Liouville theorem of the

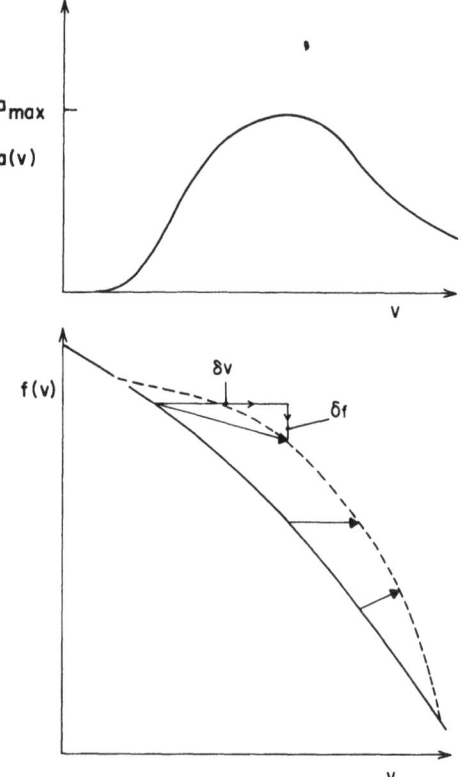

Figure 4—Deterministic representation of stochastic acceleration, showing the effect of a velocity-dependent acceleration $a(v)$ on an electron distribution $f(v)$, where v is the velocity parallel to **B**. The incremental changes to f and v are discussed in the text.

invariance of velocity-space density along a trajectory in phase space,

$$df/dt = 0$$

In the present application $\text{div}_v a$ is equal to da/dv which is generally nonzero, so the above simplification is no longer valid.

We have, since there is no source or sink and $\text{div}_c v = 0$,

$$(df/dt) + f(da/dv) = 0$$

Integration over an increment of time short enough for the change in velocity to leave da/dv unchanged gives

$$f = f_0 \exp[-(da/dv)\delta t]$$

or, for small increments,

$$\delta f = -(da/dv)f\delta t$$

The effect of this change in velocity-space density is indicated schematically in Figure 4 where, in the region of $da/dv > 0$, f is reduced; where $da/dv = 0$, f remains constant; and, where $da/dv < 0$, f is increased. The last statement is analogous to the approach to terminal velocity in a viscous medium. The validity of this correction to f in the analysis that follows has been confirmed by checking that the net current is continuous.

No attempt is made in the present approximation to adjust the energy transfer function in accordance with changes in the electron distribution. In principle, though, the method could be extended to include feedback that would clearly serve to limit the development of positive slopes.

The framework of the numerical model (Figure 5) is a flux tube along which **B** varies as R^{-3}, R being the geocentric radial distance. Electrons of energy E are incident from above with a distribution function $f(E) = 10E(\text{keV})^{-2.75}\text{km}^{-6}\text{s}^3$ and are isotropic over the pitch-angle range from 0–90°. This distribution bears a close resemblance both in shape and magnitude to the distributions observed adjacent to regions of discrete aurora [e.g., *Bryant*, 1981; *Rinnert et al.*,

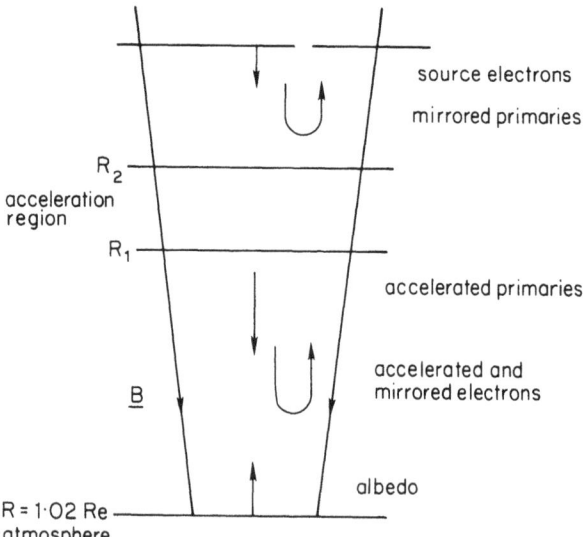

Figure 5—Framework of the numerical model. An isotropic, power-law electron distribution is incident upon a geomagnetic flux tube from a magnetospheric source at high altitude. Between geocentric radial distances R_1 and R_2, a velocity-dependent accelerator of the form shown schematically in Figure 4 acts on the electron distribution. The accelerated electrons reach the effective atmosphere at R = $1.02R_E$ (≈ 125 km altitude) where they produce an albedo of backscattered and secondary electrons. The model allows the electron energy and velocity distributions resulting from the effects of the accelerator and magnetic mirror force to be inspected at any desired radial distance.

1987; *Wilhelm et al.*, 1987], and is assumed, for present purposes, to represent the unaccelerated source distribution. Between radial distances R_1 and R_2, which define the acceleration region, an energy transfer function is allowed to operate on the electron distribution as it develops over a series of small increments in time, with the magnetic mirror force ($-E_\perp /\mathbf{B}$ grad $\mathbf{B}$, $E_\perp$ being the kinetic energy of gyromotion) acting simultaneously. The transfer function is maintained constant throughout the acceleration region. Outside, there is only the mirror force.

The effective atmosphere is reached at $R = R_0 = 1.02R_E$, an altitude of ≈ 125 km, where the accelerated electrons give rise to backscattered and secondary electrons. This albedo is evaluated from a numerical code devised by *Evans* [1974] and very generously provided by him for this analysis. The albedo electrons, which fill the loss cone isotropically over the range of pitch angles from 180° to 180° $- \sin^{-1}(R_0/R)^{3/2}$, are then allowed to enter the acceleration region from below, together with reflected accelerated electrons, where, in the bidirectional mode of operation of the model, both components experience an upward acceleration. They then constitute the upward loss cone and the remainder of the upward-moving half of the electron distribution at the source altitude. Thus the electron distribution function is fully defined at all radial distances.

3. PREDICTIONS

We show below how a number of commonly observed features in the auroral electron distribution can be recovered from the model. We illustrate first some features of the energy distribution at zero pitch angle and then continue with features of the full (assumed gyrotropic) velocity-space distribution. In all examples but one, the energy transfer function has exactly the same shape, with only its magnitude and radial distance being varied. The empirically derived $\Delta E(v)$ is shown in Figure 6 (solid curve). We emphasize that, since no feedback from changes in the electron distribution is included in the model, the results below represent only a first approximation.

Features of the Energy Distribution

An accelerated, but still monotonically decreasing energy, distribution is produced at $R \leq R_1$, as shown in Figure 7, by an energy transfer function having $\Delta E_{max} = 2.6$ keV. The original power-law distribution is added for reference.

An increase of ΔE_{max} to 5.4 keV produces a plateau (Figure 8). A further increase to 15 keV produces, as shown in Figure 9, the archetypal auroral arc distribution [*Bryant*, 1981] with (1) a peak, (2) a steep edge

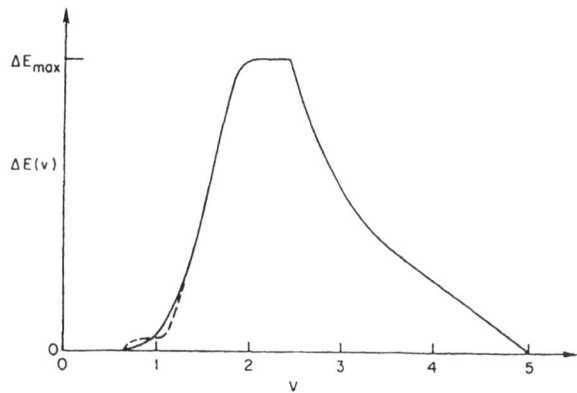

Figure 6—Empirical derivation of the energy transfer function $\Delta E(v)$ used throughout the present analysis. The dashed portion of the curve is discussed in relation to Figure 10.

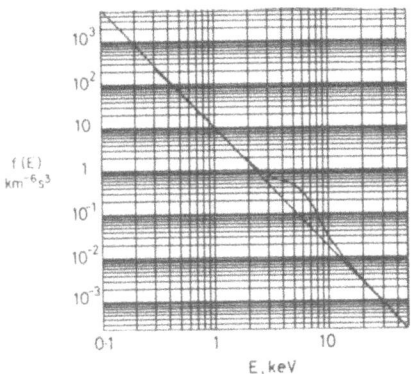

Figure 7—Monotonically decreasing electron energy distribution, seen below an accelerator with $\Delta E_{max} = 2.6$ keV. The original power-law distribution is continued for reference.

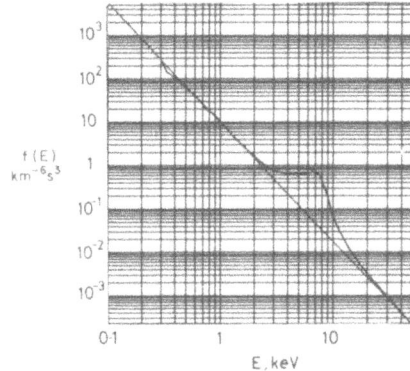

Figure 8—Energy distribution with a plateau, seen below an accelerator with $\Delta E_{max} = 5.4$ keV.

123

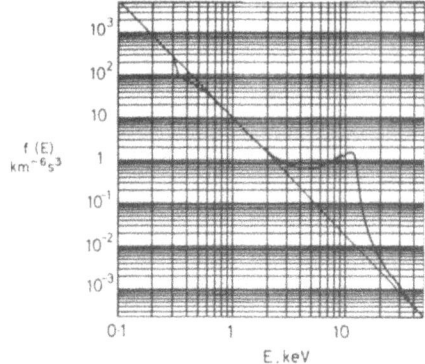

Figure 9—Archetypal peaked auroral electron distribution, seen below an accelerator with ΔE_{max} = 15 keV.

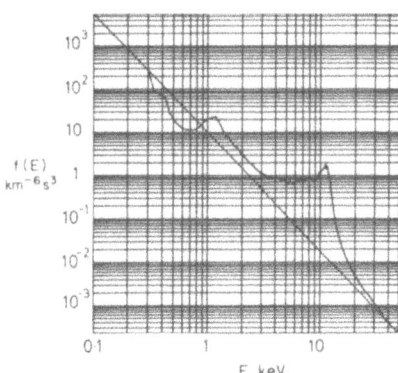

Figure 10—Energy distribution exhibiting a double peak, seen below an accelerator with ΔE_{max} = 15 keV and with a pronounced step at the low-energy threshold as indicated by the broken curve in Figure 6.

on the high-energy side of the peak [*Sandahl and Eliasson*, 1988], strongly reminiscent of an approach to terminal velocity, and (3) unchanged velocity-space densities at the lowest energies [*Whalen and Daly*, 1979; *Hall*, 1980].

When the step in the low-energy portion of the energy transfer function is made more pronounced, as in the dashed curve of Figure 6, a double peak, shown in Figure 10, is formed [*Arnoldy et al.*, 1974; *Hoffman and Lin*, 1981; *Bryant*, 1983; *Wilhelm et al.*, 1987]. This may be seen as either a simultaneous or sequential operation. The major effect introduced by the relatively small change declares a high sensitivity to the shape of $\Delta E(v)$ at low energies.

Features of the Distribution Function

As a control, we show first, in Figure 11, an electron velocity distribution at $R = 2R_E$ for the case when there is no acceleration. In keeping with the for-

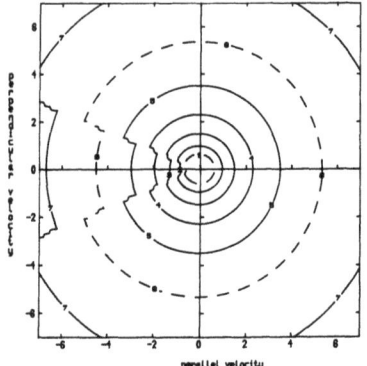

Figure 11—Control electron velocity distribution at R = $2R_E$ for the case of no acceleration. Velocities parallel to **B** (downward in the northern hemisphere) are positive to the right. Velocities perpendicular to **B** are shown along the positive and negative ordinates, in keeping with common practice in data presentation. Electrons with one unit of velocity have an energy of 1 keV. The energies range from zero at the center to ≈ 50 keV at the extremes of the axes and ≈ 100 keV at the corners (neglecting relativistic effects). Contours of velocity-space density (higher numbers representing lower f) are spaced at intervals of one order of magnitude. They are circular over most of velocity-space and vary as $v^{-5.5}(E^{-2.75})$, with the center contour representing 100 km^{-6}s^3. Albedo electrons, with relatively lower velocity-space densities than the incident and mirrored electrons, are seen within the loss cone centered on the upward velocity axis.

mat in which observations are frequently presented, the lower halves of this and other distributions below are simply mirror images of the upper halves. The downward moving electrons are isotropic and, as before, have an energy dependence given by

$$f(E) = 10E \text{ (keV)}^{-2.75} \text{ } km^{-6}s^3$$

The upward-moving electrons are composed of magnetically mirrored electrons and, in the loss cone, are composed of albedo electrons computed from the *Evans* [1974] model. By taking appropriate moments of the distribution at a range of different radial distances, we find that the net field-aligned current density varies with radial distance in such a way as to keep the total current $J_\|$ carried across the whole area of a unit flux tube (of area 1 m^2 at $R = 1R_E$) constant and equal to -0.75 μA. The downward component of energy flux incident upon the atmosphere at $R = 1.02R_E$, $\Sigma_\|$, = 2.7 mW m^{-2}.

Figure 12 shows the effect of an energy transfer function with a peak of 18 keV distributed uniformly between $R_1 = 3.0R_E$ and $R_2 = 3.5R_E$. On reaching the atmosphere at $R = 1.02R_E$, the electrons, although now exhibiting a peak in their energy distribution, are still approximately isotropic over the downward hem-

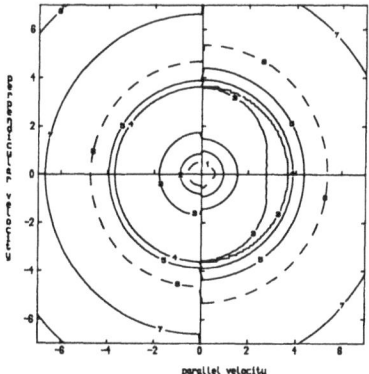

Figure 12—Peaked distribution, approximately isotropic for pitch angles of 0–90°, seen (at $|v| \approx 3$ units) at R = 1.02R$_E$, below an accelerator with ΔE_{max} = 18 keV located between R$_1$ = 3.0R$_E$ and R$_2$ = 3.5R$_E$. At this altitude the loss cone fills the whole of the left-hand side of the figure.

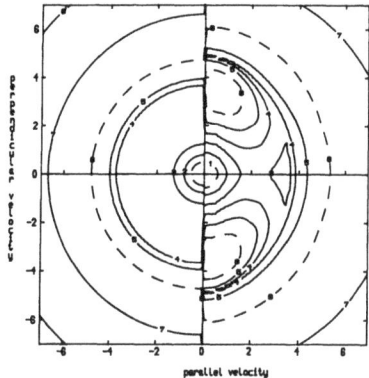

Figure 13—Peaked distribution, strongly magnetic-field aligned at velocities below the peak. Seen (in contours 3 to 6) at R = 1.02R$_E$, below an accelerator with ΔE_{max} = 18 keV located between R$_1$ = 1.1R$_E$ and R$_2$ = 1.8R$_E$.

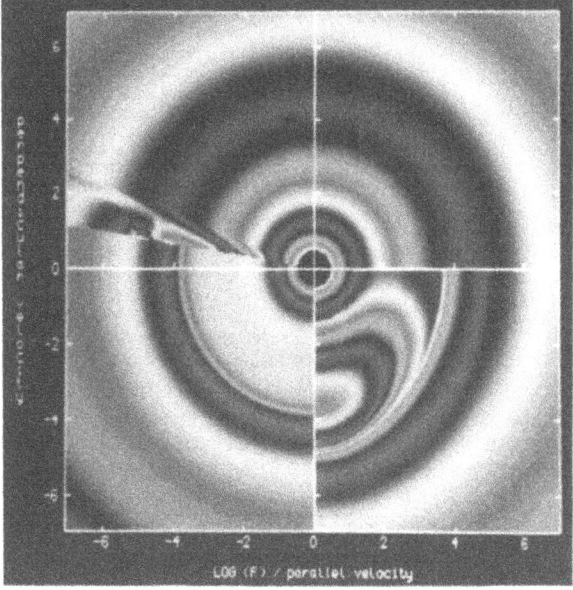

Figure 14—Strongly peaked and magnetic-field-aligned distribution shown in higher resolution than in Figure 13. The accelerator is unidirectional, with ΔE_{max} = 18 keV, located between R$_1$ = 1.1R$_E$ and R$_2$ = 1.8R$_E$. Contours of velocity-space density are color coded. In order to achieve suitable resolution, the color cycle is repeated every three orders of magnitude. An electron conic and a valley adjacent to the loss cone are prominent features of the distribution at R = 2R$_e$. Lower panels: distribution at R = 1.02R$_E$, below the accelerator. Upper panels: distribution at R = 2R$_E$, above the accelerator. (This figure also appears in color: Plate 13.)

color, the electron distributions at both $R = 1.02R_E$ (lower half) and $R = 1.6R_E$ (upper half).

The relatively weak energy gain of 2.4 keV (which, as we saw in Figure 7, produces no more than a bulge in the electron distribution function) can, if distributed uniformly between $R_1 = 1.87R_E$ and $R_2 = 1.95R_E$, produce, at $R = 2R_E$, a form known as an electron conic [*Menietti and Burch*, 1985], as shown in Figure 15. Note that this results entirely from the effects of a downward acceleration parallel to **B** combined with the magnetic mirror force. (*Temerin* drew attention to this effect in a poster paper during the International Conference on Auroral Physics.) It will be noted that the conics appear as ancillary features in several of the other examples.

In Figure 16 we show a particularly clear example of the development of a valley adjacent to the loss cone, which may be related to the observations of apparently widened loss cones [*Mizera et al.*, 1981]. In this example, observed at $R = 2R_E$, the acceleration is distributed between 1.05 and 1.4R$_E$ and has a peak energy gain of 6.3 keV.

As mentioned earlier, it is likely that the accelerator would operate on upward-moving as well as downward-moving electrons. Figure 17 shows that a

isphere, as is often the case in the center of auroral forms [*Bryant*, 1975]. The upward hemisphere at this altitude is composed entirely of albedo electrons. The net field-aligned current density again varies with radial distance, while $J_\parallel$ is again confirmed to be independent of radial distance and equal to -2.0 μA. $\Sigma_\parallel$ has increased to 31 mW m^{-2}. If the acceleration region is between $R_1 = 1.1R_E$ and $R_2 = 1.8R_E$, the resultant distribution, shown in Figure 13, is strongly magnetic-field-aligned at velocities below the peak [*Arnoldy et al.*, 1974; *Hall and Bryant*, 1974; and *Whalen and Daly*, 1979]. The current reduces to $J_\parallel = -1.0$ μA, and the downward energy flux to $\Sigma_\parallel = 13$ mW m^{-2}. These changes demonstrate that both the strength and position of the accelerator are able to modify the field-aligned current and the energy flux. Figure 14 shows, in the higher resolution afforded by

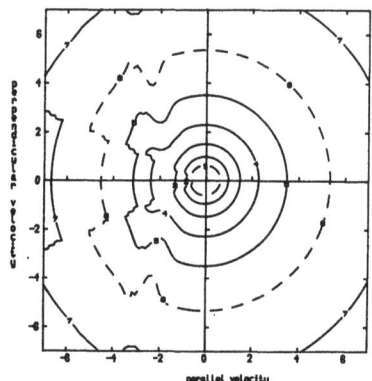

Figure 15—Electron conic, seen (especially in contours 4, 5, and 6) at R = 2R$_E$, above an accelerator with ΔE_{max} = 2.4 keV located between R$_1$ = 1.87R$_E$ and R$_2$ = 1.95R$_E$.

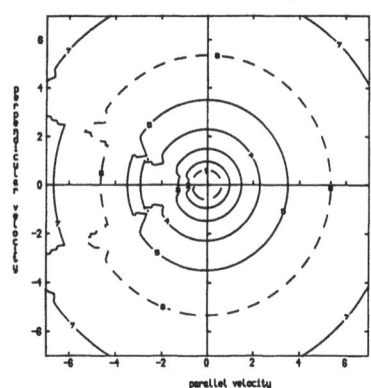

Figure 16—Valley adjacent to loss cone, seen (especially in contours 5 and 6) at R = 2R$_E$, above an accelerator with ΔE_{max} = 6.3 keV located between R$_1$ = 1.05R$_E$ and R$_2$ = 1.4R$_E$.

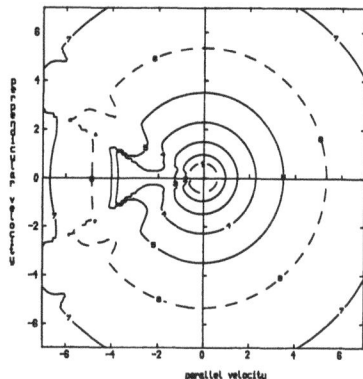

Figure 17—Upward electron beam (at velocity ≈ −4 units) at R = 2R$_E$, above a bidirectional accelerator with ΔE_{max} = 8.25 keV located between R$_1$ = 1.1R$_E$ and R$_2$ = 1.5R$_E$. Other features such as an electron conic (contours 6 and 7) and a valley adjacent to the loss cone (contours 4 and 5) are also present.

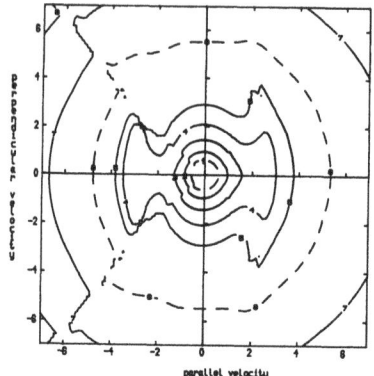

Figure 18—Counterstreaming electrons (around velocities ≈ ±3 units) at R = 1.3R$_E$, within a bidirectional accelerator with ΔE_{max} = 8.25 keV located between R$_1$ = 1.1R$_E$ and R$_2$ = 1.5R$_E$.

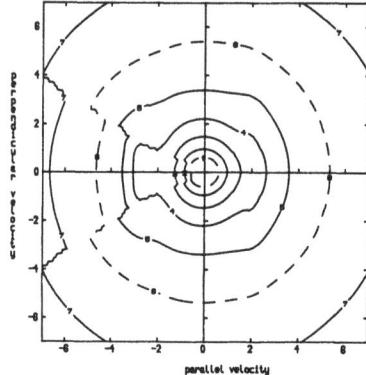

Figure 19—Horseshoe-like contours (contour 5 is further from the origin, just outside the loss cone, than from any other direction) and bulge in downward direction (in contour 4) created at R = 1.7R$_E$, within a bidirectional accelerator with ΔE_{max} = 4 keV located between R$_1$ = 1.05 and R$_2$ = 2R$_E$.

symmetric 8.25 keV bidirectional accelerator located between 1.1 and 1.5R$_E$ would create at 2R$_E$ an upward electron beam [*Klumpar and Heikkila*, 1982 and *Lundin et al.*, 1987]. Within the same accelerator at 1.3R$_E$ we observe, in Figure 18, counterstreaming electrons [*Sharp et al.*, 1980]. Within a weaker (4 keV) bidirectional accelerator distributed between 1.05 and 2R$_E$, we find at 1.7R$_E$, as shown in Figure 19, the horseshoe-like outer contours shown by *Menietti and Burch* [1985] and the bulge in the downward direction found by *McFadden et al.* [1986].

4. ENERGETICS

Bingham et al. [1984] have shown that the observed amplitudes of electrostatic turbulence near the lower-hybrid frequency, amounting to several hundred millivolts per meter [*Mozer et al.*, 1980 and *Pottelette et al.*, 1988], appear, even on the basis of quasi-linear theory, to be sufficient to produce the required doubling of the speed of electrons in the kiloelectron volt range over distances considerably shorter than one Earth radius. Nonlinear effects are anticipated, from the preliminary results of numerical simulation, to increase the effectiveness of energy transfer, therefore, the acceleration process itself seems to be a quantitative and automatic consequence of the observed wave activity.

The natural next question is, what is the source of the waves? Lower-hybrid waves could arise, as discussed by *Bingham et al.* [1988], from a number of instabilities in the ion streams regularly observed in the plasma-sheet boundary layer [*Hones et al.*, 1972 and *DeCoster and Frank*, 1979]. DeCoster and Frank have, in fact, pointed out that the energy flux of 0.5–1 mW m^{-2} seen at $10R_E$ would, if projected along flux tubes to ionospheric altitudes, become 500–1000 mW m^{-2}, substantially exceeding that of the average charged-particle precipitation over discrete auroral arcs. At $R = 3R_E$, where electron acceleration appears to begin, the power input from these ions to a unit flux tube is approximately 30 mW, by virtue of $\mathbf{B} \propto R^{-3}$. This is typical of the power carried from a unit flux tube into the upper atmosphere by electrons producing an auroral arc, appearing to indicate that for acceleration below $R = 3R_E$, essentially the whole of the ion power reaching this geocentric distance would be required. This, of course, seems unlikely. If the energy transfer took place at greater distances, $R = 4R_E$ or $5R_E$ for example, the required efficiency drops to 0.4 and 0.2, respectively, but this is inconsistent with the absence of significant acceleration at these distances. Since the ion energy flux appears to be only fractionally low, and in the absence of any other clear candidate for the power source for this or any other acceleration mechanism, as a possible explanation for the shortfall we offer the suggestion that the ions, when observed at $\approx 10R_E$, are still undergoing energization by, for example, Alfvén waves making the energy flux closer to the Earth significantly greater than 1 mW m^{-2}. This energization could be seen as a continuation or extension of the merging pro-

cess in the magnetotail generally considered to be the energy source for the ion streams. Experimental evidence in support of this hypothesis comes from the observations by *Takahashi and Hones* [1988] that, within the region of counterstreaming beams, reflected ions are typically faster than the earthward streaming ions seen at the same time, the velocity ratio at times exceeding 1.5.

5. CONCLUSIONS

We have found that many well-documented features of the auroral-electron velocity distribution (including a bulge, a plateau, or a peak in the energy distribution, magnetic-field alignment at energies below the peak, an invariant low-energy region, a steeply falling edge above the peak, a double peak, electron conics, an apparently widened loss cone, an upward beam, counterstreaming electrons, a horseshoe-like distribution, and modulations of the field-aligned current and energy flux) can all be accounted for in a deterministic model of stochastic acceleration of the electrons by electrostatic waves of the lower-hybrid type powered by ion streams incident from the magnetospheric tail or other regions of the outer magnetosphere.

More information is needed on the relative distributions of electrons, waves, and ions at all altitudes before the overall efficiency required for the process can be established with any certainty. On the basis of present measurements, which suggest a shortfall in the energy flux carried by the ion streams observed at $10R_E$, it is suggested that these ions are still undergoing acceleration, possibly through the medium of Alfvén waves, with the result that the energy flux closer to the Earth is significantly greater than 1 mW m^{-2}.

ACKNOWLEDGMENT—We are greatly indebted to D. S. Evans for most generously providing the computer code to evaluate the intensities of the secondary and backscattered electrons, and to M. A. Lauder who designed and developed the versatile computer code and graphics. It is also a pleasure to acknowledge an illuminating discussion with S. J. Schwartz on the limits of applicability of the Liouville theorem.

REFERENCES

Albert, R. D., "Nearly Monoenergetic Electron Fluxes Detected During a Visible Aurora," *Phys. Rev. Lett.*, **18**, 369-372 (1967).

Arnoldy, R. L., P. B. Lewis, and P. O. Isaacson, "Field Aligned Auroral Electron Fluxes," *J. Geophys. Res.*, **79**, 4208-4221 (1974).

Bingham, R., D. A. Bryant, and D. S. Hall, "A Wave Model of the Aurora," *Geophys. Res. Lett.*, **11**, 327-330 (1984).

Bingham, R., D. A. Bryant, and D. S. Hall, "Auroral Acceleration by Lower-Hybrid Waves," *Ann. Geophys.*, **6**, 159-168 (1988).

Boyd, D. A., F. J. Staufer, and A. W. Trivelpiece, "Synchrotron Radiation from the ATC Tokamak Plasma," *Phys. Rev. Lett.*, **37**, 98-101 (1976).

Bryant, D. A., "Local Acceleration of Auroral Electrons," in *The Scientific Satellite Programme During the International Magnetospheric Study*, K. Knott and B. Battrick, eds., D. Reidel, Dordrecht and Boston, pp. 413-423 (1976).

Bryant, D. A., "Rocket Studies of Particle Structure Associated with Auroral Arcs," in *Physics of Auroral Arc Formation*, Geophysical Monograph 25, S.-I. Akasofu and J. R. Kan, eds., American Geophysical Union, Washington, DC, pp. 91-102 (1981).

Bryant, D. A., "The Hot Electrons In and Above the Auroral Ionosphere: Observations and Physical Implications," in *High Latitude Space Plasma Physics*, B. Hultqvist and T. Hagfors, eds., Plenum Press, New York and London, pp. 295-312 (1983).

Bryant, D. A., "Wave Acceleration of Auroral Electrons," ESA Scientific Publication SP-270, pp. 273-279 (1987).

Bryant, D. A., D. S. Hall, and D. R. Lepine, "Electron Acceleration in an Array of Auroral Arcs," *Planet. Space Sci.*, **21**, 81-92 (1978).

Chernikov, A. A., R. Z. Sagdeev, D. A. Usikov, M. Yu. Zakharov, and G. M. Zaslavsky, "Minimal Chaos and Stochastic Webs," *Nature*, **326**, 559-563 (1987).

DeCoster, R. J., and L. A. Frank, "Observations Pertaining to the Dynamics of the Plasma Sheet," *J. Geophys. Res.*, **84**, 5099-5121 (1979).

Evans, D. S., "The Observations of a Near-Monoenergetic Flux of Auroral Electrons," *J. Geophys. Res.*, **73**, 2315-2323 (1968).

Evans, D. S., "Precipitating Electron Fluxes Formed by a Magnetic Field-Aligned Potential Difference," *J. Geophys. Res.*, **79**, 2853-2858 (1974).

Gurnett, D. A., and L. A. Frank, "A Region of Intense Plasma Wave Turbulence on Auroral Field Lines," *J. Geophys. Res.*, **82**, 1031-1050 (1977).

Hall, D. S., "On the Acceleration of Auroral Electrons," ESA Scientific Publication SP-152, pp. 285-288 (1983).

Hall, D. S., D. A. Bryant, and T. Edwards, "Are Parallel Electric Fields Really the Cause of Auroral Particle Acceleration?," ESA Scientific Publication SP-229, pp. 95-97 (1984).

Hall, D. S., D. A. Bryant, C. P. Chaloner, D. R. Lepine, and R. Bingham, "AMPTE UKS Electron Measurements During the Lithium Releases of 11 and 20 September 1984," *J. Geophys. Res.*, **91**, 1320-1324 (1986).

Hoffman, R. A., and C. S. Lin, "Study of Inverted-V Auroral Precipitation Events," in *Physics of Auroral Arc Formation*, Geophysical Monograph 25, S.-I. Akasofu and J. R. Kan, eds., American Geophysical Union, Washington, DC, pp. 80-90 (1981).

Hones, E. W. Jr., J. R. Asbridge, S. J. Bame, D. Montgomery, S. Singer, and S.-I. Akasofu, "Measurements of Magnetotail Plasma Flow Made with Vela 4B," *J. Geophys. Res.*, **77**, 5503-5522 (1972).

Karney, C. F. F., and N. J. Fisch, "Numerical Studies of Current Generation by Radio-Frequency Traveling Waves," *Phys. Fluids*, **22**, 1817-1824 (1979).

Klumpar, D. M., and W. J. Heikkila, "Electrons in the Ionospheric Source Cone: Evidence for Runaway Electrons as Carriers of Downward Birkeland Currents," *Geophys. Res. Lett.*, **9**, 873-876 (1982).

Lundin, R., L. Eliasson, B. Hultqvist, and K. Stasiewicz, "Plasma Energization on Auroral Field Lines as Observed by the Viking Spacecraft," *Geophys. Res. Lett.*, **14**, 443-446 (1987).

Marion, J. B., "Classical Dynamics of Particles and Systems," Academic Press, New York and London, 2nd edition, p. 232 (1970).

McBride, J. B., E. Ott, P. B. Jay, and J. H. Orens, "Theory and Simulation of Turbulent Heating by the Modified Two-Stream Instability," *Phys. Fluids*, **15**, 2367-2383 (1972).

McFadden, J. P., C. W. Carlson, and M. H. Boehm, "Field-Aligned Electron Precipitation at the Edge of an Arc," *J. Geophys. Res.*, **91**, 1723-1730 (1986).

McIlwain, C. E., "Direct Measurements of Particles Producing Visible Auroras," *J. Geophys. Res.*, **65**, 2727-2747 (1960).

Menietti, J. D., and J. L. Burch, "Electron Conic Signatures Observed in the Nightside Auroral Zone and Over the Polar Cap," *J. Geophys. Res.*, **90**, 5345-5353 (1985).

Mizera, P. F., J. F. Fennell, D. R. Croley, Jr., and D. J. Gorney, "Charged Particle Distributions and Electric Field Measurements from S3-3," *J. Geophys. Res.*, **86**, 7566-7576 (1981).

Mozer, F. S., C. A. Cattell, M. K. Hudson, R. L. Lysak, M. Temerin, and R. B. Torbert, "Satellite Measurements and Theories of Low Altitude Particle Acceleration," *Space Sci. Rev.*, **27**, 155-213 (1980).

Papadopoulos, K., "Electron Acceleration in Magnetosonic Shock Fronts: Plasma Astrophysics," ESA Scientific Publication SP-161, pp. 313-315 (1981).

Pottelette, R., M. Malingre, A. Bahnsen, L. Eliasson, K. Stasiewicz, R. E. Erlandson, and G. Marklund, "Viking Observations of Bursts of Intense Broadband Noise in the Source Regions of Auroral Kilometric Radiation," *Ann. Geophys.*, **6**, 573-586 (1988).

Rinnert, K., K. Wilhelm, H. Kohl, K. Schlegel, G. Dehmel, H. Lühr, N. Klocker, W. Oelschlagel, M. P. Gough, B. Holback, and K.-I. Oyama, "CAESAR Investigations," ESA Scientific Publication SP-270, pp. 299-303 (1987).

Rodgers, D. J., A. J. Coates, A. D. Johnstone, M. F. Smith, D. A. Bryant, D. S. Hall, and C. P. Chaloner, "UKS Measurements Near the AMPTE Artificial Comet," *Nature*, **320**, 712-716 (1986).

Sandahl, I., and L. Eliasson, "Investigation of Precipitating Electrons During an Auroral Breakup," *Physica Scripta*, **37**, 506-511 (1988).

Scarf, F. L., R. W. Fredericks, C. T. Russell, M. Kivelson, M. Neugebauer, and C. R. Chappell, "Observations of a Current-Driven Plasma Instability at the Outer Zone Plasma Sheet Boundary," *J. Geophys. Res.*, **78**, 2150-2165 (1973).

Sharp, R. D., E. G. Shelley, R. G. Johnson, and A. G. Ghielmetti, "Counter-streaming Electron Beams at Altitudes of 1 R_E Over the Auroral Zone," *J. Geophys. Res.*, **85**, 92-100 (1980).

Takahashi, K., E. W. Hones, "ISEE 1 and 2 Observations of Ion Distributions at the Plasma Sheet-Tail Lobe Boundary," *J. Geophys. Res.*, **93**, 8558-8582 (1988).

Whalen, B. A., and P. W. Daly, "Do Field-Aligned Auroral Particle Distributions Imply Acceleration by Quasi-Static Parallel Electric Fields?," *J. Geophys. Res.*, **84**, 4175-4182 (1979).

Wilhelm, K., K. Rinnert, K. Schlegel, H. Kohl, N. Klocker, H. Lühr, W. Oelschlagel, G. Dehmel, M. P. Gough, B. Holback, and K.-I. Oyama, "Co-Ordinated Auroral Experiments Using Scatter and Rocket Investigations (CAESAR Investigations), Final Report on the Scientific Aspects," MPAE-W-47-87-13 (1987).

III-5. AURORAL ION ACCELERATION AND ITS RELATIONSHIP TO ION COMPOSITION

E. G. Shelley* and H. L. Collin*

Since the early 1970's we have known that ionospheric ions are accelerated to kiloelectron-volt energies and contribute in varying degrees to the hot magnetospheric plasmas. Different acceleration processes can be distinguished. Those that accelerate ions perpendicular to the magnetic field direction are widespread at all altitudes throughout the auroral zone while those that accelerate ions parallel to the magnetic field direction occur mainly in the evening sector in association with the electric fields, which are widely believed to exist above electron inverted-V regions. In these regions the ions also show evidence of mass dependent perpendicular heating. Considerable progress has been made toward combining the theory and modeling of candidate energization mechanisms with experimental measurements. However, there are still gaps in our knowledge of the phenomena, and more sophisticated dedicated auroral missions will be needed to distinguish unambiguously between the various proposed mechanisms.

1. INTRODUCTION

In the past two decades the subject of ion composition of energetic magnetospheric plasmas has matured from a curiosity to an essential element of space plasma physics. Initially ion composition was of interest only from the standpoint of its contribution to the identification of the origins of the plasma [Axford, 1970]. Following the discovery that the Earth's ionosphere might be a significant contributor to the hot magnetospheric plasma [Shelley et al., 1972], interest in ion composition expanded beyond the simple question of plasma origin. It became obvious that velocity distributions of ions could be used to infer features of major acceleration and transport processes, for example, to distinguish processes accelerating ions parallel to the magnetic field direction [Shelley et al., 1976] and perpendicular to it [Sharp et al., 1977]. We now realize that not only is ion composition influenced by the plasma processes, but in fact can itself have a major influence on the processes themselves.

We here review some of the experimental evidence for the close relationships between aurora, auroral electron acceleration, and the subsequent outflow of ionospheric ions from the auroral zones. The observed compositional dependent characteristics of the auroral ion distributions are emphasized and the current theoretical understanding of these mass dependent processes is discussed.

2. ACCELERATED IONOSPHERIC IONS

The first measurements of composition of ions in the auroral acceleration region were made by the S3-3 satellite and were followed by several other satellite studies, ISEE 1, DE 1, Prognoz 7, and Viking. The prevalence of upflowing characteristically terrestrial O^+ at once identified the auroral ionosphere as a major route for the transport of terrestrial ions to the magnetosphere. The distributions of the upflowing ions, H^+ and He^+ as well as O^+, take several forms suggestive of a variety of acceleration or extraction processes. Narrowly field-aligned energetic ion "beams" with kiloelectron-volt energies are frequently observed, suggesting the existence of acceleration parallel to **B**. Lower energy ions, typically of hundreds of electron volts, upflowing oblique to the field line, form a conical distribution in velocity space and imply perpendicular acceleration. At lower altitudes low-energy versions of these "conics" occur almost perpendicular to the field line and are described as transversely accelerated ions (TAI). Even lower-energy upflowing ions, heated by up to tens of electron volts, are also commonly observed.

Initial surveys showed that the location of the upflowing ions is closely correlated with the auroral oval [Ghielmetti et al., 1978]. Figure 1 shows the close match in invariant latitude between the occurrence frequency of the upflows and the occurrence frequency of discrete visible aurora in the 21-24 magnetic local time sector. Later surveys that extended to lower energies [Gorney et al., 1981] and higher altitudes [Yau et al., 1984] showed that, while beams are generally restricted to the evening sector and to altitudes above about 5000 km, conics are common at lower altitudes and at all local times. At upper ionospheric altitudes the ions occasionally reach energies of a few hundred electron volts, whereas at higher altitudes more energetic ion conics and beams are common. Lockwood et al. [1985] reported a persistent region of upwelling ions, principally O^+, with energies up to about 30 eV near the dayside polar cap boundary.

*Space Sciences Laboratory, Lockheed Palo Alto Research Laboratory, Palo Alto, California 94304.

Auroral Physics, edited by C.-I. Meng, M. J. Rycroft and L. A. Frank. © Cambridge UP 1991

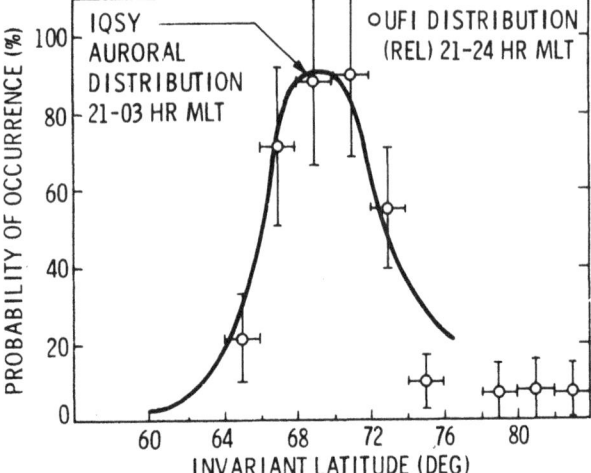

Figure 1—Comparison of visible auroral frequency distribution at solar minimum with the distribution of upward flowing ion events (from *Ghielmetti et al.* [1978]). (© American Geophysical Union.)

These observations suggest that perpendicular acceleration processes, producing conics, are widespread in local time and altitude while parallel acceleration, producing beams, is more localized and generally restricted to the evening sector above about 5000 km. These acceleration processes operate on terrestrial ions, notably O^+, which must reach the high-altitude acceleration regions from ionospheric altitudes. Ionospheric ions have temperatures of less than 1 eV, so in order for large amounts of O^+ to reach high altitudes there must be persistent low-altitude processes that heat or accelerate the O^+ enough to enable it to overcome the Earth's gravity and survive charge ex-

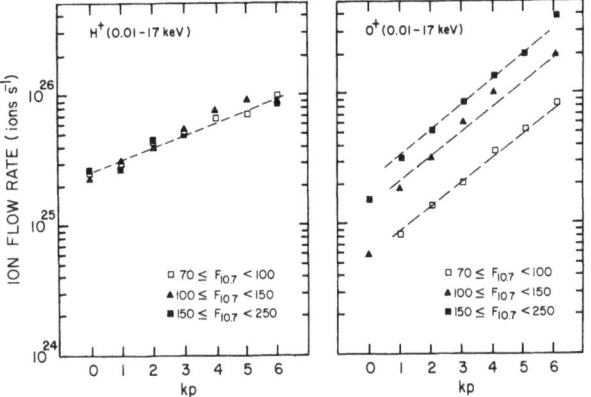

Figure 2—Ion outflow rate for H^+ and O^+ ions at 0.01–17 keV, as a function of the magnetic Kp index, for different ranges of $F_{10.7}$ values (from *Yau et al.* [1988]). (© American Geophysical Union.)

change with ambient hydrogen at higher altitudes. It would appear that ions are accelerated in stages [*Chiu et al.*, 1983]. At ionospheric altitudes they are energized to between a few electron volts and a few hundred electron volts. They can then move upward to higher altitudes where they may be additionally accelerated to form more energetic conics or beams.

The variations of the composition of the outflow of ionospheric ions provides information about the supply of ions from the ionosphere and so indirectly about the mechanisms which extract them. *Yau et al.* [1988] studied the variation of the total H^+ and O^+ outflows with magnetic activity as indicated by Kp and with solar activity as indicated by the flux of solar 10.7 cm radiation. Figure 2 summarizes their findings and shows that both H^+ and O^+ outflows increase markedly with magnetic activity and the O^+ outflow additionally increases by a factor of five between solar minimum and solar maximum. Yau et al. interpreted these results to imply that an enhancement of the O^+ outflow can result from the increase of ionospheric O^+ scale height caused by the increased solar EUV irradiation at solar maximum as suggested by *Young et al.* [1982] and that outflows of both H^+ and O^+ can be enhanced by the increased energy inputs to the ionosphere at times of high magnetic activity. A similar solar cycle variation was found in the occurrence of the minor upflowing terrestrial species, He^+ [*Collin et al.*, 1988].

In the following sections we will discuss ion acceleration at low altitudes, perpendicular ion acceleration at high altitudes and parallel ion acceleration, in that order.

3. ACCELERATION AT LOW ALTITUDE

Most of the high altitude acceleration processes operate on ionospheric ions which have already been heated enough to allow them to escape the Earth's gravitational field. At altitudes from a few hundred kilometers to about 1400 km, upward moving ions are observed which are confined to a narrow range of pitch angles close to 90°, and are characterized as TAI. There have been individual studies of low altitude ion acceleration, some of which are of a statistical nature while others are based on individual rocket experiments. Figure 3 shows an example of TAI on a rocket flight [*Yau et al.*, 1986]. The narrow pitch angle width, less than 5°, suggests a perpendicular acceleration process confined to a region with an altitude extent of 100 km or less and situated at an altitude close to 500 km. The ions are mainly low energy and have a power law energy spectrum with an exponent of about -2. Such a spectrum is suggestive of a heating or stochastic process in contrast with a bulk acceleration process. *Klumpar* [1981] found TAI to be closely associated with

130

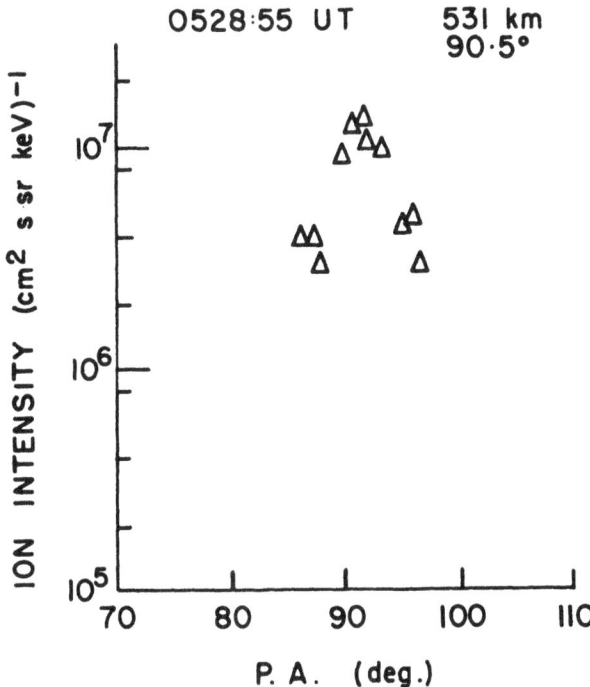

Figure 3—Pitch angle distribution of TAI during a sounding rocket experiment (from *Yau et al.* [1986]). (© American Geophysical Union.)

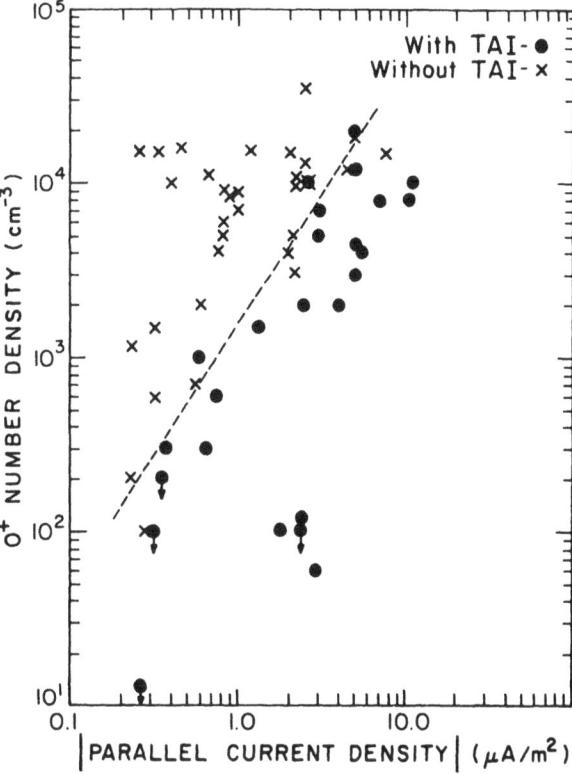

Figure 4—Scatter plot of oxygen ion number densities and field-aligned current densities for auroral oval crossings both with and without TAI. The absence of TAI in the upper left portion of this parameter space indicates ionospheric stability to the TAI acceleration mechanism (from *Klumpar* [1981]). (© American Geophysical Union.)

field-aligned currents. Figure 4 shows that the TAI appear only if the currents are sufficiently intense and that the current threshold is dependent on the local O^+ density. This association would be expected if the heating is produced by electrostatic ion cyclotron waves driven by field-aligned currents as proposed by *Ungstrup et al.* [1979]. However, this is not the only candidate process. *Kintner et al.* [1986] studied TAI and plasma waves observed on a rocket flight. They found no evidence of oxygen cyclotron waves, but did find a clear relationship (Figure 5) between the ion energy flux and the electric field amplitude of lower hybrid waves, supporting proposals of acceleration mechanisms that employ lower hybrid waves (for example *Chang and Coppi* [1981]). Another rocket flight [*Moore et al.*, 1986] showed substantial heating of the ionospheric majority species, O^+, producing a high-energy tail in the upgoing O^+ distribution (Figure 6). While the observations do not preclude the existence of tails in the H^+ and He^+, their intensities would have been at or below the sensitivity level of the instrument. On that occasion broadband waves near the lower hybrid frequency were observed in association with some of the heated O^+. However *Moore et al.* [1986] note that those particular ions appeared to have been partly thermalized rather than heated only in the perpendicular direction, and speculated that the lower

hybrid waves might have been a sink of ion energy rather than a source. Shear Alfvén waves have been proposed to accelerate the low-energy upwelling ions [*Moore*, 1986].

4. PERPENDICULAR ACCELERATION AT HIGHER ALTITUDES

Ion distributions that have symmetric peaks at an angle to **B** have been observed on auroral field lines at all altitudes up to several R_E [*Sharp et al.*, 1981] and at all local times [*Gorney et al.*, 1981]. Frequently the peaks are at large angles to **B** [*Collin and Johnson*, 1985]. If it is assumed that the acceleration had a narrow extent in altitude, was purely perpendicular to **B**, and the observed angle of the conic was the result of the ions moving adiabatically in the mirror geometry geomagnetic field, then it would appear that the location of the acceleration region in those events was within a few thousand kilometers below the satellite. This implies that perpendicular acceleration can occur anywhere throughout a wide range of altitudes.

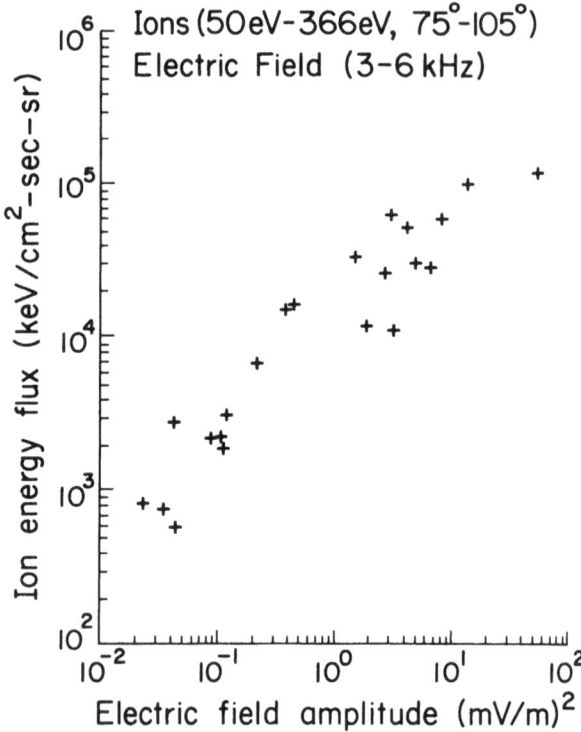

Figure 5—A comparison of transversely accelerated ion energy flux with lower hybrid electric field amplitude (from *Kintner et al.* [1986]). (© American Geophysical Union.)

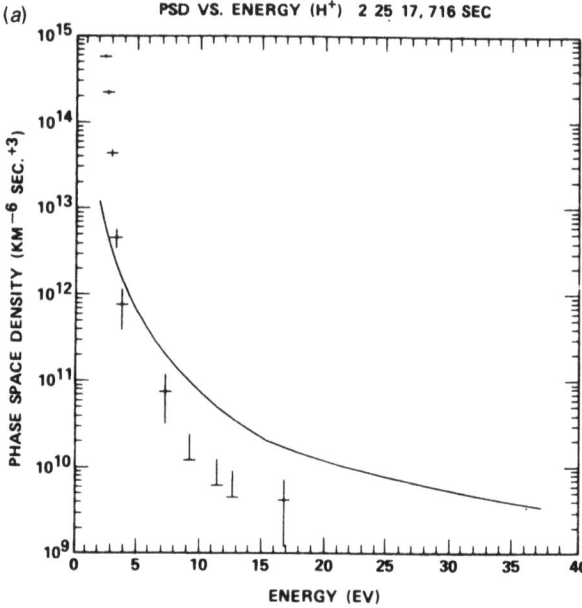

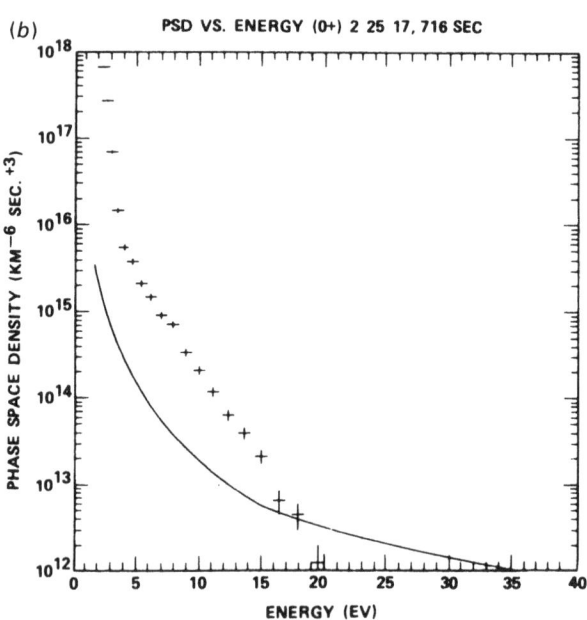

Figure 6—Detailed phase space distributions for upward moving H$^+$, and O$^+$. Error bars on the data points indicate uncertainties due to counting statistics. Points with zero counts are not shown. The solid curve indicates the 1 count/sample level. Eight samples were used for each data point (from *Moore et al.* [1986]). (© American Geophysical Union.)

Although there have been a number of studies of individual occurrences of conics and morphological studies of their average occurrence [*Gorney et al.*, 1981; *Yau et al.*, 1984], there is very little published information on the typical characteristics or mass dependence of their distributions. Their composition appears to be quite variable and some conics have been observed to be dominated by a single mass species, either H$^+$, O$^+$, or He$^+$, while others are composed of a mixture of two or more species. The available information is very limited. For example, *Collin and Johnson* [1985] found only six events in the S3-3 dataset that were suitable for mass dependence studies. Their results are consistent with the different mass components having been accelerated at the same altitude and by the same amount. The spectra of conics are generally soft and can usually be characterized by a power law with an exponent between −2.0 [*Ungstrup et al.*, 1979], and −1.3 [*Fennell et al.*, 1979], although *Collin and Johnson* [1985] found some conics to have high-energy tails. These limited results contrast with the well-known mass dependence of the heating of ion beams in which the heavier ion components are found to have higher mean energies than the H$^+$. *Collin et al.* [1981] found the O$^+$ component of beams to have a great-

er perpendicular temperature than the H$^+$ component. These beam observations can be interpreted to indicate the existence of a mass-dependent perpendicular acceleration process if, as suggested by *Chiu et al.* [1983], ion beams are formed by the acceleration of

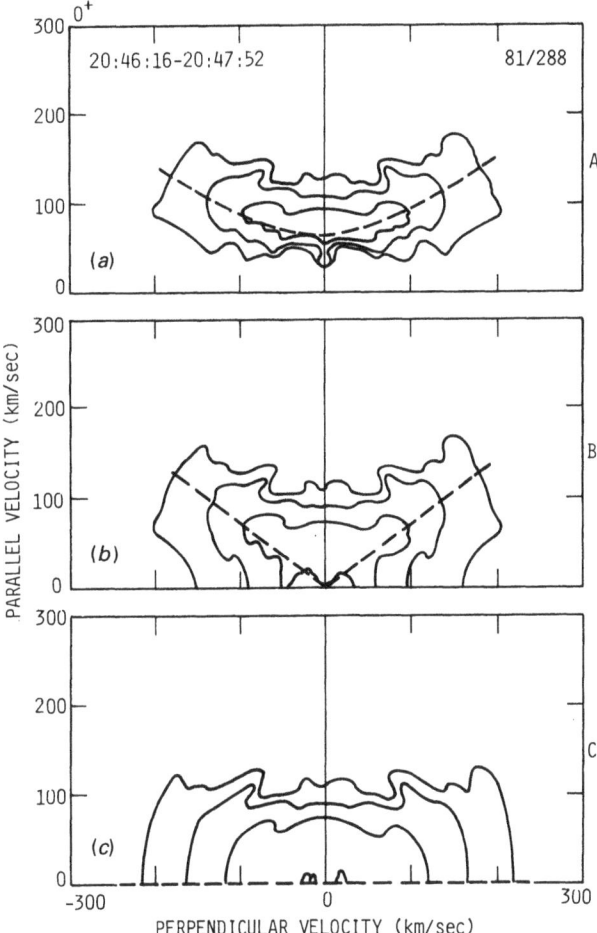

Figure 7—Phase space density contours at half-decade intervals below the peak phase space density (inner contour is at 6×10^7 s^{3}km^{-6}) for oxygen ions: (a) measured distribution; (b) transformed back through a parallel potential of 310 V; (c) further adiabatic transformation down the field line to a point where the magnetic field strength is 1.49 times that at the satellite (from *Klumpar et al.* [1984]). (© American Geophysical Union).

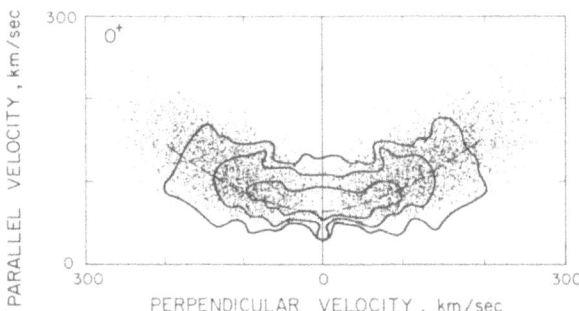

Figure 8—The points are a Monte Carlo simulation of an ion bowl distribution. The density of points has not been adjusted to take account of the smaller amount of phase space at low $v_\perp$. The contours are from the data of *Klumpar et al.* [1984] (from *Temerin* [1986]). (© American Geophysical Union).

conics by a parallel electrostatic field that would add the same energy to all ions regardless of mass. This implies that the observed mass-dependent energy had been contributed by the original conics, and thus that the conics had been produced by a mass dependent perpendicular acceleration process.

Conics are often seen in association with plasma waves of several kinds, e.g., lower hybrid mode [*Kintner and Gorney*, 1984], as well as low-frequency turbulence and electrostatic shocks [*Mozer et al.*, 1980], and much theoretical work has been done on the potential of plasma instabilities to produce perpendicular ion heating. However, as discussed by *Klumpar* [1986] and *Kintner* [1986], observations have been unable to show conclusively any firm correspondence between perpendicular ion acceleration in the auroral zone and any particular type of plasma wave. As *Klumpar* [1986] pointed out, there are substantial difficulties in confirming that the observed waves were present in the often distant acceleration region and did not propagate from elsewhere. *Peterson et al.* [1988] conducted a systematic search of the DE 1 dataset without finding an event that showed unambiguous evidence of local heating of ions by waves. They did, however, find several events in which the wave and ion observations were consistent with local ion heating. *Kintner et al.* [1979] surveyed S3-3 ion and wave observations and found one conic that was consistent with local heating by EHC waves.

Klumpar et al. [1984] have reported hybrid conics in which all ions had at least a minimum parallel velocity so that the distribution in velocity space took the form of a hollow cone or bowl displaced in parallel velocity. *Kondo et al.* [1989] examined the morphology of these hybrid conics and found that they occur most frequently on the dayside at auroral latitudes. *Klumpar et al.* [1984] attributed their displaced distributions to a two stage process, illustrated in Figure 7, in which the ions experience perpendicular acceleration together with electrostatic acceleration along **B** and adiabatic transport upward through about 3000 km. An alternative explanation was put forward by *Temerin* [1986] who modeled a stochastic process in which initially cool ions were given random perpendicular velocity increments. The mirror force partially converted these into parallel velocity. Figure 8 shows an ion distribution obtained by maintaining this process over several thousand kilometers and that gave a good match with the contours of the observed distribution. *Temerin* [1986] did not identify any particular mechanism to provide the velocity increments, and so does not necessarily incorporate any mass discrimination effects. *Retterer et al.* [1987] and *Chang* [1986] modeled an electromagnetic ion cyclotron resonance with broadband electromagnetic turbulence over a wide altitude

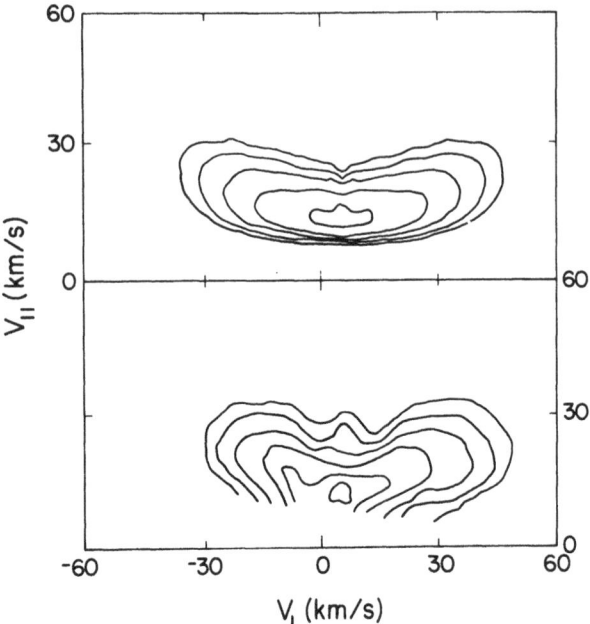

Figure 9—The bottom panel presents a contour diagram of the observed ion-conic distribution function, measured by the HAPI instrument on DE 1, while the top panel presents a theoretical ion-velocity distribution, plotted in the same way as the observed conic distribution (from *Retterer et al.*, [1987]). (© American Physical Society.)

extent. Figure 9 demonstrates the match between the modeled ion distribution and an observed ion distribution. Since the observed electromagnetic turbulence had a higher spectral density at the gyrofrequencies of heavier ions, it is these ions that can be expected to be accelerated most readily. Currently, *Crew et al.* [1988] are exploring the extent to which this process can explain a wider range of conics from the DE 1 dataset, and a quantitative comparison of the ability of resonant and stochastic models to explain observed conics is being attempted [*W. K. Peterson*, private communication 1988].

A number of plasma wave modes can be driven unstable by electron currents and have been studied to determine their potential for ion heating in mixed ion plasmas. The theory and simulations of *Ashour-Abdalla et al.* [1987] and *Schriver and Ashour-Abdalla* [1988] showed that an ion-ion hybrid (Buschbaum) mode or a modified lower hybrid mode would be driven unstable by a current flowing out of the ionosphere through a plasma containing H^+ and O^+ in various proportions. These instabilities and the resulting ion heating depend strongly on the plasma regime. *Ashour-Abdalla et al.* [1988] examined the variation of heating produced by these instabilities as the plasma conditions changed with altitude in the manner that would be expected on auroral field lines. They found differ-

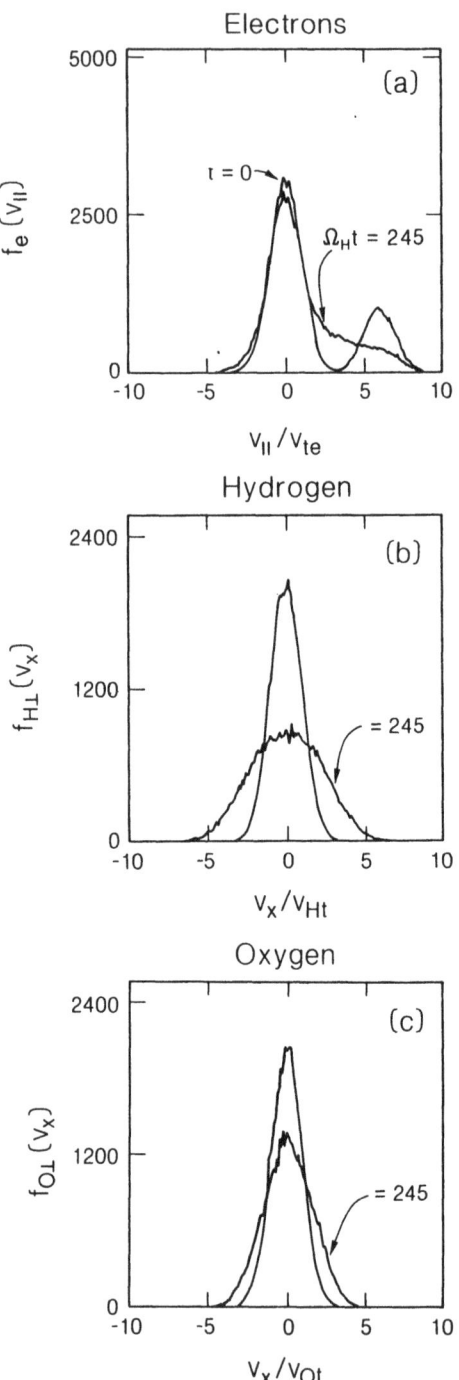

Figure 10—Simulated particle distributions. The electron parallel velocity distribution is shown at the top and the hydrogen and oxygen perpendicular velocity distributions are shown in the middle and at the bottom, respectively. The distributions for $t = 0$ are superimposed on the distributions for $\Omega_H t = 245$. The electron free energy is used up and the hydrogen is seen to be bulk heated; the oxygen forms a high energy tail and is slightly bulk heated (from *Ashour-Abdalla et al.* [1988]). (© American Geophysical Union.)

ent instabilities to be dominant in plasma regimes corresponding to different altitudes. At altitudes below 2000 km, when O^+ is the dominant species, the modified lower hybrid mode was the most unstable. Figure 10 shows the development of the particle distributions. The energy driving the ion heating is the free energy of the electron beam that is diffused out into a monotonically decreasing distribution as the instability progresses. Both the H^+ and O^+ are bulk heated and a high-energy tail develops in the O^+ distribution. In regions of plasma cavities at altitudes between 2000 km and 15,000 km where O^+ is still a significant but not dominant component, the ion-ion hybrid mode was found to be the most unstable. The source of energy for the heating is again the electron beam. There is little effect on the H^+, but O^+ is bulk heated. The extent of the O^+ heating was found to depend on the ion composition and the O^+ was more strongly heated when its density was low. If the plasma includes a small amount of a third species, He^+, both a helium-hydrogen hybrid mode and a helium-oxygen hybrid mode were expected and substantial bulk heating of the He^+ resulted.

A wave-particle interaction is not the only type of mechanism proposed to account for conics. Several authors have examined the behavior of ions that encounter a double layer oblique to the field line such as might be found at the margins of an inverted-V [*Lennartsson*, 1980; *Borovsky*, 1984; *Greenspan*, 1984]. They found that if the spatial or temporal scale of the double layer were small enough, then gyrotropy would break down and the ions would acquire energy perpendicular to the geomagnetic field direction, forming a conic distribution, and that heavier ions would be expected to acquire more energy. The accelerations contributed by a succession of weak double layers would be expected to accumulate in a fashion similar to the process described by *Temerin* [1986].

5. PARALLEL ACCELERATION AND ION BEAMS

The ion beams observed in the evening sector typically consist of energetic ions whose distributions are sharply peaked at kiloelectron-volt energies, are field aligned, and usually fill the atmospheric loss cone, or are even somewhat wider. When a beam is composed of multiple species, all species are observed to have similar energies.

The location of the beams is closely correlated with the auroral oval, but occurs predominantly in the dusk hemisphere [*Ghielmetti et al.*, 1978]. The beams are also associated with inverted-V electron distributions, field-aligned currents [*Mozer et al.*, 1980], and electrostatic ion cyclotron waves [*Kintner et al.*, 1979] that

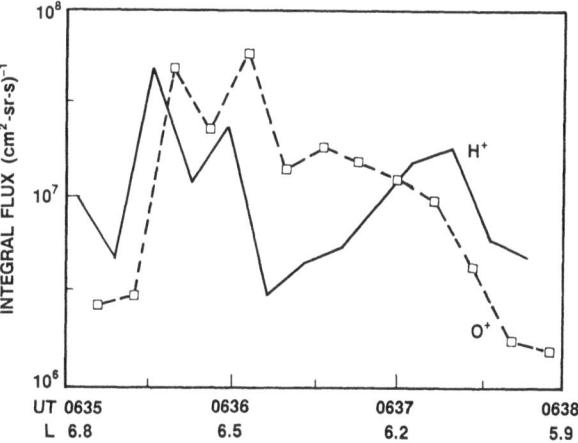

Figure 11—Integral flux of upflowing ions observed on November 5, 1981. Universal time (UT) and L shell (L) of the data are indicated below. (After *Shelley et al.* [1983].)

are believed to be generated by the beams [*Cattell et al.*, 1979]. *André et al.* [1987] found evidence that the spectrum of the accompanying waves was strongly influenced by the cold electron distribution. *Ghielmetti et al.* [1978] reported that the beams, and so presumably the acceleration region, occurred almost exclusively at altitudes above 5000 km.

When the composition of the beams was examined in detail, it was found that the species making up the beams, principally H^+ and O^+, had different and variable velocity space distributions. Their ionospheric origin was clearly indicated by the frequency with which O^+ occurred, but their composition was very variable. For example, Figure 11 shows the ion beam composition in a single transit of an inverted V. When first encountered, the beam consisted of about equal proportions of H^+ and O^+, but later the O^+ concentration increased to about five times that of H^+. *Ghielmetti et al.* [1987] found that, while the average energies of the components were well correlated, the O^+ components generally had a higher average energy than the H^+ components, as can be seen in Figure 12. They attributed the differences to a mass-dependent acceleration or energy loss process. However, the beam distributions together with their association with inverted Vs, and upward-directed parallel electric fields, indicated by the enhanced loss cones of mirroring electrons [*Sharp et al.*, 1979], implied that the ions had been accelerated by an electrostatic field parallel to **B**, which would be expected to give equal energy per unit charge to all ions regardless of species. This view was further supported by the statistical correlation between the ion beam mean energy and the simultaneous potential drop deduced from the enhanced electron loss cones [*Collin et al.*, 1986]. The existence of parallel electric fields in inverted-V regions was giv-

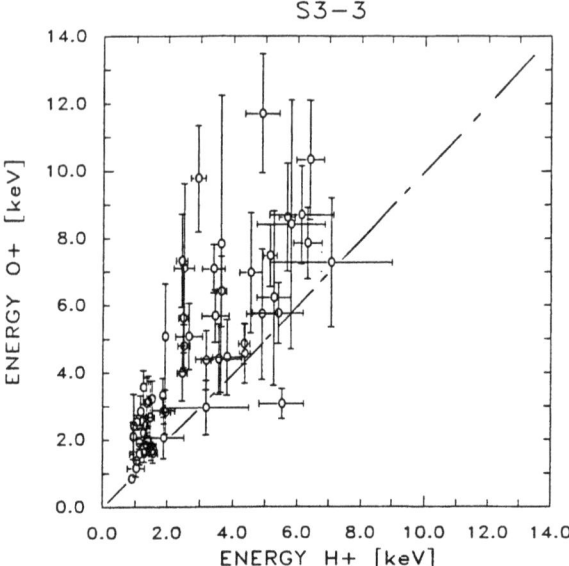

Figure 12—Scatter plot of the average energy of H+ and O+ upward flowing ions within individual acceleration structures. Dashed line indicates equal energy condition (from *Ghielmetti et al.* [1987]). (© *Physica Scripta*.)

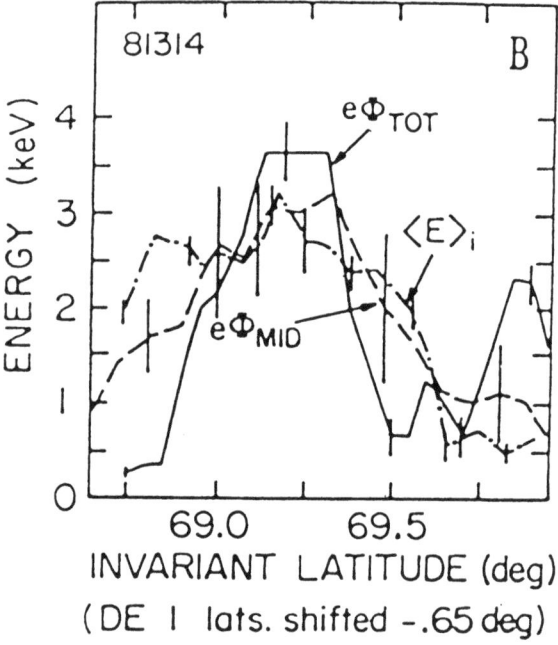

Figure 13—Comparison of the auroral accelerating energies inferred in three independent ways: from the energy of the precipitating electrons at low altitudes, $e\Phi_{TOT}$, from the enhancement of the loss cone of electrons at high altitudes, $e\Phi_{MID}$, and from the average parallel energy of upflowing ions measured at high altitudes, $\langle E\rangle_i$ (from *Reiff et al.* [1988]). (© American Geophysical Union.)

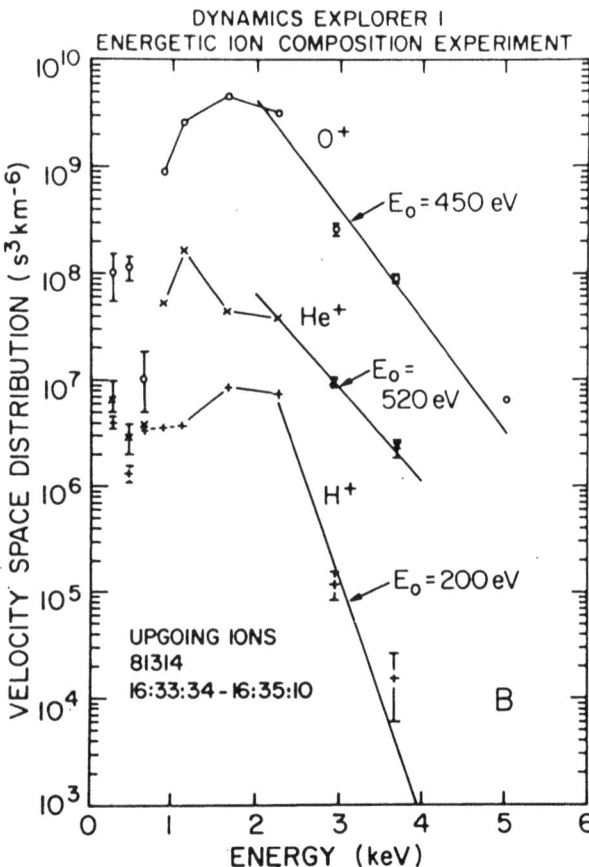

Figure 14—EICS upgoing ion distribution functions. Values for O+ are shown as o's, for H+ as +'s and for He+ as x's. The characteristic energies are estimated from the slopes of the fluxes at high energies (from *Reiff et al.* [1988]). (© American Geophysical Union.)

en additional support by *Reiff et al.* [1988], who used conjugate satellite measurements to make independent simultaneous measurements of the potential drop at two altitudes on the same field line as the inverted V. Figure 13 shows the profile of the potential drop energy inferred from the peak energy of electrons that had been accelerated by falling through the potential drop and detected by DE 2 at 670 km and that inferred from the enhanced loss cones of mirroring electrons measured by DE 1 at 11,000 km. These two measurements match quite well within their experimental uncertainties. The mean energy of upflowing ion beams measured at DE 1 also matches these estimates suggesting that the ions have acquired their energy by falling through this potential drop. However, the ion acceleration mechanism cannot be as simple as this, since the originally cold ionospheric ions have evidently been heated to temperatures of several hundred electron volts as seen in Figure 14. Had the ions been heated before being accelerated through the potential drop, their

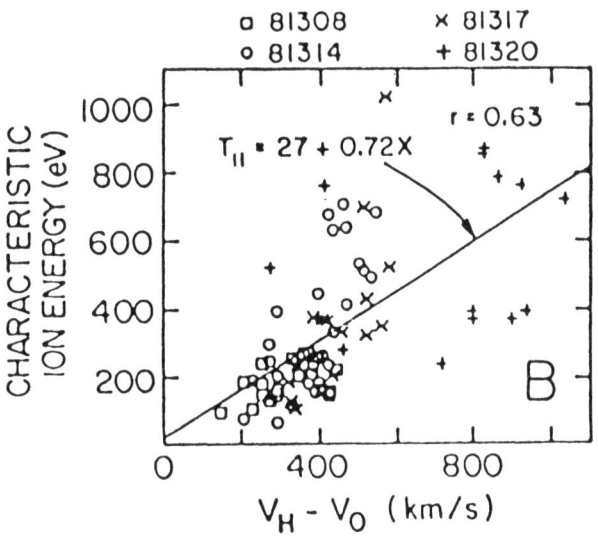

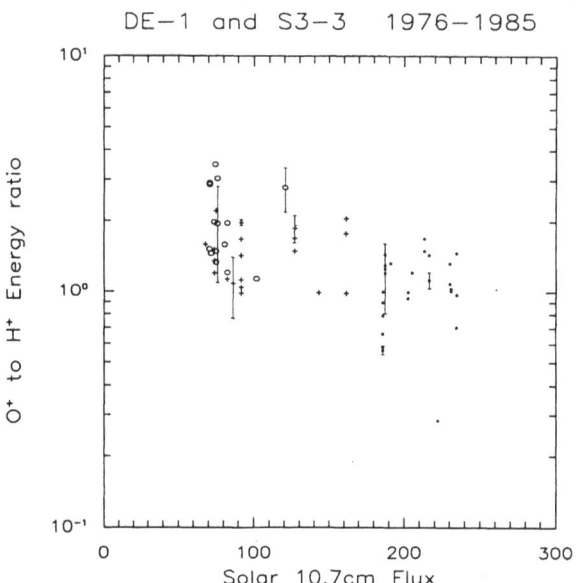

DE−1 and S3−3   1976−1985

Figure 15—Scatter plot of ion characteristic energies versus the differences between the inferred hydrogen and oxygen beam velocities. A correlation of 0.63 was found (from *Reiff et al.* [1988]). (© American Geophysical Union.)

Figure 17—The ratio of O⁺ mean energy to H⁺ mean energy versus solar 10.7 cm radio flux. Data from 1976–1977 are indicated by open circles, data from 1981–1982 by solid circles, and data from 1984–1985 by crosses. Some representative error bars are shown (from *Collin et al.* [1987]). (© American Geophysical Union.)

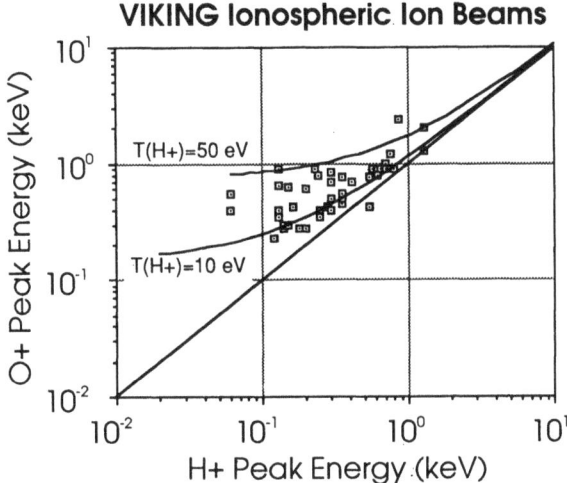

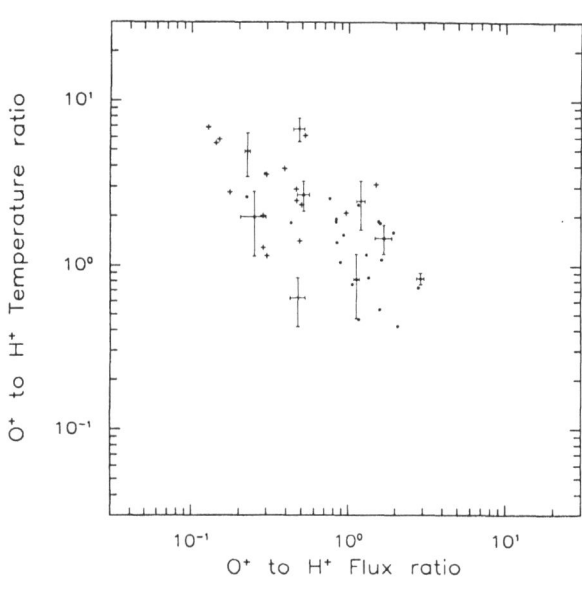

Ion Beams   1981−1985

Figure 16—Scatter plot of O⁺ peak energy versus H⁺ peak energy for upward flowing ion beams observed during a number of Viking passes through the auroral oval. The curves represent cases in which both species received the same initial velocity corresponding to H⁺ energy of 10eV (lower curve) and 50eV (upper curve) (from *Lundin and Hultqvist* [1989]). (© American Geophysical Union.)

Figure 18—The ratios of O⁺ temperature to H⁺ temperature versus the ratios of O⁺ flux to H⁺ flux. There is a strong tendency for the O⁺ temperature to decrease in relation to that of the H⁺ as the composition of the beams becomes dominated by O⁺. Data from 1981–1982 are indicated by solid circles, and data from 1984–1985 by crosses. Some representative error bars are shown. (from *Collin et al.* [1987]). (© American Geophysical Union.)

mean energy would be expected to be greater than that corresponding to the potential drop. Moreover the temperatures are mass dependent with the heavier species having gained higher temperatures.

A candidate heating mechanism is the ion-ion two-stream instability [*Kaufmann et al.*, 1986; *Bergmann and Lotko*, 1986], which is driven by the difference in velocity of the two ion species. Reiff et al. examined the dependence of the ion heating on the relative velocity of the major ion species (Figure 15). They found some support for the two-stream instability in the correlation between the ion heating, indicated in this figure by the characteristic energy, and by the velocity difference of the H^+ and O^+ components of the beams observed in four transits of inverted Vs.

Reiff et al. [1988] had found only small differences between the mean energies of the O^+ and H^+ components. Other studies that addressed the relative energization of the ion species composing beams obtained conflicting results. *Collin et al.* [1987], using data obtained by DE 1 during 1981 and 1982, found that the O^+ and H^+ components had nearly equal mean energies in contrast to the marked difference in energies found by *Ghielmetti et al.* [1987] who had used data obtained by S3-3 during 1976 and 1977 (Figure 12). Measurements made by the Viking satellite during 1986 (Figure 16) also showed beams in which the peak energy for O^+ was higher than H^+ [*Lundin and Hultqvist*, 1989].

Like the composition of the total ion outflow, the composition of ion beams was found to vary with solar activity, with O^+ becoming the major component at solar maximum [*Collin et al.*, 1987] and H^+ the major component at solar minimum. Collin et al. also examined the solar activity dependence of the relative energy and temperature of the O^+ and H^+ beam components. Figure 17 shows the O^+ energy to be greater than that of the H^+ at low solar activity and that the energies of H^+ and O^+ become more nearly equal as solar activity increases. The O^+ to H^+ temperature ratio shows a similar trend, suggesting that at least part of the excess O^+ energy could be accounted for by increased heating. The investigation of the ion-ion two-stream instability by *Bergmann and Lotko* [1986] suggested that in a beam dominated by O^+, the instability would tend to be suppressed. Indeed, Collin et al. found (Figure 18) that the temperature of the O^+ decreased relative to that of the H^+ as the composition of the beams changed to become dominated by O^+.

Further studies of the ion-ion two-stream instability have indicated more details of its expected effect on the ion beams. Linear analysis of the instability [*Bergmann et al.*, 1988] indicated that at sufficiently high beam velocities the wave growth would be oblique to

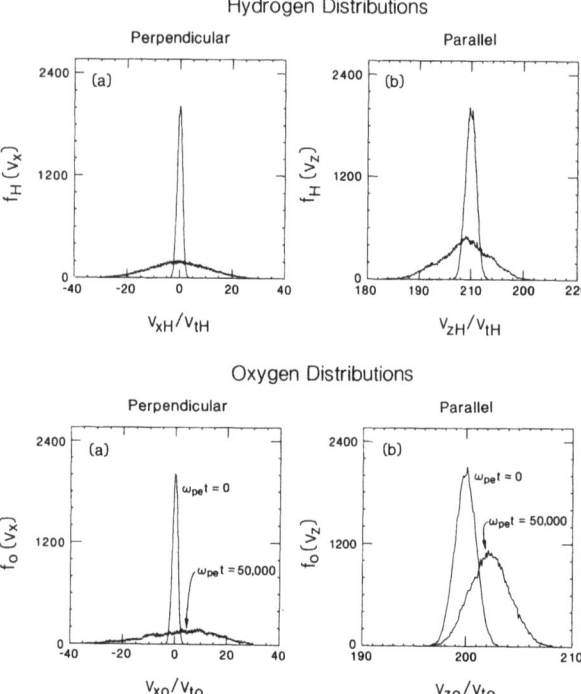

Figure 19—Velocity distributions of H^+ (*top*) and O^+ (*bottom*) before ($\omega_{pe}t = 0$) and after ($\omega_{pe}t = 50,000$) a simulated ion beam interaction. The velocity distributions are both perpendicular (left) and parallel (right) to the magnetic field shown (from *Schriver et al.* [1989]). (© American Geophysical Union.)

the magnetic field direction. They suggested that the ion heating would in consequence be at an angle to the field line. The simulations of *Roth et al.* [1989] predicted that the H^+ would be slowed down and form a low-energy tail while the O^+ would develop a high-energy tail resulting in more energetic O^+ than H^+. *Ludlow and Kaufmann* [1989] analyzed observations of an ion beam and showed that the diffusion time scales were fast enough for much of the heating to have taken place within one Earth radius below the satellite. Further numerical simulations [*Winglee et al.*, 1988; *Shriver et al.*, 1989] showed substantial heating of all species perpendicular as well as parallel to **B**. They also found the energy source for the heating was the drift velocity of the H^+, in accordance with the linear analysis predictions. Figure 19 shows the simulated heating of a beam consisting of 90% H^+ and 10% O^+. Both ions experienced heating symmetrically in the perpendicular direction, while in the parallel direction they were heated asymmetrically. This heating tended to decrease the drift velocity of the H^+ and to increase that of O^+ which gained energy at the expense of the H^+ component. Figure 20 illustrates such heating of the O^+ and H^+ components for an observed beam. The peaks of the two distributions are

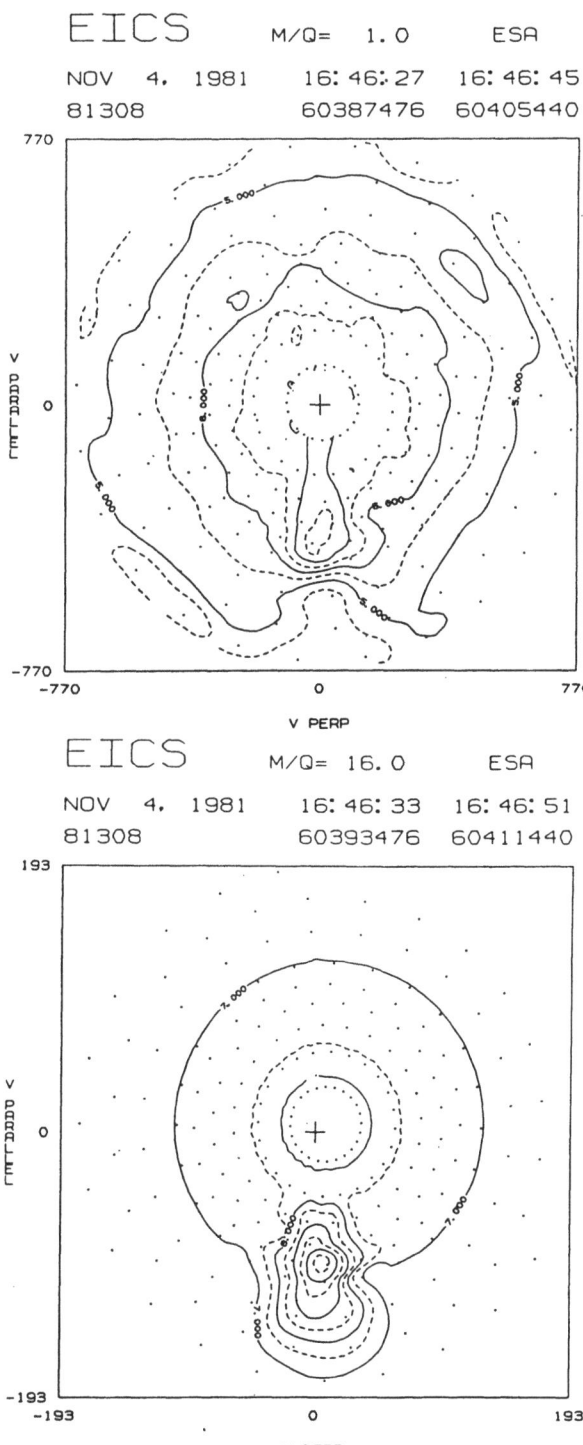

EICS   M/Q= 1.0   ESA

NOV 4, 1981   16:46:27   16:46:45
81308   60387476   60405440

EICS   M/Q= 16.0   ESA

NOV 4, 1981   16:46:33   16:46:51
81308   60393476   60411440

Figure 20—EICS velocity space contours for the H⁺ (*top*) and O⁺ (*bottom*) components of an ion beam. The velocity gained in the acceleration process is roughly four times larger for H⁺ than for O⁺, consistent with the existence of an upward electric field below the spacecraft (from *Reiff et al.* [1988]). (© American Geophysical Union.)

at about the same energy. The H⁺ can be seen to be quite asymmetric with a low temperature on the high-velocity side and an extended low-velocity tail. In contrast, the O⁺ shows much greater heating toward high velocities.

The simulations discussed above represent the interaction in only a very restricted spatial region; for a more realistic representation of the full development of the process, it would be necessary to follow the ions through the changing plasma environment of the full altitude extent of the acceleration region. Nevertheless the ion-ion two-stream interaction already shows promise of explaining the major features of ion beams.

Again, wave-particle interaction is not the only mechanism proposed to account for the high temperatures of the ion beams. An alternative scenario is the two-stage acceleration [*Chiu et al.,* 1983] described previously. *Collin et al.* [1986] found some evidence that such a sequence of events might occur. They observed an upflowing conic in a region where the downcoming electrons had a peaked spectrum and a pronounced minimum at 90°, suggestive of acceleration through an electrostatic potential drop above the satellite. Collin et al. showed that had the conic been subsequently accelerated through this potential drop, it would have emerged with beamlike properties.

Lundin and Hultqvist [1989] have suggested a process in which the ions are initially accelerated by inhomogeneities in a perpendicular electric field in a fashion similar to the oblique double-layer process referred to previously. They suggested that such a process could operate at a low enough altitude to extract cold ions from the ionosphere. The lines in Figure 16 correspond to the energies of ion beams that had acquired equal perpendicular increments of velocity before acceleration through a high-altitude parallel electric field. The statistical nature of this process appears to provide a plausible explanation for the scatter of the data points in Figure 16, which represent observations of ion beams by Viking.

These "preheating" proposals make some predictions that differ from those of the ion-ion two-stream "thermalization" model. The former would produce beams whose thermal energy was in addition to the energy acquired from electrostatic acceleration rather than the result of its redistribution. Since the heating in double layers or electric field inhomogeneities results from an interaction between the field structure and the individual ions, the extent of the heating might be expected to depend on ion mass but not on the relative composition of the plasma. While some studies such as those by *Reiff et al.* [1988] appear to favor the ion-ion two-stream thermalization model, most present observations are not adequately detailed to rule out either process.

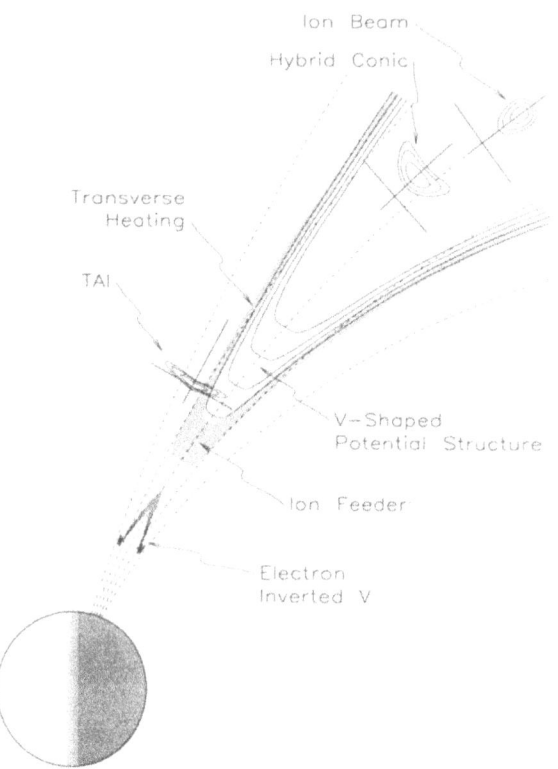

Figure 21 labels: Ion Beam, Hybrid Conic, Transverse Heating, TAI, V-Shaped Potential Structure, Ion Feeder, Electron Inverted V

Figure 21—A schematic representation of the acceleration of ions on auroral field lines. Low-energy ions from the ionosphere may receive perpendicular energy from small-scale electrostatic structures or from a plasma-wave interaction at a wide range of altitudes anywhere in the auroral zone to form conics. In inverted-V regions, the low energy ions, or possibly more energetic conics, are further accelerated, probably by a parallel electric field, to form beams. The perpendicular energy of the conics may be retained and/or a plasma interaction may redistribute the parallel energy of the beam as thermal energy. (Courtesy, D. M. Klumpar).

The field-aligned quasi-static electric potential drop, which has been considered here to be the principal energy source for the acceleration of ion beams, is at present the generally accepted interpretation of the inverted-V electron distributions. The basis for this interpretation is described by *Burch* [this volume]. An alternative interpretation of the electron distributions is discussed by *Bryant et al.* [this volume], but currently does not account for the production of ion beams.

6. SUMMARY AND STATUS

The general outline of the morphology of auroral ion acceleration is now fairly well established and is summarized in Figure 21. At low altitudes, low-energy upwelling and transversely accelerated ions are widespread. At a wide range of altitudes, up to at least several R_E, perpendicular acceleration, producing conics, occurs throughout the auroral zone. In inverted-V regions in the evening sector, substantial parallel acceleration occurs forming beams that may be narrowly field aligned or may show substantial perpendicular energization.

While considerable progress has been made toward understanding candidate energization mechanisms, it is still unclear which mechanisms are dominant in the auroral zone or how they relate to each other. In part, this is the consequence of some gaps and ambiguities in our knowledge of the phenomena. Perpendicular acceleration has been attributed to either small-scale electrostatic structures or to one of several plasma-wave interactions. However, the lack of a clear association of any particular plasma wave with perpendicular ion acceleration makes it hard to draw conclusions about which, if any, of the wave theories is important. An electrostatic field is generally accepted to be the source of the parallel ion acceleration, but the details of the velocity space distributions of the resulting beams are not yet fully explained. It appears most probable that these are the result of thermalization of the beams, but it is not possible to rule out preheating theories, since the present measurements of the details of the multi-ion distributions are of limited resolution, and since there is little direct information about the development of the distributions of ions as they move upward from the ionosphere and through the field region where they are accelerated. Furthermore, the mass-dependent flight times of the ions can lead to mass-dependent distortions of the distributions, as observed at significant distances from the acceleration or heating region.

Much progress has been made in the past few years in combining theory, modeling, and experimental measurements. However, major improvements in measurements will now be needed to distinguish the detailed predictions of the various proposed mechanisms. This will require sophisticated and dedicated auroral physics missions.

ACKNOWLEDGMENT—This work was supported by NASA contract NAS-5-29710, by the Office of Naval Research contract N00014-86-C-0159, and by Lockheed Independent Research.

REFERENCES

André, M., H. Koskinen, G. Gustafsson, and R. Lundin, "Ion Waves and Upgoing Ion Beams Observed by the Viking Satellite," *Geophys. Res. Lett.*, **14**, 463 (1987).

Ashour-Abdalla, M., H. Okuda, and S. Y. Kim, "Transverse Ion Heating in Multicomponent Plasmas," *Geophys. Res. Lett.*, **14**, 375 (1987).

Ashour-Abdalla, M., D. Schriver, and H. Okuda, "Transverse Ion Heating in Multicomponent Plasmas Along Auroral Field Lines," *J. Geophys. Res.*, **93**, 12826 (1988).

Axford, W. I., "On the Origin of Radiation Belt and Auroral Primary Ions," in *Particles and Fields in the Magnetosphere*, B. M. McCormac, ed., Reidel, Hingham, MA, 46 (1970).

Bergmann, R., and W. Lotko, "Transition to Unstable Flow in Parallel Electric Fields," *J. Geophys. Res.*, **91**, 7033 (1986).

Bergmann, R., I. Roth, and M. K. Hudson, "Linear Stability of the H^+-O^+ Two-Stream Interaction in a Magnetized Plasma," *J. Geophys. Res.*, **93**, 4005 (1988).

Borovsky, J. E., "The Production of Ion Conics by Oblique Double Layers," *J. Geophys. Res.*, **89**, 2251 (1984).

Cattell, C. A., R. Lysak, R. B. Torbert, and F. S. Mozer, "Observations of Differences Between Current Flowing Into and Out of the Ionosphere," *Geophys. Res. Lett.*, **6**, 621 (1979).

Chang, T., "Transverse Acceleration of Oxygen Ions by Electromagnetic Ion Cyclotron Resonance with Broad Band Left Hand Polarized Waves," *Geophys. Res. Lett.*, **13**, 636 (1986).

Chang, T., and B. Coppi, "Lower Hybrid Acceleration and Ion Evolution in the Subauroral Region," *Geophys. Res. Lett.*, **8**, 1253 (1981).

Chiu, Y. T., J. M. Cornwall, J. F. Fennell, D. J. Gorney, and P. F. Mizera, "Auroral Plasmas in the Evening Sector: Satellite Observations and Theoretical Interpretation," *Space Sci. Rev.*, **35**, 211 (1983).

Collin, H. L., and R. G. Johnson, "Some Mass Dependent Features of Energetic Ion Conics Over the Auroral Regions," *J. Geophys. Res.*, **90**, 9911 (1985).

Collin, H. L., R. D. Sharp, E. G. Shelley, and R. G. Johnson, "Some General Characteristics of Upflowing Ion Beams Over the Auroral Zone and Their Relationship to Auroral Electrons," *J. Geophys. Res.*, **86**, 6820 (1981).

Collin, H. L., E. G. Shelley, A. G. Ghielmetti, and R. D. Sharp, "Observations of Transverse and Parallel Acceleration of Terrestrial Ions at High Latitudes," in *Ion Acceleration in the Magnetosphere and Ionosphere*, Geophysical Monograph 38, T. Chang, ed., p. 67 (1986).

Collin, H. L., W. K. Peterson, and E. G. Shelley, "Solar Cycle Variation of Some Mass Dependent Characteristics of Upflowing Beams of Terrestrial Ions," *J. Geophys. Res.*, **92**, 4757 (1987).

Collin, H. L., W. K. Peterson, J. F. Drake, and A. W. Yau, "The Helium Components of Energetic Terrestrial Ion Outflows: Their Occurrence, Morphology, and Intensity," *J. Geophys. Res.*, **93**, 7558 (1988).

Crew, G. B., T. Chang, J. M. Retterer, W. K. Peterson, D. A. Gurnett, and R. L. Huff, "Ion Conics: Detailed Comparison of Theory and Observations," *Eos Trans. AGU*, **69**, 1374 (1988).

Dusenbery, P. B., R. F. Martin, and R. M. Winglee, "Ion-Ion Waves in the Auroral Region: Wave Excitation and Ion Heating," *J. Geophys. Res.*, **93**, 5655 (1988).

Fennell, J. F., P. F. Mizera, and D. R. Croley, "Observations of Ion and Electron Distributions During the July 29 and July 30, 1977 Storm Period," *Proc. Magnetospheric Boundary Layers Conf.*, ESA SP-148, European Space Agency, Paris (1979).

Ghielmetti, A. G., R. G. Johnson, R. D. Sharp, and E. G. Shelley, "The Latitudinal, Diurnal, and Altitudinal Distributions of Upward Flowing Energetic Ions of Ionospheric Origin," *Geophys. Res. Lett.*, **5**, 59 (1978).

Ghielmetti, A. G., E. G. Shelley, and D. M. Klumpar, "Correlation Between Number Flux and Energy of Upward Flowing Ion Beams," *Physica Scripta*, **36**, 362 (1987).

Gorney, D. J., A. Clarke, D. Croley, J. Fennell, J. Luhmann, and P. Mizera, "The Distribution of Ion Beams and Conics Below 8000 km," *J. Geophys. Res.*, **86**, 83 (1981).

Greenspan, M. E., "Effects of Oblique Double Layers on Upgoing Pitch Angle and Gyrophase," *J. Geophys. Res.*, **89**, 2842 (1984).

Kaufmann, R. L., G. R. Ludlow, H. L. Collin, W. K. Peterson, and J. L. Burch, "Interaction of Up-Going Auroral H^+ and O^+ Beams," *J. Geophys. Res.*, **91**, 1080 (1986).

Kintner, P. M., and D. J. Gorney, "A Search for the Plasma Processes Associated With Perpendicular Ion Heating," *J. Geophys. Res.*, **89**, 937 (1984).

Kintner, P. M., M. C. Kelley, R. D. Sharp, A. G. Ghielmetti, M. Temerin, C. A. Cattell, P. F. Mizera, and J. F. Fennell, "Simultaneous Observations of Energetic (keV) Upstreaming Ions and EIC Waves," *J. Geophys. Res.*, **84**, 7201 (1979).

Kintner, P. M., J. LaBelle, W. Scales, A. W. Yau, and B. A. Walen, "Observations of Plasma Waves Within Regions of Perpendicular Ion Acceleration," *Geophys. Res. Lett.*, **13**, 1113 (1986).

Klumpar, D. M., "Transversely Accelerated Ions in Auroral Arcs," in *Physics of Auroral Arc Formation*, Geophysical Monograph 25, S.-I. Akasofu and J. R. Kan, eds., American Geophysical Union, Washington, DC, p. 122 (1981).

Klumpar, D. M., "A Digest and Comprehensive Bibliography on Transverse Auroral Ion Acceleration," in *Ion Acceleration in the Magnetosphere and Ionosphere*, Geophysical Monograph 38, T. Chang, ed., p. 389 (1986).

Klumpar, D. M., W. K. Peterson, and E. G. Shelley, "Direct Evidence for Two-Stage (Bimodal) Acceleration of Ionospheric Ions," *J. Geophys. Res.*, **89**, 10779 (1984).

Kondo, T., B. A. Walen, and A. W. Yau, "Statistical Analysis of Upflowing Ions Including Hybrid Conical Pitch Angle Distributions at DE-1 Altitudes," *J. Geophys. Res.*, submitted (1989).

Lennartsson, W., "On the Consequences of the Interaction Between the Auroral Plasma and the Geomagnetic Field," *Planet. Space Sci.*, **28**, 135 (1980).

Lockwood, M., J. H. Waite, Jr., T. E. Moore, J. F. E. Johnson, and C. R. Chappell, "A New Source of Suprathermal O^+ Ions Near the Dayside Polar Cap Boundary," *J. Geophys. Res.*, **90**, 4099 (1985).

Ludlow, G. R., and R. L. Kaufmann, "Heating of Upflowing Auroral H^+ and O^+ Beams: Results From Quasi-Linear Theory," *J. Geophys. Res.*, **94**, 319 (1989).

Lundin, R., and B. Hultqvist, "Ionospheric Plasma Escape by High-Altitude Electric Fields: Magnetic Moment Pumping," *J. Geophys. Res.*, **94**, 6665 (1989).

Moore, T. E., "Acceleration of Low-Energy Magnetospheric Plasma," *Adv. Space Res.*, **6**, 103 (1986).

Moore, T. E., C. J. Pollock, R. L. Arnoldy, and P. M. Kintner, "Preferential O^+ Heating in the Topside Ionosphere," *Geophys. Res. Lett.*, **13**, 901 (1986).

Mozer, F.S., C. A. Cattell, M. K. Hudson, R. L. Lysak, M. Temerin, and R. B. Torbert, "Satellite Measurements and Theories of Low Altitude Auroral Particle Acceleration," *Space Sci. Rev.*, **27**, 155 (1980).

Peterson, W. K., E. G. Shelley, S. A. Boardsen, D. A. Gurnett, B. G. Ledley, M. Sigiura, T. E. Moore, and J. H. Waite, Jr., "Transverse Ion Energization and Low Frequency Plasma Waves in the Mid-Altitude Auroral Zone: A Case Study," *J. Geophys. Res.*, **93**, 11,405 (1988).

Reiff, P. H., H. L. Collin, J. D. Craven, J. L. Burch, J. D. Winningham, E. G. Shelley, L. A. Frank, and M. A. Friedman, "Determination of Auroral Electrostatic Potentials Using High- and Low-Altitude Particle Distributions," *J. Geophys. Res.*, **93**, 7441 (1988).

Retterer, J. M., T. Chang, G. B. Crew, J. R. Jasperse, and J. D. Winningham, "Monte Carlo Modeling of Ionospheric Oxygen Acceleration by Cyclotron Resonance with Broad-Band Electromagnetic Turbulence," *Phys. Rev. Lett.*, **59**, 148 (1987).

Roth, I., M. K. Hudson, and R. Bergmann, "Effects of Ion Two-Stream Instability on Auroral Ion Heating," *J. Geophys. Res.*, **94**, 348 (1989).

Schriver, D., and M. Ashour-Abdalla, "Linear Instabilities in Multicomponent Plasmas and Their Consequences on the Auroral Zone," *J. Geophys. Res.*, **93**, 2633 (1988).

Schriver, D., M. Ashour-Abdalla, H. L. Collin, and N. Lallande, "Ion Beam Heating in the Auroral Zone," *J. Geophys. Res.*, in press (1989).

Sharp, R. D., R. G. Johnson, and E. G. Shelley, "Observations of an Ionospheric Acceleration Mechanism Producing Energetic (keV) Ions Primarily Normal to the Geomagnetic Field Direction," *J. Geophys. Res.*, **82**, 3324 (1977).

Sharp, R. D., R. G. Johnson, and E. G. Shelley, "Energetic Particle Measurements From Within Ionospheric Structures Responsible for Auroral Acceleration Processes," *J. Geophys. Res.*, **84**, 480 (1979).

Sharp R. D., D. L. Carr, W. K. Peterson, and E. G. Shelley, "Ion Streams in the Magnetotail," *J. Geophys. Res.*, **86**, 4639 (1981).

Shelley, E. G., R. G. Johnson, and R. D. Sharp, "Satellite Observations of Energetic Heavy Ions During a Geomagnetic Storm," *J. Geophys. Res.*, **77**, 6104 (1972).

Shelley, E. G., R. D. Sharp, and R. G. Johnson, "Satellite Observations of an Ionospheric Acceleration Mechanism," *Geophys. Res. Lett.*, **3**, 654 (1976).

Shelley, E. G., H. Balsiger, P. Eberhardt, J. Geiss, A. Ghielmetti, R. G. Johnson, W. K. Peterson, R. D. Sharp, B. A. Walen, and D. T. Young, "Initial Hot Plasma Results from the Dynamics Explorer," in *Energetic Ion Composition in the Earth's Magnetosphere,*" R. G. Johnson, ed., Terra Scientific Publishing Co., Tokyo, p. 353 (1983).

Temerin, M., "Evidence for a Large Bulk Ion Conic Heating Region," *Geophys. Res. Lett.*, **13**, 1059 (1986).

Ungstrup, E., D. M. Klumpar, and W. J. Heikkila, "Heating of Ions to Suprathermal Energies in the Topside Ionosphere by Electrostatic Ion Cyclotron Waves," *J. Geophys. Res.*, **84**, 4289 (1979).

Winglee, R. M., P. B. Dusenbery, H. L. Collin, C. S. Lin, and A. M. Persoon, "Simulations and Observations of Heating of Auroral Ion Beams," *J. Geophys. Res.*, **94**, 8943 (1989).

Yau, A. W., B. A. Walen, W. K. Peterson, and E. G. Shelley, "Distribution of Upflowing Ionospheric Ions in the High-Altitude Polar Cap and Auroral Ionosphere," *J. Geophys. Res.*, **89**, 5507 (1984).

Yau, A. W., B. A. Walen, and P. M. Kintner, "Low-Altitude Transverse Ionospheric Ion Acceleration," in *Ion Acceleration in the Magnetosphere and Ionosphere*, Geophysical Monograph 38, T. Chang, ed., p. 39 (1986).

Yau, A. W., W. K. Peterson, and E. G. Shelley, "Quantitative Parameterization of Energetic Ionospheric Ion Outflow," in *Modeling Magnetospheric Plasma*, Geophysical Monograph 44, T. E. Moore and J. H. Waite, Jr., eds., American Geophysical Union, Washington, DC, p. 211 (1988).

Young, D. T., H. Balsiger, and J. Geiss, "Correlations of Magnetospheric Ion Composition with Geomagnetic and Solar Activity," *J. Geophys. Res.*, **87**, 9077 (1982).

III-6. ION PRECIPITATION AND THE TRANSPORT OF IONS ACCELERATED BY AURORAL PROCESSES

J. M. Bosqued*

It is now well established that the ionosphere is a substantial source of O^+ and H^+ ions detected in the magnetosphere. High fluxes of low-energy ions of terrestrial origin are also commonly observed in the precipitation over the auroral zone with interesting features that can be explained in terms of ion transport in the magnetosphere. While adiabatic heating applies to energetic (>5 keV) ions convected earthward from the distant plasma sheet, a new population of ionospheric origin frequently appears equatorward of the diffuse auroral zone. Recent satellite observations of energy dispersion, latitude, and local time dependences of these low-energy ions will be reviewed and compared with predictions of three-dimensional kinetic calculations of ion trajectories in a dawn-dusk electric field superimposed on the magnetic field. Detailed comparisons show a reasonable agreement between predicted drift paths as a function of mass and energy of the upflowing ions ejected from the ionosphere and soft ion precipitation patterns in the morning sector of the auroral oval.

1. INTRODUCTION

It has long been suspected that the auroral and polar ionospheres are a source of light ions [Axford, 1968]. However, as Shelley et al. [1972] have demonstrated, they are also a very significant source of energetic magnetospheric ions, essentially O^+, and to a lesser extent, H^+ and He^+ (see the review by Shelley [1985]), and even a source of minor ions such as O^{++} [Young et al., 1977; Kremser et al., 1987] and molecular ions [Chappell et al., 1982a; Craven et al., 1985; Lockwood et al., 1985a].

Since 1972, ion fluxes of magnetospheric origin have been observed in all the regions of the magnetosphere or its boundaries, including the tail lobes [Sharp et al., 1981], the central plasma sheet [Lennartsson and Shelley, 1986; Lennartsson, 1987], and the ring current [Geiss et al., 1978]. In all these regions, the ionospheric contribution is quite variable, and depends on many parameters, such as magnetic activity, season, and solar activity. As DE 1 observations have shown, there are several sources of ionospheric suprathermal plasma: a concentrated and persistent source of 10–100 eV ions at the high-altitude polar cap [Shelley et al., 1982; Chappell et al., 1982b; Yau et al., 1984, 1985a; Waite et al., 1985], and thermal upwelling ions around the polar cap [Moore et al., 1985, 1986] and near the dayside polar cap boundary [Lockwood et al., 1985a]. These upgoing ionospheric ions are dispersed in the antisunward direction according to their mass, and thus form the cleft ion fountain [Lockwood et al., 1985b]. Above the auroral zone, ions of terrestrial origin, observed over the past 10 years, are relatively more energetic, since they are accelerated in the topside ionosphere to energies from several electron volts to tens of kiloelectron volts parallel and perpendicular to the local geomagnetic field. This is demonstrated by the presence at low altitudes ($<3R_E$) of several families of pitch-angle distributions: upflowing ions (UFI) [Shelley et al., 1976; Ghielmetti et al., 1978; Gorney et al., 1981; Sharp et al., 1983; Collin et al., 1987; Yau et al., 1984, 1985a], downflowing ions (DFI) [Ghielmetti et al., 1979; Winningham et al., 1984; Bosqued et al., 1986] accelerated along the magnetic field, and conic-like distributions of transversely accelerated ions (TAI) (see Klumpar [1986] and references therein). Nevertheless, numerous problems remain unsolved, related especially to possible physical ionospheric extraction mechanisms and to parallel and transverse acceleration; considerable theoretical work has been stimulated in recent years by in situ observations (see review by Shelley and Collin [this volume]).

We do not here present a review of the morphology of the upflowing ions; this has been the subject of a large number of studies utilizing, in particular, S3-3 and DE 1, as well as of some very complete review articles [Horwitz, 1982; Shelley, 1985; Young, 1986; Klumpar, 1986; Yau and Lockwood, 1988; Peterson, 1988]. Rather, it will concentrate on the final suprathermal precipitation of O^+, H^+, and He^+ ion species in the auroral ionosphere. The energies of these ions are in the kiloelectron-volt range or lower, and thus often considerably higher than their initial temperature, if it is assumed that the majority of them are of ionospheric origin. This clearly implies several stages of transport and a recycling in the magnetosphere to the plasma sheet, followed by an acceleration and finally diffusion onto the auroral field lines. The focus here is first

---

*Centre d'Etude Spatiale des Rayonnements, Centre National de la Recherche Scientifique, Toulouse, France.

Auroral Physics, edited by C.-I. Meng, M. J. Rycroft and L. A. Frank. © Cambridge UP 1991

on recent experimental results on this plasma, that convey information on possible transport mechanisms on the polar and auroral magnetic field lines. It will be shown that several useful diagnostics exist, such as the observation of velocity and/or latitude dispersion of the flux of the terrestrial H^+ and O^+ ions, or the temporal and spatial evolution in magnetic local time (MLT) and latitude (LAT) of the detailed distribution functions of each species. Numerous two-dimensional and three-dimensional codes, developed in parallel, have proved successful in explaining the majority of the observational details and following the "life" of ions, starting with the outflow from the ionosphere. The transport of ions ejected from the ionosphere is largely dependent upon initial conditions such as emission regions, fluxes, velocities, and composition, but it is also largely controlled and dominated by the large-scale electric and magnetic fields in the magnetosphere.

2. TRANSPORT OF IONOSPHERIC IONS: EXPERIMENTAL ASPECTS

The localization, composition, intensity, and angular distribution of ion precipitation in the auroral region display quite pronounced variations that may reflect the spatial and temporal variations of the solar wind source throughout the plasma sheet and/or the ionosphere source, of the parameters that influence plasma motion (convection electric field and geomagnetic field), and, ultimately, of the losses by pitch-angle diffusion and charge exchange with upper atmospheric atoms, that increase as the particles penetrate inward. After this mixing of possible sources, there is little hope of recognizing clear and unambiguous signatures of ion transport in the distributions or precipitation morphology. However, one might hope that, in addition to the classical separation of the "trace" ions from their various sources (He^{++} from the solar wind feeding the plasma sheet, He^+ and O^+ from the ionosphere), magnetospheric transport should be separable into two steps. The first is essentially an earthward convection in the equatorial plane for ion plasma from the plasma sheet, which brings it into the quasi-trapping region, with an associated pitch-angle scattering whose consequence is the appearance of a diffuse precipitation. The second is complete single or multiple bounces along the lines of force (superposed on the earthward convection) for ions of ionospheric origin.

Here we look for the double signature of transport that might be expected in suprathermal ion precipitation. First, the diffuse precipitation should display an equatorial boundary, directly defined by plasma drift motions in the equatorial plane. Second, ionospheric precipitation, more concentrated at low energies, should occur more directly in regions near the source of the outflowing ions.

Precipitation Boundaries

The motion and transport of ions in the equatorial plane for the case of steady-state convection has been studied in great detail; as a result, the flow pattern of plasma injected from the plasma sheet has been successfully described by numerous authors [*Chen*, 1970; *Kivelson and Southwood*, 1975; *Cowley and Ashour-Abdalla*, 1976; *Ejiri*, 1978; *Ejiri et al.*, 1980; *Harel et al.*, 1981a,b; *Mauk and Meng*, 1983, among others]. In the course of this earthward convection the plasma is strongly accelerated (this follows simply from conservation of μ and J, adiabatic invariants of motion) [*Ejiri*, 1978]. However, it cannot penetrate earthward into the forbidden regions whose boundaries, or Alfvén layers, depend upon the charge of the particles, their energies, the local time, and, of course, the convection electric field. The equatorial boundary of the auroral precipitation is directly related to these Alfvén layers in the equatorial plane, at least in the stationary case. The spatial dispersion of particles as a function of energy at the inner edge of the plasma sheet and therefore at its projection into the auroral oval can be analyzed to gain insight into the adiabatic transport from the plasma sheet, if strongly disturbed periods are excluded, e.g., substorm onset [*Kivelson et al.*, 1987]. Numerous statistical studies of the low-altitude electron precipitation equatorial boundary have been carried out in recent years, for various local-time sectors [*Gussenhoven et al.*, 1981, 1983, 1987; *Sauvaud et al.*, 1983; *Feldstein and Galperin*, 1985; *Valtchuk et al.*, 1986, for most recent papers]. The position of this boundary is in reasonable agreement with equatorial transport models, including possible pitch-angle scattering losses [*Fontaine and Blanc*, 1983].

Results on ions precipitating at the low-altitude equatorward boundary are considerably rarer, essentially because the influxes are much weaker and more diffuse and display no discrete structures such as inverted Vs for electrons. This boundary, frequently difficult to locate, often fluctuates considerably with the instantaneous magnetic activity. Before examining the results, it is useful to mention that simple static models of the equatorial transport of electrons and ions under the influence of a large-scale dawn-dusk electric field predict that protons (ions) will come closer to the Earth than will electrons, by several degrees in the auroral zone, regardless of the magnetic sector. Figure 1, from *Newell and Meng* [1987], is an application of the analytic model of *Ejiri* [1978] for the 0600 MLT sector, and reflects the predictions well, especially the prediction that 1–10 keV ions (in this example) penetrate closer to the Earth than do 100 eV ions.

The relative position of the two electron and ion equatorial boundaries was recently presented by *Gussenhoven et al.* [1987] and *Newell et al.* [this volume]

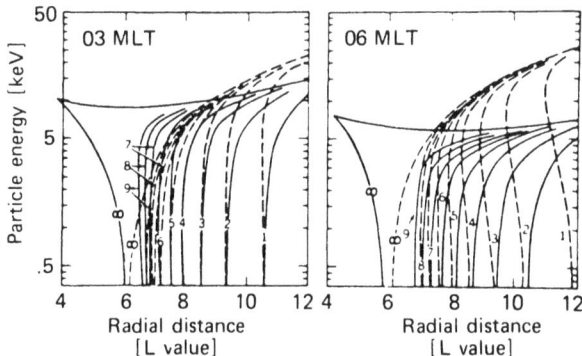

Figure 1—Time-dependent boundaries of electrons (...) and ions (—) convected in the morning sector (0300–0600 MLT) from the plasma sheet by a Stern-Volland dawn-to-dusk electric field (with $\gamma = 2$ and scaling in order to fix the stagnation point at 10 R_E in the dusk). Electric field is suddenly increased at $t = 0$ and successive boundaries are plotted at $t = 1,2,3,...$ hours after enhancement ("infinite" means steady-state Alfvén layers). Adapted from *Ejiri et al.* [1980] by *Newell and Meng* [1987].

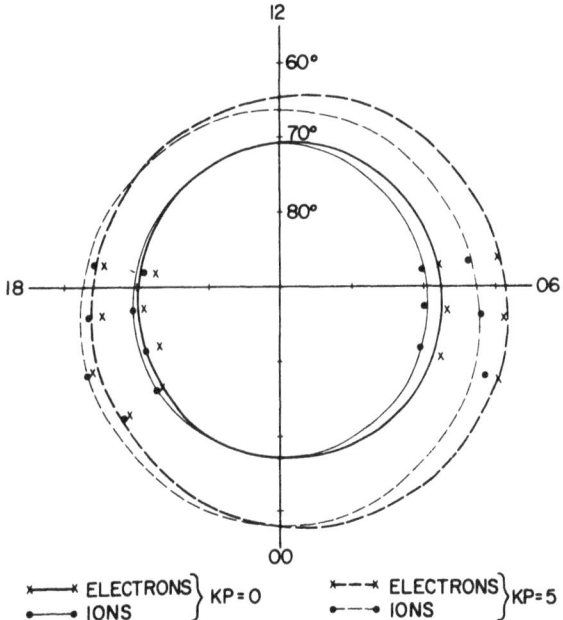

Figure 2—Electron and ion boundaries for $Kp = 0$ and 5 in the evening (1700–2100 MLT) and morning (0400–0700 MLT) sectors, given by *Gussenhoven et al.* [1987]. Ion boundaries are fitted to circles centered on the midnight meridian at 87.6° ($Kp = 0$) and 85.8° ($Kp = 5$) latitude. Electron circles slightly eccentric at 0240 MLT are a good fit to electron boundaries published by *Gussenhoven et al.* [1983].

for the evening and morning sectors, using DMSP F6 satellite data; the two boundaries (defined by the increase of the total particle number flux) correspond to two circles offset from the geomagnetic pole, centered at 87.6° and 0000 MLT (0240 MLT) for ions and electrons, respectively (Figure 2). Here we discuss the experimental results on the morning sector, in order to compare them with the result of the large-scale convection model of Figure 1. We shall see that the electron and ion boundaries, which move toward lower latitudes with increasing magnetic activity, are separated by about 3° regardless of the activity, with the equatorward electron boundary being lower, contrary to the predictions of dispersion models. To explain this, *Gussenhoven et al.* [1987] have proposed, among other things, to take into account the temporal development of the boundary positions after injection of particles from the plasma sheet toward the Earth (see also *Gorney and Evans*, 1987). The calculations of *Ejiri et al.* [1980] and *Newell and Meng* [1987] do in fact show that after 3 or 4 hours of drift in the morning sector the electrons advance closer to the Earth than do the ions, although at low energies the two boundaries should coincide. Using DMSP F6 data in addition, *Newell and Meng* [1987] have presented case studies that display results contradictory to the preceding ones, but that contain an interesting clue to the possible origin of the ions. Studying the spectral dependence of the particle equatorial cutoff, they have shown that low-energy (100 eV) ions precipitate at a lower latitude than do high-energy (1–3 keV) ions, especially in the morning sector. This soft equatorial ion population was initially noted by *Sauvaud et al.* [1981], who invoked an ionospheric origin for it.

A more detailed analysis using the energy and mass separation capabilities of the Aureol 3 (A3) French-Soviet satellite proved that this was indeed a soft population, rich in O^+ ions and, to a lesser extent, He^+ ions [*Bosqued*, 1985, 1987]. Figure 3, from A3 data, summarizes electron and ion data from spectrograms that are typical of the electrons and ions in the morning sector; the poleward boundary of the auroral oval was crossed at 0338:30 UT, and A3 was in the diffuse auroral zone until 0343 UT. The ion spectrogram is particularly interesting to examine in terms of ion transport, because several populations with intense fluxes appear successively and/or superposed:

1. In the diffuse auroral zone from 70–66° invariant latitude (0340–0341:30 UT), only the H^+ ions precipitate, with an average energy that increases as L^{-2};

2. Equatorward, mainly H^+ ions are found, whose energy decreases with latitude over about 2°. The flux of precipitating very-low-energy H^+, O^+, and He^+ ions (<300 eV) remains significant even further equatorward, to 0343 UT, which corresponds to the limit of the electrons, well dispersed in latitude over 2° (60° at 400 eV, 62° at 9 keV).

From this example (described elsewhere in detail [*Sauvaud et al.*, 1985; *Bosqued*, 1987]), it can be seen, first, that steady-state convection describes satisfactorily

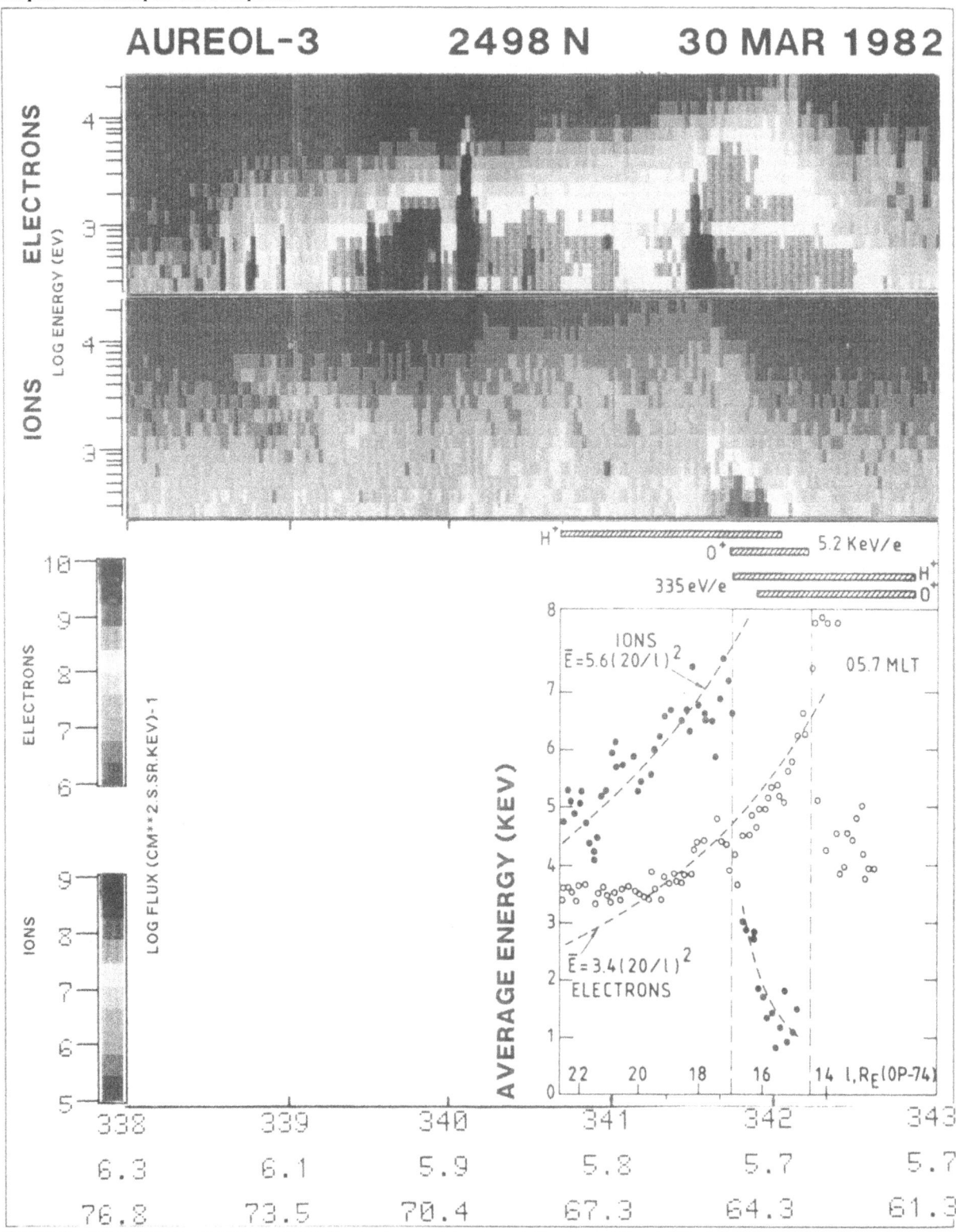

Figure 3—Precipitating electron and ion fluxes measured onboard the Aureol 3 satellite (pass 2498 north) as a function of the Universal Time (from 0338–0343 UT), magnetic local time (around 0590 MLT), and invariant latitude. Average energies are also plotted (bottom panel) and compared with the L^{-2} Fermi acceleration profile. Regions of precipitating H^+ and O^+ ions are indicated by horizontal hatched bars (extracted from *Bosqued* [1987]). (This figure also appears in color: Plate 14.)

the earthward motion of the plasma sheet that feeds the diffuse electron precipitation, and second, that electron and ion energization follow the L^{-2} law closely, as is expected for $\mathbf{E} \times \mathbf{B}$ convection with the conservation of the first two invariants of the motion [*Ejiri,* 1978]. But the most significant result is that the latitudinal position of the ion precipitation is both mass dependent and energy dependent. Soft ion precipitation (mostly O^+, <1 keV) of terrestrial origin, near the edge of the precipitation boundary, as well as the latitude dispersion of low-energy (<3 keV) H^+ ions, was interpreted by *Bosqued* [1987] as a time-of-flight dispersion of locally injected ionospheric ions. These O^+ and H^+ ions escaping from the auroral ionosphere with energies of from 10–100 eV are frequently observed by A3 with conical pitch-angle distributions [*Rème et al.,* 1985]. Injected along auroral field lines, these ions will drift under the influence of the convection field, and disperse in latitude and longitude; qualitatively, one can speculate that this dispersion will be greater for the heavier ions, O^+ in this case, and for the less energetic ones (several tens of electron volts). We will see later that three-dimensional transport models predict exactly this behavior, and that it is therefore not surprising to find soft O^+ ions at the equatorward edge of the diffuse auroral zone and at even lower latitudes.

The preceding comments and conclusions, particularly for mass-dependent dispersion, may be strengthened and generalized by stating that the morphological behavior of the ion precipitation in this morning sector is frequently found to be that presented in Figure

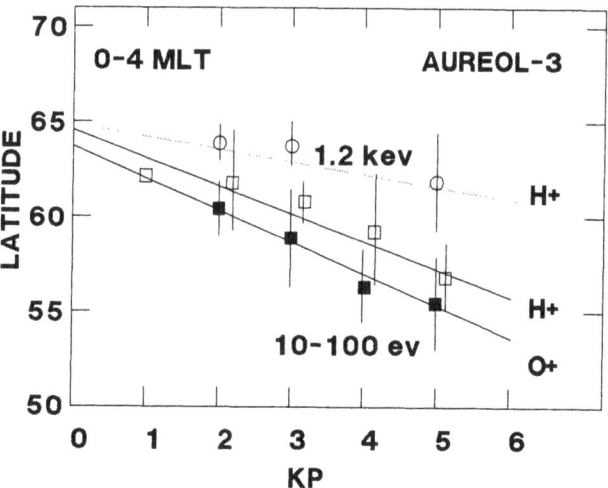

Figure 4—Latitude of the equatorial ion boundaries for the 0000–0400 MLT sector as a function of *Kp*. The boundaries were determined by using differential intensity profiles in the 10–110 eV energy channel for the H^+ and O^+ ions and 1.25 keV for the H^+ ions. About 40 Aureol 3 passes are included in this preliminary study.

3. Figure 4 shows a preliminary summary of the ion equatorward boundaries in the postmidnight sector (0000–0400 MLT) for 10–100-eV H^+ and O^+ ions, on the basis of A3 data from the high-sensitivity ion mass spectrometers. (The complete results of this analysis will be presented elsewhere.) Despite the limited dataset used here, two general tendencies may be noted:

1. Linear regressions demonstrate a systematic variation of the boundaries with *Kp*, but with a smaller slope than that found in the same sector by *Gussenhoven et al.* [1987];

2. There is a difference in the latitudes of the equatorward edges of the low-energy ions, with the O^+ ions extending over 1–2° equatorward (independent of *Kp*); the two very low energy limits display an offset with respect to the 1-keV H^+ ion boundary, which increases with the level of magnetic activity (about 6° for *Kp* = 6).

Finally we note some latitude differences of several degrees between the positions of these ion boundaries and those published recently by *Gussenhoven et al.* [1987]; these may be attributed mainly to the criteria used in the definitions of the boundaries.

Morphological Aspects of Ion Transport

As is mentioned above, ionospheric ions ejected upward by auroral acceleration processes [*Shelley,* this volume] display special initial distribution functions (conical or beam distributions) that evolve during the transport and bounce motions. Generally the transport is not totally adiabatic, since, for example, near the equatorial plane, the observed distributions, which should be highly aligned whatever their origin, are in fact broader than would be predicted. However, it is instructive to analyze the anisotropies of the distribution functions (field-aligned, "pancake," conical, etc...) since they reveal the evolution of the ions during their transport in the magnetosphere. Some recent observations that relate to the morphology of the occurrence of different types of distributions at various altitudes above the auroral oval will now be reviewed.

At altitudes above $2R_E$, the presence of H^+ and O^+ counterstreaming ions (CSI), which correspond to bidirectional pitch-angle distributions simultaneously peaked about 0° and 180° pitch angles, has been studied recently by *Horita et al.* [1985, 1987] using the ISEE 1 satellite. These counterstreaming H^+ and O^+ ions may correspond to upflowing ionospheric ions that are injected at low altitudes on closed field lines, accelerated upward along field lines at higher altitudes in one hemisphere, and that have already mirrored in the conjugate hemisphere. As mentioned by *Horita et al.* [1987], there are different possible counterstreaming ion energizing mechanisms: parallel and steady electric fields, strongly supported by recent results of *Reiff et*

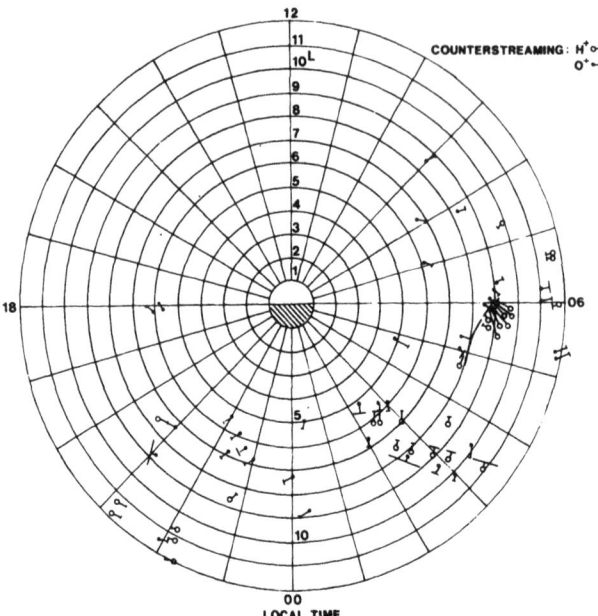

Figure 5—Counterstreaming H$^+$ and O$^+$ ion events detected by the ISEE/ion mass spectrometer and plotted in a local time-L polar diagram (from *Horita et al.* [1987]). The study contains over 60 events extended up to values of L = 12.

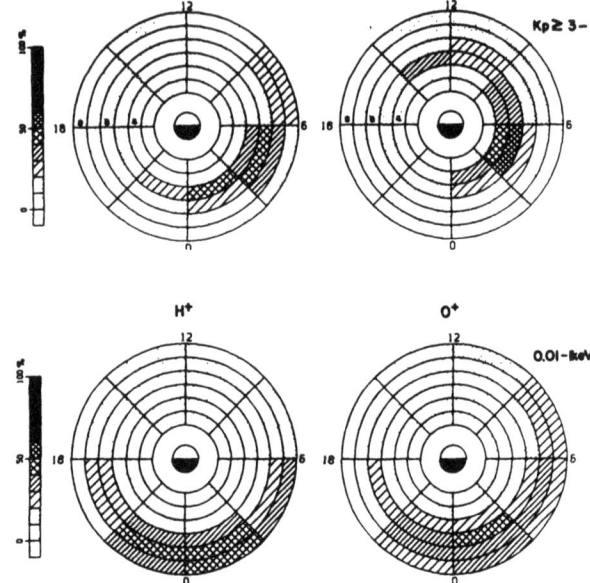

Figure 6—Occurrence probabilities of bidirectional (*top*) and unidirectional (*bottom*) H$^+$ and O$^+$ ion distributions in the polar diagram magnetic local time-L shell (between 2.5 and 8.5 by $\Delta L = 1$ bin), as measured by the DE 1/EICS instrument in the 0.01–1 keV energy range. Results are shown for magnetically disturbed periods, $Kp > 1$. Field-aligned distributions are dominated by upgoing (UFI) ion beams because DFI occurrence is relatively low (from *Sagawa et al.* [1987]).

al. [1988]; wave-particle interactions, and transient parallel electric fields, invoked by *Sharp et al.* [1980]. *Shelley and Collin* [this volume] discuss properties of ion beams and present convincing arguments that imply acceleration by electric fields parallel to *B*. However, it appears that ion heating could operate within (or below) the acceleration region. The study of these counterstreaming ions therefore provides interesting information on mass-dependent terrestrial ion acceleration, diffusion, and transport in the magnetosphere. Figure 5 shows the general characteristics of the 60 CSI events presented by *Horita et al.* [1987]: they are detected only in the 1800–0900 LT sector, and mostly in the postmidnight and predawn sectors. A latitude separation is also found: the majority of the H$^+$ CSI events are localized at L = 8–12, while O$^+$ events are lower in latitude, at L = 5–8.

At lower altitudes, a similar study has been carried out by *Sagawa et al.* [1988], but on a larger database from DE 1; it deals with low-energy (< 1 keV) ion distributions and corroborates the preceding results. The study includes H$^+$ and O$^+$ unidirectional field-aligned ions (or UFI) distributions peaked about 0°–20°, and bidirectional field-aligned (or counter-streaming) distributions. (The latter category is directly comparable to the study of *Horita et al.* [1987].) Figure 6 displays the results of *Sagawa et al.* [1987] as the probability of occurrence of UFI distributions for active conditions ($Kp \geq 3$) and energies of 0.01–1

keV (above) as well as the same distribution for counterstreaming H$^+$ and O$^+$ ions (below). In this MLT-L diagram, a direct comparison of the distributions clearly reveals that the UFI measured by DE 1 and previously by S3-3 [*Ghielmetti et al.*, 1978] are found roughly in the same region of the auroral oval centered at midnight, with a slight mass dependence: the H$^+$ UFI region is located at higher latitude (L = 7) than that of the O$^+$ ions (L = 6). These ions had been ejected only recently from the ionosphere, had not undergone multiple bounces, and thus had not scattered strongly in pitch angle while passing through the equatorial plane. One can state qualitatively that these < 1 keV upflowing ions should drift toward the morning sector in azimuth and toward lower L shells, evolving into counterstreaming beams, then, after successive pitch-angle diffusion and losses, toward loss-cone distributions. This is exactly what is observed in the CSI event distribution (Figure 6, bottom): the preferential region of occurrence is the dawn sector, always with the same mass separations.

At lower energies, *Giles et al.* [1988] and *Delcourt et al.* [1988a] have separated the < 50 eV ion distribution functions measured by DE 1 into two classes at the satellite altitude (3R$_E$): field-aligned (unidirectional) and bouncing (bidirectional). The frequency of oc-

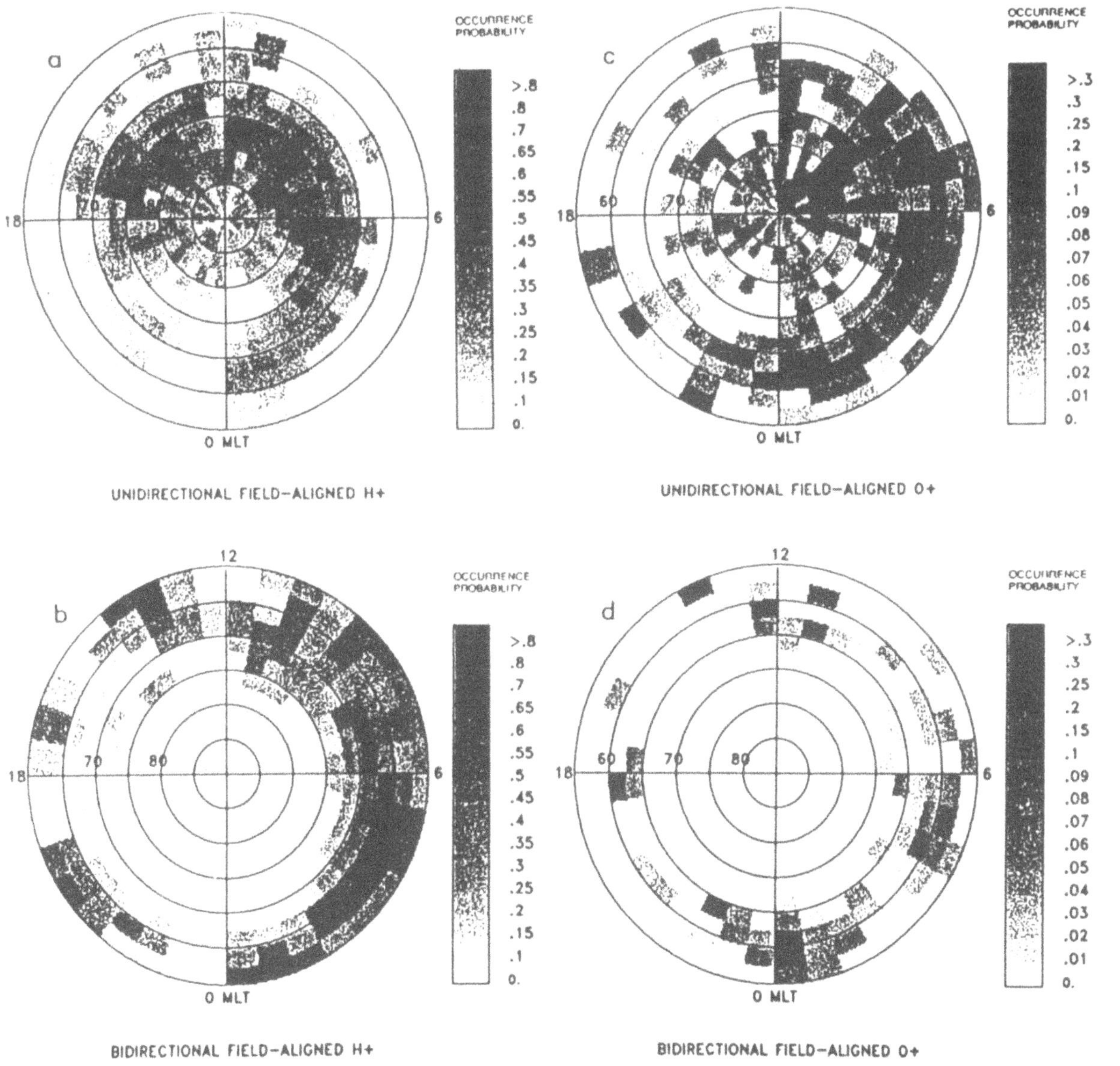

UNIDIRECTIONAL FIELD—ALIGNED H+

UNIDIRECTIONAL FIELD—ALIGNED O+

BIDIRECTIONAL FIELD—ALIGNED H+

BIDIRECTIONAL FIELD—ALIGNED O+

Figure 7—Occurrence probabilities of field-aligned low-energy (<50 eV) ion distributions measured by DE 1/RIMS at an altitude of the order of $3R_E$, plotted in MLT-ILAT diagrams. Distributions are separated in unidirectional (*top*) and bidirectional (*bottom*), for H+ (left) and O+ ions (right panels). Note the different coding scales and ILAT grids for H+ and O+ ions. (From *Giles et al.* [1987] and interpreted in terms of ion transport by *Delcourt et al.* [1988].)

curence is shown in the latitude-MLT polar diagram of Figure 7. Newly injected H+ ions are observed to have a field-aligned distribution, especially for latitude greater than 70° and in the 0600–1200 MLT sector, while the rarer O+ ions with the same distribution are spread out over lower latitudes, between 55° and 65°. If we assume that these ions have just been ejected from the auroral ionosphere at low altitudes, e.g., in the form of conics, it is clear that their transport leads rapidly to a latitude and longitude dispersion that is greater for heavier (or slower) ions, and that they have undergone multiple bounces before precipitating. This

is in fact what is observed: after several bounces, the H+ ions are found further equatorward (60° < latitude < 70°).

Observations of Latitude Dispersed Ion Beams

Recent high-latitude (>7000 km) electron and ion measurements provided by DE 1 [*Reiff et al.*, 1986, 1988] and Viking [*Block and Fälthammar*, this volume] are consistent with the parallel electric field acceleration hypothesis, although other models, i.e., acceleration by wave turbulence generated at the hybrid frequency [*Bryant*, this volume] or acoustic wave tur-

AUREOL-3 DFI BEAMS

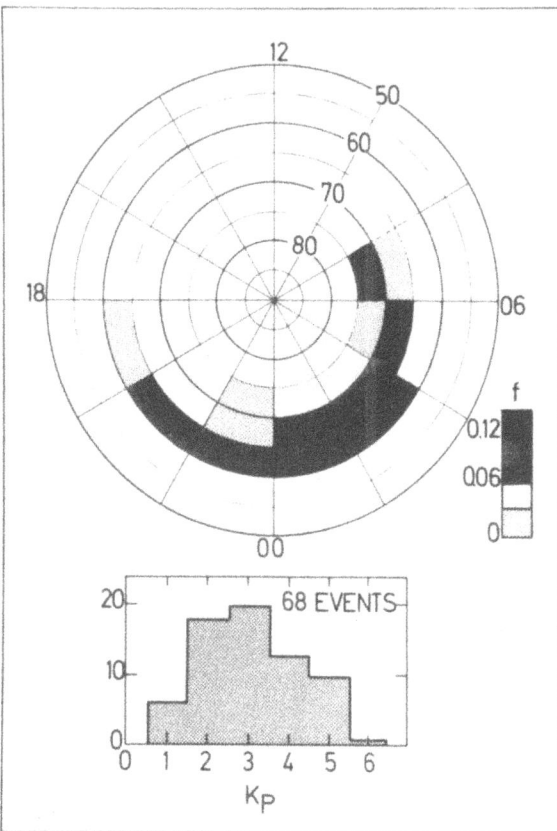

Figure 8—Frequency of occurrence of 68 events detected by Aureol 3, mostly in the morning sector. (From *Bosqued et al.* [1986].)

bulence [*Stasiewicz*, 1985], have been recently proposed. Comparing conjugate detailed plasma observations on the same field-line at two different altitudes, *Reiff et al.* [1988] were able to deduce that a parallel potential difference is required, occurring between 800 km (DE 2 altitude) and 2–3 R_E (DE 1 altitude). However, at least a part of the electric field could be strongly fluctuating and turbulent [*Bingham et al.*, 1984; *Block and Fälthammar*, this volume] as is demonstrated by electron heating within the acceleration region [*Reiff et al.*, 1988]. The counterpart to the earthward electron acceleration is the simultaneous ejection of terrestrial ions, constituting ion beams with an energy that is also peaked at several kiloelectron volts; the ions reach altitudes >5000 km in the form of O^+-rich TAI [*see Shelley and Collin*, this volume]. Ions accelerated upward out of the ionosphere are subject to convection, possibly isotropized and scattered by wave-particle interactions while passing near the equatorial plane, and should precipitate in the conjugate hemisphere. This half-bounce transport and filtering effect leads to the appearance of ion bands

in the classical energy-time spectrograms, with energy always decreasing with latitude. These structures are indeed observed by DE 1 [*Winningham et al.*, 1984; *Frahm et al.*, 1986] and A3 [*Bosqued et al.*, 1986]. The velocity-dispersed ion bands recorded by A3 in the central auroral zone are dominated by H^+ ions (with an equatorward band of O^+ ions, although having lost the detailed dispersion signature) and are dominant in the morning sector regardless of magnetic activity (Figure 8). The ion bands reported by *Frahm et al.* [1986] are, on the contrary, more related to magnetic storms in the nightside sector and to the growth of the ring current at the equatorward edge of the diffuse auroral zone. *Bosqued et al.* [1986] have discussed numerous examples in detail and have shown that the ions originate in inverted-V structures that are often poleward of the ion band.

3. MODELING ION TRANSPORT FROM THE IONOSPHERE

Initial work on suprathermal ion transport concentrated on the motion from the plasma sheet for pitch angles near 90°, for which (owing to the many bounces undergone) only a bounce-averaged correction function was introduced into the calculation of the drift velocity. However, simple equatorial transport models are insufficient for several reasons when an ionospheric source is introduced. First, for any ejection pitch angle, these ions will describe at least a complete half-bounce motion, passing through the equatorial plane practically aligned with the magnetic field. Second, heavy ions such as O^+, which are frequently ejected with small energies (<1 keV), will have bounce periods greater than the convection time and thus will move far from their original flux tube after half a bounce. The growing evidence in favor of the ionosphere as an ion source has stimulated simultaneous progress in the modeling of transport along flux tubes, in order to treat its special properties. Several models have been developed recently to calculate ion transport after the initial outflow from the ionosphere. In these models the adiabatic ion trajectories are calculated in a magnetic field with the large-scale convection and corotation electric fields superposed.

Two-Dimensional Models: Feeding the Plasma Sheet with Ionospheric Ions

The first two-dimensional kinetic trajectory models in the noon-midnight plane, developed by *Horwitz* [1984, 1987], and *Horwitz and Lockwood* [1985], followed the trajectories in the polar magnetosphere and computed the evolution of distribution functions and ion bulk parameters for <100 eV ions, injected from a narrow source located in the topside ionosphere near the cusp, as discovered by DE 1. Although magnetic

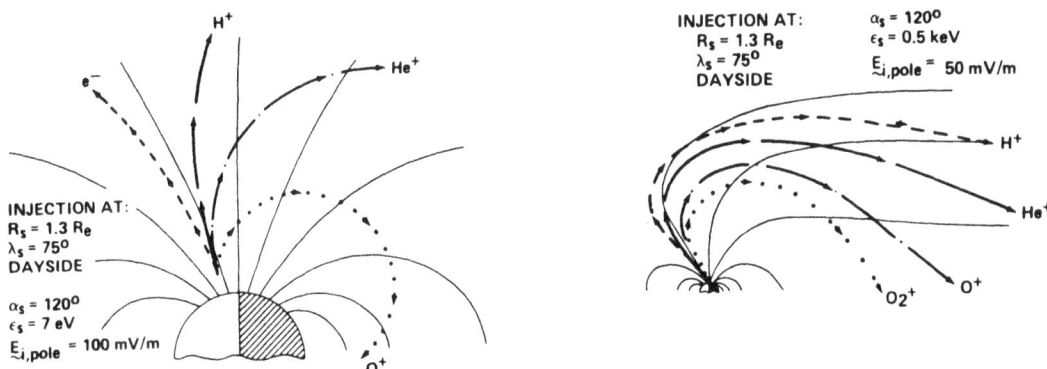

Figure 9—Two-dimensional trajectories of low-energy ionospheric ions injected near the polar cusp ionosphere, in the noon-midnight plane (from *Horwitz* [1984]): (*Left panel*) initial energy: 7 eV; dawn-to-dusk polar cap electric field: 100 mV/m (at ionospheric altitudes); O$^+$ ions are transported to the nightside auroral field-lines. (*Right panel*) initial energy: 500 eV; electric field: 50 mV/m; in that case ions are velocity-dispersed and O$^+$ ions reach the near-Earth plasma sheet.

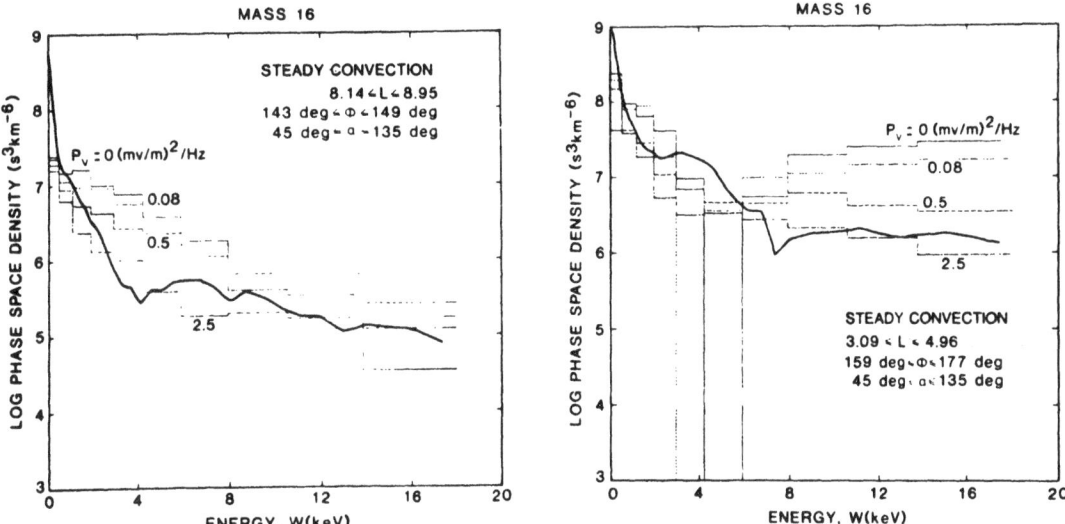

Figure 10—Comparisons between computed and experimental O$^+$ phase space densities at two radial distances: L = 8.5 (*left*) and L = 4.5 (*right*) (from *Cladis and Francis* [1985]). Observational data in the 0–17 keV/e energy range were obtained by ISEE 1 during a magnetic storm [*Lennartsson et al.* [1979]. In the calculations, ions originating from 15R$_E$ around midnight are stochastically heated perpendicular to the magnetic field during their earthward convection drift by electrostatic turbulence. Histograms are presented for different values at L = 10 of the power spectral density of the waves (measured in (mV m^{-1})^{2}Hz^{-1}) at the ion gyrofrequency.

drifts were neglected, the first interesting effect described by this type of simulation and shown in Figure 9 is the mass/charge dispersion of the plasma by the "geomagnetic mass spectrometer" above the high-altitude polar cap [*Lockwood et al.*, 1985b]. The second most obvious result is the plasma sheet trapping of ions from the polar cusp, for sufficient trapping energies. The final result is that substantial quantities of ionospheric ions, especially O$^+$, can reach the plasma sheet even in quiet periods.

Cladis and Francis [1985] have developed a two-dimensional model whose originality resides in treating energy and pitch-angle diffusion in the equatorial portion of the transport of ionospheric ions, induced

by a somewhat arbitrary broad-band transverse electric field and by losses due to charge exchange with ionospheric atoms (assumed to be composed of O and H). This is therefore for conditions that might prevail in disturbed periods. Nevertheless the results are interesting, since they demonstrate, by comparing fluxes, that during a magnetic storm, all O$^+$ ions and a large fraction of the H$^+$ ions in the ring current come from the polar ionosphere, which continuously feeds the plasma sheet. Thus even in this extreme case the ionosphere is a sufficient source of O$^+$ ions. The Cladis and Francis model can also explain the distributions measured between L = 4 and L = 8 by ISEE 1 (see Figure 10 extracted from *Lennartsson et al.*

[1979]; for this, the ionosphere must eject H^+ and O^+ fluxes of several times $10^8(cm^2s)^{-1}$, which then convect for about 2 hours. As a second step, *Cladis* [1986] studied the motion of the ions ejected from the polar cap or cusp region, and showed that they can feed the central plasma sheet (the auroral fountain) at $L > 6R_E$ and that they are sufficiently accelerated by the curvature drift component of the convection electric field to reach energies of kiloelectron-volt order. Therefore, this is a very large energy increase, particularly for O^+, when the ion approaches the equatorial plane [*Swinney and Horwitz*, 1986; *Swinney et al.*, 1988].

Three-Dimensional Models: Ion Acceleration and Drift Effects

New three-dimensional simulations of the trajectories of ions outflowing from the auroral and polar ionospheres have been presented recently, motivated by the experimental results from DE and A3 (originally by *Swinney and Horwitz*, 1985; also *Delcourt*, 1985). Here we will attempt to confront them with the experimental results accumulated over the last few years. It is clear that the dawn-dusk electric field causes latitudinal and azimuthal ion drifts that depend strongly upon the ion mass and, of course, energy (or time of flight in the magnetosphere). Thus the transport of the ions will be different as a function of their energy, their mass, the injection region, and their initial pitch angle, but what is clear is that they will be accelerated by the convection electric field and that their precipitation energies will often differ from their initial energy when they are flowing out of the ionosphere.

The transport of low-energy (< 100 eV) ions injected in the form of conics on the night and day sides was first modeled by *Delcourt* [1985], who studied the evolution of ions in the guiding center approximation by integrating the equations of *Northrop* [1963]. Various particle trajectories appear as a function of the ion mass, and depending on whether its velocity exceeds the minimum gravitational escape velocity (~ 10 km s^{-1} or ~ 10 eV for O^+ ions) or not [*Horwitz*, 1984; *Sauvaud and Delcourt*, 1987].

The "life" of ions injected with a conical distribution in the nighttime auroral oval was studied in great detail by *Sauvaud and Delcourt* [1987] using either a dipole or *Luhmann and Friesen* [1979] magnetic field, and a two-cell ionospheric convection distribution [*Volland*, 1978], with a possible dependence on the magnetic activity as measured by *Kp*. More complex patterns corresponding to northward pointing B_z have recently been included by *Delcourt et al.* [1988b]. We note here that all three-dimensional codes recently developed ignore local particle accelerations resulting from parallel potential drops and then assume equi-

potential field lines. Qualitatively terrestrial ions will undergo several bounces while drifting eastward or westward in longitude and earthward in latitude, and their energies will increase. Figure 11a, extracted from *Sauvaud and Delcourt* [1987], shows the total energy increase of H^+, He^+, and O^+ ions originating in the auroral ionosphere, as a function of their local time of injection, with realistic conditions for the potential drop (60 kV) across the polar cap. The energy gain that is found, regardless of the ion, simply reflects the fact that equipotentials of the convection field have been crossed under the effect of magnetic drift; over half a bounce the gain is directly proportional to the time of flight of the ion, and therefore is greatest for O^+. For the initial conditions of Figure 11a, detailed calculations show that the heavy ion species (O^+ or He^+, with 50-eV energy) gain enough energy to precipitate in the conjugate hemisphere in half a bounce, while an H^+ ion must bounce several times before it is lost in the atmosphere. At higher energies (350 eV in Figure 11a, bottom), since the bounce periods are smaller, all ions undergo several bounces before they obtain a sufficient incremental energy gain, 0.5, to precipitate. Simultaneous azimuthal and latitudinal drifts, associated with the energy gain, are shown in the polar diagram of Figure 11b, and also depend directly on the time of flight of the ions. To compare this with the experimental results mentioned above, we will emphasize the fact that the heavier ion, O^+, will always precipitate equatorward of the H^+ ion, whatever the initial energy and sector of origin. For the convection pattern chosen here, the associated azimuthal drift is directed westward (eastward) according to whether the ions are injected before (after) 2000-2100 MLT. As shown by *Sauvard and Delcourt* [1987] this drift pattern depends on the altitude of origin: for example, 50-eV O^+ ions injected at 4000 km and at 2000 MLT will end up precipitating eastward at 2060 MLT (and not westward of the injection point at 2000 km, as in Figure 11b).

At very low energies (30 eV), the simulations of *Delcourt et al.* [1988a] summarize perfectly the evolution of H^+ and O^+ ions injected in the form of conics in the nightside sector. Figure 12 (top panel) gives the positions of the successive mirror points of injected H^+ ions (in initial bins of $2.5° \times 0030$ MLT) in the latitude-MLT diagram and, finally, the precipitation point, when it exists. Detailed calculations show that to bounce at least once, H^+ and O^+ ions must originate at latitudes less than 72° and 68°, respectively, in the midnight sector. The general evolution described in Figure 11b for different conditions is found (i.e., an obvious motion toward lower latitudes) to be simultaneous with an azimuthal drift toward the morning sector. The main interest of this figure is that it

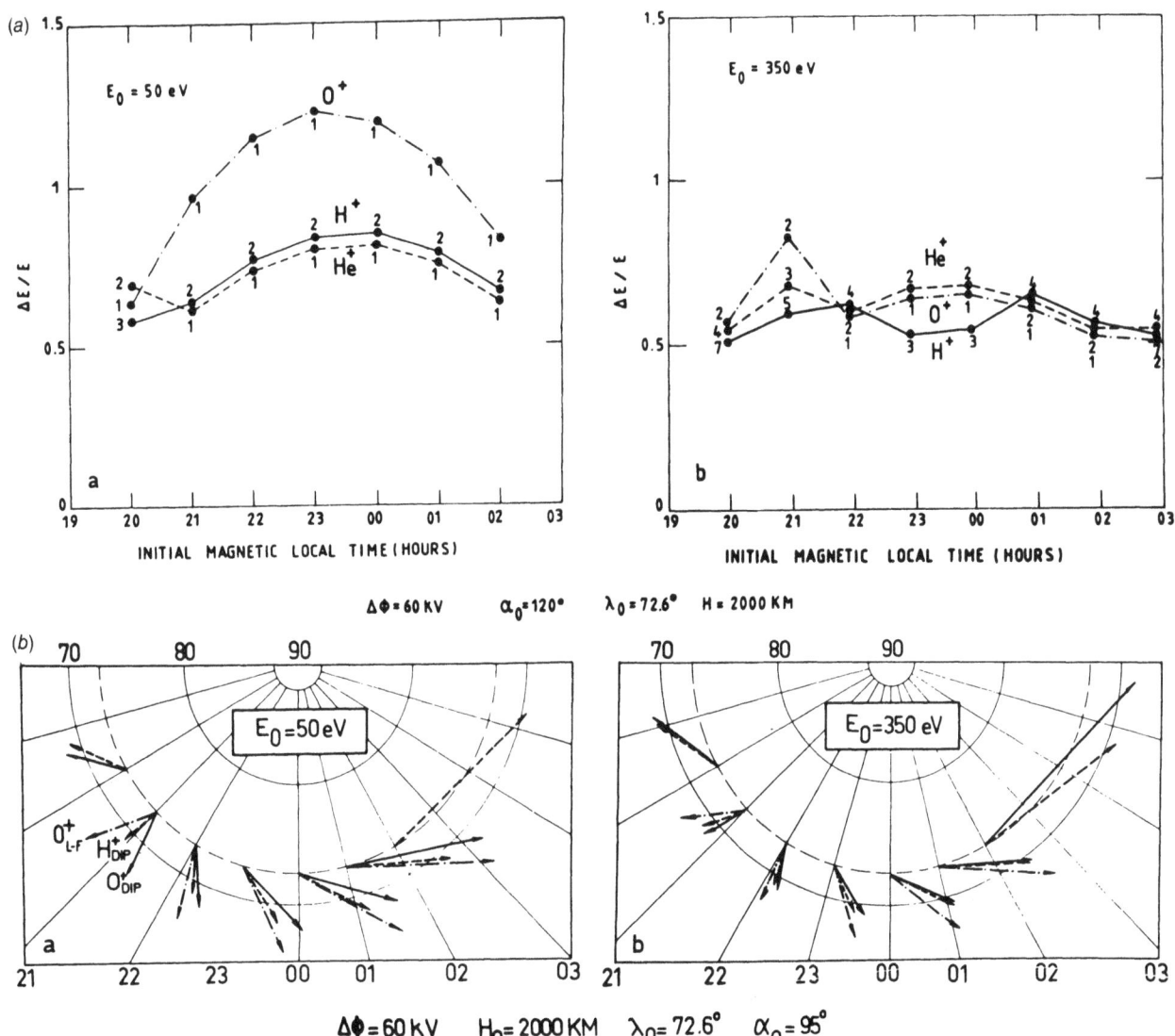

Figure 11—Transport of near perpendicularly ejected ions above the nightside auroral ionosphere (from *Sauvaud and Delcourt* [1987]): (*a*) Total energy gain normalized to the initial energy, 50 eV (*top*) and 350 eV (*bottom*) for H$^+$, He$^+$, O$^+$ species injected with similar initial conditions: altitude 2000 km, pitch angle 120°, latitude 72.6°, dawn-dusk potential drop 60 kV, and dipole magnetic field. (*b*) Simultaneous latitudinal and azimuthal drifts of the same species injected under the same conditions. Injection and final precipitation points are joined in a polar diagram by a solid (O$^+$) or dashed (H$^+$) vector; dot-dashed vectors correspond to O$^+$ ions drifting in the Luhmann-Friesen magnetic field model.

reproduces correctly the DE 1 results [*Giles et al.*, 1987] on H$^+$ ions: freshly ejected H$^+$ ions with unidirectional, field-aligned distributions (at an altitude of 3R$_E$) around 0000–2000 MLT, 70° (see Figure 7) will evolve after a residence time of about 1 hour, corresponding to several bounces, toward bidirectional distributions in the dawn sector centered at lower latitudes around 0400 MLT. A final comment is in order concerning the O$^+$ ions, for which the situation is more confused (Figure 12, bottom panel). For the most part, O$^+$ ions originating in the midnight sector do not manage to precipitate, and they reach the daytime sec-

tor after several bounces, accumulating around 55–65°.

To summarize, the most important result is that, O$^+$ and H$^+$ ions with energies <1 keV, injected with conical distributions from the auroral oval, will precipitate some 2–5° equatorward of their outflowing latitude, with a latitude separation such that O$^+$ is always equatorward of H$^+$.

4. CONCLUDING REMARKS

From the amount of data available from composition measurements at different altitudes, an overall pic-

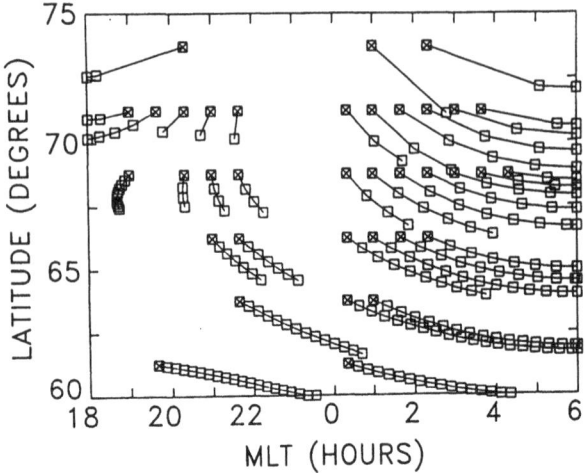

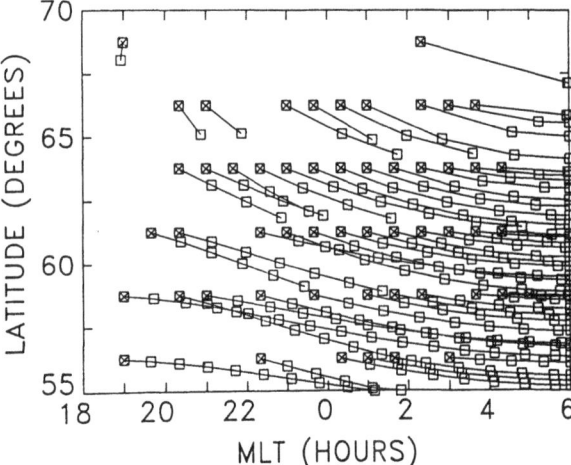

Figure 12—Simulation of the transport of 30 eV H$^+$ and O$^+$ ions injected into the nightside sector, projected in a MLT-invariant latitude diagram (from *Delcourt et al.* [1988a]). Initial position (crossed square) and successive mirror points (squares) are plotted, joined to the precipitation point (if it exists). Ionospheric ions are injected at 0.3R$_E$ altitude with a pitch angle of 60° and considered as lost when the mirror altitude is below 300 km. Three-dimensional simulations of the drift paths use the Luhmann-Friesen magnetic field model and the potential distribution of *Volland* [1978].

ture of the transport of ionospheric ions emerges. It is certain that a large fraction of the outflowing auroral ions continuously feeds the plasma sheet and is found with kiloelectron-volt energies in the auroral precipitation during the dynamic injections associated with substorms. The experimental results and the three-dimensional model calculations developed in parallel are in qualitative agreement as far as the general trans-port processes of ions accelerated by auroral processes are concerned. The following points should be emphasized:

1. Ionospheric ions outflowing from the polar cap, the auroral zone, and even the cleft, easily reach the equatorial plane while they are accelerated to kiloelectron-volt energies, and are dispersed as a function of their masses to rather large distances in the magnetotail. This feeding process was predicted even in simple models [*Horwitz*, 1984];

2. Suprathermal H$^+$, He$^+$, and O$^+$ ions are transported toward the morning sector where they are preferentially convected by the $\mathbf{E} \times \mathbf{B}$ drift. The O$^+$ precipitation region is always equatorward of the H$^+$ precipitation region whatever the magnetic activity, and this dispersion is by and large predicted by three-dimensional codes;

3. The observation of H$^+$ ion beams (and under certain quiet conditions O$^+$ ions) coming from the conjugate hemisphere and the dependence of the nightside ion boundaries on the *Kp* index, even when separated by mass, shows that the global transport is largely controlled by the large-scale convection field.

Numerical two-dimensional and three-dimensional codes (especially of *Sauvaud and Delcourt* [1987] and *Delcourt et al.* [1988a]) now provide detailed modeling of the low-energy ion transport in time and space. In particular, they demonstrate quite well that the ion energy is modified (energy gain or loss) during all of its transport across E-field equipotentials under the effect of gradient and curvature magnetic drifts. Above all, they show that the transport of ionospheric ions is strongly influenced by the drift times, which are several hours for 10–100-eV O$^+$ ions, and hence considerably greater than the time scale for field variations during a substorm. Moreover, this is where the short-comings of present models are found, since they cannot model the spatial and temporal variations of the true electric field. It will also be necessary for these models to take wave-particle interactions along the ion trajectories into account, as well as ion charge exchange with the upper atmosphere when they bounce at their mirror points. Recent calculations by *Cladis and Francis* [1985] show that these loss terms are important in the description of ion transport.

ACKNOWLEDGMENT—I express my thanks to C.-I. Meng for inviting me to give and write this paper. Stimulating discussions with Y. I. Galperin and R. A. Kovrazkhin are greatly appreciated. I would like to thank J. Cuvilo and E. Froger for their efforts in processing the Aureol 3 data. This work was supported by the Centre National d'Etudes Spatiales under grant 88-1212 and by CNRS/GRECO Plasmae.

REFERENCES

Axford, W. I., "Helium in the Atmosphere, Aurora and Solar Wind," in *Atmospheric Emissions*, B. M. McCormac and A. Omholt, eds., Van Nostrand–Reinhold, New York, p. 317 (1968).

Bingham, R., D. A. Bryant, and D. S. Hall, "A Wave Model for the Aurora," *Geophys. Res. Lett.*, **11**, 327 (1984).

Block, L. P., and C.-G. Fälthammar, "Characteristics of Magnetic-Field Aligned Electric Fields in the Auroral Acceleration Region," *this volume.*

Bosqued, J. M., "Ion Precipitation into the Ionosphere During Geomagnetic Storms," *Adv. Space Res.*, **5**(4), 179 (1985).

Bosqued, J. M., "AUREOL-3 Results on Ion Precipitation," *Phys. Scr.*, **18**, 158 (1987).

Bosqued, J. M., J. A. Sauvaud, D. Delcourt, and R. A. Kovrazhkin, "Precipitation of Suprathermal Ionospheric Ions Accelerated in the Conjugate Hemisphere," *J. Geophys. Res.*, **91**, 7006 (1986).

Bryant, D. A., and D. S. Hall, "The Auroral Electron Accelerator," *this volume.*

Chappell, C. R., R. C. Olsen, J. L. Green, J. F. E. Johnson, and J. H. Waite, Jr., "The Discovery of Nitrogen Ions in the Earth's Magnetosphere," *Geophys. Res. Lett.*, **9**, 937 (1982a).

Chappell, C. R., R. C. Olsen, J. L. Green, J. F. E. Johnson, and J. H. Waite, Jr., "Pitch Angle Variations in Magnetospheric Thermal Plasma—Initial Observations from Dynamics Explorer-1," *Geophys. Res. Lett.*, **9**, 933 (1982b).

Chen, A. J., "Penetration of Low-Energy Protons Deep into the Magnetosphere," *J. Geophys. Res.*, **75**, 2458 (1970).

Cladis, J. B., "Parallel Acceleration and Transport of Ions from Polar Ionosphere to Plasma Sheet," *Geophys. Res. Lett.*, **13**, 893 (1986).

Cladis, J. B., and W. E. Francis, "The Polar Ionosphere as a Source of the Storm Time Ring Current," *J. Geophys. Res.*, **90**, 3465 (1985).

Collin, H. L., W. K. Peterson, and E. G. Shelley, "Solar Cycle Variation of Some Mass Dependent Characteristics of Upflowing Beams of Terrestrial Ions," *J. Geophys. Res.*, **92**, 4757 (1987).

Cowley, S. W. H., and M. Ashour-Abdalla, "Adiabatic Plasma Convection in a Dipole Field: Electron Forbidden-Zone Effects for a Simple Electric Field Model," *Planet. Space Sci.*, **24**, 805 (1976).

Craven, P. D., R. C. Olsen, C. R. Chappell, and L. Kakani, "Observations of Molecular Ions in the Earth's Magnetosphere," *J. Geophys. Res.*, **90**, 7599 (1985).

Delcourt, D. C., "Circulation des Ions Ionosphériques Suprathermiques dans la Magnétosphère Terrestre," PhD thesis, Paul-Sabatier University, Toulouse, France, Dec 1985.

Delcourt, D. C., B. L. Giles, C. R. Chappell, and T. E. Moore, "Low-Energy Bouncing Ions in the Magnetosphere: A Three-Dimensional Numerical Study of Dynamics Explorer 1 Data," *J. Geophys. Res.*, **93**, 1859 (1988a).

Delcourt, D. C., J. L. Horwitz, and K. R. Swinney, "Influence of the IMF Orientation on Polar Cap Trajectories: Energy Gain and Drift Effects," *J. Geophys. Res.*, **93**, 7565 (1988b).

Ejiri, M., "Trajectory Traces of Charged Particles in the Magnetosphere," *J. Geophys. Res.*, **83**, 4798 (1978).

Ejiri, M., R. A. Hoffman, and P. H. Smith, "Energetic Particle Penetration into the Inner Magnetosphere," *J. Geophys. Res.*, **85**, 653 (1980).

Feldstein, Ya. I., and Yu. I. Galperin, "The Auroral Luminosity Structure in the High-Altitude Upper Atmosphere: Its Dynamics and Relationship to the Large-Scale Structure of the Earth's Magnetosphere," *Rev. Geophys.*, **23**, 217 (1985).

Fontaine, D., and M. Blanc, "Theoretical Approach to the Morphology and Dynamics of Diffuse Auroral Zones," *J. Geophys. Res.*, **88**, 7171 (1983).

Frahm, R. A., P. H. Reiff, J. D. Winningham, and J. L. Burch, "Banded Ion Morphology: Main and Recovery Storm Phases," in *Ion Acceleration in the Magnetosphere and Ionosphere*, Geophysical Monograph 38, T. S. Chang, ed., American Geophysical Union, Washington, DC, p. 98 (1986).

Geiss, J., H. Balsiger, P. Eberhardt, H. P. Walker, L. Weber, D. T. Young, and H. Rosenbauer, "Dynamics of Magnetospheric Ion Composition as Observed by the GEOS Mass Spectrometer," *Space Sci. Rev.*, **22**, 537 (1978).

Ghielmetti, A. G., R. G. Johnson, R. D. Sharp, and E. G. Shelley, "The Latitudinal, Diurnal, and Altitudinal Distributions of Upward Flowing Energetic Ions of Ionospheric Origin," *Geophys. Res. Lett.*, **5**, 59 (1978).

Ghielmetti, A. G., R. D. Sharp, E. G. Shelley, and R. G. Johnson, "Downward Flowing Ions and Evidence for Injection of Ionospheric Ions into the Plasma Sheet," *J. Geophys. Res.*, **84**, 5781 (1979).

Giles, B. L., C. R. Chappell, J. H. Waite, Jr., T. E. Moore, and J. L. Horwitz, "Dynamic Evolution of Low-Energy Ions in the Terrestrial Magnetosphere," in *Modeling Magnetospheric Plasma*, Geophysical Monograph 44, T. E. Moore and J. H. Waite, Jr., eds., American Geophysical Union, Washington, DC, p. 177 (1988).

Gorney, D. J., A. Clarke, D. Cooley, J. Fennell, J. Luhmann, and P. Mizera, "The Distribution of Ion Beams and Conics Below 8000 km," *J. Geophys. Res.*, **86**, 83 (1981).

Gorney, D. J., and D. S. Evans, "The Low-Latitude Auroral Boundary: Steady-State and Time-Dependent Representations," *J. Geophys. Res.*, **92**, 13537 (1987).

Gussenhoven, M. S., D. A. Hardy, and W. J. Burke, "DMSP/F2 Observations of Equatorward Auroral Boundaries and Their Relationship to Magnetospheric Electric Fields," *J. Geophys. Res.*, **86**, 768 (1981).

Gussenhoven, M. S., D. A. Hardy, and N. Heinemann, "Systematics of the Equatorward Diffuse Auroral Boundary," *J. Geophys. Res.*, **88**, 5692 (1983).

Gussenhoven, M. S., D. A. Hardy, and N. Heinemann, "The Equatorward Boundary of Auroral Ion Precipitation," *J. Geophys. Res.*, **92**, 3273 (1987).

Harel, M., R. A. Wolf, P. H. Reiff, R. W. Spiro, W. J. Burke, F. J. Rich, and M. Smiddy, "Quantitative Simulation of a Magnetospheric Substorm: 1, Model Logic and Overview," *J. Geophys. Res.*, **86**, 2217 (1981a).

Harel, M., R. A. Wolf, R. W. Spiro, P. H. Reiff, C.-K. Chen, W. J. Burke, F. J. Rich, and M. Smiddy, "Quantitative Simulation of a Magnetospheric Substorm: 2, Comparison with Observations," *J. Geophys. Res.*, **86**, 2242 (1981b).

Horita, R. E., E. Ungstrup, R. D. Sharp, R. R. Anderson, and R. J. Fitzenreiter, "Counterstreaming Hydrogen and Oxygen Ions Observed in the Magnetosphere," *Adv. Space Res.*, **5**(4), 421 (1985).

Horita, R. E., E. Ungstrup, E. G. Shelley, R. R. Anderson, and R. J. Fitzenreiter, "Counterstreaming Ion Events in the Magnetosphere," *J. Geophys. Res.*, **92**, 13,523 (1987).

Horwitz, J. L., "The Ionosphere as a Source for Magnetospheric Ions," *Rev. Geophys.*, **20**, 929 (1982).

Horwitz, J. L., and M. Lockwood, "The Cleft Ion Fountain: A Two-Dimensional Kinetic Model," *J. Geophys. Res.*, **90**, 9749 (1985).

Horwitz, J. L., "Features of Ion Trajectories in the Polar Magnetosphere," *Geophys. Res. Lett.*, **11**, 1111 (1984).

Horwitz, J. L., "Parabolic Heavy Ion Flow in the Polar Magnetosphere," *J. Geophys. Res.*, **92**, 175 (1987).

Kivelson, M. G., "Dialog on Injection-Boundary Versus Alfvén-Layer Models," in *Magnetotail Physics*, A. T. Y. Lui, ed., The Johns Hopkins University Press, Baltimore, p. 403 (1987).

Kivelson, M. G., and D. J. Southwood, "Approximations for the Study of Drift Boundaries in the Magnetosphere," *J. Geophys. Res.*, **80**, 3538 (1975).

Klumpar, D. M., "A Digest and Comprehensive Bibliography on Transverse Auroral Ion Acceleration," in *Ion Acceleration in the Magnetosphere and Ionosphere*, Geophysical Monograph 25, T. S. Chang, ed., American Geophysical Union, Washington, DC, p. 389 (1986).

Kremser, G., W. Studemann, B. Wilken, G. Gloeckler, D. C. Hamilton, and F. M. Ipavich, "Average Spatial Distributions of Energetic O^+, O^{2+}, O^{6+}, and C^{6+} Ions in the Magnetosphere Observed by AMPTE CEE," *J. Geophys. Res.*, **92**, 4459 (1987).

Lennartsson, W., "Dynamical Features of the Plasma-Sheet Ion Composition, Density, and Energy," in *Magnetotail Physics*, A. T. Y. Lui, ed., The Johns Hopkins University Press, Baltimore, p. 35 (1987).

Lennartsson, W., and E. G. Shelley, "Survey of 0.1 to 16 keV/e Plasma Sheet Ion Composition," *J. Geophys. Res.*, **91**, 3061 (1986).

Lennartsson, W., E. G. Shelley, R. D. Sharp, R. G. Johnson, and H. Balsiger, "Some Initial ISEE-1 Results on the Ring Current Composition and Dynamics During the Magnetic Storm of December, 1977," *Geophys. Res. Lett.*, **6**, 483 (1979).

Lockwood, M., J. H. Waite, Jr., T. E. Moore, J. F. E. Johnson, and C. R. Chappell, "A New Source of Suprathermal O^+ Ions Near the Dayside Polar Cap Boundary," *J. Geophys. Res.*, **90**, 4099 (1985a).

Lockwood, M., M. O. Chandler, J. L. Horwitz, J. H. Waite, Jr., T. E. Moore, and C. R. Chappell, "The Cleft Ion Fountain," *J. Geophys. Res.*, **90**, 9736 (1985b).

Luhmann, J. G., and L. M. Friesen, "A Simple Model of the Magnetosphere," *J. Geophys. Res.*, **84**, 4405 (1979).

Mauk, B. H., and C.-I. Meng, "Characterization of Geostationary Particle Signatures Based on the Injection Boundary Model," *J. Geophys. Res.*, **88**, 3055 (1983).

Moore, T. E., C. R. Chappell, M. Lockwood, and J. H. Waite, Jr., "Superthermal Ion Signatures of Auroral Acceleration Processes," *J. Geophys. Res.*, **90**, 1611 (1985).

Moore, T. E., M. Lockwood, M. O. Chandler, J. H. Waite, Jr., C. R. Chappell, A. Persoon, and M. Sugiura, "Upwelling O⁺ Ion Source Characteristics," *J. Geophys. Res.*, **91**, 7019 (1986).

Newell, P. T., and C.-I. Meng, "Energy Dependence of the Equatorward Cutoffs in Auroral Electron and Ion Precipitation," *J. Geophys. Res.*, **92**, 7519 (1987).

Newell, P. T., C.-I. Meng, and D. A. Hardy, "Overview of Statistical Global Electron and Ion Auroral Precipitation," *this volume*.

Northrop, T. G., *The Adiabatic Motion of Charged Particles*, Interscience Publishers, New York (1963).

Peterson, W. K., "Auroral Zone Ion Composition," in *Modeling Magnetospheric Plasma*, Geophysical Monograph 44, T. E. Moore, and J. H. Waite, Jr., eds., American Geophysical Union, Washington, DC, p. 145 (1988).

Reiff, P. H., H. L. Collin, E. G. Shelley, J. L. Burch, and J. D. Winningham, "Heating of Upflowing Ionospheric Ions on Auroral Field Lines," in *Ion Acceleration in the Magnetosphere and Ionosphere*, Geophysical Monograph 38, T. S. Chang, ed., American Geophysical Union, Washington, DC, p. 83 (1986).

Reiff, P. H., H. L. Collin, J. D. Craven, J. L. Burch, J. D. Winningham, E. G. Shelley, L. A. Frank, and M. A. Friedman, "Determination of Auroral Electrostatic Potentials Using High- and Low-Altitude Particle Distributions," *J. Geophys. Res.*, **93**, 7441 (1988).

Rème H., J. M. Bosqued, J. A. Sauvaud, D. Roux, R. A. Kovrazkhin, and F. K. Shuskaya, in *Results of the ARCAD 3 Project*, edited by CNES, Cepadues Editions, Toulouse, France, p. 367 (1985).

Sagawa, E., A. W. Yau, B. A. Whalen, and W. K. Peterson, "Pitch Angle Distributions of Low-Energy Ions in the Near-Earth Magnetosphere," *J. Geophys. Res.*, **92** 12241 (1987).

Sauvaud, J. A., and D. C. Delcourt, "A Numerical Study of Suprathermal Ionospheric Ion Trajectories in Three-Dimensional Electric and Magnetic Field Models," *J. Geophys. Res.*, **92**, 5873 (1987).

Sauvaud, J. A., J. Crasnier, Yu. I. Galperin, and Ya. I. Feldstein, "A Statistical Study of the Equatorward Boundary of the Diffuse Aurora in the Premidnight Sector," *Geophys. Res. Lett.*, **10**, 749 (1983).

Sauvaud, J. A., J. Crasnier, K. Mouala, R. A. Kovrazhkin, and N. V. Jorjio, "Morning Sector Ion Precipitation Following Substorm Injections," *J. Geophys. Res.*, **86**, 3430 (1981).

Sauvaud, J. A., J. M. Bosqued, R. A. Kovrazkhin, D. Delcourt, J. J. Berthelier, F. Lefeuvre, J. L. Rauch, Yu. I. Galperin, M. M. Mogilevsky, and E. E. Titova, "Positive Ion Distributions in the Morning Auroral Zone: Local Acceleration and Drift Effects," *Adv. Space Res.*, **5**(4), 73 (1985).

Sharp, R. D., A. G. Ghielmetti, R. G. Johnson, and E. G. Shelley, "Hot plasma Composition Results from the S3-3 Spacecraft," in *Energetic Ion Composition in the Earth's Magnetosphere*, R. G. Johnson, ed., Terra Scientific, Tokyo, p. 167 (1983).

Sharp, R. D., D. L. Carr, W. K. Peterson, and E. G. Shelley, "Ion Streams in the Magnetotail," *J. Geophys. Res.*, **86**, 4639 (1981).

Sharp, R. D., E. G. Shelley, R. G. Johnson, and A. G. Ghielmetti, "Counterstreaming Electron Beams at Altitudes of ~1 R_E Over the Auroral Zone," *J. Geophys. Res.*, **85**, 92 (1980).

Shelley, E. G., "Circulation of Energetic Ions of Terrestrial Origin in the Magnetosphere," *Adv. Space Res.*, **5**(4), 401 (1985).

Shelley, E. G., and H. L. Collin, "Auroral Ion Acceleration and Its Relationship to Ion Composition," *this volume*.

Shelley, E. G., R. G. Johnson, and R. D. Sharp, "Satellite Observation of Energetic Heavy Ions During a Geomagnetic Storm," *J. Geophys. Res.*, **77**, 6104 (1972).

Shelley, E. G., R. D. Sharp, and R. G. Johnson, "Satellite Observations of an Ionospheric Acceleration Mechanism," *Geophys. Res. Lett.*, **3**, 654 (1976).

Shelley, E. G., W. K. Peterson, A. G. Ghielmetti, and J. Geiss, "The Polar Ionosphere as a Source of Energetic Magnetospheric Plasma," *Geophys. Res. Lett.*, **9**, 941 (1982).

Stasiewicz, K., "The Influence of a Turbulent Region on the Flux of Auroral Electrons," *Planet. Space Sci.*, **33**, 591 (1985).

Swinney, K. R., and J. L. Horwitz, "Three-Dimensional Ion Trajectories in the Magnetosphere," *Eos Trans. AGU*, **66**, 1001 (1985).

Swinney, K. R., and J. L. Horwitz, "DE-Inspired Kinetic Modeling of the Transport of Ionospheric Ions Into the Magnetosphere," *Eos Trans. AGU*, **67**, 1138 (1986).

Swinney, K. R., J. L. Horwitz, and D. C. Delcourt, "Centrifugal Acceleration of Ions in the Polar Ionosphere," *J. Geophys. Res.*, in press (1988).

Valtchuk, E., Yu. I. Galperin, L. M. Nikolaenko, Ya. I. Feldstein, J. M. Bosqued, J. A. Sauvaud, and J. Crasnier, "The Diffuse Auroral Zone: 8, Equatorward Boundary of the Electron Precipitation Zone in the Morning Sector," *Cosmic Res. (in Russian)*, **24**, 875 (1986).

Volland, H., "A Model of the Magnetospheric Electric Convection Field," *J. Geophys. Res.*, **83**, 2695 (1978).

Waite, J. H., T. Nagai, J. F. E. Johnson, C. R. Chappell, J. L. Burch, T. L. Killeen, P. B. Hays, G. R. Carignan, W. K. Peterson, and E. G. Shelley, "Escape of Suprathermal O⁺ Ions in the Polar Cap," *J. Geophys. Res.*, **90**, 1619 (1985).

Winningham, J. D., J. L. Burch, and R. A. Frahm, "Bands of Ions and Angular Vs: A Conjugate Manifestation of Ionospheric Ion Acceleration," *J. Geophys. Res.*, **89**, 1749 (1984).

Yau, A. W., and M. Lockwood, "Vertical Ion Flow in the Polar Ionosphere," in *Modeling Magnetospheric Plasma*, Geophysical Monograph 44, T. E. Moore, and J. H. Waite, Jr., eds., American Geophysical Union, Washington, DC, p. 229 (1988).

Yau, A. W., P. H. Beckwith, W. K. Peterson, and E. G. Shelley, "Long-term (Solar Cycle) and Seasonal Variations of Upflowing Ionospheric Ion Events at DE 1 Altitudes," *J. Geophys. Res.*, **90**, 6395 (1985b).

Yau, A. W., E. G. Shelley, W. K. Peterson, and L. Lenchyshyn, "Energetic Auroral and Polar Ion Outflow at DE 1 Altitudes: Magnitude, Composition, Magnetic Activity Dependence, and Long-Term Variations," *J. Geophys. Res.*, **90**, 8417–8432 (1985a).

Yau, A. W., B. A. Whalen, W. K. Peterson, and E. G. Shelley, "Distribution of Upflowing Ionospheric Ions in the High-Altitude Polar Cap and Auroral Ionosphere," *J. Geophys. Res.*, **89**, 5507 (1984).

Young, D. T., "Experimental Aspects of Ion Acceleration in the Earth's Magnetosphere, in *Ion Acceleration in the Magnetosphere and Ionosphere*, Geophysical Monograph 38, T. S. Chang, ed., American Geophysical Union, Washington, DC, p. 17 (1986).

Young, D. T., J. Geiss, H. Balsiger, P. Eberhardt, A. Ghielmetti, and H. Rosenbauer, "Discovery of He²⁺ and O²⁺ Ions of Terrestrial Origin in the Outer Magnetosphere," *Geophys. Res. Lett.*, **4**, 561 (1977).

IV. AURORAS AND MAGNETOSPHERIC CONFIGURATION

IV-1. WHAT DETERMINES THE SIZE OF THE AURORAL OVAL?

G. L. Siscoe*

Observations reviewed here relate the size of the auroral oval to solar wind parameters (particularly IMF B_z) and the theoretical models that attempt to explain the relations. There are a diffuse, subvisual auroral oval and a discrete, visual auroral oval, giving four latitudinal borders, each of which behaves in a unique way in response to solar-wind changes. Quantitative models exist only for the borders of the diffuse auroral oval. The models can account for many of the oval's observed geometrical properties, but only with the aid of empirically determined parameters. A fully deterministic theory for the geometrical properties of the auroral oval is still in the future.

1. AN INTRODUCTION TO AURORAS AND MAGNETOSPHERIC CONFIGURATION

Whoever first used the simile, "Auroras are like TV images" must have meant similarity in both how they work and what they do. They work by magnetically focused electron beams stimulating emissions in target neutrals. What they do is to display projections of magnetospheric landscapes and actions on an atmospheric screen. In the early 1960s people used the analogy to confirm the basic away-from-the-Sun-over-the-pole-and-back-toward-the-Sun-just-equatorward magnetospheric convection pattern inferred from surface magnetometer measurements. Poleward of roughly 70° magnetic latitude, auroras drift mainly tailward, whereas equatorward they mostly go sunward. Since then the effort to view the magnetosphere through its auroral image has matured into a major research industry. Observations present an auroral ring with distinct segments and bands, which assumedly project down from distinct magnetospheric structures (like TV images of "real-life" objects). The segments and bands behave idiosyncratically, presumably mimicking the behavior of their real-life magnetospheric counterparts. The goal of the involved research industry is to identify the magnetospheric counterparts of each auroral unit and to infer their dynamics from auroral behavior.

A list of distinct auroral units includes (1) a segment of discrete dayside auroras, (2) a segment of discrete nightside auroras, (3) discrete polar cap auroras, (4) a poleward band of diffuse aurora, and (5) an equatorward band of diffuse aurora. Within these units there are subdivisions, and the distinctions between them are not always clear. The dayside and nightside segments of discrete auroras are imbedded within the bands of diffuse aurora. Nonetheless these are the principle units. The operating metaphor is that the scenes they present display the magnetosphere's major structures and the plays they enact relate the magnetosphere's main dynamical modes.

In the chapter following this paper, there are four papers from leading practitioners of auroral interpretation. Anyone dreading to find here a series of lifeless reviews of a dead subject—this auroral unit belongs to that magnetospheric structure, and when it does this it means that the structure does that—will be delighted to find instead nearly total disagreement on what structures belong to what units, and therefore necessarily on the meanings of auroral behaviors. The series presents among the most cogent and trenchant advocacy statements to date, covering a significant fraction of proposed models of mapping and dynamics. On reading these, one learns that a principle reason for disagreements is the failure of the data to capture the magnetosphere's complex behavior sufficiently to preclude multiple interpretations. There are lessons here for the design of future measurement campaigns and missions. One senses also that even with the present level of ambiguity allowed by the data, the extent of disagreement could be reduced through systematically evaluating and comparing the models in a workshop setting with broad community involvement. The following suite of papers, to which we now turn, suggests a natural program for such a workshop.

In his contribution (Discrete Auroras and Magnetospheric Processes), *L. R. Lyons* argues that the segment of discrete nightside auroras projects down through the plasma-sheet boundary layer—the tail's renowned realm of non-Maxwellian particle populations—from the tail's ion acceleration region—the neutral sheet—which lies beyond the central plasma sheet. Accepting this and bringing into play the observation that the auroral substorm erupts from a preexisting nightside discrete aurora, he concludes that the magnetospheric substorm issues from the neutral sheet—not from the central plasma sheet, where the rival near-Earth-neutral-line model initiates it. But then, in part, to appropriate an advantage of the near-Earth-neutral-line model—its potential to affect directly the middle

*Department of Atmospheric Sciences, University of California Los Angeles, Los Angeles, CA 90024.

magnetosphere, which manifests pronounced substorm responses—he postulates (with *A. Nishida*) that the neutral line in fact lies in the neutral sheet, which prior to substorm onset might impinge on the central plasma sheet and encroach close enough to the middle magnetosphere to induce the effects observed therein.

Galperin and Feldstein (Auroral Luminosity and Its Relationship to Magnetospheric Plasma Domains) disagree. In their model, the segment of discrete nightside auroras projects down from the central plasma sheet—a view for which they adduce three arguments. *First*, they report a consensus among observers that the diffuse aurora extending equatorward from the segment of discrete auroras is congruent with the magnetospheric region of stably trapped energetic electrons. They note that the trapping boundary, which must by inference map to the equatorward edge of the segment of discrete auroras, coincides with the earthward border of the central plasma sheet. Ergo the central plasma sheet maps to the segment of discrete nightside auroras. *Second*, the poleward band of diffuse auroral precipitation lies just poleward of the segment of discrete auroras. Since the polewardmost part of the plasma sheet is its boundary layer, by the principle that to every distinct auroral structure there corresponds a distinct magnetospheric structure, the plasma-sheet boundary layer evidently maps to the poleward band of diffuse auroral precipitation, and thus poleward of the segment of discrete auroras. *Third*, the segment of discrete nightside auroras encompasses more magnetic flux than the plasma-sheet boundary layer, especially during the substorm expansive phase when discrete auroras are often strewn over 5° of latitude. Therefore the segment of discrete nightside auroras cannot be the magnetic projection of the plasma-sheet boundary layer onto the ionosphere. Galperin and Feldstein's picture admits the near-Earth-neutral-line substorm model unaltered.

Mauk and Meng (The Aurora and Middle Magnetospheric Processes) extend the shift in emphasis away from the plasma-sheet boundary layer to the middle magnetosphere itself. They document one example of a discrete aurora that maps to the middle magnetosphere and refer to others. They discuss a possibly related middle magnetospheric phenomenon that McIlwain discovered in 1975—magnetic-field-aligned electron beams—and show that these are not caused by extreme distortions of the magnetic field bringing the plasma-sheet boundary layer into the domain of the middle magnetosphere. After demonstrating that the electron beams can stimulate discrete auroras and reviewing a mechanism for generating such beams on dipolar field lines—which nonetheless mimics the acceleration process at neutral sheets, albeit for short durations—they close by citing recent observations of near-Earth-

neutral-line-like phenomena in the middle magnetosphere, much closer than previously documented if not suspected.

These articles clearly describe different pictures showing the relation between the segment of discrete nightside auroras and magnetospheric structures. Some aspects of the pictures are complementary but others are incongruous. The reports set the disagreements in stark relief, and seem thereby to lay an unobstructed path for future work—verify or falsify alternative models and seek syntheses with added explanatory power. But this reviewer's impression is that, even in combination, the models fail to account for auroral phenomena during the substorm expansive phase, when the plasma-sheet boundary layer, the central plasma sheet, and the middle magnetosphere (if indeed these labels apply then) seem to produce discrete auroras simultaneously, and perhaps cooperatively. If so, we must go beyond synthesizing present models and create more structurally integrated ones.

In the fourth report (The Quiet Time Aurora and the Magnetosphere Configuration), *Lundin and Eliasson* take up discrete dayside auroras and discrete polar-cap auroras. Whereas discrete nightside auroras signify magnetospheric activity, discrete polar-cap auroras signify quiescence, and discrete dayside auroras are continual. The authors stress that the dichotomy between dayside and nightside discrete auroras is real. They attribute the former to incessant injections of plasma and energy from the solar wind into the cusp and cleft—funnel-like folds in the magnetosphere's dayside boundary—and the latter to similar but intermittent injections into the middle magnetosphere from the plasma sheet and its boundary layer—a process that is activated and suppressed by turnings of the interplanetary magnetic field (IMF), somewhat like a rheostat. They maintain that when the IMF is turned to suppress injections from the tail, thereby creating a state of relative magnetospheric quiescence, injections into the cusp and cleft energize the magnetosphere's residual dynamics (rather than, for example, magnetic merging on open field lines as advocated by others). In their version, the quiet magnetosphere's enlarged cusp and cleft feed extra plasma to the low-latitude boundary layer (LLBL) inflating it inward, while reduced boundary stresses protract it tailward. Instead of a latitudinally thin, longitudinally short strip straddling noon as during active times, the LLBL's ionospheric image becomes latitudinally thick and nearly encircles a shrunken and reshaped polar cap, leaving only a midnight-sector remnant of the formerly dominant plasma sheet. Other quiescent settings of the IMF rheostat shift the polar cap dawnward or duskward or fill it with discrete auroras. But herein lies the central controversy of the paper. According to the side favored by Lundin and Eliasson, discrete auroras actually do not form in the polar cap, but instead within a distended ionsopheric projection of the LLBL, which has insinuated itself otherwise unnoticed over polar-cap terrain. In this view, which they support with Viking observations, "discrete polar-cap auroras" is a misnomer. Polar-cap auroras are ordinary auroras masquerading in an extraordinary place. The disclosure indicts all breeds, including the famous theta aurora. The other side of the controversy, which chiefly concerns the theta aurora, also invades the polar cap to form discrete auroras, but not in the tide-usurping-the-shore manner of the LLBL. Instead thin, field-aligned walls of closed-field-line plasma rise from the plasma-sheet boundary layer and partition the tail lobe—the magnetospheric counterpart of the polar cap. The partitions' ionospheric projections impress Sun-aligned auroral arcs across the polar cap. Unfortunately we have no advocacy report supporting this side, but Lundin and Eliasson treat it fairly.

Although unrehearsed, the four reports repeat the same refrain: discrete auroras are deep magnetospheric enigmas. Perhaps the project to associate discrete auroras with magnetospheric structures is wrongheaded. The appropriate magnetospheric counterparts of discrete auroras might be processes that can occur in one structure or another or cross between them. (In the case of discrete polar-cap auroras, questions of structure and process seem to be intertwined.) The same theme recurs independently in this paper (What Determines the Size of the Auroral Oval?). The paper's opening review of material on the size of the auroral oval, as defined by discrete auroras, leads to a dead end, because no physically explicit model relates the statistical oval of discrete auroras to solar-wind parameters. Here the band of diffuse aurora (the poleward and equatorward bands are merged in the paper) enters the picture. The band's poleward border encircles the open-field-line polar cap, a physically well-defined structure whose size can be quantitatively expressed in terms of parameters that derive from the solar wind. The band's equatorward border marks the earthward limit of magnetospheric convection, which is also a physically explicit definition that allows one to locate the equatorward border in terms of parameters that derive from the solar wind. In this case, one needs information on the poleward limit of convection as well. The point of the paper is that indeed one can explicitly relate the positions of the poleward and equatorward borders of the diffuse aurora to magnetospheric parameters that derive from the solar wind, but correspondingly explicit relations between the controlling magnetospheric parameters and solar-wind parameters do not exist. To acquire the ability to predict the auroral configuration (even the diffusive part)

from solar-wind parameters using explicit physics, we must learn what determines the strength of magnetospheric convection and what fixes the capacity of the tail's magnetic-flux reservoir. These questions emerge as the cardinal problems blocking the ascendancy of geospace science to a predictive environmental science.

2. THE AURORAL OVAL AND PREDICTIVE GEOSPACE SCIENCE

"Science," proclaims Nobel Laureate Sir Peter Medewar in his *The Limits of Science*, "is, or aspires to be, *deductively ordered*." From general laws, it wants to deduce particulars. The environmental sciences prefer the synonym "predict." The maturity of an environmental science can be measured by its ability to predict the parameters of its domain from boundary or initial conditions. One might argue with this statement and say instead that the ultimate goal of an environmental science is to understand the structures and processes that define its domain. But understanding is only theory until its predictions are tested. When an environmental science achieves the ability to predict the state of its domain, it can return applications to science in the form of tested understandings and to humankind in the form of useful predictions.

How well can geospace science predict the state of its domain from its boundary conditions? The boundaries of geospace are the Earth and the solar-wind. To get a first answer to the question, we let solar-wind parameters be variable and Earth parameters be fixed. (Letting the dipole wobble and the ionospheric conductivity vary interactively would not assist predictiveness.) Then the last fully predictive theory of the geospace environment is the *Mead and Beard* [1964] solution to the original Chapman-Ferraro problem: "Find the shape of a magnetically closed, vacuum magnetosphere." The conditions "magnetically closed" and "vacuum" make the problem mathematically determined for any solar-wind ram pressure—the only solar-wind parameter that enters. That is, the solution determines all of the magnetosphere's parameters in terms of the solar-wind ram pressure. Figure 1 shows the predicted shape of the Mead-Beard magnetosphere.

Unfortunately the Mead-Beard magnetosphere has no auroral oval. The only particles that reach the ionosphere in their vacuum model leak from the solar wind through the two magnetic null points on the dayside boundary. The magnetic field then funnels them down to the ionosphere onto high-latitude spots on the noon meridian. The model lacks an auroral oval because it has the wrong kind of tail. The tail needed to make an auroral oval is bifurcated by a sheet of electric current, which (topologically) creates open field lines above and below it. Open field lines make the hole in the oval. Particles must carry the cross-tail current. Thus an

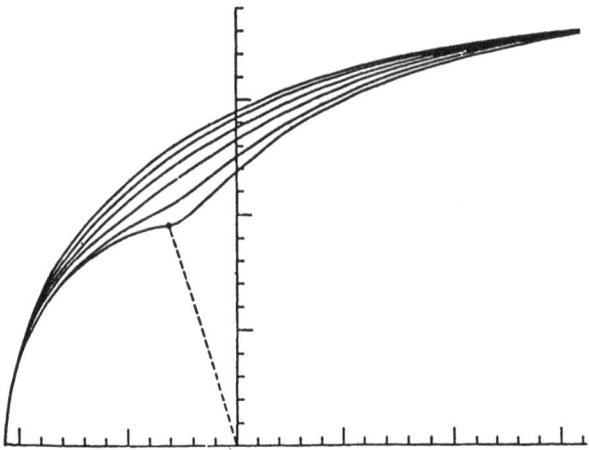

Figure 1—The *Mead and Beard* [1964] fully predictable magnetosphere has no auroral oval, because it has no equatorially partitioned tail.

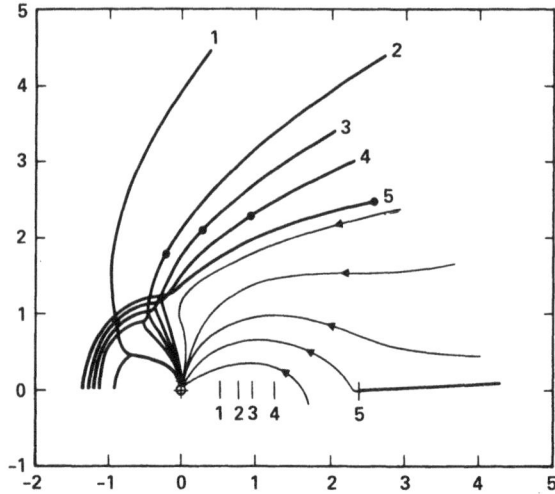

Figure 2—The *Unti and Atkinson* [1968] magnetosphere has an equatorially partitioned tail, and therefore an auroral oval, but it is not fully predictable. One must specify the size of the auroral oval.

auroral oval cannot form in a vacuum magnetosphere. The particles that carry the current are also a source for the aurora.

Unti and Atkinson [1968] published the first fully quantitative magnetospheric model with a tail. It is two-dimensional. The cross-tail current flows on an infinitesimally thin sheet through which no magnetic flux penetrates. Further, the boundary with the solar wind is magnetically closed. So constrained, the Chapman-Ferraro problem extended to include a bifurcated tail can be solved by the method of conformal mapping. Figure 2 shows boundary shapes for the Unti-Atkinson magnetosphere.

Unlike the Mead-Beard magnetosphere, the shape of the Unti-Atkinson magnetosphere is not fixed by specifying solar-wind and Earth parameters. One must also specify the amount of open magnetic flux—a purely magnetospheric parameter not predicted by the model. But the amount of open flux determines the size of the auroral oval, like a finger determines the size of a ring. In actuality then, the first quantitative model elaborate enough to include the auroral oval requires that the oval's size be specified in advance. The point warrants emphasis. The size of the auroral oval is the first unpredicted magnetospheric parameter we encounter in the progressive escalation of global magnetospheric models. The situation has not changed since 1968, although more realistic features have replaced some idealizations in the Unti-Atkinson model. (For a review of this topic, see *Siscoe* [1988].) It is fair to say that determining the size of the auroral oval from variable solar-wind parameters symbolizes the problem of achieving predictiveness in geospace science.

We here review attempts to relate the size of the auroral oval to solar-wind variables. None succeeds completely. Each invokes empirical parameters. Our

goal is to identify the empirical parameters in each case, to locate where future work should be focused. A second goal is to establish criteria to evaluate a model's effectiveness in predicting the size of the auroral oval. To begin, we summarize what is known regarding the size of the auroral oval and its dependence on solar wind parameters.

3. OBSERVED PROPERTIES OF THE OVAL OF VISUAL AURORAS

The auroral oval is not merely one thing. The more we learn about it, the more complex it becomes. The original oval is the band of visual auroras. Now we know that this band is embedded within a latitudinally broader band of diffuse, subvisual emissions. The latter is often divided into two or three latitudinal stripes defined by the spectra and composition of the particles whose impacts stimulate the emissions. However our purposes are met by resolving the auroral oval into a band containing discrete, visual auroral forms and a broader, enveloping band of subvisual, diffuse aurora. We put immediate emphasis on the poleward and equatorward borders of both bands, because their dynamics characterize the complexity of the system; each of the four borders behaves uniquely. For a running start into today's complexity, we begin historically, at a time when things seemed simpler.

In admirable pieces of scholarship, *Robert Eather* [1980] and *Asgeir Brekke* [1984] have established the history of our subject. Early in the 18th century they find explicit reference to a continuous ring of auroral light encircling the pole. By the 19th century some people knew that the ring centered on the magnetic pole rather than the rotational pole and that it changed its size with magnetic activity. But only recently have we had the technology to view enough of the ring to quantify it. Without such data Hermann Fritz in 1881 abandoned the ring itself and quantified its habitat instead. In place of the concrete picture of an actual auroral ring with a variable size and shape, he conceived the abstract picture of probable auroral rings with fixed sizes and shapes. He used a catalog, which he compiled, of auroral sightings from all times and places and drew lines of equal occurrence frequency—called isochasms—like topographic contour lines on a map. They circled the magnetic pole and peaked in value 23° from the center, thereby marking the "auroral zone."

People assumed that isochasms define the sizes and shapes of the actual auroral ring at different times, and that the number identifying a particular isochasm tells the percentage of time that the auroral ring coincides with that isochasm. If this were so, Fritz's map of isochasms would define the problem that theorists must solve: find a model that predicts an auroral ring that has an average separation from the magnetic pole of 23° and that changes its size in response to solar-wind changes (especially changes in the IMF) in a way that reproduces Fritz's auroral-occurrence-frequency-versus-geomagnetic-latitude plot. In other words, find a model that allows one to go from observed statistics of solar-wind parameters to the statistics of isochasms. Unfortunately, as we review next, the assumption that the auroral ring has the magnetic pole as its center and thus rotates with the Earth (an assumption we implicitly make if we use geographically fixed isochasms to test auroral ring models) is wrong.

With IGY all-sky camera data, *Feldstein* [1960, 1963, 1964] and, independently, *Khorosheva* [1962] discovered that the zone of highest auroral occurrence frequency does not rotate with Earth like a circle of magnetic latitude. Instead it slides as the Earth rotates under it, so that it remains closest to the magnetic pole at noon and farthest from it at midnight. To an observer on the rotating Earth, it appears to approach the pole from midnight to noon and to recede from the pole from noon to midnight. Thus Fritz's geographically fixed isochasms mismatch the auroral oval. The reason is that most entries in his auroral catalogue record sightings made around midnight. His auroral zone is the line a pen would draw on the rotating Earth if it were held on the midnight meridian at the midpoint of Feldstein's auroral oval for typical activity levels (*Kp*s from 2 to 6). When overlaid, they coincide in the midnight sector, but in the noon sector, Feldstein's auroral oval lies closer to the pole. Feldstein first called his the "oval auroral zone," but later the "auroral belt." (For an interesting, post-IGY history of the development of the auroral oval concept and

an assessment of its importance to geospace science, see Feldstein's 1986 article commemorating the silver anniversary of its real discovery.)

Whereas isochasms cover all latitudes at all times, the auroral belt has latitudinal edges fixed by local time and magnetic activity. To speak of the size of auroral oval one must specify which part—polar edge, equatorial edge, or midline. *Khorosheva* [1962] noted that the auroral oval approximates a circle offset from the magnetic pole. Apropos this observation, satellite auroral images later revealed that the quiet aurora indeed rings the pole in a rather accurate circle displaced on average 4.2° along the 0100 MLT meridian (*Meng et al.* [1977]. Because the auroral trace on the photographic image is so circular, these authors propose renaming the auroral oval the auroral circle). Using offset circles for borders, one can specify the auroral oval by merely giving the circles' radii and offsets as functions of activity. For this, the tabular data of *Bond and Thomas* [1971] for the southern auroral oval are particularly useful. A linear least-squares fit to *Kp* of the average of the noon-midnight and dawn-dusk spans of the equatorward limit of their auroral oval gives

$$\theta_e^v = 19.7 + 2.0 \, Kp \quad (r = 0.96)$$

$$(\text{B\&T data}) \tag{1}$$

$$\Delta\theta_e^v = 3.4 + 0.56 \, Kp \quad (r = 0.97)$$

$$(\text{B\&T data}) \tag{2}$$

where θ_e is the colatitude relative to the auroral pole of the equatorward boundary in degrees, $\Delta\theta$ is its offset in degrees along the midnight meridian, and r is the correlation coefficient. Similarly for the poleward border we find

$$\theta_p^v = 14.2 - 0.40 \, Kp \quad (r = 0.78)$$

$$(\text{B\&T data}) \tag{3}$$

$$\Delta\theta_p^v = 1.15 - 0.30 \, Kp \quad (r = 0.71)$$

$$(\text{B\&T data}) \tag{4}$$

(*Thomas and Bond* [1977, 1978] give more exact expressions incorporating departures from circularity. *Starkov* [1969] gives an expression for the radii of the equatorial boundary of Feldstein's northern hemisphere oval, but with a fixed offset and with Q instead of Kp as the activity index. *Holzworth and Meng* [1975] give radii and offsets of the poleward and equatorward borders of Feldstein's northern hemisphere oval as functions of Q. The Bond and Thomas statistical oval is broader at all activity levels than the standard Feldstein oval. But the borders of both ovals exhibit the same behavior relative to levels of activity, at least qualitatively. The reason we use the Thomas and Bond oval here will be immediately apparent—we can relate it to solar-wind parameters indirectly.)

Although Kp is a gauge of global geomagnetic disturbance and not a measure of solar-wind parameters (which we want ultimately to relate to the size of the oval) these formulas reveal general trends. With increasing activity, the equatorward border of the auroral oval expands and shifts tailward while the poleward border contracts and shifts sunward. In both aspects—alteration of radius and shift of center—when activity changes, the poleward border responds less and does so less obediently than the equatorward border. *Feldstein and Starkov* [1967] reported the same trends for the northern hemisphere auroral belt.

We can use an empirical relation between Kp and the "dynamo component" of the solar-wind motional electric field, VB_z (V = solar-wind speed and B_z = the meridional component of the interplanetary magnetic field, also designated IMF B_z), that *Hardy et al.* [1981] elicited by comparing how one set of auroral boundary data correlated with both parameters. To a good approximation, their relation can be written in the remarkably simple form

$$Kp = 2 - VB_z \quad (\text{mV m}^{-1}) \qquad (5)$$

This formula applies when $B_z < 1$ nT. The correlations on which it is based change dramatically and become weak when $B_z > 1$ nT, precluding a formula for this range. Because in general, B_z varies more than V, for many uses we can replace V in the formula by its long term average, 440 km s^{-1} [*Gosling et al.*, 1976], which gives

$$Kp = 2 - 0.44 \, B_z(\text{nT}) \quad (B_z < 1 \text{ nT}) \qquad (6)$$

The sizes of the Bond and Thomas auroral oval boundaries can then be expressed in terms of B_z as

$$\theta_e^v = 23.7 - 0.88 \, B_z(\text{nT}) \quad (B_z < 1 \text{ nT}) \qquad (7)$$

$$\theta_p^v = 13.4 + 0.18 \, B_z(\text{nT}) \quad (B_z < 1 \text{ nT}) \qquad (8)$$

The advent of space measurements brought three changes in specifying the auroral oval and its dependence on solar-wind variables. (1) The oval could be measured synoptically with images of major portions of its circle (and later with images of the entire circle) instead of statistically, as with the all-sky camera data on which the Feldstein-Starkov and Bond-Thomas ovals are based. (2) Solar-wind parameters could be measured in situ, replacing proxy indices such as Kp. (3) Precipitating particles whose fluxes and energies auroras gauge like proxy indices could be measured directly, and the properties of precipitation boundaries could be determined and used as more appropriate quantities for global theories to predict.

Holzworth and Meng [1975] published the first data showing how one of the auroral oval's radii responds to changes in a solar wind parameter (the meridional component of the interplanetary magnetic field, IMF B_z). They determined the radii of the circular traces that quiet auroral arcs make on DMSP images. Their least-squares fit to these data gives

$$\theta_e^v = 18.8 - 0.91 \, B_z$$

$$\text{(H\&M 1975 data)} \qquad (9)$$

They assumed that the arcs mark the poleward rim of the auroral oval, but according to *Feldstein and Galperin* [1985] they rim its equator instead. Indeed Eq. 9 agrees better with the formula for the equatorward edge of the (broader) Thomas and Bond oval (Eq. 7). We will shortly see that when $B_z < 0$ (and therefore substorms occur) the poleward edge of the statistical oval does not, in fact, mark the abode of quiet auroral arcs.

In 1984, using more cases, Holzworth and Meng published a second regression formula

$$\theta_e^v = 16.9 - 0.62 \, B_z$$

$$\text{(H\&M 1984 data)} \qquad (10)$$

Although they omit correlation coefficients, the scatter in the bigger data set is large, and the difference in the two formulas probably reflects the uncertainty in the results.

Zverev et al. [1979] arrived at a different expression of this type using all-sky camera data to determine, as a function of IMF B_z, the latitude in the midnight sector from which auroras begin their "poleward rush" at the onset of the substorm expansive phase. Normally this is the aurora's equatorward-most excursion before rushing poleward. They obtained

$$\theta_e^v = 20.0 - 0.53 \, B_z{}^m + 0.02 \, (B_z{}^m)^2$$

$$\text{(ZSF data)} \qquad (11)$$

where $B_z{}^m$ is the most negative value that the IMF B_z reaches during the substorm growth phase. The result omits the correlation coefficient, but the error bars for individual B_z values stretch typically $\pm 1°$ in latitude. The B_z dependence here is weaker than in the H&M 1984 data (above). *Feldstein and Galperin* [1985] attribute the difference to the fact that θ_e increases further after expansion phase onset, an increase that the H&M 1984 data set might have included but, by design, the ZSF dataset does not. Alternatively the difference might reflect an inherent lack of correlation, giving an intrinsic scatter whenever one plots θ_e against IMF B_z. *Zverev et al.* [1979] also showed that after substorm expansive phase onset, the poleward rushing aurora comes closer to the pole when the IMF is more strongly southward, following the empirical law

$$\theta_p^v = 17.1 + 0.24 \, B_z{}^m - 0.07 \, (B_z{}^m)^2$$

$$\text{(ZSF data)} \qquad (12)$$

The colatitude θ_p tells how near the aurora gets to the pole at the peak of the expansive phase.

Feldstein and Galperin [1985] suggest that θ_p determines the poleward border of the statistical auroral oval in the midnight sector. (Compare Eqs. 8 and 12, especially the coefficient of the linear term.) The suggestion gives the key to interpreting what the statistical auroral oval is. First, let us be clear about what it isn't. It isn't a band filled with auroral forms, a band that broadens and thins by the appearance and disappearance of auroras at its borders. The statistical oval is not "statistical" for the reason that the auroras within it move around and come and go in random ways, while maintaining a uniform, constant auroral density from border to border. The borders of the oval on satellite images are not visible as discontinuities separating space where auroras are from space where they aren't.

Instead, the statistical auroral oval is often devoid of auroral forms except for one or more quiet arcs running parallel to and near the oval's equatorward bor-

der, whose latitude is set by the activity level (as measured by Kp, Q, or IMF B_z). In the midnight sector, the arc or arcs sporadically "explode," shooting an auroral front poleward, which ultimately traverses a distance set by the activity level. Meanwhile, the explosion propagates east and west while the equatorward edge migrates equatorward, all by amounts set by the activity level. The explosion might subside and resume repeatedly at the same activity level. The oval is statistical because substorms are sporadic. Images taken at different times at the same activity level show qualitatively different auroral patterns, being taken at different phases of the substorm cycle. Like invisible barriers positioned by the activity level, the oval's equatorward and poleward borders confine the aurora's substorm expansion.

For our purposes the important point is this. While the activity level fixes the auroral oval's borders, it does not fix the aurora's gross distribution within those borders. At any but the quietest level, the poleward edge of the live aurora moves much, sweeping in fits from near the oval's equatorward border to its poleward border. By contrast, the aurora's equatorward edge moves little and more slowly at any level, shifting slightly to reach the oval's equatorward border during the substorm expansion phase. The live aurora's poleward border is much more dynamic and variable than its equatorward border.

We have learned enough about the behavior of the borders of the visual auroral oval to extract the points germane to our question: What determines the size of the auroral oval?

1. The IMF B_z and substorms play the dominant roles in determining the size of the auroral oval (no surprises to trained magnetospherists). Qualifiers: A. Besides IMF B_z, other solar-wind parameters, especially solar-wind speed, influence the auroral oval, but subordinately, in part because they vary less. (For a review, see *Reiff and Luhmann* [1986].) Since our emphasis is on the auroral oval and not the precise form of the solar-wind coupling function, we let IMF B_z represent the total solar-wind influence on the auoral oval. B. IMF B_z and substorms are not independent things, but we treat them so here because they affect the size of the oval in markedly different ways. Which brings us to the second point.

2. Substorms dominate in determining the poleward border of the statistical auroral oval. This observation has the important consequence that, until the field has advanced to where it can deduce quantitative substorm dynamics from solar-wind parameters, the poleward border of the statistical oval is inherently unpredictable from those parameters. Then what about its equatorward border?

3. IMF B_z dominates in determining the equatorward border of the statistical auroral oval. Evidently one has here a possibility to find deterministic laws that would allow one to predict the equatorward edge from solar-wind measurements. In a later section this paper reviews the directions searches for those laws have taken. Nearly without exception, the models relate primarily to the case of negative IMF B_z. In evaluating the models' predictions, it is useful to make comparisons with the constant and linear coefficients of the empirically derived dependence on IMF B_z of the size of the equatorward border of the oval. For the oval of visual auroras, this dependence is given in Eqs. 7, 9, 10, and 11. These equations present quite a scatter of coefficients. If one naively averages them with equal weights to get a representative result in a democratic sense, one finds

$$\theta_e^v = 19.9(\pm 2.9) - 0.74(\pm 0.19) \, B_z(\text{nT}) \quad (13)$$

Before turning to the models, we have still to learn how the borders of the diffuse auroral oval vary with IMF B_z.

4. OBSERVED PROPERTIES OF THE OVAL OF DIFFUSE AURORAS

On the basis of an analysis of ground-based photometer data, *Slater et al.* [1980] give a formula that shows how the invariant colatitude of the equatorward edge of the diffuse aurora in the midnight sector depends on Kp.

$$\theta_e^d = 22.5 + 2.04 \, Kp$$

$$\text{(SSK data)} \quad (14)$$

When converted to an expression in terms of B_z with Eq. 6, this becomes

$$\theta_e^d = 26.6 - 0.88 \, B_z(\text{nT}) \quad (15)$$

The borders of the diffuse aurora can be mapped more directly from polar orbiting satellites that measure the precipitating particles that produce the aurora. In this way *Gussenhoven et al.* [1981, 1983] obtain the dependence of the equatorward precipitation boundary on Kp. Averaging the results of the newer study for all local time sectors with correlations greater than 0.5, one finds

$$\bar{\theta}_e^d = 20.0(\pm 1.7) + 1.70(\pm 0.26) \, Kp$$

$$\text{(GHB data)} \quad (16)$$

or in terms of B_z

$$\bar{\theta}_e^d = 23.4(\pm 2.2) - 0.75(\pm 0.11) \, B_z(\text{nT}) \quad (17)$$

Hardy et al. [1981] correlate the equatorward precipitation boundary directly with IMF B_z. Averaging their results over all local times and including both hemispheres, one finds

$$\bar{\theta}_e^d = 23.9(\pm 1.7) - 0.89(\pm 0.21) \, B_z(\text{nT})$$

$$\text{(H+ data)} \quad (18)$$

In an independent analysis of the correlation between precipitation boundaries and the IMF B_z, *Makita et al.* [1983] report for the average of the dawn and dusk equatorward boundaries and for negative B_z

$$\bar{\theta}_e^d = 22.4 - 0.75 \, B_z(\text{nT}) \quad (\bar{r} = 0.60)$$

$$\text{(MMA data)} \quad (19)$$

A democratic average over Eqs. 15, 17, 18, and 19 gives for the equatorward edge of the diffuse auroral oval

$$\theta_e^d = 24.1(\pm 1.8) - 0.82(\pm 0.08) \, B_z(\text{nT}) \quad (20)$$

Makita et al. [1983] also give the corresponding formula for the poleward precipitation boundary. Averaging their dawn and dusk negative B_z results gives

$$\bar{\theta}_p^d = 10.9 - 0.97 \, B_z(\text{nT}) \quad (\bar{r} = 0.52)$$

$$\text{(MMA data)} \quad (21)$$

We see here that, unlike the earlier results on the behavior of the poleward border of the statistical visual auroral oval (Eqs. 8 and 12), the actual poleward precipitation boundary expands rather than contracts as B_z grows more negative. Evidently the data on which Eq. 21 is based are relatively immune to the effects of the substorm expansive phase, an inference that accords with their being taken in the dawn-dusk sector. Comparing Eq. 21 with Eq. 19 (or even 20) shows that the latitudinal width of the diffuse auroral oval shrinks as B_z becomes more negative. *Feldstein and Galperin* [1985] make a similar observation, citing scanning photometer data that show the band of "diffuse subvisual luminescence" narrowing during the growth phase of substorms.

Hardy et al. [1981] and *Makita et al.* [1983] relate that the character of the correlation between the equatorward precipitation boundary and IMF B_z changes radically between positive and negative B_z values. We have given the correlations for negative B_zs. For positive B_zs, the equatorward precipitation boundary barely moves in response to changes in IMF B_z. By contrast, *Makita et al.* [1983] find that the poleward precipitation boundary moves virtually continuously as IMF B_z crosses the division between positive and negative values. The linear equation that describes how the poleward precipitation boundary moves over the range -8nT $< B_z < 8$nT, averaged over the dawn and dusk sectors, is

$$\bar{\theta}_p^d = 11.2 - 0.79\, B_z(\text{nT}) \quad (\bar{r} = 0.70)$$

$$\text{(MMA data)} \qquad\qquad (22)$$

which is not very different from Eq. 21.

5. RELATION OF THE OBSERVATIONS TO THE MODELS

Models that relate the size of the auroral oval to solar-wind parameters take two forms: superposition and convection. Superposition models work on the ring-on-finger metaphor, in which the ring is the auroral oval and the finger is the open-field-line polar cap. Superposition models determine only the inner radius of the ring, corresponding to the polar cap perimeter, not its thickness, corresponding to the width of the auroral band, nor its ornamentation, corresponding to the discrete and diffuse bands. That being so, the observational result against which to compare the predictions of superposition models is Eq. 22 (or Eq. 21). Here we assume that poleward precipitation border is also the border between open and closed field lines. Because superposition models are steady state, substorms complicate things. As already noted, no model exists that predicts how much flux converts from open to closed during a substorm, let alone one that relates to the IMF B_z the amount of flux converted. Because substorms control the behavior of the poleward border of the visual auroral oval, one

cannot compare the information on it with the predictions of superposition models. As a decision of operational procedure, we assume that Eq. 22 (or Eq. 21) is steady state for the purpose of testing superposition models. Since the magnetopause keeps most of the IMF out of the magnetosphere, one must determine what fraction gets in to apply superposition models. Nondeterminism enters the picture here, because there is no physically explicit model that enables one to do this. The solution to the IMF penetration problem is tantamount to the physically explicit, quantitative solution to the formation of the geomagnetic tail, which has not been achieved. Not surprisingly then, the tail enters superposition models in an important way.

Convection models use the convection electric field in the form of the transpolar electrical potential to derive the equation of motion for the radius of the open-field-line polar cap, corresponding to the poleward precipitation boundary, and the equilibrium radius for the inner edge of the plasma sheet, corresponding to the equatorward precipitation boundary. Convection models are more complex than superposition models. They can accommodate substorm control of the poleward boundary of the auroral oval and IMF B_z control of the equatorward boundary. Agreement in the first instance can only be qualitative because, again, there is no quantitative model of substorm flux conversion. But quantitative tests can be made on their predictions of the size of the equatorward boundary. For this we use Eq. 20 (or its predecessors). However, we must add a step to go from the transpolar potential to the IMF B_z, Eq. 20's solar-wind parameter. This is where nondeterminism enters the picture for convection models, because there is no physically explicit model to make that step. We must use empirically determined parameters.

The other information gleaned in the previous two sections plays supporting roles in the arguments. Interestingly, there is no model for the poleward or the equatorward border of the original, visual auroral oval.

6. SUPERPOSITION MODELS

We look first at the simplest superposition model that produces an open polar cap—a uniform field, representing the IMF B_z, superposed on a dipole field, representing the geomagnetic field. We will see that it simulates the data unjustifiably well, considering that the model lacks boundary and tail currents. Adding boundary currents destroys the agreement with the data and points to the geomagnetic tail as the means for reestablishing it. The predictive power of superposition models is thereby revealed to be limited by our ability to specify explicitly and quantitatively the tail in terms of solar-wind parameters. As such ability lies in the future, we conclude our brief survey by looking at semi-

POLAR CAP RADIUS VS IMF STRENGTH
FOR SOUTHWARD FIELD

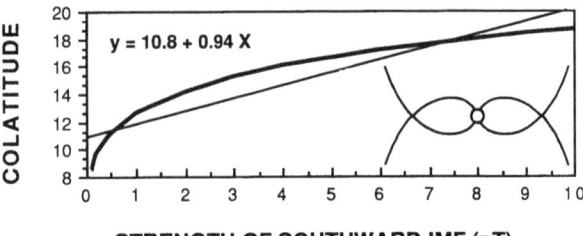

STRENGTH OF SOUTHWARD IMF (nT)

Figure 3—How the angular radius of the open-field-line polar cap increases with IMF strength when a south-pointing IMF is superposed on the Earth's dipole field.

empirical superposition models that incorporate the tail. While the models nicely illustrate qualitatively how the IMF controls polar cap geometry, we will see that as things now stand, the data and the models are not optimally suited for quantitative comparisons.

Figure 3 illustrates how, in an IMF-plus-dipole superposition model, the angular radius of the open-field-line polar cap varies with IMF strength. The figure shows the case of a purely southward IMF, for which one has an analytic relation between the two variables of interest.

$$\sin \theta_p \ = \ \left[\frac{27 \ |B_z|}{9 \ B_e} \right]^{1/6} \tag{23}$$

where B_e is the surface-equatorial strength of Earth's dipole field ($\approx 31{,}000$ nT).

Over the two-decade range of typical values, 0.1–10 nT, the curve appears nearly linear except at the small-field-strength end, where it drops abruptly as the field goes to zero. (The restriction of the nonlinearity to a region near the origin reflects the weak power dependence (1/6) in the equation.) The linear, least-squares fit to the curve shown in the figure looks remarkably like the linear, least-squares fit to the data for negative IMF B_z (x in Figure 3 is $-B_z$ in Eq. 21). However the agreement must be fortuitous, in part because the fit depends on choice of end points for the interval, but more importantly because only about 20% of the IMF penetrates the magnetospheric boundary, the rest is shielded by boundary currents. (Twenty percent is the average of the penetration efficiencies determined by *Burch* [1973], *Fairfield* [1979], and *Cowley and Hughes* [1983].) Adding boundary shielding to the model increases the numbers on the abscissa of Figure 3 fivefold, which ruins its accordance with Eq. 21. (Equivalently, the effect of the shielding currents on Eq. 23 is to replace B_z with $B_z/5$.) To restore the

agreement, something more must be added to the model to amplify the penetrating field and raise its influence on polar cap area to the level of the unshielded IMF. Of course, that something is the tail, which in effect amasses and concentrates the penetrating field.

To quantify the geotail's amplifying function, we must modify Eq. 23 to account for the fact that the solar-wind pressure confines the dipole field to a fixed magnetospheric volume, rather than to the variable volume that inheres in Eq. 23. In the IMF-plus-dipole superposition model, the area in the solar wind that connects to the open-field-line polar cap decreases as the IMF strength increases, because the null points move earthward. The offsetting tendencies of field strength increase and area decrease give Eq. 23 its weak B_z dependence. A superposition model with a magnetospheric boundary usually fixes the size of the boundary while allowing the IMF to vary. The IMF penetrates some fixed fraction of the boundary's cross-sectional area (empirically about 1/5). Equating the magnetic flux in the open-field-line polar cap to the flux of southward IMF crossing the fixed, effective penetration area, A_p, on the magnetospheric boundary gives

$$\sin \theta_p \ = \ \left[\frac{A_p \ |B_z|}{A_e \ B_e} \right]^{1/2} \tag{24}$$

where A_e is the cross-sectional area of the Earth.

The important difference between this result and the earlier one is its stronger dependence on B_z (1/2 power instead of 1/6 power). Though it is based on a more realistic model, Eq. 24 loses verisimilitude compared to Eq. 23. In a linear, least-squares fit, the stronger B_z dependence roughly halves the constant term relative to that in Figure 3 and nearly trebles the coefficient in the linear term. One can try to weaken the dependence on B_z by letting the IMF penetration contribute to the observed erosion of the dayside boundary with increasing $-B_z$ [*Hill and Rassbach*, 1975]. But the dayside boundary occupies too small a fraction of the total boundary area too help much. Another possibility emerges as we look now at the first of the superposition models with a tail.

That a model with a magnetospheric boundary also needs a tail becomes apparent when one computes how big the boundary must be to allow the flux from the open-field-line polar cap to pass through 1/5 of its area. Taking 4 nT as a typical IMF strength and 15° as the corresponding polar cap radius—consistent with Eq. 21—one finds $1.6 \times 10^4 \ R_E^2$, where R_E is the radius of the earth. Given that the dawn-dusk magnetospheric width is approximately $30 R_E$, the length parallel to the solar wind must be roughly $500 R_E$.

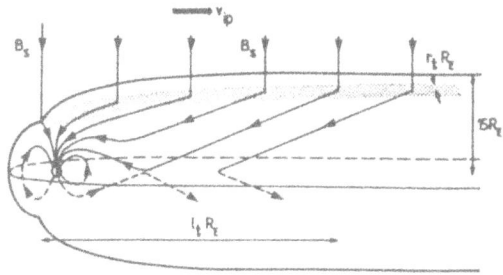

Figure 4—The *Rycroft* [1987] magnetospheric model with an open IMF window of fixed width and length.

Thus, one needs a tail. The tail captures the IMF flux from a $500R_E$ reach of solar wind and delivers it to Earth where it opens the polar cap much wider than the penetrating IMF does acting alone.

Rycroft [1987] explores the geometrical and electrodynamical properties of a superposition model in which the IMF enters the magnetosphere through a $10R_E$ slot running $400R_E$ down the spine of the tail (Figure 4). The model gives a 17° polar cap for a 4 nT IMF B_z, instead of Eq. 21's 15°; but the discord is easily fixed by changing the dimensions of the slot. More discordant is the fact that compared to Eq. 21, the polar cap expands and contracts too much in response to changes in IMF B_z. As noted in the discussion relating to Eq. 24, all (purely southward IMF) superposition models with fixed effective penetration areas exhibit excess sensitivity to IMF B_z compared to Eq. 21. Rycroft shows that around the 4 nT IMF value, the polar cap expands or contracts by 2.3° when the (southward) IMF strengthens or weakens by 1 nT. Equation 21 prescribes a significantly smaller, 1° movement for a 1 nT change in IMF strength.

One can reduce the model's sensitivity to variations in IMF strength by imitating the IMF-plus-dipole case and modulate the size of the boundary's flux-window to offset changes in IMF strength. Prevailing notions about how dayside magnetic merging works (which determines the window's width) and how nightside magnetic merging works (which determines the window's length) suggest modulating the length. Increasing $|B_z|$ might shift the tail's neutral line sunward and contract the tail segment attached to earth. To achieve the desired effect, the window's length must vary inversely as roughly the two-thirds power of $|B_z|$. For example, if B_z doubles from 4-8 nT, the window must shorten from $500R_E$ to about $300R_E$. Data neither support nor refute this hypothesis.

Though the south-IMF-plus-dipole model of Figure 3 fits the empirical Eq. 21, it fails completely to account for the polar cap's size when the IMF points northward, as given by Eq. 22. The open-field-line polar cap vanishes in the corresponding north-IMF-plus-

dipole model. Similarly, though one can modify Rycroft's structurally more complete model to agree with Eq. 21, it too is unable to account for positive B_z behavior. We look next at models that predict polar cap parameters for arbitrary IMF orientation.

The advantage of Rycroft's model is that its simplicity lets one compute analytically how, in the presence of boundary and tail currents, the polar cap's area depends on IMF strength, albeit for a strictly south-pointing IMF. To determine how, at the same structural level, the polar cap's area, shape, and location depend on IMF orientation (as distinct from IMF strength at a fixed south orientation) takes numerical models. *Akasofu et al.* [1981] and *Akasofu and Roederer* [1984] published the first results from such a model. It uses a magnified image dipole to make the boundary, and it ribs an equatorially bisected, flaring cylinder with current loops to simulate the tail. It lets 100% of the IMF into the magnetosphere (no shielding). The tail is correspondingly short. For other than a purely southward IMF orientation, the polar cap in the model distorts from circularity and its center shifts from the dipole axis. Both effects become pronounced as the IMF swings through the equator into northward orientations. *Alekseyev and Belenkaya* [1983] and *Toffoletto and Hill* [1989] report similar behavior from their models, the first without and the second with IMF shielding. Toffoletto and Hill show that for northward orientations, the topology of the polar cap depends on the (still imperfectly known) nature of dayside magnetic merging. The open-field-line polar cap can be singly or doubly connected.

Figure 5 shows how polar cap size depends on IMF orientation according to the calculation by *Akasofu and Roederer* [1984]. (The other articles cited emphasize different aspects of the problem and they omit suffi-

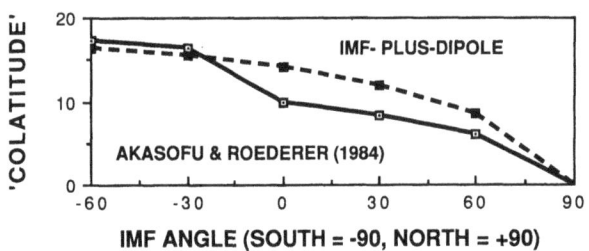

POLAR CAP 'RADIUS' VS IMF ANGLE
TWO MODELS
IMF = 5 nT

Figure 5—The *Akasofu and Roederer* [1984] polar cap sizes computed with a superposition model in which an unshielded IMF penetrates a magnetospheric boundary with a geomagnetic tail. The model treats arbitrary IMF orientations. The comparison curve shows what happens in the absence of boundary and tail currents.

cient size-vs-angle information to construct comparison plots.) The words "radius" and "colatitude" in Figure 5 appear in quotes because the cap is not strictly circular and its perimeter is not strictly a line of magnetic colatitude. The words refer to values that would result from circularizing the polar cap's area and centering it on the dipole axis.

Using a representative 5 nT IMF strength, Akasofu and Roederer give polar cap areas for the six IMF angles in the figure with data points. Note that the equatorial orientation, for which $B_z = 0$, produces a 10° polar cap, agreeing nearly with Eq. 22. If one converts the abscissa to B_z by projecting the 5 nT IMF onto the z-axis using the angles marked, one can test the model against the linear term in Eq. 22. Unfortunately, a discontinuity appears in the model between northward (including equatorial) and southward (excluding equatorial) orientations, rendering the value for the linear term nonunique. If one persists in spite of the discontinuity, one finds 0.68°/nT for northward IMF and 2.6°/nT for southward IMF, or an average of 1.64°/nT. The northward IMF value approximates the 79°/nT of Eq. 22, but comparing with the average is a better test. Then the model fails. It predicts too large a change in polar cap size as the IMF rotates from northward to southward (or vice versa).

The dashed line in the figure shows the colatitudes (again in quotations, that is, equivalent colatitudes) that result if one superposes on the Earth's dipole a uniform 5-nT field with the same six orientations chosen by Akasofu and Roederer. (*Akasofu and Ahn* [1980], first implemented this idea, but for stronger field strengths appropriate to the magnetosheath.) Considering that the model for the comparison curve has neither boundary nor tail, the agreement is remarkable, especially for southward orientations. Also it passes continuously from northward to southward IMFs, and it has a slope nearer to that of the positive IMF angles of the Akasofu and Roederer model.

These observations motivate testing the IMF-plus-dipole superposition model against Eq. 22, instead of Eq. 21, which led to Figure 3. Here we derive B_z values by rotating a constant magnitude IMF north and south of its average equatorial orientation. This is a more realistic procedure than varying the strength of a south-pointing IMF. We maintain realism by restricting the northward and southward swings of the IMF to 40°, which simulates its typical range. But within the equatorially centered 80° wedge, we assume all angles equally probable. We determine the IMF strength that gives a linear, least-squares fit over the chosen range whose linear term agrees with Eq. 22. Figure 6 shows the result. The determined IMF strength is 4.6 nT, which is close enough to the average IMF strength to claim agreement with observations. However, the

POLAR CAP RADIUS VS IMF ANGLE INTERPRETED AS BZ
IMF = 4.6 nT

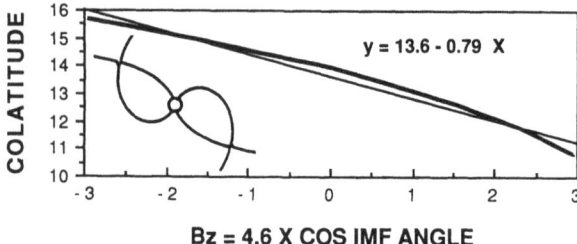

Bz = 4.6 X COS IMF ANGLE

Figure 6—How the polar cap size varies with IMF B_z when the B_z component results from rotating a fixed strength IMF northward and southward of the equator.

constant term is too large (13.6° vs 11.2°). The IMF strength that gives a linear, least-squares fit over the range whose constant term agrees with Eq. 22 is 1.9 nT, much too low, and the coefficient of the linear term in this case is 1.66, much too high.

Though not perfect, the qualitative success of the IMF-rotation-and-superposition exercise suggests that modelers should interrogate their models, as did Akasofu and Roederer, to learn how the predicted polar cap areas vary with the orientation of a constant strength IMF. Similarly, data analyzers should plot inferred polar cap size against IMF angle in separate bins of IMF strength, instead of against IMF B_z. Then the procedures would conform more closely to the way nature seems to work, and the results of models and observations could be compared directly.

What have we learned from the superposition models regarding our initial question? The bare IMF-plus-dipole superposition models simulate the empirical Eqs. 21 and 22 better than models dressed with boundary and tail currents. If this reflects some deep law of nature, it lies beyond proof by our present theoretical understanding. To test it empirically requires data from another planet. Whether the success of simple models reveals a law or is nothing but an arresting accident, the future emphasis should be on the evolution of realistic models, because they assure explicit understanding of the governing physics, and they yield answers to a greater variety of questions. The models should be able to predict polar cap parameters for all IMF orientations and strengths. Observers should correlate polar cap behavior with the IMF vector (orientation and strength), in addition to the IMF components.

Most importantly, we have learned that before superposition models can be designed to predict polar cap behavior in terms of solar-wind parameters, the field must have solved the problem of determining how much IMF leaks through the boundary and how much the tail then accumulates. More specifically, the field must first solve the problem of determining the geometry and rates of dayside magnetic merging from solar-wind parameters, and it must also solve the problem of determining the geometry and magnetodynamics of the tail from solar-wind parameters and knowledge of the geometry and rates of dayside magnetic merging. Our lack of understanding of the basic physics in two major areas of magnetospheric phenomena prevents our achieving the desired predictive capability. Since the solutions to the missing physics problems are intimately tied to the phenomenon of magnetospheric convection, another class of models has evolved which tries to circumvent the limitations of superposition models by relating polar cap size to magnetospheric convection.

7. CONVECTION MODELS

There are convection models for the poleward and equatorward borders of the auroral oval. We look at each in the stated order and then a synthesis of the two. The underlying assumption of convection models of the poleward border is this: magnetic merging at the dayside magnetospheric boundary makes the open-field-line polar cap expand; nightside magnetic merging in the tail makes it contract. If the dayside merging rate exceeds the nightside merging rate, there is net expansion. If the nightside merging rate exceeds the dayside merging rate, we have net contraction [*Siscoe and Huang*, 1985]. In the ionosphere, one measures a merging rate (that is, the amount of magnetic flux converted from closed to open or open to closed in units of Wb/s, which is the same as volts) by the electrical potential across the line onto which the magnetospheric merging line projects. The net merging rate is then the difference between the dayside and nightside merging electrical potentials. One has immediately a prognostic equation for polar cap size in terms of the net merging rate by equating the rate of change of magnetic flux in the polar cap $(2\pi B_e r_{pc}^2)$ to the net merging potential $(\Phi_d - \Phi_n)$. (r_{pc} is the equivalent polar cap radius, in the sense used earlier.) This gives

$$r_{pc}^2(t) = r_{pc}^2(t=0) + \frac{1}{2\pi B_e} \int_0^t (\Phi_d - \Phi_n)dt \tag{25}$$

or in differential form

$$\frac{dr_{pc}}{dt} = \frac{\Phi_d - \Phi_n}{4\pi B_e r_{pc}} \tag{26}$$

To compute the polar cap size for any moment, one must specify the history of the two merging potentials

up to that moment. To predict the polar cap size from solar-wind parameters, one must be able to turn them into merging potentials. As yet this cannot be done. This section reviews what has been done, by combining theory and observations relating to the net merging potential, to elucidate the behavior of the poleward precipitation boundary of the diffuse auroral oval (identified as the polar cap perimeter), and by the development of a convection theory for the closed-field-line magnetosphere, what has been done to elucidate the behavior of the equatorward precipitation boundary (or equivalently, the equatorward border of the diffuse auroral oval).

Superposition models give steady-state pictures of the magnetosphere, in which the Earth's dipole resides in the combined fields of one or more steady sources. Convection models of the polar cap are inherently time dependent. If nightside merging always balanced dayside merging, the net merging potential would always be zero. The solution of the equation for r_{pc} would be r_{pc} = constant, and there would be no way within the model to determine the constant.

Coroniti and Kennel [1973] published the first model to prescribe the coupled time dependence of dayside and nightside merging potentials. Following a sudden increase in the dayside merging potential from a previous steady-state, balanced merging condition, the nightside merging potential rises gradually to rebalance the merging rates and reestablish the steady state. During the transient interval between steady states, a net quantity of flux converts from closed to open, and the polar cap gains area. Physically their model works like an electrical RL circuit driven by a voltage source. The response of such a circuit always lags a change in the voltage source by a characteristic time given by L/R. The Coroniti-Kennel circuit links the dayside boundary and the ionosphere with field-aligned currents, which today we would designate region 1. Their voltage source is the dayside-boundary merging potential; the ionosphere provides the resistance; and the voltage drop across the ionospheric resistor is the nightside merging potential. The inductance corresponds to the reservoir of closed magnetic flux.

The quantity of open flux added to the polar cap is the transpolar potential, Φ_{pc}, times the L/R time scale, τ. To account for the entire polar cap size with this model, one equates the added flux to the total flux and solves for the angular radius of the polar cap.

$$\sin \theta_p = \left[\frac{\Phi_{pc}\tau}{2A_e B_e} \right]^{1/2} \tag{27}$$

Interpreting this expression as a formula that predicts the polar cap size in terms of solar-wind parameters, one identifies τ as a fixed model parameter and Φ_{pc} as a proxy solar-wind variable. Using typical values for model parameters, Coroniti and Kennel find a representative L/R time scale to be 40 minutes. Regarding Φ_{pc}, though no definitive MHD nor plasma physical theory exists to relate it to its solar-wind origin, there are good reasons to think its proximate antecedent is the dynamo component of the solar wind's motional electric field, VB_z. *Reiff and Luhmann* [1986] have reviewed the reasons and tabulated empirical relations that show Φ_{pc} to be essentially linearly related to VB_z. For numerical comparisons, we use their formula that contains B_z explicitly (labeled B_s in R&L).

$$\Phi_{pc}(\text{kV}) = 35 - 22\, B_z(\text{nT}) \qquad (28)$$

where in going from their expression to this one, we have assumed a standard solar-wind speed of 440 km s^{-1}, as before.

With the model value for τ and the empirical expression for Φ_{pc}, the linear, least-squares fit to Eq. 27 over the range -4 nT $< B_z < 0$, gives 6.2° for the constant term and 1.29°/nT for the coefficient of the linear term. Comparing these with 10.9° and 0.97°/nT from Eq. 21 shows that the model processes too little magnetic flux to account for the entire open-field-line polar cap, and it is too sensitive to changes in B_z to simulate the dynamics of the entire polar cap. To illustrate the first point better, note that inserting a typical 60-kV transpolar potential in Eq. 27 gives a total flux of 1.44×10^8 Wb, corresponding to a 7.8° polar cap.

The Coroniti and Kennel model seems to have more to say about how much flux converts from closed to open during a substorm growth phase than about how much open flux resides in the polar cap. The quantity of flux just calculated matches the open flux added to the polar cap during the growth phase of the one substorm for which it has been ascertained [*Frank*, 1988]. To get back on the trail of the determinants of the size of the auroral oval, we turn to convection models for the closed-field-line magnetosphere, and review their control of the equatorward precipitation boundary.

The transpolar potential drives currents in the conducting ionosphere that, following the dictates of Ohm's law for a conducting surface, flow equatorward from the polar cap's border in one local time sector, and after circuiting the auroral oval, return to the polar cap's border in the opposite local time sector. The electric field they ohmically distribute around the auroral oval map along electrically conducting magnetic field lines into the closed-field-line magnetosphere. The transpolar potential is thereby conveyed to the

magnetosphere's trapped particle populations. There it impels the particles sunward with an $\mathbf{E} \times \mathbf{B}$ drift. This is magnetospheric convection.

Most of the magnetosphere's resident particles inhabit the tail's plasma sheet. Their sunward drift carries them closer to Earth and into stronger magnetic fields. Plasma-sheet particles constitute a hot plasma, and it takes energy to move a hot plasma into a stronger magnetic field. Doing so is like compressing a hot gas. When the energy available is consumed, the particles stop their earthward advance, and drift azimuthally around to the dayside. The electric field must supply the energy. We infer that bigger transpolar potentials let the plasma-sheet particles progress nearer to Earth before drifting around.

The distance at which earthward motion stops marks the earthward edge of the plasma sheet, and by projection, the equatorward precipitation boundary. Thus we have the connection we seek between the latitude of the equatorward precipitation boundary and the transpolar potential. Working out the mathematical form of the inferred relation entails more than just recited, because drift paths of positive and negative particles cross. The resulting charge separation discharges the bulk of itself through the ionosphere as the region 2 currents, and the remainder creates a secondary electric field that profoundly alters the applied field [*Vasyliunas*, 1972].

The interactively coupled magnetosphere-ionosphere electrodynamics problem has been attacked with the full power of a numerical magnetospheric circulation model [*Wolf*, 1983]. It can follow the particle, field, and current responses to arbitrary changes in the transpolar potential. Approximate analytic models can serve our lesser need to find where the equatorward precipitation boundary lies that is in equilibrium with a given steady transpolar potential. The problem has been solved by *Jaggi and Wolf* [1973], *Southwood* [1977], and *Siscoe* [1982]. Though their approaches and approximations differ, they yield nearly identical results, which, when syncretized into a common form, become

$$\sin\theta_e = \left[1 + \frac{\Sigma^p}{\Sigma^*}\frac{\Phi_{pc}}{K_{ps}}\right]^{1/6} \sin\theta^* \qquad (29)$$

where Σ^p is the ionospheric Pedersen conductance, Σ^* is the total electric charge of the ions in a unit magnetic flux tube, K_{ps} is the energy of the plasma sheet ions (in the same units as Φ_{pc}), and θ^* is the colatitude of the circumpolar ring around which the transpolar potential is applied.

No better option exists than to identify the ring of applied potential with the region 1 current system. This gives $\theta^* \simeq 20°$ [*Iijima and Potemra*, 1976]. Measurements and estimates of the other constants vary by more than a factor of two. One can only ask whether within their ranges, Eq. 29 predicts observed values of θ_e. In this case, the comparison formula is Eq. 20. If we pick $B_z = 0$ for a test demonstration, the target number is $\theta_e = 24.1°$. Equation 28 gives $\Phi_{pc} = 35$ kV. The requirement on the unspecified constants is then $(\Sigma^*/\Sigma^p)\, K_{ps} = 19$ kV. Conductance ratios from 5–10 and particle energies from 2–4 kV fall within observed ranges and satisfy the requirement. Using the number thus determined, the linear, least-squares fit to Eq. 29 over the range $-4\,\text{nT} < B_z < 0$ gives $24.2°$ for the constant term and $1.08°/\text{nT}$ for the coefficient to the linear term, compared with the $0.82°/\text{nT}$ of Eq. 20. Although the syncretized model's linear coefficient is too large, the difference is arguably within the uncertainties in determining it. The bigger shortcoming lies in its inability to say anything about θ^*. Specifying θ^* empirically almost (but not completely) builds the right answer into the model.

What determines θ^*? In effect, θ^* is an integration constant determined by the boundary condition $\theta_e(\Phi_{pc} = 0) = \theta^*$, the tacit idea being that the circumpolar ring of applied potential is a geometrically fixed feature. It represents a rigid barrier toward which the equatorward precipitation boundary recedes as activity subsides and against which it rests when activity ceases. Structurally it must lie near the poleward edge of the auroral oval. But we have seen that the poleward edge is not stationary. It moves as much as (in fact more than) the equatorward edge. Even the heavily averaged location of the region 1 ring shifts three degrees between active and inactive averaging bins [*Iijima and Potemra*, 1978].

Identifying θ^* with the poleward edge of the auroral oval allows one to treat it as a dependent variable for which one has the determining Eqs. 25 and 26, which relate it to the difference in the dayside and nightside merging potentials. (Here $\sin\theta^* = r_{pc}/R_E$.) Now in the differential equation for θ_e with Φ_{pc} as the independent variable that led to the solution Eq. 29, θ^* is replaced by the differential Eq. 26, with Φ_d and Φ_n as independent variables. The result has too many independent variables, and another constant of integration. Nonetheless, there is a special, but realistic, solution to the problem [*Siscoe*, 1982]. It is based on the observation that much of the time either Φ_d dominates (substorm growth phase) or Φ_n dominates (substorm expansive phase). Then either $\Phi_{pc} = \Phi_d$ or $\Phi_{pc} = \Phi_n$, leaving effectively only one independent variable. The artifact of a fixed barrier to the poleward migration of the equatorward precipitation boundary is removed with the boundary condition $\theta_e(\Phi_{pc} = 0) = 0$. The remaining freedom in the problem is fixed by the prediction of Eq. 26 that during substorm growth phase, θ^*

increases much faster than θ_e (as *Meng and Makita* [1986] have observed). On time scales longer than substorms, this keeps the poleward and equatorward edges close together and allows the replacement $\theta^* = \theta_e$ as an approximation at appropriate places in the equation. The approximation carries a price, however. One loses the definiteness of the feature for which one is solving. The independent variable is a representative colatitude of the entire auroral oval, which has appreciable latitudinal width. Further, the result is valid only for southward IMF, when substorms dominate auroral oval dynamics. As expressed by *Siscoe and Crooker* [1983], the solution is

$$\sin\theta_{ao} = \left[\frac{\pi}{4}\frac{\eta\Sigma^p}{\Sigma^*}\frac{\Phi_{pc}}{K_{ps}}\right]^{3/16}\sin\theta_{ps} \qquad (30)$$

where the new variables are θ_{ao}, a colatitude representing the size of the auroral oval as a unit, η, an empirical parameter lying between 1 and 3, and θ_{ps}, the colatitude of the field line connected to the particles whose energies are K_{ps}. (In the assumed dipole field, K_{ps} varies with distance.)

The advantage of Eq. 30 is that it removes the artifact of the high-latitude barrier. The disadvantage is that as an observable, θ_{ao} is specified rather imprecisely by the condition that it lie between θ_p and θ_e. Fortunately θ_p and θ_e, as given by Eqs. 20 and 21, have virtually the same linear dependence on B_z. If we take their average to represent the auroral oval as a unit, the hardest test we can give Eq. 30 is that there exist a combination of empirically certified constants (η, Σ^p, Σ^*, K_{ps}, and θ_{ps}) for which its linear, least-squares fit in the range $-4\,\text{nT} < B_z < 0$ has a constant term lying between $10.9°$ and $24.1°$ and a linear term with a coefficient close to $0.90°/\text{nT}$. Unfortunately, this is not a severe test, and the fact that the equation passes implies more that we have failed in refuting it than that we have succeeded in validating it. Taking $\eta = 4/\pi$, $\Sigma^*/\Sigma^p = 10$, $K_{ps} = 4$ kV, and $\theta_{ps} = 14°$ and using Eq. 28, one finds a constant term of $13.8°$ and a linear coefficient of $0.92°/\text{nT}$ (Figure 7).

This is as far as convection models have evolved in the direction of specifying auroral oval size in terms of solar-wind parameters. Improvements are still possible at the present conceptual level. The vacuum dipole field should be replaced by a dipole-plus-boundary-plus-ring-current-plus-tail-field appropriate to the level of activity as given by Φ_{pc} and solar-wind ram pressure. The result will surely change the exponent in Eq. 30, but because of compensating effects, it takes detailed analysis to say which way. Equation 26 and the differential equation behind Eq. 29 can be integrated jointly instead of combined in an approximation

AURORAL OVAL RADIUS VS |Bz|
RING COUPLING MODEL

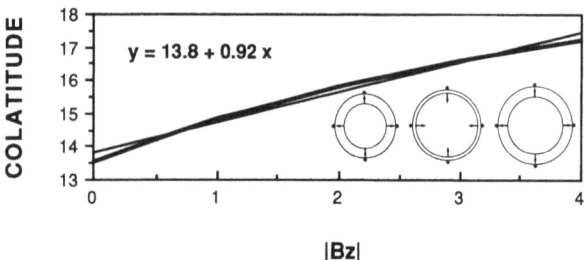

Figure 7—The *Siscoe and Crooker* [1983] auroral oval size based on a ring-coupling type of convection model. The B_z dependence is inferred from an empirical formula relating it to the convection potential.

that blurs the distinction between θ_p and θ_e.

Improvements are also needed in data presentations—the same improvements cited in connection with superposition models. All but one of Reiff and Luhmann's empirical formulas relating the IMF to Φ_{pc} involve the vector IMF (strength and angle) rather than its z-component. We need auroral oval boundary data also in terms of the IMF vector.

Once the improvements have been made, problems remain whose solutions wait upon advances in our understanding of global magnetospheric dynamics. We do not know how to extend convection models to treat the northward IMF case. When the IMF swings northward, the poleward border retreats poleward leaving the equatorward border more-or-less stranded and immobile [*Meng and Makita*, 1986], suggesting that the situation differs qualitatively from the southward IMF case. We need the physically explicit, quantitative relation between Φ_{pc} and solar-wind parameters, as distinct from physically implicit, empirical relations, which we have. Similarly we need the physically explicit, quantitative relation between K_{ps} and solar-wind parameters. As in the case of the superposition models,

the missing physics centers on the magnetospheric boundary, in this case blocking access to the relation for Φ_{pc}, and on the geomagnetic tail, keeping K_{ps} a semiempirical parameter.

8. CONCLUSION

We opened with a quote from one Nobel laureate and close with a quote from another, which stresses the same point. Richard Feynman ends book II of his *Lectures on Physics* with the following challenge to environmental sciences (in which he explicitly includes auroral physics): "The test of science is its ability to predict." We can predict the size of the magnetosphere in terms of solar-wind parameters—one of the early successes of magnetospheric physics—but not the size of the tail, and hence not the size of the auroral oval. The geometrical properties of the tail depend in part on magnetic merging at the dayside magnetospheric boundary and in part on its own internal dynamics. For predictive purposes, we are stymied at both points. We must determine quantitatively with explicit physics the geometry and rate of magnetic merging at the magnetospheric boundary, and we must determine quantitatively with explicit physics the magnetostatics and dynamics of the geomagnetic tail. As the direct creation of boundary and tail processes, albeit a reactive creation, the auroral oval remains unpredictable in terms of magnetospheric boundary conditions until we understand the governing boundary and tail processes. Also as the most visible and accessible recorder of the processes acting together globally, it represents an essential target for comprehensive data gathering and modeling as part of an integrated effort to solve the two main problems in the field of magnetospheric physics.

ACKNOWLEDGMENT—This work was supported in part by the National Science Foundation through Grant ATM 87-22962.

REFERENCES

Akasofu, S.-I., and B.-H. Ahn, "Dependence of the Amount of Open Magnetic Flux on the Direction of the Interplanetary Magnetic Field," *Planet. Space Sci.*, **28**, 545-547 (1980).

Akasofu, S.-I., D. N. Covey, and C.-I. Meng, "Dependence of the Geometry of the Region of Open Field Lines on the Interplanetary Field," *Planet. Space Sci.*, **29**, 803-807 (1981).

Akasofu, S.-I., and M. Roederer, "Dependence of the Polar Cap Geometry on the IMF," *Planet. Space Sci.*, **32**, 111-118 (1984).

Alekseyev, I. I., and Ye. S. Belenkaya, "Electric Field in an Open Model of the Magnetosphere," *Geomag. Aeron.*, **23**, 57-61 (1983).

Bond, F. R., and I. L. Thomas, "The Southern Auroral Oval," *Aust. J. Phys.*, **24**, 97-102 (1971).

Brekke, A., "On the Evolution in History of the Concept of the Auroral Oval, *Eos Trans. AGU*, **65**, 705-707 (1984).

Burch, J. L., "Rate of Erosion of Dayside Magnetic Flux Based on a Quantitative Study of the Dependence of Polar Cusp Latitude on the Interplanetary Field," *Radio Sci.*, **11**, 955-961 (1973).

Coroniti, F. V., and C. F. Kennel, "Can the Ionosphere Regulate Magnetospheric Convection?" *J. Geophys. Res.*, **78**, 2837-2851 (1973).

Cowley, S. W. H., and W. J. Hughes, "Observation of an IMF Sector Effect in the Y Magnetic Field Component at Geostationary Orbit," *Planet. Space Sci.*, **31**, 73-90 (1983).

Eather, R. H., *Majestic Lights*, American Geophysical Union, Washington, DC (1980).

Fairfield, D. H., "On the Average Configuration of the Geomagnetic Tail," *J. Geophys. Res.*, **84**, 1950-1958 (1979).

Feldstein, Ya. I., "Geographical Distribution of Aurorae and Azimuths of Auroral Arcs," *Invest. Aurorae*, **N4**, 61-78 (1960).

Feldstein, Ya. I., "On the Morphology of Auroral and Magnetic Disturbances at High Latitudes," *Geomagnet. Aeron.*, **3**, 183 (1963).

Feldstein, Ya. I., "Auroral Morphology, I. The Location of the Auroral Zone," *Tellus*, **16**, 252 (1964).

Feldstein, Ya. I., "A Quarter of a Century with the Auroral Oval," *Eos Trans. AGU*, **67**, 761 (1986).

Feldstein, Ya. I., and G. V. Starkov, "Dynamics of Auroral Belt and Polar Geomagnetic Disturbances," *Planet. Space Sci.*, **15**, 209-229 (1967).

Feldstein, Ya. I., and Yu. I. Galperin, "The Auroral Luminosity Structure in the High-Latitude Upper Atmosphere," *Rev. Geophys.*, **23**, 217-275 (1985).

Frank, L. A., "Dynamics of the Near-Earth Magnetotail—Recent Observations," in *Modeling Magnetospheric Plasma*, T. E. Moore and J. H. Waite, Jr., eds., American Geophysical Union, Washington, DC, pp. 261-276 (1988).

Gosling, J. T., J. R. Asbridge, S. J. Bame, and W. C. Feldman, "Solar Wind Speed Variations: 1962-1974," *J. Geophys. Res.*, **81**, 5061 (1976).

Gussenhoven, M. S., D. A. Hardy, and W. J. Burke, "DMSP/F2 Electron Observations of Equatorward Auroral Boundaries and Their Relationship to Magnetospheric Electric Fields," *J. Geophys. Res.*, **86**, 768-778 (1981).

Gussenhoven, M. S., D. A. Hardy, and N. Heinemann, "Systematics of the Equatorward Diffuse Auroral Boundary," *J. Geophys. Res.*, **88**, 5692-5708 (1983).

Hardy, D. A., W. J. Burke, M. S. Gussenhoven, N. Heinemann, and E. Holeman, "DMSP/F2 Electron Observations of Equatorward Auroral Boundaries and Their Relationship to the Solar Wind Velocity and the North-South Component of the Interplanetary Field," *J. Geophys. Res.*, **86**, 9961-9974 (1981).

Hill, T. W., and M. E. Rassbach, "Interplanetary Magnetic Field Direction and the Configuration of the Dayside Magnetosphere," *J. Geophys. Res.*, **80**, 1-6 (1975).

Holzworth, R. H., and C.-I. Meng, "Mathematical Representation of the Auroral Oval," *Geophys. Res. Lett.*, **2**, 377-380 (1975).

Holzworth, R. H., and C.-I. Meng, "Auroral Boundary Variations and the Interplanetary Magnetic Field," *Planet. Space Sci.*, **32**, 25 (1984).

Iijima, T., and T. A. Potemra, "The Amplitude Distribution of Field-Aligned Currents at Northern High Latitudes Observed by Triad," *J. Geophys. Res.*, **81**, 2165-2174 (1976).

Iijima, T., and T. A. Potemra, "Large-Scale Characteristics of Field-Aligned Currents Associated with Substorms," *J. Geophys. Res.*, **83**, 599-615 (1978).

Jaggi, R. K., and R. A. Wolf, "Self-Consistent Calculation of the Motion of a Sheet of Ions in the Magnetosphere, *J. Geophys. Res.*, **78**, 2852 (1973).

Khorosheva, O. V., "Diurnal Drift of the Auroral Closed Ring" (in Russian), *Geomagn. Aeron.*, **2**, 839-850 (1962).

Makita, A., C.-I. Meng, and S.-I. Akasofu, "The Shift of the Auroral Electron Precipitation Boundaries in the Dawn-Dusk Sectors in Association with Geomagnetic Activity and Interplanetary Magnetic Field," *J. Geophys. Res.*, **88**, 2744-2752 (1983).

Mead, G. D., and D. B. Beard, "Shape of the Geomagnetic Field Solar Wind Boundary," *J. Geophys. Res.*, **69**, 1169-1180 (1964).

Meng, C.-I., R. H. Holzworth, and S.-I. Akasofu, "Auroral Circle—Delineating the Poleward Boundary of the Quiet Auroral Belt," *J. Geophys. Res.*, **82**, 164-172 (1977).

Meng, C.-I., and K. Makita, "Dynamic Variations of the Polar Cap," in *Solar Wind—Magnetosphere Coupling*, Y. Kamide and J. A. Slavin, eds., Terra Scientific Publishing Company, Tokyo, pp. 605-631 (1986).

Reiff, P. H., and J. G. Luhmann, "Solar Wind Control of the Polar-Cap Voltage, in *Solar Wind—Magnetosphere Coupling*, Y. Kamide and J. A. Slavin, eds., Terra Scientific Publishing Company, Tokyo, pp. 453-476 (1986).

Rycroft, M. J., "The Electrodynamics and Magnetohydrodynamics of Geospace," *Mem. Natl. Inst. Polar Res., Spec. Issue*, **48**, 196-215 (1987).

Siscoe, G. L., "Polar Cap Size and Potential: A Predicted Relationship," *Geophys. Res. Lett.*, **9**, 672-675 (1982).

Siscoe, G. L., "Energy Coupling Between Regions 1 and 2 Birkeland Current Systems," *J. Geophys. Res.*, **87**, 5124-5130 (1982).

Siscoe, G. L., "The Magnetospheric Boundary," in *Physics of Space Plasmas (1987)*, SPI Conference and Reprint Series, Number 7, T. Chang, G. B. Crew, and J. R. Jasperse, eds., Scientific Publishers, Inc., Cambridge, MA, pp. 3-78 (1988).

Siscoe, G. L., and N. U. Crooker, "Coupling of Birkeland Current Rings," in *Magnetospheric Currents*, T. A. Potemra, ed., American Geophysical Union, Washington, DC, pp. 260-268 (1983).

Siscoe, G. L., and T. S. Huang, "Polar Cap Inflation and Deflation," *J. Geophys. Res.*, **90**, 543-547 (1985).

Slater, D. W., L. L. Smith, and E. W. Kleckner, "Correlated Observations of the Equatorward Diffuse Auroral Boundary," *J. Geophys. Res.*, **85**, 531-542 (1980).

Southwood, D. J., "The Role of Hot Plasma in Magnetospheric Convection," *J. Geophys. Res.*, **82**, 5512-5520 (1977).

Starkov, G. V., "An Analytical Presentation of the Equatorial Boundary of the Auroral Oval," *Geomagn. Aeron.*, **9**, 759-760 (1969).

Thomas, I. L., and F. R. Bond, "An Empirical Equation for the Austral Auroral Oval," *Geophys. Res. Lett.*, **4**, 411-412 (1977).

Thomas, I. L., and F. R. Bond, "A Spherical Harmonic Analysis of the Austral Auroral Oval," *Planet. Space Sci.*, **26**, 691-695 (1978).

Toffoletto, F. R., and T. W. Hill, "Mapping of the Solar Wind Electric Field to the Earth's Polar Cap," *J. Geophys. Res.*, **94**, 329-347 (1989).

Unti, T., and G. Atkinson, "Two-Dimensional Chapman-Ferraro Problem with Neutral Sheet," *J. Geophys. Res.*, **73**, 7319-7328 (1968).

Vasyliunas, V. M., "The Interrelationship of Magnetospheric Processes," in *Earth's Magnetospheric Processes*, B. M. McCormac, ed., D. Reidel Publishing Company, pp. 29-38 (1972).

Wolf, R. A., "The Quasi-Static (Slow-Flow) Region of the Magnetosphere," in *Solar-Terrestrial Physics*, R. L. Carovillano and J. M. Forbes, eds., D. Reidel Publishing Company, pp. 303-368 (1983).

Zverev, V. L., G. V. Starkov, and Ya. I. Feldstein, "Influences of the Interplanetary Magnetic Field on the Auroral Dynamics," *Planet Space Sci.*, **27**, 665-667 (1979).

IV-2. THE QUIET-TIME AURORA AND THE MAGNETOSPHERIC CONFIGURATION

R. Lundin,* L. Eliasson,* and J. S. Murphree[†]

The "quiet magnetosphere" is characterized by very low magnetic activity (low *Kp, AE*) and a correspondingly low auroral activity along the *statistical* auroral oval. However, the absence of substorm activity along the nightside oval is not necessarily associated with an absence of activity in the dayside oval and certainly not with the lack of activity over the polar cap. On the contrary, a number of studies indicate that the dayside oval and the polar cap may be quite active during times of low night-side activity. Because the nightside oval is topologically connected to the tail, one may then argue that the dayside and polar cap auroral activity is unrelated to magnetotail processes. This is only relevant for magnetically quiet periods (low *Kp, AE*). An argument for a distinction between the dayside and nightside discrete aurora is the topological connection to different boundary layers, the low-latitude boundary layer (LLBL) and the plasma-sheet boundary layer (PSBL). However, during quiet times the LLBL and PSBL may become phenomenologically (and topologically) coupled, the LLBL being the region where solar-wind energy and momentum are transferred. It will be argued here that the dynamical part of the dayside oval (here designated the dayside cleft), and during extended periods of a northward interplanetary magnetic field (IMF) also the polar region, is connected to the magnetospheric boundary layers (LLBL, PSBL). The persistent activity in the dayside cleft thus reflects the persistent appearance of a magnetospheric boundary layer. This is corroborated by satellite measurements indicating, for example, that the LLBL is wider during quiet times and a northward IMF. The widening of the LLBL is also reflected by the broadening and poleward expansion of the auroral activity during quiet times; therefore, the quiet-time aurora is, to a large extent, driven by processes in the LLBL.

1. INTRODUCTION

Throughout the years magnetospheric research has to a large extent focused on periods of high magnetic activity, on substorms, and on magnetic storms. Yet, one may argue that to understand the disturbed magnetosphere one must first understand the quiet time magnetosphere. The usage of the term quiet time is frequently misleading because it is assumed to be related to a magnetosphere in a state of complete relaxation. Rather than the expected state of quiescence, magnetically quiet periods (in terms of the *Kp* and *AE* indexes) may represent times when the magnetosphere is moderately active, yet with no significant substorm activity taking place. Discrete aurora near the noon meridian [*Murphree et al.*, 1981] appears to be a persistent feature that does not depend strongly on the nightside activity [*Evans*, 1985; *Meng and Lundin*, 1986]. Similarly, as shown in Figure 1, the region 1 field-aligned current (FAC) system displays a strong and persistent maximum around noon [*Ijima and Potemra*, 1976, 1978]. The maximum current is not very strongly dependent on the magnetic disturbance level (*AE*). *Zanetti and Potemra* [1986] also found that the total FAC, which is a good measure of the activity level, increases again for increasingly positive B_z (Figure 2). This adds some

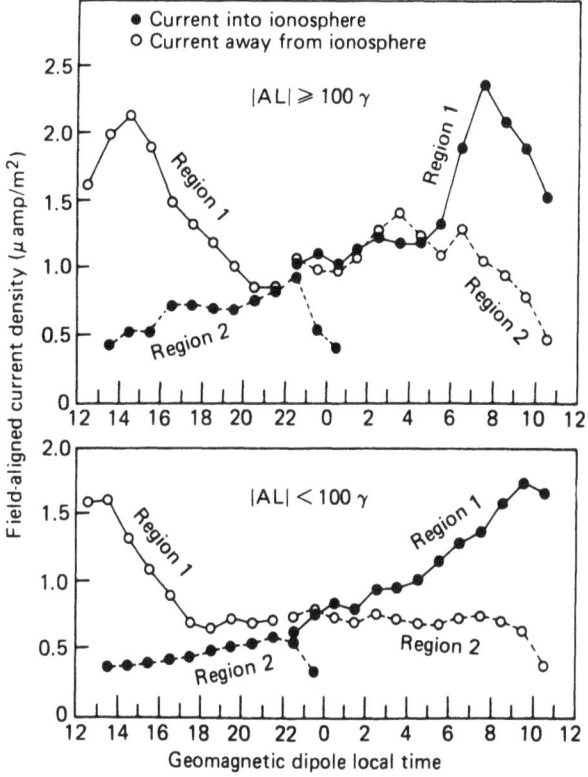

Figure 1—Local time distribution of field-aligned currents. (After *Ijima and Potemra* [1978].)

*Swedish Institute of Space Physics, P.O. Box 812, S-981 28, Kiruna, Sweden.
[†]Department of Physics, University of Calgary, Calgary, Alberta, Canada T2N 1N4.

Auroral Physics, edited by C.-I. Meng, M. J. Rycroft and L. A. Frank. © Cambridge UP 1991

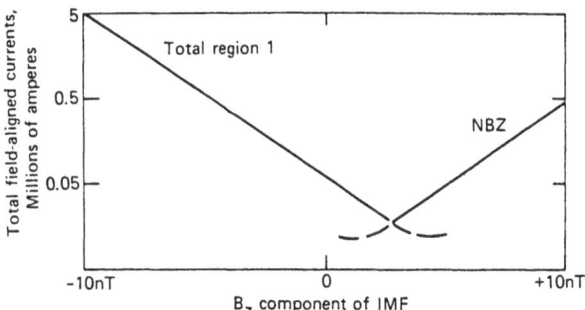

Figure 2—Total field-aligned current versus IMF B_z. (After *Zanetti and Potemra* [1986].)

complexity to the term quiet time as well as to the processes by which solar-wind energy is transferred into the magnetosphere. The "classical" picture of the merging rate into the magnetosphere [*Reiff et al.*, 1977; *Perreault and Akasofu*, 1978] evidently requires some additional modeling. Thus, with respect to the large-scale currents [*Ijima and Potemra*, 1978] and particle energy deposition into the upper atmosphere [*Brautigam et al.*, 1988], the real quiet state of the magnetosphere is rather when the IMF B_z is close to zero.

However, in what follows, we define the quiet-time aurora as the aurora occurring when the magnetic ac-

tivity (e.g., *AE* or *Kp*) is low (no substorm activity is going on) or when the IMF B_z has remained positive or close to zero for several hours or more. This includes periods when IMF B_z is strongly positive with correspondingly high activity over the polar region.

The connection between the dayside boundary layer and the midday auroral oval has been emphasized by several authors [*Eastman et al.*, 1976; *Bythrow et al.*, 1981; *Meng and Lundin*, 1986]. *Akasofu and Kan* [1980] and *Meng and Lundin* [1986] also noted that, although the dayside aurora is present along an apparently continuous auroral oval, it is not a mere extension of the discrete aurora from the nightside auroral oval. This fact is illustrated in Figures 3 and 4. Figure 3 shows a Defense Meteorological Satellite Program (DMSP) picture of the dayside aurora [after *Meng and Lundin*, 1986] which is clearly disconnected from the nightside. In Figure 4 a Viking UV image of the entire quiet-time auroral oval is shown, the dayside portion being in the sunlit area of the Earth. Although the aurora clearly forms a continuous circle along the classical Feldstein oval, Figure 4 also demonstrates that two centers of auroral activity can be distinguished—one near local midnight and one near local noon. Thus, there are two major centers for the activity along the auroral oval, one associated with plasma processes tak-

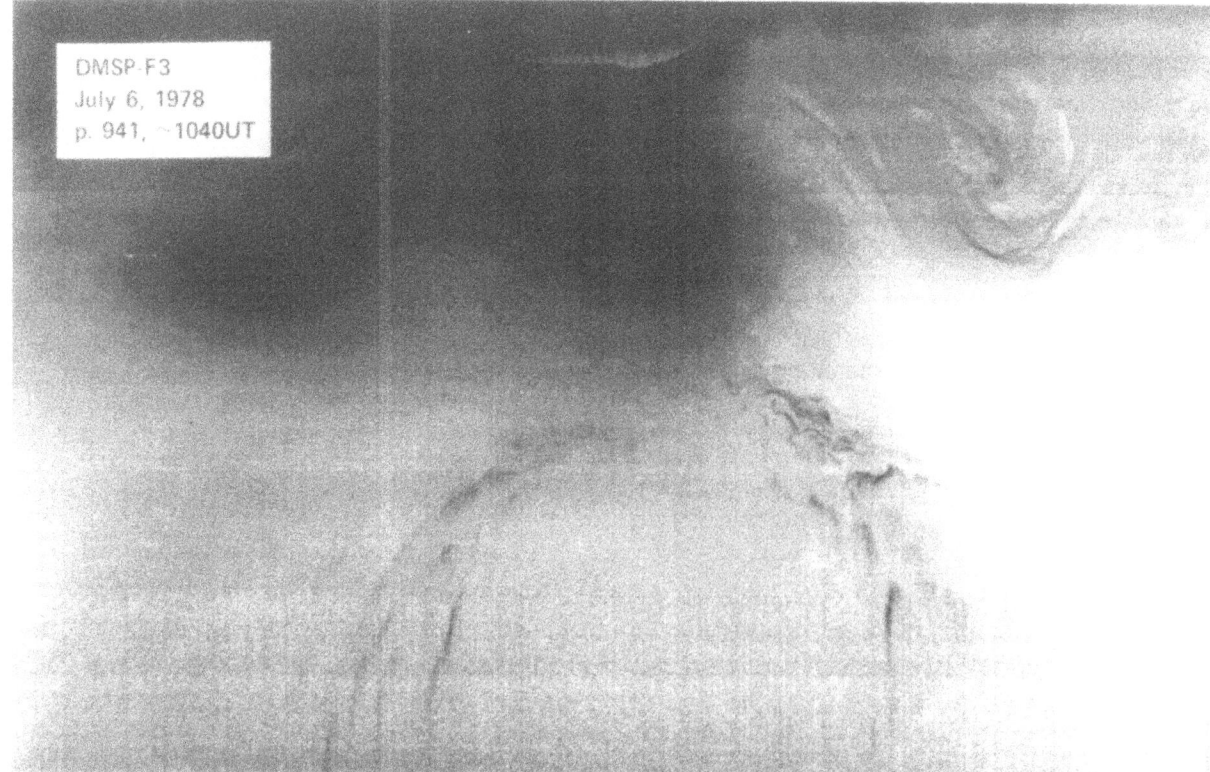

Figure 3—Dayside auroral display during medium quiet conditions (average IMF B_z slightly positive), as recorded by the DMSP F3 satellite over the southern polar region. (From *Meng and Lundin* [1986].)

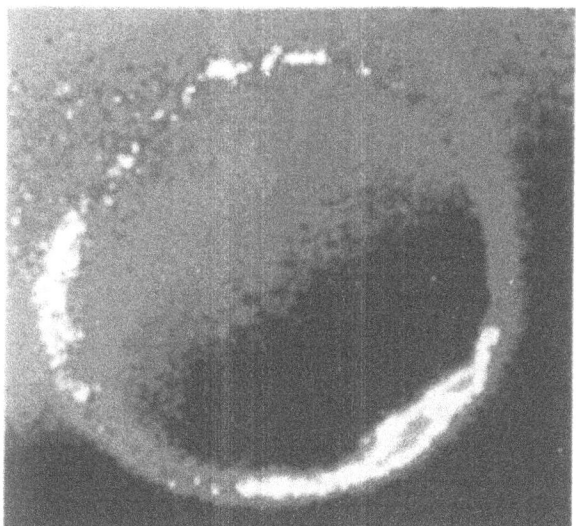

Figure 4—Viking UV image of the entire auroral oval during a time of relatively low magnetic disturbance. The image shows that the Feldstein oval is basically continuous but has two auroral activity centers, one near local midnight and one near local noon. (This figure also appears in color: Plate 15.)

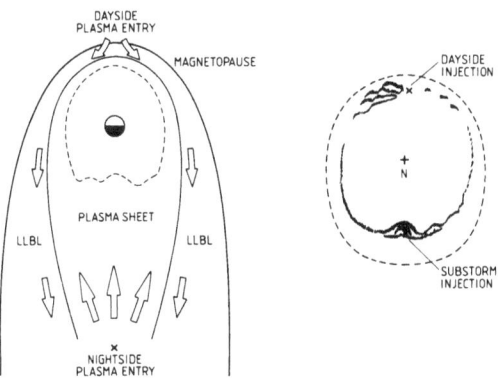

Figure 5—Diagrammatic figure showing the cause of the two activity centers along the auroral oval, the dayside and nightside plasma injection into the cusp/cleft and plasma sheet, respectively.

ing place in the tail region (plasma sheet and plasma-sheet boundary layer) and one associated with processes taking place in the dayside magnetosphere (the polar cusp and cleft).

The transient release of magnetotail energy in a substorm is, among other substorm characteristics, associated with an injection of fresh hot plasma into the inner plasma sheet. The overall increased brightness of the aurora during substorms is a consequence of an enhanced deposition of hot plasma into the upper atmosphere. This increased deposition originates from both diffusive (increased energy/pitch-angle scattering) and dynamical processes within the plasma sheet and the plasma-sheet boundary layer. Dynamical processes associated with plasma injection and discrete aurora are generally believed to result from plasma pressure or inertia effects that connect to the upper ionosphere via field-aligned currents [*Rostoker and Boström*, 1976; *Sato*, 1982; *Stasiewicz*, 1985].

Similarly, the dayside aurora can be discussed from the point of view of plasma injection, i.e., a direct injection of solar wind/magnetosheath plasma into the dayside cusp/cleft, illustrated in Figure 5. The figure shows a cross-section of the magnetosphere in the ecliptic plane (left) and its linkage to the northern auroral oval (right).

Thus, we may consider dynamical processes within the magnetosphere as plasma injection into one dayside (cusp/cleft) and one nightside (plasma sheet) entry region. The nightside injection is more transient and requires proper boundary conditions (e.g., a south-ward-turning IMF) for the release of energy into substorms to take place. Conversely, the dayside injection is persistent and depends mainly on the dynamical properties of the solar wind. This reflects also the characteristics of the aurora along the auroral oval. For instance, the dayside region is characterized by contiguous discrete aurora along the midday oval, and the nightside oval is characterized by intermittent eruptions of discrete aurora. During strong substorms, the nightside plasma injection is expected to dominate, and the injected plasma may expand well into the dayside.

During quiet times, when significant plasma injection from the nightside is lacking, the dayside injection should play the major role for dynamical processes along the auroral oval. This is corroborated by the persistency of the cusp [*Meng*, 1981], the dayside field-aligned current (FAC) system [*Ijima and Potemra*, 1978], and the midday aurora [*Murphree et al.*, 1981; *Evans*, 1985]. Inertia effects by neutral winds in the upper atmosphere [*Wygant et al.*, 1983; *Lyons et al.*, 1986] may play a role shortly after substorms, but it is not clear how long this effect continues and if it is at all significant. It will be argued here that dayside injection is quite sufficient to explain the quiet-time discrete aurora.

2. POLAR REGION AURORA

The strong decrease of particle precipitation over the polar region during substorms (the widening of the polar cap) is a well-known feature that dates back to the 1970s [*Riedler and Borg*, 1972; *Hoffman and Burch*, 1973]. The polar-cap widening is usually interpreted as an increased storage of magnetic energy in the magnetotail and enhanced merging of geomagnetic field lines with solar-wind field lines. Thus, the polar region is mostly a void of particle precipitation and auroral

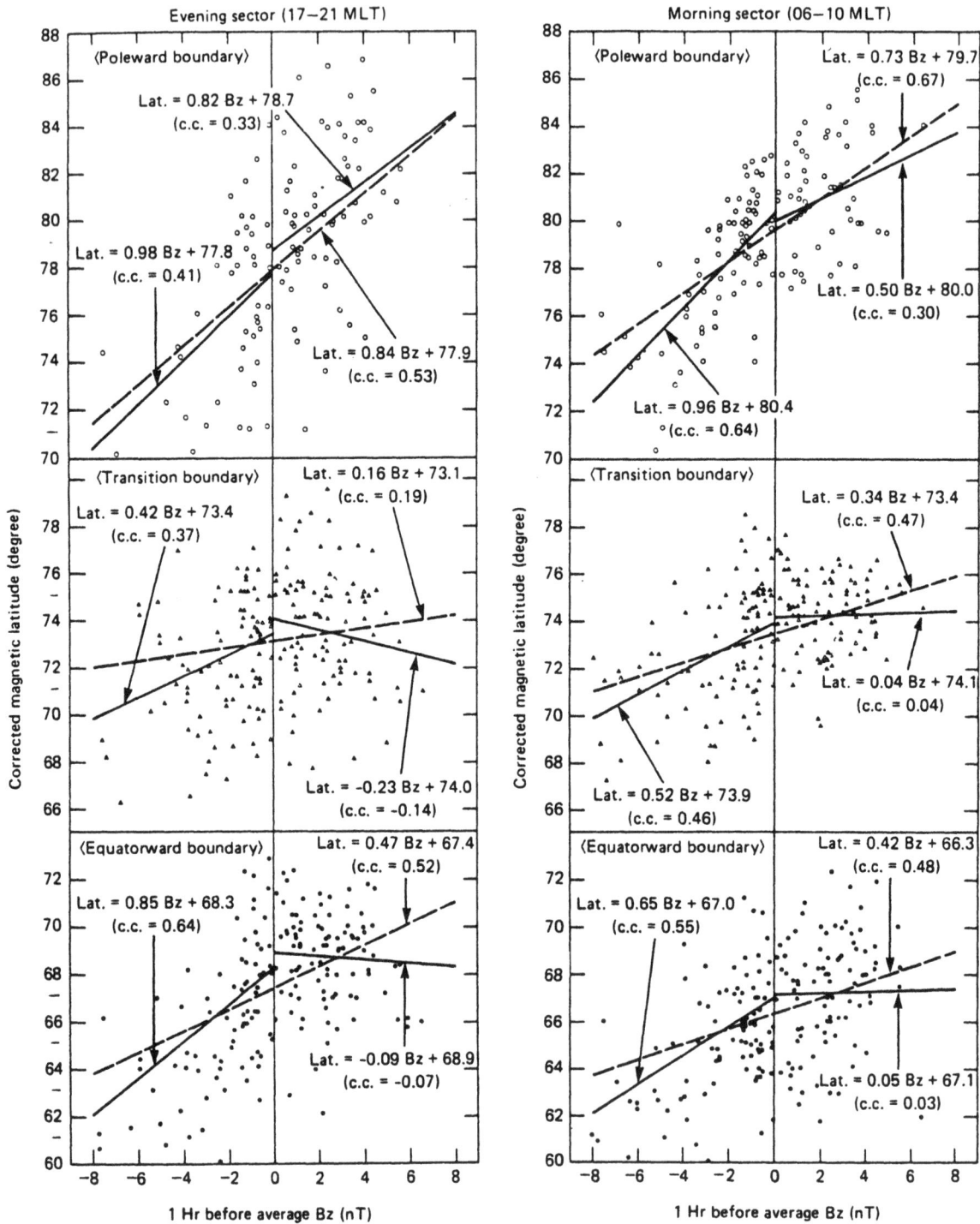

Figure 6—Scatterplot of the poleward and equatorward boundary of the electron precipitation region, illustrating the contraction of the polar cap versus IMF B_z. (After *Meng and Makita* [1986].)

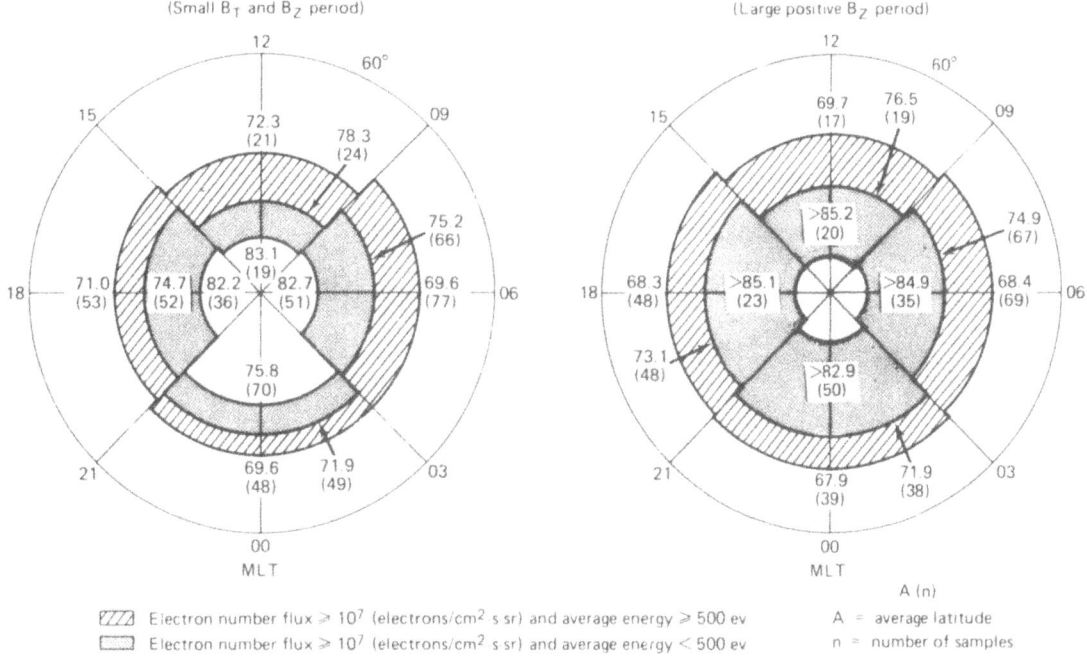

Figure 7—Average distribution of electron precipitation versus local time for small and large IMF B_z. The expanding boundary layer is illustrated by the < 500 eV precipitating electrons. (After *Makita et al.* [1988].)

activity during times of increased magnetic disturbances resulting from magnetotail activity. To the contrary, the polar region is rather active during times of low magnetic activity [*Meng*, 1981*a*; *Murphree et al.*, 1982; *Frank et al.*, 1982]. Although the intensity of this precipitation is less than that over the oval during disturbed times, the quiet-time polar-region precipitation is still considerably more intense than, for instance, the disturbed-time polar rain. This has led to speculations on the origin of the plasma responsible for the particle precipitation over the polar cap area.

The polar cap is at times characterized by a rain of very low particle fluxes, believed to originate from backstreaming solar-wind plasma on geomagnetic field lines connecting to the solar wind. An enhanced earthward flow of plasma over the polar region may thus either be due to a process with enhanced backstreaming in the tail lobe or simply a topological change permitting closed magnetic field lines to expand to higher latitudes. Recent measurements [*Frank et al.*, 1986; *Eliasson et al.*, 1987] indicate that the latter is a more likely explanation. Two different hypotheses for such an expansion of closed field lines have emerged, the plasma sheet boundary layer expansion model [*Meng*, 1981*a*] and the tail-lobe bifurcation model [*Frank et al.*, 1982, 1987; *Bythrow et al.*, 1985].

Statistical studies of the polar region performed by *Makita et al.* [1983] have shown that the polar cap size is strongly dependent on the IMF B_z component, the polar cap size being inversely proportional to the northward B_z component. Figure 6 shows a scatter plot of the poleward and equatorward boundary of the electron precipitation that displays the contraction of the polar cap versus IMF B_z [*Meng and Makita*, 1986].

The connection between the polar cap size (poleward boundary and the auroral oval) and the IMF B_z component is in favor of a coupling of the polar cap with the solar wind, in qualitative agreement with the merging hypothesis and the storage of magnetic energy in the magnetotail. Reconnection of magnetic field lines in the tail lobe has been inferred as an explanation (first by *Dungey* [1961]). However, as will be demonstrated later, only the polar cap can be considered to be on open field lines. Thus the argument of reconnection is only valid for the polar cap and not for the region of the expanded oval.

The statistical result by *Makita et al.* [1988] also suggests an expanded boundary layer during a northward IMF. Figure 7 displays the average distribution of electron precipitation versus local time for small and large IMF B_z [*Makita et al.*, 1988]. The expanding boundary layer is here represented by a widening of the zone of precipitating electrons with energies <500 eV for large positive B_z and the associated contraction of the polar cap.

ISEE 1 measurements [*Williams et al.*, 1985; *Mitchell et al.*, 1987] also indicate a widening of the LLBL for a northward IMF. Considering the topological connec-

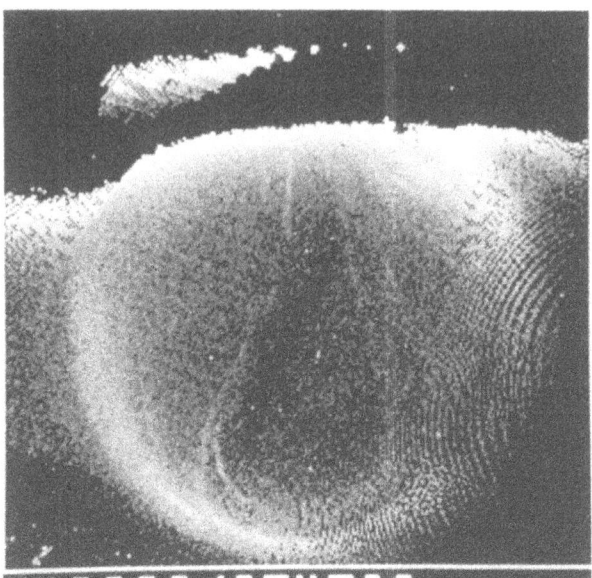

Figure 8—Isis 2 pass over the northern polar region during a period of extreme quiescence ($Kp = 0+$), illustrating the "teardrop" shape of the quiet time polar cap.

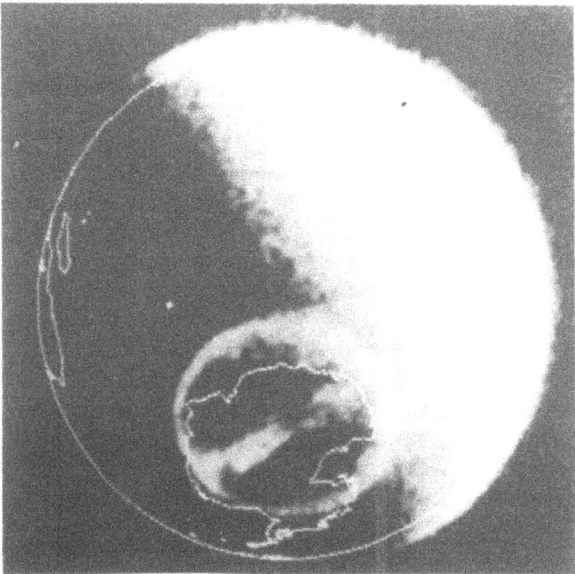

Figure 9—Example of theta aurora as observed from the Iowa imager on DE 1. (Courtesy, *L. A. Frank* [1988].) (This figure also appears in color: Plate 16.)

tion of the LLBL to the dayside-oval cleft [*Eastman et al.*, 1976; *Vasyliunus*, 1979; *Hultqvist and Lundin*, 1986], a widening of the LLBL must be associated with an expansion of the dayside cleft. Thus at least the dayside poleward boundary connected to the LLBL is expected to expand during times of persistently northward-directed IMF.

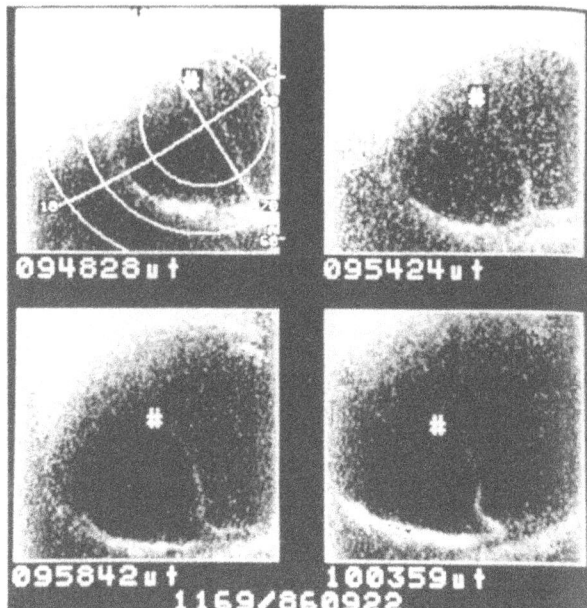

Figure 10—Viking images of a possible theta aurora over the northern polar region. The time resolution of the images in the sequence is about 1 min. (This figure also appears in color: Plate 17.)

The statistical result, for example, by *Makita et al.* [1988] gives the average distribution of the polar cap as determined from the electron precipitation (Figure 7). The shape of the instantaneous polar cap is, in general, more complex. Figure 8 shows an ISIS 2 pass over the northern polar region during a period of extreme quiescence ($Kp = 0+$). The polar cap can be recognized here by a teardrop-shaped void of auroral emissions near the pole [*Murphree et al.*, 1982]. More recent auroral images from the Polar BEAR satellite display the same teardrop-shaped quiet-time instantaneous polar cap [*Meng*, private communication]. Such a shape of the oval during extreme quiescence now appears to be well recognized [see also *Lassen et al.*, 1988; *Hones et al.* 1989]. This polar-cap topology is expected from the quiet-time expansion of the boundary layer (discussed in the next section).

The quiet-time electron precipitation, analyzed by *Meng* [1981b], implied a very structured polar region. Also, in fact, the absence of a polar cap can be inferred from measurements [*Makita et al.*, 1988]. The interpretation by *Meng* [1981a] of an expanded plasma sheet may thus be an oversimplification, considering the above arguments of the LLBL expansion.

Frank et al. [1982] found from their DE 1 images a type of polar aurora they designated the "theta aurora." They visualized theta aurora, a typical northward IMF phenomena, as a tail-lobe bifurcation expanding across the polar cap and connecting to the

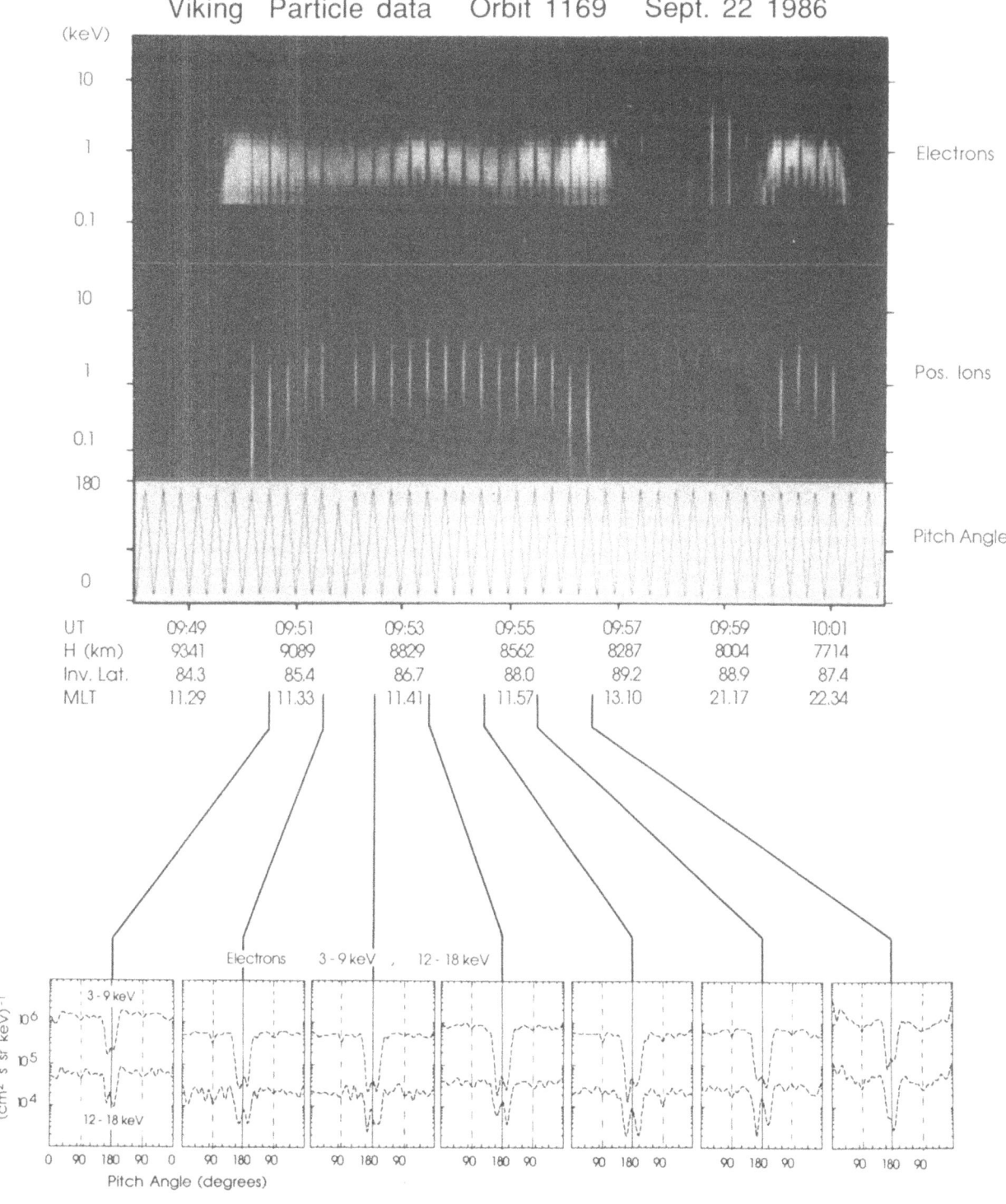

Figure 11—Viking ion and electron time energy spectrogram from the pass over the theta aurora displayed in Figure 10. The bottom panel shows the pitch angle distribution of energetic electrons (3–9 keV and 12–18 keV, respectively) with quasi-trapped "butterfly" signature at the edges of the polar arc. (This figure also appears in color: Plate 18.)

dayside oval near noon. An example of a theta aurora is shown in Figure 9 (courtesy, *Frank* [1988]). *Bythrow et al.* [1985] proposed that, as a result of the tail-lobe bifurcation, plasma from the plasma sheet may convect to the magnetopause as demonstrated in Figure 9. Simultaneous hot plasma measurements [*Frank et al.*, 1986] and ion composition measurements in a theta aurora [*Peterson et al.*, 1984] implied that the plasma within the theta aurora was of plasma sheet origin. However, it was not determined unambiguously whether the polar arc really was due to a bifurcation process or was simply isolated structures within an expanded oval as suggested by *Meng* [1981*a*].

The auroral imager aboard Viking made it possible to study aurora over the northern polar region with better temporal and spatial resolution than was done by the DE 1 satellite. An example of a theta aurora pictured by Viking with a 1-min time resolution is displayed in Figure 10. The in situ measurements (particles and fields) confirmed that Viking traversed the field lines of the polar arc between about 0950 and 1000 UT. Measurements of electrons and ions (Figure 11) show that the arc is associated with a field-aligned (upward) accelerating voltage of about 1 kV. Most of the acceleration takes place below Viking (below about 9000-km altitude), as manifested by upflowing ion beams of ~1 keV and a widened electron loss-cone. Howev-

er, the electron data also show a void of 0° pitch-angle particles below the peak energy characteristic of an acceleration of a few hundred volts above Viking. Thus, Viking traversed this polar arc within the acceleration region. One significant feature is the presence of quasi-stable trapped electrons (butterfly distribution), indicating that the arc lies on closed field lines (Figure 11). Other particle characteristics, like the period of upgoing electron beams and an isotropic plasma-sheet-like ion background around 0957–1001 UT also suggest closed field lines. In fact, there appears to be no controversy that polar arcs (theta aurora) lie on closed field lines [see also *Menietti and Burch*, 1988]. The discussion rather concerns the adjacent regions of the polar cap. However, the overall impression from most Viking polar crossings is that polar arcs lie on an expanded oval [*Eliasson et al.*, 1987] rather than being a manifestation of a bifurcated tail lobe as proposed by *Frank et al.* [1982] and *Bythrow et al.* [1985]. Ion and electron measurements both indicate that the most intense polar arc marks the poleward boundary of an expanded oval on closed field lines (see also Figure 16 and related text). Frequently, the entire expanded oval contains hot plasma with characteristics that one expects to find in the PSBL or in the LLBL. However, there are also passes when the fluxes are so low that a proper identification is impossible with the Viking

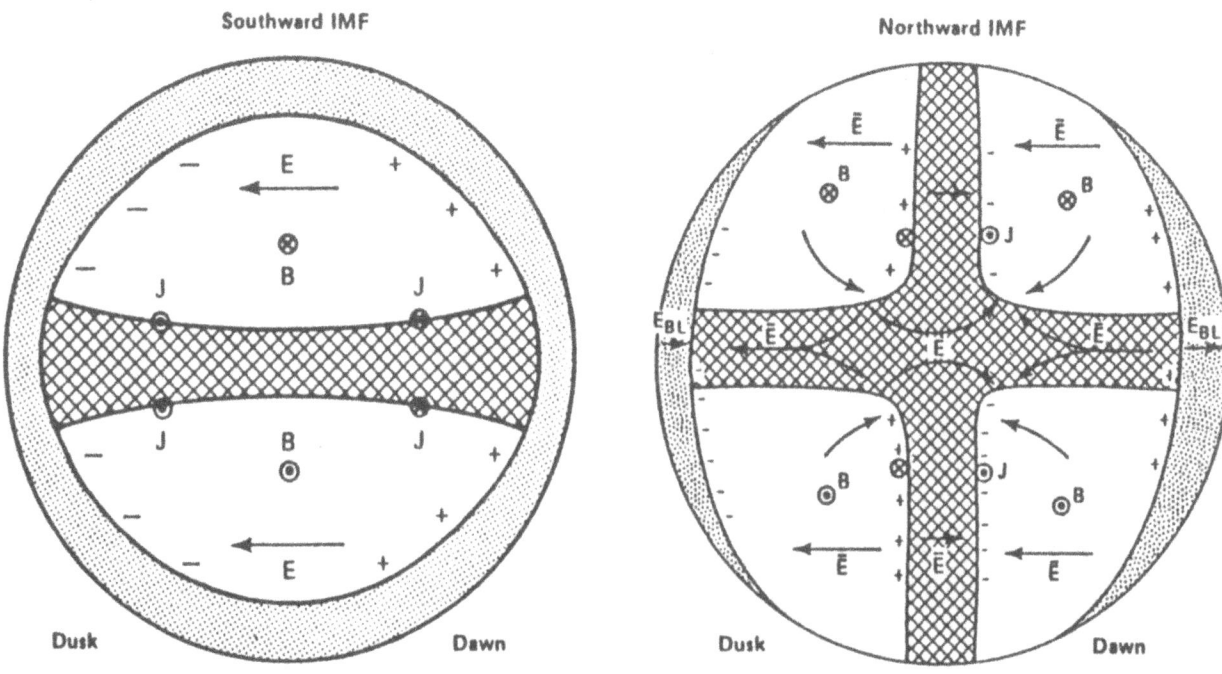

Figure 12—An illustration of the plasma sheet bifurcation expected to be responsible for the observations of theta aurora, according to *Bythrow et al.* [1985]. (After *Lui* [1986].)

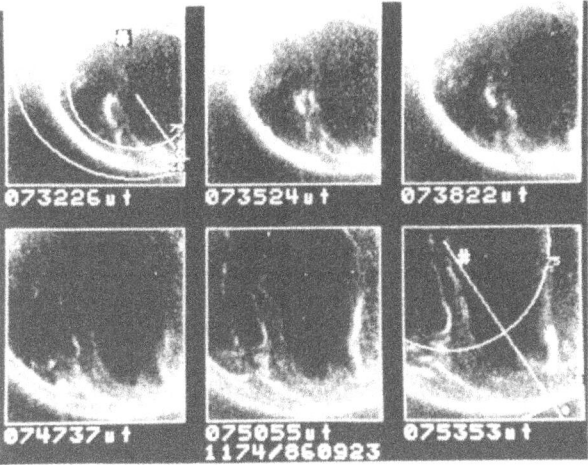

Figure 13—An example of a theta aurora formation as observed by the Viking imager, i.e., a bifurcation of the tail lobe originating in the geotail. The formation is supposed to originate from the expanding auroral "loop" on the nightside oval. (This figure also appears in color: Plate 19.)

particle experiment. In a number of these cases one may infer from the imager experiment that this was likely related to the occasional lack of particle precipitation along the field lines. Other cases simply remain ambiguous, but neither do they prove a lobe encounter. For instance, it should be kept in mind that the boundary layer may, in addition to constituting dense regions of injected plasma, also constitute a "halo" of a low-density plasma [*Sckopke et al.*, 1981; *Lundin et al.*, 1982]. Thus, what could be attributed to polar-cap traversals may also represent magnetospheric

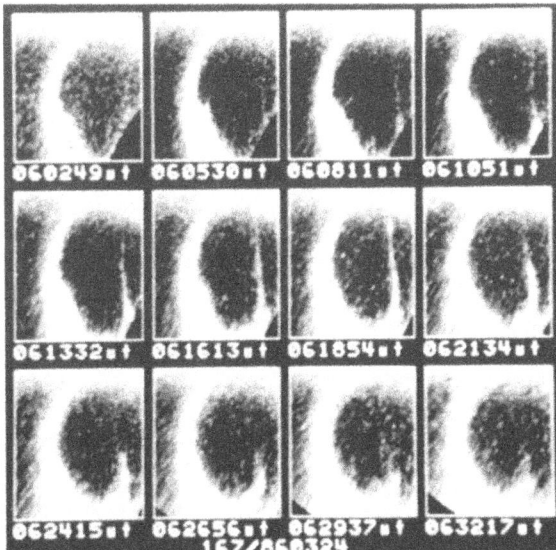

Figure 14—A time sequence of Viking auroral images of an apparently single polar arc connecting to the dayside oval with occasional "flarings" along the arc. (This figure also appears in color: Plate 20.)

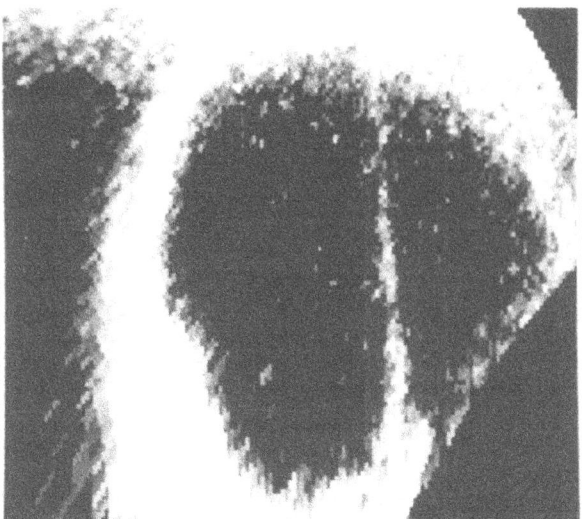

Figure 15—Detailed image of a polar arc from the same pass as that in Figure 14. The image shows that the polar arc is associated with isolated intensifications forming a "hook" at both ends in the morning oval. (This figure also appears in color: Plate 21.)

regions with relatively low particle precipitation. It is then of particular interest to understand the extension and interface of the LLBL and PSBL to the quiet-time auroral oval.

The morphology of the polar cap and the connecting points on the dayside and nightside oval of a polar arc give some clues to the interface problem. MHD simulations by *Ogino et al.* [1985] (for example) appear to be able to simulate narrow extensions of the plasma sheet into the tail lobes such as those assumed in the bifurcation hypothesis. The plasma sheet extensions are also sensitive to the IMF B_y so that the extension shifts toward dawn for $B_y < 0$ and toward dusk for $B_y > 0$. *Reiff and Burch* [1985] also discussed the polar cap from the point of view of anti-parallel merging and concluded that, during a northward IMF, similar characteristics would occur to those found in the simulations by *Ogino et al.* [1985].

The existence of a polar cap implies open field lines. Thus, a bifurcation of the tail lobe or plasma sheet, on closed field lines, suggests a split of the lobe into two parts, i.e., formation of two auroral ovals. The bifurcation of the plasma sheet with the associated field-aligned currents, as visualized by *Bythrow et al.* [1985], is depicted in Figure 12 [after *Lui*, 1986].

An example of a possible bifurcation formation as observed from Viking is shown in Figure 13 (orbit 1174). Notice that despite the apparent expansion of an auroral "loop" from the nightside to the dayside,

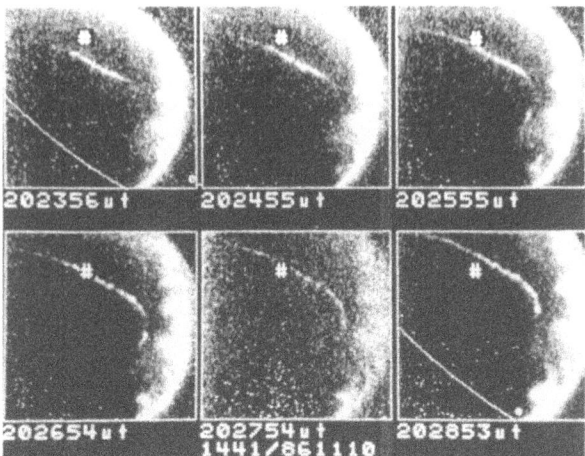

Figure 16—Series of images (≈ 1-min time resolution) displaying a "hook-shaped" arc connecting to the nightside oval. The images demonstrate the dynamics of isolated thin polar arcs with intensifications occurring within 1 min. (This figure also appears in color: Plate 22.)

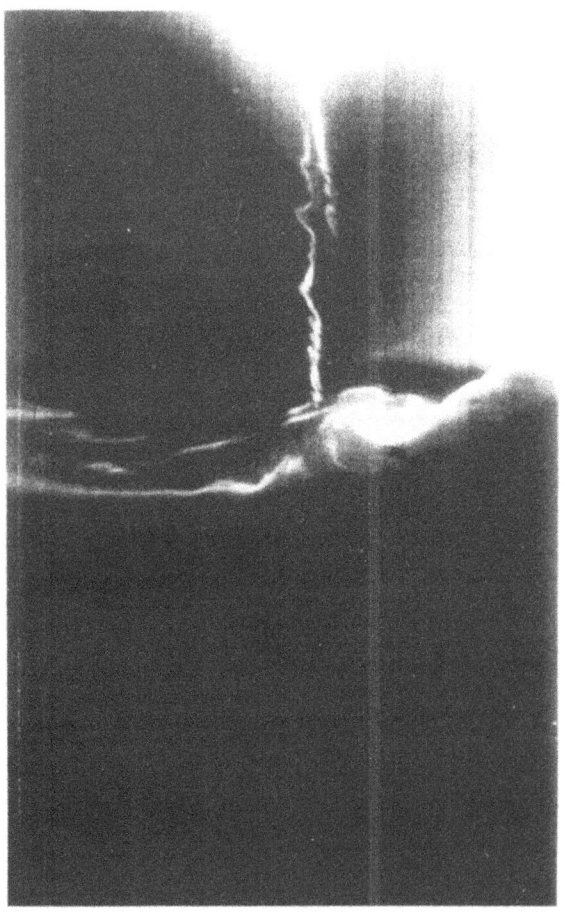

Figure 17—The fine structure of a polar arc as observed by the DMSP imager (≈ 1 km spatial resolution). The figure demonstrates that a polar arc may, in fact, consist of several "isolated" arc segments. (Courtesy, *C.-I. Meng* [1988].)

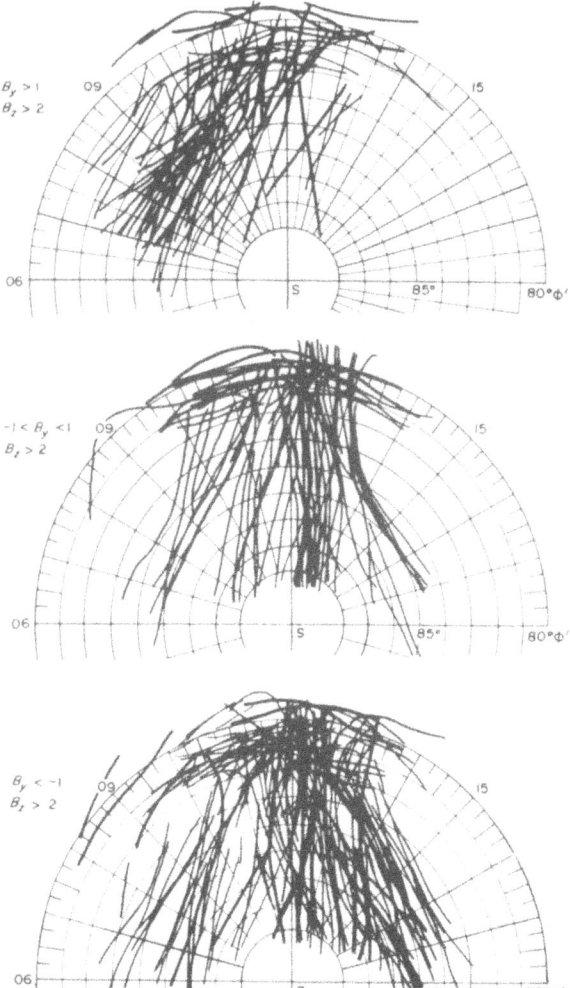

Figure 18—The IMF B_y dependence of polar arcs as observed near the south pole from the Vostok station (after *Gusev and Troschichev* [1986]).

it is not obvious that two polar caps developed. On the contrary, only one polar cap is apparent in the last two panels (0750:55 and 0753:53 UT). Thus, this example more likely represents a (dynamical) expansion of the oval in both the morning and evening sector, with vortices developing adjacent to the lobe.

The next Viking example (Figure 14) shows the dynamics of a polar arc connecting to the dayside, as also discussed by *Murphree et al.* [1988a]. The time sequence in Figure 14 demonstrates how a theta aurora appears to expand and decay along the noon-midnight meridian, from midnight to noon. Notice, however, that the flaring is along a preexisting faint arc, i.e., it may be interpreted as an expansion of a process rather than as a bifurcation of plasma over the polar cap.

A detailed picture of the connecting points is shown in Figure 15. This image shows that the polar arc is associated with a poleward (of the main oval) isolated intensification of the aurora forming a "hook" into the morning oval. Such dayside features have been shown to be dependent on the orientation of the IMF [*Murphree et al.*, 1988*b*].

Similar hooks of the polar arc are also observed on the nightside. Figure 16 (orbit 1441) shows a sequence of auroral images taken at 1-min intervals illustrating this. These Viking images also demonstrate the dynamics of isolated thin polar arcs with intensifications occurring more rapidly than the 1-min time resolution. Again the particle data from this crossing (not shown here) imply an expanded oval in the morning sector (upper part). Viking crossed the field lines of the arc around 2028 UT (fifth panel).

The isolated intensification, the hook, at both ends of the polar arc suggests that the arc either is part of an isolated structure in the magnetosphere or that it marks the boundary of a convection discontinuity (e.g., the Harang discontinuity on the nightside) at the poleward boundary of the oval. The observation of locally mirroring electrons in the polar arc, implying closed field lines there, eliminates the connection of the arc to the high-latitude boundary layer (the plasma mantle). Thus, if it is an isolated structure, it should connect to the LLBL or the PSBL, marking the edge of the polar cap (open field lines).

The fine structure of a polar arc as observed by the DMSP (~ 1 km spatial distribution) imager is shown in Figure 17. Notice from this image that what on Viking and DE 1 appear as a single continuous arc may, in fact, comprise several isolated arc structures. Even if this reflects the complexity of the discrete arc formation, it may not be significant for the overall boundary-layer morphology.

The spatial distributions of polar arcs with IMF have been studied by *Lassen and Danielsen* [1978] and *Gusev and Troschichev* [1986]. Figure 18 demonstrates the IMF B_y dependence of Sun-aligned arcs observed near the south pole by *Gusev and Troschichev* [1986]. Notice for instance the preference of finding arcs in the evening sector for $B_y > 0$ over the south pole. This should be compared with the predicted preference of observing arcs in the morning sector for $B_y > 0$ over the north pole, in conformity with the MHD simulations by *Ogino et al.* [1985]. *Gusev and Troschichev* [1986] also found that hook-shaped arcs only occurred for an IMF $B_x > 0$ over the south pole. Thus, both the IMF B_x and B_y dependence apparently favor antiparallel merging in the tail lobe. However, they also noted that polar arcs were observed inside the convection cells as well as on the "wrong" side with respect to the IMF B_y dependence. Considering what has already been noted about the rooting of polar arcs on

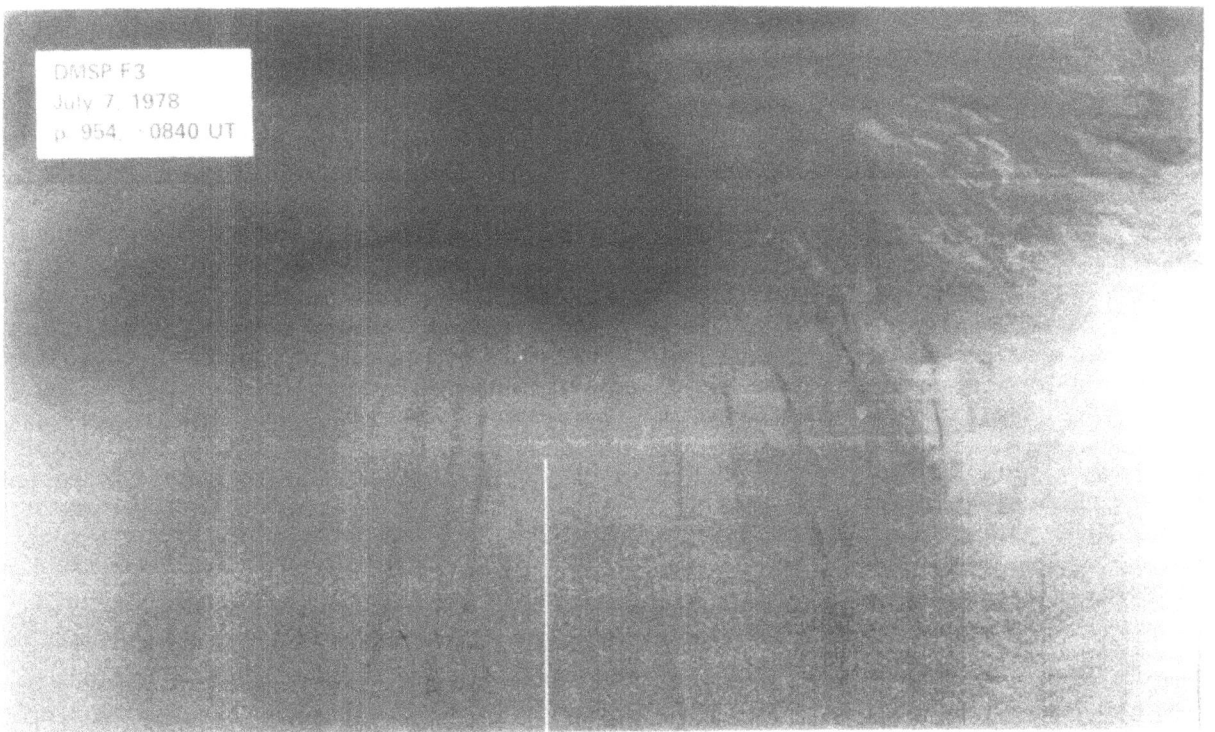

Figure 19—DMSP auroral image of multiple polar arcs over the south pole during times of extreme quiescence. (After *Meng and Lundin* [1986].)

closed field lines, one may therefore conclude that anti-parallel merging can only be relevant for polar arcs located immediately adjacent to the tail lobe.

The presence of multiple auroral structures on at least one side of the intermediate region between the northernmost polar arc and the "classical" Feldstein oval is apparent in many of the Viking images displayed here as well. A polar cap is usually identified on either the dawnside or the duskside of the polar region. Limitations due to the intensity threshold sometimes leave the auroral activity undetectable over the polar region, but this rather emphasizes the threshold problem than contradicts the boundary-layer expansion hypothesis. Particle data help to identify weak particle precipitation in the intermediate region. The frequently observed multiplicity of weaker precipitation (polar arc) structures interrupted by gaps of particle precipitation may in principle be interpreted as multiple polar cap cells. However, a more likely explanation is that small-scale convection cells (vortices) occur within an expanded boundary layer. The frequent occurrence of trapped-electron distribution within particle precipitation structures over the polar region indicates closed field lines. Thus, there are strong reasons to believe that the intermediate region is an expanded boundary layer (LLBL or PSBL) with localized dynamical structures associated with field-aligned currents and particle precipitation. In this respect, polar arcs are not different from other discrete arc structures along the auroral oval that also occur on closed field lines and are associated with field-aligned currents and particle acceleration.

Multiple polar arcs have a tendency to converge toward the dayside cusp, a fact that implies LLBL origin [*Lundin and Evans*, 1985]. However, as Figure 13 shows, polar arcs may extend from the nightside oval. Thus, it is at times difficult to distinguish between a dayside and a nightside origin of polar arcs. During the times of extreme quiescence (IMF $B_z \approx 0$, see Figure 2) when the polar cap forms a teardrop (Figure 8), the LLBL connection is more apparent. Figure 19 shows a DMSP image taken during a quiet time (*Kp* = 0+) pass over the polar region containing many weak polar arc structures. This image thus represents a closeup view of the dayside portion of the teardrop with multiple arc features resulting from a filamentation and expansion of the boundary layer.

A good terrestrial magnetic field model is important for understanding how polar arcs connect to the outer magnetosphere. It is also essential to understand the properties of the boundary layer during quiet times and how the boundary layers (LLBL and PSBL) connect along the oval. We next discuss the linkage of the boundary layer to the quiet time auroral oval.

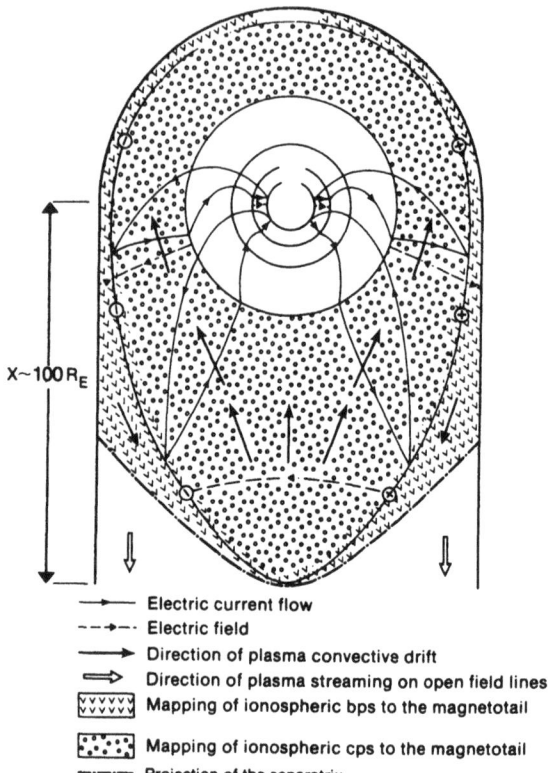

Figure 20—Magnetospheric linkage of the auroral oval in a closed magnetic field model. (After *Rostoker and Eastman* [1987].)

3. BOUNDARY LAYER LINKAGE TO THE OVAL

In a study of aurora in the midday auroral oval, *Meng and Lundin* [1986] found that the auroras they designated type 1 and type 2 were found when the IMF B_z was positive. The type 1 polar aurora was observed during times of very low magnetic disturbances. They also discussed the linkage of discrete auroral structures with polarization features resulting from freshly injected solar-wind plasma in the LLBL, on the basis of suggestions by *Lundin and Evans* [1985]. In this model the cusp proper characterizes a steady pressure center from which transient plasma injections result in polarization on closed field lines and the occurrence of discrete aurora. As a result of the draping of field lines in the boundary layer [*Eastman et al.*, 1976; *Vasyliunas*, 1979], the entire LLBL is projected down to the ionosphere in a relatively limited crescent-shaped sector near local noon—the cleft [*Heikkila and Winningham*, 1971]. During quiet times the cleft/LLBL may grow appreciably in size. Although the particle characteristics on field lines connecting to the LLBL have some similarities to those found in the cusp prop-

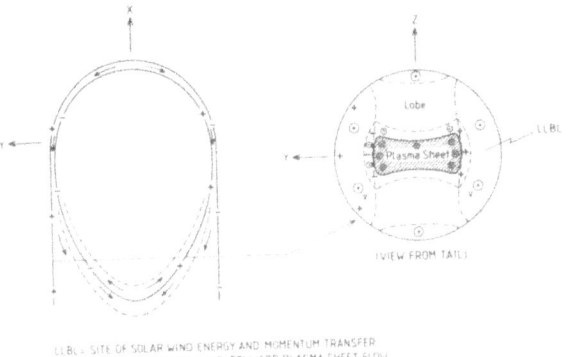

LLBL = SITE OF SOLAR WIND ENERGY AND MOMENTUM TRANSFER
⟶ POLARIZATION ⟶ EARTHWARD PLASMA SHEET FLOW

Figure 21—Modified picture of the *Rostoker and Eastman* [1987] magnetospheric linkage of the auroral oval displaying an enhanced plasma flow in the plasma sheet/PSBL induced by an increased polarization of the LLBL. The enhanced plasma sheet/PSBL flow may lead to locally strong dynamo effects and the formation of discrete arcs.

er, the cleft is characterized by significantly more particle energization [*Lundin et al.*, 1988].

Because the magnetic field draping on closed field lines is induced by currents [*Hones*, 1983; *Lundin and Dubinin*, 1985], lower draping currents expected during quiet times would imply an expansion of the LLBL footprint in the ionosphere. Consequently, a widening of the LLBL [*Williams et al.*, 1985; *Mitchell et al.*, 1987] and less draping in the LLBL both result in an expansion of the footprint in the ionosphere—the cleft. This expansion should progress in the poleward direction following the overall expansion of the magnetosphere during a northward B_z, and in the azimuth

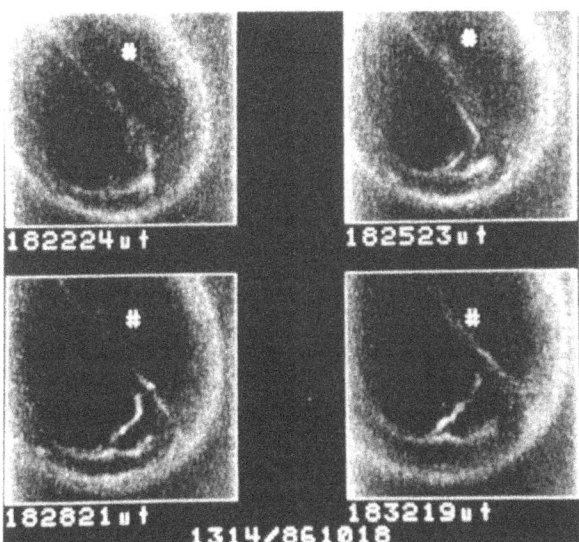

Figure 22—Viking UV image of a polar arc for a field line tracing test using the Tsyganenko magnetic field model (Figure 23). (This figure also appears in color: Plate 23.)

VIKING PROJECTIONS:

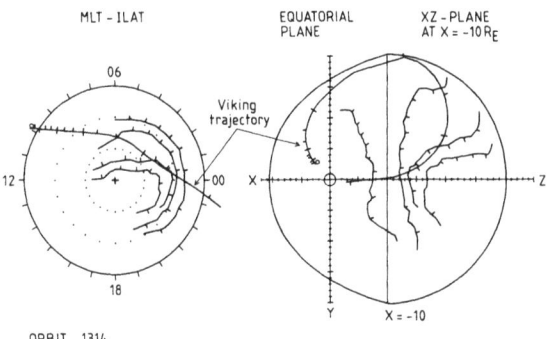

ORBIT. 1314
DATE: 86-10-18

Figure 23—Projection of the Viking UV image at 100-km altitude above the Earth in polar coordinates (left) and its magnetospheric projection using the *Tsyganenko* [1987] model (right). The projection into the distant magnetosphere is in the equatorial plane (XY) and at a distance of $-10R_E$ as a cross-section of the tail (YZ). (Courtesy, *K. Stasiewicz* [1988].)

(local time) direction as a result of decreased draping.

Rostoker and Eastman [1987] discussed the magnetospheric linkage of the auroral oval in a closed magnetic field model. It is interesting to extend their hypothesis in the frame of an expanded plasma sheet, PSBL, and LLBL. The convection boundary marks a closed loop, according to *Rostoker and Eastman* [1987], as shown in Figure 20. Notice that the loop is negatively charged at dusk and positively charged at dawn, as predicted in all models where the boundary layer plasma flow is the main generator of high-latitude currents and electric fields [*Heikkila*, 1984; *Lundin and Evans*, 1985].

A somewhat modified version of the picture by Rostoker and Eastman is illustrated in Figure 21. The YZ cross section of the tail shows the "stagnant" plasma sheet, the LLBL with a tailward plasma flow, and the PSBL with earthward plasma flow driven by the LLBL polarization. Notice that the LLBL is the dynamo for the magnetospheric convection in this model. An enhanced earthward plasma flow in the PSBL is driven by an increased polarization in the LLBL.

An interesting aspect of the Rostoker and Eastman concept [also *Heikkila*, 1984] is that it allows for a continuous auroral oval connecting with the LLBL, where the LLBL is the site of solar-wind energy and momentum transfer. Furthermore, the earthward plasma injection in the PSBL, depicted in Figure 21, may equally well result from the polarization in the LLBL as from the polarization on open field lines over the tail lobe, as suggested in the merging model.

Tsyganenko et al. [1982, 1987] have provided what is perhaps the best model hitherto of the distant magnetic field of the Earth. *Stasiewicz* [private communication] has attempted to trace the magnetic field lines

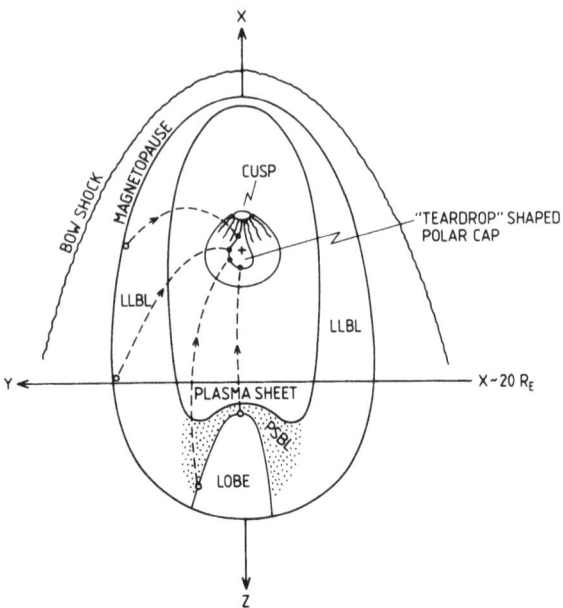

Figure 24—Diagrammatic representation of the expanded boundary layers (LLBL and PSBL) projected into the polar region during times of extreme magnetic quiescence (IMF $B_z \approx 0$) when the polar cap is expected to have a teardrop form. The magnetic connection to the polar cap boundary is illustrated by the dashed lines.

of optical and particle structures of the quiet-time aurora into the tail region, using the Tsyganenko model. An example of such a field-line tracing from a polar arc crossing by Viking is shown in Figure 22 (Viking image) and Figure 23 (model projection). The optical aurora seen by the Viking UV imaging camera and the satellite trajectory are projected into the distant magnetosphere as intersections with the equatorial plane or with the tail cut at a fixed distance from the Earth. The left-hand side in Figure 23 shows projections at 100-km altitude above the Earth in polar coordinates (magnetic local time and invariant latitude) and the right-hand side contains magnetospheric projections. The generation of polar arcs can also be described as a current-generator process driven by pressure gradients perpendicular to magnetic field gradients in the magnetotail.

It is interesting to compare the quiet-time oval topology and its magnetic projection to the outer magnetosphere with the position of the lobe/polar cap in Figure 23. The particle data as well as the magnetic field projection model suggest that the polar cap during this orbit was contracted and shifted toward dusk. The corresponding tail lobe has also contracted and shifted toward dusk. Under the assumption that the polar arc is connected to the PSBL, we may recognize from Figure 23 that the PSBL projection is only present at the nightside part of the oval (end of projection

point marked by circles). Afterward, the projection bends down into the LLBL and eventually reaches the dayside portion of the LLBL. This visualizes that the PSBL connects to the LLBL for field lines close to the magnetic pole. Thus, the polar arc can be interpreted as the footprint of processes taking place in the PSBL in one part and in the LLBL in the other, with a natural connection in the tail between the two boundary layers. The aurora thus forms a continuous boundary as visualized by *Rostoker and Eastman* [1987].

The contracted polar cap and its teardrop form during times of extreme quiescence (IMF $B_z \approx 0$) can now be understood from the expansion of the quiet-time LLBL and its connection to the PSBL, as illustrated in Figure 24, using a projection similar to that of Figure 23, but now without proper scaling. The field-line tracing of the anticipated polar-cap equatorward boundary is illustrated by the dashed curves. Notice that the nightside poleward boundary of the oval is connected to the PSBL while the dayside boundary is shaped by the expanded LLBL. The interconnection between the poleward boundary of the PSBL and the LLBL is expected to take place in the deep tail, the magnetic projection along the oval lying close to the magnetic pole near the dawn-dusk meridian.

Relatively few in situ measurements have been made at high latitudes in the distant magnetosphere that could be used for tests of either the expanded boundary layer hypothesis or the tail-lobe bifurcation hypothesis. Using data from ISEE 1/2, *Huang et al.* [1987] concluded that filamentary plasma structures in the magnetotail lobes are supportive of bifurcation processes of the magnetotail lobes. However, the ISEE 1/2 spacecraft mainly sample the central part of the near-Earth geotail ($X_{GSM} < 22R_E$, $Y_{GSM} < 15R_E$, $Z_{GSM} < 12R_E$). To identify the full extension of polar arcs to the dayside requires measurements at higher geocentric solar magnetospheric (GSM) latitudes than the ISEE spacecraft may achieve. The HEOS 2 and Prognoz 7/8 spacecraft orbits, sampling the high-latitude/high-altitude region of the magnetosphere, are more suitable for this purpose. No systematic study addressing polar arc formation has yet been published on the basis of HEOS 2 and Prognoz 7/8. *Popielawska et al.* [1988] recently reported on several tail-lobe filaments observed from Prognoz 8 data. However, there is one type of observation that may be relevant for the topic, i.e., the observation of stagnant plasma on what appears to be closed field lines in the outer magnetosphere at very high latitudes. *Sckopke et al.* [1976] from HEOS 2 measurements and *Lundin et al.* [1982] from Prognoz 7 measurements reported on several observations of plasma-sheet like populations at altitudes above $20R_E$ and GSM latitudes above $60°$. An example of such an observation is shown in Figure 25. Notice that

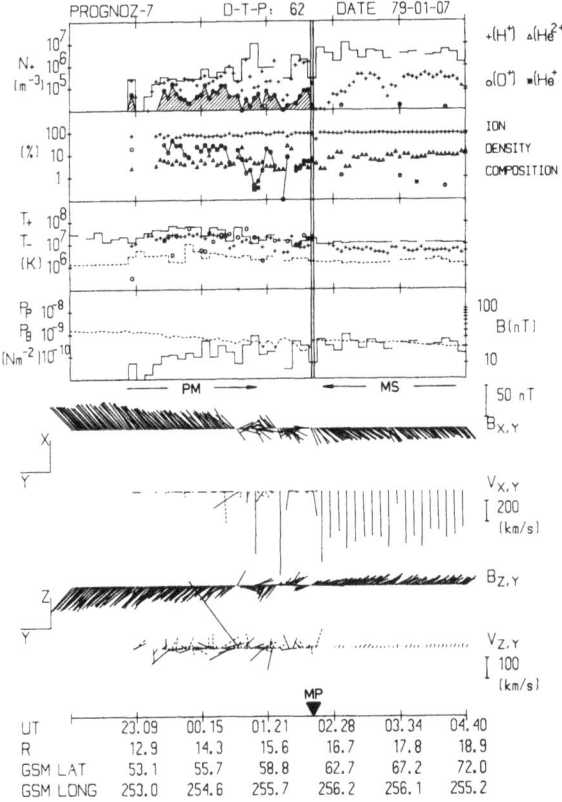

```
PROGNOZ-7        D-T-P:  62    DATE  79-01-07
```

Figure 25—Prognoz 7 outbound pass through the lobe, the plasma mantle (PM), the high-latitude magnetopause (MP), and into the magnetosheath (MS) illustrating a case of "stagnant plasma mantle." The stagnant plasma mantle corresponds most likely to plasma of plasma sheet or PSBL origin that has expanded into the high-latitude polar region. (After *Lundin et al.* [1982].)

during this outbound pass, the tail lobe was traversed before it entered the stagnant and hot plasma regime. The positive identification of a lobe was used by *Lundin et al.* [1982] as an identification of the plasma mantle. The plasma mantle, however, here lacked all the typical mantle characteristics (e.g., low-energy, low-temperature tailward-flowing ions of predominantly magnetosheath origin). In fact, the plasma characteristics are similar to those found in the plasma sheet/PSBL. Thus, cases of stagnant plasma-sheet like plasma are likely to be associated with closed field lines extending out to the high-latitude magnetopause. Whether the regions result from a tail-lobe bifurcation or are just localized plasma regions in an expanded boundary layer may be difficult to determine unambiguously from a single spacecraft. However, the fact that plasma regions remain stagnant for several hours inside the high-latitude magnetopause and may be distributed over a large GSM-latitude interval is rather in favor of an expanded boundary layer.

In a study (not yet published) based on 26 stagnant-mantle cases from HEOS 2 and Prognoz 7, it is found that about 76% of the cases show an approximately antiparallel orientation of the magnetic field inside the magnetopause compared with its outside. This is consistent with the picture of an IMF B_y dependence of open field lines occupying the magnetospheric sector with a parallel (in Y) magnetic field orientation [*Cowley*, 1981; *Akasofu and Roederer*, 1984]. Conversely, the high-latitude sector with preferentially antiparallel field lines on each side of the magnetopause is expected to be located on predominantly closed field lines.

4. CONCLUSIONS

The large-scale morphology of the quiet-time aurora and the associated magnetospheric configuration appears at present to be quite well documented. This is the result of extensive investigations performed on the basis of both ground-based observations and spaceborne auroral imagers from, for example, DMSP, DE 1, and Viking. Coincident (with auroral images) and complementary particle measurements have made it possible to determine the important boundary-layer connection of the quiet-time discrete aurora over the polar region. However, a number of uncertainties exist on how and where the polar region aurora links with the distant magnetosphere. A general consensus on the geotail-boundary layer/lobe connection must be reached first. In this report we have emphasized the expanded boundary-layer aspect of the polar region aurora, neglecting the possible influence of reconnection in the geotail.

The definition of quiet time may, to some extent, be misleading if it is interpreted on the basis of the classical energy transfer functions [*Perrault and Akasofu*, 1978]. Because of the large activity over the polar region for a northward B_z, additional terms must be added in the energy transfer function. The most quiet state of the magnetosphere apparently occurs when the IMF B_z is close to zero.

During periods of extreme quiescence ($B_z \approx 0$) the polar cap seems to contract into a teardrop shape as a result of the expansion of the boundary region between the polar cap and the average auroral oval. The aurora is very weak in this boundary region, as illustrated in Figure 19. On the dayside the boundary region represents the footprint of the quiet-time expanded LLBL, also referred to as the (expanded) cleft. The boundary region on the nightside is the (expanded) PSBL.

For a large positive IMF B_z, the aurora over the polar cap may be quite active indeed. Intense polar arcs may develop over the polar region from plasma filaments inside an expanded boundary layer (LLBL or

PSBL). Alternatively, a tail lobe bifurcation process may lead to the formation of theta aurora. Whatever the process is, it remains quite clear that discrete polar arcs are located on closed field lines, i.e., they resemble very much the discrete aurora along the auroral oval. To maintain such a magnetic field topology, it seems reasonable to assume that only one polar cap per hemisphere exists at a time, when only one well-defined polar arc is also present. This is corroborated by the frequently observed "filling in" of precipitating particles on one side of the main polar arc. Thus, a quiet-time expansion of the auroral oval boundary region is more likely than a tail-lobe bifurcation. (However, a tail-lobe bifurcation as defined from plasma expansion along preexisting closed field lines in the vicinity of the tail lobe may be considered identical to a boundary-layer process.)

A strong asymmetry of the polar-cap location may result from the direction of the IMF B_y. This agrees with both ground-based measurements of polar arcs as well as with satellite measurements of stagnant plasma at high latitudes. Further measurements from Viking, for example, are expected to shed additional light on the IMF B_y dependence.

The main conclusion from the boundary-layer dependent, quiet-time aurora is that the LLBL should play the dominant role for the generation of auroral activity. In this model, the LLBL is the major site for the quiet-time solar wind energy and momentum transfer into the magnetosphere. This is in contradiction to the reconnection model, which rather emphasizes the role of the plasma mantle instead. Clearly, the simultaneously strong activity in the polar region and a weakly developed plasma mantle during a northward B_z [*Sckopke et al.*, 1976] basically eliminate the role of the plasma mantle as a source region for polar arcs.

Future comparisons of auroral images taken over both the southern and northern hemisphere will provide conclusive information on both the expected B_y asymmetry and the tail-lobe-bifurcation/boundary-layer expansion hypothesis.

REFERENCES

Akasofu, S.-I., and J. R. Kan, "Dayside and Nightside Auroral Arc Systems," *Geophys. Res. Lett.*, 7, 753 (1980).
Akasofu S.-I., and M. Roederer, "Dependence of the Polar Cap Geometry on the IMF," *Planet. Space Sci.*, 32, 111 (1984).
Brautigam, D. H., M. S. Gussenhoven, and D. A. Hardy, "The Effects of IMF B_z and Solar Wind Speed on Precipitating Ion Morphology," Paper presented at the International Conference on Auroral Physics, Cambridge, England (1988).
Bythrow, P. F., R. A. Heelis, W. B. Hanson, R. A. Power, and R. Hoffman, "Observational Evidence for a Boundary Layer Source of the Dayside Region 1 Field-Aligned Currents," *J. Geophys. Res.*, 86, 5577 (1981).
Bythrow, P. F., W. J. Burke, T. A. Potemra, L. J. Zanetti, and A. T. Y. Lui, "Ionospheric Evidence for Irregular Reconnection and Turbulent Plasma Flows in the Magnetotail During Periods of Northward IMF," *J. Geophys. Res.*, 90, 5319 (1985).

Cowley, S. W. H., "Magnetospheric Assymetries Associated with the Y-Component with the IMF," *Planet. Space Sci.*, 29, 79 (1981).
Dungey, J. W., "Interplanetary Fields and the Auroral Zone," *Phys. Rev. Lett.*, 6, 47 (1961).
Eastman, T. E., E. W. Hones, Jr., S. J. Bame, and J. R. Asbridge, "The Magnetospheric Boundary Layer: Site of Plasma Momentum and Energy Transfer from the Magnetosheath into the Magnetosphere," *Geophys. Res. Lett.*, 3, 685 (1976).
Eliasson, L., R. Lundin, and J. S. Murphree, "Polar Cap Arcs Observed by the Viking Satellite," *Geophys. Res. Lett.*, 14, 451 (1987).
Evans, D. S., "The Characteristics of a Persistent Auroral Arc at High Latitude in the 1400 MLT Sector," in *The Polar Cusp*, A. Holtet and A. Egeland, eds., D. Reidel Pub. Co., Dordrecht, Holland (1985).
Frank, L. A., "Dynamics of the Near-Earth Magnetotail—Recent Observations," in *Modeling Magnetospheric Plasma*, Geophysical Monograph 44, T. E. Moore and J. H. Waite, Jr., eds., American Geophysical Union, Washington, DC, p. 261 (1987).
Frank, L. A., J. D. Craven, J. L. Burch, and J. D. Winningham, "Polar Views of the Earth's Aurora with Dynamics Explorer," *Geophys. Res. Lett.*, 9, 1001 (1982).
Frank, L. A., J. D. Craven, D. A. Gurnett, S. D. Shawhan, D. R. Weimer, J. L. Burch, J. D. Winningham, C. R. Chappell, J. H. Waite, R. A. Heelis, N. C. Maynard, M. Sugiura, W. K. Peterson, and E. G. Shelley, "The Theta Aurora," *J. Geophys. Res.*, 91, 3177 (1986).
Gusev, M. G., and O. A. Troschichev, "Hook-Shaped Arcs in the Dayside Polar Cap and Their Relation to the IMF," *Planet. Space Sci.*, 34, 489 (1986).
Heikkila, W. J., and J. D. Winningham, "Penetration of Magnetosheath Plasma to Low Altitudes Through the Dayside Magnetic Cusps," *J. Geophys. Res.*, 76, 883 (1971).
Heikkila, W. J., "Magnetospheric Topology of Fields and Currents," in *Magnetospheric Currents*, Geophysical Monograph 28, American Geophysical Union, Washington, DC, p. A208 (1984).
Hones, E. W., Jr., "Magnetic Structure of the Boundary Layer," *Space Sci. Rev.*, 34, 201 (1983).
Hones, E. W., Jr., J. D. Craven, L. A. Frank, D. S. Evans, and P. T. Newell, "The Horse-Collar Aurora: A Frequent Pattern of the Aurora in Quiet Times," *Geophys. Res. Lett.* (1989).
Hoffman, R. A., and J. L. Burch, "Electron Precipitation Patterns and Substorm Morphology," *J. Geophys. Res.*, 78, 2867 (1973).
Hultqvist, B., and R. Lundin, "Some Viking Results Related to Dayside Magnetosphere-Ionosphere Interactions," *Ann. Geophys.*, 5, 503 (1987).
Huang, C. Y., L. A. Frank, W. K. Peterson, D. J. Williams, W. Lennartsson, D. G. Mitchell, R. Elphic, and C. T. Russell, "Filamentary Structures in the Magnetotail Lobes," *J. Geophys. Res.*, 92, 2349 (1987).
Ijima, T., and T. A. Potemra, "Field-Aligned Currents in the Dayside Cusp Observed by Triad," *J. Geophys. Res.*, 81, 5971 (1976).
Ijima, T., and T. A. Potemra, "Large-Scale Characteristics of Field-Aligned Currents Associated with Substorms," *J. Geophys. Res.*, 83, 599 (1978).
Lassen, K., and C. Danielsen, "Quiet Time Patterns for Auroral Arcs for Different Directions of the Interplanetary Magnetic Field in the Y-Z Plane," *J. Geophys. Res.*, 83, 5277 (1978).
Lassen, K., C. Danielsen, and C.-I. Meng, "Quiet-Time Average Auroral Configuration," *Planet. Space Sci.*, 36, 791 (1988).
Lui, A. T. Y., "Solar Wind Influence on Magnetotail Configuration and Dynamics," in *Solar Wind-Magnetosphere Coupling*, Y. Kamide and J. Slavin, eds., Terra Scientific Pub. Co., Tokyo, p. 547 (1986).
Lundin, R., B. Hultqvist, N. Pissarenko, and A. Zakharov, "The Plasma Mantle: Composition and Other Characteristics Observed by Means of the Prognoz 7 Satellite," *Space Sci. Rev.*, 31, 247 (1982).
Lundin, R., and E. Dubinin, "Solar Wind Energy Transfer Regions Inside the Dayside Magnetopause—Accelerated Heavy Ions as Tracers for MHD-Processes in the Dayside Boundary Layer," *Planet. Space Sci.*, 33, 891–907 (1985).
Lundin, R., and D. S. Evans, "Boundary Layer Plasmas as a Source for High-Latitude, Early Afternoon, Auroral Arcs," *Planet. Space Sci.*, 33, 1389 (1985).
Lundin, R., L. Eliasson, and I. Sandahl, "VIKING First Results: Hot Plasma," Proceedings from the STP-symposium in Toulouse, 1986, *Phys. Script.*, 37, 482 (1988).
Lyons, L. R., and R. L. Waltersheid, "Feedback Between Neutral Winds and Auroral Arc Electrodynamics," *J. Geophys. Res.*, 91, 13,506 (1986).
Makita, K., C.-I. Meng, and S.-I. Akasofu, "The Shift of the Auroral Electron Precipitation Boundaries in the Dawn-Dusk Sector in Association with Geomagnetic Activity and Interplanetary Magnetic Field," *J. Geophys. Res.*, 88, 7967 (1983).

Makita, K., C.-I. Meng, and S.-I. Akasofu, "Latitudinal Electron Precipitation Patterns During Large and Small IMF Magnitudes for Northward IMF Conditions," *J. Geophys. Res.*, **93**, 97 (1988).

Meng, C.-I., "Polar Cap Arcs and the Plasma Sheet," *Geophys. Res. Lett.*, **8**, 273 (1981*a*).

Meng, C.-I., "The Auroral Electron Precipitation During Extremely Quiet Geomagnetic Conditions," *J. Geophys. Res.*, **86**, 4607 (1981*b*).

Meng, C.-I., and R. Lundin, "Auroral Morphology of the Midday Oval," *J. Geophys. Res.*, **91**, 1572 (1986).

Meng, C.-I., and K. Makita, "Dynamic Variations of the Polar Cap," in *Solar Wind-Magnetosphere Coupling*, Y. Kamide and J. Slavin, eds., Terra Scientific Pub. Co., Tokyo (1986).

Menietti, J. D., and J. L. Burch, "DE 1 Observations of Theta Aurora Plasma Source Regions and Birkeland Current Charge Carriers," *J. Geophys. Res.*, **92**, 7503 (1987).

Mitchell, D. G., F. Kutchko, D. J. Williams, T. E. Eastman, L. A. Frank, and C. T. Russell, "An Extended Study of the Low-Latitude Boundary Layer of the Dawn and Dusk Flanks of the Magnetosphere," *J. Geophys. Res.*, **92**, 7395 (1987).

Murphree, J. S., L. L. Cogger, and C. D. Anger, "Characteristics of the Instantaneous Auroral Oval in the 1200—1800 MLT Sector," *J. Geophys. Res.*, **86**, 7657 (1981).

Murphree, J. S., C. D. Anger, and L. L. Cogger, "The Instantaneous Relationship Between the Polar Cap and Oval Auroras at Times of Northward Interplanetary Magnetic Field," *Can. J. Phys.*, **60**, 349 (1982).

Murphree, J. S., R. Elphinstone, L. L. Cogger, and D. D. Wallis, "High Latitude Dayside Features," *J. Geophys. Res.*, submitted (1988*a*).

Murphree, J. S., L. L. Cogger, R. Elphinstone, and D. D. Wallis, "Short Term Dynamics in Polar Aurora, *J. Geophys. Res.*, submitted (1988*b*).

Ogino, T., R. J. Walker, M. Ashour-Abdalla, and J. M. Dawson, "An MHD Simulation of B_y-Dependent Magnetospheric Convection and Field-Aligned Currents During Northward IMF," *J. Geophys. Res.*, **90**, 10,835 (1985).

Paschmann, G., G. Haerendel, N. Sckopke, H. Rosenbauer, and P. C. Hedgecock, "Plasma and Magnetic Field Characteristics of the Distant Polar Cusp Near Local Noon: The Entry Layer," *J. Geophys. Res.*, **81**, 2883 (1976).

Perreault, P., and S.-I. Akasofu, "A Study of Geomagnetic Storms," *Geophys. J. R. Astron. Soc.*, **54**, 547 (1978).

Peterson, W. K., and E. G. Shelley, "Origin of the Plasma in a Cross-Polar Cap Auroral Feature (theta aurora)," *J. Geophys. Res.*, **89**, 6729 (1984).

Popielawska, B., S. Romanov, J. Blecki, S. Klimov, R. Lundin, and K. Kossacki, "Prognoz 8 Plasma Observations in the Tail Lobes vs IMF Direction: Comparison with ISEE-1 and Prognoz 7 Results," Paper 8.3.4 presented at XXVII COSPAR meeting in Helsinki, Finland (1988).

Reiff, P. H., and J. L. Burch, "IMF B_y-Dependent Plasma Flow and Birkeland Currents in the Dayside Magnetosphere, 2, A Global Model for Northward and Southward IMF," *J. Geophys. Res.*, **90**, 1595 (1985).

Reiff, P. H., R. W. Spiro, and T. W. Hill, "Dependence of Polar Cap Potential Drop on Interplanetary Parameters," *J. Geophys. Res.*, **86**, 7639 (1981).

Reiff, P. H., T. W. Hill, and J. L. Burch, "Solar Wind Plasma Injection at the Dayside Magnetospheric Cusp," *J. Geophys. Res.*, **82**, 479 (1977).

Riedler, W., and H. Borg, "High-Latitude Precipitation of Low-Energy Particles as Observed by ESRO 1A," *Space Res.*, **XII**, 1397 (1972).

Rostoker, G., and R. Bostrom, "A Mechanism for Driving the Gross Birkeland Current Configuration in the Auroral Oval," *J. Geophys. Res.*, **81**, 235 (1976).

Rostoker, G., and T. Eastman, "A Boundary Layer Model for Magnetospheric Substorms," *J. Geophys. Res.*, **92**, 12,187 (1987).

Sato, T., "Auroral Physics," in *Magnetospheric Plasma Physics*, A. Nishida, ed., D. Reidel Pub. Co., Dordrecht, Holland, p. 197 (1982).

Sckopke, N., G. Paschmann, H. Rosenbauer, and D. H. Fairfield, "Influence of the Interplanetary Magnetic Field on the Occurrence and Thickness of the Plasma Mantle," *J. Geophys. Res.*, **81**, 2687 (1976).

Sckopke, N., G. Paschmann, G. Haerendel, B. U. Ö. Sonnerup, S. J. Bame, T. G. Forbes, E. W. Hones, Jr., and C. T. Russell, "Structure of the Low-Latitude Boundary Layer," *J. Geophys. Res.*, **86**, 2099 (1981).

Stasiewicz, K., "Generation of Magnetic Field-Aligned Currents, Parallel Electric Fields, and Inverted-V Structures by Plasma Pressure Inhomogeneities in the Magnetosphere," *Planet. Space Sci.*, **33**, 1037 (1985).

Tsyganenko, N. A., "Global Quantitative Models of the Geomagnetic Field in the Cislunar Magnetosphere for Different Disturbance Levels," *Planet. Space Sci.*, **35**, 1347 (1987).

Tsyganenko, N. A., and A. V. Usmanov, "Determination of the Magnetospheric Current System Parameters and Development of Experimental Geomagnetic Field Models Based on Data from IMP and HEOS Satellites," *Planet. Space Sci.*, **30**, 985 (1982).

Vasyliunas, V. M., "Interaction Between the Magnetospheric Boundary Layers and the Ionosphere," in *Magnetospheric Boundary Layers*, SP-148, B. Battrick, ed., ESA, Paris, p. 387 (1979).

Williams, D. J., D. G. Mitchell, T. E. Eastman, and L. A. Frank, "Energetic Particle Observations in the Low-Latitude Boundary Layer," *J. Geophys. Res.*, **90**, 5097 (1985).

Winningham, J. D., and W. J. Heikkila, "Polar Cap Auroral Electron Fluxes Observed with ISIS 1," *J. Geophys. Res.*, **79**, 949 (1974).

Wygant, J. R., R. B. Torbert, and F. S. Mozer, "Comparisons of S3-3 Polar Cap Potential Drops with the Interplanetary Magnetic Field and Models of Magnetopause Reconnection," *J. Geophys. Res.*, **88**, 5727 (1983).

Zanetti, L. J., and T. A. Potemra, "The Relationship of Birkeland and Ionospheric Current Systems to the Interplanetary Magnetic Field," in *Solar Wind-Magnetosphere Coupling*, Y. Kamide and J. Slavin, eds., Terra Scientific Pub. Co., Tokyo, 547 (1986).

193

IV-3. DISCRETE AURORAS AND MAGNETOTAIL PROCESSES

L. R. Lyons*

Important information about magnetospheric phenomena associated with auroras and substorms can be inferred from low-altitude auroral observations. Satellite observations have shown that discrete auroral arcs lie within a boundary plasma sheet (BPS) region that is outside the central plasma sheet (CPS). The observations imply that arcs are generated along BPS field lines by magnetospheric processes that form large, perpendicular electric field structures. The BPS and the arc generation processes apparently lie along field lines that are in the vicinity of the boundary between open and closed field lines and cross the tail (or magnetopause) current sheet. Ground-based observations show that the first indication of a substorm onset is the brightening of a quiet, discrete arc. This suggests that substorms are initiated along the BPS field lines associated with arc generation, and not within the CPS. Finally, auroral observations have shown that the area of open, polar-cap field lines varies considerably during periods of geomagnetic activity. Expansion of the polar cap has the potential for releasing trapped plasma sheet particles along freshly open field lines. The resulting evacuation of field lines has the potential for being an important loss process for the plasma sheet and for being a source of tailward flows and energetic particle bursts in the tail.

1. INTRODUCTION

By observing auroral arcs, both optically and via measurements of charged particle precipitation and the electric fields and particle precipitation in the vicinity of arcs, we can obtain valuable information on electrodynamical processes in Earth's magnetosphere. Discrete auroral arcs, generally believed to be formed by electrons accelerated by parallel electric fields, are an important result of energy-transfer processes in the magnetosphere. Auroral arcs are often quiet and undergo only slow variations with time. However, during a substorm expansion phase arcs brighten and become active; such auroral breakups being an important aspect of substorms. The magnetospheric processes responsible for arc formation and for the auroral breakup remain important, unsolved aspects of magnetospheric physics. Here, inferences obtained from auroral observations regarding the magnetospheric processes responsible for arc formation and substorms are discussed.

The first topic involves the magnetospheric region of arc generation. The relation of the generation region to the magnetic and electric field structure is considered. This discussion leads to suggestions concerning the magnetospheric region of substorm initiation. The final topic concerns speculations on the effects of substorm-associated variations of the magnetospheric magnetic field geometry.

2. MAGNETOTAIL REGION OF ARC GENERATION

Energy associated with discrete auroral arcs is dissipated via the acceleration of particles by magnetic-field-aligned electric fields $1-2R_E$ above Earth [*Gorney et al.*, 1981] and by Joule heating in the ionosphere. The source of the arc energy is obtained from poorly understood generation processes that occur at much higher altitudes. Some theoretical analyses have addressed these processes [e.g., *Rostoker and Boström*, 1976; *Stasiewicz*, 1985; *Lotko et al.*, 1987]; but much more knowledge of arc generation processes is required. Information on these processes that can be inferred from low-altitude satellite observations is discussed here where generation processes in the tail are treated; generation of dayside auroras near noon is discussed by *Lundin et al.* [this volume].

Observations from low-altitude, polar-orbiting satellites have shown that the region of electron precipitation in the auroral zones can be divided into a high-latitude portion of structured precipitation that has been referred to as the boundary plasma sheet (BPS) and a low-latitude portion that has been referred to as the central plasma sheet (CPS) [*Gurnett and Frank*, 1973; *Winningham et al.*, 1975].

The BPS and CPS regions can be identified in the data from the polar orbiting S3-3 satellite shown in Figure 1. This figure shows energy-time spectrograms of electrons (0.17–33 keV) and ions (E/q from 0.09–3.9 keV/q). The intensity is given by a grey scale in units of differential energy flux. In addition, Figure 1 also shows intensity coded strips for 235-keV electrons and >80 keV ions, the electric potential along the satellite trajectory, and the pitch angle of the particles measured as the satellite spins. In the figure, the CPS can be seen from just before 27400 s to near 27500 s, and the BPS can be seen from the poleward boundary of the CPS to ~ 27600 s. The data in Figure 1 is from ~ 1600

*Space Sciences Laboratory, The Aerospace Corporation, Los Angeles, CA 90009.

Auroral Physics, edited by C.-I. Meng, M. J. Rycroft and L. A. Frank. © Cambridge UP 1991

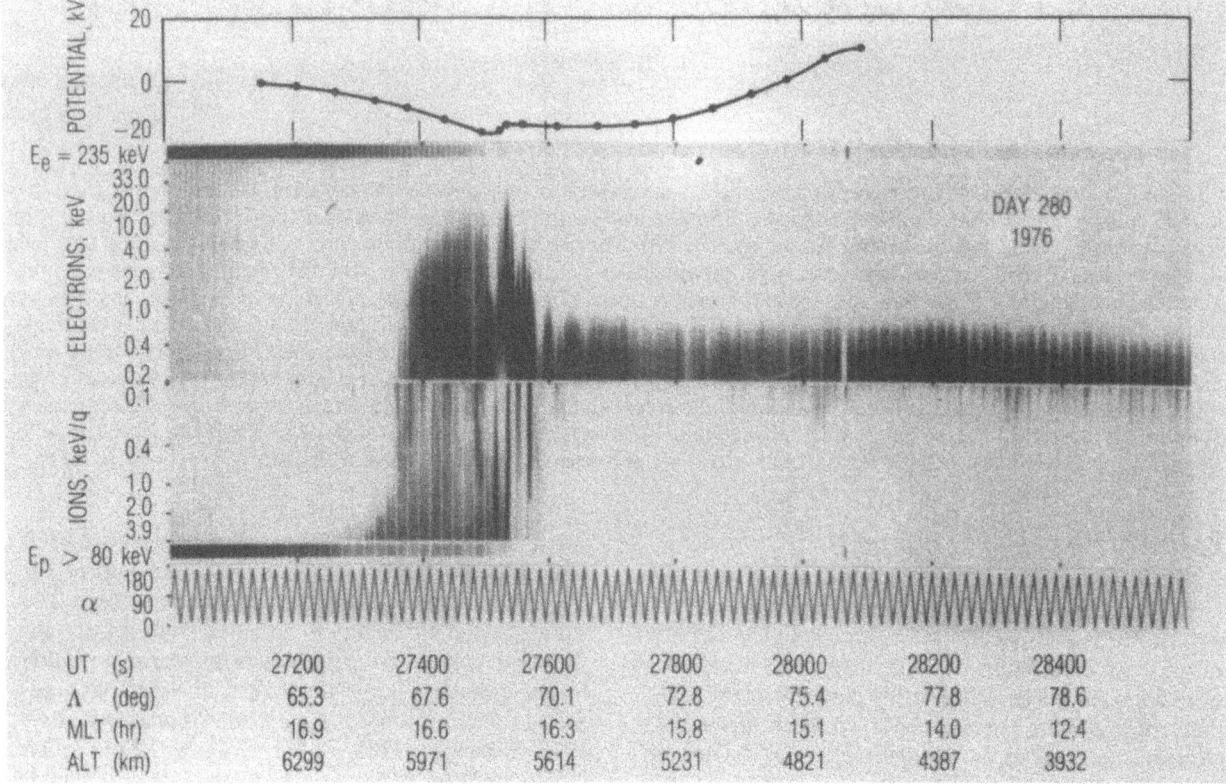

Figure 1—Spectrogram of S3-3 plasma data and plot of electric potential along satellite trajectory for 27000–28600 s UT on Day 280, 1976 (October 6). The center panels show the energy flux for 0.2–33 keV electrons and for 0.1–3.9 keV/q ions versus time. Energy flux levels are encoded in a grey scale with darker shading representing higher flux. The grey scale ranges from 7.0×10^3 to 5.0×10^6 keV/cm^2-s-sr-keV for ions and from 1.4×10^5 to 1.0×10^8 keV/cm2-s-sr-keV for electrons. Grey-scale bands at the top and bottom of the central spectrograms represent the intensities of 235-keV electrons and >80-keV protons, respectively. The pitch angle of the particle data is indicated by a line graph below the particle data. Time, invariant latitude, magnetic local time, and satellite altitude are annotated along the bottom of the figure. (From *Lyons and de la Beaujardiere* [1988].)

MLT, which maps to the nightside in most magnetic field models.

Within the BPS region, the electron spectrogram shows "inverted-V" signatures [*Frank and Ackerson*, 1971] and the ion spectrogram shows beams within the upgoing loss cones. These features have been identified as being the result of magnetic-field-aligned electric fields, as has been quantitatively shown using two-dimensional plots of the S3-3 electron and ion data [*Mizera and Fennell*, 1977; *Croley et al.*, 1978; *Fennell et al.*, 1981; *Mizera et al.*, 1981]. Regions of acceleration by field-aligned electric fields are routinely observed within the BPS, and they have been associated with discrete auroral arcs [*Johnstone and Winningham*, 1982; *Whalen*, 1983; *Lyons et al.*, 1988]. Throughout this discussion, the term "discrete auroral arc" refers to a region of precipitation of electrons that have been accelerated by parallel electric fields having a total field-aligned potential drop on the order of the electron thermal energy or greater. The unstructured precipitation

within the CPS does not show signatures of field-aligned acceleration and is referred to as the "diffuse aurora." Diffuse auroras can occur between auroral arcs within the BPS as well as within the CPS.

Observations in the tail [*Lui et al.*, 1977; *DeCoster and Frank*, 1979; *Williams*, 1981; *Spjeldvik and Fritz*, 1981; *Eastman et al.*, 1984] are consistent with the division of the plasma sheet into an outer boundary region and a CPS that has been inferred from the low-altitude observations. Significant ion flows are almost exclusively confined to the boundary layer [*Huang and Frank*, 1986], and field-aligned currents and perpendicular electric fields have been observed within the boundary layer in the tail having magnitudes appropriate for the mapping of auroral arc currents and electric fields into the tail [*Frank et al.*, 1981; *Cattell et al.*, 1982; *Levin et al.*, 1983; *Huang et al.*, 1984; *Pedersen et al.*, 1985].

On the basis of the observations described above, it is reasonable to conclude that discrete auroral arcs lie along the outer boundary of the plasma sheet within the BPS. Poleward of the BPS, the relatively uniform polar rain [*Winningham and Heikkila*, 1974] is often seen [*Winningham et al.*, 1975]. The polar rain can be seen in Figure 1 at electron energies ≤ 500 eV. It is generally believed that the polar rain results from solar-wind electrons entering the magnetosphere along open magnetic field lines [e.g., *Fennell et al.*, 1975]. Equatorward of the BPS, reduced fluxes can be seen in the downgoing loss cone, which indicates trapped particles on closed field lines. This can be seen in Figure 1 at electron energies ≤5 keV.

Observations such as those in Figure 1 show that the BPS, and thus the region of auroral arc generation, lies in the vicinity of the boundary between open and closed field lines. The convection reversal (as indicated by the minimum in the electric potential in the top panel of Figure 1) is also often associated with the BPS and the open-closed field line boundary [*Heelis et al.*, 1980], as is the case for the data in Figure 1. The relation between the BPS and the CPS is illustrated in the top panel of Figure 2. In this figure, the BPS is shown to lie only on closed field lines equatorward of the separatrix field line that reaches the tail neutral point. However, the BPS may extend poleward onto open field lines.

The association between the particle and electric field data in Figure 1 is expected for typical conditions, where convection is antisunward over the open polar cap region and sunward at lower latitudes. However, there are exceptions, as can be seen in the example of S3-3 data from a traversal of the auroral zone shown in Figure 3. In this example, the BPS lies between ~44200 s and ~44550 s Universal Time (~69° to 72° invariant latitude) and polar rain can be seen at electron energies <400 eV poleward of the BPS. Within portions of the BPS and equatorward, reduced fluxes can be seen in the downgoing loss cones at electron energies greater than a few keV. Thus, the BPS apparently lies in the vicinity of the boundary between open and closed field lines as expected.

However, the potential plot in Figure 3 does not show a reversal in the large-scale electric field in the vicinity of the BPS. Convection was approximately uniform and in the midnight-to-dusk direction as expected from sunward convection from 62° to beyond 77° latitude. While a convection reversal is often observed in the vicinity of the BPS, the lack of a reversal is not unusual [*Lyons and de la Beaujardiere*, 1988] and is possibly related to the interplanetary magnetic field (IMF). It has been shown [*Lyons*, 1980; *Chiu and Cornwall*, 1980] that the reversal of the convection electric field can form inverted-V regions containing field-aligned acceleration as are observed in the BPS. However, observations such as shown in Figure 3 (another

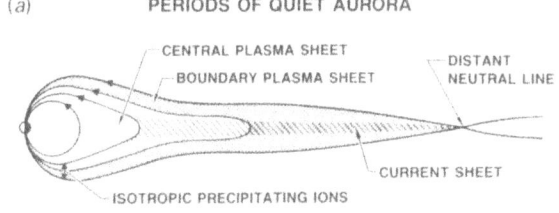

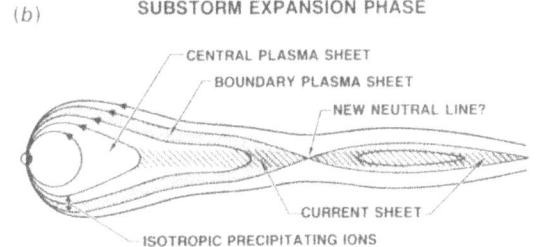

Figure 2—Schematic illustration (not to scale) of the magnetotail (*a*) for periods of quiet aurora such as might exist for a period of time prior to the onset of a substorm expansion phase, and (*b*) for the onset of a substorm expansion phase.

example has been presented by *Mizera et al.* [1981]) show that reversal of the large-scale convection electric field is not necessary for the formation of discrete auroral arcs.

The electric field phenomena that appear to be generally associated with the BPS can be seen in Figure 4 (from *Gurnett and Frank* [1973]). This figure shows the precipitating electron energy flux from the pass of a polar-orbiting satellite over the auroral zone, with the BPS and CPS identified from a more detailed data presentation. Electric field data are also shown in the figure, and large, spatially structured electric fields can be seen throughout the BPS. Such structured electric fields are commonly seen in the BPS [e.g., *Mozer et al.*, 1977] and have been seen in the absence of a reversal in large-convection electric fields [*Mizera et al.*, 1981]. These structured electric fields have also been observed in the ionosphere in association with individual auroral arcs, with the electric field consistently showing a negative divergence coincident with the arcs [e.g., *Swift and Gurnett*, 1973; *Maynard et al.*, 1977; *Evans et al.*, 1977; *de la Beaujardiere et al.*, 1977; *Heelis et al.*, 1981; *Marklund et al.*, 1983]. The structured electric fields have been shown to be able to form discrete auroral arcs as a result of the requirement for current continuity in the ionosphere [*Lyons*, 1981]; however, it is necessary to understand the formation and maintenance of the electric fields within the BPS to understand the generation of auroral arc currents, fields, and energy transfer.

Important additional information on arc generation can be inferred from data such as those in Figures 1

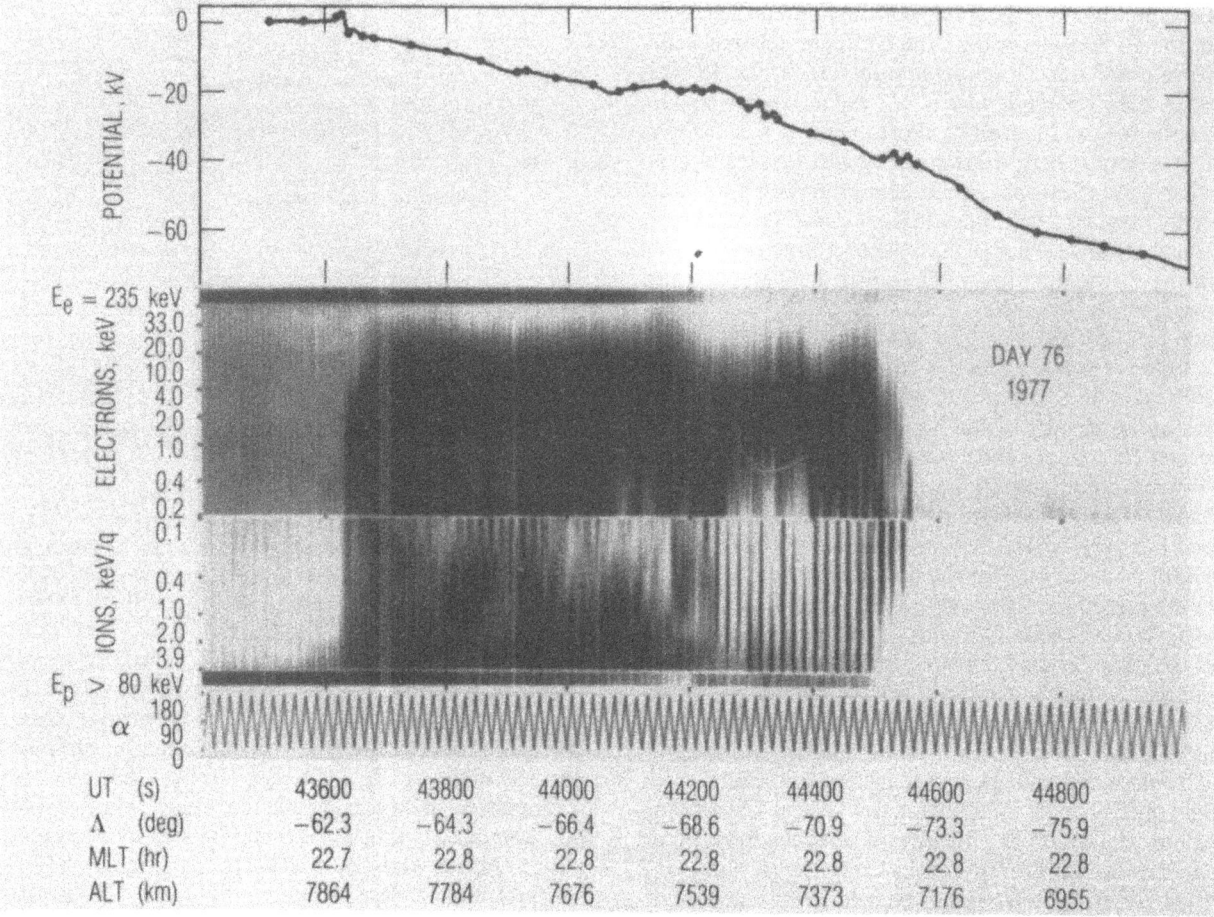

Figure 3—Same as Figure 1, but for 43400–45000 s UT on day 76, 1977 (March 17). (From *Lyons and de la Beaujardiere* [1988].)

and 3. This is that the BPS lies on field lines that cross the tail current sheet and perhaps also on field lines that cross the magnetopause current sheet.

The distribution of the flowing ions within the BPS in the tail [*DeCoster and Frank*, 1979; *Eastman et al.*, 1984] has been found to agree with what is expected if the source of the ions is the tail current sheet [*Lyons and Speiser*, 1982]. The current sheet is defined to be the region of weak magnetic field near $z = 0$ (GSM coordinates) across which B_x reverses direction. Within the current sheet, ion motion violates the guiding center approximation. As a result, ions are energized and scattered in pitch angle as they interact with the current sheet [e.g., *Speiser*, 1965; *Wagner et al.*, 1979; *Cowley*, 1980; *Lyons and Speiser*, 1982]. The particle motion within the current sheet allows the current sheet to take the place of the "diffusion region" associated with reconnection theories [*Speiser*, 1970; *Coroniti*, 1985]. However, the region of energization within the current sheet can extend over a radial distance of the order of 10–100R_E, and is not restricted to a region near the neutral line.

The scattering in pitch angle within the current sheet is expected to extend over a broad energy range ≤ 1 keV–1 MeV), and it is expected to give isotropy across the downgoing loss cone. Thus, ions within the BPS field should be observable on low-altitude, polar-orbiting satellites as ion precipitation with isotropic pitch angle distributions extending to high energies. Such isotropic precipitation has been observed over the ~1 keV–1 MeV energy range [*Hultqvist et al.*, 1974; *Lundblad et al.*, 1979; *Sharber*, 1982; *Lyons and Evans*, 1984], identifying field lines that map to the tail current sheet. This precipitation can be seen in the ion data of Figures 1 and 3 at energies >80 keV and near 3.9 keV, extending over a latitude region within and equatorward of the BPS.

Simultaneous electron and ion measurements such as those indicated in Figures 1 and 3 show that the BPS essentially always occurs on field lines containing isotropic ion precipitation from the tail current sheet. These measurements generally show the ion precipitation extending equatorward of the BPS. Occasionally the BPS extends poleward of the region of observable

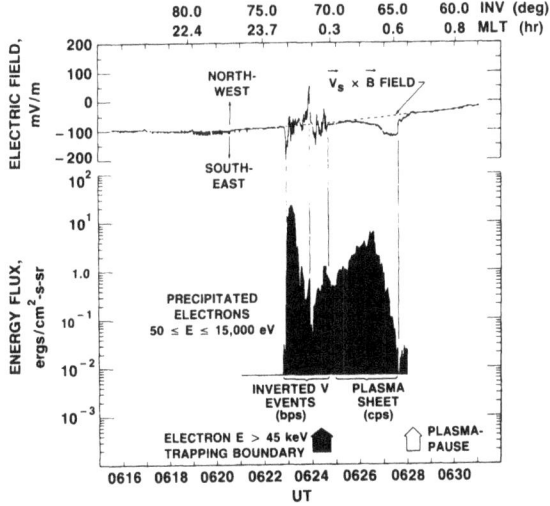

Figure 4—The electric field and precipitating electron energy fluxes from orbit 3667 (June 6, 1969) of the Injun 5 satellite. The BPS and CPS were identified from a more detailed data presentation. (From *Gurnett and Frank* [1973].)

ions. In the later cases, the poleward portion of the BPS extends either onto open polar-cap field lines that cross the magnetopause current sheet or onto more distant portions of the tail current sheet where ion fluxes are too low to be measured. Almost never does the BPS extend equatorward of the region of current sheet ion precipitation [*Lyons et al.*, 1988].

On the basis of these observations, we conclude that the BPS almost exclusively lies on field lines that thread the tail (and perhaps the magnetopause) current sheet. This implies that discrete auroral arcs are generated along field lines that thread the tail (or magnetopause) current sheet, as has been illustrated in the top portion of Figure 2 by having the BPS crossing the tail current sheet. *Mauk and Meng* [this volume] considered the possibility of field-aligned acceleration affecting particles near synchronous orbit during transient periods; however, the low-altitude observations imply that any generation of auroral arcs within the CPS would be a rare occurrence. Since the isotropic ion precipitation generally extends equatorward of the BPS, the current sheet generally extends earthward of the equatorial crossing of the BPS as shown. Sufficient energization to form the flowing ions observed within the BPS may only exist in the outer portion of the current sheet. As a result of the increase in B_z, the energization within the current sheet is expected to decrease with decreasing $|x|$. However, scattering in pitch angle can be significant even when energization during a traversal of the current sheet is small. This can cause the ion precipitation to extend equatorward of the ionospheric mapping of the flowing BPS ions, as illustrated in Figure 2.

3. MAGNETOSPHERIC REGION OF SUBSTORM INITIATION

The association of discrete auroral arcs with the BPS and the tail current sheet has important implications for substorms. Auroral observers consistently state that preexisting quiet arcs breakup at the initiation of a substorm [e.g., *Akasofu*, 1964]. Evidence does not suggest that discrete arcs form equatorward of all preexisting arcs when a substorm begins. This implies that substorms are initiated on field lines within the BPS that thread the tail (or perhaps the magnetopause) current sheet. Such an association has not been incorporated in the model involving the formation of a near-Earth neutral line within the preexisting CPS.

The most widely discussed description of magnetospheric substorms contends that a neutral line is formed in the near-Earth portion of the central plasma sheet during the expansion phase [e.g., *Hones et al.*, 1973, 1984; *Nishida and Nagayama*, 1973; *Nishida and Hones*, 1974; *Hones*, 1979; *Hones and Schindler*, 1979; *Nishida et al.*, 1981, 1983; *Cattell et al.*, 1986; *Baker et al.*, 1987]. In this description, substorm expansion phase signatures observed in the tail are a result of the neutral line formation.

More recently, substorm signatures in the tail have been described in terms of currents and ion flows that are observed in the BPS [*Rostoker and Eastman*, 1987; *Eastman et al.*, 1988]. In the boundary layer description of the substorm expansion phase, substorm phenomena observed in the tail are ascribed to features of the BPS. Formation of a near-Earth neutral line is not involved. It should be noted that this description is based primarily on ISEE 1 observations at $|x| < 22R_E$.

The apparent conflict between the two points of view on essential structural elements of the magnetotail during the substorm expansion phase has been a subject of intense debate at scientific meetings over the past few years [e.g., *Rostoker*, 1985]. However, a new study of ISEE 3 observations near $x = -80R_E$ in the tail [*Nishida et al.*, 1988] lead *Lyons and Nishida* [1988] to suggest that the two points of view are not mutually exclusive. We suggested that substorm neutral lines form within the source region for the BPS, the source region presumably being the tail current sheet.

While observations at $|x| \leq 22R_E$ consistently show a significant distinction between a quiescent CPS and an active boundary layer, such a distinction is not seen near $80R_E$. Ion flows are seen throughout the plasma sheet without being confined to the boundary region. During quiescent times the flow direction is earthward and the magnetic field has a northward polarity. On the basis of these observations, *Nishida et al.* [1988] suggested that $80R_E$ lies within the source region that supplies the flowing plasma to the BPS.

This source region lies earthward of the distant neutral line, and is presumably the tail current sheet as illustrated in the top panel of Figure 2. Auroral arcs lie along BPS field lines, and the top panel is drawn to represent a period of quiet aurora such as might exist for a period of time prior to the onset of a substorm.

At the onset of the substorm expansion phase, on the other hand, *Nishida et al.* [1988] found that the flow direction at $80R_E$ consistently turns tailward. The magnetic field polarity turns southward a few minutes later. They attributed these observations to the formation of a neutral line at substorm onset earthward of $80R_E$.

To allow for such neutral line formation, as well as for substorm initiation along BPS field lines, *Lyons and Nishida* [1988] suggest that the neutral line forms within the preexisting source region as illustrated in the bottom panel of Figure 2. Thus, in our proposal, the initiation of the substorm expansion phase occurs within the preexisting source region. Earthward of the source region, plasma flow will continue to be confined to the boundary region of the plasma sheet after the neutral line forms. Assuming the source region is the tail current sheet, we do not require the formation of a new diffusion region in the vicinity of the new neutral line. Particle acceleration is expected to continue throughout the entire source region, including the new neutral line, and would be enhanced in regions of significant induced electric fields.

This model provides a synthesis of the neutral line model and the boundary layer observations by allowing for: (1) All auroral field-aligned currents and ion flows being within the BPS at distances ($|x| < 22R_E$) sampled by ISEE 1, and not within the central plasma sheet; (2) Auroral arcs and substorm initiation occurring along field lines that thread the BPS and reach the source region for the BPS; and (3) Formation of a neutral line within the plasma sheet earthward (perhaps well earthward) of $80R_E$.

We did not give a precise estimate as to what radial distances substorm neutral lines might form. They may form anywhere earthward of $x \sim -80R_E$, and may occasionally form earthward of $x = -22R_E$. If one forms earthward of $-22R_E$, we suggested that it only does so within a source region that extends earthward of $-22R_E$. Under such circumstances, the region without a distinguishable CPS and BPS should extend earthward of the ISEE 1 apogee. This proposal is considered the key issue in the recent dispute concerning the substorm expansion phase. Other important substorm phenomena, such as variations in the plasma sheet thickness, conditions leading to neutral line formation, and injection of particles to the vicinity of synchronous orbit, were not addressed.

4. EFFECTS OF VARIATION IN THE POLAR CAP AREA

One of the most important magnetospheric phenomena that has been inferred from auroral observations is the changes that occur in the area of open (polar cap) field lines. Based on the assumption that the polar-cap boundary is located near the poleward boundary of the auroral oval, auroral observations have led to the conclusion that the polar cap is much larger during geomagnetically active times than during quiet times [e.g., *Akasofu*, 1968]. Furthermore, the area of the polar cap has been observed to increase and then decrease during individual substorms [*Craven and Frank*, 1987]. Figure 5 (from *Frank and Craven* [1988]) shows variations in the area of the polar cap during the period of a modest, isolated substorm on November 10, 1981 as obtained from the auroral imager on the DE 1 satellite. The total polar cap flux (*F*) is approximately proportional to the area of the polar cap. Figure 5 shows this flux to increase by ~50% prior to the substorm onset and to decrease following the onset by about half the amount that it had increased.

It is reasonable to assume that the increase in polar-cap area during the substorm growth phase is associated with an increase in magnetic energy in the tail (as shown by the calculated values in Figure 5), and that the decrease in the area is associated with the release of the tail energy during the expansion phase [e.g., *McPherron*, 1973]. Such changes in the geomagnetic field configuration can have significant effects on the trapped particle population in the magnetosphere. For example, the change in magnetic field configuration following the onset of the expansion phase (which returns the field to a more dipolar shape) can lead to a significant inward transport and energization of trapped particles [e.g., *Baker*, 1984; *Mauk*, 1986]. The inward motion and energization result from the electric field induced in the magnetosphere (measured by *Aggson et al.* [1983]) in association with the relaxation of the geomagnetic field.

During the substorm growth phase, on the other hand, the distortion of the geomagnetic field and associated induced electric field cause an outward motion and de-energization of trapped particles. Figure 6 shows a spectrogram of S3-3 data obtained during a period when the polar cap area was significantly larger than average and illustrates what may happen to trapped particles as the polar cap expands. The data are from a crossing of the auroral zone in the late evening local time sector on day 122, 1977.

Typically (see Figures 1, 3, and examples presented by *Lyons et al.* [1988]) energetic particle fluxes taper off rather gradually with increasing invariant latitude Λ, and particles of different energies reach detector

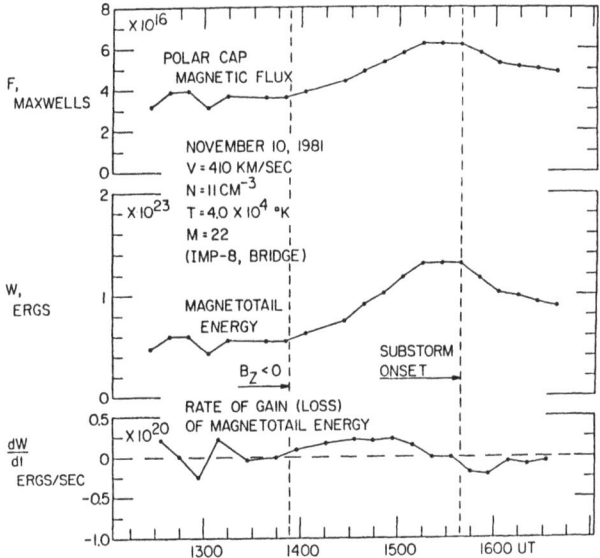

F, MAXWELLS

X 10^16

POLAR CAP MAGNETIC FLUX

NOVEMBER 10, 1981
V = 410 KM/SEC
N = 11 CM^{-3}
T = 4.0 X 10^4 °K
M = 22
(IMP-8, BRIDGE)

W, ERGS

X 10^23

MAGNETOTAIL ENERGY

$B_z < 0$

SUBSTORM ONSET

$\frac{dW}{dt}$ ERGS/SEC

X 10^20

RATE OF GAIN (LOSS) OF MAGNETOTAIL ENERGY

1300   1400   1500   1600 UT

Figure 5—The polar cap magnetic flux *F* observed from DE 1 auroral images during a modest isolated substorm on November 10, 1981. Also plotted are calculated values for the magnetotail energy and for the rate of change of magnetotail energy. (From *Frank and Craven* [1988].)

background levels at different latitudes. Occasionally, however, the S3-3 particle data show a sharp outer boundary where electron and ion fluxes at all energies fall abruptly to background at the same invariant latitude. The example in Figure 6 shows such a boundary, with the intensities of all particles abruptly decreasing to background levels near $\Lambda = 67°$. At all invariant latitudes below the boundary, 433-keV and 654-keV electron energy channels (not shown in Figure 1) showed reduced fluxes in both the downgoing and upgoing loss cones, thus indicating the presence of trapped particles on closed field lines. This was true except for the highest-latitude satellite spin equatorward of the abrupt boundary, where the energetic electron channels showed isotropy over the downgoing loss cone.

The above inferences from Figure 6 concerning regions of open and closed field lines suggest that, to within one satellite spin, the abrupt cutoff in plasma-sheet particles occurs at the boundary between open and closed field lines. The invariant latitude of the cutoff is several degrees equatorward of the typical location of the outer boundary of trapped particles, implying a significantly expanded polar cap. For example, the mean Λ of the outer boundary of observable >40-keV electrons near midnight was found by *Fritz* [1970] to be 73°.

The particle cutoff could be formed by some sort of a strong source of particles of all energies that abruptly cuts off at the outer boundary of closed field

lines, or it could be formed by a loss of particles of all energies at invariant latitudes beyond the cutoff. Recently, we suggested the second alternative [*Lyons et al.*, 1989]. If the region of open field lines expands to include field lines previously containing significant plasma sheet fluxes, we would expect particles of all energies to escape along freshly opened field lines. This would lead to an abrupt particle cutoff near the boundary between open and closed field lines, as is seen in Figure 6.

Opening of field lines containing trapped particles is possible at all local times as the polar cap expands. On the dayside, this will lead to particle escape along field lines that cross the magnetopause, so as to produce flows of magnetospheric ions into the magnetosheath. On the nightside, such escape will result in outward-flowing distributions of plasma-sheet particles until the freshly open field lines are evacuated of plasma-sheet particles. This evacuation process may be an important loss process for the plasma sheet.

For such escape to be important on the nightside, the separatrix between open and closed field lines must move toward decreasing invariant latitude at a rate greater than the rate of plasma drift toward lower invariant latitude that results from magnetospheric electric fields. Such motion of the separatrix would cause plasma in the tail to be transferred across the separatrix from closed to open field lines, which is opposite to the transfer generally associated with steady-state reconnection in the tail. In other words, the separatrix must overtake the convecting plasma. As discussed later, the electric field induced as the polar cap expands convects tail plasma in the direction required for such "reverse" reconnection to occur. However, for reverse reconnection to occur, it is necessary that $\mathbf{E} \cdot \mathbf{J} < 0$ (where $\mathbf{J}$ is current density) along the magnetic x-line in the tail [*Vasyliunas*, 1984]. This requires that other terms (such as the gradient in the off-diagonal elements of the pressure tensor) dominate the resistive term in the generalized Ohm's law (*Vasyliunas*, 1975).

Tailward plasma flows have been reported along the outer boundary of the plasma sheet as it thins in association with substorms [e.g., *Hones et al.*, 1976; *Liu et al.*, 1977b]. Such flows could result from the evacuation of freshly open field lines, the evacuation thus contributing to the observed thinning of the plasma sheet. Furthermore, since the plasma-sheet particle distribution includes energetic particles, the evacuation process should lead to bursts of energetic ions (at energies ≥ 100 keV) traveling outward in the tail during disturbed periods. Enhancements in the flux of ~0.1-1 MeV ions are often observed in the tail [e.g., *Armstrong and Krimigis*, 1968; *Sarris et al.*, 1976]. Such enhancements have been called energetic particle "bursts," and they have been associated with sub-

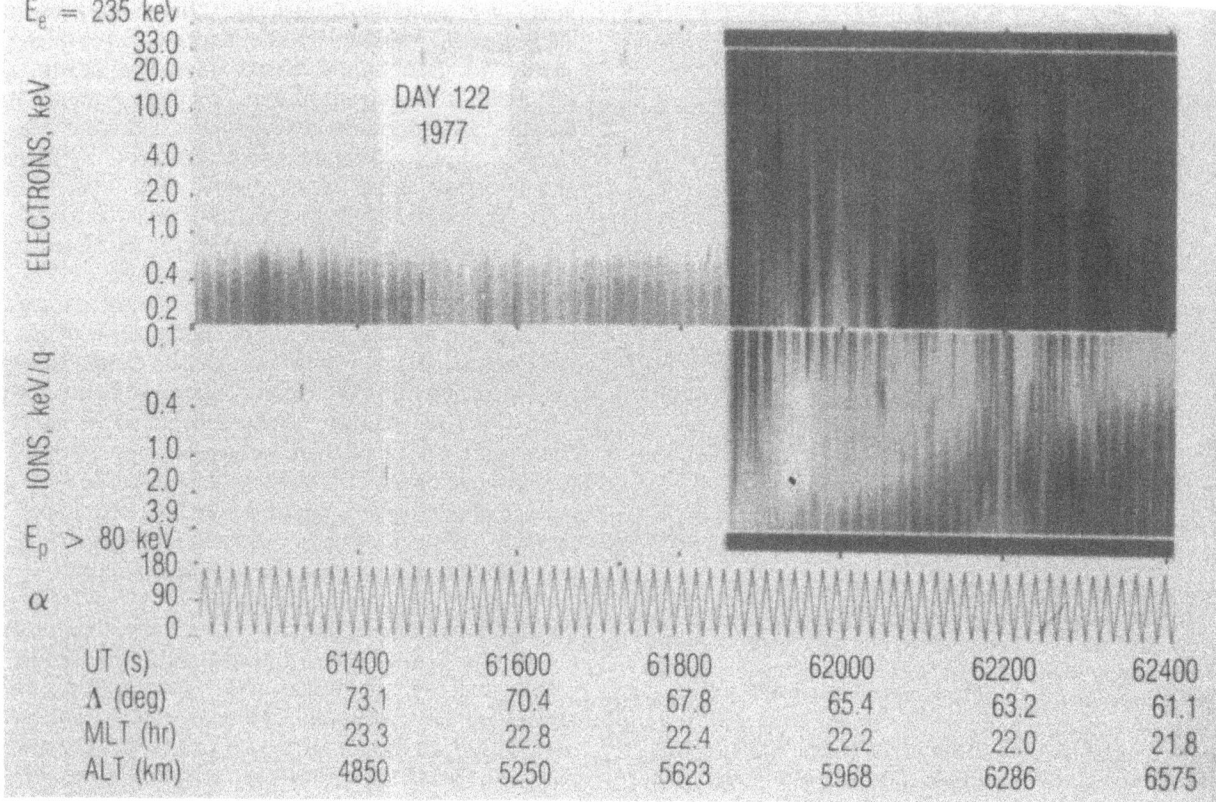

Figure 6—Same as Figure 1, but for 61200–62400 s UT on day 122, 1977 (May 2). Electric field data are unavailable for this satellite orbit. (From *Lyons et al.* [1989].)

storms [*Krimigis and Sarris*, 1979; *Kirsch et al.*, 1981]. These observations have led to the suggestion that the bursts result from impulsive acceleration by induced electric field associated with substorms [*Sarris and Axford*, 1979; *Baker et al.*, 1979]. However, *Fennell* [1970] and *Baker et al.* [1979] noted that the ion intensities within the bursts were similar to those in the outer radiation zone, as would be expected if trapped particles were the source of the bursts. Indeed, *Fennell* [1970] suggested this interpretation.

It is easy to show that trapped particles should be evacuated as the polar cap expands by considering the approximate relationship $L = \sec^2\Lambda$ between the generalized L-value [*Roederer*, 1970] and Λ. This value of L, which is inversely proportional to the magnetic flux ϕ enclosed by a particle drift shell, is conserved when ϕ is conserved. Furthermore, the above relation between L and Λ is useful, because magnetospheric processes do not significantly change B at Earth's surface.

Letting Λ^* and L^* be the invariant latitude and the L-value of the separatrix between open and closed field lines, we find that a large range of drift shells become open for modest changes in the size of the polar cap. For $\Lambda^* = 75°$, for example, trapping occurs for $L <$

$L^* = 15$. If the polar cap expands to $\Lambda^* = 70°$, field lines having $L > 8.5$ become open. A further expansion of the polar cap to $\Lambda^* = 65°$ reduces L^* to 5.5.

To evaluate the effects of polar-cap expansion on trapped particles, *Lyons et al.* [1989] applied the simple model [*Hill and Rassbach*, 1975; *Schulz*, 1976, 1980] consisting of a dipolar magnetic field to which is added a uniform southward field B_∞. We found that, as the polar cap expands, L^* decreases and the neutral line in the equatorward plane moves earthward. However, we also found that drift shells move outward toward the neutral line (the outward drift being the result of the electric field induced by dB_∞/dt) as the neutral line moves earthward. The outward motion of the drift shells as the separatrix moves inward is in the direction required for particles to be transferred from closed to open field lines. Despite the simplicity of the model, the result that particle drift shells move outward as the polar cap expands should be qualitatively correct in the outer part of the nightside magnetosphere.

The above discussion does not include the effects of the convection electric field. In general, the convection electric field will decrease particle loss on the nightside and increase dayside losses throughout the magneto-

202

pause. However, the net particle loss as the polar cap expands should not be greatly affected. For particles to be evacuated from the nightside, it is necessary that field lines open at a rate faster than the rate at which particles are carried across field lines by convective electric-field drift. This would be a signature of reverse reconnection. Using our model to evaluate the relative rates of the separatrix motion as the polar cap expands and of the convecting plasma, we found that reverse reconnection should have a good chance of occurring whenever the polar cap expands. It should be possible to check whether it actually occurs by using radar and optical measurements to infer the motion of the polar-cap boundary and the motion of the plasma normal to the boundary.

On the basis of the S3-3 observations and the model results, we suggested that expansion of the polar cap and the resulting evacuation of trapped particles along freshly open field lines may be an important cause of (1) outward plasma flow in the tail, (2) energetic particle bursts in the tail, and (3) particle loss from the tail.

We did not suggest that all energetic particle bursts and flows in the tail result from the escape of trapped particles along freshly open field lines. Energetic ions have been observed [*Spjeldvik and Fritz*, 1981; *Williams*, 1981] as part of the population of flowing ions along the outer boundary of the plasma sheet. Most of the initially earthward flowing ions mirror as they approach the Earth, so that earthward-streaming, tailward-streaming, and counterstreaming ions can be seen near the plasma sheet boundary [*Williams*, 1981]. *Spjeldvik and Fritz* [1981] suggested that the location of the boundary between the streaming ions and the lobes can vary in a wave-like manner in the direction normal to the boundary, so that a satellite can have multiple encounters with the plasma sheet boundary layer (PSBL). They found that such encounters with the boundary layer can give burst-like fluxes of energetic ions having time scales ≳ 30 s.

Figure 7a illustrates the region of flowing ions in the tail that should occur for approximately steady-state or contracting polar cap conditions. The PSBL is shown to be a result of ions ejected from the outer portion of the tail current sheet. Since the boundary layer can contain energetic ions, variations in the position of the outer boundary of the plasma sheet should give burst-like fluxes of ions near the lobe-plasma sheet boundary earthward of the distant tail neutral line.

Besides forming the flowing ions along the plasma sheet boundary, ejection of ions from the current sheet might also give tailward flowing ions tailward of the neutral line as illustrated in Figure 7a. Such tailward ejection has been invoked to explain tailward flowing ions observed at distances sufficiently far from the Earth (60–200R_E) that they might be tailward of the

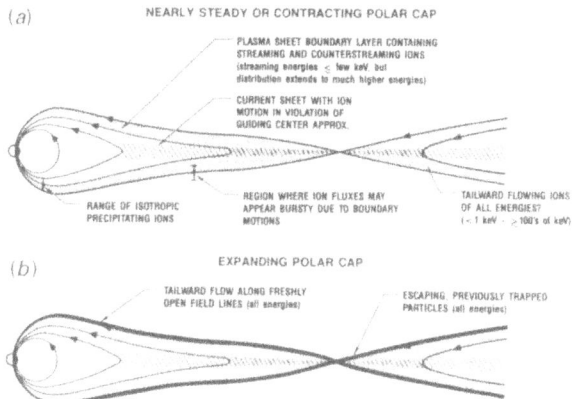

Figure 7—Schematic illustration of regions of flowing ions in the tail (*a*) for approximately steady or contracting polar cap conditions, and (*b*) for times when the polar cap expands sufficiently fast that the separatrix between open and closed field lines overtakes convecting particles trapped on closed field lines in the plasma sheet.

neutral line [*Cowley et al.*, 1984]. It must be remembered, however, that particles can only be energized by tens of kiloelectron volts or less in the current sheet [*Lyons and Speiser*, 1982], so that current sheet interactions are not responsible for energizing ions to energies ≳ 100 keV in the tail.

Figure 7b illustrates what we suggest will happen as the open field line region expands to include previously closed field lines containing trapped particles. Tailward flow should develop along the outer boundary of the plasma sheet at all energies. These tailward flowing ions should also be observable tailward of the neutral line, the time interval during which they are observable being a function of ion energy. The phase space density of escaping particles beyond the neutral line should be the same as that within the outer portion of the plasma sheet prior to the opening of the flux tubes containing the particles.

5. CONCLUSIONS

The auroral observations discussed here have led to a number of suggestions concerning magnetospheric processes associated with auroral arc generation and substorms. As of this time, most of these suggestions must be regarded as speculative. It would be extremely interesting if studies could be designed and performed to test these ideas.

Auroral particle observations from low-altitude satellites have led to the suggestion that the currents and electric fields associated with discrete auroral arcs are generated along BPS field lines that cross the tail current sheet (and perhaps also along field lines that cross the magnetopause). The generation generally occurs in the vicinity of the boundary between open and closed field lines and the generation process gives large spatially structured electric fields perpendicular to **B**. Discrete arcs are often associated with the convection reversal, but reversal of the large-scale convection electric field is not required for arc generation.

The above conclusions, together with the observation that the first indication of a substorm is the sudden brightening of one of the quiet arcs [*Akasofu*, 1964], imply that the onset of the substorm expansion phase also occurs along BPS field lines that cross the tail (or magnetopause) current sheet. If a new neutral line is formed with the plasma sheet at onset, this result implies that it must form within the preexisting tail current sheet.

Observations of auroral particles during periods when the polar cap is large suggest that field lines containing trapped particles can become open as the polar cap expands. This would lead to the evacuation of trapped particles along freshly opened field lines. On the dayside, the evacuation should occur along field lines that thread the magnetopause. On the nightside, the evacuation would lead to outward plasma flow, to energetic particle bursts, and possibly to a significant loss of plasma sheet particles during the growth phase of a substorm.

ACKNOWLEDGMENT—I am deeply grateful to my colleagues O. de la Beaujardiere, D. S. Evans, J. F. Fennell, A. Nishida, T. W. Speiser, M. Schulz, and A. L. Vampola who have coauthored the studies upon which this paper is based. I am extremely fortunate to have had the opportunity to work with these individuals. Preparation of this paper was supported by NASA grant NAGW-853, NSF grant ATM 88-00602, and the Aerospace Sponsored Research Program.

REFERENCES

Aggson, T. L., J. P. Heppner, and N. C. Maynard, "Observations of Large Magnetospheric Electric Fields During the Onset Phase of a Substorm," *J. Geophys. Res.*, **88**, 3981 (1983).

Akasofu, S.-I., "The Development of the Auroral Substorm," *Planet. Space Sci.*, **12**, 273 (1964).

Akasofu, S.-I., *Polar and Magnetospheric Substorms*, D. Reidel Publ. Co., Dordrecht, Holland, p. 5 (1968).

Akasofu, S.-I., *Physics of Magnetospheric Substorms*, D. Reidel Publ. Co., Dordrecht, Holland, pp. 71-93 (1977).

Armstrong, T. P., and S. M. Krimigis, "Observations of Protons in the Magnetosphere and Magnetotail with Explorer 33," *J. Geophys. Res.*, **73**, 143 (1968).

Baker, D. N., "Particle and Field Signatures of Substorms in the Near Magnetotail," in *Magnetic Reconnection in Space and Laboratory Plasmas*, E. W. Hones, Jr., ed., American Geophysical Union, Washington, DC, p. 193 (1984).

Baker, D. N., R. C. Anderson, R. D. Zwickl, and J. A. Slavin, "Average Plasma and Magnetic Field Variations in the Distant Magnetotail Associated with Near-Earth Substorm Effects," *J. Geophys. Res.*, **92**, 71 (1987).

Baker, D. N., R. D. Belian, P. R. Higbie, and E. W. Hones, Jr., "High-Energy Magnetospheric Protons and Their Dependence on Geomagnetic and Interplanetary Conditions," *J. Geophys. Res.*, **84**, 7138 (1979).

Cattell, C. A., M. Kim, R. P. Lin, and F. S. Mozer, "Observations of Large Electric Fields Near the Plasmasheet Boundary by ISEE 1," *Geophys. Res. Lett.*, **9**, 539 (1982).

Cattell, C. A., F. S. Mozer, E. W. Hones, Jr., R. R. Anderson, and R. D. Sharp, "ISEE Observations of the Plasma Sheet Boundary, Plasma Sheet, and Neutral Sheet, 1, Electric Field, Magnetic Field, Plasma, and Ion Composition," *J. Geophys. Res.*, **91**, 5663 (1986).

Chiu, Y. T., and J. M. Cornwall, "Electrostatic Model of a Quiet Auroral Arc," *J. Geophys. Res.*, **85**, 543 (1980).

Coroniti, F. V., "Explosive Tail Reconnection: The Growth and Expansion Phase of Magnetsopheric Substorms," *J. Geophys. Res.*, **80**, 7427 (1985).

Cowley, S. W. H., "Plasma Populations in a Simple Open Model Magnetosphere," *Space Sci. Rev.*, **26**, 217 (1980).

Cowley, S. W. H., R. J. Hynds, J. G. Richardson, P. W. Daly, T. R. Sanderson, K.-P. Wenzel, J. A. Slavin, and B. T. Tsurutani, "Energetic Ion Regimes in the Deep Geomagnetic Tail: ISEE-3," *Geophys. Res. Lett.*, **11**, 275 (1984).

Craven, J. D., and L. A. Frank, "Latitudinal Motion of the Aurora During Substorms," *J. Geophys. Res.*, **92**, 4565 (1987).

Croley, D. R., Jr., P. F. Mizera, and J. F. Fennell, "Signature of a Parallel Electric Field in Ion and Electron Distributions in Velocity Space," *J. Geophys. Res.*, **83**, 2701 (1978).

DeCoster, R. J., and L. A. Frank, "Observations Pertaining to the Dynamics of the Plasma Sheet," *J. Geophys. Res.*, **84**, 5099 (1979).

de la Beaujardiere, O., R. Vondrak, and M. Baron, "Radar Observations of Electric Fields and Currents Associated with Auroral Arcs," *J. Geophys. Res.*, **82**, 5051 (1977).

Eastman, T. E., L. A. Frank, W. K. Peterson, and W. Lennartsson, "The Plasma Sheet Boundary Layer," *J. Geophys. Res.*, **89**, 1553 (1984).

Eastman, T. E., G. Rostoker, L. A. Frank, C. Y. Huang, and D. G. Mitchell, "Boundary Layer Dynamics in the Description of Magnetospheric Substorms," *J. Geophys. Res.*, in press (1988).

Evans, D. S., N. C. Maynard, J. Troim, T. Jacobsen, and A. Egeland, "Auroral Arc Electric Field and Particle Comparisons, 2, Electrodynamics of an Arc," *J. Geophys. Res.*, **82**, 2235 (1977).

Fennell, J. F., "Observations of Proton Bursts in the Magnetotail with Explorer 35," *J. Geophys. Res.*, **75**, 7048 (1970).

Fennell, J. F., P. F. Mizera, and D. R. Croley, "Low Energy Polar Cap Electrons During Quiet Times," *Proc. Int. Conf. Cosmic Rays*, **14**, MG8-3, 1267 (1975).

Fennell, J. F., D. J. Gorney, and P. F. Mizera, "Auroral Particle Distribution Functions and Their Relationship to Inverted-Vs and Auroral Arcs," in *Physics of Auroral Arc Formation*, Geophysical Monograph 25, S.-I. Akasofu and J. R. Kan, eds., American Geophysical Union, Washington, DC, p. 91 (1981).

Frank, L. A., and K. L. Ackerson, "Observations of Charged Particle Precipitation into the Auroral Zone," *J. Geophys. Res.*, **76**, 3612 (1971).

Frank, L. A., and J. D. Craven, "Imaging Results from Dynamics Explorer 1," *Rev. Geophys.*, **26**, 249 (1988).

Frank, L. A., R. L. McPherron, R. J. DeCoster, B. G. Burek, K. L. Ackerson, and C. T. Russell, "Field-Aligned Currents in the Earth's Magnetotail," *J. Geophys. Res.*, **86**, 687 (1981).

Fritz, T. A., "Study of the High-Latitude, Outer-Zone Boundary Region for ≥ 40-keV Electrons with Satellite Injun 3," *J. Geophys. Res.*, **75**, 5387 (1970).

Gorney, D. J., A. Clarke, D. Croley, J. Fennell, J. Luhmann, and P. Mizera, "The Distribution of Ion Beams and Conics Below 8000 km," *J. Geophys. Res.*, **86**, 83 (1981).

Gurnett, D. A., and L. A. Frank, "Observed Relationships Between Electric Fields and Auroral Particle Precipitation," *J. Geophys. Res.*, **78**, 145 (1973).

Heelis, R. A., W. B. Hanson, and J. L. Burch, "AE-C Observations of Electric Fields Around Auroral Arcs," in *Physics of Auroral Arc Formation*, S.-I. Akasofu and J. R. Kan, eds., American Geophysical Union, Washington, DC, p. 154 (1981).

Heelis, R. A., J. D. Winningham, W. B. Hanson, and J. L. Burch, "The Relationship Between High-Latitude Convection Reversals and Energetic Particle Morphology Observed by Atmospheric Explorer," *J. Geophys. Res.*, **85**, 3315 (1980).

Hill, T. W., and M. E. Rassbach, "Interplanetary Magnetic Field Direction and the Configuration of the Day Side Magnetopause," *J. Geophys. Res.*, **80**, 1 (1975).

Hones, E. W., Jr., "Transient Phenomena in the Magnetotail and Their Relation to Substorms," *Space Sci. Rev.*, **23**, 393 (1979).

Hones, E. W., Jr., and K. Schindler, "Magnetotail Plasma Flow during Substorms: A Survey with IMP 6 and IMP 8 Satellites," *J. Geophys. Res.*, **84**, 7155 (1979).

Hones, E. W., Jr., S. J. Bame, and J. R. Asbridge, "Proton Flow Measurements in the Magnetotail Plasma Sheet Made with IMP 6," *J. Geophys. Res.*, **81**, 227 (1976).

Hones, E. W. Jr., J. R. Asbridge, S. J. Bame, and S. Singer, "Substorm Variations of the Magnetotail Plasma Sheet from $X_{SM} = -6R_E$ to $X_{SM} \approx -60R_E$," *J. Geophys. Res.*, **78**, 109 (1973).

Hones, E. W., Jr., D. N. Baker, S. J. Bame, W. C. Feldman, J. T. Gosling, D. J. McComas, R. D. Zwickl, J. A. Slavin, E. J. Smith, and B. T. Tsurutani, "Structures of the Magnetotail at $200R_E$ and Its Response to Geomagnetic Activity," *Geophys. Res. Lett.*, **11**, 5 (1984).

Huang, C. Y., and L. A. Frank, "A Statistical Study of the Central Plasma Sheet: Implications for Substorm Models," *Geophys. Res. Lett.*, **13**, 652 (1986).

Huang, C. Y., L. A. Frank, and T. E. Eastman, "High-Altitude Observations of an Intense Inverted-V Event," *J. Geophys. Res.*, **89**, 7423 (1984).

Hultqvist, B., H. Borg, P. Christophersen, W. Riedler, and W. Bernstein, "Energetic Protons in the keV Energy Range and Associated keV Electrons Observed at Various Local Times and Disturbance Levels in the Upper Ionosphere," NOAA Technical Report ERL 305-SEL 29, U.S. Dept. of Commerce, Boulder, CO (1974).

Johnstone, A. D., and J. D. Winningham, "Satellite Observations of Suprathermal Electron Bursts," *J. Geophys. Res.*, **87**, 2321 (1982).

Kirsch, E., S. M. Krimigis, E. T. Sarris, and R. P. Lepping, "Detailed Study on Acceleration and Propagation of Energetic Protons and Electrons in the Magnetotail During Substorm Activity," *J. Geophys. Res.*, **86**, 6727 (1981).

Krimigis, S. M., and E. T. Sarris, "Energetic Particle Bursts in the Earth's Magnetotail," in *Dynamics of the Magnetosphere*, S.-I. Akasofu, ed., D. Reidel, Dordrecht, Holland, p. 599 (1979).

Levin, S., K. Whitley, and F. S. Mozer, "A Statistical Study of Large Electric Field Events in the Earth's Magnetotail," *J. Geophys. Res.*, **88**, 7765 (1983).

Lotko, W., B. V. Ö. Sonnerup, and R. L. Lysak, "Nonsteady Boundary Layer Flow Including Ionospheric Drag and Parallel Electric Fields," *J. Geophys. Res.*, **92**, 8635 (1987).

Lui, A. T. Y., L. A. Frank, K. L. Ackerson, C.-I. Meng, and S.-I. Akasofu, "Systematic Plasma Flow During Plasma Sheet Thinnings," *J. Geophys. Res.*, **87**, 4815 (1977b).

Lui, A. T. Y., E. W. Hones, Jr., F. Yasuhara, S.-I. Akasofu, and S. J. Bame, "Magnetotail Plasma Flow During Plasma Sheet Expansions: Vela 5 and 6 and IMP 6 Observations," *J. Geophys. Res.*, **82**, 1235 (1977a).

Lundblad, J. Å., F. Soraas, and K. Aarsnes, "Substorm Morphology of >100 keV Protons," *Planet. Space Sci.*, **27**, 841 (1979).

Lyons, L. R., "Generation of Large-Scale Regions of Auroral Currents, Electric Potentials, and Precipitation by the Divergence of the Convection Electric Field," *J. Geophys. Res.*, **85**, 17 (1980).

Lyons, L. R., "Discrete Aurora as the Direct Result of an Inferred, High-Altitude Generating Potential Distribution," *J. Geophys. Res.*, **86**, 1 (1981).

Lyons, L. R., and O. de la Beaujardiere, "Critical Problems Requiring Coordinated Measurements of Large-Scale Electric Field and Auroral Distribution," in *Outstanding Problems in Solar Systems Physics: Theory and Instruments*, American Geophysical Union, Washington, DC, (1988).

Lyons, L. R., and D. S. Evans, "An Association Between Discrete Aurora and Energetic Particle Boundaries," *J. Geophys. Res.*, **89**, 2395 (1984).

Lyons, L. R., and A. Nishida, "Description of Substorms in the Tail Incorporating Boundary Layer and Neutral Line Effects," *Geophys. Res. Lett.*, **15**, 1337 (1988).

Lyons, L. R., and T. W. Speiser, "Evidence for Current-Sheet Acceleration in the Geomagnetic Tail," *J. Geophys. Res.*, **87**, 2276 (1982).

Lyons, L. R., J. F. Fennell, and A. L. Vampola, "A General Association Between Discrete Auroras and Ion Precipitation from the Tail," *J. Geophys. Res.*, **93**, 12,932 (1988).

Lyons, L. R., M. Schulz, and J. F. Fennell, "Trapped-Particle Evacuation: Source of Magnetotail Bursts and Tailward Flows?" *Geophys. Res. Lett.*, in press (1989).

Marklund, G., W. Baumjohann, and I. Sandahl, "Rocket and Ground-Based Study of an Auroral Break-up Event," *Planet. Space Sci.*, **31**, 207 (1983).

Mauk, B. H., "Quantitative Modeling of the Convection Surge Mechanism of Ion Acceleration," *J. Geophys. Res.*, **91**, 13,423 (1986).

McPherron, R. L., "Satellite Studies of Magnetospheric Substorms on August 15, 1988," *J. Geophys. Res.*, **78**, 3044 (1973).

Maynard, N. C., D. S. Evans, B. Maehlum, and A. Egeland, "Auroral Vector Electric Field and Particle Comparisons, 1, Premidnight Convection Topology," *J. Geophys. Res.*, **82**, 2227 (1977).

Mizera, P. F., and J. F. Fennell, "Signatures of Electric Fields from High and Low Altitude Particle Distributions," *Geophys. Res. Lett.*, **4**, 311 (1977).

Mizera, P. F., J. F. Fennell, D. R. Croley, Jr., A. L. Vampola, F. S. Mozer, R. B. Torbert, M. Temerin, R. Lysak, M. Hudson, C. A. Cattell,

R. J. Johnson, R. D. Sharp, A. Ghielmetti, and P. M. Kintner, "The Aurora Inferred From S3-3 Particles and Fields," *J. Geophys. Res.*, **86**, 2329 (1981).

Mozer, F. S., C. W. Carlson, M. K. Hudson, R. B. Torbert, B. Parady, Y. Yatteau, and M. C. Kelley, "Observations of Paired Electrostatic Shocks in the Polar Magnetosphere," *Phys. Rev. Lett.*, **38**, 292 (1977).

Nishida, A., and E. W. Hones, Jr., "Association of Plasma Sheet Thinning with Neutral Line Formation in the Magnetotail," *J. Geophys. Res.*, **79**, 535 (1974).

Nishida, A., and N. Nagayama, "Synoptic Survey for the Neutral Line in the Magnetotail," *J. Geophys. Res.*, **78**, 3782 (1973).

Nishida, A., H. Hayakawa, and E. W. Hones, Jr., "Observed Signatures of Reconnection in the Magnetotail," *J. Geophys. Res.*, **86**, 1422 (1981).

Nishida, A., Y. K. Tulunay, F. S. Mozer, C. A. Cattell, E. W. Hones, Jr., and J. Birn, "Electric Field Evidence for Tailward Flow at Substorm Onset," *J. Geophys. Res.*, **88**, 9109 (1983).

Nishida A., S. J. Bame, D. N. Baker, G. Gloeckler, M. Scholer, E. J. Smith, T. Terasawa, and B. Tsurutani, "Assessment of the Boundary Layer Model of the Mangetotail Substorm," *J. Geophys. Res.*, **93**, 5579 (1988).

Pedersen, A., C. A. Cattell, C.-G. Fälthammar, K. Knott, P.-A. Lindqvist, R. H. Manka, and F. S. Mozer, "Electric Fields in the Plasma Sheet and Plasma Sheet Boundary Layer," *J. Geophys. Res.*, **90**, 1231 (1985).

Roederer, J. G., *Dynamics of Geomagnetically Trapped Radiation*, Springer-Verlag, New York (1970).

Rostoker, G., "Dialogue on the Magnetotail," *Eos. Trans. AGU*, **66**, 96 (1985).

Rostoker, G., and R. Boström, "A Mechanism for Driving the Gross Birkeland Current Configuration for the Auroral Oval," *J. Geophys. Res.*, **81**, 235 (1976).

Rostoker, G., and T. E. Eastman, "A Boundary Layer Model for Magnetospheric Substorms," *J. Geophys. Res.*, **92**, 12,187 (1987).

Sarris, E. T., and W. I. Axford, "Energetic Protons Near the Plasma Sheet Boundary," *Nature*, **277**, 460 (1979).

Sarris, E. T., S. M. Krimigis, and T. P. Armstrong, "Observations of Magnetospheric Bursts of High-Energy Protons and Electrons at ~$35R_E$ with IMP 7," *J. Geophys. Res.*, **81**, 2341 (1976).

Schulz, M., "Plasma Boundaries in Space," in *Physics of Solar Planetary Environments*, D. J. Williams, ed., American Geophysical Union, Washington, DC, p. 491 (1976).

Schulz, M., "Magnetospheric and Interplanetary Electrostatics: A Simple But Explicit Model," in *High Latitude Electric Fields in the Magnetosphere and Ionosphere*, paper 18, AGU Chapman Conference, Yosemite (1980).

Sharber, J. R., "The Continuous (Diffuse) Aurora and Auroral-E Ionization," in *Physics of Space Plasmas*, T. S. Chang, B. Coppi, and J. R. Jasperse, eds., Scientific Publishers, Cambridge, MA, p. 115 (1982).

Speiser, T. W., "Particle Trajectories in Model Current Sheets, 1, Analytical Solutions," *J. Geophys. Res.*, **70**, 4219 (1965).

Speiser, T. W., "Conductivity Without Collisions or Noise," *Planet. Space Sci.*, **18**, 613 (1970).

Spjeldvik, W. N., and T. A. Fritz, "Energetic Ion and Electron Observations of the Geomagnetic Plasma Sheet Boundary Layer," *J. Geophys. Res.*, **86**, 2480 (1981).

Stasiewicz, K., "Generation of Magnetic Field-Aligned Currents, Parallel Electric Fields, and Inverted-V Structures by Plasma Pressure Inhomogeneities in the Magnetosphere," *Planet. Space Sci.*, **33**, 1037 (1985).

Swift, D. W., and D. A. Gurnett, "Direct Comparison Between Satellite Electric Field Measurements and the Visual Aurora," *J. Geophys. Res.*, **78**, 7306 (1973).

Vasyliunas, V. M., "Theoretical Models of Magnetic Field Line Merging, 1," *Rev. Geophys. Space Phys.*, **13**, 303 (1975).

Vasyliunas, V. M., "Steady State Aspects of Magnetic Field Merging," in *Magnetic Reconnection in Space and Laboratory Plasmas*, E. W. Hones, Jr., ed., American Geophysical Union, Washington, DC, p. 25 (1984).

Wagner, J. S., J. R. Kan, and S.-I. Akasofu, "Particle Dynamics in the Plasma Sheet," *J. Geophys. Res.*, **84**, 891 (1979).

Whalen, J. A., "A Quantitative Description of the Spatial Distribution and Dynamics of the Energy-Flux in the Continuous Aurora," *J. Geophys. Res.*, **88**, 7155 (1983).

Williams, D. J., "Energetic Ion Beams at the Edge of the Plasma Sheet: ISEE 1 Observations Plus a Simple Explanatory Model," *J. Geophys. Res.*, **86**, 5507 (1981).

Winningham, J. D., and W. J. Heikkila, "Polar Cap Electron Fluxes Observed with Isis 1," *J. Geophys. Res.*, **79**, 949 (1974).

Winningham, J. D., F. Yasuhara, S.-I., Akasofu, and W. J. Heikkila, "The Latitudinal Morphology of 10-eV to 10-keV Electron Fluxes During Magnetically Quiet and Disturbed Times in the 2100–0300 MLT Sector," *J. Geophys. Res.*, **80**, 3148 (1975).

IV-4. AURORAL LUMINOSITY AND ITS RELATIONSHIP TO MAGNETOSPHERIC PLASMA DOMAINS

Yu. I. Galperin* and Ya. I. Feldstein[†]

A controversy exists concerning the projection of the "typical" nightside auroral oval of discrete forms (inverted-Vs, bands, and narrow arcs) from the magnetospheric tail regions. The most popular point of view is that the oval is mapped from the boundary plasma sheet (BPS). But several researchers, including the authors, advocate the mapping from the main part of the nightside plasma sheet between say, 5 and 50 Earth radii (R_E), depending on activity. Direct experimental arguments, including new results from the ARCAD/Aureol 3 satellite and field-line tracings according to contemporary magnetospheric models for the steady near-midnight magnetosphere of moderate activity, are summarized with the following main conclusions: (1) The zone of diffuse auroral luminosity, equatorward of the oval of discrete auroral forms and of the boundary of stable trapping for energetic electrons, maps from the inner magnetosphere between the inner boundary of the plasma sheet and the steady convection boundary (the "instantaneous plasmapause"); (2) The auroral oval maps from the main, or central (or low-latitude) plasma sheet (CPS); and (3) The diffuse auroral band poleward of the most poleward arc, or inverted-V event, of the oval (narrow during disturbed times, but extended during quiet times) maps from the boundary region of the plasma sheet (BPS).

1. INTRODUCTION

The problem of mapping auroral oval features such as discrete luminous forms (arcs, bands, inverted-Vs) that are excited by accelerated and highly structured electron beams from the outer magnetosphere is still one of the key problems in the physics of geospace.

Two viewpoints can be followed since the pioneering works of the early 1960s. The first, expounded in a number of works (*Valchuk et al.* [1979]; *Feldstein and Galperin* [1985] and the references therein), claimed that the auroral oval was a mapping of an enormous magnetotail "plasma sheet," including the neutral sheet, onto the conducting ionosphere. In this case the magnetic flux threading the auroral oval at nightside ionospheric altitudes is closed through the plasma sheet, while the field-aligned currents generated in the plasma sheet are closed in the oval through the conducting ionosphere, which is ionized by plasma sheet particles. This concept is designated as VFSL (the initials of the names of its main proponents, Vasyliunas, Feldstein, Starkov, Lassen), although the concept was either discussed or used in numerous other works. We illustrate it here by Figure 1, adapted from *Feldstein and Galperin* [1985], and will discuss here only its nightside part. In terms of the alternative concept, the auroral oval of discrete forms is mapped from the BPS, which is located at the outer edge of the CPS. These designations follow the comprehensive description of

the tail plasmas by *Eastman et al.* [1984, 1985]. In this concept the extended-magnetotail CPS is mapped to the diffuse auroral zone equatorward of the oval. This concept was used extensively during the last decade after the publications by *Winningham et al.* [1975], *Lui et al.* [1977], and others (see, for example, the reviews by *Frank* [1985] and *Fairfield* [1987]). It is illustrated here by Figure 2, taken from *Eastman et al.* [1984, 1985].

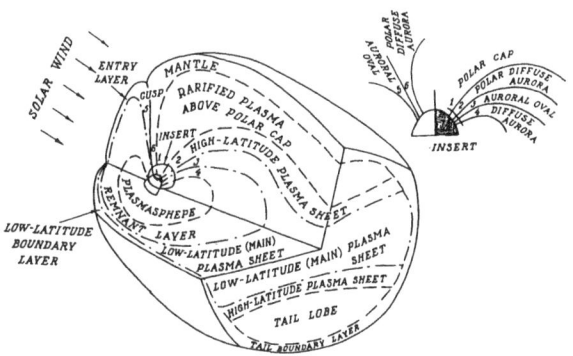

Figure 1—Three-dimensional diagram of relationships of various magnetospheric plasma domains to various regions of luminosity. The magnetosphere is cut in the meridional and equatorial planes to show its internal structure. The polar diffuse aurora is mapped from the high-latitude plasma sheet (2) and from the mantle (6). The auroral oval is mapped from the low-latitude (central, main) plasma sheet (3) and from the entry layer (5). The diffuse auroral region equatorward of the auroral oval is mapped from the remnant layer (4). (Adapted from *Feldstein and Galperin* [1985].)

*Institute of Space Research of the USSR Academy of Sciences, Moscow, 117810, Profsoyuznaya 84/32, USSR.

[†]Institute of Terrestrial Magnetism, Ionosphere and Radio Wave Propagation of the USSR Academy of Sciences, Troitsk, Moscow Region, USSR.

Auroral Physics, edited by C.-I. Meng, M. J. Rycroft and L. A. Frank. © Cambridge UP 1991

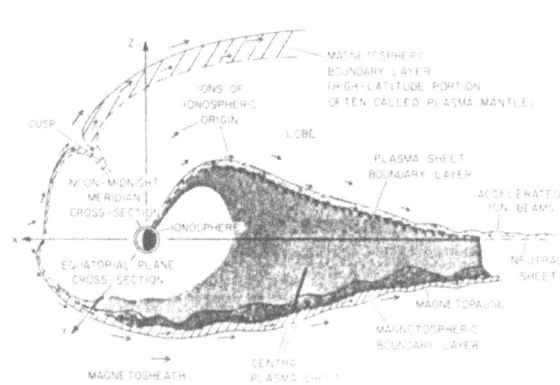

Figure 2—Plasma regimes of the Earth's magnetosphere. The arrows represent typical bulk flow velocities as observed. In the equatorial plane cross section, the tail size does not generally decrease with increasing distance, as implied in this figure. The plasma sheet boundary layer is linked via field lines to the high-latitude part of the auroral region, suggesting an association with the discrete aurora. Diffuse auroras at somewhat lower latitudes are mapped from the near-Earth central plasma sheet. (From *Eastman et al.*, 1985].)

According to *Eastman et al.* [1985]: "The plasma sheet boundary layer is linked via field lines to the high-latitude portion of the auroral region at other local times, suggesting the association with discrete aurora. Diffuse auroras at somewhat lower latitudes map-out to the near-Earth central plasma sheet." The scheme leaves it obscure as to where the band of diffuse aurora poleward from discrete auroral forms is mapped.

This mapping and the so-called "boundary plasma sheet model of auroral substorm," which is based on it, achieved wide attention and were substantiated by various, though indirect, comparisons with the plasma characteristics and dynamics, field-aligned currents at high and low altitudes, and theoretical considerations (see critical review by *Baumjohann* [1988]).

This remaining controversy in auroral mapping complicates not only descriptions of the results of particular experiments but also the interpretation of available measurements and the planning of important projects including multipoint measurements in the magnetosphere and ground-based observations of auroral events. For example, it is not yet clear where a substorm starts. Ground-based data indicate that onset is often associated with a stable extended auroral arc at the equatorward boundary of the oval in its near-midnight sector [*Akasofu*, 1968]. Can such an auroral arc be mapped from or close to, the inner boundary of the plasma sheet, as, in fact, follows from VFSL? If so, a typical substorm arises from a violation of the stability of an electrodynamic magnetospheric config-

uration deep inside the plasma sheet (in fact, rather close to the Earth, e.g., $< 10R_E$). Indeed, many independent observations are consistent with this scheme (e.g., *Takahashi et al.* [1987] and *Lui et al.* [1988]).

On the other hand, some measurements [*Frank et al.*, 1976] have shown that the violent processes of plasma heating and onset of turbulence can occur on the flanks of the plasma sheet and within its boundary layers in the magnetospheric tail. These active and localized regions in the boundary layers are called "fireballs" [*Frank*, 1976, 1985; *Frank et al.*, 1976]. If at least some substorms are caused, or in some way initiated, by the formation of fireballs, their generation is due to processes occurring not within, but at the edge of, the plasma sheet. These are but a few examples of problems that arise when any particular coordinated experiment, including measurements in the tail and in the auroral ionosphere, is being planned or analyzed.

The successful choice of optimal satellite orbits and of a measurement program in a project such as INTERBALL (a satellite and a subsatellite in the tail at the $\sim 200,000$-km apogee, together with a satellite and a subsatellite in the auroral magnetosphere at a $\sim 20,000$-km apogee in combination with low-orbiting satellites and ground-based measurements) depends on knowing when and where the particular cause-and-effect relationships can be expected along the geomagnetic field lines during a substorm.

We limit ourselves here to a steady "typical" oval of discrete auroral forms (including arcs and inverted-V structures) at Kp values from 1–4 and in the 2200–0000 MLT sector. As the mapping of the oval to the BPS has generated wide discussion and support elsewhere (see the reviews by *Lyons* [this volume] and *Lundin et al.* [this volume]), we concentrate here only on brief formulations of the arguments for, and deductions from, the alternative point of view.

We review here the concept of mapping of the nightside steady oval of discrete auroral forms from the low-latitude (main or central) plasma sheet (CPS on Figure 2), and of mapping the diffuse auroral zone equatorward of the oval onto the inner magnetosphere L-shells of the outer radiation belt. This latter region was called "the earthward edge of plasma sheet" by *Frank and Ackerson* [1971], and the "remnant layer" by *Feldstein and Galperin* [1985]. This region of the near-Earth space encompasses the inner magnetospheric domain where the large-scale plasma convection from the tail plasma sheet still proceeds, i.e., down to the instantaneous plasmapause, which is meant to be the magnetospheric large-scale convection boundary (not just the region of some particular value of the cold plasma density). We present additional arguments that the equatorial border of the diffuse auroral zone coincides with the plasmapause on the nightside and that both have the same time histories.

We then summarize the new observational data on the specific region of soft electron precipitation—the polar diffuse zone [*Eather*, 1969; *Valchuk et al.*, 1979; *Feldstein and Galperin*, 1985], located just poleward from the most poleward inverted-V structure (or auroral arc) during steady conditions. This polar diffuse zone, which is rather narrow during disturbed conditions, ~0.5–1° of latitude, adjacent to the polar oval arc, has been observed by various experimental methods. In this zone the electrons are of energies ≲ 1 keV and generally soften poleward, while in the ion spectra specific dispersed structures with energies rising up to at least 20 keV have sometimes been recently found from the Aureol 3 satellite [*Kovrazhkin et al.*, 1987]. In our opinion, this polar diffuse zone of soft electron precipitation is the ionospheric signature of the high-latitude plasma sheet (HLPS) (or the boundary plasma sheet (BPS)). The similarity of the ion dispersion structures here and those of ion beams observed in the BPS at high altitudes [*Takahashi and Hones*, 1988] suggests that this layer be identified with the tail BPS projection.

2. BASIC DATA AND DEFINITIONS OF AURORAL OVAL AND PLASMA SHEET

The plasma sheet, with a 0.1–1.0 cm^{-3} charged-particle number density, a 0.5–5.0 keV ion temperature, a 0.2–2.0 keV electron temperature, with low large-scale plasma bulk velocities ranging from 0–10^3 km s^{-1}, and a 10–30 nT magnetic field, is the main reservoir of the auroral plasma on the closed field lines in the magnetospheric nightside sector. The hot (auroral) plasma distribution in the nightside sector is delineated by an inner boundary of the plasma sheet. At this boundary, the rise of the mean electron energy prevailing inside the plasma sheet, because of adiabatic acceleration by the earthward drift, is usually replaced by a general softening of the electron energy spectrum as the radial distance decreases [*Vasyliunas*, 1968; *Schield and Frank*, 1970]. Simultaneously, the precipitating electron energy density also decreases. The region where these changes occur, between the inner boundary of the plasma sheet and the plasmapause, is several Earth-radii wide in the equatorial plane of the nightside magnetosphere. The outer boundary of the "earthward edge of the plasma sheet," as defined by *Frank* [1971, see his Figure 8] nearly coincides with the so-called trapping boundary for the >45 keV electrons. The high-altitude measurements [*Serlemitsos*, 1966; *West et al.*, 1978] have shown that this boundary is sharp through the nightside magnetosphere, but

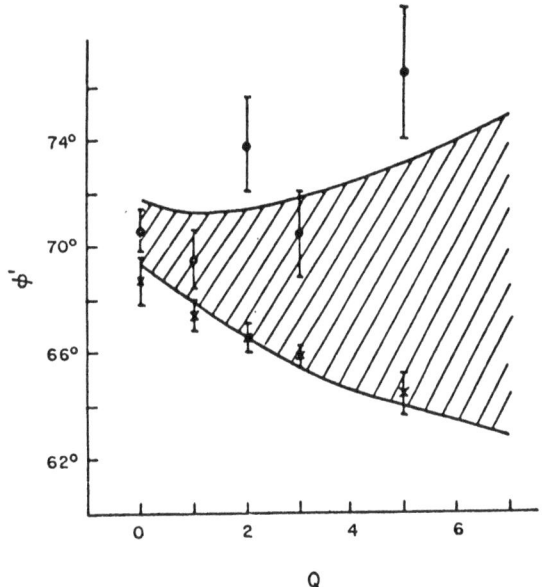

Figure 3—The region of the discrete auroral forms in the zenith, the auroral oval (in corrected geomagnetic latitude φ'), in the midnight sector, versus with the Q-index of magnetic disturbances during the IQSY period (shaded). The boundary of stable trapping of electrons with the energy >35 keV (crosses); the threshold boundary of electrons with E > 35 keV (circles). The bars show r.m.s. deviations. (From *Feldstein and Starkov* [1970].)

is noticeably extended toward the tail with respect to a dipolar magnetic field. Thus the outer radiation belt (the stable-trapping region) adjoins the inner boundary of the plasma sheet, beyond which stable trapping becomes impossible, presumably because of the curvature of the magnetic field in the current sheet [*Sergeev and Tsyganenko*, 1982; *Sergeev et al.*, 1983].

The same trapping boundary can also be easily defined at low altitudes. It is noted that this boundary at auroral altitudes is near the equatorial border of the oval [*Feldstein and Starkov*, 1970]. We reproduce here their plot as Figure 3, and two examples of the satellite measurements from *Gurnett and Frank* [1973], as Figure 4. All other figures from this paper, as well as from many others (e.g., *Akasofu*, [1974], *Deehr et al.* [1976], *Valchuk et al.* [1979], *Lyons and Evans* [1984]) are consistent with the conclusion by *Deehr et al.* [1976]: "The diffuse auroral region is associated with high-energy stably trapped energetic electrons, and the discrete aurora is poleward of the stable electron trapping boundary."

This energetic particles population boundary, where the isotropization of the angular distribution occurs, can be used as a natural field-line tracer, because it can be measured both at low and high magnetospheric altitudes. *Sergeev and Tsyganenko* [1982] related this boundary to a particular value of the ratio of the par-

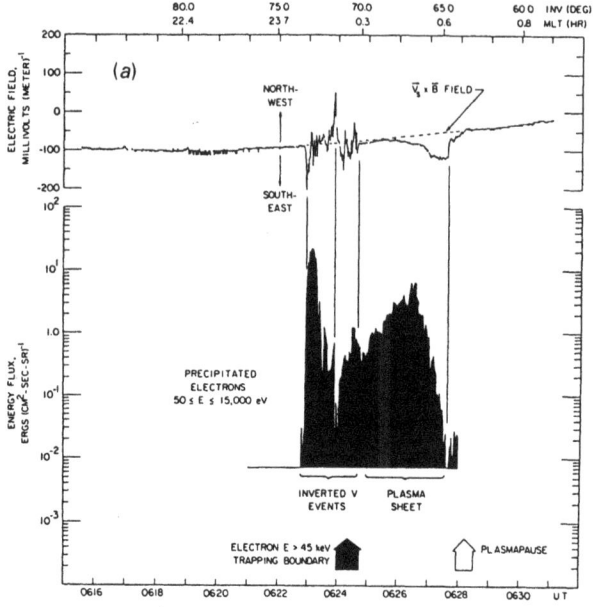

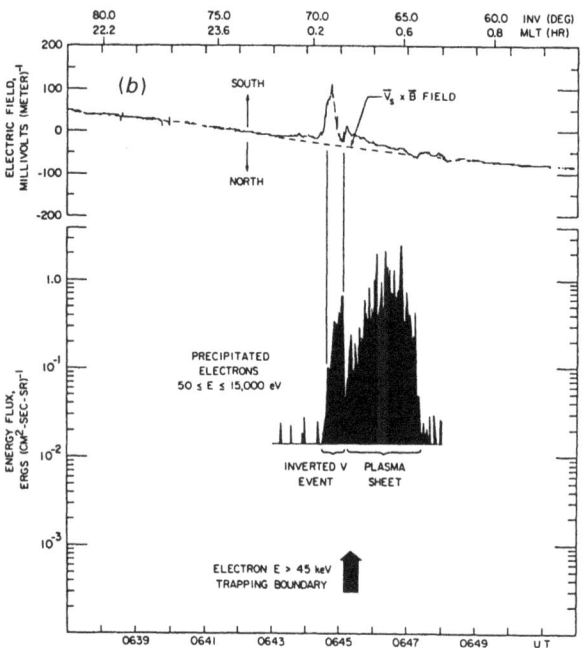

Figure 4—(*a*) The electric field and precipitating electron energy fluxes for orbit 3667 of the Injun 5 over the northern hemisphere on June 6, 1969. The equatorward termination of the convection zone is coincident with the plasmapause location as determined with simultaneous VLF electric field measurements. (*b*) The same presentation for orbit 3655 over the northern hemisphere on June 5, 1969. (From *Gurnett and Frank* [1973].)

ticle's Larmor radius to the radius of curvature of the field line at the earthward edge of the tail current sheet; that is, to the inner boundary of the plasma sheet current region. This relation for particles of various magnetic rigidities together with their respective "trapping boundaries" was used in the construction of the best available magnetospheric model by *Tsyganenko* [1987]. If we take the corrected geomagnetic latitudes from *Feldstein and Starkov* [1970] for conditions close to Kp = 1, 3, and 5+ as 67.8°; 65.5°, and 64°, respectively, then from that model we find the mapped equatorial distances in the tail as $\sim 8.9R_E$, $\sim 7.5R_E$, and ~ 6.5-$7.0R_E$—that is, in "the near-Earth plasma sheet"—which are rather close to the geostationary orbit. This is consistent with the observations of particle injections to the geostationary orbit at moderate-to-high activity levels and with the regular ground-based observations of discrete auroral phenomena at the foot of the geostationary satellites. A good example of auroral arcs that map inside the geostationary orbit is shown in Figure 2 of *Mauk and Meng* [this volume] (from *Meng et al.* [1979]).

As to the polar border of the auroral oval, a bright and active auroral arc extended in local time is often the only discrete auroral feature distinguished on the ultraviolet auroral images from DE 1 and Viking satellites with a width resolution of 50–100 km. In these conditions the width of the auroral oval of discrete forms can be misinterpreted and the diffuse zone equatorward of the oval could be misinterpreted. But the ground-based observations usually allow registration of class 1 or class 2 discrete auroral forms imbedded in the diffuse glow equatorward from the most poleward auroral arc till the equatorial border of the oval. Also, the particle measurements from satellites usually show a band of inverted-V structures of significant latitudinal width located just above the auroral discrete forms of the oval (e.g., *Deehr et al.* [1976], *Valchuk et al.* [1979]). Thus the full latitudinal width of the oval, with imbedded stationary or dynamical discrete auroral forms, must be considered in the mapping to the tail plasma sheet. According to the model by *Tsyganenko* [1987], the polar border of the oval is mapped to >40-$70R_E$; that is, to the limit of the model.

As the distance along the Z-axis from the neutral sheet increases inside the plasma sheet, the mean parameters of ions and electrons vary in the same manner as when moving away (along the X-axis) from the inner boundary of the plasma sheet, namely, the mean energy and the energy density of electrons and ions decrease. This was explained to be the result of the adiabatic compression in the plasma sheet due to the convection earthward and toward the plasma-sheet

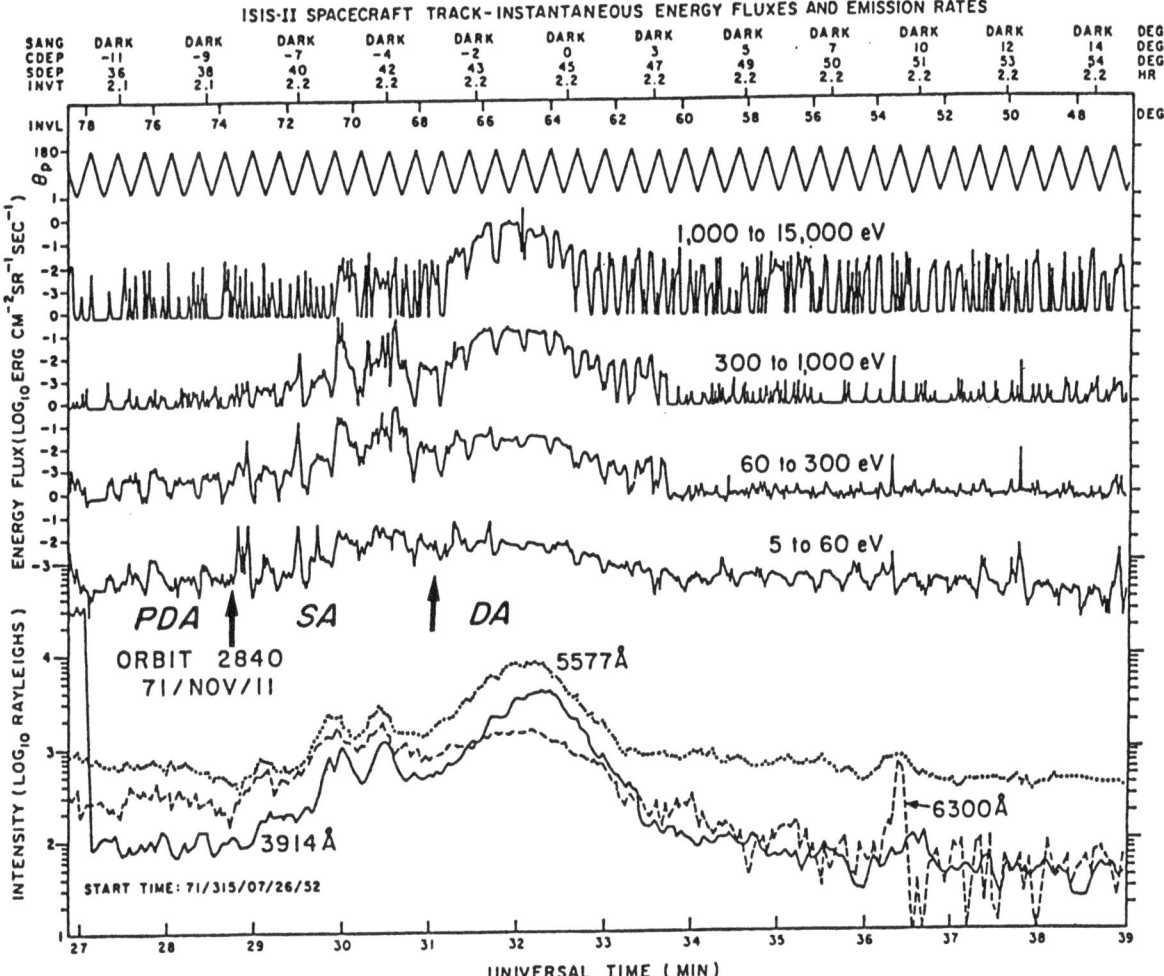

Figure 5—Intensity variations of the precipitating electrons of various energies and of the auroral intensities during an Isis 2 traversal of the night sector of the auroral oval at 0727-0739 UT on November 11, 1971 [*Shepherd et al.*, 1980]. The upper line shows the variations of the satellite pitch angle. Shown below are the intensity variations of the 6300, 5577, and 3914 Å emissions along the Isis 2 trajectory. DA, diffuse auroras; SA, structured auroras; PDA, polar diffuse auroras. The arrows indicate the approximate positions of the boundaries between auroras of various types. (From *Feldstein and Galperin* [1985].)

midplane. It was noted that similar variations of the mean electron and proton energies inside the oval from its equatorward boundary to its poleward boundary take place if variations due to the inverted-V structures are excluded [*Galperin et al.*, 1978, *Sauvaud et al.*, 1985]. These variations in the mean energies of the particles are qualitatively consistent with the proposed mapping of the nightside plasma sheet to the oval.

As has been already noted, an additional band of soft electron precipitation is usually observed poleward from the last auroral arc (or inverted-V) structure. This band gives rise to subvisual luminosity (mostly in the 630.0 nm oxygen emission) and feeble ionospheric ionization. We reproduce as Figure 5 (taken from *Feldstein and Galperin* [1985]) the results from an Isis 2 pass and have added our markings of the polar diffuse aurora (PDA), discrete, or structured, aurora (SA), and the equatorial diffuse aurora (DA). Good examples of the polar diffuse zone both in electron and ion measurements with high space/time resolution from Aureol 3 can be found in the review by *Bosqued* [this volume]. During magnetically quiet intervals (for IMF $B_z > 0$, as a rule) this high-latitude soft diffuse precipitation region expands poleward and can occupy a noticeable portion of the polar cap [*Meng*, 1981].

The polar auroral electron precipitation region was divided into harder and softer zones by *Meng and Makita* [1986] and *Makita et al.* [1985, 1988] on the

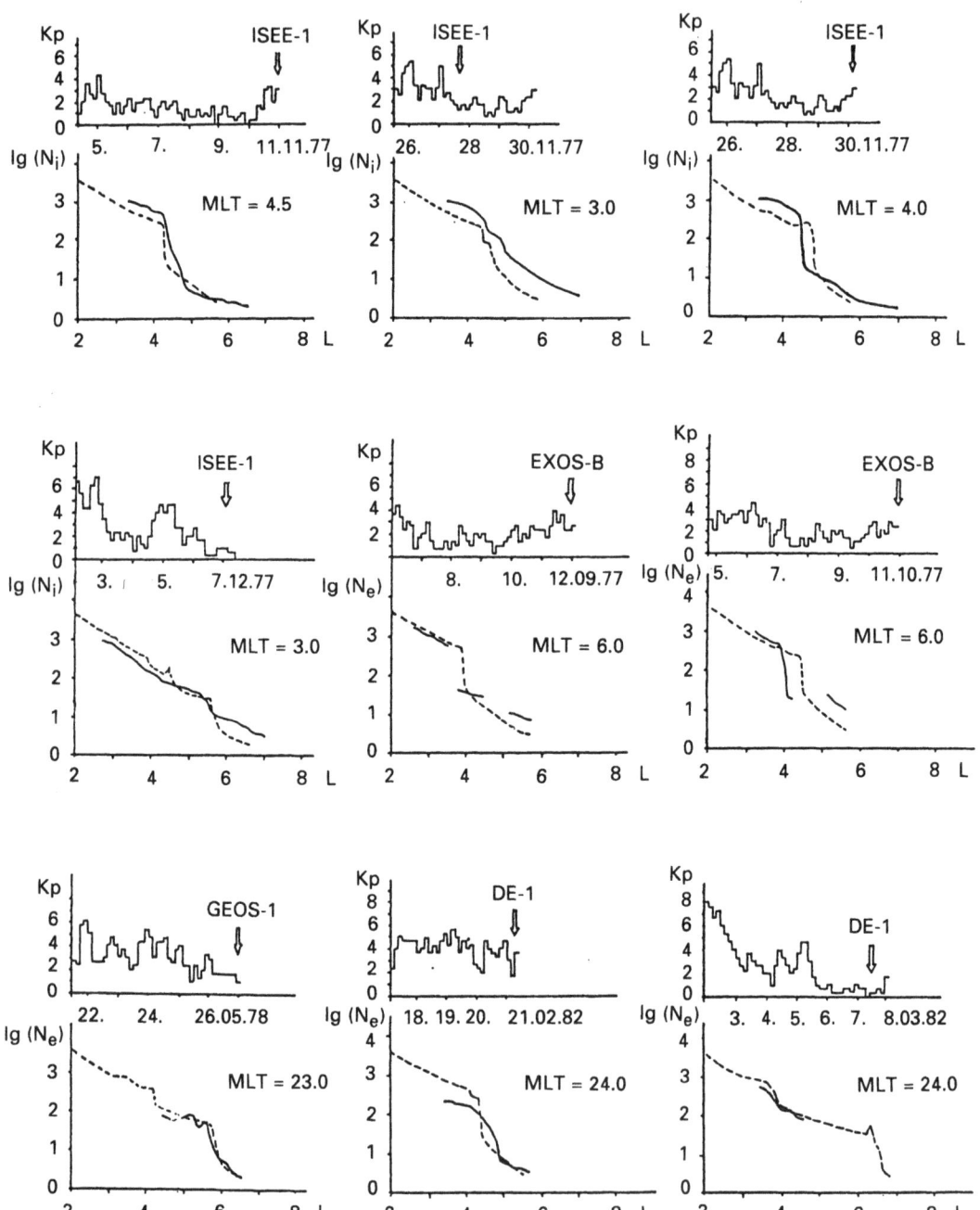

Figure 6—Three-dimensional model of the electron density in the outer plasmasphere (based on the supposition that the equatorial border of the diffuse aurora coincides with the large-scale convection boundary (i.e., with the instantaneous plasmapause), as compared with direct measurements of the electron density at high altitudes taken from published data of various satellites. (Adapted from *Soloviev et al.* [1989].)

basis of DMSP satellite data; they arbitrarily took a mean electron energy of 0.5 keV to separate the two zones. In our opinion, to identify the plasma-domain boundaries in the magnetosphere, the morphological characteristics of particle spectra and of auroral forms, and the boundaries where the plasma characteristics change systematically should rather be used for identification.

It must be admitted, however, that the physical processes occurring in the poleward boundary of the auroral oval and, particularly, in the polar soft diffusive zone during magnetic storms and substorms and during quieting conditions have been insufficiently studied. They therefore constitute one of the important present-day problems of magnetospheric physics, and the proposed mapping of this polar diffuse zone from the BPS and the distant neutral line [*Kovrazhkin et al.*, 1987], though very plausible, needs further confirmation.

In the sections that follow, we list some main aspects of the proposed mappings and provide only brief commentaries. In the last section we discuss some implications for the substorm initiation localization and some interpretation problems.

3. EVIDENCE THAT SUPPORTS MAPPING THE REGION OF DIFFUSE LUMINOSITY EQUATORWARD OF THE AURORAL OVAL TO THE REGION BETWEEN THE INNER BOUNDARY OF THE PLASMA SHEET AND THE PLASMAPAUSE

The geomagnetic field models [*Tsyganenko and Usmanov*, 1982; *Tsyganenko*, 1987], which are sufficiently accurate in the inner magnetosphere, confirm that the diffuse auroral zone equatorward of the oval of discrete forms is mapped by magnetic field lines from the inner magnetosphere, namely, from geocentric distances of 4–$12R_E$ in the equatorial plane (the remnant layer), depending on magnetic activity.

The sharp boundary of the >30 keV electron stable-trapping region observed at low altitudes coincides approximately with the boundary between the diffuse precipitation region equatorward of the auroral oval and the oval of discrete auroral forms [*Frank and Ackerson*, 1971, 1972; *Gurnett and Frank*, 1973; *Feldstein and Starkov*, 1970; *Akasofu*, 1974; *Deehr et al.*, 1976; *Lui et al.*, 1977; *Lui and Burrows*, 1978; *Valchuk et al.*, 1979]. The stable-trapping region at high magnetospheric altitudes is bounded by the flux tubes extending out to ~ 8–$12R_E$ [*Serlemitsos*, 1966; *Frank*, 1971; *Aubry et al.*, 1972; *West et al.*, 1973, 1978]. In this case the outer radiation belt at high magnetospheric altitudes adjoins the plasma sheet. According to high-altitude measurements, *Vasyliunas* [1968] states that the

plasma sheet inner boundary during quiet periods in the premidnight sector is located at $\sim 11R_E$.

In particular, the geostationary orbit in the nightside sector is mapped to the diffuse zone during magnetically quiet periods and during weak disturbances [*Eather et al.*, 1976]. The precipitating particles' energy spectra and intensities in the diffuse zone, measured simultaneously on one and the same flux tube at ionospheric altitudes and in the near-equatorial plane of the magnetosphere, at geosynchronous orbit are in practice identical, indicating that pitch angle scattering of the trapped particles under conditions ranging from weak to strong diffusion is the main reason for the precipitation of these trapped particles.

The spectrum of electrons in the diffuse auroral zone near its equatorial boundary gets monotonically softer, while the mean energy of the electrons falls with decreasing latitude to the equatorward diffuse precipitation boundary (DPB) [*Winningham et al.*, 1975, 1978; *Valchuk et al.*, 1979; *Gussenhoven et al.*, 1981; *Tanskanen et al.*, 1981]. Similarly, the mean energy of auroral electrons decreases, and their spectrum gets softer, nearer the Earth in the near-equatorial magnetosphere, between the inner boundary of the plasma sheet and the plasmapause [*Vasyliunas*, 1968, 1972; *Schield and Frank*, 1970]. These facts seem to lead to the conclusion that the diffuse luminosity located in the stable-trapping zone equatorward of the oval is mapped into the inner magnetosphere onto the outer radiation belt region adjoining, from inside, the inner boundary of the magnetotail plasma sheet (beyond which any stable trapping is no longer possible). This is one of the important differences between our scheme and the scheme proposed by *Winningham et al.* [1975]; *Lui et al.* [1977], and *Eastman et al.* [1985], which maps this region from the CPS in the tail.

The low-energy charged particles that drift earthward from the tail due to steady-state magnetospheric convection are permanently carried inside the stable-trapping zone, thereby forming a diffuse precipitation zone that reaches the convection boundary, i.e., the plasmapause. Here the plasmapause is meant to be the instantaneous boundary of the steady-state large-scale convection according to classic definitions by *Nishida* [1966] and *Brice* [1967]. For details about the relation between the diffuse auroral zone, diffuse luminosity, and the plasmapause, see *Feldstein and Galperin* [1985]. The empirical models of this precipitation boundary were constructed from satellite observations by *Galperin et al.* [1977], *Gussenhoven et al.* [1981, 1983], and *Valchuk et al.* [1986].

The hypothesis [*Galperin et al.*, 1977; *Sauvaud et al.*, 1983; *Fairfield and Vinas*, 1984] of the identification of this boundary with the instantaneous large-scale convection boundary (plasmapause) was tested by constructing the three-dimensional model of the outer plasmasphere electron density in *Soloviev et al.* [1989]. The magnetic activity prehistory as described by the *Kp* index was used to predict the prehistory of the plasmapause positions for a particular moment and place in the subauroral magnetosphere, together with simplified models of the plasma convection and filling rate due to the polar wind. It appeared that the positions of the density jumps in the equatorial electron density profiles as measured by high-altitude satellites can be reproduced by the above-mentioned model with an average error of $\pm 0.3 R_E$. Even the general shapes and density values of these radial profiles could be reasonably modeled, despite the crudeness of the convection model used. This means that the hypothesis that this plasma boundary is the convection boundary, i.e., the plasmapause, is consistent with the available information of the electron density radial profiles in the outer plasmasphere. We show in Figure 6 sample comparisons of the model by *Soloviev et al.* [1989], with published data of direct measurements of the electron density together with the *Kp*-prehistories in each particular case. So it is conceivable that the hot plasma is carried from the tail by the large-scale steady convection down to this boundary populating the diffuse auroral zone.

Much more powerful injections of the hot plasma occur during substorms into the inner magnetosphere, leading to strong enhancement of the precipitation. But the experiments show that after the injection the low-energy particles closely follow the steady convection trajectories. So the earthward/equatorward expansion of the above-described boundary is rather slow but increases by increments during a series of injections, thus making this border rather inertial, in time, except under specific conditions [*Galperin et al.*, 1977; *Sauvaud et al.*, 1983; *Feldstein and Galperin*, 1985]. If the magnetospheric electric field decreases sharply, the injected low-energy particles—"the plasma clouds"—can proceed slowly, precipitating for many hours or tens of hours still imitating, during the first several hours, the preceding precipitation boundary.

In the dusk and midnight sectors, region 2 of the field-aligned currents is mapped from the diffuse luminosity region [*Potemra*, 1977; *Feldstein et al.*, 1984] equatorward of the auroral oval. The modeling efforts were able to reproduce here the Alfvén layer as the source of the large-scale field-aligned currents of region 2, which were due to the plasma pressure gradients in the inner magnetosphere [*Harel et al.*, 1981; *Ponomarev*, 1981; *Pudovkin and Zakharov*, 1984].

We are of the opinion that the arguments presented above indicate that the diffuse auroral precipitation zone equatorward of the oval of discrete auroral forms is located in the stable-trapping zone of energetic particles (the outer radiation belt) and is mapped from the near-Earth portion of the plasma sheet (the remnant layer in Figure 1).

If, however, the equatorward boundary of diffuse auroral precipitation is defined to be the equatorward boundary of the auroral oval (as was done recently elsewhere (e.g., *Eather et al.* [1976], *Gussenhoven et al.* [1981], *Kamide and Winningham* [1977], *Lui et al.* [1982], *Horwitz et al.* [1982]), this definition means that the boundary of the oval coincides with the plasmapause, which does not obviously conform to the commonly accepted concepts concerning the oval.

The equatorward border of diffuse auroral luminosity is regarded in some cases to be associated with the inner boundary of the plasma sheet (e.g., *Kamide and Winningham* [1977], *Eather at al.* [1976], *Lui et al.* [1982], *Horwitz et al.* [1982], *Gussenhoven et al.* [1983]). In these cases, the plasma sheet should include the outer radiation zone, a concept that is at variance with the concept that any stable trapping of energetic particles is impossible in the plasma sheet, because of the magnetic field distortion by the tail current sheet.

4. EVIDENCE THAT SUPPORTS MAPPING THE AURORAL OVAL TO THE LOW-LATITUDE (CENTRAL, MAIN) PLASMA SHEET

A. That the auroral oval is located on closed field lines is indicated by the following arguments.

Argument 1: The conjugacy of discrete auroral forms in opposite hemispheres during quiet and moderately disturbed periods [*Davis et al.*, 1971; *Mizera et al.*, 1987; *Obara et al.*, 1988].

Argument 2: The pitch-angle distribution of high-energy electrons above the entire auroral oval, with two loss cones (at 0° and 180°). This agrees with the energetic particle losses into the atmosphere under repeated bounce-oscillations on closed field lines [*Burrows*, 1974; *Venkatarangan et al.*, 1975; *Lin and Hoffman*, 1979].

Argument 3: The discovery of H^+ and O^+ beams located in the immediate proximity to the typical electron inverted-V structures [*Bosqued et al.*, 1986b]. These structures were observed under the conditions of the substorm growth phase, when auroral forms generally drift equatorward. A systematic latitude-energy dispersion was detected in these ion structures that could be adequately modeled by a velocity-filter effect in the case of a radial drift in the equatorial plane of the model magnetosphere with constant average velocity V_c. Both the field-line length and the single value of V_c can be deduced in rare cases when both dispersed H^+ and O^+ structures are detected if we consider the source to be the upward ion beam above the conjugate inverted-V. These values are quantitatively consistent with the present-day magnetospheric models and with the observed drift velocities of the auroral forms during the growth phase ($V_c \sim 100$–300 m s^{-1}). According to *Bosqued et al.* [1986b] and *Bosqued* [this volume], the effect proves directly that the magnetic field lines above the inverted-V structures (i.e., above the oval of discrete forms) are closed and well described by the magnetospheric models. An inductive electric field at plasma sheet altitudes could provide the necessary nonequipotential radial drift to low latitudes.

Narrow dispersed ion beams with anisotropic distributions at high altitudes over the oval, above the accelerating regions, were also inferred from the measurements onboard DE 1 [*Winningham et al.*, 1984] and Viking [*Hultqvist*, 1988].

Argument 4: The magnetospheric magnetic field distribution and its model representations [*Mead and Fairfield*, 1975; *Tsyganenko and Usmanov*, 1982; *Tsyganenko*, 1987] lead to mapping the auroral oval equatorial border from the tail midplane at the geocentric distances ranging from X $\sim -5R_E$ to X $\sim -10R_E$ (depending on the activity) and its polar border from distances >40–$70R_E$ (to the limit of the model). This projection corresponds obviously to the low-latitude (main, central) plasma sheet and is inconsistent with mapping the whole oval width from the boundary (high-latitude) plasma sheet.

The recent simultaneous measurements from the ISEE 2 in the plasma sheet and plasma sheet boundary layer and from DE 1 (auroral oval imagery) during steady conditions allowed the mapping of the BPS to the real auroral oval that was rather extended at this time. The results (see Figure 7) have led *Frank and Craven* [1988] to the conclusion that the BPS is projected to "the poleward discrete arc that extends over a broad range of local times in the evening sector. The equatorward zone of luminosities is mapped into the ring current and the near-Earth plasma sheet." This result is entirely consistent with the mappings advocated here.

Argument 5: It has been noted already that the equatorial border of the oval closely coincides with the interface between region 1 and region 2 of field-aligned currents. In this case, at least, the equatorward portion of region 1 maps from the main (central) plasma sheet [*Potemra*, 1977; *Valchuk et al.*, 1979; *Frank et al.*, 1981, 1984; *Feldstein et al.*, 1984; *Timofeev et al.*, 1985, 1988; *Ohtani et al.*, 1988].

Also the systematic electrodynamical pattern inside the oval, in the form of encircled current loops for the visible arcs, inverted-V structures, and the region

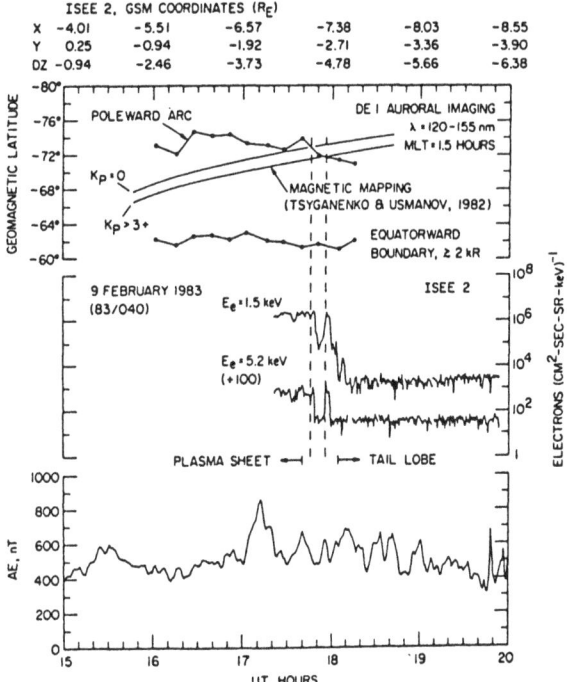

Figure 7—Energetic electron intensities for the ISEE 2 crossing of the plasma sheet boundary layer on February 9, 1983 (*center panel*). Instantaneous position of ISEE 2 relative to the boundaries of the auroral oval as determined from auroral images obtained with DE 1 (*top panel*). (From *Frank and Craven* [1988].)

1/region 2 large-scale currents found for steady conditions [*Timofeev et al.*, 1985, 1988] is difficult to reconcile with the narrow beam-like ion fluxes of the BPS.

B. That the earthward plasma convection and the adiabatic heating of particles in the plasma sheet leave their signatures in the particle spectra at the latitudes of the oval above the ionosphere is indicated by the following arguments.

Argument 1: The mean energies of plasma sheet particles (electrons and ions) rise nearer the Earth, conforming to the betatron acceleration under the general earthward plasma convection [*Vasyliunas*, 1968; *Hones et al.*, 1971; *Eastman et al.*, 1985].

Argument 2: The mean energies of particles, which excite diffuse luminosity within the auroral oval on the nightside (except inverted-V precipitation), increase with decreasing latitude [*Eather and Mende*, 1972]. According to the satellite measurements [*Galperin et al.*, 1978; *Sauvaud et al.*, 1985], the increase obeys the adiabatic acceleration law due to the earthward $\mathbf{E} \times \mathbf{B}$ drift of plasma sheet particles. The mean energy is proportional to l^{-2} where l is the length of the geomagnetic field line.

Argument 3: The plasma density at the top of an auroral magnetic field line (in the equatorial plane of the plasma sheet), as inferred from the particle spectra in the inverted-V structures observed above the oval, is 0.1–3.0 cm^{-3} [*Bruning and Goertz*, 1985; *Bosqued et al.*, 1986*a*]. These densities are characteristic of the low-latitude central plasma sheet (CPS) according to measurements [*Eastman et al.*, 1985], rather than of the BPS where the typical plasma densities are lower, 0.1–0.3 cm^{-3} [*Eastman et al.*, 1985].

5. IDENTIFICATION OF THE SOFT DIFFUSE ELECTRON PRECIPITATION AND LUMINOSITY REGION POLEWARD OF THE AURORAL OVAL AS A SEPARATE MAGNETOSPHERIC PLASMA DOMAIN MAPPED FROM THE HIGH-LATITUDE MAGNETOTAIL BOUNDARY PLASMA SHEET

The oval of discrete auroral forms (arcs, bands, rays) lies within a nearly continuous and permanently present band of weak, diffuse auroral luminosity surrounding the pole and bordering the oval on its equatorward (see above) and poleward sides (the continuous aurora) [*Akasofu*, 1974; *Whalen et al.*, 1977; *Whalen*, 1983; *Winningham et al.*, 1978].

Dividing the auroral region into hard (equatorward or the oval) and soft (poleward) precipitation zones was considered in many works (e.g., *Eather and Akasofu* [1969], *Eather* [1969], *Eather and Mende* [1972], *Isaev and Pudovkin* [1972], *Valchuk et al.* [1979], *Shepherd* [1982], *Makita et al.* [1985], *Meng and Makita* [1986], *Gogoshev et al.* [1987]).

During quiet periods in the dusk sector, the "hard zone," defined to be a region where the mean precipitating electron energy E_e exceeds 0.5 keV, nearly coincides with the oval of discrete auroras, whereas the "soft zone" (with $E_e < 0.5$ keV) extends to the poleward region [*Lassen and Danielsen*, 1987; *Lassen et al.*, 1988].

New experimental evidence concerning the polar diffuse zone was discovered from the ARCAD/Aureol 3 satellite particle measurements. Specific latitude-energy dispersed structures in the ion precipitation were sometimes found just inside the polar diffuse zone of soft electrons [*Kovrazhkin et al.*, 1987]. The ions were only H$^+$ (without traces of O$^+$), the energies increased with latitude from ~2–3 to sometimes more than 20 keV (see Figure 8 as an example adapted from *Kovrazhkin et al.* [1987] and *Zelenyi et al.* [1989]). This latitude dispersion was observed both on the poleward and equatorward passes, so the possibility of the time-of-flight dispersion from a purely impulsive source

seems to be unlikely. For some more regular events (observed usually during recovery of substorms or several hours after that) the velocity filter interpretation with a constant radial drift velocity V_c seems to be plausible. In these cases the "particle source" field line (for infinite parallel velocity) is located only at the polar edge of the polar diffuse zone, while the low-energy tail of H$^+$ ions is seen all along the width of this zone (rather narrow in these cases, about 0.5–2° of latitude). These polar oblique ion structures differ fundamentally from the oblique ion structures observed inside the oval. No electron inverted-V structure of comparable energy was observed here at the high-energy side, only some low intensity and softening electron precipitation. From the more than a thousand passes analyzed, only several tens of good examples have been found. Thus these polar ion structures are not common features at the polar border of the oval. They were interpreted in *Kovrazhkin et al.*, [1987] to be the direct consequence of the field-aligned acceleration in the distant tail, presumably at the reconnection region; the distance to it was crudely evaluated as 50–100R$_E$. There is considerable similarity between the dispersion characteristics of these low-altitude oblique ion structures and the features of the dispersed ion beams streaming along the BPS at high altitudes, observed by *Williams* [1981]. This similarity suggests their identity despite the difference in the energy ranges of the detectors used on low-altitude and high-altitude satellites. But recently the high-altitude measurements of lower energy ions in the BPS were analyzed by *Takahashi and Hones* [1988], who have demonstrated even greater similarity to the characteristics measured from the ARCAD/Aureol 3 satellite in the polar diffuse zone.

Thus we can conclude that the existence of these polar oblique ion structures makes it possible to substantiate the indirect inference made by *Feldstein and Galperin* [1985] on the projection of the polar diffuse electron precipitation zone from the BPS (at least, during the observations of such ion structures). Besides, we can observe here the change in the character of the electron precipitation spectra as compared with the adjacent inverted-V structures in the oval. In our opinion then, this region of space from auroral altitudes to the distant tail may be considered as a separate magnetospheric plasma domain. It can be identified with the BPS of which such ion beams are most characteristic. It is interesting to note that at downstream distances in the tail of about 140–220R$_E$, *Scholer et al.* [1987] were able to find only tailward ion beams, similar in characteristics to the earthward ones observed by *Williams* [1981] and *Takahashi and Hones* [1988] at smaller distances. This is consistent with the above-mentioned estimate of the distance to the ion beam

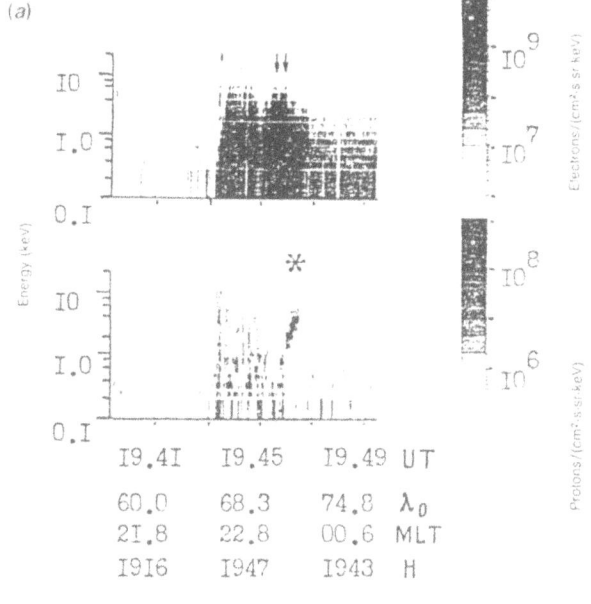

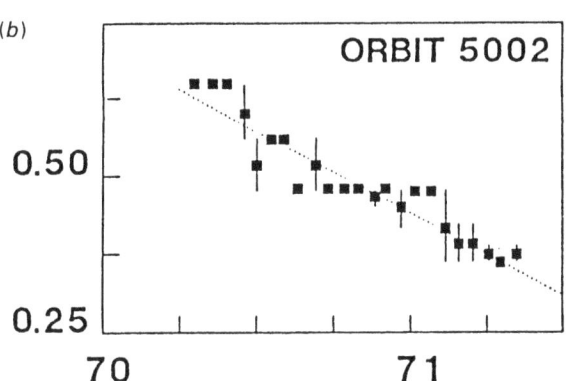

ORBIT 5002

Figure 8—An example of polar oblique ion structures in the polar diffuse auroral zone observed from ARCAD/Aureol 3 satellite. (*a*) Energy-time spectrogram for electrons (upper spectrogram) and ions (lower spectrogram) for orbit 5002, October 5, 1982. The arrows show two inverted-V structures above the electron spectrogram. The asterisk above the ion spectrogram marks a polar ion velocity-dispersed structure (from *Kovrazhkin et al.* [1987]). (*b*) The velocity-dispersed ion structure on orbit 5002. Values of $E_{max}^{-\frac{1}{2}}$ (where E_{max} is the ion energy, in kiloelectron volts, of the maximal intensity) are plotted versus the invariant latitude (in degrees). (c) Mass composition measurements at the maximum of the ion spectrum at 4.2 keV from the polar oblique ion structure on orbit 5002. Mass-channel number is indicated on abscissa axis, masses 1, 2, 4, and 8 are indicated above the graph ((*b*) and (*c*) are from *Zelenyi et al.* [1989].)

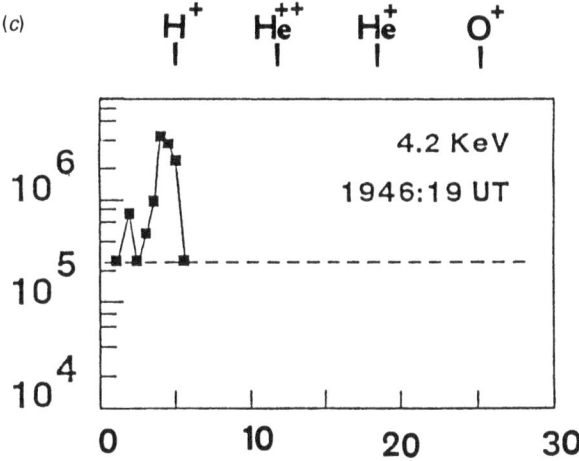

source in the tail. Another possibility, suggested by one of the referees, is the source in the low-latitude downstream magnetosphere boundary layer; then it could be similar to, or identical with, the "fireball."

We can note also that the polar region of soft diffuse electron precipitation extending to very high latitudes during quiet times can be located entirely on closed magnetic field lines. This follows from observations of the conjugacy of the polar soft precipitation regions and of the so-called theta-aurora in the two hemispheres [*Makita et al.*, 1983; *Mizera et al.*, 1987; *Obara et al.*, 1988]. But the analysis of these quiet conditions is beyond the scope of this presentation (see the review by *Lundin et al.* [this volume]).

The conclusion we draw from these data comparisons from different satellites is that the mere existence of the polar diffuse zone as the projection of the BPS is crucial for the proposed mapping of the central plasma sheet to the auroral oval of discrete forms. Then

the most poleward auroral arc in the oval can be considered to be the physical border between the plasma sheet proper and the BPS. To judge whether this auroral structure is of the same physical nature as a typical inverted-V event, or an auroral arc inside the oval, or has some singularity due to its frontier location, will be possible only from the detailed analysis of the physical processes responsible for the formation of this auroral structure.

6. SOME IMPLICATIONS OF MAPPING FOR THE SUBSTORM ONSET LOCATION AND DEVELOPMENT

Substorm Onset Development at the Inner Boundary of the Plasma Sheet

The analysis of many of the ground-based observations of aurora development, both from all-sky camera films and from visual observations (obviously with very high space/time resolution), showed that the substorm onset starts from the brightening (activation) of the most equatorward arc of the auroral oval, which then leaps rapidly poleward forming the auroral bulge [*Akasofu*, 1964]. Later, it was stated that in the night sector a substorm is initiated close to the boundary between the diffuse and the discrete aurora [*Akasofu*, 1974]. High-resolution observations provided the detail for making this important point. *Starkov and Feldstein* [1971] showed that actually, during the activation, the arc splits and the bright-rayed auroral band starts the poleward movement. The arc can stay at the place, or even move somewhat equatorward, leaving an eye-shaped region of enhanced emission between the poleward expanding front and the equatorward arc. Another important detail was stressed by *Pellinen and Heikkila* [1984], who documented the short fading of the light intensity of the arc just before the onset of the brightening. Since then, such behavior has been confirmed by many observers of auroral phenomena.

Recently two other important aspects of the auroral activations were reported. The first is the identification of the microstructure of the magnetospheric and auroral activity that was shown to be a superimposition of "microbursts" of about 1-minute duration that were observed simultaneously in the tail current sheet and at auroral altitudes at nearly conjugate locations not only during substorms [*Sergeev and Yahnin*, 1979; *Sergeev et al.*, 1986a,b], but also during low magnetic activity [*Sergeev et al.*, 1986c] and steady convection conditions [*Sergeev and Lennartson*, 1988]. Thus the local activations (but not only large-scale magnetospheric substorms) must be considered as the key explosive (burst-like) phenomena in the magnetosphere.

The second important aspect is the wealth of various morphological types of the auroral activations of medium scale ($\sim$ 100–300 km) discovered by the auroral imagers on board the DE 1 and Viking satellites [*Frank and Craven*, 1988; *Hultqvist*, 1988]. The many new types of auroral activations observed, and more detailed, and often peculiar, global substorm development sequences give new challenges to the theorists.

Rapid poleward leaps of active auroras forming the auroral bulge were observed in many intensive substorms at the boundary of the oval during the substorm expansion phase [*Pytte et al.*, 1978; *Hones*, 1985], but this was questioned by *Rostoker* [1986]. In the data from DE 1 [*Craven and Frank*, 1985] and Viking [*Anger et al.*, 1987; *Hultqvist*, 1988] the poleward leaps were absent during weak substorms and the polar oval expanded gradually both poleward and equatorward.

If only a single, bright arc is imbedded in the diffuse glow in the near-midnight sector, or if a new acti-

vation of an aurora occurs in the course of the developing substorm, it can be interpreted as a substorm onset at the poleward boundary of the auroral oval [*Hultqvist*, 1988]. But high-resolution and detailed observations from the ground at stations below the polar border of the oval (as far as we know) have not yet confirmed the initiation of an isolated substorm at the polar boundary of the oval. At the same time, local activations, which can be considered as microsubstorms, are quite common at this location.

The most poleward arc forming the leading edge of the expanding bulge often has peculiarities in respect to the auroral arcs of other types. Some cases were observed when such an arc maintained considerable brightness, long after the substorm structures inside the auroral oval faded [*Hones et al.*, 1987; *Lanchester and Wallis*, 1985]. These phenomena deserve a detailed analysis and further comparisons with the tail variations and with the more extended "typical" substorms.

The Mappings and the Substorm Theories

Above, we have put forth the arguments for mapping the polar diffuse zone and its equatorward boundary (i.e., the most poleward arc, or the inverted-V) to the BPS in the tail. If we consider the magnetic flux subtended by the BPS of the $0.5R_E$ thickness at the distance of $3R_E$ from the tail midplane with the magnetic field of 20 nT, then at auroral altitudes, with 60,000-nT field, this will correspond to a thickness of about 10 km, which is comparable to the arc width. This argument can be presented in a more quantitative way as the use of the magnetospheric model for the field-line tracing between the BPS and the oval. One very important example of this kind has been discussed above, on the basis of observations by *Frank and Craven* [1988, see Figure 7]. Still, it must be noted that the latitudinal extent of the polar diffuse zone and the magnetic flux subtended by it appear to be larger then the magnetic flux through the BPS in the tail.

From the oval mapping we have discussed, the most important implication is that a "typical" substorm starts in the tail at radial distances of not more than about $10R_E$ and just near the trapping boundary. At the auroral altitudes, this corresponds to the border between the diffuse and the discrete aurora in the near-midnight sector. It can be noted also that the average position of the so-called injection boundary [*McIlwain*, 1974; *Mauk and Meng*, 1983] at which the particle injection, or acceleration, occurs during a substorm, is also close to the oval equatorial border and is located close to the geostationary orbit during disturbed conditions. Obviously, this does not exclude other types and locations of burst activity in the tail. In particular, the active phenomena in the boundary plasma sheet can give rise to some types of substorm-like auroral activations observed along the polar border of the auroral oval [*Rostoker and Eastman*, 1987; *Eastman et al.*, 1984, 1985; *Lundin et al.*, this volume]. Aside from the definitions and semantics, it is necessary to identify what physical processes can be considered, to lead to the powerful, burst-like energy release in the near-Earth plasma sheet (for details, see *Baumjohann* [1988]).

One plausible hypothesis discussed at length (probably, with some modifications related to the above-mentioned localized microburst superimposition) is that of the "substorm neutral line" forming in the near-Earth plasma sheet (see *Hones* [1985], *Nishida* [1984], and *Nishida et al.* [1988]). Indeed, very active, hot, and turbulent plasma regions were observed at radial distances as low as 8–9R_E [*Takahashi et al.*, 1987; *Lui et al.*, 1988] and even at the geostationary orbit [*Lazutin et al.*, 1984; *Roux*, 1986; *Kozelova et al.*, 1988], but their characteristics could not be represented by a single and large-scale neutral line. Therefore it is important to explore other theoretical possibilities for the interpretation of the substorm onset phenomena inside the inner magnetosphere.

The localized bursts due to the "patchy reconnection" in the near-tail current sheet were modeled by *Sergeev et al.* [1987], who found that a minimum time scale of about 1 minute was consistent with the recent developments in the theory of the (localized) tearing-mode instability in the tail current sheet [*Buchner and Zelenyi*, 1987]. Another new approach to the substorm development due to excitation of the turbulent regime of the magnetospheric convection was described by *Trachtengertz and Feldstein* [1988].

An interesting idea of the combined Birkeland current loop and the two-sheet (Boström type) current for the prebreakup phase up to the substorm onset was developed by *Rothwell et al.* [1988]. This model, as the authors state, can explain "why breakup has been observed as low as L = 5–6," describing the inherent instability of the current configuration in certain conditions. Important modeling efforts by *Kan et al.* [1988] have led to the conclusion that the enhancements of magnetospheric convection in realistic conditions of nonhomogeneous anisotropic conductivity of the auroral ionosphere can produce localized regions of significant increase of the upward field-aligned currents elongated along the boundary of diffuse and discrete auroras and can naturally imply the equatorial arc brightening and some other characteristics of the substorm.

Also worth mentioning here are the inherent insta-

bility of the tail magnetic configuration for the increasing cross-tail current noted by *Kaufmann* [1987]; the idea of the flute instability at the inner boundary of the plasma sheet allowing for the effect of field-aligned currents on the conductivity in the auroral ionosphere [*Lyatsky*, 1987]; and the possibility of the diamagnetic outward ejection of the magnetospheric plasma cloud heated by longitudinal electric fields [*Lyatsky and Maltsev*, 1984].

These theoretical works have been mentioned to show that there are several interesting possibilities for the interpretation of various substorm-like phenomena of different scales observed in the near-Earth magnetosphere.

In our opinion, the observational evidence leaves no doubt that the inner nightside magnetosphere at distances of, say, less than 10R_E is the site of active substorm onset phenomena that develop inside of (and presumably due to the energy release of) the cross-tail plasma sheet current. These phenomena, once initiated, develop in the whole, main, or central plasma sheet. It seems quite probable that substorm-like phenomena of other types, e.g., fireballs, can originate in the BPS.

7. CONCLUSIONS

From the above it follows that the nightside auroral oval is mapped from the low-latitude (main or central) plasma sheet in the magnetospheric tail (CPS). The region of diffuse auroral luminosity equatorward of the oval is mapped via magnetic field lines from the region located between the instantaneous plasmapause and the inner boundary of the plasma sheet (the remnant layer). The region of diffuse luminosity poleward of the oval is mapped from the high-latitude portion of the plasma sheet (BPS). Magnetotail plasma dynamics during localized and large-scale auroral activity, including those recently discovered by Viking [*Hultqvist*, 1988], needs to be studied further to investigate their signatures associated with the diversity of substorms, microsubstorms, and local activations, in various discrete auroral forms. The study of the dynamics of the magnetosphere requires that active processes occurring in the tail and at the flanks of the magnetosphere be analyzed in multipoint magnetospheric projects such as INTERBALL, CLUSTER, and IMPACT. When those data are combined with coordinated ground-based geophysical measurements (including high-resolution optical auroral observations), a deeper insight into the nature of magnetospheric disturbances will be achieved.

REFERENCES

Akasofu, S.-I., "The Development of the Auroral Substorm," *Planet. Space Sci.*, **12**, 273–282 (1964).

Akasofu, S.-I., *Polar and Magnetospheric Substorms*, D. Reidel, Dordrecht-Holland (1968).

Akasofu, S.-I., "Discrete, Continuous and Diffuse Auroras," *Planet. Space Sci.*, **22**, 1723–1726 (1974).

Anger, C. D., J. C. Murphree, A. Vallance Jones, et al., "Scientific Results from the Viking Imager: An Introduction," *Geophys. Res. Lett.*, **14**, 383–386 (1987).

Aubry, M. P., M. G. Kivelson, R. L. McPherron, C. T. Russell, and D. S. Colburn, "Outer Magnetosphere Near Midnight at Quiet and Disturbed Times," *J. Geophys. Res.*, **77**, 5487–5502 (1972).

Baumjohann, W., "The Plasma Sheet Boundary Layer and Magnetospheric Substorms," *J. Geomag. Geoelectr.*, **40**, 157–175 (1988).

Bosqued, J. M., "Ion Precipitation and the Transport of Ions Accelerated by Auroral Processes," *this volume*.

Bosqued, J. M., C. Maurel, J. A. Sauvaud, R. A. Kovrazhkin, and Y. I. Galperin, "Observations of Auroral Electron Inverted-V Structures by the Aureol 3 Satellite," *Planet. Space Sci.*, **34**, 255–269 (1986*a*).

Bosqued, J. M., J. A. Sauvaud, D. Delcourt, and R. A. Kovrazhkin, "Precipitation of Suprathermal Ionospheric Ions Accelerated in the Conjugate Magnetosphere," *J. Geophys. Res.*, **91**, 7006–7018 (1986*b*).

Brice, N. M., "Bulk Motion of the Magnetosphere," *J. Geophys. Res.*, **72**, 5192–5211 (1967).

Bruning, K., and C. Goertz, "Influence of the Electron Source Distribution on Field-Aligned Currents," *Geophys. Res. Lett.*, **12**, 53–56 (1985).

Buchner, J., and L. M. Zelenyi, "Chaotization of the Electron Motion as the Cause of an Internal Magnetotail Instability and Substorm Onset," *J. Geophys. Res.*, **92**, 13456–13466 (1987).

Burrows, J. R., "The Plasma Sheet in the Evening Sector," in *Magnetospheric Physics*, B. M. McCormac, ed., D. Reidel, Hingham, MA, pp. 179–197 (1974).

Craven, J. G., and L. A. Frank, "The Temporal Evolution of a Small Auroral Substorm as Viewed from High Altitudes with Dynamics Explorer 1," *Geophys. Res. Lett.*, **12**, 465–468 (1985).

Davis, T. N., T. J. Hallinan, and H. C. Stenbaeck-Nielsen, "Auroral Conjugacy and Time-Dependent Geometry of Auroras," in *Radiating Atmosphere*, B. M. McCormac, ed., D. Reidel, Dordrecht-Holland, pp. 160–169 (1971).

Deehr, C. S., J. D. Winningham, F. Yasuhara, and S.-I. Akasofu, "Simultaneous Observations of Discrete and Diffuse Auroras by the ISIS-2 Satellite and Airborne Instruments," *J. Geophys. Res.*, **81**, 5527–5535 (1976).

Eastman, T. E., L. A. Frank, W. K. Peterson, and W. Lennartsson, "The Plasma Sheet Boundary Layer," *J. Geophys. Res.*, **89**, 1553–1572 (1984).

Eastman, T. E., L. A. Frank, and C. Y. Huang, "The Boundary Layers as the Primary Transport Regions of the Earth's Magnetotail," *J. Geophys. Res.*, **90**, 9541–9560 (1985).

Eather, R. H., "Latitudinal Distributions of Auroral and Airglow Emissions: The "Soft" Auroral Zone," *J. Geophys. Res.*, **74**, 153–159 (1969).

Eather, R. H., and S.-I. Akasofu, "Characteristics of Polar Cap Auroras," *J. Geophys. Res.*, **74**, 4794–4798 (1969).

Eather, R. H., and S. B. Mende, "High Latitude Particle Precipitation and Source Regions in the Magnetosphere," in *Magnetosphere-Ionosphere Interaction*, K. Folkestad, ed., Universitetsforlaget, Oslo, pp. 139–154 (1972).

Eather, R. H., S. B. Mende, and R. J. K. Judge, "Plasma Injection at Synchronous Orbit and Spatial and Temporal Auroral Morphology," *J. Geophys. Res.*, **81**, 2805–2824 (1976).

Fairfield, D. H., "Structure of the Geomagnetic Tail," in *Magnetotail Physics*, A. T. Y. Lui, ed., Johns Hopkins University Press, pp. 23–32 (1987).

Fairfield, D. H., and A. F. Vinas, "The Inner Edge of the Plasma Sheet and the Diffuse Aurora," *J. Geophys. Res.*, **89**, 841–855 (1984).

Feldstein, Ya. I., and G. V. Starkov, "The Auroral Oval and the Boundary of Closed Field Lines of Geomagnetic Field," *Planet. Space Sci.*, **18**, 501–508 (1970)

Feldstein, Ya. I., R. G. Afonina, B. A. Belov, A. E. Levitin, D. S. Faermark, and V. Y. Gaidukov, "High-Latitude Field-Aligned Current Patterns in Connection with Magnetosphere Structure," in *Magnetospheric Currents*, Geophysical Monograph 28, T. A. Potemra, ed., American Geophysical Union, Washington, D.C., pp. 284–293 (1984).

Feldstein, Ya. I., and Yu. I. Galperin, "The Auroral Luminosity Structure in the High-Latitude Upper Atmosphere: Its Dynamics and Relationship to the Large-Scale Structure of The Earth's Magnetosphere," *Rev. Geophys.*, **23**, 217–275 (1985).

Frank, L. A., "Relationship of the Plasma Sheet, Ring Current, Trapped Boundary, and Plasmapause Near the Magnetic Equator and Local Midnight," *J. Geophys. Res.*, **76**, 2265–2275 (1971).

Frank, L. A., "Hot Plasmas in the Earth's Magnetosphere," in *Physics of Solar Planetary-Environments*, vol. 2, D. J. Williams, ed., American Geophysical Union, Washington, DC, pp. 685–700 (1976).

Frank, L. A., "Plasmas in the Earth's Magnetotail," *Space Sci. Rev.*, **42**, 211–240 (1985).

Frank, L. A., and K. L. Ackerson, "Observations of Charged Particle Precipitation into the Auroral Zone," *J. Geophys. Res.*, **76**, 3612–3643 (1971).

Frank, L. A., and K. L. Ackerson, "Local-Time Survey of Plasma at Low Altitudes Over the Auroral Zone," *J. Geophys. Res.*, **77**, 4116–4127 (1972).

Frank, L. A., K. L. Ackerson, and R. P. Lepping, "On Hot Tenuous Plasmas, Fireballs, and Boundary Layers in the Earth's Magnetotail," *J. Geophys. Res.*, **81**, 5859–5881 (1976).

Frank, L. A., and J. D. Craven, "Imaging Results from Dynamics Explorer 1," *Rev. Geophys.*, **26**, 249–283 (1988).

Frank, L. A., C. Y. Huang, and T. E. Eastman, "Currents in the Earth's Magnetotail," in *Magnetospheric Currents*, Geophysical Monograph 28, T. A. Potemra, ed., American Geophysical Union, Washington, DC, pp. 147–157 (1984).

Frank, L. A., R. L. McPherron, R. J. DeCoster, B. G. Burek, K. L. Ackerson, and C. T. Russell, "Field-Aligned Currents in the Earth's Magnetotail," *J. Geophys. Res.*, **86**, 687–700 (1981).

Galperin, Yu. I., J. Crasnier, Yu. V. Lissakov, L. M. Nickolaenko, V. M. Sinitsin, J.-A. Sauvaud, and V. L. Khalipov, "Diffuse Auroral Zone. I. Model of the Equatorial Border of the Auroral Electron Diffuse Precipitation Zone in the Evening and Near-Midnight Sectors" (in Russian), *Cosmic Res.*, **15**, 421–434 (1977).

Galperin, Yu. I., V. A. Gladyshev, N. V. Jorjio, R. A. Kovrazhkin, V. M. Sinitsin, F. Cambou, and J. A. Sauvaud, "Adiabatic Acceleration Induced by Convection in the Plasma Sheet," *J. Geophys. Res.*, **83**, 2567–2573 (1978).

Gogoshev, M. M., G. G. Shepherd, P. V. Maglova, V. Chr. Guineva, and Ts. P. Datchev, "Structure of the Polar Oval from Simultaneous Observations of the Optical Emissions and Particle Precipitations During the Period of High Solar Activity 1981-1982," *Adv. Space Res.*, **7**, 7–10 (1987).

Gurnett D. A., and L. A. Frank, "Observed Relationships Between Electric Fields and Auroral Particle Precipitation," *J. Geophys. Res.*, **78**, 145–170 (1973).

Gussenhoven, M. S., D. A. Hardy, and W. J. Burke, "DMSP/F2 Electron Observations of Equatorward Auroral Boundaries and Their Relationship to Magnetospheric Fields," *J. Geophys. Res.*, **86**, 768–778 (1981).

Gussenhoven, M. S., D. A. Hardy, and N. Heinemann, "Systematics of the Equatorward Diffuse Auroral Boundary," *J. Geophys. Res.*, **88**, 5692–5708 (1983).

Harel, M., R. A. Wolf, P. H. Reiff, R. W. Spiro, W. J. Burke, F. J. Rich, and M. Smiddy, "Quantitative Simulation of a Magnetospheric Substorm. 1, Model Logic and Overview," *J. Geophys. Res.*, **86**, 2217–2241 (1981).

Hones, E. W., "The Poleward Leap of the Auroral Electrojet as Seen in Auroral Images," *J. Geophys. Res.*, **90**, 5333–5337 (1985).

Hones, E. W., J. R. Asbridge, S. J. Bame, and S. Singer, "Energy Spectra and Angular Distributions of Particles in the Plasma Sheet and Their Comparison with Rocket Measurements Over the Auroral Zone," *J. Geophys. Res.*, **76**, 63–87 (1971).

Hones, E. W., C. D. Anger, J. Birn, J. S. Murphree, and L. L. Cogger, "A Study of a Magnetospheric Substorm Recorded by the Viking Auroral Imager," *Geophys. Res. Lett.*, **14**, 411–414 (1987).

Horwitz, J. L., W. K. Cobl, C. R. Baugher, C. R. Chappell, L. A. Frank, T. E. Eastman, R. R. Anderson, E. G. Shelley, and D. T. Young, "On the Relationship of the Plasmapause to the Equatorward Boundary of the Auroral Oval and to the Inner Edge of the Plasma Sheet," *J. Geophys. Res.*, **87**, 9059–9069 (1982).

Hultqvist, B., "Scientific Results from the Swedish Viking Satellite: A 1988 Status Report," IRF Scient. Report No 196 (36 pp.), Kiruna (1988).

Isaev, S. I., and M. I. Pudovkin, "Auroras and the Earth's Magnetospheric Processes" (in Russian), (244 pp.), Publ. House Nauka, Leningrad (1972).

Kamide, Y., and J. D. Winningham, "A Statistical Study of the 'Instantaneous' Nightside Auroral Oval: The Equatorward Boundary of Electron Precipitation as Observed by the ISIS-1 and ISIS-2 Satellites," *J. Geophys. Res.*, **82**, 5573-5588 (1977).

Kan, J. R., L. Zhu, and S.-I. Akasofu, "A Theory of Substorm: Onset and Subsidence," *J.Geophys. Res.*, **93**, 5624-5640 (1988).

Kaufmann, R. L., "Substorm Currents: Growth Phase and Onset," *J. Geophys. Res.*, **92**, 7471-7486 (1987).

Kovrazhkin, R. A., J. M. Bosqued, L. M. Zelenyi, and N. V. Jorjio, "Observations of Evidence of Reconnection and Plasma Acceleration at a Distance of About 5×10^5 km in the Tail of the Earth's Magnetosphere" (in Russian), *Lett. ZETP*, **45**, 377-380 (1987).

Kozelova, T. V., L. L. Lazutin, M. I. Pudovkin, A. Pedersen, and K. H. Glassmeier, "The Electric Fields in the Magnetosphere in the Vicinity of the Western Edge of the Substorm Current Wedge," in *Proc. Int. Symp. Polar Geomagnetic Phenomena*, May 25-31, 1986, Souzdal, USSR, O. M. Raspopov, ed., Apatity, pp. 35-38 (1988).

Lanchester, B. S., and D. D. Wallis, "Magnetic Field Disturbances Over Auroral Arcs Observed from Spitsbergen," *J. Geophys. Res.*, **90**, 2473-2480 (1985).

Lassen, K., and C. Danielsen, "Distribution of Auroral Arcs During Quiet Geomagnetic Conditions," Geophys. Paper R-78, Danish Meteorological Institute, Copenhagen, p. 19 (1987).

Lassen, K., C. Danielsen, and C.-I. Meng, "Quiet-Time Average Auroral Configuration," *Planet. Space Sci.*, **36**, 791-799 (1988).

Lazutin, L. L., G. Gustafsson, A. A. Khrushchinsky, G. Kremser, A. O. Melnikov, W. Riedler, Ya. A. Sakharov, K. M. Torkar, and J.-P. Treilhou, "SAMBO-GEOS: Substorm Trapped Particle Boundary Movements, Particle Precipitation and Acceleration," in *Achievements of the International Magnetospheric Study*, European Space Agency, Paris, pp. 315-318 (1984).

Lin, C. S., and R. A. Hoffman, "Characteristics of the Inverted-V Event," *J. Geophys. Res.*, **84**, 1514-1524 (1979).

Lui, A. T. Y., D. Venkatesan, C. D. Anger, S.-I. Akasofu, W. J. Heikkila, J. D. Winningham, and J. R. Burrows, "Simultaneous Observations of Particle Precipitations and Auroral Emissions by the ISIS-2 Satellite," *J. Geophys. Res.*, **82**, 2210-2226 (1977).

Lui, A. T. Y., and J. R. Burrows, "On the Location of Auroral Arcs Near Substorm Onsets," *J. Geophys. Res.*, **83**, 3342-3348 (1978).

Lui, A. T. Y., R. E. Lopez, S. M. Krimigis, R. W. McEntire, L. J. Zanetti, and T. A. Potemra, "A Case Study of Magnetotail Current Sheet Disruption and Diversion," *Geophys. Res. Lett.*, **15**, 721-724 (1988).

Lui, A. T. Y., C.-I. Meng, and S. Ismail, "Large-Amplitude Undulations of the Equatorward Boundary of the Diffuse Aurora," *J. Geophys. Res.*, **87**, 2385-2400 (1982).

Lundin, R., L. Eliasson, and J. S. Murphree, "The Quiet-Time Aurora and the Magnetosphere Configuration," *this volume*.

Lyatsky, V. B., "Conductivity Waves in the Magnetosphere-Ionosphere System" (in Russian), *Geomagn. Aeron.*, **27**, 965-970 (1987).

Lyatsky, V. B., and Yu. P. Maltsev, "About the Origin of Auroral Bulge" (in Russian), *Geomagn. Aeron.*, **24**, 89-93 (1984).

Lyons, L. R., "Discrete Auroras and Magnetotail Processes," *this volume*.

Lyons, L. R., and D. S. Evans, "An Association Between Discrete Aurora and Energetic Particle Boundaries," *J. Geophys. Res.*, **89**, 2395-2400 (1984).

Makita, K., C.-I. Meng, and S.-I. Akasofu, "Comparison of the Auroral Electron Precipitations in the Northern and Southern Conjugate Regions by Two DMSP Satellites," in *Proc. Symp. Coordination and Observation of the Ionosphere and Magnetosphere in the Polar Regions*, Spec. Issue, National Institute of Polar Research, Tokyo, pp. 149-159 (1983).

Makita, K., C.-I. Meng, and S.-I. Akasofu, "Temporal and Spatial Variations of the Polar Cap Dimension Inferred from the Precipitation Boundaries," *J. Geophys. Res.*, **90**, 2744-2752 (1985).

Makita, K., C.-I. Meng, and S.-I. Akasofu, "Latitudinal Electron Precipitation Patterns During Large and Small IMF Magnitudes for Northward IMF Conditions," *J. Geophys. Res.*, **93**, 97-104 (1988).

Mauk, B. H., and C.-I. Meng, "Characterization of Geostationary Particle Signatures Based on the 'Injection Boundary' Model," *J. Geophys. Res.*, **88**, 3055-3071 (1983).

Mauk, B. H., and C.-I. Meng, "The Aurora and Middle Magnetospheric Processes," *this volume*.

McIlwain, C.E., "Substorm Injection Boundaries," in *Magnetospheric Physics*, B. M. McCormac, ed., D. Reidel, Hingham, MA, pp. 143-154 (1974).

Mead, G. D., and D. H. Fairfield, "A Quantitative Magnetospheric Model Derived from Spacecraft Magnetometer Data," *J. Geophys. Res.*, **80**, 523-532 (1975).

Meng, C.-I., "Polar Cap Arcs and the Plasma Sheet," *Geophys. Res. Lett.*, **8**, 273-276 (1981).

Meng, C.-I., B. Mauk, and C. E. McIlwain, "Electron Precipitation of Evening Diffuse Aurora and Its Conjugate Electron Fluxes Near the Magnetic Equator," *J. Geophys. Res.*, **84**, 2545-2558 (1979).

Meng, C.-I., and K. Makita, "Dynamic Variations of the Polar Cap," in *Solar Wind-Magnetosphere Coupling*, Y. Kamide and J. Slavin, eds., Terra Scientific Publ. Co., Tokyo, pp. 605-631 (1986).

Mizera, P. F., D. J. Gorney, and D. S. Evans, "On the Conjugacy of the Aurora: High and Low Latitudes," *Geophys. Res. Lett.*, **14**, 190-193 (1987).

Nishida, A., "Formation of Plasmapause, or Magnetospheric Plasma Knee, by the Combined Action of Magnetospheric Convection and Plasma Escape from the Tail," *J. Geophys. Res.*, **71**, 5669-5679 (1966).

Nishida, A., "Reconnection in the Earth's Magnetotail: An Overview," in *Magnetospheric Reconnection in Space and Laboratory Plasmas*, E. W. Hones, ed., American Geophysical Union, Washington, DC, pp. 159-167 (1984).

Nishida, A., S. J. Bame, D. N. Baker, G. Gloeckler, M. Scholer, E. J. Smith, T. Terasawa, and B. Tsurutani, "Assessment of the Boundary Layer Model of the Magnetospheric Substorm," *J. Geophys. Res.*, **93**, 5579-5588 (1988).

Obara, T., M. Kitayama, T. Mukai, N. Kaya, J. C. Murphree, and L. L. Cogger, "Simultaneous Observations of Sun-Aligned Polar Cap Arcs in Both Hemispheres by EXOS-C and Viking," *Geophys. Res. Lett.*, **15**, 713-716 (1988).

Ohtani, S., S. Kokubun, R. C. Elphic, and C. T. Russell, "Field-Aligned Current Signatures in the Near-Tail Region. 1. ISEE Observations in the Plasma Sheet Boundary Layer," *J. Geophys. Res.*, **93**, 9709-9720 (1988).

Pellinen, R. J., and W. J. Heikkila, "Inductive Electric Fields in the Magnetotail and Their Relation to Auroral and Substorm Phenomena," *Space Sci. Rev.*, **37**, 1-61 (1984).

Ponomarev, E. A., "The Model of the Nightside Magnetosphere," in *Investigations on Geomagnetism, Aeronomy and Solar Physics* (in Russian), Publ. House Nauka, Moscow, N 53, pp. 3-14 (1981).

Potemra, T. A., "Large-Scale Characteristics of Field-Aligned Currents Determined from the Triad Magnetometer Experiment," in *Dynamical and Chemical Coupling*, B. Grandel and J. A. Holtet, eds., D. Reidel, Hingham, MA, pp. 337-352 (1977).

Pudovkin, M. I., and V. E. Zakharov, "Study of Dynamic Processes in the Magnetospheric Plasma," in *Magnetospheric Researches* (in Russian), Publ. House Radio, Moscow, 3, pp. 67-85 (1984).

Pytte, T., R. L. McPherron, M. G. Kiveison, H. I. West, and E. W. Hones, "Multiple-Satellite Studies of Magnetospheric Substorms: Plasma Sheet Recovery and the Poleward Leap of Auroral Zone Activity," *J. Geophys. Res.*, **83**, 5256-5268 (1978).

Roux, A., "Generation of Field-Aligned Current Structures at Substorm Onset," in *Future Missions in Solar, Heliosphere and Space Plasma Physics*, E. Rolfe and B. Battrick, eds., ESA, Noordwijk, pp. 151-159 (1986).

Rostoker, G., "Comment on The Poleward Leap of the Auroral Electrojet as Seen in Auroral Images by Edward W. Hones," *J. Geophys. Res.*, **91**, 5879-5880 (1986).

Rostoker, G., and T. Eastman, "A Boundary Layer Model for Magnetospheric Substorms," *J. Geophys. Res.*, **92**, 12187-12202 (1987).

Rothwell, P. L., L. P. Block, M. B. Silevitch, and C.-G. Fälthammar, "A New Model for Substorm Onsets: The Pre-Breakup and Triggering Regimes," *Geophys. Res. Lett.*, **15**, 1279-1282 (1988).

Sauvaud, J. A., J. M. Bosqued, R. A. Kovrazhkin, J. J. Berthelier, A. Berthelier, and Yu. I. Galperin, "Positive Ion Distributions in the Morning Auroral Zone: Local Acceleration and Drift Effects," *Adv. Space Res.*, **5**, 73-77 (1985).

Sauvaud, J. A., J. Crasnier, Yu. I. Galperin, and Ya. I. Feldstein, "A Statistical Study of the Dynamics of the Equatorward Boundary of the Diffuse Aurora in the Pre-Midnight Sector," *Geophys. Res. Lett.*, **10**, 749-752 (1983).

Schield, M. A., and L. A. Frank, "Electron Observations Between the Inner Edge of the Plasma Sheet and the Plasmasphere," *J. Geophys. Res.*, **75**, 5401-5412 (1970).

Scholer, M., B. Klecker, D. Hovestadt, G. Gloeckler, F. M. Ipavich, A. B. Galvin, D. N. Baker, and B. T. Tsurutani, "Energetic Ion and Electron Beams at the Plasma-Sheet Boundary in the Distant Tail," in *Magnetotail Physics*, A. T. Y. Lui, ed., Baltimore, Johns Hopkins University Press, pp. 245-249 (1987).

221

Sergeev, V. A., T. Bosinger, and A. T. Y. Lui, "Impulsive Processes in the Magnetotail During Substorm Expansion," *J. Geophys.*, **60**, 175–185 (1986*a*).

Sergeev, V. A., and W. Lennartsson, "Plasma Sheet at X ~ –20R$_E$ During Steady Magnetospheric Convection," *Planet. Space Sci.*, **36**, 353–370 (1988).

Sergeev, V. A., R. J. Pellinen, T. Bosinger, W. Baumjohann, P. Stauning, and A. T. Y. Lui, "Spatial and Temporal Characteristics of Impulsive Structure of Magnetospheric Substorm," *J. Geophys.*, **60**, 186–198 (1986*b*).

Sergeev, V. A., V. S. Semenov, and M. V. Sidneva, "Impulsive Reconnection in the Magnetotail During Substorm Expansion," *Planet. Space Sci.*, **35**, 1199–1212 (1987).

Sergeev, V. A., and N. A. Tsyganenko, "Energetic Particle Losses and Trapping Boundaries as Deduced from Calculations with a Realistic Magnetic Field Model," *Planet. Space Sci.*, 30,999–31,007 (1982).

Sergeev, V. A., and A. G. Yahnin, "The Features of Auroral Bulge Expansion," *Planet. Space Sci.*, **27**, 1429–1444 (1979).

Sergeev, V. A., and A. G. Yahnin, "Mutual Location and Magnetospheric Sources of the Zones of Precipitation of Energetic Electrons, Diffusive and Discrete Aurorae at a Creation Phase of a Substorm" (in Russian), *Geomagn. Aeron.*, **23**, 972–978 (1983).

Sergeev, V. A., A. G. Yahnin, R. A. Rakhmatullin, S. I. Soloviev, F. S. Moser, D. J. Williams, C. T. Russell, "Permanent Flare Activity in the Magnetosphere During Periods of Low Magnetic Activity in the Auroral Zone," *Planet. Space Sci.*, **34**, 1169–1188 (1986*c*).

Serlemitsos, P., "Low-Energy Electrons in the Dusk Magnetosphere," *J. Geophys. Res.*, **71**, 61–67 (1966).

Shepherd, G. G., "Structure and Dynamics of the Red Auroral Oval in Quiet and Disturbed Conditions," presented at Symp. Magnetosphere-Ionosphere Processes and Airglow, Stara Zagora, pp. 132–144 (1982).

Shepherd, G. G., J. D. Winningham, F. E. Bunn, and F. W. Thirhette, "An Empirical Determination of the Production Efficiency for Auroral 6300 Å Emission by Energetic Electrons," *J. Geophys. Res.*, **85**, 715–721 (1980).

Soloviev, V. S., Yu. I. Galperin, L. V. Zinin, L. D. Sivtseva, V. M. Filippov, and V. L. Khalipov, "Diffuse Auroral Zone. IX. The Equatorward Boundary of Diffuse Precipitation of the Plasmasheet Electrons as a Convection Boundary (plasmapause)" (in Russian), *Cosmic Res.*, **27**, 232–247 (1989).

Starkov, G. V. and Ya. I. Feldstein, "Auroral Substorm" (in Russian), *Geomagn. Aeron.*, **11**, 560–563 (1971).

Takahashi, K., and E. W. Hones, Jr., "ISEE 1 and 2 Observations of Ion Distributions at the Plasma Sheet-Tail Lobe Boundary," *J. Geophys. Res.*, **93**, 8558–8582 (1988).

Takahashi, K., L. J. Zanetti, R. E. Lopez, R. W. McEntire, T. A. Potemra, and K. Yumoto, "Disruption of the Magnetotail Current Sheet Observed by AMPTE/CCE," *Geophys. Res. Lett.*, **14**, 1019–1022 (1987).

Tanskanen, P. J., D. A. Hardy, and W. J. Burke, "Spectral Characteristics of Precipitating Electrons Associated with Visible Aurora in the Premidnight Oval During Substorm Activity," *J. Geophys. Res.*, **86**, 1379–1395 (1981).

Timofeev, E. E., V. M. Smyshliaev, N. V. Jorjio, Yu. I. Galperin, J. M. Bosqued, J. J. Berthelier, M. K. Vallinkoski, and R. J. Pellinen, "Coordinated Data on Auroral Electrodynamics from Ground-Based Radar Diagnostics and Aureol-3 Satellite," in *Results of the ARCAD-3 Project and of the Recent Programmes in Magnetospheric and Ionospheric Physics*, Toulouse, France, May 1984, ed. CNES, pp. 949–971, Cepadues-Editions (1985).

Timofeev, E. E., O. M. Raspopov, Yu. I. Galperin, N. V. Jorjio, A. Berthelier, J. J. Berthelier, J. M. Bosqued, M. K. Vallinkoski, and R. J. Pellinen, "Regularities of the Birkeland Current Systems Stratification (Coordinated Experiments in the Framework of the ARCAD-3 Project)" (in Russian), *Cosmic Res.*, **26**, 709–724 (1988).

Trakhtengertz, V. Yu., and A. Ya. Feldstein, "The Substorm Expansive Phase as a Result of the Turbulent Regime of Magnetospheric Convection" (in Russian), *Geomagn. Aeron.*, **28**, 743–746 (1988).

Tsyganenko, N. A., "Global Quantitative Models of the Geomagnetic Field in the Cislunar Magnetosphere for Different Disturbance Levels," *Planet. Space Sci.*, **35**, 1347–1358 (1987).

Tsyganenko, N. A., and A. V. Usmanov, "Determination of the Magnetospheric Current System Parameters and Development of Experimental Geomagnetic Field Models Based on Data from IMP and HEOS Satellites," *Planet. Space Sci.*, **30**, 985–1007 (1982).

Valchuk, T. E., Yu. I. Galperin, J. Crasnier, L. M. Nikolaenko, J. A. Sauvaud, and Ya. I. Feldstein, "Diffuse Auroral Zone. IV. Latitudinal Distribution of Auroral Emissions and Particle Precipitation and its Relationship with the Plasmasheet and Magnetotail" (in Russian), *Cosmic Res.*, **17**, 559–579 (1979).

Valchuk, T. E., Yu. I. Galperin, L. M. Nickolaenko, Ya. I. Feldstein, J. M. Bosqued, J. A. Sauvaud, J. Crasnier, "Diffuse Auroral Zone. VIII. Equatorial Border of the Auroral Electron Diffuse Precipitation Zone in the Morning Sector" (in Russian), *Cosmic Res.*, **24**, 875–883 (1986).

Vasyliunas, V. M., "Observations of Low-Energy Electrons in the Evening Sector of the Magnetosphere with OGO 1 and OGO 3," *J. Geophys. Res.*, **73**, 2839–2884 (1968).

Vasyliunas, V. M., "Low Energy Particle Fluxes in the Geomagnetic Tail," in *The Polar Ionosphere and Magnetospheric Processes*, G. Skovli, ed., Gordon and Breach, New York, pp. 25–47 (1970).

Vasyliunas, V. M., "Magnetospheric Plasma," in *Solar Terrestrial Physics/1970*, E. R. Dyer, ed., V.3, D. Reidel, Hingham, MA, pp. 192–211 (1972).

Venkatarangan, P., J. R. Burrows, and I. B. McDiarmid, "On the Angular Distributions of Electrons in 'Inverted V' Substructures," *J. Geophys. Res.*, **80**, 66–72 (1975).

West, H. I., R. M. Buck, and M. G. Kivelson, "On the Configuration of the Magnetotail Near Midnight During Quiet and Weakly Disturbed Periods: State of the Magnetosphere," *J. Geophys. Res.*, **83**, 3805–3817 (1978).

West, H. I., R. M. Buck, and J. R. Walton, "Electron Pitch-Angle Distributions Throughout the Magnetosphere as Observed on OGO 5," *J. Geophys. Res.*, **78**, 1064–1080 (1973).

Whalen, J. A., "A Quantitative Description of the Spatial Distribution and Dynamics of the Energy Flux in the Continuous Aurora," *J. Geophys. Res.*, **88**, 7155–7169 (1983).

Whalen, J. A., R. A. Wagner, and J. Buchau, "A 12-Hour Case Study of Auroral Phenomena in the Midnight Sector: Oval, Polar Cap, and Continuous Auroras," *J. Geophys. Res.*, **82**, 3529–3546 (1977).

Williams, D. J., "'Energetic Ion Beams at the Edge of the Plasmasheet: ISEE 1 Observations and a Simple Explanatory Model," *J. Geophys. Res.*, **86**, 5507–5518 (1981).

Winningham, J. D., F. Yasuhara, S.-I. Akasofu, and W. J. Heikkila, "The Latitudinal Morphology of 10-eV to 10-keV Electron Fluxes During Magnetically Quiet and Disturbed Times in the 2100–0300 MLT Sector," *J. Geophys. Res.*, **80**, 3148–3171 (1975).

Winningham, J. D., C. D. Anger, G. G. Shepherd, E. J. Weber, and R. A. Wagner, "A Case Study of the Aurora, High-Latitude Ionosphere and Particle Precipitation During Near-Steady Conditions," *J. Geophys. Res.*, **83**, 5717–5731 (1978).

Winningham, J. D., J. L. Burch, and R. A. Frahm, "Bands of Ions and Angular V's: A Conjugate Manifestation of Ionospheric Ion Acceleration," *J. Geophys. Res.*, **89**, 1749–1754 (1984).

Zelenyi, L. M., R. A. Kovrazhkin, and J. M. Bosqued, "Velocity-Dispersed Ion Beams in the Nightside Auroral Zone: Aureol 3 Observations," *J. Geophys. Res.*, submitted (1989).

IV-5. THE AURORA AND MIDDLE MAGNETOSPHERIC PROCESSES

B. H. Mauk* and C.-I. Meng*

Much attention has been paid over the last decade to the field-aligned electromagnetic processes that give rise to the discrete aurora that often predominates at the poleward portion of the auroral regions. Conventional wisdom suggests that such processes are driven by interactions occurring within the magnetotail, with the primary transport occurring along the outer boundaries of the plasma sheet population (whether the boundaries are newly formed or preexisting). In terms of auroral processes, the middle magnetosphere ($r < 10R_E$ at the equator and corresponding to the earthward limit of the plasma sheet population) has generally been viewed as the domain of the diffuse aurora, resulting from the diffusive drizzle of magnetically trapped populations. However, observations have suggested that, during short periods, field-aligned electromagnetic processes play a prominent role in the transport of mass and energy within the middle regions. Also, there is evidence that these processes manifest themselves in discrete-like auroral behavior. Also, recent theoretical developments suggest that the physics of the field-aligned electromagnetic processes occurring within the middle "adiabatic" regions has some important similarities to that occurring within the deeper magnetotail. These and other developments are reviewed.

1. INTRODUCTION

The mapping of auroral features from the distant magnetospheric regions remains one of the highly controversial areas of auroral research. The title of this paper, as distinct from the title of the paper by *L. R. Lyons* [this volume] ("Discrete Auroras and Magnetotail Processes"), may leave the unfortunate impression that on macroscopic scales such mappings are fairly well understood. However, the *Eos* report by *Siscoe* [1988] on the Auroral Physics Conference from which this volume is derived, properly emphasizes the fact that uncertainties about such mappings, even those on macroscopic scales, lie at the very core of our ignorance concerning the processes that create the aurora. Thus, while we will in this paper be discussing phenomena observed within what we here term the middle magnetosphere ($L < 10R_E$, and in particular the geosynchronous regions), there is no guarantee that these phenomena can be cleanly separated from phenomena that conventional wisdom would place within the more distant magnetotail regions. In fact, one of the theoretical notions that are put forth here is that the physics of the deep magnetotail can be quite similar to the physics of the quasi-dipolar regions of the middle magnetosphere during dynamical periods. It remains to be determined whether the similarities are principally of academic concern (i.e., of little practical use) or are fundamental to our understanding of macroscopic auroral dynamics.

Conventionally, the middle magnetospheric regions are thought to be the domain of the diffuse aurora resulting from the diffusive drizzle of hot, magnetically trapped electrons into the magnetic loss cones and onto the atmosphere [e.g., *Lui et al.*, 1977; *Eather et al.*, 1976]. Indeed, such processes often prevail within these regions. Figure 1 shows figures from a study performed by *Meng et al.* [1979] comparing charged particle intensity spectra sampled at ~2030 local time from the geosynchronous orbit (ATS 6 satellite, University of California at San Diego (UCSD) Auroral Particles Experiment, C. E. McIlwain) with spectra sampled nearly simultaneously at very low altitudes by the DMSP 32 spacecraft near the magnetic foot point of the same field line. The right panel shows the optical auroral display (in negative) viewed by the DMSP satellite, and the triangle shows the magnetic foot point of the ATS 6 satellite as calculated using the field model of *Olson and Pfitzer* [1974]. The foot point is positioned within what is conventionally called the diffuse aurora, equatorward of the more discrete features apparent on the figure. The left panel shows the comparison between the equatorial and low-altitude intensity spectra. The comparison is obviously quite favorable for energies below 1–2 keV. (The DMSP triangles correspond to the calculated magnetic latitude of the ATS 6 foot point; however, the DMSP bars were found 0.5° equatorward.) The trapped geosynchronous particles appear to be precipitated with very little modification to their spectral shapes. The wave-particle interactions held to be responsible for the precipitations have been reviewed extensively by *Kennel and Ashour-Abdalla* [1982].

Figure 2 shows another example from the *Meng et al.* [1979] study to demonstrate that the picture is far from simple, even concerning the occurrence of the diffuse aurora. Here the calculated ATS 6 foot point lies within an auroral region that again has the appearance of being diffuse (ATS 6 local time ~1930 LT), with

*The Johns Hopkins University Applied Physics Laboratory, Laurel, MD 20707.

Auroral Physics, edited by C.-I. Meng, M. J. Rycroft and L. A. Frank. © Cambridge UP 1991

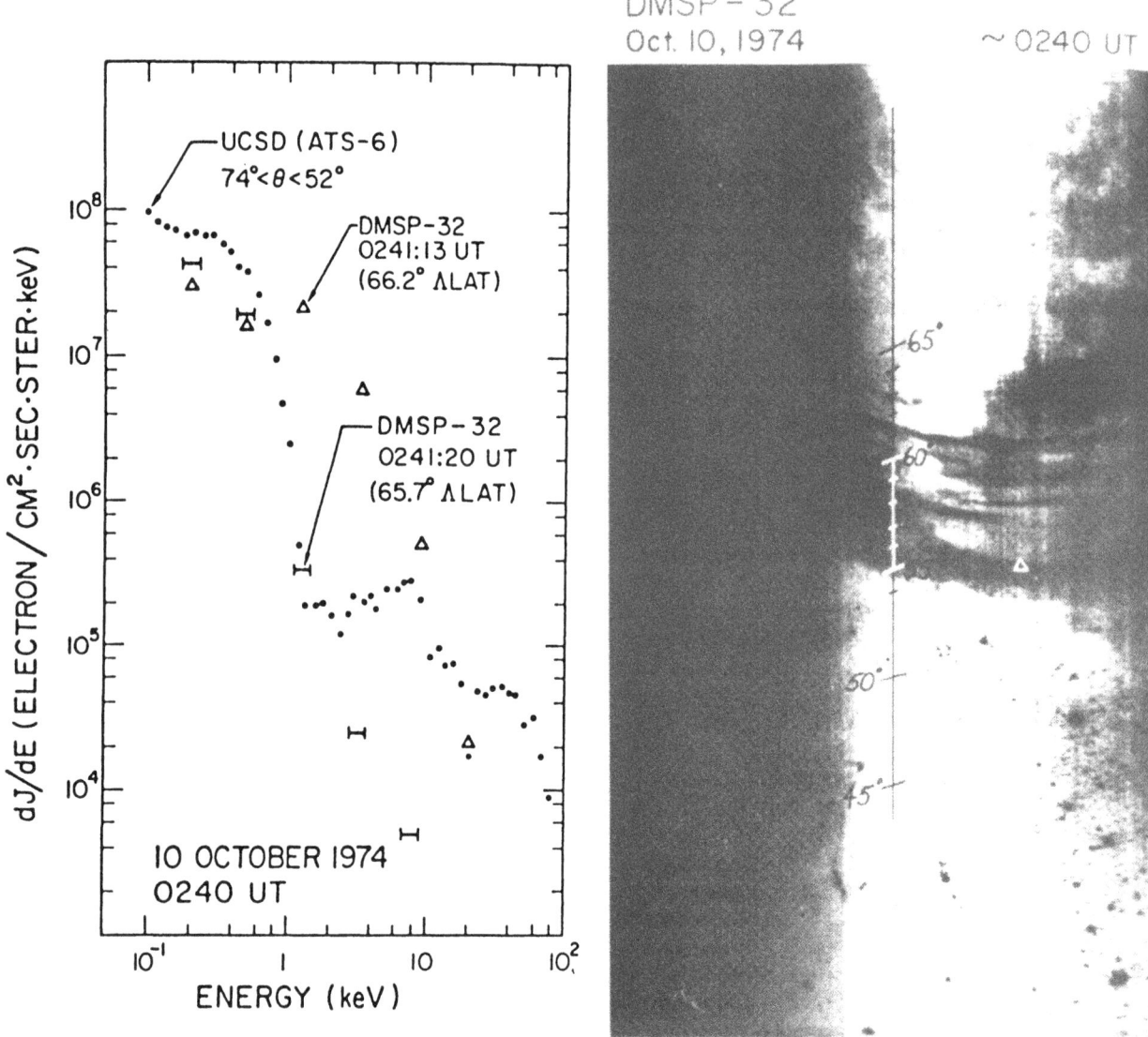

Figure 1—A coordinated observation between the plasma sheet electrons at geosynchronous orbit (ATS 6) and the auroral display and electron precipitation near its field-line conjugate observed by the DMSP 32 satellite on October 10, 1974, ~0240 UT. The right panel shows the DMSP auroral imagery (in negative) with a triangle showing the calculated ATS 6 foot point [*Olson and Pfitzer*, 1974]. The left panel shows the comparison between the conjugate ATS 6 and DMSP electron flux spectra. (From *Meng et al.* [1979].)

discrete features observed poleward. Also, the low- and high-altitude intensity spectra match each other in energy to a very remarkable degree over two orders of magnitude. (Again, a slight shift in latitude was needed to yield the "exact" match apparent with the bars.) The interesting feature of this event is that discrete auroral features are clearly observed equatorward of the ATS 6 magnetic foot point, suggesting that some discrete features map from regions earthward of the geosynchronous orbit and earthward of populations with cen-

tral plasma sheet characteristics (see *Meng et al.* [1979] for a more complete data presentation). It is of course possible that the calculated mapping is grossly incorrect in latitude (by ≥3.5°) and that the ATS 6 foot point lies equatorward of all of the discrete features. However, there exists no matching DMSP spectrum equatorward of the discrete features, and the remarkable match between the high- and low-altitude intensity spectra provides strong circumstantial evidence against such a conclusion. It is unknown whether Fig-

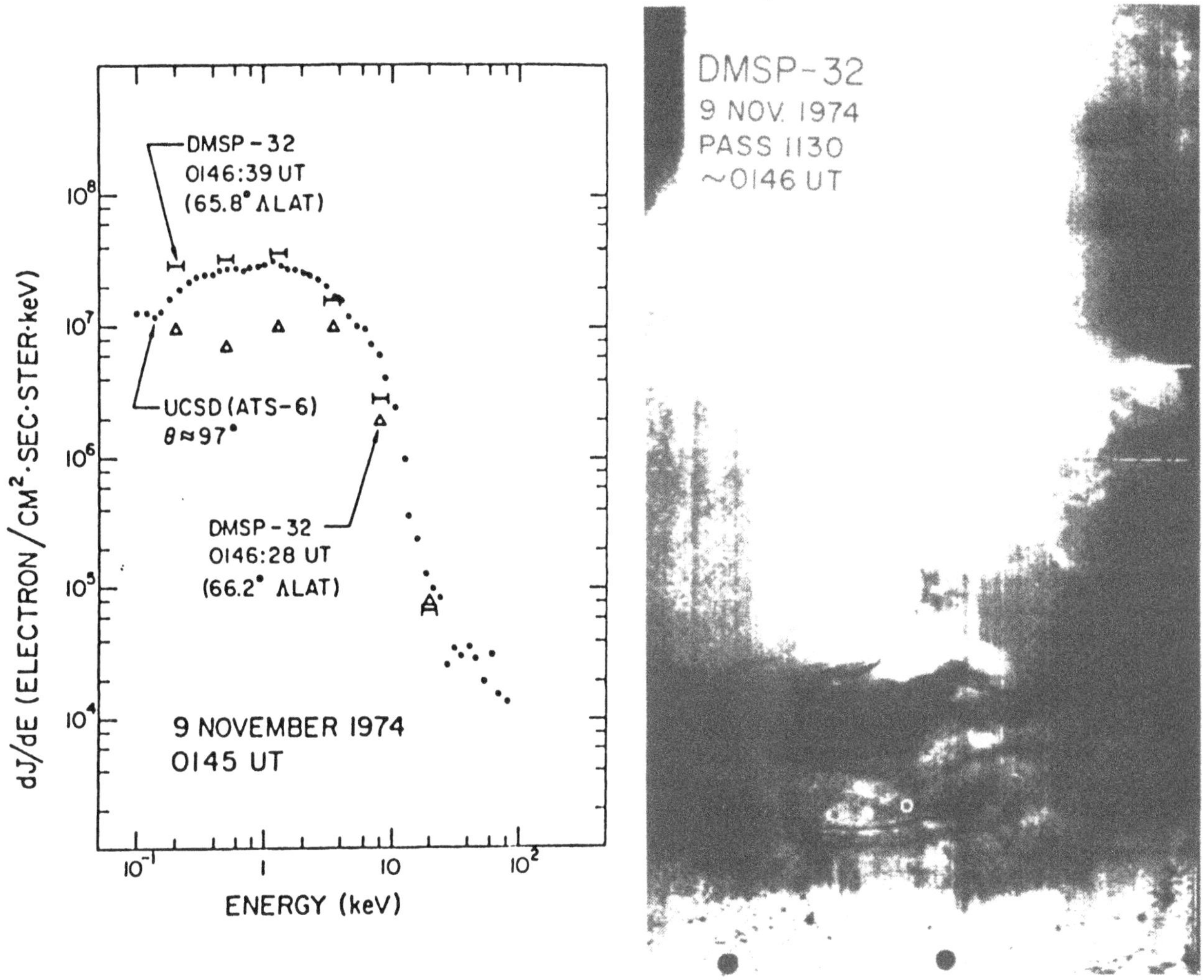

Figure 2—Same as Figure 1 for November 9, 1974, ~0146 UT, and with a circle for the calculated ATS 6 foot point. Note the discrete auroral features equatorward of the ATS 6 foot point. The mapping is confirmed by the differential flux comparison (see text).

ure 2 represents a common or a highly unusual condition. Many researchers [e.g., *Galperin and Feldstein*, 1989; *Rothwell et al.*, 1988] have argued that discrete aurora observed at very low latitudes must map to the middle (e.g., geosynchronous) regions, and the importance of very low-latitude auroral phenomena has been emphasized for many years [e.g., *Chapman* 1957]. Also, the observation of discrete auroral arc systems imbedded within the main region of the diffuse auroral emissions are common [*Murphree et al.*, 1981]. However, it is unusual to have the mapping confirmed by simultaneously obtained particle spectral characteristics.

The observations in Figure 2, showing discrete auroral phenomena equatorward of the geosynchronous foot point, offer evidence that the middle magnetospheric regions can be far more interesting in terms of auroral phenomena than the conventional picture would suggest. In the sections that follow, more direct in-situ evidence of discrete-like auroral activity occurring within the middle magnetospheric regions will be offered. This phenomenon is associated with the earthward boundaries of the hot, plasma sheet populations.

225

2. FIELD-ALIGNED ELECTROMAGNETIC PHENOMENA

The most compelling in-situ evidence for the importance of field-aligned electromagnetic phenomena within the middle or geosynchronous regions, perhaps associated with discrete-like auroral processes, is that obtained by *McIlwain* [1975]. Figure 3 shows samples of an electron distribution, replotted from *McIlwain* [1975], that was measured by the UCSD electrostatic analyzers on the ATS 6 geosynchronous satellite. The solid dots show the magnetic field-aligned distribution, and the open circles show the field-perpendicular distribution. The distribution obviously has a strongly field-aligned characteristic, and the field-aligned distribution shape has a peaked feature that, while quantitatively different, is reminiscent of electron distributions observed over discrete auroral forms at altitudes $<2R_E$ [e.g., *Arnoldy*, 1981; *Bryant*, 1981]. (It must be stressed that these similarities do not necessarily mean that the geosynchronous electron beams are associated with discrete aurora, but the distribution characteristics are suggestive of a field-aligned discharge.) As indicated by the insert to Figure 3, and as is discussed by *Fritz et al.* [1977] and *Mauk and Meng* [1987], the geosynchronous electron beams are often observed coincidently with east-west magnetic perturbations interpreted as resulting from magnetic field-aligned electric currents [e.g., *Nagai*, 1982]. The observed perturbations often do not require that the source currents be local to the spacecraft. However, independent observations have been made of localized magnetic field-aligned currents within the near geosynchronous regions [*Kelly et al.*, 1984; *Robert et al.*, 1984; *Roux*, 1985].

There have been some more recent observations of magnetic-field-aligned electron beams within the middle magnetospheric regions. *Klumpar et al.*, [1988] observed, using the AMPTE CCE spacecraft, bidirectional electron beams with energies between hundreds of electron volts to about 20 keV at radial distances of $8–9R_E$ near the magnetic equator. The authors suggest that the beams may correspond to the observations of *McIlwain* [1975]; however, little information concerning the context of the AMPTE beams was given, and a $\partial f/\partial v_{\parallel} \geq 0$ feature was not observed. (Not all of McIlwain's beams have such a feature, presumably because the beams are observed in various stages of evolution. The ATS 6 instrument had relatively poor temporal resolution, as described below.) At higher energies, *Kremser et al.* [1988] have observed, using the GEOS 2 geosynchronous satellite, bidirectional electron beams within the 16–80 keV range. The beams were rare (10/year), but their relation to substorm injection phenomena was quite similar to that

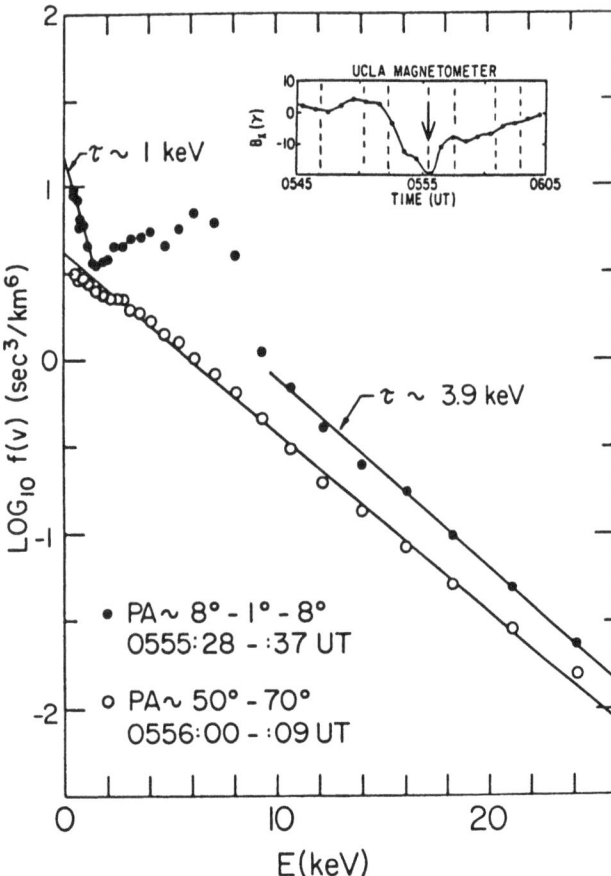

Figure 3—Electron velocity distributions, $f(v)$, measured by the ATS 6 satellite in association with a nighttime geosynchronous substorm particle injection on June 16, 1974. The intense magnetic field-aligned beam is of particular interest. The insert shows the associated magnetic signature (the spacecraft B_x is approximately minus the standard coordinate D) that is commonly attributed to field-aligned electric currents. The "parallel" (i.e., pitch angles, PA, up to 8°) particle data was sampled at the time of the vertical arrow. (Replotted from *McIlwain* [1975] and *Fritz et al.* [1977].)

of McIlwain's electron beams. The energy range of the detectors used was above the energies where one might expect a $\partial f/\partial v_{\parallel} \geq 0$ feature to be observed (and indeed none was observed); however, these rare events may well correspond to unusually high-energy tails of the beams reported by *McIlwain* [1975]. Some of the $\partial f/\partial v_{\parallel} \geq 0$ events of McIlwain do show well-developed high-energy tails extending to energies very substantially higher than the energy at the peak of the $f(v)$ spectrum (e.g., ~20 keV for an $f(v)$ peak at 1.5 keV, *Parks et al.* [1980]). However, the *Kremser et al.* [1988] events had unusually high energy, and an association with *McIlwain's* [1975] beams remains to be firmly established. As a final note on other observations, upgoing, multi-kiloelectron-volt electron beams

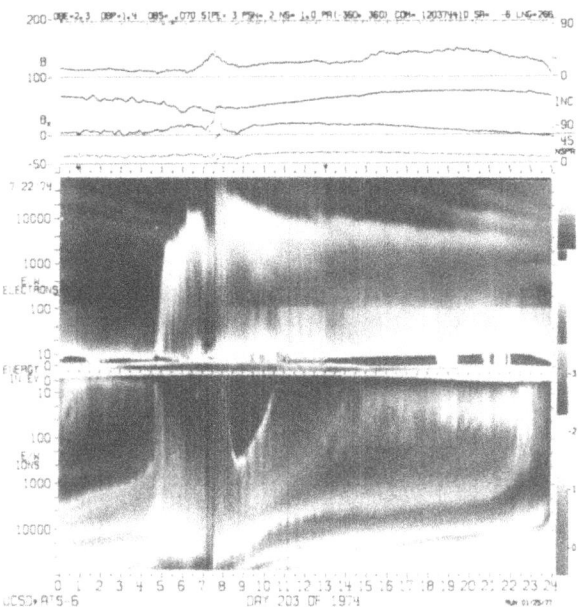

Figure 4—Twenty-four hour spectrogram from the ATS 6 geosynchronous satellite for ion and electron energies between ~ 1 eV and 82 keV (note the inverted ion energy scale). Of interest is the sudden injection at 0735 UT, and the associated electron beams of the sort shown in Figure 3. The upper three curves display the magnetic field data (From *Mauk and Meng* [1983*a,b*].)

(as opposed to the downgoing electron beams conventionally associated with auroral forms) have been observed at relatively low altitudes within the auroral regions by the Viking spacecraft [*Lundin et al.*, 1987; *Hultquist et al.*, 1988], and an association with the near geosynchronous electron beams has been suggested (private communication from a referee).

The concern was voiced in open forum at The International Conference on Auroral Physics, St. John's College, Cambridge, England, July 1988 that the geosynchronous electron beams observed by *McIlwain* [1975] could have occurred in the outer plasma sheet boundary layer regions (PSBL), rather than within the quasi-dipolar regions. (The beams have been observed at geomagnetic dipole latitudes of ~ 10°.) An understanding of the conditions under which the beams occur is important for subsequent discussions, and so this concern deserves some attention.

Context of Electron Beam Observations

The context for the beam observations was discussed by *McIlwain* [1975], and when those discussions are placed in the context of work on the convective evolutions of injected plasmas [e.g., *DeForest and McIlwain*, 1971; *McIlwain*, 1972, 1974; *Mauk and Meng*, 1983*b*] the evidence is strong that the beams exist on quasi-dipolar field lines near the inner edge of the newly ener-

gized and injected plasma sheet population. Figure 4 is presented to amplify this point.

The event of interest in Figure 4 is a nearly dispersionless substorm injection that occurred at 0735:30 UT ± 30 s. An intense magnetic field-aligned electron beam of the sort shown in Figure 3 (showing a $\partial f/\partial v_\parallel > 0$ characteristic and peaking at several kilovolts) was sampled first at ~ 0740:18 UT (5 minutes after the injection), and subsequently at ~ 0741:53 UT. The UCSD electrostatic analyzers on ATS 6 employ rotating heads that rotate back and forth with a period of 314 s (5.23 min—*Mauk and McIlwain* [1975]). The energy sweeping (from low to high energies) takes 16 seconds and, at the time of interest, is repeated every 32 seconds. It is significant that for this particular event, no electron beam (i.e., with $\partial f/\partial v_\parallel \geq 0$) was observed during a field-aligned sample that occurred about 1 minute following the time when the hot plasmas are fully established at the satellite location (near 0736:35 UT). Following the observations of the electron beam, between 0740 and 0742 UT, remnants of the beam were observed at ~ 0745 UT and for at least 15 minutes beyond, but these remnants were substantially degraded with respect to the original beam (i.e., less intense, having lower energies, and not characterized with the condition $\partial f/\partial v_\parallel > 0$).

The magnetic signatures of the ~ 0735 UT injection in Figure 4 are shown with the line plots at the top of the display. The top line plot shows that |**B**| was ≥ 130 nT before the event, and decreased suddenly by ≤ 10 nT at the initiation of the event. The second line plot shows that the inclination angle I ($\tan I = B_r/B_\theta$) increased somewhat (by ~ 12°) near the event initiation time, showing that the magnetic field becomes more dipole-like, i.e., dipolarized. Finally, the third line plot shows substantial east-west magnetic activity with a very sudden transition occurring at just the injection initiation time.

The key point in Figure 4 for the present discussions is the subsequent (several hours) convective evolutions of the injected hot particles. Note that at the highest ion energies (~ 80 keV), a so-called echo appears, starting near 0830 UT. These echo ions are ions that have drifted (principally via the magnetic gradient and curvature drifts) all the way around the Earth, only to reencounter the spacecraft on the nightside [*Lanzerotti et al.*, 1971; *DeForest and McIlwain*, 1971; *Mauk and Meng*, 1983*a*]. The point is that the injection region clearly exists on a guiding center drift shell that connects smoothly to the quasi-dipolar field lines on the dayside. What is more, the drift period is essentially identical to the purely dipolar value for ions of this energy at this L value. It is clear that the injected plasmas appear on quasi-dipolar field lines that support guiding center behavior. (Here, we will loosely use the phrase "quasi-dipolar" to refer to field lines that have dipolar symmetries and that support guiding center behavior for the bulk of the populations being considered. Such field lines could have very substantial tail-like distortions.)

With regard to how the electron beams fit into the picture, it is significant that the beams were observed well after the hot plasmas were established in the vicinity of the spacecraft (first observed 5 minutes after the hot plasmas appeared), and well after the $\Delta|\mathbf{B}|$ and ΔI activity had stabilized (the fast changing behavior apparent on Figure 4 ended by 0737:40 UT). Since the evidence is strong that, after the injection activity, the hot plasmas existed on quasi-dipolar field lines (supporting guiding-center behavior), the evidence is also strong that the field-aligned electron beams were also observed on such quasi-dipolar field lines.

The concern that has been expressed with regard to this conclusion takes the form of the following hypothesis. During the growth phase the field lines in the vicinity of the geosynchronous orbit become more tail-like, effectively causing the satellite to move from field lines that map to the middle magnetosphere to field lines that map to the more distant magnetotail regions. In principle, the satellite could be in the high-latitude outer boundaries of the plasma sheet populations (i.e., the plasma sheet boundary layer) where field-aligned particle streaming is commonly observed in the deep tail regions [*Lyons and Speiser*, 1982; *Eastman et al.*, 1984] during both substorm and nonsubstorm periods (at least for ions). Continuing this hypothesis, expansion phase dipolarization of the plasma sheet boundary layer regions could sweep the boundary outward and over the spacecraft immersing the spacecraft in the quasi-isotropic energized plasma sheet populations. The electron beams would be observed only as the spacecraft passed through the boundary regions on its way into the isotropic plasma regions. Supporting this hypothesis is the fact that, prior to the expansive phase dipolarizations (e.g., 0715–0735 UT on Figure 4), the preexisting particle intensities sometimes lessen, apparently as the spacecraft moves onto the higher latitude field lines. (In Figure 4, the peak energy fluxes dropped by almost a factor of 4 and 8 for the electrons and ions, respectively; however, quasi-Maxwellian spectra remained for both species.)

The principal rebuttal to the PSBL hypothesis is the observational fact described above, that the electron beam associated with the Figure 4 event was observed only well after the particles were injected and well after the magnetic field indicators of magnetic reconfiguration had become quiet. Just as importantly, during the one observational opportunity for the electron

beams to be observed following the satellite's immersion into the hot, energized, quasi-isotropic injected plasmas, the electron beam was not observed. (As will be seen later, the electron beams do not always lag the isotropic injections.) This evidence suggests strongly that the plasma sheet boundary layer is not the site of the observed electron beams. Again, the electron beams appear to be observed on quasi-dipolar field lines that support guiding center behavior. (Guiding center behavior could be violated during the creation of the electron beams, but the site of that creation would appear to be the hot plasma sheet populations interior to any pre- or postexisting outer plasma sheet boundaries.)

Pursuing the Figure 4 event just one step further, following the 0735 UT injection are dispersive high-energy intensity cutoffs that appear symmetrically (in a qualitative sense) in the hot electrons and hot ions. Those cutoffs slowly decrease with time, ranging from tens of kiloelectron volts 30 minutes after the injection, to below 10 keV four hours later. After the injection, the most energetic particles (tens of kiloelectron volts) feel principally the effects of the magnetic drifts rather than the electric drifts, with the electrons drifting eastward and the ions drifting westward. Given the combined electron and ion characteristics, these dispersive features have been interpreted as resulting from an earthward intrusion of energized plasma sheet populations with the most earthward intrusion occurring near midnight [*McIlwain*, 1972; *Konradi et al.*, 1975; *Mauk and Meng*, 1983a]. The dispersive high-energy cutoff features have been modeled quite successfully [*Mauk and Meng*, 1983b] using a particular form of the midnight intrusion known as the double-spiral injection boundary [*Konradi et al.*, 1975; *Mauk and Meng*, 1983b; *McIlwain*, 1985]. The point is that the higher energy ions and electrons offer a tool for remotely sensing the initial positions of sharp spatial boundaries [*McIlwain*, 1974]. Most telling for the 0735 event in Figure 4 is the extreme rapidity in which the high-energy ion cutoff forms (the fresh hot ions extend to above the ~80 keV instrument limit only sporadically and only for several minutes). Later on, ions with energies below the cutoff are observed because the postinjection convection electric field is strong enough to transport the low-energy ions radially while they are being transported azimuthally by gradient drifts. Given an earthward intrusion centered at midnight, given the 0130–0200 LT position of the satellite, and given the duskward drift motion of the ~80 keV ions, it appears that the satellite was located just at the earthward edge of the energized plasmas. The energetic electron behavior is consistent with this story. Figure 4 shows that, unlike the ions, more than 15 minutes go by before the high-energy electron cutoff is lower than the ~80 keV limit of the instrument. The east-ward drifting electrons are observed continuously until the duskward boundary of the near-midnight intrusion is sensed. It appears, then, that the field-aligned electron beams are created near the earthward boundary of the newly energized and transported plasma sheet populations, rather than at the outer boundaries.

In conclusion, there is strong, if not overwhelming, circumstantial evidence against an association between the magnetic field-aligned electron beams reported by *McIlwain* [1975] and the outer boundaries of the plasma sheet population (i.e., the plasma sheet boundary layer, PSBL). The beams appear to be created and placed onto quasi-dipolar field lines and are created near the earthward edge of the newly energized and injected plasma sheet populations.

The relationships of this earthward edge (or the injection boundary) to the conventionally characterized middle-region plasma sheet populations, and also to the ring current populations, is discussed by *Mauk and Meng* [1986, 1983b]. In particular, it is argued that dynamical injections that give rise to the impulsive earthward intrusions are an inseparable component of the processes that populate the middle regions with plasma sheet particles (i.e., quasi-time-stationary, curl-free convection never acts alone in populating these regions). The ring current populations result from subsequent transport of the injected populations. These conclusions are controversial (e.g., *Kivelson et al.* [1987]).

Possible Auroral Manifestations

The evidence of Figure 2 notwithstanding, it is likely to be difficult to identify uniquely the auroral manifestations of the above described field-aligned-electromagnetic phenomena encountered within the geosynchronous orbit. The electron beams discovered by *McIlwain* [1975] appear to be highly transient phenomena, lasting only for several minutes in their most intense forms. (Low-energy beams can last for tens of minutes, but in the kiloelectron-volt range the forms of the beams degrade rapidly from their original forms.) Thus, a low-altitude satellite, such as one of the DMSP series satellites, would have to be at the right place at the right time during the discharge, even if the high-altitude electron beams commonly propagate to, or originate from, the low-altitude regions. Space-time ambiguities would prevail for such measurements.

There is a platform that has some of the advantages of geosynchronous satellites in that it performs relatively long-term monitoring of a single region of space. That platform is the high-altitude balloon. From such a platform energetic electron acceleration and precipi-

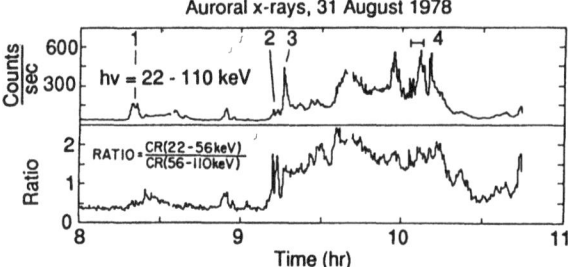

Figure 5—A three-hour summary plot (~0200 to ~0500 LT) showing bremsstrahlung auroral X-ray data from an X-ray pinhole deployed by balloon to 4–5 g/cm^2 altitudes and at a geomagnetic latitude equal to ~64°. The top panel shows total counts per second and the bottom panel shows the ratio of low-to-high energy counts per second. The two-process injection phenomena that occurred at points 2 and 3 are the focus of the interest here. (From *Mauk et al.* [1981].)

tation phenomena can be monitored via the bremsstrahlung X-rays generated by energetic electrons striking the upper atmosphere [e.g., *Anderson and Milton*, 1964; *Barcus and Rosenberg*, 1966; *Brown* 1966; *Parks* 1967).

We focus on X-ray data because phenomena with similarities to the geosynchronous phenomena have been revealed. Figure 5 shows data taken using an X-ray imaging pinhole camera deployed by balloon on August 31, 1978 [*Mauk et al.*, 1981]. The balloon was launched at 64° geomagnetic latitude. The figure shows 3 hours of data (0800–1100 UT) corresponding to local times between 0200–0500. The top panel shows the total X-ray counts/second integrated over the energy range of the instrument. The bottom panel shows a spectral index composed by taking the ratio of low-energy to high-energy counts/second. This index is qualitative only, since it is contaminated with background counts that tend to "harden" the index for low foreground rate events.

The regions of interest for the present discussions are those labeled 2 and 3 on the upper panel. Event 3 was obviously a major burst when compared with the events that preceded it. Perhaps even more revealing is the behavior of the spectral index. That index rises along with the event-3 burst, but then stays constant in its raised position as the total rates dramatically decrease. It must be recognized that measurements of precipitation fluxes are fundamentally distinct from in situ measurements of trapped fluxes in that not only must the electron populations be present within the magnetosphere, but processes must be operating to cause those particles to move into the loss cone and precipitate. The stability of the spectral index starting at point 3 suggests that the trapped populations formed or injected at point 3 are continually present within the

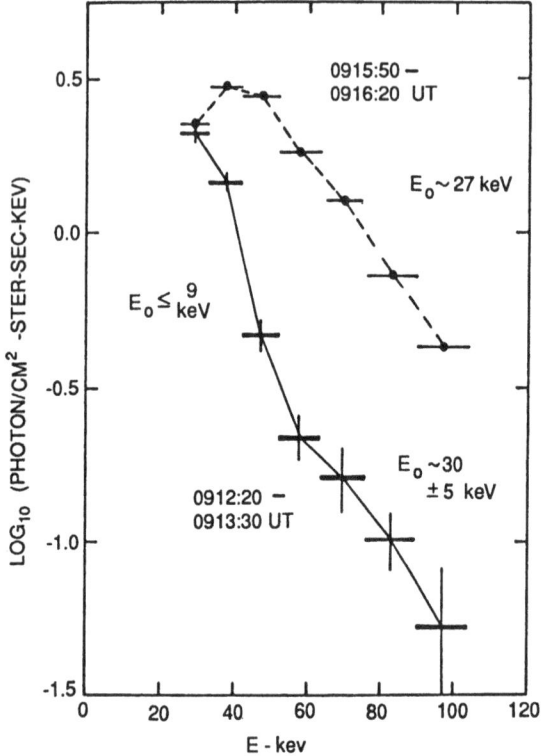

Figure 6—X-ray differential spectra (background subtracted) from points 3 (top spectrum) and 2 (bottom spectrum) in Figure 5. Corrections for atmospheric attenuation have not been made.

magnetosphere after that time. The variability of the total count rates after point 3 suggests that the loss cone emptied rapidly after point 3, and mechanisms were not continually operating to keep the loss cone full. The event-3 feature has been interpreted as corresponding to a particle injection of the sort observed in the geostationary orbit, such as the 0735 UT event shown in Figure 4 [*Mauk et al.*, 1981].

The double-humped event labeled 2 in Figure 5 is the focus of our interest in the X-ray data. This feature began about 4 minutes before the beginning of the event-3 injection. What is significant about this event is the extreme spectral softness, as partially revealed in the lower panel. Despite the background contamination in this index, the event shows itself to be substantially softer (lower in average energy) than does the event-3 injection. Figure 6 shows more details of the spectral characteristics of the Figure 5 events. The upper spectrum (dashed line) was taken during event 3 (background subtracted), and the lower spectrum (also background subtracted) was taken during event 2. Atmospheric attenuation has not been removed from these spectra, hence the strong roll-off of flux at lower energies for the upper spectrum (4–5 g/cm^2 at-

mosphere). It is traditional [e.g., *Barcus and Rosenberg*, 1966; *Brown*, 1966] to characterize the spectral characteristics with the free parameter E_0 of a fit to the high-energy portion of an X-ray spectrum of the form $j \sim \exp(-E/E_0)$. Figure 6 demonstrates how spectrally soft the event 2 fluxes were as compared with event 3 ($E_0 < 9$ keV for event 2, and $E_0 \sim 27$ keV for event 3). In fact, because of instrumental limits in spectral resolution (the spectral resolution capabilities of the X-ray camera were degraded in the process of optimizing it for spatial resolution), the $E_0 < 9$ keV value represents a coarse upper limit. It would be consistent with the characteristics of the camera if the soft component of the event-2 spectrum consisted only of X-rays with energies somewhere below 40 keV.

A final characteristic of note concerning event 2 on Figure 5 is its spatial characteristics. As documented by *Mauk et al.* [1981], the soft component of event 2 consisted of a single spot confined to about 20 km out of the ~80 km field of view of the X-ray camera. In contrast, the spectrally hard component of event 2, and the fluxes of event 3 were spatially uniform over the field of view.

All of the characteristics of the region of events 2 and 3 in Figure 5, cited above, point to the conclusion that two distinct processes were occurring during an injection event: one of these generated spatially striated and spectrally soft electron fluxes, and the other generated spatially uniform and spectrally hard electron fluxes. It has been suggested [*Mauk et al.*, 1981] that these two processes can be identified with the two processes discovered by *McIlwain* [1975] to be operating near the earthward edge of the hot plasma sheet populations injected during the expansion phase of substorms. The spectrally soft component of event 2 in Figure 5 would thus be identified with the field-aligned electron beams of *McIlwain* [1975], while the spectrally hard event 3 would be identified with the isotropic injections within the geosynchronous orbit.

Figure 7, in fact, shows an example of geostationary phenomena that has substantial similarities to the X-ray data, in terms of timing. At about 0103–0105 UT on this display (~1900 LT) the rotating UCSD ATS 6 charged particle detectors observed intense field-aligned fluxes of electrons (and ions as well). The electron fluxes were dramatically enhanced in the field-aligned directions over those in the field-perpendicular direction at energies up to 20–25 keV (i.e., at energies higher than those observed during the event shown in Figure 3). This field-aligned electron feature preceded the occurrence of a sudden isotropic electron injection by ~5 minutes. The timing similarities between this electron-beam/isotropic-injection event and the event-2/event-3 X-ray features on Figure 5 are obvi-

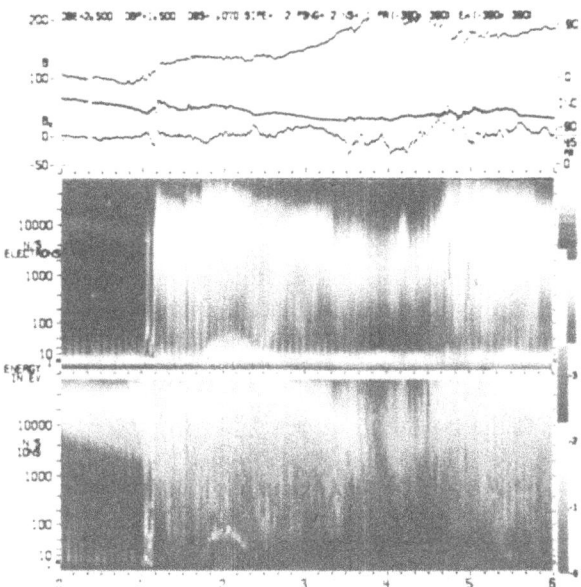

Figure 7—From the ATS 6 geosynchronous satellite (Courtesy, C. E. McIlwain), showing that the impulsive isotropic injection near 0110 UT (~ 1900 LT) is, somewhat unusually, preceded here by field-aligned particle beams after ~ 0103 UT, extending to electron energies of 20–25 keV. We focus on the similarities between this two-process geosynchronous injection and the two-process phenomena shown in Figure 5.

ous. (Care needs to be exercised since the geostationary electron beam was not characterized with a $\partial f/\partial v_{\parallel} \geq 0$ feature at this time.)

The weakest link in our attempt to associate the X-ray features with geostationary injection phenomena are the characteristic energies involved with the softer component. Given the resolution characteristics of the X-ray camera, energies up to at least ~ 35–40 keV were involved with the X-ray event, whereas the electron beams generally involve lower energies (the Figure 7 event extends only to ~ 20 keV, perhaps to 25 keV). The difference is small enough that one can hypothesize that the unique X-ray observation corresponds to an unusually energetic example. The problem would be solved if it could be shown that the *Kremser et al.* [1988] events are indeed the high-energy portion at the McIlwain beams as discussed earlier.

Thus, assuming that the associations suggested here are borne out by future observations, it appears that there is an auroral manifestation of the field-aligned electromagnetic phenomena discovered by *McIlwain* [1975] within the middle magnetospheric regions. Given the 20-km scale size observed, the label ''discrete'' would certainly be justified for this phenomenon. However, we cannot yet demonstrate that the phenomena

is discrete in the full sense utilized in modern auroral research.

3. DISCHARGE MECHANISMS

The evidence of *McIlwain* [1975], the evidence of transient and small-scale field-aligned current signatures [*Kelly et al.*, 1984; *Robert et al.*, 1984; *Roux*, 1985], and the evidence of the X-ray signatures discussed in the previous sections all suggest that transient auroral discharge-like processes, with similarities to the processes that generate discrete aurora, operate within the middle magnetospheric regions where quasi-dipolar magnetic field configurations prevail. In addition, the evidence of *Meng et al.* [1979; see Figure 2] suggests that discrete auroral processes may even be able to operate in a much less transient fashion within these same regions. We will now try to account for the more transient phenomena. The discussions will be somewhat speculative.

It is our thesis that macroscopic ion dynamics plays a prominent role in the occurrence of auroral discharge phenomena. This thesis has precedent in the works of, for example, *Stenzel et al.* [1981], *Kan* [1975], *Bingham et al.* [1988], *Bryant and Hall* [this volume], etc. We focus on ion dynamics because of recent findings showing that, even within quasi-dipolar magnetic configurations where guiding-center behavior prevails, ions can behave very nonadiabatically over macroscopic scales during dynamical periods. We begin below with a discussion of guiding-center ion dynamics and follow that with an application to observed middle magnetospheric phenomena.

Guiding Center Acceleration

Within the guiding-center approximation we may write [Northrup, 1963]

$$\frac{mdv_{\parallel}}{dt} = qE_{\parallel} - \mu\,\frac{d|\mathbf{B}|}{ds} + mv_c \cdot \frac{d\mathbf{b}}{dt} \qquad (1)$$

and

$$\mu = \frac{1/2mv_{\perp}^2}{|\mathbf{B}|} = \text{constant}, \qquad (2)$$

where $v_{\parallel}$ and $v_{\perp}$ are the velocities parallel and perpendicular to $\mathbf{B}$, m is particle mass, q is particle charge, $E_{\parallel}$ is the parallel component of the electric field E, μ is magnetic moment, s is distance along $\mathbf{B}$, v_c is the convection velocity $cE \times B/B^2$, and $\mathbf{b}$ is the unit vector in the direction of $\mathbf{B}$. The last term of Eq. 1, dubbed here the ''centrifugal term,'' will be the focus of the discussion. This term can be understood heuristically by rewriting Eq. 1 into a field-aligned energization equation. Considering only the term of interest one may write

$$\frac{d(1/2mv_\parallel^2)}{dt} \sim \left(mv_\parallel^2 \frac{\mathbf{R}_c}{R_c}\right) \cdot v_c$$
$$+ \left(mv_c^2 \frac{\mathbf{R}_1}{R_1}\right) \cdot v_\parallel \qquad (3)$$

where $\mathbf{R}_c$ and $\mathbf{R}_1$ are explained by reference to Figures 8*a* and 8*b*, respectively. One sees that two different types of terms have emerged, each with the form of a centrifugal force times a velocity.

The first term of Eq. 3 can be understood heuristically by referring to Figure 8*a*. There is a centrifugal force associated with the motion of the particle's guiding center along a curved field line, with radius of cur-

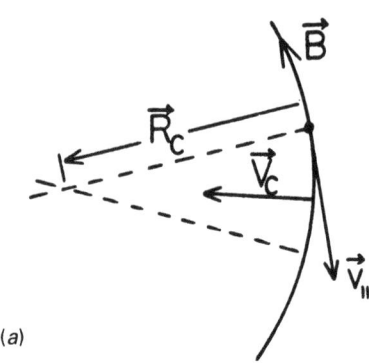

(*a*)

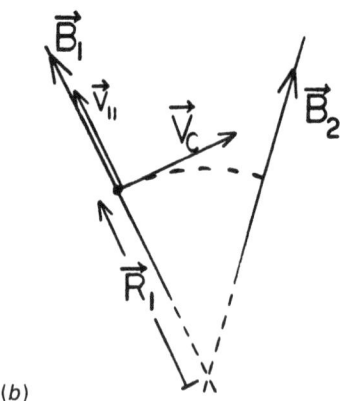

(*b*)

Figure 8—Illustrations of the "centrifugal" character of the energizations that result from the last term of the guiding-center equation (Eq. 1). (See text for explanation.)

vature R_c. Energization results when the magnetic force that balances out the centrifugal force acts in the direction of the convection velocity v_c. This energization is analogous to the resulting energization of a stone that is whirling around in a circle on the end of a rope, when the rope length is shortened.

The second term of Eq. 3 can be understood, again heuristically, by referring to Figure 8*b*. There is a centrifugal force associated with the convection of the particle from one field line to another field line that is pointing in a different direction, with an effective moment arm R_1 defined within an inertial frame of reference. Energization results when that centrifugal force (the force is not balanced in this case) acts against the particle's parallel velocity. This energization is analogous to the energization that results to a bead sliding loosely along a straight length of metal rod when the rod is, say, whirled overhead in a whip-like motion.

Until recently, the centrifugal acceleration described above has not been applied at the middle and inner magnetospheric regions because it has commonly been assumed that parallel energization can be calculated by invoking the conservation of the so-called longitudinal or second adiabatic invariant ($\oint v_\parallel ds$ = constant). Recent interest in the centrifugal acceleration term for these regions has been sparked, as described below, by the realization of the importance of fast dynamical processing on the transport of particles within these middle and inner regions (see *Cladis* [1986] and *Chapman and Cowley* [1984] for other applications).

Middle Magnetospheric Ion Dynamics

Quinn and McIlwain [1979] have reported on a phenomenon within the geosynchronous environment, and occurring during the expansion phase of substorm injections, that clearly violates the second adiabatic invariant of the ions involved. The phenomenon is the "bounce-phase-bunched" ion distributions that consist of spatially bunched clusters of ions bouncing back and forth between the northern and southern hemispheres along quasi-dipolar field lines (energies ranging from electron volts to about 20 keV). The velocity dispersion of the distributions shows them to be the result of temporally impulsive events. *Quinn and Southwood* [1982] proposed that the bunching was caused by the centrifugal acceleration mechanism acting in association with sudden magnetic reconfigurations termed convection surges.

The suggestion of *Quinn and Southwood* [1982] was modeled in detail by *Mauk* [1986]. Figure 9 provides a brief review. The top panel shows the spatial configuration. Shown is a meridional plane of the Earth's magnetosphere containing two quasi-dipolar field lines. The field line labeled T_i is the configuration before the occurrence of the convection surge. It has a tail-like distortion. The field line labeled T_f is the post-surge configuration and is purely dipolar. The transition between T_i and T_f occurs as a result of the application of intense, transient electric fields (out of the figure page) lasting for a period of 60 s. The electric field values used (20 mV m^{-1} at the equator) were constrained by the observed values reported by *Aggson et al.* [1983]. Note that the shape change of the field line (i.e., the dipolarization) is a consequence of the fact that the observed electric fields point east-to-west near the magnetic equator and west-to-east at high latitudes [*Aggson et al.*, 1983].

In these modeling efforts, the evolution of a single flux tube was considered. That flux tube was loaded by a distribution of ions that was, on the average, Maxwellian, isotropic out of the loss-cone, and empty within the loss cone. Ten thousand particles were followed and were positioned initially with weighted random numbers. The evolution of the ions of this distribution was then followed before, during, and after the convection surge or dipolarization. The second panel of Figure 9 shows a simulated spectrogram from the model. A single dot on the panel corresponds to the crossing of the magnetic equator by a single ion. Only particle pitch angles between 5° and 15° are represented. The initial Maxwellian temperature was 30 eV.

The spectrogram panel shows that the mechanism energizes the particle substantially. Also, the dramatic postsurge, dispersive streaks qualitatively match the bounce-phase-bunched ion distributions reported by *Quinn and McIlwain* [1979]. However, it is the third panel of Figure 9 that is most relevant to the present discussion. Shown are the pre- and postsurge distributions (after the bounce-phase-bunching has phase-mixed away). The postsurge distribution is dramatically enhanced in the magnetic field-aligned direction over the field-perpendicular direction. As discussed by *Mauk* [1986], such field-aligned distributions (below 1 to 10 keV in energy) are characteristic of ion measurements within the geosynchronous regions. The mechanism is strongly species dependent. The O^+ ions show even more dramatic pitch angle anisotropies, while electrons will tend toward pancake distributions.

In the following section it will be shown that one consequence of the field-aligned ion distributions generated by the convection-surge is the generation of field-aligned electric fields. We will speculate that these electric fields can give rise to a discrete-like auroral discharge. It should be noted that the mechanism by which field-aligned ion distributions are generated on these quasi-dipolar field lines is very similar to a corresponding mechanism operating within the deep tail in the vicinity of narrow neutral sheets (see *Speiser* [1965, 1967], *Lyons and Speiser* [1982], and the discussion of *Mauk*

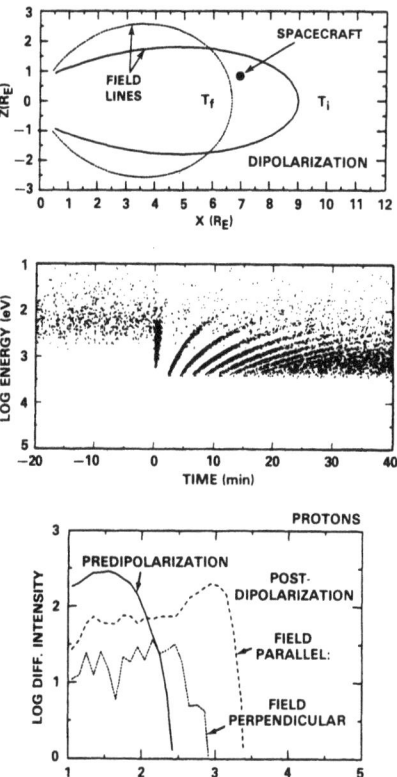

Figure 9—(Top) Meridional view of pre- and postconvection-surge field-line configurations used in the modeling efforts of *Mauk* [1986]. (Middle) Simulated ion spectrogram generated by the convection surge test particle modeling efforts of the same paper. (Bottom) Pre- and postsurge ion distributions generated with the same test particle modeling efforts. The "centrifugal" acceleration term in Eq. 1 is central to obtaining the field-aligned character of these ion distributions.

and Meng [1988]). Field-aligned ion beams are a consequence of both processes. Thus, the discharge mechanisms discussed here with regard to the quasi-dipolar regions may be applicable to the outer regions mapping along the plasma sheet boundary layer (PSBL).

Field-Aligned Electric Fields

The convection-surge process can generate strongly magnetic-field-aligned ion distributions, as Figure 9 makes clear. Hence the average ion mirrors at a much lower altitude than it did before the dipolarization. The electrons, on the other hand, will not be affected by the centrifugal term as much as the ions. Hence the electrons will largely be confined to the higher altitude regions relative to the ion positions. In other words, the action of the centrifugal acceleration term during the convection surges tries to separate charges along the field lines. The system will respond to restore quasi-

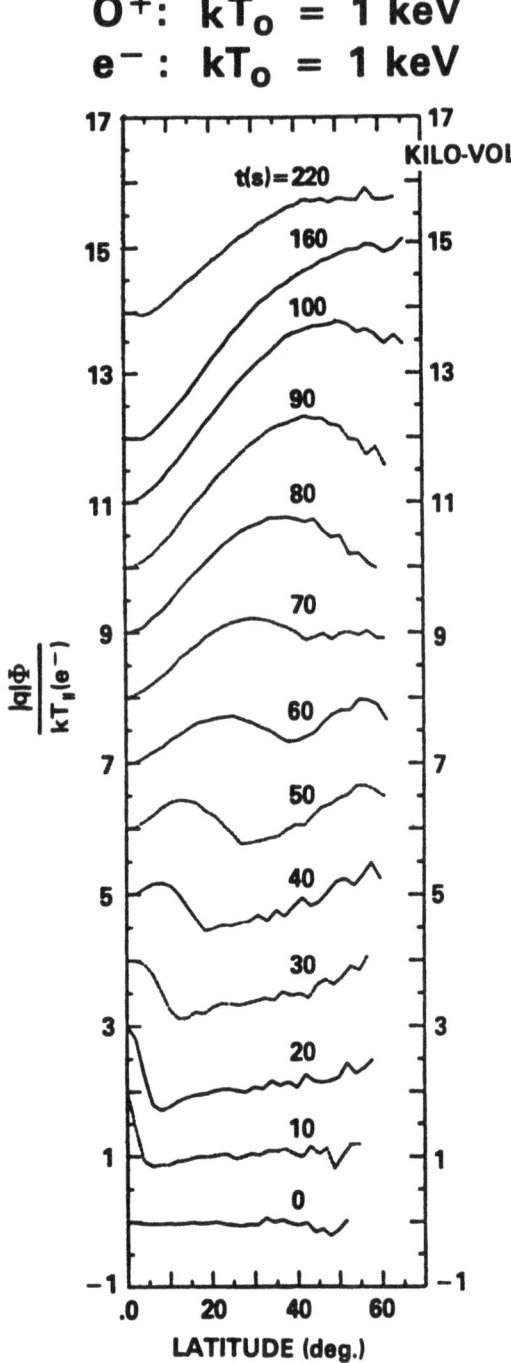

$$O^+: \quad kT_0 = 1 \text{ keV}$$
$$e^-: \quad kT_0 = 1 \text{ keV}$$

Figure 10—Profiles of magnetic-field aligned electric potential (expressed in terms of electron thermal energy) generated by the self-consistent convection surge modeling efforts documented by *Mauk* [1988]. Profiles are shown for various times into the simulation periods between $t = 0$ s and $t = 220$ s. The different profiles are displaced vertically for clarity. All initial particle temperatures were 1 keV. Oxygen ions were utilized. Guiding-center behavior is all that is required to obtain these results.

charge-neutrality by setting up magnetic-field-aligned electric fields [e.g., *Whipple*, 1977 and *Chiu and Schulz*, 1978].

This process has been modeled [*Mauk*, 1988] by expanding the previously described modeling efforts into an electrostatically self-consistent simulation. Approximated forms for the massless electron distributions were assumed (e.g., bi-Maxwellians), and the technique of quasi-neutrality was used to derive the field-aligned electric fields. The ions were forced to respond self-consistently to these electric fields.

Figure 10 shows one example of these modeling efforts. The figure shows magnetic latitude versus normalized potential for various times (labeled by seconds) during the convection surge process. For this particular simulation the predipolarization populations consisted of isotropic O^+ ions and electrons, each with Maxwellian temperatures of 1 keV. The figure shows that during the initial stages of the 60-second dipolarization, a relatively sharp potential discontinuity forms near the magnetic equator. As this field-aligned electric field front propagates from the equator to the higher latitude/ionospheric regions, it evolves into a much broader structure, finally settling down into a net equator-to-ionosphere potential drop that is several times the electron temperature. For the simulation shown, the initial front involves a potential drop of ~ 1.3 kV, and the equator-to-ionosphere potential drop at times reaches as high as 3 kV. These values are, of course, dependent on initial conditions. Potentials as high as ~ 10 kV have been reached for these geosynchronous regions using higher initial temperatures (~ 10 keV).

One of the things that has been ignored in the self-consistent modeling efforts described is the possible ionospheric interaction at high latitudes. Immediately into the simulation (e.g., see the Φ profile at $t = 20$ s) a potential barrier forms that excludes the ionospheric electrons from streaming out of the ionosphere and neutralizing the near equatorial potential structures. However, at later stages (i.e., at times > 1 minute), it is clear that the ionospheric interaction is likely to become important.

Ionospheric Interactions

It is our hypothesis that the convection surge process has a role in the generation of the electron beams of *McIlwain* [1975]. It can be stated unequivocally that without introducing additional physics, the modeling efforts described above cannot be used to explain the occurrence of electron beams of the sort shown in Figure 3. Test particle electrons have been followed through the electromagnetic fields generated by the self-consistent simulations, and only quasi-bi-Maxwellian electron distributions are observed even during the passage of field-aligned electric potential structures.

The missing physics would most likely be related to the ionospheric interaction with the simulated physics. It was argued earlier that the ionospheric interaction could probably be ignored for some period greater than 1 minute, but that switching polarities would ultimately demand a strong ionospheric involvement. Some idea of what the ionospheric interaction might consist of can be obtained by considering the laboratory studies of *Stenzel et al.* [1981] on the generation of electrostatic-shock/double-layer structures within converging magnetic field-line geometries. As diagrammed in the top panel of Figure 11, Stenzel et al. fired a spatially broad ion beam at one pole of a dipolar magnetic field structure. An important feature of the experiment was the capability to set the potential V_m of the face of the magnet's pole to arbitrary values relative to the system ground. With V_m set equal to very low positive values, broad-scale magnetic field-aligned electric field profiles were obtained that are very analogous to the final time-stationary results obtained with the convection surge modeling described here. However, as the bottom panel of Figure 11 shows, for relatively large values of V_m, narrower scale, detached double-layer structures were generated with electric fields larger than those obtained with small values of V_m. *Stenzel et al.* [1981] discovered that large V_m values were required to reflect some of the incoming ions in order to set up an electron-ion counterstreaming situation. The counterstreaming helps establish crucial electron populations trapped between the magnetic mirror and the electrostatic mirror, presumably via the scattering that results from turbulence generated by the microscopically unstable situation. It should also be noted that despite the fact that the Stenzel et al. ions generate positive spacecharge in the near magnet regions when compared to more distant regions along the same field line, the near magnet regions along the centerline of the experiment are charged negatively with respect to adjacent fieldlines, in analogy with auroral observations (i.e., a V-shape potential is obtained).

The motivation here in discussing the *Stenzel et al.* [1981] results is that the convection surge process generates pitch-angle collimated ion beams that propagate toward the field-line converging regions above the ionosphere, in direct analogy with the Stenzel et al. experiments. The question then arises concerning other requirements that must be imposed to pursue the analogy with the Stenzel et al. experiments. We speculate first that in generating detached double-layer-like structures, the need for large repelling potentials V_m might well disappear if one were to add an ionosphere at the surface of the Stenzel et al. magnet. Cold, upper ionospheric ions being accelerated out into the magnetosphere by the broad-scaled field-aligned electric field could then provide the counterstreaming conditions required to generate detached double-layer-like structures.

Magnetic field-aligned currents were also a feature of the *Stenzel et al.* [1981] results, and one might wonder how such currents might be generated and closed on quasi-dipolar field lines within the magnetosphere. The strong inductive electric fields reported by *Aggson et al.* [1983], which have been used to drive the modeled convection surges, will generate fairly large polarization currents perpendicular to **B** of the form

$$\mathbf{J}_p = \frac{1}{4\pi} \frac{c^2}{V_A{}^2} \frac{\partial \mathbf{E}}{\partial t} , \qquad (4)$$

where V_A is the local Alfvén speed. Ignoring the very large (10 second) fluctuations ($\Delta E/E \sim 1$) within the observed value of **E** and utilizing only the envelope of the *Aggson et al.* [1983] observations, $\partial E/\partial t \approx \pm 1$ mV/ms, where the signs + and − correspond to the turn-on and turn-off portions of the event. Including the large fluctuations (which characteristically last about 10 seconds; i.e., 5 seconds for turn-on) $\partial E/\partial t$ values as large as ± 6 mV/ms are observed. Utilizing the average values of V_A reported by *Moore et al.* [1987] for the appropriate region near the equator, one finds values of $|J_p|$ in the range of ~ 1-6×10^{-14} A/cm^2 or ~ 1-6×10^4 A/R$_E{}^2$, where R$_E$ is the Earth radius. These currents would flow in the plus or minus azimuthal directions. If these current densities were diverted into the field-aligned direction (say at the azimuthal boundaries of the inward hot plasma intrusion that is centered near midnight), and if the current density remains the same during that diversion, then the field-aligned current densities entering the ionospheric regions (amplified by the ratio B_{equ}/B_I) would range from 1-6 μA/m^{-2}. These values are typical of broad-scale auroral features such as inverted-V structures [*Bythrow et al.*, 1986, 1987]. If, on the other hand, the current is somewhat amplified during the diversion process because the boundaries of the inward hot plasma intrusions may be very narrow (*Moore et al.* [1981] suggests a thickness of 0.1R$_E$), even higher magnetic field-aligned current densities might be generated. Thus, because of azimuthal spatial gradients, sizeable magnetic field-aligned currents could be generated by the convection surge process on quasi-dipolar field lines. Alternatively (or in conjunction), *Roux* [1985] has suggested that a Rayleigh-Taylor-like instability at the earthward boundary of the hot plasmas on quasi-dipolar field lines can lead to charge separations and field-aligned currents. All of these ideas suggest that indeed discrete-like auroral discharges of the sort reported by *Stenzel et al.* [1981] may

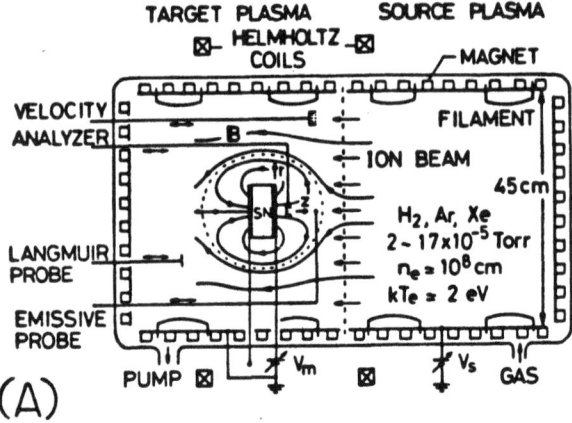

(A)

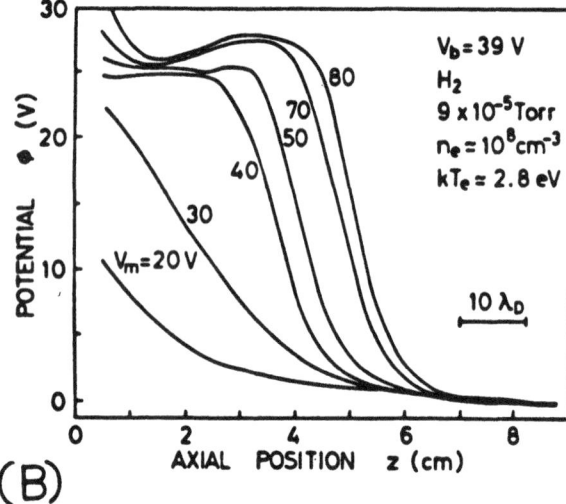

(B)

Figure 11—(a) Laboratory setup for an experimental study of the generation of electric potential double layers within a magnetic geometry of strongly converging field lines. (b) Field-aligned electric potentials measured along the symmetry axis of the dipole magnetic configuration shown in (a). V_m is a potential impressed on the face plate of the magnet (from *Stenzel et al.* [1981]). Field-aligned ion beams are key to generating the observed structures (see text).

be possible on these field lines as a result of the convection surge process.

Obviously, all of the discussions in this section are highly speculative. It must also be noted that even if it can be shown that discharges of the sort generated by *Stenzel et al.* [1981] can be generated on the quasi-dipolar field lines of the geosynchronous regions, it is still unclear how the electron distribution shown in Figure 3 would be generated. The Stenzel et al. results are promoted here because they do offer a possible scenario for generating a strong discharge-like ionospheric response to the convection surge processes.

It should be noted that a Stenzel et al. [1981] type response is not the only possible near-ionosphere response to the convection-surge process. *Bryant and Hall* [1989] argue that field-aligned ions in general, such as those to be found within the plasma sheet boundary layer (PSBL), can generate lower hybrid waves near the ionosphere. These waves can accelerate electrons via resonant scattering, and can, it is argued, generate electron spectra with characteristics matching those observed in the vicinity of discrete aurora. The convection-surge process generates field-aligned ions that converge toward the near ionospheric regions. Thus, the *Bryant and Hall* [1989] mechanism could operate on quasi-dipolar field lines and perhaps help explain the apparent discharge phenomena occuring there.

4. DISCUSSION

It was noted earlier that the guiding-center centrifugal acceleration responsible for field-aligned ion acceleration within the quasi-dipolar middle magnetosphere, is substantially similar to the neutral sheet ion acceleration that has been shown to operate within the more distant magnetosphere [e.g., *Lyons and Speiser*, 1982]. The similarities are discussed at length by *Mauk and Meng* [1988]. Both mechanisms generate magnetic-field-aligned ion beams that propagate from the equatorial regions toward the field line converging regions of the upper ionosphere. Such beams can generate magnetic-field-aligned electric fields and stimulate a discrete-like auroral discharge. It has been argued by *Lyons and Evans* [1984] (see also *Lyons et al.* [1988]; and *Lyons* [this volume]) that auroral emissions commonly called discrete arcs map magnetically to the neutral sheet acceleration regions (and generally the poleward edges of the low-altitude manifestations of those regions). We have argued here that more transient discrete-like auroral processes and emissions may operate within the middle magnetospheric regions where guiding-center behavior prevails, yielding such phenomena as the electron beams of *McIlwain* [1975]. On the basis of all of these points, we suggest that field-aligned ion dynamics is a crucial aspect of discrete-like auroral discharge phenomena and that it may unify the middle magnetospheric regions with the more distant magnetotail regions in terms of auroral processes. On the basis of this notion, the distant magnetosphere would be able to sustain discrete phenomena, whereas the similar phenomena in the middle magnetosphere is likely to be highly transient. It should be noted that there will exist a region between the middle, guiding-center regions and the distant, neutral-sheet-acceleration regions that will not yield the field-aligned ion beams needed for the above processes to operate. Here the ion distributions are isotropized on interacting with neutral sheets of intermediate thickness [*Propp and Beard*, 1984; see *Mauk and Meng*, 1988]. (Thus, the isotropization of low-altitude ion distributions as discussed by *Lyons et al.* [1988], would not provide sufficient information to know whether arcs can form.)

On the other hand, none of the above discussions address the phenomena observed by *Meng et al.* [1979; see Figure 2]. Here, relatively stable discrete auroral features are observed equatorward of the ATS 6 foot point, and thus map from positions earthward of the geostationary orbit (the mapping is confirmed by the particle spectral features). These discrete features are also equatorward of a relatively broad diffusive auroral region that is, in turn, bounded by discrete activity on the poleward boundary. The period in question was quite active ($Kp = 5+$) and one might hypothesize that the equatorward arc could be explained by an abnormally earthward motion of the neutral sheet, even at the 1900–1930 LT location of the observations. However, one must explain simultaneously the broad region of diffuse auroral behavior poleward of the arcs.

There are mechanisms that have been proposed that may generate discrete aurora within guiding-center regions of the middle magnetosphere [e.g., *Kan et al.*, 1988; *Kan and Cao*, 1988; *Rothwell et al.*, 1988; *Watanabe and Sato*, 1988; *Roux*, 1985]. These models often involve (excepting that of *Roux* [1985]) longitudinally extended regions of high ionospheric conductivity, convective accelerations (with accompanying vorticity) within the equatorial magnetosphere that magnetically map to the high conductivity regions, and the concept of ionospheric feedback. With these models, magnetic reconnection and/or neutral sheet physics would be of secondary importance to the arc formation. Given the conclusions of *Lyons et al.* [1988], whereby quasi-steady discrete arc activity is co-located with isotropic ion distributions (i.e., full loss cones), any model of quasi-steady discrete auroral activity operating on quasi-dipolar field lines would have to explain the isotropization of the ion distributions without involving neutral-sheet scattering. However, the transition from anisotropic to isotropic ion distributions (e.g., during a poleward latitudinal cut using low-altitude spacecraft) is not always in quantitative, or even qualitative, agreement with simple neutral-sheet scattering models during active periods [*Imhof*, 1988]. Searches for growth phase trends in the characteristics of that transition have failed. *Imhof* [1988] suggests that wave-particle interactions may have a strong role to play in ion isotropization for some events. In addition, the two regions of discrete activity in Figure 2 are

separated broadly enough by central-plasma-sheet-like populations, that it is easy to imagine that multiple processes could be involved. If it can be shown that the arcs do occur on quasi-dipolar field lines, we propose that the transient phenomena involved in generating the *McIlwain* [1975] electron beams could be responsible for establishing the necessary conditions for the quasi-steady arc activity to proceed. The transient activity could generate higher conductivity channels within the ionosphere. Also, the electric-field shielding across the injection boundary could result in convective accelerations and vorticity within the regions mapping to these channels. This suggestion is along the lines of one given in the review by *Siscoe* [1988] whereby energy is extracted from the distant magnetosphere to generate near-Earth and magnetospheric expansive phase phenomena. Clearly these suggestions are speculative in nature.

As a final note, there have been some recent findings by the AMPTE CCE spacecraft that could substantially alter our view of substorm auroral phenomena [*Takahashi et al.*, 1987; *Lui et al.*, 1988; *Lopez et al.*, 1988]. In association with dispersionless substorm injections for $\sim$30–400 keV ions, the spacecraft has found examples of strong current disruptions within the neutral sheet region between $R \sim 8$–$9R_E$. These events are characterized by a strong transient depression in the magnetic field strength (1 min), extreme turbulence in all magnetic field components (including reversals of the Z-component to negative values), followed by a dipolarization and an increase in the field strength over the predisruption values. It is proposed that strong particle energization results from the associated inductive electric fields in the presence of small spatial-scale scattering of particles. These events occur much closer to the Earth than has been proposed for magnetotail reconnection X-lines [e.g., *Hones*, 1984], and the responses of the ion distributions do not favor an X-line interpretation [*Lui et al.*, 1988; *Lopez et al.*, 1988]. Also, it would be difficult to associate this phenomenon with the plasma sheet boundary layer. It has not been placed into the context of phenomena described elsewhere in this work, and its significance to auroral phenomena remains to be discovered.

ACKNOWLEDGMENT—We thank D. G. Mitchell for helpful discussions. This work was supported by the Atmospheric Science Division, National Science Foundation grant ATM-8315041, by the Air Force Office of Scientific Research grant 84-0049, and by NASA and the Office of Naval Research support to The Johns Hopkins University Applied Physics Laboratory and Department of the Navy under task I2UOS1P of contract N00024-87-C-5301.

237

REFERENCES

Aggson, T. L., J. P. Heppner, and N. C. Maynard, "Observations of Large Magnetospheric Electric Fields During the Onset Phase of a Substorm," *J. Geophys. Res.*, **88**, 3981 (1983).

Anderson, K. A., and D. W. Milton, "Balloon Observations of X-rays in the Auroral Zone, 3. High Time Resolution Studies," *J. Geophys. Res.*, **69**, 4457 (1964).

Arnoldy, R. L., "Review of Auroral Particle Precipitation," in *Physics of Auroral Arc Formation*, Geophysical Monograph 25, S.-I. Akasofu and J. F. Kan, eds., American Geophysical Union, Washington, DC, p. 56 (1981).

Barcus, J. R., and T. J. Rosenberg, "Energy Spectrum for Auroral Zone X-rays, 1. Diurnal and Type Effects," *J. Geophys. Res.*, **71**, 803 (1966).

Bingham, R., D. A. Bryant, and D. S. Hall, "Auroral Electron Acceleration by Lower-Hybrid Waves," *Ann. Geophys.*, **6**, 159 (1988).

Brown, R. A., "Electron Precipitation in the Auroral Zone," *Space Sci. Rev.*, **5**, 311 (1966).

Bryant, D. A., "Rocket Studies of Particle Structures Associated with Auroral Arcs," in *The Physics of Auroral Arc Formation*, Geophysical Monograph 25, S.-I. Akasofu and J. R. Kan, eds., American Geophysical Union, Washington, DC, p. 102 (1981).

Bryant, D. A. and D. S. Hall, "The Auroral Electron Accelerator," *this volume*.

Bythrow, P. F., M. A. Doyle, T. A. Potemra, L. J. Zanetti, R. E. Huffman, C.-I. Meng, D. A. Hardy, F. J. Rich, and R. A. Heelis, "Multiple Auroral Arcs and Birkeland Currents: Evidence for Plasma Sheet Boundary Waves," *Geophys. Res. Lett.*, **13**, 805 (1986).

Bythrow, P. F., and T. A. Potemra, "Birkeland Currents and Energetic Particles Associated with Optical Auroral Signatures of a Westward Traveling Surge," *J. Geophys. Res.*, **92**, 8691 (1987).

Chapman, S., "The Aurora in Middle and Low Latitudes," *Nature*, **179**, 7 (1957).

Chapman, S. C., and S. W. H. Cowley, "Acceleration of Lithium Test Ions in the Quiet-Time Geomagnetic Tail," *J. Geophys. Res.*, **89**, 7357 (1984).

Chiu, Y. T., and M. Schulz, "Self-Consistent Particle and Parallel Electrostatic Field Distributions in the Magnetospheric-Ionospheric Auroral Region," *J. Geophys. Res.*, **83**, 629 (1978).

Cladis, J. B., "Parallel Acceleration and Transport of Ions from Polar Ionosphere to Plasmasheet," *Geophys. Res. Lett.*, **13**, 893 (1986).

DeForest, S. E., and C. E. McIlwain, "Plasma Clouds in the Magnetosphere," *J. Geophys. Res.*, **76**, 3587 (1971).

Eastman, T. E., L. A. Frank, W. K. Peterson, and W. Lennartsson, "The Plasma Sheet Boundary Layer," *J. Geophys. Res.*, **89**, 1553 (1984).

Eather, R. H., S. B. Mende, and R. J. R. Judge, "Plasma Injection at Synchronous Orbit and Spatial and Temporal Auroral Morphology," *J. Geophys. Res.*, **81**, 2805 (1976).

Fritz, T. A., J. P. Corrigan, et al., "Significant Initial Results from the Environment Measurements Experiment on ATS 6," NASA Technical Paper 1101, NASA Scientific and Technical Information Office, Washington, DC (December 1977).

Galperin, Y. I., and Y. I. Feldstein, "Auroral Luminosity and Its Relationship to Magnetospheric Plasma Domains," *this volume*.

Hones, E. W., Jr., Plasma Sheet Behavior During Substorms," in *Magnetic Reconnection in Space and Laboratory Plasmas*, Geophysical Monograph 30, E. W. Hones, Jr., ed., The American Geophysical Union, Washington, DC, p. 178 (1984).

Hultqvist, B., R. Lundin, K. Stasiewicz, L. Block, P.-A. Lindquist, G. Gustafsson, H. Koskinen, A. Bahnsen, T. A. Potemra, and L. J. Zanetti, "Simultaneous Observation of Upward Moving Field-Aligned Energetic Electrons and Ions on Auroral Zone Field Lines," *J. Geophys Res.*, **93**, 9765 (1988).

Imhof, W. L., "Fine Resolution Measurements of the L-Dependent Energy Threshold for Isotropy at the Trapping Boundary," *J. Geophys. Res.*, **93**, 9743 (1988).

Kan, J. R., "Energization of Auroral Electrons by Electrostatic Shock Waves," *J. Geophys. Res.*, **80**, 2089 (1975).

Kan, J. R., and F. Cao, "Effect of Field-Aligned Potential Drop in a Global Magnetosphere-Ionosphere Coupling Model," *J. Geophys. Res.*, **93**, 7571 (1988).

Kan, J. R., L. Zhu, and S.-I. Akasofu, "A Theory of Substorms: Onset and Subsidence," *J. Geophys. Res.*, **93**, 5624 (1988).

Kaye, S. M., E. G. Shelley, R. D. Sharp, and R. G. Johnson, "Ion Composition of Zipper Events," *J. Geophys. Res.*, **86**, 3393 (1981).

Kelly, T. J., C. T. Russell, and R. J. Walker, "ISEE-1 and -2 Observations of an Oscillating Outward Moving Current Sheet Near Midnight," *J. Geopnys. Res.*, **89**, 2745 (1984).

Kennel, C. F., and M. Ashour-Abdalla, "Electro-Static Waves and the Strong Diffusion of Magnetospheric Electrons," in *Magnetospheric Plasma Physics*, A. Nishida, ed., Center for Academic Publications Japan, Tokyo, in co-publication with D. Reidel Pub. Co., Dordrecht, Holland, p. 245 (1982).

Kivelson, M. G., J. Feynman, B. H. Mauk, R. A. Wolf, "Dialog on Injection-Boundary versus Alfvén-layer Models," in *Magnetotail Physics*, A. T. Y. Lui, ed., The Johns Hopkins University Press, Baltimore, MD, p. 403 (1987).

Klumpar, D. M., J. M. Quinn, and E. G. Shelley, "Counter-Streaming Electrons at the Geomagnetic Equator Near 9 R_E," *Geophys. Res. Lett.*, **15**, 1295 (1988).

Konradi, A., C. L. Semar, and T. A. Fritz, "Substorm-Injected Protons and Electrons and the Injection Boundary Model," *J. Geophys. Res.*, **80**, 543 (1975).

Kremser, G., A. Korth, L. Ullaland, S. Perraut, A. Roux, A. Pedersen, R. Schmidt, and P. Tanskanen, "Field-Aligned Beams of Energetic Electrons (16 keV $\le$ E $\le$ 80 keV) Observed at Geosynchronous Orbit at Substorm Onset," *J. Geophys. Res*, **93**, 14453 (1988).

Lanzerotti, L. J., C. G. Maclennan, and M. F. Robbins, "Proton Drift Echoes in the Magnetosphere," *J. Geophys. Res.*, **76**, 259 (1971).

Lopez, R. E., A. T. Y. Lui, D. G. Sibeck, K. Takahashi, R. W. McEntire, L. J. Zanetti, and S. M. Krimigis, "On the Relationship Between the Energetic Particle Flux Morphology and the Change in the Magnetic Field Magnitude During Substorms," *J. Geophys. Res.*, in press (1988).

Lui, A. T. Y., R. E. Lopez, S. M. Krimigis, R. W. McEntire, L. J. Zanetti, and T. A. Potemra, "A Case Study of a Magnetotail Current Disruption and Diversion," *Geophys. Res. Lett.*, **15**, 721 (1988).

Lui, A. T. Y., D. Venkatesan, C. D. Anger, S.-I. Akasofu, W. J. Heikkila, J. D. Winningham, and J. R. Burrows, "Simultaneous Observations of Particle Precipitations and Auroral Emissions by the ISIS-2 Satellite in the 19–24 MLT Sector," *J. Geophys. Res.*, **82**, 2210 (1977).

Lundin, R., L. Eliasson, B. Hultqvist, and K. Stasiewicz, "Plasma Energization on Auroral Field Lines as Observed by the Viking Spacecraft," *Geophys. Res. Lett.*, **14**, 443 (1987).

Lyons, L. R., "Discrete Aurora and Magnetotail Processes," *this volume*.

Lyons, L. R., and D. S. Evans, "An Association Between Discrete Aurora and Energetic Particle Boundaries," *J. Geophys. Res.*, **89**, 2395 (1984).

Lyons, L. R., J. F. Fennell, and A. L. Vampola, "A General Association Between Discrete Auroras and Ion Precipitation from the Tail," *J. Geophys. Res.*, **93**, 12932 (1988).

Lyons, L. R., and T. W. Speiser, "Evidence for Current Sheet Acceleration in the Geomagnetic Tail," *J. Geophys. Res.*, **87**, 2276 (1982).

Mauk, B. H., "Quantitative Modeling of the 'Convection Surge' Mechanism of Ion Acceleration," *J. Geophys. Res.*, **91**, 3423 (1986).

Mauk, B. H., "Generation of Macroscopic Magnetic-Field-Aligned Electric Fields by the Convection-Surge Ion Acceleration Mechanism," *J. Geophys. Res.*, **94**, 8911 (1988).

Mauk, B. H., J. Chin, and G. Parks, "Auroral X-Ray Images," *J. Geophys. Res.*, **86**, 6827 (1981).

Mauk, B. H., and C. E. McIlwain, "UCSD Auroral Particles Experiment," *IEEE Trans. Aerosp. Electron. Syst.*, **AES-11**, 1125 (1975).

Mauk, B. H., and C.-I. Meng., "Characterization of Geostationary Particle Signatures Based on the 'Injection Boundary' Model," *J. Geophys. Res.*, **88**, 3055 (1983*a*).

Mauk, B. H., and C.-I. Meng, "Dynamical Injections as the Source of Near Geostationary Quiet Time Particle Spatial Boundaries," *J. Geophys. Res.*, **88**, 10011 (1983*b*).

Mauk, B. H., and C.-I. Meng, "Macroscopic Ion Acceleration Associated with the Formation of the Ring Current in the Earth's Magnetosphere," in *Ion Acceleration in the Magnetosphere and Ionosphere*, Geophysical Monograph 38, T. Chang, ed., American Geophysical Union, Washington, DC, p. 351 (1986).

Mauk, B. H., and C.-I. Meng, "Plasma Injection During Substorms," *Phys. Script.*, **T18**, 128 (1987).

Mauk, B. H., and C.-I. Meng, "Macroscopic Magnetospheric Particle Acceleration," in *Outstanding Problems in Solar System Plasma Physics: Theory and Instrumentation*, Geophysical Monograph Series, American Geophysical Union, Washington, DC, in press (1988).

McIlwain, C. E., "Plasma Convection in the Geosynchronous Orbit," in *Earth's Magnetospheric Processes*, B. M. McCormac, ed., D. Reidel, Hingham, MA, p. 268 (1972).

McIlwain, C. E., "Substorm Injection Boundaries," in *Magnetospheric Physics*, B. M. McCormac, ed., D. Reidel, Hingham, MA, p. 143 (1974).

McIlwain, C. E., "Auroral Electron Beams Near the Magnetic Equator," in *The Physics of Hot Plasma in the Magnetosphere*, B. Hultqvist and L. Stenflo, eds., Plenum, New York, p. 91 (1975).

McIlwain, C. E., "Equatorial Magnetospheric Particles and Auroral Precipitation," in *Results of the Arcad 3 Project and of the Recent Programmes in Magnetospheric and Ionospheric Physics*, Cepadues Editions, Toulouse, France, pp. 275-280 (1985).

Meng, C.-I., B. H. Mauk, and C. E. McIlwain, "Electron Precipitation of Evening Diffuse Aurora and Its Conjugate Electron Fluxes Near the Magnetospheric Equator," *J. Geophys. Res.*, **84**, 2545 (1979).

Moore, T. E., R. L. Arnoldy, J. Feynman, D. A. Hardy, "Propagating Substorm Injection Fronts," *J. Geophys. Res.*, **86**, 6713 (1981).

Moore, T. E., D. L. Gallagher, J. L. Horwitz, and R. H. Comfort, "MHD Wave Breaking in the Outer Plasmasphere," *Geophys. Res. Lett.*, **14**, 1007 (1987).

Murphree, J. S., C. D. Anger, and L. L. Cogger, "ISIS-2 Observations of Auroral Arc Systems," in *The Physics of Auroral Arc Formation*, Geophysical Monograph 25, S.-I. Akasofu and J. R. Kan, eds., American Geophysical Union, Washington, DC, p. 15 (1981).

Nagai, T., "Observed Magnetic Substorm Signatures at Synchronous Orbit," *J. Geophys. Res.*, **87**, 4405 (1982).

Northrop, T. G., *The Adiabatic Motion of Charged Particles*, Interscience, New York (1963).

Olson, W. D., and K. A. Pfitzer, "A Quantitative Model of the Magnetospheric Magnetic Field," *J. Geophys. Res.*, **79**, 3739 (1974).

Parks, G. K., "Spatial Characteristics of Auroral Zone X-ray Microbursts," *J. Geophys. Res.*, **72**, 215 (1967).

Parks, G. K., B. H. Mauk, C. Gurgiolo, and C. S. Lin, "Observations of Plasma Injections," in *Dynamics of the Magnetosphere*, S.-I. Akasofu, ed., D. Reidel, Dordrect, Holland, p. 371 (1980).

Propp, K., and D. B. Beard, "Cross-Tail Ion Drift in a Realistic Model Magnetotail," *J. Geophys. Res.*, **89**, 11013 (1984).

Quinn, J. M., and C. E. McIlwain, "Bouncing Ion Clusters in the Earth's Magnetosphere," *J. Geophys. Res.*, **84**, 7365 (1979).

Quinn, J. M., and D. J. Southwood, "Observations of Parallel Ion Energization in the Equatorial Region," *J. Geophys. Res.*, **87**, 10536, (1982).

Robert, P., R. Gendrin, S. Perraut, A. Roux, and A. J. Pedersen, "Geos 2 Identification of Rapidly Moving Current Structures in the Equatorial Outer Magnetosphere During Substorms," *J. Geophys. Res.*, **89**, 819 (1984).

Rothwell, P. L., L. P. Block, M. B. Silevitch, and C.-G. Fälthammar, "A New Model for Substorm Onsets: The Pre-Breakup and Triggering Regimes," *Geophys. Res. Lett.*, **15**, 1279 (1988).

Roux, A., "Generation of Field-Aligned Current Structures at Substorm Onset," *Proc. ESA Workshop on Future Missions in Solar, Heliospheric and Space Plasma Physics*, Garmish-Partenkirchen, Germany, 30 April-3 May 1985, ESA SP-235, pp. 151-159 (June 1985).

Siscoe, G. L., "Meeting Reports: Auroral Physics, A Chapman Memorial," *Eos Tran. AGU*, **69**, 1596 (1988).

Siscoe, G. L., "What Determines the Size of the Auroral Oval?" *this volume.*

Speiser, T. W., "Particle Trajectories in Model Current Sheets, 1, Analytical Solutions," *J. Geophys. Res.*, **70**, 4219 (1965).

Speiser, T. W., "Particle Trajectories in Model Current Sheets, 2, Applications to Auroras Using a Geomagnetic Tail Model," *J. Geophys. Res.*, **72**, 3919 (1967).

Stenzel, R. L., M. Ooyama, and Y. Nakamura, "Potential Double-Layers in Strongly Magnetized Plasmas," in *Physics of Auroral Arc Formation Geophysical Monograph*, 25, S.-I. Akasofu and J. R. Kan, eds., American Geophysical Union, Washington, DC, p. 226 (1981).

Takahashi, K., L. J. Zanetti, R. E. Lopez, R. W. McEntire, T. A. Potemra, and K. Yumoto, "Disruption of the Magnetotail Current Sheet Observed by AMPTE/CCE," *Geophys. Res. Lett.*, **14**, 1019 (1987).

Watanabe, K., and Tetsuya Sato, "Self-Excitation of Auroral Arcs in a Three-Dimensionally Coupled Magnetosphere-Ionosphere System," *Geophys. Res. Lett.*, **15**, 717 (1988).

Whipple, E. C., T. G. Northrop, and T. J. Birmingham, "The Signature of Parallel Electric Fields in a Collisionless Plasma," *J. Geophys. Res.*, **82**, 1515 (1977).

IV-6. AURORAL PLASMA WAVES

A review is given of auroral plasma wave phenomena, starting with the earliest ground-based observations and ending with the most recent satellite observations. Two types of waves are considered, electromagnetic and electrostatic. Electromagnetic waves include auroral kilometric radiation, auroral hiss, ELF noise bands, and low-frequency electric and magnetic noise. Electrostatic waves include upper hybrid resonance emissions, electron cyclotron waves, lower hybrid waves, ion cyclotron waves, and broadband electrostatic noise. In each case, a brief overview is given describing the observations, the origin of the instability, and the role of the waves in the physics of the auroral acceleration region.

1. INTRODUCTION

The study of auroral plasma waves has a long and interesting history extending over more than half a century. The first reported observation of a plasma wave phenomenon associated with the aurora was in 1933 by *Burton and Boardman* [1933]. Using a telephone receiver and a telegraph line as a simple receiving system, Burton and Boardman discovered that bursts of very-low-frequency (VLF) radio "static" were sometimes correlated with flashes of auroral light. Following this early discovery, little progress occurred for more than twenty years, until the International Geophysical Year (IGY), which started in 1957. Because of the interest in studying whistlers and other audio frequency radio phenomena during the IGY, many of the ground stations included VLF receiving equipment. Soon a substantial number of observations of auroral radio emissions became available. These observations clearly showed that radio emissions were often detected on the ground in association with auroral displays [*Ellis*, 1957; *Duncan and Ellis*, 1959; *Dowden*, 1959; *Martin et al.*, 1960; *Morozumi*, 1963; *Jorgensen and Ungstrup*, 1962; *Harang and Larsen*, 1964]. Typically the emissions were the strongest in the VLF frequency range, from about 1–20 kHz. However, in some cases emissions were reported at frequencies as high as 500 kHz [*Ellis*, 1957]. Since the VLF auroral emissions tended to produce a hiss-like sound in the audio output of the receiver, these emissions soon became known as "auroral hiss." Among the early reports, *Ellis* [1957] has the distinction of being the first to propose a theory of auroral plasma waves. He suggested that the radio emissions were produced by Cerenkov radiation from the precipitating charged particles responsible for the aurora. As will be discussed later, elements of his

ideas still exist in modern theories of auroral hiss. For a more extensive review of the early ground-based observations, see *Helliwell* [1965].

The launch of the first Earth-orbiting satellites in the late 1950s opened an entirely new era in the study of magnetospheric plasma wave phenomena. The first satellite measurements of auroral plasma waves were reported by *Gurnett and O'Brien* [1964], using a VLF radio receiver on a low-altitude polar-orbiting satellite. These measurements clearly showed that auroral hiss is produced in regions of intense downgoing electron precipitation. Other later studies by *Gurnett* [1966], *Hartz* [1970], *Gurnett and Frank* [1972], and *Laaspere and Hoffman* [1976] showed that auroral hiss is closely related to the occurrence of a beam-like electron precipitation event called an inverted-V [*Frank and Ackerson*, 1971]. Parallel electric fields caused by space charge regions at high altitudes over the auroral zone [*Carlqvist and Boström*, 1970] soon became the accepted mechanism for producing these electron beams.

During the late 1960s and early 1970s an entirely new type of high frequency auroral radio emission was detected by eccentric orbiting satellites. The first detection of this new radio emission was by *Benediktov et al.* [1965, 1968], using data from the Elektron 2 and 4 satellites. They identified bursts of radio noise at 725 kHz and 2.3 MHz at large distances from the Earth, that were associated with geomagnetic storms. This same type of radio noise was also studied by *Dunckel et al.* [1970], at frequencies below 100 kHz, and by *Stone* [1973] and *Brown* [1973], who showed that the peak intensity occurred between 100 to 500 kHz. A short time later, *Gurnett* [1974] demonstrated that the radiation was produced at high altitudes over the auroral regions in association with discrete auroral arcs. Gurnett also found that the total radiated power was very large, up to 10^9 watts. Earth was therefore found to be an intense planetary radio source, comparable in some respects to Jupiter, which had been known for

*Department of Physics and Astronomy, The University of Iowa, Iowa City, IA 52242.

Auroral Physics, edited by C.-I. Meng, M. J. Rycroft and L. A. Frank. © Cambridge UP 1991

many years to be an intense radio emitter [*Burke and Franklin*, 1955]. The radiation is now commonly called "auroral kilometric radiation" or "AKR," a term first suggested by *Kurth et al.* [1975].

Since these early observations, many additional types of auroral plasma waves have been discovered, most of which can be detected only by satellites in or above the ionosphere. In this chapter, we give an overview of the present state of knowledge of auroral plasma waves, with the main emphasis on results obtained in the last ten years. Although the early satellite investigations revealed the main types of auroral plasma waves, it is only recently, with the launch of the DE 1, EXOS B, and Viking satellites, that measurements have been obtained in the critical altitude range from about $1-3R_E$ where the auroral acceleration occurs and many of the most intense waves are produced.

2. PLASMA WAVE MODES

Before describing the observations, it is useful to discuss first the characteristic frequencies of a plasma and their relationship to the various wave modes that can exist in the auroral ionosphere. The most important characteristic frequencies of a plasma are the cyclotron frequency f_c and the plasma frequency f_p. As discussed by *Stix* [1962], a cyclotron frequency and a plasma frequency can be defined for each species. The cyclotron frequency for a charged particle of mass m_s and charge e_s is given by

$$f_{cs} = \frac{1}{2\pi} \frac{|e_s|B}{m_s} \qquad (1)$$

and the plasma frequency is given by

$$f_{ps} = \frac{1}{2\pi} \sqrt{\frac{n_s e_s^2}{m_s}} \qquad (2)$$

where B is the magnetic field strength and n_s is the number density of the sth species.

To describe the modes of propagation that exist in a plasma, it is convenient to consider two classes of waves: electromagnetic and electrostatic. Electromagnetic waves have a magnetic field, whereas electrostatic waves do not. Usually, the propagation speed of electromagnetic waves is much higher than the thermal speed. For this reason, the propagation of electromagnetic waves can usually be described by a model in which the plasma is completely cold (i.e., zero temperature). Electrostatic waves on the other hand almost always propagate at speeds on the order of the thermal speed. Therefore, thermal effects usually must be included in the analysis of electrostatic waves.

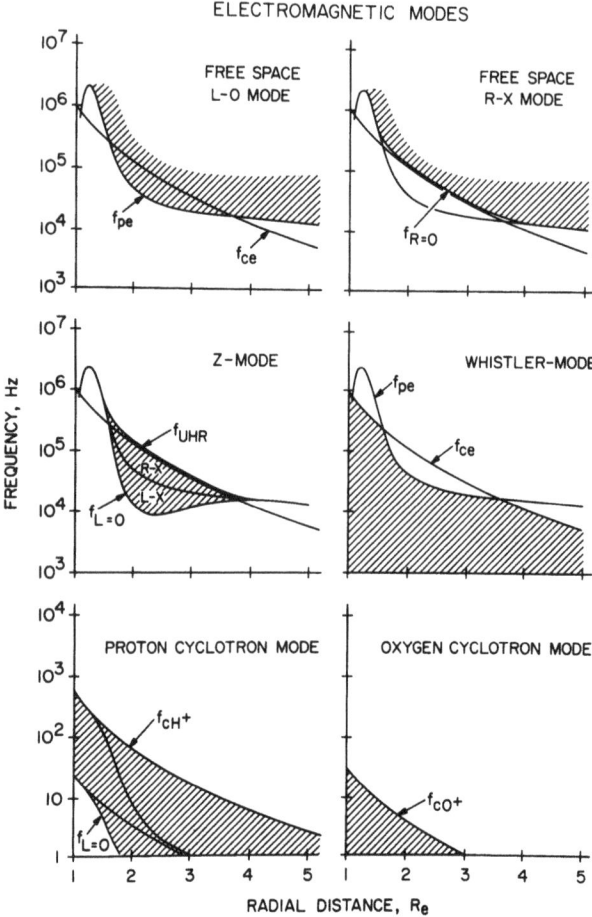

Figure 1—The frequency range of the six most important electromagnetic modes and their relationship to various characteristic frequencies. The plasma parameters used to make these plots are representative of the auroral ionosphere.

The approximate frequency ranges of the various electromagnetic modes of propagation that can exist in the auroral plasma are summarized in Figure 1. Two frequency regimes can be considered. At high frequencies, above the ion cyclotron frequencies, ion effects are generally not important. In this frequency range there are four separate electromagnetic modes of propagation [*Ratcliffe*, 1959]. These modes are shown in the top four panels of Figure 1 and are the free-space R-X mode, the free-space L-O mode, the Z mode, and the whistler mode. The term "free space" means that the mode reduces to the well-known free-space electromagnetic mode in the limit $n_e = 0$ and $B = 0$. Using the terminology of *Stix* [1962], the R and L designations indicate the polarization with respect to the magnetic field (R for right, and L for left), and the O and

X designations indicate the type of propagation perpendicular to the magnetic field (O for ordinary, and X for extraordinary). The Z mode is named after the so-called "Z trace" observed in ionograms [*Ratcliffe*, 1959], and the whistler mode is named after lightning-generated signals that propagate in this mode [*Storey*, 1953]. The L-O and R-X free-space modes have low frequency cutoffs at the electron plasma frequency, f_{pe}, and the R = 0 cutoff, $f_{R=0} = f_{ce}/2 + [(f_{ce}/2)^2 + f_{pe}^2]^{1/2}$. The Z mode is bounded at the upper limit by the upper hybrid resonance, $f_{UHR} = [f_{ce}^2 + f_{pe}^2]^{1/2}$, and at the lower limit by the L = 0 cutoff, $f_{L=0} = -f_{ce}/2 + [(f_{ce}/2)^2 + f_{pe}^2]^{1/2}$. The whistler mode has an upper frequency limit at f_{ce} or f_{pe}, whichever is lower.

At low frequencies, where ion effects are important, a new electromagnetic mode is introduced for each ion species. These modes are called electromagnetic ion cyclotron waves. The bottom two panels of Figure 1 show the electromagnetic ion cyclotron modes associated with H^+ and O^+, which are the dominant ions in the auroral ionosphere. Other ion cyclotron modes also occur in association with various minor ions, such as He^+, but for simplicity are not shown. When two or more ion species are present, the ion cyclotron modes occur in distinct bands, one associated with each ion cyclotron frequency. The low-frequency limit of each band is determined by a L = 0 cutoff. Polarization reversals and hybrid resonances also occur between each adjacent pair of ion cyclotron frequencies. For a discussion of these ion effects, see *Smith and Brice* [1964].

The electrostatic modes of propagation that can exist in the auroral plasma are summarized in Figure 2. These modes can be grouped in certain combinations, each associated with a characteristic type of resonance. First, there are the three hybrid modes, which occur at the upper hybrid resonance frequency, f_{UHR}, the lower hybrid resonance frequency, f_{LHR}, and the ion-ion hybrid resonance frequency, f_{IHR}. At these resonances the electromagnetic X-mode becomes purely electrostatic for propagation perpendicular to the magnetic field. Next there are the electrostatic electron cyclotron and ion cyclotron modes, also sometimes called the Bernstein modes. An electrostatic cyclotron mode occurs between each adjacent pair of cyclotron harmonics for each species in the plasma. These modes are indicated by the dashed lines in Figure 2. The electrostatic cyclotron modes have wavelengths comparable to the cyclotron radius of the species involved, and disappear completely in the limit of zero temperature. Finally, there is the well-known electrostatic oscillation that occurs at the electron plasma frequency, also known as the Langmuir mode [*Krall and Trivelpiece*,

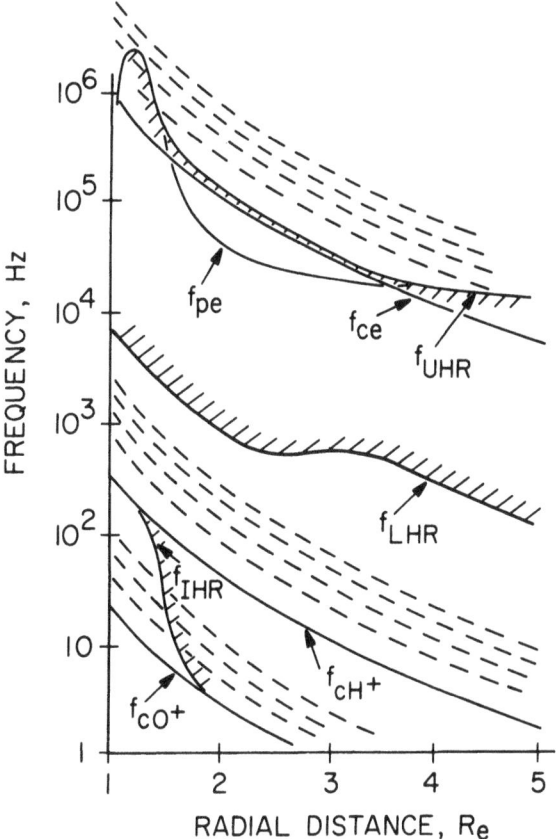

Figure 2—The characteristic frequencies of the various electrostatic modes that can exist along the auroral field lines.

1973]. The Langmuir mode has a strong dependence on the electron temperature and degenerates to a monochromatic oscillation at the electron plasma frequency in the limit of zero temperature.

In addition to the modes associated with the ambient plasma, there are also a variety of electrostatic modes introduced when beams are present, the so-called "beam modes." A good example is the electron acoustic mode [*Tokar and Gary*, 1984], which only exists in the presence of an electron beam. For a review of the electrostatic beam modes, see *Gary* [1985].

To organize the presentation, the observations are described in the same order as in Figures 1 and 2, starting first with the electromagnetic modes and proceeding to the electrostatic modes. In each section, the types of waves are described in order of decreasing frequency.

3. ELECTROMAGNETIC WAVES

Auroral Kilometric Radiation

The highest-frequency plasma wave emission generated in the auroral zone is auroral kilometric radiation (AKR). This radio emission typically has the highest

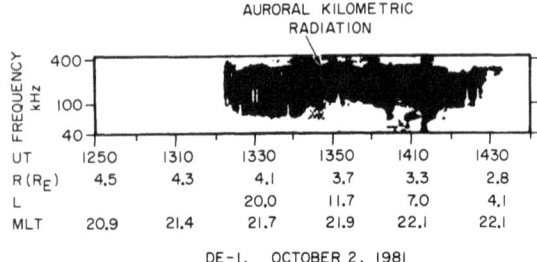

Figure 3—A frequency-time spectrogram of auroral kilometric radiation. This radio emission usually occurs in the frequency range from 100 to 400 kHz and is highly variable, both in frequency and amplitude.

intensities in the frequency range from about 50–400 kHz. A typical example of auroral kilometric radiation is illustrated in Figure 3, which shows a frequency-time spectrogram of an auroral kilometric radiation event detected during a DE 1 pass over the northern polar cap at a radial distance ranging from about 3–4R_E. The auroral kilometric radiation is the intense (dark) emission extending from about 80–400 kHz. The broadband electric field intensity of this radiation is very large, typically about 10 mV m^{-1} at a radial distance of 4R_E. The average power radiated from Earth by the auroral kilometric radiation has been estimated by *Gallagher and Gurnett* [1979] to be about 10^7–10^8

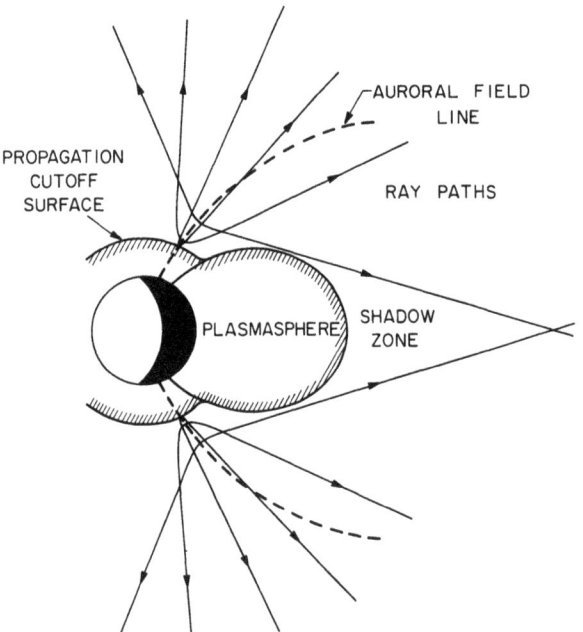

Figure 4— Representative ray paths for the auroral kilometric radiation. The propagation cutoff surface prevents the radiation from reaching the ground. This cutoff is at $f_{R=0}$ for the R-X mode, and at f_{pe} for the L-O mode.

watts, although during intense events the peak power may be as high as 10^9 watts [*Gurnett*, 1974]. The radiated power is highly variable and often changes by up to 80 dB on time scales of ten minutes or less. The intensity variations are closely correlated with auroral magnetic disturbances [*Benediktov et al.*, 1965, 1968; *Dunckel et al.*, 1970; *Voots et al.*, 1977] and with the occurrence of discrete auroral arcs [*Gurnett*, 1974; *Kurth et al.*, 1975]. Considerable fine structure is also evident in the frequency spectrum, usually consisting of narrowband tones drifting upward and downward in frequency [*Gurnett et al.*, 1979; *Morioka et al.*, 1981; *Benson et al.*, 1988].

Spatial intensity surveys [*Gurnett*, 1974; *Gallagher and Gurnett*, 1979] and direction-finding measurements [*Kurth et al.*, 1975; *Alexander and Kaiser*, 1976] clearly established that the auroral kilometric radiation is generated along the nighttime auroral field lines at radial distances from about 2–4R_E. Observations show that the radiation is generated in a broad beam directed away from Earth more or less as shown in Figure 4. High plasma densities in the lower levels of the ionosphere produce a propagation cutoff that prevents the radiation from reaching the ground. Polarization measurements by *Shawhan and Gurnett* [1982] and *Mellott et al.* [1984] using the DE 1 spacecraft show that the radiation is primarily generated in the R-X mode, although a small amount (~2 percent) of L-O mode radiation is also present. A spectrogram showing the polarization during a typical DE 1 pass over the auroral zone is shown in Figure 5. The dark black region is right-hand polarized R-X mode radiation, and the white region is left-hand polarized L-O mode radiation. The difference in the low-frequency cutoff of the two modes is believed to be a propagation effect caused by differences in the refractive characteristics of the two modes. Usually, the R-X mode is refracted outward away from Earth more strongly than the L-O mode.

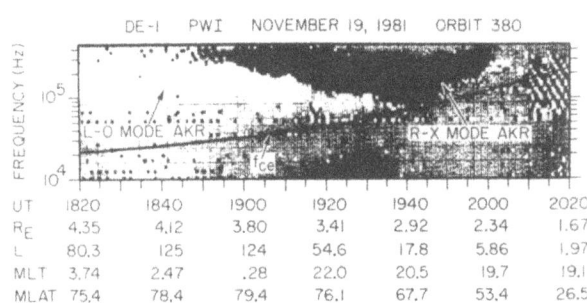

Figure 5—A spectrogram showing the polarization of the auroral kilometric radiation. The dominant polarization is right-hand with respect to the magnetic field in the source region (R-X mode). A weak left-hand component is also sometimes observed (L-O mode).

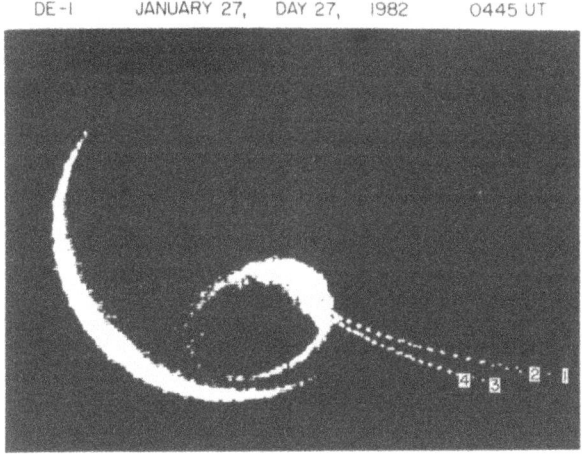

DE-I   JANUARY 27,   DAY 27,   1982   0445 UT

| AKR SOURCE # | PWI | |
	UT	WAVE FREQUENCY
1	0445	104 kHz
2	0445	136 kHz
3	0445	170 kHz
4	0445	218 kHz

Figure 6—An auroral image from DE 1 showing the occurrence of bright auroral emissions during an intense AKR event. The dashed lines show the magnetic field lines through the source. The source position is determined from the intersection of the direction of arrival and the $f = f_{ce}$ surface. (This figure also appears in color: Plate 24.)

Several factors strongly indicate that both the L-O and R-X mode radiation are generated close to the electron cyclotron frequency. The strongest evidence of this relationship comes from radio direction-finding measurements and comparisons with auroral images. By using two-dimensional radio direction-finding techniques, the direction of arrival of the auroral kilometric radiation can be determined with a high degree of accuracy. If it is assumed that the radiation is generated at the electron cyclotron frequency, then the source position can be uniquely determined from the intersection of the direction of arrival and the surface defined by $f = f_{ce}$. Once the source position is known, the magnetic field line can be traced from the source down to the atmosphere and compared with auroral optical emissions. This process is illustrated in Figure 6, from *Huff et al.* [1988]. The points labeled 1 through 4 show the source positions determined at four frequencies, 104, 136, 170, and 218 kHz. The dashed lines then show the mapping of the magnetic field lines from these source positions down to an altitude of 200 km. As can be seen, the dashed lines all terminate in a region of bright auroral optical emissions. Similar results are found in many other examples. The close agreement confirms the basic assumption used in the analysis, namely that the radiation is generated very close to the electron cyclotron frequency.

The fact that the auroral kilometric radiation is generated at frequencies very close to the electron cyclotron frequency has several implications. As can be seen in Figure 1, the low frequency cutoff of the free space R-X mode is close to the electron cyclotron frequency only in regions where $f_{pe} < < f_{ce}$. Since no radiation can be generated at frequencies below the cutoff frequency, the fact that the emission always occurs near f_{ce} implies that the condition $f_{pe} < < f_{ce}$ is a basic requirement for the generation of this radiation. The ionospheric plasma density therefore plays an important role in determining the bandwidth and conditions under which the radiation is generated.

Numerous theories have been proposed to explain the auroral kilometric radiation. For a review of some of the proposed theories, see *Grabbe* [1981]. At present, it is widely believed that the radiation is produced by a Doppler-shifted cyclotron maser mechanism. Although some elements of this theory were first introduced by *Melrose* [1973], the first really detailed cyclotron maser theory of auroral kilometric radiation was developed by *Wu and Lee* [1979]. The main elements of Wu and Lee's theory are that the electrons interact with the waves via a Doppler-shifted cyclotron resonance interaction, and that the loss cone in the auroral electron distribution provides the free energy source. By using a fully relativistic treatment of the cyclotron resonance interaction, Wu and Lee were able to show that the condition $f_{pe} << f_{ce}$ arises as a direct result of the resonance condition. Thus, one of the basic experimental observations is readily explained by the theory. Numerous other investigators have since provided further refinements of the basic theory, including growth rate calculations [*Omidi and Gurnett,* 1982] and investigations of nonlinear and inhomogeneous effects [*Wu et al.,* 1981; *Winglee,* 1985; *Zarka et al.,* 1986]. Some uncertainty still exists concerning the free energy source. For example, *Louarn et al.* [1989] have recently suggested that electrons electrostatically trapped in the auroral acceleration region provide the primary free energy source. Despite uncertainties of this type, the main features of the cyclotron maser mechanism appear to be in good agreement with observations.

Auroral Hiss and Z-Mode Radiation

Whistler-mode auroral hiss and Z-mode radiation are discussed together because they occur in overlapping frequency ranges and have somewhat similar characteristics. A spectrogram illustrating these two types of radio emissions is shown in Figure 7. This spectrogram is from a DE 1 pass over the nighttime auroral zone at a radial distance of about 3–4R_E. The auroral hiss

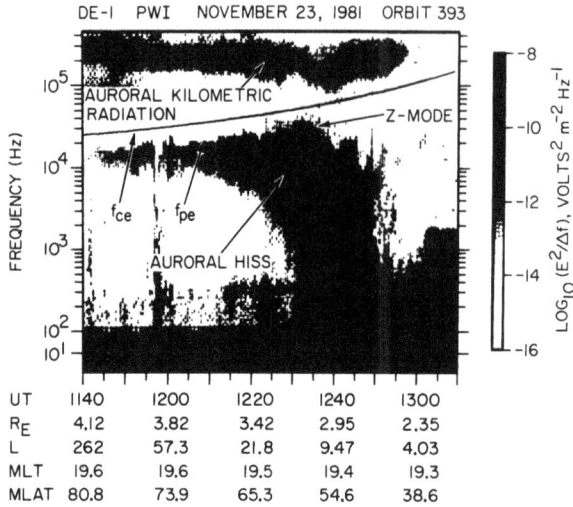

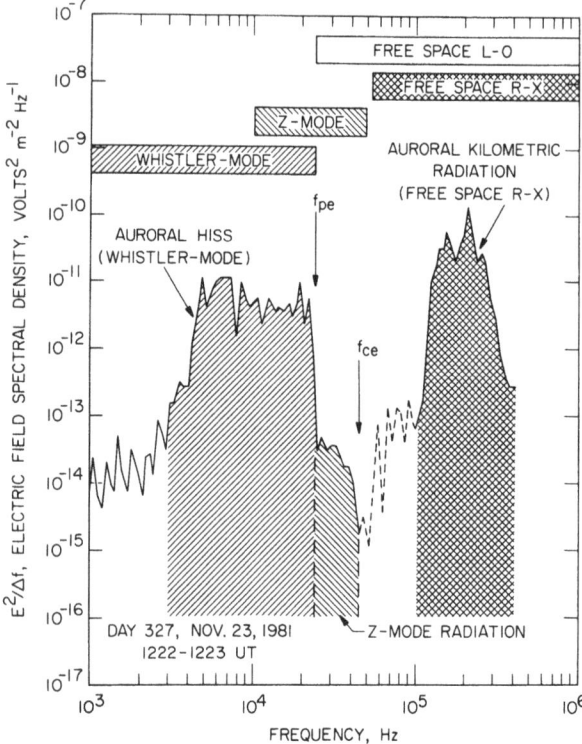

Figure 7—A spectrogram of auroral hiss and Z-mode radiation during a DE 1 pass over the auroral zone. (This figure also appears in color: Plate 25.)

Figure 8—A spectrum showing the auroral hiss cutoff at f_{pe}, and the weaker Z-mode radiation above the cutoff. The frequency ranges of the four high frequency electromagnetic modes are given at the top of the plot.

is the broad intense emission extending from a few hundred hertz up to about 20–30 kHz. At low frequencies the emission is centered on the auroral zone, which is located from about 1232–1247 UT. At high frequencies the emission spreads out over a broad region, both toward the polar cap, and to a lesser extent toward the equator. The spreading at high frequencies produces a "funnel-shaped" frequency-time signature that is a characteristic feature of this radiation [*Gurnett et al.,* 1983]. At high altitudes the auroral hiss often has a sharp high frequency cutoff. This cutoff can be seen in Figure 7, and is illustrated in greater detail in Figure 8, which shows a selected spectrum in which the cutoff is particularly sharp and well defined. The cutoff is a propagation effect that arises because the whistler mode has an upper frequency limit of either f_{pe} or f_{ce}, whichever is smaller. Since f_{pe} is typically less than f_{ce} at these altitudes, the cutoff is located at the electron plasma frequency, f_{pe}. The frequency range of the whistler mode is shown by the cross-hatched bar at the top of Figure 8.

Careful inspection of Figure 7 shows that an additional very weak emission exists above the upper cutoff of the auroral hiss. This emission is Z-mode radiation [*Gurnett et al.,* 1983]. The Z mode is unusual in that it is bounded at high frequencies by the upper hybrid resonance and at low frequencies by the L = 0 cutoff. Because of these cutoffs, Z-mode radiation is permanently trapped in the magnetosphere (see Figure 1). In most regions where the Z mode has been previously observed [*Walsh et al.,* 1964; *Gregory,* 1969; *Mosier et al.,* 1973], the electron plasma frequency is

usually greater than the electron cyclotron frequency $(f_{pe} \gg f_{ce})$. Under these conditions the Z mode is confined to a very narrow band around the electron plasma frequency. At high altitudes over the polar region, the opposite situation usually occurs, namely that $f_{pe} \ll f_{ce}$. Under these conditions the bandwidth of the Z mode becomes very large. The allowed frequency range of the Z mode is illustrated at the top of Figure 8 based on the known plasma parameters. As can be seen, the Z mode overlaps with the whistler mode. Since the auroral hiss is usually more intense than the Z mode, the lower part of the Z mode spectrum normally cannot be detected. Wideband spectrograms [*Persoon and Gurnett,* 1989] show that the Z mode is relatively smooth and continuous, very similar to auroral hiss. This is in strong contrast to the auroral kilometric radiation, which has considerable fine structure.

Several factors strongly indicate that auroral hiss is generated at wave normal angles near the resonance cone, where the index of refraction becomes infinite [*Stix,* 1962]. For wave normal angles near the resonance cone the whistler mode becomes quasi-electrostatic,

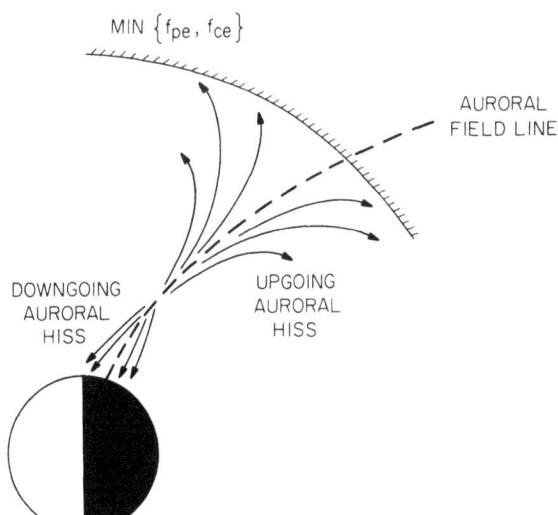

Figure 9—Representative ray paths of auroral hiss. Both upgoing and downgoing auroral hiss are observed. The upgoing auroral hiss has an upper frequency limit of f_{pe} or f_{ce}, whichever is smaller.

with $E >> cB$. The quasi-electrostatic characteristic explains why auroral hiss is easily detected by an electric antenna, but often not by a magnetic antenna. Resonance-cone propagation also explains why auroral hiss seldom extends below the lower hybrid resonance frequency. As discussed by *Stix* [1962] the whistler mode resonance cone only exists for frequencies above the lower hybrid resonance frequency. Resonance-cone propagation also accounts for the tendency of the auroral hiss to spread out at high frequencies, thereby forming the funnel-shaped frequency-time spectrum. For this type of propagation, it can be shown that the radiation is emitted in a beam around the magnetic field with a beamwidth that increases with increasing frequency. As the spacecraft approaches the source field line, the higher frequencies are detected first, thereby producing the funnel-shaped spectrum [*Gurnett et al.,* 1983].

Poynting flux measurements by *Mosier and Gurnett* [1969] show that auroral hiss propagates both upward and downward along the auroral field lines. At high altitudes, greater than 10,000 km, the radiation is usually propagating upward, and at low altitudes, less than 1000 km, the radiation is usually propagating downward, as shown in Figure 9. At intermediate altitudes both directions of propagation occur, although normally not on the same L-shell. At low altitudes the upward propagating emissions are usually very narrow, lasting only a few seconds. These upgoing low-altitude emissions are sometimes called saucers [*Gurnett and Frank,* 1972; *James,* 1976]. Comparisons with low-

energy electron measurements clearly show that the downgoing auroral hiss is correlated with downgoing, 100 eV to 1 keV, "inverted-V" electron beams [*Gurnett and Frank,* 1972; *Laaspere and Hoffman,* 1976] and that the upgoing auroral hiss is correlated with upgoing, ~ 50 eV, electron beams [*Lin et al.,* 1984]. The fact that the electrons responsible for generating the radiation are moving in the same direction as the wave, i.e., $\omega/k_{\parallel} = v_b$, provides strong evidence that the radiation is produced by a Cerenkov interaction (also called a Landau resonance). Detailed calculations of whistler-mode growth rates have been performed by *Maggs* [1976] using realistic models for the auroral electron beams and a Landau resonance interaction. These calculations show good agreement with observed auroral hiss spectrums and intensities.

Although the understanding of auroral hiss is fairly advanced, relatively little work has been done on the propagation and generation of Z-mode radiation. Because of the broad bandwidth, the Z mode can propagate long distances both horizontally and vertically over the auroral zone and polar cap. The qualitative nature of this propagation is illustrated in Figure 10, which shows the "propagation window" formed by the propagation cutoff surfaces at f_{UHR} and $f_{L=0}$. At frequencies above f_{ce} the index of refraction surface has a resonance cone, so the propagation is highly anisotropic. At frequencies below f_{ce} the resonance cone disappears, and the propagation is more nearly line-of-sight, like free space. At low altitudes the radiation tends to be refracted away from the $f_{L=0}$ surface, and once reflected asymptotically approaches the f_{ce} surface [*Gurnett et al.,* 1983]. Because of the complicated propagation geometry it is difficult to determine accurately the source of the radiation. Most likely the source is located in the auroral zone, since this is where the larg-

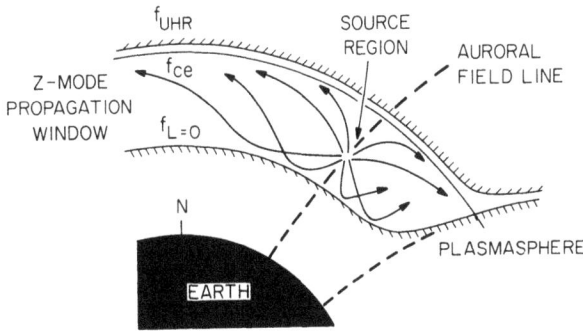

Figure 10—Representative ray paths of Z-mode radiation. This radiation is trapped between the propagation cutoffs at f_{UHR} and $f_{L=0}$, and can propagate long distances over the polar cap.

est intensities usually occur. Several generation mechanisms have been proposed. These include a beam-driven Landau resonance mechanism [*Maggs and Lotko,* 1981], a loss-cone driven cyclotron maser mechanism [*Omidi et al.,* 1984], and mode conversion from auroral hiss [*Hashimoto et al.,* 1987]. Just which of these theories provides the best mechanism for generating the Z-mode radiation remains to be determined.

ELF Noise Bands

Intense electromagnetic noise bands are frequently observed in the auroral zone by low altitude polar orbiting satellites at frequencies ranging from ten to several hundred hertz. These emissions were first reported by *Gurnett and Frank* [1972], who called them ELF (extremely low frequency) noise bands. Gurnett and Frank also showed that they are closely correlated with regions of low energy, 100 eV to 10 keV, "inverted-V" auroral electron precipitation. A frequency-time spectrogram illustrating a typical example of this type of noise is shown in Figure 11. Similar signals are observed on both the electric and magnetic antennas, so the noise clearly consists of electromagnetic waves. Typical electric and magnetic field amplitudes are 3 to 10 mV m^{-1} and 10–30 pT. The bandwidth of the emission is quite narrow, usually less than 20%. *Temerin and Lysak* [1984] have shown that the center frequency decreases with increasing altitude, from a few hundred hertz at 1000 km, to a few tens of hertz at 10,000 km. The emission frequency is bounded by the proton cyclotron frequency and the helium cyclotron frequency. The electric field is polarized perpendicular to the static magnetic field, and both left- and right-hand polarizations are observed.

Because the ELF noise bands always occur below the proton cyclotron frequency, *Temerin and Lysak* [1984] suggested that the noise is produced by electromagnetic ion cyclotron waves. The relevant ion cyclotron mode is the branch that exists between the proton cyclotron frequency, f_{cH+}, and the helium cyclotron frequency, f_{cHe+}. Once generated, the waves propa-

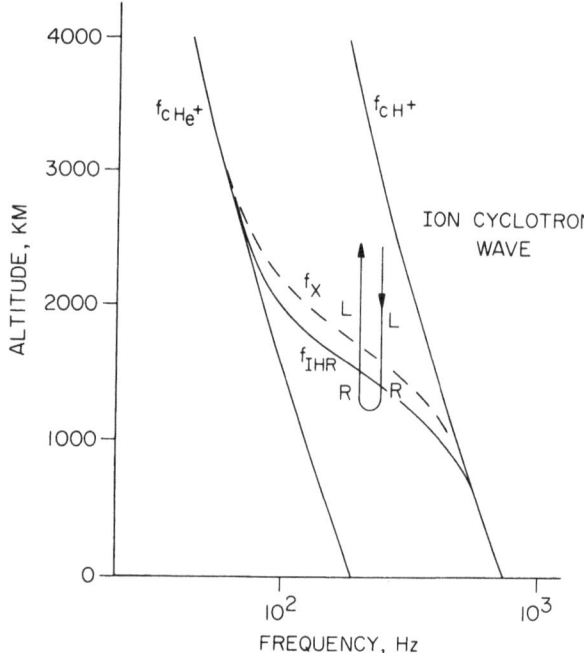

Figure 12—ELF noise bands are believed to be caused by electromagnetic ion cyclotron waves propagating at large wave normal angles between the He$^+$ and H$^+$ cyclotron frequencies. These waves undergo a polarization reversal at the crossover frequency, f_x, and are reflected slightly below the ion hybrid resonance, f_{IHR}.

gate downward as shown in Figure 12. Reflection is believed to occur as soon as the ion hybrid resonance frequency, f_{IHR}, exceeds the wave frequency. Since the polarization reverses at the crossover frequency, f_x, both left- and right-hand polarized waves can occur, in agreement with the observations. The exact mechanism for generating these waves has not been firmly established. The most likely possibility appears to be a Landau resonance interaction with low-energy electrons. In this case the free energy source would be the electron beam associated with the inverted-V electron precipitation.

Low-Frequency Electric and Magnetic Field Noise

For many years it has been known that intense low-frequency electric and magnetic field fluctuations are observed over the auroral zone by low-altitude polar-orbiting satellites. The electric field fluctuations were first reported by *Heppner* [1969] using double-probe electric field measurements on the OV1-10 spacecraft. This noise has been studied by many investigators, including *Maynard and Heppner* [1970], *Kelley and Mozer* [1972], *Kintner* [1976], *Temerin* [1978], *Curtis et al.* [1982], and *Weimer et al.* [1985]. Typically, the noise

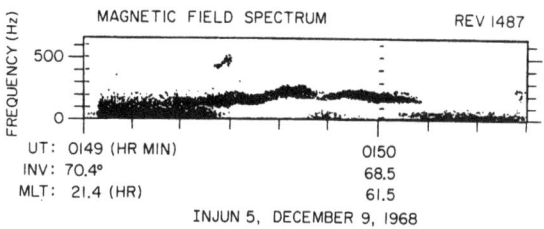

Figure 11—A frequency-time spectrogram of an ELF noise band observed over the auroral zone in association with an inverted-V electron precipitation event.

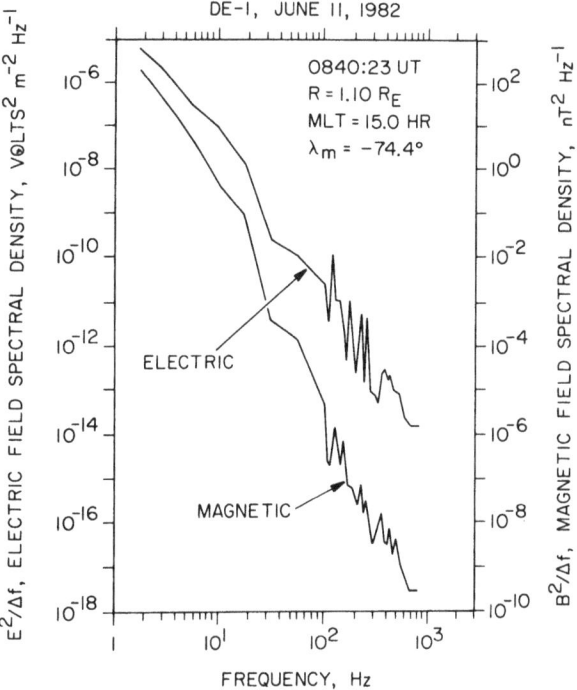

DE-1, JUNE 11, 1982

0840:23 UT
R = 1.10 R_E
MLT = 15.0 HR
λ_m = −74.4°

ELECTRIC

MAGNETIC

FREQUENCY, Hz

Figure 13—Representative spectrums of the low-frequency electric and magnetic field noise commonly observed at low altitudes over the auroral zone. The electric and magnetic fields are highly correlated at low frequencies, with the Poynting flux directed downward toward Earth.

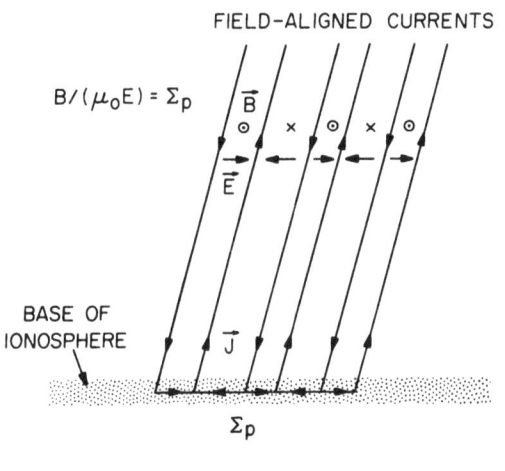

FIELD-ALIGNED CURRENTS

$B/(\mu_0 E) = \Sigma_p$

BASE OF IONOSPHERE

Σ_p

Figure 14—The low-frequency electric and magnetic field noise is believed to be caused by the motion of the spacecraft through a system of quasi-static fields generated by field-aligned currents closing through the base of the ionosphere.

is most intense at frequencies below 1 Hz and decreases more or less monotonically with increasing frequency. A typical electric field spectrum is shown in Figure 13. Enhanced levels of magnetic field noise are also observed over the auroral zone by low-altitude polar-orbiting satellites. These magnetic field fluctuations were first studied in detail by *Armstrong and Zmuda* [1973]. A typical magnetic field spectrum is shown in Figure 13. As can be seen, the general shapes of the electric and magnetic field spectrums are rather similar. Usually the magnetic field spectrum is steeper than the electric field spectrum, particularly at frequencies above 10 Hz. Typical power law spectral indices are −2.0 to −4.0 for the electric field, and −3.0 to −5.0 for the magnetic field. Both the electric and magnetic fields are perpendicular to the static magnetic field.

The initial interpretation of these observations was that the fluctuations are caused by independent quasi-static fields. The electric fields were believed to be produced by turbulent convective motions of the ionosphere [*Kelley and Kintner,* 1978], and the magnetic fields were believed to be produced by field-aligned currents [*Armstrong and Zmuda,* 1973]. However, more recently it was shown that the electric and magnetic field fluctuations are closely correlated [*Smiddy et al.,* 1980; *Gurnett et al.,* 1984]. Correlation coefficients are often as high as 80% at frequencies below 10 Hz, decreasing somewhat toward higher frequencies. The phase of the correlation is such that the Poynting flux is directed toward Earth. The net energy flow integrated over the auroral zone is substantial, ~ 10^8 watts [*Gurnett et al.,* 1984].

Two extreme views have been advanced to explain the origin of this noise. The first model, described by *Smiddy et al.* [1980], assumes that the fluctuations are caused by the motion of the spacecraft through a system of static field-aligned current structures imbedded in the ionosphere. The electric fields are generated by the closure of the field-aligned currents through the conductive layer at the base of the ionosphere, and the magnetic fields are produced directly by the currents. The basic geometry involved is illustrated in Figure 14. In this model, it can be shown that the electric and magnetic fields are related to the height-integrated Pedersen conductivity, $B/(\mu_0 E)^{-1} = \Sigma_p$. Comparisons of the $B/(\mu_0 E)^{-1}$ ratio are in good agreement with estimates of the height-integrated Pedersen conductivity [*Smiddy et al.,* 1980; *Gurnett et al.,* 1984; *Weimer et al.,* 1985]. The second model assumes that the noise is caused by Alfvén waves. Several investigators, for example, *Goertz and Boswell* [1979] and *Lysak and Dum* [1983], have pointed out that turbulent fluctuations imposed on the auroral field lines in the distant magnetosphere should be transmitted to the ionosphere as a shear Alfvén wave, or in a modified form called a kinetic Alfvén wave [*Hasegawa, 1977*]. Since large turbulent fluctuations are known to exist in the magnetopause boundary layer and in the distant magnetotail [*Coroniti et al., 1977*], there are good reasons to believe that large amplitude Alfvén waves should be excited along the auroral field lines. *Chang et al.* [1986] have suggested that these waves play an important role in auroral ion acceleration at low altitudes.

To decide between the static structure model and the Alfvén wave model, it is necessary to evaluate the Doppler shift caused by the spacecraft motion relative to the plasma. In the static structure model, the fluctuations are caused entirely by the Doppler shift, whereas in the Alfvén wave model, the Doppler shift is negligible. An evaluation of the Doppler shift requires measurements of the wavelength, which can only be obtained from multispacecraft measurements. Unfortunately, such measurements do not exist; so it is very difficult to determine which model provides the best description.

4. ELECTROSTATIC WAVES

Upper-Hybrid Waves and Electron Cyclotron Waves

As can be seen in Figure 2, there are two basic types of electrostatic waves that occur near the electron cyclotron frequency and plasma frequency. They are upper hybrid resonance waves and electron cyclotron waves. Although both types of waves are commonly found in other regions of the magnetosphere, particularly near the magnetic equatorial plane [*Kennel et al., 1970; Shaw and Gurnett, 1975*], they are seldom observed at high altitudes over the auroral zone. The absence of intense emissions at the upper-hybrid resonance frequency and near harmonics of the electron cyclotron frequency is apparently related to the low plasma densities ($f_{pe} \ll f_{ce}$) that usually occur at high altitudes over the auroral zone. When upper-hybrid and electron cyclotron waves are observed, they almost always occur when the electron density is high (i.e., $f_{pe} \gtrsim f_{ce}$). For example, an upper-hybrid noise band and a weak "$(3/2)f_{ce}$" electron cyclotron emission can be seen in Figure 9 of *Gurnett et al.* [1983]. These events occur near the polar cusp on the dayside of Earth under conditions where $f_{pe} \sim f_{ce}$. The electron density near the polar cusp is usually substantially higher than in the nightside auroral zone.

Lower-Hybrid Waves

Early spacecraft VLF measurements showed that intense narrowband electrostatic emissions frequently occurred at the lower hybrid resonance frequency [*Barrington, 1969*]. Although these emissions are a prominent feature of the plasma wave spectrum at middle and low latitudes, they are not normally observed in the auroral zone. Auroral hiss emissions are sometimes called "lower hybrid waves" [*Chang and Coppi, 1981*]. However, these emissions are actually electromagnetic whistler-mode waves and have been described in the previous section.

Electrostatic Ion Cyclotron Waves

In an early theoretical study of electrostatic instabilities, *Kindel and Kennel* [1971] predicted that electrostatic ion cyclotron waves should be generated at high altitudes over the auroral zones by field-aligned currents. Waves similar to those predicted by Kindel and Kennel were subsequently discovered by *Kintner et al.* [1978] using electric field measurements from the S3-3 satellite. Typically the electrostatic ion cyclotron waves appear as strong narrowband emissions at frequencies slightly above the proton cyclotron frequency and its harmonics. *Kintner* [1980] has distinguished between waves excited at low harmonic numbers, $n \leq 3$, which he called electrostatic ion cyclotron (EIC) waves, and waves excited at higher harmonic numbers, which he called ion cyclotron harmonic (ICH) waves. Both of these types of waves are illustrated in Figure 15. Typical electric field amplitudes are on the order of 1 mV m^{-1} [*Koskinen et al., 1987*], although field strengths of up to 25 mV m^{-1} have been reported in some cases [*Kintner et al., 1978*]. In both cases the electric field tends to be oriented nearly perpendicular to the static magnetic field. Phase velocity measurements by *Kintner et al.* [1984] and Doppler broadening analyses by *Boardsen et al.* [1989] show that the wavelengths are comparable to the proton cyclotron radius. Recently, electrostatic waves have also been discovered near harmonics of the oxygen cyclotron frequency [*Kintner et al., 1988*].

The free energy source that drives the electrostatic ion cyclotron waves and the possible role that these waves play in the physics of the auroral acceleration region is still somewhat uncertain. Comparisons with plasma distribution functions by *Kintner et al.* [1979], *Kintner* [1980], *Cattell* [1981], *André* [1986], and *Koskinen et al.* [1987] indicate that the electrostatic ion cyclotron mode can be driven unstable by ion beams and ion conic distributions as well as by field-aligned currents. The absence of adequate low energy electron measurements have made it difficult to evaluate the possible role of electrons as the free energy source. For several years it has been suggested [*Lysak et al., 1980*] that electrostatic ion cyclotron waves may play an important role in accelerating ion conic distributions. Although EIC and ICH waves are frequently observed in the region where the transverse ion heating occurs [*Peterson et al., 1988*], the role of these waves is un-

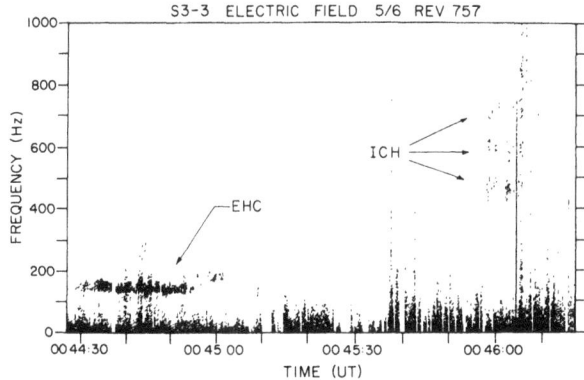

Figure 15—A frequency-time spectrogram of electrostatic ion-cyclotron waves detected by the S3-3 satellite. Two types have been identified: electrostatic hydrogen cyclotron waves (EHC), which occur at low harmonic numbers ($n \leq 3$), and ion cyclotron harmonics (ICH), which occur at high harmonic numbers.

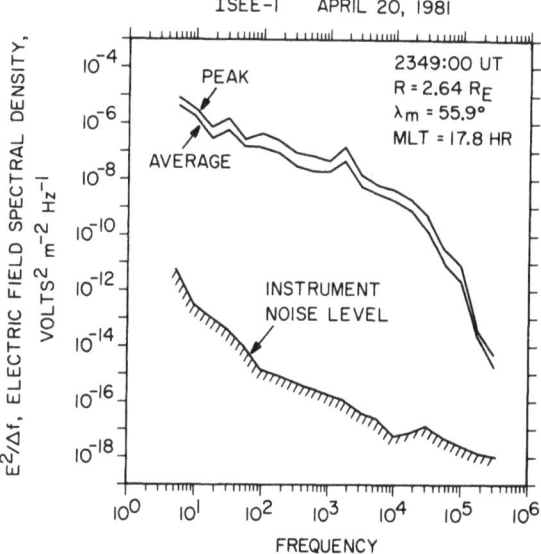

Figure 16—A spectrum of broadband electrostatic noise detected by the ISEE 1 spacecraft at high altitudes over the auroral zone. This noise is electrostatic and extends over a very broad frequency range, typically from a few hertz to more than ten kilohertz.

certain. Generally, the intensities are quite low (a few millivolts per meter). It is not clear that such low intensities can account for the observed heating rates.

Broadband Electrostatic Noise

For many years it has been known that intense broadband bursts of electrostatic noise occur along the auroral field lines. The first evidence of this noise was provided by *Scarf et al.* [1973, 1975] who identified impulsive bursts of electrostatic noise at frequencies of 1–10 kHz over the nighttime auroral region in association with field-aligned currents. A similar type of noise was later found at lower frequencies in the distant magnetotail by *Gurnett et al.* [1976]. Subsequent studies by *Gurnett and Frank* [1977] showed that the noise extends over an essentially continuous region along the auroral field lines from altitudes of a few thousand kilometers to as much as $46R_E$ in the distant magnetotail. Since the noise is electrostatic and has a very broad bandwidth, it is commonly called broadband electrostatic noise.

A representative spectrum of an intense broadband electrostatic noise event is shown in Figure 16. Typically, the noise extends over a very broad frequency range, from a few hertz to several tens of kilohertz, usually decreasing in intensity with increasing frequency. The integrated broadband electric field amplitude tends to decrease with increasing distance from Earth, from about 10 to 30 mV m^{-1} at altitudes of a few thousand kilometers, to 1 to 3 mV m^{-1} in the distant magnetotail. On a fine time scale the noise is very impulsive and spikey and shows no simple relationship to any of the characteristic frequencies of the plasma. Recent studies by *Erlandson et al.* [1987] have shown that the impulsive variations are related to fine struc-

ture in the field-aligned current distribution. At low and intermediate altitudes, up to a few Earth radii, the broadband electrostatic noise tends to be imbedded in funnel-shaped auroral hiss events. For example, in Figure 7, the intense low-frequency electric field noise from about 1232–1247 UT is broadband electrostatic noise. At high frequencies the broadband electrostatic noise is often difficult to distinguish from auroral hiss, and at low frequencies the noise merges with the electric field component of the low-frequency electric and magnetic noise described in the previous section. Isolated bursts of broadband electrostatic noise also sometimes occur over the polar cap, for example, from about 1157–1159 UT in Figure 7.

In addition to the correlation with field-aligned currents, comparisons with plasma measurements show that the broadband electrostatic noise is correlated with ion beams in the distant magnetotail [*Grabbe and Eastman,* 1984], and with ion conics at lower altitudes over the auroral zone [*Chang et al.,* 1986]. Despite the association with energetic ions, the free energy responsible for the broadband electrostatic noise has still not been clearly established. In an early theory, *Ashour-Abdalla and Thorne* [1977] proposed that the noise is caused by electrostatic ion cyclotron waves driven by field-aligned currents. The high frequencies and absence of spectral structure at harmonics of the ion cyclotron frequency were thought to be due to Doppler spreading. More recently, *Grabbe and Eastman* [1984] proposed that the noise is produced by an electrostatic ion-beam instability. Their theory accounts for the broadband nature of the spectrum without the need to consider Doppler effects. A major difficulty in evaluating all of these interpretations is that the plasma wave mode has not been clearly established. In fact, the broadband nature of the noise and the complete absence of cutoffs or resonances at any of the characteristic frequencies of the plasma suggest that the usual small-amplitude linearized mode analysis may not be applicable. Instead, the noise may represent a large amplitude turbulent state in which nonlinear effects play the dominant role. In this case, all evidence of the unstable plasma-wave mode is effectively eliminated from the electric field spectrum.

5. CONCLUSION

We have here reviewed the present state of understanding of auroral plasma waves. The observations are characterized by great diversity with many different plasma-wave modes. At present, most of the primary wave phenomena associated with the aurora are known, and well-developed theories have been advanced to explain most of them. Although much progress has been made, many details yet remain to be resolved. For example, although the auroral kilometric radiation is widely believed to be generated by the cyclotron maser instability, the detailed feature of the electron distribution function that drives this instability still has not been firmly established. Similarly, the origin of the broadband electrostatic noise is still poorly understood, and the possible role of electrostatic ion cyclotron waves and Alfvén waves in the heating of auroral ion distributions has not been clearly established. It is clear from these and other similar issues that many detailed questions are unresolved. Most likely, the study of auroral plasma waves will continue to be an active area of research for many years to come.

ACKNOWLEDGMENT—The author thanks A. Persoon, R. L. Huff, and R. R. Anderson for their help in the preparing several of the illustrations, and L. A. Frank for providing the auroral image from Dynamics Explorer 1. This research was supported by NASA through grants NAG5-310 and NAG5-1093 with Goddard Space Flight Center, and by grant NAGW-1488 with NASA Headquarters.

REFERENCES

Alexander, J. K., and M. L. Kaiser, "Terrestrial Kilometric Radiation. 1. Spatial Structure Studies," *J. Geophys. Res., 81*, 5948-5956 (1976).

André M., "Electrostatic Ion Waves Generated by Ion Loss-Cone Distributions in the Magnetosphere," *Annal. Geophys., 4*, 241-246 (1986).

Armstrong, J. C., and A. J. Zmuda, "Triaxial Magnetic Measurements of Field-Aligned Currents at 800 km in the Auroral Region: Initial Results," *J. Geophys. Res., 78*, 6802-6807 (1973).

Ashour-Abdalla, M., and R. M. Thorne, "The Importance of Electrostatic Ion Cyclotron Instability for Quiet-Time Auroral Precipitation," *Geophys. Res. Lett., 4*, 45-48 (1977).

Barrington, R. E., "Satellite Observations of VLF Resonances," in *Plasma Waves in Space and Laboratory*, Vol. 1, J. O. Thomas and B. J. Landmark, eds., Edinburgh Press, Edinburgh, p. 361 (1969).

Benediktov, E. A., G. G. Getmantsev, Yu. A. Sazonov, and A. F. Tarasov, "Preliminary Results of Measurements of the Intensity of Distributed Extraterrestrial Radio-Frequency Emission at 725 and 1525-kHz Frequencies by the Satellite Electron-2," *Kosm. Issled., 3*, 614-617 (1965).

Benediktov, E. A., G. G. Getmantsev, N. A. Mityakov, V. O. Rapoport, and A. F. Tarasov, "Relation Between Geomagnetic Activity and the Sporadic Radio Emission Recorded by the Elektron Satellites," *Kosm. Issled., 6*, 946-949 (1968).

Benson, R. F., M. M. Mellott, R. L. Huff, and D. A. Gurnett, "Ordinary Mode Auroral Kilometric Radiation Fine Structure Observed by DE 1," *J. Geophys. Res., 93*, 7515-7520 (1988).

Boardsen, S., W. Peterson, and D. Gurnett, "Double-Peaked Electrostatic Ion Cyclotron Waves," *J. Geophys. Res.*, submitted (1989).

Brown, L. W., "The Galactic Radio Spectrum Between 130 kHz and 2600 kHz," *Astrophys. J., 180*, 359-370 (1973).

Burke, B. F., and K. L. Franklin, "Observations of a Variable Radio Source Associated with the Planet Jupiter," *J. Geophys. Res., 60*, 213-217 (1955).

Burton, E. T., and E. M. Boardman, "Audio-Frequency Atmospherics," *Proc. IRE, 21*, 1476-1494 (1933).

Carlqvist, P., and R. Boström, "Space Charge Regions Above the Aurorae," *J. Geophys. Res., 75*, 7140-7146 (1970).

Cattell, C., "The Relationship of Field-Aligned Currents to Electrostatic Ion Cyclotron Waves," *J. Geophys. Res., 86*, 3641-3645 (1981).

Chang, T., and B. Coppi, "Lower Hybrid Acceleration and Ion Evolution in the Supraauroral Region," *Geophys. Res. Lett., 8*, 1253-1256 (1981).

Chang, T., G. B. Crew, N. Hershkowitz, J. R. Jasperse, J. M. Retterer, and J. D. Winningham, "Transverse Acceleration of Oxygen Ions by Electromagnetic Ion Cyclotron Resonance with Broadband Left-Hand Polarized Waves," *Geophys. Res. Lett., 13*, 636-639 (1986).

Coroniti, F. V., F. L. Scarf, L. A. Frank, and R. P. Lepping, "Microstructure of a Magnetotail Fireball," *Geophys. Res. Lett., 4*, 219-222 (1977).

Curtis, S. A., W. R. Hoegy, L. H. Brace, N. C. Maynard, and M. Sugiura, "DE-2 Cusp Observations: Role of Plasma Instabilities in Topside Ionospheric Heating and Density Fluctuations," *Geophys. Res. Lett., 9*, 997-1000 (1982).

Dowden, R. L., "Low Frequency (100 kc/s) Radio Noise from the Aurora," *Nature, 184*, 803 (1959).

Duncan, R. A., and G. R. Ellis, "Simultaneous Occurrence of Subvisual Aurorae and Radio Noise Bursts on 4.6 kc/s," *Nature, 183*, 1618-1619 (1959).

Dunckel, N., B. Ficklin, L. Rorden, and R. A. Helliwell, "Low Frequency Noise Observed in the Distant Magnetosphere with OGO 1," *J. Geophys. Res., 75*, 1854-1862 (1970).

Ellis, G. R., "Low-Frequency Radio Emission Auroral," *J. Atmos. Terrestr. Phys., 10*, 302-306 (1957).

Erlandson, R. E., R. Pottelette, T. A. Potemra, L. J. Zanetti, A. Bahnsen, R. Lundin, and M. Hamelin, "Impulsive Electrostatic Waves and Field-Aligned Currents Observed in the Entry Layer," *Geophys. Res. Lett., 14*, 431-434 (1987).

Frank, L. A., and K. L. Ackerson, "Observations of Charged-Particle Precipitation into the Auroral Zone," *J. Geophys. Res., 76*, 3612-3643 (1971).

Gallagher, D. L., and D. A. Gurnett, "Auroral Kilometric Radiation: Time-Averaged Source Position," *J. Geophys. Res., 84*, 6501-6509 (1979).

Gary, S. P., "Electrostatic Instabilities in Plasmas with Two Components," *J. Geophys. Res., 90*, 8213-8221 (1985).

Goertz, C. K., and R. W. Boswell, "Magnetosphere-Ionosphere Coupling," *J. Geophys. Res., 84*, 7239-7246 (1979).

Grabbe, C. L., "Auroral Kilometric Radiation: A Theoretical Review," *Rev. Geophys. and Space Phys., 19*, 627-633 (1981).

Grabbe, C. L., and T. E. Eastman, "Generation of Broadband Electrostatic Noise by Ion Beam Instabilities in the Magnetotail," *J. Geophys. Res.,* **89,** 3865-3872 (1984).

Gregory, P. C., "Radio Emission from Auroral Electrons," *Nature,* **221,** 350-352 (1969).

Gurnett, D. A., "A Satellite Study of VLF Hiss," *J. Geophys. Res.,* **71,** 5599-5615 (1966).

Gurnett, D. A., "The Earth as a Radio Source: Terrestrial Kilometric Radiation," *J. Geophys. Res.,* **79,** 4227-4238 (1974).

Gurnett, D. A., and L. A. Frank, "ELF Noise Bands Associated with Auroral Electron Precipitation," *J. Geophys. Res.,* **77,** 3411-3417 (1972).

Gurnett, D. A., and L. A. Frank, "A Region of Intense Plasma Wave Turbulence on Auroral Field Lines," *J. Geophys. Res.,* **82,** 1031-1050 (1977).

Gurnett, D. A., and B. J. O'Brien, "High-Latitude Geophysical Studies with Satellite Injun 3. 5. Very-Low-Frequency Electromagnetic Radiation," *J. Geophys. Res.,* **69,** 65-89 (1964).

Gurnett, D. A., S. D. Shawhan, and R. R. Shaw, "Auroral Hiss, Z-Mode Radiation, and Auroral Kilometric Radiation in the Polar Magnetosphere: DE 1 Observations," *J. Geophys. Res.,* **88,** 329-340 (1983).

Gurnett, D. A., L. A. Frank, and R. P. Lepping, "Plasma Waves in the Distant Magnetotail," *J. Geophys. Res.,* **81,** 6059-6071 (1976).

Gurnett, D. A., R. R. Anderson, F. L. Scarf, R. W. Fredricks, and E. J. Smith, "Initial Results from the ISEE-1 and -2 Plasma Wave Investigation," *Space Sci. Rev.,* **23,** 103-122 (1979).

Gurnett, D. A., R. L. Huff, J. D. Menietti, J. L. Burch, J. D. Winningham, and S. D. Shawhan, "Correlated Low-Frequency Electric and Magnetic Noise Along the Auroral Field Lines," *J. Geophys. Res.,* **89,** 8971-8985 (1984).

Harang, L., and R. Larsen, "Radio Wave Emissions in the VLF Band Observed Near the Auroral Zone. 1. Occurrence of Emissions During Disturbances," *J. Atmosph. Terr. Phys.,* **27,** 481-497 (1964).

Hartz, T. R., "Low Frequency Noise Emissions and Their Significance for Energetic Particle Processes in the Polar Ionosphere," in *The Polar Ionosphere and Magnetospheric Processes,* G. Skovli, ed., Gordon and Breach, New York, pp. 151-160 (1970).

Hasegawa, A., "Kinetic Properties of Alfvén Waves," *Proc. Indian Acad. Sci.,* **86,** 151 (1977).

Hashimoto, K., W. Calvert, and R. Huff, "On Z-mode Waves Observed by the DE 1 Satellite," Solar Terrestrial Environment Workshop, Tokyo, Japan, January 22 (1987).

Helliwell, R. A., *Whistlers and Related Ionospheric Phenomena,* Stanford University Press, Stanford, CA (1965).

Heppner, J. P., "Magnetospheric Convection Patterns Inferred from High Latitude Activity," *Atmospheric Emissions,* B. M. McCormac and A. Omholt, eds., Reinhold, New York, p. 251 (1969).

Huff, R. L., W. Calvert, J. D. Craven, L. A. Frank, and D. A. Gurnett, "Mapping of Auroral Kilometric Radiation Sources to the Aurora," *J. Geophys. Res.,* **93,** 11,445-11,454 (1988).

James, H. G., "VLF Saucers," *J. Geophys. Res.,* **81,** 501-514 (1976).

Jorgensen, T. S., and E. Ungstrup, "Direct Observation of Correlation Between Aurorae and Hiss in Greenland," *Nature,* **194,** 462-463 (1962).

Kelley, M. C., and P. Kintner, "Two-Dimensional Turbulence in a Low B Cosmic Scale Plasma," *Astrophys. J.,* **220,** 339-345 (1978).

Kelley, M. C., and F. S. Mozer, "A Satellite Survey of Vector Electric Fields in the Ionosphere at Frequencies of 10 to 500 Hz, 1, Isotropic, High-Latitude Electrostatic Emissions," *J. Geophys. Res.,* **77,** 4158 (1972).

Kennel, C. F., F. L. Scarf, R. W. Fredricks, J. H. McGehee, and F. V. Coroniti, "VLF Electric Field Observations in the Magnetosphere," *J. Geophys. Res.,* **75,** 6136-6152 (1970).

Kindel, J. M., and C. F. Kennel, "Topside Current Instabilities," *J. Geophys. Res.,* **76,** 3055-3078 (1971).

Kintner, P. M., Jr., "Observations of Velocity Shear Driven Plasma Turbulence," *J. Geophys. Res.,* **81,** 5114-5122 (1976).

Kintner, P. M., "On the Distinction Between Electrostatic Ion Cyclotron Waves and Ion Cyclotron Harmonic Waves," *Geophys. Res. Lett.,* **7,** 585-588 (1980).

Kintner, P. M., M. C. Kelley, and F. S. Mozer, "Electrostatic Hydrogen Cyclotron Waves Near One Earth Radius Altitude in the Polar Magnetosphere," *Geophys. Res. Lett.,* **5,** 139-142 (1978).

Kintner, P. M., J. LaBelle, M. C. Kelley, L. J. Cahill, Jr., T. Moore, and R. Arnoldy, "Interferometric Phase Velocity Measurements," *Geophys. Res. Lett.,* **11,** 19-22 (1984).

Kintner, P. M., J. Vago, R. Arnoldy, and T. Moore, "Simultaneous Observations of Electrostatic Oxygen Cyclotron Waves and Ion Conics," *Eos Trans. AGU,* **69,** 1396 (1988).

Kintner, P. M., M. C. Kelley, R. D. Sharp, A. G. Ghielmetti, M. Temerin, C. Cattell, P. F. Mizera, and J. F. Fennell, "Simultaneous Observations of Energetic (keV) Upstreaming and Electrostatic Hydrogen Waves," *J. Geophys. Res.,* **84,** 7201-7212 (1979).

Koskinen, H. E. J., P. M. Kintner, G. Holmgren, B. Holback, G. Gustafsson, M. André, and R. Lundin, "Observations of Ion Cyclotron Harmonic Waves by the Viking Satellite," *Geophys. Res. Lett.,* **14,** 459-462 (1987).

Krall, N. A., and A. W. Trivelpiece, *Principles of Plasma Physics,* McGraw-Hill, New York (1973).

Kurth, W. S., M. M. Baumback, and D. A. Gurnett, "Direction-Finding Measurements of Auroral Kilometric Radiation," *J. Geophys. Res.,* **80,** 2764-2770 (1975).

Laaspere, T., and R. A. Hoffman, "New Results on the Correlation Between Low-Energy Electrons and Auroral Hiss," *J. Geophys. Res.,* **81,** 524-530 (1976).

Lin, C. S., J. L. Burch, S. D. Shawhan, and D. A. Gurnett, "Correlation of Auroral Hiss and Upward Electron Beams Near the Polar Cusp," *J. Geophys. Res.,* **89,** 925-935 (1984).

Louarn, P., A. Roux, H. deFeraudy, and D. LeQueau, "Trapped Electrons as a Free Energy Source for the AKR," *J. Geophys. Res.,* submitted (1989).

Lysak, R. L., and C. T. Dum, "Dynamics of Magnetosphere-Ionosphere Coupling Including Turbulent Transport," *J. Geophys. Res.,* **88,** 365-380 (1983).

Lysak, R. L., M. K. Hudson, and M. Temerin, "Ion Heating by Strong Electrostatic Ion Cyclotron Heating," *J. Geophys. Res.,* **85,** 678-686 (1980).

Maggs, J. E., "Coherent Generation of VLF Hiss," *J. Geophys. Res.,* **81,** 1707-1724 (1976).

Maggs, J. E., and W. Lotko, "Altitude Dependent Model of the Auroral Beam and Beam-Generated Electrostatic Noise," *J. Geophys. Res.,* **86,** 3439-3447 (1981).

Martin, L. H., R. A. Helliwell, and K. R. Marks, "Association Between Aurorae and Very-Low-Frequency Hiss Observed by Byrd Station, Antarctica," *Nature,* **187,** 751-753 (1960).

Maynard, N. C., and J. P. Heppner, "Variations in Electric Fields from Polar Orbiting Satellites," *Particles and Fields in the Magnetosphere,* B. M. McCormac, ed., Reinhold, New York, 247-253 (1970).

Mellott, M. M., W. Calvert, R. L. Huff, D. A. Gurnett, and S. D. Shawhan, "DE-1 Observations of Ordinary Mode and Extraordinary Mode Auroral Kilometric Radiation," *Geophys. Res. Lett.,* **11,** 1188-1191 (1984).

Melrose, D. B., "Coherent Gyromagnetic Emission as a Radiation Mechanism," *Aust. J. Phys.,* **26,** 229 (1973).

Morioka, A., H. Oya, and S. Miyatake, "Terrestrial Kilometric Radiation Observed by Satellite JIKIKEN (EXOS-B)," *J. Geomagn. Geoelectr.,* **33,** 37-62 (1981).

Morozumi, H. M., "Semi-Diurnal Auroral Peak and VLF Emissions Observed at the South Pole, 1960," *Eos Trans. AGU,* **44,** 798 (1963).

Mosier, S. R., and D. A. Gurnett, "VLF Measurements of the Poynting Flux Along the Geomagnetic Field with the Injun 5 Satellite," *J. Geophys. Res.,* **74,** 5675-5687 (1969).

Mosier, S. R., M. L. Kaiser, and L. W. Brown, "Observations of Noise Bands Associated with the Upper Hybrid Resonance by the IMP 6 Radio Astronomy Experiment," *J. Geophys. Res.,* **78,** 1683 (1973).

Omidi, N., and D. A. Gurnett, "Growth Rate Calculations of Auroral Kilometric Radiation Using the Relativistic Resonance Condition," *J. Geophys. Res.,* **87,** 2377-2383 (1982).

Omidi, N., C. S. Wu, and D. A. Gurnett, "Generation of Auroral Kilometric and Z Mode Radiation by the Cyclotron Maser Mechanism," *J. Geophys. Res.,* **89,** 883-895 (1984).

Persoon, A. M., and D. A. Gurnett, "The High-Resolution Frequency Spectrum of Z Mode Radiation," *Eos Trans. AGU,* **70,** 434 (1989).

Peterson, W. K., E. G. Shelley, S. A. Boardsen, D. A. Gurnett, B. G. Ledley, M. Sugiura, T. E. Moore, and J. H. Waite, "Transverse Ion Energization and Low-Frequency Plasma Waves in the Mid-Altitude Auroral Zone: A Case Study," *J. Geophys. Res.,* **93,** 11,405-11,428 (1988).

Ratcliffe, J. A., *The Magneto-Ionic Theory and Its Applications to the Ionosphere,* Cambridge University Press, New York (1959).

Scarf, F. L., R. W. Fredricks, C. T. Russell, M. Kivelson, M. Neugebauer, and C. R. Chappell, "Observation of a Current-Driven Plasma Instability at the Outer Zone-Plasma Sheet Boundary," *J. Geophys. Res.,* **78,** 2150-2165 (1973).

Scarf, F. L., R. W. Fredricks, C. T. Russell, M. Neugebauer, M. Kivelson, and C. R. Chappell, "Current-Driven Plasma Instabilities at High Lati-

Gurnett—*Auroral Plasma Waves*

tudes," *J. Geophys. Res.,* **80**, 2030-2040 (1975).

Shaw, R. R., and D. A. Gurnett, "Electrostatic Noise Bands Associated with the Electron Gyrofrequency and Plasma Frequency in the Outer Magnetosphere," *J. Geophys. Res.,* **80**, 4259-4271 (1975).

Shawhan, S. D., and D. A. Gurnett, "Polarization Measurements of Auroral Kilometric Radiation by Dynamics Explorer-1," *Geophys. Res. Lett.,* **9**, 913-916 (1982).

Smiddy, M., W. J. Burke, M. C. Kelley, N. A. Saflekos, M. S. Gussenhoven, D. A. Hardy, and F. J. Rich, "Effects of High-Altitude Conductivity on Observed Convection Electric Fields and Birkeland Currents," *J. Geophys. Res.,* **85**, 6811-6818 (1980).

Smith, R. L., and N. Brice, "Propagation in Multicomponent Plasmas," *J. Geophys. Res.,* **69**, 5029-5040 (1964).

Stix, T. H., *The Theory of Plasma Waves,* McGraw-Hill, New York (1962).

Stone, R. G., "Radio Physics of the Outer Solar System," *Space Sci. Rev.,* **14**, 534-551 (1973).

Storey, L. R. O., "An Investigation of Whistling Atmospherics," *Phil. Trans. R. Soc. London,* A, **246**, 113-141 (1953).

Temerin, M., "The Polarization, Frequency, and Wavelength of High Latitude Turbulence," *J. Geophys. Res.,* **83**, 2609-2616 (1978).

Temerin, M., and R. L. Lysak, "Electromagnetic Ion Cyclotron Mode (ELF) Waves Generated by Auroral Electron Precipitation," *J. Geophys. Res.,* **89**, 2849-2859 (1984).

Tokar, R. L., and S. P. Gary, "Electrostatic Hiss and the Beam Driven Electron Acoustic Instability in the Dayside Polar Cusp," *Geophys. Res. Lett.,* **11**, 1180-1183 (1984).

Voots, G. R., D. A. Gurnett, and S.-I. Akasofu, "Auroral Kilometric Radiation as an Indicator of Auroral Magnetic Disturbances," *J. Geophys, Res.,* **82**, 2259-2266 (1977).

Walsh, D., T. F. Haddock, and H. F. Schulte, "Cosmic Radio Intensities at 1.225 and 2.0 MC Measured Up to an Altitude of 1700 km," *Space Res.,* **4**, 935 (1964).

Weimer, D. R., C. K. Goertz, D. A. Gurnett, N. C. Maynard, and J. L. Burch, "Auroral Zone Electric Fields from DE 1 and 2 at Magnetic Conjunctions," *J. Geophys. Res.,* **90**, 7479-7494 (1985).

Winglee, R. M., "Effects of a Finite Plasma Temperature on Electron-Cyclotron Maser Emission," *Astrophys. J.,* **291**, 160-169 (1985).

Wu, C. S., and L. C. Lee, "A Theory of Terrestrial Kilometric Radiation," *Astrophys. J.,* **230**, 621-626 (1979).

Wu, C. S., S. T. Tsai, M. J. Xu, and J. W. Shen, "Saturation and Energy Conversion Efficiency of Auroral Kilometric Radiation," *Astrophys. J.,* **248**, 384-391 (1981).

Zarka, P., D. LeQueau, and F. Genova, "The Maser Synchrotron Instability in an Inhomogeneous Medium: Determination of the Spectral Intensity of Auroral Kilometric Radiation," *J. Geophys. Res.,* **91**, 13,542-13,558 (1986).

V. AURORAL SUBSTORMS AND DYNAMICS

V-1. OVERVIEW OF OBSERVATIONS AND MODELS OF AURORAL SUBSTORMS

G. Rostoker*

Magnetospheric substorms represent a process in which energy from the solar wind is degraded to heat in Earth's ionosphere. The term "substorm" was originally introduced by ground-based observers to describe auroral and geomagnetic phenomena taking place in the auroral oval during episodes of transient enhanced energy dissipation. Later, researchers using in situ data in the magnetosphere adopted the term substorm to describe perturbations of the particle and field environment that they measured. These two different approaches led to conflicting views of the substorm process and, ultimately, to models that are apparently in conflict with one another. Here, we describe substorm phenomenology in a way that makes the physical processes involved apparent. A brief review of the ways in which the energy flow from the solar wind through the magnetosphere-ionosphere system can be quantitatively modeled is presented first. The key features of the near-Earth neutral line model and the boundary layer dynamics model are then reviewed and an effort is made to reconcile these two apparently conflicting pictures of the physical processes taking place in the magnetotail during substorms.

1. INTRODUCTION

Any researcher involved in the study of auroral substorms usually has the more ambitious goal of trying to understand the nature of the solar-terrestrial interaction. To acquire such an understanding, it is necessary to find a satisfactory solution to three well-defined problems that have, in the past, often been studied in isolation; these are

1. To find the physical processes through which energy from the solar wind is able to penetrate inside the magnetopause, providing the energy whose visible manifestation is the aurora (i.e., the auroral substorm).
2. To find the physical mechanisms for energy transport in the magnetosphere and for storage of part of the energy that enters from the solar wind that is not directly deposited into the upper atmosphere.
3. To determine the physical processes that regulate the magnetosphere-ionosphere interaction through which the energy from the solar wind is deposited into the upper atmosphere where it degrades to heat.

Researchers dealing with substorms normally find themselves addressing one of the aforementioned problems. However, the complexity of the overall solar-terrestrial interaction usually forces the researcher to perform quantitative treatments of any one of the three problems in isolation from the others. Ultimately, by studying the behavior of the auroras, one can hope to provide a set of constraints that will be helpful in de-

termining the boundary conditions required to distinguish between the various plausible scenarios that may be proposed in attempting to address the three major problem areas described above.

Before introducing the competing models that have been proposed to explain the auroral substorm, it is useful to review some of the knowledge that has been built up in attempts to answer the three questions above. The first of these problems addresses the question of the entry of energy from the solar wind. This topic has been studied in the context of a competition between the magnetic reconnection model of *Dungey* [1961] and the viscous interaction picture of *Axford and Hines* [1961]. The wealth of satellite data acquired over the past two decades has confirmed that solar-wind plasma can penetrate the magnetopause where it is visible in the closed-field-line region of the low-latitude boundary layer (LLBL) and the open-field-line region of the high-latitude plasma mantle as seen in Figure 1. As pointed out by *Baumjohann and Haerendel* [1986], it seems that typically about 90% of the energy entry can be attributed to the reconnection process (which is strongly dependent on the configuration of the interplanetary magnetic field (IMF)) while about 10% is due to viscous processes (which are either independent of or only weakly dependent on the interplanetary magnetic field). Thus it seems that the substorm process primarily reflects changes in magnetic field merging on the magnetopause and the responding reconnection in the magnetotail. While semi-empirical expressions for quantitatively evaluating the rate of energy input into the magnetosphere are presently in place [*Perrault and Akasofu*, 1978] the character of frontside merging is still a controversial issue. The reader is referred to *Paschmann et al.* [1986]

*Institute of Earth and Planetary Physics and Department of Physics, University of Alberta, Edmonton, Alberta, Canada T6G 2J1.

Auroral Physics, edited by C.-I. Meng, M. J. Rycroft and L. A. Frank. © Cambridge UP 1991

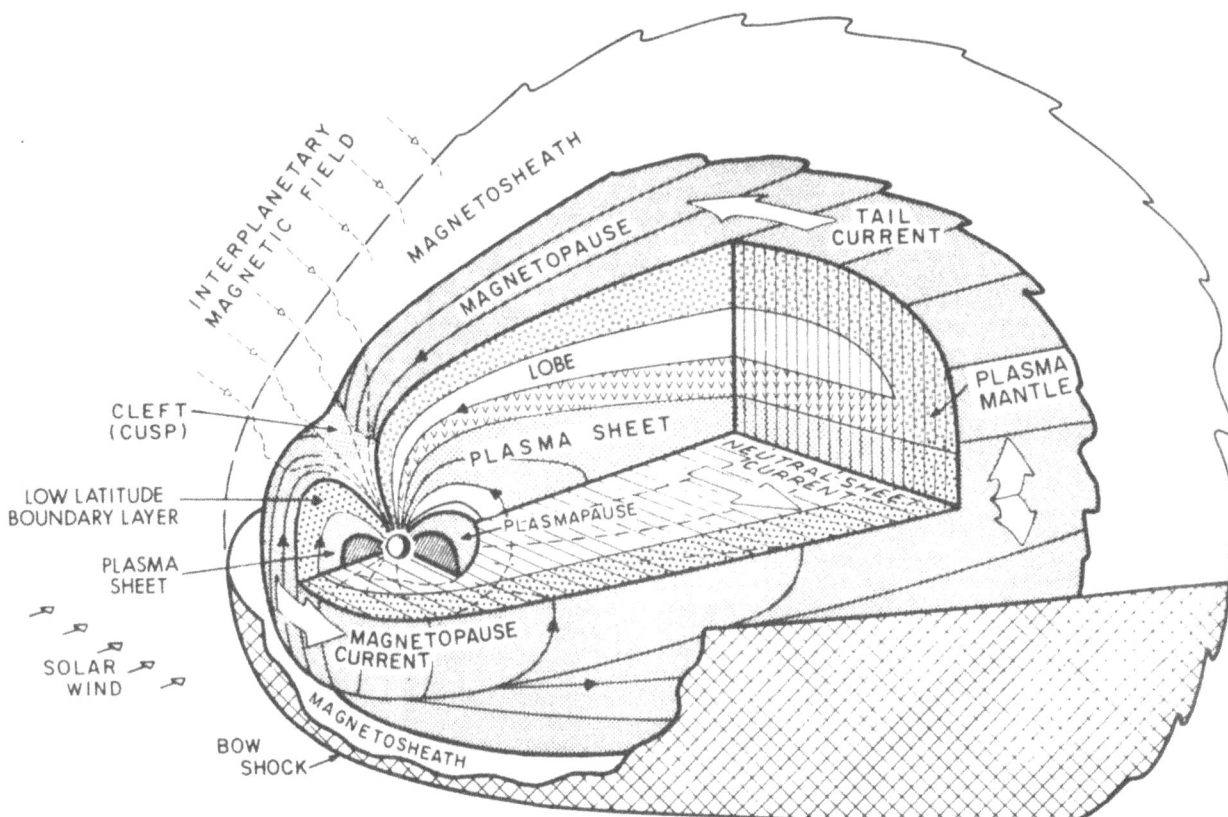

Figure 1—Three-dimensional diagram of the Earth's magnetosphere (adapted after *Heikkila* [1972]). The cut in the plane approximately perpendicular to the Sun-Earth line is normal, valid for distances out to at least $25R_E$ with the thicknesses of the central plasma sheet and the plasma sheet boundary layer being dependent on the level of the solar terrestrial interaction. The interface region between the plasma mantle, plasma sheet boundary layer, and central plasma sheet at distances above the neutral sheet is not yet well defined in terms of in situ observations.

for a modern view of how the merging rate may vary with interplanetary conditions.

While it is now quite clear that the energy that enters the magnetosphere from the solar wind can be stored both in the electromagnetic field and in the kinetic energy of drift of the magnetospheric plasma, the dynamics of the storage process during the substorm activity is not yet fully understood. It seems reasonable to expect that, before any internal magnetotail energy conversion processes come into play, the energy that has entered from the solar wind is found in three major forms:

1. Kinetic drift energy of the boundary layer plasmas as they convect antisunward on closed field lines in the LLBL and toward the center of the tail on open field lines in the plasma mantle.
2. The magnetotail magnetic field generated by the crosstail plasma sheet dawn-to-dusk current and the boundary layer/magnetopause currents that serve as closure currents for the circuit involving the crosstail current.

3. Waves generated through turbulent processes along the magnetopause that propagate from the boundary regions throughout the magnetotail.

Through processes that relate to the diversion of plasma flow by neutral lines, some of the plasma mantle population ends up convecting earthward forming at least part of the population of the central plasma sheet (CPS) [*Pilipp and Morphill*, 1976]. As the plasma drifts earthward in the CPS, it is adiabatically energized taking its energy from the electromagnetic field [*Hines*, 1963]. While adiabatic heating can energize plasma sheet particles to the order of kilovolts, rapid plasma sheet heating events observed during substorm activity [*Huang et al.*, 1988] suggest that nonadiabatic energy conversion processes are operative in the magnetotail during substorm activity. *Goertz and Smith* [1989] have suggested that the waves generated near the magnetopause can steepen into shocks and heat the plasma sheet through a resonant coupling mechanism that they have called the thermal catastrophe model. Since sudden plasma sheet warming seems to be an observed facet of substorm activity in the magnetotail, some heating mechanism that can rapidly turn on and affect a significant volume of the CPS would seem to be a necessary attribute of any comprehensive substorm model. One further way in which energy can be converted from one form to another within the magnetotail is found in the process of magnetic field reconnection. The existence of a neutral line across the magnetotail is, more or less, accepted by all researchers despite the lack of hard data taken in the expected region of this portion of the geotail separatrix (thought to be approximately $100R_E$ behind the Earth). The essence of the reconnection concept is that it leads to a conversion of magnetic field energy to the kinetic energy of drift of magnetotail plasma through the process of magnetic field annihilation. In this scenario, sunward flows out of the reconnection region are on closed field lines while antisunward flow is on open field lines threading the plasma sheet in the distant magnetotail. A series of studies by J. Birn et al. [*Birn*, 1979, 1984; *Birn and Hones*, 1981] describe simulation studies of the reconnection process in magnetotail geometry thus providing a theoretical foundation for models of auroral substorms which call on magnetic field reconnection to play an important role.

Perhaps the area of the solar-terrestrial interaction that has been studied most thoroughly and quantitatively is the topic of magnetosphere-ionosphere coupling. This involves the study of the physical processes through which the energy that enters the magnetosphere from the solar wind is deposited in the ionosphere. The understanding of magnetosphere-ionosphere coupling is crucial to the development of any model that proposes to explain auroral substorm activity. It is now evident that the coupling is effected by field-aligned electric currents that link diamagnetic and inertial currents flowing transverse to the tail magnetic field lines to ionospheric currents in the auroral oval. Some existing approaches concentrate on the interaction of ionospheric electric fields and conductivity structure to predict the nature of the field-aligned currents that provide the magnetosphere-ionosphere linkage [*Kan and Sun*, 1985; *Hori and Yamamoto*, 1987; *Kan et al.*, 1988]. Others concentrate on the drift of plasma sheet particles in the near-Earth magnetotail and couple this with constraints put on the auroral ionosphere to construct a self-consistent picture of the regions of the magnetosphere (and their ionospheric projections) that respond to the input of energy from the boundary layers and more distant regimes of the magnetotail (cf. *Harel et al.* [1981] and the many following papers dedicated to the development of the "Rice model"). Still other researchers concentrate on the breakdown of MHD conditions leading to the development of parallel electric fields in regions 1–2R_E above Earth's sur-

face on magnetic field lines penetrating the auroral oval. Their studies probe the mechanisms for generating the field-aligned electric fields (cf. *Block* [1978], *Chiu and Schulz* [1978], and *Lysak and Dum* [1983] among many others) that accelerate electrons to energies of several kilovolts leading to the bright discrete arcs of the auroral substorm and the field-aligned current that plays such an important role in the magnetosphere-ionosphere coupling process.

It is important to emphasize that any successful model for auroral substorms must incorporate an understanding of all the various problems discussed above. Yet there is a tendency to try to break the overall problem down into component parts in hope that the ultimate solution will conform to the condition that "the whole is equal to the sum of the parts." Unfortunately, the magnetosphere-ionosphere system is basically nonlinear in character, a fact that is clearly evident during the episodes of impulsive activity such as substorms. It is only recently that global models of the solar-terrestrial interaction have been developed that take, as their constraints, information from all parts of the system (viz. the entire magnetosphere-ionosphere system and the solar-wind plasma and field that surrounds it). The most sophisticated of these global models has been presented by *Ogino* [1986] and *Ogino et al.* [1986]. These three-dimensional simulation codes represent the leading edge of more comprehensive attempts to formulate a model for the solar-terrestrial interaction. The success of this approach demands the use of very large, fast computers and the implementation of the correct boundary conditions, particularly at the interface between regions of plasma and field having distinctly different properties.

Ultimately, the only way to validate any theory of the solar-terrestrial interaction is to compare the predictions of proposed models with observations of particles and fields in the magnetosphere-ionosphere system. Since in situ observations using spacecraft are only available in limited regions of space at any one time, the ionospheric signatures of the magnetospheric particles and fields (which can be remotely sensed by ground-based arrays of instrumentation or satellite-borne global imaging devices) are of crucial importance in defining the boundary conditions that represent the input parameters for any model. *Craven and Frank* [this volume] and *Shepherd and Murphree* [this volume] help to describe the observed features of auroral activity that allow one to define the boundary conditions at ionospheric levels in terms of both latitudinal and longitudinal structure. The role magnetic field reconnection plays in the reconfiguration of magnetotail plasmas is explored in the context of the formation of a second neutral line close to the Earth [*Hones and Galvin*, this volume and *Kan et al.*, this volume].

In addition, *Kan et al.* consider the implications of a quantitative treatment of the magnetosphere-ionosphere coupling problem on the formation and movement of near-Earth neutral lines. The goal presented here is to explore the key features of the near-Earth neutral line and the boundary-layer models of substorms in the context of observations of auroral substorm activity and the accompanying evolution of the three-dimensional current systems that couple the ionosphere to the distant regions of the magnetosphere.

2. EVOLUTION OF THE DEFINITION OF A SUBSTORM

Any researcher seeking to understand the nature of the solar-terrestrial interaction will inevitably encounter the term *magnetospheric substorm*. Unfortunately, in reading the literature, they will find that the term substorm is used by different people to mean different things. This problem in semantics has been instrumental in retarding the progress in our understanding of the solar-terrestrial interaction for over twenty years. In fact, it is only in the past few years that specialists dealing with substorm phenomena have reached some limited agreement on the primary processes taking place during an episode of activity that, on the basis of various magnetospheric and ionospheric signatures, would have been acknowledged to involve substorms. The reader is referred to *Rostoker et al.* [1987] for a more detailed discussion of substorm processes, and the essence of that review will be discussed later here.

It can be easily understood that, if one could not agree on the nature of the disturbance to which the term substorm should apply, the development of a universally accepted theory for the origin of substorms would encounter formidable obstacles. In fact, since the early 1970s there has been general agreement that substorm activity is driven through processes involving merging of the IMF with the dayside terrestrial magnetic field followed by subsequent reconnection of the resulting open field lines in the magnetotail [*Dungey*, 1961]. While there is some question as to whether some sort of viscous interaction at the magnetopause is responsible for a portion of the energy transferred from the solar wind into the magnetosphere [*Axford and Hines*, 1961], it would appear that at least 90% of the energy transfer can be attributed to the reconnection phenomenon [*Baumjohann and Haerendel*, 1986]. In recent years, the focus of attention has shifted to the magnetotail with important questions being asked about the processes taking place in that region of space during substorm activity. In the early stages, it was felt that a single neutral line behind the Earth was adequate to account for the transfer of magnetic flux (and accordingly of energy) in accordance with the

initial concept of *Dungey* [1961]. However, more recently a new paradigm has developed in which a second neutral line forms in the neutral sheet close to Earth during the development of a substorm. This paradigm, most carefully explored by Hones and colleagues [*Hones*, 1984], proposes that the onset of reconnection of tail lobe field lines causes the release of a significant portion of the plasma sheet into the solar wind through the formation and subsequent down-tail motion of a closed field structure called a plasmoid. This paradigm has been termed the "near-Earth neutral line" model of a substorm (hereafter called the NENL model). It has now become clear that substorm processes related to discrete auroral disruption take place on magnetic field lines that thread the boundary plasma sheet (BPS) in the auroral ionosphere and the plasma sheet boundary layer (PSBL) in the magnetotail. This observation led to the "boundary layer dynamics" model (hereafter called the BLD model) in which auroral substorm features are attributed to changes in velocity shear at the interface between the LLBL and the CPS [*Rostoker and Eastman*, 1987]. The effectiveness of these two paradigms (the NENL and BLD models) will dominate the latter part of this presentation. However, it is first necessary to adopt an acceptable definition of the substorm phenomenon. We shall attempt to do this, set in the context of the evolution of the semantic problems encountered by substorm researchers over the years.

The term substorm originally entered the vocabulary of the space scientist through a study of the dynamics of discrete auroral arcs by *Akasofu* [1964]. Using pictures from an array of widely and irregularly spaced all-sky cameras, Akasofu concluded that the development of certain highly dynamic and distorted auroral forms followed a repeatable pattern of a kind shown in Figure 2. He called this pattern of development an auroral substorm and it has served for over two decades as a framework in which many observations of particles and electromagnetic fields in ionosphere, magnetosphere, and interplanetary medium are set. As one can see from Figure 2, the auroral substorm as envisioned by Akasofu involves a brightening of auroral arcs over a significant number of local time zones across midnight, with the subsequent poleward motion involving the formation of a single S-shaped auroral structure at the western edge of the disturbed region. This latter structure was called the *westward traveling surge* because it was thought that it could propagate westward at speeds of approximately 1 km s^{-1}, while maintaining its structural identity over several time zones. The substorm disturbance was broken into two distinct sections. These were the *expansion phase* (involving poleward motion of the auroras) and the *recovery phase* (involving equatorward motion of the auroras). Near the end of the 1960s most of the other particle and field observations had been ordered in the framework of the Akasofu substorm with the entire ensemble of disturbances being given the name *magnetospheric substorm.*

Throughout the 1970s, improved observations of the substorm phenomenon through the use of coordinated networks of ground-based detectors (e.g., all-sky cameras, magnetometers) led to a more detailed picture of the magnetospheric substorm phenomenon. One of the first things that became obvious was that the poleward motion of the poleward edge of the substorm disturbed region occurred in discrete steps [*Kisabeth and Rostoker*, 1974]. This led to the substorm westward electrojet appearing as an ensemble of parallel current strips each having developed at a different time and each having its own life cycle. Shortly thereafter, *Wiens and Rostoker* [1975] demonstrated that each of these current elements had a distinctive western edge and they suggested that the western motion of the region of substorm disturbance evolved as a stepwise progression. This behavior was also inferred by *Pytte et al.* [1976], who gave the ensemble of perturbations the name *multiple onset substorm.* The entire ensemble of disturbances during a multiple onset substorm yielded an envelope of magnetic activity that stood out clearly in the auroral electrojet index *AE* (Figure 3). Plotted on this scale (which is typical of what the space researcher encounters), one has to recognize that the Akasofu substorm might occupy the time scale of only an hour or so (the disturbance starting near 0720 UT and lasting until 0830 UT, for example). Each of the elements that developed separately, having their own identifiable onset and life cycle, are not readily apparent in this format of the *AE* index. At the beginning of the 1980s, the concept of the multiple onset substorm was further refined by *Tighe and Rostoker* [1981], who noted that surge forms at the western edge of the substorm-disturbed region often moved irregularly or not at all during their lifetime. In fact, while the western edge of the substorm region often does expand westward during an episode of activity, the individual surge forms tend to show little longitudinal motion, as evidenced from the recent Viking imager data [*Rostoker et al.*, 1987]. The key change to the Akasofu paradigm is therefore that there is not one monolithic surge form that can propagate over large azimuthal distances. Rather, there is a sequence of surge activations, with the western edge of the envelope of auroral luminosity (containing the multiple surge forms) typically (but not necessarily) expanding westward. The electric currents (both ionospheric and field-aligned) that develop in conjunction with the ensemble of surge formations are responsible for the irregular magnetic perturbations that mark a rise and

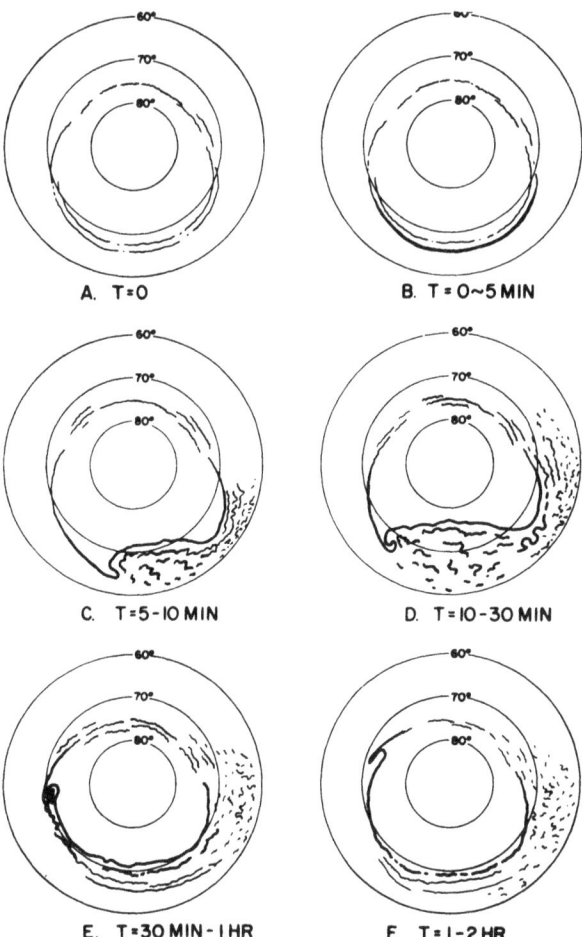

A. T=0   B. T=0~5 MIN

C. T=5-10 MIN   D. T=10-30 MIN

E. T=30 MIN-1 HR   F. T=1-2 HR

Figure 2—View looking down on the north pole of the sequence of auroral activations defined by *Akasofu* [1964] as an *auroral substorm.* Upward field-aligned current at the western edge of the disturbed region is connected to downward field-aligned current to the east by a westward ionospheric electrojet current. The magnetic field disturbance due to the growth and decay of the electric currents was called a *polar magnetic substorm.* Together, the auroral and polar magnetic substorms are the major ionospheric signatures of the *magnetospheric substorm.*

subsequent fall in the *AE* index as shown in Figure 3.

The beginning of the 1970s marked the arrival of a new definition of a substorm introduced primarily by researchers who acquired and attempted to interpret data from satellite-borne detectors. There was a tendency for these researchers to use indices of activity starting, in the early days, with the 3-hour *Kp* index. The introduction of the 1-hour *AE* index for correlation of satellite-particle and field data dates back to the study of the response of the magnetotail to substorms carried out by *Fairfield and Ness* [1970]. They viewed the rise and subsequent fall of *AE* as a substorm; that is, the envelope of the *AE* signature represented the substorm instead of the shorter time-scale perturbations that characterized the original Akasofu substorm. During the 1970s this new definition of the substorm became dominant and Akasofu's (1964) original definition was never really used by satellite experimenters. In fact, the "*AE* substorm" involved a rise and fall of that index, closely paralleling the turning toward the south and the subsequent return to its prior level of the north-south (B_z) component of the IMF.

The confusion marked by one group of researchers discussing substorms in the context of Akasofu's paradigm and a second group applying the same name to perturbations in the magnetosphere that correspond to rises and falls in *AE* impeded attempts to understand the physics of substorm processes for some considerable time. A group of researchers attempted to address this problem at a workshop in Victoria, Canada [*Rostoker et al.*, 1980]. However, their recommendations were not universally adopted. Fortunately, the solution to an old problem appeared and agreement was reached on the phenomenology of substorms. This old problem first appeared as a controversy involving equivalent current systems of geomagnetic disturbances called "bays." While *Sugiura and Heppner* [1965] contended that the equivalent current system of bays involved both an eastward auroral electrojet in the

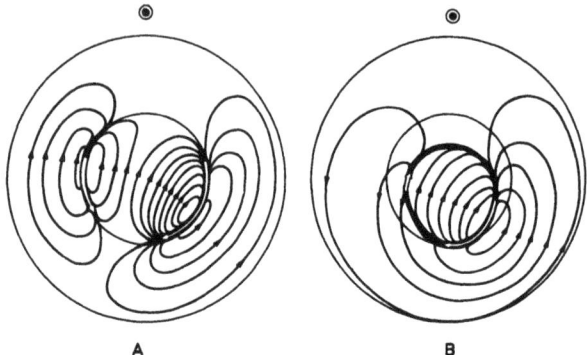

Figure 4—Equivalent ionospheric current systems proposed during the 1960s for substorms. Top panel shows the two-cell system supported by *Sugiura and Heppner* [1965], while bottom panel shows the one-cell system proposed by *Akasofu et al.* [1965] (after *Akasofu et al.* [1965]). Actually both systems exist but have different time scales. Left panel corresponds to the equivalent current one would now associate with the directly driven process, while right panel corresponds to the substorm current wedge associated with impulsive unloading of stored magnetotail energy.

afternoon sector and a westward electrojet in the morning sector (cf. Figure 4), *Akasofu et al.* [1965] argued that bays involved only a westward electrojet. Evidently, magnetic perturbations could be interpreted in such a way that both equivalent current systems could be inferred, so it is not surprising that other researchers found it necessary to define two separate types of disturbances (e.g., DP1 and DP2 introduced by *Obayashi and Nishida* [1968]). *Rostoker* [1969] moved part way toward resolving the controversy by noting that the bays whose ionospheric equivalent current systems involved both eastward and westward electrojets were of arbitrary length, often in excess of 1 hour. In contrast, shorter period disturbances, often appearing as riders on the longer period bays were thought to have the equivalent current system involving only a westward electrojet. Later on, *Perreault and Akasofu* [1978] pointed out that much of the substorm disturbance was attributable to what they called the directly driven system, which they quantified by the *AE* index. By then, a number of researchers, notably McPherron et al., had linked the shorter period bays (viz. the expansive phase of the original Akasofu substorm) to the sudden deposition of stored magnetotail energy into the ionosphere [*McPherron*, 1972]. Finally, in the early 1980s it became clear that magnetospheric activity characterized by the *AE* index included contributions from both the directly driven process (modulated closely by the rise and fall of energy input from the solar wind) and the loading-unloading process. In the latter process, energy stored in the magnetospheric tail during episodes of enhanced energy input from the solar wind was eventually deposited in the ionosphere. This deposition was triggered either by changes in interplanetary con-

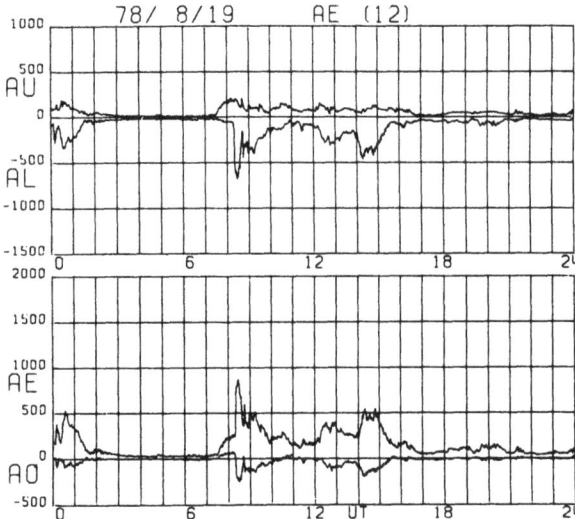

Figure 3—The auroral electrojet (*AE*), *AU* (upper), and *AL* (lower) indices derived from ground-based magnetometers in the auroral zone for a typical 24-hour interval. Researchers concentrating on the interpretation of satellite-borne detectors commonly use the *AE* index (in nanoteslas) as an indicator of the presence or absence of substorm activity. They would, for example, probably identify intervals such as 0000–0140 UT and 0730–1000 UT as substorms. Their time scale is not consistent with the time scale of the Akasofu auroral substorm, however (cf. Figure 2).

ditions or internally due to processes as yet not fully understood. The unloading of tail energy proceeds through the evolution of a sequence of intensifications, each providing an auroral signature conforming to the original Akasofu substorm.

To summarize, it has become clear that substorm activity involves both directly driven processes (responsible for the two-cell DP2 equivalent current system) and the concurrent loading and unloading of the magnetotail (with the unloading responsible for numerous azimuthally localized auroral breakups, each of which features the appearance of the single-cell DP1 current system). The reader is referred to *Rostoker et al.* [1987] for a more detailed discussion of the auroral and geomagnetic signatures of magnetospheric substorms.

3. SUBSTORM FEATURES THAT ANY MODEL MUST EXPLAIN

We have now explained how the term substorm came to mean different things to different people. We now present the phenomenology of the overall global disturbance that may deserve the name "magnetospheric substorm." In the light of the fact that the original term substorm was introduced in the context of auroral activity, we here discuss the evolution of our "substorm" in terms of its auroral signatures.

Figure 5 shows the global evolution of the auroras in response to an increase in energy input to the magnetosphere, followed (some time later) by a return of the energy input to its previous level. There are two basic behavioral patterns evident in this figure. The increase in energy input (due to a turning toward the south of the IMF) leads to an increase in polar cap area through the equatorward shift of the auroral oval. The following decrease in solar-wind energy input results in the oval contracting poleward to its presubstorm position. Even as the oval expands equatorward in response to increased energy input, the electric field and ionospheric conductivity increase in such a way as to increase the eastward and westward electrojets flowing in the oval across the dusk and dawn meridians, respectively. These electrojets reflect the contribution to the *AE* index of directly driven activity. Any model of the magnetospheric substorm must explain these auroral and geomagnetic manifestations of the directly driven activity. The second behavioral pattern involves the evolution of discrete auroral activity at the poleward edge of the oval. During the equatorward expansion of the auroral oval, one or two surges may appear at any one time at the poleward edge of the oval, either separately or coexisting with one another. Each of these surges might nearly fill the field of view of an appropriately placed all-sky camera, and each would have associated with it a filamentary westward

ionospheric electrojet directly under which a magnetometer might record a 150–200 nT "bay." If one of

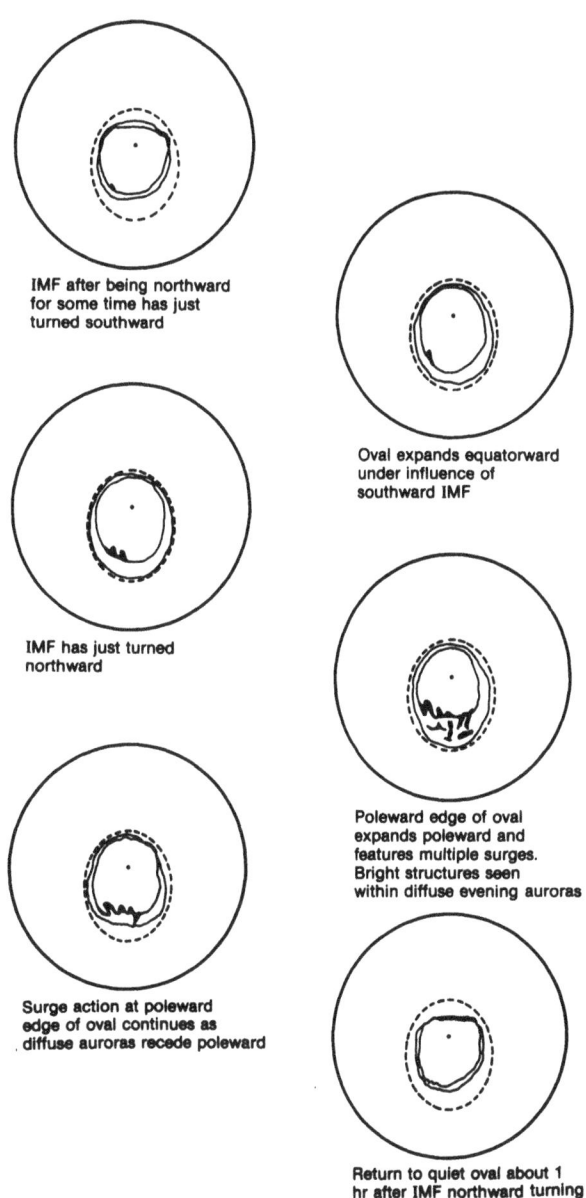

IMF after being northward for some time has just turned southward

IMF has just turned northward

Surge action at poleward edge of oval continues as diffuse auroras recede poleward

Oval expands equatorward under influence of southward IMF

Poleward edge of oval expands poleward and features multiple surges. Bright structures seen within diffuse evening auroras

Return to quiet oval about 1 hr after IMF northward turning

Figure 5—Behavior of the auroral oval (dashed line) and the discrete auroras within the oval as time progresses during the course of a magnetospheric substorm (in the same format as Figure 1). The entire substorm involves an expansion and subsequent contraction of the open-field-line region of the polar cap. Superposed on this more global behavior, localized discrete auroras undergo activations, each of which corresponds to the auroral substorm originally defined by *Akasofu* [1964].

those appropriately placed magnetometer stations happened to be part of the *AE* network, then a sharp but relatively shortlived (compared to the lifetime of the driven activity) increase in *AE* would be noted. An observer at these appropriately placed all-sky camera or magnetometer locations would claim to have detected what Akasofu originally called a substorm. These active regions would be said to experience poleward motion during the expansive phase and equatorward motion during the recovery phase—exactly opposite to the motions related to growth and decay of the driven system disturbance. While these surge structures appear sporadically during the episode of increased energy input from the solar wind, the return of the energy input to its previous level often leads to an outburst of filamentary electrojet and surge development with each surge typically (but not always) to the west of the previous one. Such an episode would have been called a multiple onset substorm in the past, but it is important to recognize that these expansive phase disturbances reach their peak even as the directly driven system is beginning to decay and the poleward edge of the oval is expanding poleward. Any model of a magnetospheric substorm must provide an explanation of the latitudinally and longitudinally localized surges, the spatial evolution of the region of multiple surge structures, and the accompanying electric currents associated with the unloading of magnetotail energy into the auroral ionosphere. Any theory that fails to provide explanations for both the directly driven and loading-unloading processes described above is not a complete theory for the magnetospheric substorm. We shall next inspect the two primary frameworks in which theoretical models of substorm activity are presently couched.

The Near-Earth Neutral Line Model (NENL Model)

The basic elements of this model follow the contention made by *Dungey* [1961] that energy from the solar wind is coupled into the magnetosphere through the process of magnetic field line merging. The physics of this process lies in the appearance of a component of the IMF antiparallel to the frontside dipole field lines of Earth. Under these circumstances, magnetic field lines on the dayside magnetopause merge wth the IMF and are swept over the poles to form a distended magnetotail. Eventually the tail field lines reconnect in the neutral sheet, separating the northern and southern hemispheres of the tail, leading to distended closed field lines that contract earthward to rejoin the dipole configuration. The energy stored in the distended field is released to energize the particle population of the magnetotail, with some of the energized charged particles precipitating into the ionosphere, leading to enhanced

electrical conductivity and auroral luminosity. The physics of the aforementioned reconnection process and its impact on energization of the magnetosphere has been discussed in detail by *Coroniti* [1985]. Detailed model studies of the reconnection process have been carried out by several researchers [e.g., *Birn*, 1984; *Walker and Sato*, 1984; and *Lee et al.*, 1985 among others].

The linkage of reconnection processes in the tail to substorm phenomenology based on ionospheric responses began in earnest with studies by *Camidge and Rostoker* [1970] and *Fairfield and Ness* [1970]. Both works pointed to specific changes in the magnetotail magnetic field associated with the appearance of auroral oval electric current previously attributed to the substorm phenomenon. About this time, the behavior of magnetotail particles was studied by E. W. Hones, Jr. et al. [e.g., *Akasofu et al.*, 1971; *Hones et al.*, 1971; *Hones et al.*, 1973] and, by the mid-1970s, the concept of the present near-Earth neutral line model had evolved. The reader is referred to *Hones* [1984] for a clear description of the NENL framework, which will now be summarized with the aid of Figure 6. It is assumed that, at the beginning of the substorm process, a neutral line already exists in the distant tail. Presumably this neutral line is involved in some quasi-steady-state process, in contrast to the transient nature of the process attributed to the "substorm" in this framework. The substorm process is initiated by enhanced reconnection on the dayside magnetopause after which a neutral line forms in the near-Earth region and reconnection of plasma-sheet field lines commences. In the auroral zone, we might anticipate that enhanced energy input through dayside reconnection would lead to a growth of the area of the polar cap in the auroral oval and a thinning of the plasma sheet over at least some portion of the tail. The substorm expansive phase does not begin until all the plasma-sheet field lines have been reconnected and tail-lobe field lines begin to be reconnected. Then the auroral substorm expansive phase (viz. an auroral breakup) commences. The plasma sheet downstream of the near-Earth neutral line is considered to be on distended solar-wind field lines; plasma is ejected into the solar wind within a closed field line structure, which has been termed a *plasmoid*. Reconnection continues at the near-Earth neutral line until some later time when that neutral line begins to retreat into the tail and the plasma sheet begins to thicken. Generally, the plasma in the thickening plasma sheet is hotter than it was before the thinning that initiated the process. Thickening of the plasma sheet proceeds until the tail returns to its original presubstorm topology.

In the NENL framework, the substorm expansive phase marks the onset of reconnection of tail-lobe field

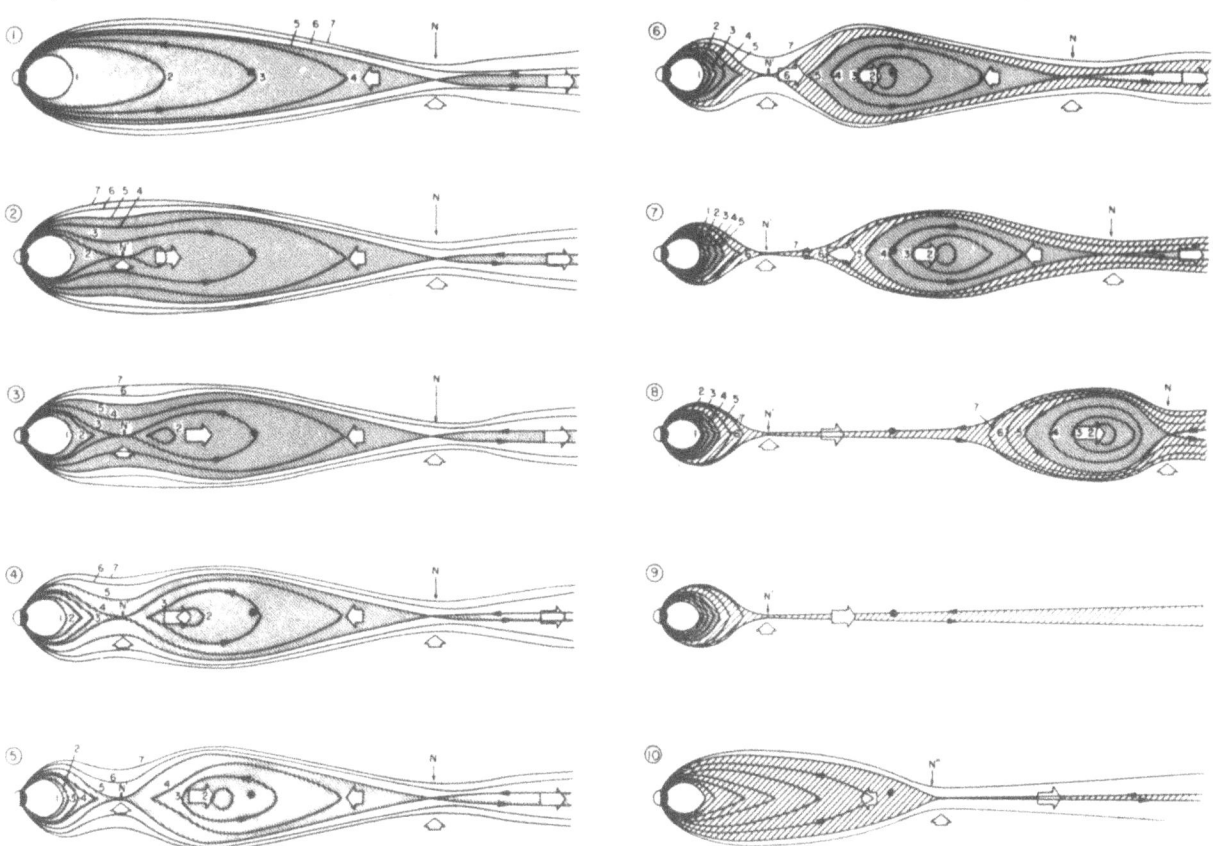

Figure 6—Evolution of the particle and field configuration in the magnetotail during the development of a magnetospheric substorm. The merging of plasma sheet field lines at a near-Earth neutral line is followed by the onset of merging of tail-lobe field lines marking the onset of the substorm expansive phase. The plasma downstream of the near-Earth neutral line is contained in a closed-field-line structure called a plasmoid that is ejected from the magnetotail. The substorm recovery is marked by the tailward retreat of the neutral line and the thickening of the plasma sheet. (After *Hones* [1984].)

lines and the poleward motion of the poleward edge of the auroral oval is linked to the thickening of the plasma sheet. Theoretical studies and numerical simulations by *Walker and Sato* [1984] have demonstrated how reconnection at a neutral line might lead to the development of a substorm current wedge on the field lines, earthward of the neutral line location. Recent studies by *Birn et al.* [1986] have also demonstrated how reconnection can lead to a boundary layer appearing at the high-latitude edge of the plasma sheet, thus accounting for the character of the tail plasma sheet described earlier by *Eastman et al.* [1985]. What is now missing from the NENL framework is an explanation of two key features: first, it does not address the origin of the directly driven activity and the colocation of electric current systems and discrete arc configurations of the directly driven and unloading elements of the substorm process; second, it does not deal with the details of auroral activity in the ionosphere, particularly in regard to the azimuthal structure of discrete auroral forms during episodes of unloading. Since the

term substorm was defined using these discrete auroral signatures, any theory of substorms must be set in a framework in which these defining attributes can be understood.

The Boundary Layer Dynamics (BLD) Model

Unlike the NENL model, which is designed to explain the substorm expansive phase, the BLD model is designed to explain the directly driven activity. As will be shown, the expansive phase signatures of a substorm then emerge as manifestations of the perturbation of the directly driven system. The essence of the BLD model is seen best in the projection of the plasma populations and plasma convective flow on the plane of the neutral sheet as shown in Figure 7. This figure is not to scale and should be viewed as representative of the topology of the magnetotail. The Birkeland currents illustrate the three-dimensional nature of the illustration. The reader is referred to *Rostoker and Eastman* [1987] and *Rostoker* [1987] for more detailed descriptions of this model.

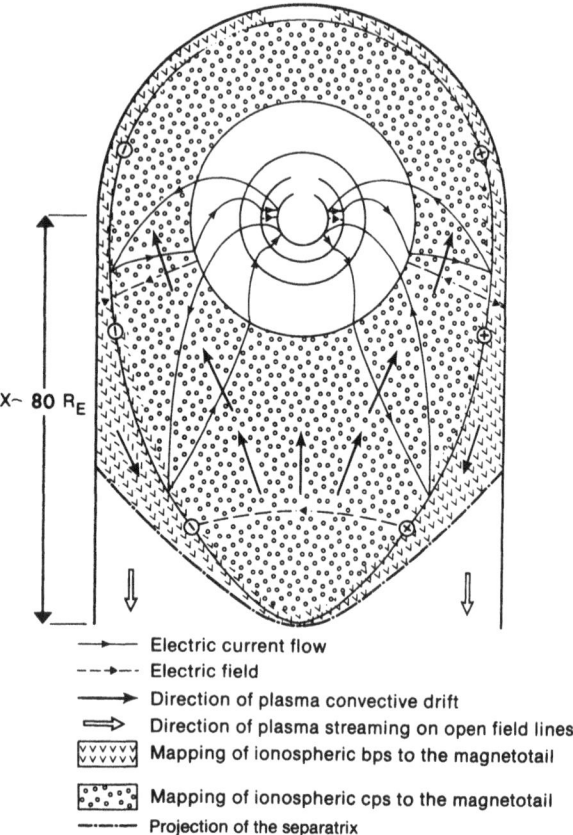

Electric current flow
Electric field
Direction of plasma convective drift
Direction of plasma streaming on open field lines
Mapping of ionospheric bps to the magnetotail
Mapping of ionospheric cps to the magnetotail
Projection of the separatrix

Figure 7—Projection of the charged particle populations of the magnetotail and velocity-shear zones on closed field lines onto the plane of the neutral sheet. Also shown is the geometry of the Birkeland currents and the mapping scheme from the magnetotail to the auroral ionosphere used in the boundary layer model of substorms (after *Rostoker and Eastman* [1987]). This figure is not to scale so as to show all important regions clearly.

The BLD model is based on the identification and mapping from the magnetosphere to the ionosphere (or vice versa) of the major identified velocity shear zones or electric field polarity transitions in the magnetosphere-ionosphere system. The major contention, insofar as the mapping is concerned, is that the ionospheric BPS, the plasma sheet boundary layer (PSBL) at the high-latitude edge of the magnetospheric CPS and the interface between the LLBL, and the midplane of the CPS further downtail are one and the same region of space extended along the magnetic field from the ionosphere to the neutral sheet. Like the NENL model, the BLD model relies on the primary energy transfer from the solar wind to the magnetosphere being achieved by magnetic-field-line reconnection. Both models also feature thinning of the plasma sheet in concert with the loading of the magnetosphere as solar-wind energy is stored in the magnetic field of the tail. In both models, the recovery of the substorm involves

thickening of the plasma sheet. However, on the BLD model there is no need for establishment of a second near-Earth neutral line. There is but one magnetotail neutral line, which is probably located about 60–$100R_E$ behind the Earth for normal magnetospheric conditions. Reconnection across this neutral line is required to provide the earthward flow of plasma observed in the CPS. The flow energy is thought to be provided to the plasma through conversion of magnetic field energy stored in the magnetotail. In the energy conversion process, the plasma can both be heated and its flow velocity enhanced. The braking of the convective earthward flow of plasma allows the kinetic drift energy to be converted to electromagnetic energy stored in the Birkeland currents through an MHD generator mechanism [*Rostoker and Boström*, 1976]. While not all the region 1 currents necessarily owe their origin to the braking of CPS plasma, the amount of drift energy available is far from inconsequential. Enhanced plasma convective flow associated with increased energy input from the solar wind leads to an enhanced convection E-field (which maps to the ionosphere) and enhanced energetic charged particle precipitation into the auroral oval through the effects of hydromagnetic energization [*Hines*, 1963]. In this way the directly driven system is closely modulated by the rate at which energy from the solar wind enters the magnetosphere.

Substorm expansive phase activity is thought to evolve from bursts of enhanced magnetic-field-line reconnection across the neutral line. As shown in Figure 8, if the momentum density contrast across the velocity shear zone between CPS and LLBL is sufficient, a Kelvin-Helmholtz instability can be excited. As has been shown by *Thompson* [1983], the growth of the instability leads to a shorting current that attempts to stabilize the instability by removing free energy from the system. Thus the growth of the Kelvin-Helmholtz instability leads directly to the upward field-aligned current, which is known to flow out of the ionosphere along field lines threading the "westward traveling" surge [*Rostoker*, 1987]. Multiple surges simply reflect the possibility of several wave crests existing simultaneously along the length of the velocity shear zone where conditions for growth of the Kelvin-Helmholtz instability are favorable. Field-aligned currents, flowing toward the source region at the LLBL/CPS interface from the ionospheric BPS, thread the PSBL where edge effects provide the negative B_z perturbation often cited as compelling evidence for the NENL model.

4. AN EVALUATION OF THE TWO MODELS

Before starting a critical evaluation of the two frameworks we have discussed, it is worth emphasizing that each model is based on a dataset that involves primarily single-point observations in a vast magnetotail volume that features rapid reconfigurations and significant spatial gradients in particle and field characteristics. The existing dataset makes it entirely reasonable that a unique framework in which substorms can be described is not yet at our disposal. All we can do at the present is ask which model best describes the magnetospheric substorm as defined by the auroral and geomagnetic signatures on which the original definition of the substorm was based.

In trying to compare the two frameworks, we might first focus on what are perceived to be the sources of disagreement in terms of the observations. The primary difference lies in the interpretation of certain magnetic field perturbations and concurrent particle flows in the magnetotail between $-10R_E > X > -20R_E$. In particular the disagreement focuses on the appearance of shortlived episodes of negative B_z and antisunward flow of plasma. Whenever this combination of effects is detected in the tail, it is associated with the satellite being on the anti-Earthward side of a near-Earth neutral line by those espousing the NENL model for substorms. In contrast, those adhering to the BLD model consider those observations to indicate the satellite position is in the PSBL duskward of a field-aligned current element flowing out of the ionosphere from a westward traveling surge in the auroral oval. Figure 9 shows an example of such a magnetic perturbation that has been studied by those in both schools of thought and interpreted in the two frameworks [cf. *Hones et al.*, 1986; *Rostoker and Eastman*, 1987]. A second major difference lies in the location of the region of the magnetotail that maps to the substorm-disturbed region of the ionosphere. The NENL model involves current disruption in the center of the near-Earth magnetotail (cf. Figure 10) with field-aligned current flowing into and out of the ionosphere at the edges of the disrupted region combining with a westward ionospheric electrojet to form the substorm current wedge. The western edge of the wedge in the ionosphere, marked by the westward traveling surge, maps to the center of the near-Earth magnetotail. In contrast, in the BLD model the surge maps to a portion of the velocity-shear zone separating the duskside CPS and LLBL. The outward field-aligned currents associated with the wedge-type current structures therefore thread the PSBL close to the Earth and only reach the plane of the neutral sheet near the flanks of the magnetotail some tens of Earth radii away from Earth.

At first sight, it may seem to be very difficult to reconcile these two frameworks with one another. In fact, it is highly unlikely that such a reconciliation can take place if typical observations of episodes of negative B_z and antisunward plasma flows in the region

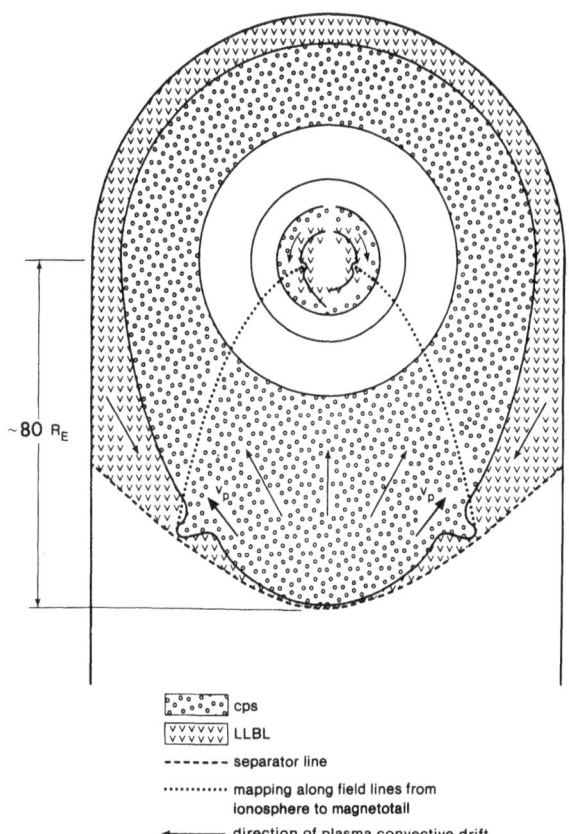

~80 R_E

⬚ cps	
⩔ LLBL	
-------	separator line
....	mapping along field lines from ionosphere to magnetotail
⬅	direction of plasma convective drift

Figure 8—Diagram illustrating the development of a Kelvin-Helmholtz instability at the interface between the low-latitude boundary layer (LLBL) and the central plasma sheet (CPS), triggered by a burst of magnetic field line reconnection at a distant neutral line. The instability develops anywhere along the boundary where the growth rate is adequate. Although only one waveform is shown on each velocity-shear zone, it is possible to have several such forms if the momentum density contrast across the velocity-shear zone is adequate for great distances along the shear zone. Field-aligned current flows from the ionosphere to the evening sector shear zone in the magnetotail, acting to stabilize the instability. (After *Rostoker* [1987].)

earthward of $X = -20R_E$ continue to be interpreted by the two schools of thought in their present fashions. It is interesting to note that some synthesis of the two frameworks is possible if we assume that the near-Earth neutral line is typically antisunward of $X = -20R_E$. Such an approach is not unreasonable in the light of the observations of *Cattell and Mozer* [1984] and, more recently, *Baumjohann et al.* [1989] who utilized ISEE and AMPTE data, respectively, to reach the conclusion that any near-Earth neutral line is typically outside $X = -20R_E$. If the near-Earth line lies outside of $-20R_E$, the BLD model can be retained while satisfying the contention of the NENL proponents that reconnection at such a neutral line

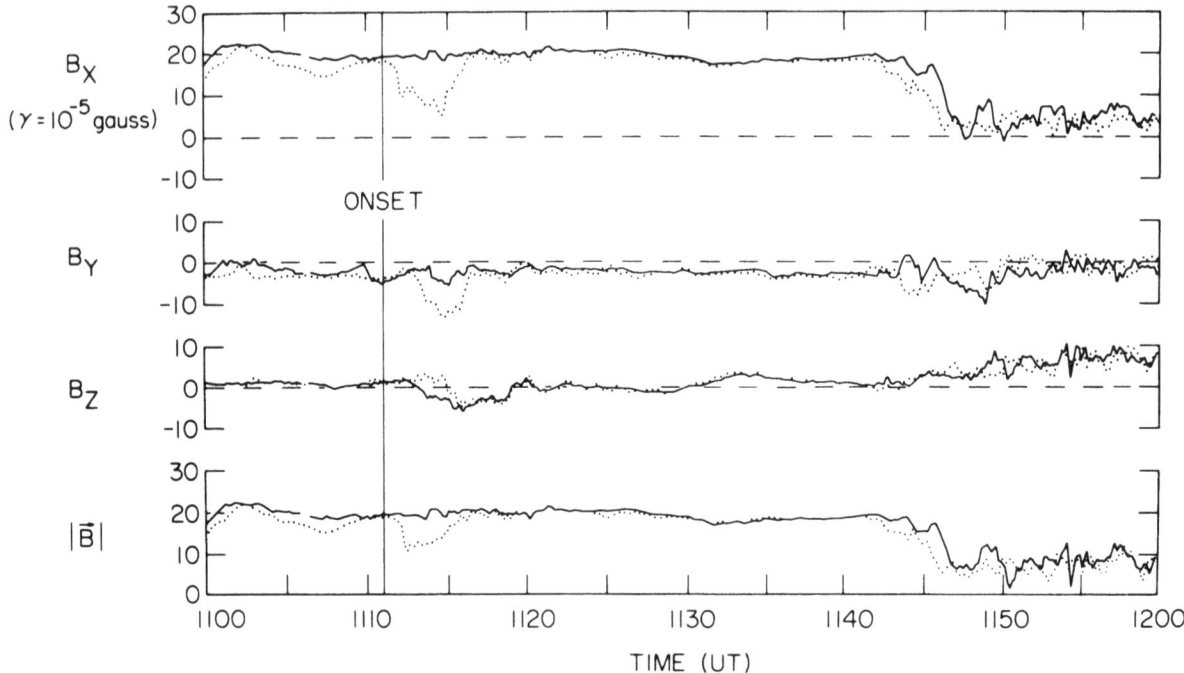

Figure 9—Magnetic field perturbations detected by the ISEE 1 and 2 satellites in the dusk sector magnetotail shortly after the onset of a substorm expansive phase at approximately 1111 UT on April 24, 1979. Despite the fact that the two satellites are separated by less than 3000 km, they detect significantly different magnetic field signatures. ISEE 2 was closer to the neutral sheet, closer to the center of the tail, and closer to the Earth than ISEE 1. The unique character of these magnetic field perturbations finds two separate explanations in the boundary layer dynamics (BLD) model [*Rostoker and Eastman*, 1987] and in the near-Earth neutral line (NENL) model [*Hones et al.*, 1986].

leads to plasmoid formation in the distant tail. Such an accommodation is possible because the satellite data on which the framework for the BLD is built is confined to inside $X = -22R_E$. While the neutral line might be considerably tailward of $-22R_E$, the data presently available do not permit that contention to be verified or denied.

The final question that one might be tempted to ask is, why should the plasmoid structure exist at all? The BLD model does require one neutral line from which fast plasma flows can emerge during episodes of reconnection, in order to provide adequate velocity shear at the CPS/LLBL interface to permit the Kelvin-Helmholtz instability to grow in an appropriate time scale (viz. a growth time of a few minutes at most). However, the BLD model does not require a second neutral line in order to meet the observational criteria. The major impetus behind the belief in a plasmoid as a typical structure in the tail during substorm activity lies in the observation of antisunward flows and bipolar magnetic signatures in the distant tail as measured by ISEE 3 [*Hones et al.*, 1984]. The observations of isotropic energetic particle fluxes during the occurrence of the magnetic signature identified as the bipolar structure [*Scholer et al.*, 1984] suggests the presence of a closed-field-line structure characteristic of a plasmoid. Thus,

on first sight, it might not seem unreasonable to accept the plasmoid hypothesis with the proviso that the near-Earth neutral line is normally located outside $X = -20R_E$. However, there is one aspect of this scenario which seems puzzling and requires some further thought.

Figure 11 shows a schematic diagram (not to scale) of the plasma populations of the magnetotail and their projection on the auroral ionosphere. In conventional views of the mapping between the magnetosphere and ionosphere, the CPS is thought to map to the auroral oval, with the inner edge of the CPS mapping specifically to the equatorward edge of the oval. The neutral line is then associated with the region of breakup arcs in the ionosphere, viz. the Harang discontinuity in the evening sector. The fact that a satellite in the CPS normally detects slow earthward convective motion even during times of demonstrable substorm expansive phase activity has led to the contention that the neutral line does not extend across the entire tail width, but rather occupies some region of limited azimuthal extent. In this manner, it is possible to contend that a satellite observing slow earthward convective drift in the CPS during a substorm expansive phase is simply not in the sector occupied by the neutral line in which the fast antisunward motion of the plasmoid particles might be

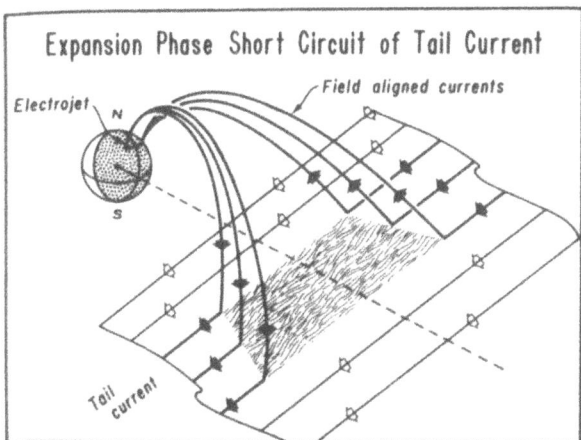

Figure 10—View of the substorm current wedge associated with the expansive phase of a magnetospheric substorm (Courtesy, R. L. McPherron).

detected. The ISEE 3 results of *Slavin et al.* [1985] suggest that, earthward of about $-100R_E$, CPS plasma normally drifts sunward. If this is the case, there should exist two velocity-shear zones at the interface between the azimuthally confined plasmoid and the earthward drifting plasma on either side of it. While the plasmoid itself is decoupled from the auroral ionosphere, the velocity shear will lead to distortion of the electric field adjacent to the interface between the plasmoid and the normal CPS. In fact, for the velocities of hundreds of km s^{-1} attributed to plasmoid ejection into the solar wind, one might well expect a Kelvin-Helmholtz instability to be excited along the interfaces between the plasmoid and the CPS. In addition, if any field-aligned current flows between the ionosphere and the magnetosphere in response to the appearance of space charge associated with the development of the plasmoid/CPS velocity-shear zone, those currents will be in the opposite sense to those expected for the substorm current wedge. That is what might be expected to occur and to have ionospheric signatures if, indeed, the plasmoid is a typical feature of the substorm-disturbed magnetotail. There is no evidence at present that these features have been detected, but demonstration of their presence would favor the existence of plasmoids.

5. CONCLUSIONS

We have first discussed the modern definition of a magnetospheric substorm, taking care to set it in the context of the evolution of the term. We then outlined the basic features of the near-Earth neutral line and boundary layer dynamics frameworks in which the physics of the substorm phenomenon is now being discussed. We have emphasized that the BLD framework evolved from ground-based observations of the auroral features and particles and field of the ionosphere complemented by some in situ spacecraft observations. In

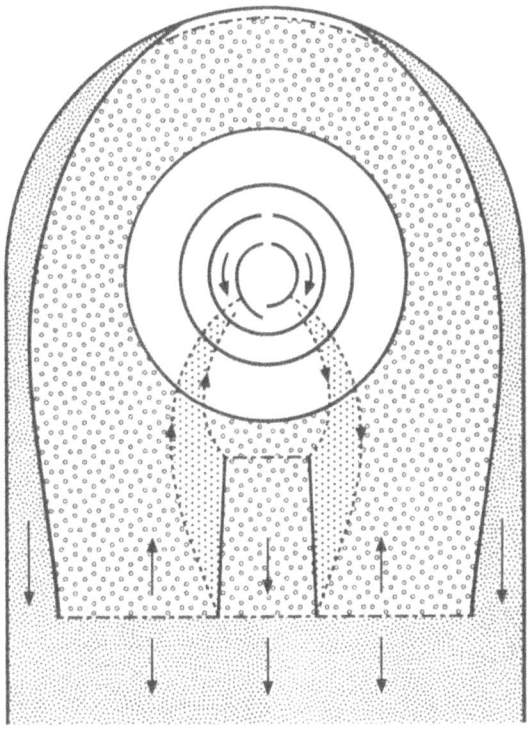

Figure 11—Schematic diagram showing the projection of a plasmoid on the plane of the neutral sheet. The velocity-shear zones at the interface between the plasmoid and the adjacent CPS ought to have some ionospheric signature ranging from field-aligned current growth to wave action that should have some north-south trace in the ionosphere. Evidence for the signatures of either of these two phenomena would be consistent with the correctness of the plasmoid hypothesis.

contrast, the NENL model evolved from spacecraft observations complemented by ground observations of geomagnetic and auroral fluctuations. It is not surprising, therefore, that the BLD model emphasizes the azimuthal structure of the auroral forms and associated currents and therefore is often portrayed using mappings onto the plane of the neutral sheet (i.e., the X-Y plane). In contrast, the NENL model is always demonstrated using projections onto the noon-midnight meridian plane (i.e., the X-Z plane) and thus this model rarely addresses the questions on the azimuthal structure of the auroral oval.

Both frameworks still require considerable study to try to find some key observational features that can lead to one framework being favored over the other. However, as we have mentioned, there is still hope that a synthesis can be found incorporating the essential features of both frameworks. At first sight, such a synthesis seems possible only if the near-Earth neutral line is normally located beyond $X = -20R_E$.

ACKNOWLEDGMENT—This research was supported by the Natural Sciences and Engineering Research Council of Canada.

REFERENCES

Akasofu, S.-I., "The Development of the Auroral Substorm," *Planet. Space Sci.*, **12**, 273 (1964).

Akasofu, S.-I., S. Chapman, and C.-I. Meng, "The Polar Electrojet," *J. Atmos. Terr. Phys.*, **30**, 227 (1965).

Akasofu, S.-I., E. W. Hones, Jr., M. D. Montgomery, S. J. Bame, and S. Singer, "Association of Magnetotail Phenomena with Visible Auroral Features," *J. Geophys. Res.*, **76**, 5985 (1971).

Axford, W. I., and C. O. Hines, "A Unifying Theory of High-Latitude Geophysical Phenomena and Geomagnetic Storms," *Can. J. Phys.*, **39**, 1443 (1961).

Baumjohann, W., and G. Haerendel, "Dayside Convection, Viscous Interaction and Magnetic Merging," in *Solar Wind-Magnetosphere Coupling*, Y. Kamide and J. A. Slavin, eds., Terra Scientific Publ. Co., Tokyo, p. 415 (1986).

Baumjohann, W., G. Paschmann, and C. A. Cattell, "Average Plasma Sheet Properties in the Central Plasma Sheet," *J. Geophys. Res.*, **94**, 6597 (1989).

Birn, J., "Self-Consistent Magnetotail Theory: General Solution for the Quiet Tail with Vanishing Field-Aligned Currents," *J. Geophys. Res.*, **84**, 5143 (1979).

Birn, J., "Three-Dimensional Computer Modeling of Dynamic Reconnection in the Magnetotail: Plasmoid Signatures in the Near and Distant Tail," in *Magnetic Reconnection in Space and Laboratory Plasmas*, E. W. Hones, Jr., ed., American Geophysical Union, Washington, DC, p. 264 (1984).

Birn, J., and E. W. Hones, Jr., "Three-Dimensional Computer Modeling of Dynamic Reconnection in the Geomagnetic Tail," *J. Geophys. Res.*, **86**, 6802 (1981).

Birn, J., E. W. Hones, Jr., and K. Schindler, "Field-Aligned Plasma Flow in MHD Simulations of Magnetotail Reconnection and the Formation of Boundary Layers," *J. Geophys. Res.*, **91**, 11,116 (1986).

Block, L., "A Double Layer Review," *Astrophys. Space Sci.*, **55**, 59 (1978).

Camidge, F. P., and G. Rostoker, "Magnetic Field Perturbations in the Magnetotail Associated with Polar Magnetic Substorms," *Can. J. Phys.*, **48**, 2002 (1970).

Cattell, C. A., and F. S. Mozer, "Substorm Electric Fields in the Earth's Magnetotail," in *Magnetic Reconnection in Space and Laboratory Plasmas*, E. W. Hones, Jr., ed., American Geophysical Union, Washington, DC, p. 208 (1984).

Chiu, Y. T., and M. Schulz, "Self-Consistent Particle and Parallel Electric Field Distributions in the Magnetospheric-Ionospheric Auroral Region," *J. Geophys. Res.*, **83**, 629 (1978).

Coroniti, F. V., "Explosive Tail Reconnection: The Growth and Expansive Phases of Magnetospheric Substorms," *J. Geophys. Res.*, **90**, 7427 (1985).

Craven, J. D., and L. A. Frank, "Diagnosis of Auroral Dynamics Using Global Auroral Imaging with Emphasis on Large-Scale Evolutions," *this volume*.

Dungey, J. W., "Interplanetary Magnetic Field and the Auroral Zones," *Phys. Rev. Lett.*, **6**, 47 (1961).

Eastman, T. E., L. A. Frank, and C. Y. Huang, "The Boundary Layers as the Primary Transport Regions of the Earth's Magnetotail," *J. Geophys. Res.*, **90**, 9541 (1985).

Fairfield, D. H., and N. F. Ness, "Configuration of the Geomagnetic Tail During Substorms," *J. Geophys. Res.*, **75**, 7032 (1970).

Goertz, C. K., and R. A. Smith, "Thermal Catastrophe Model of Substorms," *J. Geophys. Res.*, **94**, 6581 (1989).

Harel, M., R. A. Wolf, P. H. Reiff, R. W. Spiro, W. J. Burke, F. J. Rich, and M. Smiddy, "Quantitative Simulation of a Magnetospheric Substorm 1. Model Logic and Overview," *J. Geophys. Res.*, **86**, 2217 (1981).

Heikkila, W. J., "Penetration of Particles Into the Polar Cap and Auroral Regions," in *Critical Problems of Magnetospheric Physics*, E. R. Dyer, Jr., ed., IUCSTP, National Acad. Sci., Washington, DC, p. 67 (1972).

Hines, C. O., "The Energization of Plasma in the Magnetosphere: Hydromagnetic and Particle-Drift Approaches," *Planet. Space Sci.*, **10**, 239 (1963).

Hones, E. W., Jr., "Plasma Sheet Behaviour During Substorms," in *Magnetic Reconnection in Space and Laboratory Plasmas*, Geophysical Monograph 30, E. W. Hones, Jr., ed., American Geophysical Union, Washington, DC, p. 178 (1984).

Hones, E. W., Jr., and A. B. Galvin, "Poleward Motion of Auroral Structures," *this volume*.

Hones, E. W., Jr., S.-I. Akasofu, S. J. Bame, and S. Singer, "Poleward Expansion of the Auroral Oval and Associated Phenomena in the Magnetotail During Auroral Substorms," *J. Geophys. Res.*, **76**, 8241 (1971).

Hones, E. W., Jr., J. R. Asbridge, S. J. Bame, and S. Singer, "Substorm Variations of the Magnetotail Plasma Sheet from $X_{SM} \simeq -6R_E$ to $X_{SM} \simeq -60R_E$," *J. Geophys. Res.*, **78**, 109 (1973).

Hones, E. W., Jr., T. A. Fritz, J. Birn, J. Cooney, and S. J. Bame, "Detailed Observations of the Plasma Sheet During a Substorm on April 24, 1979," *J. Geophys. Res.*, **91**, 6875 (1986).

Hones, Jr., E. W., D. N. Baker, S. J. Bame, W. C. Feldman, J. T. Gosling, D. J. McComas, R. D. Zwickl, J. A. Slavin, E. J. Smith, and B. T. Tsurutani, "Structure of the Magnetotail at 220 R_E and Its Response to Geomagnetic Activity," *Geophys. Res. Lett.*, **11**, 5 (1984).

Hori, N., and T. Yamamoto, "On the Divergence of the Ionospheric Electric Field Around the Westward Traveling Surge," *Planet. Space Sci.*, **35**, 1489 (1987).

Huang, C. Y., D. G. Mitchell, and L. A. Frank, "Rapid Plasma Sheet Heating During Substorm Activity" (abstract), *Twenty-Seventh Plenary Meeting of The Committee on Space Research*, Helsinki, Finland (1988).

Kan, J. R., and W. Sun, "Simulation of the Westward Traveling Surge and Pi2 Pulsations During Substorms," *J. Geophys. Res.*, **90**, 10,911 (1985).

Kan, J. R., L. Zhu, and S.-I. Akasofu, "A Theory of Substorms: Onset and Subsidence," *J. Geophys. Res.*, **93**, 5624 (1988).

Kan, J. R., L. Zhu, A. T. Y. Lui, and S.-I. Akasofu, "A Magnetosphere-Ionosphere Coupling Theory of Substorms Including Magnetotail Dynamics," *this volume*.

Kisabeth, J. L., and G. Rostoker, "The Expansive Phase of Magnetospheric Substorms, 1. Development of the Auroral Electrojets and Auroral Arc Configuration During a Substorm," *J. Geophys. Res.*, **79**, 972 (1974).

Lee, L. C., Z. F. Fu, and S.-I. Akasofu, "A Simulation Study of Forced Reconnection Processes and Magnetospheric Storms and Substorms," *J. Geophys. Res.*, **90**, 10,896 (1985).

Lysak, R. L., and C. T. Dum, "Dynamics of Magnetosphere-Ionosphere Coupling Including Turbulent Transport," *J. Geophys. Res.*, **88**, 365 (1983).

McPherron, R. L., "Substorm Related Changes in the Geomagnetic Tail: The Growth Phase," *Planet. Space Sci.*, **20**, 1521 (1972).

Obayashi, T., and A. Nishida, "Large Scale Electric Field in the Magnetosphere," *Space Sci. Rev.*, **8**, 3 (1968).

Ogino, T., "A Three-Dimensional MHD Simulation of the Interaction of the Solar Wind with the Earth's Magnetosphere: The Generation of Field-Aligned Currents," *J. Geophys. Res.*, **91**, 6791 (1986).

Ogino, T., R. J. Walker, M. Ashour-Abdalla, and J. M. Dawson, "An MHD Simulation of the Effects of the Interplanetary Magnetic Field B_y Component in the Interaction of the Solar Wind with the Earth's Magnetosphere During Southward Interplanetary Magnetic Field," *J. Geophys. Res.*, **91**, 10,029 (1986).

Paschmann, G., I. Papamastorakis, W. Baumjohann, N. Sckopke, C. W. Carlson, B. U. O. Sonnerup, and H. Luhr, "The Magnetosphere for Large Magnetic Shear: AMPTE/IRM Observations," *J. Geophys. Res.*, **91**, 10,099 (1986).

Perreault, P., and S.-I. Akasofu, "A Study of Geomagnetic Storms," *Geophys. J. R. Astron. Soc.*, **54**, 547 (1978).

Pilipp, W., and G. Morfill, "The Plasma Mantle as the Origin of the Plasma Sheet," in *Magnetospheric Particles and Fields*, B. M. McCormac, ed., D. Reidel Publ. Co., Dordrecht, The Netherlands, p. 55 (1976).

Pytte, T., R. L. McPherron, and S. Kokubun, "The Ground Signatures of the Expansion Phase During Multiple Onset Substorms," *Planet. Space Sci.*, **24**, 1115 (1976).

Rostoker, G., "Classification of Polar Magnetic Disturbances," *J. Geophys. Res.*, **74**, 5161 (1969).

Rostoker, G., "The Kelvin-Helmholtz Instability and Its Role in the Generation of the Electric Currents Associated with Ps6 and Westward Traveling Surges," in *Magnetotail Physics*, A. T. Y. Lui, ed., Johns Hopkins University Press, Baltimore, p. 169 (1987).

Rostoker, G., and R. Boström, "A Mechanism for Driving the Gross Birkeland Current Configuration in the Auroral Oval," *J. Geophys. Res.*, **81**, 235 (1976).

Rostoker, G., and T. E. Eastman, "A Boundary Layer Model for Magnetospheric Substorms," *J. Geophys. Res.*, **92**, 12,187 (1987).

Rostoker, G., S.-I. Akasofu, W. Baumjohann, Y. Kamide, and R. L. McPherron, "The Roles of Direct Input of Energy from the Solar Wind and Unloading of Stored Magnetotail Energy in Driving Magnetospheric Substorms," *Space Sci. Rev.*, **46**, 93 (1987).

Rostoker, G., S.-I. Akasofu, J. Foster, R. A. Greenwald, Y. Kamide, K. Kawasaki, A. T. Y. Lui, R. L. McPherron, and C. T. Russell, "Magnetospheric Substorms—Definition and Signatures," *J. Geophys. Res.*, **85**, 1663 (1980).

Scholer, M., G. Gloeckler, B. Klecker, F. M. Ipavich, D. Hovestadt, and E. J. Smith, "Fast Moving Plasma Structures in the Distant Magnetotail," *J. Geophys. Res.*, **89**, 6717 (1984).

Shepherd, G. G., and J. S. Murphree, "Diagnosis of Auroral Dynamics Using Global Auroral Imaging with Emphasis on Localized and Transient Features," *this volume*.

Slavin, J. A., E. J. Smith, D. G. Sibeck, D. N. Baker, R. D. Zwickl, and S.-I. Akasofu, "An ISEE-3 Study of Average and Substorm Conditions in the Distant Magnetotail," *J. Geophys. Res.*, **90**, 10,875 (1985).

Sugiura, M., and J. P. Heppner, "The Earth's Magnetic Field," in *Introduction to Space Science*, W. N. Hess, ed., Gordon and Breach, New York, p. 45 (1965).

Thompson, W. B., "Parallel Electric Fields and Shear Instabilities," *J. Geophys. Res.*, **88**, 4805 (1983).

Tighe, W. G., and G. Rostoker, "Characteristics of Westward Traveling Surges During Magnetospheric Substorms," *J. Geophys.*, **50**, 51 (1981).

Walker, R. J., and T. Sato, "Externally Driven Magnetic Reconnection," in *Magnetic Reconnection in Space and Laboratory Plasmas*, E. W. Hones, Jr., ed., American Geophysical Union, Washington, DC, p. 272 (1984).

Wiens, R. G., and G. Rostoker, "Characteristics of the Development of the Westward Electrojet During the Expansive Phase of Magnetospheric Substorms," *J. Geophys. Res.*, **80**, 2109 (1975).

V-2. DIAGNOSIS OF AURORAL DYNAMICS USING GLOBAL AURORAL IMAGING WITH EMPHASIS ON LARGE-SCALE EVOLUTIONS

J. D. Craven* and L. A. Frank*

The spatial extent of the auroral oval and polar cap for typical auroral conditions covers an area of about 20×10^6 km^2, for which a minimum of more than 20 well-placed ground stations is required for full spatial coverage, even if significant difficulties such as sunlight, moonlight, geography, and meteorology are neglected. As shown here, auroral imaging with high-altitude spacecraft provides substantial advantages relative to ground-based techniques. Variations in the dimensions of the auroral oval with changes in the sign of the Z component of the interplanetary magnetic field (IMF) are easily observed. Auroral luminosities all along the auroral oval increase within minutes following arrival at the magnetosphere of a shock in the interplanetary medium. It appears that the dawn-dusk motion of the large-scale transpolar arc of a theta aurora observed in the northern (southern) polar cap is in the same (opposite) direction as the B_y component of the IMF. It is shown that the onsets of auroral substorms occur within a range of less than 3.5 hours of magnetic local time centered at 2250 MLT. This is nearly identical to the statistically determined location for the initial response to substorm onset at the orbits of geosynchronous spacecraft. The auroral bulge does not always expand symmetrically in the east-west direction from the position of substorm onset, but can progress preferentially into either the evening or morning sector.

1. INTRODUCTION

Ground-based auroral scientists are hampered in their investigations of the aurora by viewing limitations imposed by geography, meteorology, and season. Observations are made at night in the high-latitude, polar regions (hence a preference for the winter season), in the absence of significant cloud cover. In addition, the desire for minimum interference by moonlight can preclude observations near full moon. A single observer can view an area of about 2×10^6 km^2 at 120-km altitude when provided with an unobstructed view of the horizon above an elevation angle of 5°. This area is small in comparison to the 21×10^6 km^2 area of the auroral zone and polar cap at magnetic latitudes >67°. Hence more than 10 well-placed ground-based observing sites are required to carry out a 24-hours/day observation program within the polar cap and auroral oval in the dark hemisphere. More than 20 observing sites are required for a more reasonable minimum elevation angle of 10°.

While a campaign to uniformly monitor the polar region has never been attempted, large and well-organized campaigns have taken place in which numerous ground-based sites were distributed at polar latitudes. The International Geophysical Year (IGY) of 1957–1958 [Annals of the IGY] is an outstanding example of such an effort at the international level. Auroral observations acquired during the IGY are responsible for the significant increase in our knowledge

edge of the spatial distribution of the aurora during the early 1960s, including the description of the continuous distribution of emissions encircling the magnetic pole, i.e., the auroral oval [see Feldstein and Galperin, 1985, and references therein], and the formulation of the first global-scale description of auroral dynamical behavior, the auroral substorm [Akasofu, 1963, 1964].

With the advent of orbital observing platforms during the IGY it was quickly recognized that images from space would provide the next important step in investigations of the global distributions of aurora. Success in achieving that next step is seen in the auroral images from the spacecraft ISIS 2 [Anger et al., 1973; Shepherd et al., 1973], DMSP [Rogers et al., 1974], KYOKKO [Kaneda, 1979], Dynamics Explorer [Frank et al., 1981], HILAT [Meng and Huffman, 1984], Viking [Anger et al., 1987], and Polar BEAR [Meng et al., 1987]. A brief summary of the principal wavelengths, relevant orbital parameters, and resolutions of these imagers is provided by Frank and Craven [1988].

Here, various aspects of the large-scale spatial distribution of the aurora and its dynamic evolution as viewed from space are discussed. We concentrate on results from auroral images gained at high altitudes with the spacecraft, Dynamics Explorer 1 (DE 1) [Hoffman et al., 1981], for which the spacecraft apogee altitude is 3.65 Earth radii (R_E), and the orbital inclination and period are 90° and 6.83 hours, respectively. The DE 1 auroral imaging instrumentation comprises three imaging photometers, two for observations at visible wavelengths and one at vacuum-ultraviolet

*Department of Physics and Astronomy, The University of Iowa, Iowa City, Iowa 52242-1479.

Auroral Physics, edited by C.-I. Meng, M. J. Rycroft and L. A. Frank. © Cambridge UP 1991

(VUV) wavelengths. Effective full angle of an individual pixel is 0.29°, which provides a linear dimension for a pixel of 32 km per R_E of spacecraft altitude. Hence the images generally provide a spatial resolution of the order of a hundred kilometers. Images to be discussed here, at VUV wavelengths, are acquired with filter passbands for which the principal auroral emission features are (1) the N_2 Lyman-Birge-Hopfield (LBH) bands, (2) N_2 (LBH) and the O I multiplets at about 130.4 and 135.6 nm, and (3) the same as (2) but with the addition of a background of geocoronal Lyman-α radiation. These filters are referred to herein as 1, 2, and 3, respectively, where the number designates the increasing passband of the filter. A detailed discussion of the instrumentation is provided by *Frank et al.* [1981].

The observations discussed are presented in four sections, beginning in the next section with a summary of the basic global-scale dimensions of the auroral distribution and the variations in auroral activity with the north-south orientation of the interplanetary magnetic field (IMF). Discussions of the theta aurora, which represents the large-scale limit of Sun-aligned arcs in the polar cap during periods of northward-directed IMF, follow. The global auroral response to the arrival at Earth of shocks in the interplanetary medium is outlined in the fourth section, using examples for northward and southward IMF. Finally, there is an overview of large-scale auroral features observed during auroral substorms, which are of greatest importance during periods of southward IMF.

2. LARGE-SCALE SPATIAL DISTRIBUTIONS

The large-scale spatial distributions of the aurora at polar latitudes in the two terrestrial hemispheres are readily illustrated with the DE 1 auroral images of Figures 1 and 2. These false-color images of the aurora borealis (Figure 1) and the aurora australis (Figure 2), with overlays of the coastlines, show that, when not distinguishing between discrete and diffuse forms, the aurora can form a continuous ring of optical emissions encircling each magnetic pole. The south magnetic pole (for which the vector direction of the field is toward the pole) is located near the northwest coast of northern Greenland (Figure 1) and the north magnetic pole is positioned near the sunward side of the gap in optical emissions along the transpolar arc (Figure 2). In each case, the high-latitude boundary of the instantaneous auroral oval in magnetic coordinates is nearly represented by a circle [see *Meng et al.*, 1977]. For this northern view in Figure 1 during active auroral conditions, the latitudes of the poleward boundary are about 73° and 74°, respectively, at local noon and midnight,

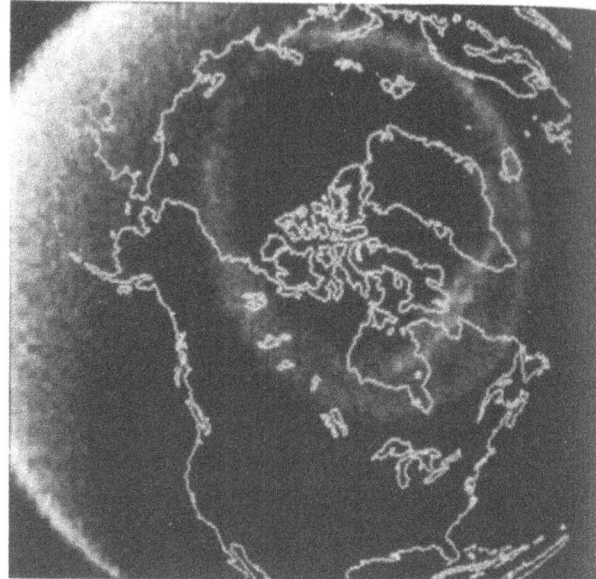

Figure 1—The distribution of auroral luminosities over the North American continent at 0241 UT on November 8, 1981 [after *Frank et al.*, 1985]. A coastline map is superposed on this false-color image of the aurora borealis at ultraviolet wavelengths 123-155 nm (filter 2). Principal emissions detected at these wavelengths are from the multiplets of atomic oxygen at about 130.4 and 135.6 nm and from the LBH bands of molecular nitrogen. For the false-color format, luminosities less than about 1 kR are coded black. For greater luminosities the code progresses from red through orange to yellow. Typical luminosities in the sunlit hemisphere are 20–30 kR, with the largest values observed near the subsolar point. (This figure also appears in color: Plate 26.)

nearly centered on the magnetic pole. The boundary is shifted by several degrees toward the morning sector. For the southern view of Figure 2, with a northward IMF, the poleward boundaries are located at about 80° and 70°, which yields a 5° antisunward offset nearly symmetric in the dawn-dusk plane. This antisunward offset of about 5° is in good agreement with the statistically derived shape of the poleward boundary as reported by *Meng et al.* [1977].

Diameters of the auroral ovals as presented here are approximately equal to the longitudinal width of Canada (5200 km) and to the cross section of the Antarctic continent (4800 km). The areas of circles of these diameters are 21×10^6 km^2 and 18×10^6 km^2, respectively. Measurement of the actual areas encircled by the high- and low-latitude boundaries of the aurora are 12.7 and 24.2×10^6 km^2, respectively, for the aurora of Figure 1, and 9.1 and 19.0×10^6 km^2 for the aurora of Figure 2. These are representative values. Hudson Bay in eastern Canada provides a convenient geological length scale for determining fields-of-views for ground-based observers, as its maximum longitudinal width of about 1000 km is equivalent to the dis-

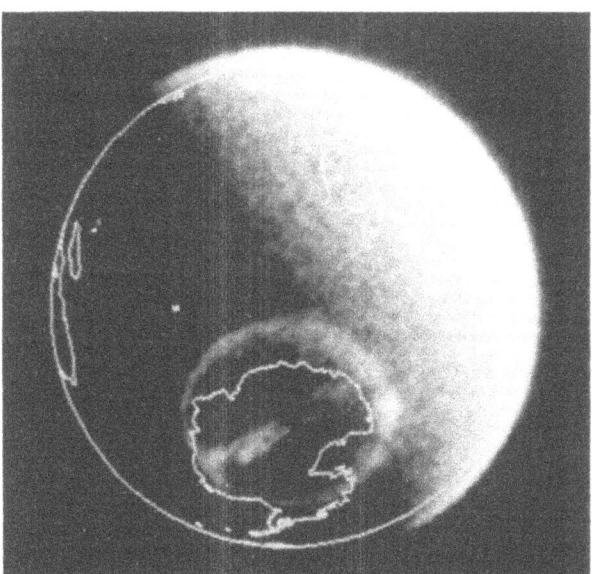

Figure 2—The distribution of auroral luminosities over Antarctica at 0022 UT on May 11, 1983 [after *Frank et al.*, 1985]. A coastline map is superposed on this false-color image of the aurora australis at ultraviolet wavelengths identified in the caption of Figure 1. This image exhibits a theta aurora that comprises the auroral oval and a transpolar arc. The transpolar arc extends into the polar cap from local midnight, traverses the polar cap, and joins with the auroral oval at local noon. (This figure also appears in color: Plate 27.)

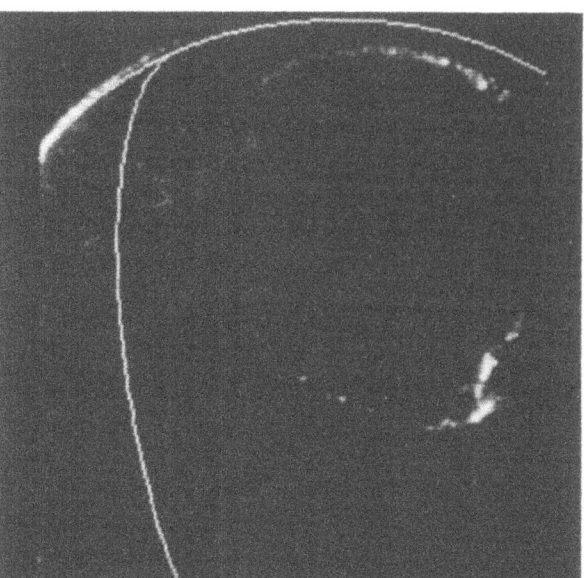

Figure 3—An image of the sunward portion of the northern auroral oval exhibits the gap in discrete aurora that can occur in the local noon sector. The gap is readily detected in the absence of bright diffuse emissions at lower latitudes, which are present in the images of Figures 1 and 2. Three auroral arcs are seen in the evening sector near the lower central part of the image, and bright, active aurora are visible at the lower right as part of the westward boundary of the expanding auroral bulge in a substorm. Contours identifying the Earth's limb and terminator are overlaid on this image (0154 UT, December 8, 1981). The sunlit hemisphere is observed in the left part of the image. Principal emissions are from the LBH bands of molecular nitrogen. (This figure also appears in color: Plate 28.)

tance at 120-km altitude that a ground-based observer views for elevation angles $>10.5°$. The area of a 1000-km diameter circle is 0.79×10^6 km^2, or about 1/25th of the area of the instantaneous oval and polar cap. Additional specific examples of the useful field-of-view for a ground-based observer are provided in Figures 2 of *Frank and Craven* [1988] and *Craven et al.* [1989a].

While the large-scale spatial distribution of these emissions is continuous, distinct local time variations seen in the images are not continuous along the entire auroral oval. In Figure 1, for example, localized, more discrete auroral forms are present in the midnight sector at the high-latitude boundary, while more uniform (diffuse) emissions at lower latitudes extend toward the local noon sector from midnight. Localized features are also noted in the early afternoon hours, where for this image in the northern hemisphere local time advances in the counterclockwise direction. In the conjugate hemisphere (Figure 2) localized, bright emissions are again seen in the early afternoon hours, where in the southern hemisphere local time advances in the clockwise direction. These localized enhancements in the early afternoon hours of local time are probably associated with enhanced electron precipitation and its effects observed at those local times [*Evans*, 1985, and

references therein]. The transpolar arc that bisects the southern auroral oval will be discussed in the next section.

The auroral distribution is not always continuous across the local noon sector, but can display a prominent midday gap [e.g., *Meng*, 1981a, and references therein]. In particular, the discrete auroral arcs can appear to be absent in the noon sector. A clear example of this midday gap in the discrete aurora is shown in Figure 3, for an observation of the northern auroral oval on December 8, 1981, from an altitude of 1.92R$_E$. The emissions are from the N$_2$ (LBH) bands. Three arcs are visible in the evening sector along the lower portion of the image, with the brightest arc at lowest latitude extending westward to a local time of 1320 MLT. A surge is visible at 1920 MLT, in the lower-right portion of the image. As emissions in the late-morning sector do not extend past 0925 MLT, the apparent gap extends over 4 hours of local time. Auroral brightnesses within the gap are at, or near, the photometer sensitivity threshold of 300 R, and the latitudinal widths of the discrete auroral forms in the evening sector are seen at the limiting spatial resolu-

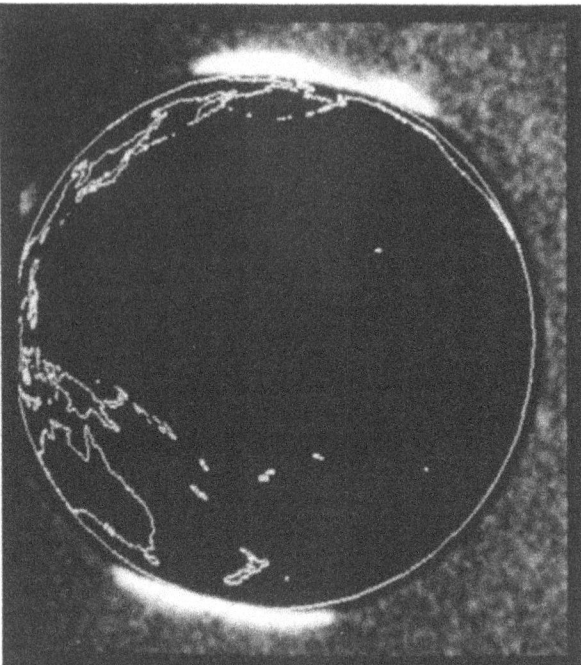

Figure 4—This unique image of Earth from Dynamics Explorer 1 at 1215 UT on March 1, 1982 records aurora in the two polar regions: The aurora borealis at northern latitudes and aurora australis at southern latitudes. Earth's limb and coastal outlines are overlaid on the image. The spacecraft is located within Earth's umbral shadow cone at an altitude of about 20,000 km above the Pacific Ocean. The active aurora in the two hemispheres rise to altitudes of about 370 km above the limb of the solid Earth. Resonantly scattered solar Lyman-α radiation from Earth's extended hydrogen atmosphere is responsible for the diffuse glow beyond Earth's limb. The dark band encircling Earth above the limb is due to absorption of the ultraviolet radiation by the atmosphere at low altitudes. The passband of the filter for this image extends from 117–165 nm. (This figure also appears in color: Plate 29.)

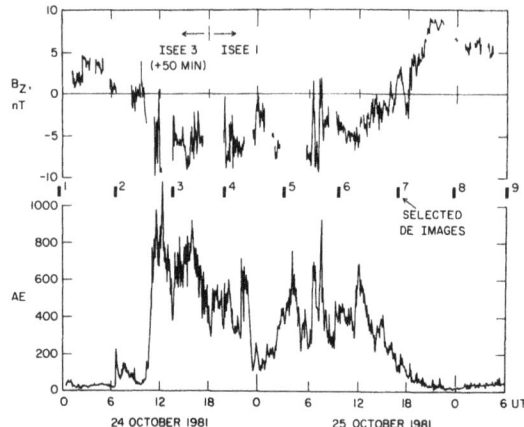

Figure 5—The Z_{GSM} component of the interplanetary magnetic field at Earth (upper panel) and the auroral electrojet index *AE* (lower panel) for the 54-hour interval from 0000 UT on October 24, 1981 through 0600 UT on October 26, 1981. The times of nine selected *DE* auroral images presented in Figure 6 are identified here by the dark vertical bars across the center of the figure.

tion of the photometer. This gap is readily detected here in the absence of bright diffuse emissions at lower latitudes, which are present in the images of Figures 1 and 2.

The auroral images presented in Figures 1 and 2 represent two typical, but not simultaneous views of the aurora in the two hemispheres. Simultaneous views of the aurora in the two hemispheres with a single spacecraft can only be obtained if the spacecraft is located at high altitudes and low latitudes. A view of the northern auroral oval from near the equator has been presented earlier by *Frank and Craven* [1988] during a brief period of spacecraft eclipse early in the DE mission. Several imaging sequences in this same several-week period of spacecraft eclipses in February 1982 have also provided simultaneous images of the two auroral ovals through a fortuitous combination of Universal Time and spacecraft position. The example

presented in Figure 4, with coastal outlines and limb of the solid Earth overlaid on the image, is a particularly outstanding example of such simultaneous imaging from a single spacecraft. The plane of the 90° inclination orbit bisects the image vertically, and acts to highlight the offset of the magnetic dipole axis aligned about 11° clockwise in this image obtained at 1215 UT on March 1, 1982.

The large-scale spatial distribution of auroral emissions varies with time in response to the north-south orientation of the IMF [e.g., *Akasofu*, 1977]. To demonstrate this, consider in Figure 5 variations in the IMF B_z component and the auroral electrojet index *AE* for the 54-hour time interval beginning at 0000 UT on October 24, 1981 (day 297). The IMF is monitored with magnetometers onboard the ISEE 3 spacecraft near the L_1 libration point about $235R_E$ upstream in the solar wind (data courtesy of E. J. Smith), and with ISEE 1 in Earth orbit (data courtesy of C. T. Russell). The correction of 50 min for the transit time of magnetic signatures in the IMF from ISEE 3 to ISEE 1 has been made by comparing limited simultaneous observations from the two spacecraft. While the acquisition of telemetry is not continuous throughout the time interval, sufficient coverage is provided to establish the overall pattern of variations in the north-south orientation. The field is oriented northward during the first six hours, and then becomes more nearly aligned in the X-Y plane (geocentric solar magnetospheric coordinates) until about 1000 UT on the same day. The field then turns and remains southward for the next 30 hours. It returns to a northward orientation after 1600 UT on October 25 (day 298).

276

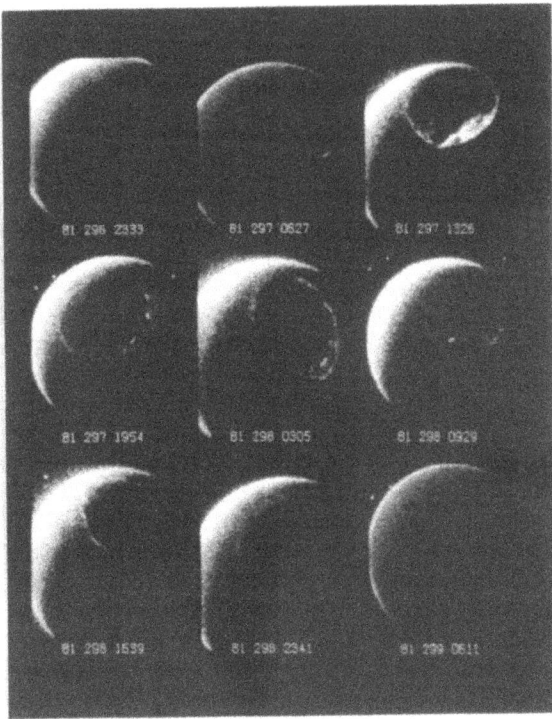

Figure 6—A series of nine *DE* auroral images taken in consecutive orbits to illustrate the gross large-scale spatial distribution of the aurora during a time interval in which the interplanetary magnetic field is first oriented northward (images 1–2), southward (images 3–7), and again northward (images 8–9) (see also Figure 5). Below each image is overlaid the year, day of year, and time (UT) of that image. Identification of the filter for each image is provided in Table 1. (This figure also appears in color: Plate 30.)

TABLE 1. Principal auroral emission features for images of Figure 6.

Image	Day	UT	Filter	Emissions
1	296	2333	2	N_2(LBH) + O I
2	297	0627	1	N_2(LBH)
3	297	1326	3	N_2(LBH) + O I + H(Lyman α)
4	297	1954	2	N_2(LBH) + O I
5	298	0305	3	N_2(LBH) + O I + H(Lyman α)
6	298	0929	2	N_2(LBH) + O I
7	298	1639	3	N_2(LBH) + O I + H(Lyman α)
8	298	2341	2	N_2(LBH) + O I
9	299	0611	1	N_2(LBH)

cal as possible for each image. Below each image is given the year, day of year, and time (UT) of acquisition. Unfortunately, the same filter is not used for each of the nine orbits, but the filters, emission lines, and other supporting information are summarized in Table 1. The first image (1) is located at the upper left in the figure, and time advances from left-to-right and top-to-bottom. The last image (9) is at the lower right. The image times are identified in Figure 5 by numbered black bars across the center of the figure.

A weak transpolar arc is visible in the first image (filter 2, 2333 UT, October 23) of Figure 6, consistent with the earlier northward orientation of the IMF (not shown) [*Frank et al.*, 1986]. Auroral brightnesses are reduced significantly for the second image with filter 1 for N_2(LBH) emissions only. Fortuitously, the time for this image corresponds to the first sharp increase in *AE* and reveals the onset near midnight of a small, localized, and brief auroral substorm shortly after the southward turning. The next five images (3 through 7) capture the auroral oval during ongoing auroral activity typical for periods of southward IMF and initial recovery from an extended period of southward orientation. The third image (filter 3, 1326 UT, October 24) presents the first observation of a substorm expansion phase that began less than 12 min earlier. Particular attention should be given to this substorm onset, because the *AE* index (Figure 5) decreases rapidly with the onset and then recovers by about 1345 UT to values of 600–700 nT observed just prior to this onset. This sharp decrease in *AE* with the onset of substorm activity is not an isolated example, and can be found in observations at other times. This further demonstrates that care must be exercised when using the *AE* index for the identification of individual substorms. It is presumed, but not yet demonstrated, that such decreases are associated with particular spatial distributions of

The response of the auroral electrojet as measured by the *AE* index during this day is also shown in Figure 5 and is typical for the observed variations in the orientation of the IMF. During the intervals of northward IMF at the beginning and end of the 54-hour time interval, the magnitude of the index is much less that 100 nT and does not vary significantly. The first large, impulsive increase in *AE* to 200 nT just after 0600 UT on October 24 is preceded by the decrease in magnitude of B_z to approximately 0 nT. The large increase in *AE* to 1000 nT is closely associated with the decrease in B_z to − 10 nT. Enhanced activity along the auroral electrojet continues for more than 25 hours. The nearly monotonic increase in B_z after 1200 UT on October 25 is accompanied by a concomitant decrease in electrojet activity.

It is not possible in a limited space to provide the full collection of auroral images obtained in the eight orbits of DE 1 during a little more than the 54-hour interval. A compact summary is provided in Figure 6 by selecting one image from each consecutive orbit, where the position along the orbit is as nearly identi-

ground magnetometers and the auroral electrojet during some substorms.

The next four images are obtained during the expansion or recovery phases of distinct substorms, as follows: image 4, late expansion; images 5 and 6, late expansion or early recovery; image 7, late recovery. Discrete auroral forms are faintly observed along the poleward boundary of the auroral oval in the evening-to-midnight sectors in each of these four examples. The last two images (8 and 9) reveal again, for northward IMF orientations, a less dynamic auroral oval similar to that observed in the first two images. Identical filters are used for images 1 and 8 (filter 2) and for images 2 and 9 (filter 1), as summarized in Table 1. The threshold luminosity was established at about 2 kR at image processing for images 3–7, and was lowered to about 1 kR for the weaker auroral luminosities of images 1, 2, 8, and 9.

3. MOTION OF THE THETA AURORA

An early perception of the polar cap in auroral research was that of a dark, relatively unimportant region poleward of the classical discrete aurora that was devoid of interesting auroral phenomena. This view has evolved with careful observations and with improving sensor sensitivities, and now it is recognized that the polar cap aurora represents the optical signature of important plasma processes within the magnetotail-ionosphere system that are most prominent when the IMF is directed northward. Physical details of these processes and an understanding of the mapping to the distant magnetotail are less well developed than those of the diffuse and discrete aurora along the auroral oval, as plasma sources are more readily identifiable in terms of the plasma sheet and the plasma-sheet boundary layer.

Sun-aligned arcs were first studied extensively with ground-based instrumentation [e.g., *Lassen and Danielsen*, 1978, and references therein]. As summarized by Lassen and Danielsen from such observations, the arcs are prominent when the IMF is northward and are observed throughout the polar cap, with a greater occurrence in the morning sector. Low-altitude DMSP spacecraft images have also been used to determine the spatial distribution of Sun-aligned polar cap arcs. For three classifications of arcs observed for quiet magnetic conditions and northward IMF, *Gussenhoven* [1982] has shown that the sign of the IMF B_y component is important in classifying the spatial distributions. Gussenhoven concludes that the arcs occur most frequently in the morning sector (P(2) classification) for $B_y < 0$ and in the evening sector (P(3)) for $B_y > 0$, with arcs near the center of the polar cap (P(1)) associated with small values of B_y. Again, arcs are observed most fre-

quently in the morning sector, but this is because B_y is negative for these observations more frequently than it is positive. A similar investigation by *Ismail and Meng* [1982], with DMSP auroral images, identifies three types of polar cap Sun-aligned arcs: type 1, distinct arcs within the central polar cap; type 2, arcs near the evening and morning sectors of the auroral oval; and type 3, arcs near local midnight that realign along the auroral oval toward the evening sector or toward both the evening and morning sector. The type-1 and type-3 arcs are observed infrequently, i.e., 4% and 0.4% occurrence frequencies, respectively. The type-2 arcs are observed more frequently in the morning sector. *Ismail and Meng* [1982] associate the type-2 arcs with an expansion of the auroral oval to higher latitudes. Arc lengths can exceed 1000 km. Correlations with IMF orientation yield greatest occurrence frequencies for northward-directed fields, and the type-1 arcs are observed more frequently in the northern (southern) hemisphere when B_x is negative (positive). It is important to recall that most DMSP auroral images do not provide a view of the entire polar cap.

With the launch of DE 1 in late 1981, global auroral imaging at 6–12 min temporal resolution became available with imaging sequences of up to 5 hours duration. Early in the analysis program large-scale auroral forms of several hundreds of kilometers width were observed within the polar cap to extend continuously from the midnight to the noon sector across the polar cap [*Frank et al.*, 1982]. There are no reports in the previous literature of the existence of large-scale transpolar arcs such as that shown in Figure 2. These unique auroral forms are observed to move in directions gener-

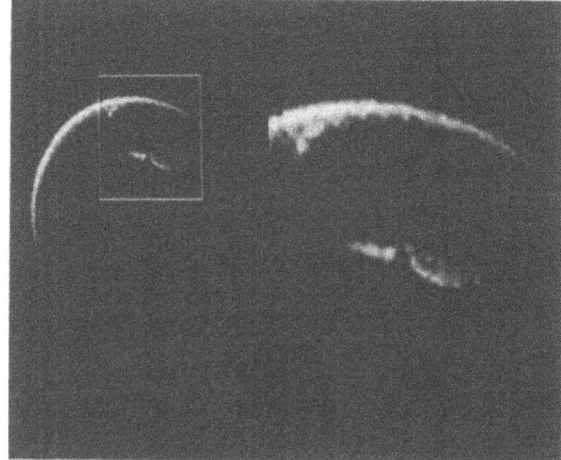

Figure 7—Image of the northern auroral oval and polar cap at 0538 UT on March 25, 1982 [after *Huang et al.*, 1989]. This image is one of 13 for the time interval 0514–0757 UT that exhibit a transpolar arc within the polar cap. Motion of the arc is toward the evening sector during a period in which the IMF B_y component is positive. (This figure also appears in color: Plate 31.)

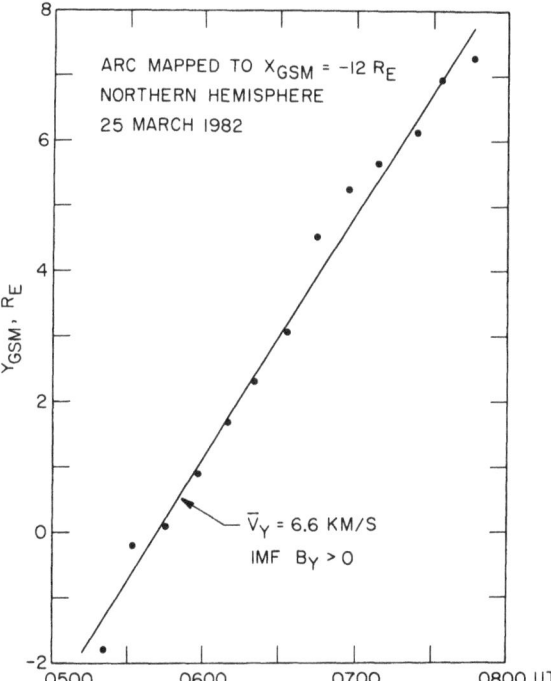

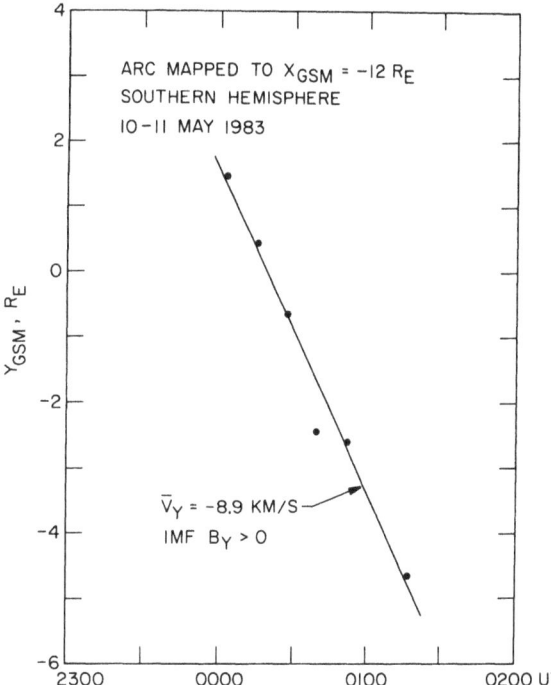

Figure 8—A position on the transpolar arc of March 25, 1982 (Figure 7) mapped to $X_{GSM} = -12 R_E$ in order to summarize its motion. The position is approximately equidistant between the two intersection points of the arc and the auroral oval. The magnetic field model of *Tsyganenko and Usmanov* [1982] is used. Average speed as determined by a least-squares fit to the data is 6.6 km s^{-1}, and the motion is in the direction of the IMF B_y component, which is positive.

Figure 9—A continuation of Figure 8 for the transpolar arc of May 10–11, 1983 observed in the southern hemisphere (see also Figure 2). Average speed is 8.9 km s^{-1} in the direction opposite to the direction of the IMF B_y component, which is positive.

ally perpendicular to the Sun-Earth line; for a negative IMF B_y component the motion in the northern hemisphere is in the dawn direction [*Frank et al.*, 1985]. A second example of motion in the northern polar cap, but with positive B_y has recently been discussed by *Huang et al.* [1989]. The position of the transpolar arc on March 25, 1982 at 0538 UT is shown in the single image at left in Figure 7, with the selected section within the rectangle enlarged at right. As a summary of the arc motion throughout the interval of observations (0514–0757 UT), the location of the arc approximately midway between the two intersections with the auroral oval has been mapped magnetically into the magnetotail using the model field of *Tsyganenko and Usmanov* [1982] for $K_p > 3+$, and the Y_{GSM} position of the arc at $X_{GSM} = -12 R_E$ is shown in Figure 8. Note that for $B_y > 0$ the motion is toward the dusk sector. The average speed is 6.6 km s^{-1}. In contrast, at southern polar latitudes a single outstanding sequence of observations associated with the image of Figure 2 demonstrates that for positive B_y the transpolar arc can move in the dawn direction, opposite to

that in the northern hemisphere. This is demonstrated in Figure 9 with the same magnetic mapping technique. The average speed of the arc is 8.9 km s^{-1}.

Simultaneous measurements of a single bright arc in the two polar caps with low-altitude, polar-orbiting spacecraft are difficult because of the trajectories of the spacecraft relative to the diurnally displaced auroral oval and the offset magnetic dipole. Nevertheless, such observations may have been accomplished by *Gorney et al.* [1986] for a bright arc observed in the northern hemisphere with DMSP F6 and NOAA 7 plasma instruments and in DMSP F6 images, and in the southern hemisphere with the NOAA 6 plasma instrument. An arc is observed in the morning sector in both hemispheres, separated from the auroral oval. However, it is not determined from the observations whether the arc is a transpolar arc. It is unclear how to interpret this observation in light of (1) the DE 1 observations that suggest that transpolar arcs move in opposite directions (hence one would not expect the northern and southern arcs, if the same feature in two polar caps, to both be in the morning sector), and (2) the DMSP observations of *Ismail and Meng* [1982], which suggest that the B_x component may select a preferred hemisphere. One suggestion is that the arc observed by

Gorney et al. [1986] really is a type-2 arc as identified by *Ismail and Meng* [1982] at the poleward boundary of the auroral oval at high latitudes [see *Meng*, 1981b]. Observations similar to those of *Gorney et al.* [1986] are reported by *Obara et al.* [1988] from Viking auroral images in the northern hemisphere and EXOS-C observations of precipitating electrons in the southern polar cap. The Viking images show clearly the presence of a transpolar arc in the morning sector of local time. *Obara et al.* [1985] conclude that an arc is also present in the morning sector of the southern polar cap.

If polar cap arcs map magnetically into the plasma sheet [*Frank et al.*, 1982, 1986; *Peterson and Shelley*, 1984], then it is expected that plasma signatures similar to those observed within the plasma sheet and its boundary layer will be intercepted with spacecraft normally within the lobes of the magnetotail. Such plasma signatures for structures aligned along the Z–X plane are reported by *Huang et al.* [1987]. In contrast to the results of Gorney et al. and Obara et al., *Huang et al.* [1989] report that observations of a transpolar arc in the evening sector of the northern polar cap appear to be accompanied by plasma structures in the morning sector of the southern lobe. Simultaneous optical observations of transpolar arcs in the northern and southern polar caps with Viking and DE 1, respectively, on August 3, 1986 [*Craven et al.*, 1989b] provide the clearest evidence to date for a dawn-dusk asymmetry in the position of transpolar arcs and their motion in opposite directions.

At the limit of weak (subvisual) 630-nm optical emissions, the principal observations continue to be made from the ground. It is concluded that these Sun-aligned arcs within the polar caps arise from sheetlike distributions of precipitating low-energy (hundreds of electron volts) electrons that maintain current continuity at shears in the plasma flow across the polar cap during magnetically quiet periods [*Reiff et al.*, 1978; *Carlson et al.*, 1984]. The Sun-aligned arcs drift in the dawn-dusk plane while the dominant plasma flow direction is in the antisunward direction. Within the velocity shear the flow can be reduced in magnitude, stagnant or sunward [*Carlson et al.*, 1984]. Simultaneous ground-based observations within two 1000-km diameter fields of view demonstrate that these weak, Sun-aligned arcs can extend across a significant fraction of the polar cap, and that they represent "the dominant optical character of the northward IMF polar cap" [*Carlson*, 1988, and as presented at the 1988 Spring AGU Meeting].

In brief review, it appears clear that Sun-aligned arcs represent the dominant discrete auroral feature of the polar cap for magnetic quiet periods of positive B_z, and are present over a wide range of brightnesses and wavelengths associated with variations in characteristic energies of precipitating electrons along velocity shears in the plasma flows. There is no reason to expect different physical processes at the extremes of arc brightness and physical dimensions. Least frequently observed of the Sun-aligned arcs is the large-scale transpolar arc. It is reasonable to associate this feature with the P(1) classification of *Gussenhoven* [1982] and with the type-3 aurora of *Ismail and Meng* [1982]. The transpolar arcs move generally transverse to the Sun-Earth line, with the direction of motion controlled by the sign of the IMF B_y component.

4. RESPONSE TO SHOCKS IN THE INTERPLANETARY MEDIUM

A shock in the interplanetary medium arriving at Earth applies a compressive force that reduces the overall dimensions of the near-Earth magnetospheric cavity and must alter the magnitudes and distributions of currents and plasmas around and within the cavity. One well-established signature of changes in the locations and magnitudes of these currents is an increase in the horizontal component of the magnetic field detected with magnetometers at low-latitude ground stations. The surface field at all local times begins to increase abruptly, in <1 min, in response to the compression [e.g., *Nishida*, 1978]. Of greater significance for the magnetosphere is the subsequent onset of a geomag-

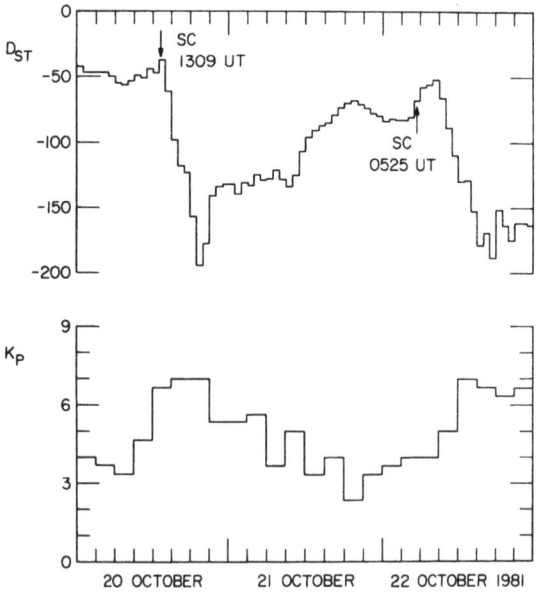

Figure 10—Hourly averages of D_{st} and the three-hour values of K_p for October 20–22, 1981. Onset times are identified for two geomagnetic storm sudden commencements (SC). The direction of the arrow for each SC specifies whether the IMF is oriented southward (down) or northward (up).

netic storm, in which the terrestrial ring current undergoes a rapid enhancement in a matter of hours, the low-latitude magnetic field decreases by hundreds of nanoteslas, and auroral activity increases dramatically at higher latitudes [e.g., *Akasofu*, 1977, and references therein]. For such cases, a preceding sudden increase in the low-latitude magnetic field accompanying the magnetospheric compression is labeled the storm sudden commencement (SC), and the subsequent rapid decrease in the magnetic field is called the main phase of the storm.

There is also an immediate magnetospheric response to the arrival of an interplanetary shock in the form of enhanced energetic (tens of kiloelectron volts) electron precipitation along the auroral oval. These increases are observed with riometers and balloon-borne x-ray detectors [e.g., *Brown et al.*, 1961; *Matsushita*, 1961; *Ortner et al.*, 1962; *Ullaland et al.*, 1970]. Within the magnetosphere a significant increase is noted in the probability of observing energetic particle bursts in the magnetotail and magnetosheath [*Tholen and Armstrong*, 1986]. Increases in auroral luminosities are also reported [*Vorob'yev*, 1974; *Craven et al.*, 1986]. The enhancement can begin simultaneously (< 1 min) with SC onset and last for 3–10 min [*Vorob'yev*, 1974] or longer [*Craven et al.*, 1986]. The magnitude of the enhancement varies from about 10% to factors of >2, depending on the method of detection, and also varies from event to event. The immediate increase in auroral luminosities is not associated with onset of an auroral substorm. The auroral electrojet is also observed to respond to a shock arrival (see the brief summary by *Craven et. al.*, [1986], and references therein).

The case study of two moderate geomagnetic substorms by *Craven et al.* [1986] with auroral images from DE 1 illustrates the immediate large-scale influence on the aurora following the arrival at Earth of a shock in the interplanetary medium. Magnetic activity for geomagnetic storms on October 20 and 22, 1981 is summarized in Figure 10, where the D_{st} and Kp indices are plotted, respectively, in the upper and lower panels. Sudden commencements are detected at 1309 UT on October 20 and at 0525 UT on October 22. The orientation of the IMF is southward before the first SC and during the main phase decrease in D_{st}. In contrast, it is northward for the second event. The main phase onset following the second SC is delayed for more than two hours until the IMF turns southward. It is only then that significant auroral activity begins, as noted coarsely with the three-hour Kp index in the lower part of Figure 10 or with the AE index (not shown). The AE indices and IMF data for these events are given by *Craven et al.* [1986]. The importance of the southward turning of the IMF for the

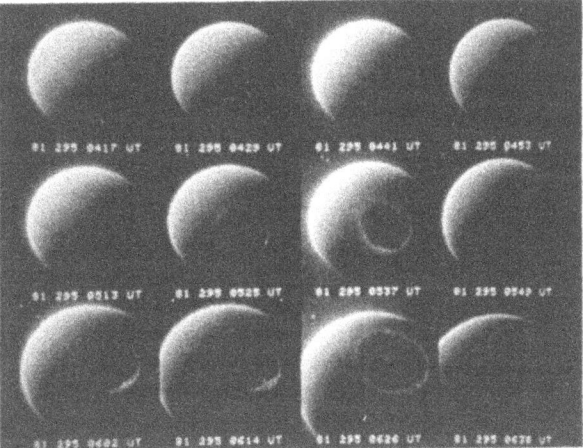

Figure 11—Sequence of 12 consecutive false-color auroral images at ultraviolet wavelengths in the time interval 0417–0650 UT on October 22, 1981 [after *Craven et al.*, 1986]. Increasing luminosities follow the SC at 0525 UT (beginning of the sixth frame). Predominant direction of the IMF B_z component is northward. Below each image is the year, day of year, and UT for the beginning of the 12-min telemetry period for the image. (This figure also appears in color: Plate 32.)

initiation of the main phase of a geomagnetic storm has been shown by *Akasofu* [1977, Figure 5.18].

The auroral response to the magnetospheric compression on October 22, 1981 can be seen in the 12-image sequence of Figure 11. The false-color format is again used in which weaker luminosities of several kilorayleighs are coded in red, and the greater luminosities of the active aurora and from the sunlit hemisphere (about 20 kR) are coded orange to yellow. For these observations the ultraviolet photometer operates in a mode for which an image is obtained with filter 3 and the next image is acquired with filter 2. The two-filter sequence repeats cyclically. The first image, at 0417 UT, is with filter 3, which passes Lyman-α radiation. The weak emissions observed above Earth's limb in the first image of Figure 11 are due to scattering of solar Lyman-α radiation by exospheric hydrogen [e.g., *Rairden et al.*, 1986]. The filter passband for the second image, at 0429 UT, excludes responses due to the Lyman-α radiation.

The IMF orientation prior to and during the shock crossing is northward. The first five images of this sequence (0417–0525 UT) are taken prior to arrival of the shock. Auroral luminosities around the oval increase noticeably in the sixth image at 0525 UT, and remain enhanced for the duration of the imaging sequence. As shown in Figure 12b (from the work of *Craven et al.* [1986]), auroral luminosities increase from about 2 kR before the shock crossing to 6–8 kR within 10 min after the crossing. The increases in luminos-

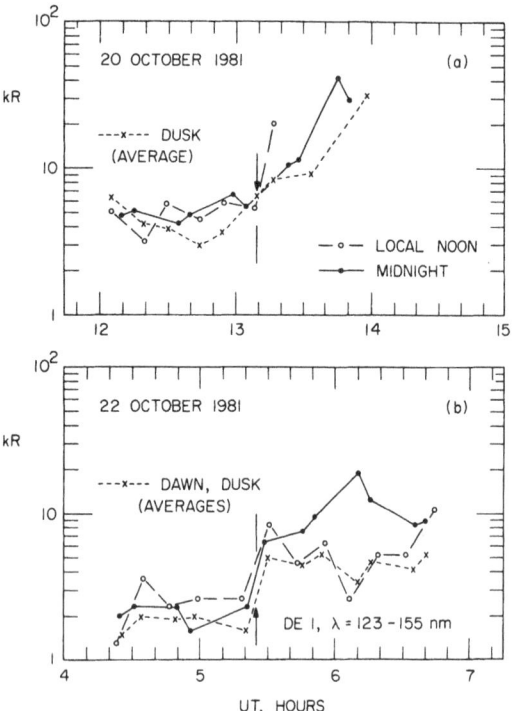

Figure 12—(a) Maximum luminosities of ultraviolet emissions along the auroral oval in the noon, dusk, and midnight sectors for the two-hour interval surrounding the SC at 1309 UT on October 20, 1981 (images not shown). (b) Continuation for October 22, 1981 and the SC at 0525 UT (images shown in Figure 11). Averages are given for the intensities in the dawn and dusk sectors.

ities after the shock crossing are observed more clearly for northward IMF due to the generally lower luminosities along the auroral oval during the period prior to the crossing. An example of this can be seen in Figure 12a for a shock crossing at 1309 UT on October 20 in the presence of a southward IMF. Note that the luminosities are about 4–6 kR prior to the shock crossing. The imaging sequence for this event is provided in the work of *Craven et al.* [1986].

The last four images of Figure 11 demonstrate that the aurora brightens at midnight about 30 min after the shock crossing and a Sun-aligned polar arc begins to form at midnight. The Sun-aligned polar arc then lengthens across the polar cap to local noon at an average speed of 1 km s^{-1}. This sunward expansion of the arc is similar in speed to westward traveling surges that move along preexisting arcs in the late-evening sector. Faint Sun-aligned auroral forms undetected with DE 1 may have been present within the polar cap in advance of the shock impact as part of the established polar cap convection pattern. No significant increases are observed in the magnitude of the AE index during this period of arc formation.

For the single example discussed here in some detail, and for the companion case with southward IMF also discussed in detail by *Craven et al.* [1986], the onset of significant auroral activity about 30 min after shock impact is independent of the sign of B_z and may be related to a time constant for reconfiguration of the geomagnetic tail in responses to changes in the interplanetary medium behind the shock.

5. AURORAL SUBSTORMS

The concept of the auroral substorm as first described by *Akasofu* [1964] provides a general description of a particular class of large-scale dynamical variations observed along the auroral oval. In this description, an auroral substorm begins near local magnetic midnight as a rapid, localized brightening along previously quiet arcs and proceeds with disruption of the ordered distribution of the arcs. A significant point is that the onset begins along an arc, and not within the diffuse auroral emissions equatorward of the arcs. Second, the onset need not occur along the most equatorward arc, though it usually does. The disruption of the longitudinally extended, stable arcs into patches and arc fragments at substorm onset is known as the auroral breakup. The region of enhanced auroral luminosities and rapidly moving auroral forms, the auroral bulge, expands quickly in longitude along the arcs, and less rapidly to higher latitudes. Diffuse emissions appear at lower latitudes. Surge activity at the westward edge of the auroral bulge can expand rapidly westward [e.g., *Akasofu et al.*, 1965, 1966c; *Craven et al.*, 1989a], with average speeds of apparent westward expansion ranging from several hundred to several thousand meters per second. The classical westward traveling surge apparently is not due to the westward motion over thousands of kilometers of a single surge, but is due to the sequential formation of a series of relatively localized surges at progressively greater distances westward along the auroral oval [e.g., *Tighe and Rostoker*, 1981; *Craven et al.*, 1989a]. The westward expansion of the auroral bulge is also observed on occasion to cease abruptly early in the expansion phase and to remain nearly stationary for much of the expansion phase [e.g., *Rostoker et al.*, 1987; *Craven and Frank*, 1987; *Craven et al.*, 1989a]. Eastward expansion of the breakup is also highly variable, and can also expand well into the morning sector [e.g., *Craven and Frank*, 1987] or can stop abruptly early in the expansion phase [*Shepherd et al.*, 1987]. The range of speeds for the poleward expansion is similar to that for westward traveling surges [see *Akasofu et al.*, 1965, 1966a], and the speed varies with time and local time in the expansion phase [e.g., *Craven and Frank*, 1987]. The expansion generally develops through the ongoing for-

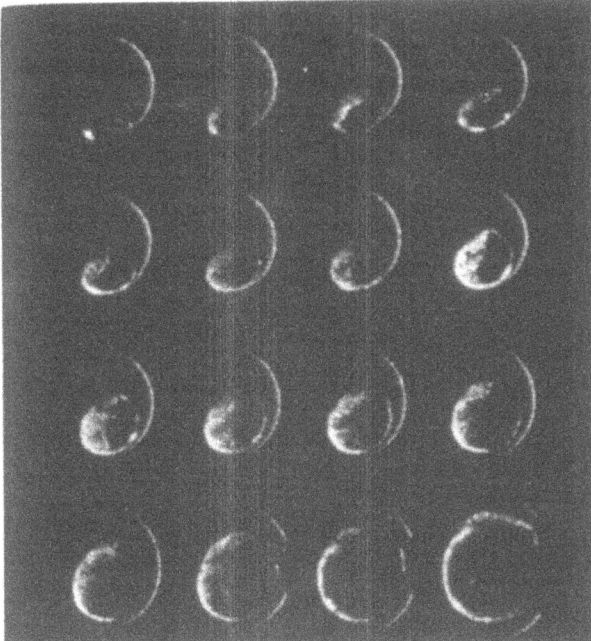

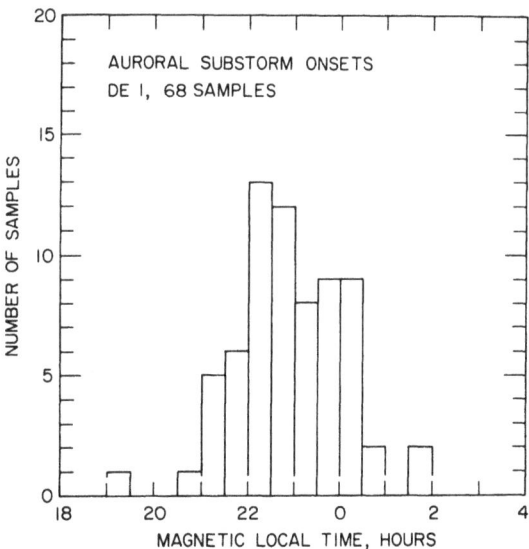

Figure 14—Histogram showing the number of substorm onsets in half-hour increments of magnetic local time along the auroral oval from a set of 68 well-defined substorm onsets identified with DE 1.

Figure 13—A sequence of 16 images of the aurora australis in the time interval 0202–0517 UT on June 13, 1983 [after *Frank and Craven*, 1988]. Intense auroral activity begins in the first image at upper left with a localized brightening (substorm onset) followed by a period of rapid expansion of the aurora in latitude and in longitude. Luminosities increase noticeably by 0326 UT (eighth image), indicating the arrival at Earth of a shock or discontinuity in the interplanetary medium. This auroral activity occurs simultaneously with the main phase decrease of the low-latitude surface magnetic field during a geomagnetic storm. (This figure also appears in color: Plate 33.)

mation of new arcs at progressively higher latitudes [e.g., *Davis and Kimball*, 1960; *Kisabeth and Rostoker*, 1974]. The formation of new arcs does not appear to occur at a uniform rate, but occurs episodically at intervals of 10–15 min [e.g., *Kisabeth and Rostoker*, 1974; *Craven and Frank*, 1987]. These individual, more-rapid poleward advances are believed to be associated with the individual substorm intensifications, or multiple substorm onsets. Equatorward of the most poleward arcs within the auroral bulge, auroral forms are observed to move equatorward [e.g., *Snyder and Akasofu*, 1972]. Equatorward expansion of the aurora takes place at rates of tens to several hundreds of meters per second [see *Akasofu et al.*, 1966b; *Craven and Frank*, 1987], and has been observed in some instances to exceed the rate of poleward expansion [*Snyder and Akasofu*, 1972]. The duration of the expansion phase of the substorm is highly variable, from tens of minutes to hours, and is terminated with the absence of additional new arcs at higher latitudes. The end of this expansion phase is followed by a recovery phase, in which the auroral luminosities diminish, the

poleward and equatorward expansions become contractions, and longitudinally extended quiet arcs begin to reform. The duration of a typical substorm cycle is 2–3 hours. Additional specific details of auroral morphology are given by *Akasofu* [1977] and references therein.

The physical location of substorm onset is difficult to identify from the ground because of the several considerations discussed previously. Low-altitude spacecraft such as DMSP are hampered significantly by the low sample rate of a single image per polar transit, usually about 100 min for the same hemisphere. High-altitude continuous imaging provided by DE 1 and Viking is not so restricted. A survey in search of images of auroral substorm onsets has been made using DE observations for the time interval from September 24, 1981 through February 10, 1982. It is important to limit the survey to those images for which the auroral bulge is still of limited spatial extent (in order to reduce the error in determining the location of onset), and for which subsequent images demonstrate the continuing development of a substorm. The first image of Figure 13 represents a good example of the localized onset, and a second example is provided by *Frank and Craven* [1988, in Figure 12]. Neither of these examples is from the original survey. From the initial survey, a total of 68 onsets that meet the two selection criteria have been identified. For each event, the geographic coordinates of onset have been measured and the corrected geomagnetic local time and latitude have been computed. The local time distribution for these onsets is shown in Figure 14 without restriction on the magnetic lati-

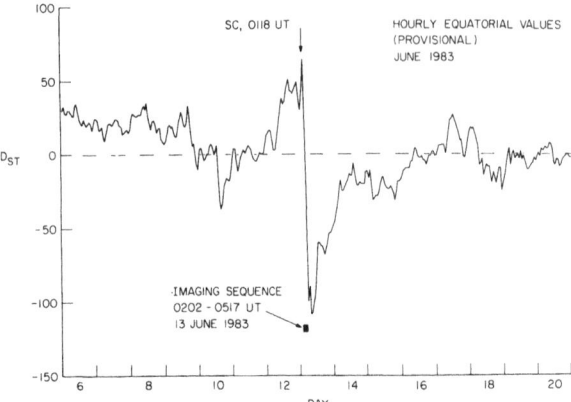

Figure 15—Hourly averages of the D_{st} index for the 15-day interval from June 6–20, 1983. The sudden commencement (SC) is at 0118 UT. The images shown in Figure 13 are from the time interval of the main phase decrease.

tude of onset, which varies from 59°–72°. The median value is 65°. The mean and median values of MLT for the histogram are the same and are 2250 MLT (22.8 hours). The width of the distribution is 3.5 hours for more than two samples per half-hour interval. The average latitudinal width of the auroral bulge for the 68 cases is 1.1°. These observations confirm the well-known but not well-documented fact that auroral substorm onsets are observed most frequently late in the premidnight sector.

An investigation of substorm-related plasma injections and magnetic field reconfigurations at geosynchronous altitudes by *Arnoldy and Moore* [1983] finds that the initial signature of a substorm at the geosynchronous altitude occurs within a limited region of local time centered just before midnight (see Figure 7 of *Arnoldy and Moore* [1983]). Their analysis shows that the time delay between arrival of substorm-related changes at the two spacecraft is dependent on the local times of the spacecraft: to the west (east) of the onset the more easterly GOES 2 (westerly GOES 3) observes the effect first. The spacecraft positions are separated by two hours in local time. At a particular pair of local times for the two spacecraft the time delay is zero and it is presumed that the two spacecraft are located symmetrically in longitude about the local time of onset. Using a simple unweighted least-squares fit of median values in one-hour increments of local time, the data of their Figure 7 yields a local time of 2300 MLT for zero time delay. This local time is nearly identical to the 2250 MLT obtained above for the DE observations, thereby confirming that statistically there is a close spatial relation in local time between auroral substorm onset and the onset of a reconfiguration of current systems within the inner magnetosphere at geosynchronous altitudes.

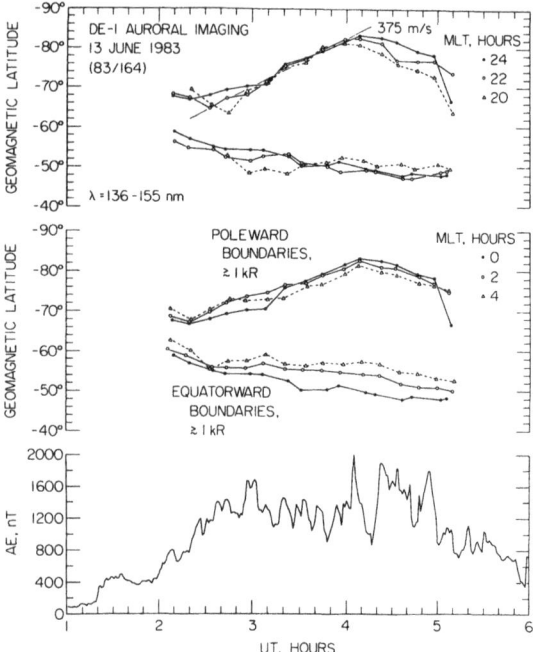

Figure 16—Latitudinal motions of the aurora for the imaging sequence presented in Figure 13 and the auroral activity index *AE* for the time interval 0100–0600 UT. The equatorward and poleward boundaries of the aurora are presented in the upper panel for the three magnetic local times 2000, 2200, and 2400 MLT, and in the center panel for 0000, 0200, and 0400 MLT. The contours for local midnight are reproduced in both panels to assist in evaluating the variations with local time. No sharp increase exists in the *AE* index that clearly is associated with the luminosity enhancement at about 0202 UT, but the index is increasing at that time.

The large-scale latitudinal and longitudinal motions of the aurora in the southern hemisphere for the particularly active period following the onset at about 0202 UT on June 13, 1983 are also shown in Figure 13. The expansion phase continues until about 0410 UT (eleventh image). The development of the terrestrial ring current and the auroral electrojet for this active period are summarized with the D_{st} and *AE* indices, which are plotted, respectively, in Figure 15 and the lower panel of Figure 16. This imaging sequence follows a SC at 0118 UT and occurs during a main phase decrease in a geomagnetic storm and development of an enhanced terrestrial ring current. The intense, brief increase in the luminosities at all local times at 0327 UT (eighth image of Figure 13) is what would be expected following a shock crossing, as discussed in the previous section, thereby suggesting an encounter with a second shock or discontinuity in the interplanetary medium. Accompanying this large-scale enhancement in the luminosities at 0327 UT is a significant, localized perturbation of the auroral oval to lower latitudes near 0540 MLT, (the eighth and ninth images of Figure 13), suggesting a large-scale perturbation in the outer magnetosphere in the dawn sector. (Local time advances in the clockwise direction for these images of the southern hemisphere.) The perturbation extends equatorward by about 4° from the 63° magnetic latitude of the auroral oval in the previous images. Also, a significant localized enhancement in luminosities is seen in the early afternoon sector at 1350 MLT, beginning with the eighth image [see *Evans*, 1985, and references therein]. Luminosities along the auroral oval decline to earlier values after the tenth image.

The latitudinal motions of the aurora during this substorm are presented in Figure 16, with motions at 2000, 2200, and 2400 MLT shown in the upper panel and at 0000, 0200, and 0400 MLT in the center panel. The measurements at midnight are reproduced in each panel to facilitate comparisons. The several significant points are (1) the poleward expansion is reasonably symmetric along the 8 hours of local time at an average speed of 375 m s^{-1}, (2) the equatorward expansion is asymmetric, with an average speed of about 97 m s^{-1} near midnight, and decreases slightly in the morning sector, (3) the recovery phase at high latitudes begins after 0410 UT with no indication of a "poleward leap" [*Hones*, 1985] separating the expansion and recovery phases, (4) the poleward boundary during the recovery phase is complex, with the beginning time for the recovery phase varying with local time [e.g., *Snyder and Akasofu*, 1972; *Kisabeth and Rostoker*, 1974; *Craven and Frank*, 1987], (5) the *AE* index remains at large values of the order of 1600 nT, well into the recovery phase, and (6) the recovery at the equatorward boundary begins in the premidnight sector at about 0440 UT, but is not detected in the postmidnight sector by the end of the imaging sequence at 0517 UT. Not resolved in this sequence at 12-min temporal resolution are the episodic poleward expansions observed in less intense substorms [e.g., *Craven and Frank*, 1987].

The symmetric poleward expansion displayed for this interval of intense auroral activity is not a consistent feature of substorms. To demonstrate this, the spatial distribution of aurora is shown in Figure 17 for a selected image during the expansion phase of each of four substorms. The first image, at 1549 UT, is from a sequence of images on November 4, 1981 for which a substorm onset is identified at about 1525 UT. The greater luminosities of a surge near the western edge of the auroral bulge identify the approximate local time of the onset. The westward edge of the bulge did not advance along the auroral oval during this substorm. The second image at 1123 UT on April 7, 1983 follows a substorm onset at about 1023 UT. The onset

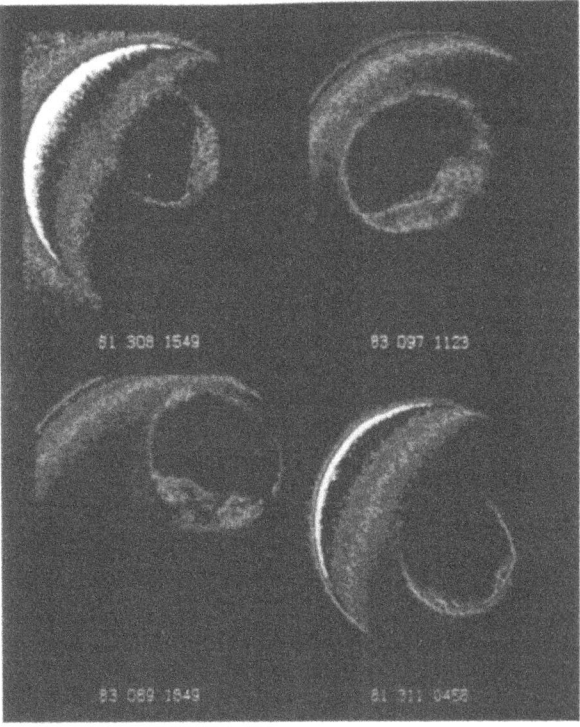

Figure 17—Four auroral images, selected to illustrate variations in the longitudinal distribution of aurora, that can be observed during the expansion phase of substorms. The first image, at upper left, features an expansion that does not proceed westward of 2200 MLT (the surge is nearly stationary) and the auroral bulge expands predominantly into the morning sector. The second image, at upper right, shows an auroral bulge more symmetric about the noon-midnight plane and a surge that advances farther into the evening sector. The auroral distribution in the third image, at lower left, is nearly a mirror image of the first image, with the expansion into the evening sector and nearly no eastward expansion into the morning sector. The last image is an example which auroral activity expands westward rapidly along the auroral oval, and there is almost no signature at midnight of an auroral bulge. For this false-color format luminosities less than about 1 kR are coded black. For greater luminosities the code progresses from blue through green, yellow, and red to a saturation value near 20 kR coded white. (This figure also appears in color: Plate 34.)

occurred near local midnight and the expanding auroral bulge is more nearly symmetric about midnight. The third image at 1849 UT on March 30, 1983 is nearly a mirror image of the first image, with the eastern edge of the auroral bulge remaining nearly stationary, and the auroral bulge expanding into the evening sector. Substorm onset is after 1825 UT. The last image, from a study by *Craven et al.* [1989a], shows that a very small substorm near local midnight can be accompanied by an intense, westward traveling surge. The presence of a westward traveling surge is verified by simultaneous ground observations, and the DE 1 im-

ILLUSTRATIVE EXAMPLES —
POLEWARD EXPANSION OF THE AURORA
IN SUBSTORMS

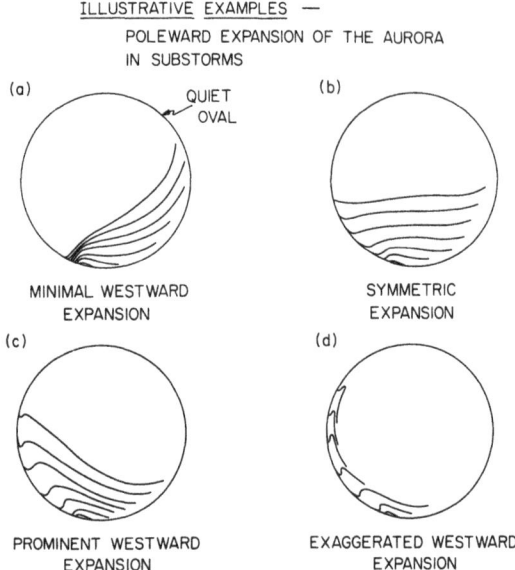

Figure 18—Four examples of poleward motion during the expansion phase of an auroral substorm, to illustrate observed local time variations.

ages show that the auroral bulge at local midnight is small. The four images show that poleward expansion of the auroral bulge need not be symmetric about midnight or the location of onset, but can be highly asymmetric. The four images presented in Figure 17 are selected to demonstrate also that asymmetries are observed in both hemispheres. The first and fourth (second and third) images of the figure were obtained in the northern (southern) hemisphere. The images are all displayed with the dusk sector to the left. It is not yet known whether the orientation of the IMF influences the asymmetric development of the bulge or how the asymmetry in one hemisphere maps to the conjugate hemisphere. Simultaneous imaging in the two hemispheres is required, as is provided by DE 1 and Viking.

Schematic illustrations are provided in Figure 18 to assist in outlining these local time variations. For each of the four examples, in magnetic coordinates, the initial spatial distribution of auroral arcs is denoted by a single, global-scale quiet auroral arc. Recall that the high-latitude boundary of the auroral oval in geographic coordinates maps to a circle in magnetic coordinates [*Meng et al.*, 1977]. Poleward and westward expansions of the auroral bulge are identified by a series of curves representing arcs. No attempt has been made to illustrate the instantaneous rates of expansion. For clarity, the magnitudes of the motions may be exaggerated from appropriate typical values, but the maximum amplitudes are not unrealistic for some extremely intense substorm expansions. Equatorward expansions are not included in these illustrations.

With these limitations in mind, an example of a minimum in westward expansion is illustrated in Figure 18a. In the midnight sector the high-latitude boundary advances poleward in discrete steps as new arcs develop rapidly at higher latitudes with substorm intensifications. Repetitive intensifications of the electrojet are observed near the westward edge of the bulge, but "the surge may grow and decay in confined longitudinal sector without suffering any significant displacement" [*Tighe and Rostoker*, 1981]. There is no reason to preclude even more exaggerated examples of eastward expansions.

The observations of *Craven et al.* [1989a] are schematically illustrated in Figure 18d, for which the prominent feature is a latitudinally confined expansion of surges along the oval. Examples of westward expansions intermediate between the more extreme cases of Figures 18a and 18d are provided in Figures 18b and 18c. A surge would be observed along the auroral oval at the intersection of the preexisting arcs of the quiet oval and the westward edge of the expanding auroral bulge. A ground-based observer at 1800 MLT would observe a surge only if the symmetric expansion of Figure 18b proceeded sufficiently far poleward. An expansion that did not extend as far poleward at midnight would also be observed at 1800 MLT if the westward motion were more prominent (Figure 18c). Repetitive intensifications and poleward advances of the high-latitude boundary of the aurora near local midnight at intervals of 10–15 min [e.g., *Craven et al.*, 1986, 1989a] are similar to the observed rates of intensifications and advances for surges as the westward edge of the bulge [*Wiens and Rostoker*, 1975]. This similarity in repetition rates further supports the inference that the poleward advance of the aurora and intensifications of the surges are directly related [see also *Rostoker et al.*, 1987].

The physical mechanism that controls the east-west asymmetry in a substorm expansion is not known, but the orientation of the IMF deserves attention as a variable of significance. Also, variations in the orientation of the IMF prior to and/or during a substorm could lead to variations in the temporal development of the substorm. As pointed out by *N. C. Maynard* [private communication, 1988], the asymmetries in our Figure 18 suggest similarities to plasma convection patterns for the polar latitudes as determined by *Heppner and Maynard* [1987], which are dependent on the orientation of the IMF.

ACKNOWLEDGMENT—Magnetic field data from the ISEE 1 and ISEE 3 spacecraft were kindly provided, respectively, by C. T. Russell and E. J. Smith. This research was supported in part by the National Aeronautics and Space Administration under grants NAG5–483 and NGL–167–001–002, and by the Office of Naval Research under grant N00014–85–K–0404.

286

REFERENCES

Akasofu, S.-I., "The Dynamical Morphology of the Aurora Polaris," *J. Geophys. Res.*, **68**, 1667-1673 (1963).

Akasofu, S.-I., "The Development of the Auroral Substorm," *Planet. Space Sci.*, **12**, 273-282 (1964).

Akasofu, S.-I., *Physics of Magnetospheric Substorms*, D. Reidel, Dordrecht, Netherlands (1977).

Akasofu, S.-I., D. S. Kimball, and C.-I. Meng, "The Dynamics of the Aurora—II, Westward Traveling Surges," *J. Atmos. Terr. Phys.*, **27**, 173-187 (1965).

Akasofu, S.-I., D. S. Kimball, and C.-I. Meng, "Dynamics of the Aurora—V, Poleward Motions," *J. Atmos. Terr. Phys.*, **28**, 497-503 (1966a).

Akasofu, S.-I., D. S. Kimball, and C.-I. Meng, "Dynamics of the Aurora—VII, Equatorward Motions and the Multiplicity of Auroral Arcs," *J. Atmos. Terr. Phys.*, **28**, 627-635 (1966b).

Akasofu, S.-I., C.-I. Meng, and D. S. Kimball, "Dynamics of the Aurora—IV, Polar Magnetic Substorms and Westward Traveling Surges," *J. Atmos. Terr. Phys.*, **28**, 489-496 (1966c).

Anger, C. D., T. Fancott, J. McNally, and H. S. Kerr, "ISIS-II Scanning Auroral Photometer," *Appl. Opt.*, **12**, 1753-1766 (1973).

Anger, C. D., S. K. Babey, A. L. Broadfoot, R. G. Brown, L. L. Cogger, R. Gattinger, J. W. Haslett, R. A. King, D. J. McEwen, J. S. Murphree, E. H. Richardson, B. R. Sandel, K. Smith, and A. Vallance Jones, "An Ultraviolet Auroral Imager for the Viking Spacecraft," *Geophys. Res. Lett.*, **14**, 387-390 (1987).

Annals of the International Geophysical Year, Vols. 1-47, Pergamon Press, New York (1959-1969).

Arnoldy, R. L., and T. E. Moore, "Longitudinal Structure of Substorm Injections at Synchronous Orbit," *J. Geophys. Res.*, **88**, 6213-6220 (1983).

Brown, R. R., T. R. Hartz, B. Landmark, H. Leinbach, and J. Ortner, "Large-Scale Electron Bombardment of the Atmosphere at the Sudden Commencement of a Geomagnetic Storm," *J. Geophys. Res.*, **66**, 1035-1041 (1961).

Carlson, H. C., Jr., "Polar Cap Sun Aligned Arcs (abstract)," *Eos Trans. AGU*, **69**, 440 (1988).

Carlson, H. C., Jr., V. B. Wickwar, E. J. Weber, J. Buchau, J. G. Moore, and W. Whiting, "Plasma Characteristics of Polar Cap F-Layer Arcs," *Geophys. Res. Lett.*, **11**, 895-898 (1984).

Craven, J. D., L. A. Frank, C. T. Russell, E. J. Smith, and R. P. Lepping, "Global Auroral Responses to Magnetospheric Compressions by Shocks in the Solar Wind: Two Case Studies," in *Solar Wind-Magnetosphere Coupling*, Y. Kamide and J. Slavin, eds., Terra Publishing Co., Tokyo (1986).

Craven, J. D., and L. A. Frank, "Latitudinal Motions of the Aurora During Substorms," *J. Geophys. Res.*, **92**, 4565-4573 (1987).

Craven, J. D., L. A. Frank, and S.-I. Akasofu, "Propagation of a Westward Traveling Surge and the Development of Persistent Auroral Features," *J. Geophys. Res.*, **94**, 6961-6967 (1989a).

Craven, J. D., L. A. Frank, J. S. Murphree, and L. L. Coggar, "Simultaneous Optical Observations of Transpolar Arcs in the Two Polar Caps with DE 1 and Viking," (abstract), *Eos Trans. AGU*, **70**, 428 (1989b).

Davis, T. N., and D. S. Kimball, "Incidence of Auroras and Their North-South Motions in the Northern Auroral Zone," Geophys. Inst. Report UAG-R100, Univ. of Alaska, Fairbanks (1960).

Evans, D. S., "The Characteristics of a Persistent Auroral Arc at High Latitudes in the 1400 MLT Sector," in *The Polar Cusp*, J. A. Holtet and A. Egeland, eds., D. Reidel, Boston (1985).

Feldstein, Y. I., and Yu. I. Galperin, "The Auroral Luminosity Structure in the High-Latitude Upper Atmosphere: Its Dynamics and Relationship to the Large-Scale Structure of the Earth's Magnetosphere," *Rev. of Geophys.*, **23**, 217-275 (1985).

Frank, L. A., J. D. Craven, K. L. Ackerson, M. R. English, R. H. Eather, and R. L. Carovillano, "Global Auroral Imaging Instrumentation for the Dynamics Explorer Mission," *Space Sci. Instr.*, **5**, 369-393 (1981).

Frank, L. A., J. D. Craven, J. L. Burch, and J. D. Winningham, "Polar Views of the Earth's Aurora with Dynamics Explorer," *Geophys. Res. Lett.*, **9**, 1001-1004 (1982).

Frank, L. A., J. D. Craven, and R. L. Rairden, "Images of the Earth's Aurora and Geocorona from the Dynamics Explorer Mission," *Adv. Space Res.*, **5**, 53-68 (1985).

Frank, L. A., J. D. Craven, D. A. Gurnett, S. D. Shawhan, D. R. Weimer, J. L. Burch, J. D. Winningham, C. R. Chappell, J. H. Waite, R. A. Heelis, N. C. Maynard, M. Sugiura, W. K. Peterson, and E. G. Shelley, "The Theta Aurora," *J. Geophys. Res.*, **91**, 3177-3224 (1986).

Frank, L. A., and J. D. Craven, "Imaging Results from Dynamics Explorer," *Rev. Geophys.*, **26**, 249-283 (1988).

Gorney, D. J., D. S. Evans, M. S. Gussenhoven, and P. F. Mizera, "A Multiple-Satellite Observation of the High-Latitude Auroral Activity on January 11, 1983," *J. Geophys. Res.*, **91**, 339-346 (1986).

Gussenhoven, M. S., "Extremely High Latitude Auroras," *J. Geophys. Res.*, **87**, 2401-2412 (1982).

Heppner, J. P., and N. C. Maynard, "Empherical High-Latitude Electric Field Models," *J. Geophys. Res.*, **92**, 4467-4489 (1987).

Hoffman, R. A., G. D. Hogan, and R. C. Maehl, "Dynamics Explorer Spacecraft and Ground Operations System," *Space Sci. Instr.*, **5**, 349-367 (1981).

Hones, E. W., Jr., "The Poleward Leap of the Auroral Electrojet as Seen in Auroral Images," *J. Geophys. Res.*, **90**, 5333-5337 (1985).

Huang, C. Y., L. A. Frank, W. K. Peterson, D. J. Williams, W. Lennartsson, D. G. Mitchell, R. C. Elphic, and C. T. Russell, "Filamentary Structures in the Magnetotail Lobes," *J. Geophys. Res.*, **92**, 2349-2363 (1987).

Huang, C. Y., J. D. Craven, and L. A. Frank, "Simultaneous Observations of a Theta Aurora and Associated Magnetotail Plasmas," *J. Geophys. Res.* **94**, 10,137-10,143 (1989).

Ismail, S., and C.-I. Meng, "A Classification of Polar Cap Auroral Arcs," *Planet. Space Sci.*, **30**, 319-330 (1982).

Kaneda, E., "Auroral TV Observation by KYOKKO," *Proc. International Workshop on Selected Topics of Magnetospheric Physics*, Japanese IMS Committee, Tokyo (1979).

Kisabeth, J. L., and G. Rostoker, "The Expansive Phase of Magnetospheric Substorms, 1. Development of the Auroral Electrojets and Auroral Arc Configuration During a Substorm," *J. Geophys. Res.*, **79**, 972-984 (1974).

Lassen, K., and C. Danielsen, "Quiet Time Pattern of Auroral Arcs for Different Directions of the Interplanetary Magnetic Field in the Y-Z Plane," *J. Geophys. Res.*, **83**, 5277-5284 (1978).

Matsushita, S., "Increase of Ionization Associated with Geomagnetic Sudden Commencements," *J. Geophys. Res.*, **66**, 3958-3961 (1961).

Meng, C.-I., R. H. Holzworth, and S.-I. Akasofu, "Auroral Circle-Delineating the Poleward Boundary of the Quiet Auroral Belt," *J. Geophys. Res.*, **82**, 164-172 (1977).

Meng, C.-I., "Electron Precipitation in the Midday Auroral Oval," *J. Geophys. Res.*, **86**, 2149-2174 (1981a).

Meng, C.-I., "Polar Cap Arcs and the Plasma Sheet," *Geophys. Res. Lett.*, **8**, 273-276 (1981b).

Meng, C.-I., and R. E. Huffman, "Ultraviolet Imaging from Space of the Aurora Under Full Sunlight," *Geophys. Res. Lett.*, **11**, 315-318 (1984).

Meng, C.-I., R. E. Huffman, F. Del Greco, and R. Eastes, "UV Images of the Dayside Auroral Oval (abstract)," *Eos Trans. AGU*, **68**, 396 (1987).

Nishida, A., *Geomagnetic Diagnosis of the Magnetosphere*, Springer-Verlag, New York (1978).

Obara, T., M. Kitayama, T. Muki, N. Kaya, J. S. Murphree, and L. L. Cogger, "Simultaneous Observations of Sun-Aligned Polar Cap Arcs in Both Hemispheres by EXOS-C and Viking," *Geophys. Res. Lett.*, **15**, 713-716 (1988).

Ortner, J., B. Hultqvist, R. R. Brown, T. R. Hartz, O. Holt, B. Landmark, J. L. Hook, and H. Leinbach, "Cosmic Noise Absorption Accompanying Geomagnetic Storm Sudden Commencements," *J. Geophys. Res.*, **67**, 4169-4186 (1962).

Peterson, W. K., and E. G. Shelley, "Origin of the Plasma in a Cross-Polar Cap Auroral Feature (Theta Aurora)," *J. Geophys. Res.*, **89**, 6729-6736 (1984).

Rairden, R. L., L. A. Frank, and J. D. Craven, "Geocoronal Imaging with Dynamics Explorer," *J. Geophys. Res.*, **91**, 13,613-13,630 (1986).

Reiff, P. H., J. L. Burch, and R. A. Heelis, "Dayside Auroral Arcs and Convection," *Geophys. Res. Lett.*, **5**, 391-394 (1978).

Rogers, E. H., D. F. Nelson, and R. C. Savage, "Auroral Photography from a Satellite," *Science*, **183**, 951 (1974).

Rostoker, G., A. Vallance Jones, R. L. Gattinger, C. D. Anger, and J. S. Murphree, "The Development of the Substorm Expansive Phase: The 'Eye' of the Storm," *Geophys. Res. Lett.*, **14**, 399-402 (1987).

Shepherd, G. G., T. Fancott, J. McNally, and H. S. Kerr, "ISIS-II Atomic Oxygen Red Line Photometer," *Appl. Opt.*, **12**, 1767-1774 (1973).

Shepherd, G. G., C. D. Anger, J. S. Murphree, and A. Vallance Jones, "Auroral Intensifications in the Evening Sector Observed by the Viking Ultraviolet Imager," *Geophys. Res. Lett.*, **14**, 395-398 (1987).

Snyder, A. L., and S.-I. Akasofu, "Observations of the Auroral Oval by the Alaskan Meridian Chain of Stations," *J. Geophys. Res.*, **77**, 3419-3430 (1972).

287

Tholen, S. M., and T. P. Armstrong, "Triggering of Magnetospheric Particle Bursts by SSC Events," in *Solar Wind-Magnetosphere Coupling*, Y. Kamide and J. Slavin, eds., Terra Publishing Co., Tokyo (1986).

Tighe, W. G., and G. Rostoker, "Characteristics of Westward Travelling Surges During Magnetospheric Substorms," *J. Geophys.*, **50**, 51-67 (1981).

Tsyganenko, N. A., and A. V. Usmanov, "Determination of the Magnetospheric Current System Parameters and Development of Experimental Geomagnetic Field Models Based on Data from IMP and HEOS Satellites," *Planet. Space Sci.*, **30**, 985-998 (1982).

Ullaland, S. L., K. Wilhelm, J. Kanges, and W. Riedler, "Electron Precipitation Associated with a Sudden Commencement of a Geomagnetic Storm," *J. Atmos. and Terr. Phys.*, **32**, 1545-1553 (1970).

Vorob'yev, V. G., "SC-Associated Effects in Auroras," *Geomagn. Aeron.*, **14**, 72-74 (1974).

Wiens, R. G., and G. Rostoker, "Characteristics of the Development of the Westward Electrojet During the Expansive Phase of Magnetospheric Substorms," *J. Geophys. Res.*, **80**, 2109-2128 (1975).

V-3. DIAGNOSIS OF AURORAL DYNAMICS USING GLOBAL AURORAL IMAGING WITH EMPHASIS ON LOCALIZED AND TRANSIENT FEATURES

G. G. Shepherd* and J. S. Murphree[†]

Observations made by the Viking Ultraviolet Imager reveal a number of processes that appear to be associated with the optical signature of substorms. Previous satellite imaging has focused on the larger scale events, but in some cases the smaller, more dynamic aspects of auroral variations are also repeatable from substorm to substorm and thus need to be included in our concepts of its morphology. Taking advantage of the temporal resolution available, examples of substorm onsets are found that produce one or more intensifications in the evening sector before the major optical expansion. While these intensifications are in the general area of the subsequent expansion, their precise locations differ. When substorm onset occurs equatorward of a preexisting arc system, its subsequent poleward expansion may be limited in latitudinal extent. Once expansion of the discrete auroral has stopped, a common feature is the rapid (< 1 min) intensification of the most poleward discrete arc. No latitudinal motion is associated with this brightening, but the arc may develop a series of vortices spread out in longitude.

1. INTRODUCTION

The discovery of the concepts of the auroral oval [*Feldstein*, 1963] and the auroral substorm [*Akasofu*, 1964] together provided a giant leap forward in our understanding of the behavior of the aurora. It is sobering to imagine now how many people (including some of us) watched the equatorward movement of the auroral in the evening and its poleward retreat in the morning without recognizing that this pattern was attached to the Sun—this brilliant separation of the auroral behavior into a static global pattern with a superimposed dynamical behavior provided a proper frame of reference for auroral studies, and firmly connected the aurora to the magnetosphere. We must remember that these major advances were made with all-sky cameras, well before the inception of satellite imaging, but it is also interesting to recall that *Sydney Chapman* [1964] was an early proponent of satellite imaging, proposing color photography from manned satellites; at that stage of space exploration one would not have thought of high-altitude imagery of the entire oval.

Since that time, the power of satellite auroral imaging has received increasing recognition; as demonstrated by recent experiments, Viking, HILAT, and Polar BEAR, as well as future missions, ISTP and Interball. These, of course, built on the success and experience of earlier auroral imagers, ISIS 2, DMSP, and Dynamics Explorer from which an excellent background to the subject has been given by *Craven* [1989]. With all of these developments in satellite imagery over the past 17 years, our knowledge both of the instantaneous auroral distribution and substorm evolution has been significantly advanced. With improvements in temporal resolution, we are now able to extend our investigations beyond the large-scale dynamical features to some of the smaller events, both preceding and following substorms. Such observations not only reveal new features, which at this point cannot be obviously fitted into the existing substorm concept, but also raise the question of the relationship between these optical measurements and more conventional ground-based measurements of substorm processes. It is a characteristic of auroral imaging that measurements made by new instruments require existing concepts to be looked at in a fresh light. Each new instrument has its own scale size, in space and time, and what it seen in the aurora is a function of that scale size. So the new information provided by the Viking Ultraviolet Imager (UVI) should allow us to extend or augment the existing well-developed substorm concept.

2. INTRODUCTION TO VIKING UVI

In Figure 1a (on September 20, 1986 at 1836:42 UT), we show a typical UVI image. The characteristics of the UVI instrument have veen described by *Anger et al.* [1987]. Here we provide a brief review. The instrument uses reflective optics of the Burch design, and employs an intensified charge-coupled-device (CCD) detector. The accumulated charges are clocked across the CCD at the same rate as the auroral image, which moves across the CCD because the satellite spins with a period of 20 seconds. With this technique, an exposure time of about 1 s can be achieved, about the time for the charges to move across the full extent of the CCD. In the interpretation of some particular im-

*Institute for Space and Terrestrial Sciences, York University, Toronto, Canada, M3J 1P3.
[†]Physics Department, University of Calgary, Calgary, Alberta, Canada, T2N 1N4.

Auroral Physics, edited by C.-I. Meng, M. J. Rycroft and L. A. Frank. © Cambridge UP 1991

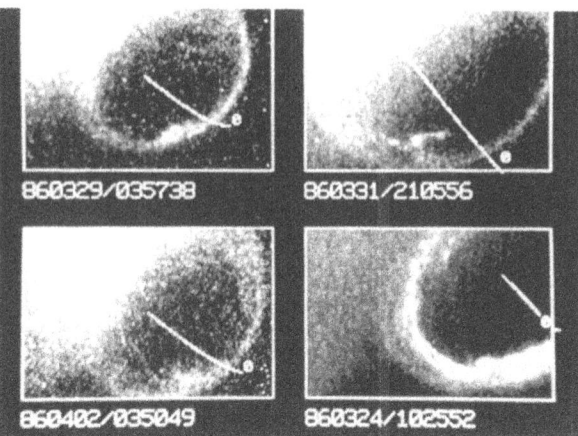

Figure 1—Examples of Viking UV imager data showing characteristics of discrete features in the auroral distribution. The data have been corrected for nonuniformity effects and background, but are otherwise unmodified. The data are from the LBH camera and dayglow emissions are apparent in the upper left part of each image. *Top left*: The field of view of an all-sky camera positioned at Tromsö is indicated by the roughly circular curve. *Top right* and *bottom left*: The 2200 MLT meridian is shown. (This figure also appears in color: Plate 35.)

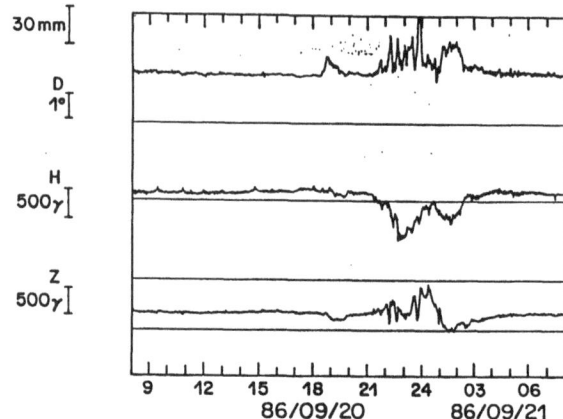

Figure 2—*D*, *H*, and *Z* components of the magnetic field acquired at Tromsö, Norway on September 20, 1986.

ages that follow, it is important to remember this point: each image is a snapshot, with all pixels recorded simultaneously for about 1 s.

The image of Figure 1*a* shows the high signal-to-noise ratio, and contains the raw data as received from the instrument with only a uniformity correction (a function of column number, which is in the horizontal dimension in this presentation) applied, which illustrates the simplification in data-processing as a result of using an array detector. The upper left corner of the image is in sunlight and the terminator is seen as the diagonal border to the upper left bright region. This image has been chosen to illustrate the appearance of the auroral distribution during a clear expansion phase of a substorm. The optical region in the night sector is significantly distorted from its quiet-time morphology to a width of 8.3° magnetic latitude at 2300 MLT. Such a distortion is not localized or transient on the scale of the Viking UV images, but represents a common substorm morphology as seen from satellite imagers [e.g., *Craven et al.*, 1989]. At the western edge of the distorted region near 2130 MLT is a localized bright region, commonly referred to as a Westward Traveling Surge (WTS) [*Akasofu et al.*, 1965]. Such features are regions of localized upward field-aligned current [*Inhester et al.*, 1981; *Opgenoorth et al.*, 1983] whose source are energetic electrons [*Meng et al.*, 1978]. Such a distorted night-sector auroral distribution fits into the common concept of how a substorm looks in two dimensions. This particular UVI observation reveals significant activity, but ground-based stations must be appropriately located in order to confirm such ac-

tivity. For example, Figure 2 shows the magnetic activity as monitored at Tromsö, whose position is noted by the circular area in Figure 1*a*. At the time of the Viking observation, Tromsö is south and west of the WTS and reveals little (< 70 nT) activity in the *H* component and just the beginning of a 100-nT depression in *Z*. A significant positive excursion in *D* is noted, however, and is just the signature for a WTS to the east of Tromsö [*Rostoker et al.*, 1980].

In Figure 1*b* is shown a more "recognizable" view of a WTS, indicating substorm expansion-phase activity [*Akasofu et al.*, 1965]. This particular view, taken from 10,200 km, emphasizes the common "S" shape of surge features, and the observations of such a feature are tacitly assumed to be that of a WTS [e.g., *Bythrow and Potemra*, 1987]. There has been much discussion concerning the actual dynamics of such features, beginning with the concept of discrete electrojet motion [*Wiens and Rostoker*, 1975] versus continuous development [*Pytte et al.*, 1976]. With the advent of Viking observations, it has become clear that such features may remain motionless for minutes at a time [*Murphree et al.*, 1989]. Figure 1*c* shows an image taken 13 min after than in Figure 1*b*, and comparison of the two indicates no motion. At typical WTS speeds of 2–3 km s^{-1} [*Opgenoorth et al.*, 1983]; this feature could have moved to 1900 MLT. In any case distortions such as in Figures 1*b* and 1*c* are classical observations of substorm expansion activity and present static two-dimensional pictures that are entirely consistent with the current ideas about substorm morphology [*Rostoker et al.*, 1980].

Figure 1*d* shows further examples of substorm activity. In this single image, there is not one, but several distortions in the night sector. Such bulges, all regions of intense electron precipitation, quite likely are all regions of substorm expansion activity [*Pytte et al.*,

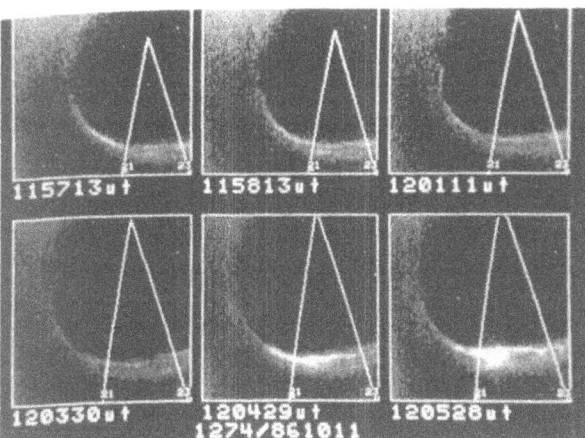

Figure 3—Viking LBH image data from orbit 1274. On each image is shown the MLT meridians of 2100 and 2300 hours. A substorm onset is apparent in the image at 1204:29 UT. (This figure also appears in color: Plate 36.)

1976]. Such multiple expansions or WTS activity have been termed substorm intensifications [*Rostoker et al.*, 1980] or microsubstorms [*Sergeev*, 1974] and their occurrence begins to reveal the complex dynamics involved in the multiple onset substorm process [*Smits et al.*, 1986].

The date presented in Figure 1 are samples or snapshots of the auroral distribution that represent the satellite image measurement of a single instant during a substorm. Historically, the optical signature of a substorm has been characterized in terms of measurements made by ground-based instrumentation [*Akasofu*, 1964; *Akasofu*, 1968]. From such measurements the paradigm for an optical substorm has evolved [*Rostoker et al.*, 1980] to describe the onset, expansion, and subsequent recovery of, primarily, the discrete aurora in the premidnight sector. Here we are particularly interested in the onset and expansion that involve an arc (lying at the equatorward edge of any preexisting system of discrete arcs) that brightens and, subsequently, moves poleward. The observation of a surge (i.e., a highly distorted, intense region of discrete aurora resulting from the longitudinal variability in the poleward motion of the substorm arc) has become a defining characteristic for substorm occurrence, and its relationship to other substorm signatures is well established [*Gledhill et al.*, 1987; *Rosenberg et al.*, 1987]. However, it has long been recognized that a single ground-based station cannot resolve whether an observation of a surge is in fact the initial onset in the magnetospheric context, or the result of dynamic development of a single expansion feature, or else of a multiple onset substorm [*Pytte et al.*, 1976].

Satellite imagers of suitable temporal and spatial resolution can unequivocally establish whether this indeed is the case in any given instance, once the general

PRE-EXPANSION OPTICAL ACTIVITY

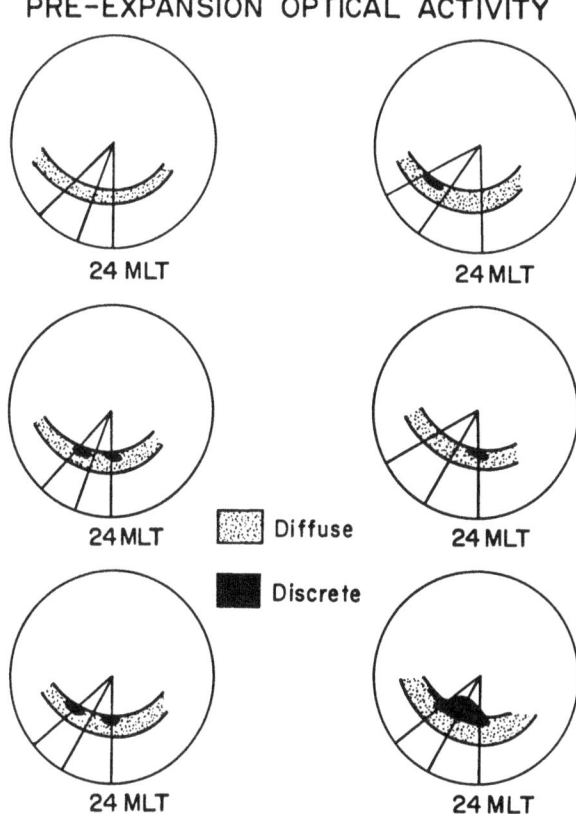

Figure 4—Schematic diagram illustrating six possible types of presubstorm expansion activity. Multiple small-scale intensifications may occur and fade away in a matter of minutes, and the subsequent major expansion may be unrelated to any single previous intensification in terms of spatial position.

character of intensifications of the optical aurora as seen by satellite imagers is understood. Previous work involving once-per-orbit images [*Lui and Burrows*, 1978; *Gelpi et al.*, 1987] has established a number of important points in the static relationship between optical features and other parameters, while the DE imager has revealed longer time scale characteristics of optical features [*Craven and Frank*, 1987]. It is now important, however, to address the dynamics of the entire auroral distribution on reasonable spatial (~ 50 km) and temporal (~ 1 min) scales in the context of substorm theory. Some initial work has been done by *Rostoker et al.* [1987a] for a particularly nice example of an isolated substorm where the large-scale relationship between the satellite image measurements and a substorm model was discussed, and by *Shepherd et al.* [1987], who reported on the observation of small-scale intensifications. The focus of this report is to present several observations made with the Viking imager of the dynamics of intensifications in the auroral distribution and place them in the context of the current phemonological description of an optical substorm.

3. AURORAL BEHAVIOR PRIOR TO SUBSTORM ONSET

Shown in Figure 3 (orbit 1274 on October 11, 1986) is a sequence of six images, covering a total of 8 min, that were acquired at intervals spaced from 1–3 min apart; the Universal Times are given under each image. The data have been corrected for background and signal uniformity. The first image in the sequence (1157:13 UT) shows a very weak auroral distribution in the dusk sector, but an extended intensification from 1900–2100 MLT at the poleward edge of the observable aurora. After 2100 MLT the aurora appears more structured, with multiple parallel arcs in the region covering several degrees eccentric dipole latitude. At about 2200 MLT there is a brightening in two of these arc systems, and at 2300 MLT (the second meridian overlaid) there is a brightening in the poleward component. None of these brightenings has a significant (>50 km) latitudinal extent and thus do not show that a substorm expansion has occurred (even a microsubstorm expansion, *Sergeev* [1974]. Such an expansion does occur (beginning at 1205:28 UT), but only after these brightenings have died away. Localized intensifications such as these are common (these are only on the order of 1 kR of equivalent 3914 Å emission), but it is interesting to note their multiplicity and spatial separation both in latitude and longitude. The subsequent substorm expansion begs the question as to the relationship between these spatially separated brightenings and the actual location of substorm onset. On the basis of this single image, one might suppose that the location of the brightest feature (the one prior to 2100 MLT) would be most probable, while the more equatorward region ($\sim$2130 MLT) would be more consistent with substorm models [*Akasofu*, 1964]. However, the next image (1158:13 UT) further indicates the transient character of these intensifications. The extended region prior to 2100 MLT is now fading and breaking into smaller regions, while a bright feature has developed at 2120 MLT. In fact such activity has been occurring for the previous 12 min with a significant brightening (to $\sim$1 kR) at 2100 MLT in an image acquired at 1154:15 UT (not shown). These intensity fluctuations are quite easy to distinguish in the satellite imagery, but the known drifting of discrete arcs equatorward prior to onset at speeds of $\sim$200 m s^{-1} [e.g., *Yahnin et al.*, 1983] is more problematic. For example, for the typical minute between images (presented here from the UVI) such drifts as seen from apogee correspond to much less than a pixel. This coupled with the common lack of stable, extended high-intensity regions as seen in the satellite imagery precludes the identification of such equatorward drifting arcs in general.

In the third image, 3 min following the previous one, the spot at 2120 MLT has practically vanished. At this time there are three candidate bright regions, one at 2000, one at 2130, and the third at 2300 MLT, all in the poleward portion of the auroral distribution. Two minutes later, in the fourth image at 1203:30 UT, everything has faded away, similar to the auroral fading observed with all-sky cameras by *Pellinen and Heikkila* [1978] although the scale of dimming observed here is much greater than can be seen by a single all-sky camera, which in turn is much greater than most of their examples of fading before breakup. In fact, one might argue that the relevant dimming is that observed between the two tiny bright spots left in the auroral distribution ($\sim$2145 MLT). One minute later, in image 5, there has been a violent explosion of plasma injected into the auroral ionosphere, and this has occurred so rapidly that, with the 1-min time resolution of the data, it is not possible to determine where it began, or how it spread. To the resolution of the data it is simultaneous over 4 hours of local time and concentrated at the poleward edge of the auroral distribution, emphasizing the longitudinal character of the plasma injection; in classical substorm descriptions it is the latitudinal expansion that is emphasized. This observation is probably related to the auroral "horn" as described by *Yahnin et al.* [1983] (see also *Pytte et al.* [1976]), where an extended auroral arc activates prior to the subsequent poleward expansion resulting from surge activity. If we make the most reasonable assumption that the onset was in the middle of the region, which happens (coincidentally?) to be near the pair of intense spots (in the image at 1203:30 UT), and that it occurred immediately after the previous image was taken, then the expansion velocity east and west is approximately 3.7 km s^{-1}, which is a little less than other arc brightenings of this type [*Steen and Collis*, 1988], but greater than typical speeds for westward traveling surges [*Opgenoorth et al.*, 1983]. It should be noted in passing that optical measurements of velocities need not correspond to source plasma motions, since loss-cone filling and energization are also a requirement for the production of optical features. In the last image, 1 min later, the brightest region has expanded equatorward with some slight poleward motion, but little further longitudinal expansion. In fact, it is interesting to note that even at the last image acquired (1206:28 UT—not shown), the substorm expansion has not extended in longitude beyond the bounds of the prior transient intensifications.

To summarize, there are a number of noteworthy features of this substorm onset. First, there is considerable activity (which means energy deposition) spread out over 4 hours of MLT prior to the onset. However, from this activity it is not evident how to predict where

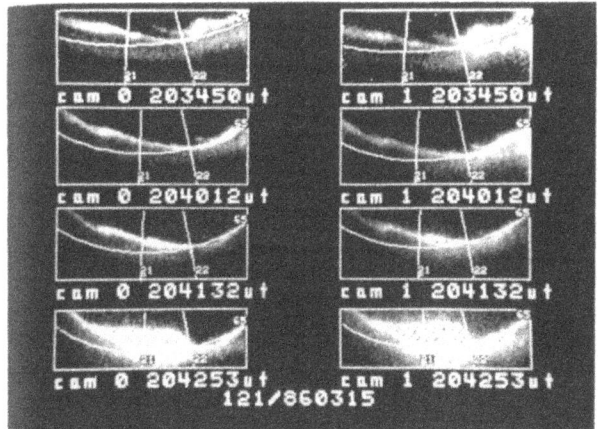

Figure 5—Four pairs of Viking UVI data, for both cameras: left at LBH wavelengths (cam 0) and right atomic oxygen (cam 1). Data are for only a small section of the late evening sector. (This figure also appears in color: Plate 37.)

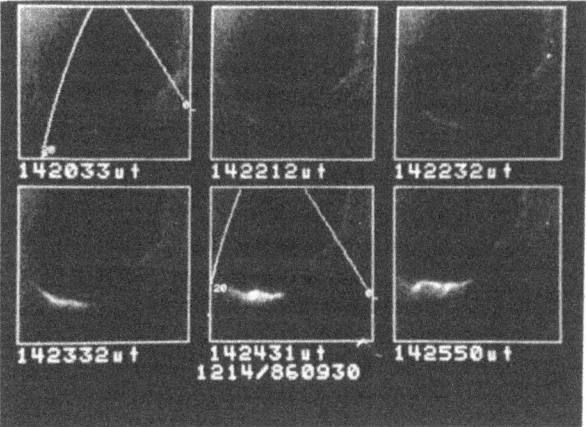

Figure 6—Viking LBH image data from orbit 1214. Three extended arc systems are noted in the evening sector. A substorm expansion occurs on the most equatorward one at 1423:32 UT just where the more poleward system begins to be observable. (This figure also appears in color: Plate 38.)

the onset subsequently occurs. A general fading of the aurora seems to occur some 2 min prior to the onset, and then the onset seems to occur perhaps in the middle of the 4-hour MLT sector in the vicinity of a pair of bright spots in the poleward component of the aurora. Within a single minute this sector brightens, encompassing the entire area of previous activity; the brightening of the arc seems not to require significant poleward expansion along its entire length. It is conceivable that the magnetosphere has been attempting to release energy by driving field-aligned currents at various locations represented by these intensifications, but conditions were not suitable for a full-blown expansion. This is consistent with the observations of *Steen and Collis* [1988] who report on observations of spiral development. In some cases an arc deformation does not proceed to full-scale substorm expansion, reportedly because of some limit on the velocity shear presumed to result in the vorticity. The data presented in Figure 3 are for a single substorm occurrence, but similar activity is evident in other UVI datasets. Figure 4 shows six schematic diagrams of what commonly happens. Here the observation of the temporal development of isolated intensifications is emphasized.

Detailed Low-Altitude View of a Substorm Onset

In Figure 5 we show a sequence (orbit 121 on March 15, 1986) in which the satellite was at low altitude, and we can explore in greater detail some of the implications raised in the previous section. Here we show, side-by-side, images taken simultaneously from both cameras, as labeled. Camera 0 (on the left) is the camera responding primarily to the N_2 LBH bands while camera 1 responds primarily to the 1304 Å and 1356 Å lines of atomic oxygen. In the first image we see for camera 0 a well-defined diffuse aurora equatorward

of 65° eccentric dipole latitude, with bright discrete arcs poleward. The brightest arcs lie before 2100 MLT and after 2200 MLT; that is, there is a break between 2100 and 2200 hours. In the camera 1 image on the right, the aurora earlier than 2100 MLT is much weaker in intensity, and the aurora after 2200 MLT is brighter. The modeling that would tell us what this means in terms of the mean electron energy has not yet been accomplished, but current understanding indicates that this is consistent with harder precipitation before 2100 MLT, and softer precipitation following 2200 MLT. In the second image down, about 5 min later, the two poleward arcs seem to have joined, and now pass smoothly through the 2100–2200 MLT region, although the brightest regions are still prior to 2100 and after 2200 MLT. The asymmetry in the two cameras still exists. In the third image pair, taken 80 s later, this weaker middle region now has become the most intense; and, in the fourth image pair, 80 s later again, a large substorm expansion has taken place. It should be stressed that there is uncertainty as to what has actually taken place between 2040:12 and 2041:32 UT. It may be that the more eastward intensification has propagated westward and then rapidly expanded poleward and equatorward at the point where it is observed at 2041:32 UT. Alternatively, the intensification at 2230 MLT may have faded and a new one appeared at 2130 MLT. Only improved temporal resolution can resolve such a question. Thus, again, there are intensified regions occurring over an extended longitudinal range prior to the substorm; however, in this case at least one of the preexisting intensifications (the westernmost one) does not disappear. This situation also is different in the sense that one could not really consider the preexisting arc to be quiet because of the presence of small-scale spiral features. Note also that there is no evidence of fading in this example—the region between 2100 and 2200 MLT remains relatively constant in intensity during the 7 min before the expansion. We note also, in the third image pair, where the intensification has occurred prior to full substorm onset, that the brightening has taken place in the diffuse aurora as well as in the brighter poleward discrete arc. Thus the level of energy deposition is widespread, and occurs here in both the diffuse as well as the discrete aurora.

The question arises as to whether this is an example of a multiple-onset substorm because of the preexisting spirals. Unfortunately, the low altitude of the satellite at this time precludes identifying the large-scale character. However, it should be noted that the width of the surge prior to 2100 MLT is only 100 km and hence of a much smaller scale than the examples presented in Figure 1. Further it has been shown that such spirals, or vortices, are common at all local times [*Murphree et al.*, 1989], particularly in the afternoon [*Lui et al.*, 1989], and reflect plasma instabilities not necessarily related to substorm processes. As *Steen and Collis* [1988] have noted in the night sector such features are common and do not always progress toward full-scale substorm expansion. Further, in the case of the data presented in Figure 5, the occurrence of the expansion does not appear to involve the expansion region westward (a small spiral already exists west of the onset region) as is found by *Wiens and Rostoker* [1975], although new expansions behind the surge front are not uncommon [*Pytte et al.*, 1976]. As is consistent with the other examples here, it seems more reasonable to discuss expansive features in the context of large-scale latitudinal and longitudinal changes.

4. THE LOCATION OF SUBSTORM ONSET

The above examples have posed the problem of where a substorm onset occurs in relation to existing regions of discrete aurora (e.g., spirals). As is noted above, it is in general not possible to establish the particular discrete arc on which the onset occurs because of the spatial resolution of the Viking imager. In fact the distinction between discrete and diffuse aurora is sometimes difficult to establish. The advantage that satellite imagers bring is the ability to view the onset region over a larger local time range and thereby compare larger scale discrete arc systems that may not extend uniformly through the night sector. In the example shown in Figure 6 (orbit 1214 on September 30, 1986) the aurora is weak, but in the original data one can identify three discrete regions of emission, seen in the first three images of the sequence. Only the most equatorward region of emission is continuous throughout the night sector, the two more-poleward, starting at approximately 2100 and 2200 MLT and extending

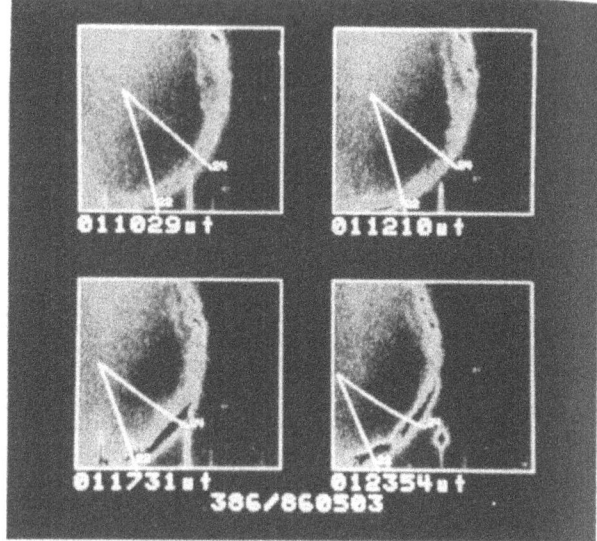

Figure 7—Image data from orbit 386. On each image are shown the MLT meridians of 2200 and 2400 hours. (This figure also appears in color: Plate 39.)

to at least midnight. Some precursor activity is noted in the poleward regions of the most equatorward region of emission at 1422:12 and 1422:32 UT. Then at 1423:32 UT, a longitudinally extended onset has begun in the poleward part of this discrete arc system. This onset occurs very near where the second discrete arc system just poleward begins to be identifiable. The farthest poleward system does not begin till later in MLT. By 1424:31 UT the expansion has extended in latitude such that there now exists no discrete features poleward of the onset region, and seems not to extend any farther poleward. This is consistent with some of the observations of *Sergeev and Yahnin* [1979] who reported that a localized expansion had a weak diffuse arc system poleward of the onset. The subsequent expansion reached, but did not extend beyond this preexisting diffuse arc. Note that in the image at 1425:50 UT two spiral forms coexist.

Another example of this type of activity is shown in Figure 7. Here image data taken at 20-s resolution is available for the auroral observations taken on May 3, 1986 between 0040:18 and 0127:55 UT. The magnetosphere is still recovering from a magnetic storm that took place on the previous day. The first image shown in Figure 7 shows considerable activity in the early morning sector, primarily in the equatorward region. In the evening sector near 2200 MLT there is some activity in the most poleward arc (the vertical streaks near the bottom of the image are due to "bleeding" of charge in several of the CCD columns, and should be ignored). At 0112:10 UT a longitudinally extended region of intensification has erupted well equatorward of the poleward edge. Because earlier

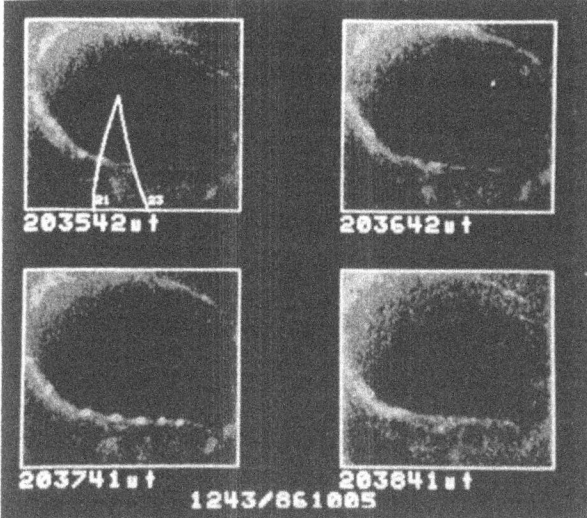

Figure 8—Viking LBH data from orbit 1243. The appearance of a vortex street is apparent at 2037:41 UT. (This figure also appears in color: Plate 40.)

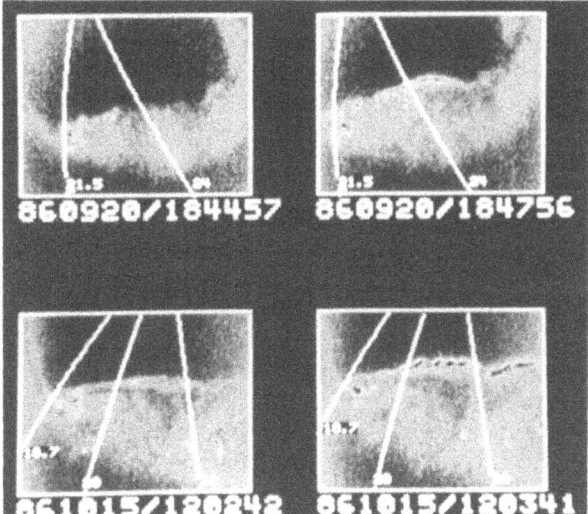

Figure 9—Examples from two orbits of the intensification of the most poleward arc after substorm expansion has been completed. In the top are data from orbit 1160 (September 20, 1986) and in the bottom from orbit 1296 (October 15, 1986). All images are from the LBH camera and have been corrected for nonuniformity. (This figure also appears in color: Plate 41.)

MLT are not in the images it is not possible to determine whether any activity is occurring westward of this expansion, but the eastward evolution is significantly progressing at a rate of over 7 km s^{-1} between 0112:10 and 0119:12 UT. Once again, however, the expansion is generally limited in latitude to the spatial position of the preexisting poleward arc. Some slight poleward expansion beyond this position is noted near 2300 MLT.

5. MULTIPLE "SIMULTANEOUS" INTENSIFICATIONS IN A POLEWARD ARC

In Figure 8 (orbit 1243) is shown an example of one of the more dramatic types of longitudinally extended dynamical behavior that has been obtained with the Viking UVI. As can be seen in the first image at 2035:42 UT considerable substorm activity has taken place. A large region of diffuse emissions extends throughout the night sector. The major poleward expansion has stopped by this time and is consistent with ground-based magnetometer observations [*Murphree and Elphinstone*, 1988]; the recovery phase is just beginning. The discrete auroral oval branches off from the diffuse aurora a little earlier than 2100 MLT, and in the first image just fades out as it extends into the midnight sector, parallel to the diffuse aurora. In the second image, taken 1 min later, this discrete arc is brighter, and thus more clearly visible, with one pronounced small intensification near 2100 MLT. In the third image, another minute later, five bright, equally spaced intensifications have appeared, while in the fourth image, one more minute later, they have virtu-

ally disappeared. In this last image, only remnants of the intensifications can be seen, with faint connections to the diffuse aurora. These structures are reminiscent of the "north-south" structures noted by *Rostoker et al.* [1987b] in the expansive phase of a substorm.

This behavior has no apparent precedent in classical substorm behavior. There is no significant latitudinal motion accompanying this intensification, and thus it seems not to be related to any possible "poleward leap" of the auroral electrojet [*Hones*, 1985]. The general pattern is that of vortex streets, or auroral curls, that have been detected with ground-based television images [*Davis and Hallinan*, 1976], but the scale size of the street here is very much larger, much too large to be observed in total with an all-sky camera on the ground.

Figure 9 illustrates two more examples of this type of behavior. In the top row are two images taken from orbit 1160. In the first image, the only discrete intensification is the WTS near 2100 MLT. Three minutes later the poleward boundary has significantly intensified and now extends past midnight. Note that in this example the poleward arc has not satisfied any instability criteria, since a vortex sheet has not been created. In the bottom of Figure 9 are two images taken from orbit 1296. On the left, at 1202:42 UT, is once again a clear substorm expansion morphology as seen in the Viking imagery. Details of this orbit have been discussed by *Rostoker et al.* [1987b]. In this case there is a poleward discrete arc that already exists, but the next image (1203:41 UT) shows that once again it has intensified and extended eastward in longitude. The striking features of the pattern are first of all the geometry, second the large longitudinal extent, and finally the rapid and virtually simultaneous (to within 1 min) turning on and off of these intensifications. It is challenging to try to visualize what magnetospheric configuration could give rise to this organized large-scale "simultaneous" energy deposition.

6. SUMMARY AND CONCLUSIONS

The high spatial and temporal resolution of the Viking UVI has allowed the instantaneous two-dimensional character of optical substorm phenomena to be observed. Some new features have been revealed that need to be incorporated in current substorm concepts. Between successive 1-s snapshots taken at 1-min intervals configurations sometimes change so drastically that it is not possible to determine whether motions are involved or whether energy is deposited simultaneously across the region. In the events presented here, it is the large longitudinal extents of the (simultaneously?) excited region that are not only most striking but also most at variance with the existing substorm concept. This longitudinally extended region may be intensified either along its full length or at a number (up to five have been observed) of discrete points along the longitudinal range. The "simultaneity" of the intensification is the same in either case.

In conclusion the following specific observations were noted.

1. Optical activation can begin in widely separated regions well before the substorm onset. It is not possible to predict, from these earlier patterns, where the ultimate onset is going to occur. There is also evidence for a fading of activity following this presubstorm activation and the substorm onset.

2. The UVI images, in general, have insufficient spatial resolution to identify the precise thin arc in which the substorm is initiated. However, in one case we are able to confirm that it occurs in the most equatorward of the discrete arcs.

3. When substorm onset occurs equatorward of a preexisting arc system, its subsequent poleward expansion may be limited in latitudinal extent.

4. A common feature during late expansion or early recovery phase is the rapid intensification of the discrete arc that defines the poleward boundary of emissions in the night sector. This arc may develop a vortex street.

ACKNOWLEDGMENT—The authors appreciate the help of R. Elphinstone and G. Enno in preparing the figures. The Viking project was managed by the Swedish Space Corporation under contract to the Swedish Board for Space Activities. The UV imager was built as a project of the National Research Council of Canada, and this work was supported under grants from the Natural Sciences & Engineering Research Council of Canada.

REFERENCES

Akasofu, S.-I., "The Development of the Auroral Substorm," *Planet. Space Sci.*, **12**, 273 (1964).

Akasofu, S.-I., D. S. Kimball, and C.-I. Meng, "Dynamics of the Aurora, II, Westward Traveling Surges," *J. Atmos. Terr. Phys.*, **27**, 173 (1965).

Akasofu, S.-I., *Polar and Magnetospheric Substorms*, D. Reidel Publishing Co., Dordrecht, Holland (1968).

Anger, C. D., S. K. Babey, A. L. Broadfoot, R. G. Brown, L. L. Cogger, R. Gattinger, J. W. Haslett, R. A. King, D. J. McEwen, J. S. Murphree, E. H. Richardson, B. R. Sandel, K. Smith, and A. Vallance Jones, "An Ultraviolet Imager for the Viking Spacecraft," *Geophys. Res. Lett.*, **14**, 387 (1987).

Bythrow, P. F., and T. A. Potemra, "Birkeland Currents and Energetic Particles Associated with Optical Auroral Signatures of a Westward Traveling Surge," *J. Geophys. Res.*, **92**, 8691 (1987).

Chapman, S., "Aurora and Geomagnetic Storms," in *Space Physics*, D. P. LeGalley and A. Rosen, eds., John Wiley and Sons, New York (1964).

Craven, J. D., "Diagnosis of Auroral Dynamics Using Global Auroral Imaging with Emphasis on Large Scale Evolutions," *this volume*.

Craven, J. D., and L. A. Frank, "Latitudinal Motions of the Aurora During Substorms," *J. Geophys. Res.*, **92**, 4565-4573 (1987).

Craven, J. D., L.A. Frank, and S.-I. Akasofu, "Propagation of a Westward Traveling Surge and the Development of Persistent Auroral Features," *J. Geophys. Res.*, in press (1989).

Davis, T. N., and T. J. Hallinan, "Auroral Spirals 1. Observations," *J. Geophys. Res.*, **81**, 3953 (1976).

Feldstein, Y., "Some Problems Concerning the Morphology of Auroras and Magnetic Disturbances at High Latitudes," *Geomagn. and Aeronomy*, **3** (1963).

Gelpi, C., H. J. Singer, and W. J. Hughes, "A Comparison of Magnetic Signatures and DMSP Auroral Images at Substorm Onset: Three Case Studies," *J. Geophys. Res.*, **92**, 2447-2460 (1987).

Gledhill, J. A., I. S. Dore, C. K. Goertz, R. Haggard, W. J. Hughes, M. W. J. Scourfield, D. P. Smits, P. H. Stoker, P. R. Sutcliffe, P. A. Wakerley, and A. D. M. Walker, "A Magnetospheric Substorm Observed At Sanae, Antarctica," *J. Geophys. Res.*, **92**, 2461-2476 (1987).

Hones, E. W., Jr., "The Poleward Leap of the Auroral Electrojet as Seen in Auroral Images," *J. Geophys. Res.*, **90**, 5333 (1985).

Inhester, B., W. Baumjohann, R.A. Greenwald, and E. Nielsen, "Joint Two-Dimensionsl Observations of Ground Magnetic and Ionospheric Electric Fields Associated with Auroral Zone Currents," *J. Geophys.*, **49**, 155 (1981).

Lui, A. T. Y., and J. R. Burrows, "On the Location of Auroral Arcs Near Substorm Onsets," *J. Geophys. Res.*, **83**, 3342 (1978).

Lui, A. T. Y., D. Venkatesan, and J. S. Murphree, "Spatially Periodic Bright Spots on the Dayside Auroral Oval," *J. Geophys. Res.*, **94**, 5515-5522 (1989).

Meng, C.-I., A. L. Snyder, Jr., and H. W. Kroehl, "Observations of Auroral Westward Traveling Surges and Electron Precipitation," *J. Geophys. Res.*, **83**, 575 (1978).

Murphree, J. S., and R. D. Elphinstone, "Correlative Studies Using the Viking Imagery," *Adv. Space Res.*, **8**, 9-19 (1988).

Murphree, J. S., L. L. Cogger, and R. D. Elphinstone, "Observations of Distortions of Optical Features in the UV Auroral Distribution," *IEEE Transactions Plasma Science*, **17**, 109-115 (1989).

Opgenoorth, H. J., R. J. Pellinen, W. Baumjohann, E. Nielsen, G. Marklund, and L. Eliasson, "Three-Dimensional Current Flow and Particle Precipitation in a Westward Travelling Surge (Observed During the Barium-GEOS Rocket Experiment)," *J. Geophys. Res.*, **83**, 3138 (1983).

Pellinen, R. J., and W. J. Heikkila, "Observations of Auroral Fading Before Breakup," *J. Geophys. Res.*, **83**, 4207 (1978).

Pytte, T., R. L. McPherron, and S. Kokubun, "The Ground Signatures of the Expansion Phase During Multiple Onset Substorms," *Planet. Space Sci.*, **24**, 1115-1132 (1976).

Rosenberg, T. J., D. L. Detrick, P. F. Mizera, D. J. Gorney, F. T. Berkey, R. H. Eather, and L.J. Lanzerotti, "Coordinated Ground and Space Measurements of an Auroral Surge Over South Pole," *J. Geophys. Res.*, **92**, 11123-11132 (1987).

Rostoker, G., S.-I. Akasofu, J. Foster, R. A. Greenwald, Y. Kamide, K. Kawasaki, A. T. Y. Lui, R. L. McPherron, and C. T. Russell, "Magnetospheric Substorms—Definition and Signatures," *J. Geophys. Res.*, **85**, 1663-1668 (1980).

Rostoker, G., A.Vallance Jones, R. L. Gattinger, C. D. Anger, and J.S. Murphree, "The Development of the Substorm Expansive Phase: The 'Eye' of the Substorm," *Geophys. Res. Lett.*, **14**, 399 (1987a).

Rostoker, G., A. T. Y. Lui, C. D. Anger, and J. S. Murphree, "North-South Structures in the Midnight Sector Auroras as Viewed by the Viking Imager," *Geophys. Res. Lett.*, **14**, 407 (1987b).

Sergeev, V. A., "The Discrete Activations of the Magnetosphere During the Substorm Explosive Phase," Proc. of STP Symposium, Sao Paulo, Brazil, ed. by Brazilian Organizing Committee, **2**, pp. 22-58 (1974).

Sergeev, V. A., and A. G. Yahnin, "The Features of Auroral Bulge Expansion," *Plant. Space Sci.*, **27**, 1429-1440 (1979).

Shepherd, G. G., C. D. Anger, J. S. Murphree, and A. Vallance Jones, "Auroral Intensifications in the Evening Sector Observed by the Viking Ultra Violet Imager," *Geophys. Res. Lett.*, **395** (1987).

Smits, D. P., W. J. Hughes, C. A. Cattell, and C. T. Russell, "Observations of Field-Aligned Currents, Waves, and Electric Fields at Substorm Onset," *J. Geophys. Res.*, **91**, 121-134 (1986).

Steen, A., and P. N. Collis, "High Time-Resolution Imaging of Auroral Arc Deformation at Substorm Onset," *Planet. Space Sci.*, **36**, 715 (1988).

Wiens, R. G., and G. Rostoker, "Characteristics of the Development of the Westward Electrojet During the Expansive Phase of the Magnetospheric Substorms," *J. Geophys. Res.*, **80**, 2109-2128 (1975).

Yahnin, A. G., V. A. Sergeev, R. J. Pellinen, W. Baumjohann, K. U. Kaila, H. Ranta, J. Kangas, and O. M. Raspopov, "Substorm Time Sequence and Microstructure on 11 November 1976," *J. Geophys.*, **53**, 182-197 (1983).

V-4. POLEWARD MOTIONS OF AURORAL STRUCTURES

E. W. Hones, Jr.,* A. B. Galvin,[†] and P. R. Higbie*

We have studied two substorms using data from the magnetotail and from synchronous orbit in correlation with auroral images from the Dynamics Explorer 1 (DE 1) satellite and from Viking. Our results lend support to the near-Earth neutral line model of substorms. They suggest that the substorm auroras map to a near-Earth ($|X_{SM}| < 18R_E$) neutral line and that the eastward and westward lengthening of the aurora during the expansive phase correspond to lengthening of the neutral line across the magnetotail. We find associations between poleward surges of the auroras and electrojet, the peak of the AL index, and recovery of the plasma sheet that are consistent with the view that a poleward leap of the auroras occurs late in substorms and may constitute the terminal stage of the substorm expansive phase.

1. INTRODUCTION

Poleward motion of auroras on the nightside of the auroral oval is a defining feature of the auroral substorm, specifically of the expansive phase of the auroral substorm. The expansive phase starts with a localized brightening of the nightside aurora, which quickly spreads in longitude, and bright, active auroral structures move rapidly poleward so that an auroral bulge of bright auroral emission, measuring hundreds of kilometers in the north-south direction and a thousand or more kilometers in the east-west direction, exists within a few minutes, or tens of minutes, after the onset of the expansive phase. The expansive phase carries auroras poleward to 75–80° magnetic latitude within an hour or so, after which they become less intense and more slowly recede equatorward in what is known as the substorm recovery phase. This description of an auroral substorm was derived through detailed study of all-sky camera observations from an array of ground stations [Akasofu, 1964]. Its general features have been confirmed, and new details have been added through observations of the whole auroral oval during substorms by DE 1 [Craven and Frank, 1985, 1987] and Viking [Anger et al., 1987].

Studies of individual auroral substorms with all-sky cameras are difficult and often unsatisfactory because of the limited view of a single camera compared to the large scale of the substorm, because of the vagaries of the weather and because of practical problems of long-term maintenance and operation of cameras. Auroral substorms have a characteristic geomagnetic signature, and a worldwide network of magnetic stations has existed for many years with which the geomagnetic signatures of substorms are recorded. Thus, information about the magnetic features of substorms is much more readily available than that about the visible features.

During the past quarter century, satellites with scientific payloads of ever-increasing sophistication have explored space around the Earth. Many of these investigations were instigated directly or indirectly to learn about the cause of the auroras and substorms. Of course such investigations often involved correlating phenomena in space, measured by a satellite, with concurrent substorm features measured at Earth. Because of the ready availability of geomagnetic data, it was usually the magnetic signatures of substorms that were used in the correlations. Thus a whole substorm science has been built up that relates phenomena in the plasma around the Earth primarily to the geomagnetic signatures of substorms and only indirectly to the auroral features by which substorms were originally defined. That is not to say that the substorm's magnetic signatures are uninteresting or unimportant. They are important because they tell about the electric currents that flow in the magnetosphere and ionosphere during substorms. Yet the auroras, being visible, offer the one possibility of presenting, for inspection, the behavior of the entire substorm at Earth. This multiplies, many fold, the opportunities for correlations with satellites. The auroral imaging capabilities of the DE 1 and Viking satellites thus represent a giant step forward in our ability to study substorms, particularly in our ability to relate them to their causative agents in outer space. It was for this reason that the PROMIS (Polar Regions and Outer Magnetosphere International Study) campaign was conceived in 1985 and carried out in the spring of 1986. PROMIS exploited the unique opportunity that existed during a three-month period for maximizing acquisition of outer magnetosphere and solar-wind data simultaneously with DE 1 and Viking auroral imaging.

*Los Alamos National Laboratory, Los Alamos, NM 87545.
[†]Institute of Physical Science and Technology, The University of Maryland, College Park, MD 20742.

Auroral Physics, edited by C.-I. Meng, M. J. Rycroft and L. A. Frank. © Cambridge UP 1991

One model views a substorm as the magnetosphere's sporadic release of energy previously acquired from the solar wind. The stored energy is partly released to the upper atmosphere, where it causes the auroral substorm, and partly returned to the solar wind. This is accomplished by magnetic reconnection at a "near-Earth" magnetic neutral line (X-line) that forms in the plasma sheet, some $15R_E$ behind the Earth. The portion of the plasma sheet tailward of the X-line becomes disconnected from the Earth, becoming a configuration of closed magnetic loops (a plasmoid) that flows down-tail and is lost. The field lines earthward of the X-line collapse (i.e., contract) earthward, somehow causing the variety of substorm effects in the polar regions. The substorm sequence in the tail ends with a rapid retreat of the near-Earth neutral lines to a location far ($\sim 100R_E$) down-tail. Important observed signatures of these proposed magnetotail features of a substorm are a plasma dropout, or plasma sheet thinning, that occurs at distances greater than $\sim 15R_E$ near the time of the expansive phase onset at Earth, and plasma sheet thickening or "recovery" that occurs about a half-hour to an hour, or more, later [*Hones*, 1979]. Recalling that the expansive phase of the auroral substorm has the auroras over much of the nightside oval advancing poleward more or less monotonically for perhaps an hour, we note the seemingly strange fact that the early stage of auroral expansion, when the auroras are just starting to move poleward, coincides with a dropout of the plasma sheet, while the late stage of its expansion, when the auroras have reached high latitudes, coincides with an opposite effect, i.e., plasma sheet thickening.

The magnetic signature of the onset of a substorm's expansive phase is the onset of a negative H-bay, caused by establishment of a westward electrojet in the ionosphere at mid-auroral latitudes, i.e., $\lambda_m \approx 65°$. This, of course, coincides with the plasma sheet thinning. It is found [*Hones et al.*, 1973; 1984] that when the plasma sheet recovers, about an hour after expansive phase onset, the westward electrojet moves rapidly poleward by perhaps 5°. This geomagnetic feature of a substorm has been called the "poleward leap" of the electrojet [*Hones et al.*, 1973]. It is natural to expect that the auroras must execute a poleward leap, too, and Hones [1985] claimed that evidence for such a feature existed in DE 1 images of a substorm on November 8, 1981, published earlier by *Craven et al.* [1984]. *Hones* [1985] proposed that the poleward leap could be regarded as the terminal feature of the substorm expansive phase, a final rapid poleward motion of the electrojet and the auroras that immediately precede the recovery phase. However, *Craven and Frank* [1987] argued that the November 8, 1981 event oc-

curred in an unusual period of auroral activity and did not support the concept of a poleward leap.

In this paper we discuss two substorms during which the ISEE 1 and 2 satellites measured magnetotail plasma and energetic particles concurrent with imaging of the polar auroras by DE 1 or Viking. These observations are relevant to the questions of the poleward leap of the auroras and electrojet, formation of a near-Earth neutral line, and configuration changes of the tail magnetic field during substorms.

2. OBSERVATIONS OF A SUBSTORM ON MAY 4, 1986

A substorm on May 4, 1986, midway through the PROMIS campaign, was uniquely well documented. The expansive phase onset was imaged at 20-second time resolution by the Viking satellite. Most of the expansive phase was also imaged by DE 1 at 6-minute time resolution. ISEE 1 and 2 were near the tail's axis $18.5R_E$ from Earth, operating at high data rate, and data were recorded by several geosynchronous satellites. Here we present a preliminary analysis of the May 4 event that utilizes only a small fraction of the available data. A much more thorough study of the event will be made in a planned PROMIS Coordinated Data Analysis Workshop (PROMIS CDAW).

Figure 1 shows H- and Z-component magnetograms from stations in Alaska and Western Canada. These stations were near 2400 MLT. Anchorage, Talkeetna, and College show negative H-bay onsets at 1155 UT. Z-component deflections at 1155 UT, negative at Anchorage and positive at College and Fort Yukon, indicate that the substorm westward electrojet was initiated

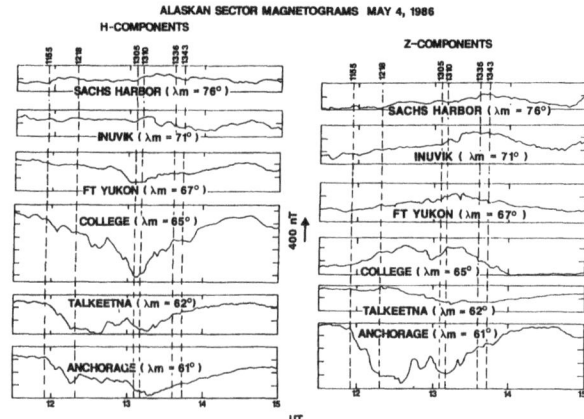

Figure 1—H- and Z-component magnetograms from several stations in Alaska and western Canada for 1130–1500 UT on May 4, 1986. Vertical dashed lines through the figures mark times of special significance that are mentioned in the text.

SATELLITE 1984-129, MAY 4, 1986

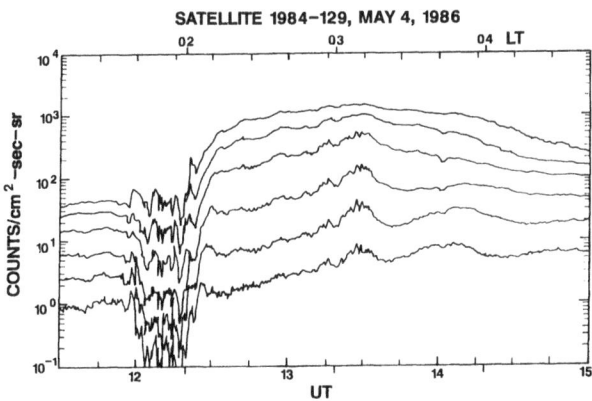

Figure 2—Electron count rates at energies greater than 30, 45, 65, 95, 140, and 200 keV measured between 1130 UT and 1500 UT on May 4, 1986 with the Los Alamos Charged Particle Analyzer on geosynchronous satellite 1984-129. The local time of the satellite is indicated at the top.

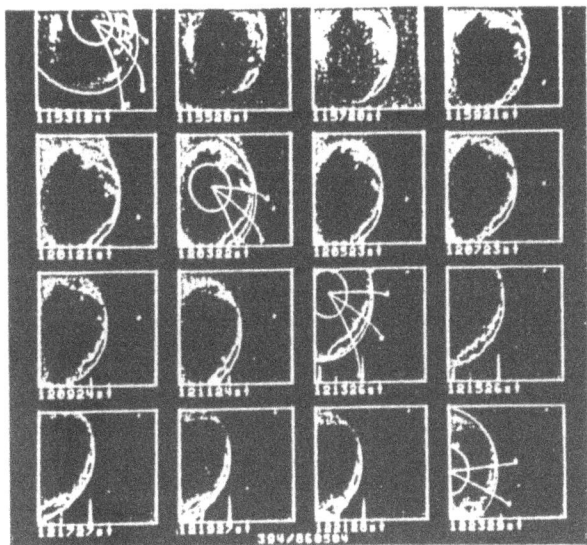

Figure 3—Images of the Earth's northern polar region showing the evolution of the auroras during the May 4, 1986 substorm. The images were recorded, each in one second, by the Viking satellite in atomic oxygen (O I) 130.4 nm emission. The dawn meridian is toward the upper right and the dusk meridian is toward the lower left in each image. The 2200-, 0000-, and 0200-MLT meridians are drawn in four of the images. (This figure also appears in color: Plate 42.)

at latitudes between 61° and 65°, perhaps near Talkeetna (62°) which showed little Z-deflection. Electron data from the Los Alamos Charged Particle Analyzer on geosynchronous satellite 1984-129, near 0130 MLT (Figure 2), shows a distinct change of character at 1155 UT, with the more energetic electrons decreasing rapidly in intensity. The combined satellite and ground data suggest the onset of a substorm expansive phase at 1155 UT.

Figure 3 shows a sequence of 16 northern hemisphere images from the Viking satellite from 1153 to 1223 UT. We show only every sixth one of Viking's images. The image at 1153:19 UT shows a localized brightening near 2200 MLT that later spreads east and west along the oval and expands poleward. The image taken 20 seconds earlier showed no such brightening, so the Viking images place the substorm onset between 1152:59 UT and 1153:19 UT. We adopt 1153 UT as the onset time rather than the later time (1155 UT) derived from the ground records and geosynchronous satellite data.

Figure 4 (top panel) shows the *AL* index for the interval 1130–1500 UT. Below that is a sequence of seventeen 6-minute images of the southern auroral region taken by DE 1 from 1202 through 1345 UT. Each image is centered under the 6-minute time interval during which it was accumulated. Below the images is a plot of the flux of 27–33 keV protons measured with the Ultra Low Energy Charge Analyzer (ULECA) on ISEE 1 [*Hovestadt et al.,* 1978]. ISEE 1 was near the tail axis about 18.5R_E from Earth. Its location is given under the second panel of Figure 4. The bottom panel shows the latitude of the poleward auroral boundary at 2100, 0000, and 0300 MLT, measured on the DE 1 images using computer graphics techniques developed for this purpose by the University of Iowa.

Also shown is the projection to the ionosphere of ISEE 1 along field lines of the magnetosphere model published by *Tsyganenko* [1987]. The model was computerized for this purpose at Los Alamos [*J. Birn,* private communication, 1988]. In this model, ISEE 1 mapped to 71° magnetic latitude and ~0110 MLT. Here we show it mapped to –71° magnetic latitude. This implies north-south symmetry of the magnetic field, which may not actually be true.

The proton flux at ISEE 1 (second panel of Figure 4) did not show any effect of the substorm onset at 1153 UT. There was a significant drop of flux at 1200 UT and then, starting at ~1217 UT, a major decrease of flux began, ending at ~1222 UT in a reduction to less than one percent of the presubstorm intensity. Figure 2 shows that synchronous satellite 1984-129, at ~0200 MLT, recorded a sudden drop of electron flux from 1215–1217 UT and then a large increase from 1217–1221 UT.

Figure 5 shows latitudinal contours of auroral intensity at 2-minute intervals at the 0200 MLT meridian. These were derived from the Viking images of Figure 3 using computer techniques developed at the University of Calgary. Although there are differences among the first six panels, there is a distinct change in the seventh (1211:24 UT) panel—that is, the appearance of the narrow spike at about 63° magnetic lati-

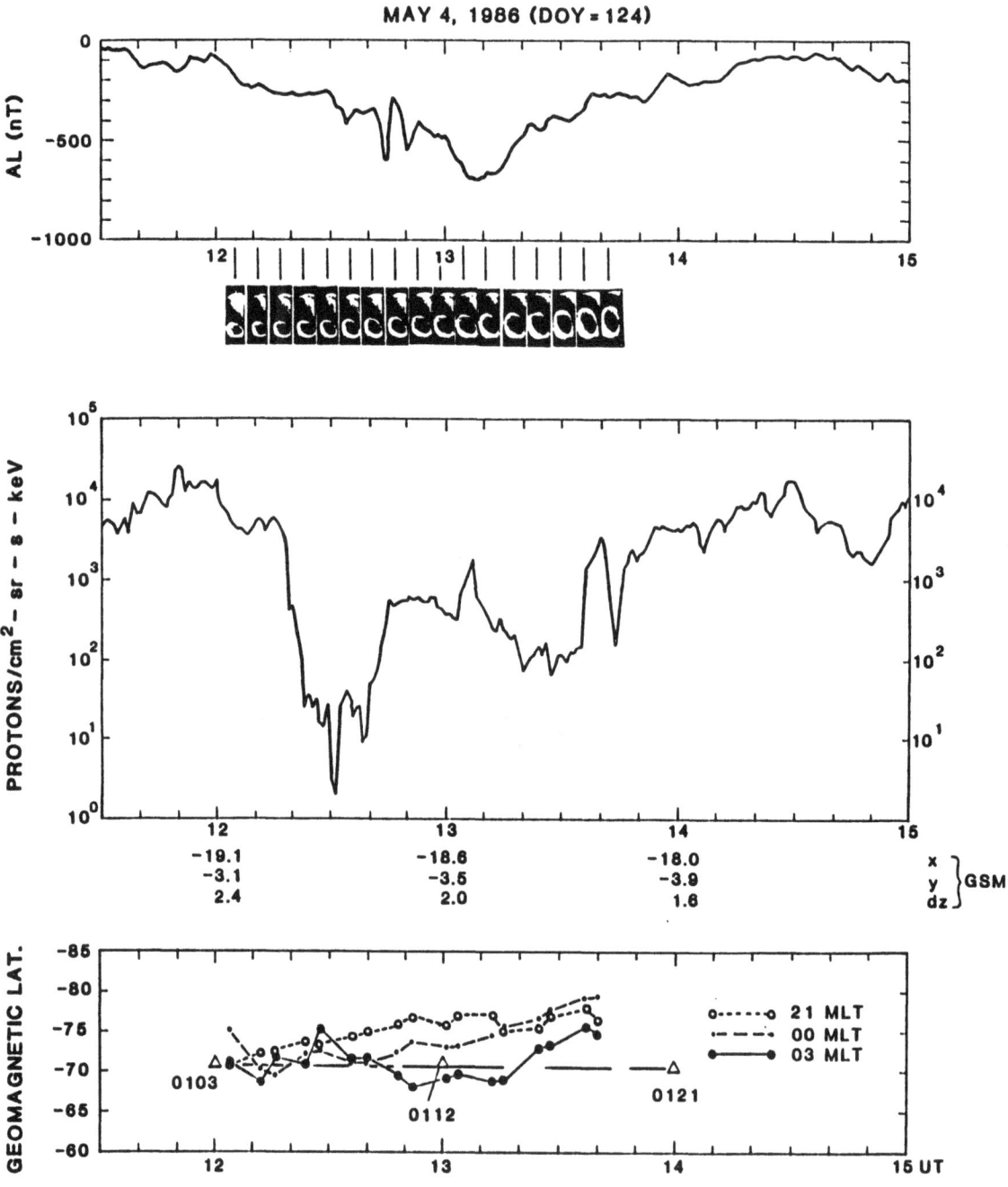

Figure 4—Data for the substorm on May 4, 1986. *Top panel:* The *AL* index. Auroral images: DE 1 images of the southern auroral oval as seen in atomic oxygen (O I) 130.4 nm emission. The dawn meridian is toward the upper left and the dusk meridian is toward the lower right in each image. The width of each image is equal to 6 minutes on the horizontal time scale of the figure and each image is centered at the midtime of its 6-minute accumulation interval.
Middle panel: Flux of 27–33 keV protons measured with the ULECA instrument on ISEE 1. The location of ISEE 1 is given under this panel.
Bottom panel: Latitude of the poleward edge of the auroras at 2100, 0000, and 0300 MLT, determined by computer graphics analyses of the DE 1 images. Also shown, by the straight long-dashed line, is the latitude of the foot-point of the field line from ISEE 1 (see text). The MLT of the foot-point is given every hour. (This figure also appears in color: Plate 43.)

LATITUDINAL PROFILES OF AURORAL INTENSITY AT 0200 MLT
VIKING AURORAL IMAGES, MAY 4, 1986

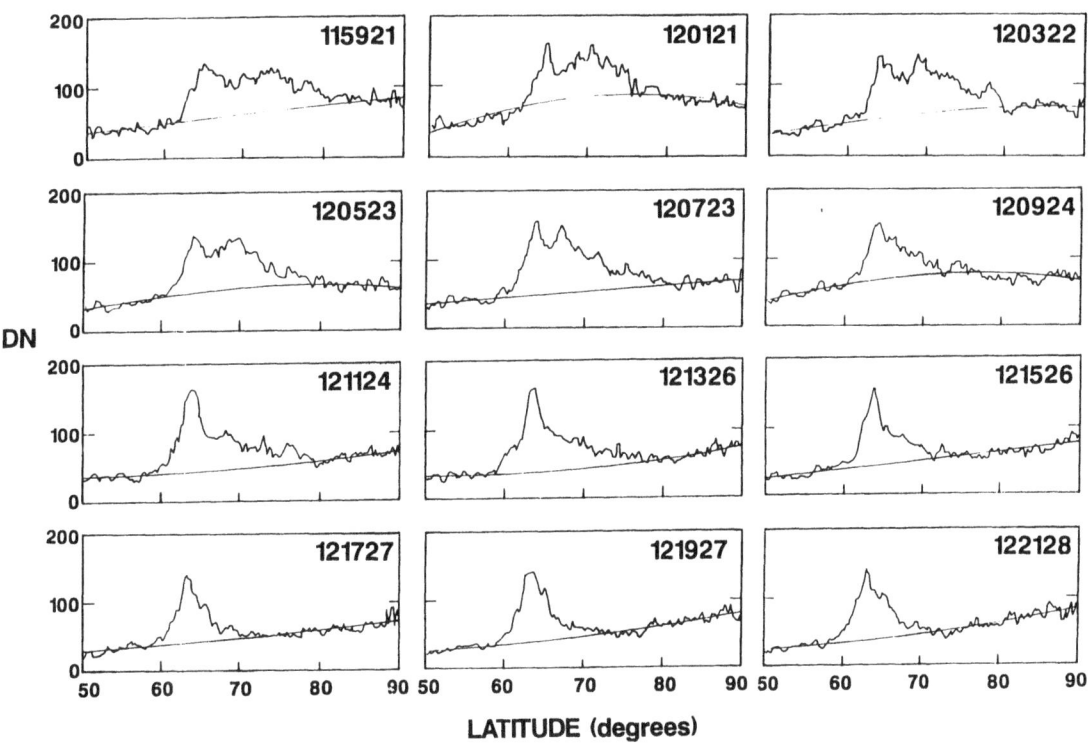

Figure 5—Latitudinal profiles of auroral intensity along the 0200-MLT meridian, derived from the Viking images. The smooth curve is the computer-derived background intensity level. The ordinate scale is in numbers of counts per pixel.

tude that signifies the arrival of the eastward-extending substorm aurora at the 0200 MLT meridian. Latitudinal profiles measured at other magnetic local times suggest that the aurora advanced from 0100–0230 MLT in the time interval 1203:22–1215:26 UT. The aurora thus spread well into the morning sector of the oval between 10 and 22 minutes after substorm onset. At 1217 UT (24 minutes after substorm onset) the proton flux at ISEE 2 started its rapid decrease (the plasma sheet thinned there) and the electron flux at synchronous orbit suddenly increased (a particle injection occurred there).

The AE index peaked at ~1308 UT and started to subside (Figure 4). At ~1320 UT the aurora at 0300 MLT surged poleward, and after 1325 UT the auroras at all three local times (2100, 0000, and 0300 MLT) moved rapidly poleward, that at 0000 MLT approaching $-80°$ latitude by 1340 UT. At 1335 UT the proton flux at ISEE 1 suddenly increased nearly to presubstorm values, marking the plasma sheet's final expansion over ISEE 1.

At 1320 UT the H-component of the field at Inuvik ($\lambda_m = 71°$) suddenly decreased and the Z-component suddenly increased (Figure 1). These signatures, together with the fact that the H-components at all the lower latitude stations were becoming less negative, are consistent with a substantial poleward movement of the westward electrojet. Note that there was also a diminution of the intensity of the electrojet—the H- and Z-components that developed at Inuvik after ~1320 UT never became as great as those that appeared earlier at the lower-latitude stations. This may imply either a reduction of the total current or a latitudinal dispersal of the current that reduced its local intensity.

3. DISCUSSION OF THE MAY 4, 1986 EVENT

The observations during the May 4 event have important relevance to several aspects of the neutral line model of substorms. We shall discuss the implications of the observations concerning (1) the formation and development of the near-Earth neutral line; (2) probable configuration changes of the tail magnetic field; (3) the substorm's late phase and the poleward leap of the auroras and electrojet.

1. Ground observations have shown that the substorm auroral brightening spreads eastward and westward along pre-existing faint auroral arcs at speeds of 10 km s^{-1} or greater [*Opgenoorth et al.,* 1980; 1983]. There have been suggestions that the spreading bright aurora maps to a magnetic X-line that is spreading across the near-tail plasma sheet [*R. J. Pellinen,* private communication, 1985]. Our observations during the May 4 substorm provide rather direct evidence in support of this view. Figure 6 depicts the idea schematically. Imagine that an X-line was initiated at the 2300 MLT meridian in the near tail (bottom of Figure 6). Spreading eastward, it brings about magnetic reconnection in successive flux regions producing closed flux loops (i.e., tail section A and then tail section B) that successively flow tailward as "elemental" plasmoids, and Earth-tied flux regions (i.e., section A′ then section B′) that successively collapse earthward, engendering eastward-spreading auroras at their feet. In Figure 6 the neutral line has just reached tail-section C, which will also experience magnetic reconnection and the resulting changes such as already occurred in sections A and B.

Briefly, the observation that conforms to Figure 6 is that the frequently observed signatures of expansive phase onset, i.e., particle injection at synchronous orbit (Figure 2) and plasma sheet thinning in the distant ($X_{SM} \lesssim -15R_E$) tail (Figure 4) occurred about simultaneously (at ~1217 UT) at satellite 1984-129 and ISEE 1, which were both in the early-morning sector of the tail. But they did so, not at substorm onset (at ~1153 UT), but more than 20 minutes later, concurrent with the eastward-extending aurora's reaching well into the morning sector (Figure 5).

An obvious question about this interpretation is how to explain the nearly immediate (1155 UT) onset effect seen by satellite 1984–129, i.e., the flux perturbations and dropouts. We suggest that collapse of the flux tubes in the ~2200 MLT sector, where the substorm onset occurred, drastically altered electron drift paths so electrons could no longer easily reach satellite 1984–129 where the collapse had not yet taken place. Another question is why the substorm signatures at 1984–129 and ISEE 1 did not occur sooner, say precisely when the advancing aurora reached the MLT of their mapping points. We suggest that this could

CROSS-TAIL LENGTHENING OF THE NEAR EARTH NEUTRAL LINE DURING SUBSTORMS

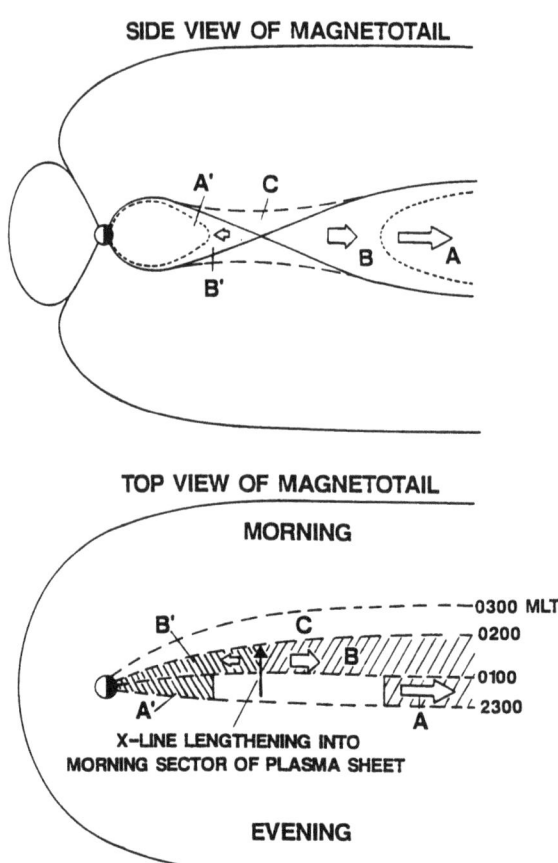

Figure 6—Side view and top view of the magnetosphere illustrating the proposed effects of progressive lengthening of the near-Earth neutral line after its initial formation at the onset of a substorm expansive phase (see text).

indicate uncertainties in projecting the satellite positions to Earth and possible delays between the local formation of an X-line and the resulting macroscopic changes in magnetic configuration that constitute the familiar substorm signatures at the satellite locations.

If our interpretation is correct, the observations imply that (i) The active substorm aurora lies at the feet of field lines that map to a magnetic X-line in the near tail; (ii) Extension of the aurora along the oval manifests the extension of the neutral line across the near-tail; (iii) In the May 4, 1986 event, the neutral line was formed earthward of 18.5R_E (and tailward of synchronous orbit).

2. It is customary to associate the auroras with closed field lines and thus their poleward border

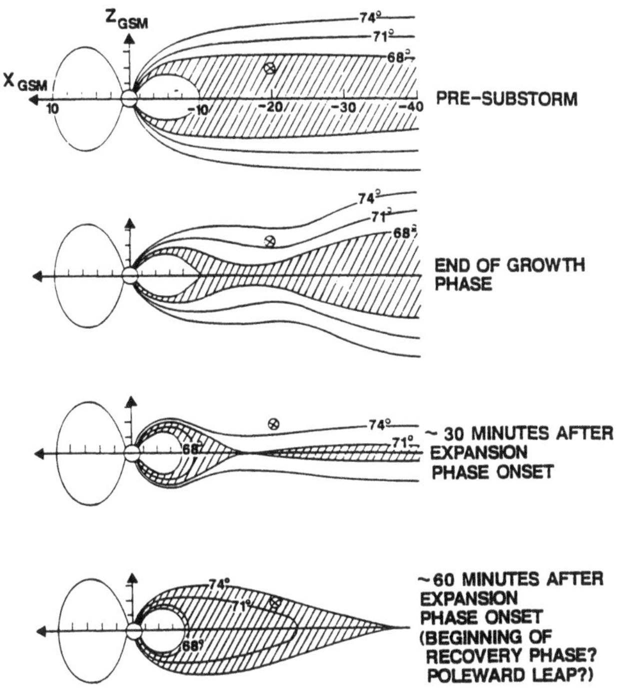

PLASMA SHEET CONFIGURATION CHANGES CAUSED BY SUBSTORMS

PRE-SUBSTORM

END OF GROWTH PHASE

~ 30 MINUTES AFTER EXPANSION PHASE ONSET

~ 60 MINUTES AFTER EXPANSION PHASE ONSET (BEGINNING OF RECOVERY PHASE? POLEWARD LEAP?)

Figure 7—Configurations of the plasma sheet (hatched region) that may occur in association with a substorm. Three representative field lines are marked in each panel. A satellite is indicated at $X_{GSM} = -20R_E$, $Z_{GSM} = 4R_E$.

with the last closed field line, i.e., with the lobe-plasma sheet interface. Thus it is reasonable to consider that a satellite that is in the plasma sheet maps magnetically to some location equatorward of the poleward border of the auroras. Also, when a satellite is crossing from lobe to plasma sheet, or vice versa, it maps to some point along the poleward border of the aurora.

The bottom panel of Figure 4 shows the geomagnetic latitude of the aurora's poleward boundary at 2100, 0000, and 0300 MLT. A nearly straight line shows also the projected magnetic latitude and MLT of ISEE 1 (as described earlier). The graphs suggest that field lines from the poleward edge of the aurora at 0000 and 0300 MLT (bracketing the projected local time of ISEE 1) extend out to ISEE 1 from 1200 UT at least until 1250 UT. That is, the poleward borders of the auroras in that MLT region were at about the same or higher magnetic latitude than the footpoint of the ISEE 1 field line ($\lambda_m \approx -71°$). The middle panel of Figure 4 suggests that this mapping is reasonable before ~ 1220 UT. But after that time and until ~ 1245 UT, ISEE 1 was

clearly outside the plasma sheet, i.e., field lines from the aurora's poleward border did not reach out to the satellite. The plasma sheet clearly returned to ISEE 1 in full intensity only after ~ 1330 UT, when the auroral border had reached latitudes of $-75°$ to $-80°$.

Figure 7 shows configurations that the plasma sheet (at least its near-midnight sector) may assume before and during a substorm. These are suggested by the neutral line model of substorms and are consistent with many observations. Field lines labeled 68°, 71°, and 74° do not apply necessarily to the situation during the May 4 substorm but are meant to be representative of field lines that ordinarily lie in the auroral zone, at low polar-cap latitude and at high polar-cap latitude. The diagrams suggest that before a substorm and during the growth phase, mid-auroral zone field lines (68°) can extend far out into the tail at large distances above the midplane. Our May 4 data tell us nothing about that because the DE 1 images start after expansive phase onset. But even then, for a few minutes, field lines from ~ 70° extended past ISEE 1. Some 30 minutes after expansive phase onset, when auroras have advanced several degrees poleward (shown as 71° in the third panel), the border field lines may reach far out into the tail, but they pass through the near-Earth neutral line, lie very close to the midplane, and are not likely to be sampled by a satellite that is well away from the midplane. In the May 4 event, field lines from the auroras, at ~ $-70°$ to $-75°$, obviously did not reach out to the ISEE 1 location, ~ $2R_E$ from the midplane.

The fourth panel of Figure 7 shows the situation late in a substorm. The auroras have advanced to high polar-cap latitude (74°) and at that time, suddenly, the satellite in the figures is re-enveloped by the plasma sheet, presumably because of the retreat of the neutral line and refilling of the plasma sheet. In the May 4 event, the final envelopment of ISEE 1 by the plasma sheet occurred as auroras reached latitudes between $-75°$ and $-80°$.

There is no existing magnetosphere model that predicts quantitatively the configuration changes shown in Figure 7 that can reasonably be believed to occur during a substorm. Attempts to map a satellite's position to Earth, or, conversely, to map the auroras into the tail, during substorms using existing quantitative models are obviously futile.

3. Recalling that the proposed poleward leap of the westward electrojet (and, presumably, of the

auroras) is visualized as a few-degree rapid pole-ward motion of the electrojet and auroras that terminates the expansive phase and that is associated with a thickening of the plasma sheet, we ask whether such a set of phenomena appeared in the May 4 substorm. The answer seems to be "yes." The peak of the AL index occurred ~1310 UT. By ~1320 UT there were signatures in the Alaskan-Canadian magnetograms of a substantial poleward displacement of the westward electrojet. A simple visual examination of the DE 1 images reveals an accelerated thickening of the nightside of the oval in the last four pictures (beginning ~1321 UT), and this impression is supported quantitatively by the curves in the bottom panel of Figure 4 that show the poleward boundary at all three local times to be moving poleward, after 1320 UT, as fast as, or faster than, at any earlier time. At 1335 UT, during this concerted poleward motion of the auroras, the plasma sheet at ISEE 1 experienced a major recovery. Thus, there was an interval of about 20 minutes (rather than the 5 minutes proposed by *Hones* [1985]) occurring about 1½ hours after expansive phase onset that displayed the features of a poleward leap of the electrojet and auroras. Unfortunately there are no images after 1340 UT to provide a look at the transition between the poleward leap and the recovery phase.

4. OBSERVATIONS OF A SUBSTORM ON MARCH 28, 1983

An isolated substorm began at ~0330 UT on March 28, 1983 following four hours of quiet AE conditions ($AE < 50$ UT). Figure 8 shows the AE index from 0230-0600 UT. DE 1 provided eleven images of the entire southern auroral oval from 0308-0512 UT. These are shown below the AE index. The center of each image lies at the center of the 12-minute interval during which it was accumulated. The two panels below the images show the fluxes of 2-keV electrons and 6-keV protons measured with electrostatic analyzers on ISEE 1 [*Anderson et al.,* 1978]. The bottom panel shows the geomagnetic latitude of the aurora's poleward boundary at three magnetic local times, 2100, 0000, and 0300 MLT, measured from the images, using the computer graphics system at the University of Iowa. The geomagnetic latitude and magnetic local time of the footpoint of the ISEE 1 magnetic field line are also shown.

The AE index suggests an onset time of 0330 UT. The first DE 1 image (0308-0320 UT) shows no auroral brightening, but the second one (0323-0335 UT) does. The images thus suggest that the substorm started af-

ter 0320 and before 0335, in agreement with the AE index. The location of ISEE 1 is shown below the middle panels. It was in the early morning sector of the magnetotail about $20R_E$ from Earth and was far (~$8R_E$) below the estimated position of the neutral sheet. Even at this large distance from the neutral sheet, however, substantial fluxes of plasma electrons and protons were measured before the substorm, suggesting that the half-thickness of the presubstorm plasma sheet was about $8R_E$. The bottom panel shows that ISEE 1 projected to 0100-0200 MLT and $-76°$ magnetic latitude.

We now note several features of the auroral development and corresponding features in the outer magnetosphere. The early poleward expansion of the auroras (second, third, and fourth images) was largely confined to the 0000 MLT sector, although the beginning of an extension into the morning sector is barely discernible in the fourth image (0347-3059 UT). A temporary enhancement of plasma intensity (0333-0350 UT) occurred during this poleward expansion. In the fifth image (0359-0412 UT) the dawnward extension of the aurora is very clear and the bottom panel of Figure 8 shows that there was a poleward expansion of nearly 5° at 0300 MLT in that image. The dawnward extension had not reached 0300 MLT in the 0347-0359 UT image, but computer analysis of that image showed that by 0352 UT the aurora at 0130 MLT, where ISEE 1 projected to Earth, had expanded poleward by 2°. At ~0350 UT there was a sharp drop of plasma intensity at ISEE 1 (middle panels of Figure 8), and by 0400 the intensity had dropped below the instrument thresholds.

The seventh image (0424-0436 UT) shows a remarkable poleward surge of the auroras that represents expansion at all three meridians, 2100, 0000, and 0300 MLT (bottom panel). The interval of this surge includes a final sharp intensification of the AL index and the beginning of its recovery, i.e., it encompasses the AL peak. The surge carries the aurora at 0300 MLT to $-75°$ latitude where it then remains through the last image (0512-0524 UT). Geomagnetic Z-component records from ground stations in the auroral zone, ranging in magnetic local time from ~2200 MLT (Great Whale River and Fort Churchill in Canada) to ~0400 MLT (Leirvogur in Iceland), registered the signature of overhead poleward passage of a strong westward current between ~0425 and ~0435 UT. This was a magnetic signature of the rapid poleward expansion of the auroras over this broad local time range.

At geosynchronous orbit (data not shown here) a >30 keV electron detector on satellite 1982-019, at ~2300 MLT, showed a gradual dropout of flux from ~0300 until 0325 UT. Between 0325 and 0330 UT the

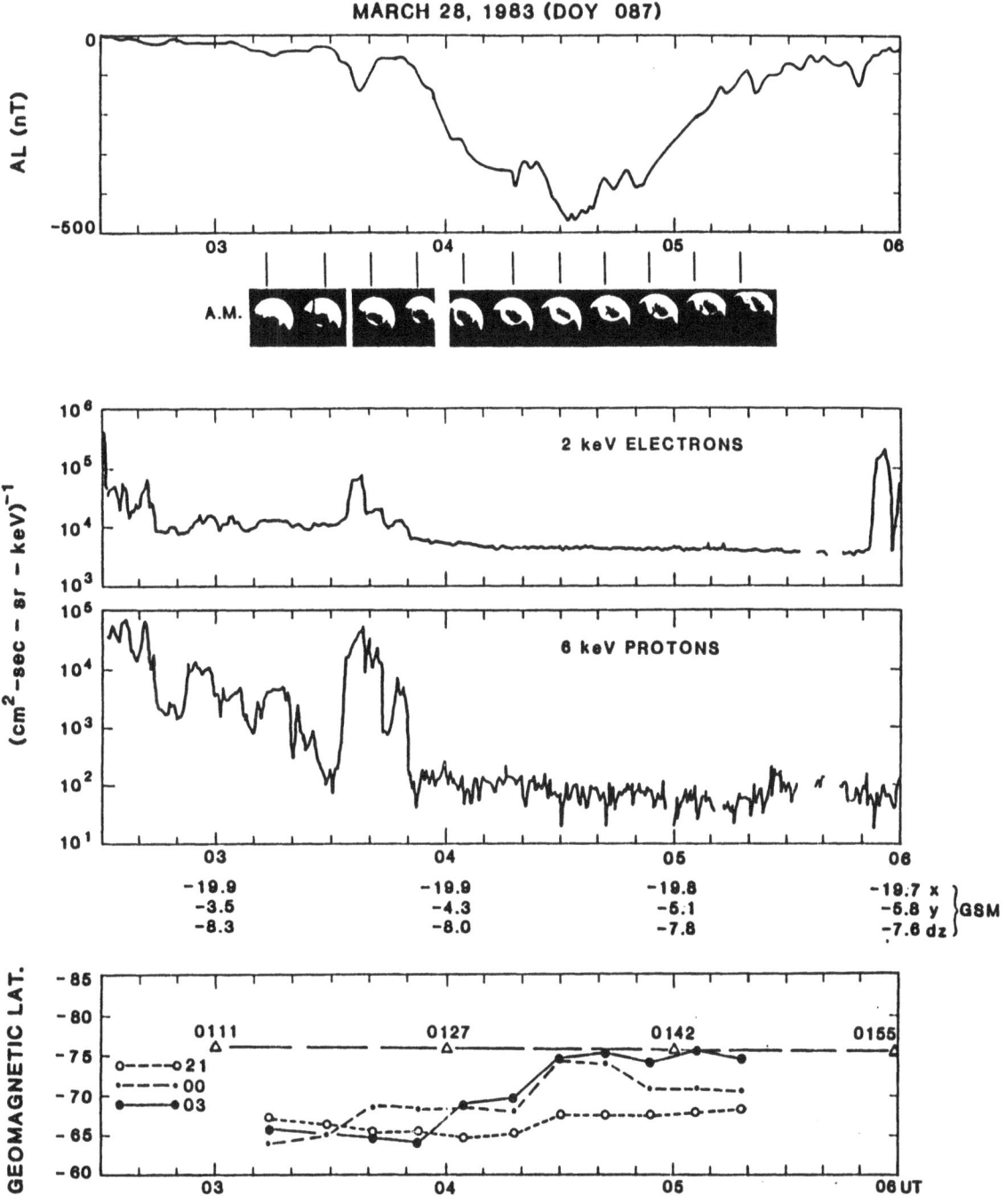

Figure 8—Data for a substorm on March 28, 1983. *Top panel:* The AL index. Auroral images: DE 1 images of the southern auroral oval as seen in atomic oxygen (O I) 130.4 nm emission. The dawn meridian is toward the upper left of each image and the dusk meridian is toward the lower right. The width of each image is equal to 12 minutes on the horizontal time scale of the figure and each image is centered at the midtime of its 12-minute accumulation interval.
Middle panels: Fluxes of 2-keV electrons and 6-keV protons measured by the University of California at Berkeley Energetic Particle Detector on ISEE 1. The location of ISEE 1 is given under these panels.
Bottom panel: Latitude of the poleward edge of the aurora at 2100, 0000, and 0300 MLT determined by computer graphics analyses of the DE 1 images. Also shown (by the long-dashed line) is the latitude of the foot-point of the field line from ISEE 1 (see text). The MLT of this point is shown every hour. (This figure also appears in color: Plate 44.)

electron flux suddenly returned to slightly higher intensity than that before the dropout. These signatures suggest a pre-expansive phase stretching of the magnetic field and its sudden dipolarization closely coincident with expansive phase onset. Later, satellite 1982-019, then at ~ 2345 MLT, recorded a large increase of electron flux beginning at 0419 UT and reaching a peak at 0431 UT. Also at geosynchronous orbit, a > 300 keV ion detector on satellite 1981-025, at ~ 1930 MLT, revealed a series of proton drift echoes [*Belian et al.,* 1978], the first in the series starting at 0425 UT and being very narrow (~ 3 minutes) and sharply peaked in time. These proton drift echoes were also observed by geosynchronous satellite 1977-007, at ~ 0900 MLT, about 9 minutes after their appearance at 1981-025, a delay appropriate for the westward drift of the proton bunches through 10 hours of local time. *Belian et al* [1978] found that proton drift echoes are seen at geosynchronous orbit in about 10% of substorms. They are thought to be produced by brief, highly localized injections of protons into the premidnight region of synchronous orbit, usually in close association with substorm onsets. The extreme narrowness of the initial proton pulse at satellite 1981-025 suggests that the satellite, at ~ 1930 MLT, was only a moderate distance to the west of the injection site. It is notable that this drift echo series was initiated, not at substorm onset (0330 UT), but about one hour later (0425 UT), at the time of the large poleward surge of the auroras that coincided with the *AL* peak.

No effect of the events at ~ 0430 UT was observed at ISEE 1, which was still outside the plasma sheet. The plasma sheet enveloped ISEE 1 again starting at 0552 UT and by about 0610 UT the plasma was hotter and more intense than it had been before the dropout at ~ 0350 UT (data not shown here).

5. DISCUSSION OF THE MARCH 28, 1983 EVENT

Although there are many differences in detail between this substorm and that of May 4, 1986, there are important likenesses in the underlying evolutions of the two events. We shall discuss these underlying features in the same sequence that we used in discussing the May 4 event.

1. The onset of this auroral substorm occurred near 0000 MLT (Figure 8). The immediate response at synchronous orbit (satellite 1982-019 at ~ 2300 MLT) seems to have been a dipolarization of the local magnetic field. Detailed analysis of the DE 1 images shows that the latitude of the poleward edge of the auroras at 2300 MLT was − 64° at 0317 UT, − 65.0° at 0329 UT, and − 66.2° at 0341 UT. Thus the substorm aurora evidently

reached 2300 MLT essentially coincident with substorm onset. The dipolarization of the field at synchronous orbit is consistent with this in the context of the interpretation sketched in Figure 6. At ISEE 1 in the early morning sector of the tail, a burst of plasma ions and electrons began at substorm onset (Figure 8) and lasted 20 minutes. Then, at ~ 0350 UT, when the auroras advanced dawnward to the neighborhood of the ISEE 1 magnetic foot-point, plasma sheet dropout occurred at ISEE 1. The observations, both at ISEE 1 and at satellite 1982-019, support the association of the advancing aurora to an advancing near-Earth neutral line portrayed in Figure 6. In this case also, the neutral line evidently formed earthward of ISEE 1 (which was 20R$_E$ from Earth) and tailward of synchronous orbit.

2. Before substorm onset ISEE 1 was within the plasma sheet despite the fact that its mapping to Earth lay nearly 10° poleward of the poleward border of the aurora. That suggests that, for the situation on March 28, the Tsyganenko model was inapplicable. But it does support the view, suggested in the top panel of Figure 7, that field lines from auroral latitudes stretch far into the tail well above the midplane before a substorm.

We again see that the effect of substorm onset is to reduce the extension of auroral field lines into the tail; i.e., even though the auroras advance to − 75°, those field lines do not reach ISEE 1 although before the substorm field lines from much lower latitudes did so.

3. In this substorm we again find a rather dramatic poleward advance of the auroras and westward electrojet along a broad range of local times occurring near the peak of the *AL* index. In this case the plasma sheet recovery at ISEE 1 occurs about 1½ hours later. We ascribe this long delay to the satellite's very large distance from the tail midplane.

6. CONCLUSIONS

This study of two substorms confirms the importance of the auroral imaging capability to the study of substorms, particularly when it is used in conjunction with in situ measurements of outer magnetosphere phenomena. Our results are quite supportive of the near-Earth neutral line model of substorms. In particular, they suggest that the substorm aurora map to a near-Earth ($|X_{SM}| < 18R_E$) neutral line and that the eastward and westward advances of the auroras manifest eastward and westward cross-tail advances of the neutral line. We found that widespread advances of the aurora and of the westward electrojet occurred late in both substorms and that these advances occurred near the times that the AL index reached its peak value and started to subside. In one substorm the plasma sheet recovered about 15 minutes after the poleward advance. In the other, when ISEE 1 was $8R_E$ above the tail midplane, plasma sheet recovery occurred about 1½ hours after the poleward advance. These results are consistent with the proposal that the expansive phase of a substorm terminates with a poleward leap of the auroras and electrojet.

Our observations suggest that before substorm onset midauroral zone field lines (i.e., $\lambda_m \approx 68°$) reach far out into the tail in the boundary of a thick plasma sheet. During the substorm expansive phase, those field lines close near the Earth and field lines from several degrees higher latitude mark the border of the plasma sheet, but a much thinner plasma sheet. After the expansive phase terminates with the poleward leap, field lines from high latitude ($\lambda_m \approx 75°$) mark the border of a much-thickened plasma sheet. There is no quantitative magnetosphere model that portrays these substorm-induced configuration changes of the outer magnetosphere.

ACKNOWLEDGMENT—We are indebted to L. A. Frank, J. D. Craven, Rae Dvorsky, and other members of the Department of Physics and Astronomy of the University of Iowa for use of the auroral images from the DE 1 satellite and for their generous assistance in computer analyses of the images. We thank J. Birn for his help and constructive suggestions in several aspects of this work. We thank N. R. Larson of the University of Washington for providing ISEE plasma data for many substorms and R. D. Elphinstone of the University of Calgary for his help in analyses of the Viking auroral image data for the May 4, 1986 substorm. We thank Jo Ann Joselyn and L. D. Morris of NOAA for digital geomagnetic data. The AE index was obtained on magnetic tape from the National Geophysical Data Center of NOAA, Boulder, CO. At the University of Iowa this research was supported by NASA under grants NAG5-483 and NGL-16-001-002 and by the Office of Naval Research under grant N00014-85-K-0404. The work at Los Alamos was done under the auspices of the U.S. Department of Energy with NASA support under order number S-56312-D.

REFERENCES

Akasofu, S. I., "The Development of the Auroral Substorm," *Planet. Space Sci.* **12**, 273 (1964).

Anderson, K. A., R. P. Lin, R. J. Paoli, G. K. Parks, C. S. Lin, H. Reme, J. M. Bosqued, F. Martel, F. Cotin, and A. Cros, "An Experiment to Study Energetic Particle Fluxes In and Beyond the Earth's Outer Magnetosphere," *IEEE Trans. Geosc. Electron.*, **GE-16**, 213 (1978).

Anger, C. D., J. S. Murphree, A. Vallance Jones, R. A. King, A. L. Broadfoot, L. L. Cogger, F. Creutzberg, R. L. Gattinger, G. Gustafson, F. R. Harris, J. W. Haslett, E. J. Llewellyn, J. C. McConnell, D. J. McEwen, E. H. Richardson, G. Rostoker, B. R. Sandel, G. G. Shepherd, D. Venkatesan, D. D. Wallis, and G. Witt, "Scientific Results from the Viking Ultraviolet Imager: An Introduction," *Geophys. Res. Lett.*, **14**, 383 (1987).

Belian, R. D., D. N. Baker, P. R. Higbie, and E. W. Hones, Jr., "High-Resolution Energetic Particle Measurements at $6.6R_E$, 2, High-Energy Proton Drift Echoes," *J. Geophys. Res.* **83**, 4857 (1978).

Craven, J. D., Y. Kamide, L. A. Frank, S. I. Akasofu, and M. Sugiura, "Distribution of Aurora and Ionospheric Currents Observed Simultaneously on a Global Scale," in *Magnetospheric Currents*, Geophysical Monograph 28, T. A. Potemra, ed., American Geophysical Union, Washington, DC, p. 137 (1984).

Craven, J. D., and L. A. Frank, "The Temporal Evolution of a Small Auroral Substorm as Viewed from High Altitude with Dynamics Explorer 1," *Geophys. Res. Lett.* **12**, 465 (1985).

Craven, J. D., and L. A. Frank, "Latitudinal Motions of the Aurora During Substorms," *J. Geophys. Res.* **92**, 4565 (1987).

Hones, E. W., Jr., "Transient Phenomena in the Magnetotail and Their Relation to Substorms," *Space Sci. Rev.* **23**, 393 (1979).

Hones, E. W., Jr., "The Poleward Leap of the Auroral Electrojet as Seen in Auroral Images," *J. Geophys. Res.* **90**, 5333 (1985).

Hones, E. W., Jr., J. R. Asbridge, S. J. Bame, and S. Singer, "Substorm Variations of the Magnetotail Plasma Sheet From $X_{SM} \approx -6R_E$ to $X_{SM} \approx -60R_E$", *J. Geophys. Res.* **78**, 109 (1973).

Hones, E. W., Jr., T. Pytte, and H. I. West, Jr., "Associations of Geomagnetic Activity with Plasma Sheet Thinning and Expansion: A Statistical Study," *J. Geophys. Res.* **89**, 5471 (1984).

Hovestadt, D., G. Gloeckler, C.Y. Fan, L. A. Fisk, F. M. Ipavich, B. Klecker, J. J. O'Gallagher, M. Scholer, H. Arbinger, J. Cain, H. Hofner, E. Kunneth, P. Laeverenz, and E. Tums, "The Nuclear and Ionic Charge Distribution Particle Experiments on the ISEE-1 and ISEE-C Spacecraft," *IEEE Trans. Geosci. Electron.*, **GE-16**, 166 (1978).

Opgenoorth, H. J., R. J. Pellinen, H. Maurer, F. Kuppers, W. H. Heikkila, K. U. Kaila, and P. Tanskamen, "Ground-Based Observations of an Onset of Localized Field-Aligned Currents During Auroral Breakup Around Magnetic Midnight," *J. Geophys.* **48**, 101 (1980).

Opgenoorth, H. J., R. J. Pellinen, W. Baumjohann, E. Nielsen, G. Marklund, and L. Eliasson, "Three-Dimensional Current Flow and Particle Precipitation in a Westward Travelling Surge (Observed During the Barium-GEOS Rocket Experiment)," *J. Geophys. Res.* **88**, 3138 (1983).

Tsyganenko, N. A., "Global Quantitative Models of the Geomagnetic Field in the Cislunar Magnetosphere for Different Disturbance Levels," *Planet. Space Sci.* **35**, 1347 (1987).

V-5. A MAGNETOSPHERE-IONOSPHERE COUPLING THEORY OF SUBSTORMS INCLUDING MAGNETOTAIL DYNAMICS

J. R. Kan,* L. Zhu,* A. T. Y. Lui,[†] and S.-I. Akasofu*

The magnetosphere-ionosphere (M-I) coupling model of auroral substorms proposed by *Kan et al.* [1988] is extended to predict changes of the plasma sheet configuration during substorms. The proposed substorm theory can explain where, when, and why an auroral arc near the poleward edge of the diffuse aurora in the premidnight sector brightens for the onset of a substorm. Substorms are powered by enhanced magnetospheric convection that must be driven by the solar wind. An enhanced magnetospheric convection forces the ionospheric convection to speed up. The information of enhanced magnetospheric convection is transmitted by Alfvén waves to the ionosphere. During the transient period, the ionospheric convection, the field-aligned current, and the ionospheric conductance are increasing with time. The most intense upward field-aligned current occurs within a narrow strip near the poleward edge of the diffuse aurora in the evening-midnight sector, when the poleward gradient of the diffuse auroral conductance belt overlaps with the convection reversal region in the evening-midnight sector. In the overlapping region, the divergence of the Hall and of the Pedersen currents are both upward to reinforce each other. Although the enhanced magnetospheric convection is the primary driving force, the ionosphere to a large extent controls the location and timing of the brightening of an auroral arc for the substorm onset. The M-I coupling theory of substorms predicts that a near-Earth X-line may form as a consequence of the onset of an intense substorm and that the location of the near-Earth X-line, when it forms in the plasma sheet after the onset of an intense substorm, should map to the poleward boundary of the diffuse auroral precipitation region in the ionosphere.

1. INTRODUCTION

The first comprehensive theoretical descriptions of magnetospheric substorms were presented by *Axford and Hines* [1961] for a closed magnetosphere and later by *Coroniti and Kennel* [1973] for an open magnetosphere. Since then, several substorm models have been proposed, each emphasizing certain aspects of the substorm phenomena. These models include (1) the X-line model [*McPherron et al.*, 1973], (2) the X-line plasmoid model [*Hones*, 1976], (3) the boundary-layer model [*Rostoker and Eastman*, 1987], (4) the magnetosphere-ionosphere (M-I) coupling model [*Kan and Sun*, 1985; *Kan et al.*, 1988], and (5) the thermal catastrophe model [*Smith et al.*, 1986; *Goertz and Smith*, 1989].

Any theory of substorm onset must explain how an auroral arc near the poleward edge of the diffuse aurora in the evening-midnight sector brightens rather suddenly. It is known that the sudden brightening arises from a chain of processes that originates from a sudden enhancement of upward field-aligned currents. In the M-I coupling theory of substorms, *Kan et al.* [1988] demonstrated that an enhanced magnetospheric convection can lead to enhanced upward field-aligned currents near the poleward edge of the diffuse aurora as the magnetosphere forces the ionosphere to respond. Alfvén waves are responsible for transmitting the in-

formation during the transient stage of enhanced M-I coupling. It is shown that as many as eight bounces of Alfvén waves are needed for the ionosphere to respond fully in the midnight sector, but only a few bounces are needed on the dayside, because of the relatively low ionospheric conductance. The most intense upward field-aligned current in their model is localized in a narrow strip along the poleward edge of the diffuse aurora in the evening-midnight sector, when the poleward gradient of the diffuse auroral conductance belt coincides with the convection reversal region in the evening sector. It is in this region where the divergence of the Hall and the Pedersen currents are both upward, reinforcing each other. Therefore, although the enhanced magnetospheric convection is the primary driving force, the sudden brightening of an auroral arc does depend on the ionospheric response to the enhanced magnetospheric convection.

The X-line model of substorms [*McPherron et al.*, 1973] and the X-line plasmoid model of substorms [*Hones*, 1976] propose that the substorm onset is caused by the formation of a near-Earth X-line from which the substorm current wedge supposedly originates. Assuming that the substorm field-aligned current can be intensified because of an X-line formation, it must form within $10 R_E$ in the magnetotail in order for the substorm current wedge to be located near the poleward boundary of the diffuse aurora, as required by observations for substorm onset [*Akasofu*, 1964]. The boundary-layer model of substorms [*Rostoker and Eastman*, 1987] proposes that the substorm onset is caused by the Kelvin-Helmholtz instability in the boundary layer from which the substorm

*Geophysical Institute, University of Alaska Fairbanks, Fairbanks AK 99775.
[†]The Johns Hopkins University Applied Physics Laboratory, Laurel, MD 20707.

Auroral Physics, edited by C.-I. Meng, M. J. Rycroft and L. A. Frank. © Cambridge UP 1991

current wedge originates. This is an extension of the boundary-layer field-aligned current model developed by *Sonnerup* [1980]. In the model of *Rostoker and Eastman* [1987], surges of multiple substorm onsets are identified with the vortices generated by the instability. It remains to be shown whether the divergence of the Hall current is nonzero at each vortex as observed in the surge [*Opgenoorth et al.*, 1983]. The thermal catastrophe model [*Goertz and Smith*, 1989] proposes that the substorm onset is caused by the absorption of Alfvénic fluctuations leading to the heating of the plasma sheet. Further study is required to show whether this can lead to the intensification of the field-aligned current in the substorm current wedge. The M-I coupling model of substorm [*Kan et al.*, 1988] proposes that the substorm onset is caused by the ionospheric response to an enhanced magnetospheric convection following a southward turning of the interplanetary magnetic field (IMF). This is the only quantitative model that shows where, when, and why the upward field-aligned current and the auroral brightness are intensified as observed at the onset of substorms [*Akasofu*, 1964].

Our purpose is to discuss an extension of the M-I coupling model of substorms [*Kan et al.*, 1988] to include the magnetotail dynamics.

2. ENHANCEMENT OF MAGNETOSPHERIC CONVECTION

It is generally accepted that a southward turning of the IMF can lead to an enhanced dayside reconnection to enhance the magnetospheric convection. The leading edge of the enhanced magnetospheric convection is propagated perpendicular to the magnetic field lines by the fast-mode wave as described by *Kan et al.* [1988]. The enhanced antisunward magnetospheric convection (on open field lines) is initiated poleward of the reconnection region on the dayside magnetopause and propagates antisunward by fast-mode compression waves. On the other hand, enhanced sunward magnetospheric convection (on closed field lines) is initiated equatorward of the reconnection region and propagates antisunward by the fast-mode rarefaction waves.

Figure 1 is a sketch of the temporal development of an enhanced magnetospheric convection projected onto the ionosphere (which is not the convection pattern in the ionosphere). Figure 1*a* shows the magnetospheric convection pattern before the enhancement of dayside reconnection. Figure 1*b* shows the magnetospheric convection pattern halfway enhanced. Figure 1*c* shows the fully enhanced magnetospheric convection pattern. The fast-mode wavefronts are indicated by shaded horizontal strips propagating at the fast-mode speed in the antisunward direction. The enhanced dayside reconnec-

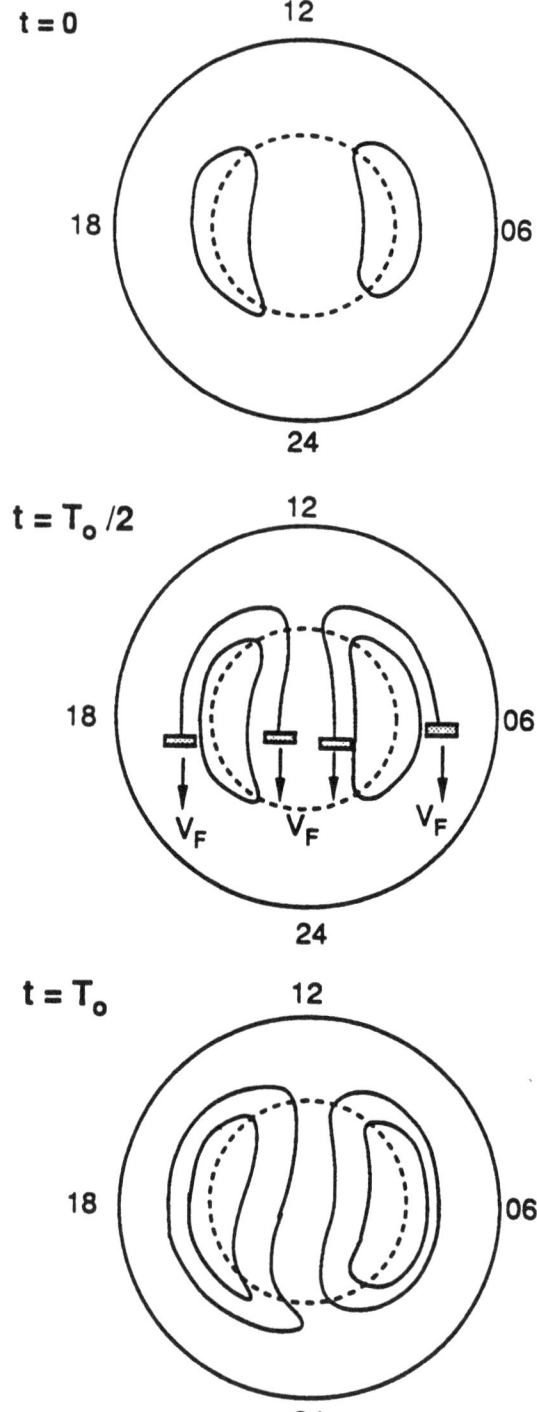

Figure 1—Temporal development of an enhanced magnetospheric convection projected onto the ionosphere following a southward turning of the IMF. The arrows indicate the fast-mode wave velocity V_F. The shaded narrow blocks denote the fast-mode wavefronts, and T_0 is the time delay of the enhanced magnetospheric convection after a southward turning of the IMF.

312

tion is communicated from the dayside reconnection site to the magnetosphere by the fast-mode wave to enhance the magnetospheric convection. The time delay between a southward turning of the IMF and the resulting enhanced magnetospheric convection can be estimated by the travel time of the fast-mode waves.

A lower limit on the time delay can be estimated by assuming that the magnetospheric convection is enhanced immediately after the passage of the fast-mode wavefront. Hence, the magnetospheric convection on auroral field lines would be enhanced as soon as the fast-mode wavefront has traveled a distance $D \approx$ 15R$_E$ (from the dayside magnetopause to the Earth) + 35R$_E$ (from the Earth to the near-Earth plasma sheet connected by field lines to auroral oval) = 50R$_E$. Taking the fast-mode wave speed in the magnetosphere to be about 2000 km s^{-1} on the average [*Moore et al.*, 1987], the wave travels 20R$_E$ in 1 min. Thus, a lower limit of the delay would be $(T_0)_L =$ 50/20 = 2.5 min.

An upper limit of the time delay can be estimated by assuming that the magnetospheric convection is enhanced after the wavefront has reached the preexisting X-line in the distant tail (~ 100R$_E$) to initiate an enhanced reconnection in the plasma sheet to drive the enhanced magnetospheric convection. The enhanced convection then propagates earthward from the enhanced tail reconnection site to the near-Earth plasma sheet (~ 35R$_E$). Hence, the total distance traveled by the fast-mode wave to set up an enhanced magnetospheric convection is approximately $D \approx$ 15R$_E$ (from the dayside magnetopause to the Earth) + 100R$_E$ (from the Earth to the X-line in the distant tail) + 65R$_E$ (from the distant X-line back to the near-Earth plasma sheet) = 180R$_E$. Again taking the fast-mode speed to be 2000 km s^{-1}, an upper limit of the time delay would be $(T_0)_U =$ 180/20 = 9 min.

The ionospheric convection is driven by and therefore lags behind the magnetospheric convection. The time lag can be one to several units of Alfvén travel time T_A depending on the ionospheric conductance, where $T_A = L_\parallel/V_A \approx 1.5$ min for $L_\parallel = 30$R$_E$ and $V_A = 2000$ km s^{-1}. In regions where the conductance is higher, the ionospheric line-tying effect is stronger so that more time is required for the ionosphere to respond to the enhanced magnetospheric convection. This is evident in the results of the M-I coupling model [*Kan and Sun*, 1985] in which the ionopspheric convection on the dayside and in the polar cap reaches a steady-state pattern after one to three T_A, while 15–17$T_A \approx$ 22–25 min are needed for the ionospheric convection in the midnight sector to reach a steady state. Thus, the ionospheric convection and the field-aligned current in the midnight sector can be fully enhanced in $\sim$25–35 min after a southward turn-

ing of the IMF. This is consistent with the observed statistical time delay [*Meng et al.*, 1973].

3. M-I COUPLING MODEL OF SUBSTORMS

The ionospheric response to an enhanced magnetospheric convection has been shown by *Kan et al.* [1988] to produce intense, upward, field-aligned currents near the poleward edge of the diffuse aurora belt in the evening sector, thus resulting in the brightening of the auroral arc for substorm onset. The upward field-aligned currents intensify most at the poleward boundary of the diffuse auroral conductance belt when the convection reversal region overlaps with the poleward gradient of the diffuse auroral conductance belt in the evening sector. Where and when the above condition is met, the divergence of the Hall current and the divergence of the Pedersen current are both upward to produce the most intense localized upward field-aligned currents. The time taken for the ionosphere in the midnight sector to respond to enhanced magnetospheric convection after a southward turning of the IMF is about 30 min. This is the first quantitative substorm model that shows where, when, and why the upward field-aligned currents should intensify for the substorm onset as observed [*Akasofu*, 1964].

Figure 2 shows the input parameters to the model. Figure 2a shows the initial ionospheric Hall conductance prior to the enhancement of magnetospheric convection. The initial Pedersen conductance (not shown) is a factor of 1.5 smaller than the Hall conductance. Figure 2b shows the distribution of the reflection coefficient at the magnetopause on open field lines and at the plasma sheet on closed field lines. Figure 2c shows the distribution of the magnetospheric reflection coefficient along the noon-midnight meridian.

The physics governing the magnetospheric reflection of Alfvén waves [*Kan and Sun*, 1985] can be summarized briefly as follows. On open field lines, the solar wind inertia is sufficiently large that the solar wind flow is more or less unchanged by the loading effect of the Alfvén wave incident on the magnetopause. In other words, the electric field on open field lines at the magnetopause can be maintained by the solar wind. Therefore, the incident wave electric field must be canceled by the reflected wave field, which then leads to the magnetospheric reflection coefficient $R_m(E) = E'/E^i \approx -1$ on open field lines at the magnetopause. This is equivalent to the input conductance of the solar wind dynamo Σ_D (on open field lines) being much greater than the Alfvén wave conductance Σ_A.

The wave reflection from the plasma sheet on closed field lines is quite different. The inertia of the $\mathbf{E} \times \mathbf{B}$ convection in the plasma sheet is somewhat limited, even including the effect of the $\mathbf{J} \times \mathbf{B}$ force due to the cross-tail current. The reflection from the plasma

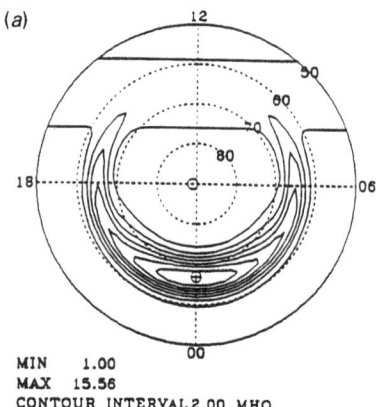

(a)

MIN   1.00
MAX   15.56
CONTOUR INTERVAL 2.00 MHO

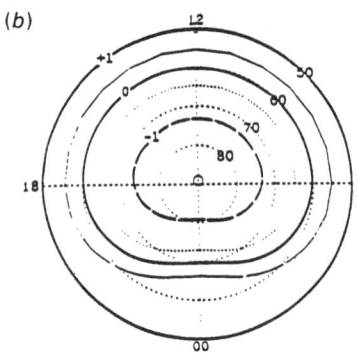

(b)

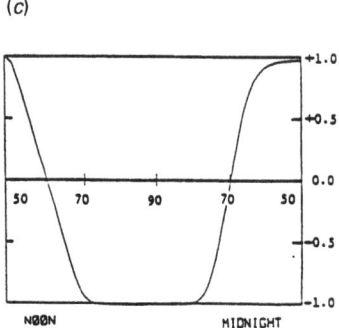

(c)

Figure 2—Input parameters for the M-I coupling model of substorms. (a) The assumed background ionospheric Hall conductance before the substorm onset, which consists of the solar conductance and the diffuse auroral conductance (for simplicity, the ratio of Hall-to-Pedersen conductances is assumed to be 1.5, independent of the particle energy); (b) and (c) the assumed distributions of the magnetospheric reflection coefficient at the magnetopause on open field lines and at the plasma sheet on closed field lines.

sheet is further complicated by the fact that the net "reflected" wave in one hemisphere consists of the reflected wave due to an incident wave from the same hemisphere and the transmitted wave due to an incident wave from the opposite hemisphere. Let $_nR_m(E)$ and $_nT_m(E)$, respectively, denote the reflection and transmission coefficients of an Alfvén wave incident on the plasma sheet from the northern hemisphere; and let $_sR_m(E)$ and $_sT_m(E)$ denote those from the southern hemisphere. The reflection and transmission coefficients are related, of course, by $_nT_m(E) = 1 + {_nR_m}(E)$. Now, consider a pair of identical but oppositely propagating Alfvén waves incident on the plasma sheet simultaneously from opposite hemispheres. By assuming that the plasma sheet is hemispherically symmetric, $_nR_m(E) = {_sR_m}(E) = R'_m(E)$ and $_nT_m(E) = {_sT_m}(E) = T'_m(E)$. The net "reflection" coefficient from the plasma sheet in this case is given by

$$R_m(E) = R'_m(E) + T'_m(E) \qquad (1)$$

The loading effect of an incident Alfvén wave can be expected to modify the electric field in the plasma sheet. In the limiting case where the inertia including the $\mathbf{J} \times \mathbf{B}$ force in the plasma sheet is negligible, such as at the inner edge of the plasma sheet, the convection can be expected to be fully modified by the incident Alfvén wave. This means the incident Alfvén wave from one hemisphere can propagate across the plasma sheet to the other hemisphere almost without reflection, i.e., $R'_m(E) \approx 0$, so that $T'_m(E) = 1 + R'_m(E) \approx +1$. Thus, the net "reflection" coefficient at the plasma sheet due to simultaneous incidence of Alfvén waves from both hemispheres is $R_m(E) = R'_m(E) + T'_m(E) \approx +1$. On the other hand, if the inertia and the $\mathbf{J} \times \mathbf{B}$ force in the plasma sheet are sufficiently large, such as in the low-latitude boundary layer, $R'_m(E)$ may be somewhat less than zero, say $R'_m(E) = -0.6$ and $T'_m(E) = +0.4$. Thus, the net "reflection" coefficient at the low-latitude boundary layer due to simultaneous incidence of Alfvén waves from both hemispheres is $R_m(E) = R'_m(E) + T'_m(E) = -0.2$. On the basis of the above discussion, the reflection coefficient for the outward-traveling Alfvén waves at the magnetospheric boundaries can be summarized by

$$R_m \geq -1 \text{ on open field lines}$$
$$-1 < R_m < +1 \text{ on closed field lines} \qquad (2)$$

It may be noted that $R_m(E) = -1$ corresponds to an idealized constant voltage source while $R_m(E) = 1$ corresponds to a constant-current source, as has been

314

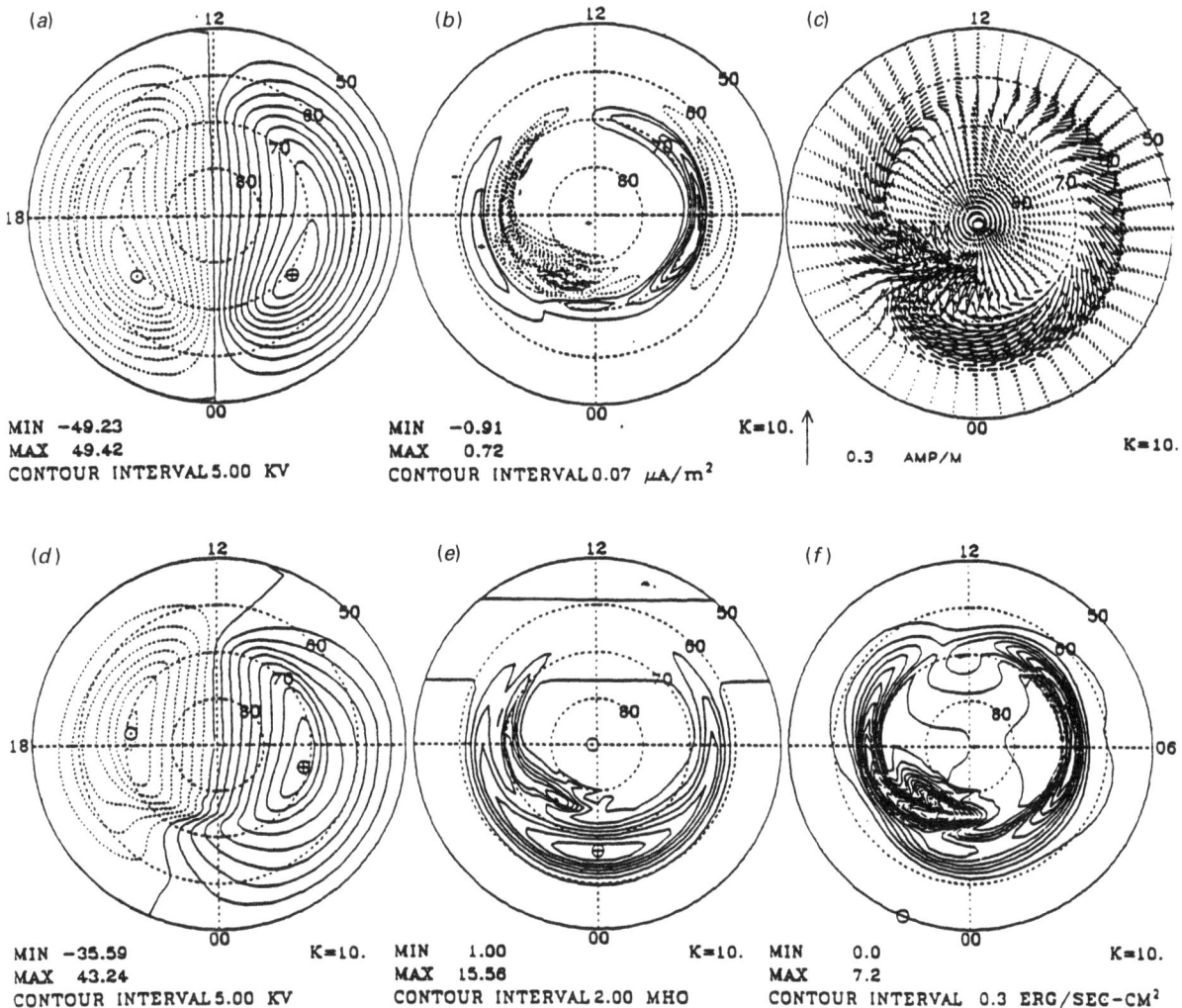

Figure 3—(a) The enhanced magnetospheric convection pattern projected (not to be taken as mapped) onto the ionosphere. All other panels are the simulation results at the 10th bounce (*K* = 10) of the Alfvén waves in the M-I coupling system. After about 8 bounces, the enhanced M-I coupling reaches a steady state; (b) the field-aligned currents (dashed contours are constant current density contours of upward currents, solid contours are for downward currents); (c) the ionospheric current vector (including both the Pedersen and the Hall currents); (d) the convection pattern distorted by the nonuniform ionospheric conductance; (e) the Hall conductance enhanced by the discrete auroral precipitation; and (f) shows the Joule dissipation.

shown by *Kan and Sun* [1985]. The value of $R_m(E)$ on closed field lines can be expected to increase from slightly greater than -1 to approaching $+1$ as one moves from the distant tail or the low-latitude boundary layer toward the inner edge of the plasma sheet. Further study of the reflection of Alfvén waves from the plasma sheet is under way to derive the reflection coefficient as a function of the plasma sheet parameters including the cross-tail current, the convection, the magnetic field, and the plasma pressure.

Figure 3 shows an example of asymptotic steady-state results of the M-I coupling model of substorms [*Kan et al.*, 1988]. Figure 3a shows the input magne-

tospheric convection pattern projected onto the iono-sphere. The Alfvén wave is partially absorbed at each reflection from the ionosphere. The steady state is reached when the Alfvén wave is completely absorbed by the ionosphere. The distortion of the convection pat-tern shown in Figure 3d leads to the formation of the Harang discontinuity [*Heppner*, 1977] in the evening sector. This characteristic distortion of the convection pattern is due to the polarization electric field produced by Alfvén wave reflections from the nonuniform and anisotropic conducting ionosphere [*Kan and Sun*, 1985]. The nonuniform Hall conductance makes the ionospheric reflection coefficient anisotropic, so that the reflected Alfvén wave electric field is rotated from the incident wave field. The rotational distortion of the convection pattern is produced by the ionosphere, bending the convection stream lines and changing the vorticity. This is an important result of the M-I coupling model of substorms, which will be discussed further in connection with the generation and closure of the region-1 field-aligned current.

Brightening of the auroral arc for substorm onset occurs at the location where the upward field-aligned current density is most intense near the poleward edge of the diffuse auroral conductance belt in the evening sector, as shown in Figure 3*b*. This agrees with the observational definition of substorm onset [*Akasofu*, 1964], keeping in mind that the diffuse auroral conductance is produced by the diffuse auroral precipitation. It has been shown [*Kan et al.*, 1988] that the maximum upward current occurs where field-aligned currents due to the divergence of the Pedersen current and the divergence of the Hall current are both upward. Figure 3*c* shows the intensified auroral electrojets. Figure 3*d* shows the distorted convection pattern with the Harang discontinuity. Note the asymmetry in the potential difference between the two convection cells. The dusk cell potential is smaller than the dawn cell because the ionospheric conductance is higher and therefore loads down the convection more on the dusk side. Figure 3*e* shows the distribution of the enhanced Hall conductance due to the discrete auroral precipitation. Figure 3*f* shows the distribution of the Joule dissipation rate at the substorm onset.

The large-scale field-aligned currents, known as the region 1 and region 2 field-aligned currents [*Iijima and Potemra*, 1976], resulting from the M-I coupling process must close in the magnetosphere. A quantitative model of the magnetospheric closure of field-aligned current is not yet available; the fundamentals of this problem is discussed next.

4. CLOSURE OF FIELD-ALIGNED CURRENTS IN THE MAGNETOSPHERE

Generation of field-aligned currents is really a part of the closure of field-aligned currents in the magnetosphere, which can be explained simply as follows. If the closure of a field-aligned current in the magnetosphere takes place in a dynamo region where $\mathbf{J} \cdot \mathbf{E} < 0$, the field-aligned current is said to have been generated from, rather than closed in, the dynamo region. If the closure takes place in a motor region in the plasma sheet where $\mathbf{J} \cdot \mathbf{E} > 0$, the field-aligned current is said to be closed in the motor region.

To meet the ionospheric demand for field-aligned currents, the magnetosphere adjusts its convection pattern $\mathbf{V}$ and plasma pressure p to meet the ionospheric demand of field-aligned currents. The magnetic field $\mathbf{B}$ in the magnetosphere must undergo self-consistent changes to satisfy Ampere's law. The flow velocity $\mathbf{V}$, the plasma pressure p, and the magnetic field $\mathbf{B}$ are coupled by the fluid equations and the Maxwell equations. The simplest fluid equations for a magnetized plasma are the MHD equations from which the field-aligned current can be written as [e.g., *Kan*, 1987]

$$\nabla \cdot \mathbf{J}_\parallel = -\nabla \cdot \mathbf{J}_\perp$$

$$= \mathbf{J} \cdot [\omega \times \mathbf{V}/V_A^2 + (2\nabla B)/B]$$

$$+ [\mathbf{V} \cdot \nabla (\omega \cdot \mathbf{B}) - \omega \cdot \nabla (\mathbf{V} \cdot \mathbf{B})]/(\mu_0 V_A^2)$$

$$- [(\omega \cdot \nabla \rho)(\mathbf{V} \cdot \mathbf{B}) + (\tfrac{1}{2})\nabla V^2 \cdot (\nabla \rho \times \mathbf{B})]/B^2 \tag{3}$$

where $\nabla \cdot \mathbf{J}_\parallel > 0$ is chosen for field-aligned currents flowing away from the equatorial plane, $\omega = \nabla \times \mathbf{V}$ is the vorticity, and $V_A = B/(\mu_0 \rho)^{1/2}$ is the Alfvén speed.

Kan [1987] demonstrated qualitatively that the first two terms in Eq. 3 are the dominant terms for the magnetospheric closure of the region 1 and region 2 field-aligned currents, respectively. The first term, $\mathbf{J} \cdot \omega \times \mathbf{V}/V_A^2$, is dominated by the convection pattern and has the proper sense for closing the region 1 current [*Kan*, 1987]. The bending of the convection streamlines around the Harang discontinuity region in the midnight sector, due to the nonuniform Hall conductance [*Kan and Sun*, 1985], is equivalent to generating vorticity in the magnetospheric convection which plays an important role in closing the region 1 field-aligned current through the $\mathbf{J} \cdot \omega \times \mathbf{V}/V_A^2$ term. The second term $(2\mathbf{J} \cdot \nabla B)/B$ is dominated by the pressure gradient, which is known to have the correct sense for closing the region 2 field-aligned current [*Vasyliunas*, 1984].

It has been pointed out by *Kan* [1987] that a large fraction of the region 1 field-aligned current is on closed field lines convecting sunward, which map into the plasma sheet. The remaining fraction of the region 1 field-aligned current is on field lines convecting antisunward, which map into the low-latitude boundary layer and the magnetopause along the tail. In the midnight sector, almost all of the region 1 current must map into the plasma sheet along the high-latitude plasma sheet boundary layer. The fact that the low-latitude boundary layer can only contribute less than 20% of the total convection potential in the magnetosphere means that it cannot be the dominant energy source for the magnetospheric substorm. This was the reason why the closed magnetosphere model driven by viscous interaction in the low-latitude boundary layer [*Axford and Hines*, 1961] was discarded in favor of the open magnetosphere model [*Dungey*, 1961].

5. MAGNETOTAIL DYNAMICS DURING SUBSTORMS

The importance of the ionosphere in initiating the substorm onset is emphasized in the previous sections. In this section we discuss the dynamic changes in the magnetotail predicted by the M-I coupling model of substorms.

Substorm Growth Phase

Observation shows [e.g., *Buck et al.*, 1973] that the plasma sheet thinning starts about 30 min before the onset of a substorm at a downstream distance of about $15R_E$ in the plasma sheet. The size of the tail lobe cross-section can be expected to increase where the plasma sheet has thinned down. The polar cap can be expected to expand equatorward starting from the middayside sector toward the midnight sector. As the polar cap size increases, the convection reversal region in the evening-midnight sector can be expected to move equatorward to overlap the poleward boundary of the diffuse auroral precipitation, shown to be a necessary condition for the substorm onset [*Kan et al.*, 1988].

The equatorward motion of the convection reversal region toward the diffuse auroral precipitation region requires that the former move equatorward faster than the latter during the growth phase. One verifiable consequence of this differential rate of equatorward motion is that the latitudinal width of the auroral oval would be reduced, especially in the midnight sector during the later part of the growth phase.

Substorm Onset and Expansive Phase

One of the predictions of the M-I coupling model of substorms [*Kan et al.*, 1988] is that the onset of a substorm occurs when the ionospheric demand for upward field-aligned currents exceeds the loss-cone limit of a few $\mu A\ m^{-2}$ [*Knight*, 1973; *Friedman and Lemaire*, 1980]. To relate this prediction to the plasma sheet dynamics, we must examine how the ionospheric demand for the substorm field-aligned currents can be delivered by the magnetosphere, i.e., how the enhanced substorm field-aligned currents are to be closed in the magnetosphere.

Figure 4a shows the scenario in which the enhanced substorm field-aligned current is enclosed in the magnetosphere by diverting the cross-tail current. In this case the cross-tail current is reduced locally where the field-aligned current is most intense. The most intense upward substorm field-aligned current occurs near the poleward boundary of the diffuse auroral precipitation

region [*Kan et al.*, 1988], which may map to a downstream distance of about $10R_E$ or further, depending on how tail-like the magnetic field configuration is in the near-Earth region. Thus, a near-Earth X-line may form as a consequence, rather than as the cause, of the onset of an intense substorm. The location of the resulting near-Earth X-line is predicted by the M-I coupling model of substorms to map to the poleward boundary of the diffuse auroral precipitation region. As the near-Earth X-line moves tailward, the plasma sheet on the earthward side of the X-line starts to recover, while the plasma sheet on the tailward side of the X-line continues to thin. The enhancement of earthward flow when the plasma sheet recovers is a well-established observational feature during the recovery phase [e.g., *Lui et al.*, 1977]. On the other hand, if the reduction of the cross-tail current is not sufficiently strong, formation of an X-line may not occur. In addition, if the poleward boundary of the diffuse auroral precipitation maps to a location closer in, say ~ 5–$6R_E$ downstream in the tail, then the dipolar field is strong enough to impede the formation of a near-Earth X-line. The criterion for the formation of an X-line due to the reduction of the cross-tail current in the plasma sheet has yet to be determined. The observed dipolarization of the field configuration [*Cummings and Coleman*, 1968], following a substorm onset, can be understood as a consequence of the enhancement of the substorm field-aligned current with or without the reduction of the cross-tail current in the near-Earth plasma sheet.

It is unlikely that the formation of the near-Earth X-line is initiated by the tearing instability. It is well-known that the tearing instability is severely inhibited by the normal field component B_n in the plasma sheet as long as B_n is large enough to magnetize the electrons [*Lembege and Pellat*, 1982]. Therefore, simulation results showing formation of X-lines in the plasma sheet due to the tearing instability is dubious. As has been shown by *Swift and Allen* [1987], the conducting ionospheric boundary cannot circumvent the stabilization effect of the normal field component on tearing in the plasma sheet. The conducting boundary can neutralize the electron space charge and electrically decouple the electrons from the ions, but the energy released by the magnetic field is not sufficient to overcome the stabilizing electron pressure that builds up during the process. This result tells us that no tearing instability can occur so long as the electrons are magnetized by the normal field component in the plasma sheet. This difficulty is even more severe in the near-Earth plasma sheet where the normal field component is comparable to the transverse component. Thus, the MHD simulation models of tearings and X-line formations in the plasma sheet are open to two considerations: (1)

the dissipation is artificially added to the MHD equations, and (2) the stabilization of tearing due to magnetized electrons is not included in the MHD models.

Figure 4b shows the scenario in which the substorm field-aligned current is enhanced due to additional current flowing from the magnetopause. In this case, the cross-tail current in the plasma sheet is practically unchanged before and after the enhancement of the substorm field-aligned current. Although the cross-tail current is unchanged, the normal field component in the plasma sheet is increased due to the enhanced field-aligned current to produce the dipolarization of the field configuration in the near-Earth plasma sheet. The earthward $J \times B$ force increases due to the increase of both J and B, resulting in the plasma injection [*Deforest and McIlwain*, 1971] and the dipolarization of field configuration after the substorm onset. As a result, the plasma sheet thickens in the near-Earth region, while thinning continues tailward, as depicted in the plasma sheet thinning model of *Chao et al.* [1977]. This scenario predicts plasma injection and dipolarization after a substorm onset without the formation of a near-Earth X-line, in contrast to the X-line plasmoid models of substorms [*McPherron et al.*, 1973; *Hones*, 1976]. However, additional plasma injection can occur because of the enhanced earthward plasma convection if the near-Earth X-line is formed as a consequence of substorm onset as described in Figure 4a.

Figure 4c shows the scenario that the cross-tail current can be redistributed to form an X-line without field-aligned currents. This is the scenario of the standard tearing-mode instability configuration where X-lines are formed at local current density minima.

The three scenarios depicted in Figure 4a, 4b, and 4c can be summarized as follows. Scenario *a* shows that the near-Earth X-line can form as a consequence of the substorm onset when the ionospheric demand for upward field-aligned current exceeds the loss-cone limit around 1 μA m^{-2} [*Knight*, 1973; *Friedman and Lemaire*, 1980]. This is in contrast to the X-line plasmoid model of substorms in which the formation of the near-Earth X-line is taken to be the cause of the substorm onset. Scenario *b* shows that the field-aligned current can be enhanced without the formation of a near-Earth X-line. The ionospheric demand for intense field-aligned currents can be provided directly by the magnetopause current without changing the cross-tail current. Scenario *c* shows that the near-Earth X-line can form without enhancing field-aligned currents. As a consequence of scenarios *b* and *c*, it is completely unnecessary to interpret the tailward motion of the near-Earth X-line as "poleward leaps" of the discrete aurora as has been suggested by *Hones* [1984]. Indeed, the lack of a one-to-one correspondence between the X-line and the enhanced field-aligned current as depicted in Figure 4 speaks against the association of auroral arcs with X-lines in the plasma sheet.

On the basis of the above discussion, the M-I coupling model of substorms predicts: (1) a near-Earth X-line may form as a consequence of the onset of a sufficiently intense substorm, and (2) the location of the near-Earth X-line, when it forms in the plasma sheet after the onset of an intense substorm, should map to the poleward boundary of the diffuse auroral precipitation region in the ionosphere.

Figure 5 is a schematic illustration to summarize the plasma sheet configurations before and after the onset of an intense substorm based on the predictions of the M-I coupling model of substorms. The model predictions are (1) a near-Earth X-line may form within a few minutes after the onset of an intense substorm, and (2) the location of the X-line, when it forms, should map to the poleward boundary of the diffuse auroral precipitation region in the midnight sector. The plasmoid depicted in Figure 5 is consistent with the observational results [*Baker et al.*, 1987] that plasmoids move down the magnetotail at 200R_E about 20–40 min after the onset of substorms at the Earth.

6. SUMMARY

The M-I coupling model of substorms has been extended to predict changes of the plasma sheet configuration during substorms. The proposed theory of substorms models quantitatively the brightening of an auroral arc near the poleward edge of the diffuse aurora in the evening-midnight sector for the onset of a substorm. The M-I coupling model of substorms starts from an enhanced magnetospheric convection, due to a southward turning of the IMF, resulting in enhanced upward field-aligned currents near the poleward edge of the diffuse auroral belt. This occurs as the enhanced magnetospheric convection forces the ionospheric convection to speed up. Alfvén waves are responsible for transmitting the information during the transient period in which the ionospheric convection, the field-aligned current, and the ionospheric conductance are enhanced with time. The most intense upward field-aligned current in this model is located within a narrow strip near the poleward edge of the diffuse aurora in the midnight sector. The condition for this to occur is that the poleward gradient of the diffuse auroral conductance belt overlaps with the reversal region of the enhanced convection in the evening-midnight sector. In this overlapping region the divergence of the Hall and the Pedersen currents are both upward to reinforce each other. Therefore, although the enhanced magnetospheric convection is the primary driving force, the ionosphere to a large extent controls where, when, and why an auroral arc is to brighten up for substorm onset.

(a) Enhancing $\mathbf{J}_{\|}$ by Reducing the Cross-Tail Current

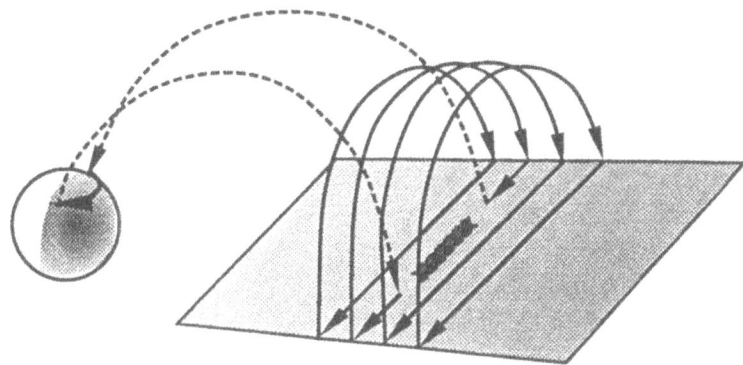

(b) Enhancing $\mathbf{J}_{\|}$ Without Reducing the Cross-Tail Current

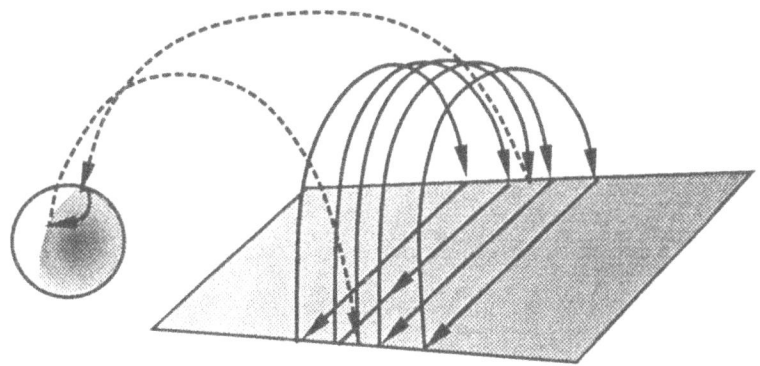

(c) Redistributing the Cross-Tail Current Without Enhancing $\mathbf{J}_{\|}$

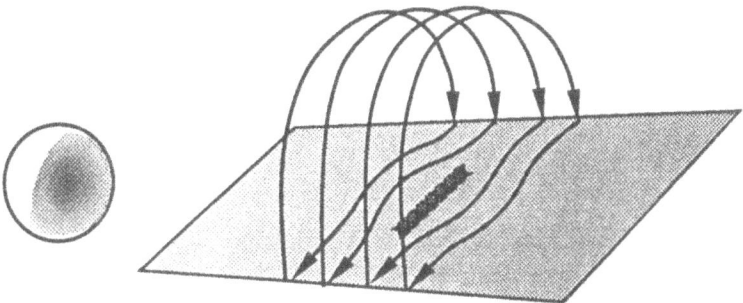

Figure 4—Schematic diagram illustrating two ways of enhancing the substorm field-aligned current. (*a*) Enhancement of the field-aligned current by reducing the cross-tail current from two units to one unit; (*b*) enhancement of the field-aligned current by increasing the magnetopause current from two units to three units and leaving the two units of cross-tail current unchanged; and (*c*) the formation of an X-line without enhancement of the field-aligned currents, as is the case of the standard tearing-mode instability configuration.

Growth Phase : T = − 20 min

15 R$_E$

Substorm Onset : T = 0

0 < T < 5 min.

new X line maps to poleward
edge of diffuse aurora

~ 400 km / s

> 150 R$_E$
(not to scale)

T > 30 min

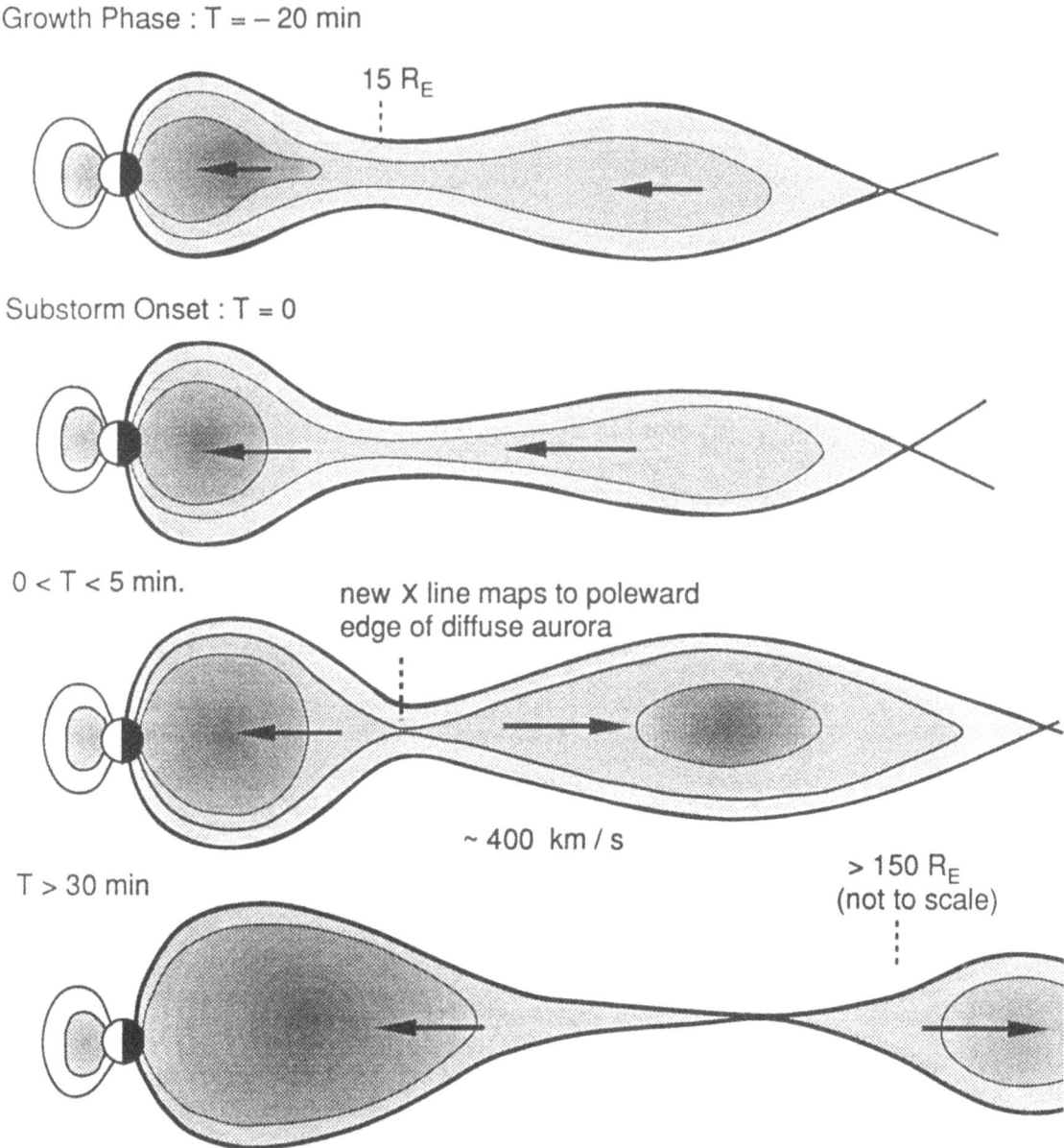

Figure 5—A schematic diagram illustrating changes of the plasma sheet associated with the various phases of a substorm as predicted by the M-I coupling theory of substorms. A near-Earth X-line may form as a consequence of the onset of an intense substorm. The location of the near-Earth X-line in the plasma sheet should map to the poleward boundary of the diffuse auroral precipitation region in the ionosphere.

The M-I coupling theory of substorms predicts that (1) a near-Earth X-line may form as a consequence of the onset of an intense substorm, and (2) the location of the near-Earth X-line, when it forms in the plasma sheet after the onset of an intense substorm, should map to the poleward boundary of the diffuse auroral precipitation region in the ionosphere in the midnight sector.

ACKNOWLEDGMENT—This work was supported by the Atmospheric Science Section of the National Science Foundation (NSF) under grants ATM 85-21194 and ATM 88-03133 and by the Air Force Geophysics Laboratory under contract F19628-88-K-0003 to the University of Alaska, Fairbanks. The effort of A. T. Y. Lui was supported by National Science Foundation grant ATM-86-11354 to The Johns Hopkins University.

REFERENCES

Akasofu, S.-I., "The Development of the Auroral Substorm," *Planet. Space Sci.*, **12**, 273 (1964).

Axford, W. I., and C. O. Hines, "A Unifying Theory of High-Latitude Geophysical Phenomena and Geomagnetic Storms," *Can. J. Phys.*, **39**, 1433 (1961).

Baker, D. N., R. C. Anderson, R. D. Zwickl, and J. A. Slavin, "Average Plasma and Magnetic Field Variations in the Distant Magnetotail Associated with Near-Earth Substorm Effects," *J. Geophys. Res.*, **92**, 71 (1987).

Buck, R. M., H. J. West, Jr., and R. G. D'Arcy, Jr., "OGO-5 Energetic Proton Observations: Spatial Boundaries," *J. Geophys. Res.*, **78**, 3103 (1973).

Chao, J. K., J. R. Kan, A. T. Y. Lui, and S-I. Akasofu, "A Model for Thinning of the Plasma Sheet," *Planet. Space Sci.*, **25**, 703 (1977).

Coroniti, F. V., and C. F. Kennel, "Can the Ionosphere Regulate Magnetospheric Convection?" *J. Geophys. Res.*, **78**, 2837 (1973).

Cummings, W. D., and P. J. Coleman, Jr., "Simultaneous Magnetic Field Variations at the Earth's Surface and at Synchronous, Equatorial Distance, 1, Bay Associated Events," *Radio Sci.*, **3**, 758 (1968).

Deforest, S. E., and C. E. McIlwain, "Plasma Clouds in the Magnetosphere," *J. Geophys. Res.*, **76**, 3587 (1971).

Dungey, J. W., "Interplanetary Magnetic Field and the Auroral Zones," *Phys. Rev. Lett.*, **6**, 47 (1961).

Friedman, M., and J. Lemaire, "Relationship Between Auroral Electron Fluxes and Field-Aligned Electric Potential Difference," *J. Geophys. Res.*, **85**, 664 (1980).

Goertz, C. K., and R. A. Smith, "Thermal Catastrophe Model of Substorms," *J. Geophys. Res.*, **94**, 6581 (1989).

Heppner, J. P., "Empirical Models of High Latitude Electric Field," *J. Geophys. Res.*, **82**, 1115 (1977).

Hones, E. W. Jr., "The Magnetotail: Its Generation and Dissipation," in *Physics of Solar Planetary Environments*, Vol. II, D. J. Williams, ed., 558–571 (1976).

Hones, E. W. Jr., "Plasma Sheet Behavior During Substorms," in *Magnetic Reconnection in Space and Laboratory Plasmas*, Geophysical Monograph, 30, E. W. Hones, Jr., ed., American Geophysical Union, Washington, D.C. (1984).

Iijima, T., and T. A. Potemra, "Large-Scale Characteristics of Field-Aligned Currents Associated with Substorms," *J. Geophys. Res.*, **81**, 3999 (1976).

Kan, J. R., "Generation of Field-Aligned Currents in Magnetosphere-Ionosphere Coupling in a MHD Plasma," *Planet. Space Sci.*, **35**, 903 (1987).

Kan, J. R., and W. Sun, "Simulation of the Westward Traveling Surge and Pi2 Pulsations During Substorms," *J. Geophys. Res.*, **90**, 10911 (1985).

Kan, J. R., L. Zhu, and S.-I. Akasofu, "A Theory of Substorms: Onset and Subsidence," *J. Geophys. Res.*, **93**, 5624 (1988).

Knight, S., "Parallel Electric Fields," *Planet. Space Sci.*, **21**, 741 (1973).

Lembege, B., and R. Pellat, "Stability of a Thick Two-Dimensional Quasi-Neutral Sheet," *Phys. Fluids*, **22**, 1995 (1982).

Lui, A. T. Y., E. W. Hones Jr., F. Yasuhara, S.-I. Akasofu, and S. J. Bame, "Magnetotail Plasma Flow During Plasma Sheet Expansions: Vela 5, 6, and Imp-6 Observations," *J. Geophys. Res.*, **82**, 1235 (1977).

McIlwain, C. E., "Substorm Injection Boundaries," in *Magnetospheric Physics*, B. M. McCormac, ed., D. Reidel Publ. Co., Dordrecht-Holland, p. 143 (1974).

McPherron, R. L., C. T. Russell, and M. P. Aubry, "Satellite Studies of Magnetospheric Substorms on August 15, 1968, 9. Phenomenological Model for Substorms," *J. Geophys. Res.*, **78**, 3131 (1973).

Meng, C.-I., B. Tsurutani, K. Kawasaki, and S.-I. Akasofu, "Cross-Correlation Analysis of the AE Index and the Interplanetary Magnetic Field B_z Component," *J. Geophys. Res.*, **78**, 617 (1973).

Moore, T. E., D. L. Gallagher, J. L. Horwitz, and R. H. Comfort, "MHD Wave Breaking in the Outer Plasmasphere," *Geophys. Res. Lett.*, **14**, 1007 (1987).

Opgenoorth, H. J., R. J. Pellinen, W. Baumjohann, E. Nielsen, G. Marklund, and L. Elliasson, "Three-Dimensional Current Flow and Particle Precipitation in a Westward Traveling Surge (Observed During the Barium-GEOS Rocket Experiment)," *J. Geophys. Res.*, **88**, 3138 (1983).

Rostoker, G., and T. Eastman, "A Boundary Layer Model for Magnetospheric Substorms," *J. Geophys. Res.*, **92**, 12187 (1987).

Russell, C. T., and R. L. McPherron, "The Magnetotail and Substorms," *Space Sci. Rev.*, **15**, 205 (1973).

Smith, R. A., C. K. Goertz, and W. Grossmann, "Thermal Catastrophe in the Plasma Sheet Boundary Layer," *Geophys. Res. Lett.*, **13**, 1380 (1986).

Sonnerup, B. U. O., "Theory of the Low-Latitude Boundary Layer," *J. Geophys. Res.*, **85**, 2017 (1980).

Swift, D. W., and C. Allen, "Interaction of the Plasma Sheet with the Lobes of the Earth's Magnetotail," *J. Geophys. Res.*, **92**, 10015 (1987).

Vasyliunas, V. M., "Fundamentals of Current Description," in *Magnetospheric Currents*, Geophysical Monograph 28, T. Potemra, ed., American Geophysical Union, Washington, D.C. (1984).

VI. AURORAL STRUCTURES

VI-1. OVERVIEW OF AURORAL SPATIAL SCALES

D. J. Gorney*

Recent theoretical works have dealt with the identification and evaluation of the physical processes that determine the characteristic scale sizes of discrete auroral arcs. It is broadly acknowledged that a characteristic spatial width of ~ 100 km (at ionospheric heights) results naturally from the resistive ionospheric mapping of the high-altitude magnetospheric convection electric field. However, recent analysis of the spatial power spectral distributions of electric and magnetic field variations has revealed structure at much smaller spatial scales. Precipitating charged particle data acquired in low-altitude polar orbit typically show monotonically decreasing inverse-wavelength spectra with slopes near unity, with no strictly "preferred" scale sizes, although the scale spectra do tend to flatten at scales larger than ~ 100–200 km. Also, typical widths of visible auroral arcs observed from the ground tend to be much smaller than the resistive scale length, and much smaller than the "average" scale size of inverted-V events observed by satellites. Here we present an overview of recent theoretical and observational works pertinent to the spatial scales that occur in discrete auroral events. The concept of a "characteristic" scale size for auroral features is elucidated and placed in perspective with the observed scale sizes and size distributions of various auroral features. Statistical and individual data results are presented. It is shown that a spectrum of scale sizes is always present in auroral structures, including both auroral particles and fields. Furthermore, it is shown that the presence of a broad spectrum of scales, including scales smaller than the resistive scale length, is nevertheless consistent with resistive mapping theory. Also, the dependence of the spatial scales on altitude and ionospheric parameters is described.

1. INTRODUCTION

Several recent theoretical works have dealt with the identification and evaluation of the physical processes that determine the characteristic sizes and size distributions of discrete auroral arcs [*Chiu and Cornwall*, 1980; *Lyons*, 1980; *Chiu et al.*, 1981; *Lotko et al.*, 1987; *Chiu*, 1986]. It is broadly acknowledged that a characteristic spatial width of ~ 100 km (at ionospheric heights) results naturally from the resistive mapping of the magnetospheric convection electric field into the ionosphere. Recent analysis of the spatial spectral distributions of electric and magnetic field variations [*Weimer et al.*, 1985] has revealed the presence of structure over a range of smaller spatial scales. These observational results have stimulated theorists to consider processes that lead to a spectrum of auroral scales [*Chiu*, 1986; 1987; *Lotko et al.*, 1987; *Gorney*, 1988] and several predictions of the expected scale spectrum and its dependence on ionospheric and plasma parameters have emerged.

Here we present an overview of recent theoretical and observational works pertinent to the spatial scales that occur in discrete auroral events. The emphasis of this present treatment is on mesoscale features; that is, scales of ~ 1 km to several hundred kilometers. Global aspects of auroral configurations such as the processes that affect the size, shape, or orientation of the auroral oval are not discussed. Also, microstructure on the scale of the warm ion gyroradius, which may be caused by a variety of plasma instabilities, is not discussed.

An important motivation for this work is the nagging discrepancy in what experimentalists consider to be the characteristic scale size of auroral features. Individual measurements of in situ particles or fields, or ground-based observations of auroral displays, each seem to yield grossly different values for the predominant scale size of auroral features. Figure 1 demonstrates the scale-size problem for a single set of satellite observations of particles and fields within the auroral region. These observations, from the S3-3 satellite at about 5500 km altitude, show electron flux on an energy-time plot and simultaneous measurements of the in situ electric field. A broad region of electron energization (commonly known as an inverted-V event [*Frank and Ackerson*, 1971]) is visible over the latitude range from ~ 75–77° (about a 200-km scale size mapped into the atmosphere). Several smaller-scale regions of electron acceleration (~ 40 km) are also apparent at slightly lower latitudes. Large-amplitude variations in the electric field are associated with each of the regions of electron energization, but the scale size of the individual electric field variations appears to be very small (~ 1–10 km). Similarly, typical observed widths of visible discrete auroral arcs tend to be very narrow (~ 1 km). Since the auroral displays are the ultimate result of the electron precipitation, it has been difficult to reconcile the differences between the observed spatial scale sizes of inverted-V events and visible auroral arcs.

Here we present a discussion in which the concept of a "characteristic" scale size for auroral features is elucidated and placed in perspective with the observed scale sizes and size distributions of various auroral features. Statistical and individual data results are present-

*Space Sciences Laboratory, The Aerospace Corporation, Los Angeles, CA 90009.

Auroral Physics, edited by C.-I. Meng, M. J. Rycroft and L. A. Frank. © Cambridge UP 1991

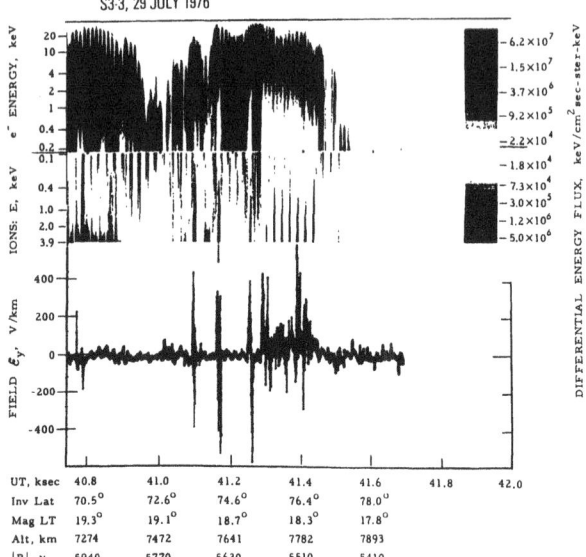

Figure 1—A composite plot of charged particle and field data from the S3-3 satellite within the auroral region. The upper panel shows electron energy flux on an energy-time (*E-t*) plot. The center panel shows energy flux similarly, but with an inverted energy scale. The lower panel shows the simultaneous measurements of the perpendicular component of the *dc* electric field.

ed. It is shown that a spectrum of scale sizes is always present in auroral structures, including both auroral particles and fields. Furthermore, it is shown that the presence of a broad spectrum of scales, including scales smaller than the resistive scale length, is nevertheless consistent with resistive mapping theory. Also, the dependence of the spatial scales on altitude and ionospheric parameters is described.

Along with an overview of previously published observational results, we present a specific observational case study to demonstrate some of the key features of spatial structures in polar discrete auroral arcs. The purpose of this particular observational case study is to perform a clear parametric test of existing theories of auroral scales. For this observational study we have used precipitating auroral electron data from the SSJ/4 instrument on the polar-orbiting DMSP F6 satellite. We have chosen data samples that allow a comparison of auroral scales under sunlit and dark ionospheric conditions. The time period chosen for study is very quiet, and auroral electron precipitation peaked at very low levels within the auroral features. Precipitating electron fluxes only rarely exceeded 1 erg cm^{-2} s^{-1} during the study interval. A period of quiet arcs was chosen to better discriminate the effects of background ionospheric conductance on the auroral spatial scale. We apply simple time-series analysis (studying the distri-

bution of auroral arcs directly), superposed epoch analysis (studying the "average" shape and width of auroral arcs within a selected interval), and power spectral analysis (to determine the distribution of spatial scales).

Discussion of these topics is organized into four sections, following this introduction.

2. THE "CHARACTERISTIC" SCALE SIZE

Theoretically, the notion that there should exist a scale size that characterizes discrete auroral features arises from the concept that the magnetic field-aligned electric potential drops that cause auroral electron acceleration are due to the inability of the magnetospheric electric field to map perfectly into the ionosphere. The imperfect mapping of the magnetospheric electric field results from the finite ionospheric conductance; thus the characteristic scale length can be thought of as the resistive scale length for the ionosphere-magnetosphere system. The existence of such a scale length can be demonstrated by considering the ionospheric Ohm's law, relating ionospheric current **j** to conductivity σ and electric field **E**.

$$\mathbf{j} = \sigma \cdot \mathbf{E} \tag{1}$$

Invoking current continuity (in three dimensions)

$$\nabla \cdot \mathbf{j} = 0 \tag{2}$$

one obtains a relationship between the parallel (vertical) current $\mathbf{j}_\parallel$, the height-integrated Pedersen conductivity Σ_p and the horizontal gradient of the perpendicular electric field.

$$\mathbf{j}_\parallel = -\Sigma_p \nabla \cdot \mathbf{E} \tag{3}$$

Expressing the electric field as the gradient of a potential, and further assuming a linear relationship between parallel current and potential drop [*Knight*, 1973], one obtains the relationship between the magnetic field-aligned current and the magnetospheric electric potential.

$$(1 - \lambda^2 \nabla^2) \mathbf{j} = \Sigma_p \nabla^2 \Phi \tag{4}$$

where λ is the characteristic (resistive) scale of the system. The quantity λ depends both on characteristics of the ionosphere and on the characteristics of the current-carrying plasma. Specifically,

$$\lambda = (\Sigma_p/e^2 n)^{1/2} (4\pi m_e K_{th})^{1/4} \tag{5}$$

where n and K_{th} are the density and temperature of the current-carrying electron population. Note that the ionospheric Pedersen conductance affects the scale size

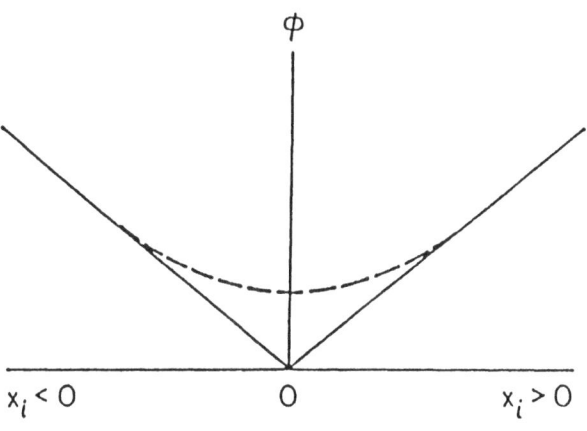

—— Assumed Outer Boundary Potential

—— Calculated Ionospheric Potential

Figure 2—An illustration of the resistive mapping of an assumed high-altitude distribution of potential into the ionosphere. The solid line shows the assumed high-altitude potential distribution, consistent with a shear in the magnetospheric plasma flow. The dashed line shows the distribution of potential at low altitudes, within the ionosphere. A magnetic field-aligned potential drop develops for scale lengths smaller than the resistive scale length (i.e., for $x < \lambda$).

fairly strongly. For typical values of n (~ 1 cm^{-3}) and K_{th} (~ 1 keV), the scale length is approximately 150 km in sunlit ionospheric conditions and about 15 km with a dark ionosphere (completely neglecting the effects of the particle precipitation on the ionospheric conductance; that is, for weak auroras). For more intense auroras, the scale size approaches 100–200 km, regardless of the amount of ionospheric insolation.

The significance of the resistive scale length λ in determining the distribution of auroral magnetic field-aligned currents and magnetic field-aligned potential drops is illustrated in Figure 2 [from *Lyons*, 1980]. Figure 2 shows the perpendicular (to the magnetic field) distribution of electric potential at high altitude, above any significant field-aligned potential drop (solid line), and at very low altitude within the ionosphere (dashed line). The diagram shows the resulting ionospheric potential distribution for an assumed distribution of potential at the high-altitude boundary. The assumed distribution at high altitude corresponds to a shear in the magnetospheric plasma convection; that is, a flow consistent with a reversal in the convection electric field such that $\nabla \cdot \mathbf{E} < 0$. This type of shear is common in the evening auroral region. The high-altitude potential distribution "maps" into the ionosphere directly for spatial scales ($x > \lambda$), but the mapping is not perfect for smaller spatial scales ($x < \lambda$). The result is a difference in electric potential between high and low

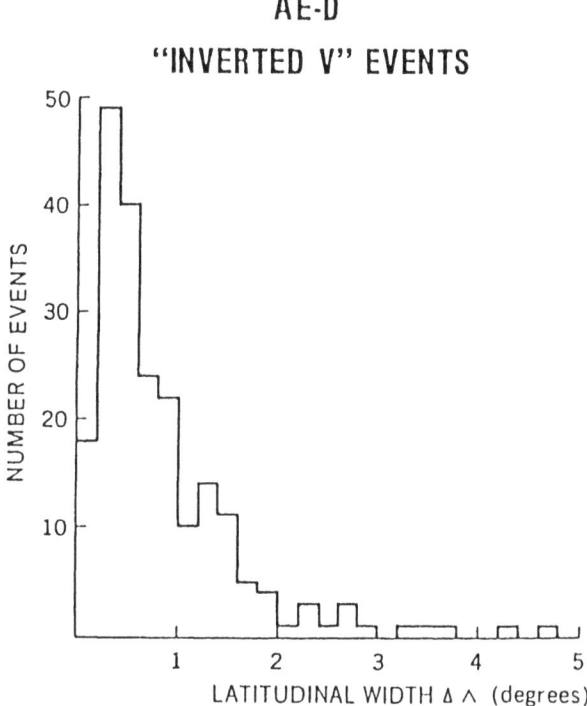

AE-D
"INVERTED V" EVENTS

Figure 3—Results of a statistical study of the latitudinal width of inverted-V events observed by the low-altitude AE-D satellite (from *Lin and Hoffman* [1979]).

altitude only for horizontal (perpendicular) scales less than the resistive scale size. The field-aligned current would have a spatial distribution similar to that of the field-aligned potential drop. Also, since the field-aligned current is carried primarily by magnetospheric electrons, the precipitating electron flux should have the same horizontal spatial distribution as well.

Thus, the resistive scale length (~ 100–200 km for most conditions) represents a horizontal scale size at which parallel potential drops are likely to occur. Several examples of observational evidence are consistent with this result. *Lin and Hoffman* [1979] performed a statistical study of the characteristics of inverted-V events using data from the low-altitude AE-D satellite. Their results for the statistical distribution of the latitudinal width of inverted-V events are shown in Figure 3. A large majority of the observed events occur with horizontal widths less than 1° of latitude (i.e., less than ~ 100 km). Note that the Lin and Hoffman study did not "resolve" or classify events with spatial widths less than 0.2° latitude. Nevertheless, the result that most events occur at spatial scales less than or comparable to 100 km seems firm.

While comparisons between the results of resistive mapping theory and actual observed events are difficult to perform (since complete observations are rare-

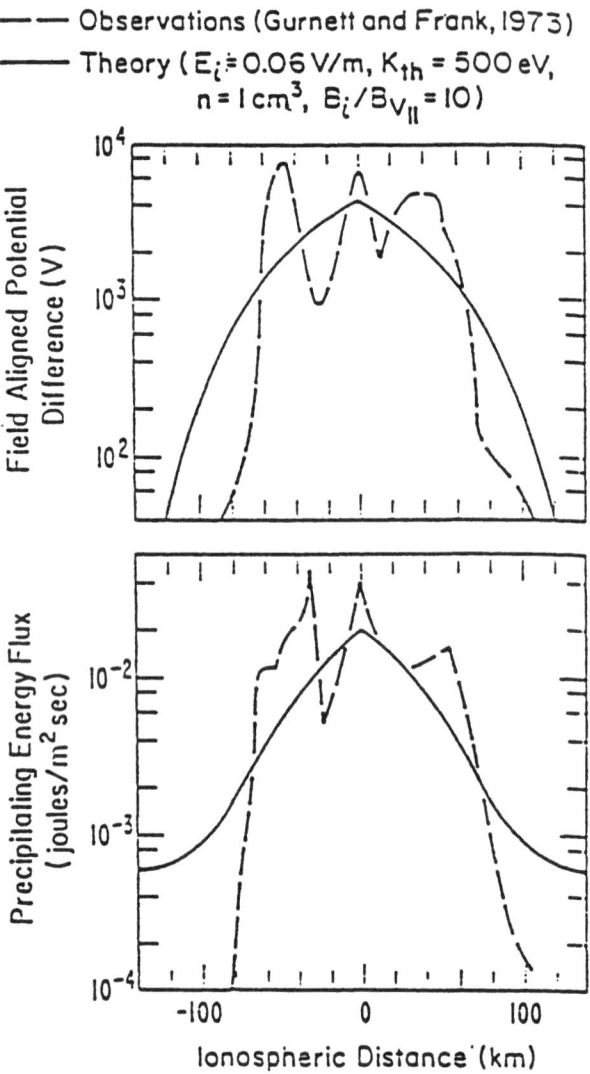

Figure 4—A comparison of theoretical (solid line) and observed (dashed line) magnetic field-aligned potential drop and precipitating electron energy flux for an inverted-V event observed in the evening sector auroral region (from *Lyons* [1980]).

ly available at both high and low altitudes), some simplified case studies have been performed with generally positive results. Figure 4 [from *Lyons*, 1980] shows a comparison between theoretical magnetic field-aligned potential differences and precipitating electron energy flux with observed values. The data are from an "inverted-V" event observed by *Gurnett and Frank* [1973] at low altitude in the evening auroral region. The calculations are based on the electric field that was measured simultaneously with the precipitating electron flux. For the theoretical results, a plasma density of $n \sim 1$ cm^{-3} and a plasma temperature of 500 eV were assumed. These are values typical of the boundary layer

of the plasma sheet [e.g., *Schield and Frank*, 1970]. The observed magnetic field-aligned potential drops were inferred from observations of the peak of the electron energy spectrum. The theoretical results clearly reproduce the overall magnitude and spatial extent of the precipitating energy flux and potential drop for the observed event. Obviously, a great deal of fine-scale structure that is not present in the theoretical result is present in the observations. Indeed, several significant (i.e., order of magnitude) deviations occur on spatial scales of less than about 10 km.

The overall spatial extent and magnitude of the precipitating electron flux and potential drop are fairly well represented by the simple theoretical model shown in Figure 4. The smaller scale observed structures do not match the theoretical result as well, in part at least because of over-simplifications of the theoretical calculation resulting from incomplete or inadequate input data. For example, electric field measurements were not available at a point above the region of the potential drop, so a simple electric field structure with a single minimum potential was assumed. Therefore, the simplified theoretical results shown in Figure 4 should not be thought of as a complete simulation of the observed event.

The presence of "unexplained" small-scale structure in data such as those shown in Figure 4 is not to be taken lightly, since the smaller scales (i.e., $x < 10$ km) are precisely those that are characteristic of visible auroral arcs viewed from the ground. It has become common to consider the existence of two distinct characteristic scale sizes; one representative of inverted V-events ($x \sim 100$ km) and another representative of discrete visible auroral arcs ($x \sim 1$-10 km).

An important outstanding theoretical issue is whether both scales (or a range of scales) can be encompassed within a single theoretical framework or whether two or more distinct physical processes act to produce the observed range of spatial scales within the aurora. A number of recent theoretical treatments has dealt with the occurrence of auroral features on spatial scales smaller than the natural resistive scale size. *Chiu* [1986] describes a process by which large-scale auroral arcs break down into narrow elements that are smaller than the resistive scale size. The process described by *Chiu* [1986] is based on the nonlinear co-attraction of individual magnetic field-aligned current elements within an initially broad auroral arc. The self-attraction of current elements leads to an evolution of spatial scales within the arc, with large-amplitude narrow features resulting from an initially broad and uniform auroral structure. This process implies that the most narrow structures should have the largest current magnitude. The co-attraction process would be significant only for very strong auroral currents. While this treatment introduces a process by which narrow auroral features can evolve from broader structures it does not predict a final state of the evolution nor does it predict a spectrum of resulting scales, since both of these quantities would depend on whatever process limits the progressive narrowing of current elements. Nevertheless, it is encouraging that the physical process outlined by *Chiu* [1986] does lead to structure at small spatial scales. (See also *Lysak and Carlson* [1981] and *Lysak and Dum* [1983] for other discussions of the development of small-scale auroral features.)

We next describe the notion of a "spectrum" of spatial scales within the framework of resistive mapping theory. (See also *Lotko et al.* [1987] for a general discussion of the temporal evolution of spatial scales in magnetosphere-ionosphere coupling.) This discussion is not meant to preclude the existence or importance of other physical processes that might also lead to filamentation or fine-scale structure in auroral events [e.g., *Chiu*, 1986, 1987; *Lysak and Dum*, 1983; *Lysak and Carlson*, 1981], but is meant to demonstrate that spatial scales smaller than the resistive scale size are nevertheless consistent with resistive mapping theory.

3. THE SCALE SPECTRUM

Scale sizes of observed individual auroral features are much smaller than even the smallest value predicted by the resistive mapping equation for reasonable values of ionospheric conductance and current-carrying plasma population. However, this should not be viewed as a failure of the theory to represent the true spatial structure of discrete auroras. Rather, the discrepancy may have to do with a common and long-standing misinterpretation of the physical meaning of the characteristic scale size predicted by resistive mapping theory. This point is best elucidated by viewing Eq. 4 in terms of its Fourier components [e.g., *Lotko et al.*, 1987; *Weimer et al.*, 1985].

$$j_{\|k} \sim \Phi_{\|k} \sim \frac{k^2\lambda^2}{1 + k^2\lambda^2} \Phi_k \qquad (6)$$

Thus, if one regards the magnetospheric convection electric field Φ in terms of its distribution in "k-space," Eq. 6 shows how spatial structure in Φ results in a spatial-scale distribution of parallel currents, $j_{\|k}$, or, equivalently, parallel potential drop, $\Phi_\|$. More importantly, Eq. 6 shows the role of the spatial scale parameter λ in determining the relationship between the spectra of electric potential and current. This relationship is plotted in Figure 5 for a simple assumed distribution of magnetospheric electric field ($\Phi_k \sim k^{-1}$). The example shows results that apply for other reasonable distributions of Φ_k as well, however. For example, for small values of k (wavelengths longer than λ) magnetospheric electric field structure does not lead

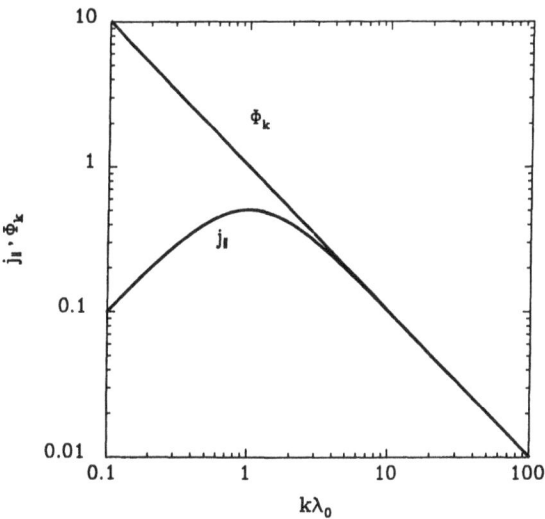

Figure 5—A plot showing the expected relationship between the field-aligned currents and magnetospheric electric potential distribution in terms of inverse wavelength, k. The results are plotted for a simple assumed distribution of magnetospheric electric potential ($\Phi_k \sim k^{-1}$).

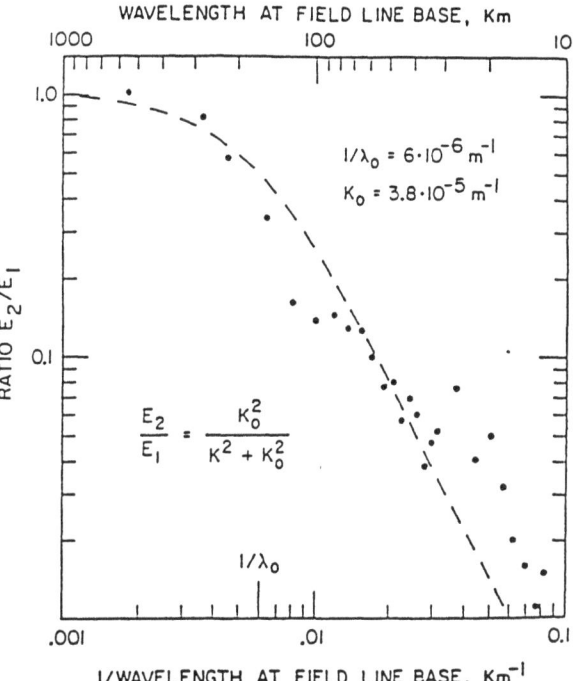

Figure 6—A plot of the ratio of electric field spectral densities measured at low (E_2) and high (E_1) altitudes (solid circles). The data are from the DE 1 and 2 spacecraft. The wavelength scale is normalized to the perpendicular scale at the base of the local magnetic field line. The dashed line represents the scaling relationship implied by resistive mapping theory, assuming a "best-fit" value for the scaling parameter k_0 (from *Weimer et al.* [1985]).

to parallel currents, while for large values of k (wavelengths shorter than λ) all structure in Φ leads to an identical spatial distribution of parallel currents and parallel potential drops. The quantity λ is simply the break point in the spatial spectrum, but by no means represents a unique "preferred" scale for the system. The formulation predicts that small-scale auroral features should be expected, provided that small-scale structure exists in the magnetospheric potential, and that auroral features should not exceed the 100–200 km characteristic scale size.

The theoretical results, presented in this manner, appear to be somewhat consistent with the data presented earlier. However, a real test of the implied relationship between field-aligned currents, potential drops, and magnetospheric electric field demands observations of (at least) the perpendicular electric field spectral distributions acquired simultaneously both above and below the altitude over which the magnetic field-aligned potential is distributed. Observations of this type have become available recently from the combined data sets of the high-altitude Dynamics Explorer 1 (DE 1) and low-altitude Dynamics Explorer 2 (DE 2) satellites [*Weimer et al.*, 1985]. Figure 6 shows an example of electric field measurements from DE 1 and DE 2 for an auroral event observed simultaneously at high altitude ($\sim 3R_E$) by DE 1 and at low altitude (below 900 km) by DE 2. The ratio of the low-altitude electric field spectral density (E_2) to the high-altitude electric field spectral density (E_1) is plotted (solid circles). The wavelength scale is normalized to the base of the magnetic field line. The ratio E_2/E_1 is approximately unity for wavelengths greater than about 200 km, but drops sharply for smaller spatial scales. This result implies that the fine-scale structure present in the high-altitude electric field does not "map" to low altitudes. This result is qualitatively consistent with the theoretical notions discussed in the preceding section.

The dashed line plotted in Figure 6 indicates a computed result for the ratio E_2/E_1 using the relationship implied by Eq. 6. The value of k_0 was chosen to provide the best visual fit of the DE data. The "best-fit" parameter used in Figure 6 corresponds to a "characteristic" scale size of 166 km (i.e., $k_0 = 3.8 \times 10^{-5}$ m^{-1}). The functional form implied by Eq. 6 is quite representative of the DE data for the case shown. Again, it is important to note that the role of the characteristic scale length is to represent the break point in the spatial scale spectrum beyond which electric fields cannot map from high to low altitudes.

Figure 6 represents data from a single case, with an assumed best-fit value for the spatial scale parameter k_0. However, the results shown in Figure 6 appear to be quite representative of auroral events generally. Figure 7 shows a composite plot of "average" electric field

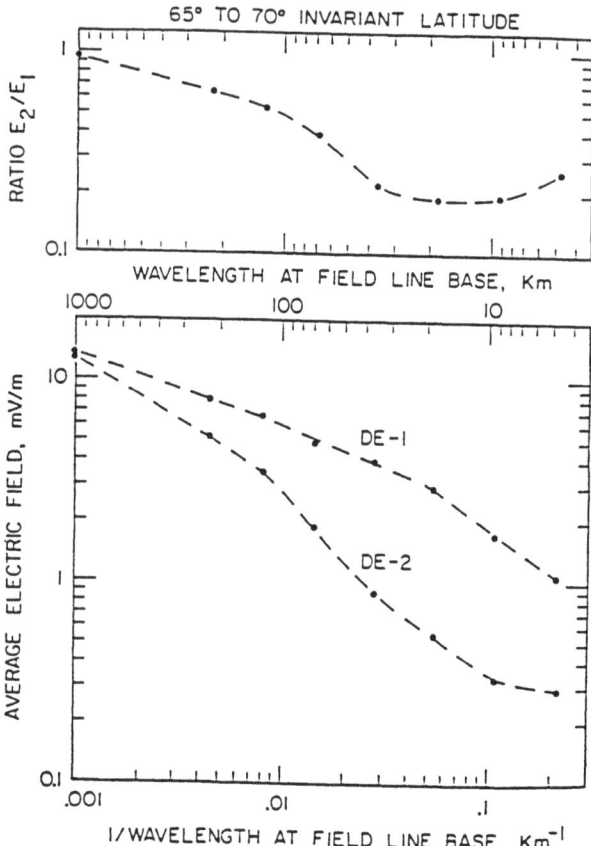

Figure 7—A composite plot of the average spatial spectra of electric fields measured by the DE 1 and 2 satellites at high and low altitudes. The upper panel shows the average ratio E_2/E_1 of the electric field spectral densities at low altitudes to those at high altitudes. The data show a clear tendency for greater electric field amplitudes at small scales for the high-altitude measurement. The ratio E_2/E_1 appears to be near unity for large-scales, but drops sharply to ~ 0.2 for scales less than ~ 100 km at the base of the magnetic field line.

spectral distributions for several auroral events observed by both the DE 1 and 2 satellites. The bottom panel shows plots of the high- and low-altitude electric-field spatial spectra, and the upper panel shows the ratio of the low-altitude field to the high-altitude field. On average, the high-altitude measurement indicates a greater amplitude of electric fields at small spatial scales than the low-altitude measurement. The ratio E_2/E_1 is unity for large-scales and drops to ~ 0.2 for small scales. On average, the break point in the spatial spectrum appears to be consistent with a wavelength of about 100 km. These results appear to be generally consistent with the results of resistive mapping theory presented in the previous section, provided one interprets the resistive mapping scale length as the appropriate spectral scaling parameter rather than as a "preferred" or "dominant" scale size. The results imply that parallel potential drops are not likely to occur for spatial scales larger than 100–200 km, but should occur at all scales smaller than 100–200 km. The results are consistent with the observations of energetic particle precipitation and the statistical distribution of inverted-V events discussed earlier. They provide an independent measure of the spatial distribution of discrete auroral phenomena.

4. A CASE STUDY

The various data examples shown in the previous sections demonstrate the importance of examining independent datasets in order to obtain a broad and thorough overview of the complex physical phenomena that determine the structure of discrete auroral phenomena. More importantly, the data examples show that incomplete or even misleading conclusions can be derived from examining only statistical average results. With this in mind, we now examine a single case study of polar discrete auroral arcs to demonstrate the results of applying different analysis techniques to a single dataset. Here, we apply simple time-series analysis (viewing the spatial distribution of auroral arcs directly), superposed epoch analysis (studying the "average" shape and width of discrete auroral arcs), and power spectral density analysis (to determine the distribution of spatial scales) to a dataset composed of precipitating electron data. The purpose of this exercise is to compare the results of different analysis procedures to better understand the concepts of "average" and "characteristic" spatial scales for auroral arcs.

The data used in this study were obtained from the Air Force Geophysics Laboratory SSJ/4 electrostatic analyzer on the low-altitude (850 km) polar-orbiting DMSP F6 satellite. The DMSP SSJ/4 instrument obtains complete precipitating electron and ion energy spectra once each second over the energy range from 30 eV–30 keV. The time period chosen for study is a period of continuous high-latitude auroral arc activity on January 11, 1983. Detailed characteristics of the auroral activity and interplanetary conditions during this time period are discussed elsewhere [see *Akasofu and Tsurutani*, 1984; *Gorney et al.*, 1986]. The period was chosen for a number of reasons. First, auroral arcs were present over a wide latitude range and the arcs occurred fairly continuously over a several-hour interval. Second, the auroral arcs, although numerous and continuous, are very weak and (presumably) do not alter the background ionospheric conditions significantly. Third, the auroral arcs are linear in configuration and are more-or-less Sun-aligned. This orientation allows thorough sampling of the arc structure from the DMSP dawn-dusk orbit. Finally, lighting conditions at this time were such that the DMSP ionospheric track

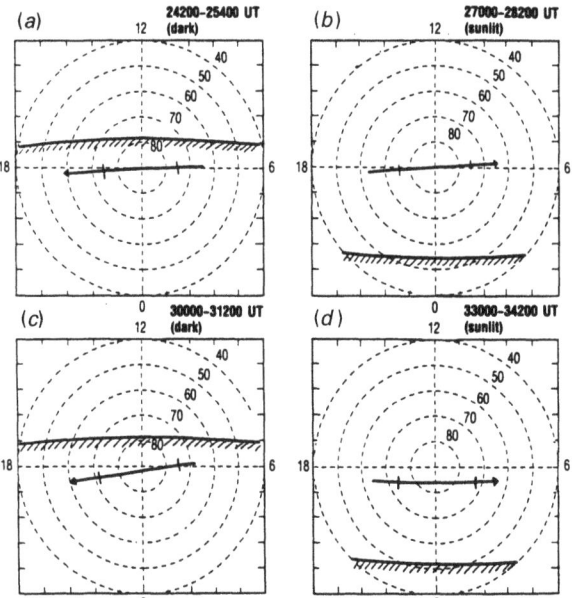

Figure 8—Plots of the trajectories of the DMSP F6 satellite in geomagnetic latitude and local time coordinates for four consecutive polar crossings. Panels *a* and *c* are northern hemisphere (dark) passes while panels *b* and *d* are southern hemisphere (sunlit) cases. The solar terminator at ionospheric altitude is shown as a thick shaded line.

was completely sunlit in the southern polar region and completely dark in the north. Since auroral arcs were present over both poles, the role of the background ionospheric conductance in determining the auroral spatial scale could be tested easily.

Figures 8a-d show the DMSP trajectories in geomagnetic latitude – local time coordinates. The orientation of the solar terminator at ionospheric altitude is indicated with the heavy shadowed line. The two left panels (*a,c*) show northern polar passes, with the trajectory completely in darkness. The right panels (*b,d*) are southern polar passes, completely in sunlight. The individual polar crossings occur about 50 minutes apart. Note that the satellite crosses the polar region from dawn-to-dusk in the north and from dusk-to-dawn in the south. During this period, DMSP passes very close to the geomagnetic pole, providing complete latitude coverage. The satellite orbital velocity is about 7 km s^{-1}, and a 500-s selected study interval is marked on each trajectory.

Figures 9a-d show linear plots of the precipitating electron number flux for each of the four intervals shown in Figure 8 (the intervals correspond to the time periods 24530–25030 s, 27440–27940 s, 30600–31100 s, and 33410–33910 s UT). Again, the left column shows data from the dark hemisphere and the right column shows data from the sunlit hemisphere. The horizon-

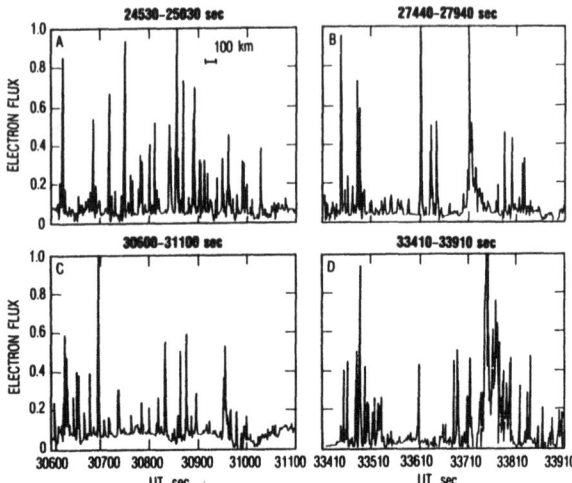

Figure 9—Linear plots of the precipitating auroral electron (30 eV–30 keV) number flux for the four polar crossings shown in Figure 8.

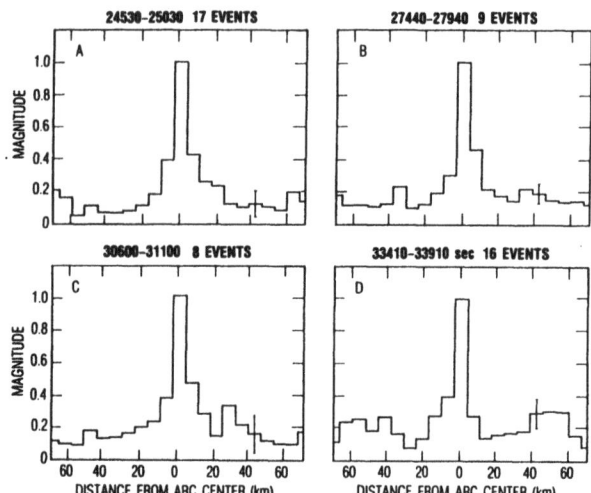

Figure 10—Normalized histogram plots of the average precipitating electron flux spatial distribution in auroral arc events relative to the point of peak flux. The vertical bracket indicates the statistical uncertainty in the average flux.

tal bracket in panel *a* shows a 100-km ionospheric scale size for reference. All four data samples show the occurrence of multiple (10–30) narrow discrete auroral arcs distributed across the entire polar region (~ 3500 km total width). The flux enhancements (sharp increases in flux by factors of ~ 5–50) have scale sizes comparable to the resolution of the instrument (~ 7 km), although some larger-scale structure is apparent as well. The separation between individual discrete features ranges from ten to perhaps hundreds of kilometers. The four data samples are qualitatively similar, although it might be argued that one or two larger scale features (~ 100 km) are present in the sunlit examples (near 27750 s and 33750 s UT).

To determine the average width of the discrete auroral features, individual events from each of the four intervals were combined and averaged to compute an average shape and width. For each interval, auroral arcs were identified by selecting discrete flux enhancements of greater than a factor of five over the background precipitation level. Using this criterion, 17 individual events were identified in case A, 9 in case B, 8 in case C, and 16 in case D. The average shape was computed by normalizing each event to its peak flux value and by centering each event about its peak flux position. The results of this averaging process are shown in Figures 10*a-d*. The vertical bracket indicates the typical statistical uncertainty for each case. In each case the average event width appears to be very narrow (~ 7 km). The flux levels approach background values within 10–20 km of the center of the structure. No significant structure is discernable at scale sizes approaching 100 km.

While statistical analysis of the auroral arc widths provides a clear demonstration of the average scale size of individual auroral arcs, it is also useful to examine the data time series using spectral methods to determine the relative distribution of scale sizes. This is particularly appropriate here, since several recent theoretical studies of the spatial scale problem utilize spectral representations. Figures 11*a-d* show amplitude spectra of precipitating electron flux as a function of inverse wavelength mapped to ionospheric height. The four spatial spectra are quite similar. The spectra show an approximate k^{-1} dependence at high wave number and are relatively flat for wavelengths greater than about 100 km. No statistically significant peaks are present. Indeed, any subsets of the study intervals also showed no evidence for individual "preferred" scale sizes.

While this single case study cannot provide sufficient information to perform a complete test of theoretical representations of auroral processes, the study is useful in demonstrating the variety of interpretations that can be made through the use of different analysis techniques. For example, the case study shows clearly that individual auroral arc elements tend to be quite narrow (< 10 km), similar to the scale sizes of visible auroral arcs viewed from the ground. However, the same dataset shows that virtually all scale sizes (< 100–200 km) are present simultaneously. This result demonstrates the role of the "characteristic" scale size in determining the distribution of scales that are present. An interesting result of this data analysis is that there is virtually no difference between the spatial scale distribution of auroral arcs in the dark and sun-

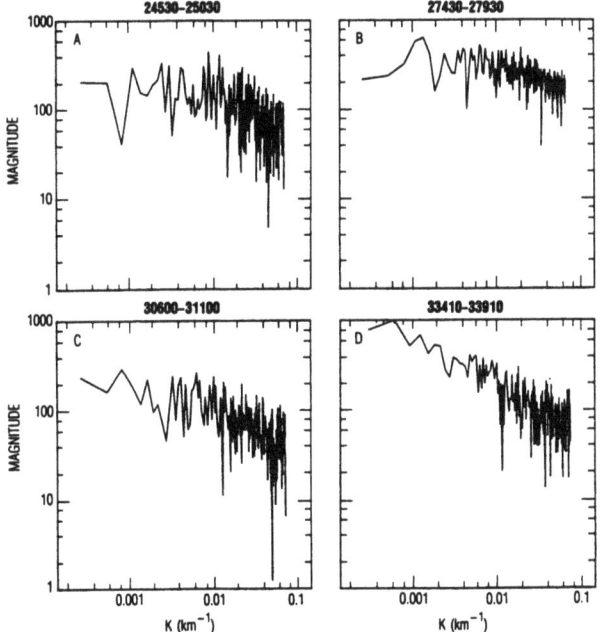

Figure 11—Power spectral density distributions of precipitating auroral electron number flux for the time series plotted in Figure 8. The power spectra are plotted with respect to inverse wavelength, normalized to the base of the magnetic field line.

lit hemispheres, even though resistive mapping theory implies that there should be as much as a factor-of-ten difference in characteristic scale size between the two hemispheres. This result indicates that further observational tests are warranted to study the dependence of the arc scale sizes on background ionospheric conditions. Coordinated space and ground-based ionospheric measurements would be useful in this regard.

5. SUMMARY

A number of recent theoretical works have dealt with the identification and evaluation of the physical processes that determine the characteristic scale sizes of discrete auroral arcs. It is broadly acknowledged that

a characteristic spatial width of ~100 km (at ionospheric heights) results naturally from the resistive ionospheric mapping of the high-altitude magnetospheric convection electric field. However, recent analysis of the spatial power spectral distributions of electric and magnetic field variations has revealed structure at much smaller spatial scales. Also, observations of visible auroral arcs from the ground seem to imply the dominance of smaller scale (<10 km) structure. Data on precipitating particles, acquired in low-altitude polar orbit, typically show monotonically decreasing inverse-wavelength spectra with slopes near unity, with no strictly "preferred" scale sizes, although the scale spectra do tend to flatten at scales larger than ~100–200 km. Also, typical widths of visible auroral arcs observed from the ground tend to be much smaller than the resistive scale length, and much smaller than the "average" scale size of inverted-V events observed by satellites.

Clearly, more observational work and comparisons with theoretical models need to be performed in order to understand thoroughly the processes that determine the spatial structures of discrete auroral features. However, it is hoped that the discussions of theoretical and observational results presented here demonstrate that several apparently contradictory observational results are, in fact, quite consistent with existing theoretical ideas, provided one performs a proper comparison. Specifically, it is important to note that a spectrum of spatial scale sizes is always present in auroral structures, including observations of both auroral particles and fields. Furthermore, the presence of a broad spectrum of scales, including scales smaller than the resistive scale length, is consistent with resistive mapping nevertheless. While the resistive scale size could be considered a "characteristic" scale size for auroral features, it should not be thought of as a "preferred" or "dominant" scale size.

ACKNOWLEDGMENT—This work was supported by NASA grant NAGW-853.

333

REFERENCES

Akasofu, S.-I., and B. Tsurutani, "Unusual Auroral Features Observed on January 10-11 and Their Possible Relationship to the Interplanetary Magnetic Field," *Geophys. Res. Lett.*, **11**, 1086 (1984).

Chiu, Y. T., "A Simple Kinetic Theory of Auroral Arc Scales," *J. Geophys. Res.*, **91**, 204 (1986).

Chiu, Y. T., "Auroral Parallel Electric Potential Structures in Magnetosphere-Ionosphere Coupling," in *Proceedings of Quantitative Modeling of Magnetosphere-Ionosphere Coupling Processes*, Y. Kamide and R. A. Wolf, eds., Kyoto Sangyo University, Kyoto, Japan (1987).

Chiu, Y. T. and J. M. Cornwall, "Electrostatic Model of a Quiet Auroral Arc," *J. Geophys. Res.*, **85**, 543 (1980).

Chiu, Y. T., A. L. Newman, and J. M. Cornwall, "On the Structures and Mappings of Auroral Electrostatic Potentials," *J. Geophys. Res.*, **86**, 10,029 (1981).

Frank, L. A., and K. L. Ackerson, "Observations of Charged Particle Precipitation into the Auroral Zone," *J. Geophys. Res.*, **76**, 3612 (1971).

Gorney, D. J., "A Study of the Spatial Scales of Discrete Polar Auroral Arcs," in *SPI Reprint Series*, Volume 8, T. Chang, ed., Scientific Publishers, Inc., Cambridge, MA (1988).

Gorney, D. J., D. S. Evans, M. S. Gussenhoven, and P. F. Mizera, "A Multiple-Satellite Observation of the High-Latitude Auroral Activity on 11 January 1983," *J. Geophys. Res.*, **91**, 339 (1986).

Gurnett, D. A., and L. A. Frank, "Observed Relationships Between Electric Fields and Auroral Particle Precipitation," *J. Geophys. Res.*, **78**, 145 (1973).

Knight, S., "Parallel Electric Fields," *Planet. Space Sci.*, **21**, 741 (1973).

Lin, C. S. and R. A. Hoffman, "Characteristics of the Inverted-V Event," *J. Geophys. Res.*, **84**, 1514 (1979).

Lotko, W., B. U. O. Sonnerup, and R. L. Lysak, "Nonsteady Boundary Layer Flow Including Ionospheric Drag and Parallel Electric Fields," *J. Geophys. Res.*, **92**, 8635 (1987).

Lyons, L. R., "Generation of Large-Scale Regions of Auroral Currents, Electric Potentials and Precipitation by the Divergence of the Convection Electric Field," *J. Geophys. Res.*, **85**, 17 (1980).

Lysak, R. L., and C. W. Carlson, "The Effect of Microscopic Turbulence on Magnetosphere-Ionosphere Coupling," *Geophys. Res. Lett.*, **8**, 269 (1981).

Lysak, R. L., and C. T. Dum, "Dynamics of Magnetosphere-Ionosphere Coupling Including Turbulent Transport," *J. Geophys. Res.*, **88**, 365 (1983).

Schield, M. A., and L. A. Frank, "Electron Observations Between the Inner Edge of the Plasma Sheet and the Plasmapause," *J. Geophys. Res.*, **75**, 5401 (1970).

Weimer, D. R., C. K. Goertz, D. A. Gurnett, N. C. Maynard, and J. L. Burch, "Auroral Zone Electric Fields from DE 1 and DE 2 at Magnetic Conjunctions," *J. Geophys. Res.*, **90**, 7479 (1985).

VI-2. MESOSCALE STRUCTURES IN AURORAL PHENOMENA

O. A. Troshichev*

Ground-based auroral observations, when coupled with spacecraft plasma and magnetic field measurements, allow the identifications of specific medium-scale auroral structures associated with processes originating at velocity shear interfaces in the distant magnetosphere. Mesoscale structures are observed in both quiet and disturbed periods, and in the dayside auroral oval as well as the nighttime oval and in the polar cap. The present review discusses a relationship between mesoscale auroral structures and the associated geophysical phenomena. Attention is focused on the features of polar cap auroral structures suggesting a closed magnetosphere topology when the IMF is northward.

1. INTRODUCTION

During the last decade the main features of large-scale, high-latitude geophysical phenomena have been derived from a number of satellite, rocket, radar, or ground magnetometer measurements. Such global structures as the auroral oval, polar cap, field-aligned current regions 1 and 2, and auroral electrojets are defined by the structure of the magnetosphere and respond to solar-wind/magnetosphere interaction.

There is another class of auroral phenomena of considerably smaller scale, from tens to hundreds of kilometers. Though some of these mesoscale phenomena, such as the auroral bulge and westward traveling surges, were singled out more than 20 years ago [*Akasofu et al.*, 1966], the study of them has been greatly advanced lately because of progress in satellite-borne instruments.

2. MESOSCALE STRUCTURES IN THE AURORAL OVAL

The auroral oval is defined [*Khorosheva*, 1961; *Feldstein*, 1963] as a region along which discrete auroral forms are distributed: these ovals are the homogeneous arcs and bands in quiet periods and bulges, surges, and patches in disturbed periods. As was shown later [*Lui et al.*, 1973], there is also the region of diffuse aurora located equatorward of the discrete auroral oval.

Figure 1 represents the generalized scheme of discrete and diffuse auroral distributions taken from *Akasofu and Kan* [1980]. It also shows some examples of typical mesoscale structures observed by spacecraft auroral images in recent years [*Lui et al.*, 1982; *Meng and Lundin*, 1986; *Lyons and Fennel*, 1986; *Bythrow et al.*, 1986; *Murphree et al.*, 1987; *Bythrow and Potemra*, 1987].

The auroral bulge and the westward traveling surge were pointed out by *Akasofu et al.* [1966] and are one of the main signatures of auroral substorms in the

nighttime auroral oval. In the early morning sector of the oval, a substorm is displayed in the omega-shaped bands appearing at the poleward edge of the diffuse aurora region [*Baumjohann*, this volume]. During nonstorm periods when the interplanetary magnetic field (IMF) is southward, discrete auroras in the early afternoon look like small (~ 100 km) isolated bright patches and then appear as multiple discrete arcs with a fan-shaped configuration converging toward the cusp. In the late morning part of the oval, the discrete arcs parallel to the auroral oval appear with lengths less than 1000 km. The midday gap of discrete auroras persistently observed in the noon sector of the oval is usually identified as the cusp region [*Meng*, 1981]. *Bythrow et al.* [1986] introduced multiple discrete auroral forms that moved poleward from the diffuse regions in the morning hours.

All these mesoscale structures are observed at the poleward border of the auroral oval. At the oval's equatorward border, wave-like structures are detected only in the evening sector, where large-amplitude undulations of the diffuse aurora boundary are seen [*Lui et al.*, 1982]. The crest-to-trough amplitude of these waveforms ranges from about 40–400 km, and the wavelength varies from about 200–900 km.

There is a clear separation between the noonside and nightside discrete auroras [*Meng and Lundin*, 1986]. Mesoscale auroral structures in the dayside oval are typical of nonstorm periods, whereas auroral bulges, westward traveling surges (WTS) and omegabands develop during substorms. Even if auroral brightenings (hot spots) occur simultaneously in both parts of the oval, their periodicity is different and they do not seem to be related to each other [*Lui et al.*, 1987].

There is a great variety in substorm auroral structures in the nightside oval. As an example, Figure 2 shows different ways of forming an auroral bulge in the course of a substorm according to satellite image data. In one case, the area of bright luminosity at 557.7 nm appearing near the midnight meridian may move toward the pole and west and, only 40 minutes later,

---

*Arctic and Antarctic Research Institute, 199226 Leningrad, USSR.

Auroral Physics, edited by C.-I. Meng, M. J. Rycroft and L. A. Frank. © Cambridge UP 1991

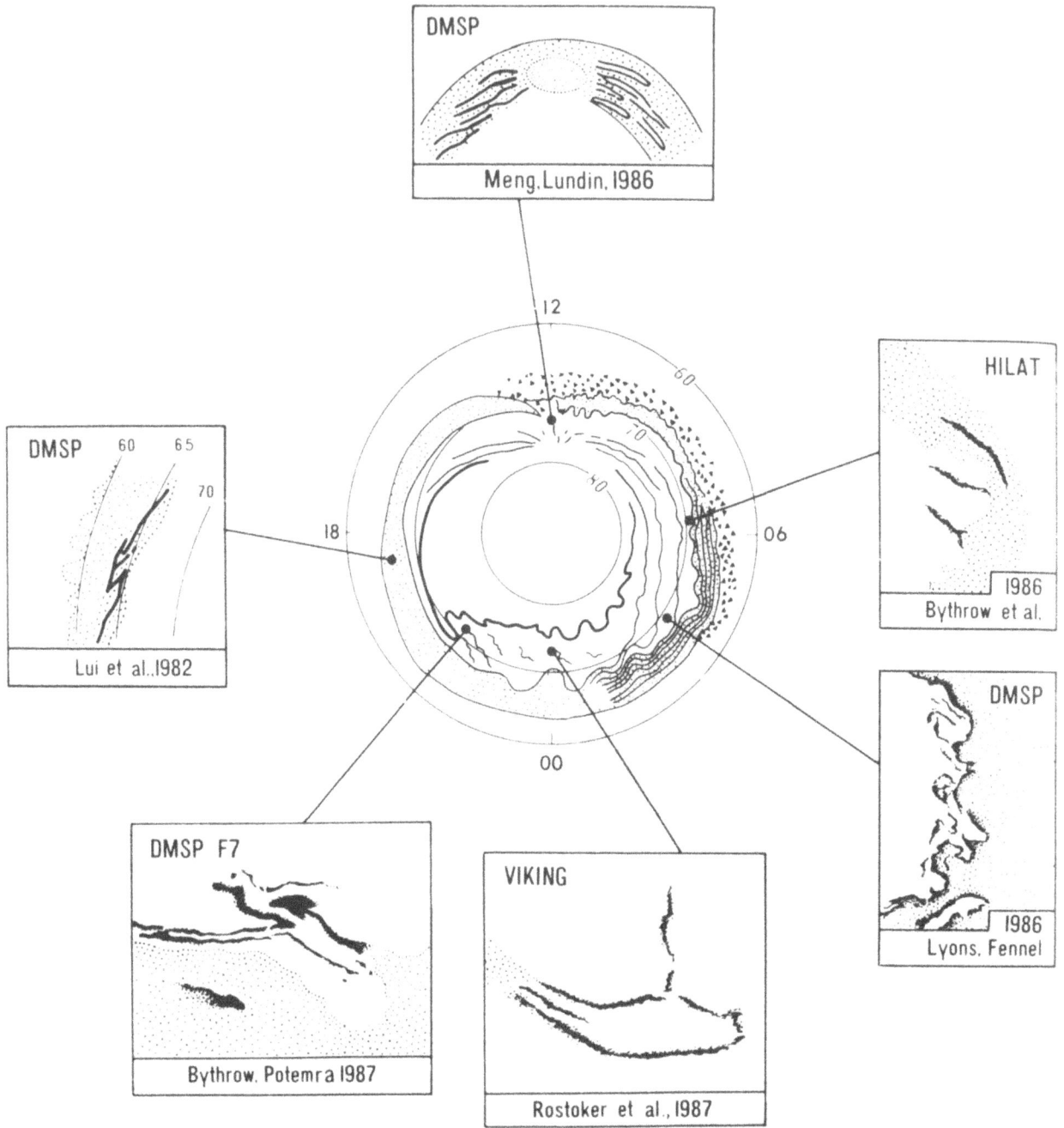

Figure 1—Diagram of the distribution of discrete and diffuse auroras given by *Akasofu and Kan* [1980], and some examples of typical mesoscale structures observed by spacecraft imagers.

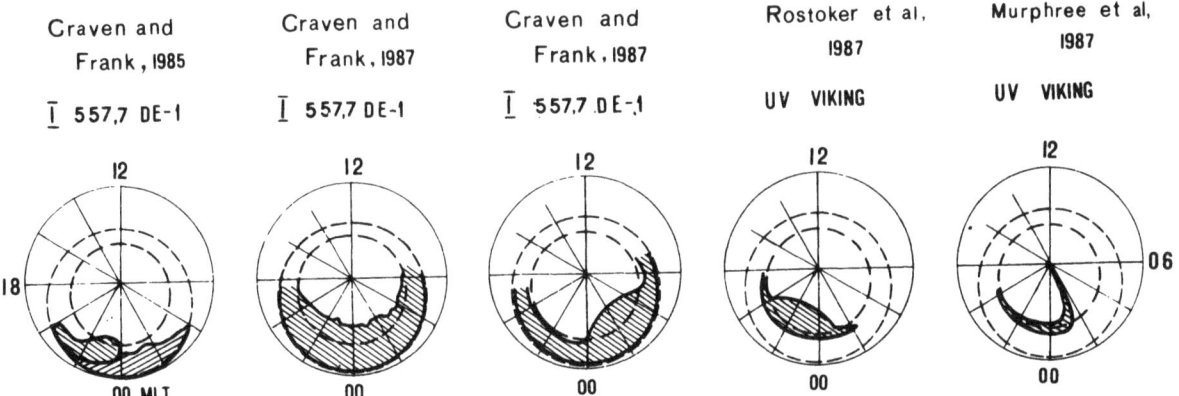

Craven and Frank, 1985   I 557.7 DE-1

Craven and Frank, 1987   I 557.7 DE-1

Craven and Frank, 1987   I 557.7 DE-1

Rostoker et al, 1987   UV VIKING

Murphree et al, 1987   UV VIKING

Figure 2—Diagram illustrating ways of building up of an auroral bulge in the course of a substorm according to spacecraft image data.

a new, more diffuse and weaker region of luminosity appears again near midnight, which spreads toward the east [*Craven and Frank*, 1987]. In another case, the bright luminosity region is expanded simultaneously toward the morning and evening meridian, but a poleward expansion begins only one hour later. In the third case, the widening of the auroral oval is observed only in the postmidnight sector, whereas the westward edge of the auroral bulge does not move to the 2200 MLT meridian in the course of the substorm [*Craven and Frank*, 1985]. Sometimes a rapid spreading of the initial auroral intensification along the oval together with a further poleward expansion of the active region results, in 5–10 minutes, in a fairly consistent lens-shaped structure that is called "the eye of the substorm" [*Rostoker et al.*, 1987a]. Each region of enhanced luminosity within the "eye" remains relatively stationary and fluctuates in intensity during the substorm. Finally, north-south oriented auroral arcs stretching from the poleward to the equatorward edges of the oval can appear. These localized structures called "fingers" [*Rostoker et al.*, 1987b] have a length up to 450 km and a width less than 100 km.

The central plasma sheet (CPS) is regarded as the source of diffuse precipitation of auroral particles related to the diffuse auroral zone [*Winningham et al.*, 1975], whereas the boundary plasma sheet (BPS) is considered to be responsible for discrete auroras [*Lui et al.*, 1977; *Frank*, 1976; *Eastman et al.*, 1984]. As for the mesoscale auroral structures, evidence regarding their origin was obtained only in recent years. *Bythrow and Potemra* [1987] carried out the first simultaneous high-resolution measurements of magnetic fields, plasma characteristics, and visible aurora associated with westward traveling surges on board the DMSP F7 satellite. Such an image shows a wide band of diffuse aurora, a pair of narrow (~20 km), parallel discrete arcs, beginning in the western part of the frame and extending eastward along the poleward edge of the diffuse

aurora and the intense auroral emission region (WTS) expanding from these arcs and poleward and westward from a lower latitude (Figure 3). The spectra presented by *Bythrow and Potemra* [1987] show that the plasma population in the diffuse zone is the same as that observed in the surge head. The charged particles that form the narrow arcs surrounding the head of the WTS are typical of beams observed in the plasma sheet boundary layer. On this basis *Bythrow and Potemra* [1987] conclude that the discrete arcs mark the poleward and equatorward edges of the boundary plasma sheet, while the WTS results from an expansion of the central plasma sheet into the boundary layer, with a subsequent deformation of the interface between the central plasma sheet, the boundary layer, and the tail lobe (Figure 4). A similar result was obtained by *Lyons and Fennel* [1986] when examining plasma spectra in the regions of omega bands in the morning sector. Precipitation in omega bands appears to be a combination of types usually found within the diffuse region and the structured region; precipitation typical of the diffuse aurora extends throughout the structure.

According to *Lundin and Evans* [1985] and *Meng and Lundin* [1986], the radial characteristics of discrete auroral arcs with a fan-shaped configuration converging toward the cusp support the idea that the dayside boundary layer is the plasma injection region for the dayside auroral oval. Figure 5 illustrates the topology of dayside plasma injection into the cusp and boundary layer, and the resulting dayside auroral display. *Evans* [1984] has reported that arc-like structures of precipitating electrons are typical in the afternoon hours.

The view that a wave propagating at the interface between the low-latitude boundary layer and the plasma sheet can be a source of multiple auroral forms was confirmed by simultaneous observations of quasiperiodic structures of small-scale Birkeland currents, energetic electrons, ion drifts, and auroral forms by the

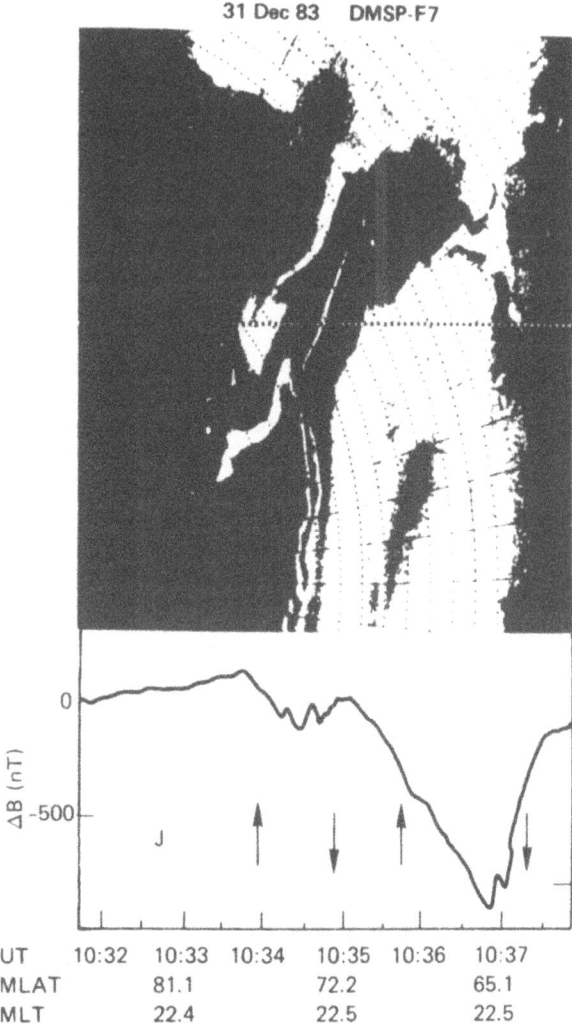

Figure 3—DMSP auroral imagery and the simultaneously measured perturbation magnetic field acquired in the premidnight sector on December 31, 1983 [*Bythrow and Potemra*, 1987].

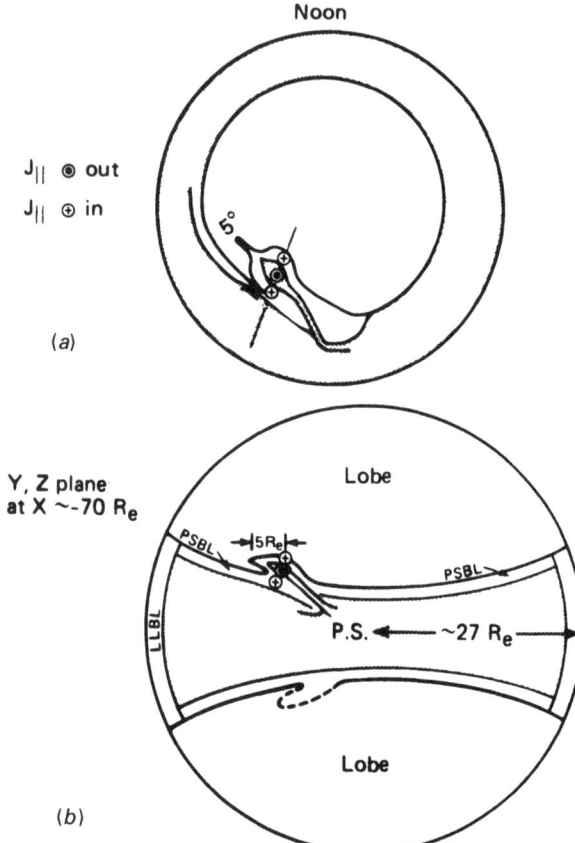

Figure 4—Diagram showing (*a*) the distribution of the aurora and associated Birkeland currents in the WTS region, and (*b*) their mapping from the magnetotail [*Bythrow and Potemra*, 1987].

HILAT spacecraft in the morning part of the auroral oval [*Bythrow et al.*, 1986].

Thus, mesoscale auroral structures at the edge of the auroral oval may be looked upon as a manifestation of some wave processes developing in the plasma sheet boundary layer. The Kelvin-Helmholtz instability driven by velocity shear is usually regarded as a possible candidate that can generate a surface wave [*Lyons and Fennel*, 1986; *Lui et al.*, 1987; *Bythrow et al.*, 1986; *Bythrow and Potemra*, 1987]. It is the plasma sheet boundary layer where velocity shear is set up by the earthward flows in the plasma sheet and the antisunward flows in the magnetotail lobes and low latitude boundary layer. Another velocity shear region may be located at the inner edge of the plasma sheet in the evening sector, where sunward convecting plasma encoun-

ters the corotating plasmasphere. Large-amplitude waves on the equatorward boundary of the diffuse aurora are evidently the result of such processes.

3. MESOSCALE AURORAL STRUCTURES AND ELECTRIC CURRENT SYSTEMS

The first data on transverse magnetic disturbances obtained by TRIAD [*Armstrong et al.*, 1975; *Kamide et al.*, 1976; *Kamide and Rostoker*, 1977] showed that auroral electrojets are confined mainly between large-scale sheets of field-aligned currents (FAC, region 1 and 2), and that discrete auroral arcs coincide with the intense upward currents. Subsequent studies based on Chatanika radar data and simultaneous TRIAD and S3-2 satellite data confirmed this [*Robinson et al.*, 1982; *Robinson et al.*, 1985a,b; *Senior et al.*, 1987]. Figure 6, summarizing the results of these studies, shows that in the morning sector the FAC region 2 coincides with the area of patchy aurora, while region 1 is located poleward of it. In the evening sector there are two systems of arcs: one lies between FAC regions 1 and 2

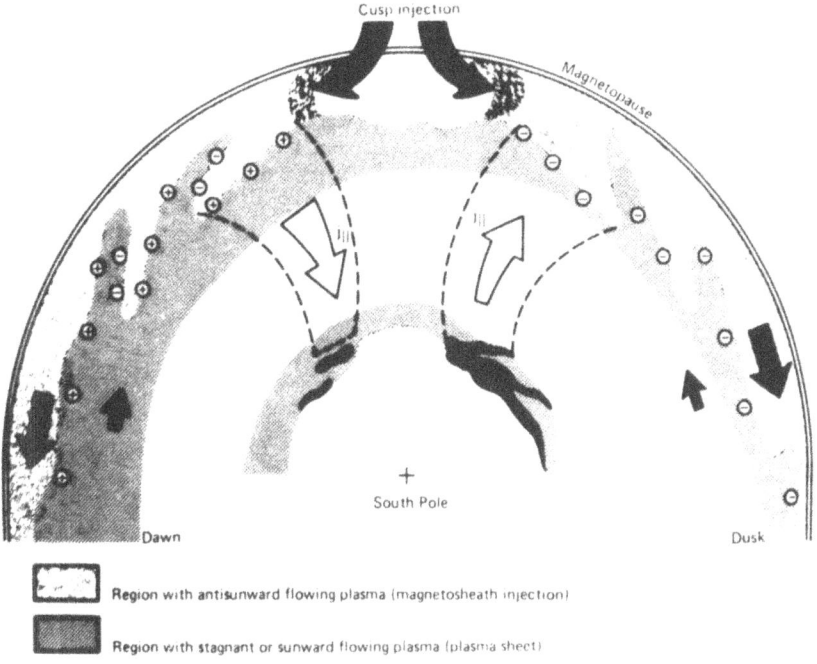

Figure 5—Topology of dayside plasma injection into the cusp and boundary layer, and the resulting dayside auroral display [*Meng and Lundin*, 1986].

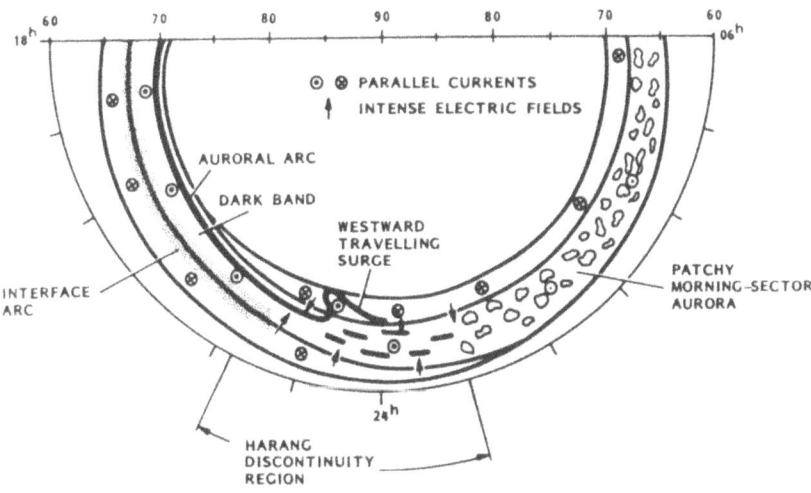

Figure 6—Schematic diagram of the relationship between field-aligned currents and visible aurora in the nightside auroral zone [*Robinson et al.*, 1985].

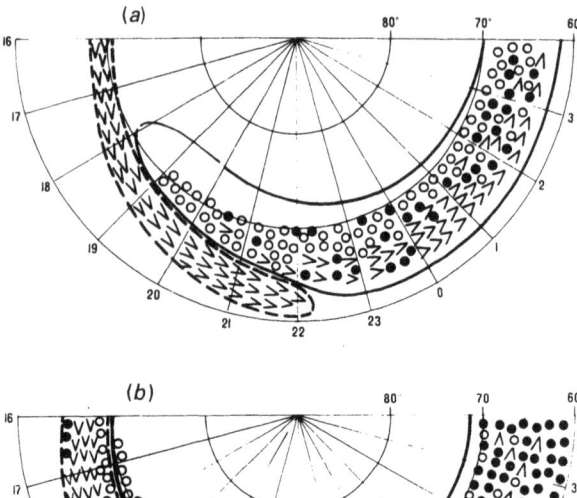

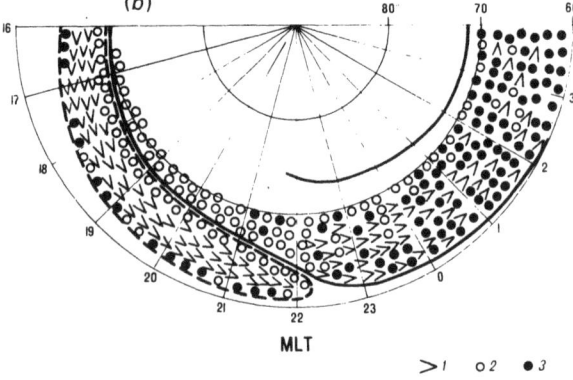

Figure 7—Schematic diagram of sporadic E layer (E_s) types in the eastward and westward electrojets for (a) isolated and (b) prolonged substorms: (1) shows an E_s layer with group delay (E_{sr}), (2) an auroral E_s layer (E_{sa}), and (3) indicates radio-wave absorption (blackouts).

and the other is located on the edge of the polar cap. Finer-scale structures of auroras and currents are not presented here.

An attempt to calculate the patterns of fields and currents for structures such as the auroral bulge, WTS, and omega bands was made by *Baumjohann et al.* [1981], *Inhester et al.* [1981] and *Opgenoorth et al.* [1983] using two-dimensional magnetic and electric field observations obtained by the Scandinavian Magnetic Array (SMA) and the STARE radars.

The results of this simulation study were presented [*Baumjohann*, 1989] for the case of a westward traveling surge that shows what strong conductivity gradients at the boundaries of the WTS lead to the generation of polarization electric fields. These enhance the westward electrojet, which is continued upward along magnetic field lines by a strong, localized field-aligned current.

However, the real distribution of field-aligned currents in the WTS region turns out to be much more complicated as compared with the calculated modeling patterns. Simultaneous imagery and high-resolution magnetic field measurements from the DMSP F7 satellite [*Bythrow and Potemra*, 1987] indicate seven sheets

of oppositely flowing Birkeland currents with a different scale in the WTS region (Figure 3). On both sides of the large-scale upward directed currents associated with the surge head there are two regions of downward flowing currents with the same scale and density (~ 0.4 μA m^{-2}). In addition to these large structures, two narrow (~ 20 km) and more intense (~ 1.5 μA m^{-2}) pairs of upward and downward field-aligned current sheets are observed on the northern and southern borders of the WTS region. The appearance of the fine field-aligned current structures in the poleward part of the auroral oval was also pointed out in other studies [*Kamide et al.*, 1976; *Kamide and Rostoker*, 1977; *Doyle et al.*, 1986; *Potemra et al.*, 1987], with the current density being as high as 30 μA m^{-2} [*Bythrow et al.*, 1987].

The regularities of the counterflowing Birkeland current sheets should give rise to interchanging northward and southward ionospheric electric fields between these sheets. These should result in filamentary structures of the westward and eastward ionospheric currents. Any manifestation of such fine-scale structures may be evident in the ground magnetic disturbance field. However, this fine structure might be hidden by the effects of larger-scale currents. Use of geomagnetic data from a meridional chain of stations could provide some indication of the splitting of the westward electrojet in the WTS region into two poleward and equatorward parts [*Kotikov et al.*, 1987].

Data from vertical incidence ionosondes operating in the Kara Sea meridian indicate different ionospheric E-layer properties in the poleward and equatorward parts of the auroral oval. In the premidnight local time sector, a sporadic E layer with a group delay (E_{sr}) is typical of a diffuse region, where the eastward electrojet flows, whereas the E_{sa} layer is more typical of a poleward region, where the westward electrojet is seen (Figure 7). The occurrence probability for each type of E_s layer depends on whether a given substorm is isolated or follows an earlier disturbance. The probability of occurrence of the E_{sa} layer in the premidnight sector increases sharply. This may be related to some change in the plasma population in the plasma sheet after the first substorm. As the study of *Baker et al.* [1985] shows, as two successive substorms develop, the leading substorm greatly perturbs and alerts the ion composition of the plasma in the outer magnetosphere. Thus the second substorm expansion phase occurs while the outer magnetospheric plasmas are largely of ionospheric (0^+) origin. A change in the plasma population affects the location and initiation of plasma sheet instabilities.

The structure of the electric current systems in the daytime cusp and oval regions is not yet clear, especially when the IMF is northward. Then additional

340

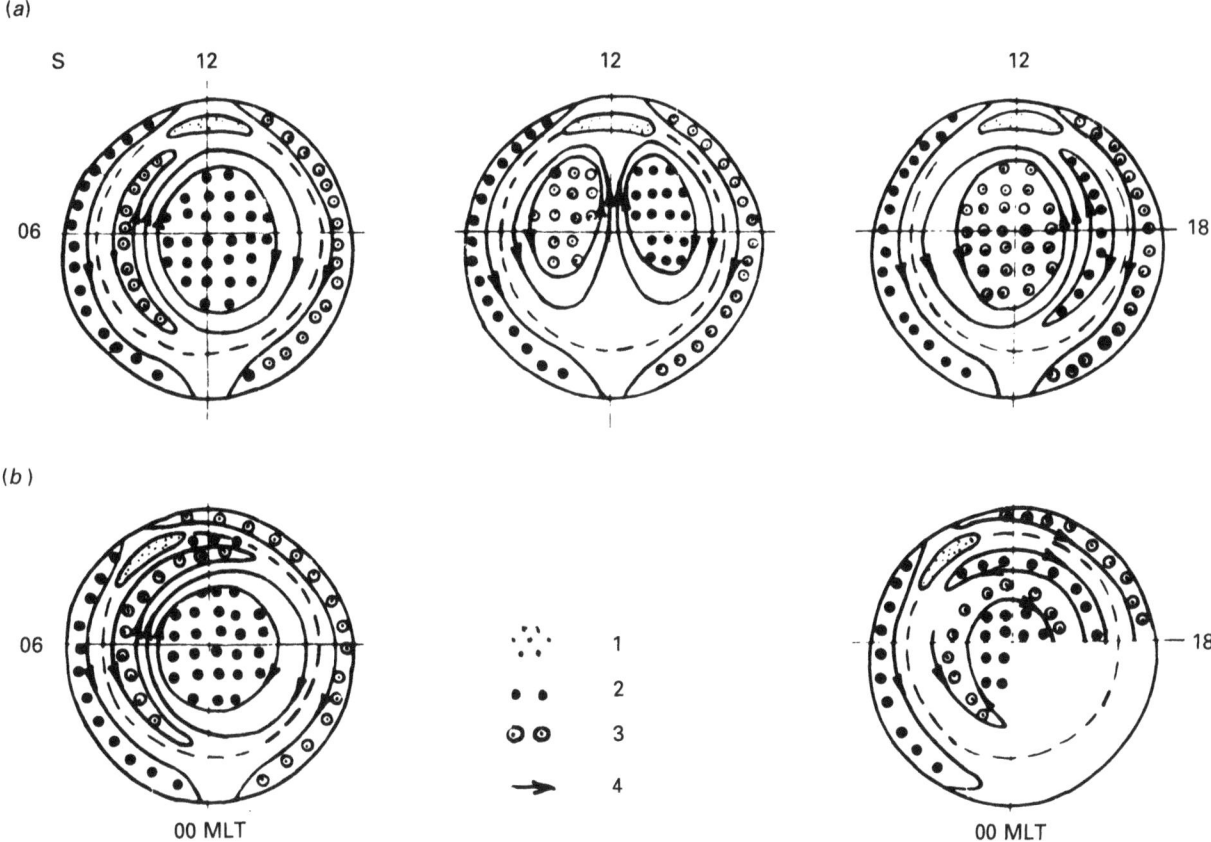

Figure 8—(*a*) Convection and field-aligned current patterns in the polar cap proposed by *Potemra et al.* [1984] for conditions of $B_z > 0$, $B_y \gtrsim 0$. (*b*) Convection and field-aligned currents obtained from meridional chain magnetic data: (1) daytime cusp region, (2) downward currents, (3) upward currents, and (4) convection flows [*Troshichev et al.*, 1988*b*].

Birkeland currents called region 3 [*Iijima and Potemra*, 1976] or NBZ currents [*Iijima et al.*, 1984] are specified in most patterns poleward of FAC region I. A relevant convection pattern, constructed taking into account the influence of the IMF B_y component is shown in Figure 8*a*. However, ground magnetic data indicate much more complicated patterns than those given in Figure 8*a* if the IMF northward component is stable and $B_y \neq 0$. Using data from the IZMIRAN chain of magnetic stations in Antarctica, *Troshichev et al.* [1988*b*] have calculated the distribution of ionospheric currents by the method of *Kotikov et al.* [1987]. They have constructed appropriate convection patterns and Birkeland currents, which are presented in Figure 8*b*. A salient feature of Figure 8*b* is a narrow crescent-shaped vortex spreading from the morning side of the southern polar cap (when $B_y > 0$) or from the evening side (when $B_y < 0$) toward the noon meridian. As a result, an additional pair of counterflowing Birkeland currents appears near the cusp region. The discrepancy between the patterns in Figure 8*a* and 8*b* is especially significant for $B_y \gtrsim 0$ when the NBZ sys-

tem decreases and Birkeland currents affected by the B_y component expand toward the pole.

In fact, the magnetic field measurements from the TRIAD, S3-2 and Intercosmos-Bulgaria-1300 spacecraft indicate the appearance of additional intense Birkeland currents in the cusp region [*Saflekos et al.*, 1982; *Doyle et al.*, 1981; *Zhuzgov et al.*, 1985]. Figure 9 presents two examples of magnetic perturbations observed in the daytime polar region, indicative of sheets of field-aligned currents. Two of them (the ones of lowest latitude) are evidently regular FAC regions 1 and 2. The poleward sheet evidently resembles the NBZ current region, and an additional pair of oppositely directed currents is located between them. It should be noted that the convection patterns obtained by *Troshichev et al.* [1988*b*] and presented in Figure 8*b* are sufficiently stable, provided that $B_z > 0$ and B_y remain unchanged. This means that these patterns are related to a steady source rather than to the wave-like motion at the interface between the low-latitude boundary layer and the plasma sheet.

(a)

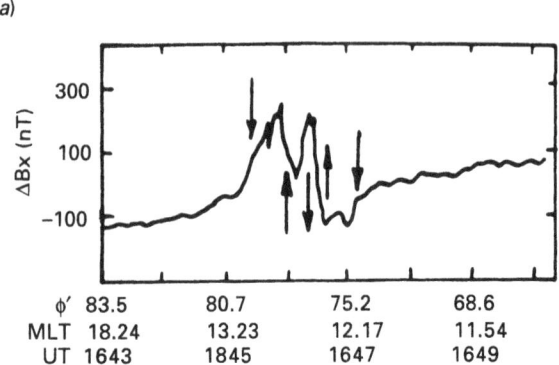

(b)

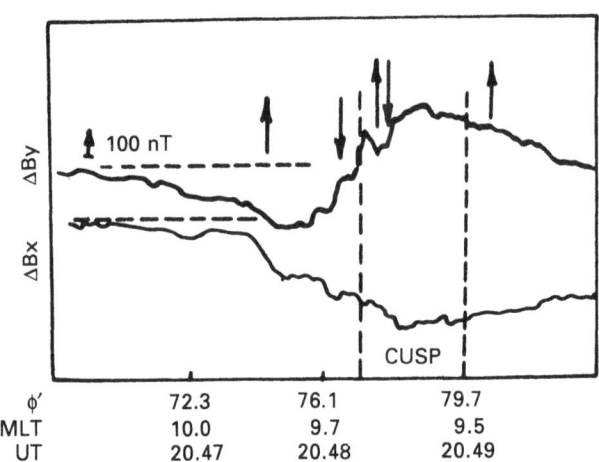

Figure 9—Field-aligned currents observed in the daytime cusp region by satellites: (a) S3-2 [*Saflekos et al.*, 1982], (b) Intercosmos-Bulgaria-1300 [*Zhuzgov et al.*, 1985].

4. MESOSCALE STRUCTURES IN THE POLAR CAPS

A typical mesoscale auroral structure in the polar cap is the Sun-aligned arcs observed only when the interplanetary magnetic field (IMF) is northward. These arcs may be single, twin, or multiple, stretched along the noon-midnight meridian as well as along the morning or evening edges of the polar cap [*Ismail et al.*, 1977; *Lassen and Danielsen*, 1978; *Yakhnin and Sergeev*, 1979; *Ismail and Meng*, 1982; *Gussenhoven*, 1982; *Gusev and Troshichev*, 1986; *Troshichev et al.*, 1988a]. Sun-aligned arcs near the edges of the dayside and nighttime auroral oval convert into hook-shaped structures. Such hook-shaped arcs are recorded by both satellite and ground-based auroral imagers in the nighttime polar cap [*Akasofu*, 1972; *Murphree and Cogger*, 1981; *Ismail and Meng*, 1982; *Gorney et al.*, 1986; *Eliasson et al.*, 1987] as well as in the dayside cap [*Gusev and Troshichev*, 1986; *Troshichev et al.*, 1988a].

In 1982 a new auroral structure called the theta-aurora was detected in the polar cap by UV imagers [*Frank et al.*, 1982, 1986]. In this structure a broad illuminated band stretching across the cap from noon to midnight is recorded simultaneously with the auroral oval arcs, provided that the IMF is northward. Taking into account the similarity in morphology between Sun-aligned arcs and the theta-aurora, *Frank et al.* [1986] and *Gusev and Troshichev* [1986] suggest that both structures relate to each other. Using data from the Antarctic station Vostok, *Troshichev et al.* [1988a] presented observations of polar cap discrete arcs and bright rays immersed in a wide band of diffuse luminosity which was seen by all-sky cameras for about three hours. Figure 10 gives two examples of the good agreement between the aurora detected from above by the DMSP auroral imager [*Ismail and Meng*, 1982] and Sun-aligned arcs observed from below by the all-sky camera at Vostok [*Gusev and Troshichev*, 1989]. The event of July 4, 1978 (Figure 10a) is of special interest because the satellite imager observed a typical theta structure. The borders of the transpolar band are clearly detected as discrete Sun-aligned arcs. And the multiple arcs seen by the Vostok camera fall within this band. These observations give a good reason to consider that the Sun-aligned arcs are in close agreement with theta-aurora, revealing the brightest auroral light arising against the background of a wide, but weaker, luminosity.

Above the polar cap arcs and theta-aurora, the plasma shows a number of unique characteristics. The plasma composition and energy spectrum differ from those in "polar rain" precipitation [*Hardy et al.*, 1982; *Burke et al.*, 1982; *Murphree et al.*, 1983; *Hardy*, 1984; *Hoffman et al.*, 1985; *Mizera and Evans*, 1986; *Greenspan et al.*, 1986; *Frank et al.*, 1986]. They are similar to those in the plasma sheet [*Murphree et al.*, 1983; *Peterson and Shelley*, 1984; *Frank et al.*, 1986; *Gorney et al.*, 1986]. The electron pitch-angle distribution indicates that the arcs occur on closed magnetic field lines [*Eliasson et al.*, 1987], and phenomena in the northern and southern polar caps may be conjugate [*Akasofu and Tsurutani*, 1984; *Gorney et al.*, 1986; *Mizera and Evans*, 1986; *Mizera et al.*, 1987]. All these peculiarities contradict the generally accepted idea that the polar caps are regions connected by open field lines to the interplanetary magnetic field and having no contact with the magnetospheric plasma sheet. To explain polar cap arcs and theta-aurora, various models have been offered [*Potemra et al.*, 1984; *Lyons*, 1985; *Chiu et al.*, 1985; *Kan and Burke*, 1985; *Reiff and Burch*, 1985; *Crooker*, 1988] taking into account magnetospheric topology and features of merging processes and magnetospheric convection for a northward directed IMF.

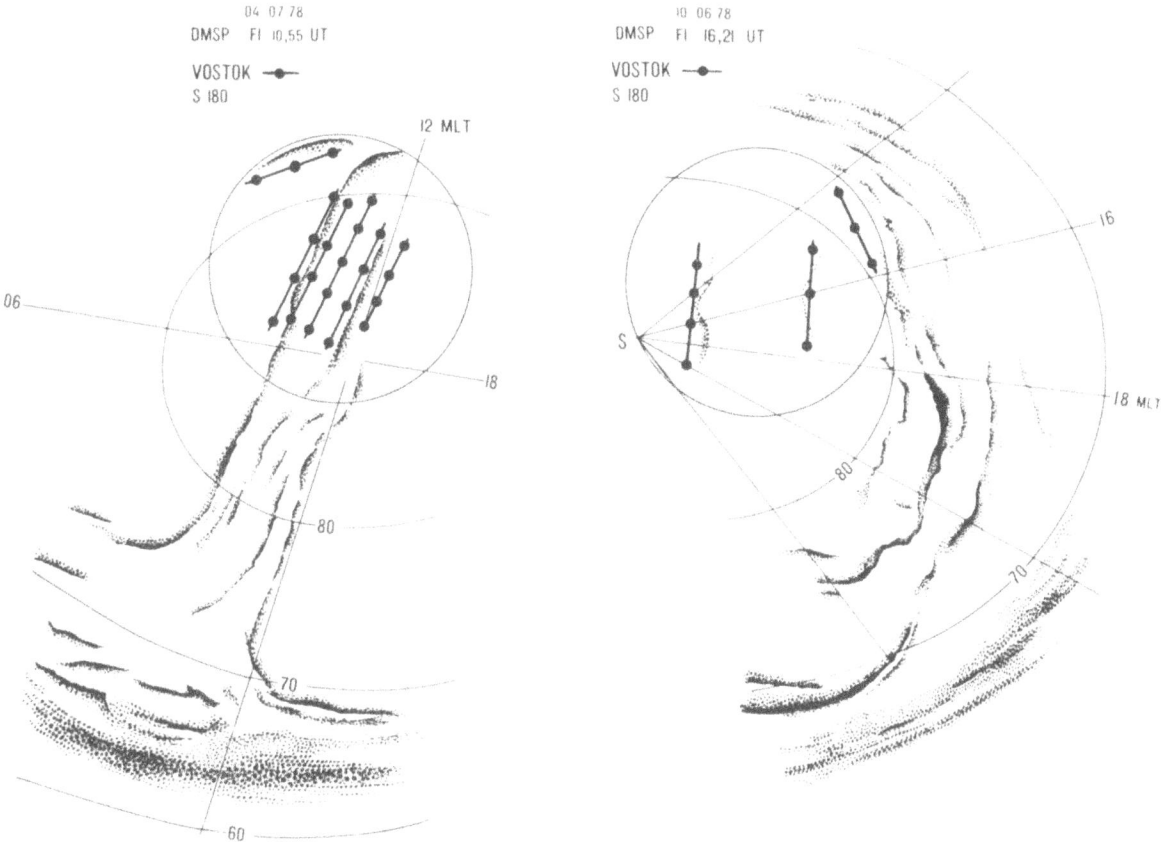

Figure 10—Two examples of good agreement between auroras detected from above by the DMSP auroral imager [*Ismail and Meng*, 1982] and Sun-aligned arcs observed from below by the all-sky camera at Vostok [*Gusev and Troshichev*, 1988].

The fact that the nature of the theta-aurora still remains obscure is attributable to some discrepancies in the experimental data. For example, *Frank et al.* [1986] point out that there is sunward convection over the polar cap arcs region and antisunward convection outside the arcs. By analyzing incoherent scatter data from Sondre Stromfjord, *Carlson et al.* [1984] came to the conclusion that the polar cap arcs lie at the boundary between oppositely directed convective flows. However, *Robinson et al.* [1987] claim that the convection over and near the polar arcs is antisunward. According to *Izrailevich et al.* [1988] and *Mende et al.* [1988] the transpolar arcs may lie in regions both of sunward and antisunward convection.

An appreciable body of information on the origin of the Sun-aligned arcs and theta-aurora may be extracted from ground-based observations. A statistical study of some peculiar features of polar cap aurora over the southern polar region was carried out by *Troshichev et al.* [1988a] using all-sky camera and scanning photometer data from the Antarctic station at Vostok. Figure 11 shows:

1. The probability of occurrence of auroral luminosity as recorded by a photometer when $B_z > 0$,

2. The occurrence of all forms of discrete aurora (Sun-aligned arcs and latitude-aligned arcs) over the Vostok station when $B_z > 0$,

3. The occurrence of latitude-aligned arcs for conditions $B_z > 0$ and $-1 < B_y < 1$ nT, and

4. The occurrence of polar cap arcs for conditions $B_z > 0$ and $-1 < B_y < 1$ nT.

Two regions of auroral luminosity seem to exist in the daytime polar cap, with a clear gap between them. Comparison of the distributions in Figure 11 *a,b,c*, and *d* shows that luminosity at invariant latitudes $\phi' = 79$–$80°$ is evidently related to the latitude-aligned arcs in the daytime auroral oval, whereas the second region at invariant latitudes of $\phi' = 82$–$84°$ is associated with polar cap arcs. It is important to note that discrete auroras most commonly occur in three areas of the polar cap, its morning and evening limbs, and a transpolar band (Figure 11*d*). That is, the distribution of polar cap arcs displays a theta-structure even if the latitude-aligned arcs are excluded from examination. Therefore a theta structure is a specified form of the distribution

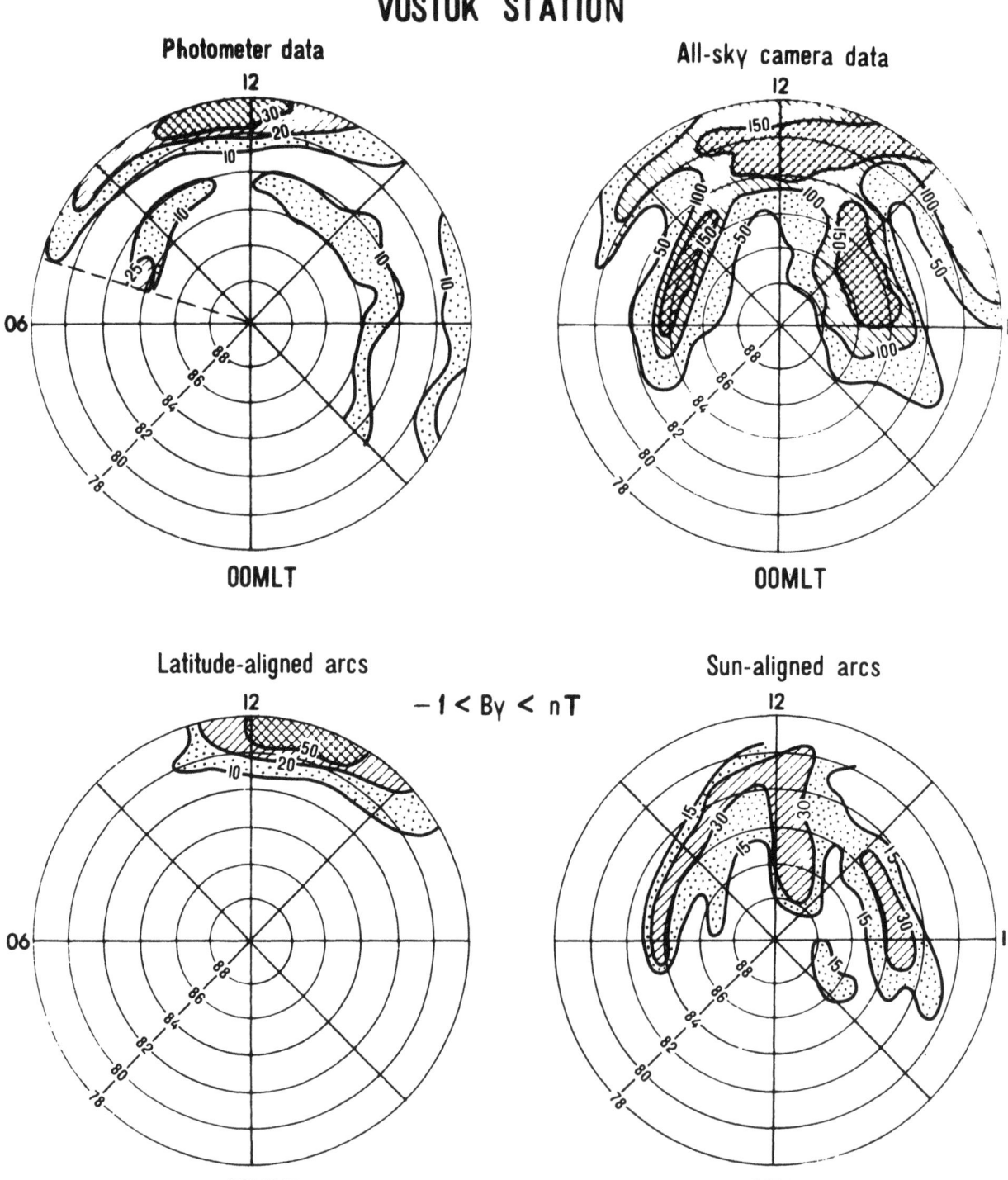

Figure 11—Distribution of auroras seen at Vostok station for $B_z > 0$ [*Troshichev et al.*, 1988a]: probability of occurrence of auroral luminosity as recorded by a photometer; distribution of all forms of discrete auroras according to all-sky camera data; distribution of latitude-aligned arcs when $-1 < B_y < 1$ nT; and distribution of polar cap arcs when $-1 < B_y < 1$ nT.

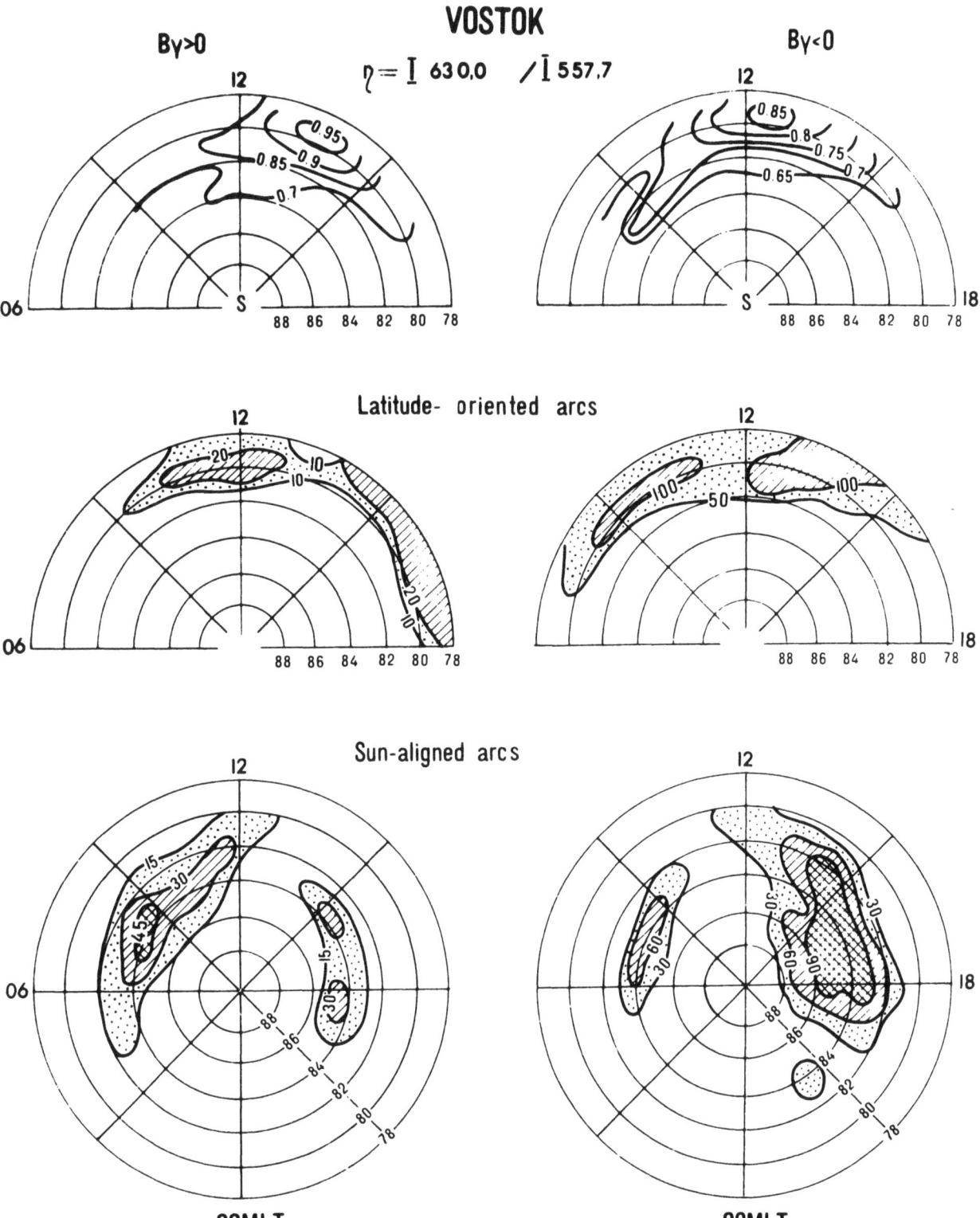

Figure 12—Influence of the IMF B_y component on auroral emissions from Vostok station data [*Troshichev et al.*, 1988a] parameter η, the ratio of auroral luminosity at 630 nm to that at 557.7 nm, characterizing the spectrum of the precipitating electrons; latitude-aligned arcs; and Sun-aligned arcs.

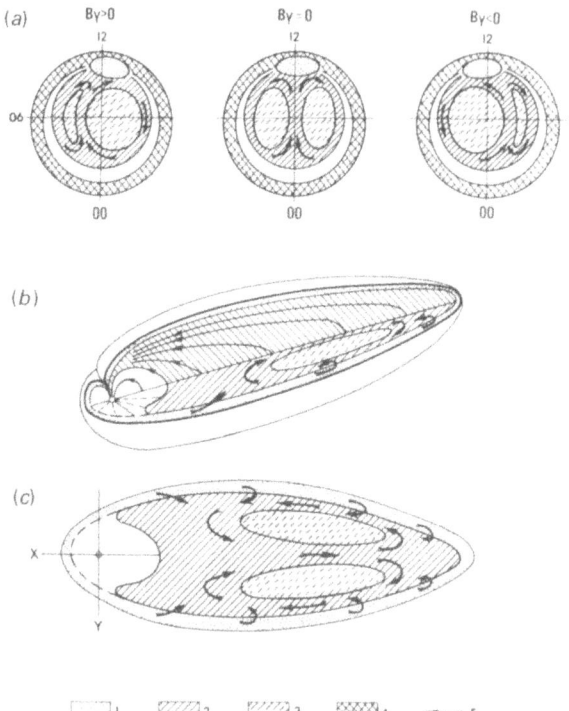

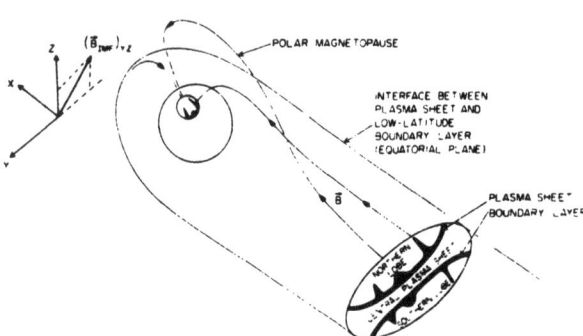

Figure 14—Interpretive sketch of lobe observations showing filaments protruding from the plasma sheet into the magnetotail lobes, and producing polar cap auroras during periods of northward IMF.

Figure 13—(*a*) Relationship between (1) the daytime cusp region, (2) auroral theta-structure, (3) regions with tenuous low-energy plasma, (4) auroral oval, and (5) convection flows. (*b*) Model of a closed magnetosphere. (*c*) Projection of the theta-structure from the equatorial cross-section of the closed magnetosphere [*Troshichev et al., 1988a*].

of polar cap arcs unrelated to auroras in the auroral oval.

The IMF B_y component affects the position of the transpolar band, shifting it toward the dawn (when $B_y > 1$ nT) or dusk ($B_y < 1$ nT), and closing it up with the appropriate limb in the theta structure (Figure 12). As a result, the polar cap arcs may be observed mainly in one half of the cap [*Meng and Lundin*, 1986] and the distribution of the polar cap auroras will show asymmetry controlled by the IMF B_y component [*Ismail et al.*, 1977; *Yakhnin and Sergeev*, 1981; *Gussenhoven*, 1982].

It is interesting to note that the influence of B_y is also displayed in the distribution of the daytime oval latitude-aligned arcs (Figure 12*b*) and the parameter η, which is the ratio of the intensity at 630.0 nm to that at 557.7 nm (Figure 12*a*). This characterizes the spectrum of the precipitating electrons: the softer the energy of precipitating electrons, the larger is the value η. The intensity of the 630.0 nm emission is a maximum where the occurrence of the auroral oval arcs is a minimum: this is at an invariant latitude of 80°, at 14.00 MLT for $B_y > 0$ and near noon for $B_y < 0$. It is known that the maximum intensity of 630.0 nm emission is

observed in the daytime cusp region [*Shepherd et al.*, 1976; *Shepherd*, 1979] where the low-energy electron precipitation is most intense and where a gap in the discrete aurora is often observed [*Meng*, 1981*a*]. Therefore the data presented in Figure 12 indicate that the daytime cusp region, coinciding with the gap in the auroral oval arcs, is displaced equatorward with respect to the theta-aurora. Figure 13*a* shows the interrelationship of the auroral oval, the theta-structure, and the daytime cusp according to the scheme of *Troshichev et al.* [1988*a*].

Since the plasma above the auroral theta structure has characteristics typical of the plasma sheet and is located on closed magnetic field lines, the conclusion is that the magnetosphere will be closed when $B_z > 0$; only the polar cusp remains open for solar-wind particles to enter. Figure 13*b,c* presents the appropriate model of the magnetosphere in which all field lines originating in the southern and northern polar caps are closed and intersect the plasma sheet [*Troshichev et al.*, 1988*a*]. Convection along the central axis of the plasma sheet is antisunward; the field lines involved move out from the inner magnetosphere toward the night termination of the magnetosphere and return along the dusk and dawn flanks of the magnetosphere. Field lines penetrating the plasma sheet would contain particles with plasma sheet characteristics. A number of plasma features observed over the polar caps when $B_z > 0$ may be explained from this point of view as follows: the similarity of the electron energy spectra in both polar cap arcs and the plasma sheets [*Murphree et al.*, 1983; *Peterson and Shelley*, 1984; *Frank et al.*, 1986], the fading of polar rain precipitation [*Hardy*, 1984; *Gussenhoven et al.*, 1984; *Riehl and Hardy*, 1986], and the conjugacy of electron precipitation above the arcs in the northern and southern polar caps [*Mizera and Evans*, 1986; *Gorney et al.*, 1986].

As evidenced by plasma composition data from the ISEE 1 and 2 spacecraft [*Huang et al.*, 1987], the protrusion of filaments containing plasma sheet particles into the magnetotail lobes is a rather common phenomenon. Only for northward IMF periods is there a greater probability of finding these filaments convecting away from the plasma sheet and their being the source of the polar cap arcs (Figure 14). A typical delay between the northward turning of the IMF and the appearance of the polar cap aurora is ~1 hour [*Troshichev et al.*, 1988a]. That is the time required for steady convection to become established when B_z > 0. In doing so, the regions with tenuous low-energy plasma on both sides of the transpolar band in the theta auroras [*Frank et al.*, 1986] must be associated with closed cavities formed in the distant plasma sheet (Figure 13b,c) where stagnation causes low thermal plasma densities.

The formation of large-scale convective vortices in the distant magnetosphere must be accompanied by the development of a large-scale pattern of field-aligned currents similar to the NBZ current system [*Iijima et al.*, 1984; *Iijima and Shibaji*, 1987]. According to *Huang et al.* [1987] the field-aligned current structures associated with plasma filaments occur in pairs of oppositely directed currents, and this is in agreement with the observations of theta-auroras [*Frank et al.*, 1986]. The current densities when mapped along the field lines are comparable with low-altitude measurements over the polar cap ($\sim\mu$A m^{-2}) and the current sheet thickness turns out to be approximately 125 km.

5. CONCLUSIONS

Mesoscale structures in auroral phenomena were detected long ago by ground imagers (all-sky cameras) as a feature of auroras in substorm periods. The use of spacecraft photometers and UV imagers with high resolving power reveals that structures with a typical size from tens to hundreds of kilometers are typical of the auroras throughout the high-latitude area. In the nighttime auroral zone, mesoscale structures (bulges, surges, and omega bands) usually appear in disturbed periods, whereas bright spots, quasi-periodic arc-like structures and bands in the daytime oval region and polar cap arcs are typical of quiet periods. Simultaneous aurora, plasma composition, and magnetic measurements from satellites show that the major part of the mesoscale structures originate at the velocity shear interfaces in the distant magnetosphere such as the boundary between the magnetotail lobes and the plasma sheet, the boundary between the low-latitude boundary layer and the plasma sheet, and the region between the plasma sheet and the corotating magnetosphere. Wave-like processes are thought to develop there as a result of the evolution of the Kelvin-Helmholtz instability.

Mesoscale structures are displayed not only in auroras but also in other geophysical phenomena such as particle precipitation, field-aligned currents, and filamentary structures in the auroral electrojets. An examination of mesoscale structures such as polar cap arcs shown that the magnetosphere is closed for long periods of the northward IMF. Auroral zone mesoscale structures can, to the best advantage, be studied by using geophysical data from a meridional chain of ground stations together with simultaneous satellite data.

REFERENCES

Akasofu, S.-I., "Mid Day Auroras and Polar-Cap Auroras," *Geofysike Publ.*, **29**, 73 (1972).

Akasofu, S.-I., and J. R. Kan, "Dayside and Nightside Auroral Arc Systems," *Geophys. Res. Lett.*, **7**, 753 (1980).

Akasofu, S.-I., and B. Tsurutani, "Unusual Auroral Features Observed on January 10-11 and Their Possible Relationship to the Interplanetary Magnetic Field," *Geophys. Res. Lett.*, **11**, 1086 (1984).

Akasofu, S.-I., D. S. Kimball, and C.-I. Meng, "Dynamics of the Aurora-IV: Polar Magnetic Substorms and Westward Travelling Surges," *Atm. Terr. Phys.*, **28**, 489 (1966).

Armstrong, J. C., S.-I. Akasofu, and G. Rostoker, "A Comparison of Satellite Observations of Birkeland Currents With Ground Observations of Visible Aurora and Ionospheric Currents," *J. Geophys. Res.*, **80**, 575 (1975).

Baker, D. N., T. A. Fritz, W. Lennartson, B. Wilken, H. W. Kroehl, and J. Birn, "The Role of Heavy Ionospheric Ions in the Localization of Substorm Disturbances on March 22, 1979: CDAW6," *J. Geophys. Res.*, **90**, 1273 (1985).

Baumjohann, W., "Electrodynamics of Active Auroral Forms: Westward Traveling Surges and Omega Bands," *this volume*.

Baumjohann, W., R. J. Pellinen, H. J. Opgenoorth, and E. Nielsen, "Joint Two-Dimensional Observations of Ground Magnetic and Ionospheric Electric Fields Associated with Auroral Zone Currents: Current Systems Associated with Local Auroral Breakups," *Planet. Space Sci.*, **29**, 431 (1981).

Burke, W. I., M. S. Gussenhoven, M. C. Kelley, D. A. Hardy, and F. J. Rich, "Electric and Magnetic Field Characteristics of Discrete Arcs in the Polar Cap," *J. Geophys. Res.*, **87**, 2431 (1982).

Bythrow, P. F., and T. A. Potemra, "Birkeland Currents and Energetic Particles Associated with Optical Auroral Signatures of a Westward Traveling Surge," *J. Geophys. Res.*, **92**, 8691 (1987).

Bythrow, P. F., M. A. Doyle, T. A. Potemra, L. J. Zanetti, R. E. Huffman, C.-I. Meng, D. A. Hardy, F. J. Rich, and R. A. Heelis, "Multiple Auroral Arcs and Birkeland Currents: Evidence for Plasma Sheet Boundary Waves," *Geophys. Res. Lett.*, **13**, 805 (1986).

Bythrow, P. F., T. A. Potemra, L. J. Zanetti, R. A. Erlandson, D. A. Hardy, E. J. Rich, and M. H. Acuña, "High Latitude Currents in the 0600 to 0900 MLT Sector: Observations from Viking and DMSP-F7," *Geophys. Res. Lett.*, **14**, 423 (1987).

Carlson, H. C., V. B. Wickwar, E. J. Weber, J. Buchau, J. G. Moore, and W. Whiting, "Plasma Characteristics of Polar Cap F-Layer Arcs," *Geophys. Res. Lett.*, **11**, 895 (1984).

Chiu, Y. T., N. U. Crooker, and D. J. Gorney, "Model of Oval and Polar Cap Arc Configurations," *J. Geophys. Res.*, **90**, 5153 (1985).

Craven, J. D., and L. A. Frank, "The Temporal Evolution of a Small Auroral Substorm as Viewed from High Altitudes with Dynamics Explorer 1," *Geophys. Res. Lett.*, **12**, 465 (1985).

Craven, J. D., and L. A. Frank, "Latitudinal Motion of the Aurora During Substorms," *J. Geophys. Res.*, **92**, 4565 (1987).

Crooker, N. U., "Mapping the Merging Potential From the Magnetopause to the Ionosphere Through the Dayside Cusp," *J. Geophys. Res.*, **93**, 1338 (1988).

Doyle, M. A., F. J. Rich, W. J. Burke, and M. Smiddy, "Field-Aligned Currents and Electric Fields Observed in the Region of the Dayside Cusp," *J. Geophys. Res.*, **86**, 5656 (1981).

Doyle, M. A., W. J. Burke, D. A. Hardy, P. F. Bythrow, P. J. Rich, and T. A. Potemra, "A Simple Model of Auroral Electrodynamics Compared with HILAT Measurements," *J. Geophys. Res.*, **91**, 6979 (1986).

Eastman, T., L. A. Frank, W. K. Peterson, and W. Lennartson, "The Plasma Sheet Boundary Layer," *J. Geophys. Res.*, **89**, 1553 (1984).

Eliasson, L., R. Lundin, and J. S. Murphree, "Polar Cap Arcs Observed by the Viking Satellite, *Geophys. Res. Lett.*, **14**, 451 (1987).

Evans, D. S., "The Characteristics of a Persistent Auroral Arc at High Latitude in the 1400 MLT Section," p. 99, Proc. NATO Conf. on the Polar Cusp, 7–11 May 1984, Lillehammer, Norway (1984).

Feldstein, Ya.I., "Some Problems Concerning the Morphology of Auroras and Magnetic Disturbances at High Latitudes," *Geomagn. Aeron.*, **3**, 227 (1963).

Frank, L. A., "Hot Plasmas in the Earth's Magnetosphere," in *Physics of Solar Planetary Environments*, vol 2, D. J. Williams, ed., p. 685 (1976).

Frank, L. A., J. D. Craven, J. L. Burch, and J. D. Winningham, "Polar Views of the Earth's Aurora with Dynamics Explorer," *Geophys. Res. Lett.*, **9**, 1001 (1982).

Frank, L. A., J. D. Craven, D. A. Gurnett, S. D. Shawhan, D. R. Weimer, J. W. Burch, J. D. Winningham, C. R. Chappell, J. H. Waite, R. A. Heelis, N. C. Maynard, M. Sigiura, W. K. Peterson, and E. G. Shelley, "The Theta Aurora," *J. Geophys. Res.*, **91**, 3177 (1986).

Gorney, D. J., D. S. Evans, M. S. Gussenhoven, and P. Mizera, "A Multiple-Satellite Observation of the High-Latitude Auroral Activity on January 11, 1983," *J. Geophys. Res.*, **91**, 339 (1986).

Greenspan, M. E., C.-I. Meng, and D. H. Fairfield, "Simultaneous Polar Cap and Magnetotail Observations of Intense Polar Rain," *J. Geophys. Res.*, **91**, 11123 (1986).

Gusev, M. G., and O. A. Troshichev, "Hook-Shaped Arcs in the Dayside Polar Cap and Their Relation to the IMF," *Planet. Space Sci.*, **34**, 489 (1986).

Gusev, M. G., and O. A. Troshichev, "Relation of Sun-Aligned Arcs to Polar Cap Convection and Magnetic Disturbances," *Planet. Space Sci.*, **38**, 1 (1990).

Gussenhoven, M. S., "Extremely High Latitude Auroras," *J. Geophys. Res.*, **87**, 2401 (1982).

Gussenhoven, M. S., D. A. Hardy, N. Heineman, and R. K. Burkhardt, "Morphology of the Polar Rain," *J. Geophys. Res.*, **89**, 9785 (1984).

Hardy, D. A., "Intense Fluxes of Low-Energy Electrons at Geomagnetic Latitudes Above 85°," *J. Geophys. Res.*, **89**, 3883 (1984).

Hardy, D. A., M. J. Burke, and M. S. Gussenhoven, "DMSP Optical and Electron Measurements in the Vicinity of Polar Cap Arcs," *J. Geophys. Res.*, **87**, 2413 (1982).

Hoffman, R. A., R. A. Heelis, and J. S. Prasad, "Sun-Aligned Arc Observed by DMSP and AE-C," *J. Geophys. Res.*, **90**, 9697 (1985).

Huang, C. Y., L. A. Frank, W. K. Peterson, D. J. Williams, W. Lennartson, D. G. Mitchell, R. C. Elphic, and C. T. Russell, "Filamentary Structures in the Magnetotail Lobes," *J. Geophys. Res.*, **92**, 2349 (1987).

Iijima, T., T. A. Potemra, L. J. Zanetti, and P. F. Bythrow, "Large-Scale Birkeland Currents in the Dayside Polar Region During Strongly Northward IMF: A New Birkeland Current System," *J. Geophys. Res.*, **89**, 7441 (1984).

Iijima, T., and T. A. Potemra, "The Amplitude Distribution of Field-Aligned Currents at Northern High Latitudes Observed by Triad," *J. Geophys. Res.*, **81**, 2165 (1976).

Iijima, T., and T. Shibaji, "Global Characteristics of Northward IMF-Associated (NBZ) Field-Aligned Currents," *J. Geophys. Res.*, **92**, 2408 (1987).

Inhester, B., W. Baumjohann, R. A. Greenwald, and E. Nielsen, "Joint Two Dimensional Observations of Ground Magnetic and Ionospheric Electric Fields Associated with Auroral Zone Currents. 3. Auroral Zone Currents During the Passage of a Westward Travelling Surge," *J. of Geophys.*, **49**, 155 (1981).

Ismail, S., and C.-I. Meng, "A Classification of Polar Cap Auroral Arcs," *Planet. Space Sci.*, **30**, 319 (1982).

Ismail, S., D. D. Wallis, and L. L. Cogger, "Characteristics of Polar Cap Sun-Aligned Areas," *J. Geophys. Res.*, **82**, 4741 (1977).

Izrailevich, P. L., I. M. Podgorny, A. K. Kuzmin, N. S. Nickolaeva, and E. M. Dubinin, "Convection and Field-Aligned Currents Related to Transpolar Arcs During Strongly Northward IMF," *Planet. Space Sci.*, **36**, 1517 (1988).

Kamide, Y., and G. Rostoker, "The Spatial Relationship of Field-Aligned Currents and Implications Regarding Acceleration of Auroral Electrons," *J. Geophys. Res.*, **82**, 5589 (1987).

Kamide, Y., S.-I. Akasofu, and G. Rostoker, "Field-Aligned Currents and the Auroral Electrojet in the Morning Sector," *J. Geophys. Res.*, **81**, 6141 (1976).

Kan, J. R., and W. J. Burke, "A Theoretical Model of Polar Cap Auroral Arcs," *J. Geophys. Res.*, **90**, 4171 (1985).

Kotikov, A. L., Yu.O. Latov, and O. A. Troshichev, *Geophysica*, **23**, 143 (1987).

Khorosheva, O. V., "The Space and Time-Distribution of Auroras and Their Relationship With High-Latitude Geomagnetic Disturbances," *Geomagn. Aeron.*, **1**, 695 (1961).

Lassen, K., and C. Danielsen, "Quiet-Time Pattern of Auroral Arcs for Different Directions of the Interplanetary Magnetic Field in the Y-Z Plane," *Geomagn. Geoelectr.*, **30**, 193 (1978).

Lui, A. T. Y., C.-I. Meng, and S. Ismail, "Large-Amplitude Undulations of the Equatorward Boundary of the Diffuse Aurora," *J. Geophys. Res.*, **87**, 2385 (1982).

Lui, A. T. Y., P. Perreault, S.-I. Akasofu, and C. D. Anger, "The Diffuse Aurora," *Planet. Space Sci.*, **21**, 857 (1973).

Lui, A. T. Y., D. Venkatesan, G. Rostoker, S.-I. Akasofu, W. J. Heikkila, J. D. Winningham, and J. R. Burrow, "Simultaneous Observations of Particle Precipitations and Auroral Emissions by the ISIS-2 Satellite," *J. Geophys. Res.*, **82**, 2210 (1977).

Lui, A. T. Y., D. Venkatesan, G. Rostoker, J. S. Murphree, C. D. Anger, L. L. Cogger, and T. A. Potemra, "Dayside Auroral Intensifications During an Auroral Substorm," *Geophys. Res. Lett.*, **14**, 415 (1987).

Lundin, R., and D. S. Evans, "Boundary Layer Plasmas as a Source for High-Latitude, Early Afternoon, Auroral Arcs," *Planet. Space Sci.*, **33**, 1389 (1985).

Lyons, L. R., "A Simple Model for Polar Cap Convection Patterns and Generation of θ Auroras," *J. Geophys. Res.*, **90**, 1561 (1985).

Lyons, L. R., and J. F. Fennel, "Characteristics of Auroral Electron Precipitation on the Morningside," *J. Geophys. Res.*, **91**, 11225 (1986).

Mende, S. B., J. H. Doolittle, R. M. Robinson, R. R. Vondrak, and F. J. Rich, "Plasma Drifts Associated With a System of Sun-Aligned Arcs in the Polar Cap," *J. Geophys. Res.*, **93**, 256 (1988).

Meng, C.-I., "Electron Precipitation in the Midday Auroral Oval," *J. Geophys. Res.*, **86**, 2149 (1981).

Meng, C.-I., and R. Lundin, "Auroral Morphology of the Midday Oval," *J. Geophys. Res.*, **91**, 1572 (1986).

Mizera, P. F., and D. S. Evans, "Simultaneous Measurements of Polar Cap Electron Distributions in Opposite Hemispheres," *J. Geophys. Res.*, **91**, 9007 (1986).

Mizera, P. F., D. J. Gorney, and D. S. Evans, "On the Conjugacy of the Aurora: High and Low Latitudes," *Geophys. Res. Lett.*, **14**, 190 (1987).

Murphree, J. S., and L. L. Cogger, "Observed Connections Between Apparent Polar Cap Features and the Instantaneous Diffuse Auroral Oval," *Planet. Space Sci.*, **29**, 1143 (1981).

Murphree, J. S., S. Ismail, L. L. Cogger, D. D. Wallis, G. G. Shepherd, R. Link, and D. H. Klumpar, "Characteristics of Optical Emissions and Particle Precipitation in Polar Gap Arcs," *Planet. Space Sci.*, **31**, 161 (1983).

Murphree, J. S., L. L. Cogger, C. D. Anger, D. D. Wallis, and G. G. Shepherd, "Oval Intensifications Associated With Polar Arcs," *Geophys. Res. Lett.*, **14**, 403 (1987).

Opgenoorth, H. J., J. Oksman, K. U. Kaila, E. Nielsen, and W. Baumjohann, "On the Characteristics of Eastward Drifting Omega Bands in the Morning Sector of the Auroral Oval," *J. Geophys. Res.*, **88**, 9171 (1983).

Peterson, W. K., and E. G. Shelley, Origin of the Plasma in a Cross-Polar Cap Auroral Feature (Theta Aurora)," *J. Geophys. Res.*, **89**, 6729 (1984).

Potemra, T. A., L. J. Zanetti, P. F. Bythrow, A. T. Y. Lui, and T. Iijima, "B$_\gamma$-Dependent Convection Patterns During Northward Interplanetary Magnetic Field," *J. Geophys. Res.*, **89**, 9753 (1984).

Potemra, T. A., L. J. Zanetti, R. E. Erlandson, P. F. Bythrow, G. Gustaffson, M. H. Acuña, and R. Lunin, "Observations of Large-Scale Birkeland Currents with Viking," *Geophys. Res. Lett.*, **14**, 419 (1987).

Reiff, P. H., and J. L. Burch, "IMF B$_\gamma$-Dependent Plasma Flow and Birkeland Currents in the Dayside Magnetosphere, 2, A Global Model for Northward and Southward IMF," *J. Geophys. Res.*, **90**, 1595 (1985).

Riehl, K. B., and D. A. Hardy, "Average Characteristics of the Polar Rain and Their Relationship to the Solar Wind and the Interplanetary Magnetic Field," *J. Geophys. Res.*, **91**, 1557 (1986).

Robinson, R. M., F. Rich, and R. R. Vondrak, "Chatanika Radar and S3-2 Measurements of Auroral Zone Electrodynamics in the Midnight Sector," *J. Geophys. Res.*, **90**, 8487 (1985).

Robinson, R. M., R. R. Vondrak, and E. Friis-Christensen, "Ionospheric Currents Associated With a Sun-Aligned Arc Connected to the Auroral Oval," *Geophys. Res. Lett.*, **14**, 656 (1987).

Robinson, R. M., R. R. Vondrak, and T. A. Potemra, "Electrodynamic Properties of the Evening Sector Ionosphere Within the Region 2 Field-Aligned Current Sheet," *J. Geophys. Res.*, **87**, 731 (1982).

Robinson, R. M., R. R. Vondrak, and T. A. Potemra, "Auroral Zone Conductivities Within the Field-Aligned Current Sheets," *J. Geophys. Res.*, **90**, 9688 (1985).

Rostoker, G., A. Vallence Jones, R. L. Gattinger, C. D. Anger, and J. S. Murphree, "The Development of the Substorm Expansive Phase: The 'Eye' of the Storm," *Geophys. Res. Lett.*, **14**, 399 (1987*a*).

Rostoker, G., A. T. Y. Lui, C. D. Anger, and J. S. Murphree, "North-South Structures in the Midnight Sector Auroras as Viewed by the Viking Imager," *Geophys. Res. Lett.*, **14**, 407 (1987*b*).

Saflekos, N. A., R. E. Sheehan, and R. L. Carvillano, "Global Nature of Field-Aligned Currents and Their Relation to Auroral Phenomena," *Rev. Geophys. Space Phys.*, **20**, 709 (1982).

Senior, C., J. R. Sharber, et al., "E and F Region Study of the Evening Sector Auroral Oval: A Chatanika/Dynamics Explorer 2/NOAA 6 Comparison," *J. Geophys. Res.*, **92**, 2477 (1987).

Shepherd, G. G., "Dayside Cleft Aurora and Its Ionospheric Effects," *Rev. Geophys. Space Phys.*, **17**, 2017 (1979).

Shepherd, G. G., F. W. Thirkettle, and C. D. Anger, "Topside Optical View of the Dayside Cleft Aurora," *Planet. Space Sci.*, **24**, 937 (1976).

Troshichev, O. A., M. G. Gusev, S. V. Nickolashkin, and V. P. Samsonov, "Features of the Polar Cap Aurorae in the Southern Polar Region," *Planet. Space Sci.*, **36**, 5 (1988*a*).

Troshichev, O. A., B. D. Bolotinskaya, A. L. Kotikov, and Papitashvilli, "B_Y Dependent Currents in the Southern Polar Region During Positive B_Z," *Planet. Space Sci.*, **36**, 6 (1988*b*).

Winningham, J. D., F. Yasuhara, S.-I. Akasofu, and W. J. Heikkila, "The Latitudinal Morphology of 10-eV to 10-keV Electron Fluxes During Magnetically Quiet and Disturbed Times in the 2100-0300 MLT Sector," *J. Geophys. Res.*, **80**, 3148 (1975).

Yakhnin, A. G., and V. A. Sergeyev, "Frequency of Occurrence of Auroras in the Polar Cap and Orientation of the IMF Vector," *Geomagn. Aeron.*, **19**, 566 (1979).

Yakhnin, A. G., and V. A. Sergeyev, "Polar Cap Aurorae and Dependence on the IMF Orientation and on Substorms, Some Features of the Morphology," *Aurora and Airglow*, **28**, 27 (1981).

Zhuzgov, L. N., L. O. Tyurmina, et al., *Space Research* (Russian), **21**, 886 (1985).

VI-3. THE PULSATING AURORA AND ITS RELATIONSHIP TO FIELDS AND CHARGED-PARTICLE PRECIPITATION

P. J. Tanskanen*

Auroral structures are reviewed with emphasis on pulsating aurora. Observations of the typical shape, size, altitude, and time variations of pulsating aurora are described. It is shown that pulsating auroras are almost always seen embedded in a diffuse background aurora at an altitude of 120–240 km. Certain pulsating forms have been observed in very thin ionospheric layers. Pulsating auroral patches seem to be subject to the $E \times B$ drift motion of enhanced plasma density that form ducts from the magnetosphere to the ionosphere. Conjugate observations of auroral structures indicate that large-scale structures are most likely controlled by magnetospheric processes whereas small-scale structures are due to local phenomena. "Black aurora" has been observed as small regions of very low luminosity embedded in a diffuse background. Electron energy spectra of pulsating aurora show a steady component extending to energies below those for which interactions with very low frequency (VLF) radio waves are expected. There is also an energetic component varying from one event to another.

1. AURORA

The aurora, a majestic multi-color polar region phenomenon, must have always been a subject of curiosity for the natives living in these areas (Figure 1). The aurora is the name given to the light resulting from the precipitation of electrons and protons from the magnetosphere into Earth's atmosphere [Omholt, 1971]. This light consists of atomic line spectra and molecular band spectra characteristic principally of oxygen and nitrogen, the chief constituents of the upper atmosphere, ionized or excited by collisions with these precipitating particles.

2. AURORAL STRUCTURES

According to their structure, auroras can be divided into three main categories [International Auroral Atlas, 1963]: quiet aurora, active aurora, and pulsing aurora. They can be characterized as follows:

1. Quiet aurora is an almost stationary aurora with an unchanging form.
2. Active aurora shows rapid (~ 1 s) changes in position and shape.
3. Pulsing aurora shows fast (≤ 1 s–100 s), often periodic, changes in intensity.

A more modern classification of auroras is based on their spatial structure, i.e., whether they are structured or unstructured. Generally the structured component is known as discrete aurora and, since it is the most visible part, it is also popularly known as "the aurora." The unstructured component is known as the continuous or diffuse aurora. Although the diffuse aurora is, due to its faintness, difficult to observe visually, the energy input associated with particles causing it is

50–80% of the total auroral energy into the polar region [Sandford, 1968]. Diffuse aurora occurs in the auroral oval and equatorward of it.

Most auroral luminosity is produced by precipitating energetic electrons. Only in rather rare cases are Doppler-shifted Balmer lines of hydrogen atoms detected in aurora, indicating the precipitation of energetic protons into the atmosphere. Every incoming proton usually catches and loses, during its passage through the upper atmosphere, a great number of electrons and produces several photons. The resulting emission is weak and barely visible to the naked eye. It occurs in homogeneous, broad forms.

3. PULSING AND PULSATING AURORA

Next, attention is paid to a special type of aurora, generally called pulsing aurora. Pulsing aurora can be divided into four classes: flaming aurora, flickering aurora, streaming aurora, and pulsating aurora. Of these, pulsating aurora is by far the most common. Some of the most pronounced characteristics of this type of optical fine structure will be discussed, together with other relevant geophysical observations.

According to Royrvik and Davis [1977] the term pulsating aurora is generally used to describe a certain class of aurora that is characterized by repetitive intensity fluctuations in the frequency range from 0.05–2 Hz, having horizontal drift velocities less than 1 km s^{-1}. This velocity differs only slightly, if at all, from the drift of adjacent forms. Usually the brightness fluctuates at least over one full cycle starting with an intensity increase. Pulsating auroras are generally seen in the midnight-to-morning sector of the auroral zone, and most frequently occur during the expansive and recovery phases of a substorm. When discrete aurora and pulsating aurora occur simultaneously, the pulsating aurora lies equatorward of the discrete aurora.

*Department of Physics, University of Oulu, SF-90570 Oulu, Finland.

Auroral Physics, edited by C.-I. Meng, M. J. Rycroft and L. A. Frank. © Cambridge UP 1991

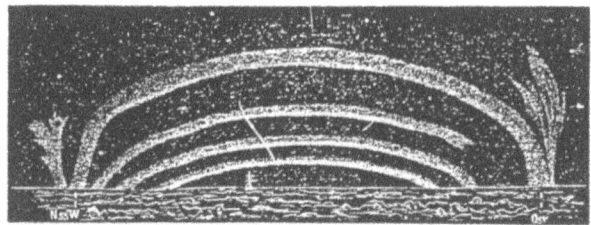

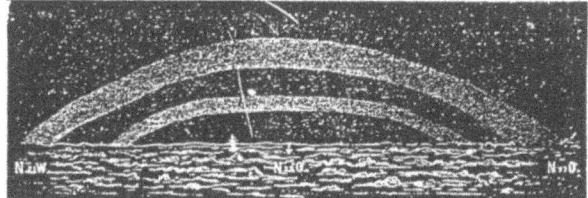

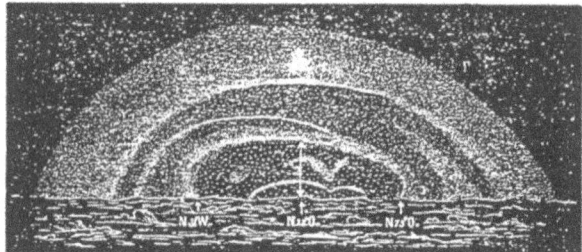

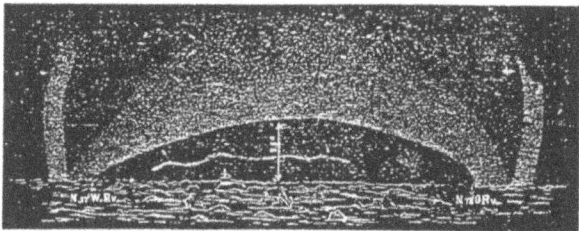

Figure 1—Drawings of aurora made by Nordenskioeld in Kolyuchin Bay (Chukchi Sea 1879 during his voyage onboard "Vega" searching for a northeast seaway passage from Europe to Asia) [*Nordenskioeld*, 1980].

Photometric studies of pulsating aurora have shown that the intensity is typically a few kilorayleighs (kR) in the nitrogen spectral line 427.8 nm but never exceeds 10 kR. The power spectral density is greatest in the period range 2–10 s [*Johnstone*, 1978]. In short, one can say that the pulsating aurora is characterized by its rapid intensity changes [*Campbell and Rees*, 1961; *Paulson and Shepherd*, 1966; *Omholt and Pettersen*, 1967; *Cresswell*, 1968; *Omholt*, 1971; *Rosenberg et al.*, 1971; *Vallance Jones*, 1974; *Pemberton and Shepherd*, 1975; *Davis*, 1978; *Sandahl*, 1986].

4. STRUCTURE AND MOTION OF PULSATING AURORA

Based on morphological features pulsating auroral forms can be divided into three main classes [*Royrvik and Davis*, 1977]: arcs, arc segments, and patches. Arcs are 1–10 km wide and of the order of 1000 km long, east-west aligned. Well-defined arcs tend to be thinner than diffuse ones. Their internal structure is uniform. Arc segments are as wide as arcs, but only some 100 km long directed in any direction. Patches vary from 10–200 km across, and are irregular in shape. They have either a uniform diffuse or well-defined internal structure.

Thomas and Stenbaek-Nielsen [1981] introduced yet another group of pulsating aurora that they called recurrent propagating forms. These types of aurora are suddenly born in a limited region of the sky from which they move in a wave-like fashion about 100 km. Their form as well as their location of origin seem to change from pulse to pulse. Their apparent velocity is 10–30 km s^{-1}, in direct contrast to stable pulsating patches with the on-off cycle described above. The periodicity of the recurrent propagating forms seems to be in the range 10–30 s, i.e., somewhat longer than the period observed in most stable auroral pulsations.

The shape, size, time variation, and drift of pulsating auroras tend to show great variability from one case to another. They can be rather stable and local, pulsating in intensity over the entire structure. Or they can show streaming features with the area growing as the pulse is switched on and decreasing as it is switched off. In arcs, the streaming is seen either as a spreading of the increased intensity eastward and westward along the arc or as the motion of a limited region of higher intensity along the arc. Sometimes there are limited northward and southward shifts when the pulse is turned on.

Patches, arcs, and segments of arcs may last for as short a time as one pulse or they may last for several minutes. A rather persistent feature of pulsating auroras is their tendency to show variations of period as a function of geomagnetic latitude, with shorter periods at lower latitudes [*Thomas and Rothwell*, 1979]. Examples of various types of pulsating aurora are given in Figures 2 through 4 [*Royrvik and Davis*, 1977; *Sandahl*, 1986].

Pemberton and Shepherd [1975] found distinct frequency bands in pulsating aurora at 2.4 and 7 s periods and at 3 and 10 s periods. Particularly, the 10 s variation tended to occur in the proximity of active auroras before and during auroral break-up. *Royrvik and Davis* [1977] also report the persistency of a 3 ± 1 Hz variation in many pulsating auroras. This is illustrated in Figure 5.

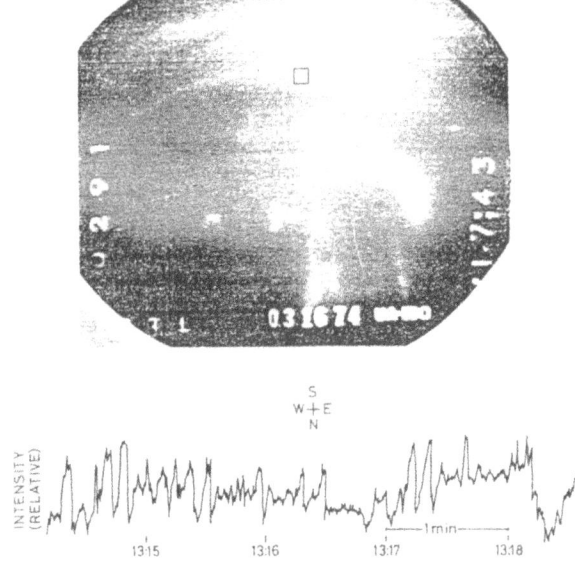

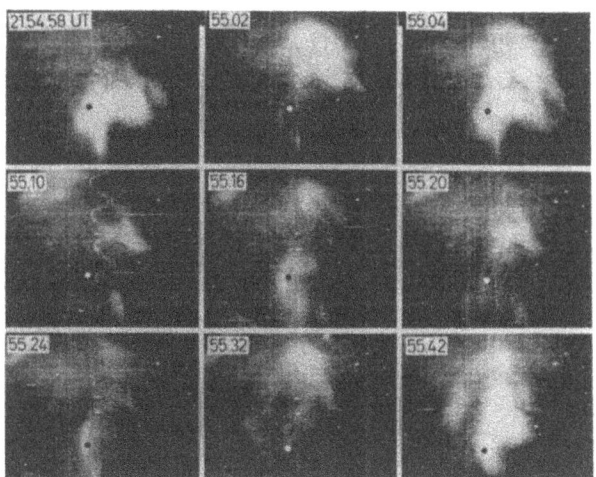

Figure 4—A typical pulsating auroral patch. (From *Sandahl* [1986].)

Figure 2—All-sky television data from March 16, 1974, 1314–1318 UT showing discrete arcs on a diffuse background. In the lower part of the figure are shown intensity recordings of irregular pulsations caused by streaming arcs and observed in the area marked by a box (from *Royrvik and Davis* [1977]).

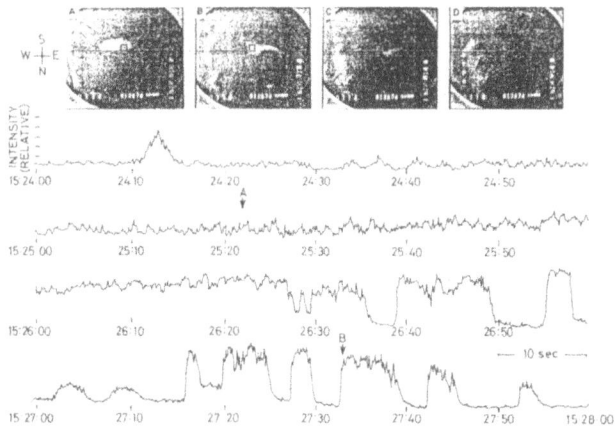

Figure 5—All-sky television data from January 26, 1974, 1524–1528 UT. Photographs show an eastward drifting pulsating patch. Note the 2–4 Hz modulation observed in the intensity around 1527:20 UT. (From *Royrvik and Davis* [1977].)

5. ALTITUDE OF PULSATING AURORA

Pulsating auroras are almost always seen embedded in a diffuse background aurora at an altitude of 120–240 km [*Brown et al.*, 1976]. The altitude of the lower border of the pulsating aurora has been located in the height interval 80–135 km. *Stenbaek-Nielsen et al.* [1979] and *Stenbaek-Nielsen* [1980] have reported very interesting stereoscopic TV observations of pulsating auroras, on the basis of which they conclude that certain pulsating auroral structures have vertical extents of less than 2 km. Sometimes these thin layers have been found in any type of aurora [*Hallinan et al.*, 1985]. The layers can be located in various places along the luminosity profile.

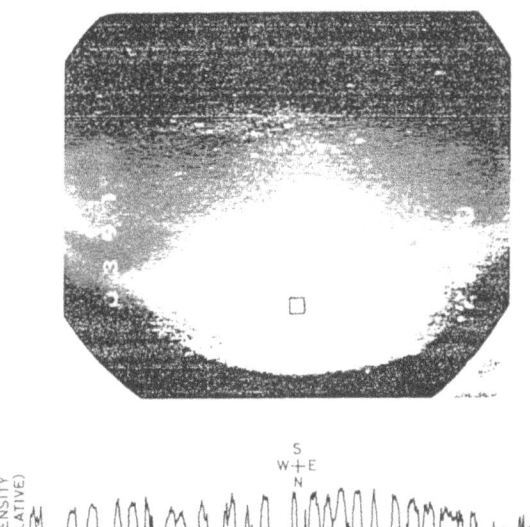

Figure 3—All-sky television data from February 24, 1974, 1220–1224 UT. The photograph shows a broad diffuse and structureless arc. The amplitude modulation in the intensity is less than 100% and variable. (From *Royrvik and Davis* [1977].)

Recently, very thin layers of enhanced electron density associated with pulsating aurora have been observed using the incoherent radar technique [*Whalund et al.*, 1988]. These layers, about 5 km thick, were observed only during the on-phase of the pulsation. It is unlikely that these observations would be linked to the so-called sporadic E-layer phenomenon.

As an explanation, localized wave-particle interaction has been suggested, which is strong enough to stop the incident electrons in the thin altitude interval of the luminous region. If, due to some instability, the waves would be enhanced as a result of the absorption of energy from the precipitating electrons and subsequently release it to the ambient electrons, then these "energized" electrons could, in turn, ionize atmospheric atoms and give rise to emissions in the lower E-region. In a qualitative sense these instabilities have been shown to exist also in laboratory experiments [*Hallinan et al.*, 1984].

Shepherd and Fälthammar [1980] have pointed out that, excluding the stopping of precipitating electrons by collisions in the upper atmosphere as being inadequate to cause auroral emissions from a very thin layer, one is left with the alternative of AC or DC electric fields. They calculated that for a DC electric field the required field strength would have to be at least 7.5 $V m^{-1}$. The required rms-value for an AC-field would be 10–40 times as large. In rocket measurements, no such narrow strong electric fields have yet been registered.

Sears and Vondrak [1981] have reported a study in which the total energy flux and characteristic energy for the precipitating particles during an intense pulsating aurora were derived by applying optical spectrophotometric techniques. These parameters were then used to model the height distribution of electron density during the pulsations. The model was compared with observations of the electron density distribution made with the Chatanika incoherent scatter radar. Radar data indicate, however, that no substantial changes in electron density profile or height-integrated conductivities occur as a result of the observed pulsating flux, probably because of the short duration of the pulsations compared with the recombination response time of the ionosphere.

It has also been pointed out that, although no verification of an optically thin layer during an intensive pulsating auroral event could be found using the radar technique, the stereo-TV observations cannot be disproved. For example, the 4.5-km transmitted pulse length is about twice as thick as the auroral luminous layer.

Just because the thin layers are in luminosity and not in ionization, one may raise the question whether the standard assumption regarding the partition of energy between optical emissions and ion-electron pair production is applicable during these anomalous processes.

6. WHERE AND WHEN DO AURORAL PULSATIONS OCCUR?

Pulsating aurora occurs during auroral substorms and particularly during their recovery and expansion phases. Sometimes pulsating aurora has been observed before the substorm onset in stable arcs, close to the equatorward border of discrete auroral arcs. Discrete arcs are well-defined bright auroral forms. In contrast to these, one sees the diffuse aurora where no borders can clearly be discerned. As has been previously mentioned, the pulsating aurora is superimposed on diffuse aurora. Thus it can be said that they constitute small-scale structures of diffuse aurora. At quiet times, however, pulsating auroras are not observed on the dark side of Earth [*Akasofu*, 1968].

A typical series of pulses in the late evening sector during the expansion phase contains pulses with rather long time intervals between the separate pulses. A westward-traveling surge often disrupts the pulsations and also the east-west aligned arcs. After the passage of the surge, the aurora becomes rather diffuse and, within about 10–15 minutes, patches or irregular arc segments of pulsating aurora may appear as superimposed structures.

In the midnight sector the pulsating forms are usually larger and rather well defined at the poleward edge of the pulsation region, whereas they tend to become diffuse and weak further to the south. In general, pulsating auroras are much more common in the midnight-to-morning sectors than in the evening sector. It has been observed that pulsating structures in the midnight and morning sectors drift with drift velocities about 1 km s^{-1}. The drift direction is always to the east in the morning sector and opposite in the evening sector with a rather undefined region (perhaps the Harang discontinuity) in between. This phenomenon has been called streaming.

Scourfield et al. [1983] have demonstrated that the pulsating auroral forms at times drift with the same velocity as that of the electron flow as seen by the Scandinavian Twin Auroral Radar Experiment (STARE). Figure 6 illustrates the use of STARE to measure the velocity of moving pulsating forms. It is suggested that auroral forms are subject to $\mathbf{E} \times \mathbf{B}$ drift in a similar way as is the background cold plasma. They further propose that a display of pulsating auroral forms is the projection on the ionosphere of an assembly of ducts of enhanced plasma density that are $\mathbf{E} \times \mathbf{B}$ drifting with the background plasma. Energetic electrons (>1 keV), with additional curvature and gradient drifts, would sweep through these ducts and precipi-

IRREGULARITY DRIFT VELOCITY

1000 M/SEC = ⊢——

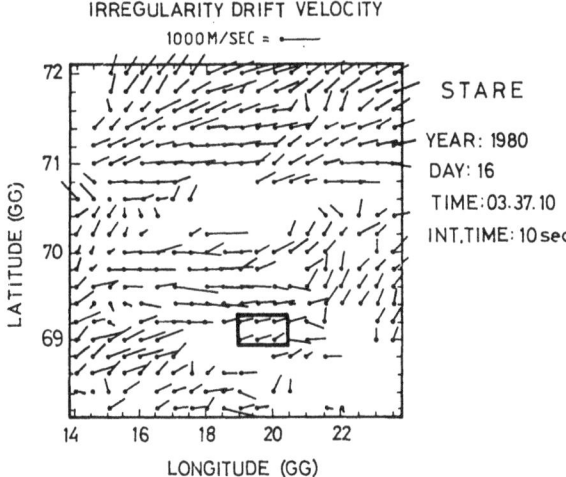

STARE

YEAR: 1980
DAY: 16
TIME: 03.37.10
INT.TIME: 10 sec

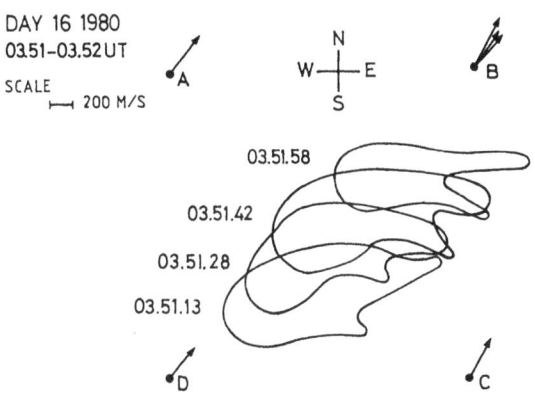

DAY 16 1980
03.51–03.52 UT
SCALE
⊢—⊣ 200 M/S

03.51.58
03.51.42
03.51.28
03.51.13

Figure 6—Four successive positions of a pulsating auroral form. Average plasma drift velocities are given for STARE cells located at A, B, C, and D. (From *Scourfield et al.* [1983].)

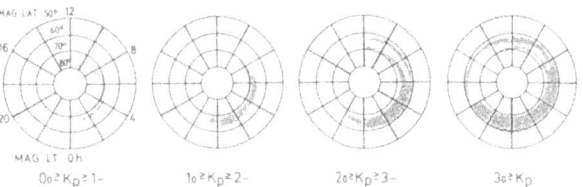

Figure 7—Schematic illustrations of the pulsating auroral regions as a function of *Kp*. (From *Oguti et al.* [1981].)

tation of these particles would cause the auroral emissions. *Oguti* [1981] has shown that the direction of the drift of pulsating auroral patches changes across magnetic midnight. This would be consistent with the general pattern of the electric field in that region. *Nakamura and Oguti* [1987] have shown that radar observations of drifting auroral forms indicate that the drift is caused by the convection electric field in the magnetosphere. This result would support the idea that the pulsating auroral patches are the ionospheric projections of cold plasma irregularities in the distant magnetosphere.

Ground magnetic pulsations under pulsating auroras in the auroral zone are believed to be due to pulsating fluctuations of ionospheric electric currents. These, in turn, are caused by conductivity fluctuations associated with the pulsating particle precipitation [*Oguti et al.*, 1984]. Pi 2 pulsations, often seen associated with substorm onsets, have been thought to be linked to the formation of oscillating field-aligned currents in a limit-

ed current wedge sector [*Tanskanen et al.*, 1987]. It has also been found that the electron temperature anisotropy in the plasma sheet plays a major role in determining the direction in which an auroral form propagates [*Rothwell et al.*, 1988].

Oguti et al. [1981] have studied the occurrence frequency of pulsating auroras, using a set of all-sky auroral TVs located between 61.5° and 74.3°N geomagnetic latitude. Their results showed that the occurrence probability of pulsating aurora is 100% after 0400 MLT. They also showed that pulsating auroras occur in the morning hours even at very low magnetic activity, i.e., when the *Kp*-value was from 0–1. For $20 < Kp < 3-$, the region of pulsating aurora in the late morning sector splits into two zones located between 64°–68° and between 61°–63° geomagnetic latitude. This is illustrated in Figure 7.

Recently *Fujii et al.* [1987] reported on conjugate measurements of rapid motions and small-scale structures of discrete auroras by all-sky TV observations made at Syowa Base in Antarctica and at Husafell in Iceland. Before the onset of a small substorm on September 26, 1984, conjugate faint discrete auroras could be identified in both conjugate areas. However, a striking dissimilarity of the conjugate auroras was noted in small-scale structures within the auroras. It was suggested that the large-scale structures and their dynamics are mainly controlled by some conditions in the magnetosphere, perhaps near the equatorial region, whereas the small-scale structures are mostly due to local acceleration processes between the magnetosphere and the ionosphere.

An interesting form of recurrent auroral pattern or pulsating aurora has been observed by *Oguti* [1976]. The most striking feature of this on-off switching aurora is that it shows a recurrence of the same pattern from several to several tens of times, with periods ranging from 1 s to tens of seconds. It is argued that the recurrence of the same pattern over and over again suggests that on-off switching auroras are due to the quasi-periodic pitch-angle scattering of trapped energetic electrons. Either the number of electrons lost in each switch-on must be relatively small in comparison with the total number of trapped electrons or the lost elec-

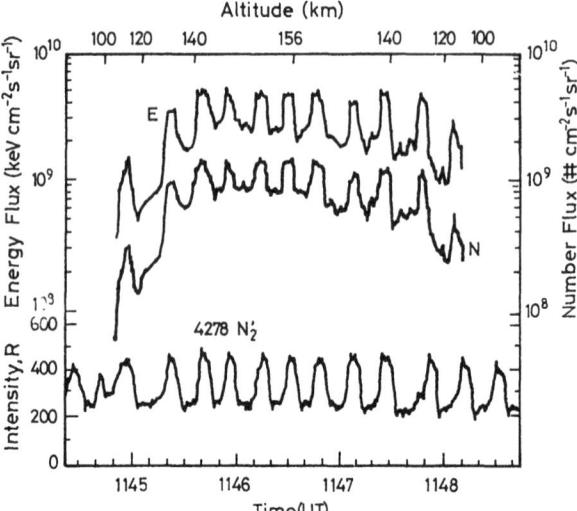

Figure 8—Rocket observations of electron energy and number fluxes together with ground-based photometer observations of auroral emissions at 427.8 nm wavelength. (From *McEwen et al.* [1981].)

trons must be rapidly replenished by the injection or drift of energetic electrons, since the same pattern is repeated many times.

7. BLACK AURORA

In the midnight and morning sectors auroral patches and arc segments may, in a short time, change into continuous arcs. This sequence of events often marks a new onset of an auroral substorm or an intensification. It is also frequently associated with the formation of east-west aligned structures that appear particularly on the northern edge of the pulsating forms. Concurrently with these phenomena another interesting observation has been made, though not so frequently reported, i.e., the occurrence of the so-called "black auroras".

Black auroras are small regions of very low auroral luminosity embedded in a diffuse aurora. It has been observed during the late recovery phase of a substorm near the poleward boundary of the diffuse aurora [*Royrvik and Davis*, 1977; *Davis*, 1978).

Observations made in 1986, with an auroral TV located in Kilpisjaervi in northern Finland, show black aurora with a vortex similar to those observed in discrete aurora. This feature could be observed to last for several minutes and showed very little, if any, motion.

8. ELECTRON PRECIPITATION, PITCH-ANGLE DISTRIBUTIONS, AND ENERGY SPECTRA

The intensity variations of pulsating aurora are directly correlated with the electron flux parameters [*Johnstone*, 1983; *Sandahl*, 1986]. Figure 8 shows var-

iations of electron energy flux (E) and number flux (N) together with intensity fluctuations of auroral optical emissions at 427.8 nm [*McEwen et al.*, 1981]. According to the report the pulsating cloud was so large that there is no doubt that the rocket observed the same region as the photometer did. It was also shown that the observed electron flux variation was sufficient to have caused the observed optical pulsations [*Johnstone*, 1983].

Electron pitch-angle distributions and spectral variations measured with a rocket-borne instrument fired into a pulsating auroral patch have been reported by *Whalen et al.* [1971]. These are shown in Figure 9. The pulsations in the early morning sector were superimposed on a relatively constant 1 kR diffuse background, and had a maximum amplitude of about 70% of the background. Observations show that during periods when the pulsations are not present the pitch-angle distributions of the lower energy electrons are highly anisotropic, peaking at 90°. The distributions are effectively isotropic at the peak of the pulsations. Electron energy spectra at 40° and 80° pitch angles show that the enhancements occur mostly at lower energies (the spectrum softens) and that the maximum relative increase occurs at small pitch angles.

However, when fitted to a Maxwellian distribution, it has been shown by *McEwen et al.* [1981] that the characteristic energy at pulsation maximum is larger than that at pulsation minimum. The intensity of the 427.8 nm nitrogen spectral line is directly proportional to the energy flux of the precipitating electrons. This explains the good correlation between these two parameters as shown in Figure 8.

Yau et al. [1981] have found that above 30 keV the electron spectrum is isotropic both during pulsation maximum and at the minimum, but for electron energies around a few kiloelectron volts the pitch-angle distribution is in accordance with that found by *Whalen et al.* [1971].

Evans et al. [1987] used coordinated observations from the NOAA 6 and S81-1 near polar-orbiting satellites to study electron spectra causing auroral pulsations. The forward-looking photometer onboard S81-1 was used to monitor the pulsating aurora from space. It provided evidence that the variations seen from NOAA 6 had a temporal component and not a spatial one only. The observations showed that the energy spectra of pulsating aurora include a steady, diffuse component, extending downward to energies well below those where resonant interactions with VLF waves are expected. The steady component is always present and the energetic part of the spectrum shows a variability from event to event. Using a time-dependent precipitation model that included the backscattered and secondary electrons, they showed that the primary

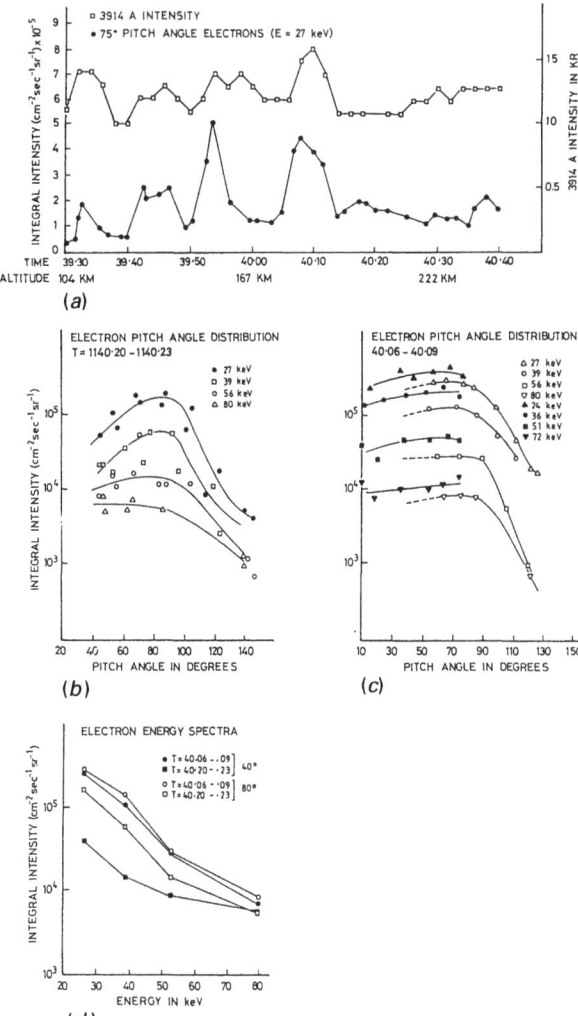

Figure 9—Optical and particle measurements recorded during a rocket flight into a pulsating auroral patch. (*a*) The 75° pitch-angle electron (>27 keV) and the 3914 Å intensities. (*b*) Electron pitch-angle distributions in a time interval after pulsations. (*c*) Electron pitch-angle distributions at the maximum of the pulsation. (*d*) Electron energy spectra recorded at the maximum (0040:06–0040:09) and minimum (0040:20–0040:23) of the pulsation at 40° and 80° pitch angles, showing spectral softening of electron precipitation during the pulsation. (From *Whalen et al.* [1971].)

precipitation need only affect electrons with energies above several kiloelectron volts. The rest of the precipitation and the diffuse component could be attributed to backscatter from the atmosphere. The only component of the observed energy spectrum that has not yet been explained or modeled is the low-energy end, the steady drizzle.

Electron density distribution measurements by the EISCAT (European Incoherent Scatter) radar facilities in northern Scandinavia have recently been compared

with optical observations of pulsating aurora within the radar beam [*Kaila et al.*, 1988]. A five-channel auroral photometer together with an auroral TV recorder were used for optical observations [*Kaila et al.*, 1987]. Based on spectroscopic emission intensity ratios (630.0 nm/427.8 nm and 557.7 nm/427.8 nm), it was observed that the characteristic energy of the background diffuse aurora was about 2 keV. For the pulsation maximum this energy was around 2.5 keV. Also the temperature sensitive emission ratio 426.0 nm/427.8 nm indicated hardening of the electron spectrum. Since the changes of electron densities were rather small during pulsations, the densities had to be integrated over the height interval 100–110 km. Electron density values were also calculated from recorded optical emissions. Measured and calculated electron densities during a pulsation event correlated well with the intensity variations of the 427.8 nm emission (Figure 10).

9. RELAXATION OSCILLATOR MECHANISM

Today, the most promising mechanism to cause optical auroral pulsations seems to be VLF (or ELF) wave-particle interactions resulting in electron pitch-angle scattering of stably trapped electrons into the loss cone [*Kennel and Petchek*, 1966; *Johnstone*, 1983; *Sandahl*, 1986; *Oguti et al.*, 1986; *Hansen et al.*, 1986]. The limit of the stably trapped magnetospheric particles is somehow exceeded, yielding precipitation of the surplus particles into the ionosphere. This occurs via VLF waves in such a way that the increased electron flux increases the VLF emission intensity and this in turn causes particle scattering into the loss cone. The interaction of waves and particles occurs as a result of cyclotron resonance between oppositely moving VLF waves and electrons that lie on the same field line. Although this mechanism, also called the relaxation mechanism, cannot explain all phenomena characteristic of auroral pulsations, it is the most advanced one. Unfortunately, correlative conjugate measurements of VLF or ELF waves in the equatorial magnetosphere and optical emissions in the auroral ionosphere are rather rare. Also, ground-based measurements of VLF waves are strongly affected by atmospheric absorption. Statistical studies of the occurrence of VLF and ELF waves show that they most likely occur after magnetic substorms at L = 5–8, and that their occurrence frequency increases after magnetic midnight (*Inan et al.*, [1978]; *Davidson* [1979, 1986 a,b]; *Davidson and Chiu* [1986]; *Helliwell et al.*, [1980]; *Koons*, [1981]; *Doolittle and Carpenter* [1983]; *Johnstone* [1983]).

Although the relaxation oscillator theory is at the moment the most common candidate to explain pulsation phenomena, there are others. One of the earliest theories for pulsating aurora was presented by

RADAR BEAM AND OPTICAL STATIONS
1 FEB 1987

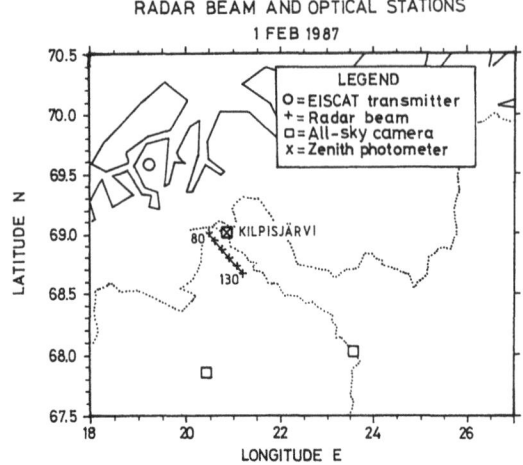

MEASURED AND CALCULATED ELECTRON DENSITIES
1 FEB 1987   ALTITUDE 110 KM

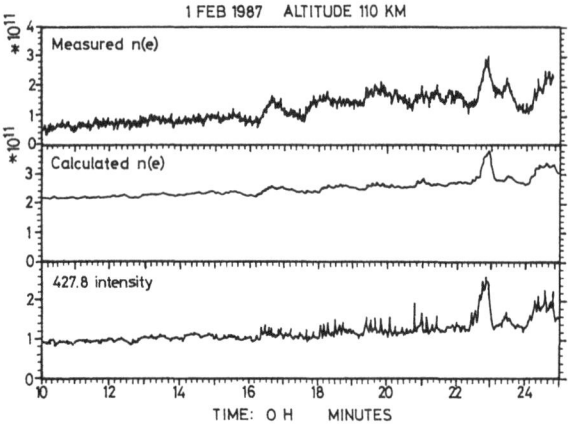

Figure 10—Upper panel: Position of EISCAT radar beam and optical stations during the coordinated radar-photometer-auroral TV experiment on February 1, 1987. Lower panel: Measured and calculated electron densities together with intensity fluctuations of 427.8 nm emissions. (From *Kaila et al.* [1988].)

Coroniti and Kennel [1970]. It is also based on wave-particle interaction in the equatorial region, but assumes a modulation of the VLF waves by magnetic pulsations. However, satellite measurements have so far failed to observe pulsations in the equatorial plane coin-cident with ongoing pulsating aurora [*Oguti et al.*, 1986]. According to *Oguti et al.* [1984] there seems to be no observational evidence for whistler mode wave turbulence being modulated by hydromagnetic waves as the scattering agent for precipitating energetic electrons as proposed by *Coroniti and Kennel* [1970].

Tagirov et al. [1986] have recently developed a so-called cyclotron maser theory that bears some resemblance to the relaxation oscillator theory.

10. SUMMARY

Auroral pulsations show great variability both in structure and in general behavior. It seems that gross features of pulsating auroral structures can be explained using existing theories and observations. However, when it comes to high resolution (both in time and space) observations of auroral structures, and their theoretical interpretation, there still exist some key questions such as:

1. Are thin luminous pulsating layers evidence for noncollisional thermalization of precipitating electrons?
2. What is the cause of recurrent pulsating forms?
3. How is black aurora explained?
4. What are the respective functions of the ionosphere and the distant magnetosphere as sources of auroral structures and their motions?
5. Is the high occurrence probability of pulsating aurora in the early morning sector due to the drift of energetic electrons and cold plasma irregularities from the tail region or to plasma injected from the ionosphere?

It is evident that coherent, high-resolution measurements of both particles and fields in the upper atmosphere, ionosphere, and the magnetosphere are needed to shed more light on these questions that are as yet unanswered or only partially answered.

ACKNOWLEDGMENT—The author expresses thanks to The Finnish Academy of Sciences and the Organizing Committee of the International Conference on Auroral Physics for financial support. He would also like to acknowledge all persons who kindly supplied material for this presentation, as well as important scientific comments on the manuscript.

REFERENCES

Akasofu, S.-I., *Polar and Magnetospheric Substorms*, Reidel, Dordrecht, Holland (1968).

Brown, N. B., T. N. Davis, T. J. Hallinan, and H. C. Stenbaek-Nielsen, "Altitude of Pulsating Aurora Determined by a New Instrumental Technique," *Geophys. Res. Lett.*, **3**, 403 (1976).

Campbell, W. H., and M. H. Rees, "A Study of Auroral Coruscations," *J. Geophys. Res.*, **66**, 41 (1961).

Coroniti, F. V., and C. F. Kennel, "Electron Precipitation Pulsation," *J. Geophys. Res.*, **75**, 1279 (1970).

Cresswell, G., "Fast Temporal and Spatial Changes in Auroras," Rep. UAG-R206, Geophysical Institute, University of Alaska, Fairbanks (1968).

Davidson, G. T., "Self-Modulated VLF Wave-Electron Interactions in the Magnetosphere: A Cause of Auroral Pulsations," *J. Geophys. Res.*, **84**, 6517 (1979).

Davidson, G. T., "Pitch Angle Diffusion in the Morningside Aurorae 1. The Role of the Loss Cone in the Formation of Impulsive Bursts of Precipitation," *J. Geophys. Res.*, **91**, 4413 (1986*a*).

Davidson, G. T., "Pitch Angle Diffusion in the Morningside Aurorae 2. The Formation of Repetitive Auroral Pulsations," *J. Geophys. Res.*, **91**, 4429 (1986*b*).

Davidson, G. T., and Y. T. Chiu, "A Closed Nonlinear Model of Wave-Particle Interactions in the Outer Trapping and Morningside Auroral Regions," *J. Geophys. Res.*, **91**, 13,705 (1986).

Davis, T. N., "Observed Characteristics of Auroral Forms," *Space Sci. Rev.*, **22**, 77 (1978).

Doolittle, J. H., and D. L. Carpenter, "Photometric Evidence of Electron Precipitation Induced by First Hop Whistlers," *Geophys. Res. Lett.*, **10**, 611 (1983).

Evans, D. S., G. T. Davidson, H. D. Voss, V. L. Imhof, J. Mobilia, and Y. T. Chiu, "Interpretation of Electron Spectra in Morningside Pulsating Aurorae," *J. Geophys. Res.*, **92**, 12,295 (1987).

Fujii, R., N. Sato, T. Ono, H. Fukunishi, T. Hirasawa, S. Kokubun, T. Araki, and T. Saemundsson, "Conjugacy of Rapid Motions and Small-Scale Deformations of Discrete Auroras by All Sky TV Observations," *Mem. Natl. Inst. Polar Res.*, (Spec. Issue) **48**, 72 (1987).

Hallinan, T. J., H. C. Stenbaek-Nielsen, and C. S. Deehr, "Enhanced Aurora," *J. Geophys. Res.*, **90**, 8461 (1985).

Hansen, H. J., M. W. J. Scourfield, and J. P. S. Rash, "Correlated Pulsations in Auroral Light Intensity and Narrow Band VLF Emissions," *Adv. Space Res.*, **6**, 227 (1986).

Helliwell, R. A., S. B. Mende, J. H. Doolittle, W. C. Armstrong, and D. L. Carpenter, "Correlation Between 4278 Å Optical Emissions and VLF Wave Events Observed at L = 4 in the Antarctic," *J. Geophys. Res.*, **85**, 3376 (1980).

Inan, U. S., T. F. Bell, and R. A. Helliwell, "Nonlinear Pitch Angle Scattering of Energetic Electrons by Coherent VLF Waves in the Magnetosphere," *J. Geophys. Res.*, **83**, 3235 (1978).

International Auroral Atlas, International Union of Geodesy and Geophysics, Edinburgh University Press, Edinburgh (1963).

Johnstone, A. D., "Pulsating Aurora," *Nature*, **224**, 119 (1978).

Johnstone, A. D., "The Mechanism of Pulsating Aurora," *Ann. Geophys.*, **1**, 397 (1983).

Kaila, K., R. Rasinkangas, E. Herrala, and P. Tanskanen, "Five Channel Auroral Photometer," Department of Physics, University of Oulu, Report no. 112 (1987).

Kaila, K., R. Rasinkangas, P. Pollari, R. Kuula, J. Kangas, T. Turunen, and T. Boesinger, "High Resolution Measurements of Pulsating Aurora by Eiscat, Optical Instruments and Pulsation Magnetometers," Proc. XXVII COSPAR Plenary Meeting, Espoo 18–29 July 1988, *Adv. Space Res.* (1988).

Kennel, C. F., and H. E. Petchek, "Limit of Stably Trapped Particle Fluxes," *J. Geophys. Res.*, **71**, 1 (1966).

Koons, A. C., "The Role of Hiss in Magnetospheric Chorus Emissions," *J. Geophys. Res.*, **86**, 6745 (1981).

McEwen, D. J., C. N. Duncan, and R. Montalbetti, "Auroral Electron Energies: Comparison of in situ Measurements with Spectroscopically Inferred Energies," *Can. J. Phys.*, **59**, 1116 (1981).

Nakamura, R., and T. Oguti, "Drifts of Auroral Structures and Magnetospheric Electric Fields," *J. Geophys. Res.*, **92**, 11,241 (1987).

Nordenskioeld, A. E., "Vegan Matka Asian ja Europan Ympaeri," reproduction of the original translation from 1881 by Wiipurin Kirjallisuus-Seura (in Finnish), vol. 2, pp. 33–38 (1980).

Omholt, A., *The Optical Aurora*, Springer, New York (1971).

Omholt, A., and H. Pettersen, "Characteristics of High Frequency Auroral Pulsations," *Planet. Space Sci.*, **15**, 347 (1967).

Oguti, T., "Recurrent Auroral Patterns," *J. Geophys. Res.*, **81**, 1782 (1976).

Oguti, T., "TV Observations of Auroral Arcs," in *Physics of Auroral Arc Formation*, Geophysical Monograph 25, S.-I. Akasofu and J. R. Kan, eds., American Geophysical Union, Washington, DC, pp. 31–42 (1981).

Oguti, T., K. Hayashi, Y. Yamamoto, J. Ishida, T. Higuchi, and N. Nishitani, "Absence of Hydromagnetic Waves in the Magnetospheric Equatorial Region Conjugate with Pulsating Auroras," *J. Geophys. Res.*, **91**, 13,711 (1986).

Oguti, T., S. Kokubun, K. Hayashi, K. Tsuruda, S. Machida, T. Kitamura, O. Saka, and T. Watanabe, "Statistics of Pulsating Auroras on the Basis of All-Sky TV Data from Five Stations. I. Occurrence Frequency," *Can. J. Phys.*, **59**, 1150 (1981).

Oguti, T., J. H. Meek, and K. Hayashi, "Multiple Correlation Between Auroral and Magnetic Pulsations," *J. Geophys. Res.*, **89**, 2295 (1984).

Paulson, K. V., and G. G. Shepherd, "Short-Lived Brightness Oscillations in Active Auroras," *Can. J. Phys.*, **44**, 921 (1966).

Pemberton, E. V., and G. G. Shepherd, "Spatial Characteristics of Auroral Brightness Fluctuation Spectra," *Can. J. Phys.*, **53**, 504 (1975).

Rosenberg, T. J., J. Bjordal, H. Trefall, G. J. Kvifte, and A. Omholt, "Correlation Study of Auroral Luminosity and X Rays," *J. Geophys. Res.*, **76**, 122 (1971).

Rothwell, P. L., M. B. Silevitch, L. P. Block, and P. Tanskanen, "A Model of the Westward Travelling Surge and the Generation of Pi 2 Pulsations," *J. Geophys. Res.*, **93**, 8613 (1988).

Royrvik, O., and T. N. Davis, "Pulsating Aurora: Local and Global Morphology," *J. Geophys. Res.*, **82**, 4720 (1977).

Sandahl, I., "Recent Developments in Pulsating Aurora Studies," Proc. 13th Annual Meeting on Upper Atmosphere Studies by Optical Methods, Rep. 86-28, pp. 141–160 (1986).

Sandford, B. P., "Variations of Auroral Emissions with Time, Magnetic Activity and the Solar Cycle," *J. Atmos. Terr. Phys.*, **30**, 1921 (1968).

Scourfield, M. W. J., J. G. Keys, E. Nielsen, and C. K. Goertz, "Evidence for the **E** × **B** Drift of Pulsating Auroras," *J. Geophys. Res.*, **88**, 7983 (1983).

Sears, R. D., and R. R. Vondrak, "Optical Emissions and Ionization Profiles During Intense Pulsating Aurora," *J. Geophys. Res.*, **86**, 6853 (1981).

Shepherd, G. G., and C.-G. Fälthammar, "Implications of Extreme Thinness of Pulsating Auroral Structures," *J. Geophys. Res.*, **85**, 217 (1980).

Stenbaek-Nielsen, H. C., "Pulsating Aurora: The Importance of the Ionosphere," *Geophys. Res. Lett.*, **7**, 353 (1980).

Stenbaek-Nielsen, H. C., and T. J. Hallinan, "Pulsating Auroras: Evidence for Noncollisional Thermalization of Precipitating Electrons," *J. Geophys. Res.*, **84**, 3257 (1979).

Tagirov, V. R., V. Y. U. Trakhtengerts, and A. Chernouss, "The Origin of Pulsating Auroral Patches," *Geomagn. Aeron.*, **26**, 502 (1986).

Tanskanen, P., J. Kangas, L. Block, G. Kremser, A. Korth, J. Woch, I. B. Iversen, K. M. Torkar, W. Riedler, S. Ullaland, J. Stadsnes, and K.-H. Glassmeier, "Different Phases of a Magnetic Substorm on June 23, 1979," *J. Geophys. Res.*, **92**, 7443 (1987).

Thomas, R. W., and P. Rothwell, "A Latitude Effect in the Periodicity of Auroral Pulsating Patches," *J. Atmos. Terr. Phys.*, **43**, 243 (1979).

Thomas, R. W., and M. C. Stenbaek-Nielsen, "Recurrent Propagating Auroral Forms in Pulsating Aurora," *J. Atmos. Phys.*, **43**, 243 (1981).

Vallance Jones, A., *Aurora*, Reidel, Hingham, MA (1974).

Whalund, J.-E., H. Opgenoorth, and P. Rothwell, *J. Geophys. Res.*, accepted for publication (1988).

Whalen, B. A., J. R. Miller, and I. B. McDiarmid, "Energetic Particle Measurements in a Pulsating Aurora," *J. Geophys. Res.*, **76**, 978 (1971).

Yau, A. W., B. A. Whalen, and D. J. McEwen, "Rocket-Borne Measurements of Particle Pulsation in Pulsating Aurora," *J. Geophys. Res.*, **86**, 5673 (1981).

VI-4. ELECTRODYNAMICS OF ACTIVE AURORAL FORMS: WESTWARD TRAVELING SURGES AND OMEGA BANDS

W. Baumjohann*

Substorm-associated active auroral structures (westward traveling surges and omega bands) in the nighttime auroral oval are associated with high ionospheric conductances in spatially confined regions. The strong conductivity gradients severely alter and distort the large-scale convection-driven electrojet system and its associated meridional electric fields, leading to small-scale and highly structured electric field and current patterns in and near westward traveling surges and omega bands. The most prominent feature are localized and intense upward Birkeland currents near the head of a surge and in the luminous tongues of an omega band.

1. INTRODUCTION

Mapping the large-scale auroral electrojet system and its associated electric field and conductivity structure is a relatively easy task, since these currents are slowly varying in time as well as with longitude. Hence, meridional magnetometer chains, meridional radar scans, or meridional satellite passes across the auroral oval have been used to get a fairly consistent and complete picture of the electrodynamics of the global auroral electrojets. However, during disturbed times, the precipitation of energetic particles associated with active aurora is often very nonuniform in the east-west direction, resulting in conductivity enhancements within relatively localized regions.

A typical distribution of substorm aurora can be seen in the DMSP photograph in Figure 1: break-up aurora and a westward traveling surge around midnight and in the late evening sector, and eastward drifting omega bands in the morning sector. The spatially confined high-conductivity regions associated with such aurora severely alter and distort the meridional electric field pattern and associated east-west current flow.

In contrast to the electrojet currents, these distortions have strong gradients along the azimuthal axis. The typical scale size of such conductivity enhancements can range from several hundred to several thousand kilometers. They are, furthermore, highly variable in time. Thus one needs simultaneous two-dimensional measurements of, for example, ionospheric electric fields and ground magnetic perturbations for a unique determination of the associated three-dimensional current system.

Such observations have first been made in northern Scandinavia, where the Scandinavian Twin Auroral Radar Experiment (STARE) radars and the Scandinavi-

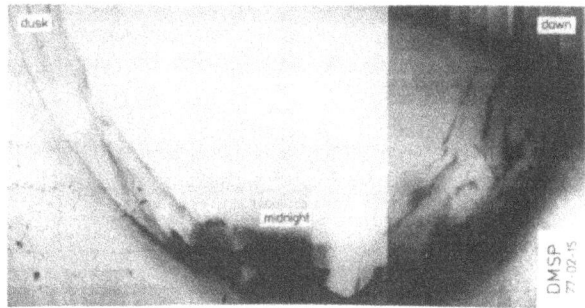

Figure 1—DMSP photograph showing a typical distribution of active auroral forms in the nighttime auroral oval.

an Magnetometer Array allowed simultaneous measurements of the two-dimensional distribution of both ionospheric electric field and ground magnetic perturbations. In addition, the Finnish All-sky Camera Chain documented visual substorm aurora with good temporal resolution.

These measurements were done at the end of the last decade and were unique. Together with modeling efforts, which basically combined the observed two-dimensional electric field distribution with an assumed conductivity distribution and varied the latter until the magnetic field of the model currents equaled the observed ground magnetic perturbations, these observations yielded a consistent picture of the electrodynamics of such auroral forms. Recent work using data from the European incoherent scatter (EISCAT) radar introduced some slight corrections to these models but otherwise verified the basic features.

Since this review concentrates on the two-dimensional synoptic measurements, it neglects earlier studies of the currents associated with active auroral forms where meridian chain magnetometer data were used. It should be pointed out, however, that especially the work done by the Alberta group served as a guideline and basis for the work described here.

*Max-Planck-Institut für extraterrestrische Physik, D-8046 Garching, Federal Republic of Germany.

Auroral Physics, edited by C.-I. Meng, M. J. Rycroft and L. A. Frank. © Cambridge UP 1991

2. WESTWARD TRAVELING SURGES

During magnetospheric substorms the ionospheric current flow is affected in two ways [e.g., *Baumjohann,* 1986; *Rostoker et al.,* 1987]. During the growth phase, the Hall current flow in the auroral electrojets increases in direct relation to the energy input from the solar wind (so-called "driven process"). In addition, sporadic release of energy previously stored in the magnetotail leads to the formation of a substorm current wedge with strongly enhanced westward current flow within the active auroral forms in the midnight sector (so-called "unloading process").

The substorm electrojet may eventually intrude deep into the evening sector along with the westward traveling surge and is superimposed on the convection electrojets [*Baumjohann,* 1983, see Figure 6]. In order to explore the electrodynamics of this substorm electrojet, *Baumjohann et al.* [1981], *Inhester et al.* [1981], and *Opgenoorth et al.* [1983a] have modeled the three-dimensional current system associated with breakup aurora around 2400 MLT and with westward traveling surges observed around 2230 and 2030 MLT.

Baumjohann et al. [1981] have analyzed two-dimensional magnetic and electric field data obtained by the Scandinavian Magnetometer Array and the STARE radars during three successive short-lived auroral breakups around magnetic midnight. All three breakups were moderate and relatively localized. In Figure 2 we summarize, for one of the events, the equivalent current vector distribution on the ground, the ionospheric electric field pattern, and the auroral structures before the breakup, at the initial brightening, and during the maximum expansion of the aurora. The observed features were alike for all three breakups.

Before the breakup, the electric field distribution exhibits a pattern typical for the Harang discontinuity region. Hardly any current flow can be seen, indicating a very low ionospheric conductance. During the initial brightening, energetic auroral electrons precipitate into a localized region south of the Harang discontinuity. In this region the westward electric field stays constant, but the northward component decreases significantly and even turns southward at some locations (especially during the maximum phase). The magnetometers observe a strong westward current developing in the same region.

These observations match in detail the features to be expected during the generation of an (incomplete) Cowling channel in the highly conducting region described first by *Boström* [1975] and summarized in Figure 3. Due to the conductivity enhancement, the westward component of the primary electric field, $E°$, drives an enhanced northward Hall current, $J_H°$. The excess Hall current deposits positive charges at the northern border of the highly conductive channel while negative charges build up at its southern boundary.

To some extent these charges are removed by Birkeland current sheets, but the remainder give rise to a southward polarization electric field, E^p. E^p drives a southward Pedersen current, J_P^p, which balances that part of $J_H°$ which is not continued via Birkeland currents. The westward currents due to the primary (convection) and secondary (polarization) electric field add up to an intense westward Cowling current.

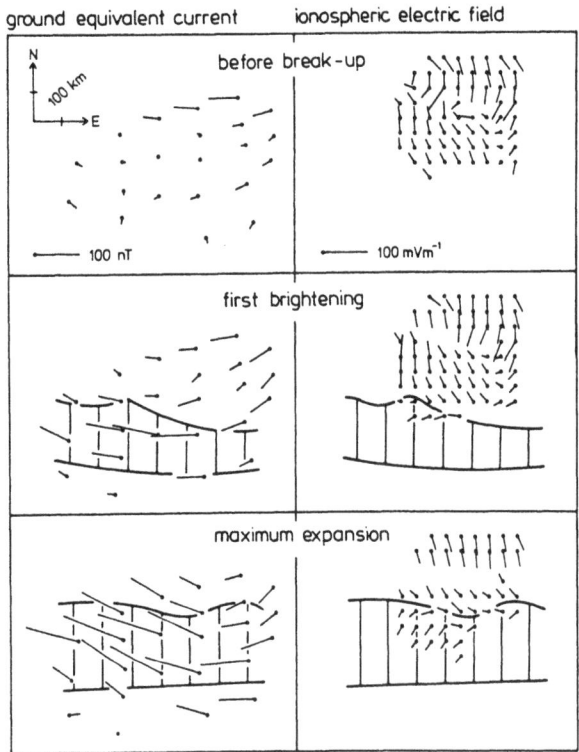

Figure 2—Spatial distribution (latitude vs. longitude) of ground equivalent current vectors, ionospheric electric field vectors, and auroral structures before, at the start of, and during the peak development of an auroral breakup. (After *Baumjohann et al.* [1981].)

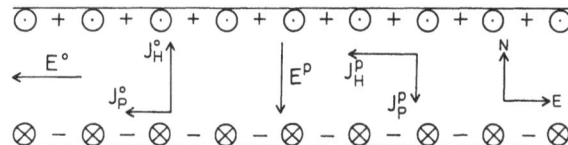

Figure 3—Generation of a polarization electric field and an incomplete Cowling channel in a region of enhanced ionization. (After *Boström* [1975].)

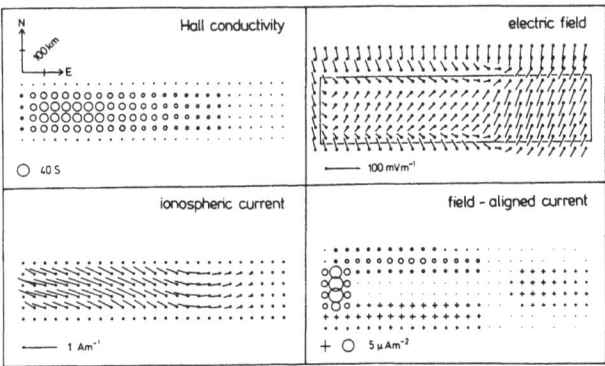

Figure 4—Spatial distribution of Hall conductance, horizontal electric field, and ionospheric and Birkeland current in the midnight auroral breakup region [after *Baumjohann et al.,* 1981]. The rectangle in the electric field panel frames the region of active aurora. Circles and crosses in the lower left panel denote upward and downward Birkeland current.

The model constructed by *Baumjohann et al.* [1981] for the maximum auroral expansion is shown in Figure 4. Within the active region, Hall and Pedersen conductances reach peak values of about 30 and 10 S, respectively, close to the western border. Outside the active region, the ionospheric conductance is close to zero. The pre-breakup electric field pattern is distorted by the superposition of a southward polarization electric field with a strength of up to 50 mV m^{-1} in the area covered by breakup aurora. The ionospheric current has sheet current densities of 600–700 mA m^{-1}, comparable with typical westward electrojet values [*Baumjohann*, 1983].

The westward component of the ionospheric current is closed by very localized and intense upward Birkeland current at the western border (about 7 μA m^{-2}) and more widespread downward Birkeland current of lower density (1–2 μA m^{-2}) in the eastern third of the breakup region. The northward current is closed by Birkeland current sheets of 1–2 μA m^{-2} at the southern and northern boundaries of the active region.

In order to see whether a similar current structure can be found when the substorm electrojet intrudes into the evening sector along with the westward-traveling surge, *Opgenoorth et al.* [1983a] have used simultaneous measurements made by the Scandinavian Magnetometer Array, the STARE radars, and the Barium-GEOS rocket payload to model the three-dimensional current system in the vicinity of a 2230 MLT westward-traveling surge. Their observations and modeling results are compared in Figures 5 and 6.

The equivalent current vector panel exhibits a clear longitudinal gradient, i.e., the westward current decreases strongly in the westward direction. This gradient is balanced by southward equivalent current at

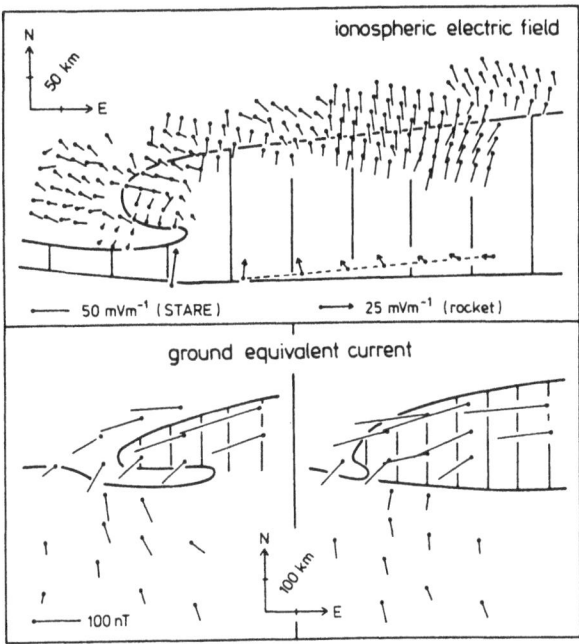

Figure 5—Spatial distribution of ionospheric electric field vectors, ground equivalent current vectors (for two moments 2 min apart), and auroral structure during the passage of a westward traveling surge in the premidnight sector. (After *Opgenoorth et al.* [1983a].)

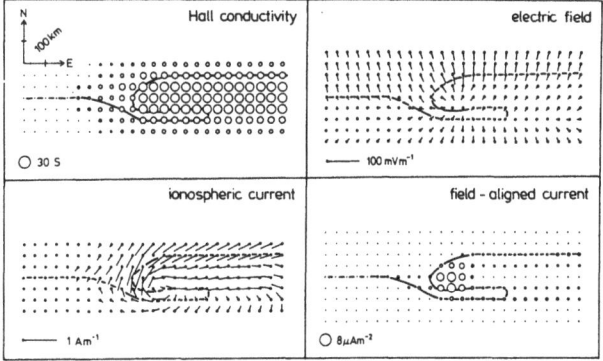

Figure 6—Same as Figure 4, but for a westward traveling surge in the premidnight sector. (After *Opgenoorth et al.* [1983a].)

subauroral latitudes, indicative of net upward Birkeland current near the head of the surge.

Electric field observations obtained by the rocket and STARE during the overhead passage of the surge are plotted with respect to their location within and around the auroral form. A radial component directed toward the center of the westward traveling surge can be seen. To the north of the aurora, the electric field is southward directed. South of the westward-traveling surge the rocket data show a northward electric-field component. Inside the surge, the rocket measured a very

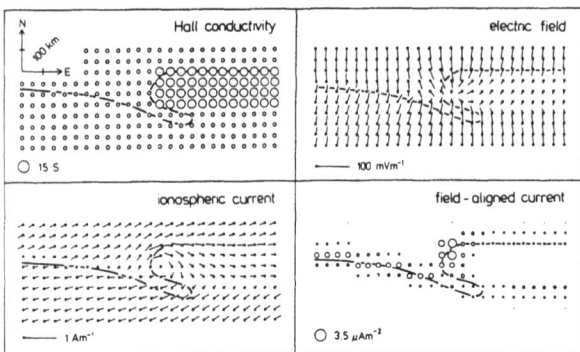

Figure 7—Same as Figure 4, but for a westward traveling surge in the early evening sector. (After *Inhester et al.* [1981].)

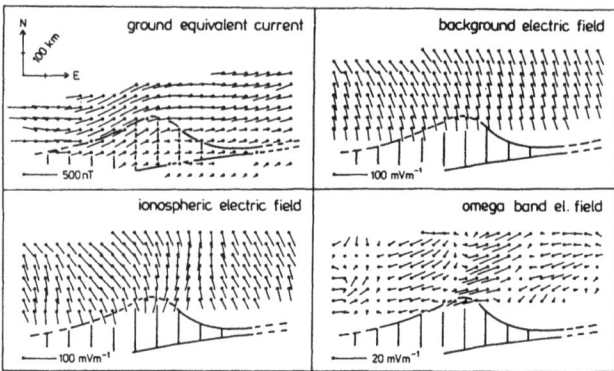

Figure 8—Same as Figure 5, but for the passage of an eastward drifting omega band in the morning sector. In addition, two different components into which the measured electric field may be decomposed are shown: a rather uniform background electric field and the omega band-associated field. (After *André and Baumjohann* [1982].)

weak electric field with a clear westward component. Apparently, here the field was below the STARE threshold (about 15 mV m^{-1}).

The modeling effort yielded that only a southwestward electric field east of the head of the surge was consistent with all observed features. This field may again be taken as a polarization electric field. Accordingly, the westward current in the westward traveling surge (about 500 mA m^{-1}) might again be a Cowling current. Continuity of this current is preserved by localized and intense upward Birkeland current in the head of the surge (about 6 μA m^{-2}). No downward current could be detected within the area of observation. Quite likely, it was located in the postmidnight region.

Comparing Figure 6 with the model results of *Inhester et al.* [1981], shown in Figure 7, which are based on observations made during the overhead passage of another westward traveling surge in the 2030 MLT sector, it becomes evident that the surge-associated features are essentially the same and that only the ambient electrojet environment is different. In the early evening sector, the surge travels along the Harang discontinuity just north of a still very prominent eastward electrojet.

In summary, the combined Scandinavian Magnetometer Array and STARE observations showed that conductivity structure, electric field distribution, and current flow associated with breakup aurora in the midnight sector and westward traveling surges in the late and early evening sector are basically the same. Strong conductivity gradients lead to the generation of polarization electric fields and enhanced westward current, which at its western border is continued via intense and very localized upward Birkeland current.

Recent observations of a westward traveling surge with the EISCAT incoherent scatter radar, the EISCAT Magnetometer Cross, and the Viking Auroral Imager [*Kirkwood et al.,* 1988] have verified the key features of the above model, but the conductivity enhancement within the aurora is apparently even more pronounced

than estimated earlier: the ionospheric conductance values measured by EISCAT within the surge are higher by a factor of 3–5. On the other hand, the electric-field strength within this region is, at least at times, much weaker than assumed in the earlier models (note that STARE could only give an upper limit in the luminous region). Hence, the model ionospheric and Birkeland current densities seem basically correct.

3. OMEGA BANDS

While auroal breakups and westward traveling surges prevail in the premidnight sector during substorms expansion phases, eastward propagating sequences of omega bands are the dominant feature in the postmidnight sector during substorm and/or storm recovery. The auroral omega bands are associated with undulations in the westward electrojet. Their magnetic signature has been denoted as Ps 6 pulsation because of the periodic nature of the magnetogram recorded when a sequence of omega bands passes overhead.

Earlier work [e.g., *Kawasaki and Rostoker,* 1979; *Gustafsson et al.,* 1981] had indicated that the current system associated with these auroral structures is truly three-dimensional. *André and Baumjohann* [1982] and *Opgenoorth et al.* [1983b] have analyzed simultaneous measurements made with the Scandinavian Magnetometer Array and the STARE radars for two postmidnight intervals during which omega bands were drifting eastward over the Finnish All-sky Camera Chain.

Figure 8 summarizes the observations of *André and Baumjohann* [1982] made around 0330 MLT. Both electric-field and equivalent current vectors were superimposed and averaged over 50 × 50 km^2 cells, using the observed feature that the electric and magnetic field pattern was actually stationary but drifting with the omega band. The undulated electrojet pattern is

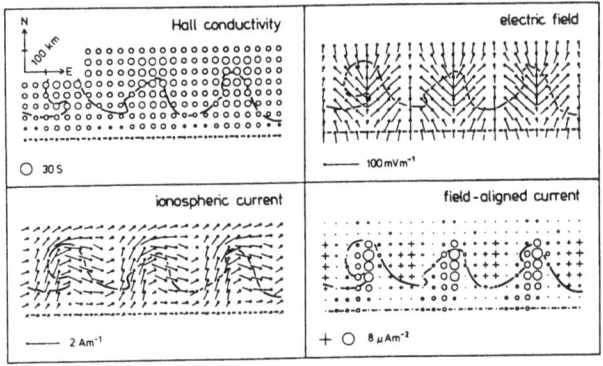

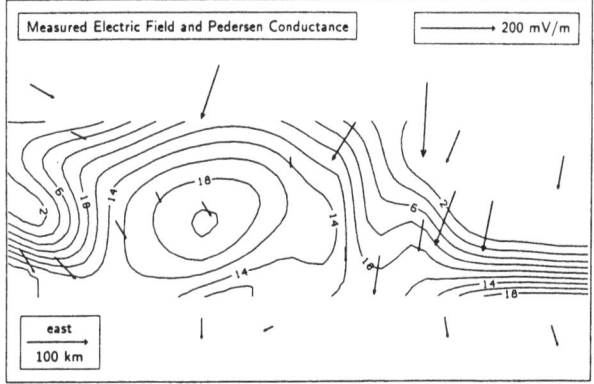

Figure 9—Same as Figure 4, but for a sequence of eastward drifting omega bands in the morning sector. (After *Opgenoorth et al.* [1983*b*].)

caused by a longitudinally alternating azimuthal electric-field component of about 10 mV m^{-1} that is superimposed on the southward convection electric field.

This becomes evident when the measured electric field pattern is decomposed into a uniform, steady background field and a second, highly structured electric-field pattern that drifts eastward with the velocity of the omega band. Moreover, in both cases studied the drift velocities of the omega band and the associated electric and magnetic field structures (typically 700 m s^{-1}) equal the **E** × **B** velocity in the uniform background electric field. Hence, the eastward motion of the omega bands is most likely caused by an **E** × **B** drift of the precipitating particles in the southward convection electric field.

Opgenoorth et al. [1983*b*] have modeled the three-dimensional current system associated with a series of omega bands. Their observations corroborate those of *André and Baumjohann* [1982]. Only the distance between the individual tongues was smaller and thus the periodic nature becomes more apparent in Figure 9. The conductivity panel shows patches of enhanced conductance, within and slightly west of the auroral tongues, on top of the ambient auroral oval conductance. The electric field panel exhibits the superposition of a southward convection electric field and an alternating azimuthal component directed toward the bright tongues.

The electric-field structure leads to a "meandering" Hall current, which may be decomposed into a westward electrojet current and counterclockwise vortices. The Pedersen current flows mainly southward, but converges somewhat toward the tongues. Hence, the ionospheric current flow has a strong southward component west of the bright tongues. On their eastern side the current flows more or less westward. Both upward and downward Birkeland currents are concentrated in local patches, in the bright tongues and the darker zone,

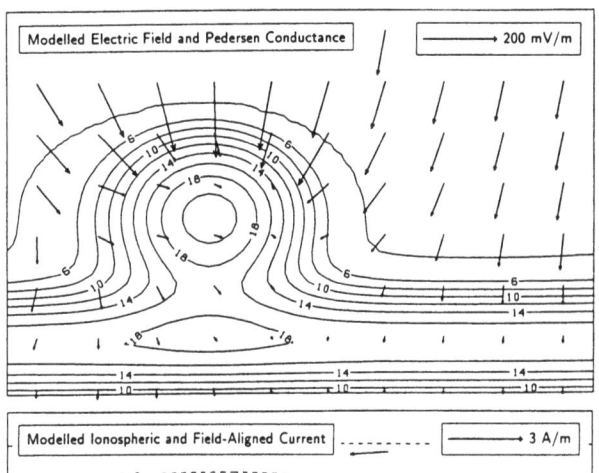

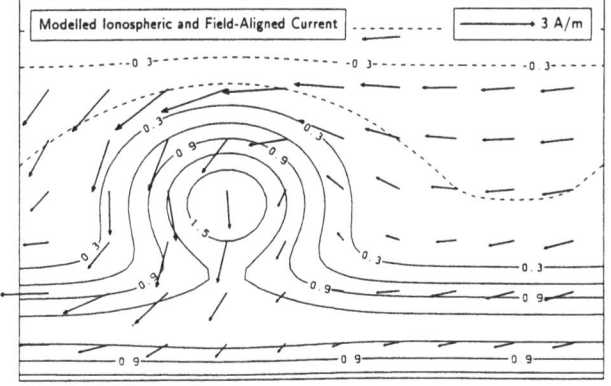

Figure 10—Measured and modeled electric field (arrows; in mV m^{-1}) and Pedersen conductance (isocontours labelled in S) patterns associated with an omega band in the morning sector. The lower panel shows the model ionospheric (arrows; in A m^{-1}) and Birkeland (solid isocontours denote upward, dashed isocontours zero or downward Birkeland current; isocontours are labelled in μA m^{-2}) current distribution. (After *Buchert et al.* [1989].)

respectively. The Birkeland current densities reach values of up to 7 μA m^{-2}.

As in the case of the westward traveling surge, recent observations of an omega band with the EISCAT incoherent scatter radar and the EISCAT Magnetometer Cross by *Buchert et al.* [1988] (Figure 10; upper panel) corroborate the earlier observations: the omega bands drift eastward with approximately the $\mathbf{E} \times \mathbf{B}$ velocity, the electric field has an azimuthal component directed toward the bright tongues, and, here measured for the first time, the ionospheric conductance is strongly enhanced within the tongues.

Buchert et al. [1989] also modeled the three-dimensional current flow associated with these auroral forms, but used an approach that is different from the trial-and-error method of *Opgenoorth et al.* [1983*b*]. Since the electric field observations were less reliable and more widely spaced than the conductance measurements, *Buchert et al.* [1989] used the well-known coupling between conductance enhancements and upward Birkeland current, i.e., the upward current is carried by precipitating electrons that create the conductivity enhancement. They varied the free parameters in the $\sqrt{j_{\parallel}} \sim \Delta\Sigma$ relation until electric fields and currents resulting as the solution of the differential equation relating electric potential to Birkeland current and conductance (basically Ohm's law and current continuity) matched the observed electric fields and ground magnetic disturbances.

This new modeling essentially verified the key features of the earlier model, but there are also differences (Figure 10, two lower panels). The most pronounced difference is that the new model does not need any localized intense downward Birkeland current in the darker zones to explain the observed electric and magnetic fields. Since the generation of concentrated downward Birkeland current is more difficult to explain than that of upward current, the new model is probably more realistic in this respect. Second, the new model has the intense southward current in the region of enhanced conductance rather than just west of it. However, *Buchert et al.* [1989] noted that the latter feature depends on the relative importance of the ΔE and $\Delta\Sigma$ variations in the auroral form and thus may vary from event to event. Finally, the new model has an asymmetric Hall current with respect to the region of enhanced conductance (the southward current being stronger than the northward current). This leads to maximum Birkeland current densities of only about

$2~\mu\mathrm{A}~\mathrm{m}^{-2}$, although the event studied by *Buchert et al.* [1989] was much more intense than that of *Opgenoorth et al* [1983*b*].

4. CONCLUSIONS

As we have shown, the distributions of conductances, electric fields, and currents in and around westward traveling surges and omega bands are reasonably well-known today. The physical processes leading to the generation of these active auroral forms are far less understood.

The generation of omega bands is least understood. At present, there is only the idea that these forms are generated by the Kelvin-Helmholtz instability. *Lyons and Walterscheid* [1985] suggested that this instability develops at the edge of an intense neutral-wind jet-stream in the highly conducting region. But *Buchert et al.* [1988] found that the neutral-wind velocities for their event are far below the threshold required for this mechanism to operate. *Rostoker and Samson* [1984] proposed that the instability develops at the interface between the low-latitude boundary layer/boundary-layer plasma sheet and the central plasma sheet. But, again, it is unclear whether the velocity shear is sufficient to excite the instability and, furthermore, if omega bands map at all to this interface.

For the westward traveling surges the situation is somewhat better. Numerical simulations by, for example, *Rothwell et al.* [1984] and *Kan and Sun* [1985] showed that bouncing Alfvén waves near a conductivity gradient can reproduce the westward expansion of the surge quite well. However, these models were recently challenged by *Hori and Yamamoto* [1987], who found that they can reproduce the convergent electric field in the surge's head only for westward electric fields below $10~\mathrm{mV}~\mathrm{m}^{-1}$.

ACKNOWLEDGMENT—The author gratefully acknowledges helpful discussions with S. Buchert and H. Opgenoorth. This work was financially supported by the Deutsche Forschungsgemeinschaft through a Heisenberg fellowship.

REFERENCES

André D., and W. Baumjohann, "Joint Two-Dimensional Observations of Ground Magnetic and Ionospheric Electric Fields Associated with Auroral Zone Currents: 5. Current Systems Associated with Eastward Drifting Omega Bands," *J. Geophys.,* **50,** 194-201 (1982).

Baumjohann, W., "Ionospheric and Field-Aligned Current Systems in the Auroral Zone: A Concise Review," *Adv. Space Res.,* **2**(10), 55-62 (1983).

Baumjohann, W., "Some Recent Progress in Substorm Studies," *J. Geomagn. Geoelectr.,* **38,** 633-651 (1986).

Baumjohann, W., R.J. Pellinen, H.J. Opgenoorth, and E. Nielsen, "Joint Two-Dimensional Observations of Ground Magnetic and Ionospheric Electric Fields Associated with Auroral Zone Currents: Current Systems Associated with Local Auroral Break-ups," *Planet. Space Sci.,* **29,** 431-447 (1981).

Boström, R., "Mechanisms for Driving Birkeland Currents," in *Physics of the Hot Plasma in the Magnetosphere,* B. Hultqvist and L. Stenflo, eds., Plenum Press, New York, pp. 431-447, (1975).

Buchert, S., W. Baumjohann, G. Haerendel, C. LaHoz, and H. Lühr, "Magnetometer and Incoherent Scatter Observations of an Intense Ps 6 Pulsation Event," *J. Atmos. Terr. Phys.,* **49,** 357-367 (1988).

Buchert, S., G. Haerendel, and W. Baumjohann, "A Model for the Electric Fields, Currents and Conductances During a Ps 6 Pulsation Event," *J. Geophys. Res.,* **94,** in press (1989).

Gustafsson, W. Baumjohann, and I. Iversen, "Multi-Method Observations an Modelling of the Three-Dimensional Currents Associated with a Very Strong Ps 6 Event," *J. Geophys.,* **49,** 138-145 (1981).

Hori, N., and T. Yamamoto, "On the Divergence of the Ionospheric Electric Field Around the Westward Travelling Surge," *Planet. Space Sci.,* **35,** 1489-1500 (1987).

Inhester, B., W. Baumjohann, R.A. Greenwald, and E. Nielsen, "Joint Two-Dimensional Observations of Ground Magnetic and Ionospheric Electric Fields Associated with Auroral Zone Currents: 3. Auroral Zone Currents During the Passage of a Westward Travelling Surge," *J. Geophys.,* **49,** 155-162 (1981).

Kan, J.R., and W. Sun, "Simulation of the Westward Travelling Surge and Pi 2 Pulsations During Substorms," *J. Geophys. Res.,* **90,** 10,911-10,922 (1985).

Kawasaki, K., and G. Rostoker, "Perturbation Magnetic Fields and Current Systems Associated with Eastward Drifting Auroral Structures," *J. Geophys. Res.,* **84,** 1464-1480 (1979).

Kirkwood, S., H.J. Opgenoorth, and J.S. Murphree, "Ionospheric Conductivities, Electric Fields and Currents Associated with Auroral Substorms Measured by the EISCAT Radar," *Planet. Space Sci.,* **36,** 1359-1380 (1988).

Lyons, L.R., and R.L. Walterscheid, "Generation of Auroral Omega Bands by Shear Instability of the Neutral Winds," *J. Geophys. Res.,* **90,** 12,321-12,329 (1985).

Opgenoorth, H.J., R.J. Pellinen, W. Baumjohann, E. Nielsen, G. Marklund, and L. Eliasson, "Three-Dimensional Current Flow and Particle Precipitation in a Westward Travelling Surge (Observed During the Barium-GEOS Rocket Experiment)," *J. Geophys. Res.,* **88,** 3138-3152 (1983*a*).

Opgenoorth, J., K. Oksman, U. Kaila, E. Nielsen, and W. Baumjohann, "On the Characteristics of Eastward Drifting Omega Bands in the Morning Sector of the Auroral Oval," *J. Geophys. Res.,* **88,** 9171-9185 (1983*b*).

Rostoker, G., and J.C. Samson, "Can Substorm Expansive Phase Effects and Low Frequency Pc Magnetic Pulsations Be Attributed to the Same Source Mechanism?" *Geophys. Res., Lett.,* **11,** 271-274 (1984).

Rostoker, G., S.-I. Akasofu, W. Baumjohann, Y. Kamide, and R.L. McPherron, "The Roles of Direct Input of Energy From the Solar Wind and Unloading of Stored Magnetotail Energy in Driving Magnetospheric Substorms," *Space Sci. Rev.,* **46,** 93-111 (1987).

Rothwell, P.L., M.B. Silevitch, and L.P. Block, "A Model for Propagation of the Westward Travelling Surge," *J. Geophys. Res.,* **89,** 8941-8948 (1984).

VI-5. LARGE-SCALE DISTRIBUTION OF DISCRETE AURORAS AND FIELD-ALIGNED CURRENTS

E. Friis-Christensen* and K. Lassen*

The system of global Birkeland currents maps regions of the magnetosphere onto the ionosphere along magnetic field lines. Similarly the pattern of auroral charged particle precipitation into the ionosphere can be regarded as a picture of the magnetospheric source and acceleration regions. Here we discuss how the field-aligned current distribution during nondisturbed conditions is related to the distribution of discrete auroras. From event studies, as well as from statistical results, we conclude that the region 1 Birkeland current system coincides with the location of discrete auroras. As a new interpretation we show that application of this view is not limited to only the gross features. Higher resolution statistical results from ground-based magnetometer chain measurements reveal that the region 1 Birkeland current system consists of two separate parts, a mainly dayside part (region 1a) and a mainly nightside part (region 1b). This separation in the pattern of field-aligned currents is accompanied by a similar separation of the dayside and nightside discrete auroral regions.

1. INTRODUCTION

The concept of the auroral oval has now been in use for more than two decades and has been widely used as a frame of reference in space physics and geophysics. While the auroral oval was initially observed visually and only in parts, using ground-based observations with a limited field of view [Feldstein, 1963], it has now been possible to prove its instantaneous existence and dynamic variation, using imagers on satellites (for example the Dynamics Explorer 1 (DE 1)). The DE 1 was the first satellite from which it was possible to observe the temporal development of the global distribution of the auroral oval [Frank et al., 1982].

The gross features of the auroral oval and its associated electric fields, currents, and particle precipitation are fairly well described. The field-aligned currents (FAC) connect the magnetosphere with the ionosphere, where they are responsible for the electric fields that drive the ionospheric currents. Figure 1 by *Iijima and Potemra* [1978] shows that the spatial distribution of field-aligned currents consists of a downward region 1 current sheet in the dawn sector and an upward region 1 current sheet in the dusk sector [*Iijima and Potemra*, 1976]. Equatorward of the region 1 current sheets there exists a system of oppositely directed current sheets, the region 2 current system. The westward and eastward ionospheric electrojets in the auroral oval are related to the region 1 and region 2 field-aligned current sheets as shown by *Kamide and Rostoker* [1977].

The statistical Feldstein auroral oval does not, however, include the dayside high-latitude distribution of

*Division of Geophysics, Danish Meteorological Institute. Lyngbyvej 100, Copenhagen, Denmark.

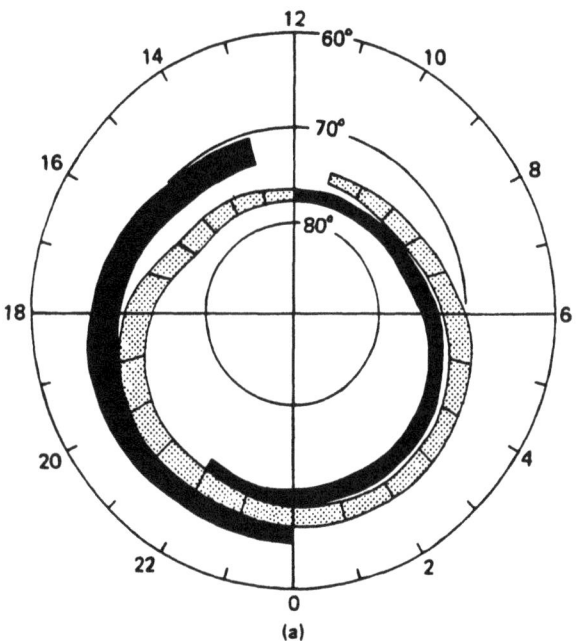

Figure 1—Statistical distribution of field-aligned currents. The filled areas indicate downward current, the dotted areas indicate upward current. (After *Iijima and Potemra* [1978].)

discrete auroras during quiet times, as described by *Lassen* [1963], using ground-based optical observations over Greenland during the IGY period (1957–59), and confirmed by independent observations by *Mishin et al.* [1970].

A more recent statistical distribution of discrete auroras has been presented by *Danielsen* [1980] and is reproduced here in Figure 2. This figure represents contour plots of the frequency of occurrence of discrete auroras for $Kp = 0$. The outermost dashed line

Auroral Physics, edited by C.-I. Meng, M. J. Rycroft and L. A. Frank. © Cambridge UP 1991

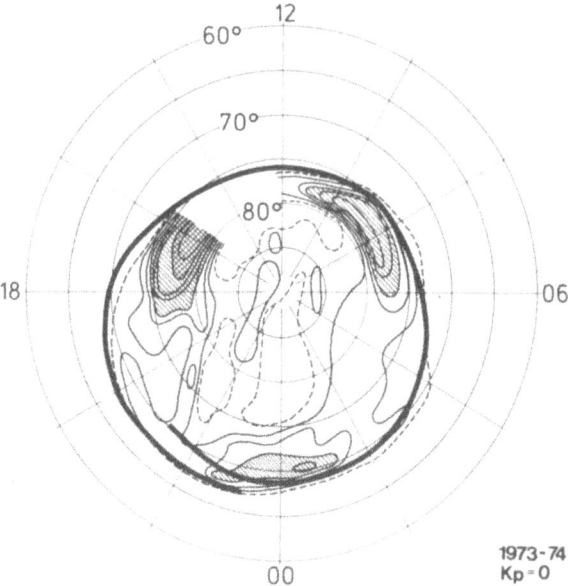

Figure 2—Distribution of the frequency of occurrence of discrete auroral arcs observed by the Greenland all-sky camera network for $Kp = 0$. The outermost contour (dashed line) corresponds to an occurrence frequency of 5%. The contour level is 5% from 5–30, and 10% for occurrence frequencies above 30%. (After *Danielsen* [1980].) Superposed on this plot is marked the equatorward boundary of the region 1 currents corresponding to Figure 1 [*Iijima and Potemra*, 1978].

corresponds to 5% occurrence rate. The heavy full line corresponds to the region 1/region 2 FAC reversal for $AL < 100$ nT presented by *Iijima and Potemra* [1978]. It is evident that the border between the region 1 and region 2 current sheets is very close to the 5% contour line of occurrence of discrete auroras. Statistically this means that the region 1 currents are located within the region of possible occurrence of discrete auroras while the region 2 currents are located in areas with a very low probability of discrete auroras.

The Feldstein oval coincides with the distribution of discrete auroras shown by *Danielsen* [1980] for the nightside from 1800–0500 MLT. On the dayside, the oval contains few or no discrete forms. According to *Lassen et al.* [1988], the discrete auroras for the dayside (0500–1800 MLT) are located poleward of the Feldstein oval, separated from the oval by the so-called transition boundary that divides the belt of auroral electrons into a poleward part of average energy <0.5 keV and an equatorward part of average energy >0.5 keV [*Makita and Meng*, 1984]. The daytime auroras are situated in the lower energy belt, poleward of the transition boundary, whereas the Feldstein oval, including the nighttime distribution of discrete auroras, is situated in the higher-energy belt equatorward of the transition boundary.

Although field-aligned currents are most directly observed using magnetometers on low-altitude polar orbiting satellites, ground-based measurements of magnetic fields and ionospheric conductivities may be used to deduce the field-aligned currents too. In such a study using a dense chain of magnetometers in Greenland, *Friis-Christensen et al.* [1985] derived field-aligned currents in the polar ionosphere that revealed a better spatial resolution of the Birkeland current pattern. Their derived region 1 current system showed a distribution consisting of two maxima, a daytime one and a nighttime one. They speculated as to whether this could be due to different region 1 current sources in the magnetosphere.

The fact that the region 1 current sheets correspond to the real distribution of discrete auroras (Figure 2), rather than to the more schematic Feldstein auroral oval, indicates the possibility that the region 1 current system may, like the auroral distribution, consist of two separate regions, one corresponding to the dayside high-latitude distribution of discrete auroras, and one corresponding to the nightside part of the auroral oval.

In this paper we will discuss the relationship between the field-aligned region 1 current system and the location of discrete auroras that has been found by studying single events. We will further show how these results may be generalized, on the basis of comparisons of statistical studies of discrete auroras as well as of field-aligned current distributions.

2. OBSERVATIONS

In the following we will briefly summarize some existing observations of electric fields and currents, together with particle precipitation information, in order to give an overview of the characteristic features for different local time sectors. Several different measurements and techniques have been used in various studies. However, owing to the capability of ground-based radars to measure a number of different geophysical parameters simultaneously, radar observations play a dominant role in the collection of observations.

Evening Sector

Klumpar [1979] used ISIS data in a comparison between diffuse auroral precipitation and the downward field-aligned current in the evening sector. In his study he concluded that the downward current generally extends 2°–3° equatorward of the central plasma sheet (CPS) precipitation, defined as a relatively stable region with precipitation of electrons above 1 keV [*Winningham et al.*, 1975]. At higher latitudes in the premidnight sector, the region 1 field-aligned current system lies within the region of particle precipitation, the boundary plasma sheet (BPS), and terminates at its high-latitude boundary. Although optical observations

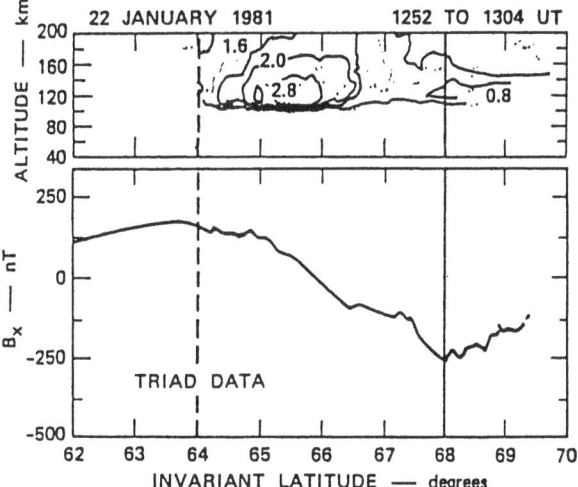

Figure 3—Relationship between field-aligned currents and ionization for the TRIAD pass of January 22, 1981. The field-aligned current reversal occurs at 68° invariant latitude. (After *Senior et al.* [1982].)

were not presented, it is obvious that this region of particle precipitation is the area in which visual auroral forms would be observed. *Klumpar* [1979] did not, however, in all cases find that the region 1 current is necessarily in the BPS. Examples were presented in which the entire CPS was located in the upward (region 1) current.

Robinson et al. [1982] used Chatanika radar measurements in conjunction with TRIAD measurements of field-aligned currents. They demonstrated the presence of a local maximum in the E-region electron density just north of the poleward boundary of the downward current, i.e., in the equatormost part of the upward current. This feature was referred to as the interface arc, because it was latitudinally localized and extended in the east-west direction.

Early Morning Sector

Senior et al. [1982] examined the relationship between the field-aligned currents deduced from the TRIAD measurements and the ionization and current pattern in the ionosphere measured by the Chatanika radar in the early morning sector. From their measurements they concluded that, characteristic of the early morning sector, there is a broad diffuse aurora extending at least 4° in latitude. The ionization is produced by precipitating electrons with energies of several kilo electron volts as deduced from the altitude of maximum precipitation. These electrons originate from the CPS according to the findings of *Klumpar* [1979]. Toward the poleward side, the altitude of the ionization increases, indicating softer precipitation and probably the poleward boundary of the low-altitude projection of

the CPS. In Figure 3, taken from *Senior et al.* [1982], is shown the relationship between the field-aligned currents and ionization. This relationship shows a broad upward region 2 current in the region of the diffuse aurora. The field-aligned current reversal takes place at the poleward boundary of the diffuse aurora, about 1° equatorward of the maximum range of the radar. The ionization is seen to maximize near the center of the current sheet and falls off toward both the poleward and equatorward boundaries.

Midnight Sector (Transition Region Between Morning and Evening)

In the transition zone between morning and evening, the field-aligned currents usually exhibit a pattern consisting of three major large-scale current sheets [*Iijima and Potemra*, 1978]. This region is called the Harang discontinuity [*Heppner*, 1972]. It is the region that separates eastward and westward convection in the auroral oval. Using S3-2 observations in conjunction with Chatanika radar measurements, *Robinson et al.* [1985] confirm the general features described by *Iijima and Potemra* [1978].

Afternoon Sector

The removal of the Chatanika radar to Sondre Stromfjord, Greenland, made it possible to study the dayside part of the auroral oval. *Robinson et al.* [1984] showed that an arc-like feature in the ionization in the early afternoon sector was colocated with the region 1 upward field-aligned current and the convection reversal region. In a study of the influence of the IMF B_y control of ionization and electric fields, *Robinson et al.* [1986] showed that the occurrence of the afternoon arc is strongly dependent on the existence of a negative B_y component. During this condition a strong convection reversal is situated in the afternoon sector and apparently favors the formation of an arc in the upward field-aligned current region. *Robinson et al.* [1986] concluded that the precipitation associated with the afternoon reversal originates from the low-latitude boundary layer.

Late Morning Sector

In the prenoon sector similar measurements have been performed using the Sondre Stromfjord radar. The radar measurements have been combined with measurements of fluxes of energetic electrons and of the magnitude of field-aligned currents using the HI-LAT satellite. *Robinson et al.* [1988] measured electron densities and electric fields during elevation scans with the incoherent scatter radar and found a broad E-region ionization extending over several degrees of latitude in the region 2 upward current region. Narrower regions of locally enhanced E-region ionization have been found to correspond to faint arc-like structures

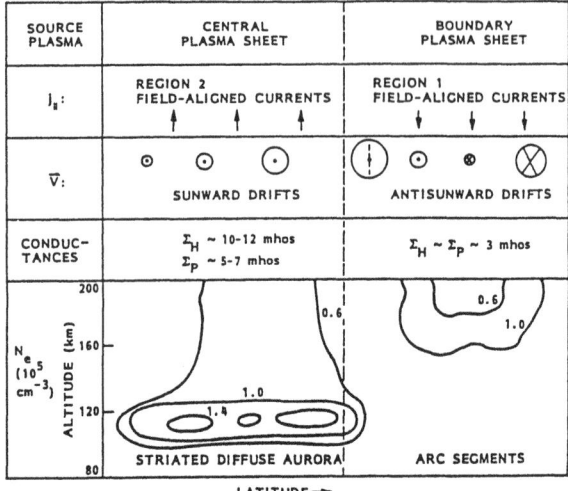

Figure 4—Latitudinal variation of electrodynamic properties across the morning sector auroral zone. (After *Robinson et al.* [1988].)

in the all-sky camera images. In this diffuse aurora the plasma is drifting sunward as deduced from the southward electric field. Poleward of this is a region of reduced electron densities at altitudes below 160 km. At altitudes above 160 km, however, enhanced electron densities are seen consistently during several elevation scans. Available DMSP photographs show that these features correspond to a system of disconnected arcs north of Sondre Stromfjord. The morphology of this population of auroral forms has been described by *Lassen* [1961, 1969]. *Lassen et al.* [1988] demonstrated that these arcs are situated in the precipitation belt of <0.5 keV electrons, separated from the diffuse aurora in the precipitation belt of >0.5 keV electrons by the so-called transition boundary [*Makita and Meng,* 1984] and also during quiescence by the trapping boundary of ≥40 keV electrons [*Lassen and Danielsen,* 1989].

Figure 4, from *Robinson et al.* [1988], shows schematically the relation between field-aligned currents, electric fields, and precipitation. The striated diffuse aurora is seen in the region of upward region 2 currents, assumed to be related to the CPS. The disconnected arc segments occur in the downward region 1 field-aligned current sheet. The spectral characteristics of the electrons in this region are similar to those measured in the cleft and are therefore speculated to originate from the magnetosheath or the low-latitude boundary layer. A similar conclusion was obtained by *Derblom* [1975] studying emissions of 6300 Å and H-alpha observed at Godhavn to the north of Sondre Stromfjord. The electric field reversal is seen to take place within the region of arc segments in the downward region 1 current.

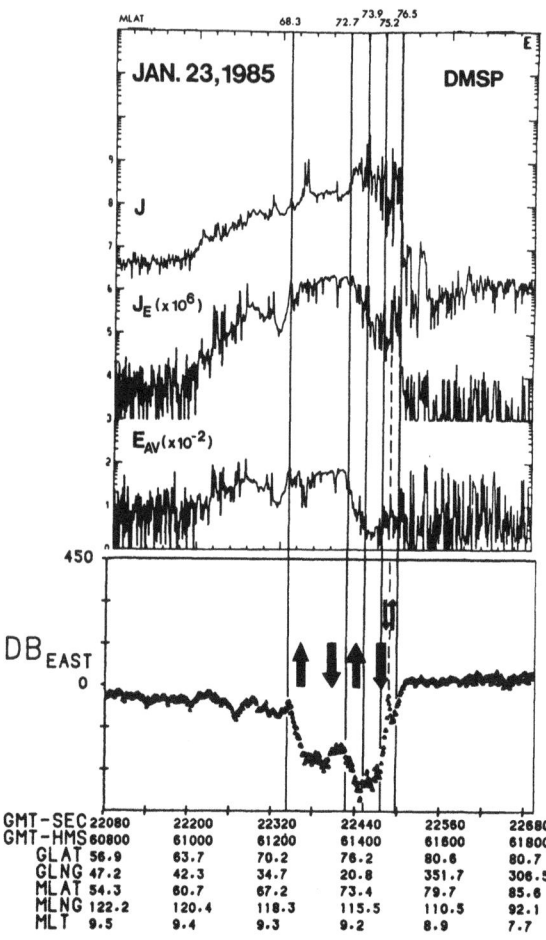

Figure 5—Electron precipitation (particle flux, energy flux, average energy) and eastward magnetic deflection during a DMSP F7 pass (0614:35 UT) on January 23, 1985. Vertical lines mark different zones of particle precipitation and Birkeland current (current direction marked by arrows). (After *Sandholt et al.* [1988].)

Figure 5 from *Sandholt et al.* [1988] show observations of electron precipitation fluxes and Birkeland currents from DMSP F7. Simultaneous ground-based optical measurements from Svalbard showed a discrete aurora associated with the small-scale structure of oppositely directed Birkeland currents that is embedded in the poleward large-scale downward current from 73.9–76.5 MLAT. Apart from the poleward set of large-scale currents that are concluded to be associated with low-latitude boundary layer particles, a second pair of large-scale current systems is observed further south. This second pair is interpreted as being the region 1 and region 2 Birkeland current sheets and the precipitating particles in these regions are characteristic of the plasma sheet.

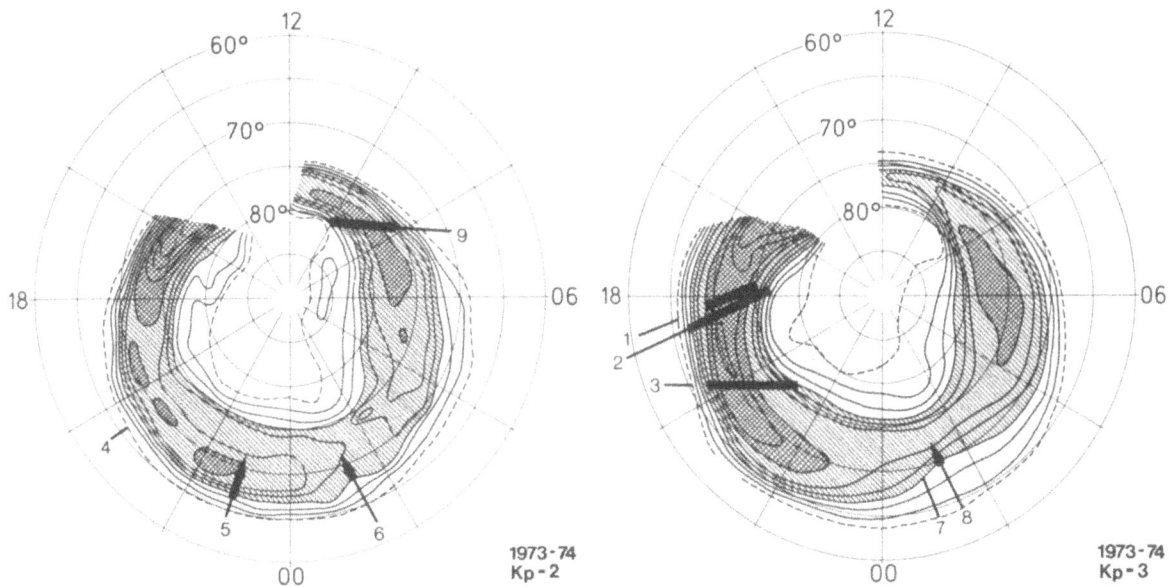

Figure 6—Summary plot showing satellite observations of field-aligned currents. A heavy line indicates region 1 current direction, a thin line indicates region 2 current direction. The observations are superposed upon a contour plot of occurrence of discrete auroras compiled by *Danielsen* [1980].

Noon

Near noon, the ionospheric current directions have been shown to depend strongly on the IMF B_y component [*Friis-Christensen et al,* 1972]. *Friis-Christensen and Wilhjelm* [1975] concluded that the ionospheric current, called the DPY current, was a Hall current flowing approximately at constant invariant latitude. Comparing the DPY current near noon with Birkeland currents measured with the TRIAD satellite, *Wilhjelm et al.* [1978] found that the DPY current was located between two oppositely directed field-aligned current sheets, the direction of which had a one-to-one correspondence with the sign of the B_y component.

Recently *Erlandsen et al.* [1988] investigated in detail the IMF B_y dependence of the region 1 currents. Using Viking measurements of magnetic fields and plasma they found that the region 1 Birkeland current system near the dayside was colocated with the region of most intense magnetosheath-like electron flux. They therefore conclude that the "traditional cusp" Birkeland current system [*Iijima and Potemra,* 1976], located poleward of the region 1 system, is associated with field lines that extend to the plasma mantle in the outer magnetosphere. Their observations are in agreement with the observations of *Wilhjelm et al.* [1978] of a systematic B_y-dependent ionospheric current sandwiched between two sheets of oppositely directed current sheets. In the model of Erlandsen et al., it is suggested that both the dawnside and duskside mantle currents (of opposite direction) may exist simultaneously, although

it is not possible to deduce this from satellite observations.

So from these events it seems to be a general feature that the region 1 currents are located in the region of discrete auroras. To summarize some of the examined events we have compared the observations of upward and downward FAC presented by *Klumpar* [1979], *Robinson et al.* [1982, 1988], and *Senior et al.* [1982] with the statistical distribution of discrete auroras for the appropriate magnetic activity level measured by Kp [*Danielsen,* 1980]. Two of the distributions, corresponding to $Kp = 2$ and $Kp = 3$, are shown in Figure 6. The observations by Klumpar are labeled 1,2, and 3, those of *Robinson et al.* [1982] are labeled 4 and 5, those of *Senior et al.* [1982] are labeled 6,7,8, and finally, the observation by *Robinson et al.* [1988] is labeled 9. The heavy lines indicate region 1 current directions, the thinner lines indicate region 2 current directions. It is obvious from this figure that the individual observations of region 1 currents match the statistical distribution of discrete auroras. Contrary to this the region 2 currents are observed equatorward of the hatched part of the statistical auroral distribution.

3. STATISTICAL DISTRIBUTION OF FIELD-ALIGNED CURRENTS AND AURORA

The average distribution of Birkeland currents by *Iijima and Potemra* [1978] shown in Figure 1 consists of two concentric rings of oppositely directed current sheets. In addition to these, but not shown here, they observed a "cusp" Birkeland current situated just poleward of the region 1 currents near noon [*Iijima and Potemra*, 1976]. This distribution of Birkeland currents, although somewhat idealized, has been shown to be a useful frame of reference for studies of auroral physics.

Despite this, we now know that particularly in the dayside, the features are much more complex than are described by the simple model. Especially, the dominant influence of the IMF B_y component on electric fields and currents is not described by the model. Several attempts have been made to modify the model to incorporate the B_y control [*Wilhjelm et al.*, 1978; *Iijima et al.*, 1978; *McDiarmid et al.*, 1978; *Zanetti and Potemra*, 1986; *Potemra et al.*, 1984; *Friis-Christensen et al.*, 1985; *Reiff and Burch*, 1985; *Erlandsen et al.*, 1988], but until now no model can be said to have gained universal acceptance, apparently because of the complex and strongly time-varying behavior of the electric fields and currents, and the lack of relevant simultaneous observations within a larger area in the noon sector.

In the Iijima-Potemra model of the Birkeland current distribution, both the region 1 and the region 2 currents form connected regions. This is a consequence of the way that the model has been formed. It is based on a large number of satellite passes across the auroral oval. The width, location, and direction of the Birkeland current has been estimated for all local magnetic times for different levels of magnetic activity determined by the AL index. The resultant system contains the qualitative features of the current systems, but is a schematic rather than a quantitative model. It is therefore difficult to compare such a model with a true average distribution of the discrete auroras.

A different way to derive the distribution of field-aligned currents has been presented by *Friis-Christensen et al.* [1985]. They inverted data from a spatially dense magnetometer array in Greenland to deduce the distribution of electric fields and currents. The resultant field-aligned current distribution for various directions of the IMF vector in the Y-Z plane is shown in Figure 7. Although *Friis-Christensen et al.* [1985] concentrated on the dayside in their data selection and in their interpretation, we may compare their distribution of field-aligned currents with the "traditional" *Iijima and Potemra* [1976] model. Although the zero line contour has been deleted for simplicity, it is evident that the results of Friis-Christensen et al. are completely missing the field-aligned currents in the region of the Harang discontinuity around midnight. This, however, is a natural consequence of the averaging technique used in the derivation rather than a feature connected

to the fact that they dealt with ground-based observations. In the region of oppositely directed FAC and ionospheric currents (the westward and the eastward electrojet) the averaging technique tends to cancel any currents unless they are systematically related to the parameters used to select the data, namely the IMF B_y and B_z components. Although the same tendency might be a concern on the dayside, where the currents are also known to be very variable, the variability here seems to be sufficiently systematically related to the IMF components that a consistent set of field-aligned currents are observed. Of course the location of the currents is smoothed out compared to the individual cases and compared to the idealized Iijima-Potemra model.

The empirical model of *Friis-Christensen et al.* [1985] is a quantitative average model in the sense that the direction as well as the magnitude of the Birkeland currents was calculated to be consistent with the average ground magnetic perturbations and an assumed conductivity model. A remarkably systematic separation near dawn and dusk of the region 1 currents is one of the prominent features of the model. *Friis-Christensen et al.* [1985] concluded that the distribution of the Birkeland currents indicated a separate source of the dayside and nightside region 1 currents and that care should be exercised in generalizing the term "region 1 current" to describe at all local times the poleward part of the double current sheet along the oval. Near noon, the region 1 current was found to be strongly controlled by the B_y component of the IMF. It was found to be closely associated with an oppositely directed field-aligned current system in the polar cap, originally referred to as the "cusp" currents by *Iijima and Potemra* [1976] but recently concluded to be associated with the plasma mantle and hence called mantle Birkeland currents [*Erlandsen et al.*, 1988].

A comprehensive dataset showing the distribution of occurrence frequency of discrete auroras for different values of Kp, but without taking into account the IMF, has been compiled by *Danielsen* [1980]. The distribution of visual, discrete auroras for different levels of magnetic activity is shown in Figure 8. In Figure 2, a subset of this figure was shown together with the region 1/region 2 current reversal according to *Iijima and Potemra* [1978] for $AL < 100$ nT. In Figure 8 we notice that during quiet conditions the distribution of discrete auroras maximizes in a prenoon and a postnoon high-latitude region. A third maximum, but less pronounced, is located near midnight in the Feldstein oval. When the activity increases, the distribution gets more ovallike, and the high-latitude maxima seem to disappear.

A similar investigation of auroral distribution was

FIELD–ALIGNED CURRENTS

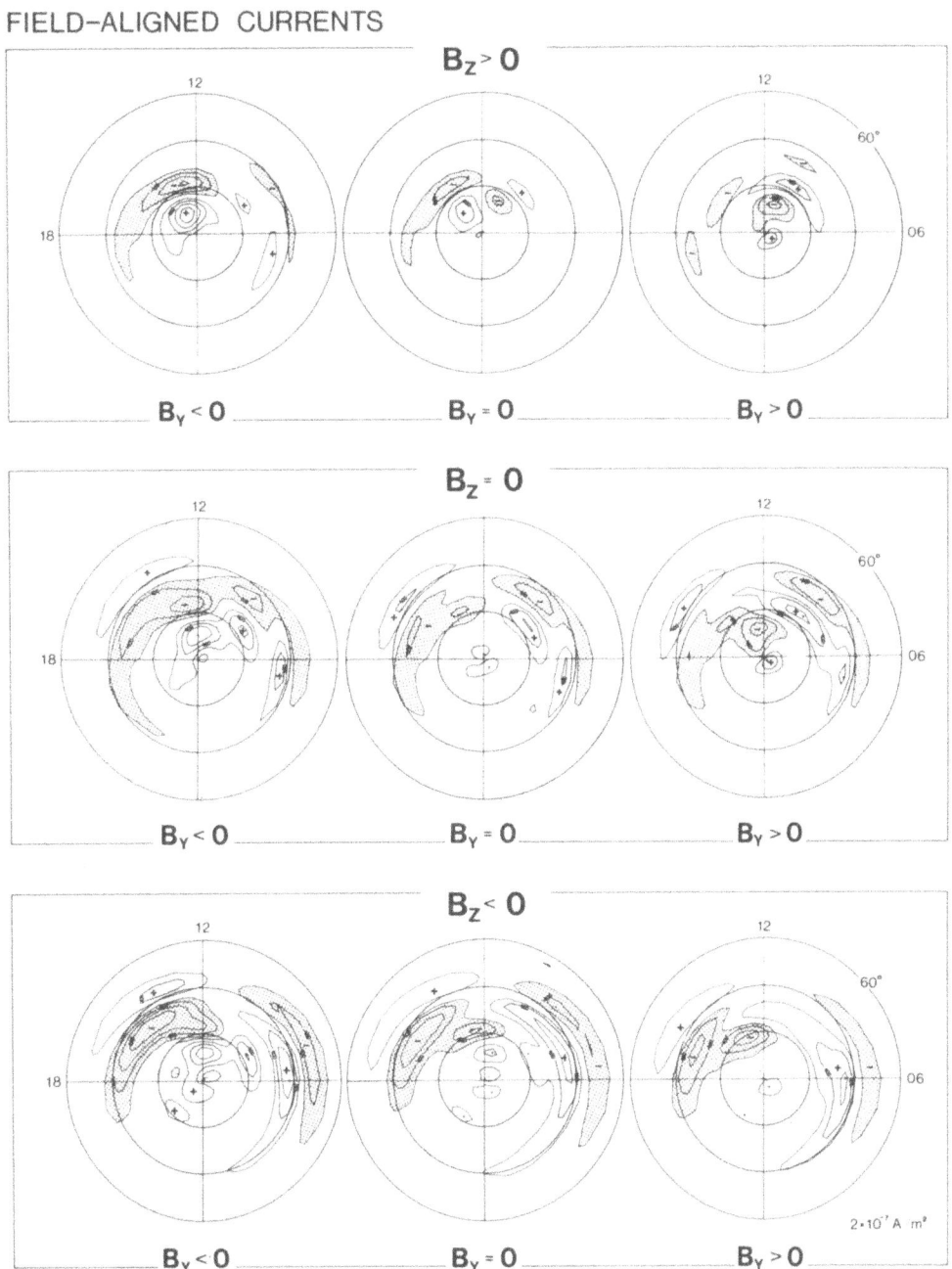

Figure 7—Distribution of Birkeland currents for different interplanetary magnetic field B_y and B_z orientations obtained by ground-based magnetometer measurements from Greenland. (After *Friis-Christensen et al.* [1985].)

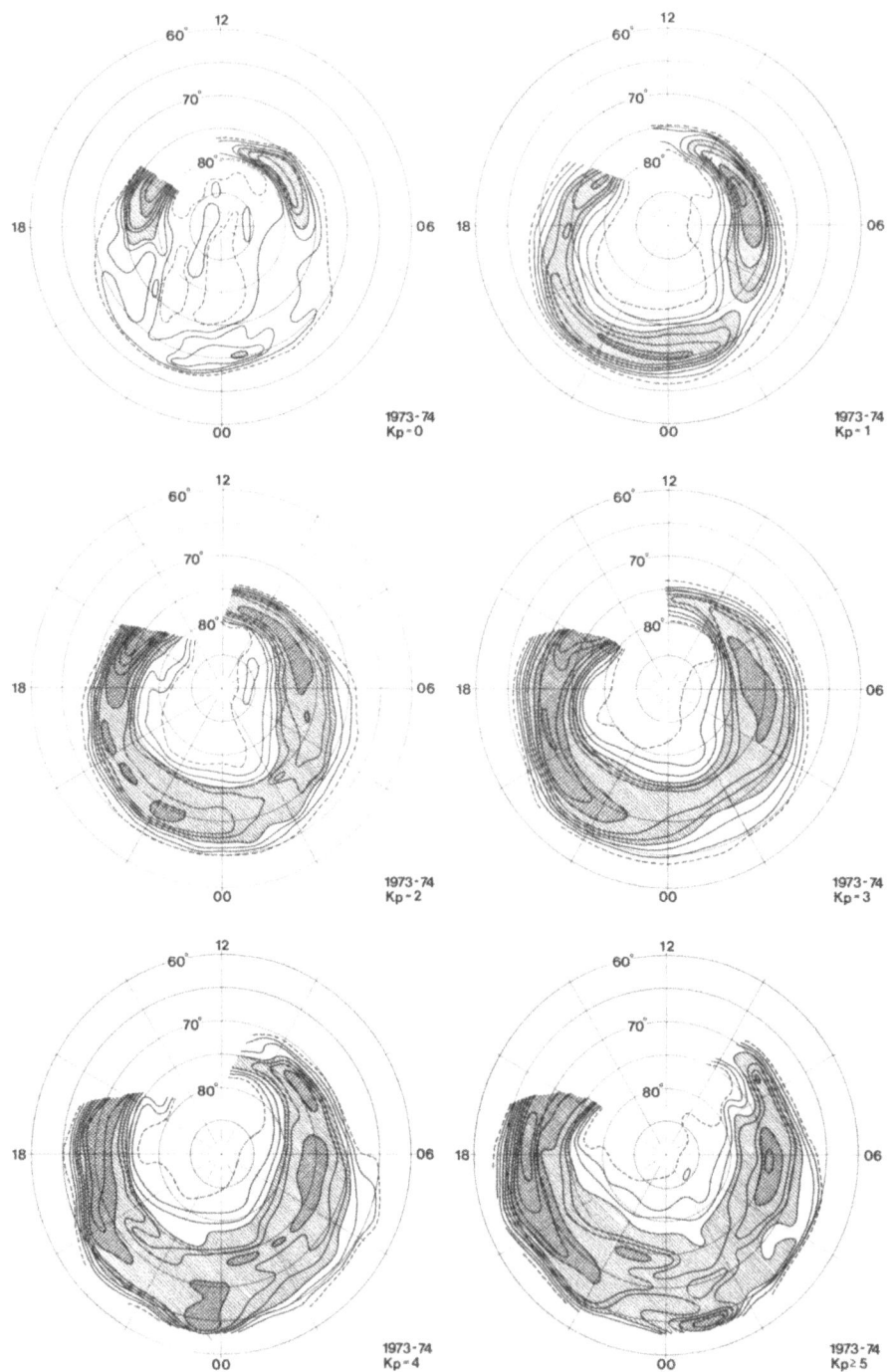

Figure 8—Distribution of the frequency of occurrence of discrete auroral arcs observed by the Greenland all-sky camera network during different levels of magnetic activity measured by *Kp*. The outermost contour (dashed line) corresponds to an occurrence frequency of 5%. The contour is 5% from 5–30% level and 10% for occurrence frequencies above 30%. (After *Danielsen* [1980].)

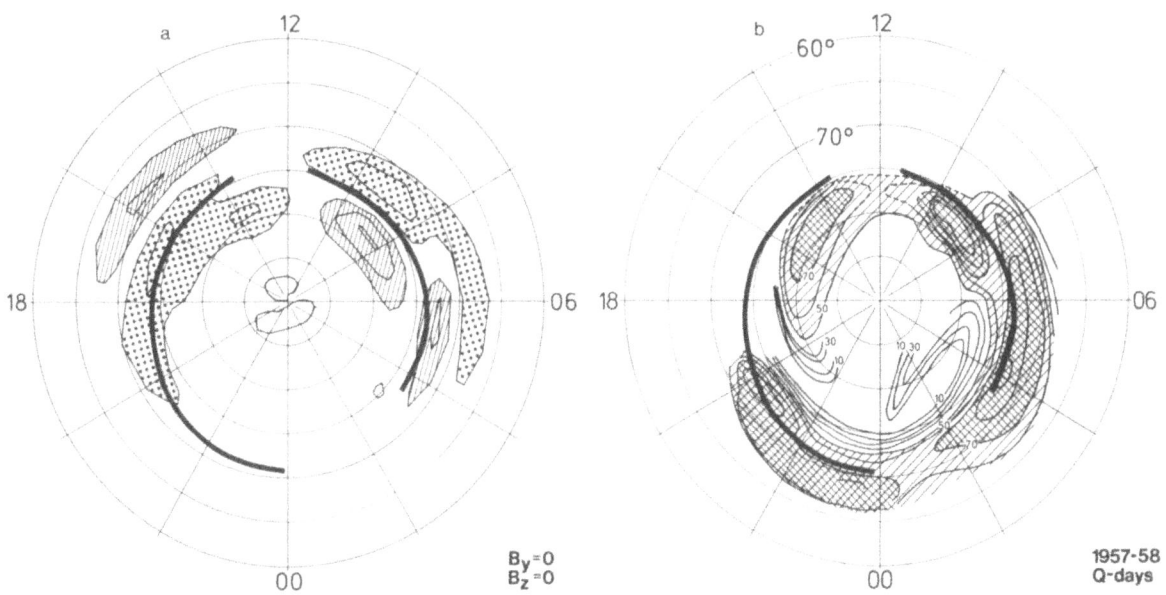

Figure 9—(a) Distribution of Birkeland currents corresponding to Figure 7, for $B_y = 0$ and $B_z = 0$. (b) Distribution of discrete auroras for Q-days 1957–58 (*Lassen*, [1963]). The transition boundary (*Makita and Meng* [1984]), that separates the belt of auroral electrons of average energy <0.5 keV from those of average energy >0.5 keV has been superposed on both plots.

made by *Lassen* [1963] for the quiet days during the IGY (1957–58) and is shown in Figure 9 together with the empirical distribution of Birkeland currents for $B_y = 0$ and $B_z = 0$ [*Friis-Christensen et al.*, 1985]. The figure reveals a nearly complete coincidence between the maximum occurrence of discrete auroras and the local maxima of Birkeland currents in the prenoon and postnoon high-latitude sectors.

We have compared the location of these maxima with the statistical position of charged particle precipitation boundaries. On top of the distributions of Birkeland currents and auroras in Figure 9, we have also drawn the transition boundary that separates the belt of auroral electrons of average energy <0.5 keV from those of average energy >0.5 keV for $Kp = 0$ or 1 [*Lassen et al.*, 1988]. The near-noon part of the region 1 current and the corresponding auroral distribution are both located in the low-energy precipitation region (average energy <0.5 keV) whereas the dawnside and duskside part of the region 1 currents and the nightside part of the auroral distribution are located in the outer belt, outside the transition line, corresponding to energies >0.5 keV.

Lassen et al. [1988] compared the location of the dayside discrete auroras with the Feldstein-Starkov oval and concluded that they are located poleward of the statistical oval, which, on the other hand, is situated in the higher energy, lower-latitude electron precipitation belt adjacent to the transition boundary. For more disturbed conditions the auroras in the statistical oval increase in intensity and extension.

It is seen when Figure 9 is compared with Figure 1 that the dawn and duskside part of the region 1 currents are colocated with the *Iijima and Potemra* [1978] region 1 currents. Comparing with Figure 7 we note that for $B_z < 0$ this distribution increases in extension, whereas for $B_z > 0$ the high-latitude dayside part of the Birkeland region 1 currents dominates. For $B_z < 0$ there is an indication that both the auroral oval/region 1 current and the high-latitude current coexist, at least in the statistical sense.

We thus have independent observations of auroras and Birkeland currents located in a particle precipitation zone with lower average energies than those characteristic of the Feldstein oval. We interpret this as an indication that these observations are related to magnetospheric regions that are different from the "nighttime" region 1 sources.

Poleward of the dayside high-latitude Birkeland currents associated with the soft particle zone (average energies <0.5 keV), there is an oppositely directed field-aligned current sheet according to *Friis-Christensen et al.* [1985], at least when B_y is different from zero. This current sheet was originally called the cusp Birkeland current [*Iijima and Potemra*, 1976]. In the study of *Friis-Christensen et al.* [1985], however, the current was concluded to be located in the polar cap. Comparing the statistical location of the currents with the statistical electron precipitation shown by *Lassen et al.* [1988], the current is found to be located in the region poleward of the low-energy belt.

A distinction between the traditional region 1/region

QUIET AURORAL OVAL

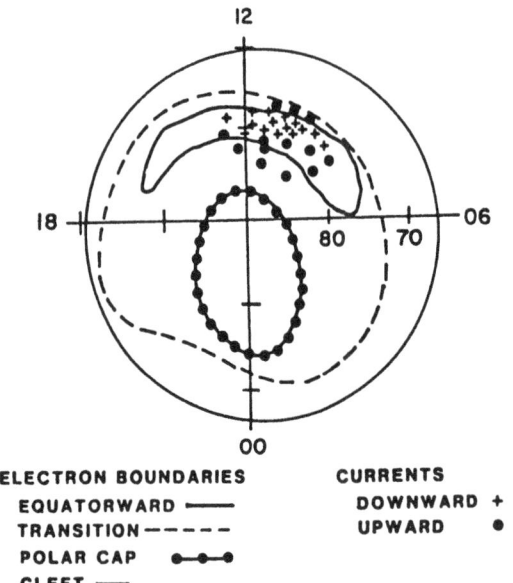

ELECTRON BOUNDARIES

EQUATORWARD ⎯⎯

TRANSITION ⎯ ⎯ ⎯ ⎯

POLAR CAP ●⎯●⎯●

CLEFT ⎯⎯

CURRENTS

DOWNWARD +

UPWARD ●

Figure 10—Sketch in mean local time (MLT)-magnetic latitude (MLAT) coordinates of the charged particle boundaries and the location of large-scale Birkeland currents for the summer South Pole with quiet magnetospheric conditions and with IMF B_y negative, on the basis of precipitating electrons and FAC observations during the quiet period of January 17–18, 1984. (After *Rich and Gussenhoven* [1987].)

2 FAC system and the noon B_y related DPY-FAC system was concluded in a study by *Rich and Gussenhoven* [1987]. They found that during prolonged quiet times the region 1/region 2 FAC system in the midnight sector was either absent or too small to be observed while, on the other hand, a clear dayside cleft FAC system is observed near noon. Figure 10 shows their sketch of particle boundaries and the location of large-scale FACs for the summer South Pole. They found that the field-aligned currents occur within the low-energy belt poleward of the transition zone.

4. REGION 1a AND REGION 1b CURRENTS

We presented observations of auroral particles and electric fields and currents above, on the basis of statistical data as well as on event studies. We now present a model of the distribution of Birkeland currents and auroras that incorporates many of these observations in as consistent a way as possible.

Individual observations indicate that the region 1 currents generally flow into or out of the ionosphere in regions where discrete auroras are present. Statistical average data of the Birkeland currents as well as of the distribution of discrete auroras both show that there is a discontinuity in these regions, statistically located

around dawn and dusk. These features are not present in the picture of a simple auroral oval as pointed out by *Lassen et al.* [1988]. The concept of a simple auroral oval is not sufficient to explain the observations, particularly in the dayside high-latitude region. The observations indicate that there may be processes that have their origin on the dayside and processes that are clearly nightside dominated. Statistically these processes do coexist, but since they seem to be influenced differently by the IMF B_y and, in particular, by the B_z component, one type of processes is likely to dominate at any given moment.

The region 1 currents are generally regarded as the "driving" currents connecting the magnetospheric regions, where dynamic processes of momentum transfer from the solar wind occur, to the ionosphere. We propose here that the "traditional" region 1 field-aligned current system consists of two separate parts, a mainly dayside region 1a system and a mainly nightside region 1b system. The region 1a system is closely associated with an oppositely directed current sheet in the polar cap or mantle, and they are both controlled by the IMF B_y component. The region 1a currents are particularly intense for strong northward IMF. In Figure 11 is a sketch of the resulting field-aligned current distribution for various B_y and B_z conditions. The region 1a currents are located in agreement with the average statistical results of *Friis-Christensen et al.* [1985], which, for $B_y = 0$, coincide with the statistical location of the high-latitude dayside discrete auroras [*Danielsen*, 1980].

The region 1b system is weak for northward IMF, but increases when the IMF becomes southward. It has its center on the nightside and extends toward the dayside along the auroral oval with increasing magnetic activity when B_z becomes more negative. In the present sketch, we have drawn the nightside part of the Birkeland current system following the distribution presented by *Iijima and Potemra* [1978], since this region cannot be properly described using only average values because of the highly variable field-aligned current intensity.

5. DISCUSSION

We have attempted to combine a variety of different observations of discrete aurora and Birkeland currents, event studies as well as statistical results, into a coherent picture. This is not a simple task because, as the single observations show, there seems not to be a simple and unique solution. There are several reasons for this. In our examinations of published observations we found that different geophysical conditions, like solar-wind and geomagnetic activity, may completely hide the basic relationships except, perhaps, during very

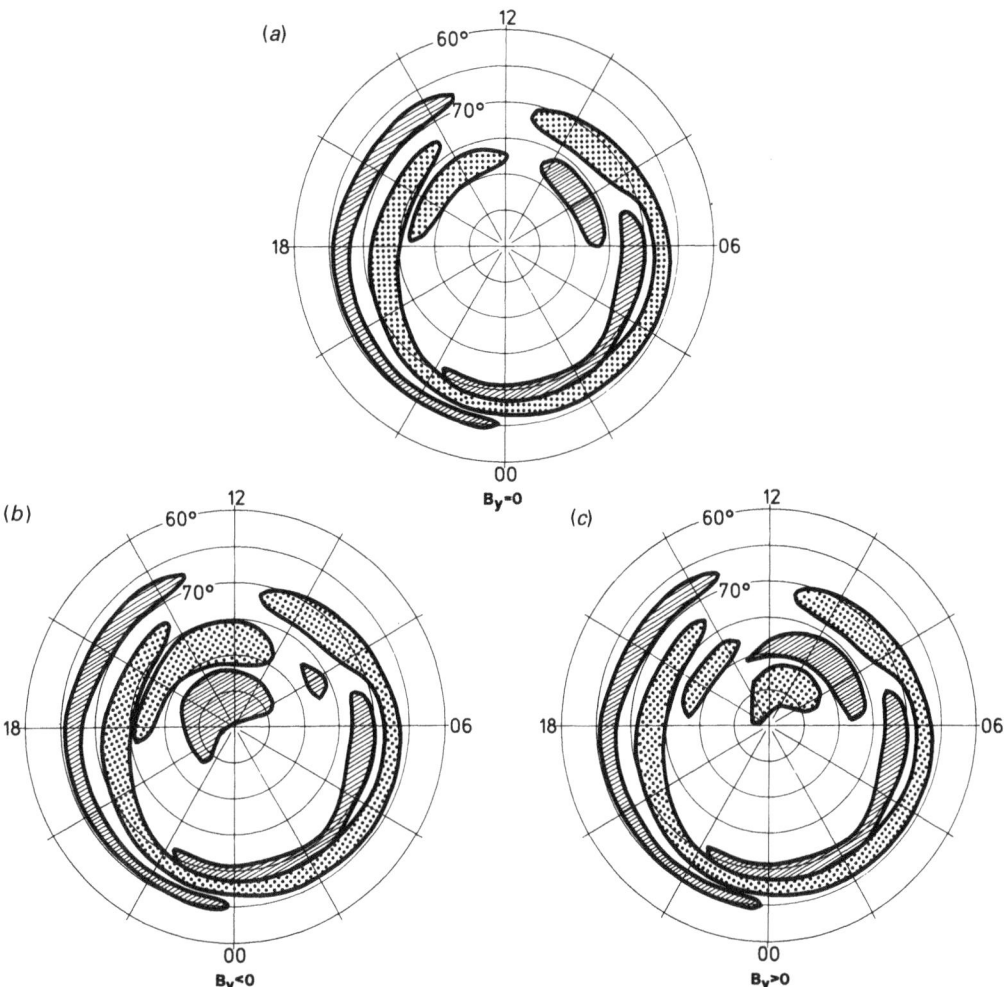

Figure 11—Model distribution of Birkeland currents on the basis of the statistical distribution obtained by satellites (*Iijima and Potemra* [1976]), as well as by ground-based measurements (*Friis-Christensen et al.* [1985]). Hatched areas indicate downward currents, dotted areas are upward currents. In (*a*) is shown the distribution for $B_y = 0$ and $B_z = 0$, (*b*) is for $B_y < 0$ and (*c*) is for $B_y > 0$. The dayside region 1 currents (region 1*a*) together with the mantle Birkeland currents on the poleward side are supposed to be strongest during positive B_z, while the nightside region 1 currents (region 1*b*) together with the region 2 currents are supposed to be strongest for negative B_z and become very weak during positive B_z.

quiet conditions. During the latter, however, very few simultaneous measurements of aurora and field-aligned currents exist. On the other hand, ground-based measurements do exist in large enough numbers to be used in statistics. Although statistical studies smooth out some of the important relationships, the general and repeatable features are thought to be conserved and any viable model should be consistent with the statistically observed average values.

A major conclusion of the present paper is that the region 1 current regions are associated with regions of discrete aurora. This is seen in the examples of observations and is confirmed also in the average statistics. A second major conclusion is that the region 1 current, previously believed to be connected to a single

topological region in the magnetosphere, namely the low-latitude boundary layer (LLBL) and the connected plasma sheet boundary layer (PSBL), probably consists of two separate regions, one (region 1*a*) connected to the LLBL and one (region 1*b*) connected to the plasma sheet. This conclusion is based on statistical results that show that during quiet conditions the distribution of discrete aurora differs considerably from the auroral oval on the dayside.

From event studies only, it is difficult to reach this conclusion since normally only a limited portion of the auroral region is encompassed in any single observation. However, in going through a number of published observations we did not find any events that obviously contradict our interpretation. The question, of course, arises: do the region 1*a* and the region 1*b* coexist. Then, if they do: is it a common phenomenon? From our investigation of published data of Birkeland currents we found only one example, the observations by *Sandholt et al.* [1988] shown in Figure 5, which we interpret as a signature of coexisting region 1*a* and region 1*b* current systems. A region 1*b*/region 2 system is located in the plasma sheet, and a region 1*a* downward current system further poleward. On the other hand, the published data may not necessarily represent an unbiased selection of passes. It is natural to imagine that a number of observations of this kind may have been regarded as too noisy and difficult to be interpreted and therefore have never been published.

Statistically we conclude that the region 1*a* current system is threaded by field-lines connected to the LLBL. As shown in Figure 9 the transition boundary corresponding to an average energy of 0.5 keV is located equatorward of the region 1*a* current distribution and also equatorward of the high-latitude discrete auroras. The region 1*b* Birkeland currents and the nightside part of the discrete auroras are statistically located in a region equatorward of the transition boundary, as shown in Figure 9. Therefore they are probably connected to the plasma sheet.

During quiescence ($AL < 100$ nT) the quiet-time region 1*b* current is, as shown in Figure 2, colocated with the maximum of auroral frequency (the auroral oval). *Winningham et al.* [1975] showed several examples of satellite passes of the midnight sector during low activity, in which they observed electron precipitation that they reference to the CPS. The location of this precipitation is coincident with the midnight maximum in Figure 2. We therefore expect the quiet region 1*b* current to be situated in the CPS region.

In the prenoon hours *Dyson and Winningham* [1974] observed electron precipitation of CPS-type between 72° and 76° latitude, immediately adjacent to what they call "cleft" precipitation, between 76° and 82° latitude. This CPS-type electron precipitation was observed at the statistical position of the Feldstein oval, whereas the cleft precipitation coincides with the high-latitude maximum of discrete auroras and the region 1*a* current. For the same orbit *Winningham et al.* [1975] reported on the occurrence of cleft electrons of lower flux in the region from the poleward border of the CPS to 79° in the early morning sector. Winningham et al. used the designation "boundary plasma sheet" (BPS) for this region. We would interpret this region to be related to the plasma sheet boundary layer, topologically connected with the low-latitude boundary layer (the cleft). In conclusion, there is observational evidence that the quiet-time region 1*b* nighttime current sheet is situated in the CPS, whereas the region 1*a* currents are situated in the low-latitude boundary layer.

Winningham et al. [1975] showed that during substorm activity active auroral forms, such as westward traveling surges and inverted-V precipitation, are located in a very disturbed precipitation belt. Since this belt is, like the quiet-time low-energy precipitation, observed immediately poleward of the CPS, the authors again used the designation BPS, now for a very disturbed region of the ionosphere. It appears from the summary plot of events (Figure 6) that the region 1*b* current is associated with the precipitation of auroral particles (discrete auroras) during moderate to high activity. Therefore, the region 1*b* is in this case associated with the BPS, rather than with the CPS.

Generally the BPS region observed in the ionosphere is thought to map into the plasma sheet boundary layer (PSBL) in the magnetotail. They are both observed poleward of the CPS, but equatorward of the polar-cap/tail-lobe region. It is, however, still an open question whether the CPS/BPS boundary maps to the CPS/PSBL boundary in the tail [*Baumjohann,* 1988].

The discrepancy between the results for low and high activity may be related to the problems with the different substorm models presently discussed in the scientific community. No single model seems to be able to explain all the observations, in the near-Earth region and in the far tail. The observations presented here indicate that, during magnetically active periods, conditions seem to be qualitatively different from conditions during quiet periods. One could imagine that the BPS, characterized by structured electron precipitation of considerable energy, is a signature solely related to the substorm situation, during which it intrudes the region previously occupied by low energy precipitating electrons characteristic of the PSBL/LLBL region. The region 1*b* current system is obviously related to the BPS region during magnetically active periods. In parallel to the increase of the BPS/region 1*b* system, the LLBL and its associated region 1*a* Birkeland current system may either withdraw from the nightside or even disappear and be replaced by the BPS/region 1*b* system.

6. SUMMARY

1. A number of independent observations indicate that the region 1 currents generally flow into, or out of, the ionosphere in regions where discrete auroras are present. Two completely independent statistical data sets both show that there is a discontinuity in these regions, statistically located around dawn and dusk.

2. The region 1 current drawn by *Iijima and Potemra* [1976, 1978] as a ring with a single discontinuity (the Harang discontinuity) near midnight is proposed to be composed of two systems, namely a current sheet (region 1*b*) coinciding with the nighttime Feldstein oval and a daytime current sheet (region 1*a*) situated poleward of the oval. Statistically the region 1*a*

and the region 1*b* systems meet and follow parallel latitude circles near dawn and dusk.

3. Like the corresponding auroral distributions, the statistical region 1*a* and region 1*b* patterns are separated by the transition boundary that separates the belt of auroral electrons of average energy <0.5 keV from those of average energy >0.5 keV.

4. The high-latitude daytime region 1*a* currents are as-

sumed to be situated on field lines threading the LLBL. The region 1*b* current system is threaded by field-lines connected to the plasma sheet. During quiet days observations indicate that the region 1*b* currents are located in the CPS, but during more active periods the region 1*b* currents are located in the BPS region.

REFERENCES

Baumjohann, W., "The Plasma Sheet Boundary Layer and Magnetospheric Substorms," *J. Geomag. Geoelectr.*, **40**, 157 (1988).

Danielsen, C., "The Dependence of Auroral Activity Upon *Kp* as Determined by the Use of an Extensive Database," *Geophysical Papers R-60*, Danish Meteorological Institute (1980).

Derblom, H., "Observed Characteristics of Polar Cleft H-Alpha and OI Emissions," *Planet. Space Sci.*, **23**, 1053 (1975).

Dyson, P. L., and J. D. Winningham, "Topside Ionospheric Spread F and Particle Precipitation in the Dayside Magnetospheric Clefts," *J. Geophys. Res.*, **79**, 5219 (1974).

Erlandsen, R. E., L. J. Zanetti, T. A. Potemra, P. F. Bythrow, and R. Lundin, "IMF B_y Dependence of Region 1 Birkeland Currents Near Noon," *J. Geophys. Res.*, **93**, 9804 (1988).

Feldstein, Ya.I., "On Morphology of Auroral and Magnetic Disturbances at High Latitudes," *Geomagn. Aeron.*, **3**, 183 (1963).

Frank, L. A., J. D. Craven, J. L. Burch, and J. D. Winningham, "Polar Views of the Earth's Aurora with Dynamics Explorer," *Geophys. Res. Lett.*, **9**, 1001 (1982).

Friis-Christensen, E., K. Lassen, J. Wilhjelm, J. M. Wilcox, W. Gonzalez, and D. S. Colburn, "Critical Component of the Interplanetary Magnetic Field Responsible for Large Geomagnetic Effects in the Polar Cap," *J. Geophys. Res.*, **77**, 3371 (1972).

Friis-Christensen, E., and J. Wilhjelm, "Polar Cap Currents for Different Directions of the Interplanetary Magnetic Field in the *Y-Z* Plane," *J. Geophys. Res.*, **80**, 1248 (1975).

Friis-Christensen, E., Y. Kamide, A. D. Richmond, and S. Matsushita, "Interplanetary Magnetic Field Control of High-Latitude Electric Fields and Currents Determined from Greenland Magnetometer Data," *J. Geophys. Res.*, **90**, 1325 (1985).

Heppner, J. P., "The Harang Discontinuity in Auroral Belt Ionospheric Currents," *Geophys. Publ.*, **29**, 105 (1972).

Iijima, T., and T. A. Potemra, "Field-Aligned Currents in the Dayside Cusp Observed by TRIAD," *J. Geophys. Res.*, **81**, 5971 (1976).

Iijima, T., and T. A. Potemra, "Large-Scale Characteristics of Field-Aligned Currents Associated with Substorms," *J. Geophys. Res.*, **83**, 599 (1978).

Iijima, T., R. Fujii, T. A. Potemra, and N. A. Saflekos, "Field-Aligned Currents in the South Polar Cusp and Their Relationship to the Interplanetary Magnetic Field," *J. Geophys. Res.*, **83**, 5595 (1978).

Kamide, Y., and G. Rostoker, "The Spatial Relationship of Field-Aligned Currents and Implications Regarding Acceleration of Auroral Electrons," *J. Geophys. Res.*, **82**, 5589 (1977).

Klumpar, D. M., "Relationships Between Auroral Particle Distribution and Magnetic Field Perturbations Associated with Field-Aligned Currents," *J. Geophys. Res.*, **84**, 6524 (1979).

Lassen, K., "Day-Time Aurorae Observed at Godhavn 1954–56," *Det Danske Meteorol. Inst. Medd.*, **15**, Charlottenlund (1961).

Lassen, K., "Geographical Distribution and Temporal Variation of Polar Aurorae," *Det Danske Meteorol. Inst. Medd.*, **16**, Charlottenlund (1963).

Lassen, K. "Polar Cap Emissions," in *Atmospheric Emissions*, B. M. McCormac, ed., Reinhold Publishing Co., New York, p. 63 (1969).

Lassen, K., C. Danielsen, and C.-I. Meng, "Quiet-time Average Auroral Configuration," *Planet. Space Sci.*, **36**, 791 (1988).

Lassen, K., and C. Danielsen, "Distribution of Auroral Arcs During Quiet Geomagnetic Conditions," *J. Geophys. Res.*, **94**, 2587 (1989).

Makita, K., and C.-I. Meng, "Average Electron Precipitation Patterns and Visual Aurora Characteristics During Geomagnetic Quiescence," *J. Geophys. Res.*, **88**, 7967 (1984).

McDiarmid, I. B., J. R. Burrows, and M. D. Wilson, "Magnetic Field Perturbations in the Dayside Cleft and Their Relationship to the IMF," *J. Geophys. Res.*, **83**, 5753 (1978).

Mishin, V. M., T. I. Saifudinova, and I. A. Zhulin, "A Magnetosphere Model Based on Two Zones of Precipitating Energetic Particles," *J. Geophys. Res.*, **75**, 797 (1970).

Potemra, T. A., L. J. Zanetti, P. F. Bythrow, A. T. Y. Lui, and T. Iijima, "B_y-Dependent Convection Patterns During Northward Interplanetary Magnetic Field," *J. Geophys. Res.*, **89**, 9753 (1984).

Reiff, P. H., and J. L. Burch, "IMF B_y Dependent Plasma Flow and Birkeland Currents in the Dayside Magnetosphere. 2. A Global Model for Northward and Southward IMF," *J. Geophys. Res.*, **90**, 1595 (1985).

Rich, F. J., and M. S. Gussenhoven, "The Absence of Region 1/Region 2 Field-Aligned Currents During Prolonged Quiet Times," *Geophys. Res. Lett.*, **14**, 689 (1987).

Robinson, R. M., R. R. Vondrak, and T. A. Potemra, "Electrodynamic Properties of the Evening Sector Ionosphere Within the Region 2 Field-Aligned Current Sheet," *J. Geophys. Res.*, **87**, 731 (1982).

Robinson, R. M., D.S. Evans, T. A. Potemra, and J. D. Kelly, "Radar and Satellite Measurements of an F-Region Ionization Enhancement in the Postnoon Sector," *Geophys. Res. Lett.*, **11**, 899 (1984).

Robinson, R. M., F. Rich, and R. R. Vondrak, "Chatanika Radar and S3-2 Measurements of Auroral Zone Electrodynamics in the Midnight Sector," *J. Geophys. Res.*, **90**, 8487 (1985).

Robinson, R. M., C. R. Clauer, O. de la Beaujardiere, J. D. Kelly, and D. S. Evans, "IMF B_y Control of Ionization and Electric Fields Measured by the Sondrestrom Radar," in *Solar Wind-Magnetosphere Coupling*, Y. Kamide and J. Slavin, eds., Terra Scientific, Tokyo (1986).

Robinson, R. M., R. R. Vondrak, D. Hardy, M. S. Gussenhoven, T. A. Potemra, and P. F. Bythrow, "Electrodynamics of Very High Latitude Arcs in the Morning Sector Auroral Zone," *J. Geophys. Res.*, **93**, 913 (1988).

Sandholt, P. E., B. Jacobsen, B. Lybekk, A. Egeland, C.-I. Meng, P. T. Newell, F. J. Rich, and E. J. Weber, "Structure and Dynamics in the Polar Cleft: Coordinated Satellite and Ground-Based Observations in the Prenoon Sector," Report 88-09, Department of Physics, University of Oslo (1988).

Senior, C., R. M. Robinson, and T. A. Potemra, "Relationship Between Field-Aligned Currents, Diffuse Auroral Precipitation and the Westward Electrojet in the Early Morning Sector," *J. Geophys. Res*, **87**, 10469 (1982).

Wilhjelm, J., E. Friis-Christensen, and T. A. Potemra, "The Relationship Between Ionospheric and Field-Aligned Currents in the Dayside Cusp," *J. Geophys. Res.*, **83**, 5586 (1978).

Winningham, J. D., F. Yasuhara, S.-I. Akasofu, and W. J. Heikkila, "The Latitudinal Morphology of 10-eV to 10-keV Electron Fluxes During Magnetically Quiet and Disturbed Times in the 2100–0300 MLT Sector," *J. Geophys. Res.*, **80**, 3148 (1975).

Zanetti, L. J. and T. A. Potemra, "The Relationship of Birkeland and Ionospheric Current Systems to the Interplanetary Magnetic Field," in *Solar Wind-Magnetosphere Coupling*, Y. Kamide and J. Slavin, eds., Terra Scientific, Tokyo (1986).

VII.   AURORA AND IONOSPHERE

VII-1. THE AURORAL ELECTROJETS: RELATIVE IMPORTANCE OF IONOSPHERIC CONDUCTIVITIES AND ELECTRIC FIELDS

Y. Kamide*

The recent availability of radar and satellite data, combined with ground magnetic records, provides us with a unique opportunity to come up with an optimized estimate of the ionospheric electrodynamic features associated with auroral displays. On the basis of our updated knowledge of the spatial relationship of ionospheric electric fields, currents, and conductivities with respect to various forms of auroral luminosities, the attempt is made to show that (1) within the so-called auroral electrojets there are two separate regions, an electric-field-dominant part and a conductivity-dominant part, each of which grows and decays with its own time constant, and (2) these two elements can be viewed as signaling two physical processes for solar-wind/magnetosphere energy coupling, i.e., the directly driven and the loading-unloading mechanisms.

1. INTRODUCTION

The ionosphere is produced primarily by the ionizing action of ultraviolet radiation from the Sun and by auroral energetic particles that strike the upper atmosphere. Electric currents flow throughout the ionosphere and are strongly influenced by the presence of the Earth's magnetic field. The motions of the ionospheric particles are affected by collisional interaction with upper-atmospheric particles, as well as by electric fields extended into the magnetosphere. Thus, the ionosphere is the physical link between the magnetosphere and the neutral atmosphere (see *Richmond* [1987] for the formation of the ionosphere).

In particular, the high-latitude ionosphere is a dynamic region, playing a key role in our understanding of the most fundamental plasma process—the magnetospheric substorm—occurring in the magnetosphere-ionosphere coupling system. It is the region where beams of auroral precipitation collide with atoms and molecules, ionizing and exciting them. Bright auroras result, intense electrojet currents flow, and significant amounts of heating are generated.

This and the next four papers in this volume present recent progress in the area of auroras and the polar ionosphere. *Nielsen* [this volume] utilizes primarily Scandinavian Twin Auroral Radar Experiment (STARE) measurements of E-layer irregularities from which the ionospheric electron drift velocity and the corresponding electric field can be estimated. One of the advantages of the STARE system is that it covers a large two-dimensional area. Nielsen demonstrates how the STARE observations combined with auroral precipitation data from a riometer are useful in discussing the location of the substorm onset region in the magnetosphere. *Röttger* [this volume] summarizes

updated knowledge on the characteristics of the auroral ionosphere, obtained from the incoherent scatter radar facility, EISCAT. It is emphasized that in addition to mapping steady-state distributions of ionospheric electrodynamic parameters, such as the convection electric field, conductivities, and field-aligned currents, coordinated observations of the radar with polar-orbiting satellites are quite important for studying polar substorms and related dynamic processes. On the basis of data from the Sondrestrom and Chatanika incoherent scatter radars, *de la Beaujardiere et al.* [this volume] stress the important of Joule heating from auroral currents as energy dissipation processes in the polar ionosphere. Three physical parameters, i.e., the electric field, the Pedersen conductivity, and the neutral wind, are evaluated in terms of their contribution to the Joule heating rate. *Iijima* [this volume] discusses the large-scale characteristics of field-aligned currents that connect the auroral ionosphere and the magnetosphere. It is shown that the field-aligned current pattern depends strongly on the sign of the interplanetary magnetic field (IMF).

We here attempt to update our knowledge about the spatial relationship of electric fields and currents and conductivities in the ionosphere with respect to auroral luminosities, and to suggest some areas that need future clarification. We focus on the concentrated ionospheric currents, called the auroral electrojects, which flow along the auroral oval. During the last several years, new datasets of ionospheric electric fields and currents relating to different auroral features at different local times under different substorm conditions have become available, providing more details on the electroject structure. In particular, the combination of radar and satellite data with ground magnetometer records permits us to determine the global configuration of ionospheric electrodynamic parameters on an individual basis. A major question to be addressed here

---

*Kyoto Sangyo University, Kyoto 603, Japan.

Auroral Physics, edited by C.-I. Meng, M. J. Rycroft and L. A. Frank. © Cambridge UP 1991

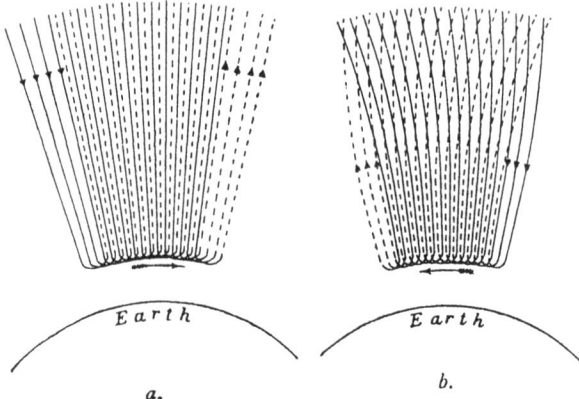

Figure 1—Birkeland's original sketch of the current system for polar elementary storms, which are now called polar magnetic substorms. (After *Birkeland* [1913].)

is: what are the auroral electrojets? Specifically, can we identify the auroral electrojets without looking at the aurora, or without using data of the auroral intensity and distribution?

2. THE AURORAL ELECTROJET

The term "auroral electrojet" was used originally and for a long time to describe a specific class of current configuration in which spatially concentrated currents flow at auroral latitudes. *How* concentrated is not defined quantitatively, but there is no doubt that the concentration of the current flow has a close connection with auroral displays. In fact, in the past, it has been tacitly assumed that the auroral electrojets are most intense in the region of bright auroral luminosity.

Figure 1 is a sketch of the current system for polar elementary storms [*Birkeland*, 1913] that are nowadays called polar magnetic substorms. The "vertical" cur-

rents, called Birkeland currents, and the "horizontal" currents, parallel to the Earth's surface, constitute the current system. Birkeland assumed that both the upward vertical current and the downward one were carried by charged particles spiraling around magnetic field lines, causing auroral forms. The horizontal portion was assumed to be carried by precipitating charged particles. It is remarkable that Birkeland proposed this current system before the discovery of the ionosphere in 1925 and before ionospheric Hall currents were known to him (see *Egeland and Kirkland* [1984] for more details).

On the other hand, *Chapman* [1927] introduced an elegant mathematical representation for analyzing the average, not individual, magnetic disturbances. Chapman derived a set of current systems consisting of the eastward and westward electrojets located in the upper atmosphere. This view was supported by the known facts about auroras (see *Sugiura* [1984] for historical details).

The Auroral Oval

Figure 2 shows the auroral oval [*Feldstein and Starkov*, 1967]. The auroral oval, which was defined as the locus of high occurrence frequency (>75%) of auroral appearance [*Feldstein*, 1966], has been used extensively to order data of electrodynamic phenomena in the polar region. As Figure 2 clearly demonstrates, the configuration of the auroral oval is not fixed in time but varies considerably with geomagnetic activity (quantified by the Q index in Figure 2). It is now known that the size of the oval is controlled also by the solar wind conditions.

The auroral oval, particularly its nightside portion, is considered to delineate the projection onto the polar ionosphere of the boundary of the inner magnetosphere (the so-called trapping region) and the outer magnetosphere. Because of the distortion of magne-

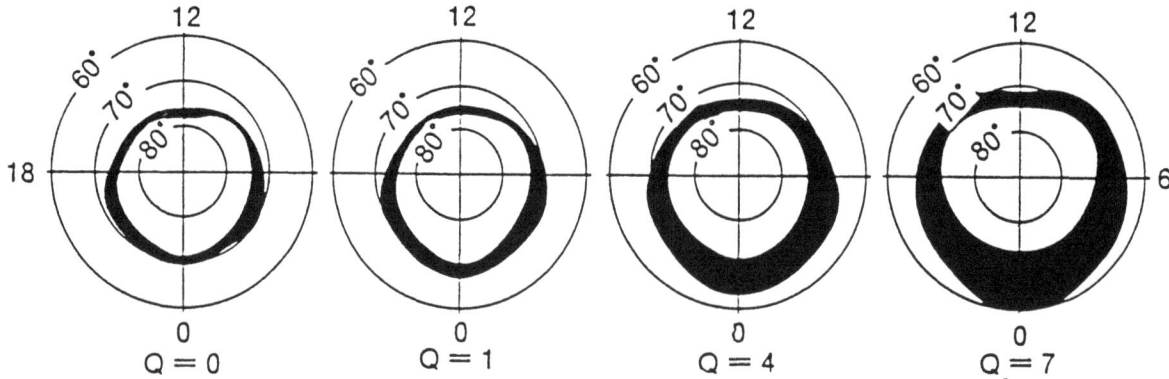

Figure 2—Changes in size of the auroral oval, in geomagnetic coordinates, for different values of the geomagnetic Q index. (After *Feldstein and Starkov* [1967].)

386

tospheric field lines, the auroral oval is not a complete circle of constant geomagnetic latitude and is eccentric with respect to the geomagnetic poles. The general pattern of many phenomena, such as precipitating charged particles and field-aligned currents, has been shown to coincide with that of the auroral oval [*Akasofu*, 1968]. This has led to the simple view that precipitating particles are in charge of enhancing ionospheric conductivities and ionospheric currents as well as causing the aurora and carrying field-aligned currents. In fact, the terminology of the auroral electrojets implies that these ionospheric currents flowing in a limited spatial region do relate directly to the aurora. If this is true, then the growth and decay of the auroral electrojets are determined by the growth and decay of the aurora. Is this really the case?

Implications of Ohm's Law

We now show, relying on the basic equations, that enhancements in the electrojet current can be established by increasing not only the conductivity but also the electric field, and that the intensity of field-aligned currents is connected in a complicated fashion to the ionospheric electric field and conductivity.

The ionosphere can be regarded as a shell of plasma surrounding the Earth. According to Ohm's law, the height-integrated ionospheric current **J** can generally be written as

$$\mathbf{J} = \Sigma_P \mathbf{E} + \Sigma_H B^{-1} \mathbf{B} \times \mathbf{E} \qquad (1)$$

where Σ_P and Σ_H are the height-integrated Pedersen and Hall conductivities, respectively; **B** is the magnetic field induction; and **E** is the electric field. Since we have dealt primarily with the global pattern of electrodynamic parameters in the high-latitude ionosphere and with the magnetogram-inversion technique from which the altitude dependence of those parameters cannot be deduced, height-integrated conductivities and currents are used throughout. The height-integrated conductivities (in siemens) are defined in the following way:

$$\Sigma_P, \Sigma_H = \int_{z_1}^{z_2} \sigma_P, \sigma_H \, dz \qquad (2)$$

where σ_P (or σ_H) is the altitude-dependent Pedersen (or Hall) conductivity (in S m^{-1}), and Z_1 and Z_2 correspond to the limits of the integration, which are taken as 90 km and 170 km, respectively, by *Brekke et al.* [1974].

Taking the two-dimensional divergence of Eq. 1 and equating it to $J_\parallel \sin\chi$, in which $J_\parallel$ is the density of the field-aligned currents, with positive being downward, and χ denotes the inclination angle of the magnetic field line, the following expression can be obtained:

$$J_\parallel \sin\chi = \Sigma_P \operatorname{div} \mathbf{E} + (\operatorname{grad} \Sigma_P) \, \mathbf{E}$$

$$+ (\operatorname{grad} \Sigma_H) \, B^{-1} \, (\mathbf{B} \times \mathbf{E})$$

$$+ \Sigma_H \, B^{-1} \operatorname{div} (\mathbf{B} \times \mathbf{E}) \qquad (3)$$

An important point to make here is that enhancements in the ionospheric current in Eq. 1 can be established enhancements either in the ionospheric conductivities (either the Pedersen or Hall conductance, or both) or in the electric field, or both. Although it is straightforward to assume that precipitating auroral electrons are responsible for active auroral displays as well as for enhancing the rate of ionization of the upper atmosphere and the ionospheric conductance, thus increasing the ionospheric currents, there is no a priori guarantee that the corresponding electric field is unchanged in the auroral region. Indeed, simultaneous observations by means of an incoherent scatter radar of the electric fields and the conductance in the auroral ionosphere indicated that there is often anticorrelation between them within auroral forms [*de la Beaujardiere et al.*, 1977]. The Pedersen and Hall conductances do not generally change in unison in bright auroral displays, reflecting the spatial energy variation of the precipitating particles.

Equation 3 is even more complicated. The intensity of field-aligned currents, the vertical currents in Figure 1, is a result of many combined effects of the ionospheric parameters. Each term represents the spatial inhomogeneity in the conductivity element and the electric field even in the thin-shell ionosphere, where the height-integrated conductivities are defined.

Substorm and Nonsubstorm Times

Birkeland dealt with individual magnetic disturbances, whereas Chapman applied his method mostly to average conditions, discussing the current system that was hoped to explain both quiet and disturbed conditions. It is now known, however, that the empirical configuration of the auroral electrojets during substorms differs significantly in character from that during steady-state conditions (e.g., *Hughes and Rostoker*, [1979]).

Figure 3 presents schematic illustrations of the eastward and westward electrojets for relatively quiet periods (on the left) and for substorm times (on the right). The Harang discontinuity is defined by the boundary between the two electrojets, or by the boundary between the regions of northward and southward electric fields [*Heppner*, 1972].

One of the differences between substorm and nonsubstorm times lies in the local time of the Harang discontinuity. During substorms, the westward electrojet penetrates deep into the evening sector. Typical values

TABLE 1. Typical values of ionospheric conductivities and electric fields during substorms and nonsubstorm times.

During Nonsubstorm Times

	Eastward Electrojet	*Westward Electrojet*
Center of electrojet	Near 1800 MLT	Near 0600 MLT
Electric field (E)	20–50 mV m^{-1} (northward)	0–30 mV m^{-1} (southward)
Pedersen conductivity (Σ_P)	2–5 S	2–5 S
Hall conductivity (Σ_H)	2–5 S	2–10 S

During Substorms

	Eastward Electrojet	*Westward Electrojet*
Center of electrojet	Near 1800 MLT	Anywhere in the dark sector
Electric field (E)	20–50 mV m^{-1} (northward)	0–100 mV m^{-1} (southward)
Pedersen conductivity (Σ_P)	5–10 S	10–20 S
Hall conductivity (Σ_H)	5–10 S	20–50 S

of ionospheric electric fields and conductivities for the two conditions are listed in Table 1, where a high degree of variability in these parameters is seen. It is also noticeable that the ionospheric conductance is nearly the same both in the eastward and westward electrojets during quiet periods, whereas it is greatly enhanced in the region of the westward electrojet during substorms. In other words, substorms accompany enhancements in the conductance.

In the next section, we show that the region of the intense auroral electrojets is not completely the same as the region of high conductivity and that of the large electric field at all local times. We also note that disturbed intervals, i.e., increase in the auroral electrojets, do not always mean substorm times, although many papers have relied on an operational definition that if the magnitude of the AE index exceeds some threshold value, say several hundred nanoteslas, that interval is a substorm interval. A substorm, however, must start with the sudden brightening of auroras and the sudden expansion of the auroral electrojet in the midnight sector, implying that the substorm onset results from plasma instabilities, not a gradual increase in current plasma.

3. CONDUCTANCE AND ELECTRIC FIELD IN THE AURORAL ELECTROJET

We now show that there are often spatial shifts among the regions of high conductivities, large electric fields, and currents. In particular, there are the electric-field-dominant part and the conductivity-rich part within the westward electrojet. The different parts tend to respond in different ways to substorms. Therefore, we present five different datasets from recent studies of electrodynamic parameters within the auroral electrojets.

Chatanika Radar Study

To assess the relative role of the electric field and the conductivity in intensifying the auroral electrojets, it is essential to measure more than one ionospheric parameter simultaneously (e.g., *de la Beaujardiere et al.* [1977]). One of the merits of incoherent scatter radars is that in certain modes of operation the radar beam is able to probe the altitude/latitude distribution of the electron density and line-of-sight plasma drifts from which the electric fields, conductivities, and currents in the ionosphere can be determined (see *Banks and Doupnik* [1975] and *Vondrak* [1983]). *Kamide and Vickrey* [1983] have presented synoptic Chatanika radar observations of the latitudinal structure of the auroral electrojets in a wide local time range.

The latitudinal profiles of electric field components, height-integrated Hall and Pedersen conductivities, ionospheric current components, and ground magnetic H and Z perturbations for three epochs are displayed in Figures 4a, 4b, and 4c, respectively. These three epochs can be considered to represent certain local times within continuous substorm activity of considerable magnitude. Figure 4a is for the eastward electrojet in the eve-

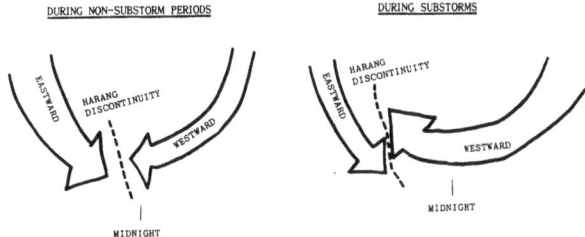

Figure 3—Sketch of the auroral electrojets during substorms (*right*) and during nonsubstorm times (*left*).

ning sector, in which the electric field features a relatively large northward component. This northward field strength tends to increase monotonically with increasing latitude up to 67.5°, and then decreases rather sharply with latitude (20 mV m^{-1} per 0.5°). Although the east-west component of the electric field is generally small (<5 mV m^{-1}), there is an indication of an increasing westward field near the poleward edge of the region containing a northward field, perhaps signifying the Harang discontinuity. The ratio Σ_H/Σ_P, a measure of the hardness of auroral particle precipitation, is not high (between 1 and 1.3), indicating also that less bright, structureless aurora and a sunlit E layer probably dominate the conductivity pattern at this time. A point of interest is that the ionospheric current, i.e., the eastward electrojet, has a northward component throughout the latitude region surveyed.

In Figure 4*b*, a similar comparison for the westward electrojet in the midnight sector is shown. Comparing this profile with that in Figure 4*a*, one can notice significant changes that occur between the evening-sector eastward electrojet and midnight-sector westward electrojet. One striking feature is that the relative contribution of the electric field and the conductivity to the auroral electrojet is reversed; in Figure 4*b* the southward electric field is less intense than the northward field in Figure 4*a*, whereas both the Hall and Pedersen conductivities are much larger than those in the eastward electrojet. The net result is a larger current density in the westward electrojet.

From a more detailed spatial comparison of the latitudinal profiles of these parameters, the following two points are noticed: (1) In the main body of the westward electrojet, the electric field is directed southward as well as westward. This corresponds to the region just eastward of the Harang discontinuity in the large-scale electric field pattern. The magnitude of the electric field is generally less than ~ 10 mV m^{-1}. (2) With the southward turning of the electric field, the Hall conductivity is enhanced significantly (maximum $\Sigma_H = 40$ S), with small-scale latitudinal variations. On the other hand, the Pedersen conductivity does not change very much along the meridian, making the Σ_H/Σ_P ratio variable as a function of latitude.

Figure 4*c* is for a period when the Chatanika radar and the Alaska magnetometer chain were observing signatures of the morning-sector westward electrojet. The latitudinal profile of the ionospheric current indicates

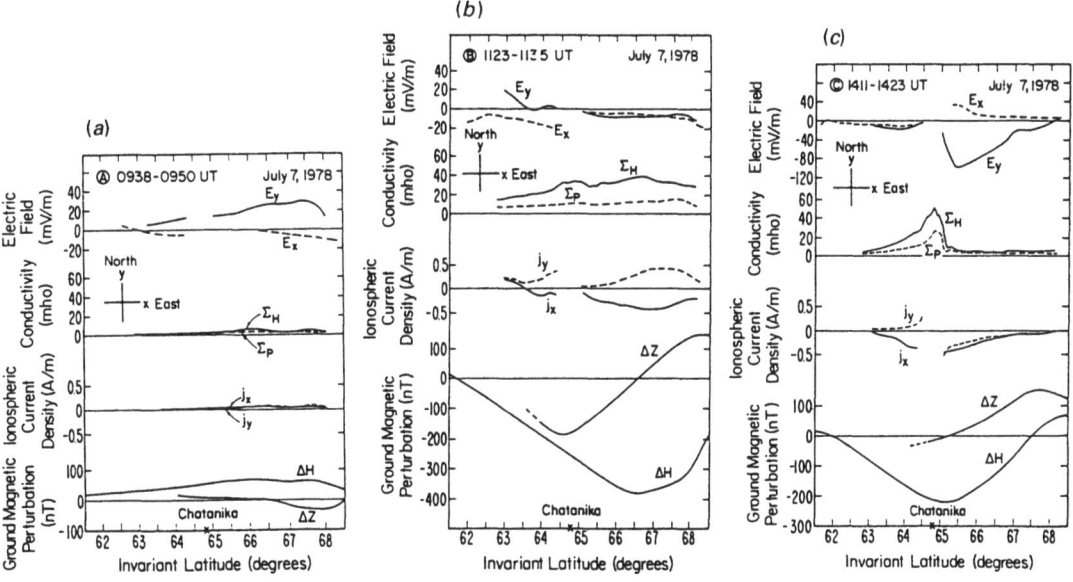

Figure 4—Latitudinal profiles of the electric field, conductivity, ionospheric current, and ground magnetic perturbations (*a*) for the typical eastward electrojet in the evening sector; (*b*) for the westward electrojet in the midnight sector; (*c*) for the westward electrojet in the late morning sector.

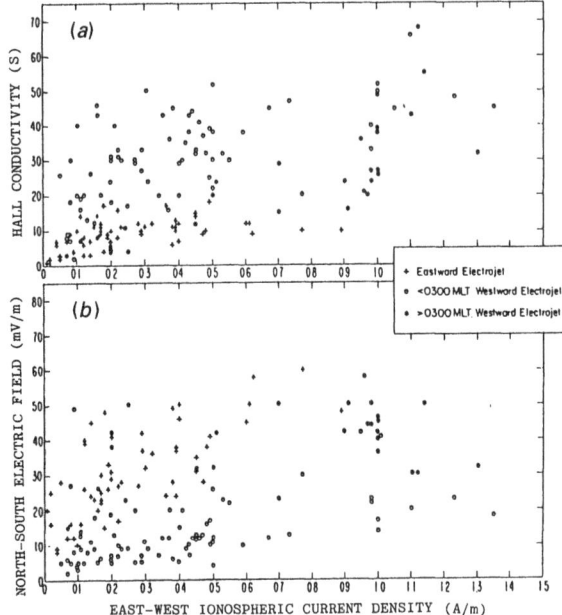

Figure 5—Influence of the east-west ionospheric current on (a) the Hall conductivity, and (b) the north-south electric field. Different symbols are used to differentiate the eastward and westward electrojets. Furthermore, the westward electrojet is split into two groups, depending on whether it occurred before or after 0300 MLT. All points represent quantities at the latitudinal center of the auroral electrojets.

that the westward electrojet was flowing in the region between 63° and 68°, with a northward (southward) component in the equatorward (poleward) half of the electrojet. In the equatorward half of the westward electrojet in this local time sector, the conductivity was high while the electric field was small. The situation is reversed in the poleward half, where the southward electric field was very intense (note the scale change between Figure 4c and Figures 4a and 4b) but the conductivities dropped drastically. This implies that bright auroras and the corresponding energetic electron precipitation were present only in the equatorward half of the morning westward electrojet [*Senior et al.*, 1982; *Kamide et al.*, 1984]. The center of the electrojet (defined by the ΔZ cross-over) seems to be located in the region sandwiched by two peaks in the southward electric field and the conductivity.

Kamide and Vickrey have further examined the local time dependence of these parameters by using a bulk dataset (a total of 160 intervals from eight days of radar operations). These 160 cases span a variety of types of substorm activity and local time sectors. For each of the cases, the latitude of the electrojet center (eastward or westward) was first identified. Then, the Hall conductivity and the north-south electric field

were determined along with the geomagnetic local time (MLT) at the electrojet center.

Figure 5a is a scatter plot of the Hall conductivity and the east-west ionospheric current density at the center latitude of the auroral electrojets. From this relationship, one may be able to gain some insight into the degree of the conductivity contribution to the auroral electrojet. Different symbols are used to distinguish the eastward electrojet and westward electrojet. The points for the westward electrojet are grouped into two categories corresponding to times before and after 0300 MLT. The choice of 0300 MLT as a dividing time is somewhat arbitrary, but best orders the points as a whole.

In Figure 5b the north-south electric field is plotted against the east-west ionospheric current in a format similar to Figure 5a. From this we can discern the different behavior of the electric field in the different temporal regimes of the auroral electrojets. In spite of the considerable scatter in Figures 5a and b, it is apparent that the eastward electrojet in the evening sector and the westward electrojet during local times after 0300 MLT have, statistically, a common character. Namely, when the current intensity is relatively small, say $|j_x| < 0.2$ A m^{-1} the increase in the current density appears to be caused by an increase in both the conductivity and the electric field. A least-squares fit to both the eastward electrojet and the late-morning westward electrojet within $|j_x| < 0.2$ gives that $\Sigma_H = 50 |j_x|$ and $|E_y| = 15 + 80 \cdot |j_x|$, where j_x, Σ_H, and E_y are in units of A m^{-1}, siemens, and mV m^{-1}, respectively.

Between the eastward electrojet and the late-morning westward electrojet, there are some differences that should also be pointed out. First, in Figure 5a the eastward electrojet does not become very large, whereas the westward electrojet can become large, say more than 1 A m^{-1}. Second, the conductivity in regions where the eastward electrojet exceeds 0.2 A m^{-1} does not increase very much with increasing current strength. This means that the increase in the current density is accomplished mainly by the increase in the northward electric field. On the other hand, in the region of the westward electrojet in the late morning sector, the conductivity as well as the southward electric field still continue to increase substantially with increasing current strength.

However, such differences are relatively minor, when contrasted with differences between the eastward electrojet and the westward electrojet in the midnight sector earlier than 0300 MLT. It is a striking feature in Figures 5a and b that the points corresponding to the westward electrojet in the premidnight and early morning sectors are well separated from those for the eastward electrojet. Conductivity values for the

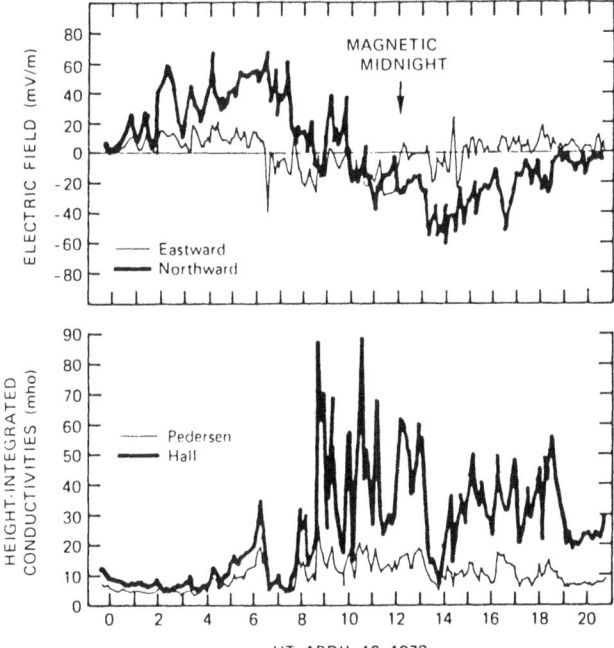

Figure 6—Variations of the electric field and height-integrated conductivities for April 16, 1973, in which intense substorms occurred successively.

midnight-sector westward electrojet are significantly higher, on the average, than those for the same intensity of the eastward electrojet and of the westward electrojet in the late morning sector.

It is also important to illustrate that the behavior described in Figures 4 and 5 is not restricted to the center of the electrojets. Figure 6 shows data from a Chatanika radar experiment during a moderate magnetic storm on April 16, 1973. This time interval was characterized by the frequent occurrence of substorms, giving us an opportunity to examine the local time dependence of ionospheric parameters under continuously disturbed conditions. We emphasize, however, that those parameters in Figure 6 do not necessarily represent the center of the auroral electrojets, since the auroral electrojets moved latitudinally relative to the radar site. The electric field was large and northward in the evening sector and the conductivity was relatively low from the beginning of the radar experiment to approximately 0830 UT or 2030 MLT (magnetic midnight is marked by an arrow in Figure 6).

The north-to-south turning of the electric field between 0830 and 1000 UT was characterized by complicated fluctuations in the field direction. This transition region corresponds to the Harang discontinuity region, discussed by *Kamide* [1978]. Between 1000 and 1300 UT, the southward electric field was relatively small, but there were large changes in the Hall

conductivity and only small changes in the Pedersen conductivity. Such a conductivity behavior suggests the sporadic precipitation of electrons with energies of several kiloelectron volts and more. This region near midnight is where the substorm westward electrojet is most intense. It should be noted that the region of the high-conductivity westward electrojet tends to penetrate into the evening sector along the poleward boundary of the eastward electrojet. This penetration is presumably associated with a westward traveling surge.

Later in the morning sector, the westward electrojet was driven by a comparatively large southward electric field. The Hall conductivity in this local time sector was lower than that in the midnight sector, although much higher than that in the evening sector. Note that the peak intensity of the morning westward electrojet was of the same order as the peak of the midnight electrojet (i.e., 1.5–2.0 A m^{-1}).

Millstone Hill and TIROS/NOAA Data

Foster et al. [1986] presented unique and very interesting patterns of ionospheric plasma convection related to auroral particle precipitation. They obtained statistically the global distribution of ionospheric convection determined from data of Millstone Hill radar for different intensities of auroral particle precipitation, using a precipitation index based on TIROS/NOAA observations. More than 2.5 million convection velocity observations over eight years were used. Diagrams showing superposed convection and precipitation patterns are particularly informative.

Kamide and Richmond [1987] pointed out that the spatial relationship between the convection electric field and the auroral precipitation has a strong local time dependence. It can be noticed in the diagrams made by *Foster et al.* [1986] that the region of the most intense precipitation in the morning sector lies equatorward of the strongest electric field, while the opposite is true in the evening sector. The implication of this is that, assuming a positive correspondence between the electron precipitation and the ionospheric conductance, the poleward half of the morning-side auroral electrojet is dominated by the electric field, while its equatorward portion is dominated by the high conductivities.

Figure 7a is a sample of the latitudinal profiles of three quantities (the electric field magnitude, the Hall conductance, and their product, i.e., the Hall current) for 0400 MLT for active times [*Foster*, 1987]. The ionospheric conductance has been calculated by the algorithm of *Fuller-Rowell and Evans* [1987]. It is demonstrated that the electric field peaks at least 5° poleward of the Hall conductances and is the stronger contributor to the poleward portion of the westward electrojet in this local time sector.

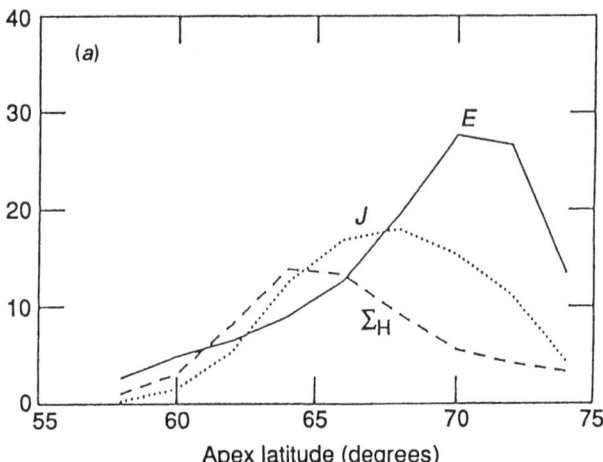

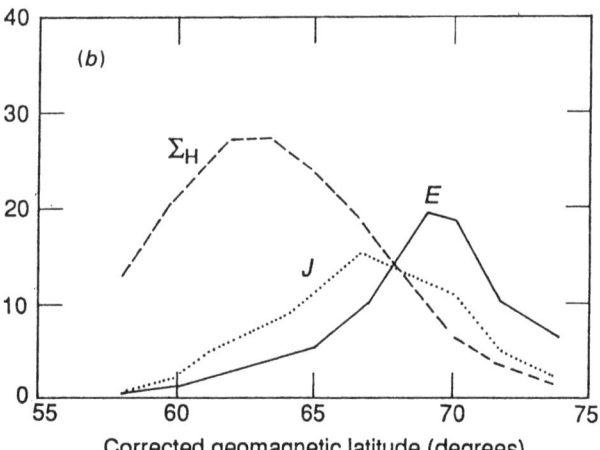

Figure 7—(*a*) Statistical latitudinal profile of the electric field *E*, the Hall conductivity Σ_H, and the Hall current *J* for 0400 MLT (after *Foster*, 1987). Vertical scale units: *E*, mV m^{-1}; Σ_H, siemens; J, 0.1 A m^{-1}. *Foster et al.* [1986] obtained the distribution of ionospheric convection determined from Millstone Hill radar data for different intensities of auroral particle precipitation, using a precipitation index based on NOAA/TIROS observations. This diagram is for index level 7. (*b*) Individual profile of the three quantities, calculated through the magnetogram-inversion technique for 1252 UT, the maximum epoch of an intense substorm, on December 12, 1981.

Conductivity from Dynamics Explorer Auroral Images

The launch of the spacecraft Dynamics Explorer 1 (DE 1) has made it possible to obtain global auroral images that can be used to estimate instantaneous, not average, conductivity distributions. The DE 1 auroral imaging instrumentation gives an ultraviolet image of auroral luminosities over the entire polar region in 12 min [*Frank et al.*, 1981, 1982]. *Craven et al.* [1983] have demonstrated that DE 1 imaging provides a means of obtaining improved conductivity models that can be used with advanced computer codes to calculate electrodynamic parameters in the polar ionosphere.

Kamide et al. [1986] have calculated the auroral enhancement component of the ionospheric conductance on the basis of images of auroral emissions observed from 20,000-km altitude, instead of employing the statistical models (e.g., *Spiro et al.* [1982]) that were used extensively in earlier studies. Assuming that the auroral emission intensity *I* is proportional to the square of the auroral electron-impact ionization density n_e (i.e., $I \propto n_e^2$) and that the height-integrated conductivity Σ can be expressed as $\Sigma \propto I^{1/2}$, the conductivities over the entire polar region can be estimated from the auroral image data. Although the absolute values of the estimated conductivity need further cross-calibrations with elaborate calculations relating to the excitation of ionospheric constituents caused by precipitating electrons [*Rees*, 1983], the region of aurorally enhanced conductivities is fairly accurately determined. This makes it possible to infer the distribution of ionospheric and field-aligned currents when the estimated conductivity distribution is combined properly with data of ground magnetic perturbations. For this purpose, *Kamide et al.* [1986] have used an improved version of the KRM magnetogram-inversion algorithm (see *Kamide et al.* [1981]).

Further, the use of simultaneous data of ion drifts from the sister satellite, DE 2, at lower altitudes (300 km) provides a unique opportunity to check the accuracy of our estimation by comparing the calculated electric fields with ion drift measurements, although such a comparison can be properly made only along DE 2 orbits. The optimum conductivity distribution has been chosen on an iterative basis so that the resultant electric fields become consistent with DE 2 ion drifts [*Kamide et al.*, 1989].

One of the characteristics seen in the deduced patterns of the global electrodynamic features is that a significant amount of ionospheric current can flow in regions where auroral activity, and thus the corresponding conductivity, are quite low. An example of this effect is found over the poleward half of the westward electrojet in the morning sector, where the southward electric field combined with relatively low conductance seems to be the main contributor to the ionospheric current. Figure 7*b* shows latitudinal profiles of three quantities (the electric field, the Hall conductance, and their product, the Hall current) for 0400 MLT. These quantities have been deduced from the best optimized result using nearly simultaneous data of DE 1 auroral imagery and of DE 2 ion drift, in addition to ground magnetometer records for the maximum epoch of an intense substorm. It is clear that the southward electric field peaks poleward of 69° latitude, whereas the

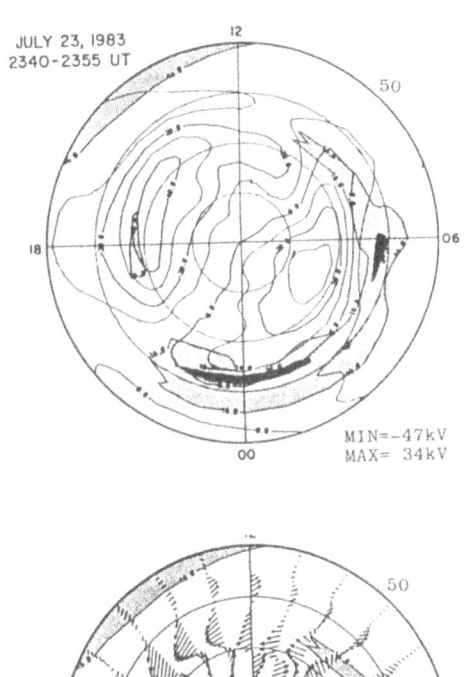

JULY 23, 1983
2340-2355 UT

MIN=−47kV
MAX= 34kV

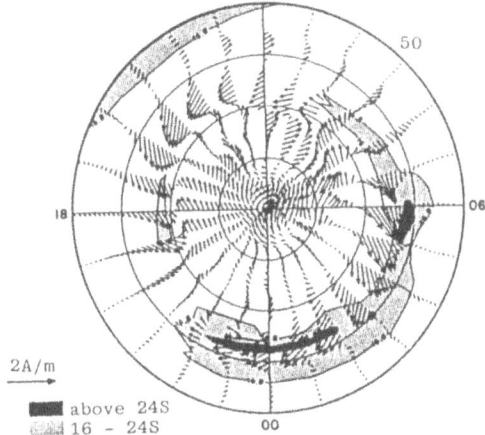

2A/m

■ above 24S
▨ 16 − 24S

Figure 8—The spatial relationships between the electric potential distributions and enhanced Hall conductance regions (*upper panel*), and between the ionospheric current vectors and enhanced Hall conductance regions (*lower panel*).

highest conductivity occurs somewhat southward of it, extending to 64° latitude in this particular example, implying that enhancements in the electric field and in auroral activity control the strength of the electrojet current separately. The peak of the westward electrojet is located between the two peaks. Comparing Figure 7a and Figure 7b, we notice that, although the exact locations of the three quantities and their magnitudes are different between the individual and average cases, the latitudinal separation by several degrees of the electric field and conductivity peaks exists consistently.

Satellite X-Ray Images

A realistic, instantaneous conductivity distribution deduced from bremsstrahlung X-ray image data of the DMSP-F6 spacecraft has been combined with the

magnetogram-inversion technique [*Ahn et al.*, 1989]. The satellite X-ray imagery has a great advantage in that the scanning detector can image a large portion of the auroral belt under both sunlit and nonsunlit conditions. In upgrading the numerical algorithm for the inversion technique, the ionospheric conductance calculated from the precipitating electron spectrum on the basis of the X-ray image data following the so-called maximum-entropy method [*Gorney et al.*, 1985] is utilized along with magnetometer data from 88 northern hemisphere observatories.

Figure 8 shows an example of overlapping plots of the Hall conductance estimated from the X-ray data with the calculated electrostatic potential (on the top) and with calculated ionospheric current vectors (on the bottom). It is evident that the equatorward portion of the westward electrojet in the morning sector is embedded in an enhanced conductance zone, while its poleward portion seems to be located in a strong electric field region. However, a somewhat different situation is found in the midnight sector, where an enhanced conductance zone is colocated with the westward electrojet. Furthermore, there is an enhanced conductance region with no significant accompanying ionospheric current, equatorward of the westward electrojet in the postmidnight quadrant. This suggests that an enhancement of conductance even in the nightside auroral latitude does not necessarily accompany an enhanced ionospheric current. On the other hand, except for a patch-like enhancement in the dusk sector, most of the eastward electrojet is located in the region of enhanced conductance of solar UV origin. This indicates that the eastward electrojet is generally dominated by the electric field. Such a finding is hardly expected from studies in which statistically derived models of the ionospheric conductivity and of the electric field are independently given in the inversion procedure.

Satellite Measurements of Electric and Magnetic Fields

Figure 9 shows a new dataset from DE 2, illustrating magnetic fields, electric fields, and energy spectra of precipitating electrons for 0500 MLT as functions of latitude (unpublished materials, courtesy of N. C. Maynard and R. Fujii). From the east-west component of the magnetic-field profile, the direction of field-aligned currents can be estimated, upward and downward on the equatorward and poleward sides, respectively. Treating the equatorward (or poleward) end of the record as the baseline near the equatorward (or poleward) boundary of the auroral oval, we notice that the downward current is more intense than the upward current, a typical feature of field-aligned currents in the morning sector [*Iijima and Potemra*, 1976; 1978]. The esti-

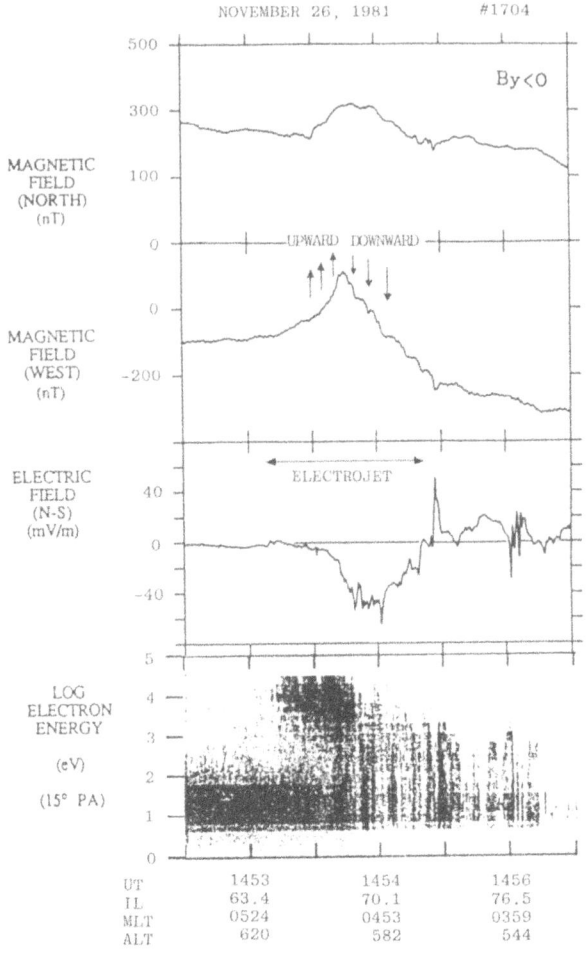

NOVEMBER 26, 1981   #1704

By<0

MAGNETIC FIELD (NORTH) (nT)

UPWARD DOWNWARD

MAGNETIC FIELD (WEST) (nT)

ELECTRIC FIELD (N-S) (mV/m)

ELECTROJET

LOG ELECTRON ENERGY (eV) (15° PA)

UT	1453	1454	1456
IL	63.4	70.1	76.5
MLT	0524	0453	0359
ALT	620	582	544

Figure 9—Electric and magnetic field data, and energy spectrogram for 15° pitch-angle electrons, from Dynamics Explorer 2 orbit 1704. The estimated width of the westward electrojets is indicated by a horizontal bar.

mated latitudinal width of the auroral electrojet is indicated in the diagram (see *Rostoker et al.* [1985]). It is clear that along this DE 2 orbit the latitudinal region of intense electron fluxes and that of large southward electric field are somewhat shifted, occupying, respectively, the equatorward half and the poleward half of the electrojet current flow. In particular, the region where the kiloelectron-volt electrons are recorded features a very weak (< 10 mV m^{-1}) electric field, while the peak of the electric field reaching 70 mV m^{-1} is embedded in < 1 kiloelectron-volt electrons.

These observations confirm that the ionospheric conductivity enhancement (assumed to be directly related to precipitating electrons) and the electric field have different roles within the electrojet current. That is, the poleward half of the westward electrojet is dominated by the electric field, while it is only in its equatorward

half that intense fluxes of keV electrons are precipitating.

4. DISCUSSION

It has been demonstrated here, by using five different datasets, that ionospheric conductivities and electric field play different roles in controlling the strength of the auroral electrojets at different latitudes and local times. To demonstrate this point, simultaneous recordings of multiple electrodynamic parameters have been essential.

Some Basic Questions

In the long history of studies of polar magnetic disturbances, magnetic field vectors observed on the "two-dimensional" Earth's surface have been the only tool with which to infer the responsible "three-dimensional" current system. It is only during the last two decades that powerful new techniques, such as satellite and radar measurements and computer simulations, have become available, providing useful information on how the current system is driven and on the extent to which the ionosphere and the magnetosphere are electrically coupled. This first-approximation model of the three-dimensional system is characterized by the eastward and westward electrojets in the regions of the northward and southward electric fields, respectively, with a considerable deformation in between at the Harang discontinuity in the premidnight sector. Crucial questions remain, however, regarding the dominant physical processes responsible for the generation of the auroral electrojets. For example, is it changes in the conductivity (σ) or electric field $\mathbf{E}$ that are most important in producing enhanced current $\mathbf{J}$ in different regions of the ionosphere? Of course, the conductivity and electric field are not totally independent in the ionosphere, in the sense that the spatial inhomogeneity of (σ) results in changes in $\mathbf{E}$.

We have here shown that the relative important of $\mathbf{E}$ and (σ) in the auroral electrojets varies considerably, depending on the location within the electrojets. In the region of the eastward electrojet in the evening sector, the northward electric field is the main contributor to the magnitude of the electrojet current. That is to say, the field magnitude is larger, compared with the southward field magnitude in the westward electrojet. Moreover, when the eastward electrojet is very weak, the electric field can maintain a reasonably large value of 10–20 mV m^{-1} (see Table 1). It is thus possible that the intensification of an already moderate eastward electrojet is mainly caused by an enhancement of the northward electric field. If the eastward electrojet current density is very small, however, any increase or decrease may depend on small changes (maximum 10

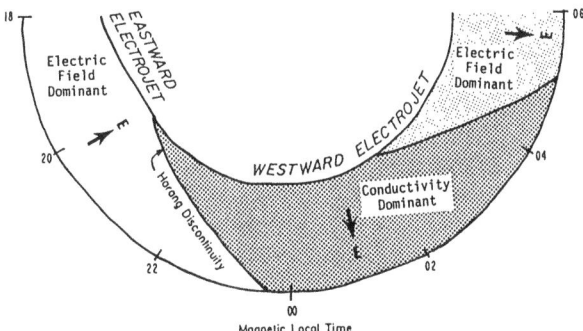

Figure 10—Schematic diagram of the auroral electrojets showing the different roles of ionospheric electric fields and conductivities in the eastward and westward electrojets at different latitudes and local times.

siemens) in the conductivity. It is important to note that in the region of the eastward electrojet, the Hall and Pedersen conductivities seem to change in unison, implying that the energies of the precipitating electrons causing the conductivity enhancements are relatively low.

Another point of interest is that there may be essentially two modes to the westward electrojet, one in which the contributions to the electrojet magnitude are "conductivity dominant," and one in which they are "electric-field dominant." The exact classification into these two modes using observed data is difficult, however, because the corresponding currents are contiguous everywhere. It has been shown statistically that the westward electrojet near midnight and in the early morning hours is characterized mainly by the relatively high Hall conductivity, whereas the westward electrojet in the late morning sector is dominated by the large southward electric field. The latter behavior is similar to that of the eastward electrojet in the evening sector, although the sense of the electric field is reversed. However, an important difference is that the Hall conductivity can become high in the late morning westward electrojet (as high as 50 S) while the maximum conductivity for the evening electrojet is less than 20 S. The high conductivity values in the morning electrojet are probably generated by kiloelectron-volt electron precipitation in patchy auroras, moving eastward from the midnight sector with the development of substorms. In other words, both the southward field and the Hall conductivity appear to be important contributors to the intense westward electrojet in the late morning sector.

To demonstrate these points, a schematic diagram is presented in Figure 10. The Harang discontinuity is clearly manifested as a switch from the eastward electrojet to the westward electrojet, and in more detail as a switch from the "electric-field-dominant" elec-

trojet to the "conductivity-dominant" electrojet. On the other hand, the boundary between the conductivity-dominant and the electric-field-dominant westward electrojet is in reality not as clear as is indicated in Figure 10, because the dominant driving mechanism changes across the boundary, by keeping the current direction unchanged. However, an important point is that, although it has been a common practice to assume that the westward electrojet is associated during substorms with a conductivity enhancement, a part of the westward electrojet in the later morning sector can be intensified without having high conductivity values. In particular, the latitudinal profile of the westward electrojet in the morning sector would indicate that its poleward portion has a relatively strong southward electric field while its equatorward portion has a relatively high Hall conductivity, consistent with what *Greenwald et al.* [1975] and *Senior et al.* [1982] have found.

In particular, using data from an auroral backscatter radar, *Greenwald et al.* [1975] found that, while there are times when the radar data showed a nearly linear dependence between the backscatter amplitude and the ionospheric current intensity, there are other times when the data exhibit little if any observable relationship. A possible explanation given by *Greenwald et al.* [1975] is that only ionospheric current variations due to electric field variations would produce the linear relationship and that in regions where particle precipitation and the corresponding conductivity enhancements are intense the electric field in the vicinity of the conductivity enhancement would be depressed and thus the radar backscatter, which is generated only when the ionospheric electric field exceeds 10–25 mV m^{-1}, would become less strong even if the current is strong. This observation indicates that the electric field and the conductivity within the auroral electrojets are not in a constant proportion everywhere.

These characteristics of the relative function of the electric field and the conductance are extremely important because many researchers have tacitly assumed that the auroral electrojets, as the terminology implies, flow in the region of high conductance where the auroral luminosity is highest. Fundamental new questions then arise: What is the auroral oval? What are the auroral electrojets? Does the auroral oval include the poleward half of the westward electrojet, where we do not necessarily expect intense auroral forms?

Two Electrojet Components

It is of great interest to combine the points demonstrated in the present paper with points made in earlier studies. As discussed in the previous sections, the auroral electrojets may have two different elements: the electric field-dominant electrojet and the conductivity-dominant electrojet. It is inferred that the auroral electrojets do not grow and decay as a whole; rather, only the portion in the midnight sector is enhanced at the time of auroral breakup, and the other portion is controlled primarily by the electric field, which does not necessarily reflect auroral activity.

It has already been shown that there are basically two types of polar current systems: DP 1 and DP 2 (see, for example, *Troshichev et al.* [1974] and *Baumjohann* [1983]). DP 2 consists of the eastward electrojet in the evening sector and the westward electrojet in the late morning sector, while DP 1 is dominated by the westward electrojet in the midnight sector. *Kamide and Baumjohann* [1985] and *Clauer and Kamide* [1985] indicated that these two modes can coexist at all times during disturbed periods and that the relative strength of these currents varies from time to time, making individual current patterns very complicated. Furthermore, the pattern of "enhanced" DP 2 has been shown to appear at times throughout substorm activity, although the DP 2 current system was originally distinguished from the DP 1 system on the basis of whether it is accompanied by a current concentration at auroral latitudes caused, presumably, by auroral enhancements (see *Obayashi and Nishida* [1968]).

The possibility of there being two types of high-latitude current sytems might have been implicit in some of the earlier studies that relied only on ground-based magnetic observations, including at least *Nishida and Kokubun* [1971], *Gizler et al.* [1976], and *Pytte et al.* [1978]. Figure 11 shows one such example [*Baumjohann*, 1983]. The two elements in Figure 11 are called the convection electrojet (DP 2 type) and the substorm electrojet (DP 1 type).

The existence of two electrojet elements can solve at least one problem that has been puzzling in the past: the complex spatial and temporal distribution of the substorm-associated auroral electrojets. *Clauer and McPherron* [1974] stressed the variability of the substorm current system by demonstrating that the westward electrojet can appear anywhere in the dark sector. Using magnetic potential contours derived from magnetic records of a densely spaced high-latitude network, *Kroehl and Richmond* [1980] indicated that the variability of individual substorms might be ordered by the preferred local time at which the westward electrojet dominates.

However, in spite of such complicated features of the current configuration during individual substorms, the westward electrojet appears to be centered in the early morning hours, say 0200–0400 MLT, in a typical distribution of magnetic disturbances at auroral latitudes. For instance, in one of the classic studies in this field, *Silsbee and Vestine* [1942] found that the westward electrojet extends from prenoon to premidnight hours with its center located at 0200–0300 MLT. *Al-*

CONVECTION ELECTROJETS   SUBSTORM ELECTROJETS

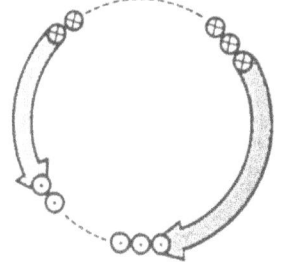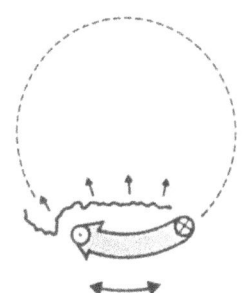

Figure 11—Schematic diagram of the location, flow direction and field-aligned current closure of convection electrojets and the substorm current wedge. (After *Baumjohann* [1983].)

len and Kroehl [1975] found that during disturbed times the most frequent times of contribution to the *AL* index by each observatory varied from 0200 to 0600 MLT, but the index was most often derived from records of stations located at about 0300 MLT. *Akasofu et al.* [1980] derived the average disturbance vector distribution and found that the maximum westward electrojet was seen again at 0300 MLT at 67° in invariant latitude. It is believed that the westward electrojet is the dominant feature of polar magnetic substorms that start most often around local midnight, not 0300 MLT, accompanied by auroral substorms. It would seem that the region of the most active aurora must move eastward, forcing the average electrojet center to move toward the morning sector. However, the most active large-scale aurora, called the traveling surge, actually moves westward into the evening sector, instead of eastward. How is it possible to explain the difference (0300 vs. 0000 MLT) between the statistical electrojet center and the generally believed substorm center?

An explanation may be that there are two essentially different current systems that coexist throughout a substorm. The first current system, whose main contributor is the ionospheric conductivity, dominates during the expansion to maximum epoch of substorms, while the other one, controlled by the electric field, exists almost all of the time. Of course, these two currents are contiguous in the ionosphere, but the main region of the first one is in the midnight sector, and the other is in the evening and the late morning hours. Simple averages of all electrojet activity without paying attention to the two physically different electrojet elements appear to lead to the 0300 MLT electrojet center found in earlier studies [*Kamide*, 1982].

Implications for Substorm Mechanisms

The implications of these two elements are interesting if they are viewed as two physical processes for solar-wind/magnetosphere energy coupling (the directly driven and loading-unloading processes). That is, the "electric field" electrojet may represent directly the effects of the solar-wind/magnetosphere dynamo, whereas the "conductivity-rich" electrojet may be a manifestation of a plasma instability process internal to the magnetosphere, relating to intense and sporadic auroral activity at substorm expansion onsets.

Baumjohann [1983] has summarized evidence showing that during periods of enhanced magnetospheric activity the ionospheric current flow is affected in two ways. First, due to direct energy input from the solar wind into the magnetosphere, and thus enhanced magnetospheric convection, the current flow in the auroral electrojet increases. This may be seen as the DP 2 current system. As the left panel of Figure 11 indicates, this system is dominated by the eastward electrojet in the evening sector and the westward electrojet in the late morning sector, both of which are controlled primarily by the electric field. Second, the sporadic release of energy previously stored in the magnetotail leads to the formation of the substorm current wedge with strongly enhanced westward current flow in the region of active breakup auroras and the westward traveling surge around midnight, i.e., to the formation of the DP 1 current system (see the right panel of Figure 11).

If we associate DP 2 currents with the directly driven substorm process and the DP 1 currents with the unloading process mentioned above, one can state that, especially on days with continuous activity and nearly persistent energy input, both processes are typically operating at the same time but in different regions of the auroral electrojets. The relative importance of the two processes varies significantly from time to time. It is conceivable that this is the reason why substorm signatures, such as the distribution of magnetic perturbation vectors or electric fields and their time changes, are quite complicated. The term "unloading" pertains to the deposition of stored energy into the auroral ionosphere as well as into the ring current. Although it has long been argued that certain substorm-associated phenomena are consistent or inconsistent with either the directly driven process or the loading-unloading process, one could resolve such controversial arguments by understanding that the magnetospheric substorm involves the two processes simultaneously.

An important point, however, is that during some interval associated with the isolated substorm, nearly pure DP 2 and DP 1 can be observed, as demonstrated by *Clauer and Kamide* [1985] for the March 22, 1979, substorms. Such an interval is particularly interesting since certain physical processes associated with each pattern can be isolated without being contami-

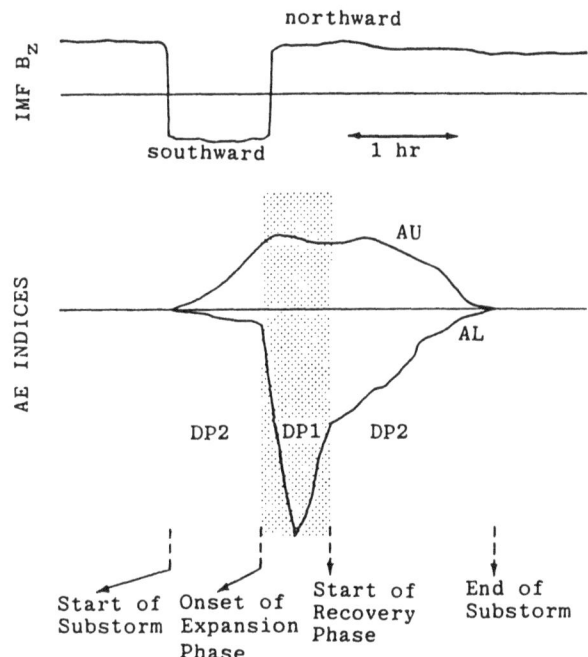

Figure 12—Schematic diagram showing the growth and decay of an isolated substorm in terms of the *AU/AL* response of the substorm phases (*bottom*) to changes in the north-south component of the interplanetary magnetic field (*top*).

nated by the other part. A typical DP 2 pattern is often seen during the period preceding the major substorm expansion onset [*Pellinen et al.*, 1982]. During such an interval, the IMF remains southward, enhancing magnetospheric convection, consistent with the dawn-dusk electric field in the polar cap and the convection-type electrojets at auroral latitudes. It is therefore plausible to assume that the increase in the ionospheric currents and the associated energy dissipation before the major substorm expansion is directly driven by the solar wind, in agreement with *Baumjohann et al.* [1981] and *Nishida and Kamide* [1983]. On the other hand, the expansion phase can occur in conjunction with the weakening or diminishing of the DP 2 component, making the DP 1 pattern dominant. Since this expansion can sometimes occur after the IMF has turned northward [e.g., *Rostoker et al.* [1982] and *Rostoker* [1983]), i.e., switch-off of the solar wind energy entry into the magnetosphere, the expansion appears to be connected to the unloading of the tail energy into the polar ionosphere [*McPherron*, 1974].

To illustrate temporal changes in the two basic processes (the directly driven and loading-unloading processes, or DP 2 and DP 1, respectively), Figure 12 shows a schematic diagram of *AU* and *AL* variations for a canonical isolated substorm, with special emphasis on the dominant physical course occurring presumably in the magnetosphere-ionosphere system. The corresponding correlation with interplanetary medium parameters is also shown. It is important to notice that the westward electrojet has two basic components, the relative predominance of which marks the start and the end of the substorm phases. In particular, the sudden increase of the DP 1 system centered in the midnight sector signals the onset of the expansion phase, whereas its peak corresponds to the maximum epoch of the substorm. On the other hand, the DP 2 system appears to increase and decrease slowly during the canonical isolated substorm, making a kind of "inflection point" in the *AL* curve. We propose that this point is the start of the recovery phase of the substorm event.

One might well wonder, since such isolated substorm events are seldom found in nature, whether one can really take the properties of isolated events and attribute them to all substorm activity. In fact, more typical substorms are marked by southward turnings of the IMF some short time prior to which the IMF had been intermittently southward. In such cases, the ionospheric conductivity could be high and the corresponding current systems of considerable strength might well be flowing before the start of the substorms. Under these circumstances, the start of the substorm in Figure 12 could be coincident with the start of the expansion phase. In other words, it is practically impossible to distinguish between the starts of DP 2 and DP 1.

It should also be emphasized that in the past we tended to simplify drastically substorm models to discuss the basic mechanisms involved. One example is the proposed three-dimensional current system for the substorm in which the westward electrojet is connected to a downward and an upward field-aligned current flow at the eastern and western edges, respectively. This itself is not a wrong approach, because we have to start with the simplest assumptions applied to the simplest examples of observations to look for the most important and fundamental cause-and-effect parameters. However, in order to unveil what governs the current system, i.e., what particles carry the field-aligned currents, how these particles are accelerated, and what is the role of these particles in creating the enhanced ionospheric conductance, it is essential to examine the space-time distribution not only of the current, but also of some key parameters describing ionospheric quantities. For example, it is very conceivable that "high conductivity and small electric field" or "low conductivity and large electric field" may carry the same electrojet currents. In those two extreme cases, we have to invoke two very different physical processes. One of our future problems is to understand the substorm generation mechanism and physical processes associated with the growth and decay of substorms by taking into account that the auroral electrojets have two components.

ACKNOWLEDGMENT—I would like to thank N. C. Maynard and R. Fujii for their permission to use their unpublished materials as Figure 9 in this paper. This work was supported in part by the Ministry of Education under grant-in-aid 62540310 and in part by the National Institute of Polar Research under a joint research program.

REFERENCES

Ahn, B.-H., H. W. Kroehl, Y. Kamide, and D. J. Gorney, "Estimation of Ionospheric Electrodynamic Parameters Using Ionospheric Conductance Deduced from Bremsstrahlung X-ray Image Data," *J. Geophys. Res.*, **94**, 2565 (1989).

Akasofu, S.-I., *Polar and Magnetospheric Substorms*, Reidel, Dordrecht, Netherlands (1968).

Akasofu, S.-I., J. Kisabeth, G. J. Romick, H. W. Kroehl, and B.-H. Ahn, "Day-to-Day and Average Magnetic Variations Along the IMS Alaska Meridian Chain of Observatories and Modeling of a Three-Dimensional Current System," *J. Geophys. Res.*, **85**, 2065 (1980).

Allen, J. H., and H. W. Kroehl, "Spatial and Temporal Distributions of Magnetic Effects of Auroral Electrojets as Derived from *AE* Indices," *J. Geophys. Res.*, **80**, 3667 (1975).

Banks, P. M., and J. R. Doupnik, "A Review of Auroral Zone Electrodynamics Deduced from Incoherent Scatter Radar Observations," *J. Atmos. Terr. Phys.*, **37**, 951 (1975).

Baumjohann, W., "Ionospheric and Field-Aligned Current Systems in the Auroral Zone: A Concise Review," *Adv. Space Res.*, **2**, 55 (1983).

Baumjohann, W., R. J. Pellinen, J. H. Opgenoorth, and E. Nielsen, "Joint Two-Dimensional Observations of Ground Magnetic and Ionospheric Electric Fields Associated with Auroral Zone Currents: Current Systems Associated with Local Auroral Breakups," *Planet. Space Sci.*, **29**, 431 (1981).

Birkeland, K., *The Norwegian Aurora Polaris Expedition 1902-1903*, vol. 1, *On the Cause of Magnetic Storms and the Origin of Terrestrial Magnetism*, sect. 2, Aschehong, Christiania, Norway (1913).

Brekke, A., J. R. Doupnik, and P. M. Banks, "Incoherent Scatter Measurements of E Region Conductivities and Current in the Auroral Zone," *J. Geophys. Res.*, **79**, 3773 (1974).

Chapman, S., "On Certain Average Characteristics of World-Wide Magnetic Disturbance," *Proc. R. Soc.*, **115**, 242 (1927).

Clauer, C. R., and Y. Kamide, "DP 1 and DP 2 Current Systems for the March 22, 1979, Substorms," *J. Geophys. Res.*, **90**, 1343 (1985).

Clauer, C. R., and R. L. McPherron, "Mapping of Local Time-Universal Time Development of Magnetospheric Substorms Using Mid-Latitude Magnetic Observations," *J. Geophys. Res.*, **79**, 2811 (1974).

Craven, J. D., Y. Kamide, L. A. Frank, S.-I. Akasofu, and M. Sugiura, "Distribution of Aurora and Ionospheric Currents Observed Simultaneously on a Global Scale," in *Magnetospheric Currents*, T. A. Potemra, ed., American Geophysical Union, Washington, DC, p. 137 (1983).

de la Beaujardiere, O., R. Johnson, and V. B. Wickwar, "Ground-Based Measurements of Joule Heating Rates," *this volume*.

de la Beaujardiere, O., R. Vondrak, and M. Baron, "Radar Observations of Electric Fields and Currents Associated with Auroral Arcs," *J. Geophys. Res.*, **82**, 5051 (1977).

Egeland, A., "Kristian Birkeland: The Man and the Scientist," in *Magnetospheric Currents*, T. A. Potemra, ed., American Geophysical Union, Washington, DC, p. 1 (1984).

Feldstein, Ya. I., "Peculiarities in the Auroral Distribution and Magnetic Disturbance Distribution in High Latitudes Caused by the Asymmetrical Form of the Magnetosphere," *Planet. Space Sci.*, **14**, 121 (1966).

Feldstein, Ya. I., and G. V. Starkov, "Dynamics of Auroral Belt and Polar Geomagnetic Disturbances," *Planet. Space Sci.*, **15**, 209 (1967).

Foster, J. C., "Reply to Kamide and Richmond," *Geophys. Res. Lett.*, **14**, 160 (1987).

Foster, J. C., J. M. Holt, R. G. Musgrove, and D. S. Evans, "Ionospheric Convection Associated with Discrete Levels of Particle Precipitation," *Geophys. Res. Lett.*, **13**, 656 (1986).

Frank, L. A., J. D. Craven, J. L. Burch, and J. D. Winningham, "Polar Views of the Earth's Aurora with Dynamics Explorer," *Geophys. Res. Lett.*, **9**, 1001 (1982).

Frank, L. A., J. D. Craven, K. L. Ackerson, M. R. English, R. H. Eather, and R. L. Carovillano, "Global Auroral Imaging Instrumentation for the Dynamics Explorer Mission," *Space Sci. Instrum.*, **5**, 369 (1981).

Fuller-Rowell, T. J., and D. S. Evans, "Height-Integrated Pedersen and Hall Conductivity Patterns Inferred from the TIROS-NOAA Satellite Data," *J. Geophys. Res.*, **92**, 7606 (1987).

Gizler, V. A., B. M. Kuznetsov, V. A. Sergeev, and O. A. Troshichev, "The Sources of the Polar Cap and Low Latitude Bay-Like Disturbances During Substorms," *Planet. Space Sci.*, **24**, 1133 (1976).

Gorney, D. J., P. F. Mizera, and J. L. Roeder, "A Maximum-Entropy Technique for Deconvolution of Atmospheric Bremsstrahlung Spectra," Space Sciences Lab. Rep. SEL-86(6940-06)-06, Aerospace Corporation, Los Angeles, CA (1985).

Greenwald, R. A., W. L. Ecklund, and B. B. Balsley, "Radar Observations of Auroral Electrojet Currents," *J. Geophys. Res.*, **80**, 3635 (1975).

Heppner, J. P., "The Harang Discontinuity in Auroral Belt Ionospheric Currents," *Geofys. Publ.*, **29**, 105 (1972).

Hughes, T. J., and G. Rostoker, "A Comprehensive Model Current System for High-Latitude Magnetic Activity, 1. The Steady State System," *Geophys. J. R. Astron. Soc.*, **58**, 525 (1979).

Iijima, T., "Large-Scale Currents Connecting Polar Ionosphere and Magnetosphere," *this volume*.

Iijima, T., and T. A. Potemra, "The Amplitude Distribution of Field-Aligned Currents at Northern High Latitudes Observed by TRIAD," *J. Geophys. Res.*, **81**, 2165 (1976).

Iijima, T., and T. A. Potemra, "Large-Scale Characteristics of Field-Aligned Currents Associated with Substorm," *J. Geophys. Res.*, **83**, 599 (1978).

Kamide, Y., "On Current Continuity at the Harang Discontinuity," *Planet. Space Sci.*, **26**, 237 (1978).

Kamide, Y., "The Two-Component Auroral Electrojet," *Geophys. Res. Lett.*, **9**, 1175 (1982).

Kamide, Y., and W. Baumjohann, "Estimation of Electric Fields and Currents from International Magnetospheric Study Magnetometer Data for the CDAW 6 Intervals: Implications for Substorm Dynamics," *J. Geophys. Res.*, **90**, 1305 (1985).

Kamide, Y., and A. D. Richmond, "Comment on Ionospheric Convection Associated with Discrete Levels of Particle Precipitation," *Geophys. Res. Lett.*, **14**, 158 (1987).

Kamide, Y., and J. F. Vickrey, "Relative Contribution of Ionospheric Conductivity and Electric Field to the Auroral Electrojets," *J. Geophys. Res.*, **88**, 7989 (1983).

Kamide, Y., A. D. Richmond, and S. Matsushita, "Estimation of Ionospheric Electric Field, Ionospheric Currents, and Field-Aligned Currents from Ground Magnetic Records," *J. Geophys. Res.*, **86**, 801 (1981).

Kamide, Y., J. D. Craven, L. A. Frank, B.-H. Ahn, and S.-I. Akasofu, "Modeling Substorm Current Systems Using the Conductivity Distribution Inferred from DE Auroral Images," *J. Geophys. Res.*, **91**, 11235 (1986).

Kamide, Y., Y. Ishihara, T. L. Killeen, J. D. Craven, L. A. Frank, and R. A. Heelis, "Combining Electric Field and Aurora Observations from DE 1 and 2 with Ground Magnetometer Records to Estimate Ionospheric Electromagnetic Quantities," *J. Geophys. Res.*, **94**, in press (1989).

Kamide, Y., R. M. Robinson, S.-I. Akasofu, and T. A. Potemra, "Aurora and Electrojet Configuration in the Early Morning Sector," *J. Geophys. Res.*, **89**, 389 (1984).

Kroehl, H. W., and A. D. Richmond, "Magnetic Substorm Characteristics Described by Magnetic Potential Maps for 26–28 March 1976," in *Dynamics of the Magnetosphere*, S.-I. Akasofu, ed., Reidel, Hingham, MA, p. 269 (1980).

McPherron, R. L., "Critical Problems in Establishing the Morphology of Substorms in Space," in *Magnetospheric Physics*, B. M. McCormac, ed., Reidel, Dordrecht, Netherlands, p. 335 (1974).

Nielsen, E., "Ionosphere-Magnetosphere Mapping of Dynamic Auroral Structures During Substorms," *this volume*.

Nishida, A., and Y. Kamide, "Magnetospheric Processes Preceding the Onset of an Isolated Substorm—A Case Study of the March 31, 1978, Substorm," *J. Geophys. Res.*, **88**, 7005 (1983).

Nishida, A., "New Polar Magnetic Disturbances, S_q^p, SP, DPC, and DP 2," *Rev. Geophys. Space Phys.*, **9**, 417 (1971).

Obayashi, T., and A. Nishida, "Large-Scale Electric Field in the Magnetosphere," *Space Sci. Rev.*, **8**, 3 (1968).

Pellinen, R. J., W. Baumjohann, W. J. Heikkila, V. A. Sergeev, A. G. Yahnin, G. Marklund, and A. O. Melnikov, "Event Study of Pre-Substorm Phases and Their Relation to Energy Coupling Between Solar Wind and Magnetosphere," *Planet. Space Sci.*, **30**, 371 (1982).

Pytte, T., R. L. McPherron, M. G. Kivelson, H. I. West, Jr., and E. W. Hones, Jr., "Multiple-Satellite Studies of Magnetospheric Substorms: Plasma Sheet Recovery and the Poleward Leap of Auroral Zone Activity," *J. Geophys. Res.*, **83**, 5256 (1978).

Rees, M. H., "Auroral Excitation and Energy Dissipation," in *Solar-Terrestrial Physics*, R. L. Carovillano and J. M. Forbes, eds., Reidel, Hingham, MA, p. 753 (1983).

Richmond, A. D., "The Ionosphere," in *The Solar Wind and the Earth*, S.-I. Akasofu and Y. Kamide, eds., Terra/Reidel, Tokyo, p. 123 (1987).

Rostoker, G., "Triggering of Expansive Phase Intensifications of Magnetospheric Substorms by Northward Turnings of the Interplanetary Magnetic Fields," *J. Geophys. Res.*, **88**, 6981 (1983).

Rostoker, G., M. Mareshal, and J. C. Samson, "Response of Dayside Net Downward Field-Aligned Current to Changes in the Interplanetary Magnetic Field and to Substorm Perturbations," *J. Geophys. Res.*, **87**, 3489 (1982).

Rostoker, G., Y. Kamide, and J. D. Winningham, "Energetic Particle Precipitation into the High-Latitude Ionosphere and the Auroral Electrojets, 3. Characteristics of Electron Precipitation into the Morning Auroral Oval," *J. Geophys. Res.*, **90**, 7495 (1985).

Röttger, J., "Incoherent Scatter Observations of the Auroral Ionosphere with the EISCAT Radar Facility," *this volume*.

Senior, C., R. M. Robinson, and T. A. Potemra, "Relationship Between Field-Aligned Currents, Diffuse Auroral Precipitation and the Westward Electrojet in the Early Morning Sector," *J. Geophys. Res.*, **87**, 10469 (1982).

Silsbee, H. C., and E. H. Vestine, "Geomagnetic Bays, Their Frequency, and Current System," *Terr. Magn. Electri.*, **47**, 195 (1942).

Spiro, R. W., P. H. Reiff, and L. J. Maher, "Precipitating Electron Energy Flux and Auroral Zone Conductivities—An Empirical Model," *J. Geophys. Res.*, **87**, 8215 (1982).

Sugiura, M., "Sydney Chapman and His Early Study of Magnetic Disturbances," in *Magnetospheric Currents*, T. A. Potemra, ed., American Geophysical Union, Washington, DC, p. 17 (1984).

Troshichev, O. A., B. M. Kuznetsov, and M. I. Pudovkin, "The Current Systems of the Magnetic Substorm Growth and Explosive Phase," *Planet. Space Sci.*, **22**, 1403 (1974).

Vondrak, R. R., "Incoherent Scatter Radar Measurements of Electric Field and Plasma in the Auroral Ionosphere," in *High Latitude Space Plasma Physics*, B. Hultqvist and T. Hagfors, eds., Plenum Press, New York, p. 73 (1983).

VII-2. LARGE-SCALE CURRENTS CONNECTING THE POLAR IONOSPHERE WITH THE MAGNETOSPHERE

T. Iijima*

Basic features of the three-dimensional electric current systems connecting the polar ionosphere with the magnetosphere have been determined from the large-scale characteristics of geomagnetic disturbances observed on the ground and above the ionosphere. They include the following:

1. Large-scale region 1, region 2, and northward interplanetary magnetic field (NBZ) Birkeland current systems.
2. The region 1 and region 2 Birkeland current systems persist quasi-permanently in the auroral-belt domain. Their latitudinal extent, current densities and intensities increase with the increase of the level of auroral electrojet activity. In particular, the density of region 1 Birkeland currents increases as the southward component of the interplanetary magnetic field (IMF) increases.
3. Ionospheric currents associated with these region 1 and region 2 Birkeland current systems are made manifest as the equivalent overhead current systems, derived from ground-based geomagnetic disturbances, with the westward current flowing in the morningside auroral belt, the eastward current flowing in the eveningside auroral belt, and the sunward current flowing in the central polar cap.
4. When the IMF turns and remains strongly northward, the NBZ Birkeland current system develops over the polar cap, poleward of the persisting region 1 and region 2 Birkeland current systems. The intensity of the NBZ Birkeland current system on the dayside increases with the increases of the IMF in the northward direction.
5. The NBZ Birkeland current system manifests the equivalent overhead current system with the antisunward current flowing across the central polar cap.
6. The region 1, region 2, and NBZ Birkeland current systems become weakest for vanishingly small IMF conditions.

1. INTRODUCTION

Polar geomagnetic disturbances observed on the ground are represented by their equivalent overhead current systems. These have been used to infer the ionospheric currents that actually flow in the polar ionosphere. The actual ionospheric currents are connected to magnetospheric currents through field-aligned currents (Birkeland currents) that flow along the geomagnetic field lines. The Birkeland currents have been observed primarily through the associated transverse magnetic disturbances that they produce above the ionosphere.

2. BASIC PATTERNS OF POLAR GEOMAGNETIC DISTURBANCES

Polar geomagnetic disturbances observed on the ground have been represented by their equivalent overhead current systems. These equivalent current systems basically conform to two distinctive patterns that depend primarily on the concurrent north-south component of the interplanetary magnetic field (IMF) B_z. One is an ordinary two-cell current system, with the current flowing sunward in the central polar cap, which is associated with a southward IMF B_z and which is

most active in the auroral belt. The other is a reversed two-cell current system, with the current flowing antisunward in the central polar cap, that is associated with a northward IMF B_z and is most active in the polar cap (poleward of auroral belt).

Southward IMF B_z, Case

Figure 1 (from *Iijima* [1973]) summarizes the characteristics of polar geomagnetic disturbances from 0000 from 2400 UT on June 20, 1970. That was a day of consistently and strongly southward IMF (i.e., the average hourly-value of the IMF is $B = 7.2$ nT, $B_x = 1.2$ nT, $B_y = -2.9$ nT, and $B_z = -5.8$ nT, respectively). The top panel represents the activity of the eastward auroral electrojet (AU), the westward auroral electrojet (AL), along with the asymmetric development of H-component disturbances at middle and low latitudes (Asy). The auroral electrojet activity continued over a full day on June 20, 1970, with a nearly constant magnitude of several hundred nanoteslas (i.e., average hourly-value of AU and AL is 272 nT and -400 nT, respectively). It has also been confirmed [*Iijima,* 1973] that the horizontal components of geomagnetic disturbance (north component X_m and east component Y_m) on this day appeared to be mainly a regular diurnal variation in the so-called polar cap region, at geomagnetic latitudes (ML) exceeding 80°. The bottom panel represents an equivalent overhead current system that was constructed from the geomagnet-

---

*Geophysics Research Laboratory, Faculty of Science, The University of Tokyo, Tokyo 113, Japan.

Auroral Physics, edited by C.-I. Meng, M. J. Rycroft and L. A. Frank. © Cambridge UP 1991

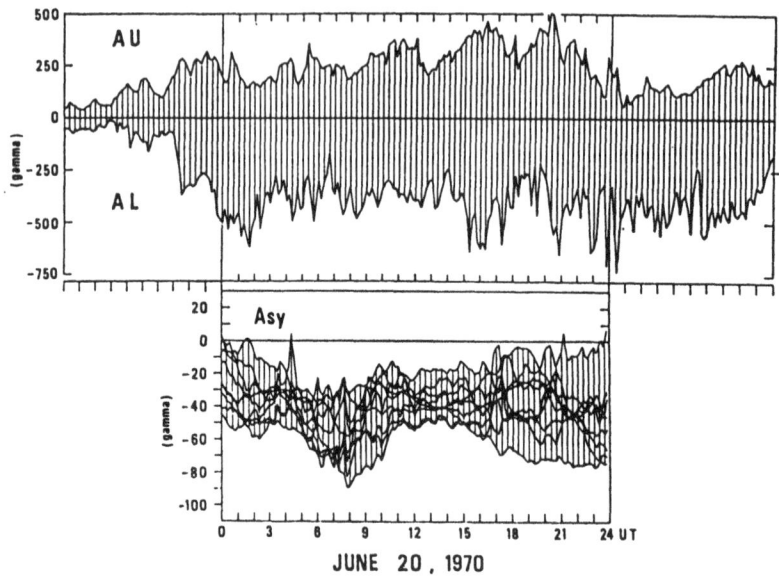

BASE OF POLAR MAGNETIC DISTURBANCE

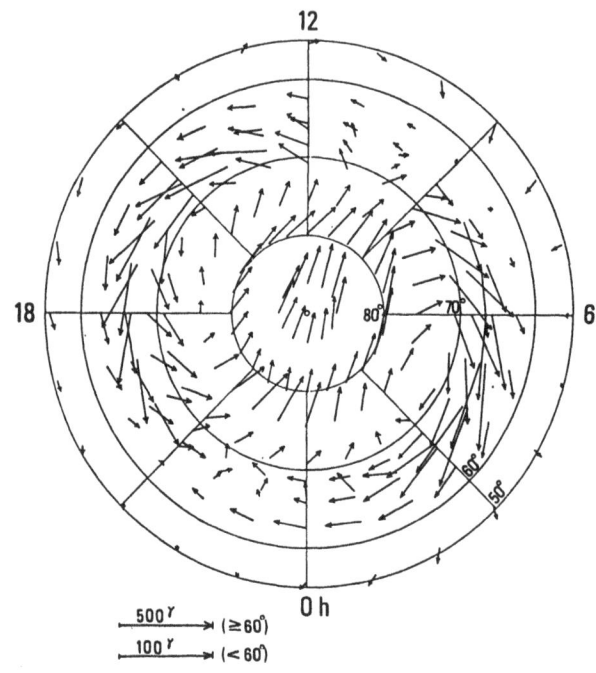

Figure 1—*Upper panel*: *AU* and *AL* values derived from 11 stations, giving the general level of auroral-zone magnetic disturbances, and the Asy showing the asymmetric development of middle- and low-latitude magnetic disturbances obtained by superimposing the H-component perturbations at 9 stations. *Lower panel*: Equivalent overhead current vectors for the polar geomagnetic disturbances on June 20, 1970, associated with stably and strongly southward IMF, in a polar map using corrected geomagnetic latitude and geomagnetic local time.

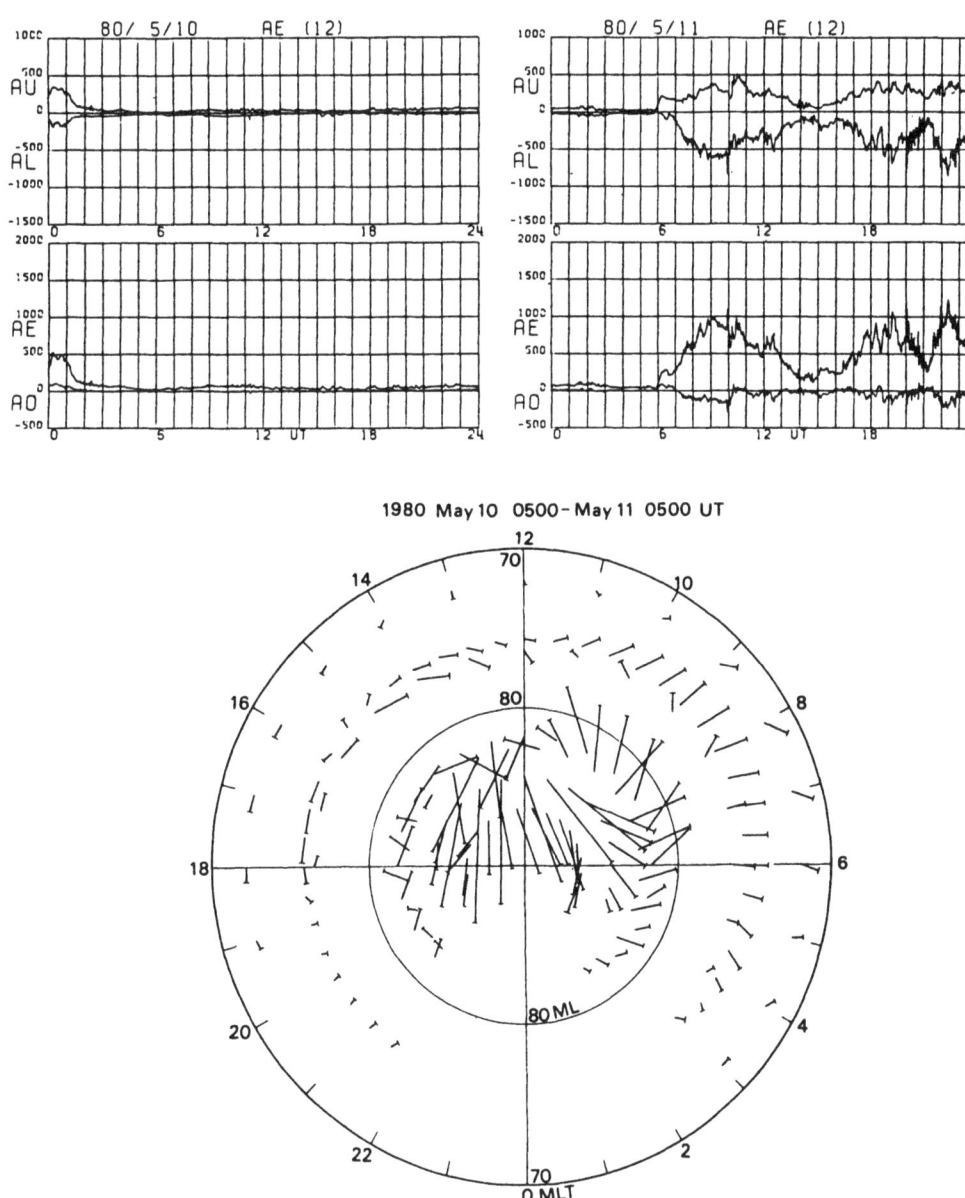

Figure 2—*Upper panel*: *AU* and *AL* values derived from 12 auroral-zone stations. *Lower panel*: Equivalent overhead current vectors on 0500 UT, May 10–0500 UT, May 11, 1980, associated with stably and strongly northward IMF. Horizontal magnetic disturbances at Alert (86.3° eccentric dipole latitude), Thule (84.3°), Resolute (82.6°), Mould Bay (81.9°), Cambridge Bay (76.7°), Godhavn (75.8°), and Baker Lake (72.2°) were used. The base-line values were defined as average values between 2100 and 0300 MLT at each station.

ic disturbances observed over a full day at 29 observatories distributed at high latitudes (ML > 50°) in the northern hemisphere. The geomagnetic effect of the hypothetical magnetospheric ring current was first subtracted (by the formula Dst (ML) = Dst (equator) × cos (ML), based upon the Dst value derived from the low-latitude stations of average ML ~34.3°). The horizontal components of the geomagnetic disturbance were then subjected to Fourier harmonic analysis. The harmonic vectors that were obtained at 29 stations were grouped onto nine latitude circles and were averaged over each circle. The first four harmonics of the horizontal components were used to construct this equivalent current system. The equivalent current system exhibits a two-cell flow pattern, with the currents flowing sunward in the central polar cap and enhanced in the auroral zone at ML ~66–69°.

Northward IMF B_z, Case

Figure 2 (from *Iijima* [1984]) summarizes the characteristics of polar geomagnetic disturbances from 0500 UT, May 10,1980 to 0500 UT, May 11, 1980 that were observed during a period of consistently and strongly northward IMF B_z (i.e., the average hourly-value of IMF is $B = 13.6$ nT, $B_x = 0.6$ nT, $B_y = -0.4$ nT, $B_z = 11.9$ nT, respectively). The auroral electrojet activity (upper panel) was persistently calm over these 24 hours (i.e., average hourly-value of AU and AL is 38 nT and -20 nT, respectively). The equivalent overhead current system (lower panel) was obtained from the geomagnetic disturbances observed at seven high-latitude observatories (ML > 70°) in the northern hemisphere by the same method as mentioned above. In contrast to the former case, the equivalent current system associated with strongly northward IMF B_z exhibits a reversed two-cell flow pattern, with the currents flowing antisunward in the central polar cap. The currents are largest at the highest latitudes (ML > 83°), particularly in the daytime hemisphere.

In the ionosphere, Pedersen currents and Hall currents flow because of the motions of ions and electrons under the electric field and their collisions with neutral particles. The equivalent overhead current system, considered to flow in a horizontal plane, is the divergence-free part of the actual ionospheric currents that cause the magnetic disturbances that are observed on the ground. The actual ionospheric currents are driven by large-scale Birkeland currents and an appropriate distribution of ionospheric Pedersen and Hall conductivities is required. For the case of Figure 1, the following model is relevant [*Crooker and Siscoe*, 1981]. The ionospheric conductivities are specially enhanced in the auroral belt. Birkeland currents exist in the auroral belt, with currents flowing into the ionosphere on the poleward edge of the auroral belt and flowing away from the ionosphere on the equatorward edge

of the auroral belt on the morning side, and with the currents flowing in opposite directions on the poleward and equatorward edges of the auroral belt on the eveningside. As for the equivalent current system in Figure 2, the model primarily requires the Birkeland currents flowing away from the ionosphere in the morningside polar cap and flowing into the ionosphere in the afternoonside polar cap region.

3. BASIC PATTERNS OF LARGE-SCALE BIRKELAND CURRENT SYSTEMS

It is now widely accepted that Birkeland currents play an essential role in transferring the plasma and electric field disturbances between the magnetosphere and the ionosphere. A number of studies have presented a variety of characteristics associated with Birkeland currents [e.g., *Potemra,* 1988]. For example, from studies using the magnetic field measurements made by polar-orbiting satellites (i.e., the Triad satellite at an altitude range of ~800 km and the MAGSAT satellite in the altitude range from ~350–550 km), the concepts of the so-called region 1 and region 2 Birkeland current systems in the auroral oval and the NBZ Birkeland current system in the polar cap have been presented to serve as a starting point for formulating the large-scale characteristics of Birkeland currents [e.g., *Iijima and Potemra,* 1976; *Iijima et al.,* 1984]. These Birkeland current systems depend basically on the associated IMF B_z component. During a period of southward IMF B_z, the dominant constituents of the Birkeland currents are the region 1 and region 2 Birkeland current systems in the auroral oval, whereas, for northward IMF B_z, the NBZ Birkeland current system in the polar cap becomes the dominant part, especially on the dayside.

Region 1 and Region 2 Birkeland Current Systems.

Figure 3 (from *Iijima and Potemra* [1978]) summarizes the characteristics of large-scale Birkeland currents over the northern hemisphere that were determined from the data of 439 Triad orbits for quiet conditions ($|AL| < 100$ nT) and 366 Triad orbits for active substorm periods ($|AL| \geq 100$). At all MLTs, Birkeland currents flow into or away from the high latitude ionosphere over a magnetic latitude range greater than 0.5° at an altitude of ~800 km. The current density is larger than 0.25 μA m^{-2}. The top diagram shows the average spatial distribution of the directions of the large-scale Birkeland currents. Below are the corresponding Birkeland current densities plotted against geomagnetic dipole local time. The principal characteristics include the following:

1. Large-scale Birkeland currents persist during relatively quiet conditions ($|AL| < 100$ nT) and during active substorm periods ($|AL| \geq 100$ nT).

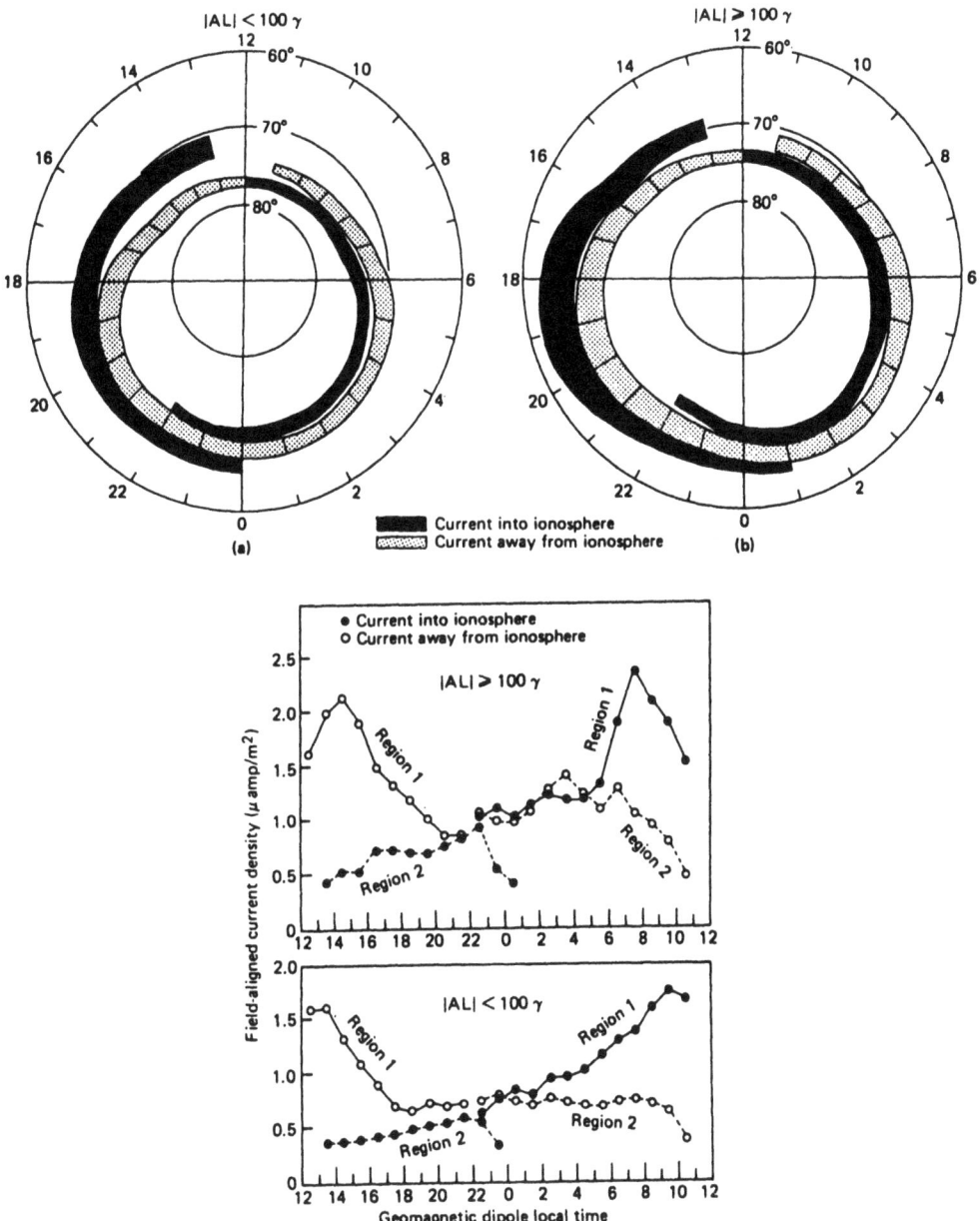

Figure 3—*Upper panel*: A summary of the distribution and flow directions of large-scale Birkeland currents determined from (*a*) data obtained from 439 passes of Triad during weakly disturbed conditions ($|AL| < 100$ nT), and (*b*) data obtained from 366 Triad passes during active periods ($|AL| \geq 100$ nT) with the current density larger than 0.25 μA m^{-2}. *Lower panel*: Diurnal distribution of Birkeland current densities for the upper diagram during active periods ($|AL| \geq 100$ nT) (*upper block*) and during weakly disturbed periods ($|AL| < 100$ nT) (*lower block*).

405

Region 1 currents are located near the poleward boundary of the Birkeland current region, with region 2 currents being located at the equatorward boundary. Birkeland currents flow into region 1 and away from region 2 in the morning sector (~2300–1100 MLT). The current flow is directed away from region 1 and into region 2 in the evening sector (~1300–2300 MLT). The Birkeland current regions appear to overlap in the midnight sector (~2100–0100 MLT). Here there are three adjacent regions, with current flow away from the auroral ionosphere and, on both equatorward and poleward sides of this region, with current flow into the ionosphere.

2. During active substorm periods ($|AL| \geq 100$ nT), the average latitudinal widths of the region 1 and region 2 currents increase by factors of 1.2–1.3. The centers of these regions shift equatorward by 2–3° in comparison with their quiet time positions($|AL| < 100$ nT).

3. The intensity and density of the Birkeland currents in region 1 are statistically larger than the currents in region 2 during all substorm phases, and in all MLT sectors except the midnight sector. In the midnight sector, the densities of region 1 and region 2 currents are nearly equal. The region 2 Birkeland currents are sometimes larger than the region 1 currents during active substorm conditions.

4. The largest net Birkeland currents are located near regions of largest region 1 Birkeland currents. These are at ~1300 MLT, with a net current flow away from the ionosphere equal to ~1.1 μA m^{-2}, and near 1000–1100 MLT with a net current flow into the ionosphere and equal to 1.1 μA m^{-2} during relatively quiet conditions ($|AL| < 100$ nT). During active periods ($|AL| \geq 100$ nT), the regions of largest region 1 Birkeland currents shift toward the nightside. The peak current region near 1000–1100 MLT shifts to 0700–0800 MLT, and the peak region near 1300 MLT shifts to ~1400–1500 MLT. There is no distinct peak in the Birkeland currents in region 2 during relatively quiet conditions ($|AL| < 100$ nT). During active periods the region 2 Birkeland currents show the largest density increase in the midnight sector, and the densities of the region 2 currents actually exceed the densities of the region 1 currents in the early morning sector.

5. The total current (summed over all regions) flowing into the ionosphere is equal to the total current flowing away from the ionosphere (within ~10% of the total amount) for all levels of substorm activity. Therefore current continuity, with

respect to the entire large-scale Birkeland current system, is preserved for a wide range of magnetospheric conditions. The total current values are $(2.5–2.8) \times 10^6$ A for $|AL| < 100$ nT, and $(4.9 - 5.4) \times 10^6$ A for $|AL| \geq 100$ nT. The total currents in region 1 are noticeably larger than the total currents in region 2 during relatively quiet conditions. During active periods the total current in region 2 becomes comparable with and sometimes exceeds the total current in region 1.

6. The density of Birkeland currents in region 1 is closely correlated with the general level of polar geomagnetic disturbance activity and persists even during quiet conditions. In fact, the densities of the dayside region 1 Birkeland currents are well correlated with an increase of the IMF in the southward direction, and in particular with a functional relationship of the form $(B_y^2 + B_z^2)^{1/2} \sin(\theta/2)$, where θ is the angle measured from the positive z direction to the IMF vector in the y-z plane in solar magnetospheric coordinates [*Iijima and Potemra,* 1982]. The density of Birkeland currents in region 2 is related to the local intensity of the auroral zone (electrojet) magnetic disturbances.

7. Ionospheric currents associated with these region 1 and region 2 Birkeland current systems during active substorm periods ($|AL| \geq 100$ nT) are related to the equivalent overhead current system derived from ground-based geomagnetic disturbances.

NBZ Birkeland Current System.

Figure 4 (from *Iijima and Shibaji* [1987]) summarizes the characteristics of large-scale Birkeland currents over the southern summer hemisphere corresponding to strongly northward IMF conditions. Here it is considered that Birkeland currents flow into and away from the ionosphere over a scale length larger than ~57 km in the altitude range from ~350–550 km, and the absolute values of their densities are larger than 0.05 μA m^{-2}. The data of 30 MAGSAT orbits were used, which covered the polar region between ~83° ML at noon and ~71° ML at midnight under the following IMF conditions: IMF $B_z > B_y > 0$, with individual B_y values being $5 > B_y > 0$ nT, individual B_z values being $8 > B_z > 2$ nT, and the average values of B_y and B_z being 2.6 nT and 4.8 nT, respectively. The top diagram shows the spatial distribution of the directions of the Birkeland currents. Below is the corresponding contour map of constant Birkeland current density, at ± 0.1 μA m^{-2} increments. The principal characteristics include the following:

1. The basic features of large-scale Birkeland currents associated with strongly northward IMF B_z

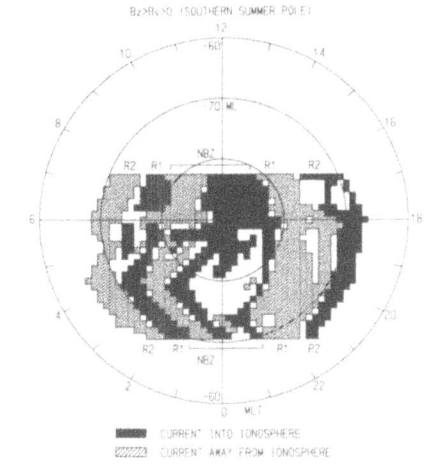

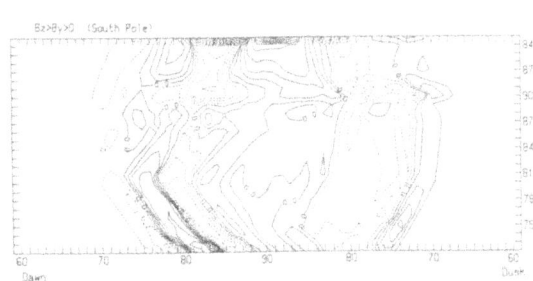

Figure 4—*Upper panel*: Spatial distribution and flow directions of large-scale Birkeland currents determined from data obtained from 30 MAGSAT passes during periods when the IMF $B_z > B_y > 0$ (average values: $B_y \sim 2.6$ nT, $B_z \sim 4.8$ nT) with the current density being larger than 0.05 μA m^{-2}. *Lower panel*: Contour maps of the Birkeland current density for the upper diagram at ± 0.1 μA m^{-2} increments. The trough and peak of the current densities are -1.7 and 1.0 μA m^{-2} located on the dayside polar cap.

are (1) the ordinary region 1 and region 2 Birkeland current systems, and (2) the polar-cap Birkeland current system flowing away from the ionosphere on the morningside (poleward of the region 1 system and with the current flowing in the opposite direction), and flowing into the ionosphere on the afternoonside (poleward of the region 1 system there and also in the opposite direction). This polar-cap Birkeland system has been referred to as the "NBZ" Birkeland current system for "northward B_z" [*Iijima et al.,* 1984].

2. The global characteristics appear to be different on the dayside and nightside. The dayside is characterized by a large NBZ Birkeland current system, which dominates the entire polar cap and is flanked by a well-defined region 1 Birkeland current system and a weaker region 2 Birkeland current system. The current densities of these dayside NBZ and region 1 systems show localized

maxima on the dayside and tend to diminish from the noon toward dawn and dusk.

3. The nightside part is characterized by a distinctive NBZ Birkeland current system that develops along with definite region 1 and region 2 Birkeland current systems. The nightside NBZ system is composed of multiple current belts that run from midnight toward both the dawn and dusk (parallel with, and poleward of, the adjoining region 1 system). The current densities of these nightside NBZ region 1 and region 2 systems also exhibit localized maxima on the nightside.

4. In the intermediate region between the localized dayside maxima and the localized nightside maxima of Birkeland current densities, the Birkeland currents are highly structured.

5. The average current density and total current of these large-scale Birkeland currents are summarized in Table 1.

6. The NBZ Birkeland current system in the polar cap intensifies and is more stable as the IMF B_z becomes more northward (reminiscent of the behavior of the region 1 Birkeland current system with increasing southward values of IMF B_z). The intensities of the dayside NBZ Birkeland currents are correlated with an IMF functional relationship of the form $(B_y^2 + B_z^2)^{1/2} \cos (\theta/2)$, where θ is the same IMF angle as defined in the previous section [*Iijima et al.,* 1984].

7. The existence of these stable NBZ Birkeland currents and the correlation of their amplitudes with the IMF show that a considerable amount of energy continues to flow into the earth's polar region even during periods of strongly northward IMF.

8. The NBZ Birkeland current system relates to Figure 2 showing the equivalent overhead current system derived from ground-based magnetograms.

4. CONCLUDING REMARKS

The basic features of the three-dimensional electric current systems connecting the polar ionosphere and the magnetosphere have been determined from the large-scale characteristics of geomagnetic disturbances observed on the ground and above the ionosphere. The most important features are the region 1, region 2, and NBZ Birkeland current systems. These currents are small when the IMF is small.

The region 1 and region 2 Birkeland current systems persist quasi-permanently in the auroral oval and increase their latitudinal extents, current densities and intensities as the general level of auroral electrojet activity increases. In particular, the density of Birkeland currents in region 1 is well correlated with the component of the IMF in the southward direction.

TABLE 1. Average current density and total current of large-scale Birkeland currents for IMF $B_z > B_y > 0$

Parameter	Morningside			Eveningside		
	Upward region 2	Downward region 1	Upward NBZ	Downward NBZ	Upward region 1	Downward region 2
Current density, μA m^{-2}						
Daytime zone	−0.12	0.21	−0.41	0.36	−0.17	0.09
Intermediate zone	?	?	?	0.10	−0.14	0.10
Nighttime zone	−0.12	0.28	−0.21	0.10	−0.13	0.17
Total current, 10^5A						
Daytime zone	−0.73	0.87	−3.11	4.17	−1.51	0.56
Intermediate zone	?	?	?	0.89	−0.83	0.29
Nighttime zone	−1.14	2.04	−1.59	1.60	−2.29	1.16

A positive value denotes current flow into the ionosphere; a negative value denotes current flow away from the ionosphere.

When the IMF turns and remains strongly northward, the NBZ Birkeland current system develops over the polar cap, poleward of the persisting region 1 and region 2 Birkeland current systems. The intensity of the NBZ Birkeland current system on the dayside is well correlated with the component of the IMF in the northward direction.

The large-scale Birkeland currents transfer momentum down the magnetic flux tubes so that large-scale plasma convection is maintained. The generation of the Birkeland currents requires the appropriate plasma stresses. The region 1 Birkeland currents develop so as to transfer the antisunward plasma stress down to the polar ionosphere. This antisunward stress is thought to be acquired in the magnetospheric low-latitude boundary layer and the plasma sheet high-latitude boundary layer [e.g., *Vasyliunas*, 1979]. The region 1 Birkeland currents also drive the sunward return plasma flow in the inner magnetosphere from the nightside to the dayside, presumably by the hot plasma pressure. The region 2 Birkeland currents are thought to be generated, and coupled with the region 1 system, to transfer the sunward plasma stress down to the auroral ionosphere. The region 2 Birkeland currents also transfer the outward-directed stress, which develops to shield the inner magnetosphere from the further intrusion of plasma [e.g., *Wolf*, 1974; *Southwood*, 1977].

The NBZ Birkeland currents are thought to flow so as to transfer the sunward plasma stress down to the polar cap ionosphere so that large-scale sunward plasma convection is maintained in the central polar cap. This arises from the interplanetary medium in the magnetotail high-latitude boundary layer, tailward of the polar cusp [e.g., *Kan and Burke*, 1985].

REFERENCES

Crooker, N. U., and G. L. Siscoe, "Birkeland Currents as the Cause of the Low-Latitude Asymmetric Disturbance Field," *J. Geophys. Res.* **86**, 11201 (1981).

Iijima, T., "Enhancement of the S_q^p Field as the Basic Component of Polar Magnetic Disturbance," *Rept. Ionos. Space Res. Japan*, **27**, 199 (1973).

Iijima, T., "Field-Aligned Currents During Northward IMF," in *Magnetospheric Currents*, Geophysical Monograph 28, T. A. Potemra, ed., American Geophysical Union, Washington, DC, pp. 115–122 (1984).

Iijima, T., and T. A. Potemra, "The Amplitude Distribution of Field-Aligned Currents at Northern High Latitudes Observed by TRIAD," *J. Geophys. Res.*, **81**, 2165 (1976).

Iijima, T., and T. A. Potemra, "Large-Scale Characteristics of Field-Aligned Currents Associated with Substorms," *J. Geophys. Res.*, **83**, 599 (1978).

Iijima, T., and T. A. Potemra, "The Relationship Between Interplanetary Quantities and Birkeland Current Densities," *Geophys. Res. Lett.*, **9**, 442 (1982).

Iijima, T., and T. Shibaji, "Global Characteristics of Northward IMF-Associated (NBZ) Field-Aligned Currents," *J. Geophys. Res.*, **92**, 2408 (1987).

Iijima, T., T. A. Potemra, L. J. Zanetti, and P. F. Bythrow, "Large-Scale Birkeland Currents in the Dayside Polar Region During Strongly Northward IMF: A New Birkeland Current System," *J. Geophys. Res.*, **89**, 7441 (1984).

Kan, J. R., and W. J. Burke, "A Theoretical Model of Polar Cap Auroral Arcs," *J. Geophys. Res.*, **90**, 4171 (1985).

Potemra, T. A., "Birkeland Currents in the Earth's Magnetosphere," *Astrophys. Space Sci.*, **144**, 155 (1988).

Southwood, D. J., "The Role of Hot Plasma in Magnetospheric Convection," *J. Geophys. Res.*, **82**, 5512 (1977).

Vasyliunas, V. M., "Interaction Between the Magnetospheric Boundary Layers and the Ionosphere," *Proc. Magnetospheric Boundary Layers, European Space Agency Special Publication* 148, p. 387 (1979).

Wolf, R. A., "Calculations of Magnetospheric Electric Fields," in *Magnetospheric Physics*, B.M. McCormac, ed., D. Reidel, Hingham, MA, pp. 167–178 (1974).

VII-3. IONOSPHERE-MAGNETOSPHERE MAPPING OF DYNAMIC AURORAL STRUCTURES DURING SUBSTORMS

E. Nielsen*

The substorm onset region in the ionosphere mapped to the (nominal) magnetospheric equatorial plane coincides closely with McIlwain's Injection Boundaries, which are located near the Earth on dipolar field lines. A good agreement is obtained both for low and high magnetic activity periods. A typical location of the substorm onset in the ionosphere is the Harang discontinuity, which just prior to onset surges equatorward. It is suggested that this equatorward surge is associated with the earthward surge of the inner edge of the plasma sheet. It is thought that this surge is caused by the formation of a neutral line in the magnetospheric tail, in which case the observations suggest the possibility that neutral line formation may trigger the substorm phenomena in the near-Earth magnetosphere. Substorm onsets can also occur on the equatorward side of the eastward electrojet, which is further evidence that physical processes in the near-Earth magnetosphere can play an important role in the substorm process. It is an important observation that during the substorm expansion phase the activated arc is moving poleward relative to the surrounding plasma, i.e., the arc is not $\mathbf{E} \times \mathbf{B}$ drifting. That result should serve as an important test of theoretical substorm models.

1. INTRODUCTION

Substorm onset characteristics as seen from the ground are reexamined to determine the regions in the magnetosphere where associated processes occur. Directly related to the ground-based observations of the onset of substorms are large energy releases in the magnetosphere.

We will concentrate mainly on two ground-based experiments: the coherent radar system, Scandinavian Twin Auroral Radar Experiment (STARE) [*Nielsen*, 1982], and a rio-imager, which is a riometer system with good spatial resolution [*Nielsen*, 1980]. An important aspect of the STARE experiment is that it can be used to obtain estimates of the ionospheric electron drift velocity [*Nielsen and Whitehead*, 1983; *Nielsen and Schlegel*, 1983, 1985]. However, the main point of the STARE system is that it allows the drifts to be measured over a large area with good spatial and temporal resolution. At a given time the system measures the spatial variations of the ionospheric flows and measures these spatial variations as a function of time. The data therefore allow a study of the temporal and spatial variations of the plasma flow patterns in the ionosphere.

While the STARE system measures the ionosphere flows, the rio-imager system measures effects caused by the precipitation of energetic charged particles. The system has unique properties that are similar to those of the STARE system. The rio imager system measures the spatial variations of particle precipitation at a given time and determines how this spatial variation changes as a function of time. This is achieved by measuring cosmic radio wave absorption in several narrow

beams pointing in different directions. The location of an arc, for example, can be determined with such a system, and its motion can be deduced from the variations of the absorption from one beam to the next. The system is particularly well suited to study the dynamics of particle precipitation as it changes as a function of longitude, latitude, and time.

We are particularly interested in determining the location of the substorm onset region, i.e., we will determine the location of the auroral arc that is activated at substorm onset relative to the ionospheric flows and energetic particle precipitation, and relative to magnetospheric features. In order to determine the physical processes that give rise to particle precipitation and the onset of ionospheric currents, it is essential to examine which regions in the magnetosphere are magnetically connected to particular phenomena in the ionosphere.

We have used electron drift-velocity measurements of large spatial coverage and good spatial and temporal resolution to show that the substorm onset location in the ionosphere is magnetically conjugate to the injection boundary [*McIlwain*, 1974] and to the inner edge of the plasma sheet [*Vasyliunas*, 1968]. This result is based on mapping ionospheric regions from the magnetosphere using a magnetic field line model [*Mead and Fairfield*, 1975]. Thus the result is subject to the usual limitations associated with the use of magnetic field line models on specific events. However, we argue that the close spatial relationship between these features may be real, in particular when it is considered that the boundaries in the magnetosphere are determined from observations in the magnetosphere. The good relationship to the features mapped into the ionosphere may therefore not be coincidental. Furthermore, it will also

*Max-Planck-Institut für Aeronomie Lindau, D-3411 Katlenburg-Lindau, Federal Republic of Germany.

be shown that the dynamics of the discontinuity are comparable with the dynamics of the inner edge of the plasma sheet, i.e., the speed at which the plasma edge moves earthward (observed in space [*Vasyliunas*, 1968]) is comparable with the speed of the equatorward surge of the discontinuity in the ionosphere.

Thus, taken at face value, the ground-based evidence indicates that the brightening arc at substorm onset is connected magnetically to the near-Earth magnetosphere (5–8R_E), probably on dipole(like) field lines, and that poleward expansion, the onset of energetic, intense particle precipitation and westward electrojet activity are all consequences of processes in that region. We have no evidence of activity of any kind poleward of the bright poleward-moving arc. There is no obvious reason, therefore, that processes tailward of the active region at the inner edge of the plasma sheet play a direct role in creating the phenomena observed from the ground. Processes in the tail may, however, have

had the effect of triggering events in the near-Earth magnetosphere.

2. OBSERVATIONS AND RESULTS

The substorm onset region is often present in the field of view of the STARE coherent radar system as shown by *Nielsen and Greenwald* [1979]. The large spatial coverage (200,000 km^2) and good spatial and temporal resolution (20 km $\times$ 20 km and 60 s, respectively), make these data particularly well suited to study the exact location of the substorm onset as well as its location relative to other phenomena, as for example auroras, and to study the ionospheric flows during the substorm expansion phase.

To determine the magnetospheric regions that are magnetically connected to specific ionospheric features, a magnetic field line model is used to map between the ionosphere and the magnetosphere. The magnetic field models used in the mapping are the quiet (MF73Q) and

STARE
FIELD OF VIEW PROJECTED ON TO
EQUATORIAL PLANE IN SM-COORDINATE
SYSTEM

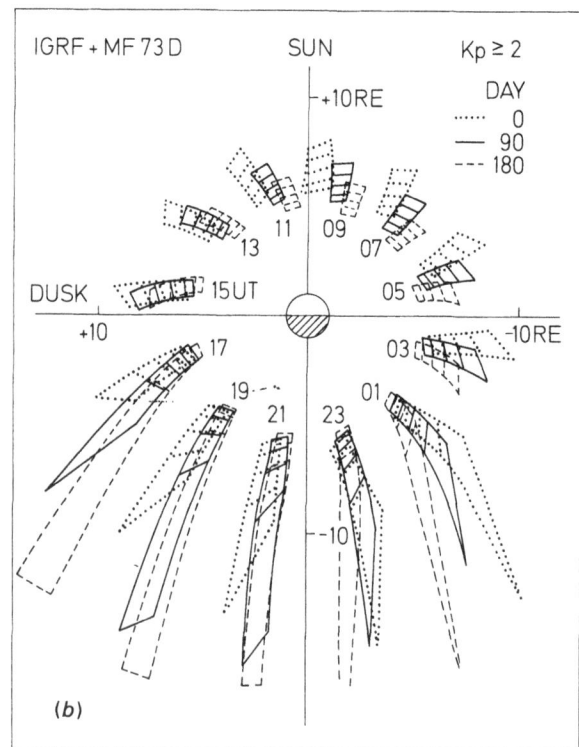

Figure 1—The field of view of the STARE system is projected onto the magnetic equatorial plane using the Mead-Fairfield model of the geomagnetic field. That equatorial plane is here defined as the collection of the points on each geomagnetic field line at the greatest distance from the Earth. The projection is in solar-magnetospheric (SM) coodinates. (*a*) Mapping at two different levels of magnetic activity; (*b*) mapping for different times of the year at higher magnetic activity.

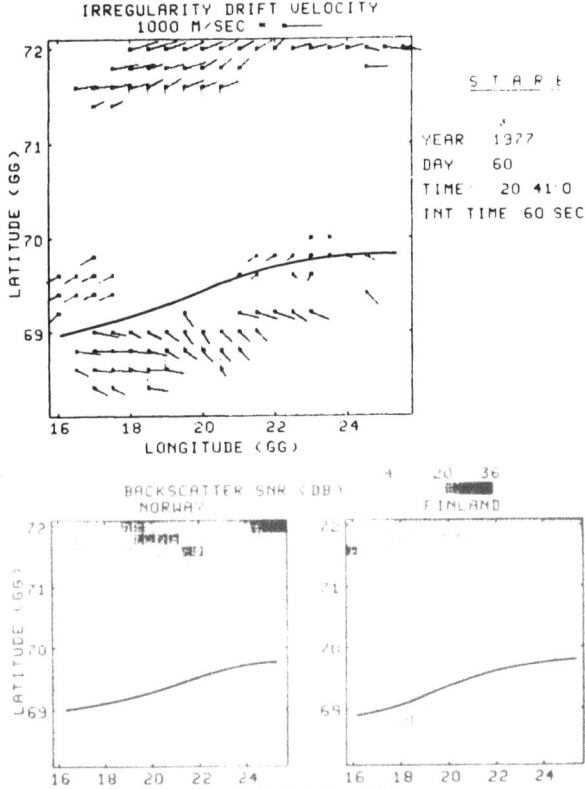

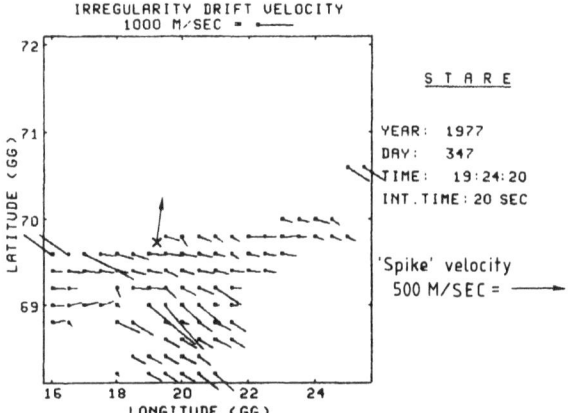

Figure 3—On the poleward border of the westward electrojet (eastward electron flows) was located a poleward expanding active arc. It has been shown to be colocated with an arc of energetic particle precipitation, observed with a rio-imager system. The velocity of the precipitation region in the ionosphere is shown as an arrow. Clearly this velocity is different from the electron flow velocities, indicating that the active arc is moving relative to the surrounding plasma. (From *Nielsen* [1980].)

Figure 2—The STARE field of view with estimates of the electron drift velocity vectors during a substorm expansion. The solid curve is the lower border of an active arc. The arc activated on the equatorward side of an eastward electrojet (westward electron flows) and left a westward electrojet (eastward electron flows) in its wake as it expanded poleward. (From *Nielsen and Greenwald* [1978].)

disturbed (MT73D) models of *Mead and Fairfield* [1975]. The Mead-Fairfield models are chosen because they are closely linked to measurements of magnetic fields in the magnetosphere and contain few assumptions. The acuracy and limitations of the models were discussed by *Mead and Fairfield* [1975] and by *Fairfield and Mead* [1975]. Figure 1 shows the STARE field of view mapped to the magnetic equatorial plane for different times of the day and the year, and for different levels of magnetic activity. The effects of tailward magnetic field lines during high magnetic activity can be clearly seen.

Figure 2 shows the STARE field of view when a substorm is in progress. The lower border of an active auroral arc is shown together with electron drift velocities in the ionosphere. The quiet arc, which eventually was activated, was initially located at the equatorward border of an eastward electrojet. The remains of this eastward current system can still be seen just poleward of the arc. When the arc activated, it

started a poleward motion through the eastward electrojet, leaving westward currents in its wake. This is the situation illustrated.

With a rio-imager experiment it has been shown that the active arc is colocated with an arc of energetic particle precipitation, and the velocity of the precipitation region could be determined. An example is given in Figure 3, where the poleward directed arrow indicates the direction and speed of motion of the arc. The electron drift velocity vectors are not pointing in the poleward direction, in the direction of the arc velocity. This means that the poleward motion of the arc is not an $\mathbf{E} \times \mathbf{B}$ velocity, but that the active arc is moving through the surrounding plasma. The importance of this finding is that the region in which the precipitating particles are accelerated is moving relative to the surrounding plasma in the magnetosphere in a manner that produces the poleward motion of the active arc.

Figures 4 and 5 illustrate different observed locations of the activated arc relative to the electron flow pattern. Shown is the latitudinal profile of the electron drift velocities versus time, with auroral arcs marked by solid lines. In Figure 4, a quiet arc is located to the north in the convection reversal, the Harang discontinuity. To the south in the field of view several successive arcs are shown to activate and expand poleward, each signaling the onset of a substorm. Clearly the arcs are intensifying on the equatorward side of the eastward electrojet. Substorms of this kind typically occur during times of low *Kp* index. During times of high *Kp* index the substorm onset is typically signaled by

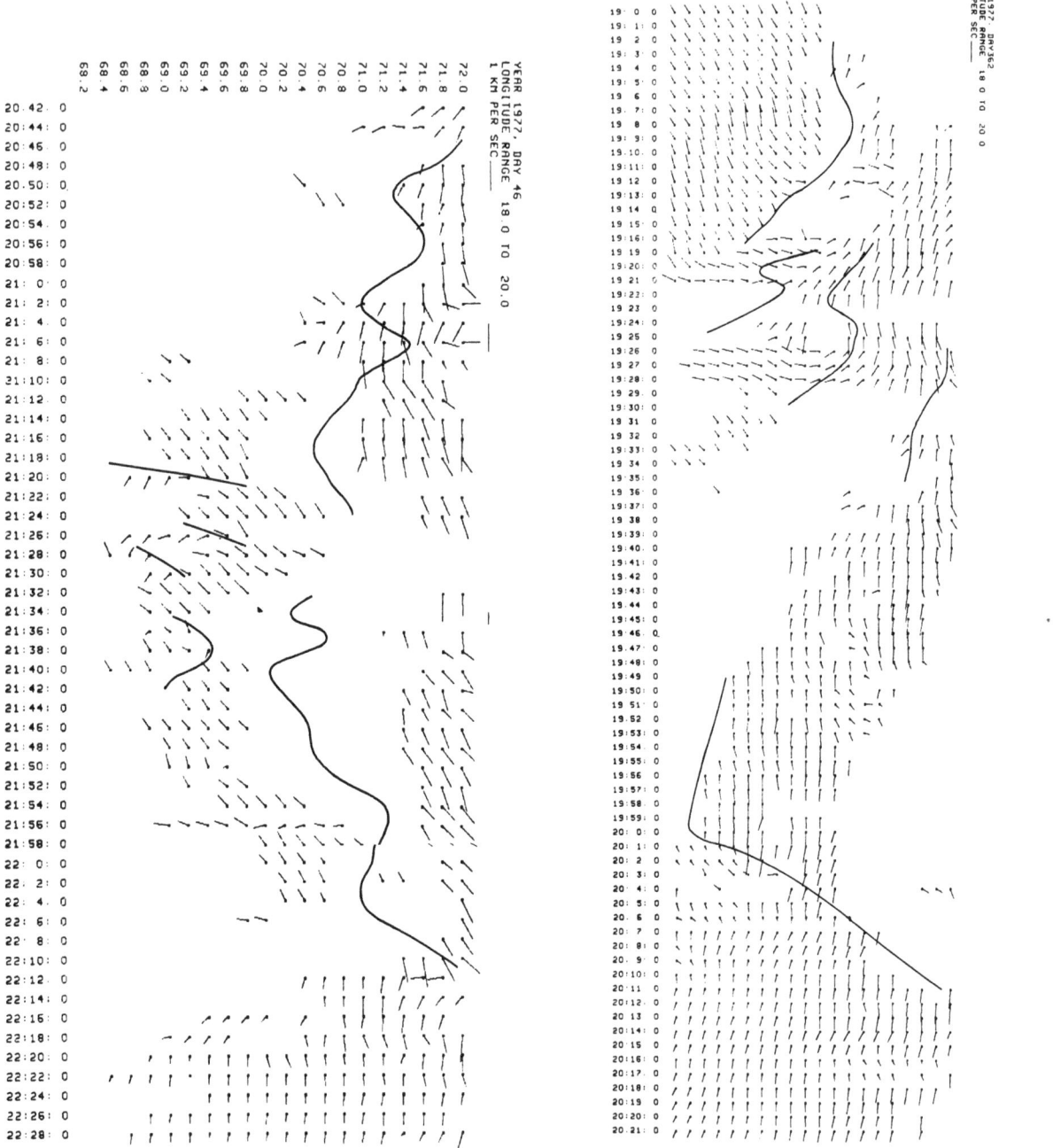

Figure 4—The latitudinal variations of the electron drift velocities versus time. The solid curves are the lower borders of optical arcs. The top-most curve is a quiet arc in the Harang discontinuity, while the lower curves are active arcs associated with substorm activity. (From *Nielsen and Greenwald* [1979].)

Figure 5—As in Figure 4, but showing the Harang discontinuity surging equatorward. An arc appears in the discontinuity (at 1949 UT). It eventually intensified (at 2000 UT) and expanded poleward. (From *Nielsen and Greenwald* [1979].)

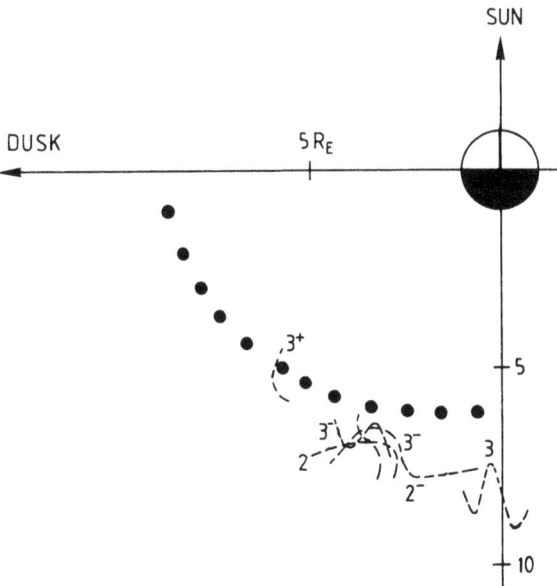

Figure 6—At times of low magnetic activity, the latitudinal position of the Harang discontinuity at 19° east longitude has been determined as a function of time and mapped onto the magnetospheric equator. The variations of this projection illustrate how the essentially east-west aligned Harang discontinuity is located in the magnetosphere. The number at each curve is the *Kp* index. The dotted curve is McIlwain's Injection Boundary for low magnetic activity. It is located equatorward of the discontinuity in the region where substorm onsets occur.

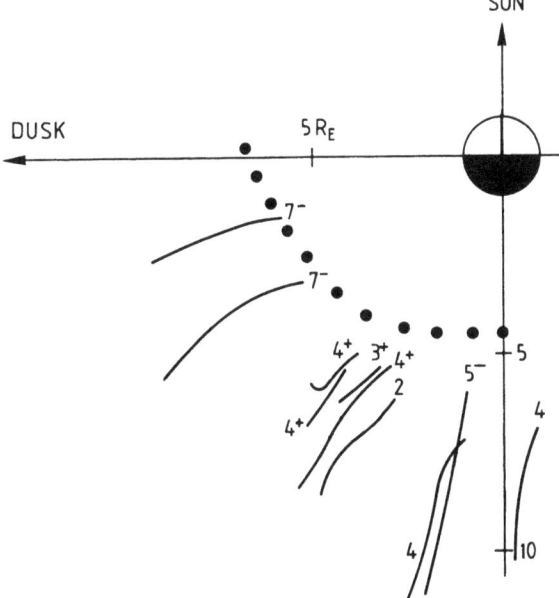

Figure 7—This plot illustrates how the essentially east-west aligned Harang discontinuity surges equatorward until it reaches McIlwain's Injection Boundary (the dotted curve) at times of high magnetic activity, at which point an arc in the discontinuity is activated and surges poleward.

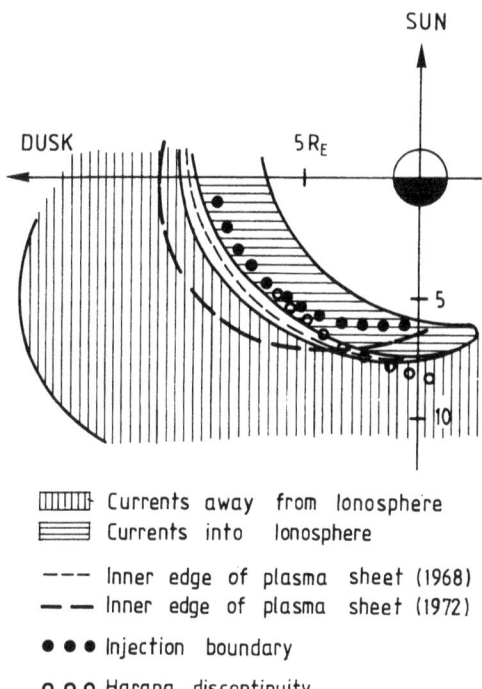

⊞	Currents away from Ionosphere
⊟	Currents into Ionosphere
---	Inner edge of plasma sheet (1968)
— —	Inner edge of plasma sheet (1972)
●●●	Injection boundary
○○○	Harang discontinuity

Figure 8—Shown are (a) the Harang discontinuity during stable conditions (from Figure 6), (b) regions associated with currents directed into or away from the ionosphere (from *Potemra* [1978]), (c) the location of the inner edge of the plasma sheet (as determined by *Vasyliunas* [1968, 1972]), and (d) the quiet time injection boundary (from *McIlwain* [1974]).

activation of an arc in the Harang discontinuity. Often the discontinuity moves equatorward prior to onset, and the arc may actually first occur during this equatorward motion. These features are illustrated in Figure 5.

In order to study the location of the substorm onset region and the dynamics of the active arc in the magnetic equatorial plane (defined as the surface that contains points on each individual field line at the largest distance from the Earth), the following procedure is followed. We choose that point of the Harang discontinuity that lies at 19° east longitude (near the center of the field of view) and follow how it moves in latitude as a function of time. The point is then mapped along a field line into the magnetospheric equatorial plane to determine the location and spatial motions of this point as a function of time. The mapping has been carried out for 16 individual events, and the results are shown in Figures 6 and 7.

The dashed curves in Figure 6 refer to events during quiet magnetic conditions when the discontinuity stayed in the field of view for an extended time. At such times the discontinuity may move slowly equatorward or poleward, but on the whole it remains nearly stationary. Mapping the velocities observed in the ionosphere into the magnetosphere yield a typical velocity during radial motions at these times of about 3 km s^{-1}. The mean location of the discontinuity is near 7R$_E$. The dotted curve marks the McIlwain injection boundary for low magnetic activity. It is clearly located near the discontinuity and on the earthward (i.e., on the equatorward) side, in the region where substorm onsets in the ionosphere occur. At low magnetic activity the magnetic field line model is probably reliable, so this result can be considered with confidence.

How is the magnetic conjugate location of the Harang discontinuity during quiet magnetic conditions located with respect to other regions or boundaries in the magnetosphere? Figure 8 shows the mean location of the Harang discontinuity as an open curve of circles. The mean location is determined using the same data as shown in Figure 6. Also shown in Figure 8 are areas in the equatorial plane that correspond to regions in the ionosphere, which, during low magnetic activity, are associated with inward or outward field-aligned currents [Potemra, 1978]. Potemra [1978] used the same magnetic field model as is used here. Thus uncertainties in comparing the spatial locations of the field-aligned currents and the discontinuity only arise from statistical fluctuations due to the limited sample size and from the mean magnetic activity represented by the samples. The figure indicates that during magnetic quiet times the discontinuity is located near the region that separates inward and outward directed field-aligned currents. Also marked in the figure is the inner edge of the plasma sheet as determined by *Vasyliunas* [1968, 1972]. These are the thin and thick dashed curves, respectively. The figure also includes the injection boundary of *McIlwain* [1974], the solid circle curve. Clearly all these curves lie within a few Earth radii in the same part of the magnetosphere, with the possibility that they may actually coincide.

The solid curves in Figure 7 refer to events during disturbed magnetic conditions, and display the equatorward motion of the discontinuity, a motion eventually terminated by the onset of a substorm signaled by the brightening of an arc in the Harang discontinuity. Typical velocities during these earthward surges are between 0.1 and 0.3R$_E$/min, close to between 10 and 30 km s^{-1}. One major difference between magnetic quiet and active times is the decisive earthward motion of the Harang discontinuity during active times. The dotted curve marks the McIlwain injection boundary for high magnetic activity. Clearly the earthward motion seems to terminate when the discontinuity reaches this boundary.

Vasyliunas [1968] found that at times associated with magnetic bay activity the inner edge of the plasma sheet surged toward the Earth with an average radial speed of

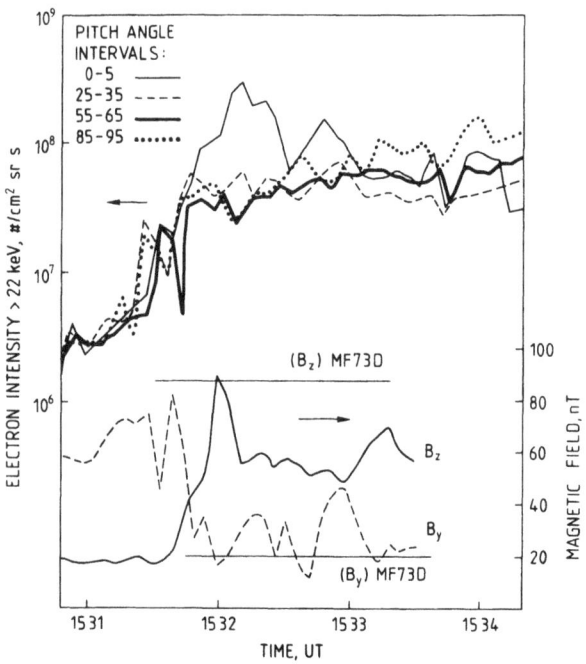

GEOS 2 OCT 30, 1978
INTEGRAL DIRECTIONAL ELECTRON FLUX
AND MAGNETIC FIELD VARIATIONS

Figure 9—Observations at geosynchronous orbit around the time of a substorm onset on October 30, 1978. At the top are shown energetic electron fluxes at several pitch angle intervals, and at the bottom the magnetic field is shown and compared with the model predictions. At the time of strong field-aligned particle fluxes (1532 UT), the magnetic field is quite close to the model values. (From *Nielsen et al.* [1982].)

12 km s^{-1}. The rapid equatorward motion of the Harang discontinuity that we have observed corresponds to a mean inward radial speed of its projection in the equatorial plane of between 10 and 30 km s^{-1}. The average of all observed speeds was 17 km s^{-1}. Thus the speed with which the inner edge of the plasma sheet moves earthward during magnetically active times is comparable with the equatorward speed of the particle precipitation arc and of the Harang discontinuity. This again supports the suggestion of the close association of these regions.

The mapping process during high magnetic activity is more uncertain owing to the increased distortion of the dipole field toward a more tail-like magnetic field configuration. Just prior to the substorm, the magnetic field may be more tail-like than predicted by the model. An example is shown in Figure 9 (from *Nielsen et al.* [1982]). Clearly, say, at 1531 UT, the B_z component is less than predicted by the model while the radial component (here B_y; the onset occurred at dusk!) is larger. The particle intensities started to increase at 1531 UT, and the magnetic field started to return toward a dipolar shape at 1531:40 UT. It is striking that the strongly field-aligned particle fluxes, which are associated with the activation of the arc in the ionosphere (i.e., with substorm expansion phase onset), occurred at 1532 UT, at a time when the observed magnetic field was quite close to the model prediction. Thus, the mapping of the onset region into the ionosphere would be quite accurate using our model.

3. DISCUSSION

We have compared the statistical locations in space of the inner edge of the plasma sheet, and of injection boundaries, as these have been reported in the literature based on measurements made in the magnetosphere, with the observed locations of the Harang discontinuity, or convection reversal, as measured with coherent radars in the ionosphere. The ionospheric observations were compared with the observations in the magnetosphere using a magnetic field line model. The result of this comparison is that these four spatial regions or boundaries are located together, to within a few Earth radii. Furthermore, these regions are at a radial distance from the Earth of between 5–8R_E. Owing to the use of statistical predictions, the results presented here cannot be conclusive, but it is unlikely that the close spatial relationship that is determined between all these independent measurements should be the result of coincidence.

The manifestations of a substorm in the ionosphere can be divided into three parts:

1. There is a gradual increase of energy input to the ionosphere through particle precipitation. It has been shown that prior to the brightening of the auroral arc (which is normally taken as the onset time of a substorm) a band of particle precipitation moved equatorward in the ionosphere with a speed of between 200 and 400 m s^{-1}, corresponding to a speed toward the Earth in the magnetosphere of between 3.5 and 7.5 km s^{-1}. The precipitating particle population was found to be energized as it moved equatorward [*Hargreaves et al.*, 1975]. *Nielsen and Greenwald* [1979] found substorm onsets to be typically preceded by an equatorward motion of the Harang discontinuity. They also found that the arc, that is eventually activated, may first appear in the discontinuity during the equatorward motion. This also implies a gradual intensification of the particle precipitation—and maybe of the field-aligned currents—during the time just prior to substorm onset (this may be referred to as the substorm growth phase).

2. There is an explosive increase in the energy input to the upper atmosphere from particle precipitation. Low-energy particles give rise to E-region ionization associated with optical aurora. *Nielsen* [1980] showed that the bright auroral arc is colocated with an arc of energetic particle precipitation, which gives rise to D-region ionization associated with riometer absorption. The equatorward motion of the precipitation arc and the Harang discontinuity with the auroral arc, mentioned under point 1, is terminated when the auroral arc suddenly brightens, signaling the onset of a substorm. The sudden activation of the arc is followed by a poleward expansion during which the arc is moving through the surrounding plasma, i.e., the arc is moving relative to the plasma. It is not $\mathbf{E} \times \mathbf{B}$ drifting with the plasma. Eventually the poleward motion stops, and the arc either immediately or after a time lapse of 10–30 min starts to move equatorward. Later this arc may again activate, signaling the onset of a new substorm, or the occurrence of a multiple-onset substorm. *Akasofu* [1964, 1968, 1974] found that the arc that is activated at substorm onset in a typical substorm is located near the poleward border of the diffuse aurora and on the equatorward side of a band of discrete arcs. *Nielsen and Greenwald* [1979] found in general that auroral arcs are located in, or poleward of, the Harang discontinuity. However, it has been observed that on occasion, typically during times of low *Kp* index, an arc can activate on the equatorward side of the eastward electrojet [*Nielsen and Greenwald*, 1978].

3. There is a large enhancement in the auroral westward electrojet current. The auroral (and particle) arc also forms the poleward border of the westward electrojet [*Nielsen*, 1980]. Thus the poleward border of the electrojet is also moving poleward through the surrounding plasma. The second and third effects are simultaneous and constitute the so-called substorm expansion phase. During the expansion phase large amounts of energy are input to the ionosphere, with Joule heating (owing to the current intensification) being typically a factor 10 larger than the heating due to particle precipitation.

It has been suggested that a fourth part may have to be considered: the poleward leap, i.e., the sudden and rapid poleward motion of the westward electrojet and auroras, which has been claimed to occur during the recovery of a substorm [*Hones*, 1985 (and references therein); *Hones*, 1986]. Some other workers have failed to determine this phase [*Rostoker*, 1986; *Nielsen et al.*, 1988].

Because of the difficulty of tracing magnetic field lines during individual events, the relationship of ground-based observations to observations in the magnetosphere is difficult to determine. However, both ground-based and magnetospheric observations have, in the past, led to suggestions of a geomagnetic connection between the active arcs in the ionosphere, the Harang discontinuity, the injection boundaries, and magnetospheric plasma boundaries.

Gustafsson [1975] indicated that the injection boundary is magnetically connected to the arc that brightens at substorm onset. *Fairfield and Mead* [1975] used their magnetic field model to map the Harang discontinuity into the magnetospheric equatorial plane, and found it had a remarkable similarity to the injection boundary. *Brekke* [1977] combined geosynchronous particle data with incoherent scatter electric field measurements, and found evidence to suggest that the Harang discontinuity is magnetically conjugate to the magnetospheric injection boundary.

McIlwain [1974, 1988] and *Mauk and McIlwain* [1974] showed that energetic particles observed at geosynchronous orbit following a substorm onset could be interpreted as having been accelerated (or injected) along a line in the magnetosphere, which they termed the injection boundary. Optical and particle observations on board a polar-orbiting satellite led *Lui and Burrows* [1978] to place the arc that brightens at substorm onset in the region that separates the boundary plasma sheet (BPS) and the central plasma sheet (CPS) [*Winningham et al.*, 1975]. The BPS coincides with regions of active auroral forms, while the CPS is associated with the diffuse aurora [*Lui*, 1975, 1978]. Since we, generally as mentioned above, have observed active arcs in or poleward of the Harang discontinuity, we propose that the discontinuity, and thus the region of substorm onset, is located near the CPS/BPS boundary. This is also in agreement with observations of ion drifts and particle precipitation made on board Atmospheric Explorer [*Heelis et al.*, 1980]. These authors found that, statistically, the Harang discontinuity and the CPS/BPS boundary are colocated in the evening quadrant.

Thus, generally, the enhanced energy input first appears near the Harang discontinuity at the CPS/BPS boundary, where it is marked by the intense brightening of an auroral arc. This subsequently executes a poleward motion through the surrounding plasma, and is followed by a complicated pattern of particle precipitation and auroral displays.

The above quoted spacecraft measurements were made either on polar-orbiting satellites or on spacecraft in geosynchronous orbit. Thus the intersection of the CPS/BPS boundary with the magnetic equatorial plane is not determined from these measurements. Since this intersection presumably is the region of substorm onset in the magnetosphere, these measurements leave open the radial distance to the onset region. However, it is probable that the CPS/BPS boundary may coincide with the "inner edge of the plasma sheet" as determined by *Vasyliunas* [1968] from observations in the magnetosphere (Figure 8). The "inner edge" is defined as the region where a tailward directed energy density gradient is present, and we suggest that it may coincide with the poleward directed gradient in the energy density on the CPS/BPS boundary [*Winningham et al.*, 1975]. Since it has been shown that this inner edge closely coincides with the mapped location of the Harang discontinuity it is suggested that the substorm onset may conceivably occur on near-Earth field lines. Thus the physical processes that are directly related to the brightening of the arc may take place in the near-Earth magnetosphere at a distance of between 5–8R_E.

Before considering the physical mechanisms that may drive an injection event, let us briefly consider the typical effects observed in the magnetotail during substorms. Observations in the magnetotail have led to a two-phase description of substorm phenomena. On the average, the plasma sheet is R_E wide and has its earthward boundary near 8–10R_E in the equatorial plane. The flow in the plasma sheet is that associated with normal convection. In the first phase of a substorm, high-speed plasma flows in the equatorial plane, toward and away from the Earth, appear. The region from which this jetting originates appears to lie from 10–15R_E. Large intensities of plasma sheet particles appear close to the Earth (at and inside geosynchronous orbit). This is normally referred to as an earthward displacement of the inner edge of the plasmasheet. It is suggested that plasma jetting earthward along magnetic field lines by intermediate processes, not fully understood, creates auroras. During the second phase the plasma sheet thickens (sometimes exceeding its presubstorm thickness), and there are strong plasma flows toward the Earth.

The first of these phases is usually associated with the substorm expansion phase. There seem to be no magnetospheric observations that are directly equivalent to the growth phase, the first substorm phase deduced from ionospheric measurements.

Owing to the limitations of using magnetic field models in individual events there are uncertainties in mapping ionospheric features from the magnetosphere. The observations in the ionosphere cannot therefore be said to be inconsistent with the substorm model that involves the formation of a neutral line in the magnetospheric tail at substorm onset. It is possible that just prior to substorm onset the geomagnetic field lines are more tail-like than the model predicts, as discussed in the previous section. This would mean that the ionospheric features would be mapped from larger radial distances, possibly from the $10-15R_E$ region where the neutral line is thought to form. In this framework one of the magnetospheric observations is not easy to reconcile with the ground-based measurements. The earthward displacement of the inner edge of the plasma sheet is thought to occur as a consequence of neutral line formation in the tail. We have associated this displacement with the equatorward displacement of the Harang discontinuity, and of the arc that eventually activates, and that happens prior to the substorm onset in the ionosphere, as marked by the explosive increase in energy input owing to particle precipitation and current intensification. If both observations are to be upheld, we must argue that the neutral line formation causes the earthward (equatorward) motion of the inner edge (discontinuity), and that this motion eventually is terminated by a substorm expansion phase onset occurring on near-Earth field lines.

If, on the other hand, the earthward surge of the inner edge of the plasma sheet is associated with the growth phase, this clear conflict between observations and the neutral line substorm model would be removed.

However, the ground-based measurements also imply that the neutral line should preferentially form in that part of the magnetosphere that is magnetically connected to the convection reversal. Substorm theories involving a neutral line have so far not accounted for this observation.

The observations of active arcs on the equatorward side of the eastward electrojet are even more severe. These observations were made at times when the magnetic *Kp* index was low, i.e., at times when the mapping was probably most reliable. We find these active arcs to lie on near-Earth dipolar field lines, in a region of the magnetosphere in which neutral line formation is not thought to be possible.

The observation that the precipitation arc, the activated arc, is not $E \times B$ drifting is probably the one observation in the ionosphere that clearly supports the neutral line model. The arc is thought to be connected magnetically to the neutral line. As magnetic field lines are convected into the merging region, this merging region, and thus the arc, connect to increasingly higher magnetic latitudes as time increases, producing a poleward motion that is nonconvective in character.

Even though the neutral line model remains a strong candidate for at least some substorm phenomenon, the model is still not fully satisfactory. Thus, theoreticians have investigated other possibilities. *Atkinson* [1979, 1984] suggested as a possible substorm mechanism "coupled ionosphere-magnetosphere instability." If an enhancement of the ionospheric conductivity is assumed, this would lead to redistribution of plasma, allowing some field lines to collapse. The precipitation of particles and the currents resulting from this collapse in turn enhance the conductivity. In this feedback system an instability is possible. In the framework outlined above, the start of this process may be the collapse of field lines following the formation of a neutral line (the "Earthward motion of the inner edge of the plasma sheet" or the equatorward motion of the Harang discontinuity, which is associated with intensifying precipitation and field-aligned currents). *Kan et al.* [1988] showed that substorm onset could occur in the convection reversal provided the region overlaps with the poleward gradient of the diffuse auroral conductance and provided the polar cap potential exceeds a certain minimum value. It is argued that intense field-aligned currents are formed in the convection reversal, and cause a potential drop along the magnetic field lines in which electrons are accelerated to cause auroral activation of the ionosphere. Both models invoke the ionospheric conductivity as a crucial factor for the substorm process, thereby accounting for the intensifications of the field-aligned currents and the precipitating particle fluxes observed during the equatorward motion of the Harang discontinuity prior to the substorm. It will be interesting to learn if these theories can also account for the poleward expansion velocity, which has a direction that is different from the $E \times B$ velocity of the surrounding plasma.

ACKNOWLEDGMENT—The STARE radar system is operated by the MPAE in cooperation with the ELAB, University of Trondheim, Norway, and the Finnish Meteorological Institute, Finland. The rio-imagers were operated by MPAE in cooperation with the University of Tromsö, Norway, and the University of Oslo, Norway.

REFERENCES

Akasofu, S.-I., "The Development of the Auroral Substorm," *Planet. Space Sci.*, **12**, 273 (1964).

Akasofu, S.-I., *Polar and Magnetospheric Substorms*, Springer, New York (1968).

Akasofu, S.-I., "A Study of Auroral Displays Photographed from the DMSP 2 Satellite and from the Alaska Meridian Chain of Stations," *Space Sci. Rev.*, **16**, 617 (1974).

Atkinson, G., "The Expansive Phase of the Magnetospheric Substorm," in *Dynamics of the Magnetosphere*, S.-I. Akasofu, ed., D. Reidel Publishing Co., Dordrecht, Holland, p. 461 (1979).

Atkinson, G., "Field-Aligned Currents as a Diagnostic Tool: Results, A Renovated Model of the Magnetosphere," *J. Geophys. Res.*, **89**, 217 (1984).

Brekke, A., "The Relationship Between the Harang Discontinuity and the Substorm Injection Boundary," *Planet. Space Sci.*, **25**, 1119 (1977).

Fairfield, D. H., and G. D. Mead, "Magnetospheric Mapping with Quantitative Geomagnetic Field Model," *J. Geophys. Res.*, **80**, 535 (1975).

Gustafsson, G., "Ionospheric/Magnetospheric Parameters Derived from Conventional Groundbased Observations," in *Radar Probing of the Auroral Plasma*, A. Brekke, ed., Universitets Forlaget, Tromsö, Norway, p. 409 (1975).

Hargreaves, J. K., H. J. A. Chivers, and W. I. Axford, "The Development of the Substorm in Auroral Radio Absorption," *Planet. Space Sci.*, **23**, 905 (1975).

Heelis, R. A., J. D. Winningham, W. B. Hanson, and J. L. Burch, "The Relationship Between High-Latitude Convection Reversals and the Energetic Particle Morphology Observed by Atmosphere Explorer," *J. Geophys. Res.*, **85**, 3315 (1980).

Hones, E. W., Jr., "Magnetic Reconnection in the Earth's Magnetotail," *Aust. J. Phys.*, **38**, 981 (1985).

Hones, E. W., Jr., Reply, *J. Geophys. Res.*, **91**, 5881 (1986).

Kan, J. R., L. Zhu, and S.-I. Akasofu, "A Theory of Substorms: Onset and Subsidence," *J. Geophys. Res.*, **93**, 5624 (1988).

Lui, A. T. Y., "Simultaneous Observations of Auroral Emissions and Particle Precipitation by Isis 2 Satellite," *J. Geophys. Res.*, **80**, 3603 (1975).

Lui, A. T. Y., and J. R. Burrows, "On the Location of Auroral Arcs Near Substorm Onsets," *J. Geophys. Res.*, **83**, 3342 (1978).

Mauk, B. H., and C. E. McIlwain, "Correlation of *Kp* with the Substorm-Injected Plasma Boundary," *J. Geophys. Res.*, **79**, 3193 (1974).

McIlwain, C. E., "Substorm Injection Boundaries," in *Magnetospheric Physics*, B. M. McCormac, ed., D. Reidel Publishing Co., Dordrecht, Holland, p. 143 (1974).

McIlwain, C. E., "Plasma Acceleration, Injection, and Loss: Observational Aspects," *Astrophys. Space Sci.*, **144**, 201 (1988).

Mead, G. D., and D. H. Fairfield, "A Quantitative Magnetospheric Model Derived from Spacecraft Magnetometer Data," *J. Geophys. Res.*, **80**, 523 (1975).

Nielsen, E., "Dynamics and Spatial Scale of Auroral Absorption Spikes Associated with the Substorm Expansion Phase," *J. Geophys. Res.*, **85**, 2092 (1980).

Nielsen, E., "The STARE System and Some of its Applications," in *The IMS Source Book*, C. T. Russell and D. J. Southwood, eds., American Geophysical Union, Washington, DC, p. 213 (1982).

Nielsen, E., J. Bamber, Z.-S. Chen, A. Brekke, A. Egeland, J. S. Murphree, D. Venkatesan, and W. I. Axford, "Substorm Expansion into the Polar Cap," *Ann. Geophys.*, **5**, 559 (1988).

Nielsen, E., and R. A. Greenwald, "Variations in Ionospheric Currents and Electric Fields in Association with Absorption Spikes During the Substorm Expansion Phase," *J. Geophys. Res.*, **83**, 5645 (1978).

Nielsen, E., and R. A. Greenwald, "Electron Flow and Visual Aurora at the Harang Discontinuity," *J. Geophys. Res.*, **84**, 4189 (1979).

Nielsen, E., A. Korth, G. Kremser, and F. Mariani, "The Electron Pitch Angle Distribution at Geosynchronous Orbit Associated with Absorption Spikes During the Substorm Expansion Phase," *J. Geophys. Res.*, **87**, 887 (1982).

Nielsen, E., and K. Schlegel, "A First Comparison of STARE and EISCAT Electron Drift Velocity Measurements," *J. Geophys. Res.*, **88**, 5745 (1983).

Nielsen, E., and K. Schlegel, "Coherent Radar Doppler Measurements and Their Relationship to the Ionospheric Electron Drift Velocity," *J. Geophys. Res.*, **90**, 3498 (1985).

Nielsen, E., and J. D. Whitehead, "Radar Auroral Observations and Ionospheric Electric Fields," COSPAR, Ottawa, *Adv. Space Res.*, **2**, 131 (1983).

Potemra, T. A., "Observation of Birkeland Currents with the Triad Satellite," *Astrophys. Space Sci.*, **58**, 207 (1978).

Rostoker, G., "Comment on 'The Poleward Leap of the Auroral Electrojet as Seen in Auroral Images'," *J. Geophys. Res.*, **91**, 5879 (1986).

Vasyliunas, V. M., "A Survey of Low-Energy Electrons in the Evening Sector of the Magnetosphere with OGO 1 and OGO 3," *J. Geophys. Res.*, **73**, 2839 (1968).

Vasyliunas, V. M., "Magnetospheric Plasma," in *Solar Terrestrial Physics*, Part III, E. R. Dyer, ed., D. Reidel Publ. Co., Dordrecht, Holland, 191–211 (1972).

Winningham, J. D., F. Yasuhara, S.-I. Akasofu, and W. J. Heikkila, "The Latitudinal Morphology of 10-eV to 10-keV Electron Flows During Magnetically Quiet and Disturbed Times in the 2100–0300 MLT Sector," *J. Geophys. Res.*, **80**, 3148 (1975).

VII-4. INCOHERENT SCATTER OBSERVATIONS OF THE AURORAL IONOSPHERE WITH THE EISCAT RADAR FACILITY

J. Röttger*

A brief description of the European Incoherent Scatter Radar Facility (EISCAT) is given and the concepts of routine operations of this facility to study the auroral ionosphere are outlined. The participation of EISCAT in international scientific programs and campaigns is summarized. We describe briefly its capabilities to study magnetosphere-ionosphere coupling, magnetospheric convection, the observation of flux-transfer events, large electric fields in the auroral ionosphere, and frictional heating. We also discuss the observations of field-aligned plasma velocities and currents. Particular emphasis is placed on the combined observations of EISCAT with satellites, especially Viking, and ground-based observations to study auroral substorms and related events, such as auroral particle precipitation, electric fields, auroral intensifications, and arcs. Furthermore, comparisons with magnetometer observations of pulsations are made from which the Joule and particle heating rates are estimated. Ionospheric conductivities and their relation to auroral activity are demonstrated. Also, E-region irregularities and plasma wave observations are elucidated. We also discuss the influence of auroral effects on D-region ionization and structure in the polar mesosphere. We consider the observation of winds and temperatures due to global thermospheric circulation, tides, and atmospheric gravity waves in the polar upper atmosphere, as well as the generation of gravity waves due to auroral Joule and particle heating and Lorentz forcing. We finally point out that EISCAT is an optimum facility to study coupling processes both downward from the magnetosphere and upward from the lower atmosphere.

1. INTRODUCTION

The incoherent scatter radar technique is an essential ground-based tool to investigate electrodynamic, plasma-physical, neutral dynamic, and aeronomical effects in the auroral ionosphere and the polar atmosphere. With this technique it is possible to measure directly the electron density, the electron temperature, and the ion temperature and composition as well as the ion velocity [*Beynon and Williams*, 1978]. From these basic parameters many related parameters are deduced, such as the electric field, currents and conductivities, Joule heating, the heat flux and energy input, and the ion-electron production rate. We concentrate here on results obtained with the European Incoherent Scatter (EISCAT) radar, operated in northern Scandinavia [*Folkestad et al.*, 1983; *Hagfors*, 1983; *Röttger*, 1983; *Baron*, 1984; *Williams*, 1985]. *Rishbeth and Williams* [1985] have described early scientific results achieved with EISCAT and *Huuskonen et al.* [1987] have outlined some applications of EISCAT for auroral research. In a review paper by *Brekke* [1988], several characteristics of the polar ionosphere are addressed that were investigated using EISCAT. Here, without claiming to constitute a complete overview or even a review, are summarized some highlights of the scientific results that were achieved in recent years, and were published in journals, proceedings, and reports, as well as in the annual reports of the EISCAT Scientific Association.

*EISCAT Scientific Association, S-98 128 Kiruna, Sweden (on leave from Max-Planck-Institut für Aeronomie, W. Germany).

2. INCOHERENT SCATTER RADAR RESEARCH OF THE POLAR UPPER ATMOSPHERE

The transfer of energy, momentum, and mass through the magnetospheric system into the polar ionosphere, the thermosphere, and the middle atmosphere can appropriately be investigated by the incoherent scatter radar (ISR) technique. Certain characteristics observed with this technique in the polar ionosphere, such as plasma convection and energetic particle precipitation, are linked from the magnetosphere and thus give ground-based support to in situ magnetospheric observations by spacecraft. The plasma flow in the polar ionosphere measured with the ISR is a manifestation of magnetospheric convection, and the interaction of the solar wind and interplanetary magnetic field (IMF) with the Earth's magnetosphere. Measurements with the ISR technique thus allow ionospheric processes to be related back to their origin in the magnetosphere and the solar wind. Precipitation of accelerated electrons and protons along the Earth's magnetic field lines and the mapping of the magnetospheric electric field into the polar ionosphere lead to the characteristic visual aurora. Because of the enhancement of the electron density (which accompanies the aurora), the electrodynamics of these phenomena and their ionospheric environment can be studied in great detail by ISR with the necessary high temporal and spatial resolution, as well as with broad horizontal coverage of a few thousand kilometers and vertical coverage up to an altitude 1000 km or more. The location of the EISCAT radars in northern Scandinavia allows these effects to be studied in the auroral oval and into the polar cap.

In principle, these observations become possible through the analysis of characteristic changes of the ion line and the plasma line of the incoherent scatter spectrum. Particular plasma-physical processes, such as instabilities and nonthermal plasma, can also be studied by specific analyses of the ISR spectra. The plasma structure of the polar ionosphere can have strong vertical and horizontal gradients; it can also travel over large horizontal distances. The ISR technique can monitor and trace these large-scale inhomogeneities. Also, the large-scale mapping of other plasma parameters (such as temperature and composition) and the electric field structure and the accompanying plasma motion can be performed. The observation of the plasma parameters and the electric field (which is deduced from the measurement of mean plasma velocities, caused by the $\mathbf{E} \times \mathbf{B}$ drift), permits the study of particular structures in the ionosphere and their originating effects in the magnetosphere as well. The relation of plasma instability processes and the resulting plasma irregularities in the polar electrojet current system to the ambient electric field, plasma gradients, and temperatures can appropriately be studied by ISRs. High plasma velocities resulting from large electric fields yield a substantial enhancement of the ionized and the neutral gas temperatures due to frictional or Joule heating. The resulting upwelling and the diffusion of atmospheric particles could cause their outflow into the polar exosphere and magnetosphere. Because ISRs can measure density, temperature, composition, and flow velocities up to altitudes of about 1000 km, they are very appropriate research tools for studying this outflow from the Earth's atmosphere, which is also called the polar wind.

There are further phenomena in the polar ionosphere that originate in the lower atmosphere and propagate upward, such as atmospheric tides and gravity waves. These deposit energy and momentum in the mesosphere and thermosphere and constitute competing effects to the input from the magnetosphere. The relative importance of the upward- and the downward-directed phenomena on the thermosphere and the middle atmosphere is not really understood and still needs to be investigated. Changes in the ionospheric plasma are mutually coupled to the background neutral atmosphere, where the temperature and wind fields become very variable. These can be studied by specially adapted ISR techniques optimized for small- and medium-scale phenomena, such as atmospheric gravity waves and the resulting traveling ionospheric disturbances, which can also be generated by auroral effects. Global-scale phenomena can be investigated in cooperation with

other radars. This is usually done in campaign mode operations and yields essential input to theoretical models of the global thermospheric circulation.

All these phenomena are frequently studied in collaboration with optical and other ground-based instrumentation and in situ observations made from rockets and spacecraft. The incoherent scatter radar technique thus comprises a powerful and often indispensable tool to study a wide range of phenomena in the polar upper atmosphere and their relation to the magnetosphere and the solar wind.

3. A BRIEF DESCRIPTION OF EISCAT

The European Incoherent Scatter Scientific Association was established to conduct research on the middle and upper atmosphere, the ionosphere, and the aurora, using the incoherent scatter radar technique [e.g., *Beynon and Williams*, 1978]. This technique is the most powerful ground-based tool for such research. EISCAT is also being used as a coherent scatter radar for studying instabilities in the ionosphere as well as for investigating the structure and dynamics of the middle atmosphere and as a diagnostic instrument to study plasma physical processes caused by artificial ionospheric modification or *Heating* experiments [e.g., *Kohl et al.*, 1987].

There exist seven incoherent scatter radars (ISR) in the world: Millstone Hill/Haystack Observatory near Boston, USA, Sondrestromfjord radar in Greenland; Arecibo Observatory in Puerto Rico; Jicamarca radar facility in Peru; the St. Santin radar in France (operation terminated), the MU radar in Japan; and the EISCAT radar facilities. The experimental sites of EISCAT are located in Scandinavia, north of the arctic circle. They consist of two independent radar systems (Figure 1). EISCAT is the only tristatic ISR facility in the world.

The UHF radar of EISCAT operates in the 933-MHz band, with a peak transmitter power of 1.5 MW, and 32-m steerable parabolic dish antennas. The transmitter and receiver are in Tromsö, Norway. Receiving sites are also in Kiruna, Sweden, and in Sodankylä, Finland, allowing tristatic measurements. The VHF radar in Tromsö operates monostatically in the 224-MHz band with a peak transmitter power of 1.5 MW (to be raised to 5 MW) and a 120 m × 40 m parabolic cylinder antenna, which is subdivided into four sectors. Each sector can be pointed into independent directions in the meridional plane from 30° south to 60° north of the zenith.

The basic data, which are measured routinely with the EISCAT ISR, are the profiles of electron density, electron and ion temperature, and ion velocity. A selection of well-designed radar pulse schemes allows the adaptation of the data taking routines to many partic-

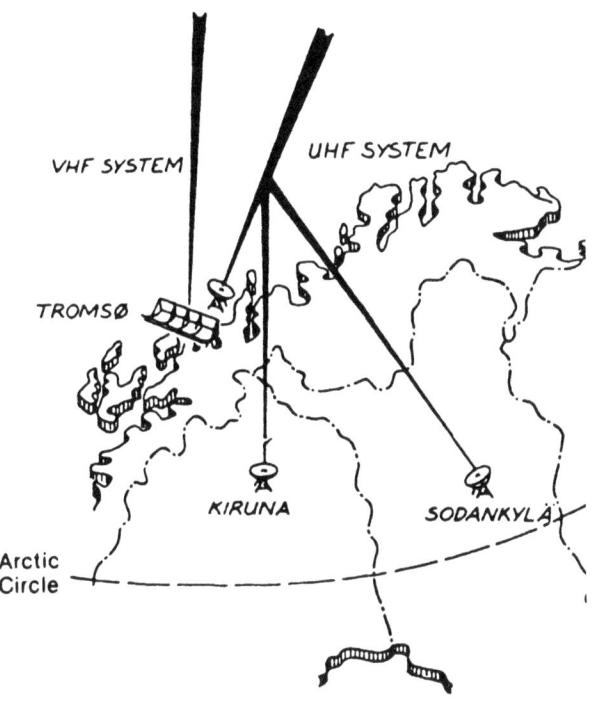

Figure 1—The EISCAT incoherent scatter radar system in northern Scandinavia, consisting of the monostatic VHF radar in Tromsö and the tristatic UHF radar with transmitter/receiver in Tromsö (Norway) and remote receivers in Kiruna (Sweden), and in Sodankylä (Finland).

4. THE EISCAT COMMON AND UNUSUAL PROGRAMMES AND THE SPECIAL CAMPAIGN OPERATIONS

One Common Programme, called CP-1, uses a fixed transmitting antenna, pointed along the geomagnetic field. The three-dimensional plasma velocity and anisotropy in other parameters are measured at certain E- and F-region altitudes by means of the remote receiving stations in Kiruna and Sodankylä. The CP-1 experiment can provide results with excellent time resolution along a fixed ionospheric profile (Figure 2). It is suitable for the study of substorm phenomena and particularly the auroral ionosphere when conditions may change rapidly. On longer time scales, CP-1 measurements also permit studies of diurnal changes (e.g., atmospheric tides) and seasonal variations. Eventually, solar cycle variability may be studied when sufficient data have been collected. The Common Programme CP-2 is designed to identify wavelike phenomena with length- and time-scales comparable with, or larger than, that of the scanning transmitter antenna (a few tens of kilometers and 10 or more min). The present version consists of a scan of four antenna positions covered in 6 min. The first three measurements in the scan form a triangle from the transmitter vertical to the south to southeast, and the final direction is parallel to the geomagnetic field line. The remote-site antennas are directed so that the three-dimensional velocities in the E- and F-regions can be measured. The Common Programme CP-3 covers a 10° wide range of latitudes in a 17-position scan up to 74°N in a 30-min cycle. The measurements are made in the plane of the magnetic meridian through Tromsö and Kiruna, and the remote-site antennas at Kiruna and Sodankylä follow the transmitter beam in the F-region. Another Common Programme, CP-4, was introduced, and covers latitudes up to almost 80°N, corresponding to about 77° invariant latitude. This Common Programme is particularly suited to studies of the plasma convection pattern at very high latitudes. A new Common Programme, CP-5, has been developed in order to suit the objectives of the international Lower Thermosphere Coupling Study (LTCS). It combines latitudinal scanning of the ionosphere with vertical sounding along the magnetic field line of Tromsö in the middle of the antenna scan. The main purpose of this experiment is to observe the dynamics of the neutral atmosphere while simultaneously exploring the electrodynamic environment. Unusual Programmes had been developed, UP-1 for D-region observations, UP-2 for auroral arc and related studies, and UP-3 for high-resolution, sporadic E-layer studies. These Unusual Programmes can be started at very short notice, as soon as suitable geophysical conditions exist. Whereas these

ular phenomena occurring at altitudes from about 60 km to more than 1000 km. Depending on geophysical conditions, a best time resolution of 1 s and an altitude resolution of a few hundred meters can be achieved, whereas typical resolutions are of the order of minutes and kilometers. The operation of a total of 2000 hours per year is distributed equally between Common Programmes (CP) and Special Programmes (SP). At present, five well-defined Common Programmes are run regularly with the UHF radar about 30 times per year for 24 or more hours to provide a database for long-term synoptic studies. VHF Common Programmes are being tested. Three Unusual Programmes (UP) can be started ad hoc during particular geophysical conditions, such as enhanced D-region ionization, auroral events, and sporadic E-layers. A large number of Special Programmes, defined individually by associate scientists, are run to study a variety of particular geophysical events. Several of the results, which are summarized here, were collected by Special Programme operations.

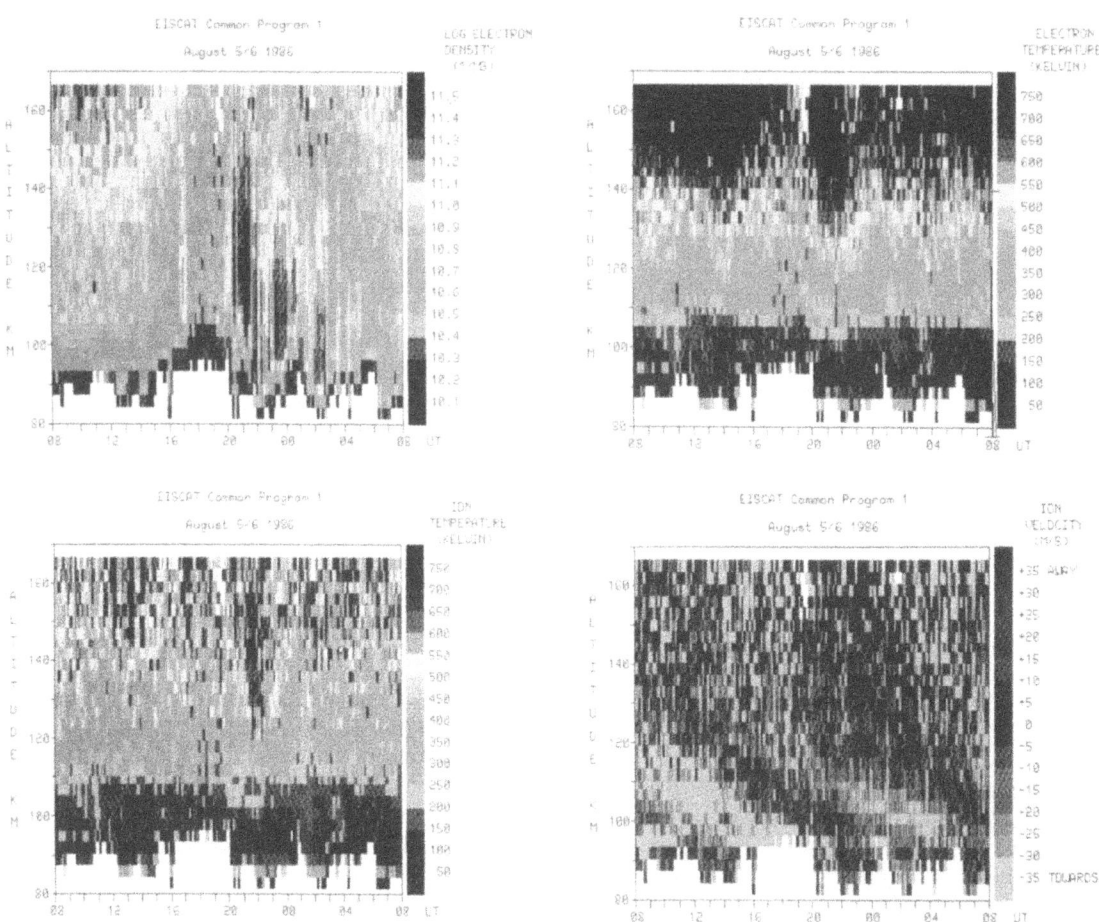

Figure 2—Color-code displays of EISCAT Common Programme observations of electron density, electron temperature, ion temperature, and ion velocity. These data were measured with the Common Programme CP-1-F from 0800 UT on 5 August to 0800 UT on 6 August 1986. In the Common Programme CP-1-F the Tromsö antenna beam is directed parallel to the Earth's magnetic profiles of the ionospheric parameters between 84 km and 168 km, which were measured at Tromsö with a special multipulse modulation scheme allowing an altitude resolution of 3 km.

The electron density plot shows the usual diurnal pattern on which are superimposed strong enhancements in the E region due to bursts of electron precipitation between 2000 UT and 0300 UT. There appears to be a concurring small increase of electron temperature above the altitude of the electron density enhancements. Between 2100 UT and 2200 UT a minor rise of ion temperature is observed above 120 km. The ion velocity shows a semidiurnal pattern, which is attributed to neutral air velocity due to atmospheric tides. (These displays are reproduced from the EISCAT Annual Report 1986.) (This figure also appears in color: Plate 45.)

CP and UP experiments are run with the UHF radar, two Common Programmes to observe the low and the high altitudes with the VHF radar are in preparation, and a high-latitude programme is planned.

Common Programme data are analyzed by EISCAT by on-line processing at the field sites in combination with the experiment operation. The analyzed data are collected at EISCAT Headquarters in Kiruna, where they are quality controlled, displayed on hardcopies and color slides, and copied on tape for distribution to the associate countries and for archiving. The World Day Common Programme data are also dispatched to the Incoherent Scatter Radar Data Base at the National Center for Atmospheric Research in Boulder, Colorado, USA (*NCAR*, 1988). The purpose of this database is to make the World Day data from all the ISRs readily accessible for scientific research by the worldwide scientific community.

In the past, EISCAT has participated in many campaign-mode operations with other radar, satellite, rocket, and balloon experiments. These experiments included the Scandinavian Twin Auroral Radar Experiments (STARE), Sweden and Britain Radar Experiment (SABRE), the French HF-Doppler radar studies of field-aligned irregularities (EDIA), the Polar Regions and Outer Magnetosphere International Study (PROMIS), the project SUNDIAL to study ionospheric processes and their role in the Sun-Earth system, the campaign Winter in Northern Europe of the Middle Atmosphere Program (MAP/WINE), the campaigns Summer in Northern Europe of the Middle Atmosphere Cooperation (MAC/SINE), and also MAC/Epsilon, Coordinated EISCAT-Balloon Experiments (CEBO), Coordinated Auroral Experiments using Scatter and Rocket investigations (CAESAR), CENTAUR and AURELD-VIP rocket campaigns, EISCAT-Heating and Partial-Reflection Radar (PRE), and cooperative, as well as coordinated, experiments with many satellites such as ARCAD 3, GEOS 2, NNSS, EXOS-C, HILAT, AMPTE/IRM, Dynamics Explorer, and Viking. The data analyses of these programs are continuing; several results have been achieved so far and are briefly outlined here; more are expected.

Together with the other incoherent scatter radars and ground-based instruments, EISCAT has participated in a number of international programs. The major ones are the Global Ionospheric Simultaneous Measurements of Substorms (GISMOS), which is a continuation of the earlier MITHRAS to study globally-simultaneous and local time effects in high-latitude ionospheric/thermospheric processes. By means of the Global Thermosphere Mapping Study (GTMS), a global view of neutral dynamical processes in the thermosphere has been obtained. The WAGS project (i.e., World Acoustic-Gravity Wave Study) is directed at the inves-

tigations of the sources and propagation characteristics of atmospheric gravity waves in the neutral atmosphere and the accompanying traveling ionospheric disturbances. There are other international programs established, such as the World Ionosphere Thermosphere Study (WITS), the Global Ionosphere-Thermosphere Coupling and Dynamics program (GIT-CAD), the Lower Thermosphere Coupling Study (LTCS). EISCAT is participating in these programs, too.

5. MAGNETOSPHERE-IONOSPHERE COUPLING

The interaction of the solar wind and the Earth's magnetosphere imposes a dawn-dusk electric field on the high-latitude ionosphere. Several consequences of this interaction are studied with EISCAT, including the distributions of electric potential and field, and of the electrojet and field-aligned currents. The plasma convection pattern is a consequence of the magnetosphere crosstail potential. At other incoherent scatter radars, techniques have been developed to derive the potential pattern from the measurements of plasma velocity as a function of space and time. In general, the previous techniques have one or both of the following limi-

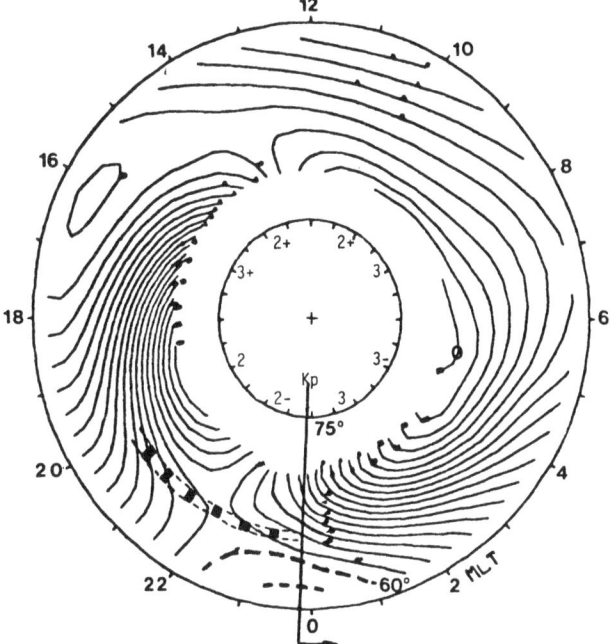

Figure 3—Ionospheric electric-field potential pattern in equipotential intervals of 2 kV. The thick line indicates the ionospheric projection of the geostationary orbit in the dusk sector. The ionospheric potential is deduced from the ionospheric plasma convection pattern measured with EISCAT. The circles indicate 75 to 60 invariant latitude, time is in MLT, and the magnetic activity index *Kp* is indicated around the inner circle. (From *Fontaine et al.* [1986a].)

tations: (1) the potential is assumed to be zero for all local times at the lowest latitude probed by the radar; (2) the convection pattern is tacitly assumed to be constant over a 24-hour period. A new approach has been made with EISCAT data in that the first assumption is not necessary, which gives information regarding the applicability of the second assumption. An example is shown in Figure 3 in which contours of constant electrostatic potential are shown as a function of invariant latitude and magnetic local time [from *Fontaine et al.*, 1986a]. The plasma flow is in the direction of the contour lines. On this rather quiet day, the cross-tail potential was in excess of 44 kV. Further work is in progress relating the potential pattern to auroral activity and solar wind parameters.

Lockwood et al. [1984] have studied plasma convection at high latitudes in connection with substorm activity, and *Farmer et al.* [1984] have studied the field-perpendicular plasma flow during a prolonged period of northward IMF. A survey of simultaneous observations of the high-latitude ionosphere and the interplanetary field with EISCAT and the AMPTE-UKS satellite has been done by *Willis et al.* [1986]. *Van Eyken et al.* [1984] investigated plasma convection up to invariant latitudes as high as 77°.

6. ANISOTROPIC AND NON-MAXWELLIAN ION VELOCITY DISTRIBUTIONS, LARGE ELECTRIC FIELDS, AND FRICTIONAL HEATING

One of the most important energy sources of the auroral upper atmosphere is Joule heating caused by friction between neutrals and the ions that are drifting due to electric fields mapped from the magnetosphere into the ionosphere. Incoherent scatter radars are extremely useful for the investigation of this process, because they can directly measure the plasma velocity and deduce the electric field and the ion temperature, which are the key parameters for this process. For electric fields larger than 40 mV m^{-1}, however, the theory predicts an increasing deviation of the ion distribution function, which would invalidate the currently applied determination of the ion temperature using standard spectrum analysis. Some first observations of this effect were reported by *Perraut et al.* [1984]. Also *Löv-haug and Flå* [1986] found an ion temperature anisotropy with EISCAT. It should be noted that EISCAT is the only incoherent scatter radar where such measurements are unambiguously possible because of the tristatic capability of the UHF radar system.

Clear examples of the characteristic spectra of radar echoes from a nonthermal plasma are shown in Figure 4 [from *Lockwood et al.*, 1987]. For a brief period during a flux transfer event, ion velocities of up to

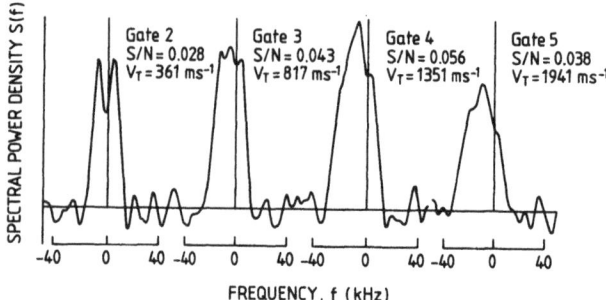

Figure 4—Incoherent scatter spectra measured with the UHF radar at four range gates at the edge of the polar cap. The gate-2 spectrum displays the usual double-hump incoherent scatter spectrum, whereas those triple-hump spectra of gates 3–5 indicate non-Maxwellian velocity distributions, which occur when the ion velocity exceeds the neutral thermal velocity. S/N stands for signal-to-noise ratio and V$_T$ for radial ion drift velocity. (From *Lockwood et al.* [1987].)

2000 m s^{-1} were measured. When the ion drift exceeds the neutral thermal speed, the familiar double-humped spectra of the incoherent scatter signal are replaced by spectra with a single central peak. These spectra have been compared with theoretical studies for a toroidal ion distribution. Similar measurements were carried out on a very disturbed day using the EISCAT tristatic radar. Among other interesting events, several exceptionally intense ion heating events took place in association with strong plasma convection. In the morning sector, ion drift velocities perpendicular to the geomagnetic field reached values up to 3000 m s^{-1} corresponding to a perpendicular electric field $E \approx 160$ mV m^{-1}. Assuming Maxwellian particle distribution functions and an enhanced fraction of molecular ions (compared with a standard model) the estimated ion temperatures, T_i, associated with the strongest convection velocities reached values of 10,000 K, about six times the temperature of the electrons, T_e. The possibility of coherent echoes from plasma irregularities in a nonthermal plasma was also discussed by *Lockwood et al.* [1988].

The validity of the incoherent scatter spectrum analysis for the case of non-Maxwellian ion distribution functions has been investigated. Theoretical incoherent scatter spectra were calculated using distribution functions for such a nonthermal plasma suggested in other works. It was found that for an increasing ratio T_i/T_e the error due to a Maxwellian interpretation of the spectra becomes less significant. *Buchert and La Hoz* [1988] found that good agreement with the measured spectra can be achieved with both Maxwellian and non-Maxwellian distribution functions. Other uncertainties such as the unknown ion composition are also important. It is concluded that within the mentioned uncertainties the measured ion temperature enhancements can fully be attributed to Joule heating.

7. FIELD-ALIGNED PLASMA VELOCITIES, MAGNETOSPHERIC CONVECTION, AND FLUX TRANSFER EVENTS

The effects of the plasma flow through the neutral atmosphere at a high velocity have also been studied using data from EISCAT Common Programmes. Near midnight the sudden reversal of the plasma velocity in the auroral zone caused an immediate increase in ion temperature, which often coincided with substantial Joule heating of the neutral atmosphere. The heating of both the ionized and the neutral atmosphere led to large upward velocities. These rapidly depleted the electron density in the ionosphere and contributed to the formation of the high-latitude electron density trough. This upwelling was observed at the latitude of Tromsö in the form of field-aligned plasma velocities of about 100 m s^{-1}. Further north, at the boundary of the polar cap, upward velocities of over 300 m s^{-1} were measured, associated with very large values of Joule heating [*Williams and Jain*, 1986; *Winser et al.*, 1988a].

Large-scale expansions of the polar cap have been observed by EISCAT. Following a southward turning of the interplanetary magnetic field, as observed by the AMPTE-UKS satellite, the weak and contracted convection pattern was replaced by an enhanced pattern that expanded over the field of view of EISCAT [*Lockwood et al.*, 1986]. On another occasion the expansion was observed jointly by EISCAT in Scandinavia and by the Sondrestromfjord radar in Greenland. The observed rate of expansion showed that the cross-cap electric field potential had increased by 200 kV, and during this expansion the Dynamics Explorer-1 satellite observed the electric field, field-aligned current, and upgoing ion-flow signature of an inverted-V structure, which moved equatorward with the polar cap boundary at the convection speed [*Lockwood et al.*, 1986]. This is a good example of a combination of ground-based and in situ observations, with the former giving the local time variations and the latter the spatial variations.

Frequently bursts of rapid poleward plasma flow were observed with EISCAT in the dayside auroral ionosphere. These were consistent with the predicted ionospheric signature of flux transfer events at the dayside magnetopause. The data revealed spatially confined (<400 km) twin-vortex flow patterns with a poleward flow of up to 2000 m s^{-1} at the center. These were transient and lasted only a few minutes, but they tended to recur within a period of 5–10 min [*Todd et al.*, 1986]. This is a convincing example of the observation of the interaction of the IMF with the Earth's magnetosphere detected remotely by the ground-based incoherent scatter radar.

The large-scale distribution of field-aligned currents, which were deduced from incoherent scatter measurements made on a geomagnetically quiet day, was found to be consistent in location and in magnitude with the usual pattern involving region-1 and region-2 current systems [*Caudal*, 1987]. The phases of the field-aligned current pattern in region 2 and of the corresponding electrostatic convection pattern can be compared. The results show a phase opposition that was found to be irreconcilable with the mechanism generally admitted as the generator of field-aligned currents in the inner magnetosphere; namely, that the charge separation is due to gradient drifts of energetic particles at the inner edge of the ring current. It is suggested that the precipitation of convected energetic electrons may produce azimuthal pressure gradients that can drive field-aligned current systems most of the time. Variations of field-aligned velocities as functions of altitude and latitude were also measured by *Winser et al.* [1988b].

8. IONOSPHERIC COMPOSITION

In the altitude region from about 130–250 km, the ion composition changes from molecular ions at the lower altitudes to atomic oxygen ions at the higher altitudes. The composition profile and its variations have been measured and studied with EISCAT. It was found that, on average, the transition altitude varies with solar zenith angle in a manner similar to that found at midlatitudes. The summer-winter seasonal variation determined at EISCAT was also in agreement with previous work at midlatitudes as well as at auroral latitudes. The variations with zenith angle and season are as would be expected from a photochemical equilibrium model. However, several unusual events were found that deviate from the equilibrium model. The discrepancies have been related to variations in other simultaneously measured ionospheric parameters of the auroral zone, such as electric fields, particle precipitation, and plasma temperature. A new result concerns the effects of particle precipitation, that can either reduce or increase the percentage of atomic oxygen ions, depending on the penetration altitude of the energetic electrons. *Lathuillere and Brekke* [1985] and *Lathuillere et al.* [1986a] have investigated the ionospheric composition and studied the accuracy achievable with early EISCAT experiments. More recent work [*Häggström et al.* 1989] using new alternating code data acquisition schemes allows much improved accuracy in determining composition profiles. *Björnä and Kirkwood* [1988] used another technique to derive the ion composition from a combined ion-line/plasma-line measurement.

9. INVESTIGATIONS OF THE AURORA WITH ISR AND OTHER GROUND-BASED INSTRUMENTS AND SATELLITES

Conventional methods of studying the aurora by optical instruments and the polar electrojet current system with magnetometers have been effectively supported by in situ rocket measurements and, in particular, by ground-based coherent and incoherent scatter radar observations as well as by satellite measurements. In the following chapters we will discuss relevant results of the electrodynamics of the aurora obtained by the combined operation of these instruments.

The EISCAT radars were involved in coordinated observations with the first Swedish satellite, Viking [e.g., *Opgenoorth and Kirkwood*, 1988]. This satellite was equipped with instruments measuring charged particle precipitation, electric and magnetic DC and AC fields, and optical auroral emissions with a temporal and spatial resolution superior to other magnetospheric satellites. An eccentric orbit was chosen such that Viking was initially positioned above or within those magnetospheric regions where auroral particle acceleration occurs. Simultaneous measurements on the same field line in the magnetosphere (by Viking) and in the ionosphere below (by EISCAT) were performed to study the nature of auroral particle acceleration and the associated electric fields and currents. Both Viking and EISCAT data complement each other in the interpretation of the observed variations of ionospheric and magnetospheric parameters in terms of temporal or spatial developments. Several combined experiments were performed to observe active auroral forms and traveling phenomena.

10. AURORAL SUBSTORMS

During a substorm the flux and the energy of precipitating particles increase rapidly and strongly. Consequently the ionization is enhanced quickly, irregularities are created, and the enhanced electric field causes (through Joule heating) the temperature of the ionosphere and neutral atmosphere to increase. During the early phases of a substorm the auroral zone moves equatorward and the polar cap plasma convection pattern changes substantially. Details of all these processes can be studied in detail by ISRs.

Lockwood et al. [1984] have investigated plasma convection during substorms. To study these events in connection with satellite observations the EISCAT dataset collected in operations coordinated with the Viking satellite has been used very effectively. The development of the substorms was studied using auroral images from the Viking satellite and from ground-based networks of all-sky cameras, magnetometers, and ri-

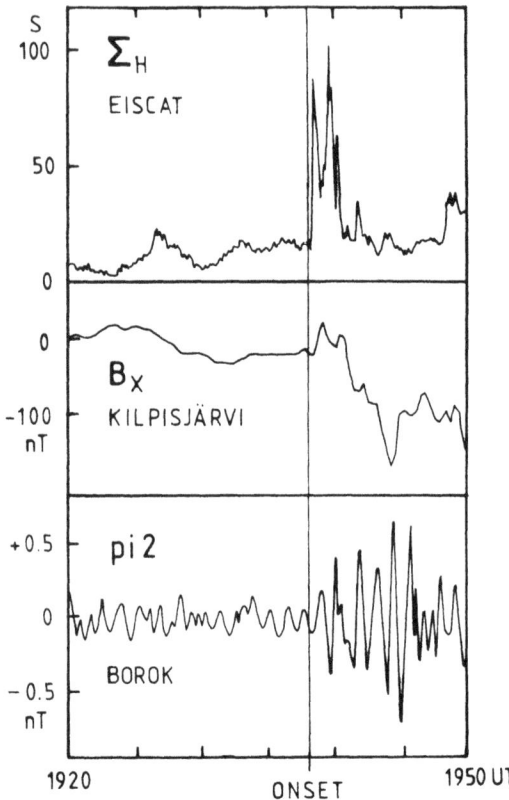

Figure 5—Signatures of substorm onset (from *Kirkwood et al.* [1988]): *Upper panel*: The ionospheric Hall conductance measured by EISCAT shows very high values. Center panel: The local magnetic perturbations (Kilpisjärvi/Finland) show Pi2 pulsations that are correlated to the conductance variations superimposed on the negative substorm bay. *Lower panel*: Simultaneous Pi2 pulsations at the midlatitude station Borok/USSR.

ometers in order to put the EISCAT measurements in context. Results obtained so far show that ionospheric conductances in the zone of diffuse aurora carrying the main substorm electrojet were close to the values that had previously been assumed in modeling work. Conductances in the active aurora leading the westward and northward substorm expansion, however, were 3–5 times higher than previously assumed. Electric field measurements showed that, at least in some cases, the electric field within a break-up arc could be very small and the ionospheric current must have been largely driven by the neutral wind. Further, strong fluctuations in conductances with periods of 1–2 min were seen in almost every case at substorm onset. In those cases where EISCAT measured under the northward expansion of the "auroral bulge," these fluctuations correlated closely with both local and midlatitude Pi2 pulsations. These effects are illustrated in Figure 5 [*Kirkwood et al.*, 1988].

From the combined EISCAT and Viking dataset a number of single substorm studies have been initiated. It has been realized that the current model for the development of the magnetospheric substorm might need some revision. The well-known features of a substorm (e.g., including distinct features and a more pronounced expansion at the westward edge of the active region) appear to be only one of several possibilities for auroral behavior after substorm onset. Cases of stationary intensifications, or pronounced eastward expansions, appear to be as common as typical substorm bulges. The set of EISCAT data coordinated with the Viking satellite data contains sufficient material to start a critical discussion of the classical substorm model [e.g., *Opgenoorth et al.*, 1988].

During the EISCAT/Viking project a campaign was carried out together with the rocket Aureld-VIP-High, which was launched during a magnetic storm to an altitude of 206 km on November 4, 1986 in coordination with a close pass of the Viking satellite. At the launch, the auroral oval was very broad, with an intrusion of the plasma sheet boundary layer into the central plasma sheet. During the rocket flight, EISCAT measured the electric field at 250-km altitude at three nearby points above the rocket trajectory. The EISCAT results, together with the rocket data, indicate that the measurements were made at the edge of a polarization field developed in intense auroral structures as a result of the increased conductivity caused by the precipitating particles [*Sandahl et al.*, 1987].

Signatures of westward traveling surges passing through the EISCAT beam were studied with high time resolution. Interesting particle spectra and electromagnetic and electric field patterns were observed in the vicinity of the surge. Comparison of EISCAT and Viking results is expected to give new insights into the mechanisms of particle acceleration during magnetospheric substorms and ionospheric reactions to the magnetospheric energy input. The fact that the surge decayed soon after the passage of the Viking satellite might help in understanding the mechanisms that limit the release of substorm associated energy in the magnetosphere [*Opgenoorth et al.*, 1988].

11. AURORAL PARTICLE PRECIPITATION

Incoherent scatter radars are efficient tools to study the impact of auroral precipitation on the ionosphere, and a few examples are described here. An auroral precipitation event lasting several hours in the dusk sector has been studied by *Fontaine et al.* [1986a]. *Collis et al.* [1988] studied the plasma convection and the auroral particle precipitation processes related to the main high latitude ionospheric trough, which was also studied in another event by *Schlegel* [1984]. *Schlegel et al.* [1984] also compared EISCAT data with the ener-

getic particle instrument on the satellite GEOS 2. Also the French-Soviet polar satellite ARCAD 3 was used for such a purpose. Electron energy spectra between about 1 and 10 keV were computed from EISCAT measurements of E-region electron density profiles by the so-called "untangle" method. They were in agreement with direct observations on board ARCAD 3 during a diffuse aurora period. They also compared well with the plasma sheet component of 3–10 keV electrons measured on board GEOS 2. The correspondence with large pitch angles suggests a quasi-isotropy of equatorial electron fluxes. The electrostatic electron-cyclotron harmonic waves, which are also observed on board GEOS 2, were not found to be intense enough to cause the strong pitch-angle diffusion of electrons of a few kiloelectron volts. These observations thus allow the conclusion that, even if these electrostatic waves contributed to the auroral precipitation, they could not be its only cause [*Fontaine et al.*, 1986b].

X-ray fluxes and electric fields were measured by *Ullaland et al.* [1985] in a balloon campaign. It was found that the build-up and decay of the X-ray event corresponded very well with the electron density variations. The height variation of the electron density profile, measured with EISCAT, was consistent with changes of the energy of the precipitating electrons. The shift in height may also represent the energy-dependent arrival of drifting electrons from a distant injection region.

12. ELECTRIC FIELDS, CURRENTS, AND INTENSIFICATIONS OF AURORAL ARCS

A special antenna scanning of the EISCAT radars covering a narrow range of latitudes is well suited to explore medium-scale structures such as auroral arcs. For example, a structure of two arcs, revealed by electron density increases in the E-region, was observed by *Girard and Senior* [1989] and is shown in Figure 6. The electric field component tangential to the structure should remain constant across the structure. This property enables the arc direction to be found. The electrodynamic parameters are then computed in the frame of reference of the arc structure. In particular, the polarization electric fields related to the structure are estimated, and the electric current circulation perpendicular and parallel to the magnetic field lines can be determined. The field-aligned current is shown in the lower panel of Figure 6 as a function of distance from the center of a two-arc structure. Pairs of adjacent current sheets were observed with densities that reached 100 μA m^{-2} above conductivity gradients in the ionosphere. Both upward and return current regions appeared to be confined over limited regions of scales of

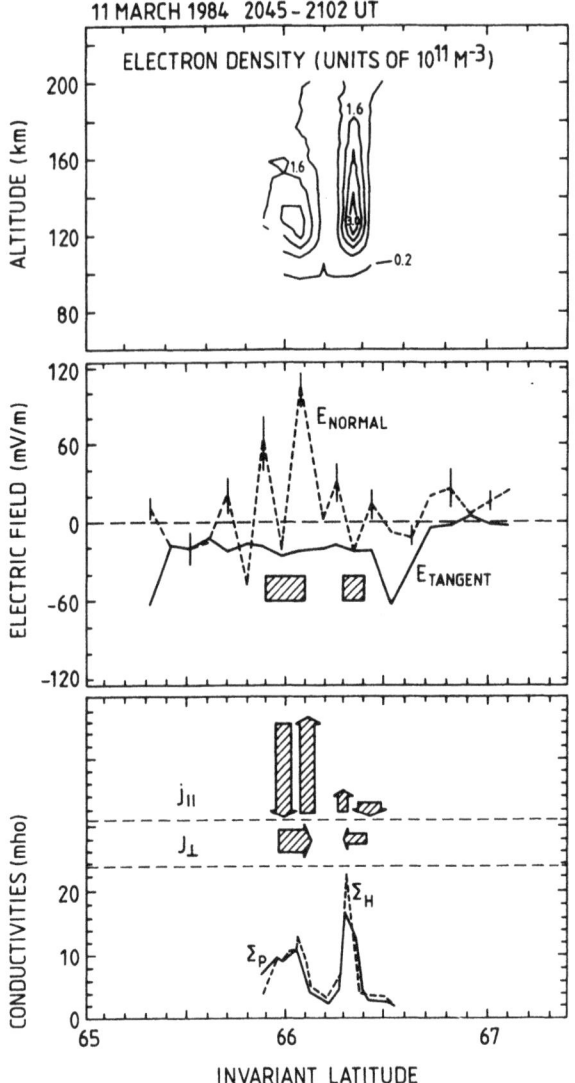

11 MARCH 1984 2045 - 2102 UT

Figure 6—The structure of two auroral arcs that are identified by two marked increases of E-region electron densities measured by EISCAT (*upper panel*). The electric field, deduced in tangential and normal direction to the arc (*center panel*), is measured with the EISCAT tristatic radar. From the conductances deduced from EISCAT data the current circulation (*lower panel*) around the arcs is deduced. (After *Girard and Senior* [1989].)

15–20 km at around 120-km altitude [*Girard and Senior*, 1989].

A special technique was applied to detect the natural plasma lines, called the chirp technique [*Hagfors et al.*, 1984]. Frequency shifts of the plasma lines can be attributed to field-aligned drift components along the magnetic field line. The preliminary data analysis has shown that apparently field-aligned current densities up to 20 μA m^{-2} could be detected. This technique offers great potential for auroral arc studies.

To understand structures and the physics of auroral arcs it is important to know how the auroral arcs move with respect to the background plasma. Experiments were carried out by *Haerendel et al.* [personal communication, 1988] to compare EISCAT radar measurements of the ion velocity with the motion of arcs. Data were obtained for one event when a quiet arc occurred above Abisko (between Tromsö and Kiruna) before the onset of a substorm. It was found that both the arc and the F-region plasma moved slowly southward in an oscillatory manner with periods between 3 and 4 min. The arc, however, was faster by an average of 24 m s^{-1}. The phase of its oscillation preceded that of the plasma by 2 min. The tangential velocity of the F-region plasma was directed westward. The results can be interpreted in the framework of oblique Alfvén waves reflected at the ionosphere. The theory provides a relation between the average relative arc motion (24 m s^{-1}), the period of the waves (3–4 min) and the thickness of the arc (a few kilometers), which is confirmed by the data. This interpretation, however, is not yet regarded as being definitive.

During a pulsating aurora, thin layers of ionization were observed with a stationary, field-aligned antenna. The observed thin layers of ionization are typically seen in no more than two radar range gates of 2.7-km separation. In spite of an apparent broadening of the observed layer by too long an integration (1 min), the derived electron density profiles can hardly be explained by normal collisional ionization by precipitating electrons. The observed direction of the field-aligned ion drift velocity in this case excludes neutral wind shear, the agent for the formation of nonauroral sporadic-E layers, as an alternative production mechanism. Hence, it is concluded that very thin layers of auroral luminosity were observed [*Wahlund et al.*, 1989].

The development of an auroral arc, from diffuse to discrete form with subsequent large-scale folding, was recorded by an all-sky camera at Kiruna and with the EISCAT radar (Figure 7). The major fold in the arc propagated westward directly over the radar. Prior to the main fold, two attempted foldings were evident in the optical data, both of which were accompanied by large increases in F-region ion temperature and the electric field. These observations were taken as evidence that the distortion of the arc was the result of a Kelvin-Helmholtz instability in the magnetosphere [*Steen et al.*, 1988a]. The transition from the leading edge of the fold was characterized by an increase of E-region electron density, which is clearly seen in the lower panels of Figure 7 as well as in the high-time resolution electron density variation shown in Figure 8. We also notice from Figure 8 the fairly rapid (time scale of 1 min) increase of the F-region electron density, which tend-

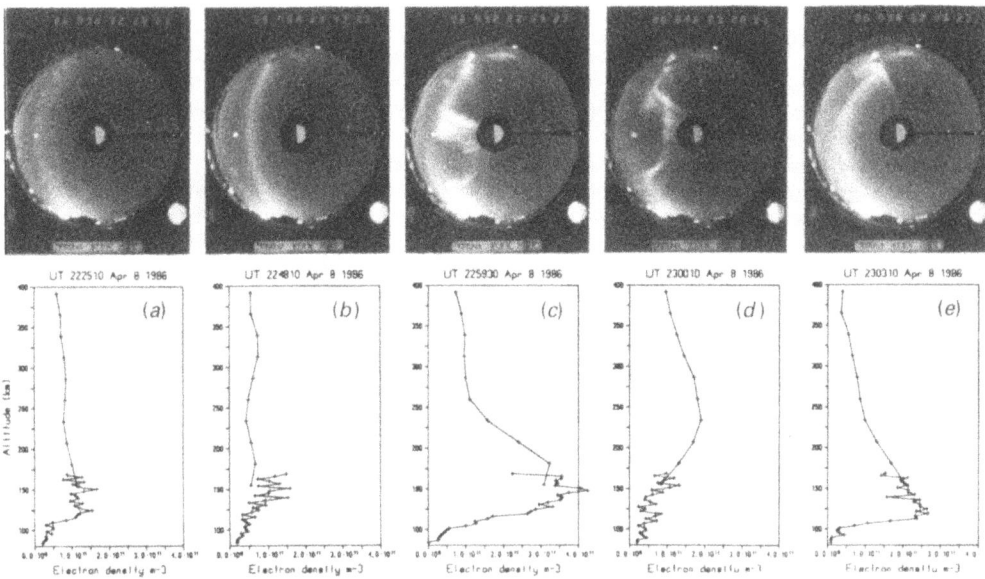

Figure 7—An eastward drifting fold in an auroral arc observed by all-sky camera in Kiruna (*upper panels*), and simultaneously measured electron density profiles (*lower panels*) from EISCAT during the passage of a fold. The leading edge of the fold (*c*) is well characterized by increased E-region densities, but the hole in the fold (*d*) is associated with a broad F-region electron density. (From *Steen et al.* [1988a].) (This figure also appears in color: Plate 46.)

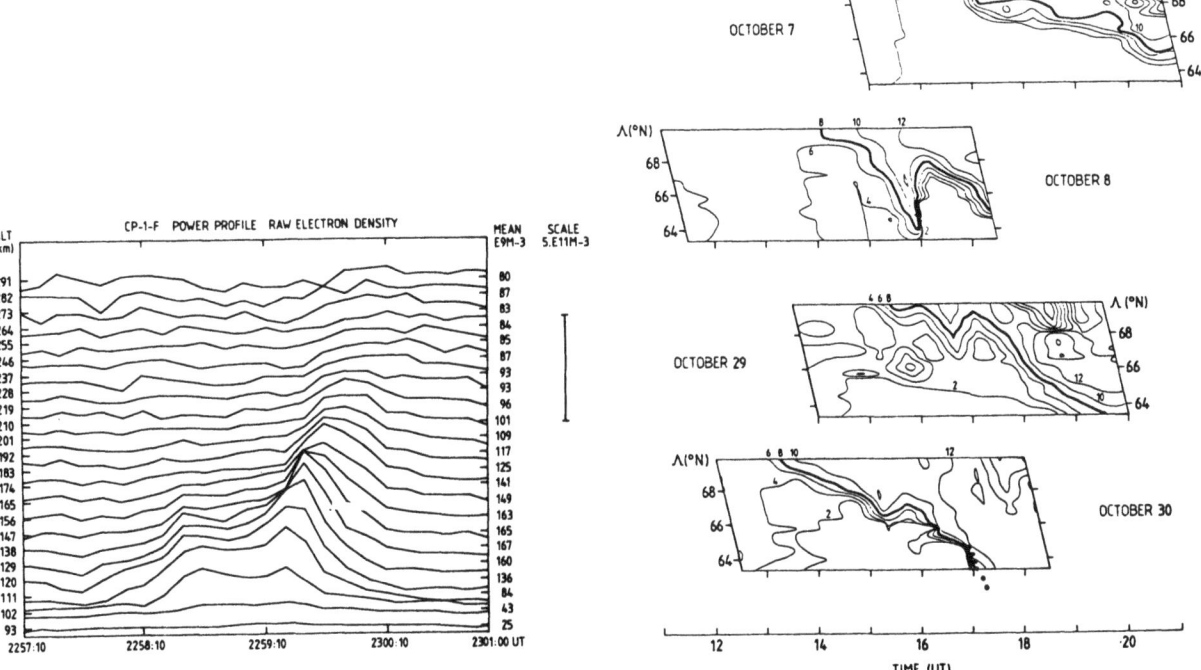

Figure 8—Electron density measured with EISCAT in the E- and F-region using a high time resolution of 10 seconds. The leading edge of the fold in the aurora structure enters the radar beam after 2259:10 UT. (From *Steen et al.* [1988a].)

Figure 9—Latitudinal distribution of electron density measured with EISCAT at 120-km altitude around times when the high-latitude trough occurs. The circles on the plot of October 30 show the location of the equatorward border of the diffuse aurora as determined from all-sky camera photographs. (From *Collis and Häggström* [1988].)

429

ed to follow the E-region increase when the hole of the fold was reached. Latitudinal distributions of E-region electron density, connected with the auroral oval, are shown in Figure 9 (from *Collis and Häggström* [1988]). From these the equatorward edge of the oval and its temporal and latitudinal variations can be investigated.

Incoherent scatter measurements were also used to study the relationship between reductions of the F-region meridional wind and auroral intensifications. In particular, the EISCAT measurements showed that the electric field and the neutral wind had consistent patterns of behavior well before any evidence of auroral brightening. It was suggested [*Steen et al.*, 1988*b*] that the increased electric field gave rise to increased Joule heating that blocked the flow of the equatorward wind. It was also speculated that these low-altitude changes could act as some sort of precursor in the ionosphere-thermosphere interactions associated with auroral intensifications.

To study auroral arc physics the CAESAR program (Coordinated Auroral Experiment using Scatter And Rocket investigations) was performed in January 1985. This combined high resolution in situ measurements using rocket-borne instrumentation and ground-based observations with EISCAT. The electric field configuration was investigated in the vicinity of an auroral arc. Therefore EISCAT was operated to measure the plasma drift vector using a scan with seven common volumes on magnetic field lines that were also on the nominal CAESAR trajectory. One scan took 15 min (the same time as the expected rocket flight duration) and these measurements were conducted before, during, and after the rocket flight. The EISCAT measurements showed that before and after the CAESAR II flight the plasma drift corresponded to the normal convection pattern. The disturbance started when the plasma drift turned to the west and increased in magnitude up to ≈ 1000 ms^{-1}. A strong eastward Hall current or electrojet was observed, in agreement with the magnetometer measurements [*Rinnert et al.*, 1986; *Wilhelm et al.*, 1987].

Figure 10 is a composite of the EISCAT and CAESAR II plasma drift measurements. The dense series of arrows represents plasma drifts, calculated from DC-electric fields measured by the rocket and assuming $\mathbf{E} \times \mathbf{B}$ drift projected along magnetic field lines down to the altitude of the EISCAT measurements at 140 km. Both datasets are in excellent agreement for latitudes below about 71.6°, which is south of the arc within the eastward electrojet. The observed difference in the plasma drifts of both datasets at latitudes above 71.6° north is real. It indicates a localized precipitation event characterized by an upward field-aligned current centered at about 14.5° east and 72° north (and probably a corresponding downward directed line current centered at

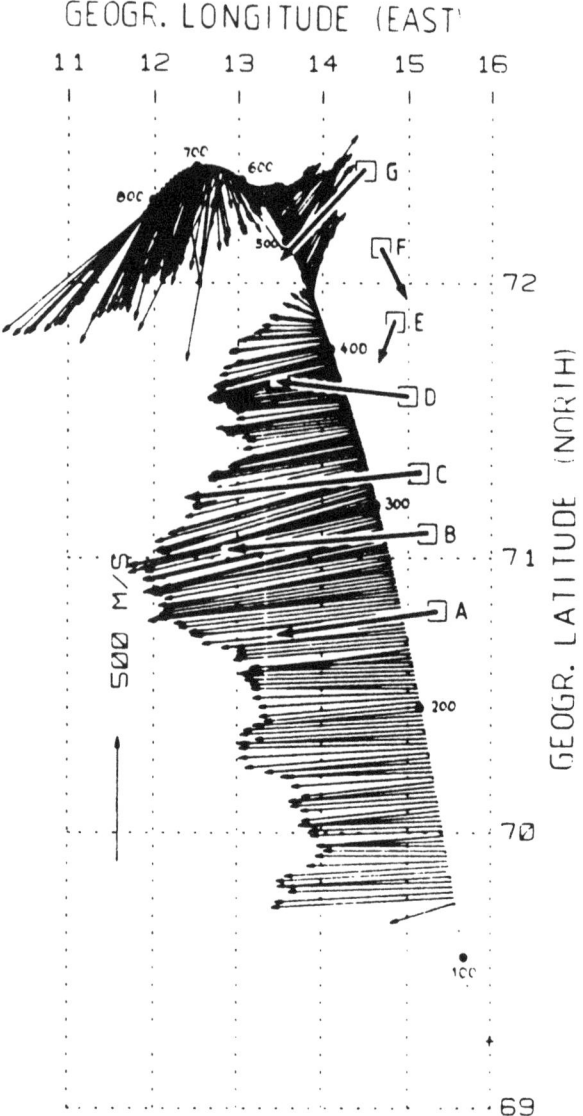

Figure 10—The comparison of plasma drifts measured by EISCAT at 140-km altitude and the plasma drifts deduced from the rocket-measured DC-electric fields along the CAESAR II rocket trajectory, projected along the magnetic field line. (From *Wilhelm et al.* [1987].)

13.5° east and 72.5° north). Other plasma parameters measured by EISCAT and on board the rocket payload were found to be in good agreement [*Schlegel and Oyama*, 1987].

13. INCOHERENT SCATTER AND MAGNETOMETER OBSERVATIONS OF PULSATIONS

During geomagnetic storms the electrojet current rapidly changes in strength and location due to the rapid changes in electron density and ionospheric conductivity. This causes variations of the geomagnetic field, called pulsations, which are measured with magnetometers at the Earth's surface. Often so-called Ps 6 pulsations, which have periods of typically 10–40 min, are observed, almost exclusively in the morning sector. On April 21, 1985 between 0300 and 0415 UT during the recovery phase of a geomagnetic substorm, which was one in a series of several strong substorms, an unusually strong Ps 6 pulsation event was recorded by the EISCAT magnetometer cross in northern Scandinavia. Simultaneously, EISCAT measured E- and F-region plasma parameters with a latitudinal scanning program. The observed four cycles of the pulsations had an amplitude of about 1200 nT and a quasi-period between 15–20 min. The integrated conductivities showed amplitudes of 50 and 20 Siemens for the Hall and Pedersen conductivities, respectively, and they had periodicities and phases that were in correspondence with those of the magnetic pulsations. By cross-correlating the measurements from different magnetometer stations, the eastward drift velocities of the pulsations fronts were calculated, yielding values between 100 and 1300 m s^{-1}. Both the direction and magnitude of the drift velocity agree roughly with the background ion drift velocities measured by EISCAT. The wavelengths of the pulsations were estimated to be about 1600 km [*Buchert et al.*, 1988].

The Ps 6 pulsations are known to be relatively stationary eastward traveling structures rather than wave phenomena. The corresponding auroral forms are the omega-bands, luminous "tongues" extending poleward from a band of diffuse aurora. Combining the magnetometer data and radar measurements, it was possible to reconstruct the two-dimensional distributions of the conductances and electric field vectors as shown in Figure 11. The conductance distribution is quite similar to that of the optical intensity of an omega-tongue. The electric field vectors tend to be parallel to the gradients of the conductances, and the electric field strength is reduced in areas of enhanced ionization. These observations are explained by an electrostatic model, which assumes that both the E-region ionization and the upward field-aligned current are caused by precipitating energetic electrons. Starting from a given distribution of field-aligned currents, ionospheric conductance, electric fields, and currents were calculated. The observed conductances as well as magnetic and electric fields are reproduced very well by this model [*Buchert et al.*, 1988].

Pulsations observed by magnetometers as well as by riometers were studied in detail and compared with rapid fluctuations in electron density measured with EISCAT. These fluctuations occurred at all ionospheric heights and were interpreted as evidence for a relative harden-

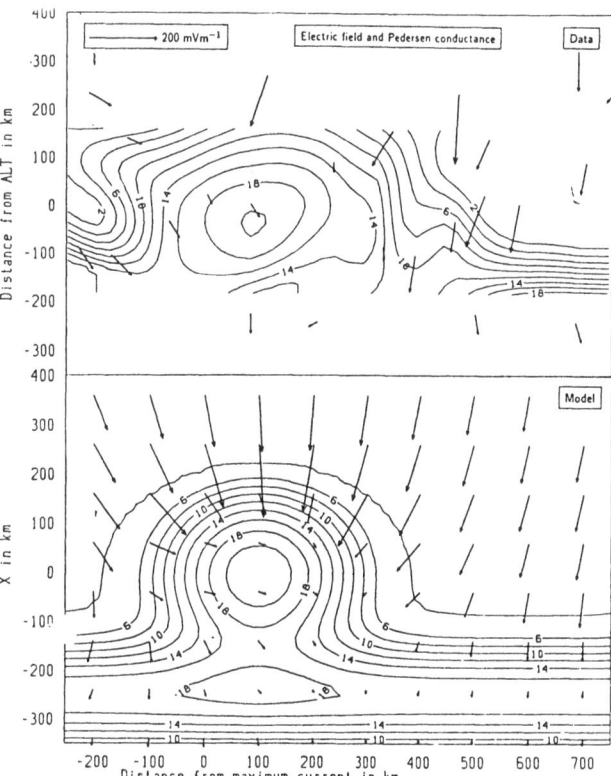

Figure 11—The two-dimensional distribution of the conductance and the electric fields in an omega-band as measured with the radar (upper panel) is compared in the lower panel with results from model computations of the omega-band. (From *Buchert et al.* [1988].)

ing and softening of charged particle precipitation accompanied by very large changes in the flux of energetic electrons [*Devlin et al.*, 1986].

14. DAMPING OF PULSATIONS BY JOULE HEATING

A study of the ion heating due to ULF (ultra-low frequency) Pc 5 magnetic pulsations with periods as low as 3 min was performed by *Lathuillere et al.* [1986b]. Ion temperature fluctuations measured by EISCAT were well correlated with the Pc 5 pulsations, the ion temperature enhancements being obtained from the ion velocity measurements. *Duboin* [1986] has determined the corresponding Joule and particle heating rates.

It has been possible to measure ion temperatures and velocity perturbations with a time resolution of 30 s during Pc 5 pulsations determined from ground-based magnetometer observations. The precise correlation between the magnetic observations and the large velocity and temperature perturbations is consistent with theoretical predictions that Joule heating is the main damping mechanism of the pulsations. Detailed quan-

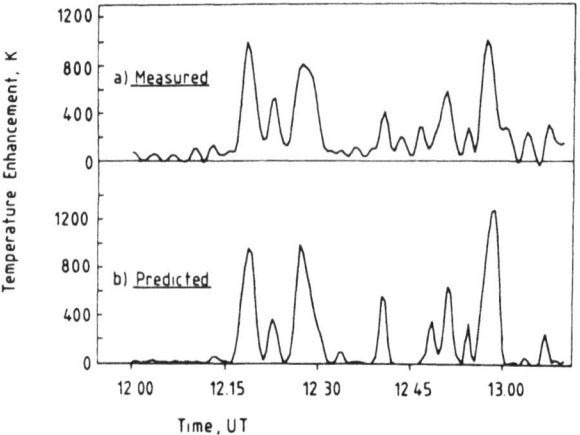

Figure 12—Temperature enhancement during pulsations, measured with EISCAT and deduced from magnetometer measurements. (After *Crowley et al.* [1988].)

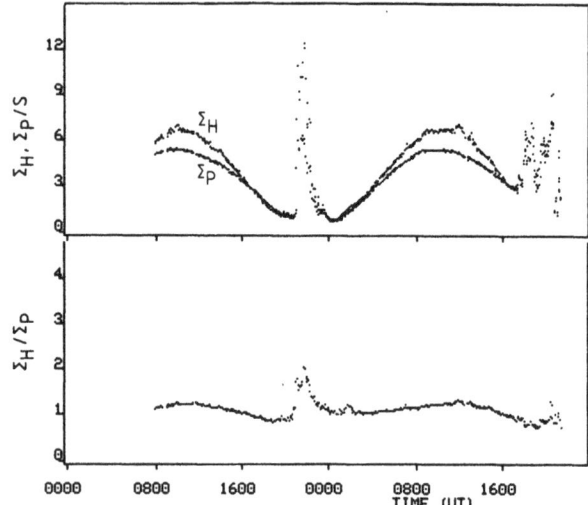

Figure 13—Hall and Pedersen conductances (upper panel) deduced from EISCAT electron density measurements on May 12, 1987 and the ratio (lower panel) of these conductances. The quasi-periodic diurnal variation is due to the changing solar zenith angle and the rapid increase is due to a superimposed particle precipitation event. (From *Brekke* [1988].)

titative analysis has shown that Joule heating is sufficient to account for all the damping. The energy dissipation by the pulsations represents about 1% of the energy deposited in the atmosphere of the northern hemisphere during a typical magnetospheric substorm. This emphasizes the importance of hydromagnetic waves in transferring energy from the outer regions of the magnetosphere to the high-latitude ionosphere. Figure 12 shows a record of the measured ion temperature perturbations over 1100 K at 312-km altitude (upper panel) and the predicted perturbation derived from ion velocity measurements (from *Crowley et al.* [1988]).

15. IONOSPHERIC CONDUCTIVITIES

Electron densities measured by EISCAT have been used to make a study of ionospheric conductivities and conductances. It is generally noticed that the Hall and Pedersen conductances are well-behaved functions of the solar zenith angle during quiet-time conditions. The Hall-to-Pedersen conductance ratio is found to be about 1.3 at quiet times but can increase to 4 during disturbed conditions (e.g., Figure 13). A method has been developed by *Brekke et al.* [1989] to derive energy spectra of the precipitating particles for electron density profiles observed during disturbed conditions. The characteristic energies are correlated with the Hall-to-Pedersen conductance ratios calculated from the same electron density profiles. An empirical relation was found that allows the characteristic energy to be derived from observations of the conductance ratios during disturbed conditions. A statistical study of E-region conductivities at low solar activity covering 2 years, 1985–1986, has been carried out by *Schlegel* [1988]. The results show quantitatively the increase of the conductivities and the decrease of the height of the conduc-

tivity maxima with increasing magnetic activity. *Brekke* [1988] has carefully reviewed all these investigations.

16. E-REGION IRREGULARITIES AND PLASMA WAVES

Strong electric fields, which drive the auroral electrojet and which can be determined by ISRs when measuring the $\mathbf{E} \times \mathbf{B}$ ion drift in the F-region, also create small-scale irregularities in the E-region, which can be studied by coherent radars (also known as auroral radars). F-region drifts measured by EISCAT and E-region drifts measured by the Sweden- and Britain-Radar Experiment (SABRE) were compared by *Robinson* [1986]. These comparisons verified that SABRE progressively underestimates the higher drift velocities. This confirms a result that was noticed earlier by *Nielsen and Schlegel* [1985] when comparing STARE auroral radar and EISCAT ISR data. New theories of current-driven plasma instabilities in the high-latitude E-region indicate that nonlinear plasma wave heating effects limit the phase speed of the plasma irregularities to the ion acoustic speed. The comparisons of EISCAT and SABRE velocities are broadly consistent with this theory.

Williams et al. [1988a] investigated the E-region volume from which auroral radars in Karelia, USSR, received coherent auroral echoes. A comparison was made between the electron concentration and electric field strength measured by EISCAT and the strength of the echoes detected by the auroral radars. This indicated that for electric fields stronger than 30 mV m^{-1} the level of plasma wave turbulence saturates at a fluctuation level of about 2.5%. A comparison was also made between the electric field strength and the enhancement of the electron temperature at the 108-km altitude. This also proved consistent with the theory of nonlinear wave heating.

Providakes et al. [1988] performed simultaneous EISCAT observations in the volume probed by the Cornell University Portable Radar Interferometer (CUPRI) 46.9-MHz radar in Lycksele. CUPRI is designed to study echoes from auroral plasma instabilities with high temporal and spatial resolution. Of particular interest are sharp-peaked spectra showing high Doppler shifts; namely, high drift velocities. These are thought to be connected to ion-acoustic waves traveling in the auroral electrojet at temporarily elevated plasma acoustic velocities, because of similarly elevated electron temperatures. The EISCAT program was specifically modified to allow high-time resolution measurements applying an eightfold 5-pulse multipulse scheme. On several occasions high electron temperatures have been observed with EISCAT at the same time and location where the CUPRI system detected high Doppler shifts. Strong plasma waves with very large phase velocities were observed at precisely the altitudes and times at which the heating was observed. It is concluded that the electron temperature increases were caused by plasma wave heating and not by either Joule heating or particle precipitation [*Providakes et*

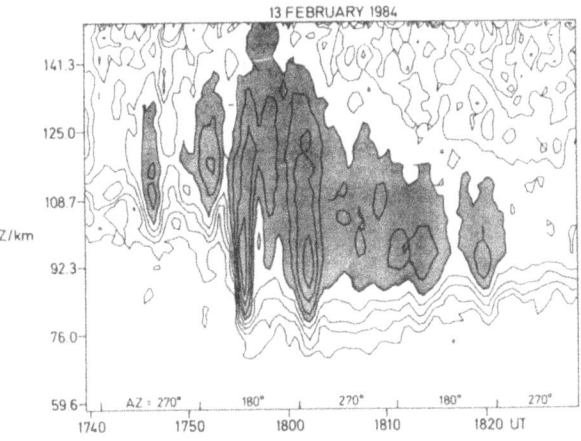

Figure 14—Electron density contour plots measured with oblique beams toward the west (270°) and the south (180°) with a time resolution of 30 s. The maximum electron density was 8×10^{10} m^{-3}. The contour lines are drawn in a logarithmic scale, where the spacing between two lines corresponds to a difference in electron density by a factor of $2^{1/2}$. The clustering of contour lines indicates the strong temporal variability of the electron density due to varying precipitation. (From *Röttger* [1984].)

al., 1988]. The current hypothesis is that these high electron temperatures are the result of wave heating by unstable plasma waves associated with auroral precipitation and electrojet currents. There is evidence of the existence of unstable ion-acoustic waves traveling at anomalously high acoustic speeds associated with these heating events. The high acoustic speeds are the result of the high electron temperatures that, in turn, are produced by the unstable waves themselves (from *Häggström et al.*, personal communication [1988]).

Almost all the observations of plasma irregularities with coherent auroral radars were done at frequencies less than a few hundred MHz. Recently *Moorcroft and Schlegel* [1988] used the EISCAT UHF radar to detect coherent echoes from the auroral E-region on 933 MHz. It appeared that these echoes, which must result from plasma waves with scales of 16 cm, occurred at times when the mean F-region plasma drift exceeded several 100 m s^{-1}.

17. EXTENSION OF E-REGION OBSERVATIONS INTO THE D-REGION AND OBSERVATIONS OF THE LOWER THERMOSPHERE AND MESOSPHERE

Precipitating particles cause substantial ionization in the E-region and frequently also in the D-region, if their energy is sufficiently high. An example of such electron density enhancements detected with the EISCAT UHF radar is shown in Figure 14 (from *Röttger* [1984]). We notice the rapid temporal changes as well as the enhancement of electron density down to 75-km altitude. The measured electron density profiles can be used to study the particle precipitation itself, to deduce the radio wave absorption, or to use it as a tracer to study the structure and dynamics of the mesosphere. *Collis et al.* [1986] have investigated the ionospheric D-region signatures of the substorm onset phase and growth phase. D-region electron densities measured during the polar cap event of February 16, 1984 have been compared with production rates computed from proton fluxes measured on the geosynchronous satellite GEOS. Profiles of the effective recombination rate were determined that show general consistency with previous estimates but differ in detail. There is evidence for a progressive change of the recombination coefficient as the polar cap absorption event proceeds [*Hargreaves et al.*, 1987]. The structure of the D-region during an auroral absorption event, caused by enhanced D-region electron density, was also studied by *Ranta et al.* [1985] and *Devlin et al.* [1986].

Measurements of the incoherent scatter spectra give information on sporadic E-layers [*T. Turunen et al.*, 1988] as well as estimates of the ion composition in these layers [*Huuskonen et al.*, 1987], which can oc-

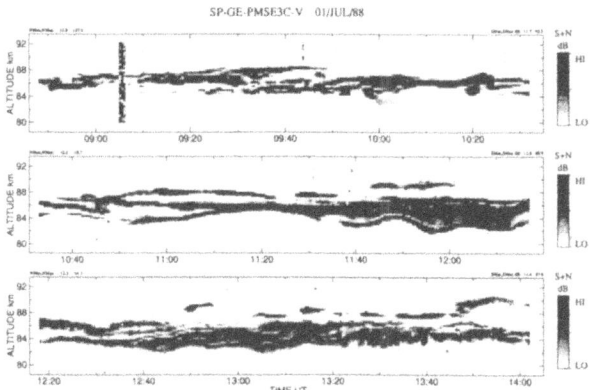

Figure 15—A grey-scale plot of polar mesosphere summer echoes observed with the EISCAT VHF radar showing characteristic filaments as well as undulations due to gravity waves. (After *La Hoz et al.* [1989].)

cur as low as 90-km altitude [*Kirkwood and Collis*, 1989]. The ion-neutral collision frequency [*Huuskonen et al.*, 1986; *Nygren et al.*, 1987] can be deduced down to the lower E-region. Indications about D-region ion chemistry can also be obtained from ISR spectra. The dependence of these parameters on auroral activity, however, still needs to be investigated in detail. *Hall et al.* [1988] have estimated the negative ion-to-electron concentration ratio. *Collis et al.* [1988] and *E. Turunen et al.* [1988] have deduced positive ion masses in the D-region. The latter appear to be very large due to clustering of ions in the cold polar mesopause region in summer.

In addition to the usual incoherent scatter echoes from the D-region, strong coherent echoes were observed in summer by the EISCAT VHF radar and later also by the UHF radar [*Hoppe et al.*, 1988; *Röttger et al.*, 1988]. Following conventional theories of radar scattering from mesospheric turbulence, such echoes should not be observable, because electron density irregularities stirred by neutral turbulence should not be detectable at the short wavelengths of the EISCAT radars. Heavy cluster ions, however, could have the effect that the ambipolar diffusion of electrons would be weakened and fluctuations in the electron gas would still be possible at very short scales to cause the unusual scatter [*Kelley et al.*, 1987]. An example of the resulting polar mesosphere summer echoes is depicted in Figure 15 (from *La Hoz et al.* [1989]). We note that these structures are completely different from the structures observed during electron precipitation events (see Figure 14). These latter structures show vertical filamentation due to the preferred direction of precipitation parallel to the almost vertical geomagnetic field lines. The structures of the polar mesosphere summer echoes, on the other hand, are almost horizontally

stratified and filamented, as well as being modulated by the occurrence of atmospheric gravity waves and turbulence. This points to their control by neutral atmospheric structure and dynamics. Although *Rishbeth et al.* [1988] claim that there is a relation of these strong polar mesosphere summer echoes to magnetic, i.e., auroral, activity, these observations need to be investigated also in terms of the dynamical and chemical structure of the polar mesosphere.

18. COUPLING WITH THE NEUTRAL ATMOSPHERE: MEAN WINDS, TIDES, AND ATMOSPHERIC GRAVITY WAVES

Mean neutral winds and temperatures in the auroral thermosphere have been studied, for instance by *Alcaydé and Fontanari* [1986], *Kofman et al.* [1986], and *Winser et al.* [1988c]. The measurements performed by EISCAT in the lower thermosphere have been compared with models. The neutral temperature was found to be 15 K less than that predicted by the MSIS model [*Kofman et al.*, 1986]. The neutral mass inferred from the temperature and scale height in winter at 100 km also differs from that of the MSIS model.

Neutral wind and temperature changes due to atmospheric tides were also studied with the EISCAT UHF radar. The dominant periodicity in wind velocity, temperature, and neutral air density is due to the semidiurnal tide [*Virdi et al.*, 1986; *Kirkwood*, 1986; *Röttger and Meyer*, 1987]. The combination of meteorological rocket and incoherent scatter measurements showed mean tidal velocities of 5 m s^{-1} down to 80-km altitude. Figure 16 shows an example of wind profiles measured with EISCAT during a proton event, which increased the D-region electron density and thus the radar signal-to-noise ratio to allow low-altitude velocity measurements. Because the neutral atmosphere tidal winds interact with the plasma drift and the electrojet current system in the lower E-region, particular normal modes of the lower thermosphere could also be excited (e.g., *Larsen and Mikkelsen* [1987]). Their periods are comparable with those of the tidal variations.

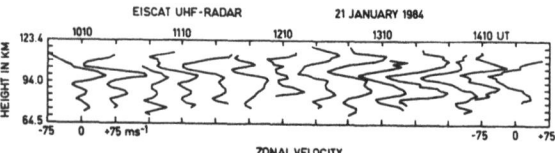

Figure 16—Time series of zonal ion velocity profiles averaged over 9.3 min in 20-min intervals, which was measured during a solar proton event. The deduced ion drift velocity comprises the neutral wind velocity up to an altitude of about 105 km and shows a long-period wave or tidal structure in the neutral atmosphere. (From *Röttger and Meyer* [1987].)

A specific mechanism of energy and momentum transfer from the magnetosphere to the ionosphere and to the neutral atmosphere is through the generation of atmospheric gravity waves. These are manifest in the ionosphere as traveling ionospheric disturbances. Sources of gravity waves are particularly found in the polar lower thermosphere through Joule and particle heating as well as the Lorentz force. By evaluating EISCAT measurements, *Crowley and Williams* [1987] showed that Joule heating is usually most effective at altitudes between 120 km and 130 km, whereas particle heating and the Lorentz force show their main contributions below 110-km altitude. All three quantities exhibit strong temporal variations due to the fluctuation in electron density. In the case of Joule heating and the Lorentz force, these are also due to electric field fluctuations. Joule heating is usually an order of magnitude larger than heating by particle precipitation. Both Joule heating and Lorentz forcing were effective in generating large-scale gravity waves, which could be detected at lower latitudes than the source region in the polar atmosphere. The auroral source often had a marked periodicity, corresponding both to intrinsic periodicities in the bursts of particles as well as in the magnetospheric electric field. These effects were intensively studied during the World Acoustic Gravity Wave Study in October 1985, when the EISCAT radar ran uninterrupted for more than 100 hours [*Williams et al.*, 1988*b*]. Gravity waves in the mesosphere were also detected because sufficient electron density was produced in the D-region by particle precipitation to allow velocity measurements (e.g., *Hall et al.* [1987]). A relation between gravity wave occurrence and auroral activity was not obvious. *Williams et al.* [1988*b*] believe that these neutral atmosphere waves were generated by atmospheric temperature variations occurring in the vicinity of the solar terminator.

The polar mesosphere and lower thermosphere are regions where processes from below, such as tides and gravity waves from the lower and middle atmosphere, and from above, such as energetic particles and electric fields from the magnetosphere, compete with each other. The EISCAT radar systems are very suitably located in the auroral zone to study these processes by means of the incoherent scatter radar technique. The EISCAT radar systems are now also used as a meso-sphere-stratosphere-troposphere (MST) radar (e.g., *Röttger* [1987]) to study the interaction region of the upper and lower atmosphere in polar latitudes.

19. CONCLUSION

An overview has been given of the capabilities of the EISCAT radars to contribute to investigations of auroral phenomena, related effects on the ionospheric plasma, and the ambient neutral atmosphere as well as the coupling from the magnetosphere to the ionosphere, thermosphere, and the middle atmosphere. It is evident that a multinational ISR facility such as EISCAT can contribute effectively to future research on solar-terrestrial physics, particularly ionosphere-thermosphere coupling and the response to energy and momentum inputs from above and below. These are priority areas of the international Solar-Terrestrial Energy Program (*STEP*, 1988) of the Scientific Committee on Solar Terrestrial Physics.

ACKNOWLEDGMENT—I appreciate the collaboration of many scientists from the EISCAT-associate countries, and with EISCAT staff who helpfully provided material and assistance for this overview. It is a summary of published papers and a compressed version of recent EISCAT Annual Reports, for which the support of S. Buchert, P. Collis, and C. La Hoz was invaluable. The EISCAT Scientific Association is funded by the Centre National de la Recherche Scientifique (France), Suomen Akatemia (Finland), Max-Planck-Gesellschaft (Federal Republic of Germany), Norges Almenvitenskapelige Forskningsråd (Norway), Naturvetenskapliga Forskningsrådet (Sweden), and the Science and Engineering Research Council (United Kingdom).

REFERENCES

Alcaydé, D., and J. Fontanari, "Neutral Temperature and Winds from EISCAT CP-3 Observations," *J. Atmos. Terr. Phys.*, **48**, 931–947 (1986).

Baron, M., "The EISCAT Facility," *J. Atmos. Terr. Phys.*, **46**, 469–472 (1984).

Beynon, W. J. G., and P. J. S. Williams, "Incoherent Scatter of Radio Waves from the Ionosphere," *Rep. Prog. Phys.*, **41**, 909–956 (1978).

Björnå, N., and S. Kirkwood, "Derivation of Ion Composition from a Combined Ion-Line/Plasma-Line Incoherent Scatter Experiment," *J. Geophys. Res.*, **93**, 5787–5793 (1988).

Brekke, A., "The Polar Ionosphere," *World Ionosphere/Thermosphere Study (WITS) Handbook*, 1, C. H. Liu and B. Edwards, eds., SCOSTEP Secretariat, Dept. Electri. Comp. Eng., Univ. of Illinois, Urbana, IL, 94–126 (1988).

Brekke, A., C. M. Hall, and T. L. Hansen, "Auroral Ionospheric Conductances During Disturbed Conditions," *Ann. Geophys.*, in press (1989).

Buchert, S., and C. La Hoz, "Extreme Ionospheric Effects in the Presence of High Electric Fields," *Nature, Lond.*, **333**, 438–439 (1988).

Buchert, S., W. Baumjohann, G. Haerendel, C. La Hoz, and H. Lühr, "Magnetometer and Incoherent Scatter Observations of an Intense Ps 6 Pulsation Event," *J. Atmos. Terr. Phys.*, **50**, 357–367 (1988).

Caudal, G., "Field-Aligned Currents Deduced from EISCAT Radar Observations and Implications Concerning the Mechanism that Produces Region 2 Currents," *J. Geophys. Res.*, **92**, 6000–6012 (1987).

Collis, P. N., and I. Häggström, "Plasma Convection and Auroral Precipitation Processes Associated with the Main Ionospheric Trough at High Latitudes," *J. Atmos. Terr. Phys.*, **50**, 389–404 (1988).

Collis, P. N., S. Kirkwood, and C. M. Hall, "D-Region Observations of Substorm Growth Phase and Onset Observed by EISCAT," *J. Atmos. Terr. Phys.*, **48**, 807–816 (1986).

Collis, P. N., T. Turunen, and E. Turunen, "Evidence of Heavy Positive Ions at the Summer Arctic Mesopause from EISCAT UHF Incoherent Scatter Radar," *Geophys. Res. Lett.*, **15**, 148–151 (1988).

Crowley, G., N. M. Wade, J. A. Waldock, T. R. Robinson, and T. B. Jones, "High Time-Resolution Observations of a Periodic Frictional Heating Associated with a Pc5 Micropulsation," *Nature, Lond.*, **316**, 354–356 (1988).

Crowley, G., and P. J. S. Williams, "Observations of the Source and Propagation of Atmospheric Gravity Waves," *Nature, Lond.*, **328**, 231–233 (1987).

Devlin, T., J. K. Hargreaves, and P. N. Collis, "EISCAT Observations of the Ionospheric D-Region During Auroral Radio Absorption Events," *J. Atmos. Terr. Phys.*, **48**, 795–805 (1986).

Duboin, M. L., "Heating Rates Measured by EISCAT: Latitudinal Variations," *J. Atmos. Terr. Phys.*, **48**, 921–930 (1986).

van Eyken, A. P., H. Rishbeth, D. M. Willis, and S. W. H. Cowley, "Initial EISCAT Observations of Plasma Convection at Invariant Latitudes 70°–77°," *J. Atmos. Terr. Phys.*, **46**, 635–641 (1984).

Farmer, A. D., M. Lockwood, R. B. Horne, B. J. I. Bromage, and K. S. C. Freeman, "Field-Perpendicular and Field-Aligned Plasma Flows Observed by EISCAT During a Prolonged Period of Northward IMF," *J. Atmos. Terr. Phys.*, **46**, 473–488 (1984).

Folkestad, K., T. Hagfors, and S. Westerlund, "EISCAT: An Updated Description of Technical Characteristics and Operational Capabilities," *Radio Sci.*, **18**, 867–879 (1983).

Fontaine, D., S. Perraut, D. Alcaydé, G. Caudal, and B. Higel, "Large Scale Structures of the Convection Inferred from Coordinated Measurements by EISCAT and GEOS 2," *J. Atmos. Terr. Phys.*, **48**, 973–986 (1986a).

Fontaine, D., S. Perraut, N. Cornilleau-Wehrlin, B. Aparicio, J. M. Bosqued, and D. Rodgers, "Coordinated Observations of Electron Energy Spectra and Electrostatic Cyclotron Waves During Diffuse Auroras," *Ann. Geophys.*, **4**, A, 405–412 (1986b).

Girard, L., and C. Senior, "Electrodynamics of Two-Arc Structure as Inferred from EISCAT Measurements," *J. Geophys. Res.*, in press (1988).

Hagfors, T., "The EISCAT Facility," in *High-Latitude Space Plasma Physics*, B. Hultqvist and T. Hagfors, eds., Plenum Publ. Corp., New York, p. 1–9 (1983).

Hagfors, T., W. Birkmayer, and M. Sulzer, "A New Method for Accurate Ionospheric Electron Density Measurements by Incoherent Scatter Radar," *J. Geophys. Res.*, **89**, 6841–6845 (1984).

Häggström, I., M. Lehtinen, T. Turunen, and M. Vallinkoski, "Incoherent Scatter Radar Measurements of Ion Composition Using Alternating Codes," *Radio Sci.*, submitted (1989).

Hall, C. M., U.-P. Hoppe, P. J. S. Williams, and G. O. L. Jones, "Mesospheric Measurements Using the EISCAT VHF System: First Results and Their Interpretation," *Geophys. Res. Lett.*, **14**, 1187–1190 (1987).

Hall, C. M., T. Devlin, A. Brekke, and J. K. Hargreaves, "Negative Ion to Electron Number Density Ratios from EISCAT Mesospheric Spectra," *Phys. Script.*, **37**, 413–418 (1988).

Hargreaves, J. K., H. Ranta, A. Ranta, E. Turunen, and T. Turunen, "Observation of the Polar Cap Absorption Event of February 1984 by the EISCAT Incoherent Scatter Radar," *Planet. Space Sci.*, **35**, 947–958 (1987).

Hoppe, U.-P., C. M. Hall, and J. Röttger, "First Observations of Summer Polar Mesospheric Backscatter Echoes with a 224 MHz Radar," *Geophys. Res. Lett.*, **15**, 28–31 (1988).

Huuskonen, A., T. Nygren, L. Jalonen, T. Turunen, and J. Silén, "High Resolution EISCAT Observations of the Ion-Neutral Collision Frequency in the Lower E-Region, *J. Atmos. Terr. Phys.*, **48**, 827–836 (1986).

Huuskonen, A., J. Kangas, and T. Turunen, "EISCAT as a Radar for Auroral Research—A Case Study," *Geophysica*, **23**, 35–46 (1987).

Kelley, M. C., D. T. Farley, and J. Röttger, "The Effect of Cluster Ions on Anomalous VHF Backscatter from the Summer Polar Mesosphere," *Geophys. Res. Lett.*, **14**, 1031–1034 (1987).

Kirkwood, S., "Seasonal and Tidal Variations of Neutral Temperatures and Densities in the High Latitude Lower Thermosphere Measured by EISCAT," *J. Atmos. Terr. Phys.*, **48**, 817–826 (1986).

Kirkwood, S., and P. N. Collis, "Gravity Wave Generation of Simultaneous Auroral Sporadic E-Layers and Sudden Neutral Sodium Layers," *J. Atmos. Terr. Phys.*, in press (1989).

Kirkwood, S., H. J. Opgenoorth, and J. S. Murphree, "Ionospheric Conductivities, Electric Fields and Currents Associated with Auroral Substorms Measured by the EISCAT Radar," *Planet. Space Sci.*, **36**, 1359–1380 (1988).

Kofman, W., C. Lathuillere, and B. Pibaret, "Neutral Atmosphere Studies in the Altitude Range 90-110 km Using EISCAT," *J. Atmos. Terr. Phys.*, **48**, 837–847 (1986).

Kohl, H., H. Kopka, C. La Hoz, and P. Stubbe, "Propagation of Artificially Excited Langmuir Waves in the Ionosphere," *Radio Sci.*, **22**, 655–661 (1987).

La Hoz, C., J. Röttger, M. Rietveld, G. Wannberg, and S. J. Franke, "The Status and Planned Developments of EISCAT in Mesosphere and D-Region Experiments," in *Handbook for MAP*, SCOSTEP Secretariat, Dept. Electri. Comp. Eng., University of Illinois, Urbana, IL, in press (1989).

Larsen, M. F., and I. S. Mikkelsen, "The Normal Modes of the Thermosphere," *J. Geophys. Res.*, **92**, 6023–6043 (1987).

Lathuillere, C., and A. Brekke, "Ion Compositions in the Ionosphere as Observed by EISCAT," *Ann. Geophys.*, **3**, 557–568 (1985).

Lathuillere, C., F. Glangeaud, and Z. Y. Zhao, "Ionospheric Ion Heating by ULF Pc 5 Magnetic Pulsation," *J. Geophys. Res.*, **91**, 1619–1626 (1986b).

Lathuillere, C., W. Kofman, and B. Pibaret, "Incoherent Scatter Measurements in the F1-Region," *J. Atmos. Terr. Phys.*, **48**, 857–866 (1986a).

Lockwood, M., A. D. Farmer, H. J. Opgenoorth, and S. R. Crothers, "EISCAT Observations of Plasma Convection and the High-Latitude, Winter F-Region During Substorm Activity," *J. Atmos. Terr. Phys.*, **46**, 489–499 (1984).

Lockwood, M., A. P. van Eyken, B. J. I. Bromage, D. M. Willis, and S. W. H. Cowley, "Eastward Propagation of Convection and Ion Temperature Enhancements Following a Southward Turning of the Interplanetary Magnetic Field," *Geophys. Res. Lett.*, **13**, 72–75 (1986).

Lockwood, M., B. J. I. Bromage, R. B. Horne, J.-P. St. Maurice, S. W. H. Cowley, and D. M. Willis, "Non-Maxwellian Ion Velocity Distributions Observed Using EISCAT," *Geophys. Res. Lett.*, **14**, 111–114 (1987).

Lockwood, M., K. Suvanto, J.-P. St. Maurice, K. Kikuchi, B. J. I. Bromage, D. M. Willis, S. R. Crothers, H. Todd, and S. W. H. Cowley, "Scattered Power from Nonthermal, F-Region Plasma Observed by EISCAT—Evidence for Coherent Echoes?" *J. Atmos. Terr. Phys.*, **50**, 467–485 (1988).

Lövhaug, U. P., and T. Flå, "Ion Temperature Anisotropy in the Auroral F-Region as Measured with EISCAT," *J. Atmos. Terr. Phys.*, **48**, 959–971 (1986).

Moorcroft, D. R., and K. Schlegel, "E-Region Coherent Backscatter at Short Wavelength and Large Aspect Angle," *J. Geophys. Res.*, **93**, 2005–2010 (1988).

The NCAR Incoherent-Scatter Radar Data Base Catalogue, National Center for Atmospheric Research, Boulder, CO (1988).

Nielsen, E., and K. Schlegel, "Coherent Radar Doppler Measurements and Their Relationship to the Ionospheric Electron Drift Velocity," *J. Geophys. Res.*, **90**, 3498–3504 (1985).

Nygren, T., L. Jalonen, and A. Huuskonen, "A New Method of Measuring the Ion-Neutral Collision Frequency Using Incoherent Scatter Radar," *Planet. Space Sci.*, **35**, 337–343 (1987).

Opgenoorth, H., and S. Kirkwood, "Ground-Based Observations Coordinated with Viking Spacecraft Measurements," *Proc. Roy. Soc.*, in press (1988).

Opgenoorth, H., B. J. I. Bromage, D. Fontaine, C. La Hoz, A. Huuskonen, H. Kohl, U.P. Lövhaug, and G. Wannberg, "Coordinated EISCAT/Viking Observations—Outline of Experiments and Description of Observations," *Annal. Geophys.*, in press (1988).

Perraut, S., A. Brekke, M. Baron, and D. Hubert, "EISCAT Measurements of Ion Temperatures which Indicate Nonisotropic Ion Velocity Distributions," *J. Atmos. Terr. Phys.*, **46**, 531–543 (1984).

Provjdakes, J., D. T. Farley, B. G. Fejer, J. Sahr, W. E. Swartz, I. Häggström, Å. Hedberg, and J. A. Nordling, "Observations of Auroral E-Region Plasma Waves and Electron Heating with EISCAT and a VHF Radar Interferometer," *J. Atmos. Terr. Phys.*, **50**, 339–356 (1988).

Ranta, A., H. Ranta, T. Turunen, J. Silén, and P. Stauning, "High-Resolution Observations of D-Region by EISCAT and Their Comparison to Riometer Measurements," *Planet. Space Sci.*, **33**, 583–589 (1985).

Rinnert, K., H. Kohl, K. Schlegel, and K. Wilhelm, "Electric Field Configuration and Plasma Parameters in the Vicinity of a Faint Auroral Arc," *J. Atmos. Terr. Phys.*, **48**, 867–878 (1986).

Rishbeth, H., and P. J. S. Williams, "The EISCAT Ionospheric Radar: The System and Its Early Results," *Quart. J. Astron. Soc.*, **26**, 478–512 (1985).

Rishbeth, H., A. P. van Eyken, B. S. Lanchester, T. Turunen, J. Röttger, C. M. Hall, and U.-P. Hoppe, "EISCAT VHF Radar Observations of Periodic Mesopause Echoes," *Planet. Space Sci.*, **36**, 423–428 (1988).

Robinson, T. R., "Towards a Self-Consistent Non-Linear Theory of Radar Auroral Backscatter," *J. Atmos. Terr. Phys.*, **48**, 417–422 (1986).

Röttger, J., "EISCAT—Das Europäische Incoherent-Scatter-Radar zur Erforschung der Polaren Atmosphäre," *Mitt. Astron. Ges.*, **58**, 67–79 (1983).

Röttger, J., "Further Developments of EISCAT as an MST Radar," in *Handbook for MAP*, **14**, S. A. Bowhill and B. Edwards, eds., SCOSTEP Secretariat, Dept. Electri. Comp. Eng., Univ. of Illinois, Urbana, IL, p. 309–318 (1984).

Röttger, J., "VHF Radar Measurements of Small-Scale and Meso-Scale Dynamical Processes in the Middle Atmosphere," *Phil. Trans. R. Soc. Lond.*, A, **232**, 611–628 (1987).

Röttger, J., C. La Hoz, M.C. Kelley, U.-P. Hoppe, and C. M. Hall, "The Structure and Dynamics of Polar Mesosphere Summer Echoes Observed with the EISCAT 224 MHz Radar," *Geophys. Res. Lett.*, **15**, 1353–1356 (1988).

Röttger, J., and W. Meyer, "Tidal Wind Observations with Incoherent Scatter Radar and Meteorological Rockets During MAP/WINE," *J. Atmos. Terr. Phys.*, **49**, 689–703 (1987).

Sandahl, I., A. Steen, A. Pellinen-Wannberg, B. Holback, F. Söraas, and J. S. Murphree, "First Results from the Viking Associated Aureld-VIP Rocket and EISCAT Campaign," in *Proc. 8th ESA Symposium on European Rocket and Balloon Programmes and Related Research*, Sunne, Sweden, 17-23 May 1987, ESA-SP-270, p. 55–60 (1987).

Schlegel, K., "A Case Study of a High Latitude Ionospheric Electron Density Depletion," *J. Atmos. Terr. Phys.*, **46**, 517–520 (1984).

Schlegel, K., "Auroral Zone E-Region Conductivities During Solar Minimum Derived from EISCAT Data," *Ann. Geophys.*, **6**, 129–138 (1988).

Schlegel, K., and K. I. Oyama, "Remote and In-Situ Plasma Measurements During the CAESAR-Flight," in *Proc. 8th ESA Symposium on European Rocket and Balloon Programmes and Related Research*, Sunne, Sweden, 17-23 May 1987, ESA-SP-270, p. 315–318 (1987).

"STEP: Solar-Terrestrial Energy Program 1990-1995—A Study of Energy Transfer Mechanisms in the Solar-Terrestrial System," Scientific Committee on Solar-Terrestrial Physics, Fairbanks, AK (November 1988).

Schlegel, K., G. Kremser, and A. Korth, "Data Comparison of EISCAT and the Energetic Particle Instrument on GEOS 2," *J. Atmos. Terr. Phys.*, **46**, 509–515 (1984).

Steen, Å, P. N. Collis, and I. Häggström, "On the Development of Folds in Auroral Arcs," *J. Atmos. Terr. Phys.*, **50**, 301–313 (1988a).

Steen, Å, D. Rees, P. N. Collis, and J. S. Murphree, "What Role Does the F-Region Neutral Wind Play in Auroral Intensifications?" *Planet. Space Sci.*, **36**, 851–868 (1988b).

Todd, H., B. J. I. Bromage, S. W. H. Cowley, M. Lockwood, A. P. van Eyken, and D. M. Willis, "EISCAT Observations of Bursts of Rapid Flow in the High Latitude Dayside Ionosphere," *Geophys. Res. Lett.*, **13**, 909–912 (1986).

Turunen, E., P. N. Collis, and T. Turunen, "Incoherent Scatter Spectral Measurements of the Summertime High-Latitude D-Region with the EISCAT UHF Radar," *J. Atmos. Terr. Phys.*, **50**, 289–299 (1988).

Turunen, T., T. Nygren, A. Huuskonen, and L. Jalonen, "Incoherent Scatter Studies of Sporadic-E Using 300 m Resolution," *J. Atmos. Terr. Phys.*, **50**, 277–287 (1988).

Ullaland, S., T. Hansen, and W. Riedler, "The Coordinated EISCAT and Balloon Observations (CEBO)—Experiment and Results," *Proc. 7th ESA Symposium*, Loen, Norway, ESA SP-229, p. 75–79 (1985).

Virdi, T. S., G. O. L. Jones, and P. J. S. Williams, "EISCAT Observations of the E-Region Semi-Diurnal Tide," *Nature*, **324**, 354–356 (1986).

Wahlund, J.-E., H. J. Opgenoorth, and P. Rothwell, "Observations of Thin Auroral Ionization Layers by EISCAT in Connection with Pulsating Aurora," *J. Geophys. Res.*, in press (1989).

Wilhelm, K., K. Rinnert, K. Schlegel, H. Kohl, N. Klöcker, H. Lühr, W. Oelschlägel, G. Dehmel, M. P. Gough, B. Holback, and K.-I. Oyama, "Co-ordinated Auroral Experiments Using Scatter and Rocket Investigations (CAESAR Investigations)—Final Report on the Scientific Aspects," Rpt. MPAE-W-47-87-13, Max-Planck-Institut für Aeronomie, Katlenburg-Lindau, W. Germany (1987).

Williams, P. J. S., "European Radar Unscrambles the Ionosphere," *New Sci.*, 46–52 (5 Dec 1985).

Williams, P. J. S., and A. R. Jain, "EISCAT Observations of the High Latitude Trough," *J. Atmos. Terr. Phys.*, **48**, 423–434 (1986).

Williams, P. J. S., B. Jones, M. Uspensky, and G. Starkov, "Multi-Radar Studies of Auroral Backscatter," *J. Atmos. Terr. Phys.*, **50**, 315–321 (1988a).

Williams, P. J. S., G. Crowley, K. Schlegel, T. S. Virdi, I. McCrea, G. Watkins, N. Wade, J. K. Hargreaves, T. Lachlan-Cope, H. Muller, J. E. Baldwin, P. Warner, A. P. van Eyken, M. A. Hapgood, and A. S. Rodger, "The Generation and Propagation of Atmospheric Gravity Waves Observed During the Worldwide Atmospheric Gravity-Wave Study (WAGS)," *J. Atmos. Terr. Phys.*, **50**, 323–338 (1988b).

Willis, D. M., M. Lockwood, S. W. H. Cowley, A. P. van Eyken, B. J. I. Bromage, H. Rishbeth, P. R. Smith, and S. R. Crothers, "A Survey of Simultaneous Observations of the High-Latitude Ionosphere and Interplanetary Magnetic Field with EISCAT and AMPTE-UKS," *J. Atmos. Terr. Phys.*, **48**, 987–1008 (1986).

Winser, K. J., G. O. L. Jones, and P. J. S. Williams, "Large Field-Aligned Velocities Observed by EISCAT," *J. Atmos. Terr. Phys.*, **50**, 379–382 (1988a).

Winser, K. J., G. O. L. Jones, and P. J. S. Williams, "Variations in Field-Aligned Plasma Velocity with Altitude and Latitude in the Auroral Zone: EISCAT Observations and the Physical Interpretation," *Phys. Script.*, **37**, 640–644 (1988b).

Winser, K. J., A. D. Farmer, D. Rees, and A. Aruliah, "Ion-Neutral Dynamics in the High Latitude Ionosphere: First Results from the INDI Experiment," *J. Atmos. Terr. Phys.*, **50**, 369–377 (1988c).

VII-5. GROUND-BASED MEASUREMENTS OF JOULE HEATING RATES

O. de la Beaujardière*, R. Johnson,* and V. B. Wickwar[†]

Joule heating in the upper atmosphere is the most important energy dissipation process between the magnetosphere and the ionosphere. Here we examine the various terms in the equation that governs Joule heating. The most notable features of the observation include the following: (1) The ionospheric electric field seasonal dependence is found to be quite significant. The shape of the ionospheric plasma convection cells and the latitude of the reversal from sunward to antisunward convection are seasonally dependent. Statistical averages of the square of the ion velocity show a maximum in fall and a minimum in summer. (2) Pedersen conductivities at F-region altitudes are examined using Chatanika and Sondrestrom radar data. It is shown that during solar minimum conditions, the F region contributes less than 20% to the total height-integrated Pedersen conductivity Σ_p. In contrast, during solar maximum conditions the contribution to Σ_p from solar-produced F-region ionization can be 60%. (3) The importance of the neutral wind term in Joule heating calculations is illustrated using a specific example. The Joule heating calculated by including the neutral wind term is 2 to 4 times smaller than that calculated without the neutral wind, but the reverse can also be true, as shown during a period when the neutral wind played the role of a dynamo in the ionosphere/magnetosphere current.

1. INTRODUCTION

It is generally accepted that the solar wind energy that reaches the magnetosphere is dissipated by three main processes: (1) injection of charged particles into the ring current and further dissipation by charge exchange, (2) Joule heating in the ionosphere and thermosphere, and (3) particle precipitation. The energy transfer rate into the ring current is often larger than that of the other two processes. At times, however, the Joule heating rate can be the largest [*Baumjohann and Kamide,* 1984]. Although particle precipitation in the auroral oval may dissipate energy at a rate comparable to Joule dissipation [*Evans et al.,* 1977; *Vickrey et al.,* 1982], globally, the dissipation by Joule heating is significantly more important than by particle precipitation [*Baumjohann and Kamide,* 1984]. The effect on the neutral atmosphere of the Joule dissipation has far-reaching consequences. Joule dissipation heats the neutral gas and causes upwelling of atomic oxygen that is then transported to lower latitudes. This process changes the global configuration of thermospheric neutral winds [*Roble et al.,* 1977; *Prölss,* 1980]. As a result, the Joule heating rate is one of the most fundamental parameters needed in studies of the coupling mechanisms between the solar wind and the thermosphere-ionosphere-magnetosphere system.

The height-integrated Joule heating rate, in the reference frame of the neutrals, is governed by the equation

$$Q_J = \int_{h_1}^{h_2} \sigma_P(h)\,(\mathbf{E} + \mathbf{U} \times \mathbf{B})^2\,dh \qquad (1)$$

where σ_P is the Pedersen conductivity, $\mathbf{E}$ is the electric field calculated from the ionospheric plasma drift $\mathbf{V}$ by $\mathbf{E} = -\mathbf{V} \times \mathbf{B}$, $\mathbf{U}$ is the neutral wind, $\mathbf{B}$ is the magnetic field, and h_1 and h_2, the limits of integration, encompass the E- and F-region altitudes.

Incoherent scatter radars are the only ground-based instruments capable of measuring all the quantities in Eq. 1. They can easily measure the electron density, electric field, and neutral wind vectors. The electron density profile is obtained from the power of the return signal, whereas the electric field is obtained from the Doppler shift [*Evans,* 1969]. In the E region, the neutral wind vector is obtained from the E- and F-region ion-velocity measurements [*Brekke et al.,* 1974; *Johnson,* 1989 (and references therein)]. In the F region, the meridional component of the neutral wind is derived from the ion velocity measured parallel to the magnetic field, combined with the electron density profile and a neutral atmosphere model [*Vasseur,* 1969; *Wickwar,* 1989 (and references therein)]. Incoherent scatter radars can also measure, although with more difficulty, the ion composition and ion-neutral collision frequency [*Lathuillère et al.,* 1983; *Johnson and Wickwar,* 1987].

Because an excellent review paper has recently been published on energetics of the thermosphere [*Killeen,* 1987], we do not review the subject of Joule heating again in this paper. Instead, we concentrate on some specific aspects of each term in Eq. 1. We first consider the electric field, and show that seasonal change

*Geoscience and Engineering Center, SRI International, Menlo Park, CA 94025.
[†]Center for Atmospheric and Space Sciences, Utah State University, Logan, UT 84322.

Auroral Physics, edited by C.-I. Meng, M. J. Rycroft and L. A. Frank. © Cambridge UP 1991

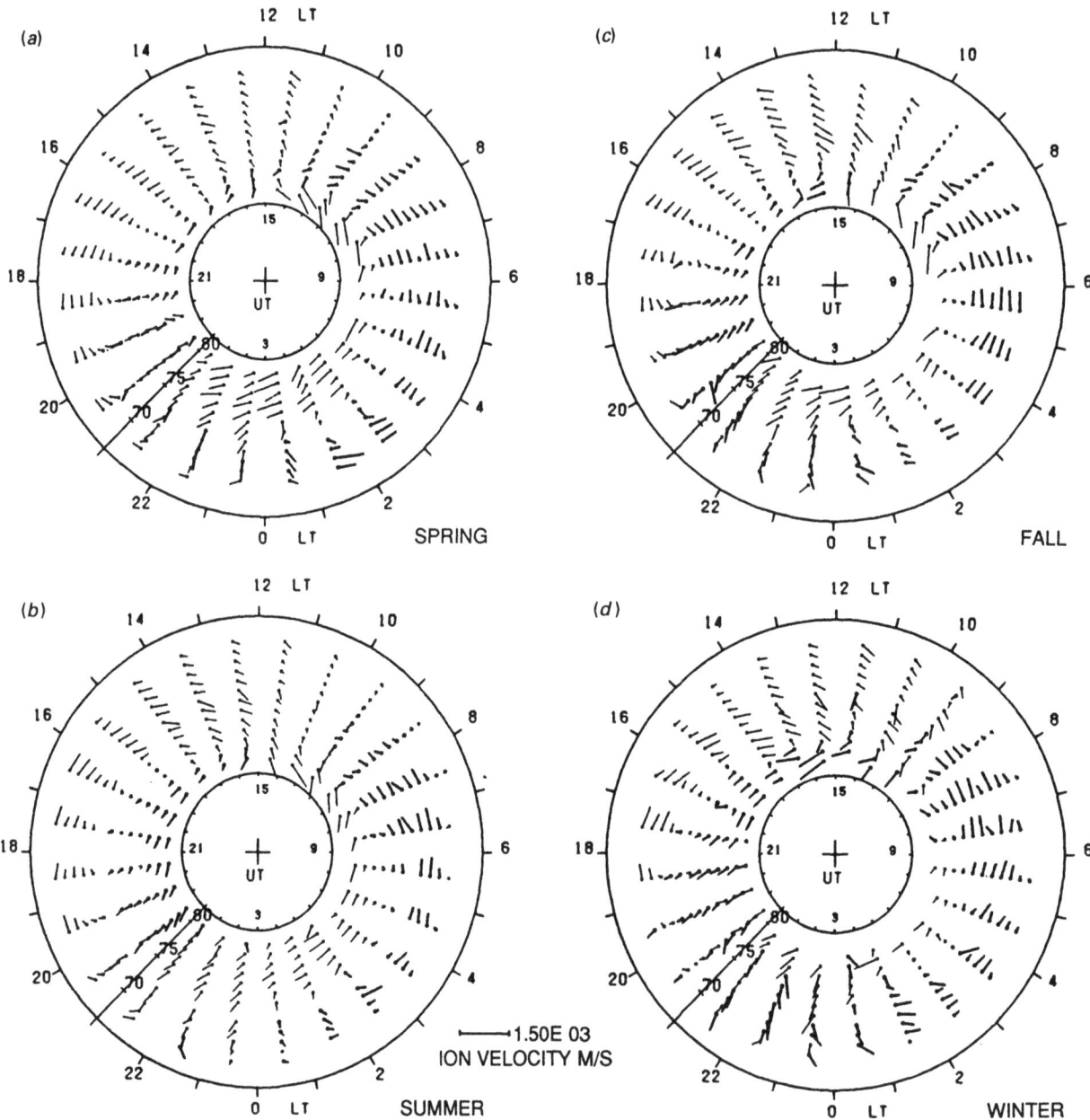

Figure 1—Average ion convection velocities (ms⁻¹) for (a) spring, (b) summer, (c) fall, (d) winter. The Sondrestrom data from March 1983–July 1988 were averaged into 1° × 1h bins. Local time is indicated on the outside and UT on the inside of the clock dial circle.

is significant in the electric field pattern and intensity observed from Sondrestrom. We then consider the conductivities, and show that during solar maximum the solar-produced F-region ionization can contribute more than 50% of the total height-integrated Pedersen conductivity. Finally, we make some remarks concerning the importance of the neutral wind contribution to Joule heating.

Here data from the Sondrestrom and Chatanika incoherent scatter radars [*Kelly*, 1983, *Leadabrand et al.*, 1972] are used. At 75° invariant latitude, Sondrestrom is usually in the polar cap during the night. It is in the cleft, cusp, and auroral zone during the day. At 65° invariant latitude, Chatanika was usually in the auroral zone during the night and equatorward of the auroral oval during the day.

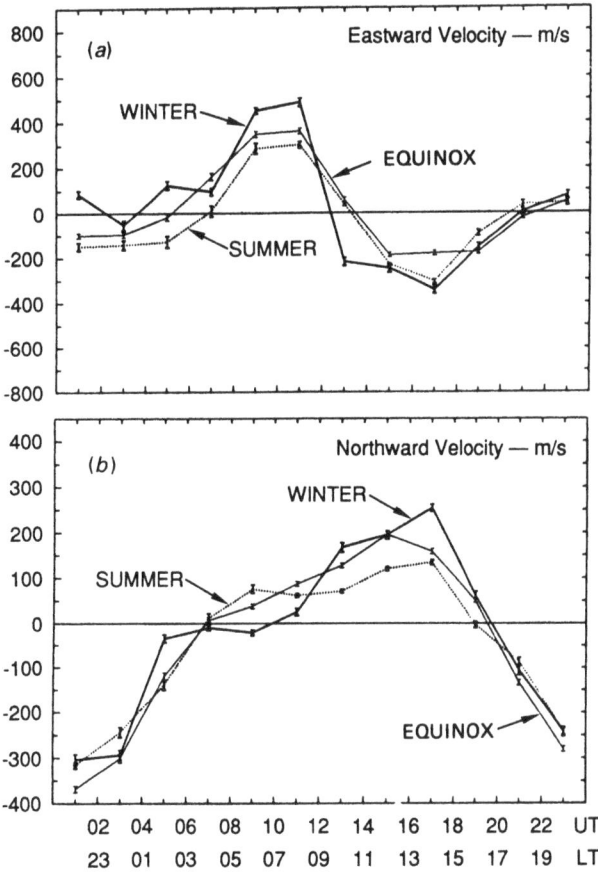

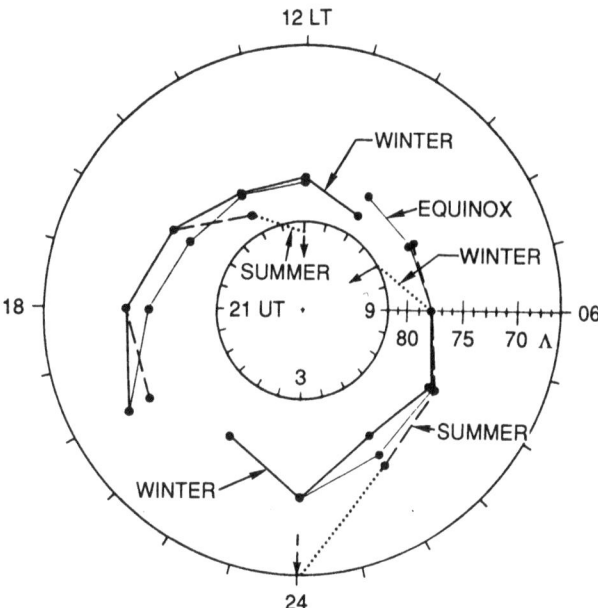

Figure 3—Latitude of the ion convection reversal for three seasons. This reversal is that of the 2° invariant latitude × 2h averaged data. Reversals in the morning and afternoon convection cells are indicated by separate curves, and the discontinuities in the noon and midnight local times show where the transition takes place. Arrows indicate where the convection reversal is located beyond the radar field of view.

Figure 2—Components of the average ion velocities (ms^{-1}) at 75° invariant latitude with (*a*) eastward and (*b*) northward velocities. Note that the velocity scale for (*a*) is half that for (*b*). The error bar indicates the mean standard deviation. Fall and spring values have been averaged together to give the equinox values.

2. SEASONAL DEPENDENCE OF HIGH-LATITUDE IONOSPHERIC CONVECTION

The dependence of high-latitude electric field on geomagnetic activity and IMF configuration has been studied for many years, using data from satellites and ground-based instruments [*Heppner*, 1977; *Holt et al.*, 1987; *Heppner and Maynard*, 1987; *Foster et al.*, 1971, 1986; *Friis-Christensen et al.*, 1985; *de la Beaujardière et al.*, 1985, 1986; *de la Beaujardière and Wickwar*, 1986; *Holt et al.*, 1987]. Little is known, however, about possible seasonal variations of electric fields. This is surprising, because it has been known for decades that the various indices of magnetic activity are seasonally dependent, with maxima in spring and fall [*Cortie*, 1912; *Russell and McPherron*, 1973 (and references therein)]. Similarly, high-latitude ionospheric currents and field-aligned currents have maxima in sum-

mer [*Matsushita and Xu*, 1982; *Campbell*, 1982; *Fujii et al.*, 1981]. To our knowledge, a systematic analysis of the seasonal dependence of ionospheric electric fields has never been done before the work described here. Seasonal variations in plasma convection are important in the context of Joule heating since the heating is approximately proportional to the square of the ion velocity.

We have binned and averaged all the Sondrestrom data taken during experiments lasting more than 7 hours. The observations spanned five years, roughly centered around solar minimum, from March 1983 to July 1988. More than 130 days were averaged. The most salient points revealed by these studies and illustrated by Figures 1 to 3 can be summarized as follows:

1. The equinox and summer convection velocities show a smooth two-cell pattern, whereas the winter convection velocity is not as smooth.
2. The winter east-west (E-W) component of the velocity is the largest in the noon and dusk sectors (1100–2100 LT). The summer E-W component is the smallest in the dawn and noon sectors (10000–1600 LT).
3. The summer north-south (N-S) component is the smallest in the noon sector (0900–1600 LT).
4. There is a phase difference in the daily variations of the E-W velocity component, the winter

months leading equinoctial and summer months. Especially on the dayside, this phase difference can be interpreted as a clockwise rotation of the convection pattern.

5. In the noon portion of the afternoon cell, the latitude of the convection reversal from sunward to antisunward is highest during the summer months and lowest during the winter months.

6. In the dawn convection cell, the latitude of the convection reversal from sunward to antisunward is highest during the winter months.

We now discuss some of these points in more detail. Figures 1a through 1d show the average velocities. Thick and thin vectors are used to distinguish west and east velocities, respectively. The bin size is 1° by 1 h, covering the invariant latitude interval 67–82°. The number of points in each bin is typically 200–300. Comparing summer and winter data (Figures 1b and d), we see that the average convection is somewhat smoother and more regular in summer. This is not unexpected because, on individual winter days, and especially on the nightside, the convection has previously been reported to be highly irregular [*Heppner,* 1977]. This is evidenced in the winter convection pattern shown, for example, by *Heelis and Hanson* [1980] and *de la Beaujardière et al.* [1985]. However, the averaging of winter data presented here smooths out much of the irregularity, and the winter drifts appear to maintain a somewhat erratic but fairly well organized two-cell convection pattern.

Figure 2 shows the individual components of the plasma drift as a function of time for all seasons. (The data for this figure and the one following correspond to a 2° × 2 h grid size.) Figure 2a shows that the E-W component (N-S electric field) is largest in winter between 1100 and 2100 LT, i.e., in the noon and dusk local-time sectors. In the summer, the E-W component is the smallest over several hours, from 0700 to 1600 LT.

The latitude at which the convection reverses from sunward to antisunward is displayed in Figure 3. For this figure, the 2° × 2 h averaged data were used. The latitude of this reversal has an opposite dependence on season in the dusk and dawn cells. In winter, the dusk cell has its convection reversal at, or equatorward of, the other seasons. In summer, the dawn cell has its convection reversal at, or equatorward of, the other seasons. In other words, the dusk convection reversal is at lowest latitudes during the winter, while the dawn convection reversal is at lowest latitudes during the summer. These latitudinal differences are most pronounced in the noon and midnight sectors.

The winter reversal is located equatorward of the summer reversal in the afternoon cell and poleward in

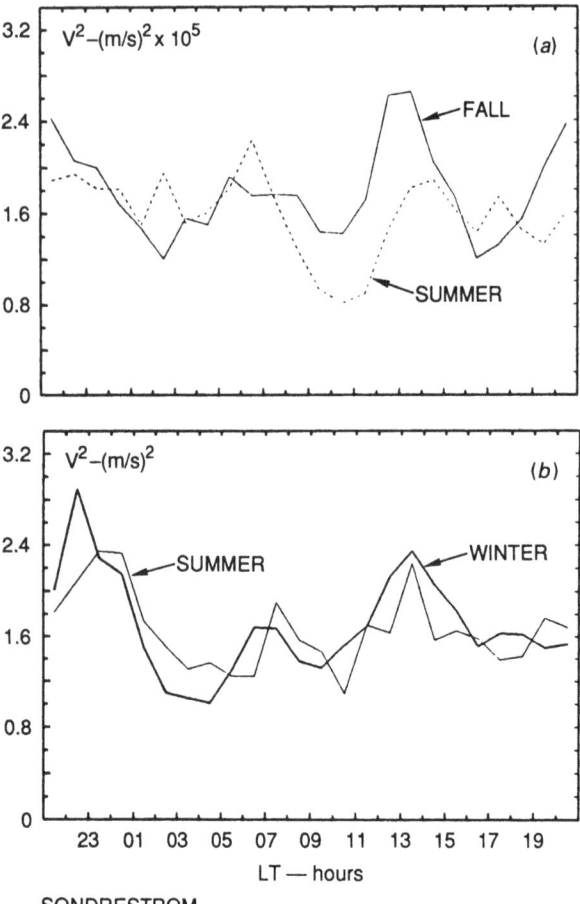

Figure 4—Average of squared ion velocity vectors as a function of local time for (a) fall and summer (solid and dotted lines, respectively), and (b) winter and spring (thick and thin lines, respectively).

the morning cell. Note that two reversals are poleward of 81°, beyond the field of view in the noon sector. These reversals poleward of 81° occur in winter on the dawnside and in summer on the duskside. At midnight, the summer reversal is also beyond the field of view, but it is equatorward of 68°.

Because Joule heating is approximately proportional to the square of the electric field or, equivalently, to V^2, the individual drift velocities were also squared and averaged according to season. The results of averaging all latitudes together are displayed in Figure 4 as a function of local time. The daily variations show three maxima that occur (1) 1–2 h before local midnight, (2) at dawn, and (3) 1–2 h after local noon. (At Sondrestrom, magnetic local time precedes local time by about one hour.) Except during summer, the largest values are those that occur before noon and before midnight. Part of these variations, however, are due

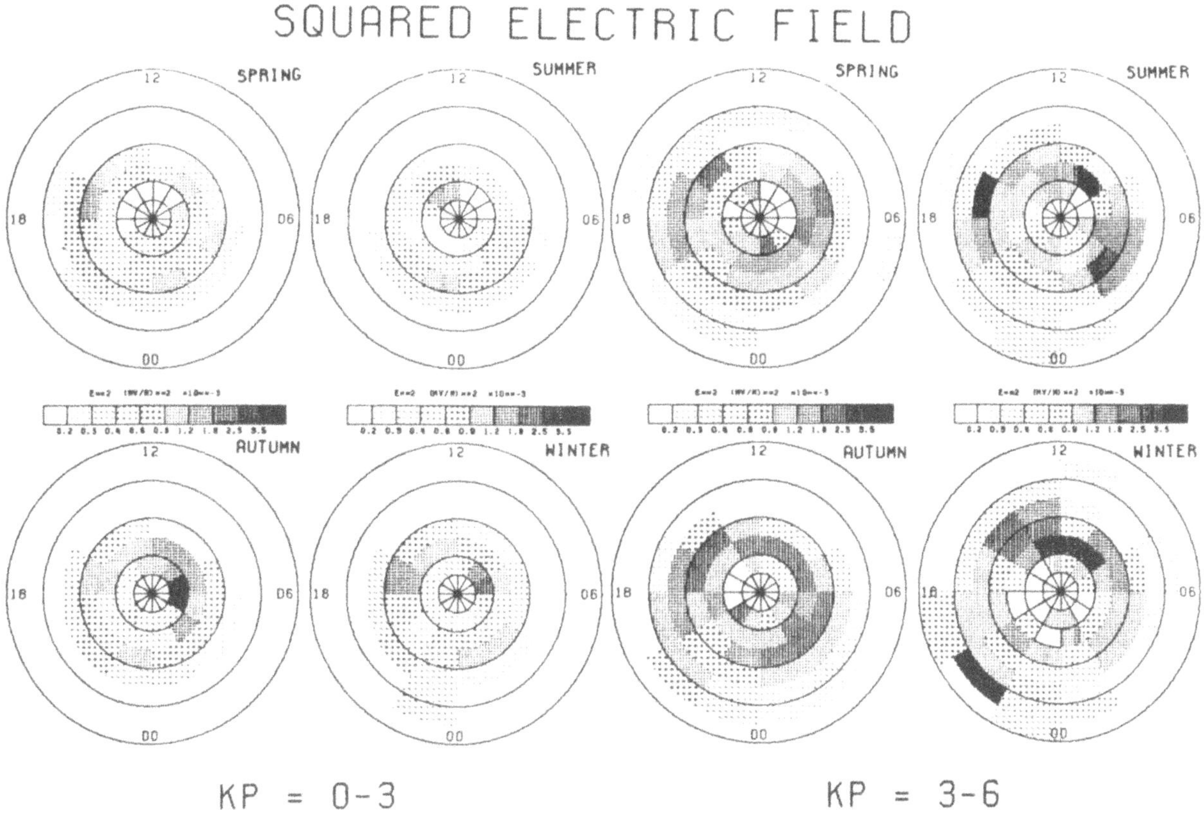

SQUARED ELECTRIC FIELD

Figure 5—Average patterns of the squared electric field magnitude measured in the four seasons for low (left) and high (right) magnetic activity conditions, obtained by binning AE-C data. (From *Foster et al.,* [1983].)

to the offset between the geomagnetic and geographic poles. Over the course of one day, the Sondrestrom radar samples differing parts of the convection.

As stated by *Foster et al.* [1983], the seasonal differences in the square of the velocity are small. However, close examination of Figure 4 reveals that the small difference between seasons is probably significant. By averaging over all local times we find that V^2 is the smallest in summer ($V^2 = 1.6 \times 10^5$ m^2 s^{-2}) and largest in fall ($V^2 = 1.8 \times 10^5$ m^2 s^{-2}). Values in spring and winter are about equal. In fact, the same seasonal dependence is also apparent in *Foster et al.* [1983], who binned the square of the electric field measured from the Atmosphere Explorer C (AE-C) satellite during two time intervals close to the previous solar minimum. By inspecting their figure, reproduced here as Figure 5, it is clear that the squared electric field (E^2) is largest in fall and smallest in summer. Also, this difference is more pronounced during the quiet periods ($Kp = 0$–3) than during the active periods

($Kp = 3$–6). And, indeed, these authors state that, overall, the E^2 is 20% lower in summer than in winter and equinox.

The F-region electron density is largest in summer and smallest in winter, due to differing solar illumination. This may bias our data since the measurement precision increases with increasing electron density. This bias, which may result in larger velocity measurements during winter, has not been taken into account in this study. However, it is probably not significant enough to affect our conclusions, which, quantitatively, agree with *Foster et al.* [1983].

The differences in E^2 have repercussions in terms of Joule heating. Neglecting the neutral wind effects, *Foster et al.* [1983, Figure 7] find that during quiet times, the Joule heating rate integrated over all latitudes is maximum in fall. The higher conductivities resulting from the larger solar illumination during summer are not sufficient to offset the fact that E^2 is smaller in summer than in the fall.

443

3. CONDUCTIVITIES FROM SOLAR-PRODUCED IONIZATION

The Hall and Pedersen conductivities are given by:

$$
\sigma_P = \frac{N_e e}{B} \left[\frac{v_{en}\Omega_e}{v_{en}^2 + \Omega_e^2} + \frac{v_O \Omega_O}{v_O^2 + \Omega_O^2} p_O \right.
$$

$$
\left. + \frac{v_{NO}\Omega_{NO}}{v_{NO}^2 + \Omega_{NO}^2} p_{N_2} + \frac{v_{O_2}\Omega_{O_2}}{v_{O_2}^2 + \Omega_{O_2}^2} p_{O_2} \right] \quad (2)
$$

$$
\sigma_H = \frac{N_e e}{B} \left[\frac{\Omega_e^2}{v_{en}^2 + \Omega_e^2} - \frac{\Omega_O^2}{v_O^2 + \Omega_O^2} p_O \right.
$$

$$
\left. - \frac{\Omega_{NO}^2}{v_{NO}^2 + \Omega_{NO}^2} p_{N_2} - \frac{\Omega_{O_2}^2}{v_{O_2}^2 + \Omega_{O_2}^2} p_{O_2} \right] \quad (3)
$$

where N_e is the electron density, e is the charge of electron, v_{en} is the electron-neutral collision frequency, v_O, v_{O_2}, v_{NO} are the ion-neutral collision frequencies, Ω_e is the electron gyrofrequency, Ω_O, Ω_{O_2}, Ω_{NO} are the ion gyrofrequencies, and p_O, p_{O_2}, p_{N_2} are the relative ion concentrations.

Recently, *Brekke and Hall* [1988] reviewed conductivity measurements and empirical models. Listing the various terms in Eqs. 2 and 3, they described the simplifying assumptions that are often made in calculating the Hall and Pedersen conductivities. In particular, these authors stressed that the collision-frequency models adopted are of great importance when deriving the conductivity.

Brekke and Hall [1988] also helped clarify the dependence of conductivities on the solar zenith angle (χ). While several authors have stated that the Pedersen and Hall conductivities (Σ_p and Σ_H) are proportional to $\cos\chi$, others have argued that conductivities are proportional to $(\cos\chi)^{0.5}$, or, more generally, to $(\cos\chi)^n$. *Brekke and Hall* [1988] showed that the conductivities are better described using an expression where the two terms in $\cos\chi$ are added. They obtained for the Pedersen and Hall conductivities an expression of the form: $\Sigma = a\cos\chi + b(\cos\chi)^{0.5}$. Adding these two terms is justified by the fact that, in the lower E region, the ionization recombination process is dominated by dissociative recombination, so the electron density is proportional to $\cos\chi$. However, in the upper E region and in the F region, charge exchange processes dominate and the electron density is proportional to $(\cos\chi)^{0.5}$.

In another recent article *Rasmussen et al.* [1988] studied the height-dependent Pedersen conductivities.

They calculated how much of the height-integrated Pedersen conductivity originates from F-region altitudes (above ~170 km) and how much originates from the E region. These authors showed that, during solar maximum conditions, the F region can contribute more than previously assumed. They derived the values of the ionospheric densities and conductivities from the combination of a photochemical equilibrium model at E-region altitudes, and the full F-region model including diffusion at higher altitudes. Under solar maximum conditions, they concluded that the contribution to Σ_p from the ionosphere above 170 km can reach 40% during daytime and 70%–80% at night and dusk. Comparison of their ionospheric density model with incoherent scatter data shows that, in the altitude range of 110–200 km, the predicted values are underestimated by 40%–50%. It is therefore important to assess, using actual data, the F-region contribution to height-integrated conductivities.

We have examined Chatanika and Sondrestrom ionospheric density measurements and calculated the conductivities (Eqs. 2 and 3) without making the approximations that have often been made in the past. The procedure was the same as used for the Chatanika data in *Rasmussen et al.* [1988], and we looked at all the days that were analyzed by these authors. All the collision terms between electrons, NO^+, O_2^+, O^+ and N_2, O_2, O were included, using the collision frequencies in *Schunk and Nagy* [1980]. The neutral atmosphere was obtained from the MSIS-83 model [*Hedin*, 1983]. The results are shown in Figures 6a through 6c where the solid and dashed lines represent Pedersen conductivity integrated up to 500 and 180 km altitudes, respectively. Figure 6a is for solar cycle minimum ($S_{10.7} = 69$); Figures 6b and 6c are for solar cycle maximum ($S_{10.7} = 197$ and 180, respectively).

These calculations do not consistently agree with *Rasmussen et al.* [1988]. During solar cycle minimum, we find that the difference between the two curves is very small, indicating that the F-region contribution to Σ_p is small, less than ~10%. These results are in agreement with *Rasmussen et al.* [1988] for the daytime, but not for dusk and nighttime when these authors found that the F-region contribution could reach 60%. During solar-cycle maximum, the difference between the E- and F-region contribution can be larger. Figure 6b shows a large F-region contribution reaching 60% during the daytime hours (i.e., on the right and left sides of the figure). Although it is daytime, the Sun has a grazing incidence and the solar zenith angle always exceeds 80°, with the result that the rate of EUV photoionization is small in the E region and large in the F region. This corresponds to the dusk conditions described by *Rasmussen et al.* [1988]

444

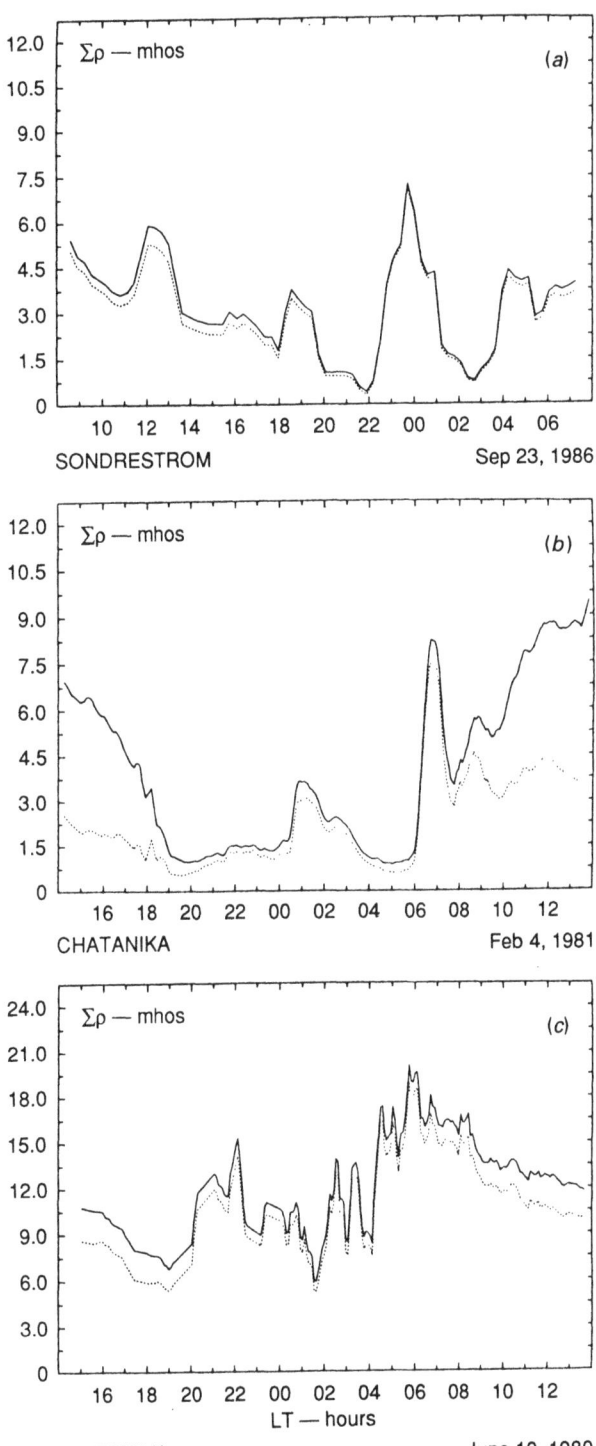

Figure 6—Pedersen height-integrated conductivity versus local time measured during (a) solar minimum at Sondrestrom and (b) and (c) solar maximum at Chatanika. The solid curve shows the Pedersen conductivity integrated up to a 500-km altitude, and the dashed curve is for a 180-km altitude limit. The solar zenith angle is indicated at the top of the plots.

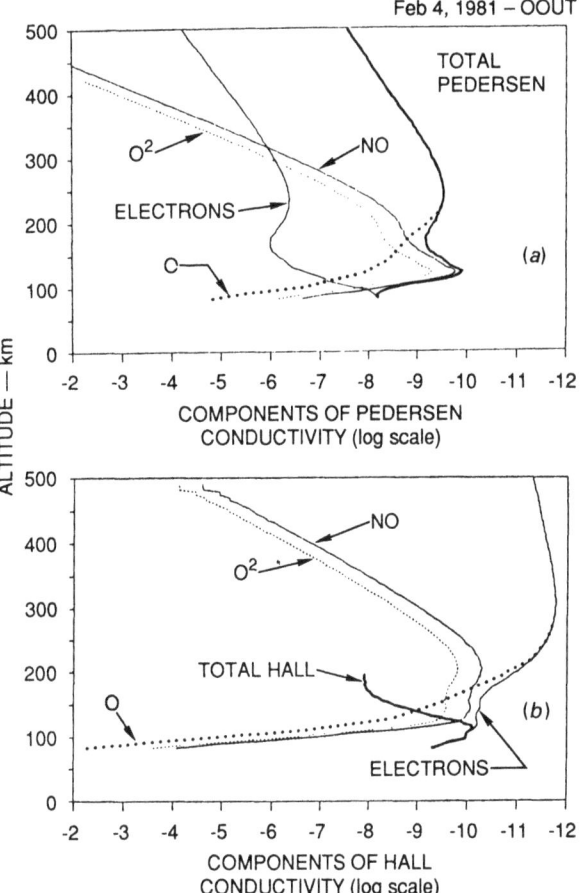

Figure 7—Height profiles of the four components of (a) Hall conductivity and (b) Pedersen conductivity (see Eqs. 2 and 3). The data correspond to 1400 LT (0000 UT) on February 4, 1981, illustrated in Figure 6b.

when the F-region contribution is largest. Thus, these results agree qualitatively with Rasmussen et al. However, our measurements show that the F-region contribution is smaller than that obtained from the model calculation, which was 80%. For the nighttime conditions, the disagreement between our observations and *Rasmussen et al.* [1988] is quite large, since we find a small F-region contribution, and these authors find a contribution around 60%–70%.

In order to assess the relative importance of each element in the conductivity calculation (Eqs. 2 and 3), we show in Figures 7a, 7b, and 8 the height profiles of the Hall and Pedersen conductivities and of the electron density. The time is local noon, and the day selected is February 4, 1981, the day illustrated in Figure 6b for which the F-region contribution to the Pedersen conductivity was large. Figure 7a shows that the contribution to the Pedersen conductivity from the electrons (the first term in Eq. 2) is negligible at all altitudes above 90 km. The contribution from the molecular ions

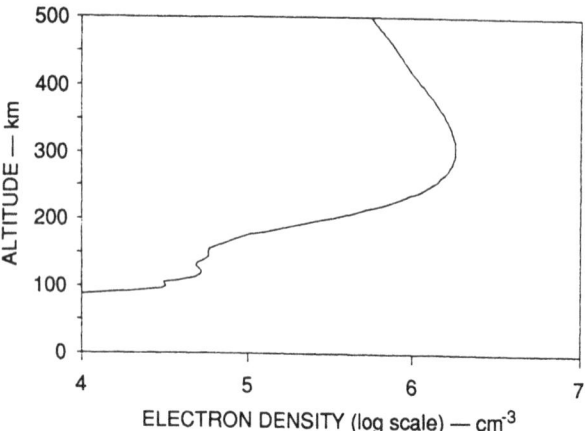

Figure 8—Electron density profile corresponding to the Figure 7 conductivities.

(last two terms in Eq. 2) dominates up to about 190 km. The contribution from atomic oxygen ions (second term in Eq. 2) dominates above 190 km. This is because the electron density, shown in Figure 8, was extremely large in the F region (1.8×10^6 el/cm^{-3}), whereas it was very small in the E region (5.3×10^4 el/cm^{-3}).

The components of the Hall conductivity are displayed in Figure 7*b*. The Hall conductivity is the difference between the contribution from the electrons (first term in Eq. 3) and from the ions (three other terms in Eq. 3). The conductivity peaks at 112 km, slightly below the E-region peak at 120 km. Below the peak, σ_H depends mostly on the electron term; above the peak, molecular ions determine the shape of the curve. The contribution from atomic oxygen is negligible below about 160 km.

4. NEUTRAL WIND

The final parameter we would like to comment on is the neutral wind, **U**. Equation 1 shows that the Joule heating of the neutral atmosphere is proportional to the square of the vector difference between the ion drift and the neutral wind. Because few techniques can provide the **V** and **U** vectors simultaneously, and because **U** is generally smaller than the **E** × **B** drift, the neutral wind is often ignored. However, its contribution can be quite important. In the F region, ion drag tends to diminish the size of **V** − **U**; in the E region, the wind can be very large during geomagnetically active conditions [*Brekke and Rino*, 1978; *Johnson and Wickwar*, 1987; *Baron and Wand*, 1983]. Depending on the magnitudes of the ion and the neutral velocities, as well as the time history of a particular event, the Joule heating rate can be either reduced or enhanced relative to the **U** = 0 value.

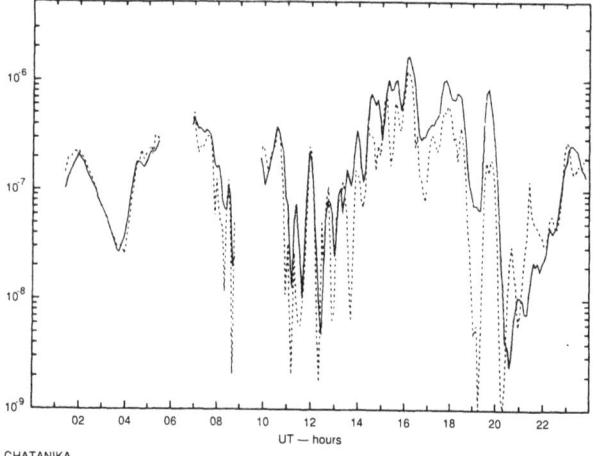

CHATANIKA

Figure 9—Joule heating rates in Wm^{-3} at 115-km altitude calculated neglecting (solid curve) and including (dashed curve) the effect of the neutral wind. The data correspond to the June 1980 period. Illustrated in Figure 6*c*.

The incoherent scatter technique provides the electric field at all heights. This technique also allows both components of **U** to be determined in the E region, and the meridional component of **U** to be determined in the F region [*Johnson,* 1989]. Joule heating rates derived from Chatanika radar measurements using Eq. 1 are displayed in Figure 9. These observations show an example of the magnitude of the neutral wind effect on a highly disturbed day characterized by large electric fields and intense particle precipitation. The two curves on this figure show the local Joule heating rates at 115 km altitude. The solid line is derived from Eq. 1, and the dashed line is computed by setting the neutral wind equal to zero. The observations correspond to the period illustrated in Figure 6*c* (June 10–11, 1980). As this particular example shows, the Joule heating rate that includes the neutral wind effect is generally 2–4 times less than the rate derived by assuming **U** = 0. The decrease is highly variable, however, and at times approaches or exceeds a factor of 10. After 2000 UT (1000 LT) the Joule heating rate derived by assuming **U** = 0 is actually exceeded by that obtained by including the neutral wind. This follows a rapid change in the direction and magnitude of ion convection, which is not immediately reflected in the neutral flow because the neutral atmosphere responds slowly to changes in the ion flow.

This result shows that the effect of neutral winds on the local Joule heating rate is nonnegligible and can at times dominate. During most of the period illustrated, the calculated Joule heating, on the average, is 2–4 times smaller when the neutral wind is included than when it is set to zero.

446

The neutral winds have seasonal variations [*Roble* 1983; *Wickwar et al.,* 1986] that have not been considered in this study. These variations must also be reflected in the seasonal variation of Joule heating.

5. SUMMARY OF OBSERVATIONS

We have focused here on the three physical parameters (the electric field, the Pedersen conductivity, and the neutral wind) that determine the Joule heating rate. Summarized below are some significant findings.

1. The data presented here indicate that the intensity of the ionospheric electric field, as well as the average convection pattern, are different for each season. The magnitude of the plasma convection is smallest in summer and largest in fall.

2. During quiet geomagnetic conditions, the Joule heating rate is maximum in the fall.

3. The F region contributes little to nighttime integrated Pedersen conductivities. During the daytime, under solar maximum conditions, the F-region conductivity can be larger than the E-region conductivity.

4. The often-neglected neutral wind can be important. In our example, the Joule heating, calculated by incorporating the actual neutral wind values, was generally 2–4 times smaller than that calculated by assuming $U = 0$. The reverse can also be true, as was illustrated during a period when the wind was larger than the $\mathbf{E} \times \mathbf{B}$ drift.

6. DISCUSSION

An inherent limitation in averaging all data by season as we have done here is that we smooth out all levels of magnetic activity and of IMF B_z and B_y, all of which profoundly affect ionospheric plasma convection. However, the present study gave enough points in each bin so that the results are statistically significant. With time we will accumulate enough measurements to calculate separate empirical models as a function of season, magnetic activity, and IMF.

As shown by Eq. 2 and Figure 7b, the F-region Pedersen conductivity depends on the collision frequency between neutrals and O^+ ions. However, it is becoming increasingly clear that the values of O^+/O collisional cross section that are generally adopted may have to be altered. As is pointed out by *Burnside et al.* [1987], the value commonly adopted may have to be multiplied by as much as 1.7. If this were the case, the F-region Pedersen conductivity would also have to be increased, since O^+ ions dominate at this altitude (see Figure 7a). This will increase even more the relative contribution from the F region to the integrated Pedersen conductivity.

The large F-region conductivity has important consequences in terms of high latitude electrodynamics: (1) Over a significant fraction of the F layer, the ions do not move in the $\mathbf{E} \times \mathbf{B}$ direction, but rather, collisions have the effect of deflecting them in the direction of $\mathbf{E}$; (2) The ionospheric current vector, $\mathbf{J}_\perp$, rotates by 90° between the E and F layers. $\mathbf{J}_\perp$ is perpendicular to the electric field in the E region, and parallel in the F region; (3) A large fraction of the field-aligned current closure by Pedersen current occurs in the F region.

Our last point concerns the role of the neutral wind in Joule dissipation. The energy dissipated to the neutrals is governed by Eq. 1, and, as we have shown, the neutral wind term should not be neglected in the calculation. When dealing with the magnetosphere, however, the actual energy dissipated by Joule heating is evaluated by considering only the electric field and by not including the neutral wind term [*Sugiura,* 1984; *Vasyliunas,* 1970].

ACKNOWLEDGMENT—This work was supported by NSF Cooperative Agreement ATM-8516436, AFGL Contract F19628-87-K006, and AFOSR contract F49620-87-K-0007.

REFERENCES

Baron, M. J., and R. H. Wand, "F-Region Ion Temperature Enhancements Resulting from Joule Heating," *J. Geophys. Res.*, **88**, 4114–4118 (1983).

Baumjohann, W., and Y. Kamide, "Hemispherical Joule Heating and the AE Indices," *J. Geophys. Res.*, **89**, 383 (1984).

Brekke, A., J. R. Doupnik, and P. M. Banks, "Observations of Neutral Winds in the Auroral E Region During the Magnetospheric Storm of August 3–9, 1972," *J. Geophys. Res.*, **79**, 2448–2456 (1974).

Brekke, A., and C. L. Rino, "High-Resolution Altitude Profiles of the Auroral Zone Energy Dissipation Due to Ionospheric Currents," *J. Geophys. Res.*, **83**, 2517–2524 (1978).

Brekke, A., and C. Hall, "Auroral Ionospheric Quiet Summer Time Conductances," *Ann. Geophys.*, **6**, 361–376 (1988).

Burnside, R. G., C. A. Tepley, and V. B. Wickwar, "The O^+ –O Collision Cross Section: Can It Be Inferred From Aeronomical Measurements?" *Ann. Geophys.*, **5A**, 342–350 (1987).

Campbell, W. H., "Annual and Semiannual Changes of the Quiet Daily Variations (Sq) in the Geomagnetic Field at North American Locations," *J. Geophys. Res.*, **87**, 785–796 (1982).

Cortie, A. L., "Sunspots and Terrestrial Magnetic Phenomena," 1898–1911, *Mon. N. R. Astron. Soc.*, **73**, 52–60 (1912).

de la Beaujardière, O., V. B. Wickwar, J. D. Kelly, and J. H. King, "IMF-B_y Effects on the High-Latitude Nightside Convection," *Geophys. Res. Lett.*, **12**, 461–464 (1985).

de la Beaujardière, O., and V. B. Wickwar, "IMF Control of Plasma Drift, Ion Temperature and Neutral Wind," *Proc. U.S.–Finland Auroral Workshop* (1986).

de la Beaujardière, O., V. B. Wickwar, and J. H. King, "Sondrestrom Radar Observations of the Effect of the IMF B_y Component on Polar Cap Convection," in *Solar Wind-Magnetosphere Coupling*, Kamide and Slavin, eds., Terra/Reidel Publishing Company, Tokyo, pp. 495–505 (1986).

Evans, J. V., "Theory and Practice of Ionosphere Study by Thomson Scatter Radar," *Proc. IEEE*, **57**, 496–530 (1969).

Evans, D. S., N. C. Maynard, J. Troim, T. Jacobsen, and A. Egeland, "Auroral Vector Electric Field and Particle Comparison. 2. Electrodynamics of an Arc," *J. Geophys. Res.*, **82**, 2235 (1977).

Foster, J. C., D. H. Fairfield, K. W. Ogilvie, and T. J. Rosenberg, "Relationship of Interplanetary Parameters and Occurrence of Magnetospheric Substorms," *J. Geophys. Res.*, **76**, 6971–6975 (1971).

Foster, J. C., J.-P. St.-Maurice, and V. J. Abreu, "Joule Heating at High Latitudes," *J. Geophys. Res.*, **88**, 4885–4896 (1983).

Foster, J. C., J. M. Holt, R. G. Musgrove, and D. S. Evans, "Ionospheric Convection Associated with Discrete Levels of Particle Precipitation," *Geophys. Res. Lett.*, **13**, 656–659 (1986).

Friis-Christensen, E., Y. Kamide, A. D. Richmond, and S. Matsushita, "Interplanetary Magnetic Field Control of High-Latitude Electric Fields and Currents Determined from Greenland Magnetometer Data," *J. Geophys. Res.*, **90**, 1325 (1985).

Fujii, R., T. Iijima, T. A. Potemra, and M. Sugiura, "Seasonal Dependence of Large-Scale Birkeland Currents," *Geophys. Res. Lett.*, **8**, 1103–1106 (1981).

Hedin, A. E., "A Revised Thermospheric Model Based on Mass Spectrometer and Incoherent Scatter Data: MSIS-83," *J. Geophys. Res.*, **88**, 10170–10188 (1983).

Heelis, R. A., and W. B. Hanson, "High-Latitude Ion Convection in the Nighttime F Region," *J. Geophys. Res.*, **85**, 1995 (1980).

Heppner, J. P., "Empirical Models of High-Latitude Electric Fields," *J. Geophys. Res.*, **82**, 1115 (1977).

Heppner, J. P., and N. C. Maynard, "Empirical High-Latitude Electric Field Models," *J. Geophys. Res.*, **92**, 4467–4489 (1987).

Holt, J. M., R. H. Wand, J. V. Evans, and W. L. Oliver, "Empirical Models for the Plasma Convection at High Latitudes from Millstone Hill Observations," *J. Geophys. Res.*, **92**, 203–212 (1987).

Johnson, R. M., and V. B. Wickwar, "Incoherent Scatter Measurements of High-Latitude Lower-Thermospheric Density and Dynamics," *Proc. Atmospheric Density and Aerodynamic Drag Workshop*, AFGL, Oct 10, 1987.

Johnson, R. M., "Lower-Thermospheric Neutral Winds at High-Latitude Determined from Incoherent Scatter Measurements: A Review of Techniques and Observations," *Adv. Space Res.* (1989).

Kelly, J. D., "Sondrestrom Radar-Initial Results," *Geophys. Res. Lett.*, **10**, 1112–1115 (1983).

Killeen, T. L., "Energetics and Dynamics of the Earth's Thermosphere," *Rev. Geophys.*, **25**, 433–454 (1987).

Lathuillère, C., V. B. Wickwar, and W. Kofman, "Incoherent Scatter Measurements of Ion-Neutral Collision Frequencies and Temperatures in the Lower Thermosphere of the Auroral Region," *J. Geophys. Res.*, **88**, 10137–10144 (1983).

Leadabrand, R. L., M. J. Baron, J. Petriceks, and H. F. Bates, "Chatanika, Alaska, Auroral-Zone Incoherent-Scatter Facility," *Radio Sci.*, **7**, 747–756 (1972).

Matsushita, S., and W.-Y. Xu, "Equivalent Ionospheric Current Systems Representing IMF Sector Effects on the Polar Geomagnetic Field," *Planet. Space Sci.*, **30**, 641–656 (1982).

Prölss, G. W., "Magnetic Storm Associated Perturbations of the Upper Atmosphere: Recent Results Obtained by Satellite-Borne Gas Analyzers," *Rev. Geophys. Space Phys.*, **18**, 183 (1980).

Rasmussen, C. E., R. W. Schunk, and V. B. Wickwar, "A Photochemical Equilibrium Model for Ionospheric Conductivity," *J. Geophys. Res.*, **93**, 9831 (1988).

Roble, R. G., R. E. Dickinson, and E. C. Ridley, "Seasonal and Solar Cycle Variations of the Zonal Mean Circulation in the Thermosphere," *J. Geophys. Res.*, **82**, 5493 (1977).

Roble, R. G., "Dynamics of the Earth's Thermosphere," *Rev. Geophys. Space Phys.*, **21**, 217–233 (1983).

Russell, C. T., and R. L. McPherron, "Semiannual Variation of Geomagnetic Activity," *J. Geophys. Res.*, **78**, 92–108 (1973).

Schunk, R. W., and A. F. Nagy, "Ionospheres of the Terrestrial Planets," *Rev. Geophys.*, **18**, 813–852 (1980).

Sugiura, M., "A Fundamental Magnetosphere-Ionosphere Coupling Model Involving Field-Aligned Currents as Deduced from DE-2 Observations," *Geophys. Res. Lett.*, **11**, 877 (1984).

Vasseur, G., "Dynamics of the F-Region Observed with Thomson-Scatter-I Atmospheric Circulation and Neutral Winds," *J. Atmos. Terr. Phys.*, **31**, 397–420 (1969).

Vasyliunas, V. M., "Mathematical Models of Magnetospheric Convection and Its Coupling to the Ionosphere," in *Particles and Fields in the Magnetosphere*, B. M. McCormac, ed., Reidel, Hingham, MA, p. 60 (1970).

Vickrey, J. F., R. R. Vondrak, and S. J. Matthews, "Energy Deposition by Precipitating Particles and Joule Dissipation in the Auroral Ionosphere," *J. Geophys. Res.*, **87**, 5184 (1982).

Wickwar, V. B., O. de la Beaujardière, and C. A. Leger, "The Analysis Phase of MITHRAS," Final Report, Air Force Office of Scientific Research Contract F49620-83-K-0005, SRI Project 4995, SRI International, Menlo Park, CA (June 1986).

Wickwar, V. B., "Global Thermospheric Studies of Neutral Dynamics Using Incoherent-Scatter Radars," *Adv. Space Res.* (1989).

VII-6. GLOBAL OBSERVATIONS: A FUTURE RESEARCH THRUST IN AURORAL AND MAGNETOSPHERIC RESEARCH

D. J. Williams*

Auroral research has benefited greatly from the availability of satellite imagery. Global auroral images have provided a unique and invaluable perspective of the temporal and spatial behavior of the entire auroral system, giving both new fundamental insights into auroral processes and an important observational framework in which to place local observations. The proven value of global observations provides the basis of an exciting future thrust in auroral and auroral-related research; namely, the provision of simultaneous global observations of the aurora and the magnetosphere. Several examples of ground-based and satellite instrumentation designed to obtain large-scale and global ionospheric, plasmaspheric, and magnetospheric observations are described. Sample results from these future measurements also are shown.

1. INTRODUCTION

Global auroral images have been prominent in auroral research for the past several years. They show in a dramatic, and often wondrous way, the power of having a global perspective of a natural phenomenon, in this case the aurora, and are a testament to the immeasurable value of being able to place local observations within a known overall framework. In a short time global auroral images have established a new era of auroral research—one that utilizes satellite auroral imagery both to observe the global behavior of the aurora and to integrate local physical processes into the total observed picture. No doubt Sydney Chapman would be pleased to know that this symposium celebrating his one-hundredth birthday would provide a forum in which to display these new and exciting avenues of auroral research.

We show in Figure 1 a representative image obtained from the first instrument to provide truly global auroral images showing the behavior of the aurora on a worldwide scale, the auroral imaging instrument on the NASA Dynamics Explorer-1 (DE 1) satellite [*Frank et al.,* 1981]. This particular UV image was obtained over the 12-minute interval 0245–0257 UT on December 13, 1981 during the late expansion phase of substorm activity at which time DE 1 was at a geographic local time of 0342 hours, a geographic latitude of 57.9°, and an altitude of 3.66 Earth radii (R_E). Such images and their time sequences present a global view of auroral dynamics previously unimagined and have become an indispensable addition to our family of auroral observations (DMSP, HILAT, Polar BEAR, DE 1, Viking). In fact, it is clear that a solution to the auroral problem will require a combination of global and local ob-

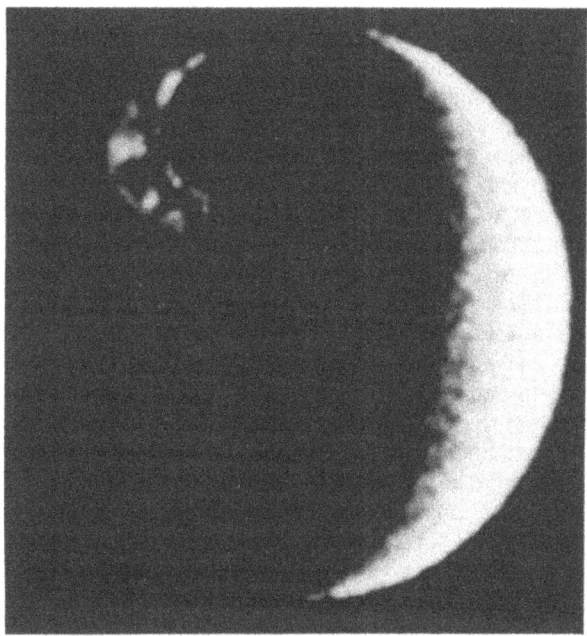

Figure 1—Aurora borealis in the late expansion phase of a substorm. This 12-minute image was obtained by the DE 1 auroral imaging instrument at 0245 UT on December 13, 1981. The observing position was at an altitude of 3.66R_E, a geographic latitude of 57.9°, and a geographic local time of 0342 hours. Principal contributions to the aurora and atmospheric dayglow in this image are from the emission lines of neutral atomic oxygen at ~ 130.4 and 135.6 nm and from the LBH bands of molecular nitrogen. (Courtesy, L. A. Frank.) (This figure also appears in color: Plate 47.)

servations in order to understand how localized physical processes combine to determine the overall behavior of the auroral system.

The need for global observations is not confined to the low-altitude auroral regions emphasized here, but is evident throughout all of geospace, that region extending from the pre-shocked solar wind through the

---

*The Johns Hopkins University Applied Physics Laboratory, Laurel, MD 20707.

Auroral Physics, edited by C.-I. Meng, M. J. Rycroft and L. A. Frank. © Cambridge UP 1991

magnetosphere and into Earth's upper atmosphere. Since the aurorae are a dynamic part of geospace with strong interactions with its many components, global observations are required to understand how auroral regions globally connect to, and interact with, the high-altitude magnetosphere.

Syndey Chapman also was interested in this linkage of the aurora to the high-altitude regions of Earth's magnetic field and spent considerable energies studying the high-altitude magnetosphere and its low-altitude signatures. Perhaps the best known of these latter works are the pioneering studies on Earth's ring current coauthored with V. C. A. Ferraro [*Chapman and Ferraro*, 1931; 1932; 1933]. As in auroral research, great strides have been made in our understanding of the ring current. However, we know neither its global characteristics nor the relationships between it and the auroral current systems—surely they are strong and will have a profound influence on our understanding of the global magnetosphere.

Thus a major thrust in future auroral research will be the measurement of global auroral and magnetospheric characteristics that, when combined with local observations, will allow us to know the dynamics of these coupled systems well enough to understand them. This thrust, global observations, forms the central theme in this discussion of future work and programs of auroral interest. Here we will consider only a few of many excellent projects as illustrative of an increasing capability of obtaining measurements over large spatial scales relating to global behavior. While local measurements are not discussed, it is emphasized that they are required not only to provide ground truth for the global observations, but also to provide an understanding of local physical processes—an understanding required in order to learn how the local physics integrates into the global picture.

2. GLOBAL OBSERVATIONS

While some of the techniques that are now available for global observations are innovative, the need for such measurements has long been recognized in the space plasma physics community. For decades, researchers utilizing ground-based observations have combined data from large numbers of stations in order to obtain a global perspective of effects seen at Earth's surface. This has led naturally to the use of multisatellite observations throughout the space age in attempts to gain a similar perspective of the magnetosphere. In fact, specific multisatellite programs have been conducted and are planned for the future in order to obtain a macroscale perspective of auroral and magnetospheric processes and regions. Past examples are the International Sun-Earth Explorer (ISEE) Program and

the NASA Dynamic Explorer (DE) Program. An example of such a future multisatellite program is the Global Geospace Science (GGS) Program, part of the International Solar Terrestrial Physics (ISTP) Program. The ISTP Program in turn is one of the core programs forming the Solar-Terrestrial Energy Program (STEP). All of these programs have been designed to maximize our capability to obtain global information from in-situ observations of solar-terrestrial and geospace characteristics and processes. For example, GGS was developed specifically as an attempt to provide an accounting of energy flow through geospace, using a minimally configured network of four satellites. These satellites are planned to be located in key regions of geospace and provide simultaneous particle and field measurements along with global auroral images in visible, UV, and X-ray wavelengths. The four key regions identified for the local plasma observations are the upstream solar wind, the high-altitude polar regions, the near-equatorial magnetospheric region, and the magnetospheric tail regions. Complementary ground-based observing programs are being developed to provide vital measurements of large-scale/global magnetospheric and ionospheric processes. For example, the Canadian CANOPUS program, consisting of an extensive ground-based array of magnetometers, riometers, radars, photometers, and all-sky cameras, will provide measurements yielding a large-scale perspective of ionospheric energy dissipation by particle precipitation and the flow of electric currents. [*A. Vallance Jones, Principal Investigator*; personal communication].

GGS, CANOPUS, and other coordinated observing programs represent an ambitious attempt to obtain an overall perspective of magnetospheric behavior and will allow an important beginning in the development and testing of global geospace models such as those envisioned by the U.S. National Science Foundation's Geospace Environment Modeling (GEM) program. However, the very size of geospace not only is testimony to the coarseness of a four-satellite grid, but also dramatizes the impracticability of implementing a sufficiently dense observational network to test various geospace models quantitatively. Other approaches are needed to obtain the required goal observations, and it is the solution to this problem that identifies a major new direction for future research into auroral and magnetospheric dynamics; namely, to obtain a global perspective of the aurora, the magnetosphere and their interactions.

3. LOW-ALTITUDE REGIONS

Global auroral images will be obtained in the future from the Japanese EXOS-D satellite, a Canadian imager on the USSR Interball satellite, and the US Polar

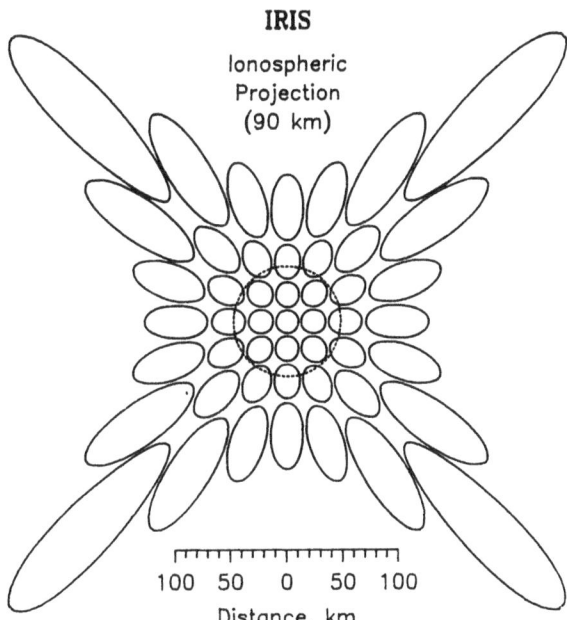

IRIS

Ionospheric
Projection
(90 km)

100 50 0 50 100
Distance, km

Figure 2—Projection to ionospheric heights of the beam patterns of the 49-beam array of IRIS (Imaging Riometer for Ionospheric Studies). This array, installed at South Pole Station during the 1987–1988 austral summer season, employs a 12° full width for each beam, operates at a frequency of 38.2 MHz and is sampled (all 49 beams) in 1 second. The dotted circle in the center of the array shows for comparison the pattern of a single wide beam riometer. (Courtesy, T. Rosenburg.)

satellite. In addition to these observations, auroral data will be obtained by the worldwide infrastructure of ground-based observing stations. To illustrate the importance of the ground-based observing program, we describe briefly two instruments that will advance greatly the capability of observing large-scale regions of the ionosphere and that will extend our global observation capability. First we consider a striking increase in the capability of observing both the spatial and temporal development of electron precipitation regions by ground-based riometers. Most measurements of cosmic radio noise absorption in the polar regions have been made with riometers using broadbeam antennas. Only limited information about the spatial structure and dynamics of energetic electron precipitation regions can be obtained by this means. Recent trends in riometry have been toward the use of multiple, narrow-beam antennas operated in a fixed-beam or one-dimensional scanning mode to examine smaller ionospheric regions. A further step in this direction has been taken with the development of a phased-array imaging riometer system by T. J. Rosenberg at the University of Maryland. The first Imaging Riometer for Ionospheric Studies, IRIS, is now installed and operating continuously at South Pole Station, Antarctica. IRIS operates at a fre-

quency of 38.2 MHz and produces 49 independent narrow beams (12° full width) arranged in a 7 × 7 square array centered on the zenith, with beam peaks out to about 45° zenith angle along the principal axes. The ionospheric sampling grid is illustrated in Figure 2 along with the area sampled by a standard broadbeam riometer. The two-dimensional viewing region of IRIS at D-region heights covers an area approximately 200 km on a side, with a resolution on the order of 20 km. The entire image is scanned once a second.

Figure 3 displays a succession of "images" obtained by IRIS in which the spatial and temporal development of an electon precipitation event is readily discernible. Each panel represents an image formed by the grid pattern shown in Figure 2 in which intensities are gray-coded in 0.25-dB steps. Each image is taken over a 10-second interval and the start of each minute is indicated on the left. The top left panel has been used to show the orientation of the images. Analysis of the data [*T. Rosenberg*, personal communication] shows a westward propagating feature traveling at a speed of ~1 km s^{-1}. Instrumentation like IRIS should prove to be very valuable in separating spatial and temporal fea-

IRIS IMAGE SEQUENCE 15 JAN 88

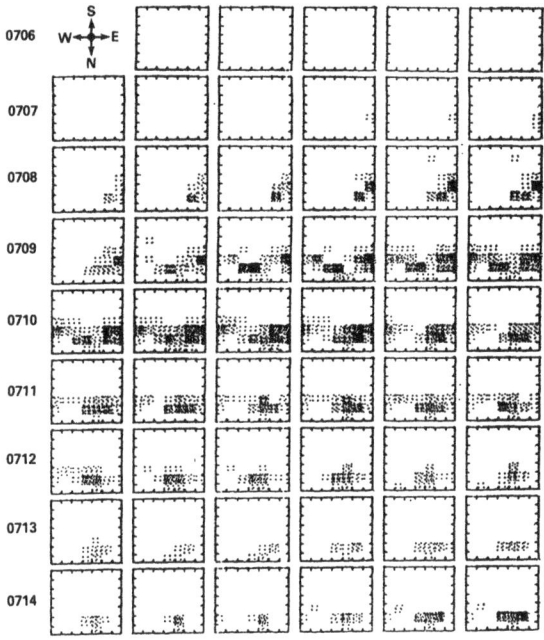

Figure 3—Sequence of IRIS images beginning at 0710:10 hours on January 15, 1988. Each panel is a 10-second sample (image) of the entire 49-beam array shown in Figure 2. Absorption is shown gray-coded in 0.25-dB increments. Image orientation is shown in the upper left and the start time for each row of 10-second images is given on the left. (Courtesy, T. Rosenburg.)

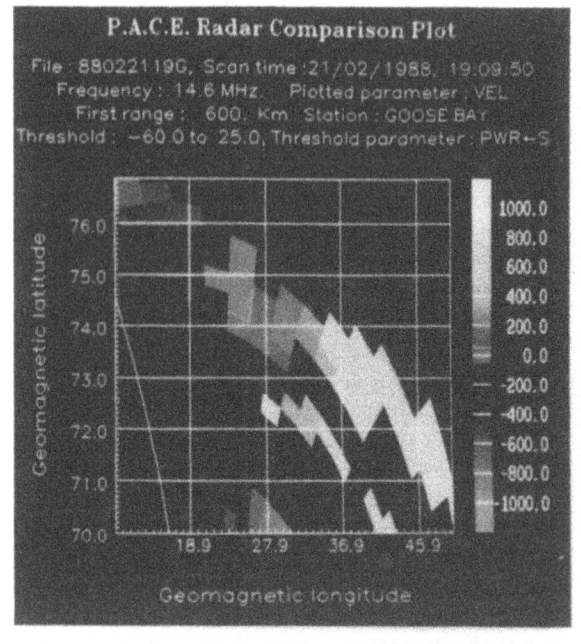

(a)

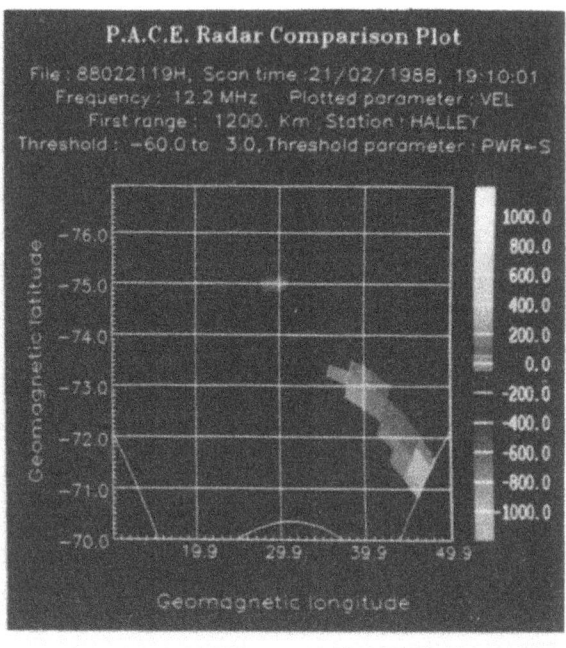

(b)

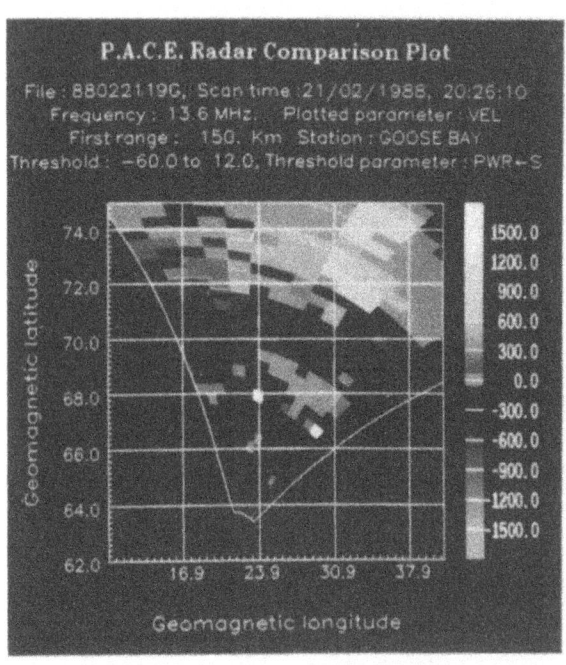

(c)

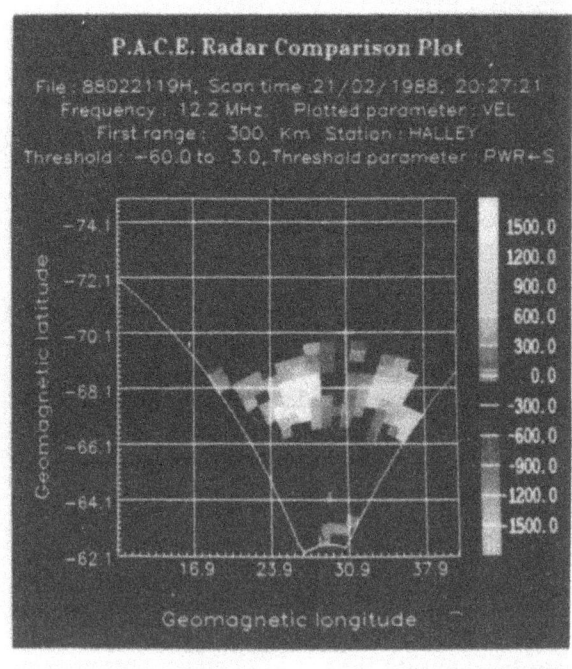

(d)

Figure 4—PACE (Polar Anglo-American Conjugate Experiment) line-of-sight velocities measured by HF (8–20 MHz) radars located in Goose Bay, Labrador (panels *a* and *c*) and Halley Bay, Antarctica (panels *b* and *d*). Velocity toward (+) and away (−) from the radar is coded in meters/second according to the color bar at the right and displayed on a grid showing geomagnetic latitude and geomagnetic longitude. The observations were made at ~2200 magnetic local time on February 21, 1988. The observations at ~1910 UT (panels *a, b*) show a velocity pattern typical of westward convection and a scattering region at the same magnetic location in both hemispheres. At ~2026 UT, two scattering regions appear and are displaced in the southern hemisphere ~5° equatorward from the corresponding northern hemisphere locations. (This figure also appears in color: Plate 48.)

452

tures of electron precipitation events over large ionospheric areas.

A second development in observing large-scale ionospheric features has come from the Polar Anglo-American Conjugate Experiment (PACE) conducted by J. R. Dudeney of the British Antarctic Survey and R. A. Greenwald of The Johns Hopkins University. PACE uses coherent scatter high frequency (HF) radars located at Goose Bay, Labrador, and Halley Bay, Antarctica, to observe ionospheric irregularities simultaneously at conjugate locations. The field of view of each radar system extends in absolute value from approximately 64° to over 85° magnetic invariant latitude.

Figure 4 shows the line-of-sight velocity obtained from the PACE measurements at ~1910 UT February 21, 1988 [*Dudeney and Greenwald*, personal communication]. The line-of-sight velocity is coded in magnitude (m s^{-1}) and direction (+, toward; −, away) according to the color bar at the right of each panel and is displayed on a geomagnetic latitude-longitude grid. Panels *a* and *c* show the Goose Bay results and panels *b* and *d* show the Halley Bay Results. The magnetic local time for these observations is ~2200 hours MLT.

Panels *a* and *b* show velocity patterns at both radars that are typical of westward convection and a scattering region that is located at approximately the same geomagnetic location in each hemisphere. Panels *c* and *d* show PACE results obtained ~1½ hours later, at which time both radars observe two distinct regions: a poleward region characterized by high speed westward convection, and an equatorward region characterized by low negative Doppler velocities. In this case, the southern hemisphere scattering regions seem displaced equatorward by ~5° relative to their northern hemisphere counterparts. Such results can provide stringent tests for global models of the magnetospheric electric and magnetic field configuration.

4. HIGH-ALTITUDE REGIONS

The preceding discussion described examples of global and/or macroscale observations to be obtained of the upper atmosphere/ionosphere regions. We now describe briefly global observations of high-altitude magnetospheric regions expected to be available in the future.

Plasmasphere

Advances in optical instrumentation, particularly in threshold sensitivities, make it feasible to carry out remote sensing of tenuous plasmas by detecting their optical emissions [*Broadfoot*, 1986]. Such measurements provide the capability of globally observing low-energy ion populations in the magnetosphere. For example, the use of intensified charged coupled devices with spectrographic and photometric instrumentation provides

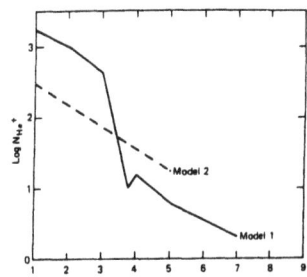

MODEL He$^+$ DISTRIBUTIONS USED TO ESTIMATE THE 304 A SIGNAL FROM THE PLASMASPHERE

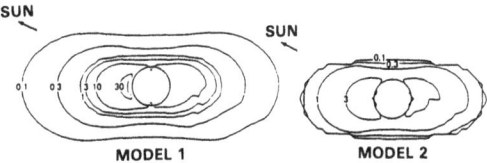

COMPARISON OF MODEL 1 (LEFT) WITH MODEL 2 (RIGHT) AS VIEWED FROM HIGH ALTITUDE. CONTOUR UNIT IS RAYLEIGHS

Figure 5—Top: Model plasmaspheric He$^+$ densities used to estimate 30.4-nm intensities. Model 1 [*Horwitz et al.*, 1984] represents magnetically disturbed conditions, and model 2 [*Waite et al.*, 1983] is an equilibrium model. Bottom: Plasmaspheric He$^+$ 30.4-nm emission maps for the two models shown above. Comparison shows that present instrumentation can readily track the plasmasphere's evolution from quiet to disturbed conditions.

sensitivities sufficient to image the plasmasphere globally by measuring resonantly scattered He$^+$ emissions at 30.4 nm. From an appropriately positioned satellite platform, it then becomes possible to observe the global morphology and dynamics of the plasmasphere.

Aside from H$^+$, He$^+$ is the most abundant plasmaspheric ion and thus is a natural candidate for remote imaging of the plasmasphere. To illustrate the capability of a plasmaspheric imager utilizing the He$^+$ 30.4 nm emission line, we show, in Figure 5, expected emission maps based on the two He$^+$ models shown. Model 1 is based on DE 1 measurements during magnetically disturbed conditions [*Horwitz et al.*, 1984] and model 2 is an equilibrium model based on ISEE measurements [*Waite et al.*, 1984]. Note that the peak intensity differs by a factor of ten for the two different models. Further, the application of standard deconvolution techniques to these line-of-sight images can be used to infer details of these (or any existing) He$^+$ distributions as a function of L.

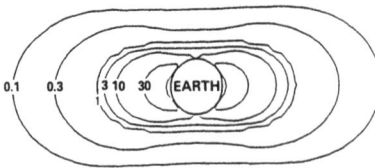

He⁺ 304 A ISOPHOTES FOR DISTRIBUTION MODEL 1. SEEN FROM A POINT ON THE EARTH-SUN LINE IN PLANE OF THE MAGNETIC EQUATOR

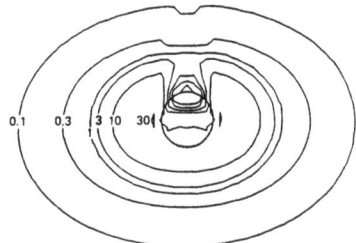

He⁺ 304 A ISOPHOTES FOR DISTRIBUTION MODEL 1. SEEN FROM A POINT IN THE NOON MERIDIAN AT A MAGNETIC LATITUDE OF 35°

Figure 6—Plasmaspheric He + 30.4-nm emission maps as observed from indicated positions.

Figure 6 shows views of the plasmasphere based on model 1 as seen from observing positions at the noon meridian both on and 35° above the magnetic equator. The brightest He⁺ 30.4 nm emission occurs near the Earth where the integrated He⁺ column density through the plasmasphere reaches its maximum value. Measurable emission should be seen out to ~5R_E based on this model. The image from 35° magnetic latitude shows a view of the full longitudinal extent of this model plasmasphere.

Imaging the entire plasmasphere, as illustrated in Figures 5 and 6, is an excellent way to monitor changes in plasmaspheric ion density and distribution globally. Since full plasmasphere images can be obtained on time scales of minutes, variations in response to geomagnetic activity can be tracked readily. For example, the refilling of newly corotating flux tubes outside the previous plasmasphere can be studied by imaging. Global images during active periods will show directly how much the plasmasphere is compressed and how much of it is detached and drifts through the magnetosphere under the influence of the ambient electric and magnetic fields. Plasmaspheric images also may allow a mapping of the inner edge of the magnetospheric convection process. These examples of the value of plasmaspheric global images show a major new capability in future research directly related to auroral studies.

Magnetosphere

To emphasize further the auroral connection, we discuss another global-magnetosphere observational technique, namely the measurement of energetic neutral atoms (ENA) to provide global images of the magnetospheric energetic particle (hot plasma) populations. The basis of this technique rests on the fact that the charged particle populations of the magnetosphere are embedded in a tenuous atmosphere of cold, neutral hydrogen atoms, Earth's hydrogen geocorona. ENA are produced by the charge exchange of the energetic ions in magnetospheric plasmas with these hydrogen atoms. The ENA travel along straight line paths from the point of charge exchange and can be detected to form an image of the source magnetospheric plasmas in a manner analogous to the imaging of aurora by the detection of photons emitted from the upper atmosphere.

The concept of measuring ENA and using their detection to image energetic ion populations in the magnetosphere has been tested successfully with measurements from instruments designed primarily for energetic charged particle observations on the IMP and ISEE spacecraft [*Roelof et al.*, 1985; *Roelof*, 1987]. *Roelof et al.*, [1985] presented the first satellite-based observations of the decay of the global ring current by means of detecting ENA (most probably oxygen atoms) generated via charge exchange. *Roelof* [1987] extended these observations and obtained the first-ever global image of the ring current by iterating ring current distributions in a model magnetic field until their convolution through charge exchange with the hydrogen geocorona produced the ENA patterns observed. Further details describing the production of ENA in the magnetosphere and the methods of obtaining images of magnetospheric energetic ion distributions are given in *Roelof and Williams* [1988], *McEntire and Mitchell* [1988], *Keath et al.* [1988], *Roelof* [1988].

Figure 7 shows the result of a simulation presented by *Roelof and Williams* [1988] in which ENA images expected to be observed by presently designed ENA cameras [*McEntire and Mitchell*, 1988] are calculated from a modeled ring current injection scenario. Panels *d*, *e*, and *f* show the resulting ENA images as observed from a position on the dusk meridian at 8R_E and 30° N geomagnetic latitude. These three panels represent a time sequence spanning the main phase asymmetric injection of the ring current (panel *d*) to the nearly symmetric recovery phase (panel *f*). Panels *a*, *b*, and *c* show the corresponding ring current ion distributions in terms of the column-integrated flux of ions whose instantaneous velocity vector points toward the viewing position. The simulation utilized by *Roelof and Williams* [1988] uses ring current intensities, pitch-angle distri-

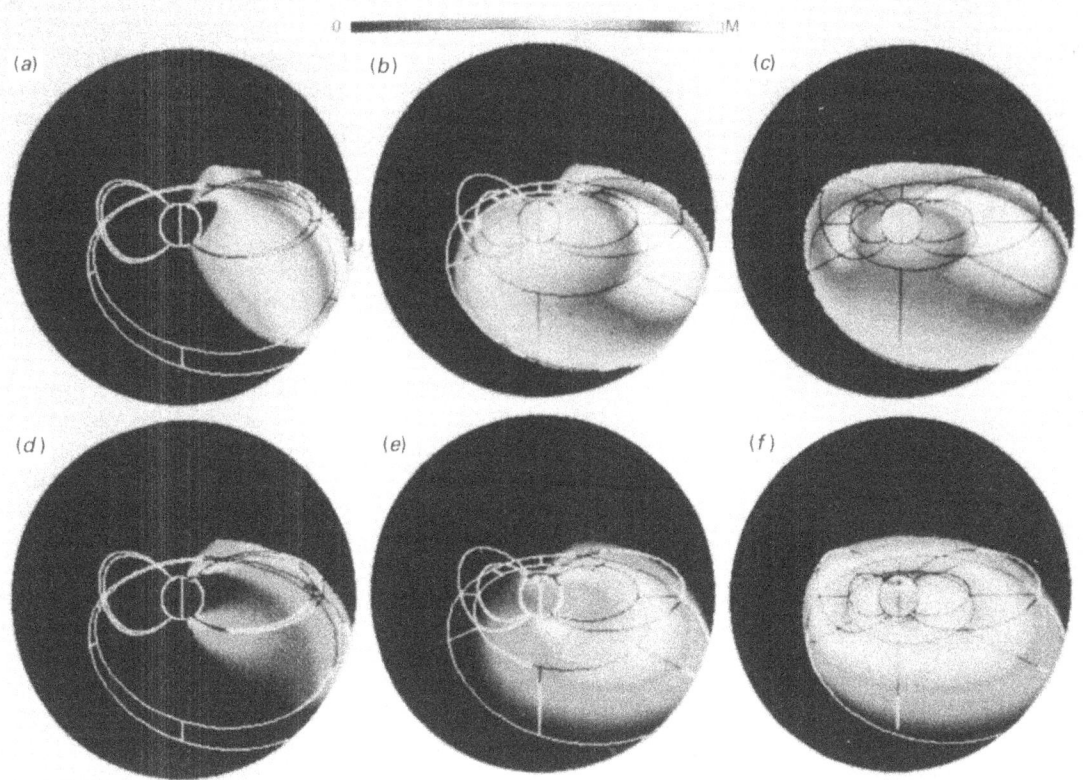

Figure 7—(*a* to *c*) False-color images of model ring current ion intensity evolution. Color represents the column-integrated flux of ions whose instantaneous velocity vector points toward the viewing point. Panels correspond to times spanning asymmetric main phase injection (*a*) through near isotropic recovery phase (*c*). Closed contours indicate the equatorial crossing locus of the inner and outer L shells containing the model fluxes. Magnetic local time cuts at every 3 hours (45°) connect the equatorial loci, with 1800 MLT being the lower vertical line; the Sun is to the left. The linear color bar runs from zero to the maximum (M). The viewing point is at a radius of 8R$_E$ and 30° geomagnetic latitude in the dusk meridian plane (*d* to *f*). False-color images of the ENA flux arising from the modeled interaction of the ions in *a* to *c* with the hydrogen geocorona. Same format as *a* to *c*. (This figure also appears in color: Plate 49.)

butions, and their evolution from main to recovery phase that are based on available observations, a hydrogen geocoronal distribution based on DE 1 observations [*Rairden et al.,*1986], and a realistic magnetic field configuration for the altitudes of interest. Thus Figure 7 can be considered to be a useful guide as to what can be expected from the ENA imaging technique, although actual ENA images may be quite different. The images shown in Figure 7 are similar to and consistent with those measured by the ISEE 1 Medium Energy Particle Instrument [*Roelof*, 1987].

The sensitivity of present instrument designs allows global images of the magnetospheric energetic particle populations to be obtained on an approximate 1-minute time scale for both substorms and magnetic storms. This time resolution and the results shown in Figure 7 indicate that the ENA imaging technique can be valuable in obtaining, for example, global assessments of the temporal and spatial evolution of (1) the ring cur-

rent, (2) magnetic field distortions, (3) the magnetospheric electric field at altitudes ≤10R$_E$, and (4) the energetic particle injection boundary. Further, the availability of global images as shown in Figure 7 will provide a measure of the particle pressure tensor for the distribution generating the ENA image. Knowing the pressure tensor allows the currents generated by this distribution, both perpendicular and parallel to the magnetic field, to be determined [*Vasyliunas,* 1970]. *Roelof* [1988] has used ISEE ENA images to obtain an estimate of the overall ring current system that includes the familiar westward current plus a radial current that agrees with statistical results based on local magnetic measurements [*Iijima et al.,* 1988] and a parallel current that appears to connect with the auroral region II current system. These preliminary results indicate that future ENA images will have a strong quantitative impact in mapping magnetospheric current systems.

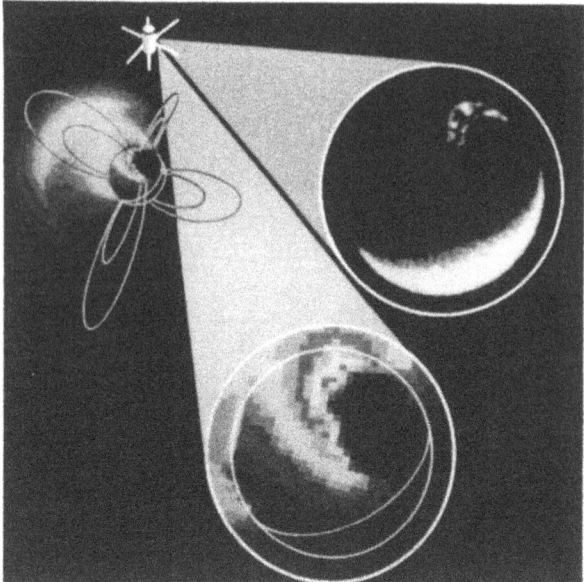

Figure 8—Global magnetospheric and auroral imaging. The DE 1 auroral image of Figure 1 is shown along with a simulated ENA image based on ISEE 1 results obtained during the main phase of the September 29, 1978 magnetic storm. The ENA images shown have been adjusted to the DE 1 position. Both an ENA low-altitude close-up view and an overall magnetospheric view are shown in comparison with the auroral image. (This figure also appears in color: Plate 50.)

We conclude by showing in Figure 8 a repeat of the DE 1 auroral image shown in Figure 1 along with simulated global ENA images as viewed from the DE 1 position. Both the global and close-up ENA images are based on ISEE 1 results obtained during the main phase of the magnetic storm of September 29, 1978 [*Roelof*, 1987]. The global ENA image on the left shows ion injection into the ring current at local midnight while close-up views of the low-altitude energetic ion distributions and the aurora are shown on the right.

The future availability of these types of magnetospheric observations (e.g., He$^+$ 30.4 nm emissions and ENA) will be important in determining the global connections and interactions between the auroral regions and the high-altitude magnetosphere.

5. SUMMARY

Great strides have been made over the past several years in our knowledge of aurora and auroral processes. Many of these advances have resulted from the availability of auroral images obtained from satellite-borne cameras. These observations not only provide an important global perspective of the aurora but also provide a known overall framework in which to cast other, local observations. We propose that a major research thrust for future auroral studies be the extension of global observations to the magnetosphere as a whole. A series of global observing techniques has been discussed that, if implemented, would provide global observations of major magnetospheric and ionospheric regions. Such observations in conjunction with auroral images will provide a first measure of the magnetosphere-ionosphere system on a truly global basis and will provide complementary observations required for understanding how local measurements fit into and determine the global picture.

ACKNOWLEDGMENT – I wish to acknowledge and thank A. L. Broadfoot, J. R. Dudeney, L. A. Frank, A. Vallance Jones, T. Rosenburg, and R. A. Greenwald for discussions and the use of data and information prior to publication. This effort has been sponsored in part by a NASA contract to The Johns Hopkins University Applied Physics Laboratory and the Department of the Navy under task I2UOS10 of Navy Contract N00039-87-C-5301.

REFERENCES

Broadfoot, A. L., in "Images of the magnetosphere and atmosphere: global effects (IMAGE)," proposal for NASA Explorer Mission Concept Studies, by D. J. Williams, L. A. Frank, A. L. Broadfoot, W. L. Imhof, S. B. Mende, D. M. Hunten, R. G. Roble, and G. S. Siscoe (1986).

Chapman, S., and V. C. A. Ferraro, "A New Theory of Magnetic Storms," *Terr. Magn. Atmos. Electr.* **36**, 77–97, 171–186 (1931).

Chapman, S., and V. C. A. Ferraro, "A New Theory of Magnetic Storms," *Terr. Magn. Atmos. Electr.* **37**, 147–156, 421–429 (1932).

Chapman, S., and V. C. A. Ferraro, "A New Theory of Magnetic Storms," *Terr. Magn. Atmos. Electr.* **38**, 79–96 (1933).

Frank, L. A., J. D. Craven, K. L. Ackerson, M. R. English, R. H. Eather, and R. L. Carovillano, "Global Auroral Imaging Instrumentation for the Dynamics Explorer Mission," *Space Sci. Instr.*, **5**, 369 (1981).

Horwitz, J. L., R. H. Comfort, and C. R. Chappel, "Thermal Ion Composition Measurements of the Formation of the New Outer Plasmasphere and Double Plasmapause During Storm Recovery Phase," *Geophys. Res. Lett.*, **11**, 701 (1984).

Iijima, T., T. A. Potemra, and L. J. Zanetti, "Large-Scale Characteristics of Magnetospheric Equatorial Currents," *JHU Applied Physics Laboratory Preprint*, **88-14** (1988).

Keath, E. P., G. B. Andrews, B. H. Mauk, D. G. Mitchell, and D. J. Williams, "Instrumentation for Energetic Neutral Atom Imaging of Magnetospheres," in *Proc. Yosemite 1988: Outstanding Problems in Solar System Plasma Physics; Theory and Instrumentation*, (1988).

McEntire, R. W., and D. G. Mitchell, "Instrumentation for the Imaging of Energetic Neutral Atoms," in *Proc. Yosemite 1988: Outstanding Problems in Solar Systems Plasma Physics; Theory and Instrumentation,* (1988).

Rairden, R. L., L. A. Frank, and J. D. Craven, "Geocoronal Imaging with Dynamics Explorer," *J. Geophys Res.*, **91**, 13613 (1986).

Roelof, E. C., "Energetic Neutral Atom Image of a Storm-Time Ring Current," *Geophys. Res. Lett.*, **14**, 652 (1987).

Roelof, E. C., "Energetic Neutral Atom Images of the Storm-Time Ring Current: ISEE-1 Observations and Computer Simulation," in *Proc. Yosemite 1988: Outstanding Problems in Solar System Plasma Physics; Theory and Instrumentation*, (1988).

Roelof, E. C., "Evidence from ENA Images for the Coupling of the Storm-Time Ring Current to Region 2 Field-Aligned Currents," *this volume.*

Roelof, E. C., and D. J. Williams, "The Terrestrial Ring Current: From in situ Measurements to Global Images Using Energetic Neutral Atoms," *Johns Hopkins APL Tech. Dig.*, **9**, 144 (1988).

Roelof, E. C., D. G. Mitchell, and D. J. Williams, "Energetic Neutral Atoms (E ~ 50 keV) from the Ring Current: IMP 7/8 and ISEE 1," *J. Geophys. Res.*, **90**, 10991 (1985).

Vasyliunas, V. M., "Mathematical Models of Magnetospheric Convection and its Coupling to the Ionosphere," in *Particles and Fields in the Magnetosphere*, B. M. McCormac, ed., D. Reidel Pub. Co., p. 60 (1970).

Waite, J. H., J. L. Horwitz, and R. H. Comfort, "Diffusive Equilibrium Distributions of He$^+$ in the Plasmasphere," *Planet. Space Sci.*, **32**, 611 (1984).

INDEX

Keyword	*Chapter and section*
Alfven layer	IV-4,IV-5
Alfven velocity	III-3
Alfven wave	III-3,V-5
Alfven wave conductance	V-5
Ampere's law	V-5
Aurora(s)	(in all chapters and sections)
Auroral arc(s) (Discrete aurora)	I-1,III-4,IV-1,IV-2,IV-3,IV-4,V-5,VII-4
Auroral bulge	I-1,IV-4,V-1,V-4,VI-1,VI-2
Auroral discharge	I-1,IV-5
Auroral dynamics (see also Dynamics)	I-1,V-1,V-3,V-4,VI-1,VII-3,VII-4
Auroral electron transport	II-2,II-4,II-5,VII-4
Auroral imaging	I-1,II-5,IV-1,IV-2,V-2,V-3,V-4,VI-1, VI-2,VII-6
Auroral ionization	II-2,II-3,II-4,II-5,VI-4,VI-5,VII-1,VII-2, VII-3,VII-4,VII-5
Auroral mapping	I-1,IV-1,IV-2,IV-3,IV-4,IV-5,V-2,V-3,V-4, VI-1,VI-2,VI-5,VII-3,VII-4,VII-6
Auroral morphology	I-1,IV-1,V-1,V-2,V-3,V-4
Auroral oval	I-1,II-4,III-1,IV-1,IV-2,IV-3,IV-4,V-1,V-2, V-3,V-4,VI-1,VI-2,VI-5,VII-2,VII-4,VII-5, VII-6
Auroral particle(s) (electrons and ions)	I-1,II-2,II-3,II-4,II-5,III-1,III-2,III-3,III-4, III-5,III-6,IV-5,VI-3,VI-4,VI-5,VII-1,VII-4, VII-5
Auroral particle acceleration	I-1,II-2,III-1,III-2,III-3,III-4,III-5,III-6,IV-3, VI-1
Auroral plasma wave	III-3,IV-6
Auroral precipitation (see also Auroral particle)	I-1,II-2,II-3,II-4,II-5,III-1,III-2,III-3,III-4, III-5,III-6,IV-3,V-5,VI-1,VI-3,VI-4,VII-3, VII-4,VII-5,VII-6
Auroral pulsations	VI-3,VII-4
Auroral spectroscopy	II-1,II-2,II-3
Auroral substorm(s)	I-1,IV-3,V-1,V-2,V-3,V-4,VI-1,VI-2,VII-3, VII-4,VII-6
Auroral thermosphere	II-3,II-4,II-5,VII-4
Auroral waves	III-3,IV-6
Beams, ion and electron	II-2,III-2,III-3,III-4,III-5,III-6,IV-5
Boundary plasma sheet	III-1,III-2,IV-1,IV-2,IV-4,IV-5,V-3,V-4,VI-2
Broadband turbulance	III-5
Central plasma sheet	III-1,III-2,IV-1,IV-2,IV-4,V-3,V-4,VI-2
Composition	II-2,II-3,II-4,II-5,III-5
Conductivity(ies)	II-4,II-5,V-5,VI-4,VI-5,VII-1,VII-2,VII-3, VII-4,VII-5
Conic, ion and electron	III-2,III-4,III-5
Convection	I-1,II-4,II-5,IV-1,IV-4,V-1,V-5,VII-3,VII-4,VII-5

Counterstreaming electrons — III-4

Cross-tail current — IV-5,V-1,V-5

Diffuse aurora — III-1,IV-1,IV-2,V-3,V-4,V-5,VI-1,VI-2,VI-3, VII-3,VII-4

Dipolarization — IV-5,V-4,V-5

Discrete aurora — III-2,III-4,IV-1,IV-2,IV-3,V-2,V-3,V-4,VI-1, VI-2,VI-5,VII-4

Double layer — III-3,III-5

Earthward edge of plasma sheet — IV-1,IV-4,IV-5,VII-4

Electric field — II-2,II-4,II-5,III-2,III-3,III-4,III-5,IV-5,V-5, VI-4,VI-5,VII-1,VII-2,VII-3,VII-4,VII-5

Electrojet — I-1,II-4,II-5,V-4,V-5,VI-2,VI-4,VI-5,VII-1, VII-2,VII-3,VII-4,VII-5

Electromagnetic ion cyclotron wave — III-5

Electron energy — II-1,II-2,III-1,III-2,III-4,VII-4

Electron transport — II-2,II-4,II-5,III-4,VI-4

Electrostatic field(s) — III-2,III-5,IV-3,IV-5,VI-1,VI-4,VII-1,VII-4

Electrostatic waves — III-4,III-5

Emission — II-1,II-2,II-3

Energy transfer function — III-4

Equatorward boundary of auroral oval — I-1,IV-1,IV-2,IV-4,IV-5,V-2,V-3,V-4,VI-1, VII-4

Excitation mechanisms — II-1,II-2,II-3

Expansive phase — I-1,V-1,V-3,V-4,V-5

Field aligned (Birkeland) current — I-1,II-4,II-5,III-4,IV-4,IV-5,V-1,V-5,VI-2, VI-4,VI-5,VII-1,VII-2,VII-4,VII-5

Fine-scale structure — VI-1,VI-2

Fireballs — IV-4

Geomagnetic (Disturbance) signatures of substorms — I-1,V-4,VI-5,VII-1,VII-2,VII-4

Geomagnetic storm — I-1,II-4,II-5

Geomagnetic substorms — I-1,II-4,II-5,IV-3,V-1,V-2,VII-3,VII-4

Gravity wave — II-4,VII-4,VII-5

Growth phase — I-1,V-1,V-3,V-4,V-5

Hall conductivity — V-5,VI-4,VII-4,VII-5

Hall current — V-5,VI-4,VII-4

Harang discontinuity — II-4,V-5,VII-3,VII-4

High-latitude electric field — II-2,VI-5,VII-1,VII-2,VII-3,VII-4,VII-5

High-latitude thermosphere — II-4,II-5,VII-4,VII-5

Hook-shaped auroral structure — VI-2

Incoherent scatter radar — VII-4,VII-5

Injection boundary — IV-4,IV-5

Inner magnetosphere — III-1,IV-1,IV-4

Inverted-V — III-1,III-2,IV-3,IV-4

Ion drag — II-4,II-5,VII-3,VII-4,VII-5

Ion-ion hybrid — III-5

Ion transport	III-6,VII-4,VII-5
Ionosphere	I-1,II-4,II-5,V-5,VI-1,VI-4,VI-5,VII-1,VII-2, VII-3,VII-4,VII-5,VII-6
Ionosphere/magnetosphere coupling	I-1,III-1,V-1,V-5,VII-3,VII-4,VII-5
Ionosphere-thermosphere general circulation model (ITGCM)	II-4,II-5
Ionospheric chemistry	II-1,II-2,II-3,II-4,II-5,VII-4
Ionospheric composition	II-1,II-3,II-4,II-5,VII-4
Ionospheric conductance	II-4,II-5,VI-4,VI-5,VII-1,VII-2,VII-3,VII-4, VII-5
Ionospheric currents	II-4,II-5,VI-4,VI-5,VII-1,VII-2,VII-4,VII-5
Joule dissipation (heating)	II-4,II-5,V-5,VI-4,VI-5,VII-1,VII-2,VII-4, VII-5
Kelvin-Helmholtz instability	V-5,VI-2
Localized plasma injection	IV-5,V-4
Loss cone	III-2,III-3
Lower hybrid wave(s)	III-4,III-5
Magnetic-field aligned (Parallel) electric field	III-2,III-3,IV-5,VI-4
Magnetic reconnection	I-1,V-1,V-4,V-5,VII-4
Magnetopause	I-1,V-5
Magnetosphere	I-1,IV-1,IV-2,V-4,V-5,VI-1,VI-2,VII-1,VII-3, VII-4,VII-6
Magnetosphere dynamics	I-1,IV-1,IV-2,IV-3,IV-5,V-1,V-2,V-3,V-4, VII-5, VII-6
Magnetospheric boundary layer(s)	IV-1,IV-2,V-5,VI-2,VI-5
Magnetospheric configuration	I-1,IV-1,IV-2,IV-4,IV-5,V-1,V-2
Magnetospheric plasma	I-1,III-1,III-6,VII-6
Magnetospheric plasma domains (regions)	I-1,III-1,IV-1,IV-2,IV-3,IV-4
Magnetotail	I-1,IV-1,IV-3,IV-4,V-1,V-2,V-4,VII-6
Magnetotail dynamics	I-1,V-1,V-2,V-4,V-5
Magnetotail lobe (Tail lobe)	IV-1,IV-2,IV-3,IV-4,V-1,V-2,V-4,V-5,VI-2
Maxwell equations	V-5
Mesoscale auroral structure	VI-2
Midday gap of discrete aurora	IV-2,VI-2
Middle magnetosphere	IV-5,VII-6
Neutral (X) line	V-1,V-4,V-5
Neutral sheet	V-4
Neutral wind	VII-4,VII-5
Neutral wind dynamo	II-4,II-5,VII-5
Nitric oxide	II-1,II-4
Nitrogen	II-1,II-2,II-3,II-4
Nitrogen bands	II-1,II-2,II-3
Omega bands	VI-2,VI-4
Open magnetosphere	I-1,IV-1,IV-2,IV-3,IV-4,V-5

461

Outer radiation belt   IV-4
Patchy aurora   VI-2,VI-3
Pederson conductivity (see also Conductivity)   II-4,V-5,VI-4,VI-5,VII-1,VII-2,VII-5
Pederson current   V-5,VI-5,VII-1,VII-2
Planetary wave   II-4,VII-4
Plasma injection   IV-3,V-1,V-5
Plasma instabilities   IV-3,IV-6
Plasma pause   IV-4,VII-6
Plasma sheet   III-1,IV-1,IV-2,IV-3,IV-4,V-1,V-2,V-5,VII-6
Plasma sheet recovery   V-1,V-4
Plasma sheet thickening   V-1,V-4
Plasma wave   IV-6,VII-4
Polar cap   I-1,IV-1,IV-2,IV-3,V-2,V-3,VI-2,VII-6
Polar cap aurora   I-1,IV-1,IV-2,VI-1,VI-2
Polar cusp region   III-1,IV-1,IV-2,VI-2,VI-5,VII-2,VII-4
Polar diffuse zone   IV-4
Polar rain   III-1,VI-2
Poleward surge (leap)   V-4
Potential drops   I-1,III-2,III-3,VII-4
Proton drift echoes   V-4
Proton gyrofrequency   III-3
Recovery phase   V-1,V-2,V-4
Reflection coefficient   V-5
Remnant layer   IV-4
Resonant acceleration   III-4
Ring current   I-1,IV-4,VII-6
Riometer absorption event   II-4,VII-4
Solar photoionization   II-4,II-5,VII-5
Solar wind   I-1,V-1,V-4,V-5,VI-2
Solar wind dynamo   V-5
Solar wind/magnetosphere interaction   I-1,IV-1,IV-2,V-1,V-2,VII-4
Stochastic acceleration   III-4
Substorm(s) (auroral, magnetic, magnetospheric)   I-1,IV-3,IV-4,IV-5,V-1,V-2,V-3,V-4,V-5,VI-5,
                                                        VII-1,VII-2,VII-3,VII-4

Substorm injection(s) (see also Localized plasma
  injection)   IV-3,IV-4,IV-5,V-4,V-5
Substorm neutral line (see also Neutral (X) line)   IV-3,IV-4,V-1,V-4,V-5
Substorm onset   I-1,IV-3,V-5,VI-4,VII-4
Sun-aligned arcs   IV-2,IV-4,VI-1,VI-2
Tearing instability   V-5
Thermosphere   I-1,II-3,II-4,II-5,VII-4,VII-5
Thermospheric composition   II-1,II-3;II-4,II-5,VII-4
Thermospheric disturbance   II-4,II-5,VII-4
Thermospheric response   II-4,II-5,VII-4
Thermospheric temperature(s)   II-5,V-3,VII-4

Thermospheric wind(s)	II-4,II-5,VII-4,VII-5
Theta aurora	IV-2,IV-4,V-2,VI-1,VI-2,VII-6
Three-dimensional time-dependent thermospheric model (GCM)	II-4,II-5
Tidal wave	II-4,II-5,VII-4
Transmission coefficient	V-5
Transpolar arcs	IV-4,V-2,VI-1,VI-2
Two-stream instability	III-5
Ultraviolet	II-1,II-2,II-3,IV-2,V-2,V-3,V-4
Upper atmosphere	II-1,II-2,II-3,II-4,II-5,VII-4,VII-5
Viscous interaction	IV-1,V-5,VII-5
Vortices	V-3,VII-5
Wave-like auroral structure	VI-2
Wave particle interaction	III-3,IV-5,IV-6
Westward travelling surge	V-1,V-2,V-3,VI-2,VI-4

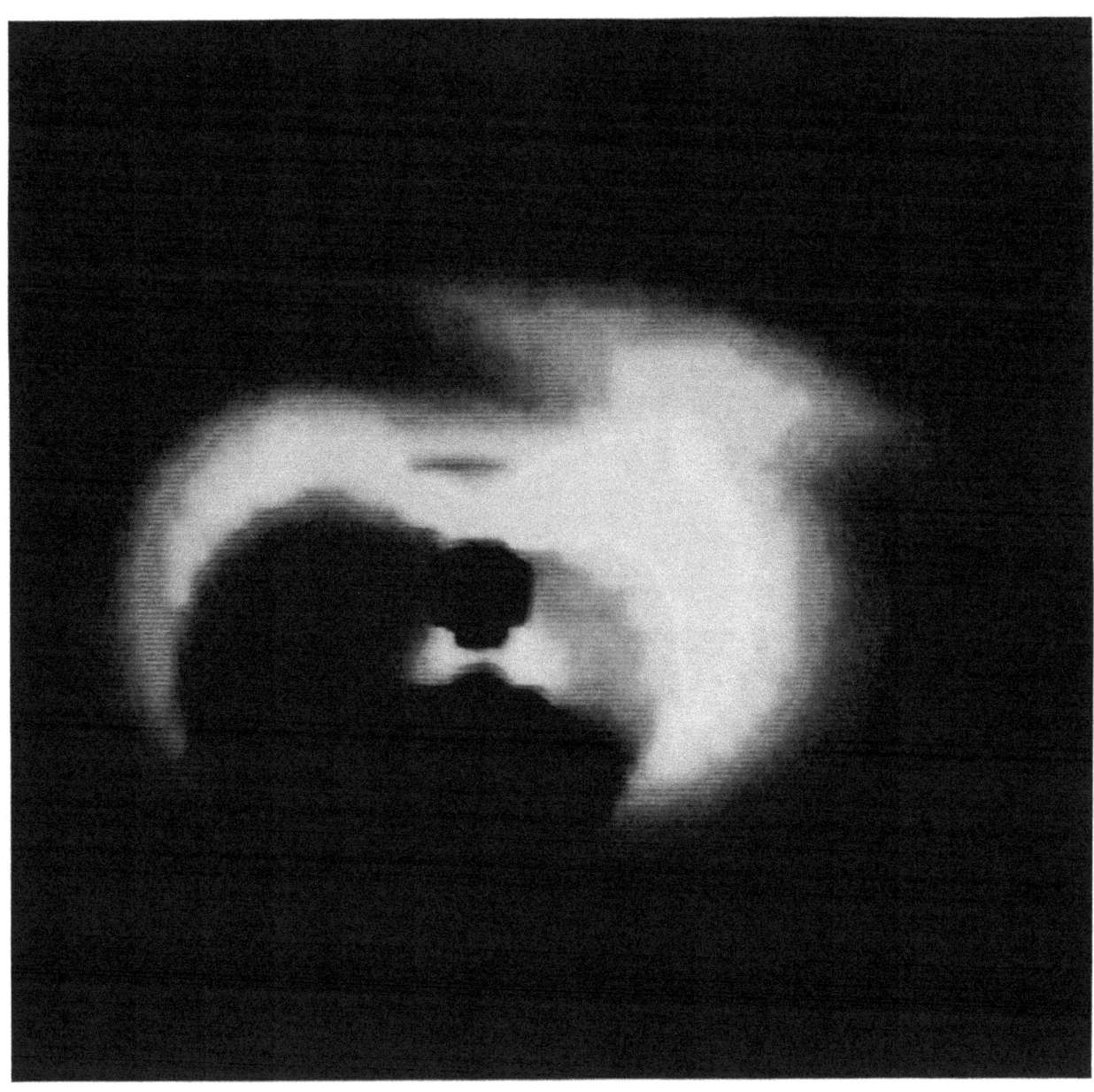

Plate 1—Results of model calculations showing an image of the magnetosphere in resonantly scattered sunlight from O^+.

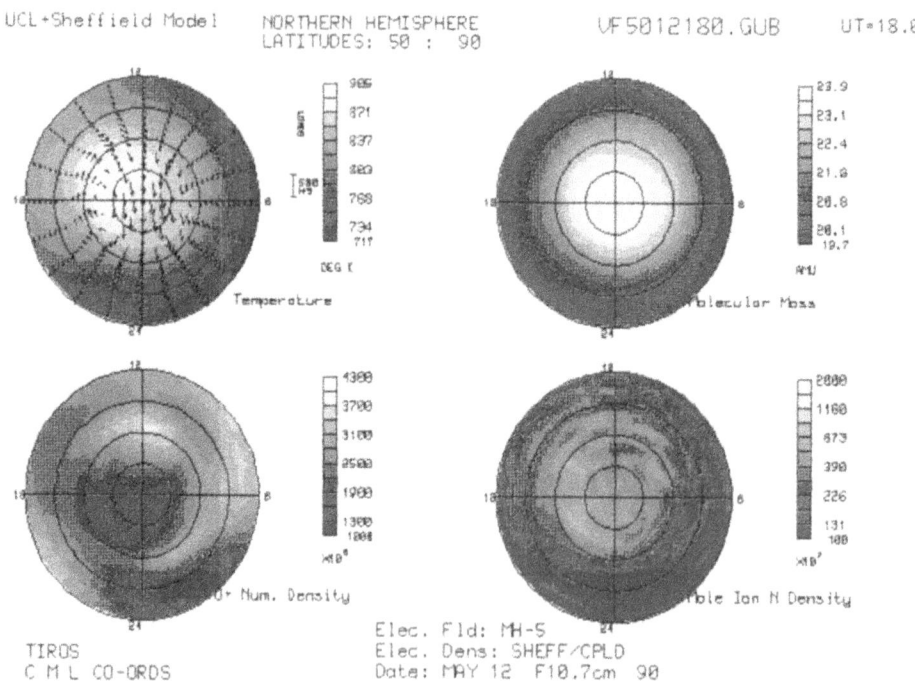

Plate 2—Northern polar distributions of upper thermospheric temperature and wind velocity, mean molecular mass, electron density, and ion temperature at pressure level 12, computed for May 12, by the coupled ionosphere-thermosphere model, using the Millstone Hill convection field, low solar activity, $F_{10.7} = 90$, low geomagnetic activity, $Kp = 2$. The simulations include the effects of lower atmospheric propagating semidiurnal tides.

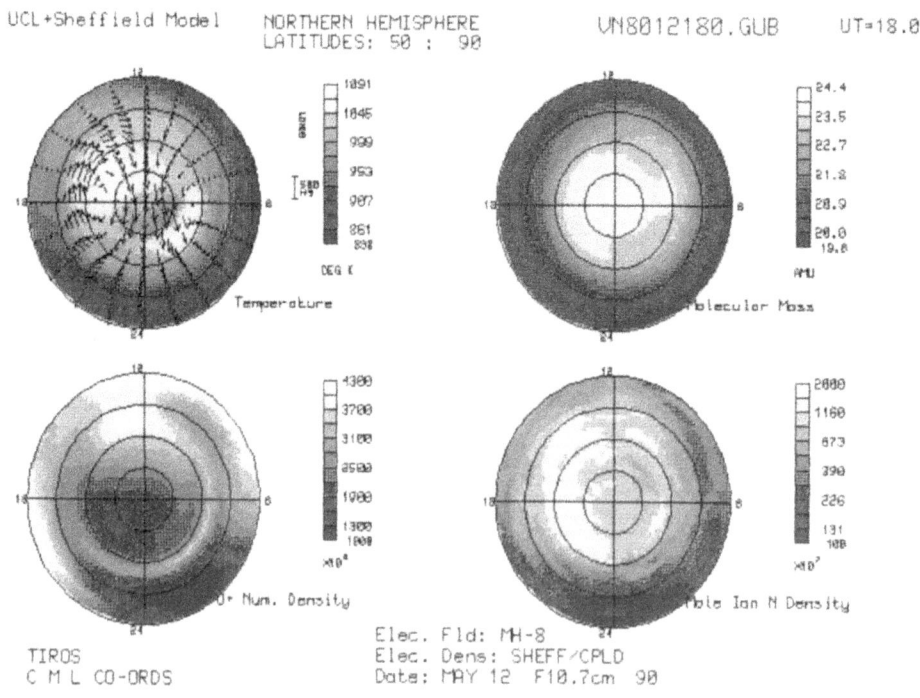

Plate 3—Same as Plate 2, for a higher level of geomagnetic activity, $Kp = 4$.

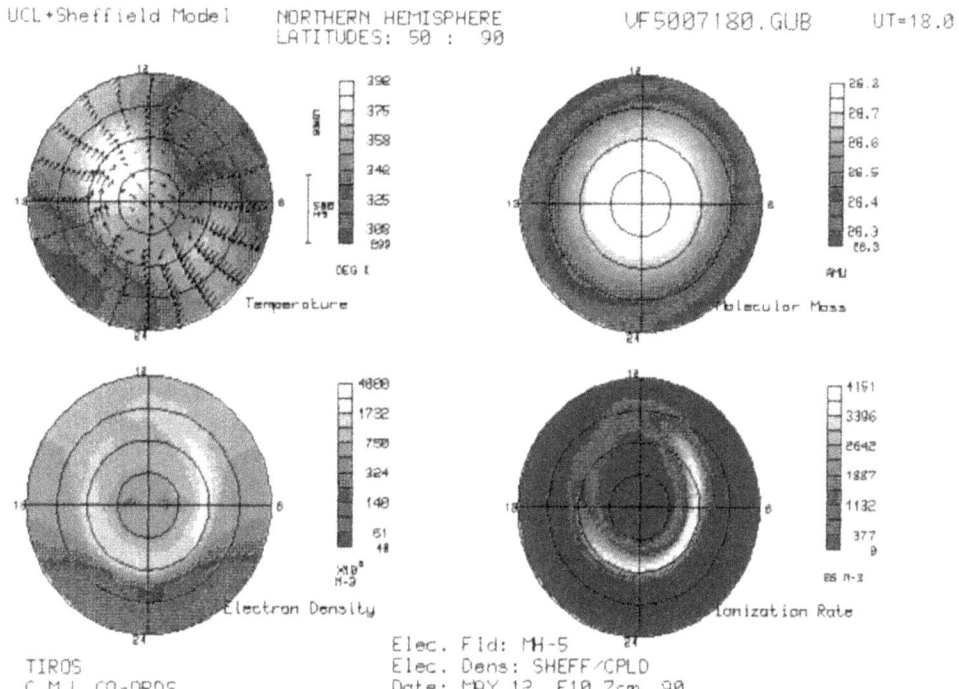

Plate 4—Northern polar distributions of lower thermospheric temperature and wind velocity, mean molecular mass, electron density and auroral ionization rate at pressure level 7 (E-region, approx. 125 km), computed for May 12, by the coupled ionosphere-thermosphere model, using the Millstone Hill convection field, low solar activity, $F_{10.7} = 90$, low geomagnetic activity, $Kp = 2$. The simulations include the effects of lower atmosphere propagating semidiurnal tides.

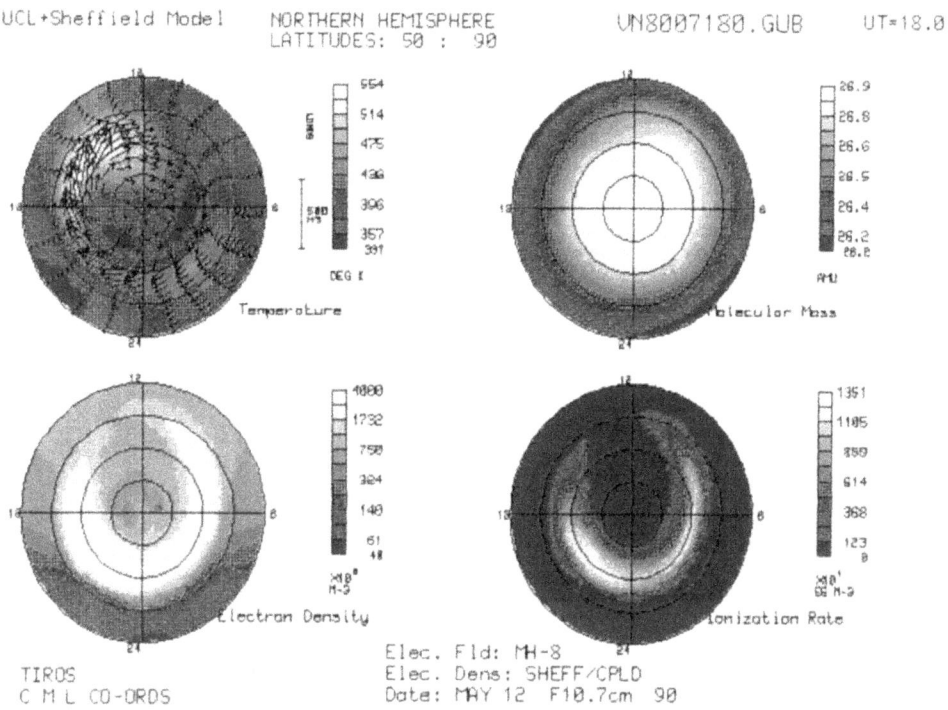

Plate 5—Same as Plate 4, for $Kp = 4$, and the data are displayed for pressure level 7, about 125-km altitude.

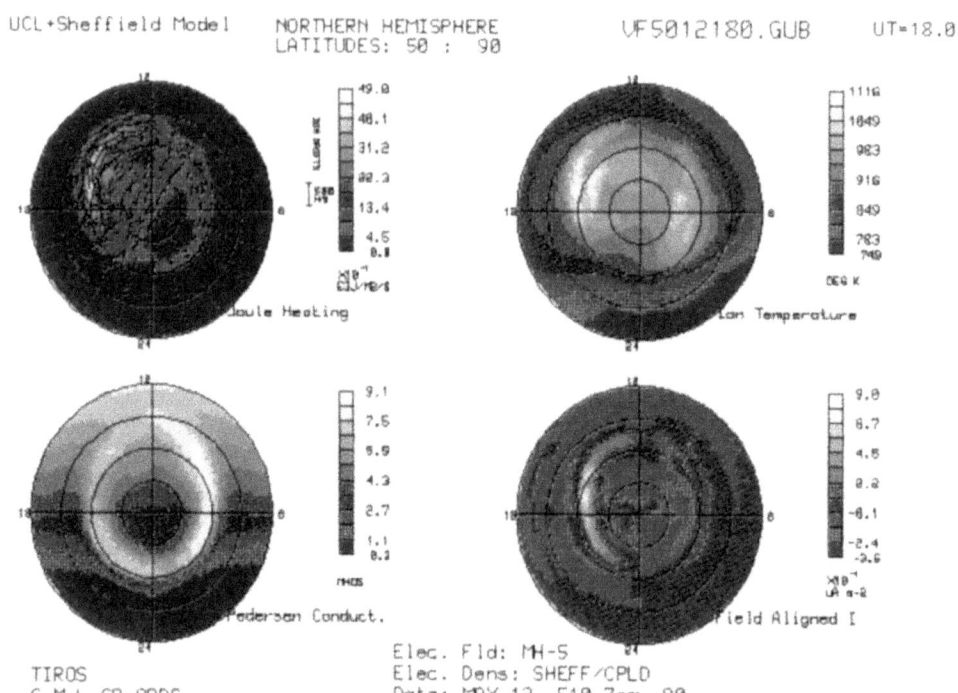

Plate 6—The northern polar region distributions are shown for the height-integrated Joule heating rate and ion flow vectors, the ion temperature (pressure level 12), the height-integrated Pedersen conductivity, and the field-aligned currents, calculated from horizontal curent convergence or divergence. The conditions are otherwise as for Plates 2 and 4.

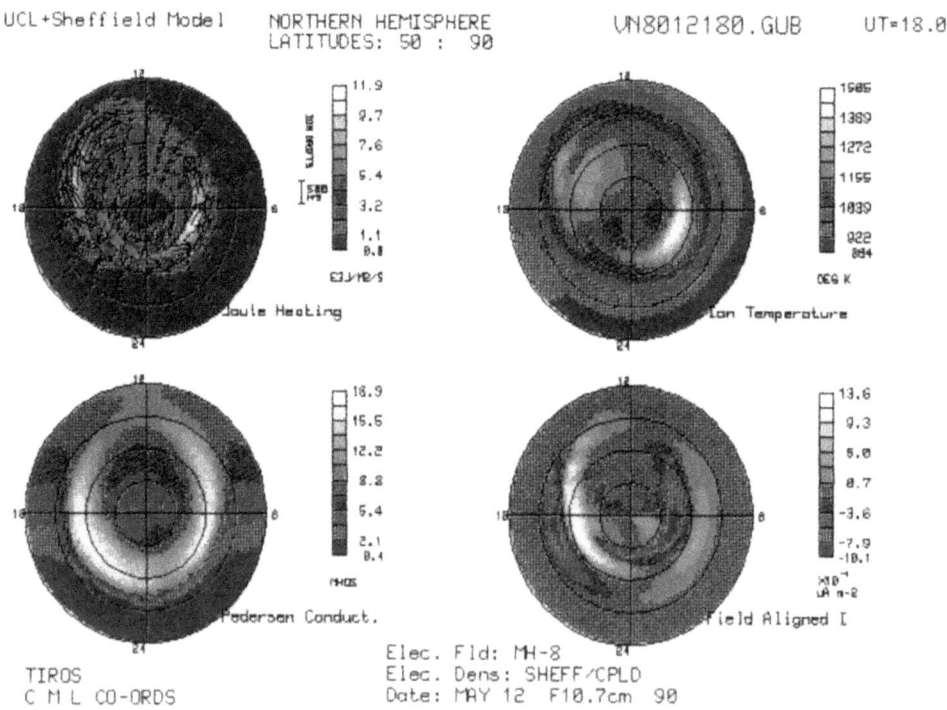

Plate 7—Same as Plate 6, for a higher level of geomagnetic activity, $Kp = 4$, corresponding to the simulations shown in Plates 3 and 5.

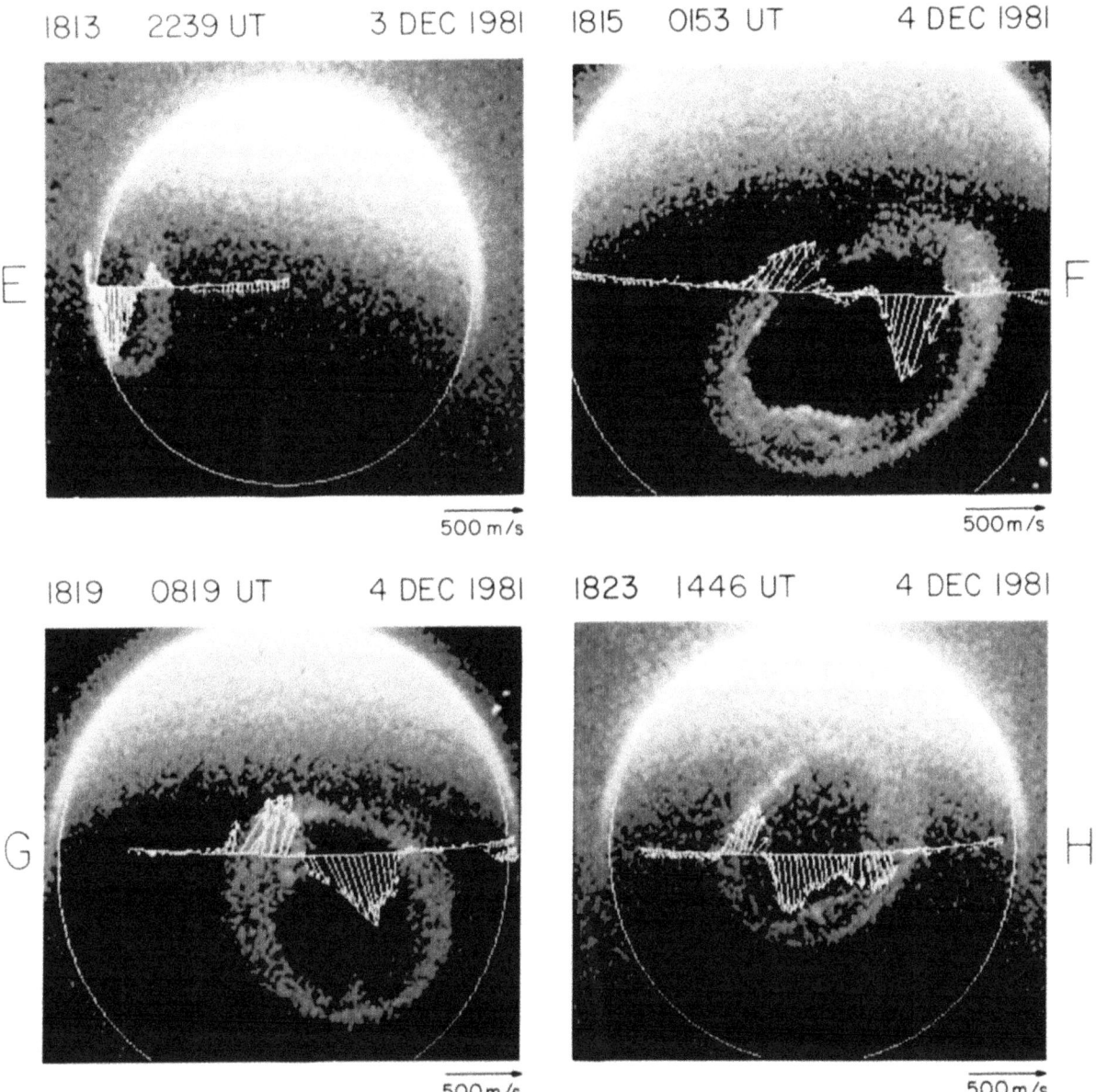

Plate 8—Simultaneously measured neutral wind vectors from DE 2 (orbit 1813) and DE 1 auroral image (courtesy L. A. Frank and J. D. Craven, University of Iowa). The images were obtained using the SAI instrument viewing at ultraviolet wavelengths. The images are oriented such that the direction towards the Sun is to the top of the figure, dusk to the left. The solar terminator is evident, running roughly horizontally across each image, as is the entire auroral oval located just to the nightside of the terminator. The neutral wind vectors are denoted by the arrows whose origins are positioned along the DE 2 orbital track. The wind scale is given at lower right. (From *Killeen et al.* [1988].)

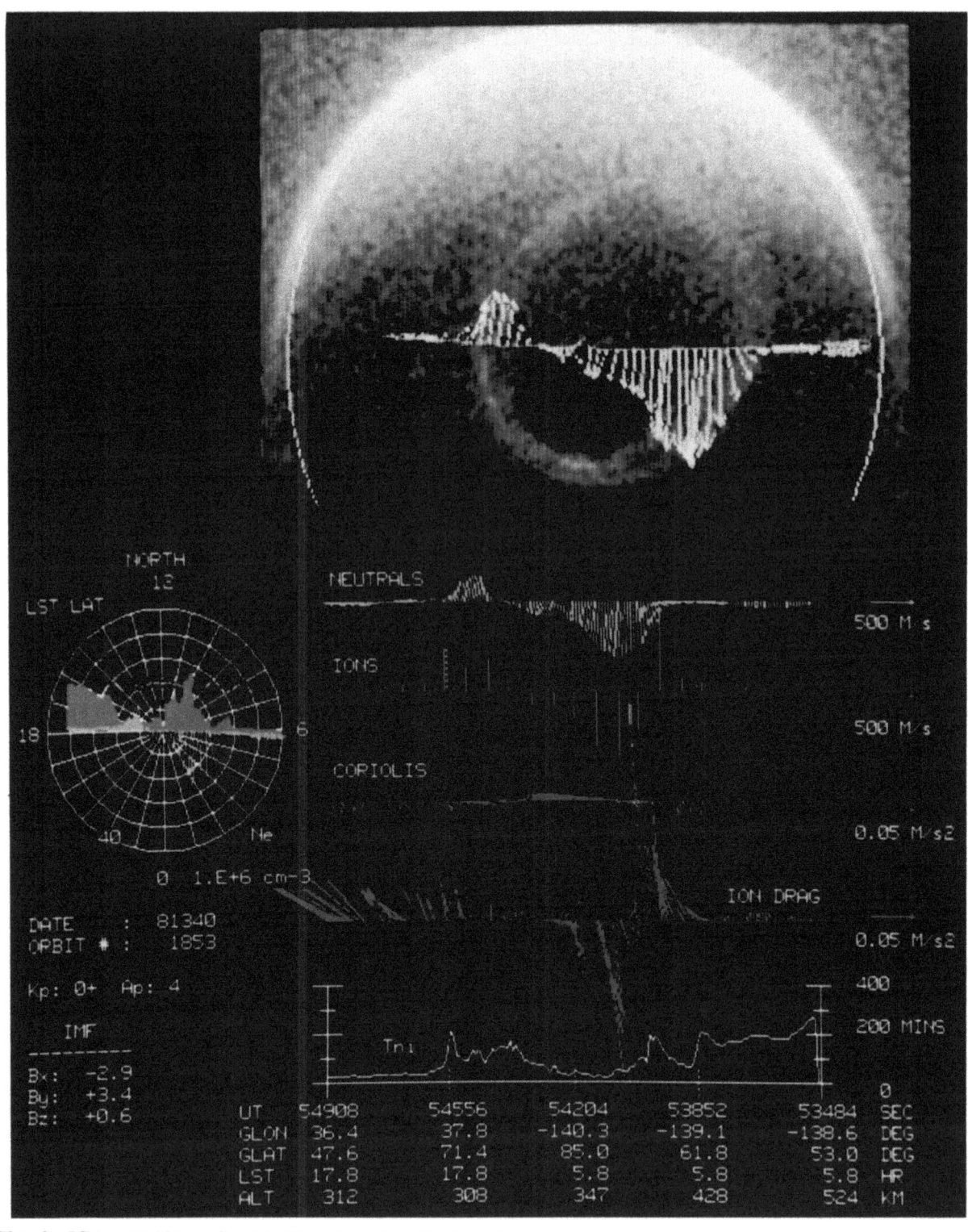

Plate 9—DE 1 auroral image for 1459 UT on December 1981 (top). Measured and derived parameters for orbit 1853 (bottom) are plotted as a function of time along the orbital track of DE 2. These parameters are from top to bottom: the neutral wind vector; the zonal (cross track) component of the ion drift vector; the Coriolis force; the ion-drag force, and the calculated ion-neutral momentum time constant. The scales for the various parameters are to the right of the diagram. The polar dial indicates the track of the spacecraft across the northern polar cap in geographic polar coordinates. The bars on the polar dial represent the neutral wind measurements and the solid fill represents the measured ion densities according to the scale given below the dial. The solar terminator is also shown on the polar dial. (From *Killeen et al.* [1988].)

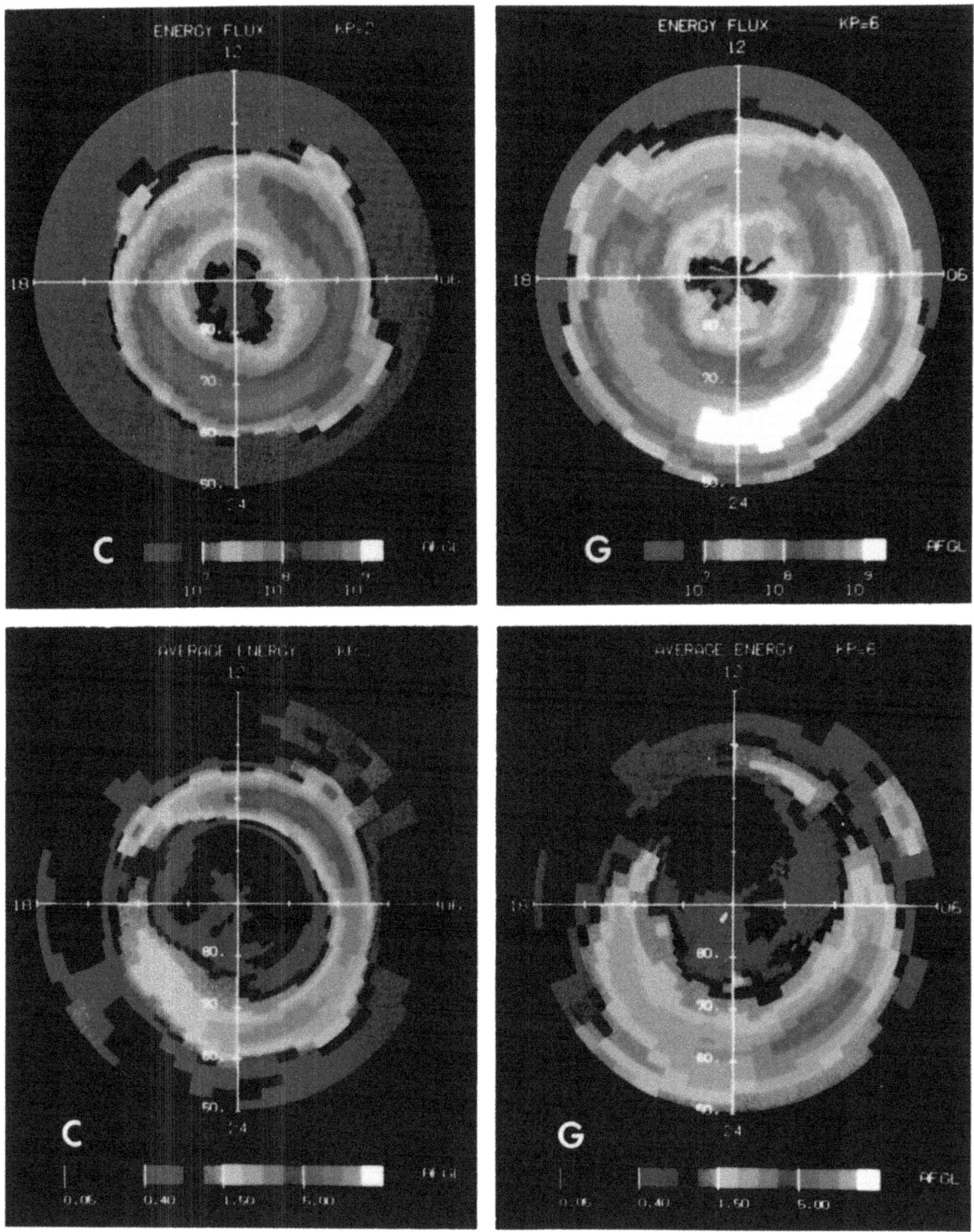

Plate 10—Polar projection map, from 50° magnetic latitude to the magnetic pole, for electron precipitations under $k_p = 2$ (left side) and $k_p = 6$ (right side) conditions of energy flux (top) and average energy (bottom). (From Hardy et al. [1985; 1989].

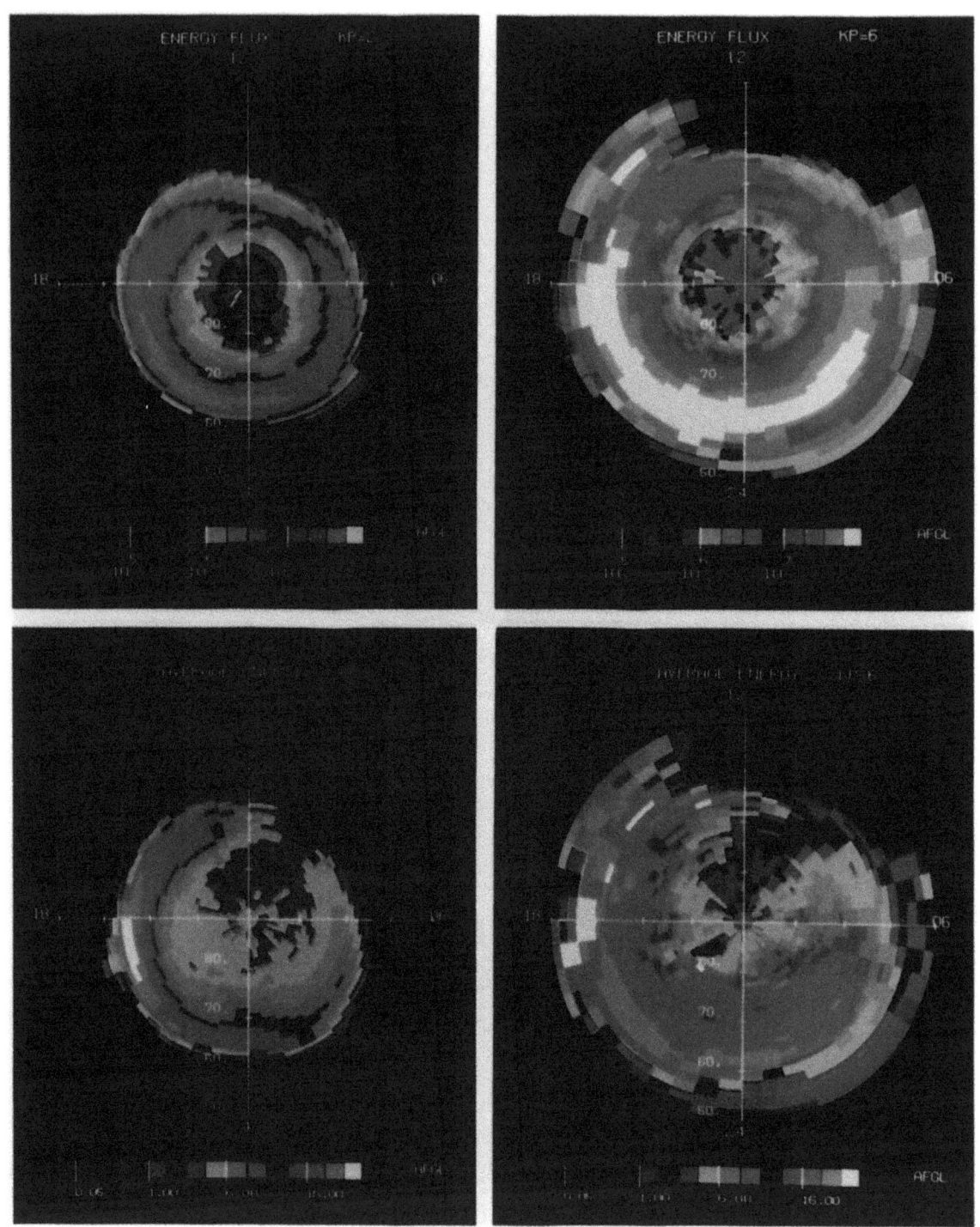

Plate 11—Same as Plate 10, except for ion precipitations.

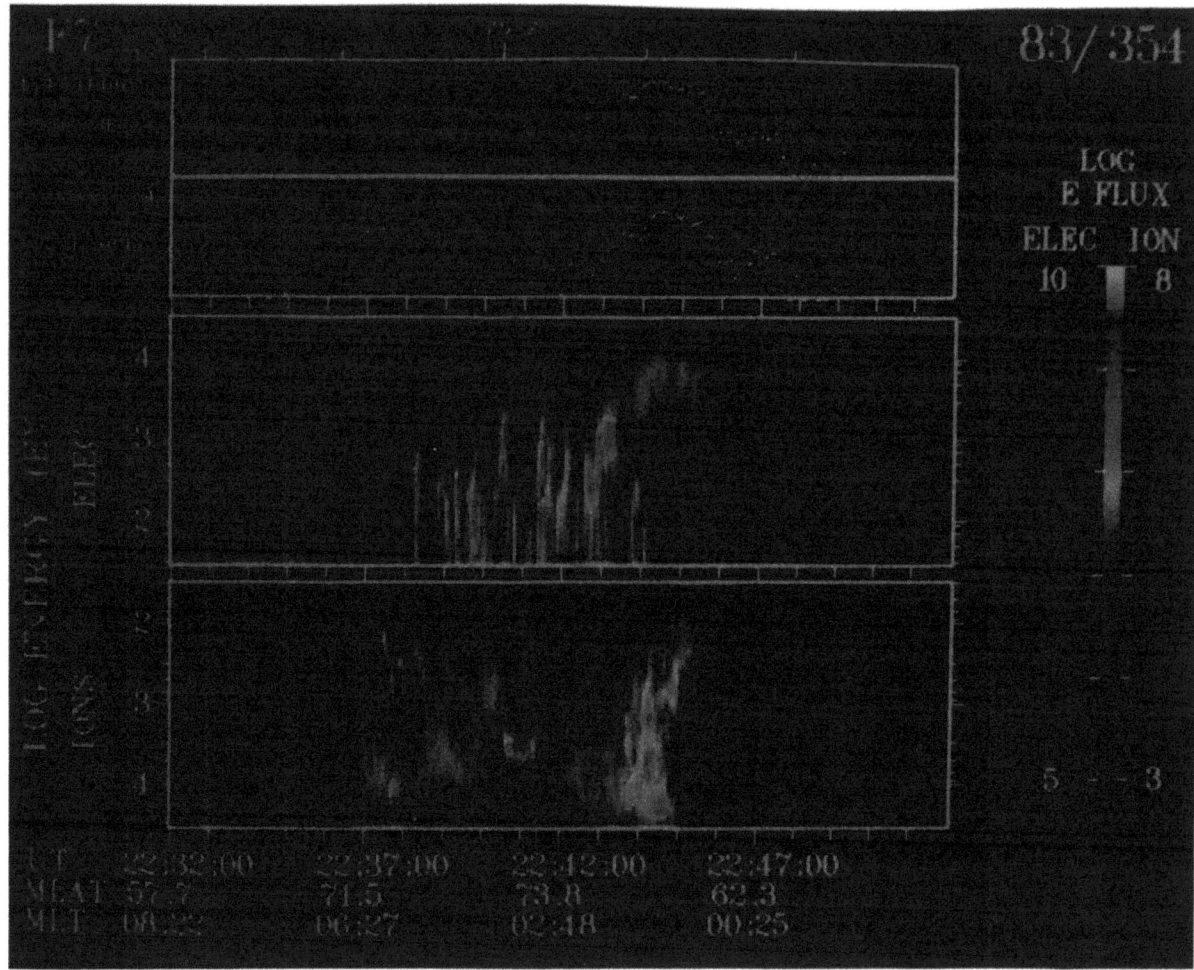

Plate 12—DMSP F7 measurements of precipitating electrons and ions for 20 min on day 354 of 1983. The top line plot is energy flux (eV/cm² s sr), and the bottom line plot is average energy (eV). The lower two spectrograms for electrons and ions, show differential energy flux values in eV/(cm² s sr eV). Notice the curve formed by the equatorward edge of the diffuse aurora on the night (right hand) side: the ions cutoffs form a "C" shape, which is the type of curve most commonly observed.

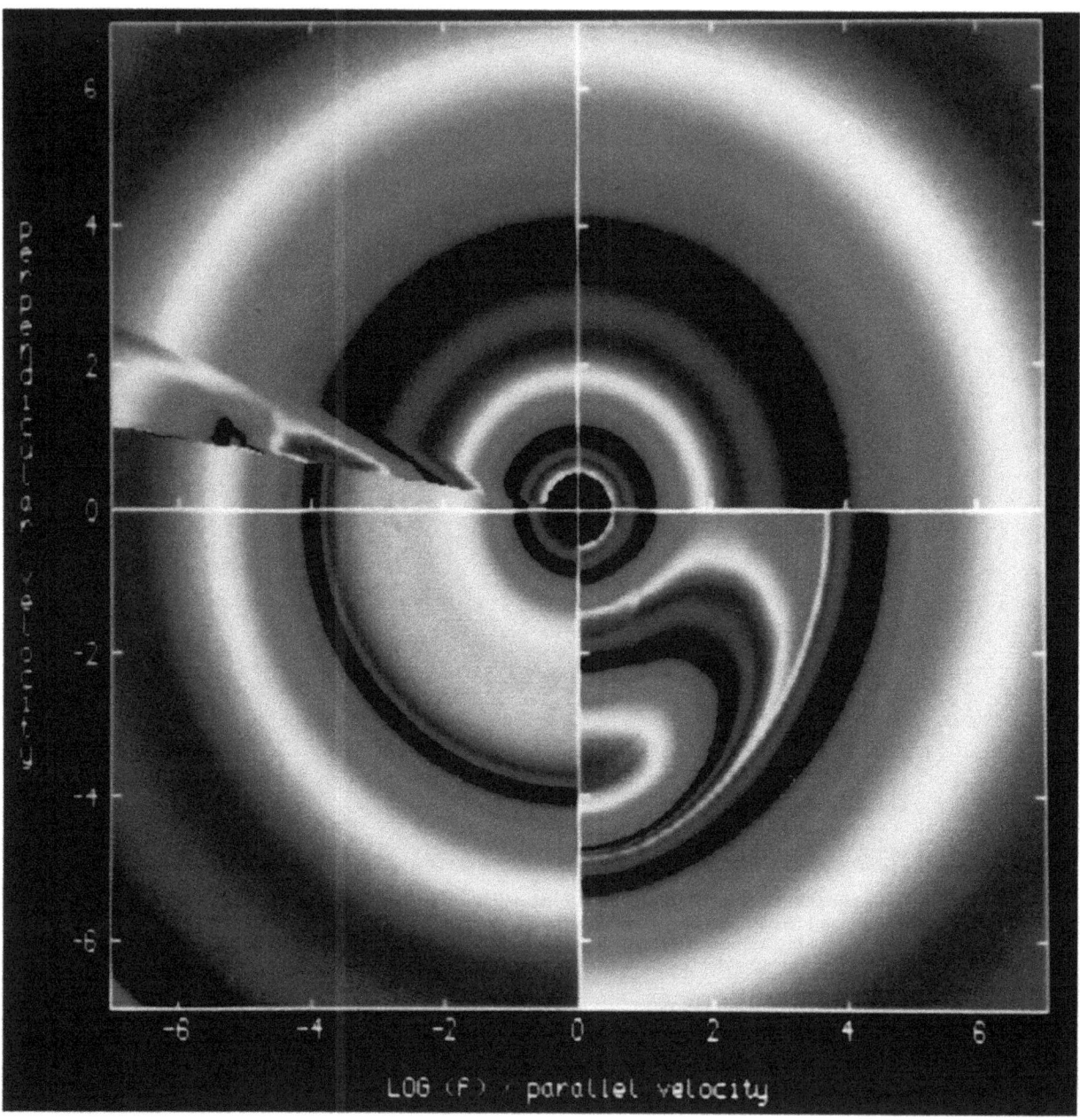

Plate 13—Strongly peaked and magnetic-field-aligned distribution shown in higher resolution than in Figure 13 of the relevant article. The accelerator is unidirectional, with $\Delta E_{max} = 18$ keV, located between $R_1 = 1.1R_E$ and $R_2 = 1.8R_E$. Contours of velocity-space density are color coded. In order to achieve suitable resolution, the color cycle is repeated every three orders of magnitude. An electron conic and a valley adjacent to the loss cone are prominent features of the distribution at $R = 2R_E$. Lower panels: distribution at $R = 1.02R_E$, below the accelerator. Upper panels: distribution at $R = 2R_E$, above the accelerator.

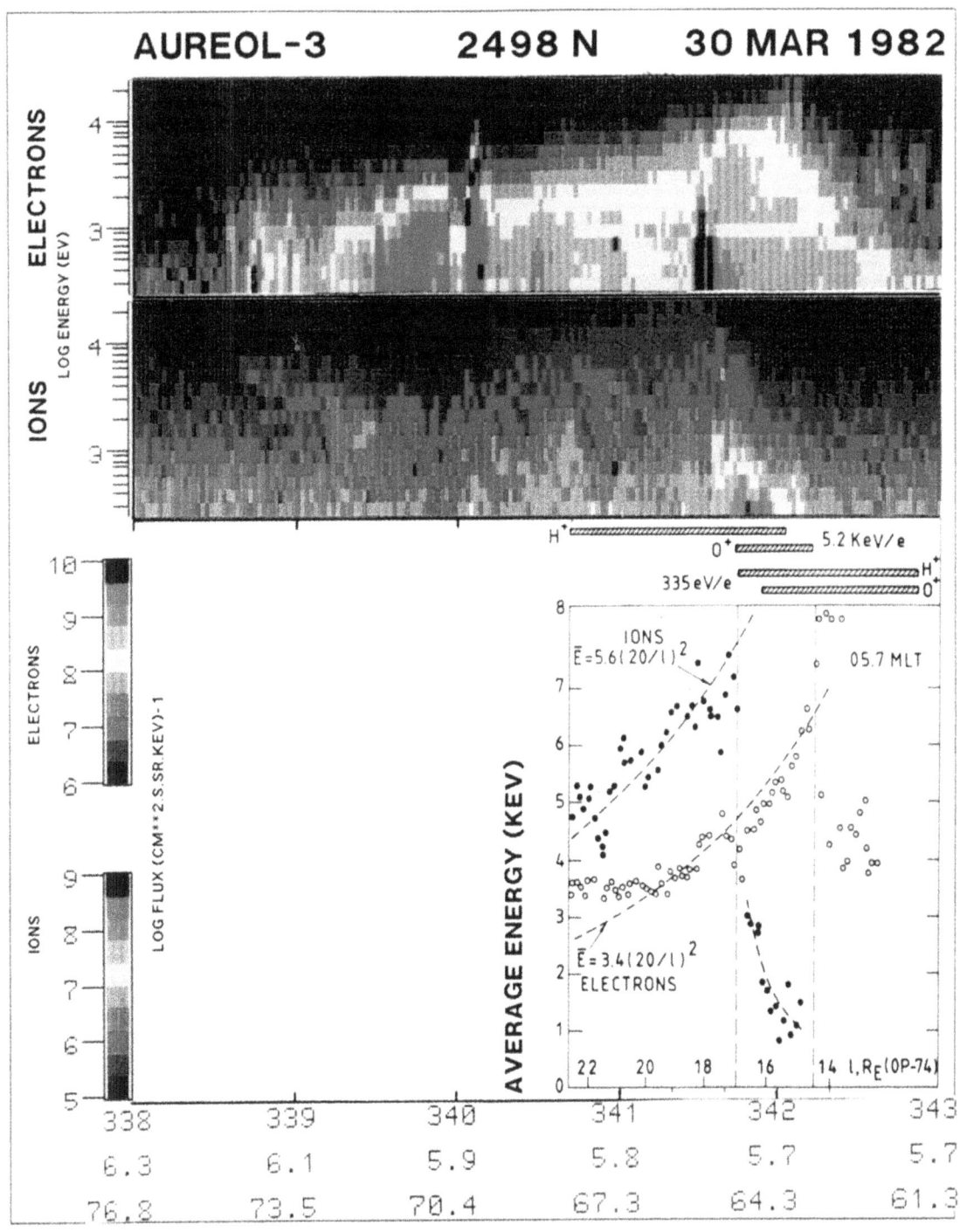

Plate 14—Precipitating electron and ion fluxes measured onboard the Aureol 3 satellite (pass 2498 north) as a function of the Universal Time (from 0338–0343 UT), magnetic local time (around 0590 MLT), and invariant latitude. Average energies are also plotted (bottom panel) and compared with the L^{-2} Fermi acceleration profile. Regions of precipitating H^+ and O^+ ions are indicated by horizontal hatched bars (extracted from *Bosqued* [1987]).

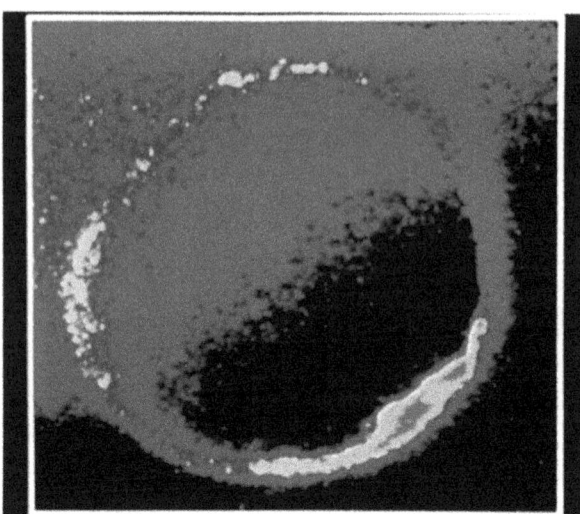

Plate 15—Viking UV image of the entire auroral oval during a time of relatively low magnetic disturbance. The image shows that the Feldstein oval is basically continuous but has two auroral activity centers, one near local midnight and one near local noon.

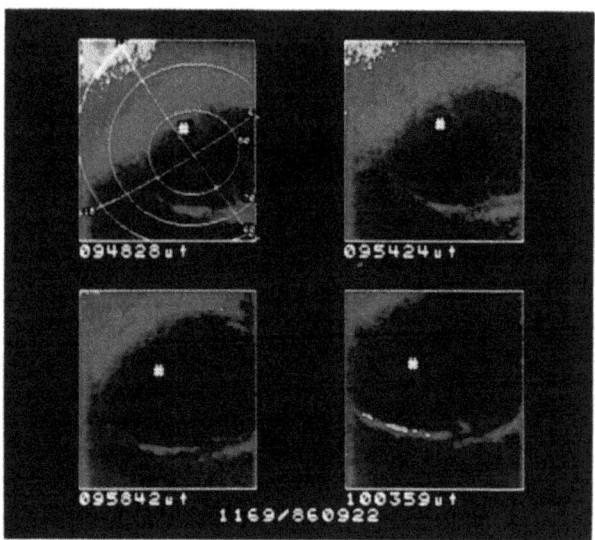

Plate 17—Viking images of a possible theta aurora over the northern polar region. The time resolution of the images in the sequence is about 1 min.

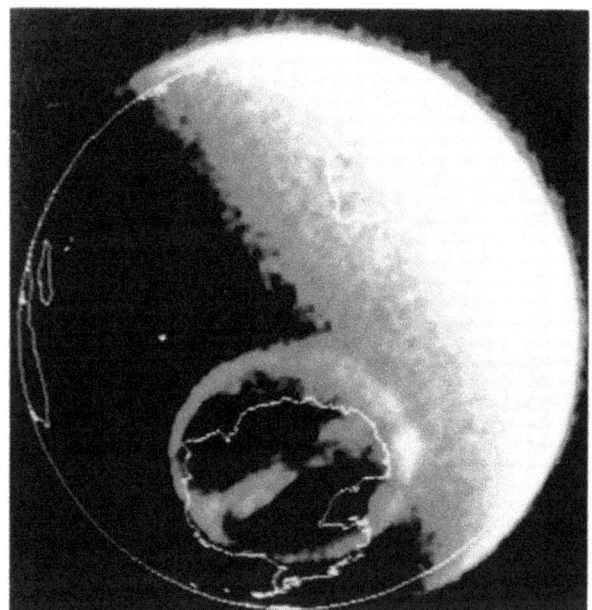

Plate 16—Example of theta aurora as observed from the Iowa imager on DE 1. (Courtesy, *L. A. Frank* [1988].)

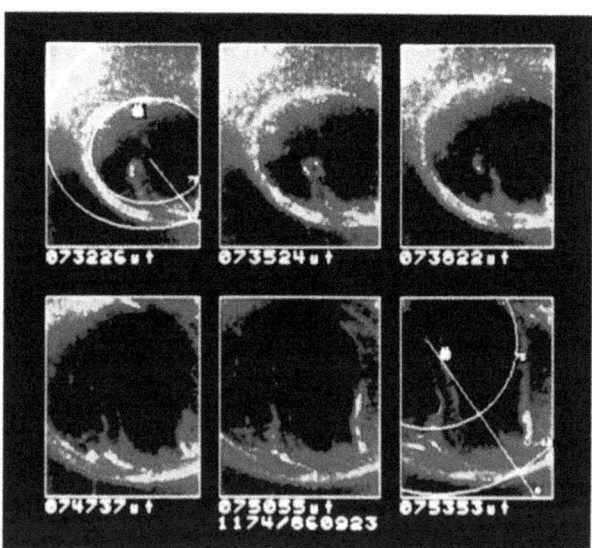

Plate 19—An example of a theta aurora formation as observed by the Viking imager, i.e., a bifurcation of the tail lobe originating in the geotail. The formation is supposed to originate from the expanding auroral "loop" on the nightside oval.

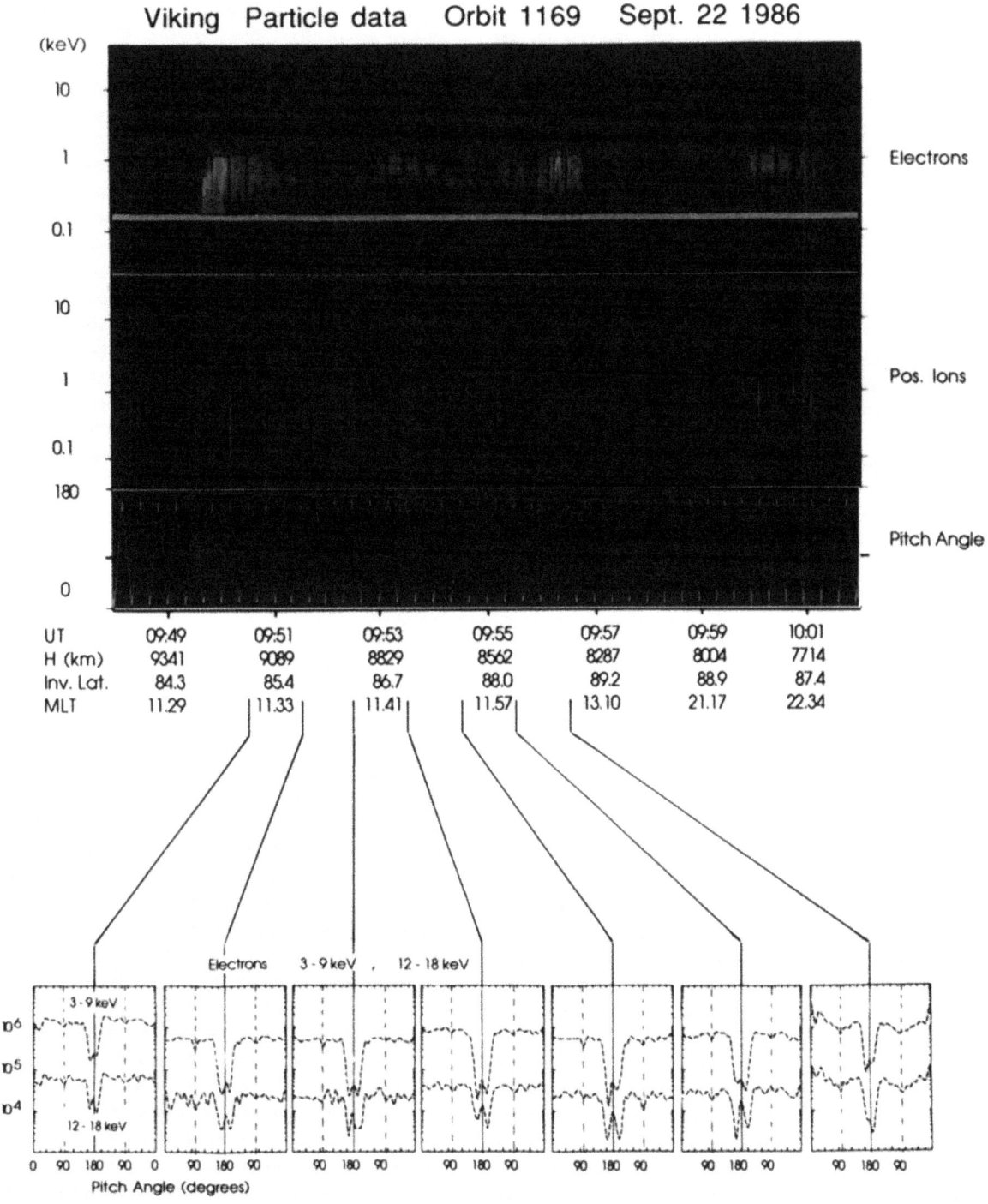

Plate 18—Viking ion and electron time energy spectrogram from the pass over the theta aurora displayed in Plate 17. The bottom panel shows the pitch angle distribution of energetic electrons (3–9 keV and 12–18 keV, respectively) with quasi-trapped "butterfly" signature at the edges of the polar arc.

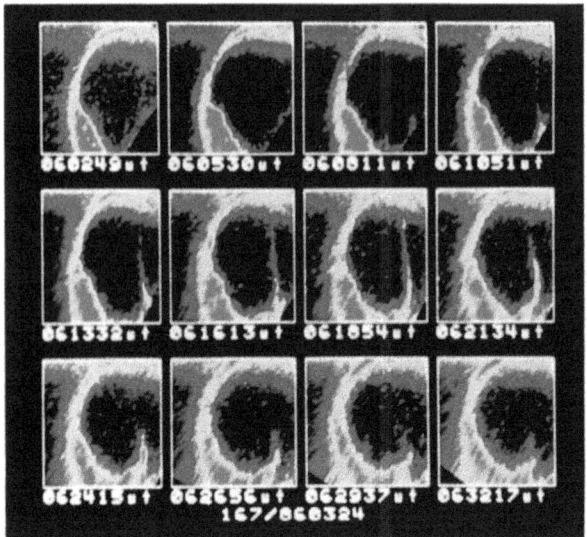

Plate 20—A time sequence of Viking auroral images of an apparently single polar arc connecting to the dayside oval with occasional "flarings" along the arc.

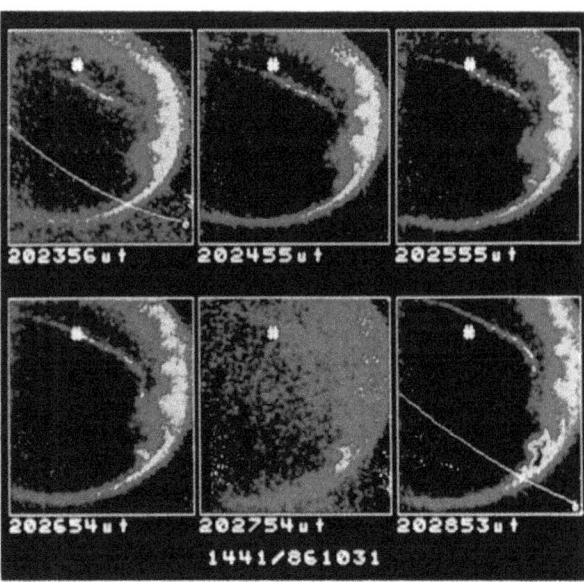

Plate 22—Series of images (≈ 1-min time resolution) displaying a "hook-shaped" arc connecting to the nightside oval. The images demonstrate the dynamics of isolated thin polar arcs with intensifications occurring within 1 min.

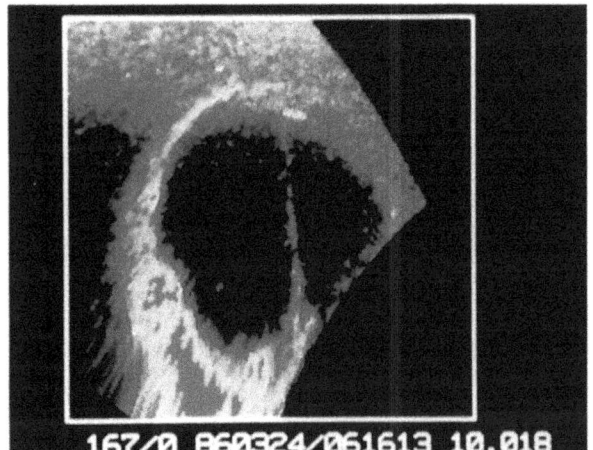

Plate 21—Detailed image of a polar arc from the same pass as that in Plate 20. The image shows that the polar arc is associated with isolated intensifications forming a "hook" at both ends in the morning oval.

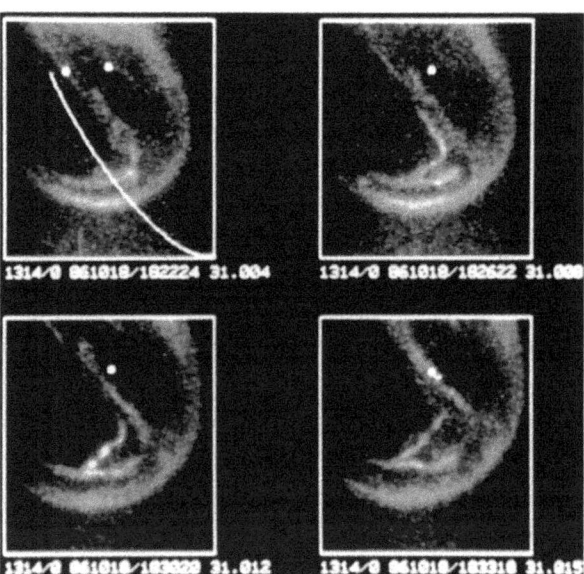

Plate 23—Viking UV image of a polar arc for a field line tracing test using the Tsyganenko magnetic field model.

DE-1   JANUARY 27,   DAY 27,   1982   0445 UT

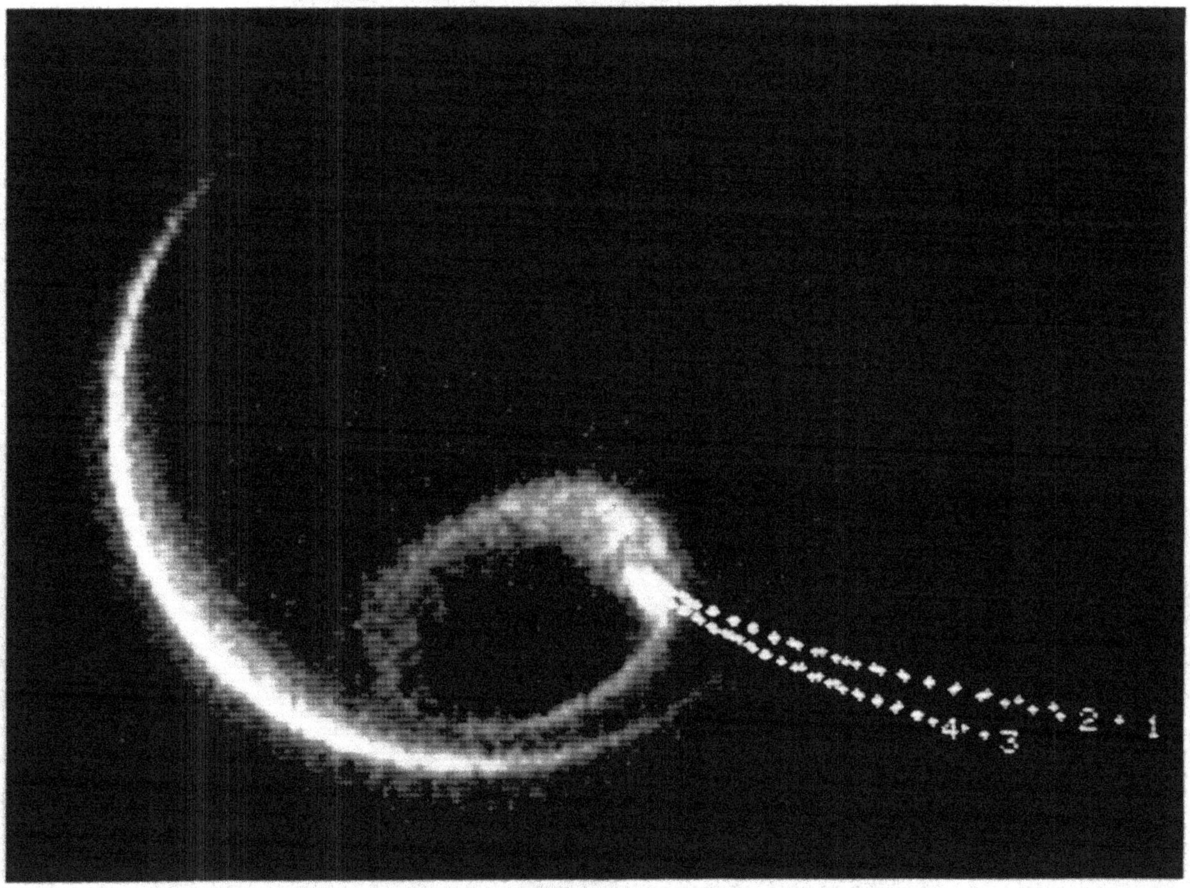

AKR SOURCE #	PWI	
	UT	WAVE FREQUENCY
1	0445	104 kHz
2	0445	136 kHz
3	0445	170 kHz
4	0445	218 kHz

Plate 24—An auroral image from DE 1 showing the occurrence of bright auroral emissions during an intense AKR event. The dashed lines show the magnetic field lines through the source. The source position is determined from the intersection of the direction of arrival and the $f = f_{ce}$ surface.

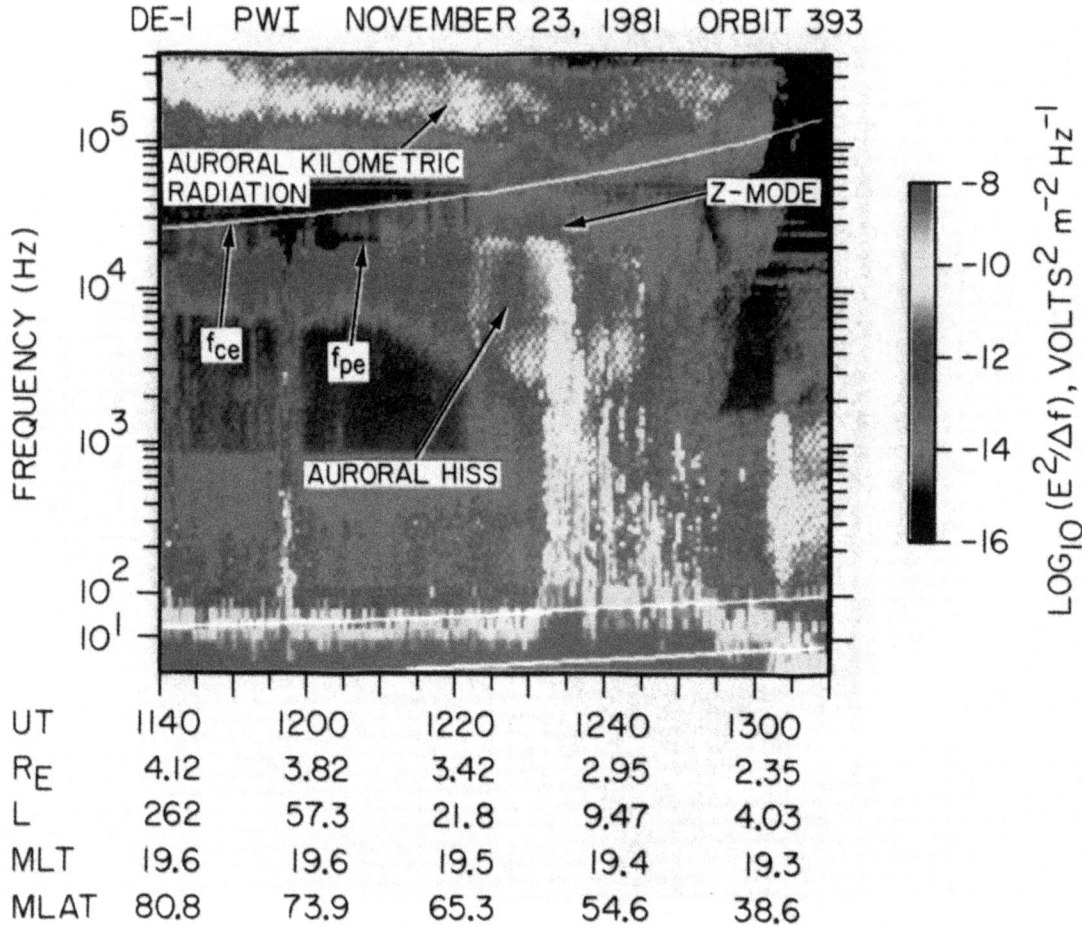

DE-1   PWI   NOVEMBER 23, 1981   ORBIT 393

UT	1140	1200	1220	1240	1300
R_E	4.12	3.82	3.42	2.95	2.35
L	262	57.3	21.8	9.47	4.03
MLT	19.6	19.6	19.5	19.4	19.3
MLAT	80.8	73.9	65.3	54.6	38.6

Plate 25—A spectrogram of auroral hiss and Z-mode radiation during a DE 1 pass over the auroral zone.

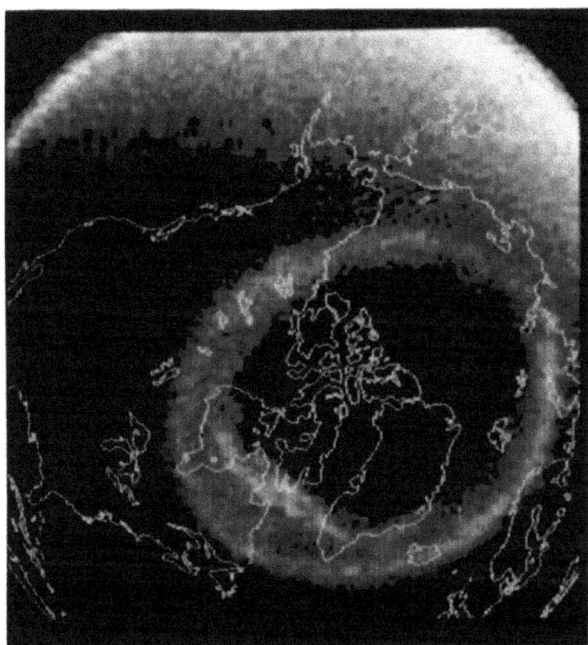

Plate 26—The distribution of auroral luminosities over the North American continent at 0241 UT on November 8, 1981. A coastline map is superposed on this false-color image of the aurora borealis at ultraviolet wavelengths 123-155 nm (filter 2). Principal emissions detected at these wavelengths are from the multiplets of atomic oxygen at about 130.4 and 135.6 nm and from the LBH bands of molecular nitrogen. For the false-color format, luminosities less than about 1 kR are coded black. For greater luminosities the code progresses from red through orange to yellow. Typical luminosities in the sunlit hemisphere are 20-30 kR, with the largest values observed near the subsolar point.

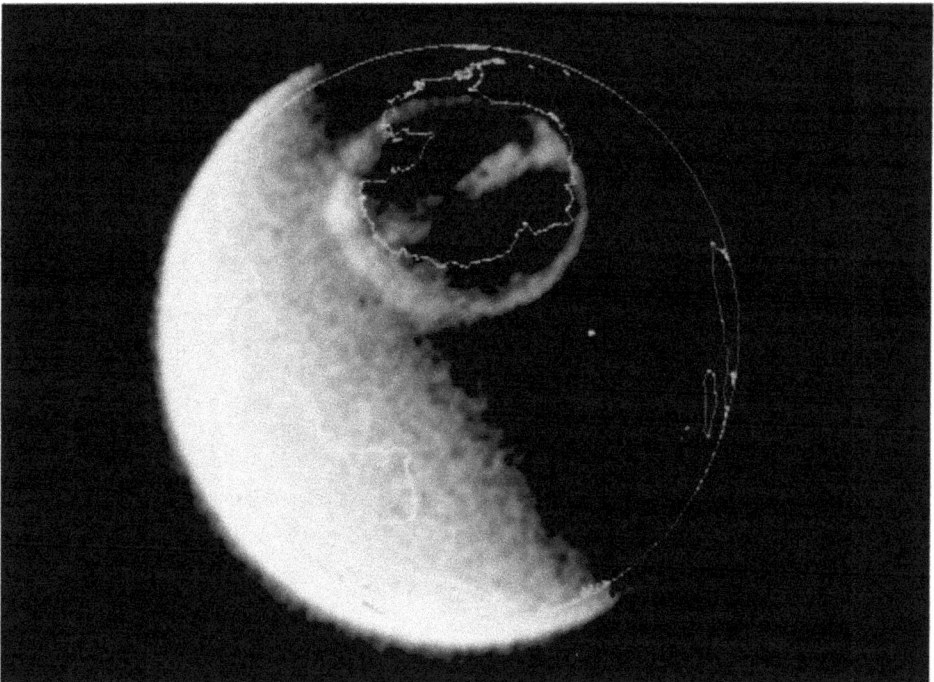

Plate 27—The distribution of auroral luminosities over Antarctica at 0022 UT on May 11, 1983 [after *Frank et al.*, 1985].A coastline map is superposed on this false-color image of the aurora australis at ultraviolet wavelengths identified in the caption of Plate 26. This image exhibits a theta aurora that comprises the auroral oval and a transpolar arc. The transpolar arc extends into the polar cap from local midnight, traverses the polar cap, and joins with the auroral oval at local noon.

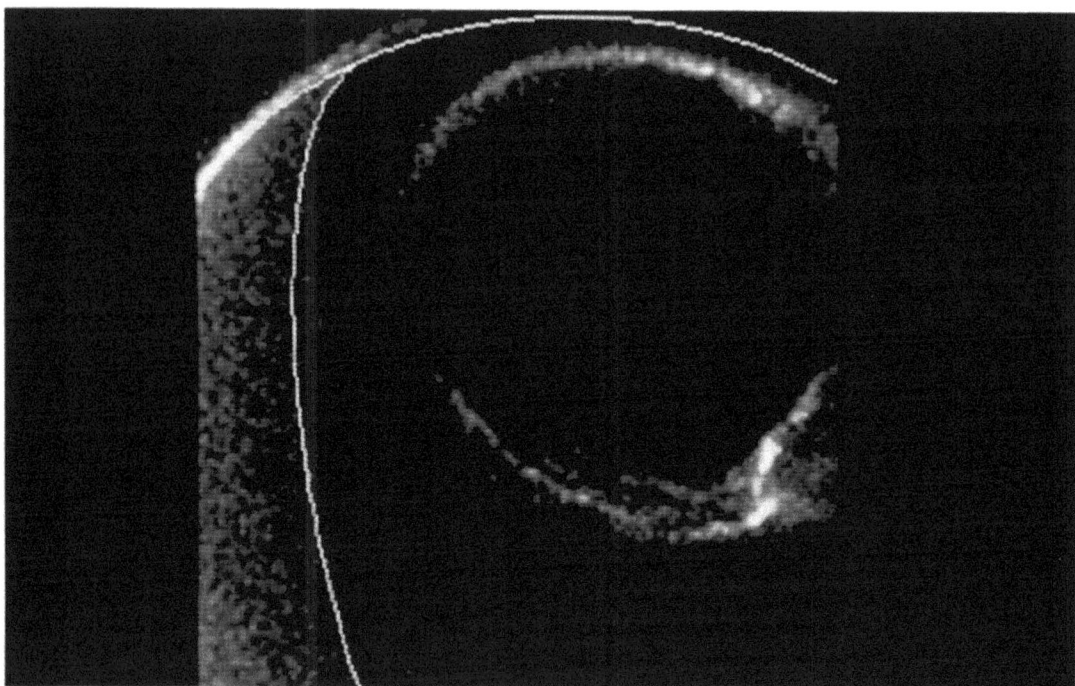

Plate 28—An image of the sunward portion of the northern auroral oval exhibits the gap in discrete aurora that can occur in the local noon sector. The gap is readily detected in the absence of bright diffuse emissions at lower latitudes, which are present in the images of Plates 26 and 27. Three auroral arcs are seen in the evening sector near the lower central part of the image, and bright, active aurora are visible at the lower right as part of the westward boundary of the expanding auroral bulge in a substorm. Contours identifying the Earth's limb and terminator are overlaid on this image (0154 UT, December 8, 1981). The sunlit hemisphere is observed in the left part of the image. Principal emissions are from the LBH bands of molecular nitrogen.

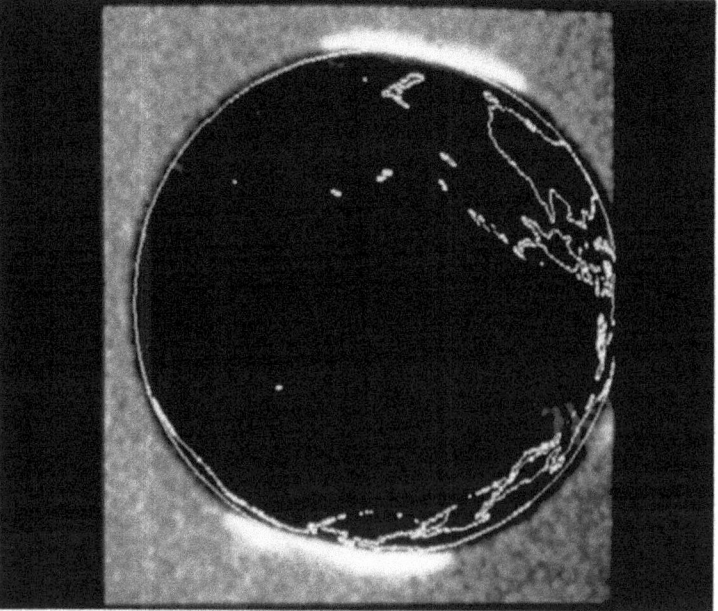

Plate 29—This unique image of Earth from Dynamics Explorer 1 at 1215 UT on March 1, 1982 records aurora in the two polar regions: The aurora borealis at northern latitudes and auroral australis at southern latitudes. Earth's limb and coastal outlines are overlaid on the image. The spacecraft is located within Earth's umbral shadow cone at an altitude of about 20,000 km above the Pacific Ocean. The active aurora in the two hemispheres rise to altitudes of about 370 km above the limb of the solid Earth. Resonantly scattered solar Lyman-α radiation from Earth's extended hydrogen atmosphere is responsible for the diffuse glow beyond Earth's limb. The dark band encircling Earth above the limb is due to absorption of the ultraviolet radiation by the atmosphere at low altitudes. The passband of the filter for this image extends from 117-165 nm.

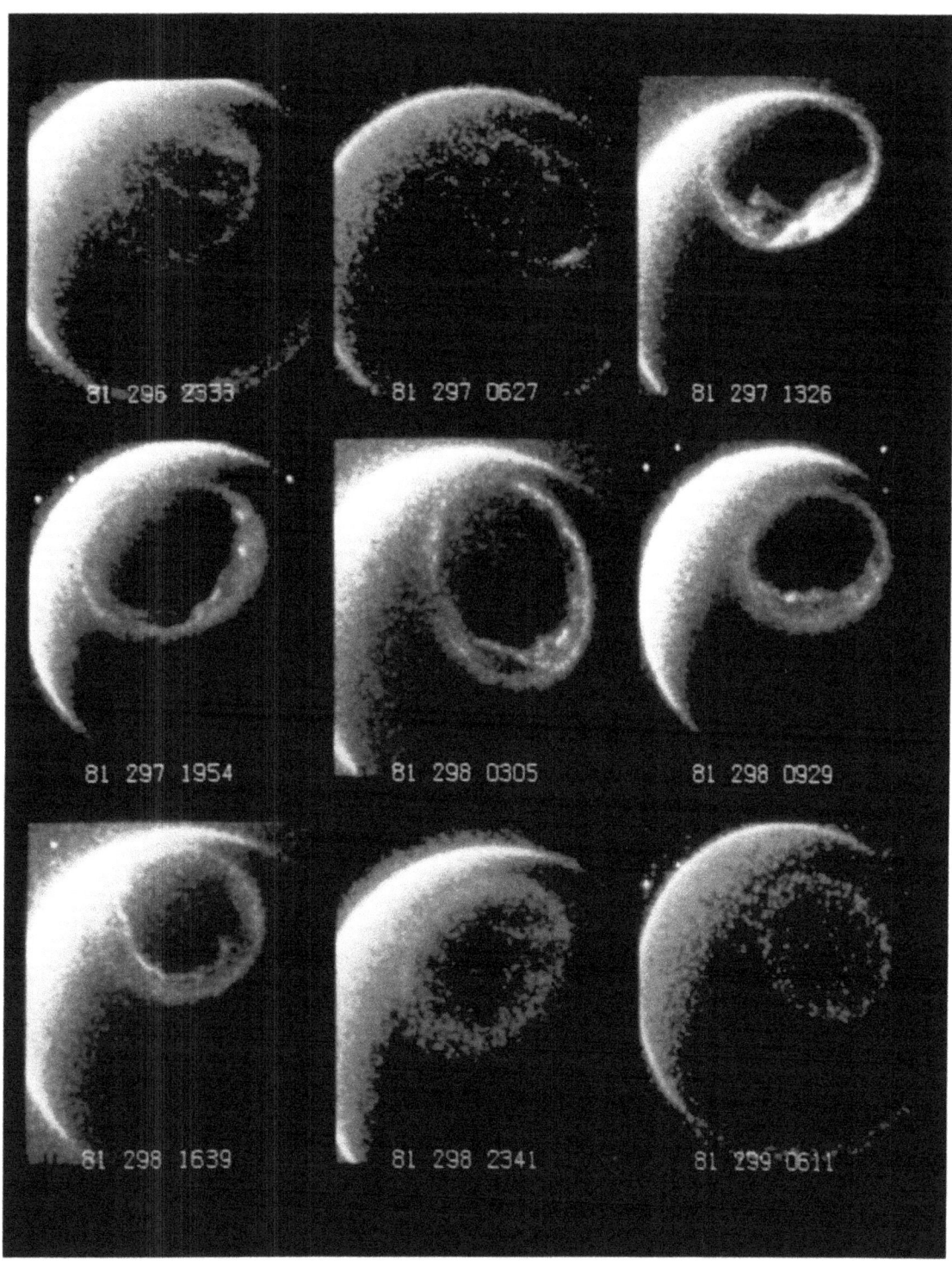

Plate 30—A series of nine *DE* auroral images taken in consecutive orbits to illustrate the gross large-scale spatial distribution of the aurora during a time interval in which the interplanetary magnetic field is first oriented northward (images 1-2), southward (images 3-7), and again northward (images 8-9) (see also Figure 5 of the relevant article). Below each image is overlaid the year, day of year, and time (UT) of that image. Identification of the filter for each image is provided in Table 1 of the relevant article.

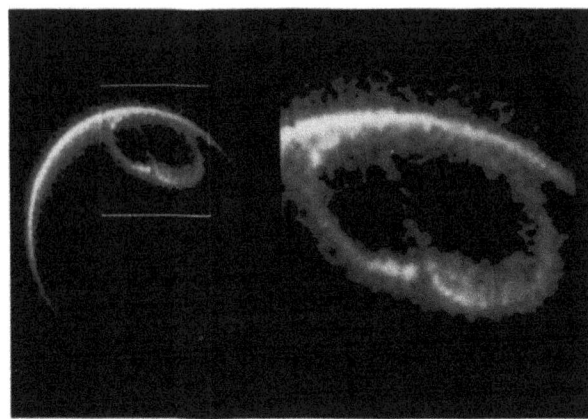

Plate 31—Image of the northern auroral oval and polar cap at 0538 UT on March 25, 1982. This image is one of 13 for the time interval 0514-0757 UT that exhibit a transpolar arc within the polar cap. Motion of the arc is toward the evening sector during a period in which the IMF B_y component is positive.

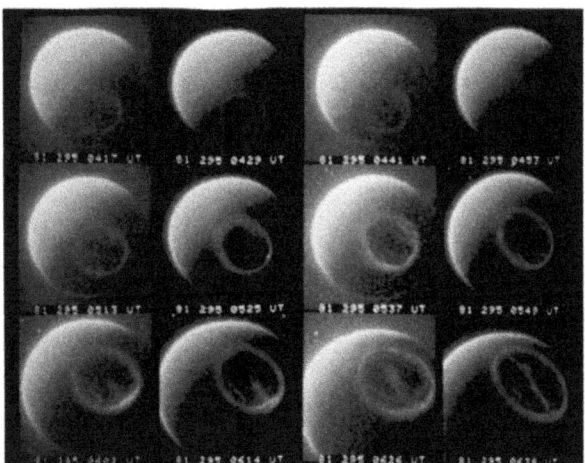

Plate 32—Sequence of 12 consecutive false-color auroral images at ultraviolet wavelengths in the time interval 0417-0650 UT on October 22, 1981. Increasing luminosities follow the SC at 0525 UT (beginning of the sixth frame). Predominant direction of the IMF B_z component is northward. Below each image is the year, day of year, and UT for the beginning of the 12-min telemetry period for the image.

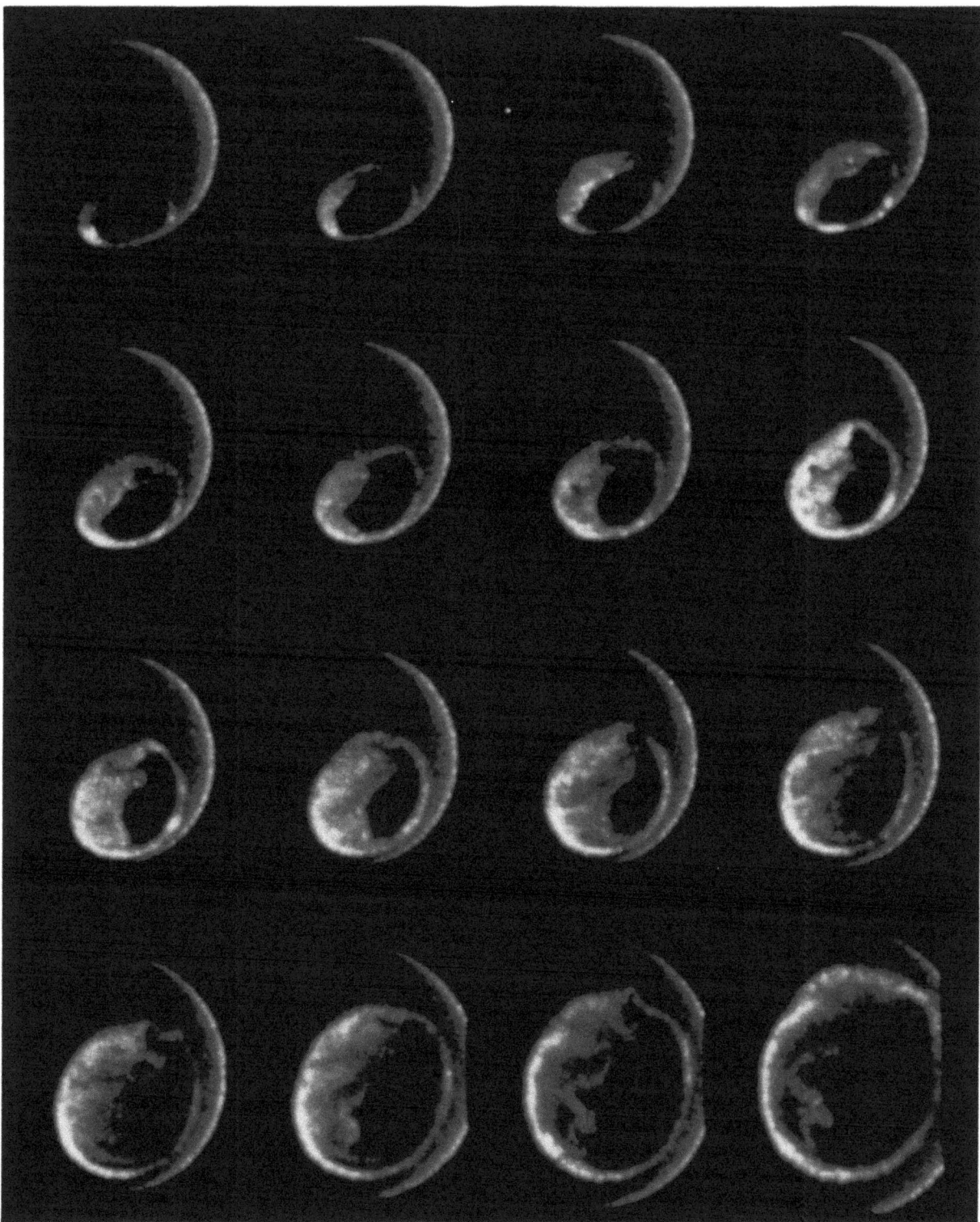

Plate 33—A sequence of 16 images of the aurora australis in the time interval 0202-0517 UT on June 13, 1983. Intense auroral activity begins in the first image at upper left with a localized brightening (substorm onset) followed by a period of rapid expansion of the aurora in latitude and in longitude. Luminosities increase noticeably by 0326 UT (eighth image), indicating the arrival at Earth of a shock or discontinuity in the interplanetary medium. This auroral activity occurs simultaneously with the main phase decrease of the low-latitude surface magnetic field during a geomagnetic storm.

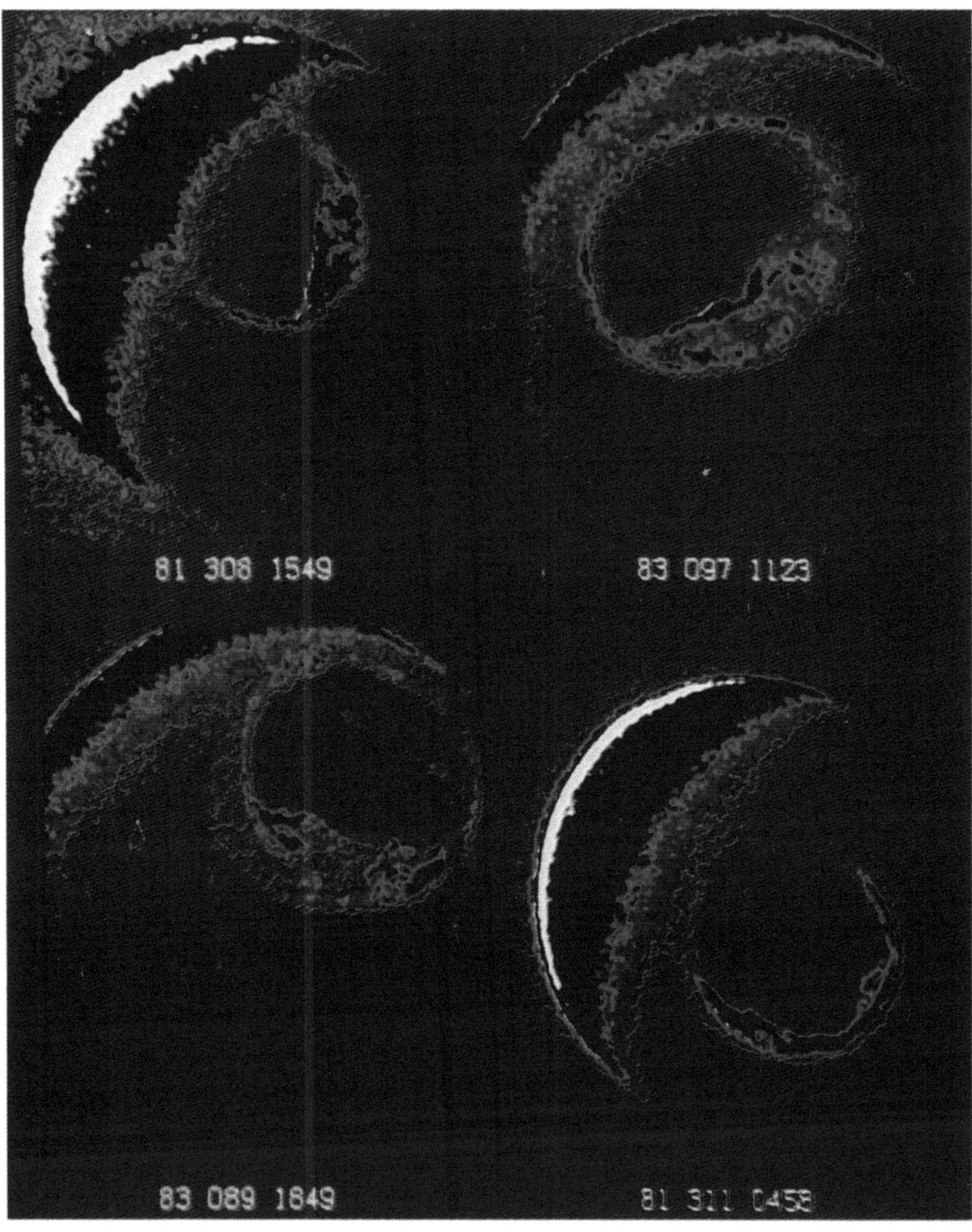

Plate 34—Four auroral images, selected to illustrate variations in the longitudinal distribution of aurora, that can be observed during the expansion phase of substorms. The first image, at upper left, features an expansion that does not proceed westward of 2200 MLT (the surge is nearly stationary) and the auroral bulge expands predominantly into the morning sector. The second image, at upper right, shows an auroral bulge more symmetric about the noon-midnight plane and a surge that advances farther into the evening sector. The auroral distribution in the third image, at lower left, is nearly a mirror image of the first image, with the expansion into the evening sector and nearly no eastward expansion into the morning sector. The last image is an example which auroral activity expands westward rapidly along the auroral oval, and there is almost no signature at midnight of an auroral bulge. For this false-color format luminosities less than about 1 kR are coded black. For greater luminosities the code progresses from blue through green, yellow, and red to a saturation value near 20 kR coded white.

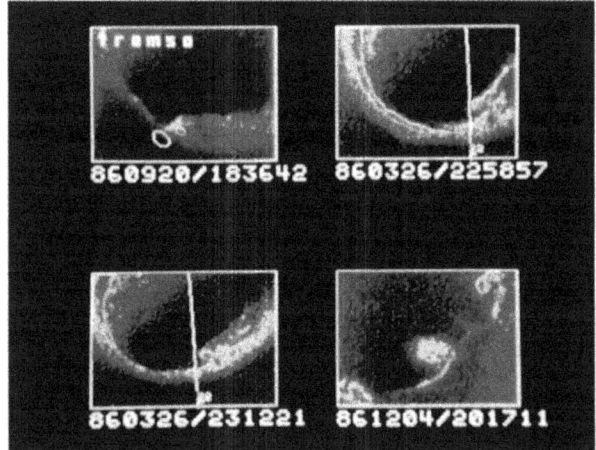

Plate 35—Examples of Viking UV imager data showing characteristics of discrete features in the auroral distribution. The data have been corrected for nonuniformity effects and background, but are otherwise unmodified. The data are from the LBH camera and dayglow emissions are apparent in the upper left part of each image. *Top left*: The field of view of an all-sky camera positioned at Tromsö is indicated by the roughly circular curve. *Top right* and *bottom left*: The 2200 MLT meridian is shown.

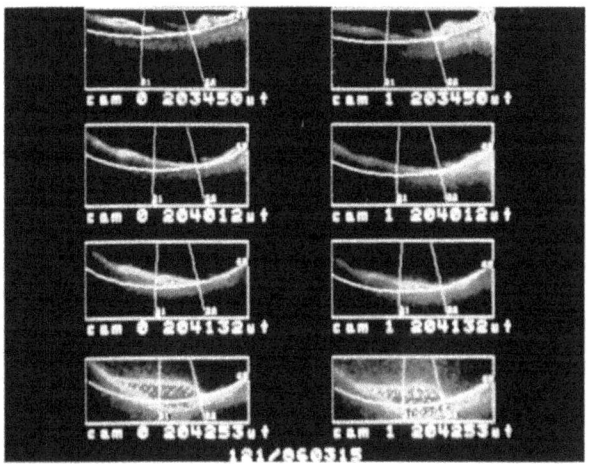

Plate 37—Four pairs of Viking UVI data, for both cameras: left at LBH wavelengths (cam 0) and right atomic oxygen (cam 1). Data are for only a small section of the late evening sector.

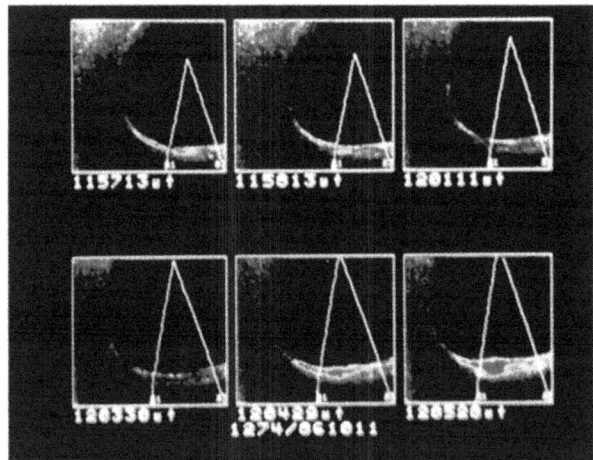

Plate 36—Viking LBH image data from orbit 1274. On each image is shown the MLT meridians of 2100 and 2300 hours. A substorm onset is apparent in the image at 1204:29 UT.

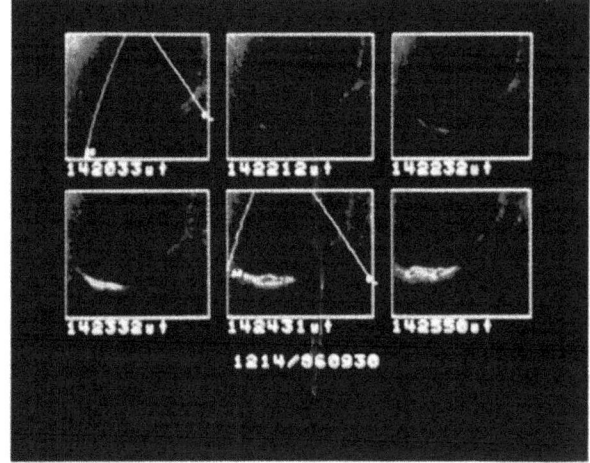

Plate 38—Viking LBH image data from orbit 1214. Three extended arc systems are noted in the evening sector. A substorm expansion occurs on the most equatorward one at 1423:32 UT just where the more poleward system begins to be observable.

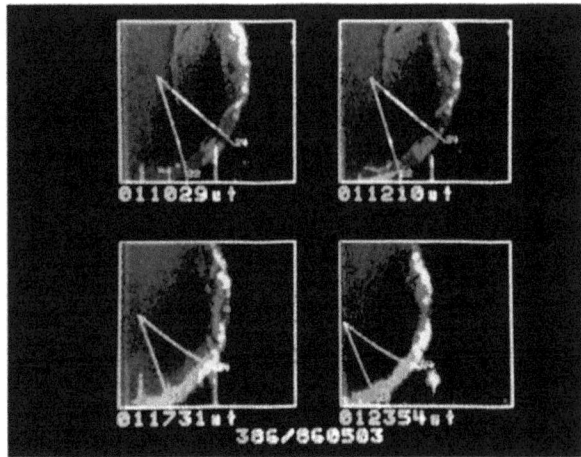

Plate 39—Image data from orbit 386. On each image are shown the MLT meridians of 2200 and 2400 hours.

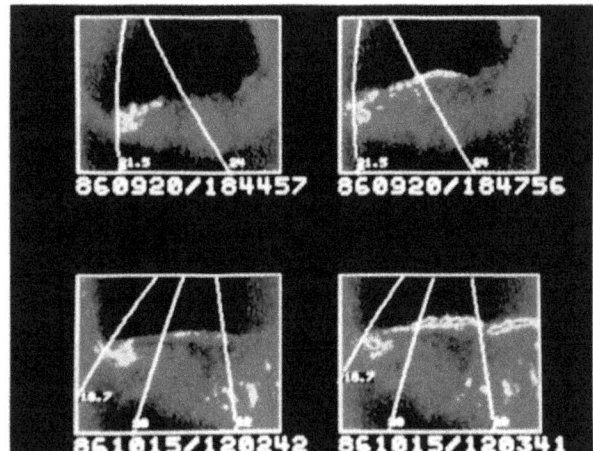

Plate 41—Examples from two orbits of the intensification of the most poleward arc after substorm expansion has been completed. In the top are data from orbit 1160 (September 20, 1986) and in the bottom from orbit 1296 (October 15, 1986). All images are from the LBH camera and have been corrected for nonuniformity.

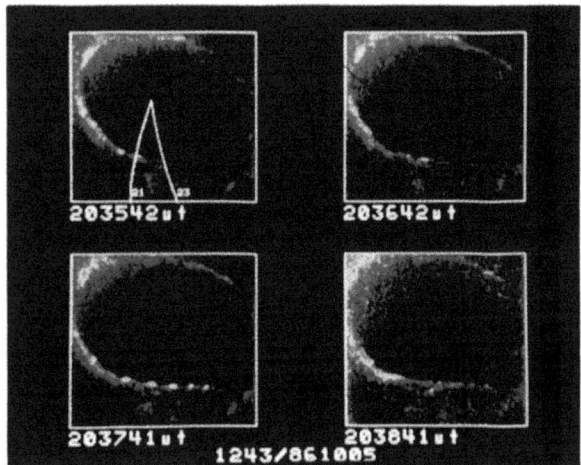

Plate 40—Viking LBH data from orbit 1243. The appearance of a vortex street is apparent at 2037:41 UT.

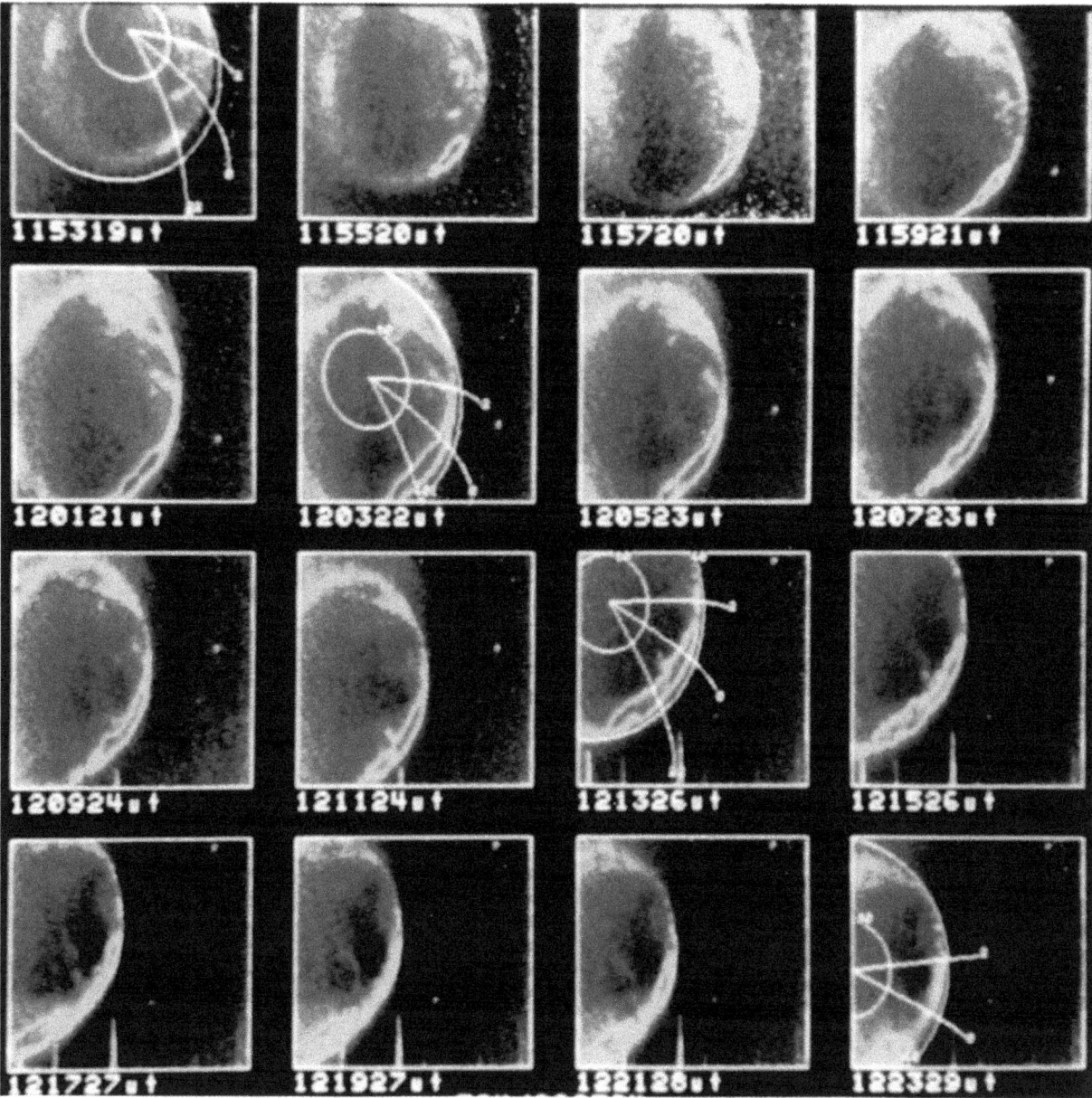

Plate 42—Images of the Earth's northern polar region showing the evolution of the auroras during the May 4, 1986 substorm. The images were recorded, each in one second, by the Viking satellite in atomic oxygen (O I) 130.4 nm emission. The dawn meridian is toward the upper right and the dusk meridian is toward the lower left in each image. The 2200-, 0000-, and 0200-MLT meridians are drawn in four of the images.

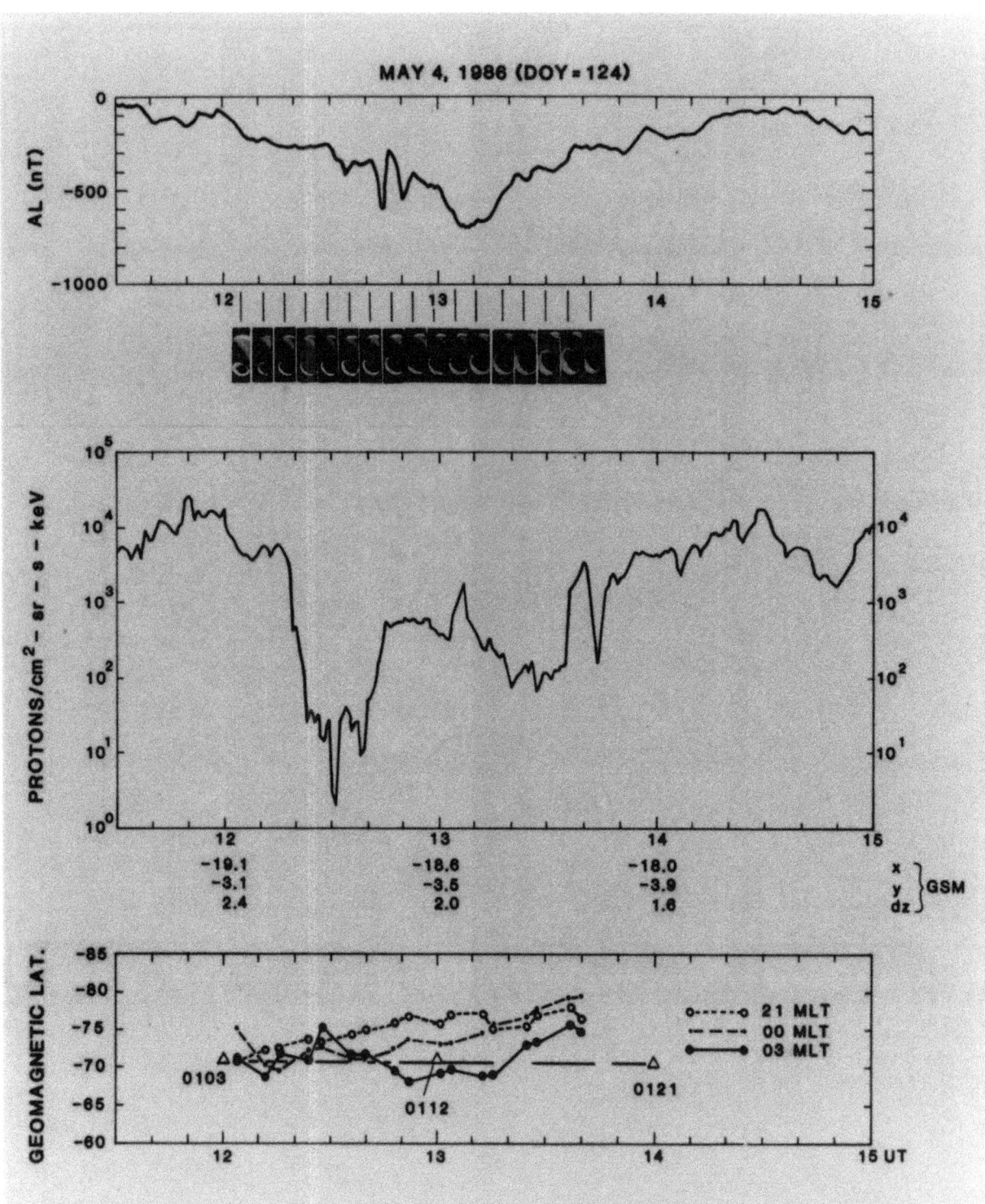

Plate 43—Data for the substorm on May 4, 1986. *Top panel:* The *AL* index. Auroral images: DE 1 images of the southern auroral oval as seen in atomic oxygen (O I) 130.4 nm emission. The dawn meridian is toward the upper left and the dusk meridian is toward the lower right in each image. The width of each image is equal to 6 minutes on the horizontal time scale of the figure and each image is centered at the midtime of its 6-minute accumulation interval.

Middle panel: Flux of 27-33 keV protons measured with the ULECA instrument on ISEE 1. The location of ISEE 1 is given under this panel.

Bottom panel: Latitude of the poleward edge of the auroras at 2100, 0000, and 0300 MLT, determined by computer graphics analyses of the DE 1 images. Also shown, by the straight long-dashed line, is the latitude of the foot-point of the field line from ISEE 1. The MLT of the foot-point is given every hour.

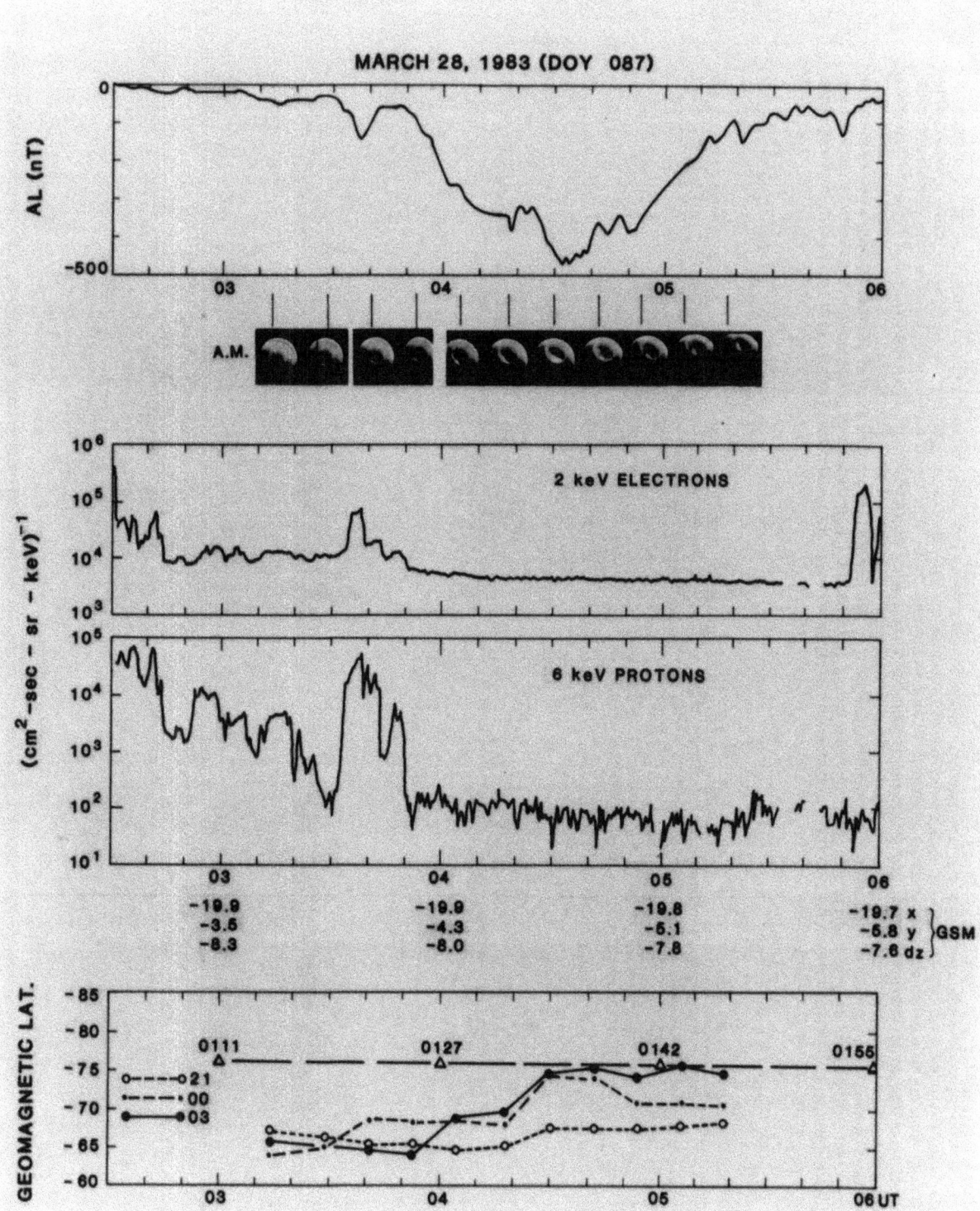

MARCH 28, 1983 (DOY 087)

Plate 44—Data for the substorm on March 28, 1983. *Top panel:* The *AL* index. Auroral images: DE 1 images of the southern auroral oval as seen in atomic oxygen (O I) 130.4 nm emission. The dawn meridian is toward the upper left of each image and the dusk meridian is toward the lower right. The width of each image is equal to 12 minutes on the horizontal time scale of the figure and each image is centered at the midtime of its 12-minute accumulation interval.

Middle panels: Fluxes of 2-keV electrons and 6-keV protons measured by the University of California at Berkeley Energetic Particle Detector on ISEE 1. The location of ISEE 1 is given under these panels.

Bottom panel: Latitude of the poleward edge of the aurora at 2100, 0000, and 0300 MLT, determined by computer graphics analyses of the DE 1 images. Also shown (by the long-dashed line) is the latitude of the foot-point of the field line from ISEE 1. The MLT of the foot-point is shown every hour.

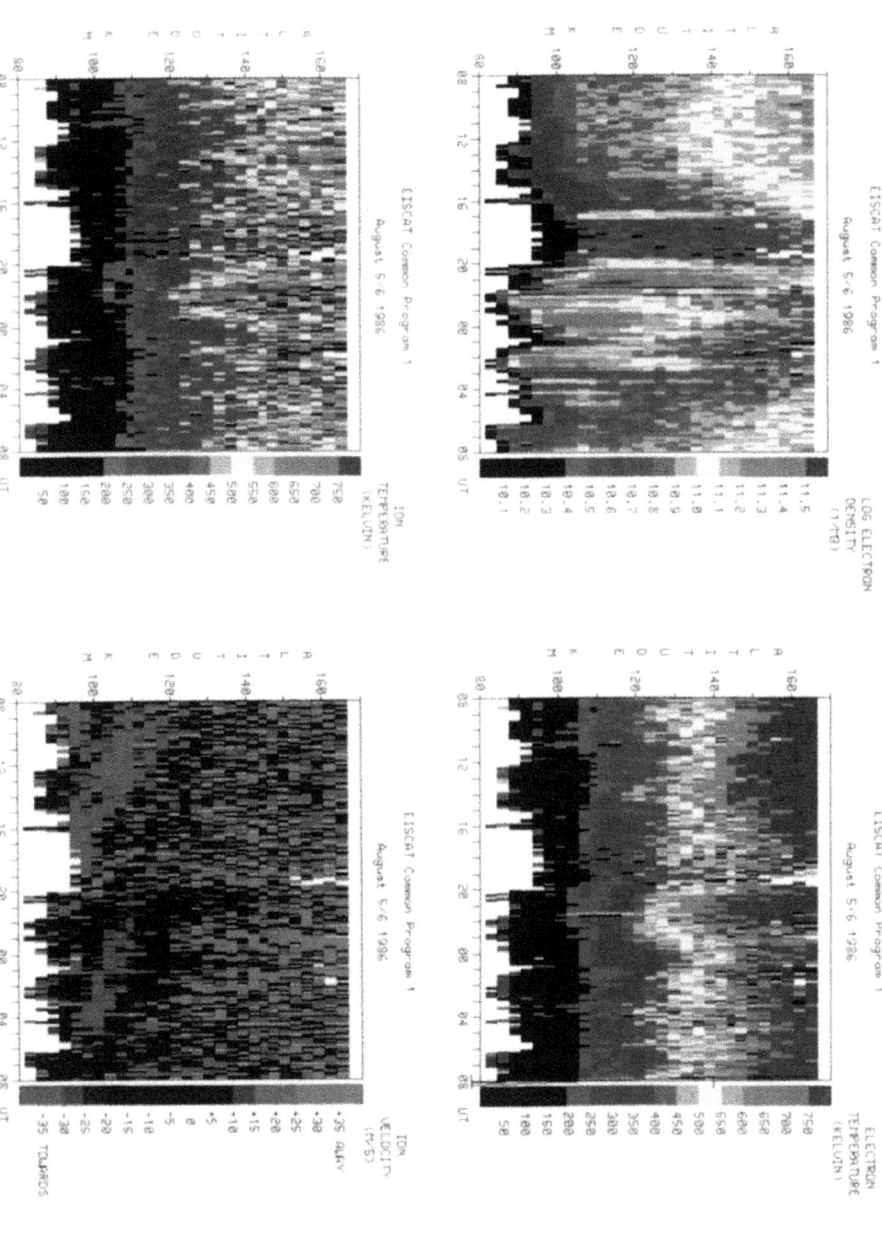

Plate 45—Color-code displays of EISCAT Common Programme observations of electron density, electron temperature, ion temperature, and ion velocity. These data were measured with the Common Programme CP-1-F from 0800 UT on 5 August to 0800 UT on 6 August 1986. In the Common Programme CP-1-F the Tromsö antenna beam is directed parallel to the Earth's magnetic field. The displays show the temporal changes of the altitude profiles of the ionospheric parameters between 84 km and 168 km, which were measured at Tromsö with a special multi-pulse modulation scheme allowing an altitude resolution of 3 km. The electron density plot shows the usual diurnal pattern on which are superimposed strong enhancements in the E region due to bursts of electron precipitation between 2000 UT and 0300 UT. There appears to be a concurring small increase of electron temperature above the altitude of the electron density enhancements. Between 2100 UT and 2200 UT a minor rise of ion temperature is observed above 120 km. The ion velocity shows a semidiurnal pattern, which is attributed to neutral air velocity due to atmospheric tides. (These displays are reproduced from the EISCAT Annual Report 1986.)

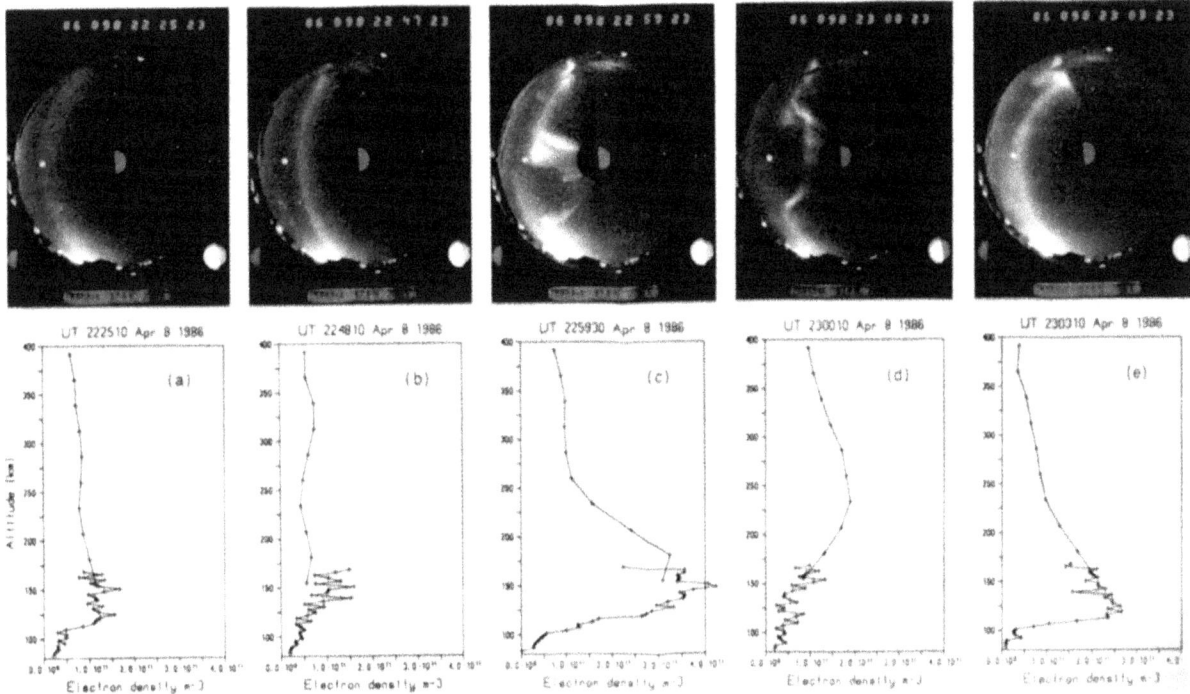

Plate 46—An eastward drifting fold in an auroral arc observed by all-sky camera in Kiruna (*upper panels*), and simultaneously measured electron density profiles (*lower panels*) from EISCAT during the passage of a fold. The leading edge of the fold (*c*) is well characterized by increased E-region densities, but the hole in the fold (*d*) is associated with a broad F-region electron density. (From *Steen et al.* [1988*a*].)

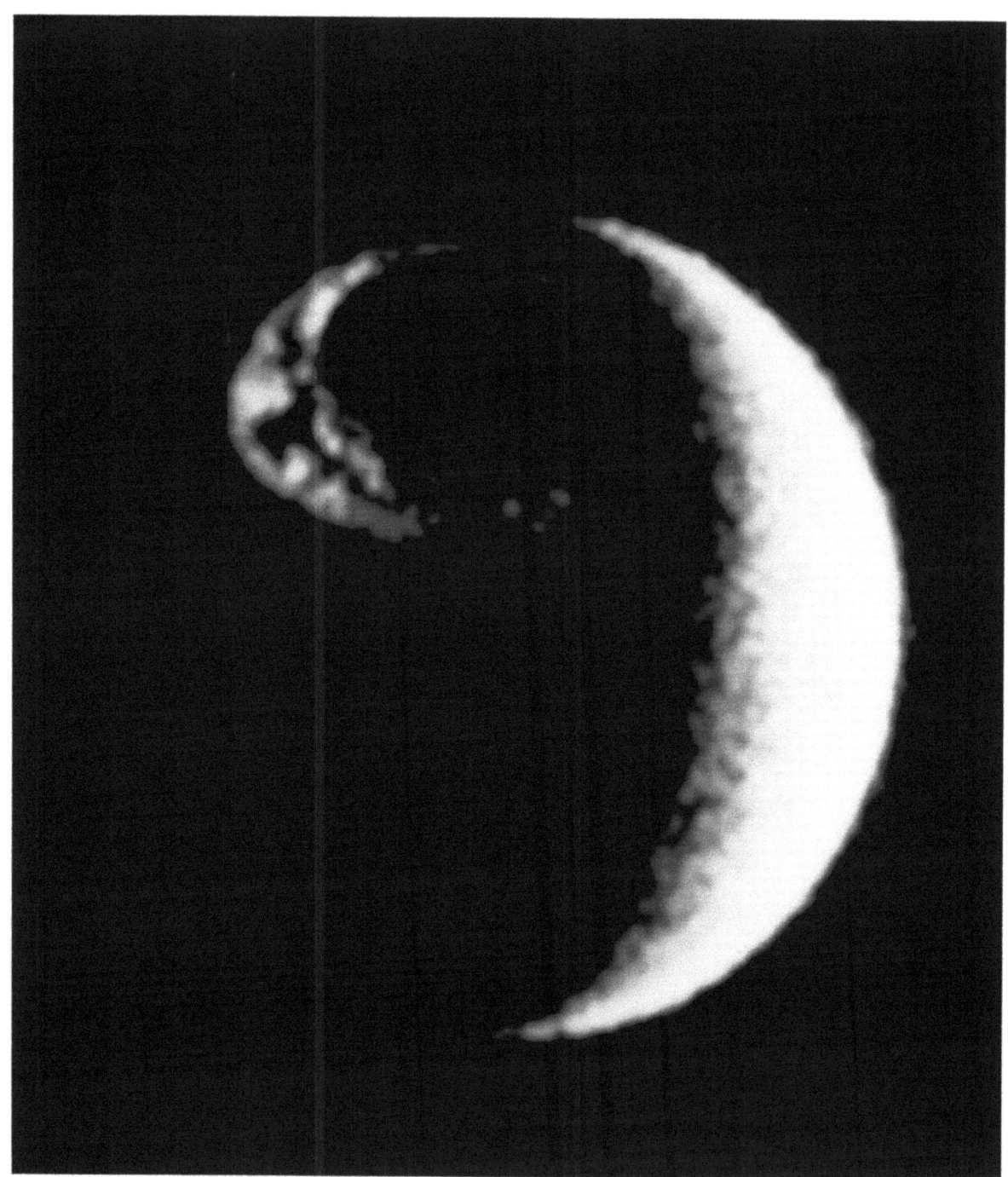

Plate 47—Aurora borealis in the late expansion phase of a substorm. This 12-minute image was obtained by the DE 1 auroral imaging instrument at 0245 UT on December 13, 1981. The observing position was at an altitude of 3.66 R_E, a geographic latitude of 57.9°, and a geographic local time of 0342 hours. Principal contributions to the aurora and atmospheric dayglow in this image are from the emission lines of neutral atomic oxygen at ~130.4 and 135.6 nm and from the LBH bands of molecular nitrogen. (Courtesy, L. A. Frank.)

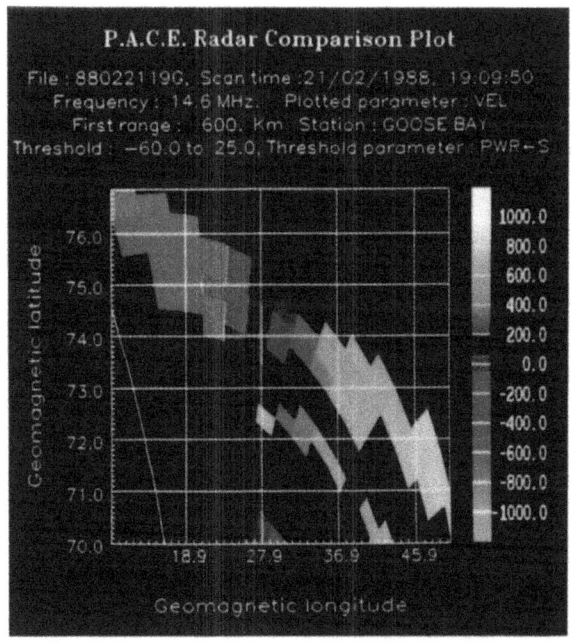

(a)

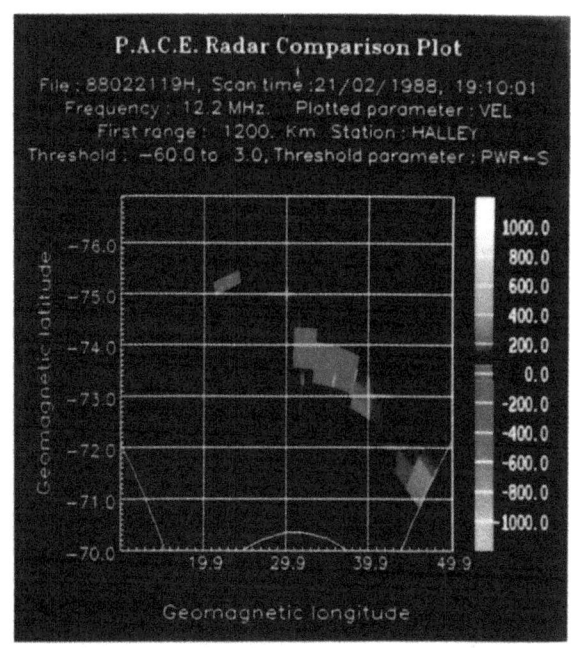

(b)

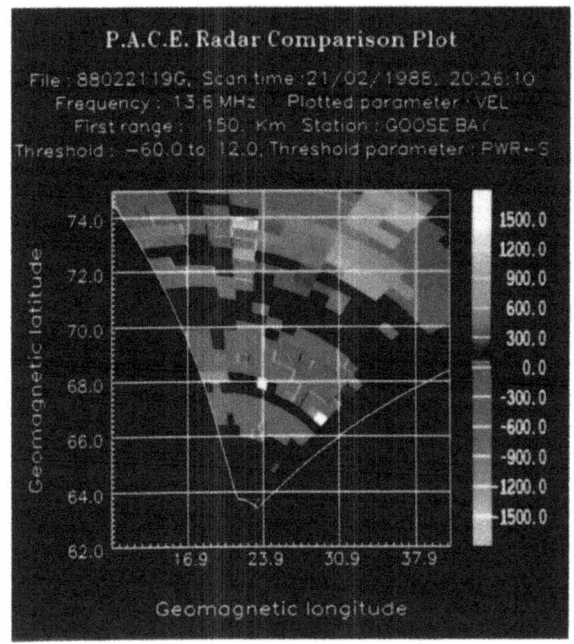

(c)

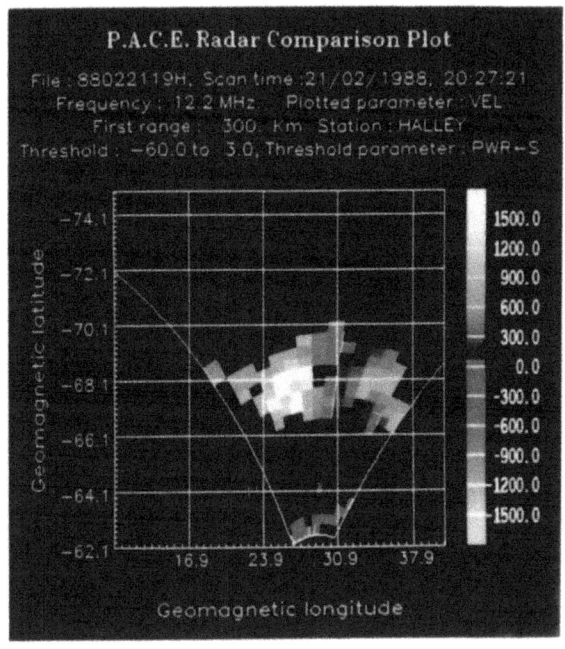

(d)

Plate 48—PACE (Polar Anglo-American Conjugate Experiment) line-of-sight velocities measured by HF (8-20 MHz) radars located in Goose Bay, Labrador (panels *a* and *c*) and Halley Bay, Antarctica (panels *b* and *d*). Velocity toward (+) and away (−) from the radar is coded in meters/second according to the color bar at the right and displayed on a grid showing geomagnetic latitude and geomagnetic longitude. The observations were made at ~2200 magnetic local time on February 21, 1988. The observations at ~1910 UT (panels *a*, *b*) show a velocity pattern typical of westward convection and a scattering region at the same magnetic location in both hemispheres. At ~2026 UT, two scattering regions appear and are displaced in the southern hemisphere ~5° equatorward from the corresponding northern hemisphere locations.

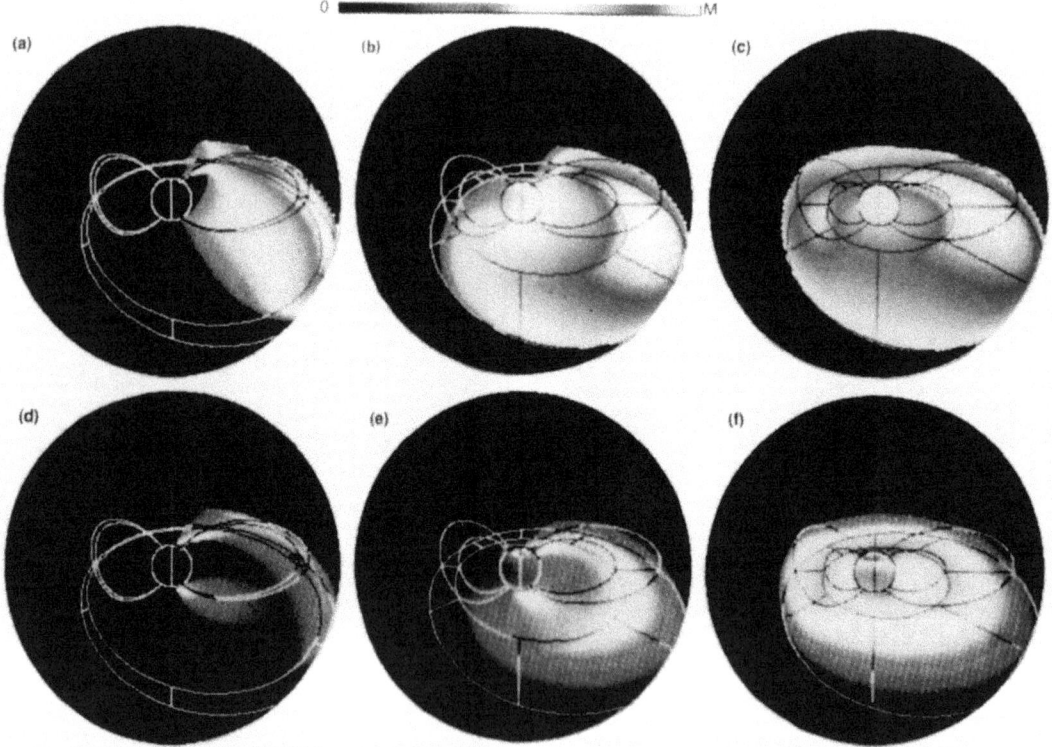

Plate 49—(*a* to *c*) False-color images of model ring current ion intensity evolution. Color represents the column-integrated flux of ions whose instantaneous velocity vector points toward the viewing point. Panels correspond to times spanning asymmetric main phase injection (*a*) through near isotropic recovery phase (*c*). Closed contours indicate the equatorial crossing locus of the inner and outer L shells containing the model fluxes. Magnetic local time cuts at every 3 hours (45°) connect the equatorial loci, with 1800 MLT being the lower vertical line; the Sun is to the left. The linear color bar runs from zero to the maximum (M). The viewing point is at a radius of $8R_E$ and 30° geomagnetic latitude in the dusk meridian plane (*d* to *f*). False-color images of the ENA flux arising from the modeled interaction of the ions in *a* to *c* with the hydrogen geocorona. Same format as *a* to *c*.

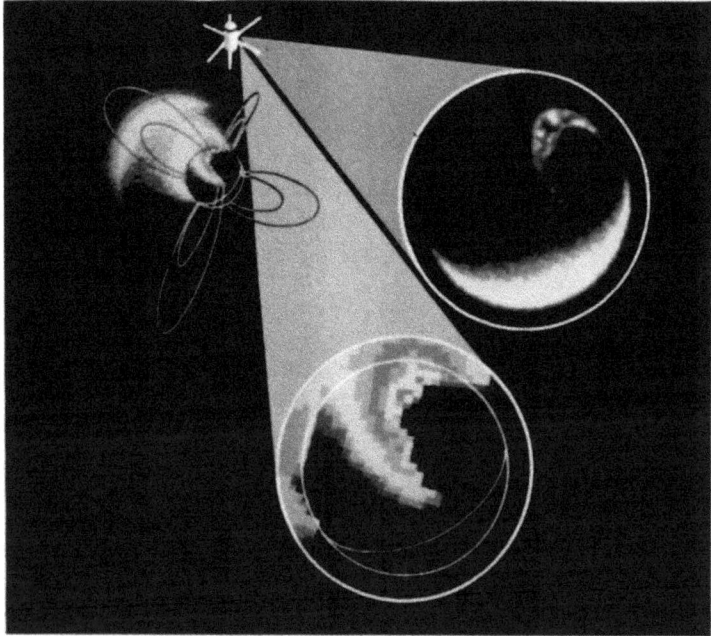

Plate 50—Global magnetospheric and auroral imaging. The DE 1 auroral image of Plate 47 is shown along with a simulated ENA image based on ISEE 1 results obtained during the main phase of the September 29, 1978 magnetic storm. The ENA images shown have been adjusted to the DE 1 position. Both an ENA low-altitude close-up view and an overall magnetospheric view are shown in comparison with the auroral image.

For EU product safety concerns, contact us at Calle de José Abascal, 56–1°, 28003 Madrid, Spain or eugpsr@cambridge.org.

www.ingramcontent.com/pod-product-compliance
Ingram Content Group UK Ltd.
Pitfield, Milton Keynes, MK11 3LW, UK
UKHW060314090126
466816UK00024B/492